DATE DUE

BRODART   Cat. No. 23-221

Glossary of the Mapping Sciences

Prepared by a Joint Committee of the

AMERICAN SOCIETY OF CIVIL ENGINEERS
AMERICAN CONGRESS ON SURVEYING AND MAPPING
and
AMERICAN SOCIETY FOR PHOTOGRAMMETRY AND REMOTE SENSING

Published by

American Society of Civil Engineers
American Congress on Surveying and Mapping
American Society for Photogrammetry and Remote Sensing

ABSTRACT

The Glossary of Mapping Sciences, a joint publication of the American Congress on Surveying and Mapping (ACSM), American Society for Photogrammetry and Remote Sensing (ASPRS), and American Society of Civil Engineers (ASCE), contains approximately 10,000 terms that cover the broad professional areas of surveying, mapping and remote sensing. Based on over 150 sources, this glossary went through an extensive review process that included individual experts from the related subject fields and a variety of U.S. federal agencies such as the U.S. Geological Survey. This comprehensive review process helped to ensure the accuracy of the document. The Glossary of Mapping Sciences will find widespread use throughout the related professions and serve as a vehicle to standardize the terminology of the mapping sciences.

Library of Congress Cataloging-in-Publication Data

The glossary of the mapping sciences.
    p. cm.
  ISBN 0-7844-0050-4. — ISBN 1-57083-011-8
    1. Surveying—Dictionaries. 2. Cartography—Dictionaries. I. American Society of Civil Engineers. II. American Congress on Surveying and Mapping. III. American Society for Photogrammetry and Remote Sensing.
TA505.G57 1994   94-30078
525'.03—dc20   CIP

© 1994 by the American Society for Photogrammetry and Remote Sensing, American Congress on Surveying and Mapping, and American Society of Civil Engineers. All rights reserved. Reproductions of this volume or any parts thereof (excluding short quotations for use in the preparation of reviews and technical and scientific papers) may be made only after obtaining the specific approval of all the publishers. The publishers are not responsible for any opinions or statements made.

Permission to Photocopy: The copyright owners hereby give consent that copies of this book, or parts thereof, may be made for personal or internal use, or for the personal or internal use of specific clients. This consent is given on the condition, however, that the copier pay the stated copy fee of $2 for each copy, plus 10 cents per page copied (prices subject to change without notice,) through the Copyright Clearance Center, Inc., 222 Rosewood Drive, Danvers, MA 01923, for copying beyond that permitted by Sections 107 or 108 of the U.S. Copyright Law. This consent does not extend to other kinds of copying, such as copying for general distribution, for advertising or promotional purposes, for creating new collective works, or for resale.

When reporting copies from this volume to the Copyright Clearance Center, Inc., please refer to the following code: ISBN1-57083-011-8 or 0-7844-0050-4.

ISBN 1-57083-011-8
ISBN 0-7844-0050-4

Published by

American Society for Photogrammetry and Remote Sensing
American Congress on Surveying and Mapping
5410 Grosvenor Lane
Bethesda, Maryland 20814

American Society of Civil Engineers
345 East 47th Street
New York, NY 10017-2398

Printed in the United States of America

Preface

In 1972, the American Congress on Surveying and Mapping (ACSM) and the American Society of Civil Engineers (ASCE) jointly published the glossary "Definition of Surveying and Associated Terms." That document contained about 1,500 terms and was revised in 1973 and reprinted in 1981. In 1989, a joint committee of ACSM and ASCE was formed to update the glossary to incorporate the many developments of that decade. The American Society for Photogrammetry and Remote Sensing (ASPRS) had been presented with a manuscript entitled "Glossary of the Mapping Sciences," prepared by one of its members, Soren W. Henriksen. This document contained approximately 10,000 terms and was reviewed and approved by the ASPRS Publications Committee, and had technical reviews by several U.S. federal agencies, including the U.S. Geological Survey, the Defense Mapping Agency, the National Oceanic and Atmospheric Administration, and the U.S. Army Corps of Engineers. When ASPRS learned of the proposed revision activities of ACSM/ASCE, the Henriksen document was presented to both societies for consideration as a three-society project.

In 1990, a joint committee of the three societies was formed and work began to conduct an intensive review and edit of the Henriksen document. The committee consisted of Mr. Neil Franklin and Mr. Jon Leverenz of the ACSM; Mr. Stephen DeLoach and Professor Earl Burkholder of the ASCE; and Mr. Don Hemenway and Mr. Soren Henriksen of the ASPRS. Professor Burkholder served as the committee chairman through May 1991 and organized a comprehensive review process. The review itself was conducted by a long list of individuals and their efforts have helped to ensure the accuracy of the document. These individuals are: Ira Alexander, Roger Anderson, Roger "Sky" Chamard, Joe Colcord, Russ Congalton, Ray Connin, Ron Cothren, Jim Crossfield, Sean Curry, Michael Falk, Bob Foster, Brian Gordon, Glen Gustafson, Galen Hart, Ray Hintz, Paul Hopkins, Chris Johannsen, Bruce Kessler, Dilwyn Knott, Jon Leverenz, Ed McKay, Alan Mikuni, Robert Mollenhauer, Glen Nave, Lewis Nichols, Dave Nystrom, Charles Ogrosky, Bob Packard, Robert Reader, James Robinson, Karen Schuckman, Bob Scruggs, Allan Shumway, Andrew Sikes, Paul Smith, Joe Ulliman, Nancy von Meyer, Don Wilson, Kam Wong, and Erich Ziebarth. Consolidation of the comments and incorporation of terms from the earlier ACSM/ASCE document was performed by Rebecca Ragon of the U.S. Army Topographic Engineering Center. Final formatting of the document was conducted by Jim Williamson of the ASPRS Publications Committee.

Over 150 sources were used in compiling the glossary. An appendix listing the principal sources of terms and definitions used, as well as many of the sources of explanatory material, is available from ASPRS for a nominal charge.

This glossary has been published to support the broad professional areas of surveying, mapping, and remote sensing. It will find widespread use throughout the related professions and serve as a vehicle to standardize the terminology used in our daily activities of the mapping sciences.

Stephen R. DeLoach, P.E., L.S.
Chairman, ASCE/ACSM/ASPRS Joint Glossary Committee

Contents

A 1
B 43
C 71
D 137
E 165
F 195
G 213
H 241
I 255
J 271
K 273
L 277
M 313
N 357
O 363
P 373
Q 427
R 431
S 461
T 515
U 553
V 557
W 565
X 575
Y 577
Z 579

A

abac - (1) SEE nomograph. (2) SEE ABAC.

ABAC - A nomograph for determining the angle, at a common origin, between the bearing of a great circle and the bearing of a rhumb line to a point. It provides the difference (t-minus-T) between the two bearings.

abandonment - Voluntary surrender or relinquishment of possession in real property without vesting this interest in another person.

Abbe number - A number ν introduced by E. Abbe for characterizing an optical medium by its dispersion, and defined by the formula $\nu \equiv (n_d - 1)/(n_F - n_c)$, in which n_d, n_F, n_c are the refractive indices of the medium at the wave-lengths of, respectively, the helium d-line, the Fraunhofer F-line and the Fraunhofer C-line. The number is used, in particular, in computations for designing achromatic systems of lenses. For example, an optical system of two lenses has the same focal length at each of the wavelengths of the F- and C-lines when the sum of the products of the Abbe numbers and the focal lengths of the individual lenses is equal to zero.

Abbe's criterion - A criterion (introduced by Abbe in 1863) for testing the hypothesis that a systematic trend exists in a set of observations. The criterion is seldom used.

Abbe prism - A prism system consisting of two double right-angle prisms and reflecting an incoming ray four times to give an erect, inverted image.

Abbe's sine condition - The equation $m = (N\sin\theta)/(n'\sin\theta')$ for the lateral magnification m by a single optical surface, as a function of the refractive indices N and N' on the incident and emergent sides of the surface, respectively; θ and θ' are the corresponding angles, at the center of curvature, between the upper and lower points of the object and of the image, respectively. SEE ALSO point, aplanatic.

Abbe's theory - The theory that for a lens to produce a faithful image, it must be large enough to transmit the entire diffraction pattern of the object.

ABC survey - SEE Airborne Control Survey.

ABC system - SEE Airborne Control System.

abeam - A bearing of approximately 90° or 270° with respect to the longitudinal axis of a vehicle.

aberration - A deviation from what is considered natural or normal. SEE IN PARTICULAR: aberration, astronomic; aberration, optical and the various kinds of aberrations listed under those entries.

aberration, annual - That apparent change in direction of a star, planet, or other celestial body which is caused principally by the Earth's revolution about the Sun. That part of stellar aberration which has an annual period. (The Sun's barycentric motion combines with the Earth's revolution to cause a very small aberration which is also covered by the term.) During the year, a star in the plane of the Earth's orbit (the ecliptic) appears to move from one end of a straight line to the other and back again. The length of this line is twice the constant of aberration. A star 90° from the ecliptic appears to move in a circle which has a diameter twice the constant of aberration. Stars at intermediate angles above or below the ecliptic appear to move in ellipses with major diameters twice the constant of aberration and with minor axes getting longer with increasing angular distance of the star from the ecliptic. Because the Earth moves in an elliptical rather than a circular orbit, the apparent paths of stars not on the ecliptic are not quite elliptical.

aberration, astigmatic - SEE astigmatism.

aberration, astronomic - The deviation of the observed direction of a source of radiation from the true direction (the direction of a straight line drawn, at the instant of observation, from observer to source), when the source and observer are moving with respect to each other. It is caused by the finite velocity of light in combination with the motions of the observer and the observed object. Also called astronomical aberration, aberration of light and aberration. Astronomic aberration is zero if the source and observer are moving directly towards each other. It is greatest when the relative motion is entirely transverse to the line joining source and observer. If the motions are considered relative to an independent coordinate system, the aberration can be considered caused by the separate motions of the source and observer in that system. The aberration caused by the observer's motion alone is then called stellar aberration or aberration of light; that caused by the source's motion alone is called light-time aberration or planetary aberration. SEE ALSO aberration, annual; aberration, differential; aberration, light-time; aberration, mensual; aberration, planetary; aberration, relativistic; aberration, secular; and aberration, stellar.

aberration, astronomical - See aberration, astronomic.

aberration, axial chromatic - SEE aberration, chromatic.

aberration, chromatic - The separation, by an optical or other system, of a single ray of polychromatic light in object-space into distinct monochromatic rays in image space, without recombining them. In the absence of other kinds of optical aberration, chromatic aberration causes white light from a point in object-space to focus into a continuous sequence of differently colored points in image-space. Two kinds of chromatic aberration (c.a.) are distinguished: longitudinal c.a. (also called axial c.a.) and transverse c.a. also called lateral c.a. or off-axis c.a.). Longitudinal c.a. produces a sequence of differently colored points along a line parallel to the optical axis of the system. Transverse c.a. produces a similar sequence of points along a line perpendicular to the optical axis. Chromatic aberration can be overcome over a moderate range of wavelengths by using a lens system composed of optical elements having different dispersive powers. (An optical system made achromatic for visual observation is not strictly achromatic for photographic work.)

aberration, comatic - SEE coma.

aberration, constant of - SEE constant of aberration.

aberration, constant of diurnal - SEE constant of diurnal aberration.

aberration, differential - The difference in apparent directions of two sources of radiation moving with respect to the observer and having the same true direction, i.e., lie on the same straight line with the observer. Also called parallactic refraction. It is a form of light-time aberration and is sometimes referred to as differential planetary aberration. It appears when, for example, artificial satellites are photographed against a stellar background.

aberration, differential planetary - SEE aberration, differential.

aberration, distortional - SEE distortion.

aberration, diurnal - The apparent change in direction of a star or other celestial body because of the observer's angular velocity about the Earth's center, combined with the finite velocity of light. Diurnal aberration is taken into account in determining astronomical azimuth and longitude for first-order surveys. It is not considered in determining astronomical latitude because an observer has practically no motion in latitude.

aberration, elliptic - SEE E-terms (of aberration of light).

aberration, lateral chromatic - SEE aberration, chromatic.

aberration, light-time - (1) Displacement of an astronomical direction from the true direction of a source, at the instant of observation, to the direction of the source at the time the observed radiation was emitted. Also called light-time correction. It is the angle through which the source of radiation has moved during the time it took for the radiation to reach the observer. (2) The sum of the annual aberration and the light-time aberration as defined in (1). When the source of radiation is an artificial satellite of the Earth, light-time aberration is called differential aberration by some geodesists.

aberration, longitudinal chromatic - SEE aberration, chromatic.

aberration, mensual - A component of stellar aberration caused by the revolution of the Earth and Moon about their common center of mass. It is approximately equal to $0".011 \cos H_M$, where H_M is the hour-angle of the Moon. This is a small quantity, taken into account when determining the deflection of the vertical from astronomical observations.

aberration, off-axis - Optical aberration at considerable angles to the optical axis. Do not confuse this with off-axis chromatic aberration. Also called extra-axial curvature and off-axis curve.

aberration, optical - Any failure of an optical system to image a point in object-space as a single point in image space or to preserve a uniform scale over the image. Also referred to simply as aberration if the meaning is unambiguous, or as lens aberration if a lens system is being discussed. An optical aberration causes either a blurring or a distortion of the image. Seven basic varieties are recognized. The five (Seidel aberrations) that affect monochromatic radiation are: spherical aberration, coma, astigmatism, curvature of field, and distortion. Radiation over a range of wavelengths is affected by two kinds of chromatic aberration: longitudinal (axial) and transverse (off-axis or lateral) aberrations. The first three cause points to be imaged as blurs. The next two image points as points but distort the shapes of lines because a point in object-space and the corresponding point in image-space do not both lie on a straight line through the center of perspective. The two chromatic aberrations cause an object seen in white light to appear as a white image with colored edges. Aberrations occur in systems other than optical e.g. radio systems and infrared systems. The above terminology applies also to aberrations in such systems. SEE ALSO aberration, chromatic; aberration, off-axis; aberration, spherical; aberration, zonal; astigmatism; aperture aberration; coma; distortion; and Seidel aberration.

aberration, planetary - The apparent angular displacement of the observed direction of a celestial body as caused by the observer's motion (stellar aberration) and the actual motion of the observed object (light-time aberration). SEE ALSO aberration, light-time.

aberration, relativistic - The tendency of a beam of charged particles moving at relativistic speeds to stay narrow and not spread. As the speed of the particles increases, the electrostatic field accompanying the particles diminishes and the tendency to spread therefore diminishes also.

aberration, secular - That difference between the apparent direction of a source of radiation and the true direction resulting from the essentially uniform and rectilinear motion of the center of mass of the Solar System with respect to the source. Also called aberration of fixed stars.

aberration, spherical - The focusing of rays from a point-source on the optical axis into points whose distance from the focal plane varies as the angle at which the rays entered the optical system, with rays entering at a small angle to the optical axis focusing closer to the focal plane than rays entering at a larger angle. The envelope of the converging rays is called the caustic of the optical system. The amount of spherical aberration varies with the location of the source the square of the distance of the outermost rays from the optical axis.

aberration, stellar - The difference between the direction of a radiating source as it would be seen by a stationary observer at a specified point, and the direction in which the source is actually seen by a moving observer at that point. In the case of an observer on the Earth, stellar aberration has three principal components: diurnal aberration, resulting from the diurnal rotation of the Earth; annual aberration, resulting from the annual motion of the Earth about the Sun; and secular aberration, resulting from the motion of the Solar System's center of mass.

aberration, transverse chromatic - SEE aberration, chromatic.

aberration, zonal - That defect of a lens system in which the focal point shifts when the aperture is decreased. It is the result of the combined effects of a number of optical aberrations.

aberration of light (astronomy) - SEE aberration, astronomic.

aberration of light (optics) - SEE aberration, optical.

aberration, off-axis chromatic - SEE aberration, chromatic. Do not confuse this with off-axis aberration.

aberration of fixed stars - SEE aberration, secular.

Abney level - A rectangular telescope tube to one side of which is fastened a vertical, graduated arc with vernier and carrying a rotatable level. The instrument is small and light enough to be held in the hand. When used as a leveling instrument, the vernier is set at zero on the graduated arc. When used to determine vertical angles, the object is sighted through the tube and the vernier rotated until the bubble is centered in the vial. The angle is then read off the graduated arc. This leveling instrument is named after W. deW. Abney, an English physicist and, with Locke's level, is one of the best for rough work.

Abney leveling instrument - SEE Abney level.

abscissa - The graphical representation of one of a pair of coordinates of a point in a two-dimensional space; the coordinate is represented as a distance along one (the abscissa axis) of two intersecting lines (axes) in the plane. The abscissa is determined by drawing a straight line from the point and parallel to the second (ordinate axis) of the two axes. It is the distance from the intersection of the axes to the intersection of the third line with the abscissa axis. SEE ALSO axis of abscissas.

abscissa (surveying) - Also called departure.

abscissa axis - SEE axis of abscissas.

absolute (adj.) - Not defined using an arbitrary origin or scale. The term has other meanings, but the present one is most common in geodesy. For specific instances of its use, SEE coordinate system, absolute; gravity, absolute; zero, absolute.

absorptance - That fraction of incident radiation which is absorbed.

absorption (of light) - The reduction in the intensity of light, on passing through a medium, because the light is transformed by the medium into some other, invisible form of energy such as heat. It should be distinguished from scattering, which diminishes the intensity of a beam by spreading the direction of propagation over a larger solid angle.

absorption band - A range of wavelengths or frequencies in the electromagnetic spectrum within which radiant energy is absorbed by a substance.

absorption filter - SEE filter, optical.

absorption spectrum - The array of absorption lines and bands that results from the passage of radiant energy from a source of continuous electromagnetic radiation thru a substance that absorbs such radiation selectively.

absorptivity - The capacity of a substance to absorb incident, radiant energy.

abstract - (1) A list of values derived directly from measurements (for a survey) and recorded in the book in which field records are kept, together with other pertinent information. Specifically, one of the following lists. (a) In triangulation, a list of horizontal angles or directions, or of differences of elevation along a base line, or of zenith angles. The measurements may be copied directly into the abstract from the book in which the measurements were recorded, or may be changed slightly (as in the case of an abstract of directions, where a constant value is added to the measured values to make a particular direction equal to zero.) (b) In leveling, a list of measured differences of elevation, with corresponding distances and other pertinent information. (2) SEE abstract of title.

abstract, dead-reckoning - A list of all courses and distances made good, together with all data to be used in plotting and adjusting the line obtained by dead-reckoning.

abstracter - The individual that takes the information pertaining to a title in its full form and puts all facts into an abbreviated form. The abbreviated form is alled an abstract or abstract of title.

abstract of title - (1) A complete summary of all information on public record relating to ownership (title) of a parcel of land. (2) A summary of all conveyances, such as deeds or wills, and legal proceedings giving the names of the parties, the description of the land in question, and the agreements, arranged to show the continuity of ownership. An abstract of title customarily cites the surveys delimiting the parcel of land, refers to plats made from surveys, and lists legal instruments affecting ownership and title, such as deeds, mortgages, liens, and the like. It usually has the information arranged in chronological order to show the history of ownership. Often referred to simply as an abstract. Rather than referring to an abstract of title as a title, as is sometimes done, the better term would be examination of title or, as is also common, title examination, because legally, an abstract and a title are not the same thing.

abstractor plat - A plat carrying a compilation of all descriptions of land within a given region, showing controlling measurements and other data. Not all dimensions in a deed are shown; those that are a controlling consideration are emphasized.

abut - (1) (v.) touch, as contiguous estates, along a border or with a projecting part; as, his land abuts on the road. (2) (v) Cause to abut (v.i.); support something by abutting; as, to abut a timber against a post. (3) (adj.) terminating or touching with an end; as, two lots abutting each other.

abutment - The structure, usually of reinforced concrete, supporting an end of an arch or bridge designed to resist thrust and holding the abutting earth in place.

abuttal - The boundary of land described in terms of the other pieces of land, highways, etc., that adjoin and bound that land. Also called buttal or butting. For example, abutted on the west by lot number 36. This term has been used to denote the boundaries on the ends as distinguished from those on the sides, as buttings and sidings.

abutter - An owner or occupant whose property adjoins or touches the land under consideration.

acceleration - The rate of change, dv/dt, of velocity v with time t. Also called linear acceleration, particularly if angular acceleration might otherwise be understood. It is sometimes called deceleration if dv/dt is negative, but such usage is unnecessary.

acceleration, absolute - The rate of change of velocity with respect to time, measured with respect to axes fixed in space.

acceleration, angular - The rate of change, $d\theta/dt$, of angular velocity θ with time t.

acceleration, linear - SEE acceleration.

acceleration error - The error resulting from a change in a craft's velocity: e.g., the apparent deflection of the vertical, as indicated by a gyroscope (artificial horizon), when the craft accelerates.

acceleration of gravity - SEE gravity; gravity acceleration.

acceleration, potential - The scaler function, if it exists, which has a gradient equal to the total acceleration of the body.

accelerometer - A device for measuring acceleration. A common type is simply a pendulum suspended from a pivot fixed in the accelerating body. As the body accelerates, the pendulum is deflected through an angle which depends on the amount of acceleration. Another common type consists of a mass held in equilibrium by springs that are fixed to the accelerating body. The amount of acceleration is determined from the amount of distortion in the springs. Masses resting on strain gages have also been used to measure acceleration. A particularly sensitive type, used in artificial satellites, encloses a spherical test-mass within a slightly larger sphere. The test-mass is kept centered by suitable means, and changes (caused by acceleration) in location of the test-mass with respect to its inclosure are measured by changes in capacitance between the two surfaces.

access, controlled - The condition existing where the right of owners, or occupants of abutting land, or other persons, to access in connection with a highway is fully or partially controlled by public authority. There are two major varieties of controlled access: fully-controlled and partially controlled. 1) Fully-controlled access (full control of access) - the authority is exercised to give preference to thru traffic by providing (access) connections with selected public roads only, and by prohibiting crossings at grade or direct connection to private driveways. 2) Partially-controlled access (partial control of access) - the authority is exercised to give preference to thru traffic to a degree that there may be, in addition to (access) connections with selected public roads, some crossings at grade and connections to private driveways.

access, legal - A right, which an owner of land that abuts a highway has, to use the highway for ingress and egress. SEE ALSO access rights.

accession to real property - The addition to real property by growth, increase, or labor. Land gradually deposited on the bank of a stream by imperceptible means. Title to real property can be acquired by accession to real property.

accessory - Any device (part, sub-assembly, or assembly) contributing to the effectiveness of a machine or piece of equipment without changing or being necessary to the equipment's basic function. SEE ALSO corner accessory.

accessory to corner - SEE corner accessory.

access right - (1) The right of entering and leaving property which abuts an existing street or highway. Also called right of access. An access right may be an easement in the street which is appurtenant to abutting property and is a private right as distinguishable from rights of the public. It is well established law, in the United States of America, that access rights cannot be denied or unreasonably restricted unless other reasonable access is available or provided, or reasonable payment made for loss of value caused by such denial or restriction. Also called right of accession. (2) The right of a riparian owner to pass to and from the waters within the width of his premises as they border upon the water.

accommodation - (1) The ability of the eye to see sharp images of objects that are at different distances. (2) The ability of the eyes to superpose two separate images. (3) The limits within which the projectors in a stereoscopic instrument can be rotated for proper orientation. For example, the Multiplex can accomodate small tilts (in the projectors) ranging from about 10° about the x-axis to 20° about the y-axis. (4) The limits within which a stereoscopic plotting instrument is capable of operating. (5) The ability of any optical instrument to adjust to differences in distance or brightness of the scene.

accretion - (1) The gradual and imperceptible addition of land by natural causes, as out of the sea or a river. Also called alluvion, although that term is also taken to mean the material itself. Accretion occurs principally by the action of water and is of two kinds: a) the deposition of solids and b) the recession of the edge of the water. The term alluvion is sometimes used for the first kind and or the material deposited (also called alluvium); the second kind is called reliction or dereliction. Some authorities take alluvion to mean only the material deposited, which makes it a synonym for alluvium. (See, however, diluvion.) Accretion applies only to the accumulation of land. Accumulation of solid matter under water is referred to as batture, as is the result of such accumulation. (2) SEE alluvion. SEE ALSO alluvion; alluvium; diluvion.

accretion topography - Surficial features, particularly meander scrolls, built by accumulation of sediment.

accuracy - (1) The difference (unsigned) between a specified value of a particular quantity and a value that has been accepted as correct for that quantity. When this difference is known and has a sign, it may be called a correction or an error, depending on the viewpoint of the user. When the difference is known only very approximately, it is usually referred to merely as high accuracy if small and low accuracy if large, without stating numerical values. (2) A measure of the closeness of a set of values, measured or calculated, to the true value. Also called outer accuracy or external accuracy by some European writers. Accuracy is sharply distinguished, in English, from precision, which is a measure of the closeness of the measurements to each other. So a very precise set of measurements can be much less accurate than a different set which is more accurate but less precise. For example, some early geodesists using a few poor measurements were able to find values of the Earth's flattening quite close to the value now accepted as correct, while later geodesists using more precise data obtained values considerably farther from the present value. The history of measurements of the speed of light provides another example of many, presumably very precise, measurements yielding quite erroneous values for the speed. Several different kinds of measure are in use. (a) The square root of the average value of the sum of squares of the differences between the values in a set and the corresponding correct or standard values. This is the most common measure and is usually referred to as the accuracy of the set of values. It is also referred to as the outer accuracy or external accuracy if the precision is referred to as the inner accuracy. (This latter terminology is rare in American usage.) If $\{x_i\}$ is the set of values (measured or calculated) and if $s is + \sqrt{[\Sigma (x_i - x_{io})^2]/M}$ then s is a measure of the accuracy if $\{x_{io}\}$ is a set of correct or standard values and M = I (the number of values in the set). It is the precision if x_{io} = x, the average value of the $\{x_{io}\}$ and if M = I - 1. (b) The average of the sum of the absolute values of the differences between the measured or calculated values and the correct or standard values. (c) The reciprocal of the s defined in (a). Accuracy cannot be calculated solely from the measured or calculated values. A standard (correct) value or set of such values must be used for comparison. The standard may be (a) an exact value, such as the sum of the three angles of a plane triangle; (b) the value of a conventional unit, such as the length of the International Meter, defined from the speed of light in a vacuum; (c) a value determined by refined methods and deemed sufficiently near the correct value that it can be used as such: e.g., the adjusted elevation of a permanent bench mark or the graticule of a map projection. (3) The standard deviation. This is a measure of precision. It should never be used as a measure of accuracy. (4) The root-mean-square error. This is approximately the same as the standard deviation and, like it, should not be used as a measure of accuracy. (5) When the set $\{xi\}$ consists of calculated numbers (e.g. those in a mathematical table), the term accuracy may mean (a) the number of significant digits in the numbers; (b) the magnitude of the least significant digit; or (c) the number of correct places in the numbers.

accuracy, absolute - Accuracy as determined by reference to a value that must be specified.

accuracy checking - Obtaining presumptive evidence that a map complies with specified standards of accuracy. It generally determines the accuracy of features on the map relative to each other rather than the accuracy with respect to some external accuracy. SEE ALSO U.S. National Map Accuracy Standards.

accuracy control system - SEE system, accuracy-control.

accuracy, external - SEE accuracy (2).

accuracy, inner - SEE precision.

accuracy method - The method used in determining the stated accuracy of a product.

accuracy, outer - SEE accuracy (2).

accuracy, positional (cartography) - The degree of compilance with which the coordinates of points determined from a map agree with the coordinates determined by survey or other independent means accepted as accurate.

accuracy, relative - (1) The square root of the average of the sum of the squares of the differences between (a) a set of measured or calculated values and a (b) set of corresponding correct values, divided by the average value of the set. I.e., relative accuracy is equal to the accuracy divided by the average value of the set. This is a useful quantity because it is dimensionless and independent of the units in which the measurements were made. (2) A quantity expressing the effect of random errors on the location of one point or feature with respect to another. In particular, (a) an evaluation of the effect of the random errors in points on a map with respect to the graticule, excluding any errors in the graticule or the coordinate system of the graticule; or (b) an evaluation of the effect of random errors in determining the location of one point on a map with respect to another point on the same map.

accuracy testing - Testing, by sampling, whether or not a map complies with specified standards of accuracy. It checks both relative and absolute accuracies of distances, lengths, and elevations shown on the map. COMPARE WITH accuracy checking.

acetate - SEE acetate film; cellulose acetate.

acetate film - (1) A nonflammable, plastic sheet used as a base for photographic film. Also called acetate (jargon).
(2) A nonflammable, plastic sheet on which drafting is done and which is used as an overlay. Also called acetate (jargon).
(3) A sheet made of plastic prepared by acetylation of cellulose.

acetate proof - An inked proof made by printing from type on thin sheets of acetate film. It can be used as a positive for making photographic negatives by contact or directly for making a positive flat.

achromat - SEE lens system, achromatic.

achromatism - SEE achromatism, actinic.

achromatism, actinic - (1) That property of a properly designed photographic lens-system which ensures that light at a wavelength near the Fraunhofer D-line (about 589 nanometers) and at the wavelength of the G-line at 430.8 nanometers, from a point source on the optical axis, are focused at the same point and produce images of the same size. Also called achromatism. SEE ALSO lens, achromatic; lens system, achromatic. (2) SEE FG achromatism.

achromatism, FG - SEE FG achromatism.

acknowledgment - A formal declaration by an individual, before some competent public officer, declaring an act or deed to be his. Usually declared before a notary public.

aclinal (adj.) - without dip; horizontal.

aclinic (adj.) - without magnetic dip.

acoustics - (1) In general, that part of physics which concerns itself with sound - its causes, characteristics, and effects. In surveying, acoustics is involved primarily in the development and use of submarine surveying by sonar and analysis and use of natural and artificially-caused seismic events. (2) In particular, that part of architecture which concerns itself with the design of buildings where the relation of the design to the quality or amount of sound in it is of first importance.

acquiescence - (1) Passive compliance or satisfaction. (2) The neglection of a legal right for a length of time sufficient to infer that that right has been waived. (3) A boundary may be established through judicial review by acquiescence of adjacent property owners. It can be described as a tacit encouragement, or a consent inferred from silence.

acquire (v.) - Detect or receive a signal from an object. The term is used in this sense in the photography of missiles and in radar.

acre - A unit of area in the English system of measure, defined as 10 square chains (1 chain is 4 rods or 66 feet). It is exactly equal to 43,560 square feet or 4,840 square yards and is approximately equal to 4046.8564224 square meters. There are 640 acres in one square mile. Although acre may be defined otherwise in some localities, it is a local definition and the word is usually qualified as builder's acre or Block Island acre. The English system of measures contains no legally defined unit of area. The size of the acre therefore depends on the length of the English yard, which is legally defined. By an ordinance of Edward I in 1303, the acre was defined as the area contained in a rectangle 40 rods long and 4 rods wide. With the rod defined as 5.5 ulnae (yards), as defined by Edward I's iron standard for the ulna, the acre is 4840 square yards.

acre, commercial - That part of an acre of newly subdivided land remaining after dedication for streets, sidewalks, parks and so on.

acreage zoning - SEE zoning, large-lot.

acre-foot - The volume of water required to cover one acre to a depth of one foot. Equivalent to 43,560 cubic feet of water, it is a convenient unit for measuring the amount of water in a reservoir, the amount needed for irrigation, or the amount of runoff.

acre, right - The share of a citizen of a New England town in the common land. The value of the acre right was a fixed quantity in each town but varied from town to town. A 10-acre lot or acre right in a certain town was equivalent to 113 acres of upland and 12 acres of meadow, and a certain exact proportion was maintained between the acre right and saleable lands.

action - The sum of the integrals of the generalized momenta of a system over their canonically conjugate coordinates. Also called a phase integral.

acuity, visual - A measure of the ability of the eye to separate details in viewing an object, and taken to be the reciprocal of the smallest angular separation, in minutes of arc, of two lines which can be seen to be separate.

acutance - An objective measure of the ability of an optical system to show a sharp edge between two contiguous regions, one of high and the other of low illuminance in object-space.

Adams' map projection - (1) A conformal map projection of the whole sphere into an area bounded by an ellipse. The map projection was devised by O. Adams but contained some singularities. These were removed by a method devised by I. Lee in 1965. (2) A conformal map projection of the sphere into a square, with the poles on a diagonal of the square and symmetrical about the center. (3) An equal-area map projection of the pseudo-cylindrical type, projecting the sphere onto a cylinder in such a fashion that the parallels of latitude go into straight lines parallel to the x-axis and spaced at the distance $2kR_a \sin\frac{1}{2}\beta$ from the origin. k is a scale factor, R_a is the radius of the authalic sphere, β is the authalic latitude and $x = [R_a(\lambda - \lambda_o)\cos\beta]/\cos\frac{1}{2}\beta$, in which λ is the longitude and λ_o is the longitude of the reference meridian. (4) A conformal projection of the sphere into a regular hexagon.

Adams' rhombic conformal map projection - A conformal map projection mapping a sphere or hemisphere into a rhombus having one opposite pair of angles equal to 60° and the other pair equal to 120°. The complex function z $(x + iy)$ is related to longitude λ and latitude ϕ on the sphere by $w(z) = \{\tan[\frac{1}{4}(\pi - 2\phi)] ei\lambda\}^{1/3}$, in which $w(z)$ is the elliptical function defined by $z = \int (1 - w_3)^{-2/3} dw$. The map projection was used by J.S. Cahill for one version of his butterfly map. It is related to λ and ϕ on the hemisphere by $u(z) = w^2(z)$, where $u(z)$ is substituted for $w(z)$ in the integral defining z. The poles will be at the apices of the 120° angles.

adaptation - The ability of the eye to adjust its sensitivity to varying intensities of illumination.

addition constant - SEE stadia constant.

address (computing machinery) - An alphanumeric code indicating the location, in a computing machine, in which an instruction or an item of information is stored.

ad filum aquae - to the central line, or middle, of the stream. Literally, to the thread of the water. SEE ALSO usque ad filum aquae.

Adindan Datum - A horizontal-control datum defined by the following location and azimuth on the Clarke spheroid of 1880 (modified); the origin is at station Zv ADINDAN:
 longitude of origin +31° 29' 21.6079" E
 latitude of origin +22° 10' 07.1098" N
 azimuth from origin to station Yu 58° 14' 28.45"
(The original origin, Station Zv ADINDAN, is now at least 10 m below the surface of Lake Nasser.) The datum was first used for adjustment of horizontal control in Ethiopia and adjoining regions. It was later extended to the entire continent. The network on Adindan datum has also been placed on Arc Datum.

adit - A horizontal or nearly passage from the surface into a mine and used for working or for removing water from the mine. Commonly called a tunnel, although a tunnel is open at both ends. SEE ALSO entry; portal.

adjacency effect - SEE Eberhard effect.

adjacent (adj.) - (1) Lying near or close to. (2) Contiguous with; touching. Adjacent, as used in land surveying, has the first meaning and implies that the two objects or parcels of land are not widely separated, although they may not actually touch, while adjoining has the second meaning and implies that no third object or piece of land lies between them.

adjoiner - The land in contact with the property of immediate interest.

adjoining (adj.) - Touching, as distinguished from being merely close to or adjacent. adjoining and abutting are at present often used as if they were synonymous. However, it is recommended that the old distinction be retained where possible. Two parcels adjoin if they have a common side; they abut if they have a common end. (2) Close to. This usage is improper but, fortunately, rare.

adjudication - The legal and demonstrative processing of casework to assure full compliance with the public-land laws and regulations and to determine rights.

adjustment - (1) The process of changing the values of a particular, given set of quantities so that the predicted values of an observable function of that set will agree best with the observed values of that function. The concept best is subjective. The most common interpretation is that one set is better than another if it leads to a lower value for the sum of the squares of the differences between the predicted and calculated values of the observable function. If the differences have a Gaussian distribution, the best set is also the most probable set; the criterion is the basis for the method of least squares. Another interpretation is that the set is best which makes the most use of the information in the observed values. This leads to the method of maximum entropy. Still a third interpretation is that the best set is one which makes the biggest difference between predicted and observed values a minimum. (2) The result of an adjustment in the above sense. Synonymous, in this sense, with results. (3) The process of finding, from a set of redundant observations, a set of best values for the observed quantities or for quantities that are functions of the observed quantities. For example, if each angle of a plane triangle is measured, it is likely that the three values will not add up exactly to the 180° required by geometry. The process of calculating three angles which do satisfy the requirement is called the adjustment of the measured values, and the three resulting angles are called adjusted values. In general, there will be N given (observed) values and N equations relating these given values to M (M≤N) unknowns. Adjustment is then the process of reducing the N equations in M unknowns to M equations in M unknowns and solving this set of equations. The M equations are sometimes called the reduced or normal equations. (4) The determination and application of corrections to observations, to reduce errors or internal inconsistencies in the derived results. (5) The changing of the values of measurements or corrected measurements of particular quantities to obtain values that conform to restrictions upon or relations among, the magnitudes of these quantities as imposed by the nature of the problem. (6) Geological changes in the Earth, under the influence of gravity, that occur because of a redistribution of mass on the Earth's surface. Among the causes of such redistribution are the erosion of land by rivers and the advance or retreat of glaciers. If the Earth's masses were in equilibrium with each other before the redistribution, they were not in equilibrium afterwards, and further changes, both in the interior and on the surface, then occur until equilibrium is reestablished. SEE ALSO isostatic equilibrium. (7) Any change in an existing situation to improve, in some sense, that situation. For example, adjustment of values, adjustment of stresses, adjustment of an instrument, etc. (8) The process of aerotriangulation. (9) Bringing the movable parts of a device into proper relationship and fit. After adjustment, the device is ready for use and is said to be in adjustment. In particular, the process of first bringing the optical elements of a telescope into proper relationship (collimation) and then bringing the optical and mechanical parts into proper relationship (alinement). (10) SEE adjustment, land-line. (11) SEE adjustment, cartographic. SEE ALSO angle method; block adjustment; Bowditch method (of adjusting a traverse); Bowie method (of adjusting triangulation); bundle method (of adjusting aerotriangulation); Chebychev adjustment; collimation adjustment; direction method; Helmert blocking method; leveling adjustment; network adjustment; peg adjustment; Schreiber's method (of adjustment); strip adjustment; track adjustment.

adjustment, analytical - Adjustment by an arithmetical process, as distinguished from a graphical adjustment, which is done geometrically. The usual procedure is to determine the discrepancies which exist among combinations of measurements, express these discrepancies in the form of condition equations or observation equations, and solve these equations to obtain corrections to the measurements (e.g., adjust the measurements) or corrections to the quantities on which the measurements depend. If there are more measurements than there are unknowns, the method of least squares is often used.

adjustment, cartographic - Placing details or symbols for control stations in their positions (on a map) relative to other details or symbols. Also called adjustment.

adjustment, free - (1) An adjustment in which the number of unknowns is greater than the number of independent equations relating the unknowns to observations. Such an adjustment has, in general, no unique solution. More or less arbitrary conditions must therefore be imposed on the unknowns to obtain a unique solution. (2) An adjustment in which the number of unknowns is greater than the number of independent equations relating the unknowns to observations, but there are just enough conditions imposed on the unknowns to produce a unique solution. (NGS) (3) The adjustment of a network containing no fixed quantities. (4) An adjustment in which the rank of the normal matrix, exclusive of that part associated with Lagrangian multipliers, is equal to the order of the normal matrix. (5) An adjustment of geodetic coordinates in which, if the number of points with unknown coordinates is N, there are 3N unknowns in 3-space (or 2N unknowns in 2-space) and there are 6 (or 3) condition equations. Also called adjustment using minimal constraints or inner adjustment.

adjustment, geodetic - The adjustment of the values of geodetic constants, particularly those (length, angle, direction, coordinates) that characterize a geodetic network.

adjustment, horizontal - The adjustment of the horizontal coordinates, or of the horizontal distances and angles or directions, in a geodetic network.

adjustment, inner-constraint - An adjustment in which conditions are imposed on certain arbitrary functions of the unknowns rather than on the unknowns themselves. For example, it may be required that the trace of the normal matrix be a minimum, or that the coordinates of all points be referred to the center of mass of the adjusted set.

adjustment, isostatic - The process whereby isostasy in the Earth's crust is maintained or, if the equilibrium is upset, is restored. The distribution of material in the outer part of the Earth is constantly being changed by erosion, sedimentation and other natural forced. The unbalanced condition which would naturally result from such changes is offset by compensating movement of material at considerable depths below the surface. SEE ALSO compensation, isostatic; depth of isostatic compensation.

adjustment, land-line - (1) The process of depicting land lines on a map to indicate the lines' location with respect to adjacent terrain and culture, using the information shown on plats and field records of the Bureau of Land Management with evidence, on the ground, of the locations of the lines to determine the location on the map. (BLM) (2) The process of depicting land lines of the public lands on a topographic map to indicate the true, theoretical or approximate locations relative to the adjacent terrain and culture.

adjustment, least-squares - The process of determining the correction δ_x to a value x_o of a vector so that the sum of the weighted squares of the differences between measured values $y_k(x)$ of a function $y(x)$ and the values $y_k(x_o + \delta_x)$ is a minimum. This is the method of least squares applied to adjustment. Also called adjustment by least squares. The adjusted values are the most probable values if the corrections have a Gaussian distribution.

adjustment, local - SEE triangulation adjustment, local.

adjustment, simultaneous - The simultaneous adjustment of all adjustable variables in a set of simultaneous linear equations representing a physical situation. In particular, the simultaneous adjustment of all adjustable elements (coordinates, lengths, angles, etc.) in aerotriangulation or in a network. SEE ALSO aerotriangulation adjustment, bundle.

adjustment, three-peg - SEE test, three-peg.

adjustment, vertical - Adjustment of the elevations or differences of elevations in a vertical network.

adjustment by conditions - Adjustment of the lengths of sides and of angles between sides in a geodetic network by formulating each observation equation as an equality between the observed value and the adjusted value plus the error (residual), and formulating each condition equation as an exact relationship between the adjusted quantities. The observation equations are usually not written down explicitly but are combined immediately with the condition equations.

adjustment by least squares - SEE adjustment, least-squares.

adjustment by variation of coordinates - Adjustment of the coordinates of points in a geodetic network by formulating each observation as a function of the adjusted coordinates and the errors (residuals). The condition equations, which are usually few, are the exact relationships between the adjusted values of the coordinates.

adjustment correction - The quantity added to an orthometric elevation to produce an adjusted elevation and to eliminate misclosures. Also called an adjustment leveling correction.

adjustment for collimation - SEE collimation.

adjustment leveling correction - SEE adjustment correction.

adjustment of a traverse - SEE balancing a traverse; traverse adjustment.

adjustment of corrections - SEE adjustment of observations.

adjustment of figure - The adjustment of the angles and lengths of sides, in a single chain of triangles, so that the sums of the angles in each triangle equals 180° and, in the case of a quadrilateral, that the sum of the angles equal 360°.

adjustment of gravity - SEE gravity adjustment.

adjustment of leveling - SEE leveling adjustment.

adjustment of observations - (1) The process of calculating, from observed values of quantities, values for these quantities that are better, in some specified manner, than the observed quantities. Also called adjustment of errors. The process of adjustment applied to the observations themselves rather than to the quantities on which the observations depend. Either the observed or the theoretical values may be considered the true values of the quantities. The numbers resulting from an adjustment are usually called the adjusted values of the observations; differences between observed and adjusted values are called the errors in the adjusted values or corrections to the observed values. This terminology is not universal, and a contrary usage is common outside of geodesy. The term adjustment of observations has the connotation that the values of the observations themselves are changed. Because an observed (measured) value should never be changed, the term is a poor one and should be avoided. (2) The process of adjustment applied to uncorrected measurements. SEE ALSO adjustment; aerotriangulation adjustment; gravity adjustment; leveling adjustment; network adjustment; and triangulation adjustment.

adjustment of triangulation - SEE triangulation adjustment.

admeasure (v.) - (1) Ascertain or set the measure or limits of. (2) Measure. (3) Determine the proper share of, or the proper apportionment of e.g., to admeasure common of pastrue.

Admiralty chart - A hydrographic chart of the British Admiralty. Sometimes written, incorrectly, admiralty chart.

admiralty law - The civil and criminal law relating to occurrences at sea and on navigable lakes and rivers, including maritime contracts.

advection - The movement of a fluid, or of a scalar quantity that is considered to move with the fluid. In particular, in oceanography, the term refers to the horizontal or vertical flow of sea water, e.g., a current. In meteorology, it refers either to the predominantly horizontal movement of large masses of the atmosphere or to the transport of an atmospheric property solely by movement of the atmosphere.

aerial (n.) - SEE antenna.

aerodynamic correction - SEE correction, aerodynamic.

aerocartography - SEE mapping, aerial.

aeroleveling - Determination of the bz values (corrections to heights) during orientation of the successive models of a strip on a stereoscopic plotting instrument, using barometric measurements of the altitudes of the air stations (recorded while photographing). Also called the aeroleveling method. Only differences in altitude are required, and these are provided by the statoscope (a differential barometric altimeter). SEE ALSO orientation.

aeroleveling method - SEE aeroleveling.

aerometeorograph - An instrument, carried on balloons or rockets, that measures and records pressure, temperature, and relative humidity of the air through which it is carried. Before the radiosonde was developed, the aerometeorograph was extensively used to investigate the upper atmosphere.

aerophotography - SEE photography, aerial.

aeropolygon method (of strip aerotriangulation) - The relative orientation of successive models of a strip, on a stereoscopic plotting instrument, without using any auxiliary data.

aerospace - The region extending outward from the solid and liquid surfaces of the Earth, i.e., the region containing the atmosphere and the space outside the atmosphere. Also called airspace. Regions interior to the Sun and other bodies of the Solar System and the universe are considered to lie in aerospace but not to be part of it.

aerostat correction - SEE buoyancy correction.

aerostatic correction - SEE buoyancy correction.

aerosurvey - SEE survey, aerial.

aerotriangulation - (1) The determination of horizontal and/or vertical coordinates of points on the ground, from measurements performed using overlapping aerial photographs and from already known coordinates of points on the ground. Phototriangulation using aerial photographs. Although the basic principles had been established as early as 1759, real progress was not made until, in the early 1900's and particularly in the 1920's, the airplane became a practical carrier for the camera. (2) The theory and procedures used for determining such coordinates. Aerotriangulation is also called aerial triangulation, aerotriangulation adjustment and phototriangulation. SEE ALSO block aerotriangulation;

independent-model method; mechanical-templet method; stereotemplet aerotriangulation; strip aerotriangulation; strip radial aerotriangulation; three-point-resection method.

aerotriangulation, analog - (1) Aerotriangulation using a mechanical, optical, or optomechanical model of the relationship between image-point and object-point. Two principal methods are used: the radial method, which obtains only the horizontal coordinates of the ground points; and stereotriangulation, which uses a stereoscopic plotting instrument to obtain horizontal coordinates, vertical coordinates or both. (2) The process of determining the location and orientation of the aerial camera at the times the photographs were taken. This meaning is rare.

aerotriangulation, analytical - Aerotriangulation in which coordinates of ground points are derived from measured coordinates of image-points using mathematical functions rather than a physical model to represent the relationship between image-point and object-point.

aerotriangulation, analytical nadir-point - Radial aerotriangulation done by computations in which nadir points are used as radial centers.

aerotriangulation, analytical radial - Radial aerotriangulation in which the unknown coordinates of ground points are determined mathematically from measured coordinates, on the image, of radial centers and directions to the images of the unknown points. If the coordinates of the principal points of overlapping photographs are determined by resection on three horizontal-control points in the overlapping region, the method is known as analytical three-point resection radial aerotriangulation. SEE ALSO aerotriangulation, graphic radial.

aerotriangulation, analytical three-point resection radial - SEE aerotriangulation, analytical radial and three-point method of radial aerotriangulation.

aerotriangulation, cantilever - (1) Determination of the coordinates of ground points in a strip of aerial photographs, using known coordinates of ground points at one end of the strip only. Commonly contrasted to bridging aerotriangulation, which uses known coordinates of ground points at both ends of the strip. Sometimes referred to simply as cantilevering. (2) The connection, by relative orientation and scaling, of a series of overlapping aerial photographs in a strip, to obtain the coordinates of points shown in the strip, relative to the coordinate system of the overlapping part of the first pair of photographs. Also called cantilever extension.

aerotriangulation, direct radial - A radial aerotriangulation done graphically by tracing the directions from successive radial centers directly onto a transparent plotting sheet, rather than by transferring the directions to templets. Also called direct radial plot.

aerotriangulation, graphic radial - A radial aerotriangulation done by other than analytical means. It is usually done directly, incorporating the ground control plotted on a map, graticule, or map grid, but it may be done first independently of such control and later adjusted to it as a unit. In that case, the scale and azimuth between all the points are not known until the coordinates have been adjusted to the ground control. A graphical radial aerotriangulation may also be done using slotted templets, spider templets, or hand templets. For definitions of various kinds of templet SEE templet (photogrammetry). SEE ALSO aerotriangulation, analytical radial and aerotriangulation, radial.

aerotriangulation, isocenter radial - Radial aerotriangulation in which the isocenter of each photograph is used as the center from which radial lines are drawn. Also called isocenter plot.

aerotriangulation, nadir-point - Radial aerotriangulation using the nadir point instead of the principal point.

aerotriangulation, radial - Aerotriangulation in which the photographs are placed in their proper relative positions and then radial lines from the center (radial center) of each vertical photograph, or from the approximate nadir point of each oblique photograph, are drawn or calculated in the directions of image-points of interest and the relative locations of these points determined by intersection and by resection between neighboring photographs. Radial aero-triangulation is also called radial plot, minor control plot, radial-line aerotriangulation, radial-line method and radial triangulation. The radial center for nearly vertical photographs may be the principal point, the nadir point, or the isocenter. A radial line (radial) is assumed to be made with the principal point as radial center unless the definitive term designates otherwise (e.g., nadir-point aerotriangulation or nadir-point plot, and isocenter aerotriangulation or isocenter plot), or unless the context states that a radial center other than the principal point was used. Radial aerotriangulation may be done graphically or analytically, but is assumed to have been done graphically unless specifically stated otherwise. If the radial aerotriangulation is not done graphically, but the process is represented by a set of equations into which measured and given coordinates are substituted, the process is called analytical radial aerotriangulation.

aerotriangulation, semi-analytical - The measurement of the x,y,z coordinates of points in a model on an analog-type instrument, and the transformation from these coordinates to geodetic coordinates or map coordinates by computation. SEE ALSO independent-model method (of aerotriangulation).

aerotriangulation, stereoscopic - SEE stereotriangulation.

aerotriangulation, vertical - Aerotriangulation done primarily for determining elevations or heights. It is of limited use because elevations or heights are generally wanted with greater accuracy than this method can give. SEE ALSO aeroleveling.

aerotriangulation adjustment - (1) Determination of corrections to the coordinates of ground points, using aerotriangulation. (2) SEE aerotriangulation. Aerotriangulation adjustment in sense (1) differs from aerotriangulation proper in that it is concerned first with corrections to coordinates, not determination of the coordinates themselves. Furthermore, it is usually done mathematically rather than mechanically. According to the number and arrangement of the photographs involved, the adjustment is referred to as individual-model, strip, bundle, or block adjustment. SEE ALSO strip aerotriangulation adjustment; strip aerotriangulation adjustment, analytical; analytical strip method; block aerotriangulation adjustment; bundle aerotriangulation adjustment; Bz-curve method; strip method, graphical.

aerotriangulation adjustment, block - SEE block aerotriangulation adjustment.

aerotriangulation adjustment, bundle - SEE bundle method of aerotriangulation adjustment.

aerotriangulation adjustment, individual-model - Aerotriangulation adjustment in which corrections to coordinates of ground points are determined separately for each stereoscopic model. Inconsistencies between coordinates from different models are removed in a second or third series of adjustments.

aerotriangulation adjustment, sequential - Aerotriangulation adjustment carried out by first adjusting the coordinates of points imaged in a strip of photographs so that the relative orientations are correct, then transferring the coordinate system of the strip to ground control. A form of strip aerotriangulation adjustment.

aesthenosphere - SEE asthenosphere.

A-factor - SEE azimuth factor.

affidavit - A statement or declaration reduced to writing and sworn to or affirmed to before some officer who has authority to administer an oath or affirmation.

affinity - (1) The ratio of the length of a segment AB to the length of the projected segment A'B', with the center of projection at infinity. (2) The same as the previous definition, the projection being orthogonal. Either meaning seems to be used by only a small number of photogrammetrists. The term scale is more generally used and is more appropriate.

age - (1) The interval of time between an astronomical event and the corresponding tidal phenomenon. (2) SEE age of phade inequality.

age, diurnal - SEE age of diurnal inequality.

age of diurnal inequality - The time interval between the time of greatest semi-monthly declination of the Moon, north or south, and the time of greatest effect of that declination upon the range of tide or speed of the tidal current. Also called age of diurnal tide and diurnal age. The age Tdi can be calculated from the formula $T_{di} = 0.911 (K_1° - O_1°)$ hours, in which $K_1°$ is the luni-solar diurnal constituent and $O_1°$ is the lunar diurnal constituent.

age of diurnal tide - SEE age of diurnal inequality.

age of parallax inequality - The time interval T_i between the time of the Moon's perigee and the time of greatest effect of the Moon's distance upon the range of tide or speed of tidal current. Also called parallax age. It is given by the formula $T_i = 1.837(M_2° - N_2°)$ hours, in which $M_2°$ is the principal lunar semidiurnal constituent and $N_2°$ is the large lunar elliptic semidiurnal constituent.

age of phase inequality - The time interval T_{pi} between the time of new or full moon and the time of greatest effect of these phases upon the range of tide or speed of tidal current. Also called age, age of the tide and phase age. It is given by the formula $T_{pi} = 0.984 (S_2° - M_2°)$, in which $M_2°$ is the principal lunar semi-diurnal constituent and $S_2°$ is the principal solar semidiurnal constituent.

age of the moon - The time elapsed since the preceding new moon. It is usually expressed in days.

age of the tide - SEE age of phase inequality.

agger - SEE tide, double (2).

agreement - A concurrence, between owners of adjoining lands, on the location of their common boundaries, defined in the exercise of good judgment by the parties to the stipulation as needed for interpretation by disinterested persons, including one qualified in land surveying, and as ordinarily required in a deed, subject to judicial opinion in the event of ambiguity or controversy.

agreement, parol - SEE parol agreement.

aiming circle - A theodolite of low accuracy with limited amount of rotation vertically, a low-power telescope, and containing a magnetic compass. Aiming circles used by the artillery have been graduated in mils. Used in horizontal surveys, such an instrument can provide an accuracy of about 1:500.

aiming line - SEE line of collimation; line of sight.

air almanac - An astronomical almanac prepared specifically for use in aerial navigation, containing the astronomical information needed and to the accuracy needed.

Air Almanac - That astronomical almanac published jointly by the U.S. Naval Observatory and H.M. Nautical Almanac Office and containing only such data as are used in aerial navigation, such as Greenwich hour angle and declination of selected celestial bodies, times of sunrise, sunset, moonrise, and moonset.

air base - (1) The line joining two different locations of the center of perspective of one camera, or the centers of perspective of a pair of cameras, at the moments of exposure. (2) The distance between two different locations of the center of perspective of one camera, or the centers of perspective of two cameras, at the moments of exposure, or the distance between two different locations of the same nodal point i.e., the length of the airbase defined in (1). (3) The distance, in a stereoscopic model, between adjacent centers of perspective as reconstructed in the stereoscopic plotting instrument. Also called model base or base. It is the analog, in the model, of the line or distance between adjacent locations of the camera.

Airborne Control Survey - A fourth-order survey carried out using a theodolite as the observing instrument and a target or beacon carried by a hovering helicopter as the aiming point to which horizontal and vertical angles are measured. Also called an ABC survey. The geodetic mark is vertically under the aiming point. Its distance beneath the helicopter is determined by using a graduated plumb line. The method was developed by the U.S. Geological Survey. A variant of this method, using distance-measuring instruments instead of a theodolite, was developed by the Royal Australian Survey Corps.

Airborne Control System - The equipment and procedures used in making an Airborne Control Survey. Also called the ABC system. SEE ALSO Airborne Control Survey.

Airborne Profile Recorder - A system that constantly measures and records the altitude of an aircraft by combining the outputs of a precise radar altimeter and a very sensitive barometric altimeter, and that has a means of determining, at the same time, the point from which the altitude was measured e.g., by means of a camera synchronized with the pulses from the altimeter. Also called Terrain Profile Recorder, APR and TPR. A variety in which the radar altimeter is replaced by an laser altimeter has been called a laser terrain profile recorder.

Airborne Profiling of Terrain System - A system for determining heights of terrain along a line of flight by using an airborne laser-altimeter for measuring height-variations in the terrain and a combination of accelerometers, gyroscopes, optical distance-measuring instrument (DMI) and computer for determining the position of the altimeter. Corner-cube reflectors are placed on geodetic markers. The DMI measures distance and direction (with respect to the gyroscopic system) to these reflectors; this information, together with information from accelerometers and gyroscopes, is sent to the computer which determines position from these data. The system was developed for the U.S. Geological Survey by C.S. Draper Laboratories.

aircraft altimetry - Determining altitudes from measurements made with an airborne altimeter.

air-density correction - SEE buoyancy correction.

air drag - Resistance of the air to motion of a body moving in it. Also called, if the intent is clear, drag. The term is applied, in particular, to (a) the resistance of the air to aircraft, projectiles, and artificial satellites, and (b) the resistance of air or other gas to the motion of a pendulum or other moving body used in a gravity-measuring apparatus. A good, approximate formula for the resistance F_a is $F_a = k \rho A (vt)^2 t/2$. k is a dimensionless coefficient whose value is determined by the shape of the moving body; ρ is the density of the air or gas; A is the greatest cross-sectional area presented by the body in the direction of motion; v is the velocity of the body relative to the gas; and t is the unit tangent in the direction of motion. In regions where the mean-free-path of the molecules of gas is much greater than the diameter of the body, k is nearly independent of the shape of the body and can be assumed to have a value between 2 and 2.5. Air drag gradually decreases the major diameter and eccentricity of an orbit, shortens the period of revolution, and increases the angular velocity of a satellite. It lengthens the period of a pendulum and the time needed for a moving body to pass over a specified distance. SEE ALSO drag, atmospheric.

air drag correction - SEE correction, air-drag.

airline distance - The distance, along a great circle, between points on a sphere taken as equivalent to the figure of the Earth.

air mile - SEE mile, international nautical.

airport zoning - Regulations (zoning) that govern land use, heights of buildings, and natural growth in the region surrounding an airport.

air rights - The rights granted by a fee simple, lease agreement, or other conveyance of an estate in real property to build upon, occupy, or use, in the manner and degree permitted, all or any portion of space above the ground or any other stated elevation within vertical planes (through the boundaries of the property).

airspace - (1) That part of the atmosphere which lines above a particular region - especially a nation or other political entity. (2) SEE aerospace.

air speed - The speed needed by an aircraft to achieve a certain air pressure under certain specified conditions. SEE ALSO ground speed.

air speed, calibrated - SEE air speed, rectified.

air speed, corrected - SEE air speed, rectified.

air speed, indicated - The uncorrected reading of pressure caused by the speed of the aircraft moving through the air, expressed in terms of the speed needed to achieve that pressure in a standard atmosphere at sea level.

air speed, rectified - Indicated air speed corrected for errors in installation and positioning. Also called calibrated air speed and corrected air speed.

air speed, true - The speed of an aircraft relative to the air around it.

air station - The point occupied by the perspective center of an aerial camera at the instant a photograph is taken.

air survey - SEE survey, aerial.

Airy disk - The pattern formed on a surface illuminated by radiation (usually light) passed through a small circular aperture. The pattern is caused by Fraunhofer diffraction and consists of a set of alternately dark and light rings. It is the form the image of a point-source would have if formed by an aberration-free optical system (sometimes called a diffraction-limited optical system). The intensity of the radiation at a distance r from the center of the pattern is proportional to $(2 J_1(z)/z)^2$, where z is $(2\pi r \sin\alpha)/\lambda$. α is the half-angle at the apex of the cone of rays forming the image and λ is the wavelength of the radiation. $J_1(z)$ is the first-degree Bessel function of the first kind.

Airy-Heiskanen gravity anomaly - An isostatic gravity anomaly Δg_{ah} calculated according to the Airy-Heiskanen theory of isostasy, by adding a correction for the variation in thickness of the crust below the geoid as well as above it. Also called an Airy-Heiskanen isostatic gravity anomaly. Equivalently, a topographic gravity anomaly from which is subtracted the Airy-Heiskanen gravity correction.

Airy-Heiskamen gravity correction - An isostatic gravity correction δg_{AH} giving the amount by which calculated gravity must be increased because the base of the crust extends below its average depth into the denser mantle in some regions (such as under mountainous regions) or lies considerably above it and is replaced by mantle in other regions (as under oceans). The isostatic theory of Airy and Heiskanen has the following postulates for use in computations: (a) the density of the crust is a constant 2670 kg/m3; (b) the density of the oceans is a constant 1030 kg/m3; (c) the density of the mantle is a constant 3270 kg/m3. The Airy-Heiskanen gravity correction is the difference between the acceleration caused by attraction of matter extending from the bottom of the crust down to the depth of compensation and having the same density as the crust, and the acceleration caused by the attraction of matter actually in the same region and having a density 3270 kg/m3. It is usually calculated by taking the sum of the effects of cylindrical shells, or sectors patterned according to the Hayford zones, and extending from the average depth of the base of the crust to the actual location of the base. For distances out to 167.7 km from a point, a cylinder of inner radius r_1 and outer radius r_2 and cut into n equal, vertical prisms contributes the amount δg_{AH} (force per unit mass or acceleration, depending on units used): $\delta g_{AH} = (2\pi G \Delta\rho/n)[(r_1^2 + D_1^2)^{1/2} - (r_2^2 + D_1^2)^{1/2} - (r_1^2 + D_2^2)^{1/2} + (r_2^2 + D_2^2)^{1/2}]$, in which D_1 and D_2 denote D + d and D, respectively, if the prism is not in an oceanic region and denote D and D - d', respectively, if the prism is in an oceanic region. In oceanic regions, the compensating layer begins at depth D - d' and ends at depth D; in other regions, it begins at depth D and extends to depth D + d. G is the gravitational constant, $\Delta\rho$ is the difference between the density of the crustal material and the density of the liquid or plastic material supporting the crust. Heiskanen used values of 20, 30, 40, 60, 80 and 100 km for D. The formula assumes that the reference surface is flat. This assumption is not satisfactory if the effect of prisms more than 167.7 km from the point of interest must be considered. This effect must be calculated using a spherical or ellipsoidal reference-surface.

Airy-Heiskamen gravity reduction - A gravity reduction applying to a measured or computed value of gravity the sum δ_{AH} of the free-air gravity correction δg_f, the complete topographic gravity correction δg_{tc}, and the Airy-Heiskanen gravity correction δg_{AH}. The quantity δ_{AH} is calculated using the quadrangles of the Hayford template.

Airy-Heiskamen isostatic gravity anomaly - SEE Airy-Heiskanen gravity anomaly.

Airy point - One of the two points on which a bar of standard length rests when in use, if it is desired that the neutral fiber be horizontal at the ends of the bar. These points are equidistant from the ends of the bar and are separated by $1/\sqrt{3}$ times the length of the bar. The neutral fiber is that surface which remains the same length before and after the bar is placed on the supporting edges. SEE ALSO Bessel point.

Airy's figure of the Earth - SEE Airy's spheroid 1830.

Airy's hypothesis - SEE Airy's theory (of isostasy).

Airy's map projection - An azimuthal map projection, approximately the arithmetic average of the stereographic and azimuthal equal-area map projections. Airy's map projection was intended to minimize the overall error of scale in the map. It is not, as is often supposed, a perspective map projection.

Airy spheroid 1830 - A rotational ellipsoid used in adjustments of horizontal control in Britain. For the triangulation of 1936 to 1950, the dimensions were taken to be:
 semi-major axis 20,923,713 feet (6,377,563.541 m)
 semi-minor axis 20,853,810 feet (6,356,257.053 m)
 flattening (derived) 1/299.325
Also called Airy's Figure of the Earth. The factor for converting from feet to meters was based on evidence that one foot of standard OI is 0.304 800 756 International Meter.

Airy's theory (of isostasy) - The hypothesis that the pressure of matter above some equipotential surface of reference such as the geoid is balanced by the pressure of matter of constant density beneath that surface but extending to depths (the depth of compensation) such that pressure of the mass above the reference surface is balanced by pressure of the mass beneath that surface. Also called Airy's hypothesis and the roots of mountains theory.

Aitoff's equal-area map projection - SEE Hammer-Aitoff map projection.

Aitoff's map projection - A map projection defined by the functions $x = 2 z \sin \alpha$ and $y = z \cos \alpha$, in which $z = \arccos[\cos \phi \cos(\lambda/2)]$ and $\alpha = \text{arccot}[\tan \phi \csc(\lambda/2)]$. x and y are the rectangular Cartesian coordinates of that point on the plane which corresponds to the point with longitude λ and latitude ϕ on the sphere. Also called Aitoff's planisphere. This map projection is neither conformal nor equal-area. It was applied by David Aitoff in 1889 to some maps in an atlas prepared by him. E. Hammer, in 1892, modified the map to make it equal area. For this reason, the name Aitoff's map projection is sometimes erroneously applied also to Hammer's modification of it. SEE ALSO Hammer-Aitoff map projection.

Aitoff's planisphere - SEE Aitoff's map projection.

albedo - The ratio of the total amount of radiation (power) reflected or scattered by a body to the total amount of radiation (power) incident on the body. In practice, albedo is measured and calculated only for the observable portion of a body. For example, the albedo of the Moon is calculated for the visible face only. The albedo of the Moon is 0.07, the albedo of the Earth is 0.39, the albedo of Jupiter is 0.51. The different albedos of different parts of a region can affect the accuracy of measurements made on photographs of that region or the ability to interpret the photograph correctly.

Albers conical equal-area map projection - SEE Albers map projection.

Albers map projection - (1) An equal-area map projection from the sphere to a cone which cuts the sphere at two latitudes ϕ_1 and ϕ_2 and characterized by the equations: $r^2 = r_1^2 + 2 k R_a^2 (\sin \phi_1 - \sin \phi)/n$, $\theta = n \Delta\lambda$, $n = \frac{1}{2}(\sin \phi_1 + \sin \phi_2)$ and $r_1 = (k R_a \cos \phi_1)/n$, in which r and $\Delta\theta$ are the polar coordinates of that point, in the plane, whose latitude on the sphere is ϕ and whose longitude is $\lambda_o + \Delta\lambda$; λ_o is the longitude of the central meridian. R_a is the radius of the sphere and k is the scale factor. r_1 is the radius corresponding to ϕ_1. (2) The map projection defined above, generalized to a mapping of the rotational ellipsoid onto the secant cone. To generalize, the differing radii at the two latitudes must be taken into account. Substitute N_1, the radius of curvature perpendicular to the meridian at latitude ϕ_1, for R_a in the fourth equation, and substitute for n the quantity $\frac{1}{2}(r_1^2 - r_2^2)/[R_a^2(\sin \beta_2 - \sin \beta_1)]$. In this equation, r_1 and r_2 are the radii of the standard parallels and the ß's are the corresponding latitudes on a sphere of equal area. The equations for transforming from the rotational ellipsoid to the plane were given in closed form by J. Snyder in 1982. Straight lines radiating from a point represent meridians; arcs of concentric circles represent parallels of latitude. The straight lines and arcs intersect at right angles. Along two selected arcs, called standard parallels, the scale is held exact. Along the other arcs, the scale varies with the latitude but is constant along any given arc. The meridional scale is too large between the standard parallels and too small between them. At any point on the graticule, the departure from exact scale along an arc is opposite in sign from the departure from exact scale along the straight line. The scale factor along an arc is the reciprocal of that along a straight line, so that an equal-area map-projection results. Passing through every point in the plane are two lines intersecting approximately at right angles and along which the scale is true; these are called isoperimetric curves. This map projection was devised by H.C. Albers in 1805 for mapping the sphere. In the United States of America, it is usually modified for mapping the rotational ellipsoid. Also called the Albers conical equal-area map projection.

Albrecht's formula - The formula $\tan A = \sin H \sec \Phi \cot \delta$ $(1/(1-a))$ where $a = \tan \Phi \cot \delta \cot H$, for the astronomical azimuth A of a celestial body in terms of the body's astronomical latitude Φ, declination δ and hour angle H. The function $1/(1-a)$ has been tabulated.

albumen process - SEE albumin process.

albumin process - A process for making photolithographic plates using bichromate albumin as the light-sensitive coating. Also called the albumen process.

Aldis signaling lamp - A signaling lamp of particular construction, used in some instances for observation of distant triangulation-stations at night.

alerts - A list giving, for specific satellites and for a particular location or locations, the Universal Times at which these satellites will rise or set, the satellites' greatest angles of elevation and azimuth at the observer.

algebra - (1) A generalization of arithmetic, in which symbols (usually letters of the alphabet) are used to represent numbers from a specified set (e.g., the set of integers or the set of real numbers) and are added, subtracted and multiplied just as if they were numbers. The principal objective of (this kind of) algebra is finding the values satisfying one or more given equations. (2) A generalization of the preceding definition, in which the symbols represent elements of a more or less arbitrary set and the operations on these elements are specified but are not necessarily those of arithmetic. See, e.g., algebra, Boolean. (3) The set and operations themselves (of definition 2 above). SEE ALSO array algebra; matrix; matrix algebra.

algebra, Boolean - SEE Boolean algebra.

algorism - SEE algorithm.

algorithm - A set of instructions for solving certain types of problems, principally mathematical ones. Also called an algorism. An algorithm is not a formula or an equation. For example, the set of instructions called Eratosthenes algorithm and used for finding all the primes less than a given number cannot be put into the form of a formula or equation. Most instructions (programs) for calculations on a digital computer are algorithms.

aliasing - (1) The effect of frequencies present in a set of data, but not represented in a Fourier series, upon the coefficients of terms that are present in the series and represent submultiples of those frequencies. Also called folding. (2) More generally, the selective effect on the remaining constants, in a mathematical representation of data, of removing certain constants in the representation.

alidade - That part of a surveying instrument which consists of a sighting device, with index, and accessories for reading or recording data. The alidade of a theodolite or surveyor's transit is the upper part of the instrument, comprising the telescope, the microscopes or verniers, the vertical circle, and accessories, all mounted on what is called the upper motion of the instrument. It is used in measuring a direction or angle on a graduated circle mounted on the lower motion. The alidade used in some topographic surveying consists of a straightedge carrying a telescope or other sighting device, and used in recording a direction on the sheet fastened to the plane-table. The movable arm of a sextant is an alidade. SEE ALSO eccentricity (of an alidade); peepsight alidade; pendulum alidade.

alidade, self-indexing - An alidade containing a damped pendulum which automatically brings the index mark of the vertical arc to the correct reading on the scale even if the base of the alidade is not quite level.

alidade, self-reducing - An alidade which provides horizontal distance and difference of elevation when the stadia interval is multiplied by the constant displayed by the alidade.

alidade, telescopic - An alidade consisting of a telescope mounted on a straightedge and used with a plane-table in topographic surveying. Also called a telescope alidade. SEE ALSO alidade.

alienation - A legal term applied to a transfer of title of a legal conveyance of property to another person.

alignment - SEE alinement.

alinement - (1) Placing points along a straight line or in a common vertical plane. (2) Locating points with respect to a straight line or to a set of straight lines. The term should be used, in surveying, only for operations associated with straight lines. It is also called lineal surveying when the distance is appreciable. (3) The ground plan showing the horizontal direction and components of the center line of a highway or railroad. (4) The horizontal position of a railroad track, as described by tangents and curves. Alinement is distinguished from profile, which shows the vertical element. Alinement is also spelled alignment.

alinement, optical - (1) The alinement of objects by the use of optical devices and instruments such as telescopes, optical micrometers, interferometers, laser beams, etc. (2) Placing the optical axes of a telescope in proper relationship to the index marks on the horizontal and vertical circles, or the right ascension and declination circles. Do not confuse with collimation, which is bringing the optical elements of the telescope into proper relationship with each other. SEE ALSO tooling, optical.

alinement chart - SEE nomograph.

alinement correction - A correction applied to the measured length of a line to allow for the tape's not being held exactly in a vertical plane containing the line. Although the two vertical planes which contain the directions of the line at its two ends do not coincide, they practically do so for all surveying over short distances. Any method of aligning a base between its end points will serve in determining an alinement correction.

alinement error - The angle between the actual line of sight of a telescope and the direction in which the horizontal or vertical circles or on an auxiliary telescope indicate that the line of sight should lie.

aliquot (adj.) - Dividing without a remainder; submultiple. The word usually appears as aliquot part, i.e., a number which is a submultiple of another. SEE ALSO description by aliquot parts.

Allen variance - The sum of the squares of the first differences in a sequence of variations in relative frequencies, divided by 2M, M being the number of first differences. The variation in frequency is usually the difference between the frequency being investigated and the standard frequency, divided by the nominal standard frequency and multiplied by 1012.

alley - (1) A narrow, dedicated street. (2) A way through the middle of a block and giving access to the rear of properties. This is the more common usage.

allodium - Land owned absolutely, without obligation to one with superior rights.

alluvion - (1) Formation of land, by imperceptible degrees, on the banks of a body of water, either by accumulation of material or by recession of the water. (2) Formation of land from the bed of a river or body of water by the gradual, natural accumulation of matter on the bed or by the gradual, natural recession of the water. Also called accretion. It is distinguished from batture by the latter's occurring beneath the water's surface and not forming land. (3) The land formed by the gradual, natural accumulation of matter on the bed of a river or by the gradual, natural recession of the water. (4) The material that is deposited along the shore of a river or the sea by gradual and imperceptible action of the water. There is some confusion, in legal and surveying terminology, between the terms alluvion, alluvium and batture. The first term is used by some as a synonym for accretion; it is used by others as a synonym for alluvium. There is also some confusion caused by the reference to land in the third definition but to material in the fourth. It is likely that material as land is meant in both. (5) SEE accretion.

alluvium - The solid material (sand, silt, gravel, or other material) deposited by running water. It may accumulate to form land (the process, and sometimes the result, being referred to as alluvion) or it may remain below the surface of the water to raise the level of the bed (the result being referred to as batture).

almanac - (1) SEE almanac, astronomical. (2) An astronomical almanac containing data useful to a navigator. SEE ALSO Air Almanac.

almanac, astronomical - An annual publication containing, for each day or other suitable fraction of the year, information on the current locations of celestial bodies, together with the times and circumstances of various astronomical events such as sunrise and sunset, eclipses, and transits of the planets - particularly such information as is useful for navigation. Also referred to simply as an almanac. It is prepared and published several years in advance of the year to which it pertains. In 1980, the almanacs The American Ephemeris and Nautical Almanac, prepared by the U.S. Naval Observatory, and The Astronomical Ephemeris, prepared by H.M. Nautical Almanac Office, were renamed The Astronomical Almanac and published as a single almanac prepared jointly by the two organizations.

almanac, nautical - An astronomical almanac prepared for use on ocean-going vessels, containing such astronomical information as is needed for navigation on the high seas and to the degree of accuracy needed for such navigation.

almucantar - Any small circle, on the celestial sphere, parallel with the horizon. Also called a parallel of altitude and circle of equal altitude.

almucantar-crossing method (of determining azimuth) - SEE equal-altitude method (of determining azimuth).

alp - A high, rocky mountain frequently covered with snow.

altazimuth - SEE alt-azimuth; instrument, alt-azimuth.

alt-azimuth (adj.) - Rotatable in altitude and azimuth. Sometimes written altazimuth. Most geodetic transits and theodolites are on alt-azimuth mountings. Satellite- tracking equipment is usually of the alt-azimuth type. SEE ALSO instrument, alt-azimuth; mounting, alt-azimuth.

altimeter - An instrument that indicates its distance above a particular surface. This distance is usually referred to as the altitude of the instrument. There are two common types of altimeters: the barometric altimeter, which indicates distance above a surface of constant atmospheric pressure, and the radar or laser altimeter, which indicates distance above the physical surface of the Earth or other celestial body. Barometers with no scale in meters and whose pressure readings must later be converted to altitudes are not altimeters. Neither are leveling instruments nor leveling systems. SEE ALSO laser altimeter; radar altimeter; radio altimeter; satellite altimeter; surveying altimeter.

altimeter, absolute - An airborne altimeter which shows, on an indicator, the distance from the altimeter to the ground below. Also called a terrain-clearance indicator. The instrument usually uses radio or sonic waves, or measures capacitance.

altimeter, aneroid - An altimeter incorporating an aneroid barometer.

altimeter, barometric - An aneroid barometer whose scale is graduated in feet, yards or meters as well as (or instead of) in units of atmospheric pressure. Also called pressure altimeter, sensitive altimeter or by surveyors, altimeter. It measures the air pressure at the altitude of the instrument but indicates its distance above a previously selected surface of constant pressure because its scale is calibrated to show altitude as a function of pressure. It is calibrated by adjusting the instrument to read the correct altitude at a point of known elevation. Mean sea level or the constant-pressure surface at the elevation of the airport is commonly used, in practice, for calibrating airborne barometric-altimeters. Altitudes are then with respect to the surface of constant pressure through the point of calibration. The relationship between altitude and pressure is obtained, in preparing the scale, from a standard atmosphere. Therefore the indicated altitude will frequently differ from the actual altitude by several meters.

altimeter, precise - A sensitive aneroid barometer used as an altimeter and accurate to within a meter when the two-base method of altimetry is carefully applied. Also called a precision altimeter.

altimeter, satellite-borne - SEE satellite altimeter.

altimeter, sensitive - SEE altimeter, barometric.

altimeter correction - One of the corrections which must be made to the readings of a barometric altimeter to obtain true altitude. Among these corrections are a correction for difference of local temperature from ground temperature and a correction for error in the zero-point of the altimeter.

altimeter setting - That value of atmospheric pressure to which the scale of a barometric altimeter is set.

altimetry - (1) Determining the distance, above the physical surface of the Earth, of a level of specified (reference) air pressure. The usual instruments are barometric altimeters or radar altimeters. A radar altimeter gives a height above the surface of the ground; the height must then be converted to altitude. An aneroid barometer indicating only air pressure may be used. The indicated pressure is then converted to altitude by subsequent calculation. SEE ALSO hypsometry. Altimetry differs from hypsometry in that the latter deals specifically with determining elevations (i.e., distances above or below the geoid). It differs from leveling in that leveling is hypsometry done with leveling instruments. (2) Determining elevations or differences of elevation (distance above the geoid) by using an altimeter. Altimeters do not measure elevations or differences of elevation directly. The accuracy of the obtained differences therefore depends not only on the accuracy of the altimeter but also on the accuracy of the assumptions made in converting to differences of elevation. SEE altimetry, barometric; and radar altimetry. (3) Determining distances above any surface of reference by any means, including leveling. For example, determining heights by radar altimeter is often referred to as altimetry. SEE ALSO fly-by method; leap-frog method; radar altimetery; satellite altimetry; single-base method.

altimetry, barometric - Determining differences of elevation with respect to a specified surface from differences of atmospheric pressure measured with a barometer or from differences of altitude determined using a barometric altimeter. By applying certain corrections and using what is sometimes called the barometric formula, a difference of atmospheric pressure at two places is transformed into a difference of altitude at those places. If the elevation of one station above a reference surface is known, the approximate elevations of other stations connected with it by barometric altimetry can be calculated. By using barometers of special design, and by including several stations of known elevation in a sequence of occupied stations, the accuracy of the elevations determined for the new stations is increased. Corrections are applied for temperature, latitude, index of barometer, misclosure of circuit, diurnal variation in atmospheric pressure, etc.

altimetry, multiple-base - Altimetry in which the variations of pressure with time are measured and recorded at fixed sites within the region being surveyed.

altimetry, thermometric - Determining elevations, or differences of elevation, by measuring the boiling point of water. The temperature at which water boils at any point on the Earth depends on the atmospheric pressure at that point, and the atmospheric pressure depends on the elevation of the point. Factors other than elevation also affect atmospheric pressure, and factors other than atmospheric pressure affect the boiling point. Hence, thermometric altimetry is less precise than barometric altimetry.

altitude - (1) The distance, along some suitable line, from a surface of reference to a point above that surface. The surface may be the physical surface of the earth, a surface of known, constant atmospheric pressure, etc. (2) The distance, along some suitable line, between two surfaces of constant atmospheric pressure, one of these surfaces being taken as the reference surface. The distance between two such surfaces (isobaric surfaces) is not constant in space or time, and one surface may be much smaller in area than the other (e.g., compare the area of the atmosphere at sea level with the area of a constant-pressure surface over land and sea). This kind of altitude, often called barometric altitude, therefore depends on the assumption or theory used in locating the surfaces as well as on the measurements themselves. A barometric altitude with an error of a meter or less is considered an accurate measurement of its kind. (3) The difference in pressure between two surfaces of constant atmospheric pressure, the lower surface being taken as referent. Also called barometric altitude. (4) The reading, in units of length, of a pressure altimeter. This definition should probably be changed to read in units of length because modern pressure altimeters usually indicate in meters rather than feet. (5) SEE elevation. (6) SEE height. (7) SEE elevation, angular. (8) The reading, in feet, of a barometer. Although this definition has appeared in glossaries, it is probably invalid; barometers are graduated in units of pressure.

altitude, absolute - Altitude above the actual physical surface, either land or water, of a planet or natural satellite. It is not the same as distance above mean sea level, which is a distance above a hypothetical surface.

altitude, angular - SEE elevation, angular.

altitude, apparent - (1) The observed angle from a horizontal plane to an object, corrected for instrumental errors, personal errors and errors in the position of the surface from which angles are measured, but not corrected for refraction, parallax, or semi-diameter of the object. Also called rectified altitude. (2) The observed angle from a horizontal plane to an object, corrected for instrumental errors and for semi-diameter of the object.

altitude, barometric - An altitude or elevation determined using a barometer or barometric altimeter.

altitude, circum-meridional - The angular elevation of a celestial object near the celestial meridian and observed to determine latitude.

altitude, ex-meridional - The angular elevation of a celestial body near the celestial meridian but not on it. A correction is applied to an ex-meridional altitude to obtain the meridional altitude.

altitude, geometric - SEE elevation, geometric.

altitude, geopotential - SEE geopotential elevation.

altitude, meridional - The arc of the celestial meridian from the horizon to a celestial body at culmination. Also called meridian altitude.

altitude, negative - Angular distance below the horizon.

altitude, observed - (1) The angular elevation of a celestial object when corrected for error in the index of the vertical scale of the instrument. (2) The angular elevation of an object as measured, without corrections of any kind. Also called sextant altitude when measured with a sextant. (3) SEE altitude, true.

altitude, orbital - The average altitude of a satellite above the surface of the primary body.

altitude, rectified - SEE altitude, apparent.

altitude, solar - The angular elevation of the Sun above the horizon.

altitude, true - (1) The angular elevation of a celestial body corrected for refraction and parallax. Also called, in American usage, observed altitude. (2) Distance above mean sea level. SEE ALSO almucantar; parallel of latitude; radar altitude.

altitude azimuth - An azimuth determined by solving the navigational triangle with altitude, declination and latitude given.

altitude circle - SEE parallel of latitude.

altitude datum - The arbitrary level from which vertical displacement is measured.

altitude difference - The difference between two angular elevations. Also called altitude intercept.

altitude hole - The blank region extending from the starting point of the sweep (on a radar's screen) to the point at which the first return from the ground shows.

altitude intercept - SEE altitude difference.

altitude tint - SEE tint, hyposometic.

altitudes, simultaneous - The angular elevations of two or more celestial bodies measured at the same time.

ambient (adj.) - Referring to a characteristic of a surrounding medium e.g., ambient pressure, ambient temperature.

ambiguity, latent - An uncertainty which does not appear upon the face of an instrument but arises from evidence aliunde - that is, one which is not involved in the words themselves but arises from outside matters. Thus, a conveyance to John Smith, living on Fifth Avenue, when it appears that there are two or more John Smiths so living, contains a latent ambiguity. The opposite of patent ambiguity.

ambiguity, patent - An ambiguity which appears on the face of the instrument and arises from the defective, obscure, or insensible language used.

ambit - A boundary line thought of as inclosing and going around a place.

ambitus - A space at least 2.5 feet wide between neighboring buildings, left for the convenience of going between them.

American foot - SEE foot.

American Meridian - SEE Washington Meridian.

American polyconic map projection - SEE map projection, polyconic.

Amici prism - A roof prism which deviates rays of light through 90° and inverts the image. It may be considered an ordinary prism having right-angled isosceles triangles as bases, in which the face forming the hypotenuse is replaced by two plane reflecting surfaces placed at right angles to each other and forming the roof of the prism.

amphidrome - SEE point, amphidromic (2).

amplifier - A device in which an input of varying voltage, current, or power is amplified. Any electronic distance-measuring instrument, angle-measuring instrument or frequency-measuring instrument depends on an amplifier to increase the strength of faint, returned signals to the point that the signals can be measured accurately.

amplifier, parametric - An amplifier which works by adding energy to the input at a frequency in resonance with the input, thus pumping in energy in the same way that the amplitude at which a pendulum oscillates can be kept constant or increased by giving a small push to the pendulum at the end of each swing.

amplitude - (1) The difference between the greatest value of a quantity and its average value. Distinguished from range, which is the difference between the greatest value and the least value. (2) The angle, at an observer, measured in a horizontal plane from east or west to a celestial body. The term is used in this sense only for bodies whose centers are on the celestial horizon, and is prefixed with E if the body is rising and W if it is setting. It is given the prefix N or S to agree with the body's declination. (3) Angular distance north or south of the prime vertical.

amplitude, magnetic - Amplitude relative to magnetic east or west.

amplitude, tidal - SEE amplitude of the tide.

amplitude, true - Amplitude relative to astronomical east or astronomical west.

amplitude of oscillation - The length of the arc passed over by a pendulum in moving from a new position of greatest displacement to the other.

amplitude of the tide - One half of the difference in elevation between consecutive high water and low water. Also referred to as tidal amplitude or tide amplitude.

amplitude of vibration - The length of the arc passed over by a pendulum in moving from its average position to the position of greatest displacement.

amplitude scintillation - That part of scintillation affecting the brightness of the image.

Amsterdams Piel, Normaal - SEE Normaal Amsterdams Piel.

anabranch - SEE ana-branch.

ana-branch - A branch (of a river) which, after diverging, rejoins the river farther down. The term is used, in particular, for a secondary channel, in a broad plain of braided channels, that maintains a semi-permanent course. Examples occur in eastern Australia, the jungles of South America, in the southern Sudan, and elsewhere. See, e.g., billabong.

anaglyph - (1) A composite picture made by superposing one picture (of a stereoscopic pair) in one color on the other picture which is in a complementary color. The usual colors are red and blue-green. When viewed through spectacles in which one lens is a red filter and the other is a blue-green filter, the anaglyph provides a three-dimensional effect to the viewer. The same kind of effect can be produced by projecting the individual pictures onto a white surface through filters of the proper (e.g., red and blue-green) colors. Anaglyphs of this kind were used in photogrammetric plotters. (2) A pair of stereoscopic pictures, one picture of the pair being in one color and the other picture in the complementary color.

analemma - A figure-eight shaped diagram showing the declination of the Sun throughout the year and also the difference between true solar and mean solar time (the equation of time). It may be shown on a plane or curved surface, such as that of an analemmatic sundial, but is most commonly shown drawn across the torrid zone on a terrestrial globe.

anallatic (adj.) - A variant spelling of anallactic.

anallactic (adj.) - Having no divergence. Also spelled anallatic. The term is applied, in optics, to the property of rendering rays from different points on an object parallel. SEE IN PARTICULAR, point, parallactic.

analog datum - SEE datum, analog.

analysis, harmonic - (1) The process of expressing a given function as a Fourier series or Fourier transform; the process of finding the coefficients in that Fourier series or evaluating the integral. Also called Fourier analysis or spectral analysis. The function is, in practice, usually given as a table of paired numbers, one of which is the time or time interval. (2) The expansion of a given function into a series of harmonic functions. (3) (of the tide) A Fourier series representing the variation in height of the tide at a particular place. The series is usually a multiple Fourier-series having the argument $(n_1T' + n_2s' + n_3h' + n_4p' + n_5N' + n_6p_1')t$, in which n_1, n_2, n_3, n_4, n_5 and n_6 are integers and the symbols T', s', h', p', N' and p_1' denote hourly rates of change of hour angle of the mean Sun, mean longitude of the Moon, mean longitude of the Sun, mean longitude of lunar perigee, mean longitude of the ascending node of the Moon's orbit and mean longitude of perihelion, respectively; t is mean solar time. Particular combinations of the quantities T', s', h', p', N', and p_1' have been assigned their own symbols. The most important combinations are given below, together with their Doodson numbers. They are arranged in descending order of amplitude of the corresponding term.

Symbol	Doodson Number	Combination
M2	255.555	$2T'-2s'+2h'$
S2	273.555	$2T'$
O1	145.355	$T'-2s'+h'$
K1	165.555	$T'+h'$
P1	163.555	$T'-h'$
N2	245.655	$2T'-3s'+2h'+p'$
K2	275.555	$2T'+2h'$
Q1	135.655	$T''-3s'+h'+p'$
V2	247.455	$2T'-3s'+4h'-p'$
M1	155.655	$T'-s'+h'+p'$
J1	175.455	$T'+s'+h'-p'$
L2	265.455	$2T'-s'+2h'-p'$

The symbols in the first column were assigned by George Darwin in his work on the tides and are still in common use to designate the combinations given. (4) The process of finding a Fourier series to represent the variation in the height of the tide at a particular place.

analysis, spectral - SEE analysis, harmonic (1).

analysis, sequential - A rule or procedure by which the number of samples or the action taken on the basis of the samples depends on the evaluation of each set of samples. SEE ALSO Kalman filtering.

analysis, tidal harmonic - SEE analysis, harmonic (3).

analysis of variance - The technique or procedure developed by R. Fisher for separating the total variance of a sample into a sum of variances traceable to specific sources.

anastigmat - SEE lens, anastigmatic.

Anblock method - A method of block aerotriangulation adjustment applied only to horizontal control (planimetry). Elevations are supposed to be satisfactorily accurate or not important. Developed about 1965 by van der Hout.

anchorage chart - A chart showing prescribed or recommended places where craft may be anchored.

Anderson's scale-point method - SEE scale-point method.

anemometer - An instrument for measuring the speed of the wind. There are two common types: the cup anemometer, which uses freely-rotating cups spun by the wind, and that which uses the air-pressure produced by the wind in a restricted passage.

angle - (1) A geometric figure formed by (a) a pair of intersecting, straight half-lines terminated at the point (called the vertex) of intersection; (b) a pair of intersecting half-planes terminated at the line of intersection (called the axis); (c) the conical surface generated by moving a straight half-line about a fixed point (called the apex, the line returning upon itself once and only once; (d) a pair of singly-intersecting, curved segments terminating at the point of intersection. The figure defined by (a) lines in a plane and is called a plane angle. The figure defined by (b) is called a dihedral angle. The figure defined by (c) is called a solid angle or, if the surfaces generated by the straight half-line are planes, a polyhedral angle. That defined by (d) is called a spherical angle if formed by segments of great circles. (2) A number expressing the size or measure of the rate at which two intersecting half-lines, two intersecting half-planes, the sides of a conical surface, or two intersecting curved segments diverge. The number is usually dimensionless and in any case is distinct from the geometric figure which it describes. To avoid confusion, the distinction will be made explicit when this makes an explanation or definition clearer. [Figurate angles will be identified by appending (f), numerical angles by appending (n)]. SEE ALSO plane angle; angle, dihedral; angle, solid; angle, spherical for definitions of these angles (n) in terms of the method used to obtain them, and for definitions of the units in which the angles are measured. The angle (n) between two intersecting curved segments is the angle (n) between the tangents to these curves at the point of intersection, and therefore does not require a separate definition. Each particular science and technology has also its own symbolism and nomenclature for angles that occur frequently. (a) Astronomic angles (f) are classified according to where the vertex or apex of the angle is located. SEE angle, geocentric; angle, heliocentric; angle, topocentric. There are two kinds of angle (n): those measured in dimensionless units such as radians or degrees, and those measured in units of time [hours (h), minutes (m), and seconds (s) of time]. There are 24 h in 360°, 15 m in 60', and 15 s in 60". (b) Geodetic angles (f) are classified according to whether they lie in a horizontal plane (horizontal angles) or a vertical plane (vertical angles). Geodetic angles (n) are classified in the same way as astronomical angles (n), according as they are in dimension-less units or in units of time. (c) Photogrammetric practice generally follows that of geodesy. However, certain angles that occur frequently in dealing with aerial photographs are given special names. SEE tilt (pitch); roll; yaw (swing). SEE ALSO angle of reflection; angle of refraction; azimuth angle; break angle; Brewster's angle; Cardan angle; conversion angle; crab angle; deflection angle; depression, angular; direction angle; distance angle; elevation, angular; exterior angle; grivation; hour angle; interior angle;

look angle; plane angle; position angle; screen angle; traverse angle; zenith angle.

angle, acute - A plane angle both of whose sides can be included within the sides of a right angle. Its size is less than 90° (100g or π/2 radians).

angle, adjusted - An adjusted value of an angle. An adjusted angle may be derived either from an observed angle or from a concluded angle.

angle, ascending vertical - SEE elevation, angular.

angle, break - SEE break angle.

angle, centesimal - An angle expressed in centesimal units rather than decimal or sexagesimal units. An example of such a unit is the gon, which is 1/100 of a right angle. SEE ALSO gon and grad.

angle, central - (1) The angle, at the center of a circular arc, included between the radii which pass through the beginning and end of the arc. This angle is equal in size to the change in direction of the tangents at the beginning and end points. In alinement surveys, it is commonly called the delta angle and denoted by Δ. (2) The angle, in curved lines containing segments of compound curves, or spirals and circular curves, between the radii to the beginning and end points or between the tangents at these points. The central angle of a spiral is called theta (θ) or delta (Δ).

angle, clockwise - (1) An angle (f) which is described in the direction in which the minute hand of a normal clock is seen to move. (2) SEE angle to right.

angle, complementary - (1) The angle formed by erecting a half-line perpendicular, at the vertex, to the initial side of the given angle and then removing the initial side. Its size is 90° minus the size of the original angle. (2) The size of the angle defined in (1). (3) The same as (1) but applicable only when the given angle is an acute angle. Also called the complement (of an angle).

angle, concluded - An interior angle between adjacent sides of a closed figure, which is obtained by subtracting the sum of all the other interior angles of the figure from the theoretical value of the sum of all the interior angles. The concluded angle is most frequently met with in triangulation, where it is obtained by subtracting the sum of two known angles of a triangle from 180° plus the spherical excess of the triangle.

angle, conjugate - SEE angle, explementary.

angle, counterclockwise - An angle described in a direction opposite to that described by the hand of a clock.

angle, critical - The smallest angle at which a wave passing from a substance of higher refractive index to one of lower refractive index can be incident on the surface between the substances and be totally reflected at that surface. At the critical angle, some of the waves may propagate along the surface. At angles less than critical, some or all of the electromagnetic waves propagate into the surface. The angle is expressed by the formula sin(critical angle) = N'/N, in which N is the higher refractive index and N' is the lower.

angle, crossing - The angle at which two lines of position, course lines, etc., cross.

angle, delta - SEE angle, central.

angle, descending vertical - SEE depression, angular.

angle, dihedral - (1) The figure formed by two intersecting half-planes and the line (called the axis) of intersection on which they terminate. The half-planes are called the sides of the dihedral angle. (2) The number expressing the size of the plane angle formed by intersecting the two sides with a plane perpendicular to the axis. The dihedral angle is termed acute, right or obtuse according as the size of the corresponding plane angle is less than, equal to or greater than 90°, respectively.

angle, direct - An angle measured directly between two lines, as distinguished from a deflection angle such as is encountered in traverses. SEE angle to left; angle to right.

angle, double-zenith - The angle obtained by pointing the telescope of a transit at an object, reading the angle on the vertical circle, then reversing the instrument to put the vertical circle on the other side of the observer, redirecting the telescope at the object, and reading the new angle on the vertical circle. The difference of the two readings is the double-zenith angle and is twice the zenith angle. Double-zenith angles are used in trigonometric leveling and in astronomy because they are little affected by inclination of the vertical axis of the instrument used. Only certain types of transit, however, allow double-zenith angles to be measured.

angle, eccentric - The angle ε such that a pair of coordinates x and y can be found, on an ellipse with major diameter 2a and minor diameter 2b, satisfying the equations $\cos\varepsilon = x/a$ and $\sin\varepsilon = y/b$.

angle, explementary - (1) The figure formed from a given angle by using the initial side of the given angle as the terminal side of a new angle and the terminal side of the given angle as the original side of the new angle. (2) The size of the figure (angle) formed as described in (1): 360°, 2π radians, or 400 gon minus the size of the given angle. (3) That angle, of a pair of angles, which results on subtracting the other angle from 360°. Also called the conjugate angle, conjugate (of an angle) or explement (of an angle).

angle, external - (1) One of the angles lying outside a pair of parallel lines intersected by a third line. Also called an exterior angle. (2) SEE exterior angle (1).

angle, geocentric - An angle whose vertex is at a specified center of the Earth. Centers commonly specified are the center of mass, the center of an ellipsoid representing the geoid, and the center of an ellipsoidal solid representing the Earth.

angle, heliocentric - An angle whose vertex is at the center of the Sun. The Sun's center of mass is usually specified or understood.

angle, horizontal - An angle between two directed lines in a horizontal plane. Equivalently, the dihedral angle between two planes intersecting in a vertical line. The angles measured on the horizontal circle of a theodolite are horizontal angles if the standing axis of the theodolite is vertical and if the horizontal circle is perpendicular to the standing axis.

angle, included - The angle, inside a polygon, between adjacent sides.

angle, interlocking - The angle between the optical axes of any two rigidly connected cameras. Also, in particular, the angle between the optical axes of the vertical and oblique cameras of a multi-camera system, or the angle between the planes of the vertical and oblique photographs.

angle, internal - SEE interior angle.

angle, locking - The complement of the interlocking angle. The angular depression of the oblique photograph when the tilt of the vertical photograph is zero.

angle, mapping - SEE gisement.

angle, measured - An angle as measured, without any correction for local conditions. Also called an observed angle. SEE ALSO angle, observed.

angle, observed - (1) SEE angle, measured. (2) A measured angle which may or may not have been corrected for local conditions only, at the point of observations.

angle, obtuse - (1) An angle (f) whose measure is greater than one right angle (90°, π/2 radians, or 100 gons). (2) An angle whose measure is greater than one right angle but less than two right angles (180°, π radians, or 200 gons).

angle, parallactic - The angle, on the celestial sphere, between a great arc from a heavenly body to the pole and a great arc from the celestial body to the observer's zenith. It should not be confused with parallax.

angle, planar - SEE plane angle.

angle, polyhedral - The solid angle bounded by a set of planes that intersect at a common point in such a way that each plane is bounded by the lines of intersection and intersects exactly to other planes. The solid angles at the apices of a pyramid are polyhedral angles.

angle, pseudo-Cardan - One of a set of three angles (θ_1, θ_2, θ_3) specifying the orientation of the axes (x,y,z) of one rectangular, Cartesian coordinate system with respect to the corresponding axes (X,Y,Z) of another. Rotations generating the angles are about each of the x,y,z axes, successively, in their original positions. In general, the amounts of rotation required to obtain the needed angles are not simply described. If, however, the axes of the two systems are close together, the pseudo-Cardan angles are approximately the same as the Cardan angles. Pseudo-Cardan angles have been used in photogrammetry.

angle, reciprocal vertical - One of a pair of vertical angles of the same line measured close together in time and from opposite ends of the line, in trigonometric leveling, to reduce the effects of refraction. Reciprocal observations are most effective when they are made simultaneously.

angle, right - A plane angle such that exactly four rotations in the plane, initial side to former position of terminal side, about the vertex bring the angle back to its original position. Its measure is therefore exactly 1/4 that of a full circle - π/2 radians, 90°, or 100 gons.

angle, skew - (1) The acute angle between a line perpendicular to one center-line and to another center line. (2) The amount of angle by which that two intersecting lines deviate from a right angle.

angle, solar - The angular elevation of the Sun.

angle, solid - (1) The figure (surface) generated by moving a straight half-line about a fixed point at which it terminates. The line is returned to coincidence with its initial position once and only once, and the generated surface does not touch or intersect itself. An alternative definition is: the locus of all straight half-lines joining a fixed point to a given, closed curve that does not intersect itself. Also called a cone. If the straight line generates a set of planes, the solid angle is called a polyhedral angle. The fixed point is called the apex of the solid angle. A straight line lying entirely on the cone is called an element of the cone. (2) A number (ratio) indicating the rate at which the cross-sectional area of the cone increases with distance from the apex. This is called the size or magnitude of the solid angle. It is determined by drawing a sphere of arbitrary radius about the apex and dividing the area of that portion contained within the cone by the area of the entire sphere. In geodesy and in some other sciences, the above ratio is multiplied by 4π (or, equivalently, the included portion is divided by the square of the radius). The resulting value is said to be in steradians. SEE ALSO degree, square.

angle, spherical - (1) An angle (f) formed by the intersection of segments of two great circles on a sphere. The two segments in the vicinity of the point of intersection are called the sides of the angle. The point itself is called the vertex of the angle. (2) A number expressing the rate at which the two sides of a spherical angle (f) diverge in the neighborhood of the vertex, and defined as the plane angle between the tangents to the two segments at the vertex.

angle, spheroidal - (1) An angle (figure) formed by the intersection of segments of two curves on a spheroid. The two segments are called the sides of the spheroidal angle and the point of intersection is called the vertex. (2) A number indicating the rate at which the sides of the spheroidal angle (f) diverge in the neighborhood of the vertex, and defined to be the plane angle between the tangents to the two segments at the vertex.

angle, straight - An angle equal or equivalent to two right angles.

angle, supplementary - (1) The figure formed by extending the initial side of an angle beyond the vertex and then removing the initial side. (2) That number which is obtained by subtracting the given angle from the numerical equivalent of two right-angles, e.g., π radians, 180°, 200g or whatever units are used for the given angle.

angle, theta - SEE theta angle.

angle, topocentric - A plane angle with vertex at the surface of the Earth (or, more usually, at a point on an ellipsoid representing the surface of the Earth).

angle, trihedral - The figure formed by three lines intersecting at a common point, and those portions of the planes included between these lines.

angle, vectorial - (1) A vector which has a length is equal to the magnitude of a given plane angle and which is directed so that it is perpendicular to the plane of that angle and in a sense so as to form a right-handed system with the sides of the angle. (2) The angle between the fixed line to which a direction is referred and a radius vector.

angle, vertical - An angle in a vertical plane. Also called a gradient (Brit.). In surveying, one of the sides of a vertical angle is usually either (a) a horizontal line in the vertical plane or (b) a vertical line in that plane. In case (a), the angle is then called, also, the angular elevation or angular depression or altitude. In case (b), it is also called the zenith angle. The vertical angle between two lines neither of which is horizontal or vertical is usually obtained a combination of two vertical angles as defined previously.

angle correction - SEE correction, arc-to chord.

angle equation - A condition equation expressing the relationship between the sum of the measured angles of a closed figure and the theoretical value of that sum, the unknowns being the corrections to the measured angles used in the adjustment. Sometimes called a triangle equation. An angle equation is used to make the sum of the three observed angles of a triangle, with corrections applied, equal to 180° plus the spherical excess of the triangle.

angle instrument - A theodolite in which the horizontal circle is fixed in position while the angle between two points is being measured, but is rotated with the telescope when preparing to measure the angle between a different pair of points. SEE ALSO theodolite, repeating; angle measuring instrument.

angle measurement, Schreiber's method of - SEE Schreiber's method.

angle measurement in all combinations - SEE Schreiber's method.

angle measuring equipment - An angle-measuring instrument together with such other devices as are used when measuring angles. Also called an angulation instrument. It could, for example, denote the instrument, the support for the instrument, the centering devices, recording instrument and so on.

angle measuring instrument - Any instrument used for measuring angles. Sometimes called an angulation instrument. Examples are the goniometer, the theodolite, the transit, and electronic devices which measure angles by interference of radio waves. Although photogrammetric cameras and plotters are used for determining angles, they do not indicate the angles directly and hence are not classified as angle-measuring instruments.

angle method (of traverse adjustment) - The method of adjusting a survey traverse by using angles between successive courses as the measured quantities.

angle method (of triangulation adjustment) - A method of adjusting a triangulation network, using adjustment by conditions, in which the measurements are of angles and the unknowns are corrections to the measured angles. The angle method is similar to the direction method, in which the corrections are to measured directions (azimuths). The angle method should be used where a chain of single triangles is to be adjusted. For an extensive network with overlapping triangles, the direction method of adjustment is preferred.

angle of aberration - SEE aberration.

angle of commutation - The difference between the celestial longitudes of the Sun and a planet, as observed from the Earth.

angle of coverage - The angle at the apex of the cone of rays passing through the front nodal point of a lens system. SEE ALSO angle of view; field of view. This term is commonly used in photogrammetry when lens systems are classified according to the size of the area they image. The following categories are generally recognized: narrow-angle - an angle between 0° and 60°; normal-angle - an angle between 60° and 75°; wide-angle - an angle between 75° and 100°; super-wide-angle - an angle greater than 105°.

angle of coverage, effective - The angle at the apex of the largest cone of rays usable throughout the photogrammetric process. Also called the working value of the angle (of coverage). Its value is necessarily less than that of the full angle of coverage.

angle of crab - The smallest angle between the line of flight and that side of the photograph most nearly in the direction of the line of flight. Also called crab and crab angle.

angle of current - In stream gaging, the angular difference between 90° and the angle made by the current with a measuring section.

angle of cut - The angle at which two lines of position intersect. It is preferably a right angle.

angle of depression - SEE depression, angular.

angle of distortion - The angle by which a ray passing from a point P through the projection center is deviated from a straight line.

angle of elevation - SEE elevation, angular. SEE ALSO angle, vertical.

angle of field - (1) The widest angle at which light entering a lens system a system will produce, at its focal plane, a circle of good definition. (2) The angle subtended by lines through the center of a lens system and the diameter of the largest area on the image having a specified definition. Also called angular field.

angle of incidence - The angle between an electromagnetic ray incident at a point on a surface and the normal to the surface at that point, measured from the normal to the ray. Also called the incidence angle.

angle of inclination - An angular depression or angular elevation.

angle of reflection - The angle between an electromagnetic ray leaving a point on a surface and the normal to the surface at that point, measured from the normal to the ray. Also called the reflection angle.

angle of refraction - (1) The angle between an electromagnetic ray passing from one substance into another and the normal to the interface at the point of emergence, measured from the normal (in the substance entered) to the ray. Also called a refraction angle. (2) That portion of an angle measured from the zenith which is caused by atmospheric refraction. Also called a refraction angle. Alternatively, the difference between the angle, measured from the zenith, at which a source of radiation would be observed in the absence of the atmosphere, and the angle, measured from the zenith, at which it actually is observed. (3) The difference between the direction to a source of radiation and the direction from which the radiation arrives at the observer. Also called the refraction angle.

angle of tilt - SEE tilt.

angle of view - (1) The angle, at the rear nodal point of an optical system (or at the rear perspective center of such a system), between the two rays to two points determining a characteristic dimension of the image formed by that optical system. Also called covering power or field of view, although this latter term is more often used for the angle, at the front nodal point or perspective center, between the rays from two points in object-space that determine a characteristic dimension of the image. Characteristic dimensions are, for example, the length of side or length of diagonal of a square image, or the length of longer or shorter side of a rectangular image. (2) In photogrammetry, twice the angle whose tangent is half the length of the diagonal of the image, divided by the calibrated focal length.

angle of yaw - (1) The angle between the direction in which a craft is moving and the longitudinal axis of the craft. (2) The angle between the projection, onto a horizontal plane, of the direction in which a craft is moving the projection, onto the same plane, of the longitudinal axis of the craft. The angle is usually small enough that either of the definitions gives the same angle within the tolerances allowed by the situation. (3) The angle through which a coordinate system fixed in a photograph must be rotated about the z-axis in object space to make the x-axis and y-axis of the photograph parallel to the x-axis and y-axis in object space. Yaw is usually the tertiary rotation, preceded by pitch and roll, so that the photograph's z-axis is presumed already to have been made parallel to the z-axis in object space. (4) The angle through which a coordinate system fixed in a photograph is rotated about a specified line through the origin of the coordinate system and nearly perpendicular to the photograph. The angle is measured in a plane perpendicular to the specified axis of rotation between a specified line in that plane and the projection of the x-axis of the coordinate system of the photograph onto that plane. In all the above definitions, the angle, also called crab is usually denoted by κ and called kappa.

angle point - (1) A monument marking a point, on an irregular boundary line, reservation line, boundary of a private claim, or a re-established, non-riparian meander line, at which a change in azimuth occurs. (2) A stake driven into the ground at a point on a traverse, to indicate a change in direction of the traverse at that point. (3) A point, in a survey, at which the alinement or boundary deflects from its previous course. (4) A marker placed to indicate a point at which there is a change in the direction of a surveyed line.

angle radar, side-looking - Side-looking radar operating at decimeter wavelengths and measuring the angle at which the radiation is emitted and returned. Most side-looking radar operates at millimeter wavelengths. Such radiation is scattered by vegetation and does not reach the ground if the vegetation is dense, and the radar sees only the top of a densely forested region. Radar operating at decimeter wavelengths is able to see through the vegetation to the ground.

angle reduction to sea level - SEE reduction (of an angle) to the geoid.

angle reduction to the ellipsoid - SEE reduction (of an angle) to the ellipsoid.

angle reduction to the geoid - SEE reduction (of an angle) to the geoid.

angle reduction to the spheroid - SEE reduction (of an angle) to the spheroid.

angle-to-left - The horizontal angle, measured counterclockwise, from the preceding leg of a traverse to the following leg.

angle-to-right - The horizontal angle, measured clockwise, from the preceding leg of a traverse to the following leg.

Ångström - A unit of length equal to 10^{-10} meter. Also called the Ångström unit. Also spelled Angstrom and angstrom. Its symbol is Å or A. It was invented to measure wavelengths in the optical (8000 - 4000 Å) part of the spectrum and below (ultraviolet and X-ray). It is not an accepted part of the SI but is still used in spectroscopic literature. The designation preferred by the SI for, e.g., the visible part of the spectrum would be 800 - 400 nm (nanometers).

Ångström unit - SEE Ångström.

angulation - The measurement of angles.

angulation, vertical - The process of obtaining differences of elevation by means of observed vertical angles together with lengths of lines. In geodesy, vertical angulation is called trigonometric leveling.

angulation instrument - SEE angle measuring instrument.

angulator - A device for converting angles measured on an oblique plane to their corresponding projections on a horizontal plane. Also called a rectoblique plotter. SEE ALSO equiangulator; photoangulator; topoangulator.

annexation - (1) The act of attaching, adding, joining, or uniting one thing to another; generally spoken of the connection of a smaller or subordinate thing with a larger or principal thing. Attaching an illustrative or auxiliary document to a deposition, pleading, deed, etc., is called annexing it. So the incorporation of newly-acquired territory into the national domain as an integral part thereof is called annexation, as in the case of the addition of Texas to the United States of America. (2) In the law relating to fixtures, actual annexation includes every movement by which a chattel can be joined or united to the freehold.

annexation, constructive - The union of such things as have been holden parcel of the realty but which are not actually annexed, fixed or fastened to the freehold.

annex point - A point selected, in the overlap between a vertical photograph and its corresponding oblique photograph, about half way between the pass points.

annotation, cartographic - The placing of additional data, drawing of new features, or deletion of destroyed or unimportant features on a photographic mosaic to show only current details. It may include elevations of airfields, cities, and large bodies of water; new structures and destroyed or dismantled bridges, dams, or railroads.

anomaly - In general, any quantity whose values differ from those which are expected or which are predicted by simple theory. The following particular usages are common. (1) (astronomy) The angle from the line of apsides of an elliptical orbit to the radius vector from the center or focus to the revolving body. In particular, the angle, at the Sun, from the line of apsides of a planetary orbit, taken at perihelion, to the radius vector to the planet. The angle determined by the radius vector drawn from the focus at which the attracting body is located is the true anomaly. The angle determined by the radius vector from the center of the ellipse is the eccentric anomaly. [The term was applied because orbits were at one time supposed to be circular and any departure from circularity was anomalous.] (2) (geodesy) A deviation of an observed value from a theoretical value because of an abnormality in the observed quantity. (3) (oceanography) The difference between some particular characteristic of sea water and the corresponding characteristic either of sea water under standard conditions or of a standard sample of sea water. (4) SEE gravity anomaly. SEE ALSO Bouguer gravity anomaly; deflection anomaly; depth anomaly; Faye gravity anomaly; gravity anomaly; height anomaly.

anomaly, eccentric - The angle from the line of apsides of an elliptical orbit to a radius vector drawn from the center of the ellipse to that point on the circumscribed circle which lies on the same perpendicular (to the line of apsides) as the moving body. The point on the circle and the moving body therefore project onto the same point on the line of apsides. The eccentric anomaly is usually denoted by E; less often by η. It is related to the mean anomaly by Kepler's equation.

anomaly, free-air - SEE gravity anomaly, free-air.

anomaly, geodetic - The deviation of the measured value of a geodetic quantity from the theoretical value of that quantity, because there is a corresponding deviation of the Earth's actual structure at the point of measurement from the theoretical structure.

anomaly, geomagnetic - SEE anomaly, magnetic.

anomaly, geopotential - SEE depth anomaly, dynamic.

anomaly, isostatic - SEE gravity anomaly, isostatic.

anomaly, local magnetic - An abnormal or irregular variation of the Earth's magnetic field extending over a relatively small region, because of local geomagnetic influences. Also called anomalous magnetic variation, local attraction, local magnetic disturbance and magnetic anomaly.

anomaly, magnetic - The difference between the intensity of the magnetic field at a particular place and the intensity predicted, for that place, by a standard formula such as that for a magnetic dipole. There is a theoretical relationship between magnetic anomaly and gravity anomaly. (2) SEE anomaly, local magnetic.

anomaly, mean - (1) The angle from the line of apsides of an elliptical orbit to the radius vector from the attracting focus to a point moving on the ellipse with angular velocity equal to the average velocity of the actual, moving body. It is related to the eccentric anomaly by Kepler's equation. It is related to the period P of the body and to the time t since the body passed through pericenter by the formula $(M) = 2\pi t/P$. (2) In undisturbed elliptical motion, the product of the average angular speed times the interval of time since the body passed pericenter-center. Thus the mean anomaly is the angle from pericenter to a hypothetical body moving with a constant angular speed which is equal to average angular speed.

anomaly, orbital - An angle, in the plane of an orbit, identifying the direction of a satellite. SEE ALSO anomaly, eccentric; anomaly, mean; anomaly, true.

anomaly, point-mass - One or more point-like masses taken to be of such a size and distribution as to account for deviations of the Earth's gravity field from that of an ellipsoid having the same mass, center of mass, rate of rotation as the Earth, and the same gravity potential on its surface as on the geoid.

anomaly, pseudo-gravimetric - A quantity g(x,z,M) introduced by Baranov (1957) to take into account not only the variation of gravitational attraction with horizontal (x) and vertical (z) distance, respectively, but also the variation with magnetization, M.

anomaly, specific-volume - SEE anomaly, steric.

anomaly, steric - The difference between the reciprocal of the actual density $\rho(S,T,P)$ of a sample of water and the reciprocal of sea water having a salinity S of 35 o/oo, a temperature T of 0° C, and the given pressure P. Also called specific-volume anomaly.

anomaly, thermosteric - That steric anomaly which would be reached if the pressure of sea water were changed isothermally to a standard pressure of one atmosphere. I.e., the steric anomaly with pressure terms omitted.

anomaly, true - The angle from the line of apsides of an elliptical orbit to the radius vector from the attracting focus to the center of mass of the moving body. It is commonly denoted by v or by ν.

anomaly of geopotential difference - SEE depth anomaly, dynamic.

anomaly of gravity - SEE gravity anomaly.

anomaly of the specific volume - SEE volume anomaly, specific.

Anschutz compass - A particular and very common make of gyroscopic compass, named after its inventor and manufacturer.

Ansermet test - The sum of the products of the weights of the unadjusted observations and the reciprocal weights of the adjusted observations must be equal to the number of observations for the problem to have a unique solution.

Antarctic Circle - The parallel of latitude, in the Southern Hemisphere, at which the latitude is equal to the complement of the declination of the winter solstice. Also called the South Polar Circle. Because the obliquity of the ecliptic is changing steadily, the winter solstice is not a point of fixed declination. Thus, the Antarctic Circle is not a line of fixed position. When the Antarctic Circle is shown on a map, it should be treated as a line of fixed position; a conventional value, 66° 33' South has been adopted for its latitude. This is the complement of 23° 27' South adopted for the latitude of the Tropic of Capricorn. The region inclosed by the Antarctic Circle is called the Antarctic Zone or South Frigid Zone. At the time of the winter solstice, the Sun is above the horizon at local midnight at all points in this region.

Antarctic Zone - The region between the Antarctic Circle (66° 33' S.) and the South Pole. SEE ALSO Antarctic Circle. Also called the South Frigid Zone. Because the Antarctic Circle is not a fixed limit, different values will be found in the literature for its latitude. The most common value is 66° 30'.

antenna - (1) A conductor at which portions of radio waves traveling through space are intercepted and converted into current which is then detected by a radio receiver, or, the conductor in which a radio transmitter terminates and at which currents generated by the transmitter are sent off into space. Also called aerial (now chiefly Brit.) (2) A component, in an electronic circuit, which receives or transmits radio waves. SEE ALSO dipole antenna; reference antenna.

antenna, dipole - SEE dipole antenna.

antenna, helical - An antenna in the form of an open-ended, loosely-coiled helix; the distance around one turn is approximately equal to the wavelength for which the antenna was designed. It has appreciable gain in the direction of its axis. Because of its high gain and small size (at the wavelength for which it was designed), it has been much used for receiving long-wavelength signals such as those of the TRANSIT satellites.

antenna, Mills-cross - An antenna consisting of two mutually perpendicular, horizontal arrays of dipoles arranged in a cross. Also called a Mills-cross antenna. A Mills-cross antenna produces a very narrow beam (main lobe). The direction of this beam can be controlled by properly phasing the currents in the dipoles of the array. It has been used for tracking artificial satellites as well as for radio astronomy. The antenna usually has very long arrays and therefore produces a very narrow field of view which can be directed by properly phasing the currents in each arm separately.

antenna, paraboloidal - A paraboloidal reflector with an antenna at its focus. SEE ALSO reflector, paraboloidal.

anti-cyclonic (adj.) - Clockwise, as seen from above in the Northern Hemisphere, or counterclockwise as seen from above in the Southern Hemisphere.

antihalation - The reduction of that blurring or fogging which takes place in photographic film at the border between a highlighted region and a shadowed region. SEE ALSO antihalation coating; halation.

antihalation coating - A light-absorbing coating applied to the back of the support of a film or plate (or between the emulsion and the support) to suppress halation.

antimeridian - That meridian which is 180° of longitude from a given meridian. A meridian and its antimeridian constitute a complete great circle.

antipode - Anything exactly opposite to something else. In particular, that point on the Earth which is 180° from a given place.

A1 time - An atomic-time scale established by the U.S. Naval Observatory, with the origin on 1 January 1958 at zero hours UT2 and with the unit (second) equal to the occurrence of 9,192,631,770 cycles of cesium-generated radiation of specified wavelength at zero field.

apastron - SEE apoastron.

aperture - (1) In general, any material part of an optical system which is specifically intended and designed to let some light pass through and to stop the rest. Equivalent, in this sense, to stop. (2) In particular, that part (element) of an optical system which determines the amount of light (power) reaching the detector. Also called the aperture stop. (3) A measure of the light-gathering power of an optical system. In simple refracting optical systems, this is approximately the diameter of the front (objective) lens. More exactly, it is the diameter of that stop (the aperture stop) which determines the angular size of the axial cone of rays from the object. In reflecting optical systems, the diameter of the primary mirror is usually taken as the aperture. If the mirror is small, the square-root of the unobscured area of the primary mirror is often used. In catadioptric systems, the diameter of primary mirror or lens may be used. Note that aperture is not the same as numerical aperture. (4) A surface, at or near an antenna, on which the values of the field are conveniently assumed for the purpose of computing fields at external points.

aperture, angular - The angle subtended, at an object point on the optical axis, by the radius of the entrance pupil.

aperture, critical - That aperture (2) for which the unsharpness caused by diffraction most nearly balances the unsharpness caused by aberration.

aperture, effective - (1) The ratio of the received power available at the terminals of an antenna to the power per unit area in the incident radiation. Also called effective area. For all antennas, effective aperture A is related to the gain G at a given wavelength λ by the equation $A/G = \lambda^2/4\pi$. The effective aperture of an ideal antenna is equal to its actual area a. In practice, the ratio A/a is always less than 1, a representative value for paraboloidal reflectors being 0.6 at microwave wavelengths. (2) That part of the opening, in an optical system, which determines the useful amount of light reaching the emulsion or other detector, or the diameter of that part of the opening. The diameter of the effective aperture is less than the diameter of the real aperture because light passing close to the edge of the real aperture is diffracted and does not contribute usefully to the image.

aperture, numerical - The product NA of the refractive index N of the medium in image space by the sine A of the half-angle of the cone of illumination there. In microscopic optics, this definition with object-space substituted for image-space is preferred.

aperture, relative - The ratio of the equivalent focal length of an optical system to the diameter of the entrance pupil. Also called f-number, f-stop or aperture stop. When objects other than point-sources such as stars are viewed or photographed, the illumination in the image is determined by the relative aperture rather than by the aperture alone. The diameter of the entrance pupil limits the power arriving at an element of the image, while the focal length determines the area over which this power is spread.

aperture, working - The largest opening of the diaphragm of a lens system with which the system gives satisfactory definition over that part of the film or plate covered by the image.

aperture aberration - An optical aberration occurring because rays from points at different distances from the optical axis do not come to the same focus.

aperture diaphragm - That aperture, in a telescope, which determines which parts of the light entering through the objective lens will pass through the rest of the optical system.

aperture ratio - The ratio of the effective aperture of a lens system to its focal length.

aperture stop - (1) That opening, in an optical system, which determines the largest apex-angle of a cone of rays from a point in object space, for which all rays within the cone pass entirely through the optical system. (2) That opening, in a lens system, which limits the area of the lens through which light can pass to the field stop. Also called a stop when the meaning is clear. Adjusting the size of the aperture stop of a given optical system regulates the brightness of the image without necessarily affecting the size of the image. (3) SEE aperture. (4) SEE aperture, relative.

apex - (1) The highest point of something, as of a cone or triangle, or the greatest latitude (vertex) of a great circle. (2) The point at which all the elements of a cone or solid angle intersect.

aphelion - That point, in an elliptical orbit of a body around the Sun that is farthest from the earth.

aplune - SEE apolune.

apo - (prefix) Farthest from the attracting body. E.g. apogee, that point, in the elliptical orbit of a terrestrial satellite, at which the satellite is farthest from the Earth. But aphelion, not apohelion, is customary.

apoapsis - SEE apofocus.

apoastron - That point, in the orbit of a body about a sun, in which the body is farthest from the sun.

apocenter - SEE apofocus.

apochromat - SEE lens, apochromatic.

apocynthion - SEE apolune.

apofocus - That point, in an elliptical orbit, at which the moving body is farthest from the focus where the attracting mass is located. Also called apoapsis and apocenter. It is at the opposite side of the orbit from the perifocus.

apogee - That point, in the elliptical orbit of a terrestrial satellite, at which the satellite is farthest from the Earth's center of mass.

apolune - That point, on the elliptical orbit of a lunar satellite, which is farthest from the Moon. Also called apocynthion and aplune. Opposite of perilune.

aposphere - A surface of rotation whose meridional section is defined by the equation $r = a\ \text{sech}(b(\psi + c))$, where a, b and c are constants, ψ is the isometric latitude and r is the perpendicular distance from the axis of rotation to the surface. The constants are chosen to make the aposphere osculate the spheroid (with which it has a common axis or rotation) along some parallel that passes through the center of the region for which a transformation is required.

a posteriori (adj.) - Obtained from observation or experience; obtained after an event. In science, it is usually said of data. In geodesy, the term is often used to denote a solution obtained after data are available. SEE ALSO a priori; Bayes' theorem.

apothegm - SEE apothem.

apothem - The perpendicular from the center of a regular polygon to one of the sides. Also spelled apothegm.

apparatus - A device of considerable complexity, usually composed of several parts that must be assembled before the device can be used.

apparatus, bar-check - An apparatus consisting of a bar or disk, with associated machinery for lowering and hoisting, which is lowered from the side of a ship to an accurately known depth. The depth is then measured independently by the ship's depth-sounder.

apparatus, iced-bar - SEE base line apparatus, iced-bar.

apparatus, stretching - SEE tape stretcher.

apparatus, taut-wire - A 100-meter long, stranded, sounding-wire, graduated at 25 meter intervals, and used to measure the distances between offshore-control buoys during a hydrographic survey.

apparatus, two-pendulum - A pendulum apparatus having two working pendulums hanging from the same support. In use, the two pendulums are set swinging in opposite directions. The effect of flexure in the support is thereby almost entirely canceled, as is the effect of microseisms. The Cambridge pendulum-apparatus is an outstanding example.

apparent (adj.) - (1) As observed; as seen. (2) Measurable or observable, referring to characteristics of real or visible bodies. The term, common in the older scientific literature, is now used mostly by astronomers.

appearance ratio - SEE hyperstereoscopy.

Appleton layer - (1) The F-layer in the ionosphere. It occurs in the general region between 150 and 300 km above the ground. (2) The F_2-layer in the ionosphere. It occurs at about 300 km above the ground.

apportionment - The division of rights or liabilities among several persons entitled or liable to them, in accordance with their respective interests. E.g. as where a contractor is given partial payment in return for partial performance, or where rents are divided according to some scale of interest, or as between seller and purchaser in the middle of a term.

appraisal - (1) An estimate and opinion of value. (2) Usually, a written statement of the market value or the value as defined by the appraisal, of an adequately described parcel of property as of a specific date. (3) A conclusion that results from an analysis of facts.

appreciation, cartographic - The assessment, in a fair and objective manner, of the accuracy and usefulness of a map.

approach chart - An aeronautical chart providing information essential for approaching an airfield under either visual or instrumental conditions of flight.

approach zone - All of the air-space lying within the boundaries of the approach-zone district at an airport and lying above the floor of the district.

approach zone district - SEE district, approach-zone.

appropriation - (1) The taking of a public thing for private use, particularly personal use. (2) SEE doctrine of appropriation. (3) Used incorrectly as a synonym for condemnation or expropriation.

approximation - (1) A value believed to be close to the correct value of a quantity. The terms first approximation, second approximation, etc., are used for sequentially calculated values that are successively closer to the correct value. (2) The process of obtaining a value believed to be close to the correct value of a quantity. Two different processes are recognized: direct approximation, in which an approximation is calculated only once and to a closeness that is considered satisfactory; and (b) successive approximation, in which values successively closer to the correct value are calculated in sequence, the approximation calculated in one step being used as a given value in the next step. The process is repeated until either a satisfactory value is obtained or no change of value occurs or the process begins to diverge or oscillate. This process is also known as an iterative process. Direct approximation provides an answer faster but the formula used are generally more complicated. Successive approximation takes longer to provide the final answer but uses much simpler formula.

approximation, asymptotic - A method of successive approximation in which successive values converge toward some final value but then, after a certain number of steps, begin to diverge.

appulse - The near approach of one celestial body to another on the celestial sphere, as in conjunction or occultation.

appurtenance - That which belongs to another thing as principal and passing as incident to it. As, e.g., a right of way or other easement to land, a right of common to pasture, an outhouse, barn, garden, or orchard. In a strict legal sense, land cannot pass as an appurtenance to land.

a priori (adj.) - Obtained solely through reasoning from axioms or assumed principles, without appeal to experience. Also used as an adverb. Often used, erroneously, of computational data or conclusions reached before an experiment or event takes place, e.g., a priori information as a pretentious term for data. In geodesy, the term is often used to denote a solution obtained before experimental or observational results are available. The term prior data, earlier data or previous data is greatly preferable, particularly if an appeal is to be made to Bayes's theorem.

apse - A point, on a curve, at which the radius vector is a maximum or a minimum. SEE ALSO apsis.

apse, line - SEE line of apsides.

apsides - SEE apsis.

apsides, line of - SEE line of apsides; apsis.

apsis (plural, apsides) - Either of the two points, in an orbit, at which the distance of the moving body from the center of attraction is an extremum. Also called apse. For an elliptical orbit, these two points lie on the major axis of the ellipse and the center of attraction lies on a focus of the ellipse. The point closest to that focus is called the periapsis and that farthest from the focus is called the apoapsis. The line joining the two points is called the line of apsides.

aqua - (1) Water. (2) In the civil and old English law, sometimes, a stream or water-course. (3) In Roman antiquity, a site of a mineral spring, usually in the plural, as Aquae Sextiae. (4) An aqueduct, as, the Aqua Cladia at Rome.

aquatone - A printing process using a zinc plate which is gelatin coated, hardened, and sensitized to print type, line drawings, and halftones by an offset process.

arbitrary method (of adjusting a traverse) - Corrections are applied to the measured quantities (or to the calculated components of these) according to the surveyor's analysis of conditions that prevailed along the traverse. I.e., the corrections needed to eliminate the misclosure are weighted according to the surveyor's judgement and not according to fixed rules.

arc - (1) A continuous portion of a mathematically defined curve - in particular, of a circle. A circular arc is part of a circle; an elliptical arc is part of an ellipse, etc. (2) The graduated scale of an instrument for measuring angles. SEE ALSO arc, graduated. (3) SEE triangulation and arc of triangulation. (4) SEE arc, graduated. SEE ALSO Beaman arc; Jordan arc.

arc, astronomic - The apparent arc described above (diurnal arc) or below (nocturnal arc) the horizon by the Sun or other celestial body.

arc, diurnal - SEE arc, astronomic.

arc, graduated - A piece of metal curved to conform to a circular arc and graduated to indicate angles with respect to a point at the center of the circle. Also called, simply, an arc e.g., Beaman stadia arc; arc of a sextant. SEE ALSO quadrant.

arc, great - An arc of a great circle.

arc, great elliptic - An arc (1) defined by the intersection of an ellipsoid with a plane through the center of the ellipsoid and two specified points on the surface.

arc, nocturnal - SEE arc, astronomic.

arc, short - A small portion, usually less than half, of one complete revolution of a satellite about its primary body. SEE ALSO short-arc method.

arc, simple - SEE Jordan arc.

arc correction - The quantity added to the period of a pendulum to account for the pendulum's departure from simple harmonic motion. The correction can be kept small by limiting the amplitude of oscillation.

Arc Datum - A horizontal-control datum in Africa, based on Cape datum and having presumably the same origin with the same coordinates and azimuth there. However, the vast longitudinal and latitudinal extent of the networks have caused complex modifications to take place during the calculations, and it is difficult to say exactly what datum or datums govern the networks. The Clarke 1880 spheroid (modified) is used for the most part, but triangulation in some former French territories has been on a datum using Clarke's values for a and b rather than his values for a and 1/f. There is a series of datums called Arc datum 1950, Arc datum 1960, etc. It is not clear to what extent these are different datums, different networks, or both.

Archimedes' spiral - The spiral given by the equation $r = k\theta$, in which r and θ are polar coordinates and k is a constant. Geometrically, it is the locus of a point which moves about a fixed point in such a manner that the distance from the fixed point is proportional to the angle through which the moving point has revolved. SEE ALSO spiral, Archimedean.

arc length - Length measured along a curved line. The line could be circular, parabolic, hyperbolic, spiral, a catenary, etc.

arc measurement - (1) Determination of the lengths of arcs of triangulation and the astronomic coordinates of the ends of the arc in order to derive the dimensions of an ellipsoid best fitting the Earth's surface or the geoid. Also called grade measurement, from the German grad for degree. One customarily differentiates between meridional-arc measurement and latitudinal-arc measurement. (2) The methodology of the procedure defined in (1).

arc method of adjusting triangulation - A method of adjusting a triangulation network, in which the network is separated into sets of connecting arcs and the corrections in each arc are determined separately. The results are then modified so that there is no disagreements between coordinates at the intersections or junctions of the arcs. The Bowie method of adjusting triangulation is an arc method.

arc navigation - Navigation in which the prescribed course is a circular arc and the vessel or aircraft is kept on this course by reference to radio signals from a transmitter at the center of the arc.

arc of parallel - A part of an astronomic or geodetic parallel of latitude.

arc of triangulation - (1) A triangulation network forming a band or belt on the surface of the Earth or on the reference ellipsoid. Also called a triangulation arc, chain of triangulation and triangulation chain. The axis (a line running down the center of the band and giving its general direction) of an arc of triangulation may approximate, in position, an arc on the ellipsoid following a meridian of longitude, a parallel of latitude, or an oblique arc. It may follow a natural feature such as a river, or it may follow an artificial feature such as a civil boundary. It is usually given a name identifying its general or particular location or having some other geographic significance: e.g., the Mississippi River Arc, the Ninety-eighth Meridian Arc, and the Arc of the 30th (meridian). (2) The geodesic joining the initial and terminal points of an arc of triangulation as defined in (1).

arc of visibility - The arc through which a beacon, usually a lamp, is visible. It is designated by its limiting bearings as observed from the sea.

arc second - A second of arc. It is denoted symbolically by "; e.g., 11 seconds of arc is written as 11". SEE ALSO second.

Arctic Circle - The parallel of latitude, in the Northern Hemisphere, having a northern latitude equal to the complement of the declination of the summer solstice. Because the obliquity of the ecliptic is steadily changing, the summer solstice is not a point of fixed declination, and the Arctic Circle as defined above is not a line of fixed position. The Arctic Circle is generally shown on a map as a line of fixed position, and a conventional value should be adopted for its latitude. The value 66° 33' N. has been proposed. This is the complement of 23° 27' N. proposed for the latitude of the Tropic of Cancer. SEE ALSO Antarctic Circle. The region inclosed by the Arctic Circle is called the Arctic Zone. At the time of the summer solstice, the Sun is above the horizon at all points in this region at local midnight. Actually, when an observer sees the Sun at a solstice just touch the horizon at local midnight without setting, he is probably much nearer to the 66th parallel of latitude than to the one proposed above for cartographic use. This is an effect of atmospheric refraction. It is modified by local conditions such as dip of the horizon.

arc triangulation - (1) Triangulation designed to progress in a single, general direction and to produce a network of limited width but of considerable length. Arc triangulation is done to connect independent and widely separated survey networks, to coordinate and correlate local networks along the arc, to furnish data for the determination of a geodetic datum, to provide a network of control points for a country-wide survey, and the like. (2) SEE ALSO arc of triangulation.

are - A unit of area equal to 100 square meters. The unit is used primarily in agriculture and related arts.

area - (1) A numerical measure of the amount of surface contained within a closed curve on a surface. It is usually referred to as the area of the figure if the surface is flat. Area is a function of the shape of the surface, the shape of the curve, and the location of the curve on the surface. On a plane, the area within a rectangle is defined as the product of the lengths of two adjacent sides of the rectangle. The area within a triangle can be shown to be equivalent to the area within a rectangle having the base of the triangle as one side and a line half as long as the height of the triangle as the other side. The area within a more complex, straight-sided figure can be found by dividing the figure into rectangles and triangles and taking the sum of the areas of these smaller figures. The planar areas within some simple closed curves such as the circle and ellipse can be found by direct integration (quadrature), but areas of planar surfaces within more complex curves must usually be found by numerical integration. Numerical integration consists basically of dividing a figure geometrically into rectangles and triangles whose sides are nearly straight, summing the areas, and finding the limiting value of the sum as the subdivision becomes finer and finer. The area of a curved surface is similarly found, by direct integration if the surface is a simple one, or by numerical integration if the surface is complex. Again, this is equivalent to subdividing the surface into small three-sided and four-sided figures bounded by curves, estimating the areas of these curved figures, summing the areas, and finding the limit as the subdivision becomes finer and finer and the small figures come closer and closer to becoming (planar) rectangles and triangles. This mathematical process has been mechanized, and instruments such as the planimeter allow one to find the area contained by a closed curve by tracing the perimeter of that curve. For many purposes, however, the area of a small region can be found well enough by covering a map of the region by a rectangular grid and summing over the areas of the squares and triangles formed by the grid and the boundary of the region. Area is almost always given as the area of a planar rectangle having the same area as the surface in question, and the area of a planar rectangle is given as the number of squares of unit length on a side that it can contain. The area of any surface is therefore given in terms of the areas of unit squares. The two units of area in almost universal use today are the square meter (the planar area contained within a square 1 meter on a side) and the hectare (the planar area contained within a square 100 meters on a side). In the United States of America, the square foot and the square inch are used for the measures of small regions. For large regions, the acre or square mile is used. For areas of small, circular figures, the circular inch (the area of a circle 1 inch in diameter) and the circular mil (the area within a circle 0.001 inch in diameter) are sometimes used. (2) That portion of a surface bounded by a closed curve. It differs from region in that the latter does not have the connotation of numerical value. (3) Some clearly defined space. Approximately equivalent to region. (real estate) - A parcel assumed to be level and at mean sea level. These assumptions are used to get consistent descriptions of land.

area, effective - For any aerial photograph that is one of a series in a strip of photographs taken during a single flight, that central part of the photograph delimited by the bisectors of overlaps with adjacent photographs. On a vertical photograph, all images within the effective area have less displacement than their conjugate images on adjacent photographs.

area, metropolitan - The region in and around a large city.

area coverage - SEE coverage, areal.

area of error - A figure within which the true location of a point whose location has been determined by measurement is believed to lie, with stated degree of probability.

area survey - SEE survey, areal.

area symbol - A continuous and distinctive shading, tone, or repetitive pattern used on a map to cover a part representing a region where a particular phenomenon occurs, or to which a particular value is assigned.

area triangulation - SEE triangulation, areal.

Argelander catalog - SEE Bonner Durchmusterung.

argument - (1) (astronomy) SEE angle; argument of latitude; and argument of perigee. (2) (mathematics) An independent variable (q.v.). The term is also used for the discrete values indexing rows or columns in tables of functions. (3) (tides) The angular variable in the representation of tidal variation by Fourier series. SEE ALSO equilibrium argument; Greenwich argument.

argument number - SEE Doodson number.

argument of latitude - The angle, measured in the orbital plane and at the center of mass of the attracting body (or at the center of mass of both bodies), from the ascending node to the orbiting object. The sum of the argument of perigee and the true anomaly.

argument of perigee - The angle, at the center of attraction, from the ascending node to perigee, measured in the direction of motion of the orbiting body.

Aries, First Point of - SEE First Point of Aries.

arithmetic, fixed-point - SEE calculation, fixed-point.

arithmetic, floating-point - SEE calculation, floating-point.

arm of the sea - A comparatively narrow extension of a larger body of water.

arpent - (1) An old French unit of area. The size of the arpent depends on its origin and on local custom. The value 0.8507 acres has been used for surveys in Arkansas and Missouri. For surveys in Louisiana, Mississippi, Alabama, and northwestern Florida, when no alternative definition for the arpent has been known, the value 0.84725 acres has been used. The arpent was used, in the U.S.A, in surveys of land granted by the French crown. (2) An old French unit of distance, taken as the length of a side of a square region one arpent in area. Its values in other units, corresponding to the values given in (1) for area, are:
 1 arpent (0.8507 acres) = 192.500 feet or 58.674 m
 1 arpent (0.84725 acres) = 191.995 feet or 58.5198 m In Canada, the arpent is exactly 180 French feet, which is about 191.85 English feet. (SEE foot).

arpentator - A land surveyor.

array - A set of numbers (or symbols representing numbers), objects, or points placed at regular intervals. The numbers, etc., are called elements of the set and of the array. Note that matrix and array are not the same thing. A matrix is an array whose elements obey certain rules. An array, in general, does not have any special rules associated with it.

array, linear - A set of points or objects placed at regular intervals in a straight line. In particular, a linear array of detectors such as charge-coupled devices (CCDs) or antennas. SEE ALSO push-broom method.

array algebra - A set of arrays of mathematical quantities, together with a set of rules for manipulating the arrays in such a way that the results are the same as if the quantities had been elements of matrices. Array algebra has no particular advantage over matrix algebra in a mathematical sense. It does, however, allow the elements to be grouped more efficiently for computation using electronic computers.

arrow, surveying - SEE chaining pin.

arrow, taping - SEE chaining pin.

Arrowsmith's map projection - SEE Nicolosi's map projection.

arroyo - A deep gully cut by an intermittent stream.

artillery atmosphere, standard - A set of values describing atmospheric conditions on which ballistic computations are based, namely: a surface temperature of 15 ° C; a surface pressure of 1000 millibars; a relative humidity at the surface of 78%; a lapse rate which yields a prescribed relation of density to altitude; and no wind.

artillery criterion - The criterion used by the artillery of some countries for accepting or rejecting data, requiring that measurements farther from the average than 4 standard deviations be rejected.

artillery map - SEE map, fire-control.

artillery survey - A survey made to determine the horizontal and vertical locations of cannon with respect to each other and their targets.

art-work prediction - SEE experience radar prediction.

Arundel method - A method of mapping from aerial photographs, proceeding point by point by a combination of graphical and analytical methods of aerotriangulation.

ascendant - A vector representing the rate of increase of a quantity.

ascension - A motion from a lower to a higher location. In particular, a point on the celestial equator at which such a motion occurs.

ascension, oblique - The arc of the celestial equator, or the angle at the celestial pole, between the hour circle of the vernal equinox and the hour circle through the intersection of the celestial equator and the eastern celestial horizon at the instant a point on the oblique sphere rises. It is measured eastward from the hour circle of the vernal equinox through 24 hours.

ascension, right - The angle between the plane of the hour circle passing through a celestial body and the plane of the hour circle passing through the vernal equinox. It is measured eastward from the vernal equinox through 24 hours (1 angular hour equals 15°) or 360°. It may also be defined as the angle, at the celestial pole, between the tangents to the hour circles of the celestial body and of the vernal equinox. Right ascension and declination form a pair of coordinates that conveniently specify the location (direction) of a body on the celestial sphere. It is conventionally represented by α.

aspect - (1) The apparent position of a (cartographic) graticule in relation to the longitudes and parallels of latitude that they represent. Also called a case of a map projection. The term is derived from, and relates to, the concept that a graticule is produced by actually projecting the meridians and parallels on an ellipsoid onto a developable surface. The appearance of the graticule then depends on how the pattern of meridians and parallels is positioned with respect to the developable surface. Terminology is not uniform. However, that aspect which produces the simplest graticule (e.g., meridians and parallels are represented as straight lines) is commonly termed the normal aspect of the map projection and of the graticule. The aspect produced by rotating the ellipsoid through 90° from its position in normal aspect is then the transverse aspect, and any rotation between 0° and 90° produces an oblique aspect. (2) The appearance of a graticule according as certain meridians or parallels of latitude are represented as lines of zero distortion (this appearance is then called the normal aspect), the line of zero distortion is perpendicular to the line which would be the line of zero distortion in the normal aspect (this appearance is called the transverse aspect), or neither relation holds (this appearance is then called an oblique aspect.) This set of rules for naming aspects is often applied to graticules not produced by true projection. (3) The appearance of a graticule according as the center of the graticule represents a pole of the rotational ellipsoid (polar aspect), a point on the equator (meridional aspect or equatorial aspect) or neither (oblique aspect). If the oblique aspect of a map projection is not symmetric about the central meridian, the aspect is called a skew oblique aspect. There is, however, no standard or generally accepted terminology for the concept of aspect. (4) The orientation of a set of directed axes associated with a scene, with respect to the image of that set of axes. For example, an image is said to be in an inverted aspect if the imaged set of axes is inverted with respect to the set of axes associated with the scene. SEE ALSO image.

aspect, equatorial - (1) That aspect, produced by an azimuthal map projection, in which the origin is at some point on the equator. (2) SEE aspect (3).

aspect, meridional - (1) The transverse aspect of a graticule. Also called a meridional map projection. (2) SEE aspect.

aspect, normal - The mathematically simplest aspect of the graticule (typically, the principal directions in the graticule coincide with those of the meridians and parallels of latitude.) SEE ALSO aspect (1).

aspect, oblique - SEE aspect (1).

aspect, skew oblique - SEE aspect (3).

aspect, transverse - SEE aspect (1).

assembly, multiple-camera - A combination of two or more cameras mounted with a fixed angle between their respective optical axes.

assembly language - A mnemonic code close to the code (machine language) in which the computer actually does its work.

assessment - (1) A levy of tax against property. (2) The amount of tax levied against property.

assessment map - A map used by taxing authorities, showing size and location of properties and improvements.

assessment work - Work required to be performed annually by the claimant in order for him to maintain a possessionary right to a mining claim for which a patent has not been issued.

assign (v.) - In law, transfer or make over to another; especially, transfer to, and vest in, certain persons, called assignees, for the benefit of creditors.

assumption, principal-point - The assumption that radial directions are correct if measured at the principal point. The assumption is satisfactory only for nearly vertical photographs.

assurance - In law, the act of conveying, or the instrument or other legal evidence of the conveyance of, real property. It is usually called common assurance.

assurance, common - SEE assurance.

A-station - (1) A subsidiary station established between principal stations of a traverse for convenience in measuring the distance between the principal stations. A-stations are established along a curved route, e.g. along a curved section of a railroad, where the measured lengths must be carried through a series of short, straight lines even tho azimuth control may be carried through widely-spaced stations. The A-stations form a loop with the (main) line connecting the principal stations, the distance between which is obtained by projecting the measured lengths of the short lines onto the main line. SEE ALSO equation, perpendicular. A-stations are so called because, in a given series, they are designated by the name of a principal station followed by the letters A, B, C, etc., in order of distance from that station. They are sometimes referred to as A,B,C stations. (2) SEE B-station.

asteroid - A small, planet-like body of the Solar System having its orbit largely in the region between Mars and Jupiter. Also called a minor planet or planetoid. Asteroids are important geodetically because they can be used for determining the mass of the Earth, the Earth's distance from the Sun, and other quantities. About 100,000 asteroids have been photographed with the 100 inch reflecting telescope at Mount Wilson, and undoubtedly a vastly larger number exist but are too faint to be detectable. Only about 1000 asteroids have been observed closely enough to allow their orbits to be calculated and other properties to be determined. The largest asteroid known (Ceres) has a diameter of about 1020 km, while some of the most recently discovered ones have diameters of less than 2 km. Most of the asteroids have orbits that are more strongly elliptical than the orbits of the principal planets and that lie in planes with higher inclinations to the plane of the ecliptic. A few of these asteroids have orbits so elliptical that they get comparatively close to the Earth - Adonis (0.44 A.U.) and Icarus (0.08) A.U. are among the closest. Observations of these planets are used to determine the distance of the Earth from the Sun (the solar parallax), the mass of the Earth-Moon pair, etc.

asthenosphere - (1) A hypothetical region, composed of that part of the Earth which comprises the lower part of the crust and the upper part of the mantle, is plastic, and allows continental drift. The name was proposed by Barrel, in 1914, for the zone immediately beneath the crust. Also called aesthenosphere or rheosphere. Conjectures as to the depth and thickness of the asthenosphere vary greatly. One estimate is that it is 340 km thick. (2) A region, in the upper mantle, characterized by low speed and high attenuation of seismic waves. (3) SEE mantle.

astigmatism - (1) The quality of focusing light from a point in object-space into a line in image-space. (2) The distance between the points at which at which two fans of rays from a common point in object-space focus in image-space, if the planes of both fans contain the optical axis but are perpendicular to each other. Also called astigmatic aberration. Astigmatism is one of the five Seidel aberrations. It is zero if the point-source is on the optical axis (of a symmetrical optical system) and increases with distance of the source from the optical axis. Thus, a single, off-axis point-source is imaged as two short, mutually perpendicular lines at different distances from the lens.

astigmatizer - A lens which introduces astigmatism into an optical system. An astigmatizer is usually mounted so that it can be moved into or out of the optical path at will. In a sextant, an astigmatizer may be used to elongate the image of a celestial body into a horizontal line.

astre fictif - A fictitious star assumed to move along the celestial equator at a uniform rate corresponding to the frequency of one of the several harmonic constituents of the tide-producing force. Each astre fictif crosses the meridian at a time corresponding to the maximum of the constituents it represents.

astrocompass - A direction-determining machine into which the coordinates of a celestial body can be set, together with the observer's latitude, and which then indicates astronomic north, azimuth and heading.

astrodynamics - (1) The application of celestial mechanics, ballistics, the theory of rocket propulsion, and allied subjects to the problem of planning and establishing the trajectories of spacecraft. (2) SEE mechanics, celestial.

astro-geodetic (adj.) - Involving the combined methods and/or data of astronomy and geodesy. For example, astro-geodetic leveling and astro-geodetic coordinates.

astro-geodetic datum orientation - SEE datum orientation, astro-geodetic.

astro-geodetic method (of determining the geoid) - A method of determining geoidal heights by measuring astronomic longitude and latitude at closely spaced points at which the geodetic longitude and latitude are also known. Also called astro-geodetic leveling and astronomic leveling. The deflections of the vertical determined for the Earth's surface are reduced to corresponding values on the geoid, and the changes in these deflections from point to point are converted, using the measured distances between points, into changes in geoidal height.

astrograph - (1) A device for projecting a set of curves showing stellar or solar angular elevations onto a chart or plotting sheet; the curves move, if the machine is properly adjusted, so as to remain in the correct position on the surface. (2) SEE telescope, astrographic.

Astrographic Catalog - One of the star catalogs published as part of the international project for determining the photographic coordinates of all stars down to the 11-th magnitude and with an r.m.s. error of 0.3".

astrograph mean time - A form of mean time used particularly in setting an astrograph (navigational device); 1200 hours mean time is set to occur when the local hour angle of the vernal equinox is 0°.

astro-gravimetric method of determining the geoid - A method of determining geoidal heights from deflections of the vertical calculated from gravimetric data, or from a combination of gravimetric and astronomic data.

astrolabe - (1) An instrument for measuring the angle between a celestial object and the horizontal plane at the observer. The term is derived from Greek words meaning take a star. It has been used to designate a great variety of instruments. An astrolabe which allows observation only at a 60° angle is sometimes called an equiangulator. (2) SEE astrolabe, planispheric. SEE ALSO Claude and Driencourt astrolabe; Danjon astrolabe; pendulum astrolabe.

astrolabe, circumzenithal - An astrolabe designed for observations on stars at a fixed distance (typically 30 - 40 degrees from the zenith). Also called a circumzenithal instrument. In one model, the light is split into two parts by a pair of internal mirrors in such a way that the two resulting images appear to approach each other as the star moves through the field of view. Also called a circumzenithal.

astrolabe, 45° - A prismatic astrolabe in which the prism directs the line of sight at a 45° angle to the horizontal.

astrolabe, impersonal - An astrolabe so constructed that the personal error of the observer does not affect the measurements. The Danjon astrolabe is such an instrument.

astrolabe, planispheric - An astrolabe, of ancient origin, consisting of a full, graduated circle with a centrally mounted alidade, and with adjustable plates on which are engraved graticules (on the stereographic projection) of the heavens and of the sphere for local latitudes. Also called, simply, an astrolabe if the type is obvious. The instrument is held suspended in a vertical plane, and the angular elevation of a star is observed with the alidade. The projection-bearing plates are so adjusted that graphical solutions of astronomical problems are obtained.

astrolabe, prismatic - An astrolabe consisting of a telescope with a prism and artificial horizon attached at its objective end so that only objects near one specific angular elevation can be observed. The usual form of this instrument has a horizontal telescope and contains a 60° prism, with the face nearest the objective being perpendicular to the line of collimation of the telescope, and a small pool of mercury attached beneath the prism to give an artificial horizon. In observing, two images of a star are seen. One image has been reflected directly into the telescope by the lower face of the prism. The other has been reflected first by the mercury's surface and then into the telescope by the upper face of the prism. The two images move in opposite directions either to or from coincidence. At the instant of coincidence, the star is at the apparent angular elevation imposed by the prism's angle. Prismatic astrolabes may be made with the angle at other that 60°. An English instrument uses a prism with 45° angle and has accessories allowing one to obtain more than one observation for a given setting. SEE ALSO Danjon astrolabe.

astrometry - SEE astronomy, positional. The term is sometimes restricted to mean the positional astrometry of stars.

astronomic (adj.) - of, or pertaining to, astronomy. It is completely equivalent to the form astronomical. Although Merriam-Webster (2nd edition) gives precedence to astronomical, it is the compiler's experience that surveyors use astronomic while astronomers prefer astronomical.

astronomic-direction method (of determining azimuths) - Determining the astronomic azimuth of a line by measuring, with a direction theodolite, the horizontal angle between a selected star and a suitable mark, and applying that angle to the azimuth of the star, computed for the epoch of the observation. In the horizontal-control surveys of continental United States, this method is preferred over others. A circumpolar star is observed at any hour angle, the mark being a lamp-signal on a station of the main scheme or at a station (called an azimuth mark) established for the purpose. A correction for inclination of the horizontal axis, depending upon the angular elevations of the star and of the mark, is applied to the observed angle. Corrections for curvature of the apparent path of the star, for polar motion, and for diurnal aberration are also applied. SEE ALSO azimuth determination from rate of change of zenith distance.

astronomy - (1) The science dealing with bodies and phenomena occurring in regions outside the Earth's atmosphere, and with meteors, cosmic rays, and other bodies or particles entering the atmosphere from outer space. This is the most commonly accepted definition. (2) The science of matter in outer space. This is a narrower definition, excluding interstellar radiation and similar subjects from astronomical consideration. The major categories of astronomy are: positional astronomy (the locations of celestial bodies); (b) celestial mechanics (the motions of celestial bodies); (c) descriptive astronomy (descriptions of celestial bodies; and astrophysics (intrinsic properties such as temperature, constitution, and radiation of celestial bodies). Some astronomers give the name astronomy to what is called positional astronomy above, and apply the term positional astronomy to what concerns the locations and kinematics of celestial bodies. Various categories are also distinguished according to the wavelength at which observations are made (these categories may of course overlap those enumerated before), e.g., infrared astronomy, X-ray astronomy and radio astronomy. Astronomy is related to geodesy principally through the use of observations of the Sun, Moon, or stars for determining coordinates of points on the Earth's surface and for determining the rotation of the Earth. SEE ALSO radar astronomy; radio astronomy.

astronomy, descriptive - The part of astronomy which is concerned merely with describing the features (size, shape, mass, relief, etc.) of celestial bodies like planets, without concern for the locations or physics of such bodies.

astronomy, geodetic - SEE geodesy, astronomical.

astronomy, positional - The part of astronomy which is concerned with the locations of celestial bodies such as planets, comets and stars, and with the changes (kinematics) in location. Also called spherical astronomy, but often referred to simply as astronomy. Astrometry is often used as a synonym for positional astronomy but is more often considered to concern the positional astronomy of stars only. Celestial mechanics is not usually considered a part of positional astronomy. Positional astronomy is the basis for stellar triangulation, very-long-baseline interferometry, determining the geoid, providing orientation to geodetic networks, establishing independent datums, and many other applications.

astronomy, spherical - The theoretical part of positional astronomy, concerned with the techniques for determining directions to celestial bodies. Its mathematical basis is spherical trigonometry.

astronomy, statistical - That part of astronomy which is concerned with the statistical distribution of the locations, motions, and other characteristics of stars and galaxies. It finds its subject matter both in positional astronomy and in astrophysics, and is very important to cosmology.

astrophysics - The part of astronomy which is concerned with the physical structure of celestial bodies - their temperatures, radiation, composition, and so on.

astro-triangulation - Surveying done by photographing, against a stellar background, a beacon carried by a balloon or satellite. Also called stellar triangulation. The method was proposed by Y. VÄisÄlÄ in 1946 and has been used with great success by various geodetic organizations.

asymmetry of object - A lack of symmetry in the appearance of an object as seen from a particular point of observation, and caused by an actual asymmetry in the object or its aspect. It causes an object to appear differently when viewed from different points and can result in different points of the object being sighted at from different points of observation. A square or rectangular pole may face the observer in such a way that a line bisecting the angle made by lines of sight tangent to the pole's edges does not pass through the center of the pole. If, instead of a pole, a square cupola or tower is sighted on, the error may be quite large. The error caused by observing on such objects is of the same character as the error caused by observing an eccentric object. Asymmetry of object differs from phase in that the former is caused by an actual asymmetry of the object as viewed; the latter is caused by an asymmetric illumination of the object.

asymptote - (1) The limit approached by the tangent to a curve as the point of tangency approaches infinity. (2) The line approached by a curve as the distance along the curve becomes very large.

Atlantis map projection - SEE Mollweide map projection.

atlas - A collection of maps designed to be kept (bound or loose) in a volume.

atlas, regional - An atlas depicting different aspects of a specific part of a country or continent.

atlas grid - A grid in which the lines of one of the two families is labeled numerically while the lines of the other family are labeled alphabetically. Also called an alphanumeric grid.

atmosphere - (1) The envelope of gas surrounding a planet or other celestial body. (2) The envelope of gas (air) surrounding the Earth. The Earth's atmosphere is considered to consist of a sequence of indistinctly bounded layers having different properties. Meteorologists generally recognize the following layers, based on the temperature and rate of change of temperature in each: the troposphere, extending from the surface of the Earth to about 11 km; the stratosphere, extending from about 11 km to about 50 km; the mesosphere, continuing from the top of the stratosphere to about 80 km; and the thermosphere, extending from there to about 500 km. The outermost region, where there is still gas caught in the Earth's gravity field but some molecules may escape, is called the exosphere and may or may not be considered part of the atmosphere. Geodesists and radio engineers accept the same troposphere and stratosphere as the meteorologists, but place the ionosphere next above those two. The ionosphere continues from the top of the stratosphere up to 300 or 400 km and is characterized by the presence of enough ions and electrons to detectably affect radio waves. The atmosphere affects the values of gravity at the surface, and the mass of the atmosphere must therefore be taken into account in precise work. The total mass is about 5.136×10^{18} kg. (3) A unit of pressure equal to the pressure of 760 mm of mercury at mean sea level and at 0° C. Also called a standard atmosphere. (4) A unit of pressure equal to 101.325 pascals. Also called a standard atmosphere. SEE ALSO ICAO atmosphere; ICAO Standard atmosphere.

atmosphere, ambient - (1) The air in the immediate vicinity of a surveying instrument or surveying station. The temperature and pressure of this air are referred to as the ambient temperature and the ambient pressure. (2) That part of the air which affects a geodetic measurement. This usage is rare and inappropriate.

atmosphere, model - Any set of values (formula or tables) representing the atmosphere, particularly the vertical distribution of temperature.

atmosphere, standard - (1) A set of values (formula or tables) relating temperature, pressure, and density (and sometimes height) within an atmosphere, and generally accepted as satisfactorily representing the actual atmosphere for many purposes. (2) A hypothetical, vertical distribution of temperature, pressure and density which, by international agreement, is taken to represent the actual distribution for many purposes. The standard atmospheres of most geodetic importance are: (a) the ICAO Standard Atmosphere (1964); (b) the COSPAR CIRA 1965 and CIRA 1972; (c) the ISO Standard Atmosphere (1973); (d) GOST 1964; and (e) the U.S. Standard Atmosphere (1976). The first three are international standards, the fourth is standard for Russia, and the fifth is standard for the U.S. Air Force and the U.S. National Oceanic and Atmospheric Administration. Atmosphere (e) agrees with the ICAO Standard Atmosphere up to 32 km and with the ISO Standard Atmosphere (1973) up to 50 km. Above 50 km and up to its limit of 1000 km, atmosphere (e) is based on data from rockets and artificial satellites. Also of geodetic importance is (f) Table 51, Smithsonian Meteorological Tables (5th edition, 1939). This standard atmosphere was used as the basis for the scales on various barometric altimeters such as those made by Wallace and Tiernan. (3) SEE atmosphere (3). (4) SEE atmosphere (4).

atmosphere, supplementary - A set of values (formula or tables) representing the temperature, pressure, and other characteristics of the Earth's atmosphere for a specific region or part of the year. It is derived from observations in the region or season involved and depends less on theory than does a standard atmosphere.

atoll - A ring-shaped, organic reef inclosing a lagoon. SEE ALSO coral atoll.

Atomgewichtsilber - A pure silver used in the 1938 redetermination of atomic weights. It is specified as the silver to be used in the chemical determination of chlorinity of sea water.

attachment, solar - A device which can be attached to an engineer's transit or other kind of theodolite, to permit the instrument's use as a solar compass.

attachment, valley-crossing - An attachment to the telescope of a leveling instrument, consisting of a wedge-shaped prism rotatable about an axis parallel to the optical axis of the telescope and deflecting the line of sight. It is used in certain techniques devised particularly for leveling across valleys, rivers, and large bodies of water.

attenuation - Any process in which the flux density, power, amplitude, intensity, or illuminance, etc., of radiation decreases with increasing distance from the source.

attenuation coefficient - A measure of the rate at which flux density, power, amplitude, intensity, or illuminance, etc., of radiation decreases with distance from the source. This quantity may be identified with k in a form of Bouguer's (Beers') law: $dI = k I dr$. Here I is the flux density, r is distance from the source and k is the attenuation coefficient. Also called attenuation constant, attenuation factor or decay constant. Extinction coefficient may be considered a synonym, but usually refers to visible radiation.

attenuation constant - SEE attenuation coefficient.

attenuation factor - SEE attenuation coefficient.

attitude - (1) The orientation of a body, as specified by the angles made by a set of axes fixed in the body with a set of axes used as reference. In particular, in photogrammetry, (2) the orientation of a camera, or of the photograph taken with that camera, with respect to some external reference system. Usually expressed as tilt, swing and azimuth, or as roll, pitch and yaw.

attraction, local - SEE magnetic anomaly, local.

augmentation - (1) The number of times the apparent topocentric, angular diameter of a celestial body is greater than the geocentric, angular diameter of that body. (2) The apparent increase in diameter of a celestial body as its angular elevation increases, because the body is closer to the observer. This should not be confused with the apparent increase in size of the Moon or Sun at the horizon as compared to the size when well above the horizon. This latter phenomenon seems to be psychological rather than physical in nature.

augmetation correction - A correction for the effect of augmentation on a measurement, and particularly that correction made to an angular elevation measured with a sextant.

augmenting easement - SEE easement, augmenting.

August's map projection - A conformal map projection of the whole sphere into a two-cusped epicycloid. It was published by F. August in 1874. The sphere is first mapped conformally onto a disk. A two-cusped epicycloid is generated by rolling a circle with radius ½ on the circumference of the disk within which the sphere was mapped. The equation of the epicycloid is found in terms of the x, y coordinates of the circumference, thus mapping the 180°-meridian onto the epicycloid. These same equations therefore map the interior of the circle into the interior of the epicycloid.

austral - Southern.

Australian Geodetic Datum - A geodetic datum in use before 1965 and based on a rotational ellipsoid with semi-major axis 6 378 165 meters, flattening 1/298.3. The origin was at CENTRAL (GRUNDY) with coordinates
longitude 134° 32' 46.457" E.,
latitude 24° 54' 11.078" S.,
elevation 1304 feet above the geoid.
The axis of rotation was defined to be parallel to the Earth's own axis of rotation.

Australian Geodetic Datum, 1965 - A geodetic datum replacing, in 1965, the earlier Australian Geodetic datum and based on a rotational ellipsoid with
semi-major axis 6 378 160 meters,
flattening 1/298.25 .
The origin is at Johnston Geodetic Station, with coordinates
longitude 133° 12' 30.0771" E.
latitude 25° 56' 54.5515" S.
elevation 571.2 m above the geoid
height of geoid - 6 meters.
The minor axis of the ellipsoid was defined to be parallel to the average position of the Earth's axis of rotation in 1961. In 1970, the definition was changed to make the minor axis parallel to the CIO axis. The datum was adopted in 1965.

Australian Height Datum - A vertical-control datum in Australia, based on a defined geoidal height of -6 meters at Johnston Geodetic Station.

Australian Map Grid - A grid system used with Australian maps on the transverse Mercator map projection and having its origin on the equator. Eastings are given the value 500

000 plus the value given by the formula for distances from the central meridian; northings are given the value 10 000 000 plus the negative value given by the formula for distances from the equator. Each grid covers a sector 6° wide, with ½° overlap on adjacent grids. The central scale factor is 0.9996 in all directions. Officially called the Australian Map Grid (AMG).

Australian Mag Grid System - SEE Australian Map Grid.

Australian National Spheroid - SEE spheroid, Australian National.

auto-centering - The interchange of instrument with target, sub-tense bar, etc., on a tripod in such a manner that centering over the marker is maintained automatically.

autocollimation - (1) The process of using an autocollimator. (2) The procedure used to determine or transfer azimuth to an instrument or device by using an autocollimator.

autocollimator - A telescope containing illuminated crosshairs placed so that they are imaged, by the objective lens-system, at infinity (i.e., light from it is collimated by the objective lens-system). When the telescope is aimed at a flat reflector placed approximately perpendicular to the telescope's optical axis, the crosshairs and their reflected image can be viewed together through the telescope's eyepiece. The reflected image will be displaced from the direct image by twice the angle between the optical axis and the perpendicular to the reflector.

autocorrelation - The correlation of the members of a time series with each other.

autocorrelation function - The limit, as T approaches infinity, of $(1/2T)$ times the integral of $f(t) f(t - \tau)dt$ from $-T$ to $+T$; τ is a parameter.

auto-covariance - SEE covariance; covariance function.

automecoic (adj.) - True scale.

autopositive - (1) A black-and-white reproduction, as of a map, plan, or photograph, on transparent film in positive form and made directly from a positive. (2) A material which gives a positive copy from a positive transparency (or a negative from a negative transparency) by direct development. Also termed a direct copy or direct positive.

autoregression - A relationship connecting the members of a time series in such a manner that the value at one instant is expressible in terms of the values at previous instances, plus a random value.

autoscreen film - SEE film, autoscreen.

average - A single number calculated from a set of numbers and used to represent the set. Also called the mean. The most common types of average are the arithmetic average, the geometric mean, the median, and the mode. If the term average is used alone, the arithmetic average is usually meant. The average of a continuous function is the integral of that function with respect to the independent variable, divided by the difference between the upper and lower limits of integration. This is a generalization of the arithmetic average, but is always referred to simply as the average.

average, arithmetic - The sum of all the numbers in a set, divided by the number of numbers in the set. Also called the average, mean, arithmetic mean or expectation. The term expectation should not be used unless the set of numbers being averaged is known to have a Gaussian distribution. Thomas Simpson (1755) was the first to consider taking the average of a number of measurements to reduce the effect of random errors.

average, geometric - SEE mean, geometric.

average, harmonic - SEE mean, harmonic.

average, moving - (1) One of a set of averages over subsets of fixed size n in a set of size N>n of ordered numbers, the numbers in the subsets being in the same sequence as the numbers in the full set and each subset advancing by the same amount in the sequence. Also called a consecutive mean, overlapping mean or running average. The subsets usually overlap, but they may be disjoint. The numbers are usually ordered at fixed intervals of time. (2) One of a set of averages, between variable limits but over a fixed interval, of a continuous function. The lower and upper limits are of the form $a + n \delta$ and $a + \Delta + n \delta$, respectively, where a, Δ and δ are constants and n is an integer going from 0 to some suitable upper limit. The quantity being integrated is usually a function of time. A graph of the moving average against time is a smooth version of the graph of the original function. The narrower the interval, the closer the derived graph is to the original.

average, running - SEE average, moving.

average, weighted - The sum of the products of each number in a set {X} multiplied by a corresponding number in a set (w), this sum being divided by the sum of the numbers in set (w). I.e., the weighted average X(W) is given by $(\Sigma wX)/(\Sigma w)$. The quantities w are often defined so that they lie between 0 and +1 and $\Sigma w = 1$. Also called the weighed arithmetic mean. (2) If X is a continuous function with the known probability density p(X), then the weighted average of X is the integral of $(p(X)X)dX$.

average-end-area method - The volume of earth to be added or removed between two cross-sections is assumed to be the average area of the two cross-sections, times the distance between them.

aviation map - SEE chart, aeronautical.

avoirdupois - SEE avoirdupois weight.

avoirdupois weight - That system of units based on the pound (weight) for measurement of weight and mass. Also called avoirdupois. At present, avoirdupois weight is used almost exclusively in the United States of America. It includes the ounce, the pound, and the ton. It is not used for weights of precious stones or precious metals.

avulsion - (1) The sudden removal of land from the estate of one person to another. (2) The breaking of a stream through its banks in a sudden and unexpected manner and in such a way as to form another channel. The term is of legal significance when the avulsion results in cutting off a large amount of land from one owner and adding it to another's land. Also called revulsion. (3) The rapid erosion of a shore by waves during a storm. Some courts assign only meaning (1) to the term. (4) The loss of lands bordering on the seashore by sudden or violent action of the elements, perceptible while in progress; a sudden and rapid change in the course and channel of a boundary river. None of the above results in a change in the boundary.

axis - (1) Any line along which measurements are made in determining the coordinates of a point, or any line from which angles are measured for the same purpose. (2) Any line about which a body rotates or revolves. In geodetic and astronomic instruments, the line usually coincides with the axis (sense 2) of a cylindrical rod or tube carried in a bearing, so the term axis is also applied to this rod or tube. However, the term axle for the solid is more common and is preferred. (3) A straight line connecting two related points as, for example, the poles of a magnet. (4) (of an aircraft) - One of the three lines fixed in an aircraft which pass (usually) through the centroid of the aircraft, are mutually perpendicular, and are directed fore and aft, up and down, and from side to side. The fore-and-aft axis (longitudinal axis) lies in the aircraft's vertical plane of symmetry and is horizontal when the aircraft is flying level. The up-and-down axis also lies in the vertical plane of symmetry. (5) (of a spirit level) - The line tangent to the top inner surface of the tube (vial) of a (toroidal) spirit level, at the center of the graduated scale of the level, and in the plane of the tube and its center of curvature. Also called axis of the level, axis of the level bubble, spirit-level axis and level axis. When the bubble is in the center of the graduations and the plane through the center-line of the vial and the center of curvature is vertical, the axis of the spirit level will be horizontal. (6) SEE axis of symmetry. SEE ALSO axis of abscissas; collimation axis; axis of ordinates; topple axis.

axis, bubble - SEE bubble axis; axis of the level bubble.

axis, equatorial - (1) A diameter of the Earth, drawn between two points on the Equator. (2) A line drawn thru the center of an ellipsoid, in a plane perpendicular to the shortest axis of the ellipsoid, and extending from one side of the ellipsoid to the other. (3) An axis, in the mounting of a telescope, which is parallel to the plane of the equator.

axis, fiducial - The line joining two opposite fiducial marks on a photograph.

axis, horizontal - The axis (line) about which the telescope or alidade of an instrument rotates when rotated in a vertical plane. For an instrument in perfect adjustment, it is perpendicular to the standing axis of the instrument and to the collimation axis of the telescope. It should coincide with the line through the centers of the pivots that support the telescope. For an instrument in perfect adjustment and properly leveled, this axis is horizontal; when the telescope is rotated around it, the collimation axis will define a vertical plane. Deviations from proper instrumental adjustment are measured with a striding level or a hanging level.

axis, level - SEE axis (5).

axis, longitudinal - An axis generally along the main body of a craft, in the direction of motion. Its precise definition is usually set by the constructor of the craft. It frequently serves as a reference for directions in navigation and in surveying.

axis, main - SEE axis, standing.

axis, major - (1) The longest line that can be drawn thru the center of an ellipse or ellipsoid to the figure. (2) Any one of the set of lines of greatest length that can be drawn thru the center and from one side to the other of a spheroid. It is usually referred to as a major axis or an axis in the equatorial plane if the set contains more than one such line. Note that the major axis may also be the axis of rotation.

axis, mechanical - The central line of an optical device as defined by the surfaces on which the device rests, i.e., where the optical axis of the device ought to be.

axis, minor - The shortest line that can be drawn thru the center and from one side to the other of an ellipse or ellipsoid.

axis, normal - An axis perpendicular to the longitudinal axis and generally in the plane of symmetry of the craft. It is oriented downwards or as defined by the constructor of the craft.

axis, optical - (1) A prolongation of the line passing through the nodal point of a lens system. (2) The line connecting the centers of curvature of all the surfaces in a lens system. (3) The line perpendicular to all surfaces of a centered lens system.

axis, perspective - SEE axis of homology.

axis, polar - (1) The axis about which the Earth, or an figure representing the Earth, rotates. (2) The shortest diameter of an ellipsoid. (3) The shortest diameter of an ellipse. (4) That axis, in a polar or spherical coordinate system, which is the primary axis to which directions are referred (i.e., from which angles are measured.

axis, primary - SEE axis, standing.

axis, principal - (1) Any one of the three (two) perpendicular lines through the center of an ellipsoid (ellipse) that are the shortest and longest of all lines through the center (and terminated by the figure). If the ellipsoid is represented by a general polynomial of the second degree in x,y,z, all terms except those in x^2, y^2 and z^2 can be removed by a sequence of rotations and translations. This is called a reduction to principle axes. (2) The line perpendicular to both surfaces of a lens. In a well-centered lens-system, in which the principal axes of the components are in the same straight line, that line is called the optical axis or, simply, the axis of the system. The principal axis should never be called the optic axis, which is a term used in crystallography for something completely different. (3) SEE moment of inertia.

axis, semimajor - The line from the center of an ellipse to the extremity of the longest diameter. I.e., one of the two longest lines from the center to the ellipse. The term is also used to mean the length of the line.

axis, semiminor - The line from the center of an ellipse to the extremity of the shortest diameter. I.e., one of the two shortest lines from the center to the ellipse. The term is also used to mean the length of the line.

axis, standing - The axis, in a surveying instrument such as theodolite or leveling instrument, that is intended to be placed in a vertical position when the instrument is in use. This axis is also commonly called the vertical axis or, less frequently, the main axis or primary axis. However, the standing axis is never exactly vertical and may, in automatic leveling instruments, be more than a degree from the vertical while leveling. To call it the vertical axis can therefore be misleading.

axis, transverse - (1) An axis perpendicular to the longitudinal axis of a craft, generally in the plane of symmetry of the craft, and oriented positively to the right looking forwards. (2) The line joining the apsides of an orbit. It is identical with the major axis of elliptical orbits.

axis, vertical - The line about which the telescope or alidade of an instrument rotates when moved horizontally. For an instrument in perfect adjustment and properly leveled, this axis is vertical, passes through the center of the horizontal circle, and is perpendicular to the plane of that circle. The term standing axis or azimuth axis is preferred.

axis of abscissas - The x-axis of a rectangular coordinate (x,y) system in which the x- and y-axes are mutually perpendicular. SEE ALSO coordinate system.

axis of camera - An imaginary line through the optical center of the lens-system and perpendicular to focal plane of the camera.

axis of collimation - SEE line of collimation. It is not the same thing as collimation axis. Hence, the term is best avoided altogether.

axis of homology - (1) The line of intersection of two projectively related planes i.e., of two planes containing projectively related figures. (2) The intersection of the plane of a photograph with the horizontal plane of reference at the ground or with the plane of a horizontal map. Also called axis of perspective and perspective axis. The ground parallel and map parallel are special cases of the axis of homology. Corresponding lines in the planes of photograph and map intersect on the axis of homology.

axis of inertia - SEE axis, principal.

axis of lens - SEE axis, optical.

axis of level - SEE axis (5).

axis of ordinates - That one, of a pair of directed straight lines intersecting at a point (the origin), along which distance is measured to determine the coordinate called ordinate of a point in the plane determined by the two lines. Also called ordinate axis and y-axis. The two lines are usually perpendicular and, in geodesy, graduated to the same scale. SEE ALSO coordinate system.

axis of perspective - SEE axis of homology.

axis of sighting - The line drawn through the optical center and centers of curvature of the lenses of a telescope.

axis of spirit level - SEE axis (5).

axis of symmetry - A straight line through a figure bisecting all straight lines drawn perpendicular to it and terminating on the figure. I.e., a straight line with respect to which a geometric figure is symmetrical.

axis of the level bubble - SEE axis (5).

axis of tilt - A line through the perspective center of a photograph and perpendicular to the principal plane of the photograph.

azimuth - (1) A horizontal angle reckoned clockwise from the meridian. Thebasic control surveys of many countries have been based upon azimuth from the southern USA. In military control surveys of most countries, including the USA, azimuths are reckoned clockwise from north. Beginning in 1985, azimuths for basic control surveys in the USA are measured clockwise from north. (2) A horizontal angle measured clockwise from a specified reference point (usually north) to the intersection with the great circle drawn from the zenith through a body on the celestial sphere. SEE ALSO grid azimuth; Laplace azimuth; Laplace formula; Laplace point.

azimuth, assumed - An azimuth chosen as a matter of convenience, usually by assigning north-south line in a survey as the meridional direction, or by defining an east-west line as being at right angles to the meridional direction.

azimuth, astronomic - (1) The angle, measured in a horizontal plane, from the vertical plane through the celestial pole to the vertical plane through the observed object. Also called an astronomical azimuth, true azimuth and zenith angle. This last usage is rare and may result from a misunderstanding of the word's meaning (an angle measured from the zenith, not at the zenith). The astronomic azimuth is the azimuth which results when the reference direction is established directly from observations on a celestial body. It is measured in the plane of the horizon and is usually reckoned clockwise from north through 360°. In navigation, it is sometimes reckoned clockwise or counterclockwise through 180° from south in the southern hemisphere and from the north in the northern hemisphere. The National Ocean Survey of the USA reckons azimuth clockwise from north (0°) through east (90°), south (180°) and west (270°), back to north (360° or 0°). In recording an astronomic azimuth, both the initial direction and the direction of reckoning must be indicated. Astronomic azimuths differ in value from geodetic azimuths because of the local deflection of the vertical (station error) which, in the USA, produces differences of up to 10" in the eastern regions or as much as 26" in the mountainous western regions. In extreme cases, the difference may be as much as 1'. (2) The angle, at the point of observation, measured from the vertical plane through the celestial pole to the vertical plane through the observed object. This is a dihedral angle and is not properly an azimuth.

azimuth, astronomical - SEE azimuth, astronomic.

azimuth, back - The azimuth of a line at the end (the end point) opposite that used for reference. If, on an ellipsoid, the azimuth of point B from point A is given, the back azimuth is the azimuth of point A from point B. Because of the meridional convergence, the forward and back azimuths of a line do not differ by exactly 180° except where A and B have the same geodetic longitude or where the geodetic latitudes of both points are 0°. If the two points are on a plane, in a rectangular Cartesian coordinate system, the same relationship between forward and back azimuths holds, but the two azimuths differ by exactly 180°.

azimuth, forward - The azimuth of a line as seen from the end (the initial point) used for reference. SEE ALSO azimuth, geodetic.

azimuth, geodetic - The angle, at a point A on a surface, between the tangent to the meridian at A and the tangent to the geodesic from A to the point B whose geodetic azimuth is wanted. The angle is measured from the direction of north clockwise. (Until 1985, the U.S. National Geodetic Survey (NGS) and some other organizations measured the angle from the direction of south clockwise.) The angle is called the forward azimuth for the line AB. The azimuth at B between the tangent to the meridian and the tangent to the geodesic is called the back azimuth or reverse azimuth of the line AB. The geodesic is not, except in special cases, a plane curve. However, for short lines of the length commonly used in triangulation, the small difference between the tangent to the geodesic and the tangent to the actual line of sight between A and B can be neglected.

azimuth, inertial - An azimuth measured with respect to lines whose directions are established by a gyrocompass. In particular, an azimuth measured using a gyrotheodolite (gyroscopic theodolite).

azimuth, magnetic - The angle, at the point of observation, between the vertical plane through the observed object and the vertical plane in which a freely suspended, symmetrically magnetized needle, influenced by no transient or artificial magnetic disturbance, will come to rest. Magnetic azimuth is generally reckoned from magnetic north (0°) clockwise through 360°. Such an azimuth should be marked as being magnetic, and the date of its applicability be given.

azimuth, normal-section - The angle at a point A, on the ellipsoid, between the meridional plane through A and the normal section which passes through A and B.

azimuth, primary - A astronomic azimuth taken as the astronomic azimuth of a side in first-order (primary) triangulation.

azimuth, reverse - SEE azimuth, geodetic.

azimuth, secondary - An astronomic azimuth taken as the astronomic azimuth of a side in second-order (secondary) or lower triangulation.

azimuth, true - SEE azimuth, astronomic.

azimuth angle - (1) The angle between the plane of the celestial meridian and the vertical plane containing the observed object, reckoned from the direction of the elevated pole and of size less than 180°. It is the spherical angle at the zenith in the astronomical triangle with vertices: pole, zenith, star. (2) The angle, taken clockwise, from the vertical plane through the celestial pole (usually the north pole) to the vertical plane through the observed, terrestrial object. Probably synonymous with azimuth. Not to be confused with azimuth angle as used in traverse and triangulation. Hence, azimuth should be used instead of azimuth angle if the meaning is the same. SEE ALSO (3) following. (3) An angle, in traverse or triangulation, through which the computation of azimuth is carried. In a simple traverse, every angle is an azimuth angle. Sometimes, to avoid carrying azimuths over very short lines, supplementary observations are made over comparatively long lines, the angles between these lines forming azimuth angles. In triangulation, certain angles, because of their size and position in the figure, are selected for use as azimuth angles and enter into the formation of the condition equation (azimuth equation) governing azimuths.

azimuth bar - SEE azimuth instrument.

azimuth by altitude - An azimuth determined by solving the astronomical triangle with latitude, declination and angular elevation given.

azimuth circle - A ring made to fit snugly over a compass or compass repeater and provided with a means for observing bearings and azimuths.

azimuth coordinate system - SEE coordinate system, horizontal.

azimuth determination - The determination of astronomic azimuth. Although the term properly should include determination of geodetic azimuths, it is used only for astronomic azimuths. The process consists essentially of measuring the angle between the object whose azimuth is wanted and north established by observations on some celestial body under known conditions (such as at equal distances from the zenith before and after transit of the meridian, etc. For specific methods of determining azimuth, see for example, astronomic-direction method; Black's method; equal-altitude method; micrometer method; Polaris hour-angle method; solar-observations method; and azimuth determination from rate of change of zenith distance. SEE ALSO Black's method; solar-observation method.

azimuth determination by solar observation - SEE solar-observation method (of determining azimuth).

azimuth determination by the astronomic-direction method - SEE astronomic-direction method (of determining azimuth).

azimuth determination by the method of equal altitudes - SEE equal-altitude method (of determining azimuth).

azimuth determination by the method of repetitions - SEE repetition method.

azimuth determination by the micrometer method - SEE micrometer method.

azimuth determination from the hour angle of Polaris - SEE Polaris hour-angle method.

azimuth determination from the hour-angles of crossing of the alumcantar - SEE equal-altitude method.

azimuth determination from the rate of change of zenith distance - A variant of the astronomic-direction method, in which the astronomic azimuth of a star is calculated from the rate of change of zenith distance with time.

azimuth determination from the Sun's angular elevation - Determination of astronomic azimuth by measuring the Sun's angular elevation, the angle of the marker from the Sun at that instant and the time of measurement. The formula is $\cot^2(\frac{1}{2} A_s) = [\sin\frac{1}{2} (S - \phi_a) \sin(S + \theta)]/[\cos S \cos (S + \delta)]$. The Sun's azimuth is A_s, the astronomic latitude of the observer is ϕ_a, the zenith distance of the Sun is θ (90° - measured angular elevation), the Sun's declination at the instant of observation is δ and S is the quantity $(\phi_a - \theta - \delta)$. The azimuth of the Sun is added to the measured angle of the marker with respect to the Sun to get the azimuth of the marker.

azimuth equation - A condition equation expressing the relationship between the fixed azimuths of two lines connected by traverse or triangulation. When traverse or triangulation connects two lines whose azimuths are fixed by direct observations or by previous surveys, an azimuth equation is used to make the azimuth of either line (as computed through the adjusted survey) from the other line agree with its azimuth as previously fixed.

azimuth error - (1) The angle, in a horizontal plane, from a vertical plane through the horizontal axis of a meridian telescope to the prime vertical through the center of that axis. Alternatively, the angle by which the east-west axis of the telescope deviates from being perpendicular to the plane of the meridian. It is considered positive if the nominally west end of the horizontal axis points south of west. (2) The angle defined similarly (to 1) for any telescope with an alt-azimuth mounting when the horizontal axis should be perpendicular to the meridian. (3) For any instrument for which a nearly vertical plane can be defined, the difference between the azimuth of that plane and the azimuth in which it is supposed to lie. This definition is often applied to the calibration of radio telescopes.

azimuth error of closure - SEE error of closure in azimuth.

azimuth factor - The quantity a, in Bessel's formula or Mayer's formula, giving the amount by which the vertical plane perpendicular to the horizontal axis of a telescope diverges from a vertical plane through North. Also called azimuth error.

azimuth instrument - An instrument designed for measurement of azimuths. Also called an azimuth bar or a bearing bar. In particular, such a device which fits over a central pivot in the cover of a magnetic compass.

azimuth line - (1) A radial line from the principal point, isocenter, or nadir point of a photograph, representing the direction to a similar point of an adjacent photograph in the same flight line. It is used extensively in radial triangulation. (2) SEE azimuth plane.

azimuth mark - The marked point, on an azimuth marker, for which the azimuth(s) to one or more distant points is (are) given. SEE ALSO azimuth marker.

azimuth mark, astronomic - SEE azimuth marker, astronomic.

azimuth mark, geodetic - A marked point established in connection with a triangulation (or traverse) station to provide a starting azimuth for dependent surveys. The geodetic azimuth from the station to the azimuth mark is determined instrumentally; its distance is often determined only approximately but should be accurate enough that the effect, on measured angles, of the ordinary errors made in horizontal control of the USA. These markers are usually located so they are readily available, without special construction to elevate either instrument or centering instrument and target, can be ignored. Since 1928, azimuth markers consisting of bronze tablets set in concrete or stone have been established in connection with the basic target. At a station having an established azimuth mark, both the geodetic azimuth and the grid azimuth of the mark on the State Plane Coordinate System have been computed and published.

azimuth mark, Laplace - SEE mark, Laplace-azimuth.

azimuth marker - (1) A geodetic monument carrying a mark whose azimuth from a given point is known either by measurement or by definition. (2) A marker set at a significant distance from a triangulation or traverse station to mark the end of a line for which the azimuth has been determined and to serve as a starting or reference azimuth for later use.

azimuth marker, astronomic - A signal or target whose astronomic azimuth from a survey station is determined by direct observations on a celestial body. The marker may be a lamp or illuminated target placed especially for the purpose or may be a signal lamp at another survey station.

azimuth misclosure - SEE misclosure in azimuth.

azimuth plane - Any vertical plane. Such a plane cuts an ellipsoid in an azimuth line or in an azimuth section. These terms are rarely encountered and should not be used. Although the vertical planes containing the prime vertical and the meridian are sometimes thought not to be azimuth planes, they are actually special cases of that term as here defined. Also called azimuth section.

azimuth section - SEE azimuth plane.

azimuth station, Laplace - SEE Laplace station.

azimuth transfer - The process of connecting, with a straight line, the nadir points of two vertical photographs selected from overlapping strips.

azimuth traverse - A survey traverse in which the direction of a measured course is determined by azimuth and verified by taking a back azimuth. To make this kind of traverse, an initial meridian usable as reference must be established.

B

Babinet's map projection - SEE Mollweide's map projection.

Bache-Wurdeman base line apparatus - A compensating base-line apparatus composed of a bar of iron and a bar of brass, each a little less than 6 meters in length, held together firmly at one end, with the free ends so connected by a compensating lever as to form a compensating base-line apparatus.

back - The top of a drift, cross cut or stope. SEE ALSO pressure back; vacuum back.

back, locating - A flat surface, in an aerial camera, located parallel to but out of the focal plane by an amount equal to the thickness of the photographic film. The film is held against the locating back by a vacuum or air pressure so that the emulsion lies on the focal plane at the instant of exposure. Locating backs are usually of metal. They are perforated or slotted to allow a differential pressure to be built up or air to be removed to form a vacuum.

back fill - Waste rock or other material filling a mined-out stope to prevent caving in.

background - The effects always present along with a particular phenomenon and from which the phenomenon must be distinguished in order to be measured. Sometimes restricted to denote only effects in an apparatus. Also called background noise, although this term should not be used unless the effects are random.

background noise - Noise from sources other than the source of the desired signal.

backscatter - (1) That energy which is scattered from a surface on which energy is incident and which lies within a hemisphere having its center on the surface and including the source of the radiation. (2) That part of the energy incident on a surface which is neither transmitted nor reflected. The term may or may not include energy absorbed at the surface and re-radiated.

backshore - (1) That zone of the land lying between mean (ordinary) high water and the place, landward, where there is a marked change in material or physiographic form, or to the line where permanent vegetation begins. (2) That part of a beach which is usually dry, being reached only by the highest tides and waves.

backside - In English law, a yard at the back part of or behind a house and belonging thereto. The term was formerly used in conveyances and also in pleading.

backsight - (1) A sight to a previously established point of a survey, or the reading or measurement obtained by that sight. (2) In leveling from an initial point to a final point through a sequence of intermediate points, a sight to (or the reading on) a leveling rod held on a preceding point. The sight to, or reading on, a leveling rod held on the succeeding point is called the foresight. In leveling, a backsight is sometimes called a plus sight because its value is added to the elevation of the point on which the leveling rod is placed, to obtain the elevation of the leveling instrument. But if the sight is to a mark on a wall or is in the roof of a mining tunnel, with the instrument at a lower elevation than the mark, the backsight will be subtracted from the known elevation to obtain the height of the instrument. The term backsight is preferred over plus sight. Neither of these definitions requires that the point or leveling rod to which the sight is made be one whose coordinates have been defined or determined previously. If the sequence consists of only two points, it is a matter of indifference which is called the backsight and which the foresight. (3) The visual intersection of the horizontal crosshair of a leveling instrument with a staff or leveling rod held vertically over or under the preceding point when the line of sight is horizontal or when the vertical angle of the line of sight is known or will become known from actual measurement. (4) The angle measured on the horizontal circle of a surveying instrument (set or to be read and recorded) when the vertical crosshair of the instrument coincides visually with the preceding point and the center of the instrument.

backsight method - A method of traversing by plane-table, in which the alidade is first placed on a selected line on the map which corresponds to an established line on the ground. The table is then rotated until the line of sight through the alidade coincides with the corresponding line on the ground.

backstaff - An instrument for measuring the Sun's angular elevation by determining the location, on a graduated staff, of the shadow of a horizontal vane. It was invented by John Davis in the 16th century and comes in two forms: one for angular elevations less than 45° and one for angular elevations of greater than 45°.

backstep - A method of determining the offsets for the line representing the lowest latitude of a map projection by measuring the appropriate distances down from the line representing the highest latitude.

backup - (1) An image printed on the reverse side of a map sheet which already has printing on the obverse side. (2) The printing of such images.

backwater - (1) A body of water held back from its natural course by some obstruction such as a contrary current, dam, etc. (2) A region affected by water held back from its natural course by some obstruction.

badlands - Terrain characterized by a very fine network of drainage, with short, steep slopes and narrow ridges between valleys. Vegetation is usually sparse.

Baker-Nunn camera - A telescope having a spherical primary mirror and a specially-figured correcting plate, and a special mounting for tracking artificial satellites. The optical system was designed by J.A. Baker and the mounting by J. Nunn. It is a variation of the Schmidt type, with the focal surface having a toroidal shape instead of a spherical shape. The correcting plate contains three elements, giving an aperture of 50 cm and a focal length of 50 cm. The field of view is 5° by 30°. The mounting is of the alt-azimuth type but allows tracking in right ascension and declination also. The optical system and mounting together allow a 12-th magnitude star to be photographed when the direction of pointing is fixed, or allow an object of 8.5 magnitude moving at 0.1° per second to be photographed when the telescope is following the motion.

balance - An instrument for measuring the force of gravity or for comparing the force of gravity on two masses. The common form of balance is a rigid beam balanced at its mid-point and having a means for adding weights to both ends. It was used in some early measurements (e.g., by Poynting in 1891) of the gravitational constant, but has been replaced by the torsion balance, which is much more sensitive. The spring balance consists of a spring suspended at one end and having the weight to be measured attached at the other. In its common form, it is too insensitive to be used for precise measurements. However, specially designed versions are used as gravimeters. SEE ALSO gravimeter; spring balance; torsion balance.

balancing a survey - SEE balancing a traverse.

balancing a traverse - Calculating, according to some reasonable hypothesis of how the errors of a traverse behave, a set of values for the angles and lengths in the traverse which eliminate the mathematical misclosure that exists if the measured values are used. The term is equivalent to traverse adjustment, although the latter term is sometimes restricted to mean only traverses adjusted by the method of least squares. Also called balancing a survey. In the older literature, the process is sometimes described as distributing corrections through a traverse. A traverse is usually a loop beginning and ending at a known point, or is a line beginning and ending at known points. The misclosure is then the only datum available for making corrections to the several courses of the traverse. There are always fewer misclosures than there are corrections to be made. The lack of sufficient misclosures must therefore be made up for by assumptions about how the errors behave from course to course in the traverse. The rules or methods used in adjusting a traverse therefore differ according to the according to the kinds of assumptions made. SEE ALSO Bowditch's method; compass rule; Crandall's rule; Jolly's method; least-squares method (of adjusting a traverse; transit rule.

balancing backsight and foresight distances - SEE balancing sights.

balancing lines of sight - SEE balancing sights.

balancing sights - Making observations, in leveling, in such a way that distances to the two leveling rods are approximately equal. Also referred to as balancing lines of sight and balancing backsight and foresight distances. The difference in distances is sometimes called the imbalance.

baldio - (1) Waste lands. (Spanish) (2) Unappropriated public domain, not set apart for municipalities. (Texas)

Baldwin solar chart - A diagram designed for orienting a planetable by means of the Sun's shadow.

balk - (1) A low ridge of earth which marks a boundary. (2) A low ridge of earth left unplowed to mark a boundary or thru carelessness.

ball (mathematics) - A sphere plus all points inside the sphere.

ballast - Selected material, such as crushed stone, placed on the roadbed to hold the railroad tracks in place.

ballistics - The science of projectiles. It is usually divided into three parts: interior ballistics, which deals with motion of the projectile while it is within or on the launching apparatus; exterior ballistics, which deals with the motion of the projectile after it leaves the apparatus; and penetration ballistics, which deals with the motion of the projectile after hitting the target or other material. Photogrammetry and surveying have been extensively applied, in exterior ballistics, to determining the trajectory of the projectile.

ballistics, celestial - Celestial mechanics applied to the motion of artificial satellites.

balloon - SEE radiosonde balloon; sounding balloon.

Bamberg broken-telescope transit - SEE transit, broken-telescope.

Bamberg transit - SEE transit, broken-telescope.

Banachiewicz-Cholesky-Craut method - SEE Cholesky's method.

banco - A portion of the flood plain or channel of a river cut off by the shifting of the course of the river. Also called a cutoff. In particular, land along an international boundary and remaining under the jurisdiction of the country from which it has been separated. Elimination of bancos means, in this context, transfer of sovereignty over the land affected.

band - (1) (geography) SEE zone. (2) (electromagnetism) SEE frequency band.

band, latitudinal - SEE zone (3).

band, spectral - A sequence of wavelengths or frequencies lying between two given values.

banding, cone-angle - In analytical photogrammetry, the separation of images into annular zones defined by specific bands contained within, usually, 5° of arc. Regions read then lie only in certain outer bands depending on, the calibration of the lens cone.

band inversor, Pythagorean - SEE Pythagorean band inversor.

bandwidth - The range of frequencies required to transmit information at a specified rate.

bank - (1) The rising ground bordering a sea, lake, or river. (2) (surveying) The continuous margin, along a stream, where all vegetation characteristic of dry land ceases. The right bank of a stream is to the right and the left bank to the left as one faces downstream. (3) A more or less flat-topped, elevated feature over which the depth of the water is relatively small, but is sufficient for navigation at the surface.

bank, left - That bank of a stream or river which is to the left of the observer when he is facing downstream.

bank, right - That bank of a stream or river which is to the right of the observer when he is facing downstream.

bank of a stream - SEE bank; bank, left; bank, right.

bar - (1) A unit of pressure equal to 106 dynes per square centimeter. Formerly called the barye. It was first suggested by Bjerknes in 1911. In SI units, it is 105 newton/m² or 105 Pa. The unit is used in both meteorology and oceanography. One standard atmosphere is 1.013250 bars. The pressure exerted by a column of water 1 m deep is very nearly 1 decibar (1 dbar), which makes the unit convenient for oceanographers. (2) A long object of uniform but arbitrary cross-section and of considerable mass. The width and depth of the solid are about the same. If the width is considerably greater than the depth, or vice versa, the object is called a sheet. (3) A ridge or mound of sand, gravel, or other unconsolidated material below the level of high water, especially at the mouth of a river or estuary, or lying a short distance from and usually parallel to a beach, and which may obstruct navigation.

bar, standard - A bar (2) on which representative units of length are marked off and used to calibrate other measures of length or distance.

bar, subtense - A horizontally-held bar carrying two easily visible marks a fixed, known distance apart, and used for determining the distance from an observer to the bar by means of the angle subtended at the observer by the marks on the bar. The marks are often placed on fittings mounted at the ends of the bar and connected by invar wires under slight but constant tension. This counteracts the effect of temperature on the distance between marks. In use, the subtense bar is mounted horizontally on a tripod and centered over the mark to which distance is to be determined. The angle subtended at the observing instrument is measured and automatically converted by the instrument to distance. The observer reads off the distance, not the subtended angle.

Baranov elevation - The value H^B_N found for the orthometric elevation H_N of a point P_N by dividing the value of the potential function W_N per unit mass at P_N (considering the value of the potential function on the geoid to be 0) by an approximation $\frac{1}{2}(\tau_\phi + g_N)$ to the average value of the acceleration of gravity between the geoid and P_N. g_N is the value of the acceleration at P_N, τ_ϕ is the value of the acceleration (calculated by a standard gravity formula) on the geoid at the latitude ϕ of P_N: $H^B_N = 2W_N/(\tau_\phi + g_N)$. W_N is calculated from the measured differences ΔH_n of elevation between points P_{n-1} and P_n along a route from point P_o on the geoid to P_N, and the measured or otherwise determine values gn of the acceleration at P_n: $W_N = \Sigma g_N \Delta H_n$, (n = 1 to N) or, for infinitesimal separations dH between points along the route, $W_N = \Sigma \int_{P_{n-1}}^{P_n} g_N \, dH$. Also called the Baranov height.

Baranov height - SEE Baranov elevation.

bar chart - SEE bar graph.

bar check - Checking the accuracy of or calibrating a depth sounder by suspending a bar or disk at various measured distances beneath the sound generator and measuring the depths independently with the depth sounder.

bar graph - (1) A set of rectangles of uniform width but varying heights, placed in a rectangular Cartesian coordinate with one side (the base) on the axis of abscissa and with regular intervals between the centers of the bases. Also called a bar chart, barograph and histogram. The heights of the rectangles (bars) are the values of the functions that the bars represent. The rectangles are usually filled in with black or with some suitable color, or each bar may be in a different color. Bars may be placed on top of or on the bases of other bars to give the values of particular components of the primary function. Bar graphs are used primarily for their visual effect and ease of understanding as compared with graphs of functions. They are therefore used mostly for showing the variations in simple things like density of population, costs of projects, and so on. However, they are also used to show the variation of elevations over a region, the relative performances of surveying instruments, etc. (2) A sheet on which are printed black, rectangular figures of equal length but of different widths and which is used for testing the resolution of lens systems, photographic systems, or other imaging systems. The figures are generally placed in groups of three or more, the figures within any one group being uniform in size and placed at uniform intervals from each other with the shorter sides on a common baseline. The interval between figures is usually equal to the length of the shorter side of the figures.

Barker's equation - The equation for the true anomaly of a body in parabolic orbit, as a function of time elapsed since passage of the body through perigee. Also called Barker's formula.

Barker's formula - SEE Barker's equation.

barleycorn - An old measure of length, equal to the average length of a grain of barley; the third part of an inch.

Barlow leveling rod - A self-reading leveling-rod on which graduations are marked by triangles each 0.02 foot in height. The Barlow leveling-rod is no longer in general use.

bar map scale - SEE scale (of a map); scale, graphic.

baroclinicity - That state of stratification, in a fluid, in which surfaces of constant pressure intersect surfaces of constant density.

barograph - (1) A recording barometer. (2) SEE bar graph (1).

barometer - An instrument for measuring atmospheric pressure. There are two basic types of barometers: the type in which atmospheric pressure is balanced by the weight of a column of liquid (usually mercury); and the type in which atmospheric pressure is balanced by a spring or other elastic device. Barometers of the latter type are called aneroid (without liquid) barometers. SEE ALSO base barometer; cistern barometer; hypsometer; manometer; mountain barometer; siphon barometer.

barometer, aneroid - A barometer which balances the atmospheric pressure against a mechanically elastic device. The usual form is a thin box of flexible, corrugated metal almost exhausted of air. When the atmospheric pressure increases, the box contracts; when the pressure lessens, the box expands. By a mechanical linkage, these movements are amplified and communicated to an index hand which registers on a graduated dial. The graduations on the dial are usually such as will show the atmospheric pressure in terms of the height of an equivalent column of mercury (a column exerting the same pressure). The aneroid barometer may be equipped with an auxiliary dial graduated to show the elevations, in feet or meters, corresponding to atmospheric pressures in some standard atmosphere. Another form is a thin, closed tube (Bourdon tube) of flexible metal wound into the form of a spiral and exhausted. The central end is fixed; the other end moves out or in as atmospheric pressure varies and tries to force the tube into a straight line or allows it to coil up. The free end is coupled by a linkage to the indicator on the dial. Also called a Bourdon gage or Bourdon-type barometer.

barometer, Bourdon-type - SEE barometer, aneroid.

barometer, mercurial - A barometer in which atmospheric pressure is balanced against the weight of a column of mercury. There are two types of mercurial barometer, differing in the form of apparatus used but not in the underlying principle - the cistern barometer and the siphon barometer (q.v.).

barometer, roving - That barometer, in the single-base method of barometric altimetry, which is moved to those points whose elevations are to be determined.

barometer elevation - The vertical distance between mean sea level of the ivory point (zero point) of a station's mercury-barometer. The term is denoted by the symbol Hz in international usage. The value of atmospheric pressure with respect to this elevation is termed actual pressure. The requirement that the elevation be referred to mean sea level may be replace by the requirement that it be referred to the geoid.

barotropy - The state of stratification of a fluid in which surfaces of constant pressure coincide with surfaces of constant density. It is the state of zero baroclinicity.

barrel distortion - Optical distortion in which the magnification in the image decreases radially (toward the edges, away from the optical axis. Also called negative distortion.

barrier - A partially emergent, bar-like ridge of sand or coarse sediment lying off a shore or a shoal, usually approximately parallel to the shore, and projecting from the flank of a headland or connecting two headlands. SEE ALSO barrier island.

barrier island - A narrow, sandy island exposed to the open ocean on one side and separated on the other side from the mainland by a narrow, shallow lagoon or embayment. There is a prominent chain of barrier islands bordering the mainland along the east coast of the United States, from Virginia on southward.

barrier reef - A partially submerged coral reef on the seaward side of a lagoon. A prominent example is the Great Barrier Reef of Australia.

bar scale - SEE scale (of a map); scale, graphic.

barycenter - The center of mass of a body or collection of bodies. Also called center of gravity, although this latter term has a slightly different meaning. E.g., the barycenter of the Earth and Moon lies on the line connecting the barycenter of each and is about 1700 km inside the Earth's surface. The coordinates x_{ci} of the center are given by $x_{ci} = (1/M) \int x_i \, dm$ where the integral is taken over all the elements dm of the total mass M and x_i is the ith coordinate of the element dm. In astronomy, the symbol S8 refers to the center of mass of the Sun and all the planets except Pluto; S4 refers to the center of mass of the Sun, Mercury, Venus, Earth, and Mars.

barye - (1) A pressure of 1 dyne per square centimeter. It is equivalent to one one-millionth of a bar. The unit was established in 1900 but has since been replaced by the Pascal. (2) SEE bar (1).

base - (1) An important point in a survey. A point used as origin or having a given coordinate, or having associated with it a value to which measurements elsewhere in the survey are referred. (2) A location or camp used as a center for surveying operations. (3) The material on which a map to be kept permanently for reference is drawn. (4) The material on which a map is drawn or which supports a photographic emulsion. SEE ALSO film base. (5) The plate, pier, column, or pedestal to which an instrument is firmly fixed while measurements are being made. A base not only provides a stable and steady support for the instrument but often provides also a point or surface to which measurements can be referred. (6) SEE base line. The term base is common in British practice; base line or baseline in American. (7) SEE air base; model base. SEE ALSO comparator base; film base; model base; photo base; platform, inertial.

base, broken - SEE base line, broken.

base, inertial - A base (5) (on which an instrument may be mounted) whose orientation with respect to the distant galaxies is constant. Also called inertial platform. The term is also applied (improperly) to a base whose orientation is known but is not necessarily constant. Note that an inertial base need not have an inertial coordinate system associated with it; the base may be accelerating. SEE ALSO mounting, inertial; platform, inertial.

base, leveling - SEE quadribrach; tribrach.

base, stabilized - A base (on which an instrument may be mounted) which has a constant orientation in a particular coordinate system. Also called a stabilized platform.

base, stereoscopic - The distance and direction between complementary image points in a stereoscopic pair of photographs.

base-altitude ratio - SEE ratio, base-height.

base apparatus - SEE base-line, apparatus.

base barometer - That barometer, in the single-base method of barometric altimeter, which is left at the point of known elevation.

base color - On a polychromatic map, the first color printed to which succeeding colors are registered.

base construction line - The bottom line of a graticule, at right angles to the line representing the central meridian and along which the distances to other meridian-representing lines are established.

base correction - The adjustment made to measured values to refer them to the value established at the base station.

base density - The photographic density of the material supporting a photographic emulsion.

base direction - The direction of the vertical plane containing the air base. It may be expressed as an azimuth or as a bearing.

base film, topographic - A photographic film for use in aerial photography done for mapping purposes. The film has a dimensionally stable base.

base-height ratio - SEE ratio, base-height.

base in - That situation, in using some types of stereoscopic plotting instrument, in which the x-component of the air base, translated parallel to itself in the direction of a projecting ray down to the plane of projection, lies inside the parallelogram of Zeiss. Also called base inside. The situation in which the translated air base lies outside the parallelogram is called base out.

base inside - SEE base in.

base length, photographic - The distance, on a photograph, between the principal point of that photograph and the point which corresponds to the principal point on another photograph. The principal point on each photograph coincides with the image of a different corresponding point in object-space. Each of these points in object-space is imaged into a point, called the conjugate image or conjugate point on the other photograph. Each photograph therefore has a base line corresponding to the base line on the other, but usually of different length. Also called a photobase.

base level - A level surface from which vertical distances are measured or calculated.

base line - (1) An imaginary line connecting two points and used as a referent for measurements. Also spelled baseline. (2) A surveyed line established with more than usual care, to which surveys are referred for coordination and correlation. Also spelled baseline. Base lines are established for specific purposes. The more important ones are defined as follows. (a) The side of one of a series of connected triangles, the length of which is measured with prescribed ratio of standard deviation of the average to the length, and from which the lengths of the sides of the other triangles are obtained by computation. Base lines are classified according to the character of the network in which they are to be included. The instruments and methods to be used in their measurement are such that the prescribed ratios of standard deviation of the average to the length are not exceeded. The ratios specified by the U.S. Federal Geodetic Control Committee (1974) were as follows: for inclusion in

    first-order class I triangulation   1 : 1 000 000
    second-order class I triangulation 1 :   900 000
    second-order class II triangulation 1 :   800 000
    third-order class I triangulation   1 :   500 000
    third-order class II triangulation 1 :   250 000

A new set of standards for geodetic networks was published by the National Geodetic Survey in 1984. That set contains no standards for the accuracy of base lines. It does contain specifications for the errors allowable in measurements of distance using various kinds of instruments (other than tape or wire). Because the new set supersedes the set of 1980, there are no federal standards of accuracy for base lines per se. (b) A line serving as basis for measurement of other lines or the angles between them. This definition is clearer than that given in (a) and is recommended for geodetic purposes. (c) The center line (central line) of location of a railway or highway. It is often called the base line of location or baseline of location. (d) A line used as reference for distances and directions in the construction of a bridge or other structure. (e) A line extending east and west along the astronomic parallel passing through the initial point, along which standard township, section, and quarter-section corners are established. (USPLS) It serves as the starting line for the survey of the meridional township boundaries and section lines. Auxiliary governing lines, known as standard parallels or as correction lines, are established along the astronomic parallel - usually at intervals of 24 miles north or south of the base line. In some of the very early surveys, the base line is referred to as the basis parallel. (f) The line from which the outer limits of the territorial sea and other offshore zones are measured; the dividing line between inland waters and the territorial sea. This definition is part of the international Law of the Sea. (g) SEE baseline (1a). SEE ALSO Caithness base line; calibration base line; base line (2a); comparison base line; Loernermark base line; VÄisÄlÄ base line.

baseline - (1) An imaginary line connecting two points and used as a referent for measurements. More particularly, (a) The line joining the antennas of a pair of transmitting stations in a radio-navigation system. Also spelled base line. (b) The line joining the phase centers of the ends of an array of antennas. For example, the line joining the phase centers of the antennas at the two ends of a very-long-baseline interferometer. (c) SEE base line (2c). (d) SEE base line (2e). (2) SEE base line (2a).

base line, broken - A base line(2b), for triangulation, consisting of two or more lines that form a continuous line and have the same general direction.

base line, geodetic - SEE base line (2a) and (2b).

base line, normal - The line following the sinuosities of the low-water mark, except where indentations are encountered that fall within the category of true bays, when the base line becomes a straight line between headlands.

base line, photographic - That line, on a photograph, joining the principal point on that photograph to the point, on the same photograph, which corresponds to the principal point on another (overlapping) photograph.

baseline, standard - A line whose length is considered so accurate that it can be used for calibrating all kinds of distance-measuring equipment or apparatus (other than the kind used in establishing the baseline). In particular, a line established using the Väisälä base-line apparatus. SEE ALSO Väisälä base line.

base line apparatus - Any apparatus designed for accurate and precise measurement of the length of a base line in triangulation or of the length of a line in the first- or second-order traverse. The various kinds of base-line apparatus include the following. (a) Apparatus having a length unit composed of simple bars or tubes of wood, metal or other material, suitably constructed for determining alinement, contact and temperature. The earliest bars used in geodetic work were of wood. The most effective apparatus of this type was the iced-bar base-line apparatus. (b) Apparatus consisting of bars or tubes or different metals having unequal coefficients of thermal expansion, so arranged that either the effects of temperature on the lengths of the component parts are largely neutralized and the effective length of the apparatus remains almost constant, or the component parts form a metallic thermometer, by means of which the temperature of the apparatus is obtained and its effect on the length of the apparatus determined. (c) Apparatus consisting of tapes or wires of metal (usually steel or a nickel-steel alloy). Because of its large coefficient of thermal expansion, a steel tape or wire can be used for measuring base lines precisely only in favorable temperature. Although tapes (bands) are used in the USA and in Canada, wires have been used in Mexico and in Europe. (d) Apparatus consisting of a source of light, a collimator, a set of reflectors, and a receiver in which the phase of the received light is compared with the phase of the transmitted light by interference. (e) Apparatus consisting of a source of electromagnetic radiation (light, infrared radiation, or radio waves), a modulator of the radiation, a collimator, a set of reflectors, and a receiver in which the phase of the modulation on the returned modulation is compared to the phase of the modulation on the transmitted radiation. (f) Apparatus consisting of a source of radio waves, a modulator, a collimator, a receiver-transmitter which returns the signal to the original transmitter, and a receiver in which the phase of the modulation on the transmitted radiation is compared to the phase of the modulation on the returned radiation. Apparatus of types (a) and (b) is no longer used, while apparatus of type (c) is gradually falling out of use and being replaced by apparatus of types (d), (e), and (f). Types (e) and (f) are the most common, being used also for measuring distances in general. SEE ALSO Bache-Wurdeman base line apparatus; Bessel's base line apparatus; contact base line apparatus; Hassler base line apparatus; Jäderin base line apparatus; Repsold base line apparatus; Schott base line apparatus; Väisälä base line apparatus.

base line apparatus, compensating - A base-line apparatus with a length element consisting of two metals having different coefficients of thermal expansion, so arranged and connected that the differential expansion of its components maintained a constant length of the element at all temperatures encountered in use. SEE ALSO Jäderin wire.

base line apparatus, contact-slide - A modified contact base-line apparatus consisting of two steel measuring bars (rods), each 4 meters in length, so mounted that contact was effected by making lines on a rod and a contact-slide coincide. Each rod formed a metallic thermometer with two zinc tubes, one on each side of the bar; opposite ends of the bar were fastened to one end of each tube while the other end was free to move as temperature changed.

base line apparatus, duplex - A contact base-line apparatus consisting of two disconnected bars, one of brass and the other of steel, each 5 meters in length and so arranged as to indicate the accumulated differences of length of the measures from the brass and steel components. Similar in concept to Bessel's base-line apparatus.

base line apparatus, iced-bar - A base-line apparatus consisting essentially of a steel bar maintained at a constant temperature by being surrounded with melting ice. Also called the Woodward base apparatus. The bar was rectangular in cross-section and was carried in a Y-shaped trough which was filled with melting ice and mounted on a cart that moved on a track. The apparatus was used in the laboratory for calibrating tapes; it was also used to some extent for establishing base-line apparatus to be used in the field.

base line apparatus, optical - (1) A base-line apparatus consisting of bars whose lengths were defined by distances between lines at or near their ends, these lines being observed by suitably mounted and adjusted microscopes. (2) Any base-line apparatus by which distance is measured in terms of the speed or wavelength of light. The earliest and most accurate is the Väisälä base-line apparatus, which measures the base line in terms of the length of a standard bar, the difference between the unknown length and an integral multiple of the bar's length being measured in terms of the wavelength of light. Most other optical base-line apparatus uses the group velocity of light as an essential constant. SEE ALSO distance-measuring instrument, optical.

base line extension - SEE triangulation network, base-extension.

baseline extension - (1) The continuation, in both directions, of the line connecting the centers of the antennas of a pair of radio stations that are used in determining a line of position in radio-navigation. (2) SEE triangulation network, base-extension.

base line level - SEE level, base-line.

base-line measuring apparatus, duplex - SEE base line apparatus, duplex.

base line of location - SEE base line.

base line reduction - Conversion of the measured length of a base line to the length of a line lying on some other surface but having its end points on the same verticals or normals that pass through the end points of the base line. The surface chosen is usually the reference ellipsoid (base-line reduction to the ellipsoid) but may be to an ellipsoid approximating an equipotential surface through mean sea level at some point (base-line reduction to mean sea level) or to an ellipsoid approximating the geoid (base-line reduction to the geoid).

base line reduction to sea level - Conversion of the measured length of a base line to the length of the vertical projection of the base line onto a specified surface defined by reference to mean sea level. Theoretically, the specified surface should be an equipotential surface. Because the shape of the surface must be known for the definition to be practical and because the shape of an equipotential surface is known only approximately, if at all, a sphere or rotational ellipsoid is usually specified instead. This works because a base line is so short that a sphere or rotational ellipsoid is a very good fit to an equipotential surface over the length of the base line. Base lines are reduced to sea level when the location of the geoid and the location of the reference ellipsoid are not known.

base line reduction to the ellipsoid - Conversion of the measured length of a base line (or the equivalent length on the geoid) to the length of that line which is the orthogonal projection of the base line onto the reference ellipsoid. The quantity Δl is added to the measured length l; a formula often used is $\Delta l = -lh/(R + h)$, where h is the average geodetic height of the base line and R the average radius of the ellipsoid along the length of the base line. An error of 6 m in h causes an error of about 1 part in 10.6 in the reduced length. By adding, to the measured length L_m, corrections for refraction, calibration of the tape, etc., L_m is converted to the length L_c of the chord between the ends of the base line. L_c is then converted to the length L_{ce} of the chord joining corresponding points on the ellipsoid (points on the same normals as the end points). Finally, L_{ce} is converted to the length of the geodesic joining the two points on the ellipsoid. In the past, it was customary to do the reduction first from the measured length to the length of the corresponding line on an ellipsoid approximating the geoid and then to go to the length of the arc on that ellipsoid. Also called base-line reduction to the spheroid if, as is almost always the case, the ellipsoid is an ellipsoid of revolution.

base line reduction to the geoid - Conversion of the measured length of a base line to the length of a line that is the vertical projection of the base line onto the geoid. Sometimes referred to, erroneously, as base-line reduction to sea level. The correction ΔL is calculated from the formula $\Delta L = LH/(R + H)$ and added to the measured length L. H is the average elevation of the base line and R is the radius of the geoid at the midpoint of the base line. This formula is sufficiently accurate for most purposes.

base line reduction to the spheroid - Base-line reduction to the ellipsoid when the ellipsoid is a spheroid (ellipsoid of rotation).

base manuscript - SEE compilation manuscript.

base map - (1) A map on which information can be placed for comparison or for geographical correlation. The term base map was at one time applied to a class of maps now known as outline maps. It may be applied to topographic maps which are used in the construction of many special types of maps by adding particular data. (2) A map from which other maps are prepared by adding information. In particular, a planimetric map used in preparing topographic maps.

basement contour - A contour on the surface of the lower-most rock-complex or basic metamorphic and volcanic rocks underlying a region.

base net - SEE triangulation network, base-extension.

base network - SEE triangulation network, base-extension.

base out - That situation, in using some types of stereoscopic plotting instrument, in which the x-component of the air base, translated parallel to itself in the direction of a projecting ray down to the plane of projection, lies outside the parallelogram of Zeiss. Also called base outside.

base outside - SEE base out.

base plat - A plat showing only those details essential for identifying the legal subdivisions shown on it. A base plat may also show such cultural and topographic features as are thought desirable. However, it is now usual to show culture, topography, etc., by transparent overlays to the base plat.

base plate - That part of a theodolite which carries the lower ends of the foot screws and attaches the theodolite to the tripod for observing.

base sheet - A sheet of dimensionally stable material, upon which the graticule and ground control are plotted and upon which aero-triangulation or compilation is done.

base side - A side, of a triangulation network, which is measured with a JÄderin baseline apparatus or with a geodimeter type of instrument.

base station - (1) A station on a baseline or base line. (2) A station at which enough accurate measurements have been made that instruments can be calibrated there or that measurements made elsewhere can be referred to the value obtained at the base station. For example, a point at which gravity has been measured is a base station if gravity is found accurately at other points by measuring the difference between gravity at these other points and at the initial point. (3) A radio receiver or radio transmitter which is used for navigation, locating, positioning or measurement of distance. The particular method of locating, etc., involved is often indicated by prefixing the name of the method to station; e.g., Hiran station. SEE ALSO master station; station, fixed; station, mobile. (4) The point at which a survey begins.

base tape - A surveyor's tape or band of metal or alloy so designed and graduated, and of such excellent workmanship, that it is suitable for measuring the distances between points in first- and second-order traverses. In used, a base tape is subject to certain physical conditions which influence its effective length and must be taken into account in calculating the length of a measured base line. These conditions include tension, temperature, method of support, sag, grade, alinement and calibrated length. (SEE tape correction). At one time, base tapes were made of steel, which has a high coefficient of thermal expansion. They are now made of invar, nilvar or some other alloy having a very small coefficient of thermal expansion. SEE ALSO invar.

base terminal - One of the two ends of a base line.

base tilt - SEE orientation, basal.

base unit - One of the units of length (or the speed of light in vacuo), mass, time, temperature, amount of substance, electrical current and luminous intensity.

basin - (1) The entire region drained by a given stream and its tributaries. (2) A low-lying region, wholly or largely surrounded by higher land. It may be as small as a single, nearly inclosed valley or as large as an extensive depression rimmed by mountains. (3) SEE basin, oceanic.

basin, closed - A basin draining to some depression or point within its region from which water is usually lost only by evaporation or percolation. (2) A basin without a surficuial outlet for precipitation falling on it.

basin, oceanic - (1) An extensive, depressed region of an ocean's bottom, of more or less oval or circular form. (2) That portion of an ocean's bottom which lies below the continental shelf, slope and rise, but exclusive of oceanic ridges, volcanoes, trenches and island arcs. Also called an ocean basin.

basin, turning - That part of a harbor or channel which is constructed or set aside for the turning of ships.

basis meridian - SEE meridian, principal.

basis parallel - SEE base line (2e).

batch data processing - Data processing in which all the original data are submitted to the computer at the same time and operated on by the computer as a single set. Also called batch processing.

batch processing - SEE batch data processing.

bathymeter - An instrument used for determining depths of the oceans or other deep bodies of water. There are a number of designs of such instruments. The principal ones are the lead line and the echo sounder.

bathymetry - The art or science of determining depths of oceanic or other deep waters.

bathythermograph - An instrument for obtaining and recording temperature as a function of depth (strictly speaking pressure) in the ocean, from a ship under way.

batter board - A horizontal board nailed horizontally to two substantial posts, with the upper edge of the board preferably at grade or at some whole number of feet above or below grade. Batter boards are commonly erected, during construction of a building or in laying sewers, etc., to help in alinement of the foundations, determine the depth of the necessary excavation, etc.

batter pile - A pile driven at an inclination from the vertical. Also called a brace pile or spur pile.

battle map - A tactical map at a scale, usually, of 1:25000.

batture - (1) The gradual raising of the bed of a river or body of water by the accumulation of solid matter such as silt, soil, or gravel. (2) That portion of the bed of a river or body of water which has been raised by the gradual accumulation of solid matter. When accumulated matter has reached the point where it extends above the surface of the water and is therefore no longer part of the bed, it is still sometimes referred to as batture. However, the terms alluvion (for the process) and alluvium (for the land) are more specific and appear to be preferred.

baud - One element of code per second. The qualifier binary is usually understood. Baud is used to mean, in particular, one binary digit per second. It is also used, erroneously, to mean one binary digit. SEE ALSO bit.

Baudet code - A code (1) which uses five binary digits to represent one character.

Bauernfeind four-sided prism - A four-sided prism containing two right angles, one 45° angle, and one 135° angle, and used in surveying to establish a right angle. Also called a Bauernfeind prism.

Bauernfeind prism - SEE Bauernfeind four-sided prism; prism, three-sided.

bay - (1) In general, an unmistakable and pronounced indentation of a coast. According to the Geneva Convention of 1958, a bay is: a well-marked indentation whose penetration is in such proportion to the width of its mouth as to contain landlocked waters and constitute more than a mere indentation of the coast. An indentation should not be regarded as a bay, however, unless its area is as large as or larger than that of a semicircle whose diameter is a line drawn across the mouth (i.e., from headland to headland, or from extremity to extremity) of that indentation. For the purposes of measurement, the area of an indentation is that lying between the low-water mark around the shore of the indentation and a line joining the low-water marks at its natural entrance points. (2) The body of water contained within a bay as defined in (1) above.

bay, closed - (1) A coastal indentation which encloses an area greater than that inclosed by a semi-circle whose diameter is a line drawn between the extremities of the indentation. (2) A bay indirectly connected with the sea through a narrow pass.

bay, historic - (1) Any coast and contiguous body of water which, for historical reasons, has become known as a bay. (2) In international law, a bay over which there has been an exclusive assertion of sovereignty by a coastal nation, and an acquiescence by foreign (i.e., other) governments. It is sometimes required also that the bay have historically been known as such. Historic bays are well recognized exceptions to the rules applicable to ordinary bays and neither the semicircular rule nor the 10-mile limitation applies. Delaware and Chesapeake Bays are examples of historic bays in the USA.

bay, hooked - An open bay having only one headland.

bay, juridical - A bay that conforms to the requirements of the law.

bay, open - (1) A coastal indentation that is part of the open sea. (2) A coastal indentation which encloses an area less than that inclosed by a semi-circle whose diameter is a line drawn between the extremities of the indentation.

Bayesian least squares - The method of least squares, using already available information about the covariances of the unknowns to derive, from new measurements, new values of the unknowns and their covariances.

Bayes' theorem - The theorem that if an event B is known to have occurred as the result of some one event in a set $\{A_k\}$, then the probability that it was the particular event A_j that caused B is given by $P[A_j\ B] = P[B\ A_j]\ P(A_j)/\ \Sigma k\{P(A_k)\ P[B\ A_k]\}$, where A_j is from a set of J members. The numerator is the product of the probability of event A_j, regardless of whether B occurred or not, times the probability that B would have occurred if A_j did. The denominator is the sum of the products of the probability of each event A_k times the probability that B would have occurred if that A_j had. If all the conditions of the theorem are fulfilled, the theorem is certainly correct. The difficulties in applying the theorem, however, are caused by insufficient knowledge of the a priori probabilities. The theorem has gained considerable notoriety through its misuse, e.g., by mistaking prior data for a priori knowledge.

bayou - (1) The outlet of a lake. (southern USA) (2) A small river or creek. (southern USA).

beach - (1) The zone of unconsolidated material which extends landward from the low-water line to the place where there is a marked change in material or topography, or to the line of permanent vegetation (usually the effective limit of storm waves). (2) The loose material which is in more or less active motion in the immediate vicinity of the shoreline. This meaning is rare. (3) The zone extending from the upper and landward limit of the effective action of waves to low water level. SEE ALSO shore.

beach, transient - A beach whose sand is removed by storm waves but is quickly restored by longshore currents.

beach berm - A nearly horizontal portion of the beach, formed by the material left by the waves. Some beaches have no beach berms, while others may have several.

beach face - SEE foreshore.

beach ridge - A continuous mound of material, behind the beach, which has been heaped up by the action of waves carrying the material of the beach farther inland.

beach scarp - An almost vertical slope along the beach, caused by erosion of the beach by waves. Its height may vary from a few centimeters to a meter or more, depending on the kind of waves acting on the beach and on the kind of material composing the beach.

beach survey - The collection of data describing the physical characteristics of a beach (a region bounded by a shoreline, a coastline, and two natural or arbitrarily assigned flanks).

beacon - (1) Any artificial device which makes itself and its location known through the emission of light, infrared radiation, or radio waves, either continuously or according to a schedule. This last requirement distinguishes a beacon from a transponder, the latter being turned on and off by the observer. (2) Anything serving as a signal or conspicuous indication either for guidance or warning. In particular, such a device used in navigation. For example, the lamp in a lighthouse and the transmitter in a TRANSIT satellite are beacons in this sense. (3) SEE monument. SEE ALSO radio beacon.

beacon, geodetic - A beacon designed specifically for use in geodetic surveying. SEE ALSO signal, geodetic.

beam - The direction or opposite direction of the transverse axis of a craft.

beam (of light) - SEE light beam.

Beaman arc - SEE Beaman stadia arc.

Beaman stadia arc - A specially graduated, metallic arc fitted to the vertical circle of a transit or plane-table alidade for the easy reduction of observations made with the stadia. Also called Beaman arc and stadia circle.

beam compass - A drafting instrument for drawing a circular arcs with a long radius. The point to be placed at the center of the arc's defining circle and the pen (or pencil tip) for drawing the arc are separate units, mounted to slide along and clamp to a long bar or beam so that the distance between them is equal to the desired radius.

beam splitter - A device or optical element that converts a single beam of electromagnetic radiation into two or more beams going in different directions. In the optical and near-optical region of the electromagnetic spectrum, a common type of beam splitter is a semi-reflecting mirror set with its normal at an angle to the direction of the beam. Another type is a completely-reflecting mirror with its normal set at an angle to the direction of the beam, but obscuring only a part of the beam. In a third type, a prism may be used instead of a mirror.

beam width - The angle between half-power intensities in the lobes of an antenna pattern.

bearing - (1) The horizontal angle between a line from an initial point (e.g., an observer) to a given point, and a line from the initial point in a specified direction - as, for example, north. Also called bearing angle. (2) The horizontal angle, measured clockwise from a given direction at a point to a line from that point to a second point. Bearing and azimuth are often used interchangeably in geodesy, although they obviously have different meanings. (3) The same as bearing in sense (2), but with the angle being measured either clockwise from North (and then called a bearing East) or counter-clockwise from North (and then called a bearing West). The bearing then never exceeds 180°. (4) The same as bearing in sense (3), but with the angles being measured either from north clockwise (eastward) or counterclockwise (westward) up to maximum of 90° or from south clockwise (westward) or counterclockwise (eastward) up to a maximum of 90°. Also called a forward bearing. In hydrography, bearing in the sense of (2) or (3) is used only for angles between points on the Earth's surface. Angles to points on the celestial sphere are called azimuths. (5) The horizontal angle which a line makes with the meridian of reference adjacent to the quadrant in which the line lies. Bearings of this kind are classified according to the meridian used as true bearings, grid bearings, or magnetic bearings. A bearing is identified by naming the end of the meridian (north or south) from which it is reckoned and the direction (east or west) of that reckoning. Thus, a line in the northeast quadrant making an angle of 50° with the meridian will have a bearing of N.50° E. Bearings are often used in land surveying, but in most surveying it is preferable to use azimuths rather than bearings. (6) The bearing as defined in any one of the above definitions, together with the two directed lines and point of intersection that forms the corresponding geometrical figure. (7) The direction of one point from another, measured in the horizontal plane at the first point, from true, grid, or magnetic North. (8) The direction of a line or the direction to a point. The terms bearing and heading have similar meanings. However, the former is more appropriate when dealing with static objects like lines or aiming points, while the latter is more appropriate when dealing with the direction in which something moves. SEE ALSO heading; compass bearing; grid bearing; Lambert bearing; Mercator bearing; transit bearing.

bearing, assumed - A bearing referred to some meridian chosen as a matter of convenience i.e., a bearing referred to an assumed meridian.

bearing, astronomic - The horizontal angle between the astronomical meridian and a line on the Earth. This term is preferred to the term true bearing which was used in early surveys.

bearing, back (USPLS) - (1) The bearing at the opposite end of a line from the observer, as measured from the astronomic meridian at the opposite end of the line. Back bearings to all lines (other than north-south lines) differ from the bearings at the observer's station by the amount of the convergence of the meridians between the points. (2) The direction of a line from point A to point B as determined at point B and as corrected for the curvature of the line from A.

bearing, direct - The bearing measured at the point at which the observer is located. SEE ALSO bearing, back.

bearing, electronic - A bearing obtained using an electronic instrument.

bearing, false - The difference between the direct bearing and the back bearing, caused by meridional convergence.

bearing, forward - SEE bearing (4).

bearing, geodetic - SEE azimuth, geodetic; bearing.

bearing, great-circle - The initial direction of a great circle through two terrestrial points, expressed as angular distance from a reference direction. It is usually measured clockwise from the reference direction taken as 0°.

bearing, magnetic - (1) In navigation, the bearing with respect to magnetic north i.e., compass bearing corrected for deviation. (2) In surveying, the angle with respect to magnetic north or magnetic south and stated as east or west of the magnetic meridian e.g., N 15° E. SEE ALSO heading.

bearing, polar - The angle, in a polar or spherical coordinate-system, formed by the intersection of the meridional plane used as reference and the meridional plane through the point. This is a dihedral angle but is commonly thought to be a plane angle.

bearing, reciprocal - The direction opposite to a bearing (bearing ± 180°).

bearing, reverse - A bearing taken with a magnetic compass and directed oppositely to the direction of the point whose bearing is desired.

bearing, rhumb - SEE bearing, rhumb-line.

bearing, rhumb-line - The initial bearing of one point from another, taken as the bearing of the rhumb-line between the two points with respect to a reference line.

bearing, true - The horizontal angle between the meridian line and a line on the Earth. The term true bearing was used in many of the early descriptions of land boundaries in the U.S.A. It is associated with astronomic north, i.e. the direction of the north point as determined by astronomical observations. If an astronomically determined bearing is used, the term astronomic bearing is now preferred over true bearing.

bearing angle - SEE bearing (1).

bearing bar - SEE azimuth instrument.

bearing circle - A ring designed to fit snugly over a compass.

bearing line - A line drawn in the direction of a bearing.

bearing method (of adjusting triangulation) - A method of adjusting a triangulation network which combines some of the characteristics of the direction method and the method of variation of coordinates. The unknowns are a corrections at each station to orient the set of measured directions there, and the azimuths, with respect to north on the grid used, of all observed lines. To this set of observations equations is added the required number of side conditions with zero absolute terms. Also called the gisement method (of adjusting triangulation). The bearing method is applicable only to adjustments in plane rectangular coordinates.

bearing object - A corner accessory that can be readily identified by its distance and direction from the corner being recorded.

bearing of line - SEE bearing (5).

bearing tree (USPLS) - A tree acting as corner accessory, its distance and direction from the corner being recorded. Also called a witness tree. Bearing trees are identified by prescribed marks cut into their trunks; the species and sizes of the trees are sometimes recorded.

beat - In some of the southern states of the U.S.A., (such as Alabama, Mississippi and South Carolina), the principal legal subdivision of a county, and corresponding to towns or townships in other states or a voting precinct.

bed - (1) That region which contains the water of a stream or river and which lies within the high-water lines of that stream or river. It is the region which is kept practically bare of vegetation by the wash of the waters of the stream from year to year. (2) The bottom of a water course, or any body of water, such as a pond or lake. Also called a river bed when referring specifically to the bottom of a river. (3) That region which forms the bottom of a body of water (i.e., contains that body) and which lies between the low-water mark. (3) A stratum, in the Earth's crust, which has been formed in an approximately horizontal position. If the stratum is of rock, it is called bedrock. (4) A layer of rock.

bed of water - SEE bed (2).

bedrock - (1) A hard layer of rock covered by soil, sediment, etc., and more or less continuous over a large region. It is distinguished from silt, clay, etc., by being hard, and from sand, pebbles, and boulders by being continuous. Bedrock which has been cleared of its covering is referred to as exposed bedrock. Large, deeply-imbedded boulders, with upper surfaces planed and leveled through exposure, are sometimes erroneously reported as bedrock in descriptive writing. (2) A stratum of rock in the Earth's crust which has been formed in an approximately horizontal position.

Behrmann's map projection - A cylindrical equal-area map projection, described by Behrmann in 1910, mapping the sphere onto a cylinder secant at 30° latitude north and south. This map projections has less average-maximal- angular distortion than any other equal-area map projection, according to Behrmann.

belt, marine - SEE sea, marginal.

belt, maritime - SEE sea, marginal.

bemetist - A surveyor who measures distances by pacing them off.

bench - (1) A level or gently sloping, flat surface produced by erosion and inclined seaward. (2) A nearly horizontal region at about the elevation of highest high water or the sea side of a dike.

bench mark - A relatively permanent, material object, natural or artificial, bearing a marked point whose elevation above or below an adopted surface (datum) is known. Sometimes written benchmark. Also called a vertical control point. Usually designated as a B.M., such a mark is sometimes further identified as a permanent bench mark (PBM) to distinguish it from a temporary bench mark (TBM) or supplementary bench mark, a mark of less permanent character intended to serve only for a comparatively short period of time. SEE ALSO instability, bench-mark.

bench mark, first-order - A bench mark connected to the surface of reference by continuous first-order leveling.

bench mark, intermediate - SEE bench mark, temporary.

bench mark, junction - SEE junction bench mark.

bench mark, monumented - SEE bench mark, permanent.

bench mark, nonmonumented - SEE bench mark, temporary.

bench mark, permanent - A bench mark of as nearly permanent character as it is practicable to establish. Usually designated simply as a bench mark or BM (or sometimes PBM), a permanent bench mark is intended to maintain its elevation with respect to an adopted surface of reference without change over a long period of time. Concrete or stone posts, with suitable marks in their tops or with imbedded, inscribed disks, are used. (The disk used by the National Geodetic Survey is made about 9 cm in diameter and is made of brass or bronze; it has a spherical top-surface to support the foot of a leveling rod.) A permanent bench mark is set in firm ground deeply enough to be free from action of frost and in a location where disturbing influences are believed to be negligible or, preferably, in outcropping rock or buried rock. SEE ALSO rod mark. Also called, without justification, a monumented bench mark.

bench mark, primary - A bench mark close to a tide station to which the zero of the tide gage and tidal datum were originally referred.

bench mark, second-order - A bench mark connected to the reference surface by continuous second-order leveling or by a continuous combination of first-order and second-order leveling.

bench mark, supplementary - A temporary bench mark placed at the junction between two sections of a line of levels. The term is obsolete.

bench mark, temporary - A bench mark established to hold, temporarily, the end of a completed section of a line of levels and serve as a starting point from which the next section is run. Spikes and screws in poles, bolts on bridges, and chiselled marks on masonry have been used as temporary bench marks and have lasted for years. Also called an intermediate bench mark or non-monumented bench mark and sometimes designated as TBM.

bench mark, tidal - A bench mark the elevation of which with respect to mean sea level at a nearby tidal gage has been determined and which is used as reference for that tidal gage and for tidal datums. National geodetic organizations follow two different policies with respect to tidal bench marks. (a) The elevation of a tidal bench mark, once determined, is kept fixed. Changes in mean sea level at the tidal gage will then not affect elevations of points in the leveling network connected to that gage. (b) The elevation of the tidal bench mark is redetermined at intervals. Elevations of points in the leveling network then change with the location of mean sea level on the tidal gage, but mean sea level itself does not change. A tidal bench mark may be included in a geodetic line of levels.

bench mark, tide-gage - A stable bench mark near the tidal gage to which tide-gage datum is referred. SEE ALSO bench mark, tidal.

bench mark, vertical-angle - A bench mark the elevation of which has been established by trigonometric leveling.

Benday process - A process for printing shadings consisting of lines, dots, stipples, etc., by inking a Benday screen (a rectangle of hardened gelatin with the pattern in relief), printing with it on portions of the metallic plate on which an outline has been photographically printed, and then etching the metal as a line plate. Also spelled benday process and Ben Day process. It is named after Benjamin Day, the originator of the process.

Benioff zone - A nearly flat region of seismic activity (hypocenters) dipping, at continental margins or in continents, into the crust at an angle of about 45° and extending to depths of up to 700 km.

Benoit's ratio - Any one of several ratios, determined by Benoit, between various units of length and the meter. The more important ratios are:
Indian foot : meter = 1 : 0.304 798 41 (38, 270.4)
British foot : meter = 1 : 0.304 800 34 (270.4)
toise : meter = 1.949 034 : 1 (270.4)

berm - (1) That part of the section of graded road adjacent to and outside of the paved portion. (2) A horizontal strip of shelf built into an embankment to break the continuity of an otherwise long slope. (3) SEE beach berm.

Bernoullian flow - Flow in the direction of the difference of horizontal component of pressure between two points in a fluid.

Bernoulli's distribution - SEE distribution, binomial.

Bernoulli's theorem - In an inviscid fluid in steady motion, $v^2/2 + p/D + gz =$ constant, where v is the rate of flow, p the pressure, D the density, z the depth and g the acceleration of gravity. It is equivalent to saying that the sum of the per-mass kinetic energy plus the external pressure plus the gravitational pressure is constant everywhere in the fluid.

Bertrand's problem - The problem of finding the average length of that segment, of a straight line laid at random on a rectangle of specified dimensions, which is contained within the rectangle. Alternatively, the average distance between two points placed at random on different sides of a rectangle of specified dimensions.

Bessel ellipsoid of 1841 - SEE Bessel spheroid 1841.

Besselian date - The time of an event, designated by giving the Besselian year in which the event occurred and the fraction of the Besselian year that had passed since the beginning of that year. Because the Besselian and tropical years are of nearly the same length, it is customary to give the fraction of a tropical year rather than the fraction of a Besselian year.

Besselian day-number - Any one of the 7 quantities A,B,C,D,E,J,J' which appear in the two equations for right ascension α and declination δ: $\alpha = \alpha_o + \mu\alpha\Delta t + A(\sin \alpha_o \tan \delta_o + m/n) + B \cos \alpha_o \tan \delta_o + C \cos \alpha_o \sec \delta_o + D \sin \alpha_o \sec \delta_o + J \tan^2 \delta_o + E$ and $\delta = \delta_o + \mu\delta\Delta t + A \cos \delta_o - B \sin \alpha_o + C(\tan \epsilon \cos \delta_o - \sin \alpha_o \sin \delta_o) + D \cos \alpha_o \sin \delta_o + J' \tan \delta_o$. These rotate the equatorial coordinate system from mean equinox at time t_o to true equinox at time $t_o + \Delta t$. In the equations, α_o and δ_o are the right ascension and declination at time t_o, α and δ the corresponding quantities at time $t_o + \Delta t$; ϵ is the obliquity of the ecliptic; m and n are precessional constants. The numbers A,B,E involve first-order precession and nutation and are given by $A = n\Delta t + \Delta\psi \sin\epsilon$, $B = -\Delta\epsilon$, E

= $L'\Delta\psi/\psi'$ where ψ' is the annual luni-solar precession, L' is the annual planetary precession in the equator and $\Delta\psi$ and $\Delta\epsilon$ are the total nutation in longitude and obliquity, respectively. The numbers C and D involve the annual aberration of light and are given by C = + κ cos ϵ (cos L_s - e_1 cos θ) and D = + κ(sin L_s - e_1 sin θ). Here, e_1 is the eccentricity of the Earth's orbit, θ is the celestial longitude of the perihelion of the Earth's orbit, κ is the constant of aberration and L_s is the Sun's geocentric longitude. $\mu\alpha$ and $\mu\delta$ are the proper motion in right ascension and in declination, respectively. The numbers J and J' provide second-order corrections to the precession and nutation. They need not be used in calculations unless considerable accuracy is required. Besselian day numbers are commonly listed in ephemerides for every day of the year, together with Δt, the fraction of a year from the nearest beginning of a Besselian year. Also called a Besselian star number.

Besselian element - One of a set of numbers and a system of coordinates, introduced by Bessel, for predicting the progress of solar eclipses and of occultations. The coordinate system has its origin at the center of mass of the Earth. The z-axis is drawn in the direction of the Moon and parallel to the line joining the centers of the Sun (or other star) and the Moon. The y-axis lies in the plane of the z-axis and the North Pole and is positive northwards. The x-axis is drawn to complete a right-handed, rectangular Cartesian coordinate system. The Besselian elements are then the following: (a) the coordinates (x,y) of the center of the Moon's shadow on the (x,y) plane (eclipse) or the right ascension and declination (α^*, δ^*) of the star (an occulation); (b) the rates of change (x',y') of x and y per hour; (c) the Universal Time or Ephemeris Time To of conjunction in right ascension of the Moon and star (occultations only); (d) the hour angle μ of the z-axis at the Greenwich meridian (eclipses) or the hour angle H of the star at Greenwich at the instant To (occultations); (e) the declination d of the z-axis (eclipses) or the value Y of the y-coordinate at the instant To (occultations). The quantities α^*, δ^*, x', y', To, H and Y are the Besselian elements of an occultation. The quantities x, y, x', y', μ, sin d and cos d, together with the following quantities peculiar to eclipses, are the Besselian elements of an eclipse; (f) the quantity dμ/dt (μ'); (g) the radii l_1 and l_2 of the umbral and penumbral cones on the (x,y) plane; (h) the angles f_1 and f_2 (half the vertex angles of the penumbral and umbral cones, respectively. The Besselian elements both for eclipses and for occultations are usually given in astronomical almanacs. However, many almanacs, because interest in eclipses and occultations has decreased and because calculating machines can easily calculate some of the elements, no longer list all the elements. (Some authors consider only x, y, x', y', μ, sin d and cos d to be Besselian elements of an eclipse.)

Besselian pendulum - A pendulum consisting of a heavy sphere suspended at the end of a very long wire. Also called a filar pendulum. This kind of pendulum was used by Bessel in Königsberg in 1826. Wires as long as 200 m have been used. SEE ALSO Besselian pendulum-apparatus.

Besselian pendulum-apparatus - A pair of filar pendulums differing in length by 1 toise, as used in Bessel's pendulum-apparatus of 1826-1827, and having a special kind of suspension.

Besselian solar year - SEE Besselian year.

Besselian year - The interval of time beginning when the right ascension of the fictitious mean sun, as affected by aberration and measured from the mean equinox, is 18h 40m (280°) and ending when the right ascension of the same sun is again 18h 40m. Also called a fictitious year and Besselian solar year. The beginning of a Besselian year is denoted by the notation .0.(e.g., 1950.0) The Besselian year is shorter than the tropical year by 0.148xT seconds, where T is given by (Date -1900)/100. The difference is usually ignored. For example, the Besselian year 1966.0 began on 1966 January 0.799d Greenwich mean sidereal time. At this instant, the date was 1966 January 0.00d at points on the meridian at 72.36° W.

Bessel point - One of the two points on which a standard bar rests when in use if the distance between lines engraved on the ends of the neutral fiber is to be a maximum. The distance between symmetrically placed supports is approximately 0.5994 times the length of the bar. The neutral fiber is that surface, in the bar, which has the same length before and after the bar is placed on the supports. Note that although the length between marks remains the same, the distance diminishes because the bar is distorted by its own weight when placed on the supports.

Bessel's base line apparatus - A set of four rods, each composed of one iron rod 2 toises (about 3.898 meters) long on which is laid a zinc rod of the same length but half the width. The zinc and iron rods are fastened together at one end only. Base lines were measured by placing the iron rods end to end. The gaps between ends of the zinc rods were measured to determine corrections for thermal expansion. Some of the most important base lines in European triangulation were measured with this apparatus.

Bessel's formula - (1) The formula which relates the time (sidereal) T_{true} of passage of a star through the astronomical meridian to the observed time T_{obs} of passage through the telescope's vertical cross-hair, to the declination δ of the star and to the constants m, n and c which are functions of instrumental errors a and b, as follows:

$T_{obs} - T_{true}$ = m + n tan δ + c sec δ
tan m = sin Φ tan a + cos Φ sec a tan b
sin n = sin Φ sin b - cos Φ sin a cos b
-cos c = -sin δ sin n + cos δ cos n sin (T_{true}- T_{obs} - m).

The quantity a is the azimuth error (azimuth constant) of the instrument, b is the level error (level constant) and c is the collimation error (collimation constant), i.e., the angle by which the horizontal axis of the instruments fails to point exactly east-west, the angle by which the horizontal axis fails to lie in the horizontal plane, and the angle by which the line of sight fails to be perpendicular to the horizontal axis, respectively. Φ denotes the astronomic latitude of the instrument. These equations apply to transit at upper culmination. They apply to lower culmination if (180° - δ) is substituted for δ. In practice, a, b and c are usually not measured. Instead, observations are made on stars for which the times of transit are known, and m, n, c are determined from the observations by the method of least squares. (2) The same as (1) but with the right ascension α of the star substituted for T_{true} and with a term T, the clock correction, added to the right-hand side. It is usually written as α = T + ΔT + m + n tan δ + c sec δ, in which T is the observed time of transit, $m = a \sin \Phi + b \cos \Phi$ and $n = b \sin \Phi - a \cos \Phi$ and the other symbols have the meanings given above. SEE ALSO Hansen's formula; Mayer's formula.

Bessel's formulae - The formulae for the standard deviation s of a set of N numbers and the standard deviation s_a of the average: $s = \sqrt{[\Sigma v^2/(N-1)]}$ and $s_a = \sqrt{[\Sigma v^2/(N-1)N]}$. The v's are deviations from the average.

Bessel's method - SEE triangle of error method.

Bessel spheroid 1841 - A rotational ellipsoid having the dimensions:
  semi-major axis   272 077.14 toise (6 377 397.16 m)
  semi-minor axis   3 261 139.33 toise (6 356 078.97 m)
The flattening (derived) is 1/(299.152 813). This is usually the ellipsoid meant when the Bessel spheroid is referred to. It was used extensively for computation of triangulation in many parts of Europe and has been used for Tokyo datum and other Asiatic datums. The metric values were computed using A.R. Clarke's value, 0.513 074 074, for the ratio of the toise to the meter.

beta plane - A plane, taken to represent the Earth, whose rate of rotation, corresponding to the Coriolis parameter, varies linearly with the north-south direction.

b-factor - SEE level factor.

Bianchi identity - The statement that the antisymmetric first covariant derivative of the Riemann tensor vanishes identically.

biangle screen - SEE screen, biangle.

bias - The departure, from a reference value, of the average of a set of values. SEE ALSO error, constant; error, systematic.

bid - (1) An offer by an intending purchaser to pay a designated price for property which is about to be sold at auction. (2) An offer to perform a contract for work and labor or to supply materials at a specified price.

bienes - In Spanish law, goods, property of every description, including real as well as personal property; all things (not being persons) which may serve for the uses of man.

bienes communes - In Spanish law, common property: those things which, not being the private property of any person, are open to the use of all, such as rain, water, the sea and its beaches.

bienes genaciales - In Spanish law, a species of community in property enjoyed by husband and wife, the property being divisible equally between them on the dissolution of the marriage. The property does not include what they held as separate property at the time of contracting the marriage.

bienes publicos - In Spanish law, those things which, as to property, pertain to the people or natio n and, as to their use, to the individuals of the territory or district, such as rivers, shores, ports, and public roads.

biens - (1) In French law, real and personal property. (2) In English law, property except estates of inheritances and freeholds.

bight - A long, gradual bend or recess in the coastline which forms a large, open, receding bay.

BIH System 1968 - A method of deriving the polar motion from a weighted set of observations from 34 instruments for latitude and 17 for time (longitude) at 29 observatories in existence in 1966. Corrections are applied each year so that the system does not change.

Bilby steel tower - SEE Bilby tower.

Bilby tower - A tower used in triangulation and consisting of two braced tripods, one within the other, constructed, according to the design of J.S. Bilby, so that an observer on a platform at the top of the outer tripod can move about without disturbing an instrument mounted on the top of the inner tripod. The Bilby tower was put into use in 1927 and is standard equipment in triangulation done by the U.S. National Geodetic Survey. The tower can easily be put up, taken down, and moved to a new location. It is also durable. Because the original Bilby tower and most subsequent ones of that design have been built of steel, the tower is often referred to as the Bilby steel tower.

billabong - An ana-branch of a watercourse usually filled with water.

bimargin format - SEE format, bimargin.

binocular (adj.) - Referring to the use of both eyes simultaneously.

binocular (n) - (1) An optical instrument providing for the use of both eyes simultaneously. (2) A pair of eyepieces or telescopes placed side by side so that both eyes can observe simultaneously. The term pair of binoculars is often used instead. Note that observing through a binocular does not necessarily give binocular vision. Microscopes are often provided with binocular eyepieces but only a single objective. Both eyes therefore look at the object from exactly the same viewpoint.

binocular comparator - SEE comparator, binocular.

binormal - That vector which, at a point on a curve, forms a right-handed set with the positive tangent and the positive normal, in that order.

biome - A region classified according to the kinds of plants and animals living there. Types used as standard by the U.S. Bureau of Land Management are: tundra, boreal forest, coniferous forest, woodland-brushland, grassland and desert.

biosphere - That transitional zone which lies between earth and atmosphere and within which most forms of terrestrial life are commonly found. That part of the Earth commonly inhabited by plant or animal life. It comprises the outer portion of the lithosphere and hydrosphere, and the inner or lower portion of the atmosphere.

Birch discontinuity - A seismic discontinuity at a depth of about 900 km.

bit - One of the two characters of a two-valued or binary number system. A digit in the binary number-system. Also called, erroneously, a baud.

blackbody - SEE body, black.

black-body - SEE body, black.

Black's method - A method primarily of determining geodetic azimuths from astronomical observations, by measuring the angle θ' between a star and the marker of interest, the star and marker being at large zenith distances z^* and z_m, respectively. The difference between measured angle θ' and the corresponding angle θ in a geodetic coordinate system is $\theta' - \theta = \cot z^* (y \cos A^* - x \sin A^*) - \cot z_m (y \cos A_m - x \sin A_m)$ where x and y are the deflections of the vertical in the meridian and in the prime vertical, respectively; A^* and A_m are the astronomic azimuths of the star and marker, respectively. The astronomic azimuth of the star is given by $A^* = \sin\phi \cot H - \tan\delta \cos\phi \csc H$, where ϕ is the latitude of the observer and H and δ are the hour angle and the declination, respectively, of the star. At large zenith distances, $\theta'-\theta$ is very small, and the geodetic azimuth is given closely by the second equation. If it is not small, three or more observations are made on different stars to determine x and y also. The method is particularly useful at high latitudes. The method has also been used to determine astronomic latitude and longitude.

blaze - (1) A mark made upon a tree trunk at about breast height by cutting out a piece of bark, leaving a flat scar upon the tree's surface. (2) A blaze, in the same sense as (1), with the proviso that a small amount of the live wood be removed also.

bleed - (1) That edge of a map or chart on which cartographic detail is extended to the edge of the sheet. Also called bi-margin format, bleeding-edge or extended color. SEE ALSO format, bimargin. (2) A condition wherein ink is dissolved by the press's fountain solution, causing a light film of ink (scum) on the plate and impression.

bleeding edge - SEE bleed.

blink comparator - An instrument for comparing two photographs by allowing the operator to view the photographs alternately at such a high rate that the two pictures seem to be superposed on each other. The two photographs (taken of the same scene and at the same scale) are placed side by side in the comparator and viewed simultaneously through a single microscope. They are adjusted until their images overlap as exactly as possible. Illumination is switched rapidly from the one photograph to the other. An object which has moved in the scene between photographs will show up in different places on the photographs and its image will appear to flicker or move back and forth. The instrument is used to detect movement of stars, satellites, etc. Also called a blink microscope or flicker comparator.

blip - SEE echo (3).

blister - (1) A small bubble formed under an emulsion because the emulsion has there become detached from the base. (2) SEE border break.

block - (1) A square, or a portion of a city or town inclosed by streets but not containing any streets. (2) A subdivision of a town site. (3) The platted portion of a city, surrounded by streets. The term need not be limited to blocks platted as such but may designate a region bounded on all sides by streets or avenues. It must be surrounded on at least three sides by streets which must be marked on the ground and not simply indicated on a plat. (4) Two or more strips of overlapping photographs.

block adjustment - (1) Determination of corrections to the coordinates of a set of points distributed over a large area, the solution being obtained simultaneously for all the points. The term is used to distinguish this process from that in which the points are distributed along strips or arcs and coordinates are obtained first for coordinates of points within each strip or arc, and the results then modified so that there are no inconsistencies between coordinates of the same points in different strips or arcs. SEE ALSO block aerotriangulation adjustment; Bowie method; bundle aerotriangulation adjustment; Helmert's blocking method. (2) The same as definition (1), except that the adjustment need not be done for all points simultaneously. (3) The adjustment of coordinates of points shown on a strip of photographs, using the data from several laterally-overlapping strips. In general, the combination of a set of such strips into a block and the application of this process to all the strips. (4) The process of improving an approximate solution of a large set of equations.

block aerotriangulation - (1) Aerotriangulation in which blocks of photographs, rather than pairs or strips, are treated as a unit. (2) Aerotriangulation involving the combination of individual strips of photographs into a block. SEE ALSO block adjustment (3).

block aerotriangulation adjustment - (1) An aerotriangulation adjustment in which the ground points whose coordinates are to be adjusted are imaged throughout a block of photographs, and the adjustment is carried out without special regard to the arrangement of the photographs into pairs or strips. The photographs, when assembled, generally form a rectangular, block-like array. (2) The same as the previous definition, except that the corrections are determined more or less simultaneously for all the photographs. Also called bundle aerotriangulation adjustment. (3) The same as the first definition, except that the photographs occur in strips and the coordinates of points imaged in each strip are adjusted first for each strip and then adjustments are made between coordinates of points in different strips. This is essentially strip aerotriangulation adjustment.

block correction - A corrected reproduction of a small region of a nautical chart, which is pasted to the chart for which it is issued.

block method (of adjustment) - SEE block adjustment.

block method (of aerotriangulation) - SEE block aerotriangulation.

block method (of aerotriangulation adjustment) - SEE block method (of phototriangulation adjustment).

block method (of phototriangulation) - Phototriangulation in which measurements of coordinates of points on a large number of photographs are converted to the corresponding coordinates of the ground points for all the points at one time rather than for selected subsets of the points, separately. Also called block photo-triangulation and phototriangulation by blocks.

block out (v.) - SEE opaque (v.).

block triangulation - SEE block method (of phototriangulation adjustment).

blooming - (1) Localized overexposure caused by radiant energy coming in at a level which exceeds the tolerance of the emulsion and thereby causing the image to lack definition. (2) A process for increasing the amount of light transmitted by a lens by coating the lens with anti-reflective coating.

blowup - Enlargement of a photograph. (jargon)

blueline - A nonreproducible blue image or outline usually printed photographically on paper or plastic sheeting, and used as a guide for drafting, stripping or layout. Sometimes called blind image.

blueline guide - A drafting surface on which a blueline has been placed.

blueprint - A copy made by placing the original, in black lines on a transparent material, in contact with a paper sensitized with ferrocyanic salts and illuminating the original. The exposed paper is developed in a bath of water and shows the original pattern in white lines on a blue background.

blunder - (1) A mistake or error caused by mental confusion, carelessness, stupidity, or ignorance. (2) A numerical mistake or error caused by mental confusion, carelessness, stupidity, or ignorance. Examples of blunders are: reading a horizontal circle wrong by a whole degree; neglecting to record a whole tape length in a traverse; and reversing the numerals in recording a measurement. It would also apply to the number recorded as the result of observing on the wrong target or from the wrong control point. (3) A large numerical mistake or error caused by mental confusion, carelessness, stupidity, or ignorance. By this definition, a small error would not be a blunder. (4) An error, made by a machine, of the same nature as that defined in (1). For example, an error caused by a poorly working card-punch or paper-tape punch would be such a blunder.

blur circle - SEE circle of confusion.

Board on Geographic Names - A board established by the act of 25 July 1947, (61 Stat 456, 43 U.S.C. 365) to provide uniformity in geographic nomenclature and orthography throughout the Federal Government. Membership is comprised of a group appointed by various governmental departments and agencies. Subject to approval by the Secretary of the Interior, the Board formulates principles, policies, and procedures on domestic and foreign geographic names and their spelling.

boat sheet - SEE field sheet.

body, black - A hypothetical body (object) absorbing all electromagnetic radiation falling upon it. Also written black-body and blackbody. A black body reflects none of the incident radiation but it does emit just as much energy as it absorbs. However, the emitted radiation is usually at different wavelengths from that absorbed.

body, celestial - Any material object which is observable over a reasonable interval of time and which moves as an individual object along a path outside the immediate vicinity of the Earth. Also called a heavenly body. Specifically, all members of the Solar system other than the Earth, all stars, nebulae, galaxies and clouds of gas in intergalactic space are celestial bodies. Meteors and meteorites and not usually considered celestial bodies. Neither are the individual particles and grains dispersed throughout space. Artificial satellites and spacecraft are celestial bodies for as long as they remain outside the Earth's atmosphere.

body, elastic - (1) A body which returns to its original size and shape after being deformed slightly. (2) A body which yields, under stress, according to Hooke's law. Do not confuse an elastic body with a rigid body. A rigid body does not yield at all.

body, heavenly - SEE body, celestial.

body, pear-shaped - A body which, described in terms of zonal spherical harmonics, contains a third-degree term of appreciable size and no higher-degree terms of comparable size. The Earth has for this reason been called pear-shaped.

body, primary - The body about which another body revolves under the force of the first body's gravitational attraction and which is so massive that the barycenter of the binary system is close to it. Also called the primary of the system.

body, rigid - A body which keeps its shape regardless of the forces imposed upon it. No actual bodies are completely rigid, but the concept is convenient in mechanics. For example, a good, approximate theory of oceanic tides can be derived by assuming that the lithosphere is a rigid body and that only the oceans yield to the attractions of the Moon and Sun.

body, rotating - A body which rotates about an axis within the body. Not to be confused with a revolving body, which moves about an axis outside the body. The theory of rotating bodies is quite extensive because it is fundamental to an understanding of the Earth's figure and internal structure.

Bogg's map projection - SEE map projection, eumorphic.

bolometer - A device which measures radiant energy by using a thermally sensitive electric resistor. Also called an actinic balance. One form absorbs the radiation on a blackened, metallic foil and records the resultant change in resistance by means of a Wheatstone Bridge. This form, as invented by Svending and Langly, was composed of blackened strips of platinum. Other forms are composed of metals that undergo very large changes of resistance at very low temperatures. Bolometers used for measuring the power in radio waves are called barretters or thermistors. COMPARE radiometer.

Boltz's method - A method devised by Boltz for solving the normal equations occurring in the adjustment of a triangulation network; it allows a large set of equations to be solved in one straight operation by the method of least squares without biasing the solution towards any particular unknown. Initially, the normal equations were solved using the Gaussian method of successive elimination. This method, however, causes the last-determined value to contain larger errors than the first-determined value. Boltz's method treats all unknowns equally and is particularly suitable for solving large systems of equations.

Bomford's geoid 1963 - An approximation to the geoid, prepared by Study Group Number 10 (International Association of Geodesy) under Bomford and presented at the meeting of 1963 in Berkeley, California. It is based on triangulation and astro-geodetic data on Europe and adjoining regions, with the backbone extending from Scotland through Switzerland and Italy to Turkey and Iran, and thence to Malaya. It was adopted as the geoid to be used in calculating the European triangulation.

Bomford's map projection - SEE Hammer-Aitoff map projection.

Bonne map projection - (1) The equal-area map projection from the hemisphere onto a tangent cone and defined by the equations $x = k (N \cot\phi_o - \Delta S) \sin\theta$, $y = k (N \cot\phi_o (1 - \cos\theta) + \Delta S \cos\theta$, $\theta = \Delta\lambda \cos\phi/(\cot\phi_o - \Delta S/N)$, in which x and y are the rectangular Cartesian coordinates of that point whose longitude is $\lambda = \lambda_o + \Delta\lambda$ and whose latitude is $\phi = \phi_o + \Delta\phi$, λ_o is the longitude of the central meridian, ϕ_o is the latitude at which the cone is tangent, k is the scale factor, N is the length of the principal normal and ΔS is the distance, along the meridian, from ϕ_o to ϕ. (2) An equal-area map projection similar to that defined above, but mapping the rotational ellipsoid onto a tangent cone. Also called the Projection de la Carte de France and Projection du Depot de la Guerre. The line representing the central meridian is straight and the scale along it is exact. All parallels of latitude are represented by arcs of concentric circles at their truly-scaled distances apart. All lines representing meridians, except the central meridian, are curved lines connecting corresponding points on the lines representing parallels of latitude. This map projection is strictly equal-area, and its invention is attributed to Petrus Apianus in 1520. It was introduced later by Rigobert Bonne in 1752. The Sanson-Flamsteed or sinusoidal map projection is a particular variant of the Bonne map projection; the equator is used as the standard parallel.

Bonner Durchmusterung - A star catalog, compiled by F.W.A. Argelander, of over 324,000 stars, accompanied by a chart and having location and approximate magnitude of each star. Usually referred to as the BD (sometimes as the Argelander catalog or the Bonner Durchmusterung star catalog). The catalog shows practically every star brighter than 10th magnitude and north of 2° declination, and has served as the basis for many later catalogs, as well as having been used as a source of stars to be used in programs for determining latitude or azimuth. Various astronomers have extended the catalog southward. It is now superseded by the Smithsonian Catalog.

Bonner Durchmusterung star catalog - SEE Bonner Durchmusterung.

Boolean algebra - A set of symbols with two binary operations denoted as (logical) addition and (logical) multiplication, such that for all symbols A, B, C in the set the following rules apply.
  A + B   and   AB   are in the set.
  A + B = B + A   and   AB = BA
  A + (B + C) = (A + B) + C   and   A(BC) = (AB)C
  A(B + C) = AB + AC   and   A + BC = (A + B)(A + C)
  A + A = A   and   AA = A
  A + B = B   if and only if   AB = A.
The set contains elements 0 and I such that
  A + 0 = A   and   A0 = 0
  A + I = I   and   AI = A.
For every element A, there is an element A' such that
  A + A' = I   AA' = 0.
A relationship of inequality between elements is also often included:
  For all X, X ≤ X
  If X ≤ Y and Y ≤ Z, then X ≤ Z
  If X ≤ Y and Y ≤ X, then X = Y
It is a branch of logic, named after the 19th-century logician George Boole. It allows statements to be represented symbolically and manipulated algebraically, and has been extensively used in designing algorithms, in designing computers, and in analyzing geodetic networks.

Boole's rule - The Newton-Cotes formula for N = 4. Simpson's rule is preferred to Boole's rule for simplicity; Weddle's rule is preferred for both accuracy and simplicity.

Borda scale - A metallic thermometer composed of two metals having different coefficients of thermal expansion. The scale was devised by the French scientist Jean Charles Borda, who placed a strip of copper on a strip of platinum, fastened the two together at one end, and by measuring the relative movement of their free ends determined their temperature. He used this as a means of determining the temperatures of bars used in baseline apparatus. Later, Bessel used a combination of zinc and iron for the same purpose.

border - (1) The outer part or edge of anything; margin, verge. (2) A boundary. (3) A frontier of a state or of the settled part of a country; a district on the frontier.

border break - A cartographic technique whereby detail of a map or chart is extended beyond the sheet lines into the margin. This makes it unneccessary to produce an additional sheet for the small amount of detail involved. Also called a blister.

borderland - SEE borderland, continental.

borderland, continental - An oceanic region adjacent to a continent, usually occupied by or bordering on a continental shelf, that is very irregular, with depths much greater than those typical of the region over a continental shelf. Also called borderland.

bore - SEE bore, tidal.

bore, tidal - A tidal wave progressing up a relatively shallow and sloping estuary or river as a solitary wave. Also called, simply, a bore. The leading edge of the tidal bore rises abruptly.

boresighting - The process of aligning a directional antenna, often by optical means.

borrow pit - SEE pit, borrow.

Boss star catalog - A compilation star-catalog compiled by Benjamin Boss and six collaborators and published in 1937 as the General Catalog of 33342 Stars for the Epoch 1950. Also called the General Catalog and the Boss catalog. It is a fundamental star-catalog and contains the right ascension, declination, and proper motions, for 33 342 stars, referred to the epoch of 1950. It was at one time widely used for determinations of astronomic latitude and longitude. However, it is now little used in geodesy because many of the proper motions in it are so uncertain that coordinates calculated using them contain unacceptably large errors.

Boston rod - SEE Boston leveling rod.

Boston leveling rod - A two-piece, target leveling-rod with the target fixed on one end. Also called a Boston rod. The target is adjusted in height by moving one or the two pieces on the other. The scale on the rod is read against a vernier. For heights greater than 5½ feet, the target end is up; for lesser heights, the target end is down. The design is little used.

bottle, water-sampling - A container used for sampling sea water and consisting of a brass cylinder fitted with a plug valve at each end and capable of being reversed in position. The two valves close simultaneously. The cylinder holds about 1200 cc. In use, the bottle is lowered on a wire with the valves open. A weight sliding down the wire is let go from the surface when the bottle has reached depth. The weight, called a messenger, closes the valves and reverses the bottle. There are usually several bottles, fastened at different depths along the wire, so the messenger closes each bottle in succession. The Nansen bottle (also called the Nansen water- bottle) is the one most widely used in oceanography. The Ekman bottle has plates seated in rubber gaskets as valves. SEE ALSO Ekman bottle; Nansen bottle.

bottom - SEE sea bottom.

bottom, deep-sea - Those regions of the sea bottom which lie at depths between about 2440 and 5750 meters. Also called the floor, ocean floor or sea floor. Regions below 5750 meters are called deep- sea depressions and trenches. The deep-sea bottom covers, in all, an area of about 284 000 000 square km and lies at an average depth of 4420 meters.

bottom, oceanic - SEE sea bottom.

Bouguer anomaly - SEE Bouguer gravity anomaly.

Bouguer correction - SEE Bouguer gravity correction.

Bouguer cylinder - A fictitious, vertical cylinder whose upper surface passes through a point P on the Earth's surface, whose axis passes through P and is perpendicular to the terrestrial ellipsoid, and whose height is equal to the height of P above the ellipsoid. The radius is arbitrary but is usually between 40 km and 100 km. The cylinder is assumed to be filled with matter, a density of 2.67 g/cc being usual.

Bouguer gravity - (1) A value of gravity at a point P, on the ground, obtained by adding, to a hypothetical value of gravity on a reference surface, (a) the free-air gravity correction δg_f and (b) the Bouguer gravity correction δg_B. Also called simple Bouguer gravity or incomplete Bouguer gravity. In this calculation, one assumes that there is no mass outside the reference surface. The first correction calculates the decrease in gravity in going from the reference surface to the point in question; the elevation H_p of the point is usually used in making this correction. The second correction assumes that the effect of matter actually outside the reference surface can be represented by the effect of a flat, infinite plate (the Bouguer plate) of thickness Hp lying between the point and the surface and of standard density (usually 2.67 g/cc). The hypothetical value of gravity is usually given by a standard gravity formula and is the value on a particular ellipsoid at constant gravity potential. The Bouguer gravity is therefore in error because (a) the ellipsoid is not the same as the one specified by the datum in which the coordinates of P are given; (b) the ellipsoid is not the surface (the geoid) to which the elevation Hp refers; (c) gravity, on the Earth, does not vary as the inverse square of the distance; (d) the matter actually present is not correctly represented by a Bouguer plate; and (e) the standard gravity formula used does not correctly represent gravity on the geoid. Nevertheless, Bouguer gravity (or, more usually, measured gravity minus Bouguer gravity) is much used in geophysics because it gives a rough idea of the difference between expected and actual masses present within a region. (2) The value of gravity, at a point Q on the geoid, obtained by subtracting, from a measured value of gravity at point P at elevation Hp above Q, (a) the Bouguer gravity correction δg_B and (b) the free-air gravity correction δg_f. Also called simple Bouguer gravity or incomplete Bouguer gravity. The matter between geoid and P is represented by an infinite, flat plate (the Bouguer plate) of thickness Hp and density ρ (usually 2.67 g/cc) placed horizontally between m P and the geoid. The Bouguer gravity is erroneous because (a) the matter in question is not well represented by the Bouguer plate; (b) gravity between the geoid and P does not, on the Earth, vary exactly according as the inverse square of the distance, and (c) the upward attraction of this matter on the point on the geoid, after the two corrections have been subtracted, is not considered. A frequently used formula for Bouguer gravity is $gB = g(measured) + 0.3086 H_p - 0.1119 H_p$, in which H_p is in meters. Note that this definition is not the converse of definition (1). (3) Gravity calculated as specified in either of the definitions (1) or (2), but using a cylindrical Bouguer gravity-correction δg_{Bc} for the attraction of a disk instead of the Bouguer gravity- correction δg_B for the attraction of an infinite plate. Also called cylindrical Bouguer gravity. (4) Gravity calculated as either of the definitions (1) or (2), but using a spherical Bouguer gravity-correction δg_{Bs} for the attraction of a spherical shell instead of the Bouguer gravity correction δg_B for the attraction of an infinite plate. Also called spherical Bouguer gravity. (5) SEE gravity, free-air.

Bouguer gravity, complete - SEE Bouguer gravity, expanded.

Bouguer gravity, cylindrical - SEE Bouguer gravity (3).

Bouguer gravity, expanded - (1) The value of gravity at a point P at elevation H_p, obtained by adding to a hypothetical value of gravity on a reference surface containing the entire mass of the Earth: (a) the free-air gravity-correction δg_f (to take account of the decrease of gravity with distance above the reference surface); (b) the Bouguer gravity-correction δg_B for the approximate attraction of matter between the reference surface and P; and (c) a topographic gravity-correction δg_t which takes into account the departure of the terrain from the plate assumed in adding the Bouguer gravity-correction. Also called Bouguer gravity, complete Bouguer gravity and Bouguer gravity with topographic correction. The topographic gravity- correction is usually calculated as if the terrain between the geoid and elevation H_p has its base on the lower surface of the Bouguer plate, i.e., as if the geoid were flattened and placed, with the matter above it, on the lower surface of the Bouguer plate. An additional correction (δg_c) is therefore sometimes added to account in part for the curvature of the geoid. Bouguer gravity calculated with this correction is sometimes called Bouguer gravity with topographic and curvature corrections. (2) The value of gravity obtained for a point Q on the geoid, at a distance H_p below a point P on the ground, by subtracting, from the measured value at P (a) the Bouguer gravity-correction δg_B, to account for the attraction of matter between P and Q; (b) the free-air gravity-correction δg_f to account for the increase in gravity in going from P to Q; and (c) a topographic gravity-correction δg_t which takes into account the difference between the actual distribution of matter above the geoid and that assumed by the Bouguer gravity-correction. Also called Bouguer gravity, complete Bouguer gravity and Bouguer gravity with topographic correction. The terrain is assumed to have its base on the lower surface of the Bouguer plate, so a correction $\delta g_c'$ is often introduced to take account of the curvature of the geoid. The resulting value is then sometimes called the Bouguer gravity with topographic and curvature corrections. (3) Gravity calculated as specified in either of the definitions (1) or (2), but using a cylindrical Bouguer gravity-correction δg_{Bc} for the attraction of a disk instead of the Bouguer gravity-correction δg_B for the attraction of an infinite plate. (4) Gravity calculated as either of the definitions (1) or (2), but using a spherical Bouguer gravity-correction δg_{Bs} for the attraction of a spherical shell instead of the Bouguer gravity correction δg_B for the attraction of an infinite plate.

Bouguer gravity, incomplete - SEE Bouguer gravity.

Bouguer gravity, simple - SEE Bouguer gravity.

Bouguer gravity, spherical - SEE Bouguer gravity (4).

Bouguer gravity anomaly - (1) The result of subtracting from a gravity anomaly g the following gravity corrections: (a) the free-air gravity correction δg_f, which gives the decrease in gravity when moving from the reference surface to the point of observation (ignoring the mass outside that surface); and (b) the Bouguer gravity correction δg_B, which approximates the effect of the mass ignored when applying the free-air gravity correction (an infinite, horizontal plate of thickness equal to the elevation of the point of observation is substituted for the ignored mass.) $\Delta g_B = \Delta g - \delta g_f - \delta g_B$. Also called the simple Bouguer gravity anomaly. It can be considered a first approximation to the difference between the value found for gravity as measured at a point and the value predicted for gravity at that point. It can also be considered a first approximation to difference between the value of gravity on the geoid, as calculated from a measured value of gravity, and the value of gravity on a reference ellipsoid such as the terrestrial ellipsoid. The following approximations are among those implicit in computing the Bouguer gravity anomaly and similar quantities. (a) The elevation H_p of the point of observation P is used instead of the geodetic height of P above the reference surface. (b) The point Q for which the value of gravity was first calculated is on the normal through P instead of on the vertical. (c) The thickness of the plate is taken as H_p instead of the value needed to take the ignored mass into account correctly. (d) The mass of the atmosphere is ignored. The Bouguer anomaly is named after the French mathematician Pierre Bouguer (1698 -1758). He did not develop the method itself but did propose using the effect of a plate of indefinite extent to make the first approximation. Bouguer's rule has also borne the name of Young (Thomas Young, 1773 -1829) and Poisson (Simon Denis Poisson, 1781 -1840). About the beginning of the nineteenth century, pendulums began to be used for measuring gravity as a means of determining the geoid. Bouguer's rule for the reduction of gravity data followed immediately. The method continued in favor until Clarke and Helmert, applying the theories of Pratt and Airy which are now known as the principle of isostasy, showed that Bouguer's rule did not come anywhere near being a good approximation to the value of gravity in many places. At present, the rule is used principally in theoretical studies and in geophysical prospecting. Bouguer gravity anomaly has proved of little or no value in reducing gravity measurements to obtain an even approximately satisfactory, theoretical value of gravity. It ignores isostatic compensation and creates large, fictitious gravity anomalies; it neglects large amounts of matter existing between the geoid and the terrestrial ellipsoid; and it takes no account of actual topography. (2) The quantity obtained by subtracting a topographic gravity correction from the Bouguer gravity anomaly and adding Bullard's term to the result. By adding Bullard's term, a spherical cap of the same thickness as the Bouguer plate and extending out to Hayford's zone O2 is substituted for the Bouguer plate. (3) The gravity anomaly obtained by subtracting from a measured value of gravity, the value assumed for gravity on a reference surface, the free-air gravity correction, the correction for an infinite plate between the two points (the Bouguer plate) and a correction for such of the terrain as was not accounted for by the introduction of the plate. In this process, no correction is made for terrain below the lower surface of the Bouguer plate. (4) The difference between the value of gravity on the geoid of the actual Earth, after masses outside the geoid have been removed, and the value of gravity (calculated from a gravity formula) on the reference surface. (5) SEE Bouguer gravity anomaly, cylindrical. (6) SEE Bouguer gravity anomaly, spherical. (7) SEE gravity anomaly, expanded.

Bouguer gravity anomaly, complete - SEE Bouguer gravity anomaly, expanded.

Bouguer gravity anomaly, cylindrical - The quantity Δg_{BC} obtained by subtracting, from a gravity anomaly Δg, (a) the free-air gravity anomaly δg_f, and (b) the cylindrical Bouguer gravity correction δg_{BC}: $\Delta g_{BC} = \Delta g - \delta g_f - \delta g_{BC}$. The cylindrical Bouguer gravity correction introduces the mass of a finite disk instead of an infinite plate.

Bouguer gravity anomaly, expanded - (1) The quantity gBe obtained by subtracting from a gravity anomaly Δg the free-air gravity correction δg_f, the spherical Bouguer gravity correction δg_{Bs}, and the topographic gravity correction δg_t: $\Delta g_{Be} = \Delta g - \delta g_f - \delta g_{Bs} - \delta g_t$. Equivalently, the quantity obtained by subtracting from the spherical Bouguer gravity anomaly the topographic gravity correction. The topographic gravity correction is introduced because of the difference between the actual distribution of mass ignored in applying the free-air correction and that assumed by introducing the spherical shell. (2) The result of subtracting, from a Bouguer gravity anomaly with topographic correction, a further gravity correction to account for the curvature of the Earth. This is practically, if not exactly, equivalent to using a spherical Bouguer gravity anomaly with topographic correction. The expanded Bouguer gravity anomaly is also called the complete Bouguer gravity anomaly and the refined Bouguer gravity anomaly. The latter term, however, seems to be applied also to the Bouguer gravity anomaly with topographic correction.

Bouguer gravity anomaly, incomplete - SEE gravity anomaly, free-air (2).

Bouguer gravity anomaly, modified - The difference gBm between a measured value g of gravity at a point P and a corresponding, calculated value of gravity, τ' is calculated by adding, to the value τ obtained from a standard gravity formula, a gravity correction $\delta g'$. $\delta g'$ is an estimate of the gravity caused by matter between the reference surface and an equipotential surface through P and uses a suitably sized a disk, shell or spherical cap to represent that matter. Alternatively, a Bouguer gravity anomaly modified by substituting a disk, shell, or spherical cap for the Bouguer plate. A modified Bouguer gravity anomaly is more commonly called by the name of the approximating body used, e.g., cylindrical Bouguer gravity anomaly if a disk is used, and spherical Bouguer gravity anomaly if a shell is used. SEE ALSO Bullard's term.

Bouguer gravity anomaly, refined - The result Δg_{Bf} of subtracting from a Bouguer gravity anomaly Δg_B the topographic gravity correction δg_t: $\Delta g_{Br} = \Delta g_B - \delta g_t$ or $\Delta g_{Br} = g - (\tau + \delta g_B + \delta g_f + \delta g_t)$, in which g is the value of g measured at a point P, τ is the value of gravity calculated for a point Q vertically below P on the reference surface, δg_B is the Bouguer gravity correction, δg_f is the free-air gravity correction and δg_t is the topographic gravity correction. Also called the complete Bouguer gravity anomaly and expanded Bouguer gravity anomaly. While strict interpretation of the definitions indicates that the terms have different meanings, much of the distinction is blurred if the terms Bouguer gravity correction and topographic gravity correction are not defined precisely. This lack of precision is common in geodetic writings, so that the three terms can often be considered synonymous. The context must be used as a guide.

Bouguer gravity anomaly, simple - SEE Bouguer gravity anomaly (1).

Bouguer gravity anomaly, spherical - The result Δg_{Bs} of subtracting from a gravity anomaly δg (a) the free-air gravity correction δg_f, moving the point for which gravity is predicted from a point on the reference surface to the corresponding point at which gravity was measured, but ignoring the effects of intervening masses outside the reference surface, and (b) the spherical Bouguer gravity correction δg_{Bs}, which represents the effect of masses outside the reference surface by the effect of a spherical shell horizontal at the point of measurement, of thickness equal to the elevation of that point, and of outer radius equal to the radius of curvature at that point (or of a suitably close approximation to that radius). $\Delta g_{Bs} = \Delta g - (\delta g_f + \delta g_{Bs})$. Alternatively, it can be defined as a Bouguer gravity anomaly in which the Bouguer gravity correction has been replaced by the spherical Bouguer gravity correction. That is, the flat plate assumed in using the Bouguer gravity anomaly is replaced by a spherical shell of the same thickness. SEE ALSO Bouger gravity anomaly, expanded.

Bouguer gravity correction - (1) A quantity which approximates the amount by which matter outside a reference surface and below a point on the ground increases the value of gravity at that point, by the gravity due to the mass of an infinite slab horizontal at the point and having a thickness equal to the distance of the point from the reference surface. Also called a Bouguer gravity reduction and simple Bouguer gravity correction. The Bouguer correction is given by the formula $\delta g_B = 2\pi G h \rho$, in which G is the gravitational constant, h is the thickness of the slab and ρ is the density. The slab is called the Bouguer plate. The density is constant and is usually taken to be 2670 kg/m3, so that the formula is, approximately, $\delta g_B = 0.0418\ h\ \rho$ newton/kg or $0.0418\ h\ \rho$ m/s^2. The correction was invented and used by Bouguer for analyzing the results of the 1735-1745 expedition to Peru. (2) A quantity which approximates the amount by which matter outside the geoid and below a point on the ground increases the value of gravity at that point, by the gravity due to the mass of an infinite slab horizontal at the point and having a thickness equal to the elevation of the point above the geoid. Also called a Bouguer gravity reduction. The value of gravity on the geoid is usually approximated by a value calculated using a gravity formula. This definition differs from that above essentially in substituting for the height h the elevation H. (3) The same as either of the preceding definitions above but using, instead of the surfaces used there, some other reference surface and using a corresponding distance of the given point above that surface. This definition is more common among geophysicists than among geodesists. (4) A gravity correction similar to those defined in (1) and (2) but substituting some other solid such as a spherical shell, a spherical cap, or a disk. Such a gravity correction is also, and more exactly, named for the shape involved, e.g., spherical Bouguer gravity correction and cylindrical Bouguer gravity correction. (5) The sum of the free-air, Bouguer (2), and topographic gravity corrections, i.e., a topographic gravity reduction. (6) The same as definition (5) but without the free-air gravity correction. (7) The sum of the free-air and Bouguer gravity corrections (2). (8) The difference between gravity as measured and an average value of gravity.

Bouguer gravity correction, cylindrical - The quantity δg_{Bc} (= $2\pi Gh^2 \rho/2a$) representing the force per unit mass, or acceleration, caused by the attraction of a cylindrical solid of density ρ, radius a and height h on a unit mass located at the intersection of the cylinder's axis with the top surface. G is the gravitational constant. The cylindrical Bouguer gravity correction is used in calculating Faye's gravity anomaly.

Bouguer gravity correction, spherical - A gravity correction δg_{Bs}, given by $\delta g_{Bs} = (4\pi \rho G/3)[(R+h) - R^3/(R+h)^2]$, which is the gravity caused by a spherical shell, density ρ, inner radius R and thickness h. G is the gravitational constant. The radius R is usually chosen to be the average radius of the Earth, although it is more properly chosen so that the outer surface passes through the point of measurement, P at an elevation H = h. This gravity correction is frequently used instead of the Bouguer gravity correction, leading to the spherical Bouguer gravity anomaly instead of the Bouguer gravity anomaly.

Bouguer gravity field - The formula $\tau = \tau_o[1-2h/a+ \rho'h/(2a \rho)]$ for the value of gravity at a height h above the surface of reference, τ_o being the value of gravity on the surface, a the radius of a spherical body with mass equal to the Earth's, ρ the average density of the Earth and ρ' the density of material between the surface and height h.

Bouguer gravity reduction - (1) The process of applying to a measured or computed value of gravity the sum δ_B of the free-air gravity correction δg_f and the Bouguer gravity correction δg_B. Also called the simple Bouguer reduction, incomplete Bouguer gravity reduction, complete Bouguer gravity reduction and Bouguer gravity correction. The incomplete Bouguer gravity reduction is also the process of applying the Bouguer gravity correction only. (2) The quantity δ_B defined in (1). (3) The process of applying, to a measured of computed value of gravity, the sum of the free-air gravity correction δg_f and the complete topographic gravity correction δg_{tc}. This meaning is rare and is more usually denoted as Bouguer gravity correction with topographic correction or topographic gravity reduction. (4) The process of applying, to a measured or calculated value of gravity, the sum of the free-air gravity correction δg_f, the Bouguer gravity correction δg_B, the topographic gravity correction δg_t, and the condensation gravity correction δg_c. (5) SEE gravity correction, Bouguer (1).

Bouguer gravity reduction, complete - SEE Bouger gravity reduction.

Bouguer gravity reduction, extended - (1) The process of applying, to a measured or calculated value of gravity, the sum of a spherical Bouguer gravity correction and a topographic gravity correction. (2) The sum of the gravity corrections mentioned in the previous definition.

Bouguer gravity reduction, incomplete - (1) The process of applying, to a measured or computed value of gravity, the Bouguer gravity correction. (2) SEE Bouger gravity reduction (1).

Bouguer plate - An imaginary, cylindrical plate of infinite radius and of thickness equal to the elevation of a particular gravity station, placed with its upper surface horizontal at the gravity station. Its density is usually taken to be 2.67 g/cc. The attraction of the Bouguer plate at the gravity station is roughly equal to the attraction there of all matter having elevations between zero and the elevation of the gravity station. The gravitational acceleration attributable to the plate is $2\pi GdH$, where G is the gravitational constant, d is the average density of the matter in the plate and H is the thickness of the plate. Its attraction corrects (approximately) for the error made in computing the free-air anomaly when the mass between the point and the geoid is ignored.

Bouguer reduction - The process of obtaining (calculating) the Bouguer gravity anomaly. SEE ALSO gravity reduction; Bouguer gravity reduction.

boulder - A large, rock which has been separated from the mass of which it was originally a part. Deeply imbedded boulders may show an exposed surface which, through weathering, has the appearance of bedrock.

boulder clay - SEE till.

boulevard - A major road into which other streets and roads feed their traffic and from which other streets and roads receive their traffic. It is generally shorter than a highway, occurs in cities, and usually serves thru traffic on a continuous route. It often has a median strip or traffic islands.

boundary - (1)(mathematics) An N-1 dimensional surface separating two N-dimensional regions (A and B) so that (a) for any two points lying entirely within a region, a line can be found that connects these two points and lies entirely within that region; and (b) for any two points lying in different regions, no connecting line can be found which does not contain a point of the N-1 dimensional surface. E.g., a point on a straight line is a boundary separating the line into two 1-dimensional regions, a closed line is a boundary separating the surface on which it lies into two 2-dimensional regions, and a closed surface separates the space in which it lies into two 3-dimensional regions. (2)(surveying) The (imaginary) line in which a vertical surface intersects the Earth's surface, or the (real), continuous object or discontinuous set of marks identifying such a line. Also called a mere. The terms boundary and boundary line are usually used with the same meaning. For example, a fence or wall running along a boundary is often referred to as either the boundary or the boundary line. However, it is more precise to refer to the fence or wall, etc., as marking the boundary. (3) Every separation, natural or artificial, that marks the confines or the line of division between two continuous estates. In a legal sense, a boundary between two pieces of property is usually considered to extend indefinitely both upwards and downwards from the surface unless there exist explicit legal restrictions against the extension. The boundary is then a surface, not a line. The surface need not be vertical. A boundary may be horizontal, separating one layer of ground from another. SEE ALSO exterior boundary; gradient boundary; land boundary; law of boundaries; state boundary.

boundary, coastal - A line (or a boundary measured from the line or points thereon) used to depict the intersection of the ocean's surface and the land at the elevation of a particular (vertical) datum. SEE ALSO boundary, tidal.

boundary, free - A boundary not limited by a call for a monument. E.g., thence N 12° E, 120 feet. Thence N 12° E, 120 feet to Brown's south line is not a free boundary.

boundary, international - A boundary fixed by an international agreement.

boundary, lateral - (1) A boundary fixing one side of a piece of land. (2) A boundary, under Public Law 31, between adjacent States and extending from shore to their seaward boundaries. (3) A boundary between adjacent nations through the marginal sea and the contiguous zones.

boundary, marine - A line of demarcation between adjoining regions at sea. The boundary may be at the surface of the sea or at the bottom.

boundary, national - A boundary between countries. Such a boundary is established by the sovereign powers concerned.

boundary, riparian - A boundary formed by the waters of a river, lake, sea or other body of water. Also called a water boundary, although the meaning of that term is doubtful. The general rule is that riparian boundaries shift with changes due to accretion or erosion. Most law states that they also change if the waters change their location by avulsion or by artificial causes.

boundary, tidal - The intersection of a tidal datum with the land. This definition permits many intersections to be included which may not be legal boundaries. The intersection should be specified, otherwise, as one which is also legally a boundary.

boundary line - A line separating two regions. In specific cases, the word boundary is sometimes omitted, as in State line; and sometimes the word line is omitted, as in International boundary, county boundary, etc. The term boundary line is usually applied to boundaries between politically defined territories, as State boundary line, between two states. A boundary line between privately owned parcels of land is preferably called a property line, or, if a line of the U.S. public land surveys, is given the particular designation of that survey system, as section line, township line, etc. In geodesy, the term is used as equivalent to boundary.

boundary map - A map prepared specifically for delineating a boundary line and adjacent territory.

boundary monument (USPLS) - A material object placed on or near a boundary line to preserve and identify the location of the boundary line on the ground. Where it is impracticable to establish a monument on or very close to a boundary line, the position of the boundary line on the ground is preserved by means of reference marks. The term monument is sometimes used to include both the mark on the boundary line and the reference mark. SEE ALSO monument.

boundary precedence - The ranking of boundaries of civil units to indicate which boundary should be shown when two or more boundaries overlap on a map. For example, where a national boundary coincides with a state boundary, only the national boundary is shown on the map. In general, the boundary of one political unit takes place over the boundary of a lower political unit and is preceded by the boundary of a higher political unit.

boundary survey - (1) A survey made to establish or to re-establish a boundary line on the ground, or to obtain data for constructing a map or plot showing a boundary line. The term boundary survey is usually restricted to surveys of boundary lines between political territories. For the survey of a boundary line between privately owned parcels of land, the term land survey is preferred (although property survey is also used); for official surveys of the public lands of the United States of America, cadastral survey is used. (2) SEE land survey.

boundary-value problem - SEE problem, boundary-value.

boundary-value problem of geodesy - SEE problem, boundary-value (2).

boundary vista - A lane cleared along a boundary line which passes through a wooded region. A boundary vista is used to make it easy to identify a boundary line and to help civil administration relating to the boundary line.

bounds - (1) The external or limiting lines of any object or space. For example, in to set bounds, as of a property.
(2) That which limits or restrains, or within which something is limited or restrained. Terms of similar meaning are limits, confines and boundaries.

bounds description - A description of the boundary of a piece of land in terms of adjoining land or waters. For example, bounded on the north by the Potomac River, on the west by Turkey Run Park, on the south by the land of Joseph Taylor, and on the east by Turkey Run.

bounty land - A portion of the public domain given to soldiers for military service by way of bounty, in place of monetary payment.

Bourdon gage - SEE barometer, aneroid.

Boussinesq approximation - (1) An approximation (Boussinesq 1903), to the exact equations of motion of water, which neglects, in the inertial term, variations in the density of water. (2) An approximation (Boussinesq 1903), to the exact equations of motion of a fluid, which assumes that the fluid is incompressible except insofar as thermal expansion produces buoyancy.

Bowditch adjustment - SEE Bowditch's method (of adjusting a traverse).

Bowditch-Baarda method (of adjusting a traverse) - SEE Bowditch's method (of adjusting a traverse).

Bowditch hypothesis - SEE Bowditch's method (of adjusting a traverse).

Bowditch's method (of adjusting a traverse) - (1) The adjustment of a traverse by calculating corrections to measured distances and directions, using the assumption (the Bowditch hypothesis) that the end of a line in the traverse is likely to be located with the same precision in the direction along the line as in the direction at right angles to it, and that the angular error is therefore inversely proportional to the square root of the distance. Bowditch's method is practically equivalent to the application of the compass rule. It has been modified by Baarda. Also called Bowditch's adjustment and the Bowditch-Baarda method. (2) SEE compass rule (for adjusting a traverse).

Bowie correction - SEE Bowie gravity correction.

Bowie effect - (1) SEE Bowie gravity correction. (2) SEE effect, indirect.

Bowie gravity correction - The quantity $2H\tau'/R$, denoted by $\delta^2 g$, in which τ' is the value of gravity, at point P on a given surface, corrected for removal and addition of masses between that surface and the geoid and for isostatic compensation. H is the distance (at P) between that surface and the position to which the geoid is moved by redistribution of the masses. Also called the Bowie correction, Bowie effect, indirect effect and Bruns term. The correction may be calculated with an isostatic gravity correction added to the value of gravity calculated from a standard gravity formula, as well with a gravity correction obtained by adding matter to the oceanic basins to bring them up to the density assumed for land. It is applied to the theoretical value of gravity, with the sign reversed. Values of the Bowie correction are tabulated in the Fundamental Tables: Deformation of the Geoid and its Effect on Gravity. The symbol $\delta^2 g$ has been used to represent the Bowie effect. (2) The equivalent value for gravity potential.

Bowie's gravity table - One of a set of tables calculated by W. Bowie (1917) giving combined topographic and isostatic gravity corrections, according to the Pratt-Hayford theory of isostatic compensation, for depths of 56.9, 85.3, 127.9, 156.2, and 184.6 km.

Bowie method (of adjusting triangulation) - A variant of the arc method of adjusting a triangulation network, in which the length and azimuth of one side of a triangle at every junction between arcs (chains) are assumed correct and carried into a suitable figure at the junction. Directions or angles in arcs between these figures are then calculated, the corrections in the individual chains calculated, and the misclosures passed into the longitudes and latitudes of the initially fixed sides in the junctions by an adjustment of the entire network, using the method of least squares. The method was devised in the Coast and Geodetic Survey, under the direction of William Bowie, for the adjustment of large triangulation-networks.

box, black - Any device changing the input to it in a predictable manner but whose inner workings are unknown to the user.

box compass - A combined straightedge and magnetic compass, suitable for use on a planetable to mark the magnetic meridian. Also called declinatoire or trough compass. A common form is a magnetic needle enclosed in a long, narrow box, with a scale having a range of only a few degrees on either side of zero.

boxing the compass - Naming the points and quarter-points of the compass in order clockwise around the circle, beginning with north.

box sextant - A cylindrical box containing two vertical mirrors, one of which can be rotated through 60° about a vertical axis. The movable mirror is turned by a knob at the top of the box and, as it rotates, so does an external pointer which indicates, on a metallic, graduated arc, the angle turned. The largest angle measurable is 120°. The instrument is best used for approximate but rapid measurement of horizontal angles in a reconnaissance for a later survey.

brace pile - SEE batter pile.

brachyte - A telescope whose physical length is made a fraction of the focal length by folding the optical path with plane mirrors.

branch, lower - That half of a meridian or celestial meridian which passes from one pole to another and through the antipode or nadir of the observer.

branch, upper - That half of a meridian or celestial meridian which passes from one pole to another and through the zenith of the observer.

branch of the sea - SEE arm of the sea.

brayer - A hand-held roller for spreading ink over type or over a plate.

break angle - The smaller dihedral angle between the two vertical planes passing through the common exposure station and the principal points of the right and left oblique photographs taken by a trimetrogon camera.

breakaway method - SEE breakaway strip method.

breakaway strip method - A method of preparing a photomosaic composed of two or more sheets. An extra-wide strip of masking tape is placed along the neatline of one sheet before the photographs are assembled. The portion of photographs extending beyond the edge is then cut along the neatline and transferred to the adjoining sheet. Also called the breakaway method.

breaker zone - (1) The region extending from the farthest line, on shore, to which the water reaches except under extreme conditions, to the line marking the outer boundary at which breakers form. Also called the surf zone and littoral zone. (2) That offshore region in which breakers form.

breaking tape - The procedure, when measuring horizontal distances on sloping ground, of holding only short lengths of tape horizontal and marking the end points of the length by placing taping pins beneath them in the ground. The tape is usually extended to its full length from the starting mark and successive, partial lengths held horizontal.

breakwater - An offshore barrier erected to break the action of the waves of the open sea and thereby create calm water inside the barrier.

Breusing's map projection - (1) A map projection transforming parallels of latitude into circles with radii which are the geometric averages of those of the stereographic and the azimuthal equal-area map projections. The radius r of the circle representing a parallel of latitude ϕ is given by $r = 2 k R^{1/2} [\tan^{1/4} (\pi - 2\phi) \sin^{1/4} (\pi - 2\phi)]$, in which k is the scale factor and R is the radius of the sphere. The meridians are represented by the radii of these concentric circles and have the same directions as the radii of the parent graticules. (2) A somewhat similar map projection, based on the harmonic average rather than on the geometric average of the radii, and with the scale changed to give the least total error for a map of given radius. This variant was introduced by A.E. Young in 1920.

Brewster's angle - That angle at which a ray incident upon a surface is completely polarized upon reflection. It is given by $\arctan (n_s / n_i)$, where n_i is the index of refraction of the medium on the side at which the ray is incident and n_s is the index of refraction of the medium on the other side.

Brewster's law - When unpolarized light incident on the flat surface of a transparent substance is most polarized in a plane upon reflection, the tangent of the angle of incidence is equal to the index of refraction of the substance. The reflected and refracted rays are just 90° apart.

bridge - A structure, including its supports, erected over a depression or an obstruction, having a track or passageway for carrying traffic or other moving loads, an having an opening measured along the center of the roadway of more than 20 feet between undercopings of abutments or springlines of arches, or extreme ends of openings for multiple boxes. It may include several pipes, where the clear distance between openings is less than half of the smaller contiguous openings.

bridging - (1) The extension and adjustment of a photogrammetric survey between regions containing ground control. The process contrasts with cantilevering, which proceeds from a region containing ground control into a region not containing ground control. (2) The production of maps by using stereoscopic plotting instruments. Also called stereocompilation.

bridging, analog - The photogrammetric measurement of points, in a stereoscopic model, and their adjustment to points of horizontal and vertical control. SEE ALSO control point, bridged supplemental.

bridging, horizontal - SEE bridging.

bridging, vertical - Photogrammetric bridging done primarily to extend vertical control. SEE ALSO bridging.

Briesemeister's map projection - An equal-area map projection of the sphere into an ellipse and defined by $x = 2\sqrt{2} kRm(\cos \phi' \sin\frac{1}{2} A_z)/D$, $y = \sqrt{2} k R (\sin \phi')/(mD)$, $D = [1 + \cos \phi' \cos\frac{1}{2} A_z]^{1/2}$, $\sin \phi' = \frac{1}{2}\sqrt{2} (\sin \phi - \cos \phi \cos \Delta\lambda)$ and $\cos A_z = \frac{1}{2} \sqrt{2} (\sin \phi + \cos \phi \cos \Delta\lambda)/\cos\phi'$, x and y are the rectangular Cartesian coordinates of that point, in the plane, corresponding to the point, on the sphere, with transformed latitude ϕ' and azimuth A_z from an origin on the central meridian situated 90° from the center of the projection. m is a constant; it was chosen by Briesemeister to be $1.75/\sqrt{2}$. This map projection has been used by the American Geographical Society for its maps of the world. The sphere is shown inclosed within an ellipse whose axes are in the ratio 1: 1.75 in length. It is an oblique version of the Hammer-Aitoff map projection, modified by a change of scale along the x- and y-axes.

brightness (of a planet) - A measure of the amount of reflected sunlight which reaches the observer. It depends on the planet's location relative to the Sun and Earth and on the albedo of the planet's surface.

brightness scale - The ratio of the brightness of highlights to the deepest shadow in the actual terrain, as measured from the camera stations, for field of view under consideration.

brightness temperature - (1) The temperature of a black-body radiating the same amount of energy per unit area at the wavelength under consideration as the body observed. Also called the effective temperature of the body observed. (2) That temperature of a radiating body not a black-body which is measured with an optical pyrometer or radiometer.

Brillouin's geoid - The level surface passing through the summit of Mount Everest (Chomulgunga). This is not the geoid; nor is it an approximation to the geoid.

Brillouin's gravity reduction - Calculation of the value of gravity on a surface lying 100/g kilometers above the Earth's surface.

British foot - SEE foot.

bronze pendulum - A quarter-meter pendulum, used by the U.S. Coast and Geodetic Survey, made of aluminum-bronze consisting of one part of aluminum and nine parts of copper. Beginning in 1920, bronze pendulums were replaced by invar pendulums in the gravity work of the U.S. Coast and Geodetic Survey.

brook - A rapidly flowing watercourse of little depth.

Brosche's method - A method of determining corrections α and $\Delta\delta$ to right ascension α and declination δ of one star catalog with respect to the right ascension and declination in another by representing the corrections as series of associated Legendre functions, e.g.: $\Delta\alpha = \Sigma\ Pnm(\delta)(Cnm \cos m\alpha + Snm \sin m\alpha)$.

Browne's correction - A gravity correction $-ga\theta^2/2$ applied to the value ga of gravity measured by a horizontally accelerated gravimeter to obtain the value of gravity that would have been measured by a stationary gravimeter. θ is the angle through which the axis of the gravimeter is deflected from the vertical. The measured value is in error because the vehicle's horizontal acceleration causes the sensitive axis of the gravimeter to make, on the average, an angle θ with the vertical.

Browne's gravity correction - SEE Browne's correction.

Brownian motion - Frequent, random movements caused by the randomly spaced impacts of sub-microscopic particles (atoms, molecules, or elementary particles) in the fluid surrounding a more massive and visible body. The motion is most noticeable when the body a small, suspended particle such as a mote of dust in the air. Brownian motion is a particular case of the motion called random walk. It is one factor limiting the accuracy achievable with some kinds of measuring instrument such as accelerometers. It should not be confused with the motion caused by random motions of the molecules in the body itself.

Brown's gravity apparatus - SEE Brown's pendulum apparatus.

Brown pendulum - SEE Brown's pendulum apparatus.

Brown's pendulum apparatus - A pendulum apparatus comprising the Mendenhall pendulum, a special kind of clamp for holding the pendulum in the receiver while the apparatus is being transported, and an electrical detecting and amplifying device for measuring and recording, on a chronograph, the oscillations of the pendulum and the corresponding times. Also called the U.S. Coast and Geodetic Survey pendulum-apparatus. The earlier pendulum-apparatus of Mendenhall was improved by placing leveling screws near the top of the receiver. This reduced flexure of the pendulum's support; the earlier form had the leveling screws at the bottom of the receiver and thus much farther from the point at which the pendulum was supported.

brown-print process - SEE VanDyke process.

Bruns' equation - The equation $Dng = 4\pi Gd(H) - 2g/R - 2\omega^2$, where Dng is the directional derivative (gradient) of the acceleration of gravity, G is the gravitational constant, d(H) the average density of matter at elevation H above the reference surface (equipotential), R the average radius of curvature at elevation H and ω the rate of rotation of the Earth. The equation holds if ω does not depend on the distance from the axis of rotation. Also called Bruns' theorem.

Bruns' formula - The formula $\Delta H = (W - U)/g_o$, where W is the potential on a particular equipotential surface of the Earth and U is the potential on an equipotential ellipsoid; ΔH is the difference of elevation between the two surfaces at the point where the gravity is g_o.

Brun's gravity reduction - The process of applying the term $2Ng_o/r$ (Bruns term) to a calculated value g_o of gravity on the geoid, to reduce gravity to the ellipsoid at a distance N below the geoid; r is a suitable radius of the Earth. It amounts to about 0.031 cm/s².

Bruns' spheroid - The surface defined by the zero-th and second-degree terms and a term representing centrifugal force, in the representation of the Earth's gravity potential by a series of Legendre functions.

Bruns' term - SEE Bowie gravity correction.

Bruns' theorem - SEE Bruns'equation.

Brunton compass - A small instrument combining the features of both the sighting compass and the clinometer. Also called a Brunton pocket transit. It can be held in the hand or supported upon a Jacob's staff or light tripod when being used to determine horizontal or vertical angles, for leveling, or for determining the magnetic bearing of a line.

Brunton pocket transit - SEE Brunton compass.

B-station - (1) A particular station of a pair of transmitting LORAN stations; the signal from the B-station always occurs more than half a repetition period after the next succeeding signal from the other station of the pair (the A-station) and less than half a repetition period before the next preceding signal from the A-station. (2) SEE A-station (1).

bubble axis - The horizontal line tangent to the upper surface of the centered bubble, in a spirit level, which lies in the vertical plane through the longitudinal axis of the vial.

buble level - SEE spirit level.

bubbler tide gage - SEE pressure gage, gas-purged.

bubble sextant - A sextant in which the bubble of a spirit level serves as the horizon. A simple form of bubble sextant was designed by E.E. Byrd and used by him for aerial navigation, on expeditions to the North and South Poles and on other arctic and antarctic expeditions. It has a spirit level with a circular bubble attached to the sextant quite close to the horizon glass. The angular elevation of a celestial body is measured by bringing its image into contact with the image of the bubble as seen in the horizon glass.

bubble tube - SEE level.

buck - A support for a surveyor's tape. Probably synonymous with saddle, tape support, etc. Also called a trestle (mainly Brit.). For very precise work, intermediate supports are made so that the tape (or wire) is supported on rollers. This avoids introducing frictional tension into the readings of the tensiometers at the ends of the tape.

buffer zone - A strip of land separating land devoted to one kind of use from land devoted to another kind.

building, accessory - A building located on a lot and used for a purpose other than that of the principal building on the same lot.

building line - SEE line, set-back.

building sewer - A pipe carrying sewage from the plumbing system of a building to a larger sewer serving several buildings or to a disposal point.

bulge, equatorial - (1) The excess of the Earth's equatorial diameter over its polar diameter. (2) That part of the Earth lying outside a sphere having the Earth's polar diameter as its diameter.

bulkhead - (1) A structure of wood, stone, or concrete erected along shores of bodies of water to arrest the action of waves or along steep embankments to control erosion. (2) The wall of a partition or structure in a mine.

bulkhead line - A line, in a harbor or navigable waters, defining the channelward limit of solid fills or bulkheads. Also called a pierhead line.

Bullard method (of gravity reduction) - SEE Hayford-Bullard method of gravity reduction.

Bullard method (of isostatic reduction) - SEE Hayford-Bullard method of gravity reduction.

Bullard's term - A quantity B giving the difference between the acceleration caused by the gravitational attraction of a spherical cap and the acceleration caused by the gravitational attraction of a Bouguer plate. I.e., of an infinite plate of the same thickness and composition as the cap. The spherical cap has a radius equal to the average radius of the Earth, a thickness equal to that of the Bouguer plate (usually taken as the elevation at the point of measurement) and an apex angle of 2° 59'56" (the outer diameter of Hayford zone O). Bullard's term is subtracted from a Bouguer gravity anomaly to obtain the equivalent, modified Bouguer gravity anomaly.

Bumstead tripod - A tripod similar to the Johnson tripod but simplified and made lighter. In the United States of America, it replaced the Johnson tripod about 1900.

bundle adjustment - SEE bundle method of aerotriangulation adjustment.

bundle method of aerotriangulation adjustment - (1) An aerotriangulation adjustment based on the principle of collinearity - that is, the geometry on which the adjustment is based is that of bundles of rays passing through perspective centers and joining ground-points to image-points. (2) SEE aerotriangulation adjustment (2). (3) SEE aerotriangulation adjustment, bundle.

buoy - A structure made to float on the surface of the water or at a predetermined depth below the surface, and to carry instruments or serve as an aid to navigation. Buoys used as aids to navigation are moored to the bottom; those made to carry instruments may be moored or free to move.

buoyancy correction - The amount subtracted from the period of a pendulum to account for the period's being lengthened by the upward pressure of the gas on it and the pull of gravity being thus partly countered. Also called the aerostat correction and air density correction.

Burckhardt's formula (for resection) - A formula used in solving for the location of a point P, in the plane, from the two angles measured at that point between three other points, using the distances and angle from the central point of the three to the other two. Denoting the measured angles by A and B, the corresponding subtended distances by a and b, and the angle between a and b by C, three auxiliary angles M, N, and Z are determined by the formula: $M = 180 - (A+B+C)/2$, $\tan Z = (a \sin B)/(b \sin A)$ and $(\tan N)/(\tan M) = (1 - \tan Z)/(1 + \tan Z)$. Then the angles D, E at the ends of the known lines are found from $D = N + M$ and $E = N - M$, and the problem is reduced to finding P by intersection.

Bureau International de l'Heure (B.I.H.) - The organization which is responsible, by international agreement, for coordinating the measurements of time by national observatories and for providing an internationally acceptable, common time; it is also responsible for maintaining the international atomic second, i.e., providing to users a unit of time, the second, against which other standards can be calibrated. As part of its function, it calculates the position of the Earth's axis of rotation with respect to points on the Earth (SEE motion, polar), and changes in the Earth's rate of rotation. The Bureau was founded in 1919 and its offices since then have been at the Paris Observatory. By an action of the International Astronomical Union, the BIH ceased to exist on 1 January 1988 and a new organization, the International Earth Rotation Service (IERS) was formed to deal with determination of the Earth's rotation. The time-keeping portion of the BIH was transferred to the Bureau International des Poids et Mesures (BIPM).

Bureau International des Points and Mesures (BIPM) - The BIPM was established in 1875 in accordance with the Convention du Metre. In 1988, it took over, from the Bureau

International de l'Heure, the responsibility for establishing and maintaining an international standard for the unit of time.

Bureau of Land Management - That agency of the U.S. Department of the Interior created by Executive Order effective 16 July 1946 and superseding the General Land Office in administering the general land-laws, with supervision of the subdivisional survey, and disposal of title to the public lands of the United States, and of the leasing, grants of right-of-way, and other surficial uses thereof, including all rights incident to the extraction of minerals; coal, phosphates, potash, oil, and gas on the public lands; withdrawals for irrigation, dams, and reservoirs, and for grazing and settlement rights of every description.

burn - The process, in lithography, of exposing a press-plate.

burn, double - The intentional exposure of two or more negatives in succession and in register on the same sensitized surface. Not to be confused with double exposure, which has the same nature but is usually unintentional. Also called double shooting.

Burt compass - SEE compass, solar.

Burt solar compass - SEE compass, solar.

buttal - SEE abuttal.

butte - A high, isolated rock outcrop from surrounding flat land.

butterfly map - A interrupted map constructed by mapping the sphere onto an octahedron and unfolding the octahedron into a butterfly-like arrangement of faces. Also called, erroneously, the butterfly map projection. The butterfly map was first described by B. Cahill in 1929. He devised four variants, using a different map projection for each. The most common variant is that using O. Adams's rhombic, conformal map projection.

butterfly map projection - SEE butterfly map.

butterfly shutter - A shutter shaped like a filled-in figure 8 and rotating about a pivot through the waist of the figure in such a way that light passing through the system can be interrupted by one or the other of the two branches. The shutter is located between the elements of the camera's lens-system.

butting - SEE abuttal.

butts and bounds - SEE metes and bounds.

Buys-Ballot's law - Facing the wind in the northern hemisphere, atmospheric pressure decreases toward the right and increases towards the left; the reverse is true in the southern hemisphere.

Bx - The x-component of the representation, in a photogrammetric model, of the air base - the line joining two camera stations. Also written bx.

Bx curve - A graphical representation of the errors, in the x-direction, of horizontal coordinates, with the errors plotted as ordinates against the x-coordinates as abscissae.

By - The y-component of the representation, in a photogrammetric model, of the air base - the line joining two camera stations. Also written by.

By curve - A graphical representation of the errors, in the y-direction, of horizontal coordinates, with the errors plotted as ordinates against the y-coordinates as abscissae.

byroad - (1) A road which in its earlier existence was obscure or local to neighborhood and not used to any great extent by the public, yet so far a public road that the public has the right of access to it at all times. (2) The statue law of New Jersey recognizes three kinds of roads: the public road, the private road, and the byroad. A byroad is a road used by the inhabitants and recognized by statute but not laid out. Such roads are often called driftways. They are roads of necessity in newly-settled countries.

byte - A set of consecutive, binary digits, usually eight in number. One byte usually represents one character.

Bz - The z-component of the representation, in a photogrammetric model, of the air base - the line joining two camera stations. Also written bz.

Bz-curve - A graphical representation of the vertical errors in the aerotriangulation of a strip of photographs. The x-coordinates of the vertical-control points, referred to the initial nadir point as origin, are plotted along the axis of abscissas and the differences between the known elevations of the control points and their elevations as determined by aerotriangulation are plotted along the axis of ordinates. The Bz curve is a smooth curve drawn through the plotted points. The elevation of any pass point in the strip is adjusted by the amount of the ordinate of the Bz curve for the corresponding x-coordinate of the point.

Bz-curve method of aerotriangulation adjustment - A method of orienting the stereoscopic models in a strip of photographs by using the Bz-curve to find the difference between the true photographic nadir point and that indicated by a Multiplex type of stereoscopic plotting instrument. The strip of models can also be leveled by this method if the aircraft's altitude, as given by the altimeter, is used.

C

cable length - 120 fathoms or 720 feet.

cadaster - (1) An official register of the location, quantity, value and ownership of real estate, compiled to serve as a basis for taxation. (2) An official register of real estate, with details of boundaries, area, value, ownership, and other rights associated with the real estate. (3) An official register of the quantity, value, and ownership of parcels of land (real estate) within the legal jurisdiction and administrative boundaries of a governmental unit. Also spelled cadastre.

cadaster, metes-and-bounds - A cadaster specifying the boundaries of a region by giving their metes and bounds. A type of numerical cadaster. SEE ALSO metes and bounds.

cadaster, multipurpose - A cadaster containing a) a geodetic network; b) a series of current, accurate, large-scale maps; c) an overlay delineating all parcels; d) a unique, identifying number assigned to each parcel; and e) a series of land-data files each including a parcel identifier used for retrieving the data about the parcel in that and other files.

cadaster, numerical - A cadaster specifying the boundaries of a region numerically. In particular, a cadaster specifying the boundaries by giving their lengths, directions, and locations with respect to certain identified monuments of marks. In this sense, it is a cadaster made in terms of metes and bounds, and may therefore be called a metes and bounds cadaster.

cadastre - SEE cadaster.

caging - Orienting and mechanically locking the rotational axis of a gyroscope to an internal reference-position.

cairn - A mound of rocks, stones or masonry constructed to mark a surveyed point or a point of importance to a survey. It is usually conical or pyramidal. SEE ALSO monument.

Caithness base line - A base line (2b), in the primary triangulation of Great Britain, between Spital Hill and Worth Hill. It is 24 828.000 meters long, measured with tape. It has been used extensively for the testing and calibration of electronic distance-measuring instruments.

calculation - The processes of addition, subtraction, multiplication and division. Arithmetic is a more general term; it includes calculation and the theory behind calculation.

calculation, fixed-point - Calculation in which each quantity used is represented by a single number (e.g., 395.28 and 10110.1) and the calculation is done according to the usual rules of arithmetic. In fixed-point calculations, the decimal point is not moved. Contrasted with floating-point calculation, in which each quantity is represented by two numbers and the decimal point is moved so that it immediately precedes or follows the most significant digit.

calculation, floating-point - Calculation in which each quantity used is represented by two numbers: a coefficient and an exponent (e.g. 3.9528-2 and 1.01101-3); the rules of arithmetic are applied separately to coefficient and exponent. Quantities are added or subtracted by first multiplying the coefficients by suitable powers of 10 to make the exponents the same and then doing the calculations on the coefficients. Quantities are multiplied or divided by multiplying or dividing the coefficients and at the same time adding or subtracting their exponents.

calculator - (1) A machine for automatically doing simple mathematical operations like addition, subtraction, multiplication and division. Most calculators can also determine the values of simple functions such as square roots, exponentials and logarithms. They differ from computers in not being able to carry out, automatically, predetermined sequences (algorithms) of calculations. Also, and preferably, called a calculating machine. (2) SEE computer.

caldera - A large, volcanic depression more or less circular or cirque-like and having a diameter many times that of the volcanic vent it contains.

calendar - (1) A numbered sequence of days, each sequence being called a year or calendar year and containing, usually, 365 or, every fourth year, 366 days; the days in a year are usually organized further into shorter subsequences of 7 days each, called "weeks" and into twelve subsequences of varying length but generally between 28 and 31 days, called months. Alternatively, a dividing of time into days and longer intervals such as weeks, months, and years, and the systematic numbering of such intervals. (2) A division of time into periods based on astronomical principles. The number of days in a year is usually referred to simply as the year for example, the year 2000 in the Gregorian calendar. The days within each year may be numbered sequentially or, more usually, may be organized into subsets, called months, of approx-imately the same size, the days within each month being numbered sequentially. The two calendars of particular civil importance are the Julian calendar and the Gregorian calendar. To change from the Julian calendar, which was in use up to 1581, to the Gregorian calendar, ten days were omitted from the calendar in 1582, the day after 1582 October 4 being redesignated 1582 October 15. SEE ALSO ephemeris calendar.

calendar, intercalary - A calendar in which days are inserted or removed from time to time to make the length of the calendar year agree better with the length of the tropical year. Tropical year and the mean solar day are incommen-surable, so any calendar year containing an integral number of mean solar days must therefore fall short of or exceed, by a fraction of a day, the tropical year.

calendar, proleptic - A calendar applied to dating events that occurred outside the period for which the calendar was originally defined. In particular, a calendar extended to date events that occurred before the officially established starting date of the calendar. E.g., the date, 49 B.C., of Caesar's crossing the Rubicon is a date in the proleptic Gregorian calendar.

calendar day - The period from midnight to midnight. It is 24 hours of mean solar time long and coincides with the civil day unless a change in time occurs during the day.

calendar month - An interval of time containing from 27 to 31 days, the exact number of days being a whole number and dictated by law or custom but approximating the interval of time from full moon to full moon (about 29.5 mean solar days). Also called a calendarial month.

calendar year - The interval of time containing an integral number of days, the exact number being set by law or custom but usually 365 days plus or minus a few days, and designated as a year in a particular calendar. A calendar year always contains an integral number of days; an astronomical year never does. The round of seasons follows the astronomical calendar. The number of days in a calendar year is therefore changed from time to time according to some rule set by law, custom, or religion, so that the average length of the calendar year and the astronomical year will remain approximately equal. SEE ALSO year, tropical.

calibrate (v.) - Determine, by measurement or comparison with a standard, the correct value of each division on a scale.

calibration - (1) Determining the systematic errors in an instrument by comparing measurements made with the instrument with the markings or measurements of a device which is considered more nearly correct. The correct value is established either by definition or by measurement with a device which has itself been calibrated against a device considered correct. Frequently-used instruments are usually calibrated using a measuring device (a working standard) which has itself been calibrated. Working standards are calibrated by comparing them with another calibrated set called laboratory standards, and these in turn are calibrated by comparison with the primary standard. Also called standardization. However, that term is now used to mean the imposition of a standard on otherwise diverse processes and products. Also called verification (Brit.) (2) The determination, in terms of an adopted unit and by mechanical interpolation based on values obtained by comparison with a standard, of the values of supplementary marks on a measuring device. The determination of the values of the divisions of a circle as proportional parts of a circumference. When the length of a surveyor's tape has been determined by comparison with a standard, the values of intermediate marks on the tape may be determined by calibration, the assumption being that the tape has been divided into parts which are strictly proportional to its length. Also called standardization. (3) The physical correction or adjustment of a measuring device to make its units of measurement corre-spond to those of a standard. Also called standardization.

calibration (of a camera) - Determination of the following quantities: (a) calibrated focal length; (b) location of the principal point with respect to the fiducial marks; (c) location of the point of symmetry; (d) the distortion effective in the focal plane of the camera and referred to the particular focal length; and, sometimes, (e) the resolution of the lens system, (f) the degree of flatness at the focal plane, and (g) the opening and closing cycle of the shutter as a function of time. Sometimes the locations of fiducial marks and intersections in a reseau are included in a calibration. If the camera has more than one lens system, calibration includes determining the angles between the optical axes of the individual lens-systems. Setting the fiducial marks and positioning the lens system are ordinarily considered adjust-ments, although these actions are sometimes done during calibration. Unless calibration is stated specifically to be of a camera, distortion and other optical characteristics of a lens system are determined in a focal plane located at the equivalent focal length and the process is called lens calibration.

calibration (of a camera system) - Calibration of camera and associated equipment and accessories such as range finder, exposure meter, intervalometer and photographic film or plates.

calibration (of a distance-measuring instrument) - SEE Schwendener method.

calibration (of a lens system) - In photogrammetry, determination of the equivalent focal length, resolution, distortion and other aberrations in a lens system. Also called lens-system calibration.

calibration base line - A line on which markers are placed at intervals so accurately measured that they can be used for calibrating distance-measuring instruments or equipment. Also called a field comparator. The specific characteristics depend on the kind of distance-measuring instrument to be calibrated. Lines on which surveyor's tapes are to be calibrated are commonly short and may be paved between markers. In the United States of America, lines on which electronic distance-measuring instruments are to be calibrated have the markers placed, typically, at distances of 150 meters, between 400 and 430 meters, and between 1000 and 1400 meters from the marker marking the starting point (zero distance). The term base line has the meaning given under base line (1) and under baseline (1), not that given under base line (2). Most calibration base lines are built specifically for calibration and are not part of a geodetic network. The spelling calibration base line is therefore better.

calibration card - A card having a list of calibration corrections or calibrated values.

calibration constant - One of a set of numbers obtained by calibrating a camera and giving the calibrated focal length and the relationship of the principal point to the fiducial marks.

calibration constant, angular - In a camera having several lens systems, or in an assemblage of several cameras, the values of the angles between the optical axes of the several lens-systems or cameras and a common reference line. For example, in a trimetrogon camera, the angles between the optical axes of the oblique cameras with respect to the optical axis of the central (vertical) camera.

calibration correction - (1) The value to be added to or subtracted from the reading of an instrument to obtain the correct reading. (2) The correction, to the nominal length between two graduations on a surveyor's tape, needed to account for the difference (calibrated-minus-nominal) between the nominal length and the length found by calibration. When the graduations mark the full length of the tape, the calibration correction and the length correction have the same value.

calibration error - That error, in an instrument, caused by imperfect calibration of maladjusted parts. Also called scale error and instrument error.

calibration plate - A photographic plate of glass exposed, when calibrating a camera, with its emulsion side in the same position in the camera as is the emulsion side of the photographic film during normal use. Also called a master glass negative or flash plate. The calibration plate provides a record of the distances between the fiducial marks of the camera.

calibration table - A list of calibration corrections or calibrated values.

calibration templet - (1) A sheet of glass, plastic, or metal made to show, according to the data provided by calibration of a camera, the relationship of the principal point to the fiducial marks. A calibration templet is used to rapidly and accurately mark principal points on a sequence of photographs. (2) A sheet of glass, plastic, or metal made to show, according to the data obtained by calibration of a multiple-lens camera, the relationships of the principal points to the fiducial markers and used to assemble the photographs from the individual photographs into one composite photograph.

call (USPLS) - (1) A clause, phrase, or statement, in a description of property, which identifies a particular characteristic of the boundary of the property. For example, a call for a monument identifies a monument on the boundary; a call for a survey identifies an earlier survey of the property; etc. (2) A reference to, or a statement of, an object, course, distance, or other matter of description in a survey or grant requiring or calling for a corresponding object, or other matter of description, on the land. SEE ALSO directory call.

call, locative - A call to specifically determine location.

call, passing - A call for a topographic or cultural feature along a surveyed line.

Callippic cycle - A period of four Metonic cycles, equal to 76 Julian years or 27 759 days. It was introduced by Callippus, a Greek astronomer, about 350 B.C. as suggested improvement to the Metonic cycle for a period in which new and full moon would recur on the same day of the year. Taking the length of the synodical month as 29.530 588 days, there are 940 lunations in the Callippic cycle, with about 0.25 day remaining.

calotte - A small region, on a rotational ellipsoid, which is bounded by a small circle.

camballeria (Spanish) - (1) A quantity of land of variable size, depending on the locality. (2) An allotment of land given to a horse soldier as a result of conquest.

Cambridge pendulum-apparatus - A pendulum apparatus constructed in 1926 by Lennox-Coyningham and consisting of three synchronous pendulums of the Sterneck type, made of invar, with knife edges of stellite resting on plates of agate. Any two of the pendulums are swung simultaneously in opposite directions. Two Cambridge pendulum-apparati have been made. They were designed to be able to measure gravity-acceleration to a precision of about 0.2 mgal. Both were used extensively in measuring the intensity of gravity for the International Gravity Standardization Net 1971.

camera - Any device for converting electromagnetic radiation from an object into an image of that object and projecting that image onto a surface. Camera without any modifier is understood to mean a camera that works with light - either a photographic camera or a television camera. Infra-red cameras and X-ray cameras use infrared and X-radiation, respectively. There are no cameras as such using radio waves. Radio waves are converted to images by first being captured by an antenna, then amplified, and then used to control the motion of a cathode-ray beam which, directly or indirectly, exposes photographic film to form the image. A photographic camera records the incoming radiation directly on radiation-sensitive emulsion. SEE ALSO camera, photographic. Cameras designed specifically for photographing the Earth's surface from the air are called aerial cameras; cameras designed for photographing the Earth from the ground are called terrestrial cameras. SEE ALSO Baker-Nunn camera; calibration (of a camera); copying camera; Moon camera, dual-rate; exposure camera; fan camera; frame camera; horizon camera; mapping camera; Nenonen camera; pinhole camera; positioning camera; process camera; restitution camera; satellite camera; strip camera; surveying camera; zenith camera.

camera, aerial - A camera designed particularly for use in aircraft.

camera, astrographic - SEE telescope, astrographic.

camera, axis of a - SEE camera axis.

camera, ballistic - (1) A camera used for photographing rockets or missiles, and producing photographs with sufficient metric fidelity that the object's trajectory can be determined photogrammetrically. Erroneously called a tracking camera. A ballistic camera may be of the fixed or the tracking type. A fixed ballistic camera keeps its optical axis pointed in the same direction while taking many pictures, and a photograph made by the camera shows the missile as a thin streak interrupted by spacings made by opening and closing the camera's shutter. A tracking ballistic camera keeps its optical axis pointed in the general direction of the object being photographed, and a photograph made by such a camera shows the object as a dot; the mounting often carries one or two graduated circles to show the direction in which the camera is pointed (this may be photographed or transmitted by selsyn motors to a recorder). (2) Any camera used in ballistics.

camera, CCD - A camera in which the light-detecting elements are charge-coupled devices (CCDs). SEE ALSO device, charge-coupled.

camera, charge-coupled-device - SEE camera, CCD.

camera, continuous-strip - A camera in which the film moves continuously past a slit in the focal plane in synchronism with the motion of the scene past the camera and producing a photograph in one unbroken length. In particular, an aerial camera in which the film moves at a rate proportional to the speed of the aircraft and producing one continuous photo-graph of the terrain. Also called a Sonne camera.

camera, convergent - An assemblage of two cameras mounted so as to have a fixed angle between their optical axes, and taking pictures simultaneously. The angular coverage is greater than that obtainable by a single camera of the given type - usually along the axis of the aircraft carrying the assemblage.

camera, copying - SEE copying camera.

camera, direct-scanning - A panoramic camera in which the lens system swings or rotates about the rear nodal point at a given rate.

camera, double - An assemblage of two equal cameras placed so that their optical axes are parallel to each other and are perpendicular to the line joining their inner focal points. If the cameras are both capable of taking pictures that can be treated photogrammetrically to give accurate results, they are often called a stereoscopic camera.

camera, geodetic stellar - A terrestrial camera used to photograph lighted objects or lights at considerable distances above the ground and against a background of stars. The object may be the Moon, a flare, a light carried on an artificial satellite, etc. Glass plates are usually used instead of film.

camera, metric - A camera constructed so that its geometric characteristics do not change from photograph to photograph and the image is as little distorted as possible. I.e., a camera whose calibration constants remain constant over long periods of time and which gives minimal distortion. Also called a precise camera. Metric cameras are essential for most photogrammetric projects. Non-metric cameras require, at best, frequent recalibration.

camera, multiband - A camera able to take several different photographs at the same time, each photograph being taken in a different part of the spectrum (e.g., in the visible, near infrared and far infrared parts). A common form has several objective lenses close together, each being followed by its own filters and focussing onto its own appropriate kind of film. Some, however, have only a single lens; the radiation is directed to the proper compartment by a beam splitter. Also called a multispectral camera.

camera, multiple-camera-assembly - SEE assembly, multiple-camera.

camera, multiple-lens - A camera having two or more objective lenses and shutters arranged at fixed angles to cover, together, a wide field of view when the shutters are opened and closed together.

camera, multispectral - A camera designed for taking a number of pictures simultaneously, each in a different part of the spectrum. SEE ALSO camera, multiband.

camera, non-metric - A camera not designed specifically for use in photogrammetry.

camera, panoramic - A camera which takes a partial or complete panoramic photograph of the terrain. Some designs contain a lens system and slot-like aperture which rotate about an axis perpendicular to the optical axis, the photographic film remaining fixed in position during the exposure. Other designs place a rotating prism in front of the lens system proper, the lens system itself remaining fixed. Still other designs rotate the entire camera, the photographic film being exposed as a sequence of frames. Only the first kind is used at present for taking aerial photographs.

camera, photogrammetric - A camera used in any of the several branches of photogrammetry. It is usually considered the same as a metric camera.

camera, photographic - A camera designed specifically for recording images directly on radiation-sensitive emulsion. A photographic camera consists of four parts, usually distinct and often separable: the lens assembly, consisting of the lens system and the lens; one or lens cylinder which keeps the lens system in place; the shutter mechanism, which determines the total amount of light admitted to the camera for taking a picture; the magazine, which stores, moves, and positions the light-sensitive material on which the images are recorded; and the body, which holds the other three subassemblies in place and also keeps out all light but that admitted by the shutter mechanism.

camera, precise - A camera whose optical qualities are good enough that the camera can be used photogrammetrically. Also called a "precision camera". There is no clear distinction between a precise camera and a metric camera.

camera, rectifying - SEE rectifier.

camera, rotating-prism - A class of panoramic camera in which a double dove prism in front of the lens system is rotated while the lens system itself remains fixed. This configuration allows photographing through 180° or more.

camera, satellite-tracking - A camera so mounted that its optical axis can be kept pointed in the direction of an artificial satellite. The Baker-Nunn camera, for example, is a satellite-tracking camera.

camera, solar - SEE camera, stellar.

camera, split - An assemblage of two cameras having their optical axes including such an angle that the fields of view overlap. Also called a split-vertical camera.

camera, split-vertical - SEE camera, split.

camera, stellar - A camera designed for photographing stars. The term is used in geodesy to designate (a) a camera designed for photographing artificial satellites against a stellar background and (b) a cameras designed photograph the night sky from an aircraft or artificial satellite so that the craft's orientation can be determined. A camera designed for photographing the Sun is called a solar camera.

camera, stereometric - An assemblage of two similar cameras mounted on opposite ends of a short, rigid bar and having their optical axes parallel. It is used in terrestrial photogrammetry for taking stereoscopic pairs of photographs.

camera, terrestrial - A camera designed particularly for taking pictures on the Earth, as distinguished from an aerial camera, which is designed for taking pictures from above the Earth. Also called a ground camera, although that term is also used for any camera used from the ground. This is a very general term and is not ordinarily applied to specific types of terrestrial cameras e.g., ballistic cameras, satellite cameras, or portrait cameras. Used alone, the term usually means a terrestrial camera designed for mapping.

camera, tracking - A camera placed on a mountings which allow the optical axis to be kept pointed in the direction of a moving object such as an automobile, a missile or an artificial satellite. Small tracking cameras may be directed manually by the operator, who looks through an auxiliary optical system (guiding telescope) at the object being tracked. Larger tracking cameras are usually directed automatically by an optical system in which photocells take the place of an operator's eye, or by an auxiliary radar.

camera, trimetrogon - An assemblage of three cameras equipped with wide-angle, Metrogon lens-systems and placed with their optical axes in the same vertical plane; the axis of the central camera is vertical and the axes of the two flanking cameras intersect the axis of the central camera at 60° to provide overlapping fields of view. Trimetrogon cameras are less used since very-wide-angle lens systems were introduced.

camera, variable-perspective - SEE camera system, variable-perspective.

camera axis - An imaginary line passing through the optical center of the lens system of a camera and perpendicular to the focal plane. Also called axis of camera.

camera calibration - SEE calibration (of camera).

camera cone - SEE lens cone.

camera constant - SEE length, calibrated focal.

camera lucida - (1) An optical device which projects a virtual image onto a flat surface and used for sketching the image or comparing that image with another. (2) A monocular instrument using a half-silvered mirror (or its equivalent) to superpose the image of an object onto a flat surface. Called, incorrectly, a camera obscura.

camera magazine - SEE cassette; magazine.

camera mount - SEE camera mounting.

camera mounting - A structure which holds a camera firmly in position for photographing and which itself is fastened to the permanent pier, base, or vehicle. Also termed a camera mount. Mountings for fastening cameras to aircraft are commercially available for most aerial cameras. Such a mounting may be designed to hold the camera in a desired, fixed position in the aircraft or it may be designed to keep the camera's optical axis vertical. (This second type is also called a gyro-stabilized camera mounting or inertial camera platform. Mountings for terrestrial cameras are usually designed to affix the camera to a tripod and to allow pointing the camera in any desired direction.

camera obscura - A darkened chamber or box into which light is admitted through a small hole, to form an image on the wall or side opposite the hole. The hole may or may not contain a lens. An example of the lensless type is the pinhole camera; an example of the type with lens is the ordinary box camera. The solar telescope, which projects an image of the Sun onto a screen, is also a camera obscura. The camera obscura should not be confused with the camera lucida; the former creates a real image, the latter a virtual image.

camera port - The window of optical quality, in an aircraft, through which aerial photographs are taken. The quality of the camera port has a significant effect on the quality of the photographs taken.

camera shutter - SEE shutter.

camera station - That point, in space or on the ground, occupied by the lens system of a camera at the moment of exposure. Also called an exposure station. The perspective center of the lens system is usually taken to represent the lens system.

camera system - A camera and all the equipment used with it for taking pictures. A camera system consists, typically, of the camera itself, a range-finder, an exposure meter, the photographic film or plates and the operator. An aerial-camera system also includes a timing device or inter-valo-meter for controlling the shutter, a vacuum pump if the camera holds the film in place by vacuum, and a source of electric power for driving the camera and other items. The aircraft may be considered a part of the system if it is used only for aerial photography. The items other than the camera are called accessories.

camera system, calibration of - SEE calibration (of camera system).

camera system, variable-perspective - A camera system which allows photographs to be taken of an object from different viewpoints even though camera and object occupy fixed positions. The system consists of a camera, a tiltable easel on which is mounted the object to be photographed, and, between the camera and the easel, a spherical mirror of large diameter to provide a variable perspective. A flat secondary mirror is often added to lengthen the optical path. Such a camera system can be used for rectifying highly oblique photographs or for making images with arbitrary perspective from vertical photographs or from maps.

camera transit - SEE phototheodolite.

camino (Spanish) - Road or highway.

cam inversor - An inversor embodying a cam which follows a specially-shaped track and used in rectifiers and similar instruments to keep the photographic negative, the lens system's perspective center, and the easel (onto which the photographic image is projected) in proper relationship when the distances of projection are changed.

Camp Colonna Datum - The horizontal-control datum defined by the following location and azimuth (clockwise from South) on the Clarke spheroid of 1866; the origin is at Camp Colonna:

longitude of origin	140° 59' 13.50" W.
latitude of origin	67° 25' 05.11" N.
azimuth from origin to South Meridian	0° 00' 00.000" .

The values of the coordinates and the azimuth are based on astronomical observations made in 1890, the longitude being determined by the lunar-culmination method.

Canadian Geodetic Datum - A vertical-control datum defined, by Order in Council 11 March 1935, as: mean sea level as determined at coastal points by the Canadian Hydrographic Survey and extended inland by the Canadian Geodetic Survey.

Canadian grid - SEE grid, perspective.

Canadian Initial Principal Meridian - The meridian of 97° 27' 28.4" west longitude. All Dominion Land Surveys are based on this meridian.

canal - An artificial, open channel or waterway constructed for one or more of the following purposes: transporting water, connecting two or more bodies of water and serving as a waterway for watercraft. SEE ALSO ditch.

candela - A unit of radiation, equal to 1/60th of the luminescent intensity from one square centimeter of a black body at 2042° K (the temperature of solidifying platinum). Formerly called a candle. The unit was approved by the International Committee on Weights and Measures on 1 January 1948. It was adopted by the U.S.A. as a legal standard on 21 July 1950.

candle - SEE candela.

cantilever (adj.) - Overhanging or projecting beyond a support. Used in photogrammetry to denote a method or process of computing data into a region not containing control from a region in which control is available e.g., cantilever aerotriangulation and cantilever extension.

cantilever (structural element) - A projecting beam or overhanging portion of a structure, supported at one end only.

cantilever extension - SEE extension, cantilever.

cantilevering - SEE aerotriangulation, cantilever; extension, cantilever.

cantilever method - SEE extension, cantilever.

canyon - A long, deep and narrow valley with steep sides and confined between very high and steep walls in a plateau or mountainous region. It often has a stream at the bottom. It is similar to but larger than a gorge.

canyon, submarine - (1) A steep-sided, broad-floored valley with V-shaped cross-section cutting across the continental shelf or continental slope and resembling a river-cut, unglaciated canyon on land. (2) A small gully that cuts the submerged slope of a great delta. Such a gully usually ex-tends no deeper than 70 meters over the continental slope.

capacity curve - A graph of the volume of a reservoir, tank, etc., as a function of the height of the water referred to some horizontal surface.

Cape Datum - A horizontal-control datum in South Africa defined by the following location and azimuth (clockwise from South) on the Clarke spheroid of 1880 (modified); the origin is at Buffelsfontein:

longitude of origin	25° 30' 44.622" E.
latitude of origin	33° 59' 32.000" S.
azimuth from origin to Zurrberg	184° 15' 26.311" .

On extension of the triangulation on this datum northward to the arc of the 30th meridian, Cape datum was applied to entire network and to extensions from it, but the name for the extended network was changed to Arc Datum. SEE ALSO Arc Datum.

card, control-data - A card on which is given locations and descriptions of individual horizontal and/or vertical control points. Also called a geodetic-data sheet and control-data sheet.

Cardan angle - (1) One of a set of three angles (θ_1, θ_2, θ_3) specifying the orientation of the three axes (x, y, z) of one rectangular Cartesian coordinate system with respect to the three axes (X,Y,Z) of another. The angles are generated by rotating the first system about the x-axis until the xy plane is parallel to the Y-axis (θ_1), then rotating about the new y-axis until the xy plane is parallel to the XY plane (θ_2), and finally rotating about the new z-axis until the three axes of the first system are all parallel to the corresponding axes of the second (θ_3). There is no generally accepted symbolism for the Cardan angles. In photogrammetry and aerodynamics, they are usually denoted by ω, ϕ, and κ (or sometimes by α, β, τ) and are called, respectively, roll, pitch, and yaw. These same terms are also used, however, for a slightly different set of angles that are not Cardan angles. (2) SEE angle, pseudo-Cardan.

Cardan coupling - A device for connecting two shafts so that rotation of the one shaft will be passed on to the other even though the axes of the shafts are at an angle to each other and may even vary during the rotation. It consists basically of two separate but otherwise identical rings each carrying a pair of trunnions. These are connected by a single intermediate member consisting of a pair of axles fastened together in such a way that the one axle is always perpendicular to the other (to which it may be fastened rigidly or in such a way as to allow the one axle to rotate in a plane perpendicular to the other). Also called a Cardan joint, Cardan link, universal coupling and universal joint. SEE ALSO Cardan suspension. In use, the connected rings are fastened to the ends of the shafts to be connected. The device does not transmit rotation efficiently if the angle between shafts is greater than about 15°. The form in which the intermediate member is a pair of axles rigidly fastened together to form a cross is sometimes called Hooke's coupling. The rate of rotation of the driven shaft then depends on the angle between the shafts.

Cardan joint - SEE Cardan coupling.

Cardan link - SEE Cardan coupling.

Cardan suspension - A Cardan coupling connecting a suspended object and a supporting framework. Cardan suspensions are used to keep instruments level regardless of the motion of the supports, e.g., on board ships or other craft, or in gyroscopic theodolites.

cardinal (surveying) - SEE direction, cardinal.

Carpentier inversor - An inversor consisting of a long rod pivoted a fixed distance below the plane of the lens system's perspective center and having sliding sleeves, on the upper and lower arms, through which pass rods rigidly connected to the planes of negative photograph and easel.

carrier sheet - A transparent sheet to which material to be copied or used as masking is fastened.

carrying contour - SEE contour line, carrying.

carrying contour line - SEE contour line, carrying.

Carte de France ellipsoid - SEE Plessiss ellipsoid.

Carte du Ciel - A photographic atlas of the whole sky, intended to show the locations of all stars of magnitude 13.5 or less (about 3 - 4 million stars in all) and produced by international cooperation according to an agreement reached among astronomers in Paris in 1887. A number of observatories have telescopes designed particularly for this work, but the project is not complete.

Cartesian coordinate - A coordinate in a Cartesian coordinate system. In 2-dimensional space, the coordinates are commonly designated by (x, y) or (x_1, x_2). In three-dimensional space, they are commonly designated by (x,y,z), (χ, η, ζ), (x_1, x_2, x_3) or (x_1, x_2, x_3). The last of these is the most usual if tensors are being used.

Cartesian coordinate, rectangular - SEE Cartesian coordinate system, rectangular.

Cartesian coordinate system - A coordinate system consisting of N straight lines (called the axes) intersecting at a common point (called the origin) and determining N-1 distinct, N-1 dimensional hyperplanes. The n-th coordinate ($1 \leq n \leq N$) of a point is the distance between that point and the hyperplane determined by all axes but the nth, and measured parallel to the n-th axis. Alternatively, a set of N families of N-1 dimensional hyperplanes such that members of the same family have no line in common, while members of different families intersect in one and only one line. The coordinates of a point are then the set of values of the parameters determining the N hyperplanes passing through that point. The units in which distances are measured need not be the same along all the axes, and the axes need not intersect at right angles. A Cartesian coordinate system (CCS) for which the units of distance are different in different directions is sometimes erroneously called an affine CCS. If all the axes intersect at right angles, the system is called a rectangular CCS or simply a CCS. Otherwise, it is called an oblique CCS. Distances are usually measured from the hyperplane to the point and are assigned a positive or a negative value according to some specified convention.

Cartesian coordinate system, oblique - SEE Cartesian coordinate system.

Cartesian coordinate system, rectangular - A coordinate system consisting of three (or two) straight lines (called the axes) intersecting at a common point and perpendicular to each other; the coordinates of a point are the distances to the point from the three planes defined by the three pairs of axes (or from the two axes). This is equivalent to specifying a coordinate system which is rectangular and Cartesian.

Cartesian coordinate system, square - Rectangular Cartesian coordinate system in which the divisions along both axes have the same length.

cartogram - (1) A small diagram, on the face of a map, showing quantitative information. (2) An abstracted and simplified map the base of which is not true to scale.

cartographer - One who practices the art of cartography.

cartographic - Associated with cartography; having to do with mapping in a general sense.

cartography - The art of expressing graphically, by maps, the known physical features of the Earth or of another celestial body. It usually includes showing the works of man and his varied activities. (2) The science, art and technology of making maps, together with the study of maps as scientific documents and works of art. (3) The theory and practice of making maps. (4) The process of making maps. The process is generally considered to include all steps from conception to finished product. The usual steps in the process are: (a) establishing the coordinate system for the map; (b) plotting the data onto the map from surveys or other sources (including other maps); (c) filling in where data are insufficient and generalizing where data are too plentiful; (d) comparing the resulting manuscript form of the map with new data (field checking) gathered specifically for this purpose; (e) making plates or other printing intermediates; and (f) printing copies of the map. In the context of the above definitions, the term may be regarded as including all types of maps, plans, charts, plots, and section; three-dimensional models and globes, etc. Although some authorities consider the term to include land surveying, hydrographic surveying, and photogrammetric surveying as part of cartography, this viewpoint is not common.

cartouche - A panel on a map, often with decoration, inclosing the title or other legends, the scale, etc.

cartridge - A light-tight, cylindrical container for standard lengths of film; the cartridge can be placed in a camera, the film exposed, and another cartridge containing the exposed film removed, all without exposing the film to daylight. SEE ALSO cassette (1).

cartway - A way or road for carts; usually, a rough road used by or passable only by carts.

carucata - A measure of land and a unit for the measurement of land, equivalent to the plowland or hide and, like the hide, typically reckoned at 120 acres. It was formerly in use in parts of England. An alternative spelling is carucate. SEE ALSO hide.

carucate - SEE carucata.

case (of a map projection) - SEE aspect (of a map projection).

case law - Law derived from cases, as opposed to law based on statutes or other sources of law.

Cassegrainian focus - In a Cassegrainian telescope, that point behind the primary mirror at which light from the secondary mirror is brought to a focus. The term is sometimes applied to a focal point behind the primary mirror in any kind of telescope.

Cassegrainian telescope - A reflecting telescope which reflects incoming radiation from the paraboloidal primary mirror to a small, hyperboloidal secondary mirror and thence back through a hole in the primary mirror to the detecting device or observer. Cassegrainian telescopes have been made for use at radio wavelengths through the spectrum to the far ultraviolet. The combination of paraboloidal and hyperboloidal mirrors makes the telescope much shorter than a telescope of the same focal length and with a flat secondary mirror.

cassette - (1) A container for holding short lengths of exposed or unexposed film and intended to be attached to or placed in the body of a camera. It may be cylindrical in shape (in which case it is often called a cartridge), with a slit through which the film is pulled out or pulled in. The cassettes used in cameras of the kind attached to theodolites for photographing the horizontal and/or vertical circles take standard 35-mm film. (2) A flat, rectangular case containing magnetic tape which is rolled off one reel in the cassette and rolled up on another, and on which information is recorded. In surveying, this information is recorded in digital form. Also called tape cassette or tape cartridge.

Cassini coordinate - One of a pair of coordinates in the rectangular spherical coordinate system. Also called a Soldner coordinate.

Cassini's aid figure - SEE Cassini's method.

Cassini's map projection - The meridional (transverse) form of the Plate Carrée map projection. It plots to scale, in a rectangular Cartesian coordinate system: (a) the geodetic distance on the rotational ellipsoid, from the central meridian to a specified point, as the x-coordinate; and (b) the distance along the central meridian, from the center to the point of intersection with the geodesic, as the y-coordinate. For the sphere of radius R, the rectangular Cartesian coordinates x and y of that point on the plane which corresponds to a point, on the sphere, with longitude $\lambda_o + \Delta\lambda$ and latitude ϕ, are given by $\sin(x/kR) = \cos\phi\sin\Delta\lambda$; $\cot(\phi_o + y/kR) = \cot\phi\cos\Delta\lambda$. ϕ_o is a specified latitude and λ_o is the longitude of the central meridian. k is the scale factor. Also called the Cassini-Soldner map projection, equal-coordinate map projection and transverse Plate Carrée map projection. The map projection is not conformal but scale is exact along the line representing the central meridian and along lines representing geodesics perpendicular to the central meridian. It was introduced in 1745. It can easily be modified to map a rotational ellipsoid onto the plane.

Cassini's method (of resection) - A method of determining the location of a point P in the plane from the known locations of three points A, B, C and the angles at P to each of these three points, by introducing two auxiliary points Q and R. Points Q and R lie on a line perpendicular to the line PB. Q lies on the circle through points B, P, C; R lies on the circle through points B, P, A. The equations of the coordinates of P, given the coordinates of A, B, C, are simple. Also called Cassini's aid figure.

Cassini-Soldner coordinate system - (1) A coordinate system, on the sphere, whose geometric elements consist of two great circles intersecting at right angles. One of these great circles is called the central meridian or x-axis; one of the two points of intersection is singled out and called the origin. The y-coordinate of a point is the perpendicular distance along a great circle from the central meridian to the point. The x-coordinate is the distance from the origin to the foot of the perpendicular. Suitable conventions are adopted for assigning positive and negative values to the distances. Also called the Soldner coordinate system. However, that term has acquired at least one other meaning, so it is best to use the full name to avoid confusion. The Cassini-Soldner coordinate system is associated with the map projection of the same name. (2) The same as the preceding definition, but with rotational ellipsoid or spheroid substituted for sphere and segment of a geodesic for great circle.

Cassini-Soldner map projection - SEE Cassini's map projection.

Cassinis's gravity formula - The formula adopted in 1930 by the International Association of Geodesy and called the International Gravity Formula.

Cassinis's gravity table - One of a set of tables calculated by G. Cassinis, P. Dore, and S. Ballarin (1937) giving the gravity correction, for each Hayford zone, for a complete cylindrical shell of unit density whose bottom (or top) is at the same elevation as the point of measurement and which extends upwards (or downwards) to all possible elevations (or depths of compensation) to 200 km. The total topographic and isostatic gravity corrections can then be found by simple arithmetic. SEE ALSO gravity table, fundamental.

catalog (astronomy) - A compilation of coordinates of celestial objects such as stars, galaxies, and radio sources, whose directions are relatively fixed. The analogous compilation for rapidly-moving objects such as the Moon, planets, and comets is termed an ephemeris or an almanac. SEE, in particular, star catalog. SEE ALSO Astrographic Catalog; zone star catalog.

catalog, stellar - SEE star catalog.

catalog equinox - The zero point of right ascension for stellar catalogs.

catchment area - SEE watershed.

catch point - (1) A point at the intersection of the side slope of a road with the natural surface of the ground. (2) A stake marking a point on the intersection of the side slope of a road with the natural surface of the ground.

catenary - The curve whose equation is $y = k \cosh(x/k)$, where k is a constant. A homogenous, perfectly flexible, inelastic cable, tape, or wire will, when suspended from two fixed points at its ends, assume this curve. If the density of the cable is d and the force of gravity is g, then the distance y' of a point on the cable from the ground is $y' = h + (k/dg) \cosh(xdg/k)$, where h and k are constants.

catenary correction - SEE sag correction.

cathode-ray tube - SEE tube, cathode-ray.

caveat emptor - Let the buyer beware. I.e., let the buyer examine the article he is buying and act on his own judgment and at his own risk.

C-band - That portion of the electromagnetic spectrum having frequencies between 3.90 and 6.20 gigahertz (4.8 and 7.7 centimeters wavelength, approximately).

C-constant - SEE C-factor.

Ceccini's origin - SEE Conventional International Origin.

cellulose acetate - A material prepared by acetylation of cellulose. Cellulose acetate is widely used in transparent sheets of plastic, as a base for photographic film, in which form it is also called acetate film. It burns very slowly, in contrast to the nitrated cellulose (nitrate film) used previously.

Celsius scale - SEE Celsius temperature scale.

Celsius temperature scale - (1) A temperature scale on which 0° denotes the boiling point of water under standard atmospheric pressure and 100° denotes the melting (triple) point of ice. On this scale, absolute zero (the state) is at approximately ± 273.16°. The scale was invented by Anders Celsius. (2) A temperature scale on which 0° denotes the melting (triple) point of ice and 100° denotes the boiling point of water at standard atmospheric pressure. Also called the Centigrade temperature-scale. This scale was devised as early as 1710 and was used by Linnaeus before 1737. Its invention is mistakenly attributed to Anders Celsius, but under the S.I. is nevertheless called the Celsius (temp-erature) scale. However, the term Centigrade (temperature) scale is still used by the Meteorological Office and is in common use in America and England.

centares - SEE meter, square.

center - (1) That point, in an instrument, at which angles, or from which distances, are assumed to be measured. (2) That point, in an instrument having a telescope, which lies on the vertical axis of rotation (the standing axis) and which is at the same elevation as the line of sight when that line is horizontal. In a transit or theodolite, the center is close to or at the intersection of the horizontal and standing axes of the instrument. (3) The spindle or spindles which, in a theodolite or leveling instrument, are in a vertical position when the instrument is in use, and about which the instru-ment, or a part of the instrument, rotates. This is a manufacturer's term. The engineer's transit (repeating theodolite) has two such centers; an inner center to which the alidade is attached, and an outer center to which the horizontal circle is attached. It is hollow and rotates on the inner center. The rotation of the alidade is spoken of as the upper motion and the rotation of the horizontal circle as the lower motion of the instrument. (4) SEE center of photo-graph. SEE ALSO photograph center; reduction to center.

center, anallactic - SEE point, anallactic.

center, exterior perspective - SEE center of perspective, external.

center, geographic - The point on which a geographical region would balance if it were a plate of uniform thickness and density. In other words, it is the center of gravity of the plate. The geographic center of continental United States of America (exclusive of Alaska and Hawaii) is in the eastern part of Smith County, Kansas, at longitude 98° 35' West, latitude 30° 50' North. The geographic centers of various states are given in the U.S. Geological Survey Bulletin 817, Boundaries, Areas, geographic Centers, and Altitudes of the United States and the several States.

center, interior perspective - SEE center of perspective, internal.

center, optical - That point, on the optical axis of an optical system, which is midway between the front and rear nodal points of the system. For a thin lens in air, the optical center, front and rear nodal points, and first and second principal points are all one and the same point. An oblique ray, even if it passes through the optical center, still undergoes a longitudinal displacement which increases with the thickness of the lens.

center, perspective - SEE center of perspective.

center, radial - SEE radial center.

center, substitute - A point, on a photograph, which is used as a radial center (because of its ready identifiability) instead of the principal point, the nadir point, or the isocenter.

centering - (1) Bringing the center of an instrument vertically above a specified point. SEE ALSO auto-centering. (2) Placing the optical centers of all elements of an optical system on a single straight line. This is a very delicate operation best done in the laboratory. It is similar to collimation, except that the latter deals with components and not with individual elements (unless these are also components).

centering, double - A method of prolonging a line from a fixed point. The transit and a foreword target are set up approximately on the line to be prolonged and a backsight taken with the telescope in the direct (normal) position. The telescope is then inverted and a foresight taken. The point where the vertical cross-hair intersects a horizontal line on the target is marked on the target. The telescope is then rotated 180° horizontally and, still in the inverted position, used for taking a backsight. The telescope is then restored to its direct position and another foresight taken. The new point of intersection of the vertical crosshair with the target is again marked. The prolonged line passes through a point midway between the two marked points. Also called double-sighting, double reversing, double reversion, reversing in altitude and azimuth, wiggling-in on a line and working-in on a line. Note, however, that this method is not the same as wiggling-in. The latter involves shifting the instrument to put it on a line.

centering, forced - Replacement of one device or instrument by another in such a way that if the original object was vertically over a mark, then so is the replacement. Some mountings are designed to hold a number of different kinds of instrument and targets, and are therefore made to allow forced centering. For example, a theodolite may be centered over the mark and, when observations are completed there, be replaced by a target which is then also automatically centered.

centering, optical - Bringing the center of an instrument vertically above a specified mark, using an optical device that allows the observer to view the point with the instrument or its mounting in place. The device itself is called an optical plummet, because it replaces the leaden plummet formerly used for centering instruments.

centering device - A part of, or an attachment to, the leveling base (tribrach) which permits the instrument mounted on it to be moved a short distance in each of two orthogonal directions. The standing axis of a surveying instrument can thus be placed above a mark on the ground without moving the tripod or putting the standing axis out of the vertical. SEE ALSO tripod, centering.

centering device, optical - A centering device used for optical centering. Also called an optical plummet. It usually consists of a telescope having a mirror or prism attached at the reticle so that the line from the center of the eyepiece to the center of the reticle is horizontal, while the line from the center of the reticle to the center of the objective is vertical.

centering error - An error caused by inaccurate pivoting in an instrument. Also called eccentric error.

center line (of a section) (USPLS) - The line connecting opposite quarter-section corners, or opposite sixteenth-section (quarter-quarter section) corners.

center line (of a street) - The line, along a street, midway between the sides of the street. The term usually applies to the center line of the original street, as built before its widening or closing. To avoid ambiguity, if the street has been narrowed or widened on one side or unequally, the center line would be defined as the center line as existing on (a specified date). closing; i.e., the center line of the original street midway between the sides.

center line (of a strip of land) - The line midway between the side lines of any strip of land of uniform width as, e.g., a street or right of way.

center of circle - That point, in the plane of a circle, which is equidistant from all points of the circle. Lines drawn from the center of circle to points on the circle are called radii.

center of figure - SEE centroid.

center of gravity - That point, in a body, at which the body is balanced under the force of gravity regardless of the position of the body. That is, the body, if supported at that point, will remain at rest no matter what its position may be. It is the same as the center of mass if the gravity field is uniform and if the ratio of gravity to mass is constant. SEE ALSO barycenter. If an element dV of volume V has a density d and coordinates x, y, z, then the coordinates x_o, y_o, z_o of the center of gravity are given by

$$x_o = \int_V x\, g d\, dV,$$ with similar formulae for y_o and z_o. g is the force of gravity.

center of instrument - SEE center (2) and center (3).

center of level - A manufacturer's term for the spindle or spindles which are vertical when the leveling instrument is in used, and about which the leveling instrument, or a part of it, rotates.

center of mass - SEE barycenter.

center of oscillation - (1) That point, in an actual pendulum or other swinging body, which lies on the line joining the point of suspension and the center of mass, and which is at a distance l_o from the point of suspension; l_o is the length of a simple, hypothetical pendulum having the same frequency of oscillation as the actual pendulum (or other body). If l is the distance from the point of suspension to the center of mass and k is the radius of gyration of the actual pendulum or body, then $l_o = (l^2 + k^2)/l$. (2) The point, in a compound pendulum, occupied by the particle which corresponds to the point-mass of an equivalent simple pendulum. The center of oscillation and of suspension are interchangeable. If the center of oscillation is made the center of suspension, the former center of suspension becomes the new center or oscillation. This principle is the basis for design of compound reversible pendulums.

center of percussion - That point, in a suspended bar, which is located at such a distance below the point of suspension that a force applied perpendicular to the line joining the two points causes no reaction at the point of suspension.

center of perspective - (1) The given point, in a perspective projection, from which straight lines are drawn through points on one surface to determine, by intersection, corresponding points on a second surface. Also called a perspective center. In the elementary theory of map projections, one surface is an ellipsoid (usually a sphere or rotational ellipsoid) and the other is a plane, cone, or cylinder. (2) The point of common intersection of lines drawn between corresponding points in object-space and image-space. Also called a perspective center and point of perspective. No such point exists for real optical systems, but it is usually convenient to describe the performance of an optical system in terms of the differences between the actual image and the hypothetical image created by a hypothetical center of perspective. (3) One of the two points on the optical axis, in a camera or projector, such that a ray approaching it from object-space emerges from the other point, in image-space, in the same direction. The two points may coincide. Also called center of projection.

center of perspective, exterior - SEE center of perspective, external.

center of perspective, external - In a surveying camera, the terminal point, outside the camera, of bundles of perspective rays. Also called the exterior center of perspective and external perspective center. In a perfectly-adjusted lens-system, this corresponds to the front nodal point of the system.

center of perspective, interior - SEE center of perspective, internal.

center of perspective, internal - In a surveying camera, the point of origin, within the camera, of bundles of perspective rays. Also called the internal center of perspective and interior perspective center. In a perfectly-adjusted lens-system, this corresponds to the rear nodal point of the system.

center of photograph - SEE photograph center.

center of projection - SEE center of perspective (3).

center of radiation - SEE radial center.

center of section - The point formed by lines connecting opposite quarter corners in a section of land. It is also called the center quarter corner.

center of suspension - The fixed point about which a pendulum or other hanging body oscillates. Note that the center of suspension exists, strictly, only for an ideal pendulum. An actual pendulum oscillates about a point or line whose location shifts as the pendulum oscillates. For example, the knife edge on which a pendulum rests is round, not sharp, with a radius of from 5 to 25 micrometers. If the situation is examined in still more minute detail, it will be found that the pendulum actually is in contact over a small area, not at a point or along a line. SEE ALSO center of oscillation; and point of suspension.

center of the Moon, mean - (1) The point at which the Moon's surface is intersected by the lunar radius directed towards the Earth's center when the Moon is at the mean acending node and the node coincides with mean apogee or mean perigee. (2) A central point, on the Moon, selected as an origin for a coordinate system for the Moon's surface.

center of transit - A manufacturer's term for the spindle or spindles which are upright when the transit is in use and about which the transit or a part of the transit rotates. The engineer's transit has two such spindles: an inner spindle to which the alidade is attached, and an outer spindle to which the horizontal circle is attached and which is hollow and rotates on the inner spindle.

center of volume - SEE centroid.

center point - SEE radial center.

center-to-center method (of mosaicking) - Assembling a strip mosaic from aerial photographs with more than 50% overlap by matching a point near the center of one photograph with corresponding points in the overlap of adjacent photographs.

center-wire method of leveling - Leveling in which only the reading at the line of sight through the central line of the three horizontal lines on the reticle is taken. For precise work, readings are taken along the line of sight through all three horizontal lines and the average of the three readings taken. The center-wire method is faster than the three-wire method but does not provide as many checks of the readings.

centiares - SEE meter, square.

centimeter-gram-second system of units - SEE system of units, c.g.s.

Centigrade temperature scale - SEE Celsius temperature scale (2). This temperature scale is called the Celsius temperature scale in the SI, but should not be confused with the temperature scale actually invented by Celsius.

centroid - (1) That point, in a geometric figure, whose coordinates are the average values of the coordinates of the points of the figure. (2) In three dimensions, that point which is at the average distance of all point of the figure from any plane. Also called center of figure and center of volume. It is the geometric analog of the center of mass of a body, and it coincides with that center if the density of the line or surface is not a function of location. The coordinates (x_o, y_o, z_o) of the centroid are given by the formulae $x_o = \int [x \, ds]/s$, in which x is the x-coordinate of a point in the figure, s denotes arc-length, area or volume and ds indicates an infinitesimal element of arc-length, area or volume. The integration is to be extended over the region for which the quantity s is defined. The formulae for y_o and z_o are similar. The formulae can be applied in one, two or three dimensions by making two, one or none of the variables x, y, z constant.

centrosphere - The core of the Earth, composed of dense material and making up most of its mass.

centuriatio - The Roman system of land surveying, carried out as rectangular surveys.

Ceplecha's method - A method of plate reduction developed by Z. Cephlecha and applied to determining the locations of meteors and satellites.

certificate, final - A document which evidences that an entryman is entitled to a patent provided that no irregularities are found in connection with his entry.

certificate of title - (1) A document, usually given to the buyer of a home with the deed, stating that the title to the property is clear. It is usually prepared by an attorney or qualified person who has examined the abstract of title for the property. It is only an opinion that title is good and is not to be confused with title insurance. (2) A certificate issued to show title registered on lands covered by Torrens title.

cesium clock - An atomic clock in which the characteristic frequency of 9 192 631 770 cycles per second of the cesium atom is used to determine the basic time interval for the clock. Because the second is defined by international agreement as the length of time needed for this number of cycles to occur, time intervals given by a cesium clock operating under the conditions specified by the definition have zero error. A cesium clock is therefore used whenever the highest possible accuracy over long intervals is needed. However, over short intervals, the frequency may vary appreciably and clocks with better stability over short intervals may be used to give time over such intervals. Cesium clocks are then used to monitor the time over longer intervals.

cession - (1) The act of ordering a yielding or giving up; surrender; relinquishment of property or rights. (2) In the Civil Law, an assignment; the act by which a party transfers property to another; the surrender of assignment of property for the benefit of ones creditors. (3) In public law, the assignment, transfer, or yielding up of territory by one state or government to another.

cesspool - An underground pit with porous walls, used to catch and temporarily contain sewage and other liquid refuse. The refuse decomposes, filters out from the pit, and is absorbed into the soil.

C-factor - (1) A quantity C given by one of a number of different formulas and expressing the rate at which the line of sight in a leveling instrument diverges from the horizontal. The commonly used definitions of the C-factor C are as follows. (a) The quantity $-[\Sigma N_r - \Sigma F_r]/[\Sigma N_i - \Sigma F_i]$, in which ΣN_r is the sum of the average readings on a leveling rod placed near the leveling instrument, ΣF_r is the sum of the average readings on a leveling rod placed far from the leveling instrument, and ΣN_i and ΣF_i are the sums of the averages of the corresponding intervals read on the rod between the top and bottom horizontal lines on the reticle. Corrections for curvature and for refraction must have been applied to the F_r's before using them in the formula. Also called C-constant, level constant, collimation correction, collimation correction factor, level collimation correction factor, error of the level, altitude-contour ratio, etc. However, most of these terms are used also for similar but different concepts. The C-factor is defined by a formula; the level constant, etc., are defined as physical quantities (or the tangent of a physical quantity). In American practice, the leveling instrument is usually placed between the two leveling rods, about 10 meters from the one and about 50 meters from the other. (The distance to the farther leveling rod may be greater - sometimes even as much as 100 meters. The shorter distance is preferred). The C-factor is expressed as millimeters per millimeter of stadia interval or as milliyard per milliyard stadia interval. (b) The quantity $-[(N_{rb} + N_{rf}) - (F_{rb} + F_{rf})]/[(N_{ib} + N_{if}) - (F_{ib} + N_{if})]$. The leveling instrument may be placed between the two leveling rods as required by definition (a) preceding, but first at a point about 10 meters from leveling rod b and about 50 meters from leveling rod f; after taking the readings and intervals on the two rods, the leveling instrument is moved to about 10 meters from leveling rod f; readings and intervals are recorded as before. The readings are corrected for curvature and refraction and the formula evaluated. Also called level collimation correction factor, collimation correction factor, level constant, etc. The C-factor is expressed as rod units per rod unit. (c) The same as definition (b) preceding, but with the leveling instrument placed first halfway between the two rods and then outside the interval and about 10 meters from the closer leveling rod. This is the present practice of the National Geodetic Survey. (d) SEE collimation correction factor. (2) The ratio of the altitude of a camera above the ground to the smallest contour interval that can be plotted accurately using the resultant photographs and a particular photo-grammetric plotter. The optical axis of the camera is assumed to be vertical and the altitude constant. A part-icular camera system and stereoscopic plotting system must be specified because the C-factor varies considerably from camera system to camera system and from plotting instru-ment to plotting instrument. The C-factor is often used to determine the altitude at which photographs should be taken, for a particular mapping project, given the smallest contour interval required and the camera system and stereoscopic plotting instrument available. The practicable altitude is the contour interval multiplied by the C-factor. It is also called the altitude-contour ratio.

c.g.s. system of units - SEE system of units, c.g.s.

chain - (1) The unit of length prescribed by law for the survey of the public lands of the United States of America and equal to 66 feet or 100 links. One acre equals 10 square chains. The chain derives its name from the Gunter's chain, which was widely used in early surveys and had the form of a series of links connected together by rings. (2) A Gunter's chain or similar measuring device. SEE ALSO engineer's chain; Gunter's chain; throwing the chain.

chain, four-pole - SEE Gunter's chain.

chain, nautical - A unit of length equal to exactly 15 feet.

chain, two-pole - A chain 33 feet long.

chain gage - SEE tape gage.

chaining - Measuring distances on the ground with a graduated tape or with a chain. The term taping is now preferred if a tape is used. SEE ALSO taping. Although the chain has been superseded by the graduated tape for making land and other surveys, the term "chaining" has continued in use in some surveying organizations and in places where reference is to surveys of the public lands of the United States of America. For the corresponding operation in other surveys, the term taping is preferred. In chaining, the persons who mark the chain's are called chainmen. SEE ALSO drop chaining; slope chaining.

chaining, slope - SEE slope chaining.

chaining pin - A short, metallic rod one end of which is pointed for driving into the ground and the other end of which is bent into a loop used in pulling the pin out of the ground. Also called a chaining pin, surveyor's arrow, taping arrow and taping pin. It is used in sets of 11 pins for marking the ends of taped (measured) intervals on the ground. A measuring tape can be laid down end-to-end 10 successive times before the set is exhausted.

chainmaker - A person who assembles a chain of title.

chainman - The person who marks the locations of the tape's ends in measuring distance with a graduated tape.

chain of title - A chronological list of documents which comprise the historical record of title of a specific parcel of real estate.

Chandlerian motion - A nearly periodic variation in the astronomic latitudes of points on the Earth's surface, with a period of about 1.2 years and an amplitude of about 0.3" (10 meters). Also called wander and polar wander. Chandlerian motion is nearly the same thing as the motion of the Earth's instantaneous axis of rotation about the Earth's average axis of greatest moment of inertia. It is not quite what is gotten from astronomical observations because (a) the Earth's crust moves with respect to the mantle and core, and (b) the method of measurement used does not refer to the instantaneous axis of rotation.

Chandlerian wobble - SEE Chandlerian motion.

change, annual - SEE change, annual magnetic

change, annual magnetic - The amount of secular change (i.e., change in one direction) undergone by the geomagnetic field in one year. Also called annual change, magnetic annual change, annual rate and annual rate of change.

change, imperceptible - Change which is so slight, gradual, or subtle that it can not be mentally perceived or discriminated.

change, secular - SEE change, secular magnetic.

change, secular magnetic - (1) An increase or decrease in the intensity and/or direction of the total magnetic field over a period of many years. Also called secular change when it is obvious that the magnetic field is meant. (2) The change, from year to year and usually extending for many decades in the same direction, of magnetic declination, inclination, or intensity. Also called magnetic secular change.

change detection - The detection of a change in object space by looking for changes in the images made of that object-space.

change of grade - The difference between the elevation of a newly constructed highway and the elevation of the land, street or highway previously existing at the site.

channel - A natural or artificial waterway which either periodically or continuously contains moving water or which forms a connecting link between two bodies of water.

channel, main - As a boundary between states or nations, the main channel of a navigable stream is the deepest and most navigable channel as it was at the time the boundary was surveyed. The main channel of a non-navigable stream was decided to be, in a particular case (boundary between Oklahoma and Texas, 1923) to be the median line between gradient lines on each bank.

character (cartography) - The distinctive trait, quality, property, or behavior of manmade or natural objects that is portrayed by the cartographer in representing such objects. The more character applied to detail, the more closely will the detail resemble the actual features. SEE ALSO generalization.

character, alphanumeric - All characters used by an electronic computer, including letters, numerals, punctuation marks, and such symbols as @, #, $ and %.

chart - A map designed for use in navigation. The term chart is applied chiefly to maps made primarily for nautical or aeronautical navigation, and to maps of the heavens. It is sometimes used, however, to describe other types of special-purpose map. SEE ALSO Admiralty chart; alinement chart; anchorage chart; approach chart; Baldwin solar chart; bar chart; current chart; Lambert conformal chart; map chart; Marsden chart; Mercator chart; Mercator chart, transverse; Ney's chart; pilot chart; portolan chart; sailing chart; star chart; test chart; World Aeronautical Chart.

chart, aeronautical - A chart designed for use in navigating aircraft. An aeronautical chart shows airports and aids and hazards to aerial navigation, as well as some of the natural features and man-made objects shown on general maps. Also called a navigational chart.

chart, aeronautical pilotage - SEE pilotage chart, aeronautical.

chart, azimuthal - A chart drawn using an azimuthal map projection. Also called a zenithal chart.

chart, azimuthal equidistant - A chart on the azimuthal equidistant map-projection.

chart, bathymetric - A topographic map of the bed of the ocean or other body of water. Bathymetric charts usually indicate depths by contour lines, with meaningfully tinted zones between curves. They are designed especially for geophysical (oceanographic) studies, but are also useful for navigation.

chart, coastal - A nautical chart intended for coastwise navigation inshore when a vessel's course may carry her inside outlying reefs and shoals. It is intended for use in entering or leaving bays and harbors of considerable size or for use in navigating large, inland waterways.

chart, conformal - A chart on a conformal map-projection.

chart, conic - A chart on a conic map-projection.

chart, co-tidal - (1) A chart displaying lines that show approximate locations of high water at hourly intervals, measured from a reference meridian (usually Greenwich). (2) A chart combining lines along which the range of the tide is the same at all points with lines along which the average times of high and low waters, or both, occur simultaneously. Co-tidal charts may deal with the tide as a whole or with one or more harmonic constituents of the tide.

chart, enroute - An aerial chart showing air routes in specific regions and the location of such aids to radio navigation as radio-direction-finder stations, radio and radar marker-beacons, and radio-range stations. Also called a radio-facility chart.

chart, equatorial - (1) A chart of equatorial regions. (2) A chart on an equatorial aspect of a map-projection.

chart, equatorial cylindrical orthomorphic - SEE Mercator chart.

chart, fair - SEE sheet, smooth.

chart, first-approximation - SEE chart, historical.

chart, general - A nautical chart intended for coastwise navigation offshore. A general chart is of smaller scale than a coast chart but of larger scale than a sailing chart.

chart, gnomonic - A chart on the gnomonic map-projection.

chart, great-circle - Any chart on which great circles appear as straight lines. A chart on the gnomonic map projection.

chart, historical - A chart based on data from previous years to determine the probable oceanographic patterns for a specified time. Also called a first-approximation chart.

chart, hydrographic - Also called a navigational chart, hydrographic chart, hydrographic map or marine map. A nautical chart shows aids and hazards to navigation, critical depths of the water, shorelines when present, and similar information. SEE chart, nautical.

chart, hypsographic - SEE map, hypsographic.

chart, hypsometric - SEE map, hypsometric.

chart, instrumental-approach - An aeronautical chart designed for use when flying by instrument, for making instrumental approach and letdown-to-contact landing in the vicinity of an airfield. Also called an instrument approach chart.

chart, inverse - SEE chart, transverse.

chart, inverse cylindrical orthomorphic - SEE Mercator chart, transverse.

chart, isobaric - A chart showing isobars. Also called a constant-pressure chart.

chart, isoclinic - A chart of which the chief feature is isoclinic lines each for a different value of the magnetic inclination along the corresponding line on the Earth.

chart, isogonic - A chart on which lines (isogons) of constant declination from geodetic north to magnetic north are drawn.

chart, isogriv - SEE isogriv chart.

chart, isomagnetic - A chart showing the shape of the Earth's magnetic field by isogonic, isoclinic, or isodynamic lines.

chart, isoporic - A chart on which lines connecting points of equal angular magnetic change are drawn.

chart, local - A large-scale aeronautical chart designed for contact flight in a congested region.

chart, long-range - SEE chart, long-range navigation.

chart, long-range navigation - A small scale (1:3,000,000 or less) aeronautical chart designed for navigation on long flights and relying principally on dead reckoning and celestial navigation.

chart, magnetic - (1) A special-purpose map showing the distribution of one of the magnetic elements such as local deviation of magnetic north from geodetic north. (2) A special-purpose map showing the distribution of the secular change in one of the magnetic elements.

chart, mean - A chart on which isopleths of the average value of a given oceanographic quantity are drawn. Also called a mean map.

chart, meteorological - A chart giving information about the weather.

chart, nautical - A chart designed for use in navigating ships, boats, or other watercraft.

chart, navigation - SEE chart, aeronautical; chart, nautical.

chart, oblique - A chart in the oblique aspect of a map projection.

chart, oblique cylindrical orthomorphic - SEE Mercator chart, oblique.

chart, oblique Mercator - A chart drawn using the oblique Mercator map projection.

chart, orthographic - A chart on an orthographic map-projection.

chart, orthomorphic - A chart on a conformal map-projection.

chart, perspective - A chart on a perspective map projection.

chart, polar - (1) A chart of polar regions. (2) A chart drawn using a polar map projection. The map projections most used for polar charts are the azimuthal equidistant, the gnomonic, the modified Lambert conformal, the stereo-graphic, and the transverse Mercator.

chart, polyconic - A chart on a polyconic map-projection.

chart, radio-facility - SEE chart, enroute.

chart, rectangular - A chart on a rectangular map-projection.

chart, search-and-rescue - A chart designed primarily for directing and conducting searches and rescues.

chart, secant conic - A chart on a conic map projection having two standard parallels.

chart, sectional - One of a series of aeronautical charts covering the United States of America at a scale of 1:500,000 and suitable for contact or visual navigation.

chart, simple conic - A chart on a simple conic map projection.

chart, smooth - SEE sheet, smooth.

chart, solar - A diagram graduated according to the principles of the sun dial, designed to be placed on a plane-table or other flat and level surface, to determine the direction of astronomic north, relative to the direction of a shadow cast by a gnomon. It is much more accurate than a sun dial.

chart, spherical - A section of a globe with a radius of 1 meter, representing without distortion a part of the surface of the Earth. It was invented by the Dutch mathematician Adraien Veen and patented in 1594.

chart, synoptic - A map showing the distribution of selected meteorological elements over a large region at a specified instant of time. It is approximately the same as a weather map.

chart, tidal-current - A chart showing by arrows and numbers the average direction and speed of tidal currents at a particular part of the currents' cycle. A number of such charts, one for each hour of the currents' cycle, are usually published together.

chart, time-zone - A small-scale map of the world showing the legal time zones.

chart, transverse - A chart made using a transverse map projection. Also called an inverse chart.

chart, transverse cylindrical orthomorphic - SEE Mercator chart, transverse.

chart, zenithal - SEE chart, azimuthal.

chart accuracy - SEE map accuracy.

chart accuracy, relative - SEE map accuracy, relative.

chart datum - (1) The tidal datum to which depths (soundings) in a hydrographic survey or on a chart are referred. I.e., a tidal datum used as a referent for depths only. It is usually taken to correspond to low water, and its distance below mean sea level is represented by the symbol Zo. According to the National Tidal Datum Convention of 1980, all tidal datums of the U.S.A. should refer to mean lower low water. (2) The permanently established surfaced from which soundings or tidal heights are measured (usually low water). The surface is called a tidal datum when referred to a certain phase of the tide. In order to provide a factor of safety, some elevation lower than mean sea level, such as mean low water or mean lower low water, is generally selected. Also called datum, datum plane, hydrographic datum, plane of reference, reference plane and tidal datum.

charter - A (legal) instrument emanating from the sovereign power, in the nature of a grant, either to the whole nation or to a class or portion of the people, or to a dependency, and assuring to them certain rights, liberties or powers.

chart depth - The actual depth below the reference surface. This is the depth shown on a smooth sheet.

charting - The preparation of maps of the bottoms of navigable bodies of water and of land at the edges of these bodies. Also called hydrographic charting.

charting photography - SEE mapping photography.

chartlet - A small chart, like those annexed to Notices to Mariners.

chart reference - The indication, on a hydrographic chart, of the chart or plan of next larger scale for some particular region.

chart with two standard parallels, conic - A chart on a conic map-projection having two standard parallels.

chasm - A large, gaping hollow or gash in rock, with steep sides.

chattel - An item of tangible, personal property.

Chauvenet's criterion - The numerical statement of the limits in Chauvenet's rule.

Chauvenet's problem - The problem of finding all practical operations for determining the inclination of a transit's horizontal axis of rotation as a function of the observed zenithal distance, taking the irregularity and inequality of the pivots into account.

Chauvenet's rule - A single, doubtful observation should be rejected if it is greater in absolute value than the value of a for which $\sigma\sqrt{(2\pi)[(2N-1)/2N]} = \int_0^a \exp(-x^2/2\sigma^2)dx$
in which σ is the standard deviation of x and N is the total number of observations. Chauvenet's rule is most useful when N is so large that there is little difference between the Gaussian distribution and Student's-t distribution but not large enough that $(2N-1)/(2N)$ is effectively unity.

Chebychev adjustment - The process of changing the values of a set of quantities so that the predicted values of a function of the set shall differ less, in absolute value, from the corresponding observed values of the function than for any other set.

Chebychev's map projection - The most general map projection producing the least distortion in representing longitudinal distances. It was invented by Chebychev in 1856.

check, diagonal - SEE diagonal check.

checkerboard method - A method of determining details of the terrain, in a small topographic survey, by covering the region with a rectangular grid marked by stakes at 15, 20 or 30 meter intervals, determining the relative elevations at nodal points, and locating all irregularities or significant features from the closest pair of intersecting lines of the grid. Contours are drawn by interpolation.

checking positive - A composite printing, on glass, of the drawing of contour lines and drainage used on the shadow projector for checking the horizontal accuracy of landforms to be shown on relief models.

check point - One of a set of points selected, in an oblique aerial photograph, in the vicinity of each tie point and distant point, for the purpose of checking the identification of those points.

check profile - A profile plotted from data obtained by surveying, and used to check a profile plotted from measurements on a topographic map.

Chezy level - A wye-type leveling instrument having a level attached under the telescope. The invention of this instrument contributed in part to making the spirit-level type of leveling instrument popular.

Chicago leveling rod - A four-piece, target leveling-rod divided into numbered feet and tenths of feet, with each tenth of a foot divided into unmarked graduations. Also called a Chicago rod. Each piece is 3½ feet long, but the bottom half-foot of each piece above the first is not marked since it slips behind the next lower piece.

chlorinity - (1) Originally, a number giving the amount of chlorine, Cl, in rams per kilogram, in a sample of sea water. (2) It was later defined to be the number $Cl(o/oo)$ giving the mass, in grams, of atomic-weight silver just necessary to precipitate the halogens in 0.3285233 kg of a sample of sea water. (3) It is now related to salinity, $S(o/oo)$ by the defining formula $S(o/oo) = 1.80653\ Cl(o/oo)$ in which the salinity is expressed in parts per thousand. In practice, chlorinity is seldom determined chemically (e.g., by titration against silver nitrate) but is determined by measuring the conductivity of sea water and calculating the chlorinity from this.

Cholesky's method - The method of solving a set of equations when the matrix of coefficients is square. The matrix is converted row by row, by applying recursive formulae to elements, to a pair of matrices of which one has only zeros below the main diagonal and the other has only zeros above the main diagonal. The same operations are applied to the vector of observations (coefficients of dependent variables) and the unknowns are determined by successive substitutions. Also called the Banachiewicz- Cholesky-Crout method. It requires less recording of intermediate results and is particularly simple if the original matrix is symmetric (e.g., if the original matrix is a normal matrix). In that case, the two resultant matrices are transposes of each other and only one need be calculated.

chopping (v.) - Periodically interrupting, by means of a shutter, that photographic image of the track of a star or artificial satellite which is produced when the camera's position is held fixed. Chopping is used to provide points on the photograph whose coordinates and be measured precisely and whose times of creation are known.

chord (USPLS) - (1) In general, a straight line joining two points on a curve or surface. (2) (land surveying) On a great circle, the arc connecting any two corners on a baseline, standard parallel or township's latitudinal boundary. (3) (route surveying) A straight line between two points on a curve, regardless of the distance between them.

chord, long - (1) On a simple curve, the chord, or the length of the chord, that extends from the point where a straight line ends and the curve begins (point of curvature) to the point where the curve ends and its extension in a straight-line continues (point of tangency). (2) On a curve composed of two circular arcs, the chord that extends from the point where the two arcs meet either to where the first arc is met by the straight line leading into it, or to where the second arc ends and its extension in a straight-line continues. In descriptions of a circular boundary of land, the length and direction of the long chord are important features.

chord, nominal - The degree of curve for chords less than 100 feet long.

chorogram - A continuous color, shading, or pattern applied to a part of a map and indicating quantitatively some characteristic of the region depicted.

choropleth - A continuous color or shading applied to parts of a map other than those parts bounded by isopleths and indicating quantitatively some characteristic of the region depicted.

chromaticity - (1) That quality of color which is definable by its chromaticity coordinates. (2) That part of a color's specifications which does not involve luminance.

chromaticity coordinate - The quantity $P_i / (\Sigma P_i)$, where P_i (i = 1,2,3) is the amount of a primary color required to give, by additive mixture with the other two primary colors, a match to the color desired.

chromaticity diagram - A diagram made by plotting any one of the chromaticity coordinates against another.

chromatism - A false coloring of an image, usually caused by chromatic aberration.

chronograph - An instrument for producing a graphical record of the time, shown by a clock or other device, at which an event occurs. In use, a chronograph produces a double record. The first is made by the associated clock and forms a continuous line with significant marks indicating periodic beats of the clock. The second is made by some external agency, human or mechanical, and records the occurrence of an event or series of events (which may be the beats produced by a second clock). The times, as shown by the clock, of such occurrences are read from the record made by the chronograph. In observations for time and longitude, the occurrences of stellar observations are recorded on the chronograph either manually by pressing a key at the instant a star is bisected by a line of the reticle of the telescope used in the observing, or automatically by keeping a star bisected by a movable wire as it travels across the field of view. (SEE transit micrometer). In determining longitudes, the chronograph also records the time at which signals are received from the station of known longitude. The chronograph has been replaced, by many organizations, by equipment which records digitally, on magnetic tape, the time at which an event occurs.

chronometer - A portable, mechanical clock with compensated balance wheel, capable of showing time with high precision and accuracy. Chronometers have been used in scientific and engineering work (astronomy, geodesy, geophysics, navigation, etc.) where accuracy and precision in the timing of observations, in addition to portability, are demanded. Chronometers used in surveying are usually constructed with a special type of balance and escapement and beat to half-seconds. For scientific work, the chronometer has been largely replaced the crystal clock or the atomic clock. SEE ALSO hack chronometer.

chronometer, break-circuit - A chronometer equipped with a device which automatically breaks an electric circuit to indicate a specific time or interval of time.

chronometer correction - SEE clock correction; correction, chronometric.

chronometer error - SEE error, chronometric.

chronometer rate - SEE clock rate.

chronometric method (of determining longitude) - A clock is set to local time at a point whose longitude is known. It is then carried from that point to the point whose astronomic longitude is to be determined. Its time at the new point is compared to the local time there. The difference in times, converted to angular measure is (with certain corrections) the difference in longitude. Also called the clock method. It is still in use, atomic clocks being used to carry the time. With the advent of artificial satellites and the availability of inexpensive receivers for the signals from such satellites, other methods have largely replaced the chronometric method.

chronometer time - Time determined by reference to the indications of a chronometer. I.e., clock time for which a chronometer is used.

Church's method - (1) Any one of three methods of spatial resection, developed by Earl F. Church, that have been used to determine tilt. (2) In particular, a rigorous, mathematical method of determining all three angles (tilt, swing and azimuth) of orientation of an aerial photograph by solving first for the coordinates of the perspective center (given the coordinates of at least three ground points and the coordinates of the corresponding images) and then solving for the angles; the process is repeated, using Newton's method of successive approximation, until the result is satisfactory.

chute - A channel or trough underground or an inclined trough aboveground, through which ore falls from a higher to a lower level. Also spelled shoot.

cine-theodolite - (1) A theodolite which takes pictures of moving objects and which simultaneously photographs the readings on the horizontal and vertical circles of the instrument. Also called a kine-theodolite. The object is kept in the telescope's field of view by one or two operators tracking the object through a separate telescope. The cine-theodolite is used principally in ballistics but has also been used for flare triangulation. Newer versions of the instrument have horizontal and vertical circles on which the angles are marked in special magnetic or optical codes; the codes are read by optical or magnetic sensors and recorded on magnetic tape. (2) A theodolite which photographs the readings on the horizontal and vertical circles. This is more properly called a photo-theodolite.

circle (geometry) - (1) The locus of a point moving in a plane so as to remain at a constant distance from single, fixed point in that plane. Alternatively, the set of all points, in a plane, equidistant from a single, fixed point in that plane. Algebraically, a plane curve having the equation (in rectangular Cartesian coordinates): $(x-a)^2 + (y-b)^2 = r^2$, in which a, b and r are constants. (a,b) identify the fixed point, called the center of the circle. r is the constant distance, called the radius of the circle. In polar coordinates (r, θ) the equation is $r^2 - 2r_1 r \cos(\theta - \theta_1) + r_1^2 = r_o^2$. The center of the circle is at (r_1, θ_1); the radius is r_o. SEE ALSO aiming circle; finder circle; hour circle.

circle (geography) - A curve on the Earth, or a circle of constant latitude on an ellipsoid of rotation representing the Earth, whose latitude marks the extreme angular elevation of the sun at a certain date. SEE Antarctic Circle; Arctic Circle; Tropic of Cancer; Tropic of Capricorn.

circle (instrument) - (1) A telescope having motion only about a horizontal axis. E.g., a meridian circle is a telescope having movement only in the plane of the meridian. (2) A circular disk or ring whose perimeter is graduated in angular units. A circle in a surveying instrument is perpendicular to an axis of rotation and has its center on that axis. It is called a horizontal circle if it is intended for measuring horizontal angles and is called a vertical circle if it is intended for measuring vertical angles. Many instruments do not require a circle's full 360° and use only a sector of a circle. These sectors and often the instruments containing them, are then called quadrants, sextants, or octants according as they comprise 1/4-th, 1/6-th of 1/8-th of a circle.

circle (road) - (1) A circular roadway having only one point of access to the adjoining street. (2) A circular road terminating other roads or streets and into which or from which traffic flows to those roads or streets.

Circle, Antarctic - SEE Antarctic Circle.

Circle, Arctic - SEE Arctic Circle.

circle, critical - That circle, in the plane, which passes through three known points. The accuracy with which a fourth point in the plane can be located depends on the location of that point with respect to the critical circle. If it lies on the circle, its location cannot be found by resection.

circle, diurnal - The apparent daily path of a celestial body, approximating a parallel of declination.

circle, divided - SEE circle, graduated.

circle, equatorial - That circle, on a rotational ellipsoid, in which a plane through the center of the ellipsoid and perpendicular to the axis of rotation intersects the surface. If the rotational ellipsoid is a sphere, a particular diameter of the sphere is specified as the axis of rotation, or a specific great circle is designated as the equatorial circle.

circle, fundamental - SEE circle, primary great.

circle, geodesic - A curve, on a given surface, all of whose points are at the same distance, measured along a geodesic, from a central point. Also called a geodetic circle. On the plane and on the sphere, the curve is a circle. On a rotational ellipsoid, the curve is in general not a circle. The difference in length of arc from the meridian is approximately $(s_4 \varepsilon^2 \sin \phi_o \cos \phi_o \sin \phi)/[3N_o 3(1-\varepsilon^2)]$, where ϕ_o is the geodetic latitude of the central point, N_o is the length of the radius of curvature in the prime meridian at the central point, ε is the eccentricity of the ellipsoid, ϕ is the geodetic latitude of a point on the geodesic circle and s is the length of the geodesic from that point to the central point.

circle, geodetic - SEE circle, geodesic.

circle, graduated - (1) A circular disk marked at equal intervals on the perimeter. The graduated circles in surveying instruments were formerly made of metal and were viewed through microscopes by reflected light. The circles in modern instruments are made of glass and are viewed through the main telescope or through an auxiliary telescope by light transmitted through the disk. Also called a divided circle. (2) SEE circle, proportional.

circle, great - The circle in which any plane which passes through the center of a sphere intersects the sphere. The shortest distance between any two points on a sphere is along the arc of a great circle connecting the two points. Great circles having particular designations on the celestial sphere are the equator, the ecliptic, meridians, hour circles, prime verticals and horizons. The shortest distance between two points on a rotational ellipsoid is a geodesic, but this is not a plane curve unless the points lie on a plane through the rotational axis or on a plane perpendicular to that axis and through the center of the ellipsoid. Of all map pro-jections, the gnomonic map projection is the only one that transforms all great circles into straight lines. Also called an orthodrome.

circle, horizontal - A ring with its perimeter graduated in angular measure and placed perpendicular to a vertical axis of rotation of an instrument, with the axis passing through its center. Horizontal circles of theodolites were formerly made of metal and were read through two microscopes placed 180° apart. In modern theodolites, the horizontal circle is made of glass and is read through a single microscope which, by a prismatic train, lets the observer see the two opposite sides of the circle simultaneously.

circle, osculating - A circle which moves along a specified curve so as to be tangent to that curve at each point. In particular, in route surveying, the osculating circle to a spiral.

circle, polar - Either the Arctic Circle or the Antarctic Circle.

circle, primary - SEE circle, primary great.

circle, primary great - (1) A great circle as one of the reference figures for the origin of a coordinate system. (2) In particular, such a circle 90° from the poles of a spherical-coordinate system, such as the Equator.

circle, prime vertical - The circle containing the vertical at a point and passing through the east and west points of the horizon. SEE ALSO vertical, prime.

circle, principal vertical - That vertical circle through the North and South points on the horizon, which coincides with the celestial meridian.

circle, proportional - A disk-shaped symbol placed on a map and proportional, in actual area or appearance, to the amount of the phenomenon being mapped, relative to other similarly-shaped symbols. Also called a graduated circle.

circle, secondary - SEE circle, secondary great.

circle, secondary great - A great circle perpendicular to a primary great circle, such as a meridian other than the prime meridian. Also called a secondary circle and secondary.

circle, small - (1) The intersection of a sphere with a plane not passing through the center of the sphere. The formulae of spherical trigonometry generally do not apply to arcs of small circles but only to arcs of great circles. Circles of latitude are small circles. (2) The intersection of a sphere representing the Earth with a plane which does not pass through the center of the sphere. Some definitions specify a circle on the Earth; such definitions are empty.

circle, solar - A circle, placed on the reticle of a transit or theodolite, with an angular radius of 15' 45" and centered on the intersection of the cross-hairs. The Sun's radius on 1 July is 15'45" and is approximately the minimum for the year; there is a slight, noticeable overlap of the image from about September to April. In making observations, the image and the circle are taken concentrically to result in horizontal and vertical angles to the Sun's center.

Circle, South Polar - SEE Antarctic Circle.

circle, vertical - (1) A circle having one of its diameters vertical. (2) Any great circle of the celestial sphere passing through the zenith (or nadir). It is the line of intersection of a vertical plane with the celestial sphere. (3) A graduated circle mounted on an instrument in such a manner that the plane of its graduated surface can be placed in a vertical plane. It may be an auxiliary, attached to a theodolite or transit, or it may be the major feature of an instrument intended primarily for measuring vertical angles in astronomical and geodetic work. The latter type is called a vertical circle. (4) A ring with its perimeter graduated in angular measure and placed perpendicular to a horizontal axis of rotation passing through its center. It is similar in construction to a horizontal circle. However, the vertical circle of a theodolite is usually smaller than the horizontal circle for mechanical reasons and because less accuracy is demanded for vertical measurements than for horizontal measurements. (5) An instrument used in astronomy for measuring angular distances from the zenith. SEE (3) above.

circle left - That position of a transit in which, when the telescope is pointed at a target, the vertical circle is to the left of the observer. Also called direct sighting, circle to the left, face left (Brit.) and vertical circle left.

circle of confusion - The circular image of a distant point, formed in a focal plane by an optical system. Also called a blur circle. A distant, point-like object (e.g., a star) is imaged by an optical system as an approximately circular spot of finite size for one or more of the following reasons: (a) the focal plane is not at the point of sharpest focus; (b) certain aberrations such as astigmatism and coma are present; (c) the optical system causes diffraction; (d) the photographic emulsion is grainy; or (e) the lens is of poor quality.

circle of declination - SEE hour circle.

circle of equal altitude - SEE almucantar.

circle of equal declination - SEE parallel of declination.

circle of error - SEE error circle.

circle of latitude - (1) A great circle on the celestial sphere, through the poles of the ecliptic and perpendicular to the plane of the ecliptic. (2) A meridian along which latitude is measured.

circle of least confusion - That image, of a pointlike source, whose largest diameter is the least of those of the other images of that point, for different distances of the focal plane. I.e., the smallest circle of confusion, as measured by the largest diameter of the image. The plane on which the circle of least confusion lies is often taken as the focal plane.

circle of longitude - (1) A circle, on the celestial sphere, parallel to the plane of the ecliptic. (2) A circle, on a sphere representing the Earth parallel to the plane of the Equator; a parallel along which longitude is measured.

circle of perpetual apparition - That circle of the celestial sphere, centered on the polar axis and having an angular distance from the elevated pole approximately equal to the latitude of the observer, within which celestial bodies are not seen to set at the observer's latitude. Contrasted to a circle of perpetual occultation. SEE ALSO star, circumpolar.

circle of perpetual occultation - That circle of the celestial sphere, centered on the polar axis and having an angular distance from the depressed pole approximately equal to the latitude of the observer, within which celestial bodies are not seen to rise at the observer's latitude. Contrasted to a circle of perpetual apparition.

circle of position - A small circle on a sphere representing the Earth, at every point of which, at the instant of observation, the observed celestial body (Sun, star, planet) has the same angular elevation and therefore the same zenith distance. At the instant of observation, the center (on the sphere) of the circle of position is that point which has the same longitude and latitude as the celestial body. The radius of the circle is the observed zenith angle of the body. The point representing the observer is therefore somewhere on that circle. A second observation on the same object at a different time (or on a different object at the same time) will determine a second and different circle of position, and if the observer has not moved, his representing point will be at the intersection of these circles. If the observer has moved, as may occur if he is on a ship or aircraft, allowance is made for the direction and amount of movement and a location determined for either point of observation. In navigation, a short portion of the circle of position is plotted as a straight line and termed a line of position or a Sumner line. A Sumner line, however, is merely a particular variety of line of position and the term should not be used as a synonym for line of position.

circle of right ascension - SEE hour circle.

circle of the sphere - A circle upon the surface of a sphere specifically, on the surface of a sphere representing the Earth or the heavens. It is called a great circle when its plane passes through the center of the sphere; otherwise, it is called a small circle.

circle of uncertainty - A circle having as its center a point whose location has been determined and as its radius the probable error of the location.

circle position - SEE circle.

circle right - That position of a transit in which, when the telescope is pointed at a target, the vertical circle is to the right of the observer. Also called reversed sighting, circle to the right, face right (Brit.) and vertical circle right.

circle sheet - A field sheet or smooth sheet on which intersecting families of curve are shown, each curve corresponding to the locus of points at
which some constant angle exists between two stations.

circle to the left - SEE circle left.

circle to the right - SEE circle right.

circuit - (1) A line of levels, a series of lines of levels, or a combination of lines or parts of lines of levels that, together with a continuous series of measured differences of elevation, extends continuously from a starting point through a sequence of intermediate points back to the starting point. (2) The set of consecutive measurements of distances and directions from a starting point to a remote point and back to the starting point.

circuit closure - SEE circuit misclosure.

circuit misclosure - (1) The amount by which the initially and finally determined values of some measured function of location differs when the measurements or calculations are carried out for points along a line that begins and ends at the same point. Also called circuit closure, loop misclosure and loop closure. (2) (leveling) The amount by which the algebraic sum of the measured differences of elevation fails to equal zero. Also called circuit misclosure. SEE ALSO misclosure.

circulation map - SEE map, traffic-circulation.

circumferentor - A surveyor's compass which has attached to it a radial arm with a vertical, slotted strip at each end. The surveyor sights through the two slots at an object and then, by raising his eye slightly, sees and reads off the direction shown on the compass.

circumpolar (adj.) - Remaining above the observer's horizon during a complete rotation of the Earth, i.e., for 24 hours. The term is commonly applied only to stars and other fixed objects in the sky. A celestial body is circumpolar when the complement of its declination is approximately equal to or less than the observer's latitude. SEE ALSO circle of perpetual apparition.

circumzenithal - SEE astrolabe, circumzenithal.

cirque - An armchair-like valley having a steep, nearly vertical wall at the back (head), a concave floor meeting the back in a sharp break of slope, and a lip, at the front end (entrance), which may be of bedrock, glacial moraine or both.

cislunar (adj.) - (1) In the region between the Moon and the Earth. (2) In the region between the Moon's orbit and the Earth's orbit.

cistern - An artificial reservoir or tank designed for storing rain-water collected from a roof.

cistern barometer - A mercury barometer in which a column of mercury is enclosed in a vertical glass tube, the upper end of which is sealed and exhausted of air; the lower end is placed in a cistern or reservoir of mercury which is exposed to atmospheric pressure. The amount of atmospheric pressure on the free surface of the mercury in the cistern determines the height to which the mercury will rise in the vertical tube. This height may be measured, and the pressure reported as so many centimeters, millimetersor inches of mercury. Also called a Fortin barometer.

city, inner - An urban local generally recognized as the central commercial or residential part of a city even tho it does not have definite political, geographic, or economic boundaries.

city map - A map of a city. Also called an urban map.

city planning - That phase of urban development in which plans are made for extensive changes in the structures and in use of land within a city. Also called urban planning.

city survey - SEE survey, urban.

city surveying - SEE surveying, urban.

claim - (1) A right or title. (2) A challenge of property or ownership of a thing which is wrongfully withheld. (3) The means by or through which a claimant obtains possession or enjoyment of privilege or thing. (4) Under land-laws, a tract of land taken up by a pre-emptioner or other settler (and also his possession of it).

claim, adverse - An assertion of title, right, or special privilege of beneficial use of the surface, or an underground land-or-water region by one party in opposition to that of another, wherein judicial opinion may be required on the conformity with state law (or, if public land of the United States of America are involved, an adjudication by the Secretary of the Interior).

claim, private land - (1) A claim, to a tract of land, that is based on the assertion that title thereto was granted to the claimant or his predecessors in interest by a foreign government (before the territory in which it is situated was acquired by the United States of America). (2) The land so claimed.

Clairaut-Darwin formula - SEE Darwin's modification (of Clairaut's formula).

Clairaut-Darwin gravity formula - SEE Darwin's modification (of Clairaut's formula).

Clairaut-Levallois equation - SEE Levallois' modification (of Clairaut's equation).

Clairaut-Molodensky equation - SEE Molodensky's modification (of Clairaut's equation).

Clairaut-Moritz equation - SEE Moritz's modification (of Clairaut's equation).

Clairaut's differential equation - (1) The second-order differential equation $\rho[(d^2f/dR^2) - (6f/R^2)] + (6\rho/R)[(df/dR) + (f/R)] = 0$ relating the flattening f of a rotational ellipsoidal solid composed of homogeneous, ellipsoidal shells and of average radius R and density $\rho(R)$. Sometimes referred to as Clairaut's equation. It has been generalized (e.g. by Jeffreys) by replacing f by fn, where fn is the n-th coefficient in an expansion of r/R in zonal harmonics. The factor 6 in the equation is then replaced by n(n+1). (2) The first order differential equation $y = x(dy/dx) + f(dy/dx)$. Its solution is $y = Cx + f(C)$, where C is a constant of integration.

Clairaut's equation - (1) The equation $(a-b)/a + (\tau_p - \tau_e)/\tau_e = \omega^2 a/2\tau_e$, in which τ_p and τ_e are the values of acceleration of gravity at the poles and at the equator, respectively, of a rotational ellipsoidal solid (representing the Earth) whose surface is equipotential, rotating with angular speed ω and having the lengths 2a and 2b for major and minor diameters, respectively. Also called Clairaut's theorem, Clairaut's third theorem and the third equation of Clairaut. It was stated by Clairaut in 1738 and is often written in the form $f + ß = 5m/2$. It is not a true equation, the left and right sides not being exactly equal. Numerous better approx-imations to the expression on the left side have been derived, as have actual equalities. Upon first consideration, Clairaut's equations may seem paradoxical: an increase in the flattening f of the ellipsoid apparently diminishes the increase in the acceleration from equator to pole. The paradox is explained by the fact that the equation applies to values on an equipotential surface and not necessarily to values on the surface of the solid itself. If the ellipticity of the solid's surface is assumed to change but that surface to still remain an equipotential surface, the equipotential surface must change accordingly and Clairaut's equation will still hold. (2) Any of several equations resemble Clairaut's equation and giving either a better value of $(f + ß)$ of that equation or an exact formula for it. All these equations, as well as the various forms of and improvements on Clairaut's formula, must be used carefully because they often use the same symbols but differ in the definitions of these symbols and in the meanings of the terms used in the definitions. SEE ALSO Levallois' modification; Molodensky's modification; and Moritz's modification. (3) SEE Clairaut's theorem.

Clairaut's equation, extended - The equation $(a-b)/a + (\tau_p - \tau_e)/\tau_e = (5\omega^2 a/2\tau_e) - f(f + \omega^2 a/\tau_e) + aC_4/4$, in which τ_p and τ_e are the values of gravity acceleration at the poles and at the equator, respectively, of an equipotential, rotational- ellipsoidal solid having equatorial diameter 2a and polar diameter 2b and rotating at the rate ω. f is the flattening, (a-b)/a and C_4 is the coefficient of the fourth-degree term in the representation of the geopotential by a Legendre series.

Clairaut's formula - The formula $\tau = \tau_e[1 + 5m/2 - f) \sin^2\phi']$ giving the value of gravity-acceleration on an equipotential, rotational-ellipsoidal solid as a function of geocentric latitude ϕ', with parameters f [the flattening (a-b)/b], $m = \omega^2 R/\tau$, in which ω is the rate of rotation, R the average radius of the ellipsoid and τ_e the average value of gravity-acceleration. Also called the Clairaut formula, Clairaut's gravity formula and Clairaut's theorem. There are other versions of the formula, these versions differing in the way they define f and m. SEE ALSO Darwin's modification (of Clairaut's formula).

Clairaut's problem - The problem of relating the shape or an inhomogeneous, rotating body in hydrostatic equilibrium to the distribution of density in the body.

Clairaut's theorem - (1) For every point P on a geodesic on a surface of rotation, the product of the distance r_o of that point from the axis of rotation by the sine of the angle θ between the meridian through P and the tangent to the curve at P is a constant: $r_o \sin\theta$ = constant. Also called Clairaut's equation. (2) SEE Clairaut's equation.

Clairaut's theorem, extended - SEE Clairaut's equation, extended.

Clairaut's third theorem - SEE Clairaut's equation.

Clarke-Rainsford formulae - SEE Clarke's formulae.

Clarke's formulae - A set of three formulae for the longitude λ, latitude ϕ and back azimuth A of the normal section (to P_o) at a point P on a rotational ellipsoid, given the longitude λ_o and latitude ϕ_o of an initial point P_o and the distance s and forward azimuth A_o (of the normal section) of P from P_o, as derived by A.R. Clarke (1880). The formulae are correct to about 1 part in 106 at 150 kilometers if 7 significant figures are used in the computations. The useful distance can be increased to about 750 km by using one of several modifications of Clarke's formula by, e.g., Rainsford (the Clarke-Rainsford formula).

Clarke's map projection - One of a set of perspective map projections devised by A.R. Clarke and projecting a truncated sphere onto a plane from a point between 1.35 and 1.65 radii from the center (the plane is not necessarily tangent to the sphere.) In particular, that perspective map projection having the perspective center at a distance of 1.4 radii from the center of the sphere and the plane at a distance of 0.3537 radii on the other side of the center. If the graticule is centered at the sub-solar point for any given instant, the boundary of the map will indicate the limits of the illuminated part of the Earth, including those parts influence by astronomical twilight. The map projection is therefore sometimes referred to as the twilight map projection or twilight projection.

Clarke spheroid 1858 - One of three rotational ellipsoids calculated by Clarke in 1858, of which the following two are particularly important (the factor 0.30480047 was used to convert from feet to meters). (1) A rotational ellipsoid having the dimensions

semi-major axis	20 926 348 feet	(6 378 361 m)
semi-minor axis	20 855 233 feet	(6 356 685 m)
flattening (derived)		1/294.26.

This is the ellipsoid referred to as the Clarke 1858 spheroid. It was used in the Austrian Land Survey and in the triangulation of many British Crown colonies in Africa. The factor 0.30480047 was used in converting from feet to meters). (2) A rotational ellipsoid having the dimensions

semi-major axis	20 927 005 feet	(6 378 558 m)
semi-minor axis	20 852 372 feet	(6 355 810 m)
flattening (derived)		1/280.40 .

This spheroid was calculated from topographically reduced observations in Great Britain and Ireland. The metric values were calculated using A.R. Clarke's value, 0.513 074 074, for the ratio of the toise to the meter and 0.304 800 47 for the ratio of the foot to the meter.

Clarke spheroid 1866 - A rotational ellipsoid having the dimensions

semi-major axis	20 926 062 feet	(6 378 206.4 m)
semi-minor axis	20 855 121 feet	(6 356 583.8 m)
flattening (derived)		1/294.978 .

This ellipsoid has been in use since 1880, in the United States of America, for calculating triangulation. The metric values were calculated using Clarke's legal meter of 1866, and were used in the past in tables based on this ellipsoid. Values based on the value 0.304 800 47 for the ratio of the foot to the meter are

semi-major axis	6 378 274 m
semi-minor axis	6 356 650 m.

Clarke spheroid 1880 - A rotational ellipsoid having the dimensions

semi-major axis	20 926 202 feet	(6 378 316 m)
semi-minor axis	20 854 895 feet	(6 356 582 m)
flattening (derived)		1/293.465 .

The metric values were calculated using 0.304 800 47 for the ratio of the foot to the meter. Note that the value given by Clarke for the flattening does not agree with the values given for the lengths of the semi-major and semi-minor axes. Those values yield a flattening of 1/293.466. A modification of this ellipsoid, with a value of 6 378 249 m for the length of the semi-major axis, has been used for triangulation in South Africa.

Clarke spheroid 1880, modified - The Clarke 1880 spheroid modified by taking the flattening and the length, in feet, of the semi-major axis as given and converting the length to meters (6 378 249.145 m) using the factor 1/(3.28086933). Also called the Clarke 1880 spheroid (modified). The modified Clarke 1880 spheroid was used in the adjustment of the triangulation extending from the Cape of Good Hope to Cairo. SEE ALSO Cape datum.

classification - The separation of a collection of things, called elements - objects, features, concepts, phenomena, etc. - into sets, called categories, according to a set of rules such that: given any element, the set in which that element belongs can be determined from the rules; and given any set, the elements in that set can be found. The number of elements may be finite or infinite; the number of sets may be finite or infinite. The sets themselves may be collected into larger groupings which are sets of sets, and these again collected into larger grouping which are sets of sets of sets, etc. The different stages of collection are usually given distinctive names such as phylum, class, order, family, and so on, or may be numbered as 1st level, 2nd level, and so on. A complete classification assigns each element to one and only one set of a particular level; all the elements in one set have at least one common characteristic (which is given by the rules); two elements from different sets (of a particular level) differ in at least one characteristic (as specified by the rules). A partial classification assigns each element to one or more sets of a particular level, i.e., an element may belong to two or more sets at the same level at the same time.

classification (of base lines) - SEE base line (2a).

classification (of control) - (1) The classification of the coordinates of marked, identifiable points used in surveying as the basis for further surveying. Control is commonly classified either by the kind of coordinates involved as, for example, horizontal control or vertical control, or by the precision or accuracy as, for example, first-order control or second-order control. The USA has, from time to time, reclassified the control established by the Coast and Geodetic Survey and its successors. The most recent classification is that of 1984. (2) SEE classification (of survey).

classification (of leveling) - (1) Classification of the procedures used in determining elevations or differences of elevation. The category in which a particular procedure is placed may depend on the accuracy of the results obtained by the procedure or on the instruments and methods involved. If the latter criterion is used, the specifications for these criteria are usually such as to ensure that the results of a leveling carried out according to them will produce a leveling network of the same category or better. The U.S. National Geodetic Survey classifies leveling as first-order class I, first-order class II, second-order class I, second-order class II and third order. It uses, as criteria, not only the instruments and methods employed but also the geometric characteristics of the leveling network produced, i.e., the density of control points in the network. (2) SEE classification of vertical control.

classification (of leveling networks) - Classification of leveling networks according to the nature and/or accuracy of the elevations or differences of elevation of points in the network. Also called classification of leveling. The U.S. National Geodetic Survey classifies only leveling networks produced by differential leveling. It classifies such networks both by the linear density of the differences of elevation determined and by a number roughly indicative of the accuracy of the differences as a whole. The categories are the same as those used for classifying the differences of elevation (vertical control) and for classifying the leveling used in producing the networks, although the criteria are somewhat different. SEE ALSO classification of vertical control.

classification (of map projections) - The separation of map projections into sets according to criteria uniquely characterizing each set. Map projections are commonly classified either by the geometric quality preserved (i.e., left unchanged) by the transformation or by the kind of surface onto which the ellipsoid is mapped as an intermediate step in mapping onto the plane. The geometric qualities used are distances between points, angles, and areas within closed curves (or distances and areas scaled to the size of the map produced.) A map projection keeping distances from a particular point or line unchanged except for scale is called an equidistant map-projection; one keeping the angles between lines unchanged is called a conformal map projection; one keeping the azimuths from a particular point unchanged is called an azimuthal map-projection; and one keeping areas within closed figures unchanged except for scale is called an equal-area map-projection. There are no generally-accepted names for other sets. Map projections usually map the ellipsoid either directly onto the plane or first onto a sphere, cone, or cylinder and then from that surface onto the plane. If the map projection is from a cone or cylinder, the mapping onto the plane is done by the simple algebraic analog of unrolling the cone or cylinder. A spherical surface cannot be unrolled, but the algebraic transformations from ellipsoid to sphere and from sphere to plane are usually very much simpler than the single-step process from ellipsoid to plane. A map projection using the cone as an intermediate surface is called a conical map-projection; one using the cylinder is called a cylindrical map-projection. There is no particular name for the set of map-projections using the sphere as an intermediate surface. Except for the aposphere, which has been used as an inter-mediate surface for the transverse Mercator map projection, choice of intermediate surfaces has been pretty much limited to the sphere, the cone, and the cylinder. Map projections are also classified into two sets according as they are true projections (i.e., are equivalent to drawing straight lines from a given point, through the points on the ellipsoid, and onto the plane or intermediate surface) or are not. Those which are true projections are then further classified according to where the center of projection and the ellipsoid are placed with respect to the plane or intermediate surface. Many other schemes of classification exist but are comparatively little used. The best known are those of Maurer (1935) and Tobler (1962); there are more recent ones by Wray and Chovitz.

classification (of map projections), parametric - A classification into 8 classes, as proposed by Tobler (1962), based on the functional relations required to determine either the Cartesian or polar coordinates in any given graticule.

classification (of maps) - The separation of maps into sets according to criteria uniquely characterizing each set. Maps are commonly classified according to (a) what geometric quality of the mapped region is preserved unchanged in the map; (b) the ratio (scale) between distances in the mapped region and lengths of corresponding lines in the map; (c) the region represented by the map; (d) the intended use of the map; (e) the type of information shown; or (f) some measure of the accuracy of the map. (a) Maps are classified as equidistant, conformal, azimuthal, or equal-area according as they keep unchanged, except for scale, distances from a given point or line, angles between lines, azimuths from a given point, or areas within closed curves. An additional category is that of maps which introduce the smallest possible error (distortion) into the map under specified conditions (minimal distortion maps). (b) Maps are classified as large-scale, medium- or intermediate-scale, or small-scale maps according as the average ratio between distances on the map and the corresponding distances on the ellipsoid is larger than some number used as criterion for the set, lies between that number and some smaller number selected as criterion, or is smaller than that second number. The criterion for large-scale maps is usually 1/75 000 or larger; that for medium-scale maps is usually larger than 1/1 000 000 - 1/500 000 is a common number. (c) Maps are classified as world maps, national maps, State maps, etc., according to the size and location of the region represented. A world map (also called a global map) represents the entire Earth, a map of North America represents the continent of North America, etc. (d) There is no limit to the variety of uses to which a map may be put or for which it may be intended. However, some categories based on intended use are quite large: aeronautical charts and nautical charts take many different forms and the set contains many subsets; military maps, for tactical or strategic planning; route maps, for traveling by various means; and so on. (e) Classification according to the type of information shown is almost equivalent to classification according to intended use. In fact, it differs from that classification principally in having a general- purpose category called topographic maps which show, symbolically, just about everything one would see if viewing the ground from a point well above it. Maps which are not topographic maps, i.e., which show only selected details or give, graphically, non-topographic information are called thematic maps. Typical are maps showing the geologic structure of the crust (geologic maps), density of population (population-density maps), the use to which land is being put (land-use map), or the ownership of land (cadastral maps). (f) Classification of maps according to their accuracy is fairly standard for topographic maps, although the scheme varies in detail from country to country and from application to application. In the United States of America, a map of standard accuracy has been defined as one (a) meeting the National Map Accuracy Standards, (b) representing all features by symbols and treatments approved at the time of publication, (c) portraying cultural detail with reasonable completeness at the time of appraisal, and (c) having a contour interval representing all significant hypsometric detail (considering the scale of the map.) Each actual map is then classified according to the extent to which it meets these standards, as follows: class 1 maps meet all four criteria; class 2 maps meet three of them, class 3 maps meet two of them, class 4 maps meet one or none of them, and class 5 maps may meet any or all of these criteria but have been superseded by maps of better quality.

classification (of surveys) - Classification of the coordinates of a set of points (the survey) produced by a single surveying project. It differs from the classification of control in that the set as a whole rather than the individual points in the set is classified. The practice is to place the survey in the same category as would have been placed a single control point considered characteristic of the points actually produced by the survey. The largest error produced by the survey may be used as a basis, or the average error, or some more complicated function of coordinates and their errors may be used.

classification (of traverses) - A control traverse is commonly classified according to the accuracy of the control established by it; in the United States of America (USA), this is the same as the classification of that control. SEE ALSO classification (of control). Horizontal- control networks in the USA contain some traverses which established control of much higher accuracy than that required for first-order control. These traverses are called high-precision traverses by the national Geodetic Survey. Other countries have also felt the need for a term to designate networks significantly more accurate than those commonly considered first-order, but there is as yet no agreement on any one designation.

classification (of triangulation) - The scheme used for classifying a triangulation (process or network) is the same as that used for classifying the control established by that triangulation or contained in that network. The classification adopted by the USA recognizes the following categories of geodetic control and, therefore, the same categories of project and network.

CATEGORY	RELATIVE ACCURACY NOT LESS THAN
First order	1 : 100 000 and
Second order, Classes I and II	1 : 50 000 and
	1 : 20 000
Third order, Classes I and II	1 : 10 000 and
	1 : 5 000 .

The ratios given are the standard deviations in the distances between directly connected, adjacent points, divided by those distances. The practice has been to name categories in order of increasing size of the relative error involved. Before 1921, the categories were: primary, secondary, and tertiary triangulation. From 1921 to 1924, they were: precise, primary, secondary, and tertiary triangulation. From 1925 to 1957, they were: first-order, second-order, third-order, and fourth- order triangulation. In 1957, they were changed to: first-order, second-order and third-order triangulation, and each of these categories was subdivided into classes. In 1974, a different scheme of subdivision into classes was adopted.

classification (of trilateration) - The category into which a trilateration or trilateration network is placed is the same as the category of the horizontal control established by or contained in that trilateration or trilateration network. Criteria for trilateration were first published in 1974, and trilateration was therefore not included in earlier classifications.

classification (of trilateration network) - A trilateration network is placed in the same category as the control it contains. SEE ALSO classification (of trilateration).

classification (of vertical control) - Classification of the elevations (or differences of elevation) in a leveling network according to their accuracy or other characteristic. Classification of an elevation (vertical control) according to its accuracy is inseparable from the classification of the leveling network in which the elevation occurs, because the values of all the differences of elevation in a particular network are solved for simultaneously - usually by the method of least squares. The U.S. National Geodetic Survey uses the same categories for classifying vertical control as it uses for classifying leveling and leveling networks networks. The criterion used for each category is derived in a complicated fashion but is roughly as follows:

category	largest standard deviation of difference of elevation with respect to any other point in the network
first-order class I	$0.5 \sqrt{K}$ mm
first-order class II	$0.7 \sqrt{K}$ mm
second-order class I	$1.0 \sqrt{K}$ mm
second-order class II	$1.3 \sqrt{K}$ mm
third-order	$2.0 \sqrt{K}$ mm,

in which K is the distance, in kilometers, along a level line between the points.

classification copy - Source material used as a guide by the compiler and/or draftsman in preparing a map or chart and consisting, usually, of details on roads, railroads, cities, and the like that have been developed from field surveys. It is usually furnished in the form of overlays, annotated maps, drawings, photographs or field sheets.

classification survey - SEE field inspection.

Claude and Driencourt astrolabe - SEE Danjon astrolabe.

clause, habenum - A clause, in a real-estate document, that specifies the extent of the interest (as life or fee) to be conveyed. This clause usually includes the words "to have and to hold".

clay - (1) Any wet, adhesive earth such as mud. (2) An earthy, highly plastic sediment consisting of a considerable amount of hydrous silicates in the form of finely-ground colloidal or clay particles.

clay particle - (1) Any mineral particle occurring in sedimentary rock and derived from pre-existing rock, and having a diameter less than 4 micrometers. (2) The same as the previous definition, except that the upper limit on the diameter is 3 micrometers.

clearing y-parallax - SEE orientation, relative.

clinometer - (1) An instrument combining a spirit level, vertical circle, and sighting device for coarse but rapid measurement of vertical angles, and made small enough that it can easily be held in one hand while in use. One type consists of an open tube with cross-hairs at one end (for sighting). The tube is pivoted on a flat piece to which are attached a vertical circle and spirit level. The inclination of the line of sight can be read on the circle by holding the flat piece so that the bubble in the spirit level is centered at the instant of observation. By means of a prism, the bubble is viewed while looking through the tube at the cross-hair and object observed. A clinometer is usually combined with a compass such as the Brunton compass. SEE ALSO hand level. (2) An instrument for indicating the degree of slope or angle or pitch of a craft.

clock - (1) An instrument for indicating time or measuring intervals of time. (2) Conventionally, any device for indicating time and too large to be carried on the person. Clocks can conveniently be distinguished according to the kind of mechanism (oscillator) that provides the basic interval of time: balance wheel, pendulum, crystal, molecule, or atom. Any clock of high accuracy was formerly called a chronometer; this usage has almost disappeared. Precise clocks incorporating balance wheels and used in astro-nomical surveying are still referred to as chronometers, but they are not accurate enough for use in more demanding work such as satellite surveying. For such work, crystal clocks are commonly used if accurate time signals are regularly available. If such time signals are not available, atomic clocks are used. SEE ALSO cesium clock; crystal clock; pendulum clock; rubidium clock; Shortt clock.

clock, astronomical - (1) A clock whose indicated time is determined by the rotation or revolution of the Earth. (2) The rotation or revolution of the Earth, considered as a timekeeping mechanism.

clock, atomic - Any clock in which the basic intervals of time are determined by reference to the resonant frequency of radiation absorbed or emitted by atoms or molecules. If the period is governed by the resonant frequency of molecules, the clock is preferably referred to as a molecular clock. Atomic clocks are commonly classified as active or passive according as they use radiation emitted by or absorbed by atoms. Typical passive atomic clocks are cesium clocks and rubidium clocks. Cesium clocks use cesium atoms for determining the basic intervals; rubidium clocks use rubidium atoms. Because the second is defined by international agreement as the length of time needed for a specified number of cycles of a particular frequency of radiation emitted by the cesium atom, cesium clocks are primary standards of time and frequency. Rubidium clocks have show smaller variations in frequency over short periods of time than do cesium clocks. The first complete and operating atomic clock was developed by H. Lyons and his associates at the U.S. National Bureau of Standards (NBS) in 1948 - 1949. It was actually a molecular clock; it used ammonia molecules for determining the basic frequency and interval of time. True atomic clocks came later and resulted from the work of L. Essen and his associates at National Physical Laboratory (England) and of Lyon and his associates at NBS from about 1952 onward.

clock, crystal - SEE crystal clock.

clock, molecular - A clock in which the intervals of time are determined by reference to the frequencies radiated or absorbed by rotating molecules. The first clock of this kind was the ammonia clock, using the molecules of gaseous ammonia to provide the basic frequency. Radio waves generated in an attached circuit were fed into the gas and the frequency of the circuit adjusted until meters showed maximal absorption of the waves by the gas; this was the desired frequency of resonance. A more recently developed clock uses methane as the gas but the basic principles are the same. Molecular clocks are frequently referred to as atomic clocks but this usage is erroneous. They are also called masers if their principal output is frequency, not time.

clock correction - The quantity added algebraically to the time shown by a clock to obtain the correct time at a given meridian. If the clock is slow, the correction is positive; if it is fast, the correction is negative.

clock method (of determining longitude) - SEE chronometric method (of determining latitude).

clock paradox - The statement, resulting from Einstein's special theory of relativity, that if one clock is moving with respect to another, the moving clock will be found to be running slower than the other. The paradox is of little importance in surveying except when clocks at widely separated locations are to be synchronized by moving a clock from the one location to the other. In an experiment in which four cesium clocks were carried around the world, it was found that moving the clocks in the direction of rotation of the Earth slowed the moving clocks up, while moving them in the opposite direction speeded them up.

clock rate - The rate of change of a clock correction; the amount gained or lost by a clock in a unit of time. When applied to a chronometer, the quantity is called chronometer rate. Clock rate is usually expressed as the increase or decrease of the clock correction per day, hour, or minute as shown on the clock's face. If the clock correction is decreasing algebraically, the clock is gaining and the rate is negative; if the correction is increasing, the clock is losing and the rate is positive.

clock stability - SEE stability.

clock star - One of a set of certain selected stars, in the vicinity of the celestial equator, whose right ascensions are accurately known and which are used for the exact determination of longitude or of time.

clock time - Time determined by reference to the indications of a clock. It is convenient to differentiate between atomic time, which is time indicated by clocks using the frequency of radiation from particular species of atoms or molecules, and clock times indicated by clocks using the frequency of mechanical vibrations such as the swinging of a pendulum, the vibration of a quartz crystal, or the oscillations of a balance wheel. SEE ALSO chronometer; clock, atomic; crystal clock; Shortt clock.

closing the horizon - Measuring the last of a series of horizontal angles, at a station, such that the sum of the series is a multiple of 360°. At any station, the sum of all horizontal angles between adjacent lines should equal 360° (400g). The amount by which the sum of the observed angles fails to equal 360° is the misclosure. This is distributed as a correction among the observed angles to bring their sum to exactly 360°.

closure - SEE misclosure.

closure, relative error of - SEE misclosure, relative.

closure error, horizontal - SEE misclosure of traverse; misclosure of the horizon.

closure in azimuth - SEE misclosure in azimuth.

closure of horizon - SEE misclosure of the horizon.

closure of triangle - SEE misclosure of triangle.

clothoid - A spiral having the parametric representation
$$x(t) = \int_0^t \cos(\pi u^2/2)\, du \; ; \; y(t) = \int_0^t \sin(\pi u^2/2)\, du$$
Also called the Cornu spiral, spiral of Cornu and Cornu's spiral, particularly in physics where the curve is used for determining the pattern of diffraction by holes; and Euler's spiral, Euler spiral or clothoid in surveying, where it is used as a transition between two other curves on the same path, as in laying out roads or tracks. It can also be defined as a spiral along which the curvature $1/\rho$ is directly proportional to the distance s, along the spiral, from the point of zero curvature. The usual form of the equation is then $2s\rho = a^2$, in which a is a constant. In surveying, the spiral has a radius of infinite length where the spiral begins and a radius R_c at the point where it joins a circular curve of radius R_c.

cloud on title - (1) An outstanding claim or encumbrance which, if valid, will affect or impair the owner's title. (2) A judgement; a dower interest.

cluster analysis - A statistical procedure for identifying those points in the data about which the data seem to cluster.

cluster development - A development in which related housing-units are grouped or clustered with common, open space surrounding each group, such open space serving as a buffer between groups or used for common recreation, walks, pathways for bicycles or other such informal activity.

cluster zoning - Zoning whereby a specific density of residence or dwelling units is prescribed for an entire locale.

co-altitude - SEE zenith angle.

coast - The strip of land, of indefinite width, that extends inland for perhaps 1 to 5 km from the shore to the first major change in type of terrain. SEE ALSO littoral; seaboard.

coastline - (1) The line that forms the boundary between the coast and the shore and marks the seaward limit of the permanently exposed coast. (2) The U.S. National Ocean Survey uses the terms coastline and shoreline as synonyms and defines them as being the line of mean high water. (3) The U.S. Submerged Lands Act, 43 U.S.C. 1301(c), states that the term coast line means the line of mean low water along that portion of the coast which is in direct contact with the open sea and the line marking the seaward limit of inland waters. (4) A narrow strip of land adjacent to the line separating the surfaces of land and water at the edge of a sea or ocean. This last definition is used in civil engineering and is best not used at all. Use coast instead.

coastline, physical - The line where the land and water meet along the open coast, irrespective of coastal indentations. This term is used to distinguish this kind of coastline from a political coastline.

coastline, political - The limits of inland waters in the vicinity of islands. This term is used to distinguish this kind of coastline from a physical coastline.

coastlining - The process of obtaining data from which the coastline can be drawn on a chart. (Jargon)

Coast-Survey method - SEE triangle-of-error method.

coating - SEE antihalation coating; scribe coating.

cocked hat - SEE triangle of error.

code - (1) A set of symbols, together with rules for using these symbols to carry information. All alphabets, for example, are codes. The set of symbols that make a computing machine add, multiply, print, start, stop, etc., is a code. Numbers are frequently used as code for computing machines. The machine translates the numbers into instructions. (2) Any set of symbols related by formal rules to another set of symbols in such a way that all information contained in the first set is contained in the second set. Such codes are used principally to hide information from those not entitled to it, or to transform information into a form suitable for transmission and/or computation. The Morse code and Gray code are examples of codes designed for transmitting information. Binary-coded decimal code and sequenced codes are examples of codes designed for mechanical calculating. (3) The set of rules used to convert data from one form of representation to another. SEE ALSO Baudet code; Gray code.

code, alphabetic - A code (1) which uses the letters of an alphabet to convey information.

code, alphanumeric - A code (1) which uses a combination of letters and numbers to convey information.

code, binary - A code (1) which uses combinations of only two digital characters (usually 0 and 1) to represent other characters. For example, the number 5 may be represented by 101.

code, numeric - A code (1) which uses numbers to convey information.

code, pseudo-random - A code (1) based on a pseudo-random sequence of numbers, i.e., based on a sequence (usually periodic) of numbers generated according to definite mathematical rules but having approximately the statistical properties of a random sequence of numbers. A common method of generating a pseudo-random sequence containing numbers $x_1, x_2, \ldots x_N$ uses the method of congruences: $x_{n+1} = ax_n " b \pmod{T}$, where b and T are relatively prime integers. a and b are chosen to make the sequence as long as possible before it repeats, while keeping the numbers as random as possible. T is chosen to fit the base of the number system used by the computer. Pseudo-random codes have been used in cryptography, in radio-communications systems, distance-measuring instruments operating at radio frequencies. The purpose of the first application is secrecy; that of the second two applications is lessening the effect of extraneous noise on the signals.

co-declination - The complement of the (astronomic) declination, i.e., 90° minus the declination. Also called the polar distance.

coded-decimal system, binary - SEE system, binary coded-decimal.

coefficient, harmonic - The coefficient of a term in a series of harmonic functions.

coefficient of refraction - (1) The ratio of the angle of refraction at the point of observation to the angle, at the center of the Earth, subtended between the point of observation and the point observed. (2) The ratio of the average radius of curvature of the Earth to the average radius of curvature of the path of the electromagnetic ray. (3) Half the ratio specified in definition (2).

coefficient of thermal expansion - The relative change (expansion or contraction), expressed as a ratio, in a linear dimension of a body, corresponding to a change of 1° in the body's temperature. The coefficient of expansion, as it is generally called, may be in terms of the Celsius (SI), Fahrenheit, or other thermometric scale. To a very great extent, its magnitude is peculiar to the material' a steel tape, for example, has a coefficient of expansion about 25 times as great as has an invar tape. The coefficient is usually expressed as a decimal fraction. In a measure of length such as the length of a base line, it enters as a correction which is the product of the coefficient of expansion of the apparatus used, the length measured, and the difference between the temperature at which the measurement was made and the temperature (standard or calibration) at which the length of the apparatus is known.

coelostat - A mirror or arrangement of mirrors rotated at such a rate by a motor as to direct light from the stars into a fixed direction on the ground, e.g., into a stationary telescope.

cofferdam - A barrier built in the water so as to form an enclosure from which water can be pumped to permit free access to the space bounded by the dam.

co-geoid - (1) That surface which lies below the Earth's surface a distance equal to the elevation of the Earth's surface as determined by leveling. The distance is measured along the vertical to the local level surface. This kind of co-geoid is the easiest to determine and the most closely related to leveling. (2) Any surface derived from observed values of gravity by using Stokes's formula or a similar formula for the height of the geoid. (3) An equipotential surface obtained from the geoid by removing all matter external to the geoid either permanently or with replacement inside the geoid. (4) That equipotential surface which is reached by repeatedly transferring matter from above the geoid to below the geoid (thereby repeatedly lowering the geoid to a new position) until no change in the position of the geoid results. This final position is then the co-geoid. Also called a regularized geoid or a compensated geoid. (5) Any surface defined in terms of the values of gravity or of gravity potential on it. A geop is a co-geoid, in this sense. (6) The bounding surface of a theoretical solid constructed so that the surface is at a constant gravity potential and the gravity field agrees, to a satisfactory extent, with that of the actual Earth where observations are possible. The co-geoid is not, according to the first two definitions, an equipotential surface nor is it defined in terms of potential.

coherence - The amount of correlation between the intensity and/or phase of electromagnetic fields at points separate in space and time.

coherent (adj.) - having the field's amplitude varying in unison at all points. While the relative phases at different points may be different, the absolute phases vary with time in identical fashions. Also called "spatially coherent".

coherent, spatial - SEE coherent.

coincidence (in viewing a bubble) - SEE coincidence bubble.

coincidence (of pendulums) - SEE coincidence method (of determining period).

coincidence bubble - The image presented when the bubble in a spirit level is viewed through an optical system that shows the bubble split in half lengthwise, one of the halves being then reversed end for end and placed alongside the unreversed image. If the bubble is not exactly centered in the vial, the ends of the two halves will appear separated, longitudinally. When the bubble is centered, the ends of the image are aligned and one seems to be viewing one end of the entire bubble.

coincidence method (of determining period) - The intervals of time between coincidence (in the same direction) of the position of a freely swinging pendulum and a clock pendulum are measured, usually at the point of lowest descent. The period of the freely swinging pendulum is then $m P_c/n$, where $m P_c$ is the length of the interval ($2mP_c$ if coincidences in the opposite direction are measured also) and n is the number of swings of the freely swinging pendulum.

col - A high, sharp-edged pass in a mountain ridge. It is usually produced by the headward erosion of opposing cirques.

co-latitude - The complement of the latitude; 90° minus the latitude. Co-latitude forms one side, zenith to pole, of the fundamental astronomical triangle. It is opposite the celestial body which is at one of the vertices of the triangle, when such a triangle exists.

collateral (adj.) - Designating or pertaining to an obligation or security attached to another to secure its performance.

collateral (n.) That which is used as collateral security or obligation.

collation - (1) The assembling of pages of publications, in sequence. (2) The verification of the order, number and date of maps.

collimate (v.) - To bring the optical elements of an optical system into proper relationship with each other.

collimation - (1) The process of bringing the optical elements of an optical system into proper relationship with each other. The process of bringing the collimated system into proper relationship with the pointing mechanism is called alinement. (2) SEE collimation adjustment. (3) Adjusting the fiducial marks in a camera so that lines through them intersect at the principal point. Also called adjustment for collimation. SEE ALSO error of collimation; line of collimation.

collimation adjustment - The process of bringing the line of collimation of a surveying instrument into close agreement with the collimation axis. I.e. the act of collimating an instrument. Also called adjustment for collimation or ambiguously, collimation. The process itself should always be referred to as a collimation adjustment.

collimation axis - The line passing through the second nodal point of the objective of a telescope and perpendicular to the axis of rotation of the telescope. In a surveyor's transit or theodolite, the collimation axis is perpendicular to the horizontal axis of the telescope. In a leveling instrument, it is perpendicular to the standing axis (vertical axis) of the instrument. When the telescope of a transit or theodolite is rotated about its horizontal axis, the collimation axis describes a plane called the collimation plane.

collimation correction - SEE C-factor; level correction.

collimation correction factor - (1) The negative of the collimation factor (2). This is not the same as level correction, with which it is sometimes confused. Also called the level collimation correction factor. (2) The tangent of the angle between the actual line of sight in a leveling instrument and a horizontal plane. Also erroneously called the C-factor. Equivalently, it is the tangent of the collimation error when applied to a leveling instrument. It is practically the same as the collimation factor, which measures an error; the name is therefore inappropriate and misleading. (3) SEE C-factor.

collimation correction factor, level - SEE collimation correction factor.

collimation error - (1) The angle between the actual line of sight through an optical instrument and the position the line would have in a perfect instrument. There are several different interpretations of this definition. (a) The angle between the actual line of sight and a line which passes through the (rear) nodal point of the objective lens system and is perpendicular to the axis of rotation. This interpretation is commonly accepted in surveying, is applied to transits, theodolites, meridian instruments, and leveling instruments, and is synonymous with error of collimation if the line of sight is through the center of the reticle. (b) The angle between the actual line of sight and the optical axis of a perfectly adjusted optical system. Divergence of the optical axis from perpendicularity to the axis of rotation is then called alinement error. This interpretation is used in astronomical optics and other sciences. Collimation error in sense (a) above is then the sum of alinement error and collimation error in sense (b) if the line of sight is through the center of the reticle. (c) The line of sight and the axis of rotation are generally skew to each other. The collimation error may therefore be defined as the angle between two planes, one through the line of sight and parallel to the axis of rotation and the other through the rear nodal point (of the objective lens system) and the axis of rotation. This is the projection, onto a plane perpendicular to the axis of rotation, of the collimation error as defined in (a). The other component of direction between the two lines does not have a common name in English, although the term skew angle or skew error has been used. Collimation error in leveling instruments may be removed or reduced by the two-peg method. (2) By extension, in radio telescopes, the angle between the plane perpendicular to the antenna baseline and the surface of symmetry of the central lobe. Because the angle varies with distance from the antenna, a cone asymptotic to this surface is sometimes used instead. Also called squint angle. (3) The amount by which the angle between the optical axis of an optical telescope and its east-west mechanical axis, or between the axis of the main lobe of a radio telescope and the east-west axis, deviates from 90°. (4) SEE collimation factor (1). (5) SEE error of collimation.

collimation error, level - SEE level collimation error; collimation error.

collimation eyepiece - An eyepiece containing a prism for use with a collimator.

collimation factor - (1) The factor c^* or c in Mayer's formula or Bessel's formula. It is also called the collimation error and is equal to the angle between the optical axis of a telescope and a vertical plane through North. (2) The tangent of the angle between the line of sight and a horizontal line through the center of the instrument when the bubble of the leveling vial is centered. This quantity is the negative of the C-factor when the stadia constant is equal to 1. (One writer denotes it by C.) Multiplying the collimation factor by the distance to the object viewed gives the linear amount, in the same units, by which the point actually sighted is above or below the point that should have been sighted. It is therefore an error. SEE ALSO collimation correction factor.

collimation plane - The plane described by the collimation axis of the telescope of a transit or theodolite when the telescope is rotated around a horizontal axis.

collimation position - The ideal position of the line of sight of a telescope - that is, the optical axis. This definition may be self-contradictory.

collimator - (1) Any device for lining up (collimating) the optical axes of the various elements of an optical system. The usual collimator consists of a reticle and an arrangement of lenses for passing light through the reticle perpendicularly to the plane of the reticle e.g., as by placing the observer or a light source at the principal focus of the lens system. A source of light is often included. The mark on the reticle may be viewed from very short distances with the same effect as if it were at an infinite distance. It may therefore be used in place of a distant mark when making any adjustment of the line of sight or line of collimation of an instrument. In adjusting a surveying instrument, the telescope of another surveying instrument may be used as a collimator, the reticle furnishing the mark; or the telescope or a discarded instrument may be placed on a special mounting to form a permanent installation. (2) A lens or lens system for producing a planar wavefront of light from a spherical wave-front. Equivalently, it may be considered a lens system for producing parallel rays of light. Also called a collimating lens.

collimator, auto - SEE autocollimator.

collimator, vertical - A telescope so mounted that its collimation axis may be made to coincide with the vertical (or direction of the plumb line). The vertical collimator serves as an optical plumb line. It may be designed for use in placing a mark on the ground directly under an instrument on a high tower or for use in centering an instrument on a high tower directly over a mark on the ground.

collinearity condition - The condition that a point in object-space, the corresponding point in image space, and the corresponding perspective center must be collinear (i.e., all lie on the same straight line). This is similar to the co-planarity condition, which states a relationship between three lines and one plane. The two types of condition are theoretically equivalent, but it has been found that in practice collinearity conditions are easier to use for solving photogrammetric problems than are co-planarity conditions.

Collins' auxiliary point method - SEE Collins' method.

Collins' method (of resection) - A method of determining, in the plane, the location of a point A from the known locations of three points B, C, D and the two measured angles BAC and CAD. One auxiliary point Q is determined as the vertex of the triangle whose base is BD and whose opposite angles are equal to BAC and CAD. The center of the circumscribed circle is determined from the perpendicular bisectors of BQ and DQ. A lies on the circumscribed circle and on the lie through Q and C. Also called the Italian method, Italian section and Collins' auxiliary point method. It was used extensively by the U.S. Coast and Geodetic Survey.

Collins' point - One of the auxiliary points established in using Collins' method or Cassini's method (of resection).

collocation - (1) The locating of two different geodetic instruments at the same control point. By extension, the location of different instruments near but at the same point, with all instruments connected to the control point by surveying. Also written co-location. (2) The exact agreement of a function with the boundary values of a boundary-value problem. (3) A method of solving, approximately, a system of differential equations by expressing the solution as a sum $y = y_o(x) + k_1 y_2(x) + ... + k_n y_n(x)$, where $y_o(x)$ is any function satisfying the boundary conditions. (4) SEE collocation method (of least squares).

collocation, least-squares - SEE collocation method (of least squares).

collocation method (of least squares) - (1) A special case of the method of least squares, in which the set of unknowns is divided into two subsets with different weights. (2) Bayesian least-squares applied to particular problems of geodesy.

co-location - SEE collocation (1).

color, complementary - One of a pair of colors whose spectra, combined, give the spectrum of white light.

color, extended - SEE bleed.

color, primary - The color of one of the projected visible spectrum bands; the additive mixtures of which all other colors of the spectrum can be produced. The colors are red, green and blue.

color composite - A composite image in which the seperate bands making up the composite are shown in different colors.

color emulsion - A photographic emulsion capable of producing colored pictures. The emulsion usually consists of several layers of emulsion each sensitive to a different part of the spectrum (not necessarily the optical portion).

color enhancement - The use of a wide range of hues or shades to exaggerate otherwise small variations of photographic density on black-and-white images, or of hue and shade on colored images.

color equation - The equation $Cl = r_R + g_G + b_B$, which characterizes a color Cl of luminosity l by matching it against a mixture of r units of color R (red), g units of color G (green) and b units of color B (blue).

color filter - An optical filter which allows only light within a certain narrow range of frequencies (wavelengths) to pass through. Color filters have been used in the telescopes of both leveling instruments and of theodolites to improve the accuracy with which an object can be sighted. E.g., a

Wratten #26 Stereo Red has been found best for use in leveling, a Wratten #8 K-2 orthochromatic has been found best for use in theodolites and transits.

color gradient - SEE tinting, hypsometric.

colorimeter - An instrument for measurement of color, i.e., for identifying in some unique and standard way the color of an object. One form, used for opaque objects, compares the object with a set of standard colors. Another, used for light-emitting objects such as lamps and stars, compares the light from the object with light of known composition passed through colored filters. Yet another, used for translucent substances such as filters or colored solutions, passes light of known composition through the substance and deter-mines the amount of light absorbed in various parts of the spectrum. A colorimeter differs from a spectrometer primarily in that it gives the color of an object (or the light from it) by specifying how that color can be made up from a small number (from 1 to 3; usually fewer than 9) other, standard colors. A spectrometer gives the color of the light from an object in terms of the intensity of the light as a more or less continuous function of wavelength.

color index - The difference between the photographic and photovisual magnitudes of a star or, more generally, the difference between the magnitudes of a single star in two separate and specified regions of the spectrum. By convention, the color index (C.I.) is then defined as: magnitude in short-wavelength region minus magnitude in long-wavelength region. The C.I. for an A0 star is defined in the Johnson-Morgan UBV system of stellar classification to be zero. The C.I. of the Sun is 0.62. The color index appears in geodesy when the coordinates of stars are to be determined very accurately from stellar photographs. It is one of the quantities appearing in the equations used in reducing the measurements of the photographs.

color mixture, additive - A combination of light of different chromaticities. I.e., a single color produced by mixing lights of different colors.

color mixture, subtractive - A single color produced by subtracting from white light a mixture of colored lights.

color of title - A person is said to have color of title if a claim to a parcel of real property is based upon some written instrument, although it may be a defective one. The title appears good but in reality may not.

color plate - A general term for the printing plate which prints any particular color. Normally, the term is modified to indicate a special color or type of plate e.g., brown plate; contour plate. SEE ALSO plate (2).

color process, additive - A method for creating all colors by adding various intensities of lights of the three primary colors (red, green, blue). The lights are produced by three projectors each provided with a filter that transmits only one of the primary colors.

color proof - (1) A multicolored print made by coating a sheet with sensitizer and making consecutive exposures through the scribed sheet. Each color requires a separate coating, which is applied over the previous color print to form the composite, multicolored proof. (2) A printed, colored picture which is made specifically to let the user check that all elements of the picture are in the correct positions and that the colors register. Color proofs come in two forms - pre-press proofs and press proofs.

color separation - (1) The process of preparing a separate drawing, engraving, or negative for each color required in printing a map. For instance, one drawing would be prepared showing only streams, lakes, and so on; another would show only contour lines; still another would show only man-made features. (2) A photographic process or electronic scanning procedure using color-filters to separate multicolored copy into separate images in each of the three primary colors.

color-separation plate - SEE plate, color-separation.

colure - An hour circle through the equinoxes or the solstices. Equivalently, a great circle, on the celestial sphere, perpendicular to the ecliptic at the equinoxes or solstices.

colure, equinoctial - The hour circle through the equinoxes.

colure, solsticial - The hour circle through the solstices.

coma - That aberration, of an optical system, which causes rays from an off-axis point in object space and passing through a given circular zone of the system, to be imaged in a circle rather than a point. Also called comatic aberration. The circles through different zones are of different sizes and are located at different distances from the optical axis. The resulting pattern, in the focal plane, looks like a falling drop of water and is called the comatic patch. The tip of the comatic patch is the intersection of the principal ray with the focal plane.

comb - A notched scale placed at right angles to the movable wire of a filar micrometer, and designed so that one turn of the micrometer screw will move the micrometer wire across one notch of the scale. The central notch of the comb, in conjunction with the zero of the micrometer head, furnishes a fiducial point from which all readings are reckoned. The comb is used for keeping count of whole turns of the micrometer screw; parts of turns are read on the graduated head of the micrometer.

combat chart - A chart showing the coastal region in which a military landing is to take place, having the characteristics of a map in the land part of the region and of a chart in the marine part, with such special characteristics as to make the chart useful in military operations. Also called a map chart.

combination plate - (1) A printing plate having both halftone and line work on its surface. (2) A printing plate having images of two or more objects on its surface.

commission, planning - SEE planning commission.

Committee Metre - The iron bar of one metre length brought to America in 1805 by Ferdinand R. Hassler, the first superintendent of the U.S. Coast Survey, for use by that survey as a standard of length. The Committee Metre was one of sixteen such bars calibrated by the Committee on Weights and Measures in Paris in 1799 against the Prototype Metre. It served as the standard of length for geodetic surveys in the USA until 1889 or 1890, when it was replaced by the National Prototype Meter. The Committee Metre was presented by Hassler to the American Philosophical Society in Philadelphia.

common establishment - SEE establishment of the port.

communications satellite - A satellite acting as a relay for electromagnetic radiation carrying signals between stations on the ground and between other satellites and the ground. Communications satellites are of two kinds: active communications satellites, which receive, amplify and re-transmit signals, and passive communications satellites, which act merely as reflectors for radiation aimed at them. Both kinds have been used as beacons for geodetic surveying.

community development, planned - A subdivision of land planned to consist primarily of a variety of residential structures but probably also of commercial and service structures and possibly some industry, the subdivision not being controlled initially, during construction, by specific restrictions imposed by zoning. Also called a planned unit-development.

community development ordinance, planned - SEE ordinance, planned-community development.

community property - Assets accumulated through joint efforts of husband and wife living together. Community property is equally owned by the husband and wife. It does not have a legal existence in some states.

comparator - (1) An instrument or apparatus for measuring a length or dimension in terms of a standard. In surveying, a comparator may be an instrument for comparing standards of length, for subdividing such standards, or for determining a standard length of a measuring device such as a bar, tape or wire. Special types of comparators are used in astronomy and in other sciences and in engineering. SEE, e.g., Väisälä comparator. (2) A precise optical instrument used to determine the coordinates of a point with respect to those of another point on a flat surface such as a photographic picture. This usage is common in photogrammetry. In other disciplines, this type of instrument is referred to as a measuring engine; the term comparator is reserved for instruments of the time defined in (1). (3) An instrument which measures a quantity by comparing that quantity with a scale. SEE ALSO blink comparator; field comparator; Väisälä comparator.

comparator, binocular - A measuring engine having a separate eyepiece for each eye but a common objective lens-system. A binocular comparator allows both eyes to see the object (e.g., a photograph) under study, but does not provide a stereoscopic view of the object.

comparator, blink - SEE blink comparator.

comparator, flick - SEE blink comparator.

comparator, longitudinal - An instrument which allows two standards of length to be compared by translating the standards parallel to each other. The standards are placed parallel to each other on an bench one part of which can move in the longitudinal direction of the standards.

comparator, monocular - A measuring engine having only a single eyepiece and objective lens-system for viewing the object (e.g., a photograph). The monocular comparator should not be confused with the monoscopic comparator (monocomparator), which may be monocular or binocular.

comparator, monoscopic - SEE monocomparator.

comparator, stereoscopic - SEE stereocomparator.

comparator, transverse - An instrument which allows two standards of length to be compared by substituting the one standard (bar) for the other in the instrument.

comparator, vertical - A device used for measuring short vertical lengths or distances, and consisting of a base, an upright, calibrated bar fixed to the base, and two horizontal microscopes that can be slid up and down the bar. It has been used in gravimetry for measuring the lengths of pendulums.

comparator base - SEE field comparator.

comparison base line - (1) The distance between two marked points, determined with highest possible accuracy, and used for calibrating baseline apparatus. (2) Two marked points, the distance between which has been so accurately determined that base-line apparatus can be calibrated by using it to measure the distance.

comparison viewer - An optical device permitting two separate pictures to be viewed separately or together in any of a variety of fashions, for comparison. The principal applications are to the comparison or maps or aerial photographs.

compass - An instrument which indicates the direction of north and, usually, other directions with respect to north. The magnetic compass indicates the direction of magnetic north; the gyroscopic compass indicates the direction of astronomic north. The solar compass and the sun compass also indicate astronomic north but can be used only during daylight and require that the Sun's position be known. SEE ALSO Anschutz compass; box compass; Brunton compass; declination (of compass); declination, magnetic; declination, vernier; declination arc; deviation of compass; Rittenhouse compass; sighting compass; sun compass; surveyor's compass; variation of compass.

compass, aperiodic - A compass whose indicator, after being deflected, returns by one direct movement, without oscillation, to its proper reading. Literally, a compass without a period. Also called a deadbeat compass.

compass, deadbeat - SEE compass, aperiodic.

compass, earth-inductor - A compass depending for its indications on the current generated in a coil rotating in the Earth's magnetic field.

compass, eccentric - A compass having the line of sight deliberately placed tangent to the rim of the compass.

compass, gyro - SEE compass, gyroscopic.

compass, gyromagnetic - (1) A compass that uses a gyroscope to smooth the indications of directions obtained from magnetically sensitive components. I.e., a magnetic compass with a gyroscope added to smooth the readings. (2) A gyroscopic compass so aligned, by applying torques derived from components detecting the Earth's magnetic field, that the zero point of the azimuth scale is kept on the magnetic meridian.

compass, gyroscopic - (1) A compass whose principal component is a gyroscope suspended so that its axis of rotation points to astronomic north. Gravitational torque and the Earth's rotation combine to make the gyroscope's axis of rotation precess and trace out an elliptical cone if motion is not constrained. A dampening force is therefore imposed on the precessing motion to make the axis move along a surface that spirals inward and ends at north. At high latitudes, gyroscopic compasses become undependable. (2) A magnetic compass fixed to a mounting stabilized by gyroscopes. This usage is misleading and undesirable.

compass, hand - SEE compass, hand-held.

compass, hand-held - A compass small enough and light enough that it can be held in the hand while in use. Also called a hand compass. SEE ALSO various types such as compass, prismatic; liquid hand compass, etc.

compass, lensatic - A compass equipped with a lens which permits the user to read, conveniently, the far side of the movable dial.

compass, liquid-damped, hand-held - A hand-held compass containing a liquid that damps the motion of the needle.

compass, magnetic - A device which indicates magnetic north by means of a magnetic needle suspended at its midpoint in such a way that the magnet aligns itself with the local magnetic field. That end of the magnet which points in the general direction of local magnetic north is marked. The magnetic compass used in surveying generally has the shape of a large, flat pillbox with a glass top. Inside the box, the bottom is marked with a graduated circle centered on the pin supporting the magnet and showing the cardinal directions and intermediate directions. The magnet itself is a long, thin needle suspended on a vertical pin rising from the center of the bottom. Some magnetic compasses have the directions marked on a disk affixed to a flattened needle. A mark is then placed on the rim of the box to indicate north.

compass, peepsight - SEE peepsight compass.

compass, prismatic - A small, hand-held magnetic compass equipped with a peep-sight and a glass prism arranged so that the magnetic bearing or azimuth of a line can be read while a sight is being taken over the line. Also called a pocket compass, although this term is also applied to small magnetic compasses without prism or lens.

compass, solar - A compass for determining astronomic North, the astronomic meridian, or bearings from astronomic North by observations on the Sun. The observer sets his latitude and the Sun's hour-angle and declination on accurately divided circles on which two sighting devices move. When one of the sighting devices is pointed at the Sun, the other sighting device points toward astronomic North. One such instrument was designed by W. A. Burt, a surveyor for the U.S. Government and was used extensively in surveying public lands of the U.S.A. It is often referred to as the Burt solar compass or Burt compass. SEE ALSO declination arc.

compass amplitude - Amplitude relative to compass east or compass west. SEE ALSO amplitude.

compass bearing - The horizontal angle from North, as indicated by a magnetic compass, east or west to the object being sighted, whichever gives the smaller angle. It differs from magnetic bearing in that the latter is corrected for deviation.

compass direction - Direction as indicated by a compass without any allowance for compass error. It may differ by a considerable amount from the astronomic or magnetic direction.

compass error - The angular difference between the direction indicated by a compass and the corresponding, true direction. Compass error includes the effect of declination of the compass.

compass face - SEE compass rose.

compass index error - The instrumental error in the magnetic bearing given by readings of the needle.

compass method (of adjusting a traverse) - SEE compass rule.

compass north - The direction indicated by the north-seeking end of the needle or other magnetic component of a magnetic compass. It differs from magnetic north, the direction of a magnetic line through the point of observation, by the amount that the supports and housing of the magnetic compass affect the needle's movement.

compass point - One of the 32 equal divisions into which the circle surrounding the needle of a compass is divided. Each division is 11.25° and may be further subdivided into quarters called quarter point.

compass rose - A circle drawn on a chart and graduated in degrees clockwise from 0° at the direction used for reference to 360° in both degrees and points or in bearings. Also called compass face. A navigator can obtain the direction of a course by transferring the line of the course to the compass rose, using an instrument like the parallel rulers for creating parallel lines.

compass rule - The rule which states that the correction to be applied to the departure (or latitude) of any course in a traverse has the same ratio to the total misclosure in departure (or latitude) as the length of the course has to the total length of the traverse. Also called Bowditch's rule. The compass rule is used when it is assumed that the misclosure results as much from errors in measured angles as from errors in measured distances. Care must be used in applying the rule because there is no general agreement on the sign to be used for the misclosure; the corrections must be applied with signs determined by the particular conventions in use. SEE ALSO Bowditch's method.

compass survey - A traverse which relies on the magnetic compass for orienting the traverse as a whole of for determining the directions of individual lines.

compensation - (geology) The hypothetical, geological process by which the Earth's crust and interior adjust themselves to stay in or close to static equilibrium in spite of the constant shift, by geological forces, of matter from one part of the surface to another. A familiar example of such shifting of matter is the erosion of mountains by water and the deposition of that matter on the bottom of lakes or seas.

compensation, forward-motion - SEE compensation, image-motion.

compensation, image-motion - Keeping an image motionless on a camera's focal plane even though the camera is moving with respect to the scene being imaged, to prevent blurring of the image during exposure. Also called forward-motion compensation. In particular, keeping the image motionless on an aerial camera's focal plane while the camera is moving forward with respect to the during exposure. In this case, true image-motion compensation must be done after the camera is oriented to the aircraft's direction of flight and the camera is fully stabilized. Strip photography is an extreme form of image-motion compensation, the shutter being open all the time. Another form of image-motion compensation occurs in the dual-rate Moon camera, in which the motion of the Moon's image during exposure is halted by a slowly rotating glass disk interposed between objective lens-system and the focal plane.

compensation, isostatic - (1) According to the theory of isostasy, the balancing of masses above and below the geoid so that vertical pressure in regions close to the surface is approximately constant regardless of the elevation or depth of the physical surface. Such regions are said to be in isostatic equilibrium. If all parts of the region are in isostatic equilibrium, compensation is said to be local; if isostatic equilibrium holds for the region as a whole but not for parts of the region, compensation is said to be regional. The terms completely compensated and undercompensated are applied to compensation in such regions. (2) The quantity: [1 - (average isostatic anomaly)]/(average Bouguer anomaly).

compensation, just (law) - Reasonable redress. The term is often included in constitutional provisions and statutes authorizing the taking of private property by the power of eminent domain. SEE ALSO domain, eminent.

compensation, regional - Isostatic compensation occurring within a large region.

compensation error - The angle from the vertical through the compensator in a self-leveling instrument to the line normal to the reflecting surface of the compensator's moving part.

compensation plate - A glass plate inserted into the optical system of a diapositive printer or a stereoscopic plotting instrument and having a surface ground to a predetermined shape to compensate for radial distortion introduced by the lens system of the camera that took the picture. Also called a compensating plate.

compensator - An optical component interposed in the line of sight or line of collimation of a leveling instrument to keep the line of sight horizontal regardless of the tile (within limits) of the rest of the instrument. It consists of at least one optical element (prism, mirror, or train of prisms or mirrors) suspended in such a way as to keep its orientation with respect to the direction of gravity fixed. It may be suspended by fine wires of invar or by flexible tapes, or it may be attached as the bob of a pendulum. Then gravity keeps the component from rotating with the rest of the instrument when the instrument is tilted slightly. The arrangement of rotating and non-rotating components is such that a horizontal ray entering the objective lens of a tilted telescope is deflected by the compensator just enough that it passes through the center of the reticle and thence through the ocular to the observer. The compensator of most self- leveling instruments can compensate for about 10' of tilt in the instrument. It replaces the sensitive spirit level formerly attached to the telescope and thereby does away with the need to compensate manually for tilt. A leveling instrument equipped with a compensator is frequently referred to as an automatic level. However, the terms compensator leveling instrument or self-leveling instrument are preferred. (2) SEE micrometer, optical.

compensator, image-motion - (1) A device installed on certain aerial cameras to keep the image stationary on the focal plane while photographing objects on the ground. (2) A device installed or used in conjunction with a camera to keep the image stationary during exposure even though the camera is moving with respect to the object being imaged.

compensator instrument - An optical instrument containing or consisting of a device (compensator) for keeping a line of sight or line of collimation level even though the optical axis of the instrument may not be level.

compensator level - SEE leveling instrument, compensator; leveling instrument, automatic.

compensator leveling - Leveling carried out with a leveling instrument containing an optical element (the compensator) which keeps the line of sight level even though the axis about which the telescope rotates is not exactly vertical. SEE ALSO leveling instrument, automatic.

compensator leveling instrument - A leveling instrument in which the line of sight is kept horizontal by a set of prisms and/or mirrors (the compensator; pendulous component) free to swing in response to gravity. Also called a self-leveling level, self-leveling leveling-instrument, self-adjusting leveling instrument, automatic level, automatic leveling instrument, etc. The range of motion of the pendulous component is quite limited, so the instrument must first be made nearly level (to within 1°) using a level (usually a circular level).

compilation - (1) The production of a new or revised map or chart, or portion thereof, from existing maps, aerial photographs, surveys, new data, and other sources. (2) The production of a map, or portion thereof, from aerial photographs and geodetic control data by means of photogrammetric instruments. It is called stereocompilation if stereoscopic plotting instruments are used. (3) Selection, assembly, and graphic presentation of all relevant information needed to prepare a map. SEE ALSO Klimsch-Variomat method of compilation; map compilation.

compilation, photogrammetric - SEE compilation.

compilation, primary - The depiction, on a specially prepared, matte, sheet of plastic, of sounding data corrected to true depths.

compilation, secondary - In bathymetry, a compilation on a specially-prepared, matte, plastic sheet, depicting uncorrected or discrete soundings.

compilation manuscript - The original drawing, or group of drawings, of a map as compiled from various data, and consisting of cartographic and related information delineated in colors on a stable material. Sometimes called a manuscript. However, that general term should not be used without adequate qualification or identification. The compilation manuscript may consist of a single drawing, called a base manuscript or because of the cluttered appearance if the entire drawing were made on a single sheet, it may consist of a basic drawing with several overlays showing, separately, vegetation, relief, names, and other information. Because a single drawing alone is usually inadequate, the base manuscript and its appropriate overlays are collectively called the compilation manuscript. Also called a base manuscript.

compilation scale - The scale at which a map is compiled, as distinct from the expressed scale of the map.

compilation star catalog - A star catalog prepared by combining stellar coordinates given in more than one independent stellar catalog. The most important compilation catalogs for geodetic purposes are (a) the General Catalog of 33342 Stars for the Epoch 1950, by Benjamin Boss and six collaborators (1937); and (b) the Smithsonian Astrophysical Observatory's Star Catalog: Position and Proper Motions of 258,997 Stars for the Epoch and Equinox of 1950 (1966). The General Catalog is a fundamental star-catalog; the Smithsonian Astrophysical Observatory's star catalog is a compilation star-catalog but is not a fundamental star-catalog. SEE ALSO Boss star catalog.

compiler - (1) A program converting a program written in a code (language) other than that directly usable by the computer into the code used by the computer (machine language). (2) One who produces maps using photogrammetry.

complement (of an angle) - SEE angle, complementary.

completion survey - (1) A survey made to finish a partially subdivided township or section, or to finish parts of boundaries of townships or sections which are unsurveyed. SEE ALSO extension survey. (2) SEE survey, field completion.

component - (1) One of the parts into which a vector may be divided and whose vectorial sum is the vector. For example, the vector representing the intensity of the Earth's magnetic field at a point can be considered the sum of the horizontal intensity and the vertical intensity, which are then the components. (2) One of the parts of a complete system. (3) SEE component, optical.

component, harmonic - In the representation of a periodic phenomenon by Fourier series, one of the terms in that series.

component, optical - One or more optical elements treated as a unit, for example, a cemented doublet or a Nicol prism.

composite - (1) A reproduction from a series of images. (2) A printer's proof made by exposing color-separation negatives one after the other in register on a single sheet of photographic paper. It is used in checking and editing. It is also called a composite print.

composite, final - A composite of the principal color-separations made after all corrections have been made.

composite photograph - SEE photograph, composite.

composite print - SEE composite.

comprehensive - A detailed sketch showing how a sheet or page will look when printed.

compression of the Earth - SEE flattening.

compression of the geoid - SEE flattening; flattening of the Earth.

computation (of geodetic coordinates), direct - Computation of the geodetic longitude and geodetic latitude of a point, from the azimuth and distance of that point from a second point whose geodetic coordinates are known. Also called direct position computation.

computation, inverse - Computation of the length, forward azimuth, and backward azimuth of a geodesic between two points whose geodetic longitude and geodetic latitude are known. Also called an inverse position computation.

computation, inverse-position - SEE computation, inverse.

computation of inverse - SEE computation, inverse.

computed-data method - A method of rectification by first computing the amount of tilt present in an aerial photograph and then, using this result to compute the settings needed on an autofocus rectifier. The photograph is then rectified without further comparison with a templet or other guide.

computer - A machine for automatically carrying out a predetermined sequence of mathematical or logical operations. Also called a computer. The terms computing machine and calculating machine may be considered synonyms. In modern usage, however, the terms are often used with different, if undefined, meanings. (a) A calculating machine can do only simple arithmetic operations like addition, subtraction, multiplication, division and so on. A computing machine can do complicated arithmetic, algebraic and logical operations. (b) A calculating machine must have the data to be used and the operations to be performed entered by the user in the sequence as the calculation progresses. A computing machine can have the data and instructions inserted separately; the machine will then, upon command, do the indicated operations on the supplied data without further intervention by the user. (c) A calculating machine is a small machine which does arithmetical or logical operations. A computing machine is a large machine which does mathematical or logical operations.

computer, digital - A computing machine that represents the numbers involved by sequences of discrete states within the machine and manipulates these sequences by discontinuous operations. Also called a digital computer. Most computing machines of this type represent numbers as specific sequences of positive and negative voltages or as well-separated levels of voltage. They operate on these sequences by means of switching circuits that can combine sequences to form the sequences representing sums, products, etc. An analytical plotter contains a digital computing machine as an essential component.

computing machine, analog - A computing machine in which quantities are represented by continuous mechanical, electrical, or optical equivalents and mathematical operations are done by analogous mechanical, electrical or optical processes. Also called an analog computer. Among the well-known analog type are the gun-control mechanisms (based on cams and linkages) used during World War II. Many early machines for predicting tides were analog computing machines. Stereoscopic plotters which connect the photographic image to the plotter through mechanical or optical systems are analog computing machines.

concentric-circle theory - SEE theory, concentric-circle.

condemnation - (1) The process by which property is acquired, through legal proceedings under the power of eminent domain, for constructing highways. (2) The act of a federal, State, county or city government or district or public-utility corporation vested with the right of eminent domain to take private property for public use when a public necessity exists. It is the act of a sovereign in substituting itself in the place of the owner and/or the act of taking all or parts of the rights of the owner. (3) The acquisition of property by the exercise of the right or power of eminent domain. Pursuant to the right or power, the sovereign, whether it is the federal or State government, or an agency to which there has been delegated this right or power, may, upon payment of just compensation, acquire property for the benefit of the public. SEE ALSO condemnation, excess.

condemnation, excess - The policy, on the part of the condemner, of taking by right of eminent domain, more property than is actually necessary for the public improvement. The same rights may be exercised by procedure or by direct purchase. SEE ALSO condemnation.

condemnation, inverse - The legal process by which a property owner may claim and receive compensation for the taking of, or payment for damages to, his property as a result of work on a highway.

condensation - The theoretical replacement of matter between geoid and Earth's surface by an equivalent layer of mass on the geoid. If the average density of matter between geoid and Earth's surface is D and the elevation is H, then the average amount m of mass per unit area on the geoid is m = DH.

condensation gravity anomaly - The quantity gc obtained by subtracting from a gravity anomaly Δg the following gravity corrections: (a) the condensation gravity correction δg_c; (b) the free-air gravity correction δg_f; and (c) a Bouguer gravity correction δg_B, a modified Bouguer gravity correction δg_{Bm} or an expanded Bouguer gravity correction δg_{Be}. Applying the condensation gravity correction is the mathematical analog of replacing all the mass between the reference surface and an equipotential surface through the point P at which gravity was measured by a thin layer of mass condensed onto the reference surface (or at a predetermined distance below it). Applying the free-air gravity correction then changes gravity from its value on the reference surface to the corresponding value at elevation Hp, and applying the Bouguer gravity correction replaces the missing mass by the Bouguer plate or some modification thereof. If the distance (called the depth of condensation) of the layer of condensed matter below the reference surface is zero, i.e., if the mass is applied as a coating to the reference surface, the gravity anomaly is sometimes referred to as Faye's gravity anomaly, because Faye proposed putting the layer on the reference surface. However, because the term Faye's gravity anomaly has also been used for other kinds of gravity anomaly, the term is best not used at all. Note that to some extent, depending on how the mass to be condensed is estimated and on what kind of Bouguer correction is applied, the condensation gravity is a variant of the Bouguer gravity anomaly with topographic correction.

condensation gravity correction - (1) The change (δg_c) in the calculated value of gravity, on the reference surface, when matter in the form of a thin disk, radius r at depth D below that surface is removed. The matter in the disk is used, after applying a free-air gravity correction, to fill the Bouguer cylinder. The mass M in the Bouguer cylinder is given by $M = \rho\pi r^2 h$, in which h is the height of the cylinder, r the radius and ρ the density of the matter in it. The change δg_c (in newtons per kilogram or meters per second^2) is $\delta g_c = 2GM\{[(h+D)/\sqrt{r^2+(h+D)^2}] - 1\}/r^2$, in which G is the gravitational constant. (2) The same as the preceding definition but with elevation H being used instead of height h, the geoid being used as the reference surface. (3) The same as either of the preceding definitions, but with the condensed mass lying in a thin layer on a surface parallel to the reference surface and equal to the topographic masses outside that surface.

condensation gravity reduction - (1) The process of applying, to a measured or calculate value of gravity, the sum δ_c of the following gravity corrections in the order listed: (a) the condensation gravity correction δg_c; (b) the free-air gravity correction δg_f; and (c) the cylindrical Bouguer gravity correction δg_{Bc}. The advantage of the condensation gravity reduction over, e.g., the Bouguer gravity reduction, is that the former changes the location of the geoid (or other reference surface) by less than 3 meters, whereas the Bouguer gravity reduction may change that surface's location by hundreds of meters. (2) The same as the previous definition but with a topographic gravity correction added and with the condensation gravity correction modified to allow for the effect of topography. (3) The quantity δ_c defined in the first definition. (4) The sum of the gravity corrections applied according to the second definition. The condensation gravity reduction is also called Faye's gravity reduction and the Helmert gravity reduction.

condenser - A lens system which concentrates the light from a source onto a limited area. An ellipsoidal reflector having the source at one of its foci concentrates the light at the other focus and is optically equivalent to a condenser.

condition - A limitation on the values or range of values that can be taken by a set of variables or parameters. Also called constraint. The limitation may be expressed as values between which the set or a function of the set is allowed to vary, or may be given as a value that the set or a function of the set must have. E.g., the coordinates of a control point are often subject to the condition that they have certain defined or previously determined values. The angles measured within a small triangle may be required to add up to 180° or, if the triangle is large, to 180° plus the spherical excess.

condition, holonomic - A condition equation which can be solved for one of the variables in terms of the others, or which can be integrated to allow such a solution. Also called a holonomic constraint.

condition, non-holonomic - An equation expressing a relationship between (a condition on) the differentials of the independent variables and which is not integrable. Also called a non-holonomic constraint.

condition equation - (1) An equation in which the unknowns may take on those, and only those, values which allow the equality to hold exactly, regardless of the results of measurements. Also called conditioned equation, conditional equation and constraint equation. For example, in measuring the angles of a spherical triangle, no relationship exists between the measured angles until all three angles have been measured. The condition that the three measured angles (plus certain corrections) must be equal to 180° plus the spherical excess of the triangle is then expressed as a condition equation. The various condition equations used in geodesy are defined under the terms that indicate the kind of quantity involved, e.g., angle equation, azimuth equation, latitude equation, length equation, longitude equation, and side equation. (2) An equation expressing a condition that must be satisfied by the unknowns, given certain measured values. This type of condition equation is commonly formed by combining a condition equation (definition 1) with an observation equation. Equations of this kind are generally solved using correlate equations. Do not confuse condition equation with equation of condition. The latter term was formerly used for what is now called an observation equation.

conditions, standard - SEE standard temperature and pressure.

condominium - (1) A form of ownership less than the whole. (2) The fee ownership of separate portions of multi-storied buildings by statute which provides the mechanism and facilities for formal filing and recordation of a divided interest in real property, where the division is vertical as well as horizontal. (3) A building held in condominium.

cone - That part of an aerial camera which includes the lens, the between-the-lens shutter, the diaphragm, and the metallic section corresponding to the bellows in other cameras.

cone-angle banding - SEE banding, cone-angle.

confidence interval - The upper and lower limits of the values, of a random variable, within which there is a pre-assigned probability (confidence) that a measurement or observation will be found. If the value lies outside those limits, there is only a small probability that the measurement occurred by chance.

configuration, critical - An arrangement of points such that the set of equations for the location of an unknown point in terms of the distances and angles between known points and the unknown point does not have a unique solution. For example, the location of an unknown point in a given plane can be determined, in general, by measuring the angles from that point to three points of known location in that plane. A critical configuration is present if all four points lie on the same circle, or if the three known points lie on a straight line. Many critical configurations are possible in the arrangement of stations observing an artificial satellite.

configuration, planetary - The apparent locations of the planets relative to each other and to other bodies of the Solar system, as seen from the Earth.

configuration of terrain - SEE expression, topographic.

confluence - A junction or place where streams meet.

confluence zone, coastal - A region lying between the shoreline and either (a) a line 50 nautical miles from shore or (b) the contour at a depth of 100 feet.

conformality - That property of a transformation from one surface to another surface which ensures that the angle between any two curves on one surface is reproduced in magnitude and sense between the corresponding curves on the other surface. This property is equivalent to the condition that the coefficients of the first fundamental form on one surface be proportional to the corresponding coefficients of the form for the other surface, both surfaces having the same coordinate system. Hence, isometric mappings are also conformal.

conjunction - (1) The configuration taken by the Sun, the Earth, a second planet when the apparent geocentric (celestial) longitude of Sun and second planet are the same. (2) The phenomenon in which two bodies have the same apparent celestial longitude or right ascension as viewed from a third body. Conjunctions are, however, usually tabulated as geocentric phenomena.

conjunction, geocentric inferior - The conjunction of a planet whose orbit lies within the Earth's, when the planet is between the Earth and the Sun. Usually referred to simply as inferior conjunction.

conjunction, geocentric superior - The conjunction of a planet whose orbit lies within the Earth's, when the planet is on the opposite side of the Sun from the Earth. Usually referred to simply as superior conjunction.

conjunction, inferior - SEE conjunction, geocentric inferior.

conjunction, superior - SEE conjunction, geocentric superior.

connection (geodesy) - The joining of two separate triangulation networks into a single network without internal discrepancies.

connection line - A surveying line connecting a monument with a permanent reference-mark.

Conrad discontinuity - A region, at depths of 17 to 20 km under some parts of the continents, in which the speed of seismic P-waves increases abruptly from about 6.2 km/s to about 6.6 km/s. Unlike the deeper, Mohorovicic discon-tinuity, the Conrad discontinuity seems to be absent over large regions. It may mark the contact of granitic and basaltic layers.

conservation of energy - SEE law of the conservation of energy.

consideration, just - A price, right, or benefit recognized by the law as sufficient to form a binding contract.

Consol - A hyperbolic radio-navigation system with a baseline only 5-6 km long, operating at a frequency of about 300 kHz. It uses three radiators to generate two daisy-shaped patterns which are slowly rotated and also switched alternately. The signal consists of coded dots and dashes. Because the baseline is so short, Consol should not be used at distances of less than about 50 km from the baseline. The usable range is about 900 to 1800 km during daytime; it is more at night. During the day, the error in direction is about $0.6°$ at an angle of $60°$ to the baseline. The error in location is a few kilometers, depending on distance from stations, diurnal effect, propagative constants, etc. The system was originally developed under the name Sonne and is still in limited used, mainly by shipping (in 1977 there were 13 stations in operation).

constant - (1) A quantity which, in an equation or function, has a fixed value. It is distinguished from a parameter in that the latter is itself considered a variable quantity which is held fixed while the independent variables progress through their allowed ranges. A constant is thought of merely has having a fixed value; the effect of changes in the constant on the equation or function is not considered. SEE ALSO constant, absolute. (2) SEE constant, physical. SEE ALSO stadia constant.

constant, absolute - A single, unchanging number. Usually referred to simply as a constant.

constant, anallactic - SEE stadia constant.

constant, arbitrary - SEE parameter.

constant, astronomic - SEE constant, astronomical.

constant, astronomical - One of a set of constants which are used in the calculation of ephemerides and which are astronomical in nature. Also called an astronomic constant. The set itself is sometimes referred to as a system of astronomical constants. That term, however, should be used only for a set which is necessary and sufficient. Not all astronomical constants are geodetically important. Those of particular importance, with the values adopted by the International Astronomical Union in 1976, are given in the Appendix.

constant, essential - A constant (1) which cannot be eliminated from a function without changing the set of values taken by the function, regardless of what changes are made in the other constants in the function. In the equation $y = x+b$, b is an essential constant. However, in the equation $y = (ax+b)/(cx+d)$, b is not an essential constant, as can be seen by dividing numerator and denominator by b.

constant, fundamental - SEE constant, fundamental physical.

constant, fundamental physical - (1) One of a set of quantities which, together with a theory of the properties of matter and radiation, are sufficient to predict these properties, both microscopic and macroscopic. Also referred to as a fundamental constant. Customarily included as fundamental physical constants are the gravitational constant G, the speed c of light, the charge e on the electron and the Boltzmann constant k. It is still not known whether or not the fundamental physical constants are really constant or they change with time. (2) SEE constant, physical.

constant, geocentric - SEE constant, geocentric gravitational.

constant, geocentric gravitational - The product of the gravitational constant G and the mass M of the Earth. Also called the geocentric constant and gravitational parameter. Its value is approximately $3.986\ 005 \times 10^{14}\ m^3/s^2$.

constant, geodetic - (1) A number having geodetic meaning and used frequently in geodetic calculations. Examples are the numbers defining a reference ellipsoid, the length of one degree of arc at the equator of a rotational ellipsoid, the rate of rotation of the Earth, the coordinates of the origin of a geodetic network, the force of gravity at Potsdam and the stadia constant. Sets of such numbers can be found in textbooks and professional journals. (2) One of a set of numbers adopted by the International Union of Geodesy and Geophysics (IUGG), on the recommendation of the International Association of Geodesy, and recommended for general use in geodetic calculations. The general use of such a set makes it easy to compare the results of various geodetic organizations and for one organization to use the results of another. If the numbers are changed frequently, they lose their usefulness. The IUGG therefore, except in recent years, has opposed frequent changes. Geodetic constants as of 1986 are given in the Appendix.

constant, gravitational - The constant of proportionality G or k appearing in formula for the force of attraction between two bodies. If the bodies can be considered points with masses M_1 and M_2, the formula is $f = G\ M_1\ M_2\ (r_2 - r_1)/r_3$, in which r_1 and r_2 are the vectors from the origin to masses M_1 and M_2, respectively, and r is the distance between the two masses. The value of the numerical constant depends upon the units in which the force f, the vectors r_1 and r_2 and the distance r, and the masses M_1 and M_2 are expressed. The value adopted for G by the International Astronomical Union in 1976 is $6.672 \times 10^{-11}\ m^3/(kg\ s^2)$. Recent measurements of the variation of gravity with depth in the Earth, and data from nuclear experiments, suggest that a fifth force, similar to gravitation, may exist or that the inverse square law for the variation of gravitation with distance is not quite correct. The uncertainties in the data, however, are such that no definite conclusions are possible at present. The symbol k^2 is reserved for the gravitational constant when the masses are expressed in units of the mass M_s of the Sun, distances are expressed in terms of the astronomical unit (A.U.), and time is expressed in ephemeris days (E.D.; 86 400 ephemeris seconds). The quantity k, also called the Gaussian constant or Gaussian gravitational constant, has the defined value $0.017\ 202\ 098\ 950\ (A.U.)3/[M_s\ (E.D.)^2]$. This value then specifies the astronomical unit through Kepler's third law $k^2\ M_s\ (1 + M_1/M_s) = (2\pi/T_o)^2\ a_3$, in which T_o is the period of an orbiting point-mass of mass M_1 and a is the radius of the circular orbit. Note that the astronomical unit is not the same as the average semi-major axis of the Earth's orbit. The difference arose because Gauss, in calculating k, used a value for the mass of the Earth about 7% too small. If k is calculated instead of being defined, it can be determined to many more significant figures than can G because, with the units adopted, k is equal to the average angular speed of the Earth in its orbit, divided by (1 + mass of Earth in units of solar mass). The mass of the Earth in these units is a very small quantity and is known to better than 6 significant figures; the average angular speed is immediately calculable from the period of revolution of the Earth, which is known to better than 10 significant figures.

constant, harmonic - The amplitude or epoch of a harmonic constituent of the tide or tidal current at a place.

constant, harmonic tidal - A tidal constant which is the amplitude or epoch of a tidal constituent.

constant, leveling-rod - On a leveling rod having two scales in the same units, the difference in readings between marks at the same height.

constant, non-harmonic tidal - A tidal constant other than a harmonic tidal constant, e.g., the range of an interval derived directly from measurements of high water and low water.

constant, physical - Any number which is accepted by physicists as being characteristic of a measurable or calculable quantity; a constant quantity or value having physical significance. Also referred to as a fundamental constant, although this is a misnomer. Those physical constants occurring frequently in the equations of physics and from which all other physical constants can be derived are called fundamental physical constants.

constant, solar - The amount of energy received, on a planet or satellite, per unit area per unit time from the Sun, at the average distance of that planet or satellite from the Sun. In particular, the amount of energy received at the Earth per unit area per unit time from the Sun, at the average distance of the Earth from the Sun (i.e., at the distance of one astronomical unit). A well-determined average value (1982) for the Earth's solar constant is 1368 watts/meter² ± 0.2%. The actual value of the energy varies by about 7% between aphelion and perihelion.

constant, tidal - Any characteristic of the tide which remains practically constant at a particular locality. Tidal constants are classified as harmonic or non-harmonic.

constant of aberration - The largest amount (annual aberration), in theory, by which a star's apparent direction varies during the year solely because of the Earth's revolution about the Sun. It is commonly denoted by κ and is given by the formula $\kappa = 2\pi\ a/[cT\ (1 - e^2)^{1/2}]$, in which a is the length of the semi-major axis of the Earth's orbit, e is the eccentricity of that orbit, T is the length of the sidereal year and c is the speed of light in space. For the epoch 2000 A.D., k has the value 20.495 52 (I.A.U. 1977).

constant of diurnal aberration - The largest amount, theoretically, by which a star's apparent direction varies, as seen from a fixed latitude, during one day solely because of the Earth's rotation. It is commonly denoted by k and is given by the formula $k = (2\pi R\ \cos\ \phi')/cT$, in which R is the radius of the Earth at the observer's geocentric latitude ϕ', T is the length of the sidereal day and c is the speed of light.

constant of general precession - The value of the constant term in the formula for general precession as a function of time.

constant of gravitation - SEE constant, gravitational. Either term is acceptable and both are in common use, with no distinction of meaning. Geophysicists seem to prefer gravitational constant.

constant of nutation - The amplitude of that term, in the equation for nutation in obliquity, which has the period of the longitude of the Moon's ascending node. Alternatively, the coefficient of $\cos\Omega$ in the equation $\Delta\varepsilon = a \cos \Omega + b \cos 2\Omega + c \cos 2L'_s + d \cos 2L_m$, in which $\Delta\varepsilon$ is the nutation in obliquity of the ecliptic, Ω is the longitude of the Moon's ascending node, L_s is the mean longitude of the Sun and L_m is the longitude of the Moon. The value adopted by the International Astronomical Union in 1976 is 9.2109" at epoch 2000.

constant of precession - (1) The quantity $[A+B\mu/(1+\mu)]/H$, in which μ is the ratio of the mass of the Moon to the mass of the Earth, H is the dynamical ellipticity of the Earth, and A and B are functions of the orbital elements of the Earth and of the ratio of the mass of the Sun to the combined masses of Earth and Moon. (2) The rate of motion of the general precession in longitude. Use of this meaning is discouraged.

constant of the cone - In a conical map projection of normal aspect, the ratio of the angle $\Delta\theta$ between meridional lines on the graticule to the angle $\Delta\lambda$ between corresponding meridians on the rotational ellipsoid.

constant-pressure chart - SEE chart, isobaric.

constellation - (1) A precisely defined region of the celestial sphere, associated with a stellar grouping, which the International Astronomical Union has designated as a constellation. (2) A stellar group, usually with pictorial or mythical associations, which serves to identify a region of the celestial sphere. (3) A set of satellites placed in simultaneous orbits to form a particular pattern.

constituent, astronomical tidal - SEE constituent, tidal.

constituent, diurnal - Any constituent of the tide whose period approximates that of a lunar day (24.84 solar hours).

constituent, long-period - (1) A tidal constituent having a period longer than one day. (2) A tidal or tidal-current constituent having a period independent of the rotation of the Earth but dependent on the orbital movement of the Moon or of the Earth. The principal lunar long-period constituents have periods approximating the month and half-month, and the principal solar long-period constituents have periods approximating the year and half-year.

constituent, lunar diurnal - A tidal constituent which represents the diurnal effect of the Moon's declination on the tides. Its period is 25.82 hours and its angular speed is 13.9430 degrees per hour. The Doodson number is 145.555.

constituent, luni-solar diurnal - A tidal constituent which, together with the lunar diurnal constituent, expresses the effect of the Moon's declination on the tides. The two constituents together account for the diurnal inequality and, at extremes, diurnal tides. With constituent P1, it expresses the effect of the Sun's declination. Its period is 23.92 hours, its angular speed is 15.0411 degrees per hour. Its symbol is K1; the Doodson number is 165.555.

constituent, luni-solar semidiurnal - A tidal constituent which modulates the amplitude and frequency of the principal lunar and principal solar diurnal constituents for the declinational effects of the Moon and Sun, respectively. It has a period of 11.97 hours and an angular speed of 30.0821 degrees per hour. Its symbol is K2; the Doodson number is 275.555.

constituent, meteorological - SEE tide, meteorological.

constituent, principal lunar semidiurnal - The tidal constituent which represents the effect of the rotation of the Earth (with respect to the Moon) on the tides. Its period is 12.42 hours and its angular speed is 28.9841 degrees per hour. Its symbol is M2; the Doodson number is 255.555.

constituent, principal solar semidiurnal - The tidal constituent which represents the effect of the rotation of the Earth (with respect to the Sun) on the tides. Its period is 12.00 hours and its angular speed is 30.0000 degrees per hour. Its symbol is S2; the Doodson number is 273.555.

constituent, semidiurnal - A tidal constituent that has two maxima and two minima each constituent day.

constituent, species of - SEE species (of constituent).

constituent, tidal - (1) A term in the Fourier-series representation of the variation of tidal height with time at a particular tidal station. (2) One of the terms in a Fourier-series representation of the tide-producing force and its corresponding formula for the tide or tidal current. A single constituent is usually written in the form $y = A \cos (at + \alpha)$, in which y is a function of the time t and is reckoned from a specified origin. The coefficient A is called the amplitude of the constituent and is a measure of the relative importance of the constituent to the amplitude of the tide as a whole. The angle $(at + \alpha)$ changes uniformly and its value at any time is called the phase of the constituent. The speed of the constituent is the rate of change of its phase and is represented by the symbol a. The period of the constituent is the time required for the phase to change through 360° and is the cycle of the astronomical condition represented by the constituent. Also called constituent and astronomical tidal constituent. SEE ALSO tide.

constituent, tidal potential - A term in a Fourier-series representation of the variation of the tide-producing potential.

constituent day - A period of rotation of the Earth with respect to a fictitious star and representing one of the periodic elements in the tidal forces. It approximates in length the lunar or solar day and corresponds to the period of a diurnal constituent or twice the period of a semidiurnal constituent. The term is not applicable to the long-period constituents.

constituent hour - One twenty-fourth part of a constituent day.

constraint - SEE condition; condition equation.

constraint, holonomic - SEE condition, holonomic.

constraint, inner-adjustment - A condition placed on the constants defining a coordinate system and resulting in the smallest sum of the elements along the main diagonal in the inverse matrix of a set of normal equations.

constraint, non-holonomic - SEE condition, non-holonomic.

constraint equation - SEE condition equation.

construction leveling instrument - A particularly sturdy leveling- instrument sufficiently accurate for leveling over short distances such as those involved in construction of minor roads, buildings, etc.

construction survey - A survey made, prior to and while construction is in process, to control elevation, horizontal location and dimensions, and configuration; to determine if the construction was adequately completed; and to obtain dimensions essential for calculating quantities used in paying for construction.

construction surveyor - A surveyor employed in construction. Also called a layout engineer.

construction theodolite - A theodolite sufficiently accurate and sturdy that it can be used in construction projects. The accuracy of a construction theodolite is usually sufficient that the instrument can also be used for third-order surveying.

contact base line apparatus - A base-line apparatus composed of bars whose lengths were defined by the distances between their end faces or points. In use, the bars were laid end to end. One bar was kept in position while another bar was moved ahead.

contact correction - A quantity added to the time of the chronographic signal of a stellar transit observed using an electrically recording, impersonal micrometer, to allow for the time required for the contact spring to cross half the width of a contact strip in the head of the micrometer. In order to ensure a satisfactory signal, the contact strips are given an appreciable width. As the micrometer wire travels from different sides of the instrument at upper and lower culminations, and also before and after reversal of the instrument, the contact spring produces a signal sometimes from one edge of a contact strip and sometimes from the other. The contact correction is intended to produce the time that would have been recorded if the electrical signal had occurred only when the contact spring crossed the middle of the contact strip.

contact diapositive printer - SEE contact printer (2).

contact exposure - The process of copying an image by placing the copying emulsion directly in contact with the material carrying the image. If the image itself is on film, two processes are recognized: emulsion-to-base exposure in which the base of the copying material is in contact with emulsion holding the image to be copied; the emulsion-to-emulsion exposure, in which the emulsion of the copying material is in contact with the emulsion holding the image.

contact glass - SEE plate, focal-plane.

contact mark - A permanent mark, at a tide gage, from which the level of the water inside the float well can be measured directly. A contact mark is used in checking the accuracy of the elevations displayed on the recorder and is located, preferably, on the frame supporting the recorder. The zero mark of the permanent tide-scale and/or the fiducial mark on a sounding probe are established by reference to this mark. The elevation assigned to a particular contact mark is referred to as tide-gage datum and may be adjusted according to the results of periodic leveling from the tide-gage bench mark.

contact plate - SEE plate, focal-plane.

contact print - A photographic copy of a transparent photographic negative or positive, made by placing a light-sensitive emulsion on an opaque base (the backing) in contact with the negative or positive and exposing directly to light.

contact printer - (1) A photographic apparatus for making prints by contact, and comprising a lamp(s) and a mechanism for holding the photographic negative and the sensitized material together during exposure. (2) An apparatus designed specifically for producing copies of diapositives at the same scale as that of the original photographic negative. Also called a contact diapositive printer.

contact printing - The process of making contact prints by placing the original and a sensitized sheet in contact.

contact-printing frame - SEE frame, contact-printing.

contact screen - (1) A transparent sheet carrying a pattern of fine, gray or magenta, vignetted dots of different densities. It is used to produce a half-tone negative by contact-printing. (2) A half-tone screen made on a film and used in direct contact with another film to make a half-tone image from a continuous-tone original. (3) A patterned image on a film used in contact with another film or photographic plate to obtain a patterned image from an open-window negative.

contact screen, magenta - A contact screen composed of dots of magenta hue and variable density. It is used for making halftone photographic-negatives.

contact vacuum printing frame - SEE frame, contact-printing.

contact vernier - A vernier lying in physical contact, or nearly so, with the primary scale. This is the type usually meant by vernier.

conterminous (adj.) - Having a boundary in common. Also called coterminous. The term is particularly useful when referring to that part of the United States of America before the admission of Alaska and Hawaii to statehood. For example, the northernmost point in the conterminous United States is in Minnesota near the Lake of the Woods.

continent - (1) A large mass of land rising abruptly from the bottom of the oceanic deep and including marginal regions that are shallowly submerged. (2) One of the land masses identified as continents and named Africa, Antarctica, Asia, Australia, Europe, North America and South America. Greenland is sometimes included. There is disagreement over the validity of classifying Europe and Asia as separate continents if a geological basis is used for the classification. Modern evidence indicates, however, that at one time the two continents were up to 3000 km apart.

continuation - The extension of a curve beyond its end point and in the same direction. If the curve is circular, the continuation has the same curvature. Continuation is the extension of a curve; prolongation is the extension of a straight line.

continue (v.) - To extend a curve beyond its end point, in the same direction.

continuous (adj.)(law) - Close; adjoining or abutting; near; coterminous.

contour - An imaginary line on the ground, all points of which are at the same elevation above or below a specified surface of reference. The definition is illustrated by the shore line of an imaginary body of water whose surface is at the elevation represented by the contour. A contour forming a closed loop round lower ground is called a depression contour. Contour should not be confused with contour line; the latter is the term for a line drawn on a map. However, the distinction is ignored by some writers. SEE ALSO depression contour.

contour, accurate - A contour line whose error lies within one-half the basic vertical interval between contour lines. Also called normal contour line and normal contour.

contour, approximate - SEE contour line, approximate.

contour, bathymetric - A line joining points, on the bed of the sea or other body of water, situated at equal vertical distances beneath the surface. Also known as a depth contour or isobath. It is different from a submerged contour. Although the definition merely specifies surface of the water, it can be inferred that an average surface such as mean sea level or the geoid is meant.

contour, carrying - SEE contour line, carrying.

contour, geoidal - A line, on the surface of the geoid, which is at a constant height above the rotational ellipsoid used as reference. Geoidal contours represent differences in height between the geoid and the ellipsoid of reference. They depend not only on the shape of the geoid but also on the dimensions and location of the ellipsoid to which the heights are referred. The same geoid referred to different ellipsoids will give different sets of geoidal contours. Although the definition specifies a rotational ellipsoid (spheroid), it will remain valid if the word rotational is dropped.

contour, illuminated - SEE contouring, illuminated; contour line, illuminated; relief, illuminated.

contour, inclined - A line representing a constant perpendicular distance in a constant direction above or below a plane representing the geoid. SEE ALSO orthogonal method (of depicting relief).

contour, intermediate - (1) A contour lying between two other contours of known elevation. (2) SEE contour line, intermediate.

contour, normal - SEE contour, accurate.

contour, sea-level - (1) A contour at mean sea level. (2) SEE contour line, sea-level.

contour, shadowed - SEE contour line, shadowed.

contour, submerged - A line joining points of equal elevation on the bed of a lake or reservoir where the elevation is related to a datum used for mapping adjacent land. This is not the same as a bathymetric contour.

contour, supplementary - SEE contour line, supplementary.

contour accuracy - SEE contour-line accuracy; contour-line error.

contour finder - A stereoscope of simple design for tracing contour lines from stereoscopic pairs of photographs. It does not compensate for changes of scale resulting from differences of elevation or from tilt in the optical axis of the camera.

contouring, illuminated - A method of showing relief, in which contour lines appear lighter when indicating illuminated slopes and darker when indicating shadowed slopes. It has also been called illuminated contour, although this usage was probably not intentional.

contouring, logical - A method of drawing contour lines - spacing them at intervals proportional to the differences of elevation between points at which the slope changes. The method is based on the fact that contours are equally spaced along a uniform slope. It permits contour lines of adequate accuracy to be drawn from carefully located elevations and makes it unnecessary to run a level line for every contour.

contour interval - (1) The difference of elevation represented by adjacent contour lines. (2) The difference of elevation between two adjacent contours.

contour interval, variable - A non-uniform contour interval. It may result because cartographic sources were used which contained different contour intervals, or it may result because the contour interval was adapted to best show the differences of elevation in different types of terrain.

contour line - A line, on a map, representing equal elevation. Frequently but incorrectly abbreviated to contour or used to mean contour. SEE ALSO index contour line.

contour line, approximate - A contour line substituted for the usual contour line when that line may not be reliable. Reliable here means accurate to within half the contour interval. Also called, erroneously, an approximate contour.

contour line, carrying - A single contour line representing two or more contours. It is used to show vertical or nearly vertical, topographic features such as cliffs, cuts and fills. Also called, incorrectly, a carrying contour.

contour line, illuminated - A contour line which is made lighter to indicate illuminated slopes and darker to indicate shadowed slopes.

contour line, index - SEE index contour line.

contour line, intermediate - A contour line drawn between those contour lines which are labeled with the elevation. Depending on the contour interval, there are three or four intermediate contour lines between the labeled contour lines. Sometimes incorrectly called an intermediate contour.

contour line, normal - SEE contour, accurate.

contour line, sea-level - A contour line representing points at mean sea-level. Also called a sea-level contour. Although the term says sea-level, mean sea-level or its equivalent is meant.

contour line, shadowed - A contour line whose thickness varies to give the effect of a shadow thrown by oblique illumination on a plate having the thickness of the contour interval. Incorrectly called a shadowed contour.

contour line accuracy - A measure of the range of differences, on a topographic map, between the elevations indicated by the contour lines and the elevations of the corresponding contours. SEE ALSO contour line error; Koppe's formula.

contour line error - The difference between the actual location of a contour line and the correct location of a contour line having the same assigned value of elevation. The relationship between the standard deviation σ_1 of the error in the contour line and the standard deviation σ_h of the error in elevation of the contour is $\sigma_1 = \sigma_h \cot \beta$, where ß is the angle of slope of the ground. SEE ALSO contour line accuracy.

contour map - (1) A topographic map portraying relief by means of contour lines. (2) A map showing the elevations and configuration of the ground by contour lines and lacking any other details except notations and the elevations of contours.

contour point - A point, on the ground, which lies on a specified contour.

contour sketching - Sketching relief onto a map as seen in perspective but controlled by locations on the map that correspond to salient points on the ground.

contour value - A number placed on a contour line denoting the elevation of the contour relative to a given datum - usually the geoid.

contract - (1) A legally enforceable agreement between two or more competent persons to do or refrain from doing something. (2) The same as the preceding definition, except that the agreement is for a consideration.

contract for deed - A contract to sell real estate, payments being made in installments; upon payment of the last installment, the deed is delivered to the purchaser.

contra solem (adj.) - SEE rotation, counter-clockwise.

contrast - (1) The actual difference in photographic density between the highlights and the shadows on a negative or positive. Contrast indicates not the magnitude of photographic densities but the differences of such densities. (2) The rating of a photographic material, corresponding to the relative difference in density which it is capable of reproducing.

contrast enhancement - Exaggeration of the ratios between densities of elements of an image.

contrast index - A number expressing the average gradient of a graph of photographic density versus exposure in terms of the slope of a straight line joining the two points, on the characteristic curve, which represent the minimal and maximal densities used in practice. For materials of high contrast, the contrast index is almost the same as gamma.

contrast transfer curve - A curve showing the relative contrast in the photographic image of a square-wave or progressive square-wave pattern as a function of the nominal spatial frequency of the pattern. I.e., a curve showing the ratio of the contrast in the image to the contrast in the pattern used as object.

contrast transfer factor - The ratio of the modulation in the image of a sinusoidal pattern at any spatial frequency to the ratio at zero frequency.

contrast transfer function - A function giving the variation of contrast transfer factor with frequency.

control - (1) Anything which is assumed to be correct and to which other similar things are compared or on which such other things are made to depend. (2) The set of previously fixed points (the framework) used to establish the position, scale, and orientation of the detail in a map. (3) SEE control, geodetic. (4) SEE control station. (5) SEE control point. In general, the exact meaning of the term must be determined from the context. SEE ALSO ground control; Laplace control; level control; Shoran control.

control, airborne-electronic-survey - Control put in by using airborne electronic-surveying equipment such as Hiran or Shoran.

control, astronomic - A network consisting of control stations whose locations have been determined by astronomic measurements. The astronomic longitudes and latitudes will normally differ from the geodetic longitudes and latitudes of the same stations by amounts corresponding to components of the deflections of the vertical.

control, basic - The data associated with a set of control stations, giving the locations of these stations and used as the basis for detailed surveys. Basic control is not changed by the detailed surveys nor by their subsequent adjustment. It may contain horizontal coordinates, vertical coordinates, or both. The basic control for the Topographic Map of the United States consists of the data from first-order and second-order triangulation and traverse, and from first-order and second-order leveling.

control, cadastral (USPLS) - (1) Control established specifically for use in a cadastral survey. (2) Lines established, and marked on the ground by survey monu-ments that are used as starting and closing points in surveys of the public domain of the United States of America. The fundamental cadastral control of the public-land surveys of the USA consists of base lines, standard parallels (correction lines), principal meridians and guide meridians.

control, electronic - Control established using electronic devices. (Jargon)

control, geodetic - (1) The geometric data associated with a collection of control stations, such as coordinates, distances, angles, and/or directions between control stations. It is practically equivalent, in this sense, to basic control. The data of geodetic control consist, first, of the distances, directions, and angles between control stations. These are converted to geodetic coordinates and azimuths. These in turn may be converted into other kinds of coordinates such as plane coordinates in a State Plane Coordinate System. This is the form in which the data are used for local surveys in the USA. (2) A set of control stations. (3) A set of control stations and the geometric data associated with them. It is equivalent, in this sense, to control network. (4) A set of control stations established by geodetic methods, i.e., methods in which the Earth's curvature has been taken into account. (5) The same as (4), but with the proviso that the geoid have been considered in calculating coordinates. I.e., the height of the geoid above the ellipsoid has been considered in calculating heights. (6) The coordinates of a control station. (7) A control station and its coordinates. (8) SEE control station. (9) SEE ground control. The exact meaning of the term geodetic control must be determined from the context.

control, horizontal - (1) Control points which have accurately known horizontal coordinates (longitude and latitude), which can be identified with physical points on the ground, and which can be used to provide horizontal coordinates for other surveys. (2) A control station or stations whose horizontal coordinates have been determined. (3) The geometric data relating to the horizontal coordinates of a control station. For definitions of specific categories of horizontal control, see control, photogrammetric; traverse; triangulation; trilateration. The main purpose of lower-order horizontal control is providing a transition from major geodetic networks to the control required for large-scale mapping or for land surveying and engineering.

control, minor - SEE control, photogrammetric.

control, photogrammetric - (1) Control (geodetic control in particular) established to provide scale and position for a set of points whose coordinates are to be determined photogrammetrically (i.e., for a photogrammetric network). Also called minor control. Locations for photogrammetric control are marked on the ground before the photographs are taken so that they can be identified in the photographs. Also called ground control. (2) Geodetic control established by photogrammetric methods. (3) Control established by photogrammetric methods. (4) Any point whose coordinates are known and whose image can be positively identified.

control, selenodetic - The set of points, on the Moon's surface, to which coordinates have been assigned either by definition or be determined by measurement, and to which the locations of other points are referred.

control, starting - Photogrammetric control available for absolute orientation of the first pair of photographs along a line of flight covering ground for which control is to be extended.

control, supplemental - Control established by surveys depending on already established geodetic control to relate aerial photographs used in mapping to the system of geodetic control. Also called supplementary control.

control, supplementary - SEE control, supplemental.

control, vertical - (1) Control points which have accurately known elevations, which can be identified with physical points on the ground, and which can be used to provide elevations for other surveys. Elevations are referred, by definition, to the geoid. However, horizontal surfaces through selected points at mean sea level and project datums have been used as references, as have non-horizontal surfaces defined by a combination of leveling surveys and points at mean sea level. (2) The elevations, or approximations thereto, associated with control points. SEE ALSO datum, vertical-control; sea level, mean.

control base - A surface upon which ground control and the graticule of a map are plotted and upon which templets have been assembled or aerotriangulation has been done and the control points thus determined have been marked.

control classification - A system by which control is labelled according to certain criteria that place control of like kinds under the same labels. The sets defined by the labeling are called categories. E.g., the kinds of control used in surveying are commonly placed in categories such as horizontal control, vertical control, photogrammetric control and gravity control. A separate classification places control in categories according to some arbitrary criterion for the accuracy or precision of that control; labels such as first-order, second-order class I and third-order. The USA has used various control classifications according as the number of different kinds of control or the attainable precision has increased, or as the organization of the surveying organization has changed. The most recent control-classification was promulgated in 1984.

control data card - SEE card, control-data.

control densification - The addition of control throughout a region in which control has already been established. Unless otherwise specified, geodetic control may be assumed to be meant.

control document - Any legal document that affects the ownership or use of public lands or resources.

control document index - SEE index, control-document.

control extension - SEE control densification.

control flight - SEE control strip.

controlling-point method - A method of determining details of the terrain, in a topographic survey, by locating those points (controlling points) of the terrain at which there are significant changes in the contours or direction of the contours. The relative elevations and planimetric coordinates of these points are then determined by survey, and the contours drawn by logical contouring. The controlling point method is extremely efficient, since it requires surveying to fewer points than does any other method. However, considerable experience is required of the surveyor to identify the controlling points.

control net - SEE control network.

control network - (1) A survey network in which the stations are control stations. (2) Geodetic control together with the measured and/or adjusted values of distances, angles, directions and/or heights used in determining the coordinates of the control.

control photography - Photography dealing with the taking of aerial photographs for locating and extending control.

control point - (1) A point to which coordinates have been assigned, these coordinates then being used in other, dependent surveys. The term is sometimes used as a synonym for control station. However, a control point need not be perpetuated by a marker on the ground. (2) A point which is identifiable on a photograph, to which coordinates have been assigned, and which is used in determining the scale and absolute orientation of the photograph. The term is usually modified to indicate the type or purpose of the point, as, for example, ground control point, photocontrol point and vertical control point.

control point, horizontal - A point whose horizontal coordinates, or whose distance and direction to other points of like kind, are known accurately enough that they can be assumed known in further surveys.

control point, photo - SEE photocontrol point.

control point, photogrammetric - A point of horizontal control that has been established by photogrammetric triangulation.

control point, secondary - A photographically identifiable point whose location has been determined photogrammetrically or by use of an accurate positioning, locating, or navigation system. Locating systems such as HIRAN and SHIRAN have been used for this purpose.

control point, supplemental - A point of supplemental control.

control point, vertical - (1) A point whose elevation or geodetic height is known accurately enough that it can be assumed known in further surveys. (2) SEE bench mark.

control station - (1) A point, on the ground, whose location (horizontal or vertical coordinates) is used as a basis for obtaining locations of other points. (2) A survey station whose coordinates are accepted as being sufficiently accurate that the coordinates of other survey stations can be determined with reference to it. In other words, a survey station used for control. (3) A point on the ground whose location (horizontal or vertical or both) is used as a basis for a dependent survey, and from which observations were made.

control station, horizontal - SEE station, horizontal-control.

control strip - (1) A strip of aerial photographs taken to aid in planning and doing aerial photography later. (2) A strip of aerial photographs taken to serve as control in assembling other strips. Also called control flight, tie flight and tie strip.

control survey - A survey which provides coordinates (horizontal or vertical) of points to which supplementary surveys are adjusted. The fundamental control survey of the United States of America provides the geodetic coordinates and plane coordinates of thousands of triangulation and traverse stations, and the elevations of thousands of bench marks. These are used in hydrographic surveys of the coastal waters, for control of the national topographic survey, and for control of surveys by State, city and private organizations.

control survey, airborne - SEE Airborne Control Survey.

control survey, first order - SEE survey, first-order.

control survey, second order - SEE survey, second-order.

control survey classification - A system by which control surveys are labelled to indicate their approximate precision or accuracy. The highest prescribed order of control survey, in the USA, is designated first order; the next lower is designated second order; the lowest third order. The second order and third order categories are further subdivided into classes I and II, with class II being in each case the category containing the less precise surveys. The term fourth order control survey is also used but its definition is not as precise as the definitions of the higher order control surveys. In general, a fourth order control survey is one which fails to meet the criteria necessary for inclusion in one of the higher orders.

control survey net, national - SEE survey network.

control traverse - A survey traverse made to establish control.

Conventional International Origin - The origin of a coordinate system and defined by the following original latitudes of five observatories (degrees and minutes are the same for all observatories):

Misuzawa	N 39° 08' 3.602"
Kitab	1.850"
Carloforte	8.941"
Gaithersburg	13.202"
Ukiah	12.096"

The Conventional International Origin (CIO) is approximately the average location of the point at which the Earth's axis of rotation intersected the Earth's surface during the years 1900-1905. It is often used as the origin for the coordinates of the instantaneous pole of rotation. In 1967, the International Union of Geodesy and Geophysics recommended that the CIO be used in defining the direction of a geodetic north pole. The acronym OIC is used in French publications.

convergence - (1) SEE convergence, meridional. (2) SEE convergence, oceanic. SEE ALSO projection convergence.

convergence, linear - The amount, in linear units, by which two meridians approach each other when extended from one parallel of latitude to another. (16) Also called lineal convergency and linear convergency.

convergence, meridional - The difference between the two corresponding angles formed by the intersection of a geodesic with two different meridians. Sometimes abbreviated to convergence or convergency. It may also be defined as the difference in forward and back azimuths at the two ends of a geodesic. At the equator, all meridians are parallel - they make the same angle, 90°, with the equator. Passing north or south from the equator, two meridians converge until they meet at the poles, intersecting at angles equal to their differences of longitude. If the surface involved is spherical, the quantity is called spherical convergence; if it is a rotational ellipsoid (spheroid), it is called spheroidal convergence.

convergence, oceanic - A region in which two currents converge.

convergence, spherical - The difference between corresponding angles made by a great circle drawn on a sphere and intersecting two meridians.

convergence, spheroidal - The difference between corresponding angles made by a specified kind of curve, on a rotational ellipsoid (spheroid), with two meridians on that ellipsoid. The curve is usually specified to be a geodesic. The resulting convergence has been termed the spheroidal geodesic convergence.

convergence angle - (1) The difference in direction between a rhumb line drawn between two points and the initial or final direction of a great circle between those two points. (2) SEE correction, arc-to-chord.

convergence of the meridians - (1) The angular drawing together or convergence of the meridians in passing from the equator to the poles. (2) SEE convergence, meridional.

convergency - SEE convergence, meridional.

convergency, lineal - SEE convergence, linear.

convergency, linear - SEE convergence, linear.

convergent position - SEE position, convergent.

conversion angle - (1) The difference in direction between a rhumb line between two points and the initial or final direction of a great circle between those two points. (2) SEE correction, arc-to-chord.

conversion factor - The number by which the numerical value or a quantity in one system of units must be multiplied to arrive at the numerical value in another system of units. E.g., The conversion factor for converting measurements in feet to the equivalent values in yards is 1/3. These measurements are in different units but are in the same system. The conversion factor for converting from feet to meters is 12/39.37; these units are in different systems.

conversion scale - A scale indicating the relationship between two different units of measurement. E.g., meters to feet.

convey (v.) - (1) Pass or transmit the title to property from one to another. (2) Transfer property or the title to property by deed or instrument under seal. The term applies properly to the disposition of real property, not of personal property. To convey real estate, by an appropriate instrument, is to transfer the legal title to it from the present owner to another.

conveyance - A written instrument which passes an interest in real property from one person to another. It may be a deed, mortgage or lease, but not a will. SEE ALSO conveyance by division line; conveyance by exception; strip conveyance.

conveyance, mesne - SEE mesne conveyance.

conveyance by division line - A conveyance that separates the land to be conveyed from that not to be conveyed by specifying a line or natural feature separating the two.

conveyance by exception - A conveyance that expressly states what land is not to be conveyed by the instrument.

Conybeare leveling rod - A self-reading leveling rod on which graduations are marked by alternately black and white, tenth of foot divisions. Each division consists of three hexagons in either black or white. The rod is no longer used.

coordinate - (1) One of an ordered set of N numbers (q_1, q_2, q_3, ... q_N) which designates the location of a point in a space of N-dimensions. In geodesy, N is usually 2, with the space a plane, sphere or rotational ellipsoid, or 3, with the space an ordinary three-dimensional Euclidean space. Time may be considered a fourth coordinate but there is no particular advantage, in geodesy, to doing so except in those few cases where relativistic effects are important. The single letter x is commonly used to designate a coordinate in 1-dimensional space. The letters x and y are commonly used for coordinates in 2-dimensional space (i.e., on surfaces) although a subscripted letter (e.g., x_1, x_2) is also common. In 3-dimensional space, the letters x, y, z, χ, η, ζ, and x_1, x_2, x_3 are commonly used for the three coordinates. The special theory of relativity commonly uses x, y, z or a subscripted letter for the space-like coordinates and t or x_4 (= it) for the time-like coordinate. A space of 6 dimensions, the so-called phase space is frequently used in dynamics. Three of the coordinates are the ordinary space-like ones, usually denoted by x,y,z as usual. The other three coordinates are the velocities or the momenta in the directions of the x-axis, y-axis, and z-axis. Coordinates are always associated with coordinate systems. The coordinate system provides the means of finding the point designated by the set of coordinates or a means of assigning coordinates to a point. The coordinates are said to be in the associated coordinate system. The point designated by the coordinates may also be said to be in that coordinate system. But the point and its location do not depend in any way on the coordinate system, while the coordinates of the point do. (2) More generally, one of a set of N numbers designating a point in a space of M dimensions, where M may be less than N. In this case, N-M coordinates can be assigned arbitrary values unless conditions are imposed on the coordinates. SEE ALSO Cassini coordinate; chromaticity coordinate; Gauss-Krüger coordinate; grid coordinate; Lamé coordinate; machine coordinate; Mercator coordinate; model coordinate; plane rectangular coordinate; Plücker coordinate; Riemann homogeneous coordinate; Soldner coordinate; space coordinate; space polar coordinate, space rectangular coordinate; State coordinate; State plane coordinate, strip coordinate.

coordinate, angular - An angle used as a coordinate, i.e., an angle which, alone or together with other coordinates, specifies the location of or direction to a point. Astronomers use two angular coordinates for specifying the directions to stars and similar celestial bodies. These coordinates are called the position of the body; the third coordinate, distance, is usually absent.

coordinate, assumed plane - A coordinate in an assumed plane coordinate system.

coordinate, astronomic - (1) One of a pair of numbers indicating the direction of the zenith at a point on the Earth's surface, the direction being given (a) with respect to a selected celestial meridian to give an angle called the astronomic longitude and (b) with respect to the celestial equator to give an angle called the astronomic latitude. Also called astronomical coordinate, geographic coordinate, gravimetric coordinate and terrestrial coordinate. Astronomic longitude and astronomic latitude are denoted, usually, either by λ_a and ϕ_a or by λ and Φ. To within the accuracy of present-day instruments, the direction of the zenith is unique at every point on the Earth's surface. Each point on the surface therefore has a unique pair of coordinates. (2) For a point on the Earth's surface, one of a pair of numbers indicating the direction of the zenith as determined at that point on the geoid which is vertically below the point on the surface; the coordinate system, terminology, and symbolism are otherwise the same as for the preceding definition. Also called an astronomical coordinate. Some organizations define astronomic coordinates according to the second definition but put in their records the coordinates according to the first definition. The user of such data should ascertain which definition actually applies. (3) A coordinate in a coordinate system defined wholly or in part by reference to celestial bodies, e.g., by right ascension and declination.

coordinate, astronomical - SEE coordinate, astronomic (1).

coordinate, barycentric - A coordinate of a point in a barycentric coordinate system.

coordinate, bipolar - (1) One of two coordinates (r_1, r_2) of a point in the plane, indicating the distances of that point from two given, fixed points in the plane. The coordinates do not give a unique location for the point. To each point with a given pair of bipolar coordinates, there corresponds a second point on the other side of the line joining the two fixed points and an equal distance from that line. Hence a convention is needed to distinguish between the two locations. Bipolar coordinates are often used in electrostatic theory. They are also used in defining Nicolosi's map projection. (2) One of three coordinates (β_1, β_2, β_3) related to rectangular, Cartesian coordinates (x,y,z) through the equations $x = a \sinh \beta_2 / (\cosh \beta_2 - \cos \beta_1)$, $y = a \sin \beta_1 / (\cosh \beta_2 - \cos \beta_1)$ and $z = \beta_3$, where a is a constant (length).

coordinate, Cartesian - SEE Cartesian coordinate.

coordinate, celestial - A coordinate in any set used to specify the location or direction of a point on a celestial sphere.

coordinate, circular cylindrical - SEE coordinate, cylindrical.

coordinate, curvilinear - A coordinate in a curvilinear coordinate system. Symbolism for curvilinear coordinates generally varies with the particular system in question. E.g., (x, y, z) is commonly used for Cartesian coordinates, (r, θ, ϕ) for spherical coordinates, etc. If curvilinear coordinates in general are being used, common symbols are (q_1, q_2, q_3) or, on a surface, (u, v).

coordinate, cylindrical - One of the three coordinates of a point in a cylindrical coordinate-system.

coordinate, ellipsoidal - (1) One of a set of three coordinates of a point in an ellipsoidal coordinate system. Also called a Lame coordinate. Algebraically, an ellipsoidal coordinate is defined as one of a set of three numbers p_i (i = 1 to 3) satisfying simultaneously the three equations $x^2/(a^2 + p_i^2) + y^2/(b^2 + p_i^2) + z^2/(c^2 + p_i^2) = 1$ (i = 1,2,3), where x, y, z are rectangular Cartesian coordinates and a, b, c are constants. (2) SEE coordinate, spheroidal (2). (3) SEE coordinate, geodetic.

coordinate, elliptic - (1) One of that set of coordinates which results when one of the semi-axes of an ellipsoidal coordinate system is infinite. (2) SEE coordinate, spheroidal.

coordinate, elliptical - One of the two angles, celestial longitude and celestial latitude, in an ecliptic coordinate system.

coordinate, equatorial - One of two angles, right ascension (α) and declination (δ), in an equatorial coordinate system.

coordinate, equatorial spherical - (1) One of three coordinates specifying: the distance of a point from the origin; the angle, at the origin, from a given plane through the origin to the point; and the angle, from a second plane through the origin and perpendicular to the first, to the line from origin to point. The same as the first definition except that the complement of the angle from a given plane to the point is used instead of the angle itself.

coordinate, galactic - A coordinate in a galactic coordinate system. Galactic coordinates are usually designated as galactic longitude, L; galactic latitude, b; and galacto-centric or heliocentric distance, r. Because the constants defining the system have been changed several times, numeric subscripts are usually attached to the letters to identify the particular version with which the symbols are to be used.

coordinate, Gaussian - SEE Gaussian coordinate.

coordinate, generalized - One of a set of coordinates just sufficient for a unique characterization of the position of a mechanical system. The coordinates need no all be of one kind. One may be a distance, another an angle, a third a potential, etc.

coordinate, generalized spherical - One of a set of N coordinates, in N dimensional space, consisting of a distance r, and azimuth θ, and N-2 latitude-like angles $\phi_1, \phi_2, \phi_3, \ldots \phi_{N-2}$. If the rectangular, Cartesian coordinates of a point are $x_1, x_2, x_3, \ldots x_N$, then the corresponding generalized spherical coordinates are given by

$$x_1 = r \cos \phi_1$$
$$x_2 = r \sin \phi_1 \cos \phi_2$$
$$x_3 = r \sin \phi_1 \sin \phi_2 \cos \phi_3$$
$$\cdots$$
$$x_{N-1} = r \sin \phi_1 \sin \phi_2 \sin \phi_3 \ldots \sin \phi_{N-2} \cos \theta$$
$$x_N = r \sin \phi_1 \sin \phi_2 \sin \phi_3 \ldots \sin \phi_{N-2} \sin \theta$$
$$r = \sqrt{(x_1^2 + x_2^2 + x_3^2 + \ldots + x_N^2)}.$$

coordinate, geocentric - (1) One of a set of three coordinates, of a point, in a geocentric coordinate system. The coordinates are commonly represented by (λ, ϕ', r) where λ, the longitude, is the angle from a reference plane through the polar axis to the plane containing the polar axis and the point; ϕ', the geocentric latitude, is the angle from the equatorial plane to the radius vector to the point; and r, the geocentric distance or geocentric radius, is the distance of the point from the center. The letter ψ is frequently used instead of ϕ'. Note that the origin (intersection of polar axis and equatorial plane, or center of reference ellipsoid if one is used) need not be at the Earth's center of mass or geometric center. The complete set of three geocentric coordinates is also called a geocentric position. (2) One of a set of coordinates designating the location of a point by means of (a) the angle from the plane of the celestial equator to a line from the center of the Earth to the point and (b) the angle from the plane of a selected, initial geodetic meridian to that line. The first angle is called the geocentric latitude. The term geocentric longitude is not used for the second angle because that angle is the same as the geodetic longitude. This definition properly applies only to the exceptional case where the plane of the equator passes through the center of the Earth and the polar axis of the reference ellipsoid coincides with the Earth's polar axis. (3) The longitude and latitude of a point (on the Earth's surface) relative to the center of the Earth. The Earth's center of mass is usually meant.

coordinate, geodesic - (1) One of a set of coordinates in the neighborhood of a point at which the gradient of the metric tensor is zero. (2) One of a pair of parameters (u,v), on a surface, such that the curves v = constant are a singly-infinite family of geodesics on that surface and the curves u = constant are the family of curves orthogonal to the geodesics.

coordinate, geodetic - One of a set of three coordinates designating the location of a point with respect to an ellipsoid of reference, the plane of ellipsoid's equator, and the plane of a selected meridian on that ellipsoid. Called, by one author, an ellipsoidal coordinate; called, by another author, a geographical coordinate. One coordinate, the geodetic height, is the distance from the ellipsoid to the point in question; it is usually denoted by h. Another, called the geodetic longitude and usually denoted by λ, is the angle from a specified plane through the polar axis to a plane containing the polar axis and the point. The third, called the geodetic latitude and usually denoted by ϕ, is the angle from the equatorial plane of the ellipsoid to the line through the point and perpendicular to the ellipsoid. SEE ALSO coordinate system, geodetic. The geodetic coordinates (λ, ϕ, h), when defined using a rotational ellipsoid, are related to the coordinates (x,y,z) of a rectangular Cartesian coordinate system whose axes coincide with those of the ellipsoid by the equations $x = (N + h) \cos\phi \cos\lambda$, $y = (N + h) \cos\phi \sin\lambda$ and $z = (N + h) \sin\phi - e^2 N \sin\phi$. N is the length of that part of the perpendicular (the normal) to the ellipsoid from the point lying between the surface and the minor axis: $N = a / \sqrt{(1 - e^2 \sin^2\phi)}$. e is the eccentricity of the ellipsoid and a half the length of the major axis. The equations must be modified if a general ellipsoid is involved.

coordinate, geographic - (1) A general term for either a geodetic or an astronomic coordinate. Also called a terrestrial coordinate. The pair of geographic coordinates is also called a geographic location or, less fitly, a geographic position. (2) One of a pair of coordinates which specify the angle between a specified line and a meridional plane and the angle between that line and an equatorial plane.

coordinate, geographical - (1) Longitude and latitude on the sphere, as contrasted to rectangular coordinates (on the plane). (2) SEE coordinate, geodetic.

coordinate, geomagnetic - A coordinate in a geomagnetic coordinate system.

coordinate, gravimetric - SEE coordinate, astronomic.

coordinate, heliocentric - A coordinate in a coordinate system having its origin at the Sun's center.

coordinate, homogeneous - In three-dimensional space, one of a set of four coordinates w_1, w_2, w_3, w_4 assigned to each point in such a way that the points in Cartesian coordinates x,y,z are given by $x = w_1/w_4$; $y = w_2/w_4$; $z = w_3/w_4$. Homogeneous coordinates have the desirable feature that all algebraic equations in homogeneous coordinates are homogeneous. This avoids the necessity of considering the special cases that arise with inhomogeneous equations and can make programming easier.

coordinate, horizontal - (1) One of a pair of coordinates in a two-dimensional coordinate system lying on a level surface such as the geoid. (2) One of a pair of coordinates in a coordinate system lying on an ellipsoid taken to represent the Earth's surface. (3) One of a pair of coordinates in a horizontal plane. (4) One of the two coordinates azimuth and zenith distance (or angular elevation) in a horizontal coordinate system (astronomy).

coordinate, isometric - The quantity $\ln \{[\tan ¼(\pi-2\phi)][(1 + e \sin\phi)/ (1 - e \sin\phi)]e/2\}$ in which ϕ is the geodetic latitude and e is the eccentricity of the rotational ellipsoid. This quantity is also called the isometric latitude.

coordinate, military - SEE coordinate system, military.

coordinate, oblate spheroidal - A coordinate in a spheroidal coordinate system which is based on oblate ellipsoids of rotation. Oblate spheroidal coordinates $(r, \psi, \theta, \lambda)$ are related to rectangular Cartesian coordinates (x, y, z) in a coordinate system having the same origin and axes by the equations: $x = r \cosh \psi \cos \theta \cos \lambda$, $y = r \cosh \psi \cos \theta \sin \lambda$ and $z = r \sinh \psi \sin \theta$. SEE ALSO coordinate, spheroidal.

coordinate, oblique - One of a set of coordinates in an oblique Cartesian coordinate system.

coordinate, photo - SEE photograph coordinate.

coordinate, photographic - One of the coordinates of a point in a coordinate system associated with a photograph. The system may be intrinsic in the photograph, as in the case of a numbered reseau imprinted on the photograph, or may be extraneous to the photograph, as in the case of the coordinate system of a measuring device used in measuring a particular photograph. SEE ALSO photograph coordinate.

coordinate, plane rectangular - SEE plane rectangular coordinate.

coordinate, polar - One of the two quantities: distance r from a central point of reference or direction θ from a specified line or plane of reference. Polar coordinates are related to plane rectangular coordinates x, y by $x = r \cos \theta$ and $y = r \sin \theta$. The point of reference is called the pole, center or origin. The line joining the center to the point whose coordinates are wanted is called the radius vector. The angle from the fixed line or plane of reference and the radius vector is the vectorial angle, central angle or polar angle. In surveying, observations are usually put in the form of polar coordinates as a first step in computing coordinates in another coordinate system. For example, geodetic coordinates (longitudes and latitudes) are derived from measurements of distances and directions.

coordinate, rectangular - A coordinate in any coordinate system whose axes intersect at right angles. Also called a rectilinear coordinate.

coordinate, rectangular Cartesian - SEE Cartesian coordinate system, rectangular.

coordinate, rectangular spherical - One of a pair of coordinates in a rectangular spherical coordinate system. One of the coordinates of a point is the distance from the origin along one of the axes to the intersection with the great circle through the point and perpendicular to the axis. The other coordinate is the distance along that great circle from the intersection to the point. SEE ALSO Soldner coordinate.

coordinate, rectilinear - SEE coordinate, rectangular.

coordinate, reduced astronomic - SEE coordinate system, astronomic (3).

coordinate, selenocentric - A coordinate related to the center of the Moon; i.e., a coordinate in a selenocentric coordinate system.

coordinate, space polar - SEE space polar coordinate.

coordinate, spatial - One of the three coordinates of a point in a three-dimensional coordinate system.

coordinate, spherical - (1) One of a set of three coordinates in a spherical coordinate system. (2) A coordinate in a coordinate system on a sphere. Usually, two coordinates, in angular or linear units, fix the location of a point with respect to two great circles that intersect at right angles. However, the coordinates may be the great-circle distances from two points on the sphere or may be the distance along one great circle and distance from another. The term is sometimes used to designate a coordinate on any surface approximately spherical.

coordinate, spheroidal - (1) One of a set of three coordinates (u, ψ, λ) of a point in a spheroidal coordinate system of given distance 2k between foci of the rotational ellipsoids of the system. Spheroidal coordinates are related to the rectangular Cartesian coordinates (x, y, z) in a system whose axes coincide with those of the spheroidal coordinate system, by the equations $x = [\sqrt{(u^2 +k^2)}] \cos \psi \cos \lambda$, $y = [\sqrt{(u^2 +k^2)}] \cos \psi \sin \lambda$ and $z = u \sin \psi$, in which u is the length of the semi-major axis of the rotational ellipsoid through the point, ψ is the reduced latitude of the point with respect to the same ellipsoid and λ is the angle from a reference plane through the minor axis to the plane through the minor axis and the point. Called an elliptic coordinate or ellipsoidal coordinate by some geodesists. Other version of the above set of coordinates and of the transformation equations are in use. E.g., the form $x = r\cosh \Gamma \cos \psi \cos\lambda$, $y = r\cosh \Gamma \cos \psi \sin\lambda$ and $z = r\sinh \Gamma \sin \psi$ has been used in satellite geodesy. (2) A coordinate in a coordinate system on a rotational ellipsoid (spheroid). (3)

A coordinate in a coordinate on a surface which is approximately spherical.

coordinate, standard - One of a pair of coordinates in a standard coordinate system.

coordinate, terrestrial - SEE coordinate, astronomic; coordinate, geographic.

coordinate, topocentric - A coordinate in a topocentric coordinate system. Unless otherwise stated, astronomic coordinates of celestial bodies are topocentric coordinates.

coordinate, topocentric equatorial - One of the three coordinates in a topocentric equatorial coordinate system.

coordinate, Universal Polar Stereographic - One of the two coordinates of a point in the Universal Polar Stereographic Grid System. The particular grid must be specified. That designation may be considered an additional coordinate.

coordinate, Universal Transverse Mercator - One of the two coordinates in a Universal Transverse Mercator Grid System.

coordinate, vertical - The vertical distance (i.e., distance measured along a vertical) of a point above or below a surface of reference (datum). The vertical coordinate of a point may be positive or negative, depending on whether the point is above or below the surface of reference. This surface may be assigned a large positive value so that all vertical coordinates referred to it will be positive. The terms elevation and height are often used for vertical coordinate. The term elevations is best reserved, however, for vertical coordinates with respect to the geoid. The term height is best reserved for distance measured along a perpendicular, e.g., geodetic height.

coordinate cadaster - A cadaster specifying the boundaries of a region by the coordinates of points on these boundaries.

Coordinated Universal Time - (1) The time kept by a clock whose rate was controlled by atomic clocks so as to be as uniform as possible for one year but the rate, chosen by the Bureau International de l'Heure, could be changed at the beginning of a calendar year. Also called Universal Time Coordinated which shortens to UTC. This time was used from 1960 through 1971 for broadcasting time-signals. (2) The time kept by a clock controlled by atomic clocks and running at the correct rate (zero offset in frequency) but changed by the occasional addition or deletion of 1 second (the leap second) to keep the time within 0.90s of Universal Time 1. This time was adopted by the International Radio Consultive Committee in Geneva in February 1971 and became effective 1 January 1972.

coordinate protractor - A square protractor having graduations on two adjacent edges, with the center at one corner, and ruled with a square grid having intervals of one centimeter or other linear unit. It has a movable arm turning with the center as pivot and graduated in the same units and at the same scale as the ruled grid. In using the protractor, the arm is set to a given angle, azimuth or bearing, and the length of a line is marked on the arm; inspection of the grid at this point will give the components of that length along the axes marked by the edges of the protractor.

coordinate system - A set of points, lines, and/or surfaces (the geometrical elements) and a set of rules, whereby each point in a given space can be identified uniquely by a set of numbers. The numbers identifying a point are called the coordinates of that point. There are two different but equivalent ways of assigning coordinates to points. The first gives rules for measuring distances and/or angles from each geometrical element to the point in question. These measured quantities are the coordinates of the point, and there must be at least as many coordinates as there are dimensions of the space. (There may be more, as with homogeneous coordinates or Plücker coordinates.) If the system contains a point from which distances are measured, that point is called a center. A line parallel to which or perpendicular to which distances are measured is called an axis, as is a line with respect to which angles are measured. A plane parallel to which or from which distances are measured, or from which angles are measured, is called a reference plane or plane of reference. The point in which two or more axes intersect is called the origin. The second way of assigning coordinates is to define families of lines and/or surfaces is such a way that through each point of the space passes one and only one line or surface from each family. Each line or surface in a family is assigned a unique number; that number is the value of the parameter generating the family and is called a coordinate. The coordinates of a point are then the coordinates of all the lines and surfaces passing through that point. It is possible that the system defined by either way is defective it that certain points in the space may have more than one set of coordinates. E.g., in a spherical coordinate system, points on the polar axis have unique values of latitude and of radial distance, but not unique values of longitude. In geodesy, only coordinate systems of 2, 3, 4 and 6 dimensions are important, and of these only the 2-dimensional and 3-dimensional systems are common. The principal 2-dimensional coordinate systems are the horizontal coordinate systems such as the State plane coordinate systems, the astronomic coordinate system and, in a sense, the system of geodetic longitudes and latitudes. The principal three-dimensional coordinate systems are the geodetic coordinate system and those extensions of the horizontal and astronomic coordinate systems which are obtained by adding vertical axes. Three-dimensional, rectangular, Cartesian coordinate systems are used primarily for computing, the values obtained being converted to one of the other systems for final use. The coordinate system in which elevations or vertical coordinates are given could be considered an 1-dimensional system. However, most elevations have associated horizontal coordinates. The principal 2-dimensional coordinate systems used in astronomy are the equatorial, the ecliptic, and the galactic coordinate systems. These are converted to 3-dimensional coordinate systems by adding radial distance as another coordinate. Coordinate systems may be classified according to (a) where the origin is located (topocentric, geocentric, heliocentric, etc.); (b) the kinds of surfaces used for reference (plane, spherical, spheroidal, etc.); or (c) the orientation of the axes (horizontal, equatorial, etc.). A coordinate system is sometimes called, by physicists and engineers, a "reference frame". However, that term has several other meanings besides that of coordinate system and is best not used with that meaning. SEE ALSO Cassini-Soldner coordinate system; GEOREF coordinate system; grid coordinate system; Luneburg bipolar coordinate system; Marussi's coordinate system; origin (of coordinate system); plane coordinate system; Soldner coordinate system; space coordinate system; State coordinate system; State plane coordinate system; Universal Space Rectangular Coordinate System.

coordinate system, absolute - A coordinate system associated with a physical system having its origin fixed in the Earth and its axes non-rotating with respect to the distant galaxies. Also called an absolute reference frame.

coordinate system, assumed plane - SEE plane coordinate system, assumed.

coordinate system, astro-geodetic - A coordinate system which has its origin at a point with known geodetic coordinates and its axes oriented with respect to astronomic north and the astronomic meridian of Greenwich. This is equivalent to defining the system by giving the geodetic and astronomic coordinates of a point in the system.

coordinate system, astronomic - (1) A coordinate system consisting of the celestial equator and of a selected celestial meridian; or, equivalently, of one plane perpendicular to the Earth's axis of rotation and another plane parallel to that axis and passing through the vertical at a specified point. The coordinates of a point on the Earth's surface are determined by the meridional arc between the equator and the point, and by the equatorial arc between the selected meridian and the meridian through the point. (2) A coordinate system the same as that of the preceding definition except that the coordinates of a point on the surface are those of that point which is vertically above or below it on the geoid. (3) A coordinate system the same as that in (2) except that an average position of the Earth's axis of rotation is used. This coordinate system is sometimes referred to as a reduced astronomic coordinate system; the coordinates are then referred to as reduced astronomic coordinates. (4) A coordinate system in which the locations of celestial bodies (Sun, planets, stars, etc.) are given. An astronomic coordinate system is also called an astronomical coordinate system.

coordinate system, astronomical - SEE coordinate system, astronomic.

coordinate system, azimuthal - SEE coordinate system, horizontal.

coordinate system, baricentric - SEE coordinate system, barycentric.

coordinate system, barycentric - A coordinate system associated with a set of massive points, particles, or bodies, with the origin of the system at the barycenter (center of mass) of the set. Also spelled baricentric coordinate system. SEE ALSO barycenter.

coordinate system, bipolar - (1) A coordinate system, in the plane, consisting of two points a specified distance apart, and the rule that the coordinates of a point are its distances from the two given points; a plus or minus sign indicates on which side of the line joining the two given points the point lies. Alternatively, the system consists of two families of circles concentric about two given points (centers) in the plane. The coordinates of the point are then the values of the two parameters defining the two families of circles. Usually the distance between the two centers is constant. However, some map projections (such as the van der Grinten) are best described in a bipolar coordinate system with variable distance between centers. (2) SEE Luneburg bipolar coordinate system.

coordinate system, (B,L) - A coordinate system devised for locating a satellite's location in (B,L) space, rather than in a spherical coordinate system; B is the local value of the total magnetic field, L is calculated through the use of some relationships among the adiabatic invariants valid in a static dipole field.

coordinate system, Cartesian - SEE Cartesian coordinate system.

coordinate system, curvilinear - Any coordinate system in which at least one of the geometric elements (lines or surfaces) used for reference is curved. Coordinates are determined by the intersections of curved lines or surfaces rather than by intersections of straight lines or planes only. In geodesy, the most important curvilinear coordinate systems are the astronomic, ellipsoidal, geocentric, geodetic, spherical and spheroidal.

coordinate system, cylindrical - A coordinate system in which the location of a point is determined by three coordinates: (a) its perpendicular distance from a given line; (b) its distance from a selected plane perpendicular to this line; and (c) its angular distance from a selected line when projected onto this plane. Also called a circular cylindrical coordinate system and a cylindrical polar coordinate system.

coordinate system, Earth-fixed - A coordinate system the axes of which are fixed with respect to the Earth. Because the Earth not rigid but (a) is imperfectly elastic and (b) moving irregularly in all its parts, the above definition is inadequate for specifying a unique, time-invariant system. A valid definition would be, for example, a coordinate system the axes of which are parallel to those of another coordinate system defined as follows: (a) a rectangular Cartesian coordinate system in three-space, with the three axes fixed with respect to each other, is defined; (b) two of the axes lie in a plane passing through three given points in the Earth (one of these points is usually the Earth's center of mass), and one of these axes passes through two of the points; (c) the direction of the third axis with respect to the plane is specified; (d) a (possibly curvilinear), time invariant coordinate system is specified by a transformation from the Cartesian coordinate system to this second one; and (e) the Earth-fixed coordinate system is then defined by parallelism with respect to the second coordinate system. However, the fact that all parts of the Earth are probably in motion with respect to each other makes it unlikely that one Earth-fixed coordinate system, however precisely defined, will serve all needs. Passing through two given points in the Earth (one of these points is usually taken as the Earth's center of mass).

coordinate system, ecliptic - (1) A right-handed, rectangular Cartesian coordinate system which has the z-axis perpendicular to the plane of the Earth's orbit and the x-axis parallel to the line of intersection of the plane of the Earth's orbit with the plane of the celestial equator. Only the system's orientation is defined. The origin may be anywhere in the Solar system but is usually placed in the Earth or in the Sun. (2) A coordinate system whole geometric elements are a straight line (called the polar axis or z-axis), a straight line (called the x-axis) intersecting the z-axis at right angles, and the point (called the origin) at which the z-axis intersects the x,y-plane. The coordinates of a point are the angle (called the longitude) from the plane of the z- and x-axes to the plane through the z-axis and the point in question; the angle (called the latitude) which is the complement of the angle from the z-axis to a line joining the origin to the point in question; and the distance from the origin to the point in question.

coordinate system, ellipsoidal - (1) A coordinate system defined by three families of confocal quadric surfaces, as follows: (a) a family of confocal ellipsoids; (b) a family of hyperboloids of one sheet and (c) a family of two-sheet hyperboloids. The families are so defined that three surfaces, one from each family, intersect orthogonally at every point of the space. The coordinates of a point are the values of the parameters identifying the three surfaces intersecting at that point. (2) SEE coordinate system, spheroidal. Use of ellipsoidal coordinate system for spheroidal coordinate system is inconsistent with the usage of mathematics and physics. It may be limited to gravimetric geodesy and should be eschewed.

coordinate system, elliptical - A coordinate system, in the plane, in which the coordinates are given as the inter-sections of a family of numbered confocal ellipses and numbered confocal hyperbolae.

coordinate system, equatorial - (1) A coordinate system, on the celestial sphere, consisting of the great circle (called the celestial equator) in which a plane perpendicular to the Earth's axis of rotation intersects the celestial sphere, and the great circle (called the reference meridian in which a plane through the Earth's axis of rotation intersects the celestial sphere. The equatorial coordinates are (a) the arc of the great circle included between the reference meridian and a great circle through the point in question and perpendicular to the celestial equator; and (b) the arc from the celestial equator along the perpendicular great circle to the point in question. (2) A right-handed, Cartesian coordinate system in which the z-axis is parallel to the Earth's axis of rotation and the x-axis is parallel to the line of intersection of the planes of the celestial equator and the ecliptic. The x-axis points toward the First Point of Aries (vernal equinox) at some specified epoch and the z-axis points to the North Celestial Pole. Only the orientation of the system is defined. The origin may be anywhere in the Solar System. If it is at the observer, the system is called a topocentric equatorial coordinate system; if it is at the Sun, the system is called a heliocentric equatorial system. (3) A coordinate system in which one axis is parallel to the Earth's axis of rotation and another axis is perpendicular to the first axis and points in the direction of the First Point of Aries. (4) A coordinate system which has as geometric elements (a) a plane perpendicular to the Earth's axis of rotation and (b) a plane parallel to the Earth's axis of rotation. Also called celestial-equator coordinate system, celestial-equator system of coordinates, equator system, equatorial system and equinoctial system of coordinates.

coordinate system, equinoctial - SEE coordinate system, equatorial.

coordinate system, Eulerian - SEE Eulerian coordinate system.

coordinate system, galactic - A coordinate system with two intersecting, orthogonal axes one of which (the x-axis) lies in the galactic plane and the other (the z-axis) is perpendicular thereto. One coordinate (called galactic longitude) is the angle from the plane of the two axes to the plane containing the z-axis and the point in question; the second coordinate (called galactic latitude) is the complement of the angle from the z-axis to the line between the point in question and the point of intersection of the two axes. Two such systems have been in general use. The first, established by Ohlsson's conversion tables of 1932, placed the pole of the system at 12h 40m right ascension and +28° declination. The origin for galactic longitude was the ascending node of the intersection (the x-axis) of the galactic plane with the plane of the celestial equator. Coordinates in this system are labelled with a subscript I. In 1958, the International Astronomical Union adopted a system in which the galactic pole is at 12h 49m right ascension, +27.4° declination, and the origin of galactic longitudes is at a position angle of 123° with respect to the pole of the equator. The equator and equinox of the old system are those of 1900.0; those of the new system are of 1950.0 (and are labelled with the subscript II). Galactic longitudes in the new system are approximately 32° greater that the longitudes in the old system.

coordinate system, geocentric - (1) Any coordinate system with origin at a specified and defined center of the Earth - as, for example, the center of mass or the geometric center. (2) A coordinate system with origin at some point called the center of the Earth but otherwise left poorly or incorrectly defined. (3) A spherical coordinate system using the complement of the angle from the z-axis instead of the angle from the z-axis. It consists of a plane (called the equatorial plane) and a line (called the polar axis) perpendicular to the equatorial plane at a point (called the origin). The geocentric coordinates of a point are the distance of the point from the origin, the complement of the angle between the polar axis and the line joining the point to the origin, and the angle from a specified plane through the polar axis to a plane through the polar axis and the point in question.

coordinate system, geocentric geodetic - A geodetic coordinate system with origin at the Earth's center of mass.

coordinate system, geodetic - A coordinate system whose geometric elements consists of an ellipsoid, the least diameter (called the polar axis) of that ellipsoid, a plane (called the equatorial plan) perpendicular to the least diameter at its mid-point and a plane (called the plane of the zero meridian or the reference plane) perpendicular to the equatorial plane and containing the normal at a specified point on the ellipsoid. The geodetic coordinates of a point are (a) the perpendicular) distance (called the geodetic height) of that point from the ellipsoid, (b) the angle (called the geodetic longitude) from the reference plane to a plane perpendicular to the equatorial plane and containing the normal to the ellipsoid from the point in question. If the ellipsoid is a rotational ellipsoid (spheroid), the two planes used in defining longitude pass through the least diameter and may, in the definition of the coordinate system, be defined as passing through that diameter. Also called, improperly, a geoidal coordinate system, which is properly a coordinate system on the geoid.

coordinate system, geoidal - (1) A coordinate system on the geoid. (2) SEE coordinate system, geodetic.

coordinate system, geomagnetic - A spherical coordinate system which has its polar axis coincident with the axis of the dipole representing the Earth's magnetic field. The zero meridian is usually selected to pass through the astronomic or geodetic North Pole.

coordinate system, ground-space - A coordinate system assigning two coordinates, such as distance and azimuth or rectangular Cartesian coordinates to points on the ground.

coordinate system, horizontal - A coordinate system, on the celestial sphere, consisting of two poles (zenith and nadir) defined by (a) a line through a specified point and drawn in the direction of gravity at that point, (b) a specified great circle drawn through the poles, and (c) a great circle (called the horizon) 90° from the poles. One coordinate is the arc drawn from the zenith to the point in question; the other is the arc along the horizon from the specified great circle to a great circle through the poles and the point. Also called an azimuth coordinate system and azimuthal coordinate system.

coordinate system, hour-angle - An equatorial coordinate system (1) which has the plane of the equator and the plane of the local meridian as its geometric elements. A point in both planes is specified as the center to which directions are referred. The coordinates of a celestial body are the angle (called the hour angle) from the plane of the local meridian to a plane perpendicular to the equatorial plane and passing through the celestial body and the center, and the angle from the equatorial plane to the line from the center to the body. The hour angle is accounted positive when measured in a clockwise direction.

coordinate system, inertial - A coordinate system fixed in a material body which is neither accelerating nor rotating. Also called a Galilean coordinate system and an inertial frame of reference. The latter term is, however, also used to denote only the geometric elements of the system.

coordinate system, isothermal - A coordinate system with coordinates (u,v) such that, in the first fundamental form for the element ds of length, $ds^2 = E\,du^2 + F\,du\,dv + g\,dv^2$, the conditions $E = G$ and $F = 0$ are satisfied.

coordinate system, Lagrangian - SEE Lagrangian coordinate system.

coordinate system, left-handed - SEE coordinate system, right-handed.

coordinate system, local - A coordinate system which has its origin within the region being studied and which is used principally for points within that region.

coordinate system, material - SEE Lagrangian coordinate system.

coordinate system, military (USA) - A coordinate system based on a polyconic map projection and used formerly by the United States Army for fire control.

coordinate system, natural - An orthogonal, curvilinear coordinate system associated with a physical system and consisting of an axis t tangent to the instantaneous-velocity vector and an axis n normal to this vector to the left in the horizontal plane. A vertically directed axis z may be added to describe three-dimensional motions.

coordinate system, oblique Cartesian - SEE Cartesian coordinate system.

coordinate system, orthogonal - A coordinate system defined by families of curves or surfaces intersecting at right angles. Although this kind of coordinate system is also referred to, sometimes, as a rectangular coordinate system, the latter term is more usually applied to coordinate systems composed of straight lines or planes.

coordinate system, photographic - (1) A rectangular, Cartesian coordinate system having its origin at the rear nodal point of the camera's lens system and its z-axis perpendicular to the plane of the image. (2) SEE photograph coordinate system.

coordinate system, plane - SEE plane coordinate system.

coordinate system, plane-rectangular - A Cartesian coordinate system in the plane, the axes intersecting at right angles.

coordinate system, polar - A coordinate system in the plane, consisting of a straight line (called the base line) and a point (called the center) on that line; the coordinates of a point are the distance of that point from the center and the angle, at the center, from the base line to a line through the center and the point. SEE ALSO coordinate, polar.

coordinate system, rectangular - (1) Any coordinate system in which surfaces or lines from the separate families constituting the system intersect at right angles. More often called an orthogonal coordinate system or rectangular coord-inate system. (2) A Cartesian coordinate system in which the lines or planes intersect at right angles. To avoid confusion, the term should not be used in this sense. Use rectangular Cartesian coordinate system instead.

coordinate system, rectangular Cartesian - SEE Cartesian coordinate system, rectangular.

coordinate system, rectangular spherical - (1) A coordinate system, on the sphere, consisting of: (a) two mutually perpendicular great-circles forming the axes and called the central meridian and the equator; (b) one of the points of intersection, called the origin; and (c) the two families of great circles through the poles of the central meridian and the equator. It is called the Cassini-Soldner coordinate system if one of the meridians is a north-south meridian. SEE ALSO coordinate, rectangular spherical. (2) The same as the preceding definition except that, instead of the family of great circles through a pair of poles, it has a family of small circles parallel to the equator. Also called a spherical rectangular coordinate system.

coordinate system, rectilinear - SEE coordinate system, rectangular.

coordinate system, reduced astronomic - SEE coordinate system, astronomic (3).

coordinate system, relative - A coordinate system which is moving with respect to a fixed (inertial) coordinate system.

coordinate system, right-handed - A rectangular Cartesian coordinate system in three dimensions which has the positive directions on the three axes (x-, y-, z-axes) defined in such a way that if the thumb of the right hand is imagined to point in the positive direction of the z-axis and the fore-finger in the positive direction of the x-axis, then the middle finger, extended at right angles to the thumb and forefinger, will point in the positive direction of the y-axis. If the coordinate system is left-handed, the middle finger will point in the negative direction of the y-axis. While two oppositely oriented, rectangular Cartesian coordinate systems can be defined on a plane, neither is intrinsically right-handed or left-handed. However, if the plane has a positive and negative side defined, then a z-axis with positive and negative directions could be considered present implicitly and right-handed and left-handed coordinate systems could be defined on the plane.

coordinate system, selenocentric - (1) A coordinate system having (a) its origin at the center of mass of the Moon, (b) a reference plane through the origin and parallel to the plane of the ecliptic, and (c) a reference plane through the origin, perpendicular to the first reference plane, and parallel to the line of equinoxes. Selenocentric longitude is measured counterclockwise (as seen from the north pole of the ecliptic) from the second reference plane to a plane through the origin, perpendicular to the first reference plane, and through the point of interest. Selenocentric latitude is measured positively northward from the first reference plane to a line joining the origin to the point of interest. (2) A coordinate system having its origin at a central point (either the center or mass or the geometric center) of the Moon. This usage is deprecated.

coordinate system, selenodetic - A coordinate system similar to a geodetic coordinate system but with origin at the center of mass of the Moon and fixed in the Moon.

coordinate system, selenographic - (1) A spherical coordinate system with origin at the Moon's center of mass, a plane of reference (called the equatorial plane) perpendicular to the Moon's average axis of rotation and passing through the average center of the lunar disk, and a plane of reference containing the average axis of rotation and the average center of the lunar disk. SEE ALSO latitude, selenographic; longitude, selenographic. (2) A coordinate system for points on the lunar surface, specifying (a) the selenographic longitude of a point as the arc, in the lunar equator, between the first lunar meridian and the lunar meridian through the point; and (b) the selenographic latitude as the arc of the meridian intercepted between the lunar equator and the point.

coordinate system, space polar - SEE space polar coordinate system.

coordinate system, spherical - A coordinate system consisting of a plane of reference called the equatorial plane, a plane of reference (called the zero-meridian plane) perpendicular to the equatorial plane, and a point (called the center) on the line of intersection of the two planes. The coordinates of a point are (a) its distance from the center, (b) the angle at the center between the equatorial plane and a line joining the center to the point in question and (c) the angle from the zero-meridian plane to the plane which contains the center, the point in question and a perpendicular to the equatorial plane at the center.

coordinate system, spherical rectangular - SEE coordinate system, rectangular spherical.

coordinate system, spheroidal - A coordinate system defined by three families of surfaces, as follows: (a) longitude is defined by a family of numbered planes intersecting in a common line (the polar axis); (b) latitude is defined by a family of numbered, confocal hyperboloids with common axis on the polar axis; and (c) distance is defined by a family of numbered, confocal spheroids (rotational ellipsoids) having one of their common axes on the polar axis and their common center on the center of the family of hyperboloids. If the spheroids are oblate, the latitudes are given by single-sheet hyperboloids with common axis of rotational symmetry on the polar axis. If they are prolate, latitudes are given by two-sheet hyperboloids.

coordinate system, standard - A rectangular, Cartesian coordinate system in a plane tangent to the celestial sphere at the point to which the optical axis of a telescope is directed. The positive η-axis points towards the north pole and the positive ξ-axis points eastward.

coordinate system, topocentric - A coordinate system having its origin at the point of observation or, by inference, on the surface of the Earth or the ellipsoid representing the Earth. State coordinate systems are topocentric, as are military grid coordinate systems. Many astronomic coordinate systems are topocentric, having their origin at the observatory using the system.

coordinate system, topocentric equatorial - A coordinate system with its origin the observer's location on the spheroid representing the Earth, and with its other defining axes and planes parallel to those of an equatorial coordinate system having its origin at the center of the spheroid.

coordinate transformation - SEE transformation.

coordination - The placing of control on a common coordinate system or datum. Coordination does not imply the adjustment of data to remove discrepancies. Two surveys over the same region may be coordinated by computation on the same datum, but there may remain between them discrepancies which can be removed only by correlation.

coordinatograph - A device used to plot points from their plane coordinates. Also called a rectangular-coordinate plotter. A common variety consists of a table along one edge of which is fixed a rigid scale divided into millimeters and multiples thereof. A second, rigid scale similarly divided is placed perpendicularly to the first and can be moved parallel to itself over the length of the table. A pen, punch, or other marking tool is attached to the second scale so as to be able to slide along it. The coordinatograph may be an integral part of a stereoscopic plotter by means of which the planar coordinates (x,y) of the floating mark are plotted directly, or it may be operated by means of signals sent from a computer or magnetic tape in which the coordinates to be plotted are stored.

Copenhagen curve - SEE curve (4).

coplanarity condition - The condition placed, in photogrammetry, on three lines - the line joining the perspective centers at the times of exposure of two photographs, and the two lines joining the point in object-space to the corresponding points in image-space - that these lines be co-planar. Also spelled coplanar. Denoting the lines by vectors y, x_1, and x_2, condition is $x_1 \times x_2 \cdot y = 0$. Although this condition has been used by several organizations as the basis for computer programs, it is not as versatile nor as widely used as the co-linearity condition.

coplane, basal - The plane common to a pair of photographs when coplanarity exists and the plane is parallel to the air base (the line joining the perspective centers at the times of exposure). Also called a horizontal coplane. If the air base is horizontal, the two photographs are said to be exposed in horizontal coplane.

coplane, horizontal - SEE coplane, basal.

copy - (1) A facsimile, duplicate, or other reproduction of an original. (2) SEE copy, original.

copy, lithographic - A copy produced by a lithographic process. Also called litho copy. (jargon)

copy, original - The photographs, artwork, scribed material or other materials furnished for reproduction by printing or other methods. Also called original and copy.

copying camera - A precise camera used for making photographic copies of documents, pictures, etc.

coral atoll - An atoll the principal material of which is coral. Geodetic markers cannot be emplaced on coral atolls in the same way as they are emplaced in igneous or metamorphic rock because the coral is much softer and is often impregnated with salt water.

coral reef - A rocky eminence composed of calcium carbonate created principally by corals, and extending upward from the sea floor to the tidal limit.

co-range line - SEE line, co-range.

cord - A measure of volume defined legally, in the United States of America, as a pile or stack equivalent to 128 cubic feet of wood and air-space (3,625 cu.m.) closely ranked, usually 8 feet long, 4 feet high and 4 feet wide. A one-foot length of such a pile is called a cord-foot, which is 16 cubic inches.

cordillera - An entire mountainous province, including all the subordinate ranges and groups and the plateaus and basins in the interior.

Cordoba Durchmusterung - A star catalog giving the approximate coordinates of 613 954 stars between declinations -21° and -90° and brighter than 10.0m. Usually referred to as the C.D. It complements the Bonne Durchmusterung, which gives approximate coordinates for stars from +90° to -23° declination.

core - That portion of the Earth extending from the lower boundary of the mantle to the center of the Earth. The radius of the core is about 3500 km; the core's surface is about 2800 km below the Earth's surface. Only longitudinal (P) seismic waves will propagate through the core; transverse (S) seismic waves will not. Because liquids will transmit longitudinal but not transverse seismic waves, the core (or at least that part adjoining the mantle) is thought to be liquid. There is some seismological evidence that a region from the center of the Earth out to about 1400 km is partly or entirely solid. This portion is called the inner core; the other, liquid portion is then called the outer core.

corer - A tube driven into the ocean's floor to collect a sample of the sediment at the bottom. SEE ALSO gravity corer.

Coriolis acceleration - The acceleration produced by the Coriolis force (q.v.).

Coriolis correction - A correction applied to an assumed or calculated location or component thereof to allow for apparent acceleration caused by the Coriolis force.

Coriolis effect - Either the Coriolis acceleration or the Coriolis force.

Coriolis force - A fictitious force introduced by an observer in a rotating coordinate system to account for the deviation of a body from the straight-line motion it would have in a non-rotating coordinate system. A body following a straight line at constant velocity in the non-rotating system appears to follow a curved path with variable velocity in the rotating system. An observer in the rotating system ascribes a Coriolis force $-2m(\Omega \times v)$ to the motion of a body of mass m moving with velocity v in non-rotating coordinate system; Ω is the angular rate of rotation of the observer. The quantity $-2(\Omega \times v)$ is called the Coriolis acceleration. The force amounts to about 0.01 dynes/cm3 in mid latitudes. Also called compound centrifugal force and deflecting force.

Coriolis parameter - The quantity $2\Omega \cos \phi'$, where Ω is the Earth's rate of rotation and ϕ' is the geocentric latitude. It is usually denoted by f and has a value of about 0.0001/s in mid latitudes.

corner - (1) A point, on a boundary of land, at which two or more boundary lines meet. It is not always the same position as a monument, which is the physical evidence of the corner's location on the ground. (2) A point on the earth, its location determined by surveying, marking an extremity of a boundary of a subdivision of the public lands, usually at the intersection of two or more surveyed lines. The term is often used incorrectly to denote the physical structure, or monument, erected to mark the corner. Corners are described in terms of the points they represent, as follows: township corner - a corner at the extremity of a township boundary; section corner - a corner at the extremity of a section boundary; quarter-section corner - a corner at an extremity of a quarter-section boundary, midpoint between or 40 chains from the controlling section corners, depending on location within the township. SEE ALSO meander corner; meander corner, auxiliary; meander corner, special; witness corner.

corner, closing - A corner at the intersection of a surveyed boundary with a previously established boundary line. In surveying the public lands of the USA, when the line connecting the last section corner and the corner to be reached on an established township boundary departs from the astronomic meridian by more than the allowable amount, the line being surveyed is extended in a cardinal direction to an intersection with the township boundary, where a closing corner is established and a connection made to the previously established corner. A closing corner is established at the intersection of a line being surveyed with a previously established township boundary to avoid excessive deviation from a cardinal direction which might be required to connect with the intended corner on that boundary. A closing corner is also established at the intersection of a township, range, or section line with the boundary of a previously surveyed and segregated tract of land such as a private land claim, mineral claim, etc.

corner, double - (1) The two sets of corners along a standard parallel, i.e., the standard township, section, and quarter-section corners placed at regular intervals of measurement. This is the standard meaning. (2) The closing corners established on the line at the points of intersection of the guide meridians, range, and section lines of the surveys brought in from the south. This is a less common meaning.

corner, existent (USPLS) - A corner which has a location that can be identified by verifying the evidence of the monument, or its accessories, by referring to the description contained in field notes, or by an acceptable record of a supplemental survey, some physical evidence, or testimony as to where the point can be located. Even though its physical evidence may have disappeared entirely, a corner will not be regarded as lost if its position can be recovered through the testimony of one or more witnesses who have a dependable knowledge of the original location.

corner, found - (1) A corner (a) of which the original or restored monument or mark is recovered, or (b) the location of which was definitely established by one or more witness corners or reference monuments. (2) An existent corner of the public-lands surveys or others which have been recovered by investigation in the field.

corner, indicated - A corner of the public-land surveys whose location can not be verified by the criteria necessary to class it as a found or existent corner, but which is accepted locally as the correct corner and whose location is perpetuated by such marks as intersections of fence lines, piles of rock, or stakes or pipes driven into the ground which have been recovered by investigation in the field.

corner, lost (USPLS) - A corner whose location cannot be determined beyond reasonable doubt either from traces of the original marks or from acceptable evidence or testimony that bears upon the original location, and whose location can be restored only by reference to one or more interdependent corners.

corner, obliberated (USPLS) - (1) A corner where there are no remaining traces of the monument or its accessories but whose location has been perpetuated. (2) A corner which may be recovered beyond reasonable doubt, by the testimony of the interested landowners, competent surveyors or other qualified local authorities or witnesses or by some acceptable recorded evidence. (This definition is undesirable because it applies also to corners for which the monuments or other physical evidence still exist.) A location which depends upon the use of collateral evidence for verification can be accepted only insofar as it is supported, generally through proper relation to known corners and agreement with the field notes on distances to natural objects, stream crossings, line trees and off-line blazes, etc., or unquestionable testimony.

corner, quarter-quarter-section (USPLS) - A corner at an extremity of a quarter-quarter section; the midpoint between or 20 chains from the controlling corners on the section or township boundaries. Also called a sixteenth-section corner. Written as 1/16-section corner.

corner, quarter-section (USPLS) - A corner at the extremity of a boundary of a quarter-section, midway between or 40 chains from the controlling section corners, depending on location within the township.

corner, sixteenth-section - SEE corner, quarter-quarter-section.

corner, standard (USPLS) - (1) A corner, on a standard parallel or base line, which serves as a starting point for surveys to the north. (2) A corner on a standard parallel or base line. (3) A senior corner on a standard parallel or base line.

corner, theoretical - A corner, on a map, for which no marks have been identified on the ground. The locations are determined by adjustment and are indicated on the map only by the intersections of the subdivisions' lines.

corner accessory - A physical object close to a corner and which is referred to in describing the location of the corner. Examples are bearing trees, mounds, pits, ledges, rocks and other natural features to which distances or directions, or both, from the corner or monument are known. Corner accessories are considered part of the monument of the corner. As a witness to the corner, it carries the same weight, legally, when the latter is found to be missing or destroyed.

corner contiguity - The property of having a corner but not a boundary line in common.

corner cube - (1) A reflector consisting of two planes intersecting at right angles. Also called a dihedral reflector. (2) SEE reflector, cube-corner.

corner mark - SEE register mark.

corner post - (1) A post set as a monument to mark the corner of a section, quarter-section, township, etc. (2) A post set to mark a corner on any boundary, public or private.

corner reflector - SEE reflector, cube-corner.

corner stake - A stake set by a surveyor, in running a survey by metes and bounds, set at every change of direction to assist the surveyor and to fix the survey to the ground.

corner tick - SEE register mark.

Cornu spiral - SEE clothoid.

Cornu's spiral - SEE clothoid.

correction - (1) That quantity which, added to an incorrect value, gives the correct, or a more nearly correct, value. Which is the correct and which is the incorrect value in any particular instance usually depends on the viewpoint taken. For example, in communication theory it is assumed that the message being sent (the signal) is correct and that the received message is incorrect because errors (noise) have been added during transmission. However, it is also possible to consider the received message as the correctly received signal and to apply various corrections to it to derive hypothetical originals. In gravimetric geodesy, gravity can be measured at the Earth's surface. Its value can also be predicted for a different location and under different conditions, using a gravity formula. To compare measured values with predicted values, corrections must be added to the values given by the gravity formula. On the other hand, one could also consider the values given by the gravity formula as being correct. One would then apply corrections to the measured values. (2) A quantity added to or subtracted from a measured quantity to compensate for known deviations from ideal conditions of measurement and for non-negligible effects of variations in uncontrolled variables. This definition is practically equivalent to the first definition, above. (3) The theoretical value of a quantity minus the observed value. (4) The observed value of a quantity minus the average (or other theoretical) value. This usage makes correction equivalent to error in its most usual sense and is rare. SEE ALSO the specific name of the correction: e.g., leveling correction, orthometric; slope correction; gravity correction. SEE ALSO adjustment correction; collimation correction; gravity correction; tension correction; timing correction.

correction, aerodynamic - The amount subtracted from the period of a pendulum to account for its lengthening because the total mass of the pendulum is increased by gas absorbed on the surface.

correction, air-drag - The amount subtracted from the period of a pendulum to account for its lengthening by the resistance of the gas surrounding the pendulum. Also called the damping correction, although this term has also a different meaning.

correction, anamorphic - The change made in the scale of a picture in one direction, while leaving the scale in the other direction unchanged; this corrects for differences in scale introduced when the original picture was taken. The correction is usually made by an optical system interposed between the original picture (film) and the projected image. It may be used, for example, to aid photo-interpretation by providing some correction for tilt of the camera or for systematic differences of elevation in the ground.

correction, arc-sin - A small correction made, in a computation, to allow for the difference between the true value of an angle and the sine (used in the computation) of that angle.

correction, arc-to-chord - The difference between the geodetic azimuth of one point from a second point, on the rotational ellipsoid, and the azimuth from grid north of the line joining the two corresponding points on a map. Also called the convergence angle, conversion angle, (t-minus-T) and t-minus-T correction.

correction, astronomic - A correction applied to measured differences of elevation to remove the tidal effect.

correction, bolometric - The quantity, usually denoted by BC, added to an absolute (visual) magnitude M_v or to an apparent (visual) magnitude m_v to get the corresponding absolute bolometric magnitude M_b or apparent bolometric magnitude m_b: $BC = M_b - M_v = m_b - m_v$.

correction, chronometric - A correction added to the observed time of an event to obtain the true time of the event.

correction, curvature-of-the-Earth - SEE correction for curvature of the Earth.

correction, dynamic - The quantity that must be added to the observed or calculated elevation at a point to obtain its dynamic elevation (dynamic height) or dynamic number. Also called a dynamic leveling correction when confusion with other kinds of dynamic correction is possible.

correction, flexural - The amount subtracted from the period of a pendulum to account for the period's lengthening because of that flexure in the pendulum's supports caused by the pendulum's motion. The pendulum's swinging in one direction causes the support to sway in the opposite direction. The amount of sway is not constant. It can be determined by actually measuring it or by hanging and additional pendulum from the same support. When the first pendulum is set in motion, the second pendulum, at first motionless, will appear to swing with respect to its support. The amount of apparent swing is a measure of the amount of flexure in the support. A flexural correction can be made unnecessary if the second pendulum is set swinging in opposition to the first; the induced flexures then cancel each other. SEE also apparatus, two-pendulum.

correction, free-air - See gravity correction, free-air.

correction, height-of-eye - That correction, to angular elevation measured with a sextant, required to take into account the height of the observer above the horizontal plane (causing the dip of the horizon). Also called the dip correction.

correction, ionospheric - The quantity subtracted from a distance measured, using radio waves, between a point on the surface and a point in or above the ionosphere, to obtain the distance that would have been measured had the ionosphere not been present. The correction is significant for distances measured using radio waves having frequencies lower than about 0.9 GHz, e.g., for distances measured using the radio waves emitted by TRANSIT satellites (150 MHz and 400 MHz).

correction, isostatic - The adjustment made to measured values of gravity or to calculated deflections of the vertical to take account of the assumed deficiency of mass under topographic features for which a topographic correction is also made.

correction, knife-edge - The amount by which the observed period P of a pendulum must be increased because the knife edge on which the pendulum swings has a circular rather than V-shaped cross-section. The correction amounts to about $P/2s$, in which s is the quantity $\sqrt{[1 + (d/r)^2 - 2(d/r) \cos \alpha]}$. r is the radius of curvature of the cross section, d is the distance from the center of gravity to the center of curvature and α is half the angle through which the pendulum swings from side to side. In practice, the radius r may not be measurable; the correction must then be determined empirically.

correction, light-time - SEE correction for light time.

correction, orthometric - The quantity added to a difference of elevation to correct for the error introduced when level surfaces at different elevations are not parallel. Also referred to as the orthometric leveling correction.

correction, rotational - A correction made to the period of a pendulum to account for effects of the Earth's rotation: centrifugal force, Coriolis acceleration, tilting of the pendulum's plane of motion and differing angular accelerations of different supports.

correction, semi-diameter - The quantity added to an angle measured from a reference line or point to a limb of a celestial body, to make the angle refer to the center of that body.

correction, tidal - (1) A correction to the period of a pendulum made because passage of the Moon or Sun through the zenith or nadir not only affects the value of gravity directly but, through the tide, increases the elevation of the station and thereby decreases gravity farther. If the Earth were rigid, passage of the Moon or Sun through the zenith would decrease gravity by about 0.08 and 0.16 mgal respectively, relative to the value when these bodies are on the horizon. Owing to the yielding of the lithosphere and hydrosphere, the effect is somewhat greater and its magnitude is not entirely predictable. The effect is usually ignored. (2) SEE gravity correction, tidal.

correction, t-minus-T - SEE correction, arc-to-chord.

correction code - A code consisting of letters, numbers, and symbols used to indicate corrections on maps or on overlays attached.

correction for curvature of the Earth (surveying) - (1) That offset from the tangent to the curve resulting from curvature of the Earth. The offset amounts to about 8 cm at 1 km distance. (2) That offset which is given in surveying tables as correction for curvature of the Earth. The values given in such tables usually combine the effects of both curvature of the Earth and atmospheric refraction. Under normal conditions, the atmospheric refraction compensates partially for the curvature of the Earth, so the correction is about 6 cm for a point 1 km away.

correction for datum - A factor used, in the prediction of tides, to resolve the difference between chart datum of the reference station and chart datum of a secondary station.

correction for elasticity of tape - SEE tension correction.

correction for flexure - SEE correction, flexural.

correction for inclination of the horizontal axis - A correction applied to a measured horizontal angle to eliminate any error that may have been caused by the horizontal axis of the instrument not having been exactly horizontal. If the horizontal axis of the instrument is not exactly horizontal, the line of collimation will not cut the horizon at a point directly underneath (or above) an observed point, that is a plane described by the line of collimation when the telescope is rotated about the horizontal axis will not be vertical. The inclination of the horizontal axis may involve one or both of two conditions: the horizontal and vertical axes of the instrument may fail to meet exactly at a right angle, and the vertical axis may deviate from the direction defined by the plumb line. So, determining the correction requires that both the inclination of the axis and the angular elevation of the observed point be known.

correction for inclination of the tape - SEE grade correction.

correction for light time - The apparent angular displacement of the apparent direction of a body from its geometric direction, resulting from the motion of the body during the interval in which light traveled from the body to the observer. Also called the light-time correction. SEE ALSO aberration, planetary.

correction for run of micrometer - A correction applied to a reading observed with a micrometer-equipped microscope, to compensate for the difference between the nominal distance the moving line is moved by one turn of the micrometer screw and the actual distance is moved.

correction for slope of tape - SEE grade correction.

correction line (USPLS) - SEE parallel, standard.

correction overlay - A transparent sheet on which corrections to the underlying drawing or map are noted.

correction plate - (1) A glass plate having a surface ground to a predetermined shape, for insertion in the optical system of a diapositive printer or stereoscopic plotter to compensate for radial distortion introduced by the lens of the camera that took the photographs. Sometimes called a compensation plate. (2) SEE corrector plate.

corrector plate - A glass plate having a surface ground to a predetermined shape and placed ahead of the primary mirror in a reflecting telescope to correct for spherical aberration or other distortion. Also called a correction plate.

correlate - One of a set of factors introduced as additional unknowns called Lagrangian multipliers and used in the method of correlates for solving a set (the observation equations) of simultaneous, linear equation. A set of equations (the normal equations) is first formed to solve for the factors; another set of equations is then formed to solve for the original unknowns in the observation equations. Also called a correlative.

correlate equation - An equation derived from observation and/or condition equations by including undetermined multipliers (Lagrangian multipliers) and expressing the condition that the sum of the squares of the residuals (or corrections) resulting from the application of these multipliers to the condition equations shall be a minimum. In adjusting triangulation by the method of least squares, correlate equations are formed directly from the observation and condition equations, there being as many correlate equations as there are corrections to be determined, but only as many undetermined multipliers as there are condition equations. From these correlate equations, new equations are formed equal in number to the number of undetermined multipliers which constitute the unknowns. Solving these equations determines values for the multipliers which, when substituted into the correlate equations, give values for the corrections which satisfy the condition equations, make the measurements and their functions consistent among themselves, and make the adjusted values the most probable that can be derived from the given measurements.

correlate method of adjustment - SEE method of correlates.

correlation - (1) Determining the extent to which one randomly varying quantity can be expressed as a function of another or both can be expressed as functions of a third, non-random quantity. This definition is extended to situations involving several variables by applying it to each different pair of variables in the set. Correlation is linear, quadratic, etc., if the function is linear, quadratic, etc. The extent is measured by a quantity called the correlation constant or correlation coefficient. (2) The removal of discrepancies that may exist among survey data so that all data agree without apparent error. The terms coordination and correlation are usually applied to the harmonizing of surveys of adjacent regions or of different surveys over the same region. Two or more such surveys are coordinated when they are computed on the same datum; they are correlated when they are adjusted together.

correlation, optical - (1) The calculation of correlation coefficients by an optical method. (2) The determination of the correlation between the image on a photograph taken some time previously, and the image on a new photograph, on the screen of a television set, or in a telescope. The method is used to provide information on location of an aircraft, to check an aerial guidance system, and for other purposes.

correlation function - (1) The average value of the quantity $x(r+nh)$ and $y(nh)$, where x and y are randomly varying functions of the quantity nh and n being an integer which varies from $-N$ to $+N$ and r and h are constants. (2) The limit of the quantity defined in (1) as N becomes very large. (3) The average value $\phi(r)$ defined by $(1/2T) \int x(r+t)y(t)\,dt$. (4) The limit of the quantity $\phi(r)$ defined in (3) as T goes to infinity. If x and y are different variables, the correlation function is called the cross-correlation function of the two variables. If they are the same variable, it is called the auto-correlation function of x.

correlative - SEE correlate.

correspondence - (1) The condition that exists when corresponding images on a pair of photographs lie in the same epipolar plane; the absence of y-parallax. (2) SEE function.

coterminous - SEE conterminous.

coudé focus - That point to which light entering a telescope is focused by bringing the light through the declination axis and the polar axis.

coulee - (1) A small stream, often intermittent, or the bed of such a stream when dry. (2) The valley, gulch or wash of an intermittent stream, often of considerable extent. (3) A small valley or a low-lying region.

Coulomb force - SEE force, electrostatic.

count, least (on a micrometer or vernier) - The finest reading that can be made directly (without estimation) on a micrometer or vernier.

countercurrent, equatorial - A current moving eastward between the North and South Equatorial Currents of the Atlantic, Pacific, and Indian (in northern winter) Oceans. In the Atlantic and Pacific Oceans, its axis lies about latitude 7° North, and in the Indian Ocean, about 7° South.

counter-etch (v.) - Remove, with certain diluted acids, impurities from a lithographic plate, thus making it receptive to an image.

course - A line having a specific direction. Many variations of this basic definition are in use in different disciplines.

course (surveying) - In surveying generally, the bearing (or azimuth) of a line, the length of a line or both. The term should not be used in describing control surveys, but if it is used, it should be with the inclusive meaning, azimuth and length of line. However, it is better to use azimuth and length of line as separate terms.

course (land surveying) - (1) In general, the bearing or azimuth of a line. (2) In law, the direction of a line run with a compass or transit and referred to a meridian.

course (traversing) - In describing a traverse, the azimuth and length of a single line of traverse considered together.

course (navigation) - The azimuth or bearing of a line along which a ship or aircraft is to travel or does travel for a specified period of time. This definition is also given as the azimuth or bearing of a line along which a ship or aircraft is to travel or does travel, without change of direction, or as the direction of intended motion of a craft in the horizontal plane, measured relative to the water in the instance of a watercraft but relative to the earth in all other craft. It is also given as the direction a navigator wants his craft to travel for a given period of time, or hopes his craft has followed. The direction of a course (d) is always measured in degrees from the astronomic meridian, and that direction is always meant unless it is otherwise qualified as, e.g., a magnetic or compass course.

course (of a river or stream) - (1) The route along which a river or stream flows. (2) The river or stream itself.

course, relative - A course measured relative to a moving object such as a moving craft.

course made good - The average course actually achieved.

court - A round or square passageway with only one entrance.

covariance - The average value of the quantity $(x_1 - \mu_1)(x_2 - \mu_2)$, where x_1 and x_2 are random variables with average values μ_1 and μ_2, respectively. If x_1 and x_2 are the same variable, the covariance is called the variance. If x_1 and x_2 are continuous functions, the name covariance function is sometimes given to their covariance.

covariance function - The covariance of two continuous functions.

covariance function, deterministic - A function which satisfies the geometric requirements to be a covariance function but does not necessarily have an interpretation as a random function.

covariance matrix - A matrix whose element in row i and column j is the average value of $(x_i - \mu_i)(x_j - \mu_j)$, where x_i and x_j are random variables with average values μi and μj, respectively. The elements along the main diagonal are called the variances of the corresponding variables; the elements off the main diagonal are called the covariances.

covenant - A promise, incorporated in a trust indenture or other formal instrument, to perform certain acts or to refrain from performing certain acts. SEE ALSO covenant, restrictive.

covenant, protective - A contract between a subdivider and the purchaser of a lot, intended to assist the developer in marketing his lots and to assure the purchaser that the value of his property and the character of the neighborhood will be maintained, and that nuisances will not be created by other purchasers.

covenant, restrictive - A private agreement restricting the use and occupancy of real estate which is a part of the conveyance and is binding on all subsequent purchasers. Such covenants may deal with control of the size, set-back and/or placement of buildings, architecture or cost of improvements.

cover, stereoscopic - SEE coverage, stereoscopic.

coverage - (1) A measure of the amount of surface represented in an image or set of images. (2) The set of images depicting a particular region. E.g., photographic coverage would denote the set of photographs depicting a particular region. (3) The extent of a region shown on a map, set of photographs or other representation of the region. (4) SEE ground coverage.

coverage, aerial - (1) A set of aerial photographs completely depicting a region, having stereoscopic overlap between exposures along the line of flight and sidelap between photographs on adjacent lines of flight. (2) A set of aerial photographs, for each of which recorded SHORAN distances at the time of exposure are available. (3) A set of images or other graphic material completely depicting a region.

coverage, stereoscopic - A set of aerial photographs of a region, providing stereoscopic views of the entire region.

crab - (1) That condition in which the sides of an aerial photograph are not parallel or perpendicular to the planned direction of flight. This condition may result if the camera is at an angle to the longitudinal axis of the aircraft. More commonly, it results when the aircraft's longitudinal axis does not coincide with the direction of flight. This may occur if the photographs are taken when cross-winds are strong. (2) That condition in which the direction in which an aircraft is moving is not the direction in which it is pointed. SEE ALSO yaw. (3) SEE angle of crab.

crab angle - SEE crab; angle of crab.

Cracovian - SEE krakovian.

Crandall's method (for adjusting a traverse) - The misclosure in azimuth or angle is first distributed in equal portions to all the measured angles. The adjusted angles are then held fixed and all remaining corrections distributed among the measurements of distance through the method of weighted least-squares.

Crandall's rule (for adjusting a traverse) - The corrections ΔD_i and ΔL_i to the x-component D_i and the y-component L_i, respectively, of course i (of length l_i) are given by
$\Delta D_i = (D_i^2 B/l_i) + (D_i L_i A/l_i)$
$\Delta L_i = (D_i L_i B/l_i) + (L_i^2 A/l_i)$, where
$B = -[(\Sigma \Delta L_i)\Sigma(L_i D_i /l_i) + (\Sigma \Delta D_i)\Sigma(L_i^2/l_i)]/D$
$A = -[(\Sigma \Delta D_i)\Sigma(L_i D_i /l_i) + (\Sigma \Delta L_i)\Sigma(D_i^2/l_i)]/D$
$D = \Sigma(D_i^2/l_i)\Sigma(L_i^2/l_i) - [\Sigma(L_i D_i /l_i)]^2$.

Craster's parabolic map projection - SEE map projection, parabolic.

crater, volcanic - A cone or ring formed around the vent through which materials from below the Earth's surface are ejected.

creek - (1) A watercourse along which water is usually flowing. (2) A rivulet smaller than a river but larger than a brook.

creep (of the period) - A gradual change in the period of a pendulum. The term is nearly synonymous with drift.

creep meter - SEE extensometer; strain gage. Also spelled creepmeter.

crest - (1) The summit of any eminence. (2) SEE crest, military. (3) SEE crest, topographic.

crest, false - SEE crest, military.

crest, military - A line, along a ridge, hill or mountain, from which all points lower down on one side of the line are visible. Also called a crest and false crest. The location of the military crest will depend on the height of the observer. The term "military" crest is preferred if confusion with the "topographic" crest is possible.

crest, topographic - A line, along a ridge, hill, or mountain, on both sides of which the ground slopes downward and is at a lower elevation. Also called the crest, topography crest and true crest. Topographic crest is preferred if confusion with military crest is possible.

crest, true - SEE crest, topographic.

crib - The space between successive ties of a railroad.

criterion, least-squares - The requirement that the sum of the squares of the differences between the measured values in a set and the corresponding calculated values be a minimum the values being a function of a set of variables which can be given values more or less arbitrarily.

critical-path method - A method of selecting, from among a number of possible schedules for a project, that schedule which best satisfies certain criteria such as efficiency, time for completion, etc. It involves separating the project into sub-projects or steps and identifying those steps (the critical steps) whose place or timing in the sequence cannot be changed, and adjusting the rest of the schedule to accommodate these critical steps as well as possible.

Crone method - SEE principal-plane method.

crop (v.) - Cut off or trim unwanted parts of a photograph. It is usually done by masking the image during printing.

cross-coupling - In mechanics, the interaction of two independent events to produce another event. If a way can be found to prevent the original events from acting on each other, the events are said to be decoupled. E.g., when a gravimeter is used on board a ship, cross-coupling takes place between vertical and horizontal motions of the ship and the vertical reaction of the gravimeter. This causes the gravimeter to give erroneous readings of gravity.

cross cut - A horizontal passage driven from a shaft to a vein across the course of the vein in order to reach the zone of the vein.

cross-hair - One of a set of hairs, filaments, or etched lines placed on a reticle held in the focal plane of a telescope. It is used as an index mark for pointing the telescope when pointings and readings must be made on a leveling rod or stadia rod, when objects must be aligned, etc.

cross-hatching - A combination of two sets of closely-spaced, parallel lines, the one set crossing the other at a predetermined angle. It is used on charts and maps to designate or distinguish particular kinds of terrain, etc.

crossing, Shoran-line - SEE Shoran line-crossing method.

crossline glass screen - SEE screen, halftone.

crossover - Two turnouts with the track between the frogs arranged to form a continuous passage between two nearby, generally parallel, tracks.

cross-profile method - A method of determining details of the terrain in a topographic survey carried out together with a traverse. Cross-profile stations are established at 100 meter intervals along the line of the traverse, and profiles run at right angles to that line, distances and differential elevations being measured to successive breaks in the terrain along the profile.

cross-ratio - SEE ratio, cross.

cross-section - (1) The intersection of a solid with a developable surface (usually a plane). For example, a vertical section of the ground or of underlying strata. A cross-section, in this sense, should not be confused with a profile, which is the intersection of a surface with a developable surfaced. (2) The trace, in a vertical plane perpendicular to a given, horizontal line or route, of the terrain through which the plane passes. (3) A horizontal grid laid out on the ground for determining contours, quantities of earthwork, etc., using the elevations of the intersections of the grid. (4) A section taken perpendicularly to the center line of a proposed construction such as a road, canal or dam.

crown - (1) The vertical rise, expressed in inches, between the edge and the center line of the base or surface. (2) The rate of rise between the edge and the center line of the base or surface. (3) The inside top of a sewer.

Crown lands - (1) In general, land belonging to the reigning sovereign. (2) Those regions, especially within the original colonial states and also within all public-land states and territories, wherein title granted by a foreign government was passed before the acquisition of sovereignty by the USA.

crust - (1) The outermost layer or shell of the Earth, defined according to various criteria including speed of seismic waves, density, and composition. (2) That portion of the Earth having considerable rigidity, sustaining both transverse and longitudinal seismic waves, and allowing speeds of up to about 78.2 km/s for seismic waves. Also called the lithosphere, although this latter term is best reserved for the solid Earth as a whole. (3) That part of the Earth above the Mohorovicic discontinuity. The crust extends from the surface of the Earth down about 5 to 10 km in oceanic regions, 35 to 40 km under much of the continents, and perhaps 60 to 80 km under mountainous regions. It is separated from the mantle by a thin region, called the Mohorovicic discontinuity (after the Yugoslavian geologist Mohorovicic, 1857-1936), in which the speed of seismic waves jumps from about 7.2 km/s to over 8 km/s. SEE ALSO M crust.

crust, elastic - SEE crust, rigid.

crust, oceanic - A mass of gabbroic material, approximately 5 kilometers thick, which lies under the bottom of the ocean and which may be more or less continuous beneath the continental crust.

crust, rigid - That part of the Earth which reacts elastically to all differences of stress below a specific elastic limit. Vening Meinesz set the limit at over 1000 kg/cm². The term's definition is inconsistent with the usual definition of rigid. A better term would be elastic crust.

crystal clock - A clock in which the intervals of time are determined by reference to the frequency at which a crystalline solid (crystal) oscillates when held between the plates of a condenser in a resonant circuit. The crystal is actually a flat plate cut from a quartz crystal at such an angle to the crystal's axes as to give the least variation in frequency of oscillation. The crystal is usually cut to give a frequency of 1 Mc/s or 5 Mc/s at resonance, although other frequencies that relate better to the sidereal day or the mean solar day have been used. The resonant frequency of the quartz plate is divided down to one cycle or pulse per second by electronic circuits. A good crystal clock has a long-term stability of about 1 part in 1012.

C-shot - A sighting made in determining the C-factor in the field.

C-station - SEE A-station (1).

C-test - SEE test, two-peg.

cubit - A measure of length, in its origin the length of the forearm from the elbow to the extremity of the middle finger, but of uncertain length in standard units such as inches or centimeters; in English measure, 18 inches (45.72 cm) [22 inches]. The ancient Egyptian cubit was 20.7 inches (52.5 cm) [20.63 inches]. It later gave rise to the Saxon Elne. The Northern cubit is of uncertain origin, but is thought to be of non-semitic origin and to have been in use from about 3000 B.C. to 1850 A.D. in one form or another. It was about [26.6 inches]. (Values given between [] are from the second reference).

Cucunutti method (of resection) - A method of resection in which lines A,B,C are drawn through the plotted and observed points, on a plane table, to give intersections 1 of A with B and 2 of B with C. The horizontal motion of the plane table is loosened and the table rotated slightly. The process is repeated to give points of intersection 1 and 2'. The desired point is at the intersection of lines 11 and 22'. The method was devised by Sergeant Cucunuitti, of the Mexican Army, in 1592.

cul-de-sac (French) - Dead end; blind alley.

culmination - (1) The location of a heavenly body at the instant it crosses an observer's meridian above or below his horizon. (2) For a heavenly body which is continually above the horizon, the location of lowest apparent angular elevation. Culmination occurs when the body transits the local meridian: upper culmination at the branch of the meridian above the celestial pole; lower culmination at the branch below the celestial pole. As an observer approaches a pole of the earth, culmination of the fixed stars becomes less noticeable, disappearing when the pole is reached. Culmination of bodies within the Solar System may, under some conditions, be obscured by changes in declination. At one time, lunar culminations were used extensively in determining astronomic longitude.

culmination, lower - SEE culmination.

culmination, upper - SEE culmination.

culture (cartography) - Features of the terrain that have been constructed by man. Also called manmade features. Included are canals, roads, buildings, boundary lines and, often, all names and legends on a map. The term is used, with this meaning, in relation to what appears on maps.

culvert - A covered channel or pipe of large diameter taking a watercourse below the level of the ground or under an obstruction such as a road or railway.

cum solem - SEE rotation, clockwise.

curb - A barrier, made of stone, concrete, or wood, paralleling the side limit of a roadway to guide the movement of vehicles' wheels and to safeguard trusses, railings or other construction existing outside the side limit and pedestrians on sidewalks.

current - (1) A horizontal movement of water in a general direction which is maintained over a great distance. Also called a stream. Currents may be classified as tidal or non-tidal. Tidal currents are caused by cross-coupling between the rotation of the Earth, the gravitational attractions of the Sun, Moon and other members of the Solar System on the Earth, the shape of the seas' bottoms, and meteorological factors. Non-tidal currents are currents not caused primarily by gravitational actions of other bodies. Eddies (bodies of water in rotary motion) are usually not classified as currents because they are only a few hundred kilometers in diameter at most. (2) The non-tidal, horizontal movement of the sea. The movement may be in the upper or lower layers of the water or in all layers. In some regions, this movement may be nearly constant in rate and direction, while in others it may vary seasonally or may fluctuate with changes in the weather. SEE ALSO current, littoral; ebb current; flood current; gradient current.

current, equatorial tidal - A tidal current occurring semi-monthly as a result of the Moon's being over the equator.

current, geostrophic - (1) A current whose motion is controlled principally by geostrophic forces. (2) A current defined by assuming that an exact balance exists between the horizontal component of pressure and the Coriolis force per unit area. Also called a gradient current. It is usually derived by preparing a chart which is based on observations of temperature and salinity at various depths at various points in a network or along a line. The direction of the current is indicated by the contours of dynamic topography and its speed by the spacing of the contours.

current, hydraulic - A current that results from a difference in water level. Most natural currents are of this type.

current, inertial - That part of a current, disturbed from steady motion in equilibrium, which has the characteristic features of purely inertial movement.

current, littoral - A current in the littoral zone, e.g., a long-shore or rip current.

current, longshore - A current paralleling the shore and occurring largely within the surf zone. It is caused by the excessive water brought to the zone by the small, net transport of water by wind waves. Longshore currents feed into rip currents.

current, relative - The current which is a function of the dynamic slope of an isobaric surface and which is determined from an assumed layer of no motion. The current flows along the contours of dynamic topography; the speed is inversely proportional to the distance between contours.

current, reversing - A tidal current which flows alternately in approximately opposite directions, with slack water at each reversal. Such currents occur principally in regions where flow is largely restricted to narrow channels.

current, rotary - A tidal current which flows continually, with the direction of flow changing through all points of the compass during the tidal period.

current, tidal - A horizontal movement of water caused by motion of the Earth in the gravitational fields of the other members of the Solar System. More particularly, it is the horizontal component of the particular motion of a tidal wave. It is part of the same general movement of the sea which is manifested in the vertical rise and fall of the waters.

current chart - A chart on which data relative to currents are depicted.

current cycle - The complete set of tidal currents occurring during a specified cycle such as a tidal day, the Metonic cycle, etc.

current diagram - A diagram showing the speed of the currents occurring during ebb and flood, and the times of slack water and strength water over a considerable extent of the channel of a tidal waterway, the times being referred to phases of the tides or currents at some reference stations.

current ellipse - A graphic representation of a rotary current in which the velocity of the current at different hours of the tidal cycle is represented by vectors. The cycle may be completed in one-half tidal day or in one tidal day according as the tidal current is semi-diurnal or diurnal.

current hour - The average interval between the transit of the Moon over the meridian of Greenwich and the time of strength of flood, modified by the times of slack water (or least current) and strength of ebb.

current line - A graduated line attached to a pole and used in measuring the speed of the current. Also called a log line.

current meter - An instrument for measuring the speed or velocity of a current. SEE ALSO pygmy current meter; Savonius current-meter.

current-meter, acoustic - An instrument which measures rate of flow in rivers and oceans by transmitting acoustic pulses in opposite directions parallel to the flow and measuring the difference in the time it takes a pulse to travel between transmitter and receiver.

current meter, Savonius-rotor - SEE Savonius current meter.

current pole - A pole used in measuring the velocity of the current.

current rose - A graphic representation of currents for a specified region; arrows at cardinal and intercardinal points of the compass show the directions toward which the prevailing current flows and the frequency with which this direction occurs for a given period of time. The thickness or pattern of the arrows may be varied to designate categories of speeds of currents.

current station - (1) The geographic locations at which measurements of current are made. (2) The facilities used to make measurements of current at a particular site. These may include a buoy, ground tackle, current meters, recording devices and radio transmitter.

current station, subordinate - (1) A current station from which a relatively short series of observations is reduced by comparison with simultaneous observations from a control current-station. (2) A station listed in the Tidal Current Tables for which predictions are to be obtained by means of differences and ratios applied to the full predictions at a reference station.

current table - A table giving daily predictions of the times of slack water and of the times and velocities of the flood and ebb maxima.

curtesy - (1) The right which a husband has in his wife's estate at her death. (2) The estate to which, by common law, a man is entitled, on the death of his wife, in the lands or tenements of which she is seized in possession in fee-simple or in tail during her coverture, provided they have had lawful issue born alive which might be capable of inheriting the estate. It is a freehold estate for the term of his natural life.

curtilage - A yard, courtyard, or piece of ground, included within the fence surrounding a dwelling house.

curvature - (1) The rate at which a curve deviates from a straight line. (2) The vector dt/ds, where t is the vector tangent to a curve and s is the distance along that curve. SEE ALSO curve. (3) The rate at which a curved surface deviates from a flat surface in a particular direction. SEE ALSO Petzval curvature.

curvature, extra-axial - SEE aberration, off-axis.

curvature, Gaussian - SEE Gaussian curvature.

curvature, mean - The average value of the two principal curvatures of a surface.

curvature, mean radius of - SEE radius of curvature, mean.

curvature, normal - The curvature, at a point P on a surface, of the curve which is the intersection of a plane through the normal to the surface and a tangent to the surface. For any point, the value of the normal curvature changes as the direction of the tangent changes. For any regular surface, there is one direction in which the normal curvature is a maximum and another direction in which the normal curvature is a minimum. These two directions are called the principal directions and the corresponding curvatures are called the principal curvatures.

curvature, normal radius of - SEE radius of curvature, normal.

curvature, principal - The greatest or least value of the normal curvature at a point on a surface. SEE ALSO curvature, normal.

curvature correction - (1) A correction applied to the average of a series of observations on a star or planet, to take account of the divergence of the apparent path of the star or planet from a straight line. Any star which is not an equatorial star does not strictly run along the horizontal wire of a transit instrument as it crosses the meridian of the observer. This phenomenon was first observed by Cassini and was explained by him in 1719. Observations on a star for the purpose of calibrating a micrometer are subject to correction for the apparent curvature of the star's path during observation, as also are some observations for azimuth. The correction to reduce an observation for latitude on a star close to but not on the meridian to what it would be if the star were on the meridian may be considered a curvature correction. (2) The correction applied to some geodetic data to take into account the divergence of the surface of the Earth (or its representing ellipsoid) from a plane. In geodetic leveling, the curvature correction and the effect of atmospheric refraction are considered together, and tables have been prepared from which combined corrections can be taken. In aerotriangulation, the curvature correction is also usually combined with the correction for refraction.

curvature in the meridian - The quantity $a(1 - e^2)/(1 - e^2 \sin^2\phi)^{3/2}$, where a is the length of the semi-major axis of a rotational ellipsoid, e is the eccentricity and ϕ is the geodetic latitude; often denoted by M or μ.

curvature in the prime vertical - The quantity $a/(1 - e^2 \sin^2\phi)^{1/2}$, where a is the length of the semi-major axis of a rotational ellipsoid, e is the eccentricity and ϕ is the geodetic latitude. It is often denoted by N or ν. Geometrically, it is the distance, along the normal at latitude ϕ, from the ellipsoid to the minor axis.

curvature of field - (1) The condition whereby objects in a plane perpendicular to the optical axis are imaged in a curved or dish-shaped surface. Equivalently, an aberration affecting the longitudinal position of images off the axis so that objects in a plane perpendicular to the axis are imaged on a curved surface. Some telescopes used in astronomy and geodesy have pronounced curvature of field. This is particularly true of Schmidt telescopes, in which the focal surface is spherical and photographic film must be cut in the form of a Maltese cross in order to lie flat on the surface. (2) The distance, parallel to the optical axis, from an off-axis image-point to the plane through an on-axis image point and perpendicular to the optical axis.

curvature of the Earth - The curvature of an ellipsoid chosen to represent the Earth. In surveying over short distances, if the Earth's curvature needs to be taken into account at all, representing the Earth by a sphere is adequate.

curvature of the ray-path - The curvature of the path taken by a narrow beam of electromagnetic radiation in passing through the atmosphere. The curvature is not constant but varies with altitude h. At each point of the path, the radius R of curvature is given, as a function of h, the index of refraction n at the point, and the angle θ that the path makes with the horizontal these, by R = -(n secθ)/(dn/dh).

curve - (1) Any one-dimensional figure having no abrupt changes of direction except at cusps. (2) Any one-dimensional figure not a straight line and having no abrupt changes of direction except at cusps. (3) A continuous, explicit function of one variable or a pair of continuous, explicit functions of two variables, etc. In general, a curve in N dimensions is specified by N continuous, explicit functions of N variables. (4) A curved, stiff piece of material, usually plastic or sheet metal, flat on one side and used for drawing curved lines other than circular arcs. Also called an irregular curve, French curve or Copenhagen curve. A curve designed particularly for plotting the layout of railroad track is called a railroad curve, while curves intended for use in shipbuilders' drafting offices are called shipbuilders' curves. A spline is used for the same purpose but is not rigid and can be made to follow an infinite variety of geometric curves. SEE ALSO distortion curve; easement curve; tide curve; transition curve.

curve, algebraic - (1) The set of points in N-space satisfying a polynomial equation in N variables. In particular, the set of points, in the plane, satisfying a polynomial equation in two variables.

curve, analytic - A curve whose parametric equations are real analytic functions of the same real variable.

curve, Bz - SEE Bz-curve.

curve, catenary - SEE catenary.

curve, characteristic - A curve showing the relationship between exposure and resulting density in a photograph, usually plotted as the density (D) against the logarithm of the exposure (log E) in candle-meter-seconds. It is also called a D-logE curve, Hurter and Driffield curve, H and D curve, sensitometric curve, time-gamma curve and density-exposure curve.

curve, circular - (1) A curve of constant radius. An arc of a circle. (2) A plane curve having the equation $(x^2 + y^2)f(x) + g(x,y) = 0$, where $g(x,y)$ does not contain $(x^2 + y^2)$ as a factor.

curve, clothoid - SEE clothoid.

curve, compound - A curve composed of two or more simple curves which deflect in the same direction and are tangent at the points where they join.

curve, continuous - SEE Jordan arc.

curve, D-logE - SEE curve, characteristic.

curve, easement - SEE easement curve.

curve, fictitious loxodromic - SEE rhumb line, fictitious.

curve, foot-point - A curve which gives the standard deviation in any arbitrary direction. It is tangent to the error ellipsoid at the ends of the major and minor axes, and bows outward at intermediate points.

curve, H & D - SEE curve, characteristic.

curve, Hurter and Driffield - SEE curve, characteristic.

curve, hypsographic - A diagram showing the frequency of occurrence of groups of elevations over the entire world.

curve, isoperimetric - A line, on a map, along which there is no variation from the exact scale of the map. There are two isoperimetric curves passing through every point on a map drawn using an equal-area map projection. This characteristic gives that class of map projections some preference for maps used in engineering.

curve, latitudinal - An easterly and westerly property line adjusted to the same average bearing from each monument to the next one in regular order, as distinguished from the long chord or great circle that would connect the initial and terminal points.

curve, lead - The curve between the switch and the frog in a turnout.

curve, loxodromic - SEE rhumb line.

curve, meizoseismal - A line, on a map, representing points at which destruction was greatest in the region around an earthquake's epicenter.

curve, off-axis - SEE aberration, off-axis.

curve, point of tangency of a - SEE point of tangency.

curve, reverse - A curve composed of two circular arcs with a common tangent at their point of junction and lying on opposite sides of the tangent, that is with their centers of curvature on opposite sides of the tangent.

curve, sensitometric - SEE curve, characteristic.

curve, spiral - (1) A transition curve of uniformly varying radius and connecting a circular arc and a tangent, or two circular arcs whose radii are, respectively, longer and shorter than its own extreme radii. Also called an easement curve or transition curve. (2) SEE spiral; curve, transition.

curve, time-gamma - SEE curve, characteristic.

curve, vertical - (1) A parabolic curve used as a transition curve between different grades or slopes. (2) A Bz curve. (3) A transition curve in the vertical direction. It is usually parabolic and symmetrical about the highest point. (4) A parabolic curve used to connect grades of different slopes to avoid the sudden change in direction in passing from one grade to the other. This method of changing grade is usually used when there is an algebraic difference of more than 0.2 percent in the initial and final grades.

curved-path error - SEE error, curved-path.

curve fitting - Representing a set of points by a curve. Also called filtering, graduation and smoothing. The term is also, though inappropriately, applied to the representing a set of points by a surface. Intuition or experience plays an important part in choosing one particular curve from the infinite number of different curves that could represent a particular set of points, but the deciding criterion is usually that the curve be the simplest that agrees with physical theory while passing as close as possible to the points of the set. The closeness of the curve to the points is determined by requiring, for example, that the sum of the squares of the distances of the points from the curve be as small as possible. SEE ALSO adjustment; smoothing; Vondrak's method.

curve of alinement - A line connecting two points on the surface of the rotational ellipsoid in such a manner that at every point, the azimuths of the two end-points of the line differ by exactly 180°. A curve of alinement is a line of double curvature slightly less in length than the normal-section lines connecting its end points.

curve of equal bearing - The curve connecting all points, on a sphere, at which the great-circle bearing of a given point is the same.

curve spiral point - SEE point, curve-spiral.

curve to spiral - SEE point, curve-spiral.

cusp - A curved segment containing a point at which the normal to the curve changes direction by 180°. If the tangent is a vector, the tangent also changes direction by 180°. The point is called the point of cusp.

cut - (1) The depth to which material is to be excavated (cut) to give the surface a predetermined slope. It is the difference in elevation of a point on the surface and a point on the proposed subgrade vertically below it. (2) The volume of material to be removed to give the surface a predetermined slope.

cut line - A line drawn on a photograph to indicate where the photograph should be cut or torn for best matching it to the adjacent photographs in a mosaic.

cutoff, avulsive - (1) A river's action when an avulsion takes place. (2) The region of land circumscribed by the old and new channels which are the result of an avulsion. One should consult the laws of the particular state in the matter of the effect on the titles and boundaries, including the interpretations through the leading judicial opinions. It appears in all cases that (a) the title to the land within the region cut off and the rules applicable to the old bank-line boundary remain what they were before the avulsive cutoff took place; (b) boundaries described as the center of the main channel and those as along the medial line between the former banks remain as situated immediately prior to the avulsion; and (c) the rules of navigability and new-bank-line riparian right apply to the new channel the same as to the old channel.

cutoff line - A survey line run between two or more stations on a traverse to produce a closed traverse of that part of the survey.

cut tape - SEE tape, subtracting.

cut tide - An oceanic tide which fails to reach its predicted height at high water.

cutting positive - A printing, on glass, of the contour drawing used to make the etched zinc-plate.

cycle (astronomy) - A set of events which recurs regularly; also, the time between recurrences. SEE names of particular cycles such as: cycle, Callippic; cycle, Metonic; Saros.

cycle, anomalistic tidal - The average period, of about 27½ days, measured from perigee to perigee, during which the Moon completes one revolution around the Earth.

cycle, lunar - Any cycle related to the Moon's orbit, particularly the Callippic cycle or Metonic cycle. The synodic month, etc., are also considered lunar cycles, although they are rarely referred to as such.

cycle, nodal - The period of about 18.61 Julian years required for the Moon's nodes to regress through a complete circuit of 360° of longitude. Also called the node cycle. It is accompanied by a corresponding cycle of changing inclination of the lunar orbit relative to the plane of the Earth's equator, with resulting variations in the rise and fall of the tides and speeds of tidal currents.

cycle, tidal - A period which includes a complete set of tidal conditions or characteristics such as a tidal day, a lunar month, or the Metonic cycle. Also called a tide cycle.

cyclonic (adj.) - Having a rotary motion which is counter-clockwise, as viewed from above, in the Northern Hemisphere and clockwise, as viewed from above, in the Southern Hemisphere. Cyclonic motion is undefined at the equator.

cylinder, cut-off - A device consisting essentially of a short, rigid bar so made and so mounted that it forms a direct connection between the apparatus being tested and a permanent monument on the ground. Its length and inclination provide the means for determining the relative locations of the fiducial marks on the base tape or bar and on the monument. It is used to refer the mark at the end of a surveyor's tape or bar standard to a mark on the ground.

D

Dalby's theorem - A formula for the reverse azimuth A_{21} from point P_2 on a spheroid (rotational ellipsoid) to point P_1 on that spheroid, in terms of the forward azimuth A_{12} from P_1 to P_2 and the forward and back azimuths B_{12} and B_{21} of points P_1' and P_2' on a sphere. Points P_1 and P_1' have the same coordinates (λ_1, ϕ_1) and points P_2 and P_2' have the same coordinates (λ_2, ϕ_2); λ and ϕ denote longitude and latitude, respectively. The formula is $A_{21} = A_{12} + (B_{21} - B_{12}) - (e_4/4)(\lambda_2 - \lambda_1)(\phi_2 - \phi_1)^2 \cos 4\phi_1 \sin \phi_1$, where e is the eccentricity.

dam, ground-water - A body of material that is impermeable or has only low permeability and occurs below the surface of the ground in such a position that it impedes horizontal movement of ground water and consequently causes a pronounced difference in the level of the water-table on opposite sides of it.

damages - In the application of eminent domain to taking part of a property, the loss in value of the remainder. Generally, the difference between the value of the whole property before the taking and the value of the remainder after the taking is the measure of the value of the part taken and the damages to the remainder. There are recognized two types of damages: consequential and severance. SEE ALSO plottage damages.

damages, consequential - Damages to property arising as a consequence of a taking and/or construction on other lands. In many States, the owner may be compensated for damage as a consequence of a change in grade of a street which adversely affects ingress to and egress from the affected property. The owner may not be compensated for damage to business, for frustration, or for loss of good will which result as a consequence of a taking or construction by the government.

damages, indirect - SEE severance damages.

damages, plottage - SEE plottage damages.

damage, severance - SEE severance damages.

damages to remainder - The loss in value of the remainder of a piece of property, resulting from the acquisition of a part of the property.

damping correction - SEE correction, air-drag.

dancing - SEE shimmer.

Danjon astrolabe - An astrolabe designed by A. Danjon and based on the double-image astrolabe of Claude and Driencourt, but having the images side by side and having a motor-driven prism. The Claude and Driencourt astrolabe splits light from a star nearing 30° zenith distance to give two images in the focal plane. As the star approaches 30° zenith distance, the two images approach each other along diagonal lines intersecting on the optical axis, and coincide when the zenith distance is exactly 30°, at which point the time is recorded. In the Danjon astrolabe, a double Wollaston prism is placed at the focus. This, together with some screens, produces two images side by side if the prism is placed at the proper distance from the focal plane. The images are kept side by side by a motor that moves the prism uniformly along the optical axis. The zenith distance over a short interval about 30° is therefore measured by the distance of the prism from the focal plane. The observer need merely make small corrections to the distance from time to time.

dark-plate technique - A method of hill-shading by drawing, on a transparent sheet of plastic which has been coated lithographically with a gray tone, sloping regions by darkening or lightening those regions using black or white pencils.

darkroom - A room or alcove devoted to the development of pictures from exposed photographic film and therefore kept dark (or, at least, free from radiation at those wavelengths to which the film is sensitive).

Darwin's constant - The constant κ in the equation $R = a[1 - f \sin \phi' - (4\kappa - 3f^2/2) \sin^2\phi' \cos^2\phi']$, in which a is the equatorial radius of a rotationally symmetric, ellipsoidal, liquid body, f is the flattening of the ellipsoid and ϕ' is the geocentric latitude. The body has the same mass as the Earth and is composed of concentric layers of constant density bounded by confocal ellipsoids. It rotates at the same rate as the Earth and the minor axis is the axis of rotation. Darwin used Roche's law for the variation of density ρ from center to outside: $\rho = \rho_o(1 - kr^2)$, in which ρ_o is the density at the Earth's center, r is the average radius of a layer and k is 0.764. Darwin obtained +0.000 008 20 for κ. de Sitter, later, obtained 0.000 002. Darwin's constant is a fundamental constant of geophysics.

Darwin's modification (of Clairaut's formula) - The formula $\tau = \tau_e [1 + (5m/2 - 17 mf/14 + 15m^2/4) \sin^2\phi + (1/8)(f^2 - 5mf) \sin 4\phi]$ for the numerical value τ of gravity on an equipotential, rotational ellipsoid at geodetic latitude ϕ, given the theoretical value τ_e of gravity at the equator, the flattening f and the constant m. The flattening is $(a - b)/a$, where 2a is the length of a major diameter of the ellipsoid and 2b the length of a minor diameter. m is the quantity $\omega^2 a/\tau_e$, where ρ is the rate of rotation of the ellipsoid. Also called the Clairaut-Darwin formula and Clairaut-Darwin gravity formula.

data - The plural form of datum(1).

data, analog - The charting record of a continuous physical quantity representing the corresponding magnitude of a mathematical variable or measurements. E.g., the voltage between two points of an electric circuit may represent a varying distance or angle. One single point from that record is called an analog datum.

data, a priori - SEE data, prior.

data, digital - A sequence or collection of numbers giving the magnitude of a mathematical variable or measurements. The term is usually applied to numbers obtained by using electrical circuits to convert the amplitude of a voltage, current or other physical quantity to digital form. A single number from the sequence is called a digital datum.

data, gridded - Data arranged in an array.

data, prior - Data obtained before those currently under consideration or being used. Also called, erroneously, a priori data.

data bank - (1) A specific collection of data such as, e.g., the material in the National Archives of the USA. (2) A place of organization containing or holding data such as, the U.S. National Archives of the USA or the memory of a computing machine. (3) Data. Also called a data base. Using data bank or data base to mean data is unnecessary.

data base - (1) Data collected for a particular project. It differs from a data bank in that the latter need not be for a particular project. (2) The same as data base (1) but with the condition that the data be organized for a particular purpose. (3) Data. Also called data bank. Using data base or data bank to mean data is unnecessary and pretentious.

data chamber - The part of a camera wherein data can be recorded on the margin of the film or plate. Data placed on photographs taken for mapping usually include the time of exposure, altitude, frame number and identification number. They are usually recorded automatically.

data processing - Any sequence of operations on data, usually in accordance with a specified or implied set of rules.

data processing, batch - SEE batch data processing.

data processing, sequential - Data processing in which the data are submitted to the computer as separate sets. The computer operates on the sets in sequence, combining the results of operating on a particular set with the data of the next set and operating on the combination of results and new data. If the amount of data is very large, sequential processing may have to be used.

data reduction - The transformation of measured values into some other form such as another order, or into averaged or adjusted values.

data smoothing - This is a pleonasm; SEE curve fitting; smoothing.

date - (1) A specified instant. The instant is specified by giving its time within the day and by identifying the day. (2) A specified day. Also called civil date. The day may be specified by giving its ordinal number within a month, the name or ordinal number of the month within the year, and the number of the year in some calendar. (In civil matters, the instant is not usually considered part of the date. SEE ALSO date, astronomical. In scientific literature, the number of the day in the month is usually given first, then the name or number of the month, and then the number of the year, e.g., 29 February 1980. This order is reversed in some sciences such as seismology. Most countries adhere to this same usage in civil matters. In English-speaking countries, however, the practice in civil matters is to give first the name of the month, then the number of the day followed by a comma, and then the number of the year, e.g., February 29, 1980. SEE ALSO Greenwich sidereal date; Julian ephemeris date.

date, astronomic - SEE date, astronomical.

date, astronomical - A specified instant designated by year, month, day, and fractional part of a day. The year is defined to begin at 0h (i.e., the starting midnight) of 31 December of the previous year; this day (31 December) is then day 0 of the reckoning. Also called an astronomic date. For example, the astronomical date corresponding to 21 December 1966, at 18 hours Universal Time is 1966 December 21.75 U.T. The astronomical date 1980 January 0.0 U.T. corresponds to 0h on 31 December 1979. There is not general agreement on the order in which the parts of an astronomical date are to be given or on the manner in which they are to be given. SEE ALSO date.

date, Besselian - SEE Besselian date.

date, civil - SEE date (2).

date, double - SEE dating, double.

date, Julian - SEE Julian date.

date line - A hypothetical line separating two neighboring time-zones in which the civil dates differ by one day. The term is often used to designate the international date line. It is written, on charts of the U.S. Hydrographic office, as Dateline.

date line, international - A particular, hypothetical line on the Earth separating neighboring regions in which the civil dates differ by one day. There is no international date line which has been formally adopted by the various nations of the world. In 1884, the International Meridian Conference, held in Washington, D.C., established the meridian of Greenwich as the meridian from which time was to be reckoned. The meridian 180° from the Greenwich meridian thus became the international date line. The international date line in use by the U.S. Hydrographic Office is defined as follows: it is drawn from the North Pole due south along the 180th meridian to 75° N.; thence southeastward, to the east of Herald Island, to 68° N and the longitude of the meridian passing between the Diomede Islands (approximately 168° 58' 22" W.); thence due south through the Bering Straits to 65° 30' N.; thence southwestward to 53° N., 170° E.; thence southeastward to 48° N. and the 180th meridian; thence due south to 5° S.; thence southeastward to 15° 00' S., 172° 30' W.; thence due south to 45° 00' S., 172° 30' W.; thence southwestward to 51° S. and the 180th meridian; thence south to the South Pole.

dating, double - Dating an event according to both the Julian and the Gregorian calendars. During 1752, Great Britain and her colonies officially replaced the Julian calendar, with 25

March as the beginning of the year, by the Gregorian calendar, with 1 January as the beginning of the year. The difference between the two calendars having increased from 10 days in 1582 to 11 days, the change was made by having 2 September 1752, followed immediately by 14 September 1752. To avoid confusion, dates according to the Julian calendar were marked old style or (O.S.); dates according to the Gregorian calendar were marked new style or (N.S.). In double dating early American records, not only the day and the month but also the year are subject to change. In England, at the time of the change, the official year commenced on 25 March, so that 24 March 1730 (O.S.), corresponded to 4 April 1731 (N.S.), but the next day was 25 March 1731 (O.S.), or 5 April 1731 (N.S.). This practice was not universal, for there was some popular use of 1 January as the beginning of the year even before the official change.

datum - (1) In general, a single, isolated piece of information. A collection or set of such pieces of information is referred to as data. E.g., geographic data refers to a list of longitudes and latitudes. The term is sometimes used to denote only information in numerical form or even only to values obtained by measurement or observation. Where confusion might otherwise result, it is convenient to refer to numerical information as numerical data and to measured or observed values as observational data, measurement data or raw data. The values obtained from numerical data by calculation are referred to as results or calculated data. (2) Any quantity, or set of such quantities, which may serve as a referent or basis for calculation of other quantities. In particular, a geodetic datum, project datum, photogrammetric datum, chart datum or tidal datum. The plural form is datums. SEE ALSO Adindan Datum; altitude datum; Arc Datum; Australian Geodetic Datum; Australian Geodetic Datum 1965; Australian Height Datum; Camp Colonna Datum; Canadian Geodetic Datum; Cape Datum; Cape Kennedy Datum; chart datum; correction for datum; Doppler Satellite Datum; elevation datum; European Datum 1950; European Datum 1979; Golofnin Bay Datum; Gulf Coast Low-Water Datum; Imbitube Datum 1959; Indian Datum; International Great Lakes Datum 1955; Kripniyuk-Kwiklokchun Datum; Kronstadt Datum; leveling datum; Luzon Datum; springs datum, low-water; Mercury Datum 1960; Mercury Modified Datum 1968; model datum; National Geodetic Vertical Datum; New England Datum; Newlyn Datum; North American Datum; North American Datum 1927; Old Hawaiian Datum; Ordnance Datum; Ordnance (Liverpool) Datum; Ordnance Newlyn Third Geodetic Leveling Datum; origin (of a datum); Panama-Colon Datum; Port Clarence Datum; Puerto Rico Datum; Pulkova Datum 1932; Pulkova Datum 1942; St.George Island Datum; St. Michael Datum; St.Paul Island Datum; South American Datum 1969; South American Provisional Datum 1956; Standard (California) Astronomical Datum 1885; state plane coordinates datum; Tsingtao Datum; Tokyo Datum; Transcontinental Triangulation Datum; Unalaska Datum; United States Standard Datum; Valdez Datum; world datum; Yakutat Datum; Yof Astro Datum 1967; Yolo Base Datum; Yukon Datum.

datum, absolute - SEE datum, absolute geodetic.

datum, absolute geodetic - (1) A geodetic datum whose ellipsoid is specified to have its center at the Earth's center of mass and its least axis parallel to the Earth's axis of rotation. Also called an absolute datum. (2) A geodetic datum defined solely by reference to definable and accessible physical objects.

datum, adjusted - A reference surface (for plane coordinates), above or below the reference surface on which the plane coordinate system was originally placed by mathematical procedures. The new surface is always parallel to the original surface. Its location above or below the original surface is governed by the degree to which the inherent difference can be reduced between (a) distance measured on the ground and (b) distance computed using plane coordinates of points and features delineated on maps or surveyed distance computed and/or measured on the adjusted datum.

datum, analog - SEE data, analog.

datum, astro-geodetic - A geodetic datum defined in terms of astronomic and geodetic quantities.

datum, complete - A datum containing exactly all the constants needed for specifying the coordinate system used. By definition, all geodetic datums (in sense (1)) are complete datums. However, the term datum is sometimes applied to sets of constants insufficient to specify the coordinate system. Then usually certain constants are assumed but are not specified. E.g., one or more of the constants specifying the orientation of the coordinate system are often omitted if a conventional orientation can be inferred.

datum, convergent-model - SEE model datum.

datum, digital - SEE data, digital. However, number or value is better.

datum, floating - SEE network, independent.

datum, geocentric - Any datum in which the Earth's center of mass is involved in specifying the coordinate system. The center may be involved directly, e.g. by specifying that the origin of the coordinate system be at the center of mass. It may be involved indirectly, e.g., by including specifications for values of gravity or gravity potential. A geocentric datum is sometimes referred to as an absolute geodetic datum.

datum, geocentric geodetic - (1) A geodetic datum which specifies that the center of the ellipsoid shall be located at the Earth's center of mass. It is customary to specify also that the shortest axis of the ellipsoid be parallel to the Earth's axis of rotation (instantaneous, average or conventional) or to some other similar line of known orientation. Another axis may be required to be parallel to plane of the Greenwich meridian or, for some purposes, to the plane of the zero meridian of the Bureau International de l'Heure. More often, the ellipsoid is placed to provide a satisfactory fit to the geoid. A rectangular coordinate system is often associated with a geocentric geodetic datum by placing the origin of that coordinate system at the center of the ellipsoid, making the x- and z-axes coincide with the longest and shortest axes, respectively, of the ellipsoid, and requiring that the coordinate system be right-handed. (2) A geodetic datum specifying that the center of the ellipsoid be at the Earth's center of mass, that the shortest axis be parallel to the Earth's conventional axis of rotation, and that one of the other two axes be parallel to the plane of the meridian of Greenwich.

datum, high-water - The high-water plane (surface) to which elevations of features on land are referred. Also called datum for heights.

datum, horizontal - SEE datum, horizontal-control.

datum, horizontal geodetic - SEE datum, horizontal-control.

datum, hydrographic - (1) The vertical-control datum for depths (soundings), depth contours, and elevations of foreshore and offshore features. (2) SEE chart datum.

datum, independent - (1) A datum established independently of any other datum. (2) SEE network, independent.

datum, local - A datum defining a coordinate system which is used only over a region of very limited extent. Use project datum for projects of limited extent.

datum, lower low-water - An approximation to mean lower low water adopted as a tidal datum for a limited region and retained for an indefinite period. It is used primarily for civil engineering in rivers and harbors. Columbia River lower low water datum is an example.

datum, low-water - (1) An approximation to mean low water that has been adopted as a standard tidal datum for a specific region although it may differ slightly from a later determination. (2) The dynamic height for each of the Great Lakes and Lake St.Clair, and the corresponding sloping surfaces of the St.Mary's, St.Clair, Detroit, Niagara, and St.Lawrence Rivers, to which are referred the depths shown on charts and the authorized depths for improvements to navigation. Elevations of these datums are referred to International Great Lakes Datum (1955).

datum, low-water springs - SEE springs datum, low-water.

datum, major - SEE datum, preferred.

datum, marine - A point, line or surface used as reference from which to reckon elevations or depths. It is called a tidal datum when defined by a certain phase of the tide.

datum, mean lower low-water - The average level of the lower of two successive low waters. It is frequently specified as the average during a particular part of the month or year, for example, mean lower low-water springs, mean lower low water at spring tide near the solstices, etc. Also a datum based on extensive observations of tidal lower lows.

datum, mean sea-level - A determination of mean sea level that has been adopted as a datum for elevations.

datum, photogrammetric - (1) A surface, usually flat, to which elevations in the region being studied are referred. The surface is usually taken as nearly horizontal as possible. (2) The plane, corresponding to the surface in the first definition, in a stereoscopic model or photogrammetric instrument, to which heights or elevations are referred. Sometimes referred to simply as model datum or datum.

datum, photographic - A horizontal plane at the average elevation of the terrain shown in a photograph and on which distances measured will be approximately proportional to corresponding distances measured on the photograph.

datum, preferred - A geodetic datum selected as the single datum for a region containing several independent datums. Also called a major datum.

datum, quasi-geodetic - A datum which does not meet all the requirements for a geodetic datum. A quasi-datum is usually either under-defined or, more rarely, over-defined, that is, it contains either too few constants or too many constants.

datum, sea-level - (1) An equipotential surface passing through a specified point whose elevation above mean sea level is known and which is used as a referent for elevations. (2) A surface passing through mean sea level at certain specified points and through other specified points whose elevations are known, and to which vertical distances determined by leveling are referred. Because these points do not lie on a single equipotential surface, the surface they and the leveling are used to define is neither an equipotential surface nor a simple geometrical surface. A datum of this kind was defined by the U.S.C.& G.S. in 1929 and called Sea Level Datum of 1929. The name, but not the definition, was changed in 1976 to National Geodetic Vertical Datum. (3) Mean sea level used as a datum for elevations or depths. This datum is not valid for points on land, since sea level cannot be extended very far inward from the shore.

datum, sounding - SEE sounding datum.

datum, tidal - (1) A surface with designated elevation, from which heights or depths are reckoned and which is defined by a certain phase of the tide. A tidal datum is a local datum, usually valid only for a restricted area about the tide gage used in defining the datum. For permanency and for convenience, a bench mark is emplaced in stable ground close to the tide gage. The elevation of the bench mark with respect to the tidal datum is determined by the tide gage, not the other way around. Also called a datum surface. When used as reference surface for hydrographic surveys, tidal datums have been termed datum planes; however, they are not planes and are treated as curved, level surfaces. The tidal datum in most general use in geodesy is mean sea level. In land surveying, where boundaries and riparian rights are involved, mean high water, mean low water and mean lower low water are sometimes used as tidal datums. Also called datum level. (2) SEE chart datum.

datum, tide-gage - A horizontal plane defined at a particular, arbitrary distance below a tide-gage bench mark and from which tidal heights at a tide gage are, in theory, measured. In the period between successive levelings to a tide-gage bench mark, the plane is established at the tide gage by reference to the contact mark.

datum, vertical - SEE datum, vertical-control.

datum, vertical-control - (1) The set of constants specifying the coordinate system to which elevations are referred, or (2) the coordinate system specified by the constants. When vertical control is understood to refer only to elevations determined by geodetic leveling, the constants consist of the elevations which are known by definition (that is, independent of measurement by leveling). The vertical-control datum (2) then consists of the points specified by the constants and the surface through these points as determined by leveling and adjustment. If only one elevation is specified, the datum is the geoid or an approximation thereto. If several elevations are specified, the surface is complicated and is neither the geoid nor mean sea level. When vertical control is understood to refer to elevations or heights determined by methods not involving the geoid (as, for example, heights determined by satellite geodesy), the datum is either the complete datum used for the survey or that portion of the datum sufficient for specifying elevations or heights. (3) Any level surface taken as a surface of reference from which to reckon elevations.

Datum Control Network, National Tidal - SEE National Tidal Datum Control Network.

Datum Convention of 1980, National Tidal - SEE National Tidal Datum Convention of 1980.

Datum Epoch, National Tidal - SEE National Tidal Datum Epoch.

datum for heights - SEE datum, high-water.

datum level - SEE datum, tidal.

datum line - SEE reference line.

Datum of 1929, National Geodetic Vertical - SEE National Geodetic Vertical Datum of 1929.

Datum of 1929, Sea-level - SEE Sea-level Datum of 1929.

datum orientation, astrogeodetic - Adjustment of the ellipsoid of reference of a particular datum so that the sum of the squares of the deflections of the vertical at selected points throughout the geodetic network is minimized.

datum orientation, gravimetric - Adjustment of the reference ellipsoid, for a particular geodetic datum, so that the differences between the geoidal heights determined from gravimetric and astrogeodetic surveys are (usually in a least-squares sense) minimized.

datum orientation, single astronomic-station - The definitions of a geodetic datum by accepting the astronomically determined coordinates of the origin and the azimuth to one other station without any correction.

datum origin - SEE origin (of datum).

datum plane - A tidal datum. The term is obsolescent; it is also inappropriate, because a plane is not involved. SEE ALSO datum, tidal. Also called datum level; reference level.

datum plane, standard - An imaginary surface containing all points at which the barometric pressure is 29.92 inches of mercury at a temperature of 15° C.

datum plane, tidal - SEE datum, tidal.

datum point - Any point of known or assumed coordinates used as a reference which measurements or calculations may be taken.

datum reduction - The quantity added to a depth, after all other corrections to the measured depth have been made, to refer each actual depth to the reference surface for the particular region.

datum surface - SEE datum, tidal.

datum tick - One of a pair of marks placed on opposite edges of a chart to indicate the location of a line of longitude or of latitude with respect to a particular datum.

datum transformation - (1) The systematic elimination of discrepancies between adjoining or overlapping triangulation networks on different datums by moving the origin, rotating, and stretching the networks until they fit each other. (2) The set of constants used to transform the coordinates of a station from one datum or ellipsoid to another, together with the coordinates on the two datums.

Davidson meridian instrument - SEE meridian telescope.

Davis quadrant - SEE backstaff.

day - (1) The period during which the Earth rotates through 360°; the interval between successive passages of a fixed point on the Earth's surface through a specified plane in space. Because the direction of this plane is not usually fixed in inertial space, the length of the day will depend on how the plane is specified. For example, the reference plane may be required to pass through the center of the Sun, giving rise to the apparent solar day; or the plane may be required to pass through the vernal equinox, giving rise to the sidereal day. (2) A period containing 86,400 seconds of time. If the second is defined in terms of a spectroscopic frequency, the length of the corresponding day is constant. The length is also constant if the average value of the second during some past period is used. The Ephemeris second is defined in this way. If the second is defined by reference to current positions of celestial bodies, the length of the day is not constant and the definition of day may become equivalent to the definition (1) of day. (3) A period which is a specified fraction of the period of motion of some celestial body such as the Earth, Moon, etc. This definition is a generalization of the definition (1) of day. The term is often used, together with the term for a longer period, as a way of expressing a date: for example, the 23rd day of May, or the 2nd day of the fourth week of May. This mode of expression is still used in legal practice, where it originated (and should remain). SEE ALSO calendar day; constituent day; Ephemeris day.

day, apparent sidereal - The interval of time between two successive, upper transits of the true vernal equinox over some meridian.

day, apparent solar - The interval of time between two successive upper or lower transits of the Sun across a given meridian. Because the revolution of the Earth is not uniform, apparent solar days vary in length throughout the year. The greatest deviation from a mean solar day is not quite a half-minute in either direction.

day, astronomic - SEE day, astronomical.

day, astronomical - (1) A solar day beginning at noon. Also called an astronomic day. It may be based either on apparent solar time or on mean solar time. It begins 12 hours later than the civil day of the same date. Before 1925, the astronomical day was used in the American Ephemeris and Nautical Almanac. Beginning in 1925 and since, the civil day has been used instead. (2) A mean solar day beginning at noon, 12 hours after the midnight of the same civil day.

day, civil - A solar day beginning at midnight. The civil day may be based either on apparent solar time or on mean solar time. It begins 12 hours earlier than the astronomical day of the same date.

day, equinoctial - SEE day, sidereal.

day, half-pendulum - The period T_d, in sidereal hours, at a point at geodetic latitude ϕ, as given by the formula $T_d = 12h/\sin\phi$. A period of 12 pendulum hours.

day, intercalary - A day introduced into the calendar in certain years to make the calendar agree with the tropical year. The 29th of February is such a day.

day, Julian - SEE Julian day.

day, lunar - (1) The interval of time between two successive upper or lower transits of the Sun through a meridian of the Moon. (2) The interval of time from one upper transit of the Moon through a meridian of the Earth to the next upper transit. It is about 24 hours 50 minutes 24.9 seconds long, on the average, but varies as much as 10 minutes from this value.

day, mean sidereal - SEE time, solar.

day, mean solar - The interval of time between two successive upper or lower transits of the mean Sun across a given meridian. It is the average length of the apparent solar day throughout the year. SEE ALSO time, mean solar.

day, modified Julian - SEE Julian date, modified.

day, sidereal - The interval of time between two successive upper or lower transits of the (true) vernal equinox across a given meridian. Also called an equinoctial day. The length of the sidereal day is subject to slight irregularities on account of small differences between the positions of the true equinox, which is affected by precession and nutation, and the mean equinox, which is affected by precession only. A sidereal day contains 24 sidereal hours each consisting of 60 sidereal minutes each of which consists of 60 sidereal seconds. It therefore contains 1 440 sidereal minutes and 86 400 sidereal seconds. It contains 23h 56m 04.091s of mean solar time, which is 23.934 469 722 hours, 1 436.068 183 333 minutes, or 86 164.091 seconds.

day, solar - (1) The interval of time between two successive upper or lower transits of either the Sun or the mean Sun across a given meridian. SEE ALSO day, apparent solar; day, mean solar. (2) The length of time it takes the Sun to make one rotation.

day, tidal - (1) The interval of time between two successive occurrences of higher high water at a particular place. It is about 24h 50m long and is approximately equal to the lunar day (meaning 2 below). (2) The interval of time between two successive upper transits of the Moon over the meridian at a place. This is approximately the same interval as that defined in (1). Tidal day and lunar day are not the same. Lunar day is sometimes used instead of tidal day but is also used to denote the interval between two successive upper or lower transits of the Sun over a meridian on the Moon.

daylight-saving time - SEE time, daylight-saving.

day number, Besselian - SEE Besselian day-number.

day number, independent - Any one of the six independent quantities f, g, G (related to precession and nutation), and h, H, i (related to annual aberration) in the two equations for rotation of the coordinate system from right ascension α_o and declination δ_o at mean equinox at time t_o, to right ascension α and declination δ at equinox of date at time $t_o + \Delta t$: $\alpha = \alpha_o + \mu\alpha_o + f + g \sin(G+\alpha_o) \tan\delta_o + h \sin(H+\alpha_o) \sec \delta_o + J \tan^2\delta_o$; $\delta = \delta_o + g \cos(G+\alpha_o) + h \cos(H+\alpha_o) \sin \delta_o + i \cos \delta_o + J'\tan \delta_o$. $\mu\alpha$ and $\mu\delta$ are the proper motions in right ascension and declination, respectively. The quantities f, g, G, h, H, i are given by

$$f = m\Delta t + \Delta\psi \cos \varepsilon$$
$$g \sin G = -\delta\varepsilon$$
$$g \cos G = n\Delta t + \Delta\psi \sin \varepsilon$$
$$h \sin H = +\kappa (\cos L_s - e \cos \theta) \cos \varepsilon$$
$$h \cos H = +\kappa (\sin L_s - e \sin \theta)$$
$$i = -\kappa (\cos L_s - e \cos \theta) \sin \varepsilon .$$

In these formulae, m and n are precessional constants, $\Delta\psi$ and $\Delta\varepsilon$ are the total nutation in longitude and in obliquity, respectively, Ls is the Sun's geocentric longitude and κ is the constant of aberration. ε is the obliquity of the ecliptic. J and J' are small numbers providing second-order corrections to precession and nutation; they need not be considered for most geodetic calculations.

day number, Julian - SEE Julian day-number.

day number, modified Julian - SEE Julian day-number, modified.

deadbeat compass - SEE compass, aperiodic.

dead reckoning - SEE reckoning, dead.

debug (v.) - Remove the errors present in an algorithm intended to be used by a computing machine. (jargon)

December solstice - SEE winter solstice.

decentering distortion - Distortion caused by the failure of one or more elements of an optical system to have their optical axes lie on the same line as the other optical axes.

decentration - Failucal axes lie on the same line as the other optical axes. re of the optical center of one element of an optical system to lie on the optical axis of a preceding or succeeding element. Decentration causes distortion, tangential distortion in particular, similar to that caused by introducing a thin wedge into a perfectly centered optical system. Decentration is therefore sometimes called the wedge effect.

decimal (adj.) - Using the number 10 as a unit in counting. Also called denary.

decimal, binary coded - SEE number, binary coded-decimal.

decimal, coded - SEE number, binary coded-decimal.

decimal system, binary coded - SEE system, binary coded-decimal.

decimeter, dynamic - The quantity 1 m^2/s^2. It is not, even approximately, the same as the amount of work done in lifting a unit mass one decimeter.

declination - (1) The angle, at the center of the celestial sphere, between the plane of the celestial equator and a line from the center to the point of interest (on a celestial body). Declination, conventionally denoted by δ, is measured by the arc of hour circle between the celestial body and the equator; it is positive when the body is north of the equator and negative when south of it. It corresponds to latitude on the Earth and with right ascension forms a pair of coordinates which defines the position of a body on the celestial sphere. (2) SEE declination, magnetic. SEE ALSO grid declination; parallel of declination.

declination, magnetic - (1) The direction (in angle east or west from the north branch of the celestial meridian) of magnetic north as determined by the positive pole of a freely suspended magnetic needle which is subject to no transient, artificial disturbance. (2) The angle, at any point, between the direction of astronomic north and the direction of the horizontal component of the Earth's magnetic field. Also called declination of compass. In nautical and aeronautical navigation, the term variation is used instead of declination, and the angle is called variation of the compass or magnetic variation. Except in navigational usage, magnetic declination is not synonymous with magnetic variation, which refers in other disciplines to regular or irregular change with time of the magnetic declination, dip or intensity.

declination arc - (1) A graduated arc, on a surveyor's solar compass or on the solar attachment of an engineer's transit, on which the declination of the Sun (corrected for refraction) is set off. (2) A graduated arc, attached to the alidade of a surveyor's compass or transit, on which the magnetic declination is set off. A reading of the needle will give a bearing corrected for that declination.

declination axis - The axis, on a telescopic mounting, perpendicular to the polar axis (i.e., in a plane perpendicular to the Earth's axis of rotation) so that the telescope can be rotated in declination.

declination of compass - SEE declination, magnetic (2).

declination of grid north - SEE gisement.

declinatoire - SEE box compass.

declinometer - A magnetic compass similar to a surveyor's compass but constructed so that the line of sight can be rotated to conform with the direction of the needle or be directed to any desired setting on the horizontal circle. It is used in determining the magnetic declination. Some models are self-registering or recording.

decree - The court's decision in equity. A decree usually directs a defendant to do or not do some specific thing, as opposed to a judgment for damages in a court of law. The same court may ordinarily sit either as a court of equity or a court of law.

dedicate (v.) - Appropriate and set apart land from one's private property to some public use.

dedication - The act of dedicating something. The dedication may be either express or implied. It is express when there is an express manifestation, on the part of the owner, of his purpose to devote the land to a particular, public use such as e.g., the streets in platted subdivisions. It is implied when the owner's acts and conduct manifest an intention to devote the land to public use. To make the dedication complete, there must not only be an intention on the part of the owner to set apart the land for the benefit of the public, but there must be an acceptance by the public.

deed - A legal instrument which, when executed and delivered, reports to an estate in real property or an interest therein.

deed, corporate - A deed transferring title to lands by a corporation.

deed, general-warranty - A deed in which the grantor warrants the title against defects arising at any time either before or after the grantor became connected with the land.

deed, mortgage - SEE mortgage deed.

deed, quitclaim - A form of conveyance whereby whatever interest the grantor possesses in the property described in the deed is conveyed to the grantee without warranty of title.

deed, special-warranty - (1) A deed in which the grantor warrants the title against defects arising after he acquired the land but not against defects arising before that time. (2) A deed wherein the grantor limits his liability to the grantee to anyone claiming by, from, through or under him, the grantor.

deed, unrecorded - A deed which is not properly recorded in a registry of deeds or other depository of public records. An unrecorded deed is binding on the grantor, his heirs and devisees, and on all other persons having actual notice thereof, but is not valid and effectual against any other persons.

deed, void - A deed whose provisions cannot be enforced.

deed in trust - A deed which establishes a trust. Also called a trust deed. It generally is a legal instrument which conveys legal title to property to a trustee and states his authority and the conditions binding upon him in dealing with the property held in trust. Trust deeds are frequently used to secure lenders against loss. In this respect, they are similar to mortgages. Also called a trust deed. Do not confuse deed in trust with deed of trust (q.v.).

deed of trust - A deed in which title to property is transferred to a third party as trustee to be security for an obligation owed by the trustor (borrower) to the beneficiary (lender). Also called a trust deed, but not to be confused with deed in trust.

deep (hydrography) - The deepest part of a submarine depression.

definition - (1) The degree of clarity and sharpness of an image. The term, in this sense, does not have the objective and qualitative connotations of resolution. However, when used in defining precise concepts such as focal plane or focal length, it is often used in the sense of resolution and particularly in the sense of resolution on the optical axis or area-weighted average resolution in the focal plane. SEE ALSO resolution. (2) The subjective impression of clarity and sharpness made on an observer by a photograph and resulting from the combined effects of resolution, sharpness, contrast, graininess and tonal reproduction. It should not be confused with or used for resolution, which denotes an objective concept and is only one of the elements producing definition. Also referred to as photographic definition.

definition, photographic - SEE definition (2).

deflection, astrogeodetic - SEE deflection of the vertical, astrogeodetic.

deflection, plumb-line - SEE deflection of the plumb-line.

deflection, relative - SEE deflection of the vertical, astro-geodetic.

deflection, residual - The unexplained difference between the direction of gravity and the normal to the ellipsoid best representing the geoid, after corrections for the effects of topography and compensation have been applied to the former. It is usually specified by its components in the meridian and in the prime vertical.

deflection, topographic - SEE deflection of the vertical, topographic.

deflection angle - (1) A horizontal angle measured from the prolongation of the preceding line, right or left to the following line. Only directed polygons, such as traverses, have deflection angles. (2) The angle, measured in that vertical plane containing the flight line, between the plane $z = 0$ for one model in a strip and the plane $z = 0$ for the preceding model of the strip.

deflection anomaly - The difference between a deflection of the vertical calculated from astronomical observations and the deflection calculated from data on gravity. The term is seldom used. Its meaning is, besides, inconsistent with that given to anomaly in other terms and is often expressed, instead, by residual.

deflection of the plumb line - (1) The angle, at a point, from the negative direction of the normal to the spheroid of reference, to the direction of gravity (i.e. of the plumb-line.) SEE ALSO deflection of the vertical, which is the angle opposite to the deflection of the plumb line. The magnitudes and directions of both kinds of deflection are, however, the same. When the deflection of the plumb line is resolved into components ξ in the meridional plane through the point and η in the prime vertical, ξ is usually considered positive southwards and η positive westwards. (2) The negative of the quantity defined in (1).

deflection of the vertical - (1) The angle ζ, at the point P on the Earth's surface, between the vertical at that point and the normal there to the reference ellipsoid. In modern usage, the normal and the vertical are considered to be pointing upwards and the angle is measured from the normal to the vertical. The deflection is usually resolved into two components: ξ, in the meridional planed through P and positive northwards; and η, in the prime vertical through P and positive eastwards. (Some authors interchange ξ and η.) Also called Helmert deflection of the vertical, astro- geodetic deflection of the vertical, astronomic-geodetic deflection of the vertical, ground-level deflection of the vertical, normal deflection of the vertical, relative deflection of the vertical and station error. Unless specifically stated to be otherwise, deflection of the vertical always has the meaning given above. (2) SEE deflection of the vertical, reduced. (3) SEE deflection of the vertical, gravimetric. (4) The angle ζ', at a point P on the Earth's surface, between the theoretical direction of the vertical there and the actual direction of the vertical. (5) The angle ζ' at the point P' which is on the geoid and vertically beneath a point P on the Earth's surface, between the normal to the equipotential surface used as reference and the normal to the geoid. SEE ALSO Helmert deflection of the vertical; Vening Meinesz' formula.

deflection of the vertical, absolute - SEE deflection of the vertical, gravimetric.

deflection of the vertical, absolute gravimetric - SEE deflection of the vertical, gravimetric.

deflection of the vertical, astro-geodetic - (1) The deflection of the vertical referred to a datum whose coordinate system has been oriented by astro-geodetic methods. Also called relative deflection and astrogeodetic deflection. (2) The angle, at a point, between the vertical there and the normal to the ellipsoid of reference of an astrogeodetically oriented datum. Also called relative deflection. (3) SEE deflection of the vertical (1).

deflection of the vertical, astronomic-geodetic - SEE deflection of the vertical (1).

deflection of the vertical, external - The angle, at a point P on the Earth's surface, between the theoretical direction of the vertical (opposite to the direction of gravity) and the actual direction of the vertical.

deflection of the vertical, gravimetric - (1) An angle calculated from gravimetric measurements using Vening Meinesz's formula for the deflection of the vertical or some variant of that formula. This angle is theoretically related to the deflection of the vertical (1), and is nearly the same as that given in the following definition. (2) The angle ζ', at a point P' which is on the geoid and is vertically below the point P on the Earth's surface, between the actual vertical through that point and the vertical through it from an equipotential surface of reference. The gravimetric deflection of the vertical is sometimes called the absolute deflection of the vertical or absolute gravimetric deflection of the vertical. Helmert called it the divergence of the vertical.

deflection of the vertical, ground-level - SEE deflection of the vertical.

deflection of the vertical, normal - SEE deflection of the vertical (1).

deflection of the vertical, reduced - The angle ζ', at the point P' which is on the geoid and vertically below the point P on the Earth's surface, from the normal at the reference ellipsoid to the vertical through P'. Because P' is not accessible, the directions of both normal and vertical must be calculated.

deflection of the vertical, relative - SEE deflection of the vertical (1).

deflection of the vertical, topographic - That part of the deflection of the vertical which is caused by the gravitational pull exerted by topographic masses. Topographic deflection of the vertical is not the same as deflection of the vertical or station error but is the theoretical effect produced by the resultant gravitational pull of the unevenly distributed, topographic masses around the station, no allowance being made for isostatic compensation.

deflection of the vertical, topographic-isostatic - That part of the gravimetric deflection of the vertical which is caused by the gravitational pull of topographic masses and by isostatic compensation.

deflection-of-the-vertical anomaly - SEE deflection anomaly.

deformation, affine - A deformation, of a coordinate system, resulting in a coordinate system such that the scale along one axis or reference plane is different from the scale along the other axes or planes. SEE ALSO transformation, affine.

deformation, horizontal - In the relative orientation of two photographs, the cumulative warpage of the models affecting the horizontal datum because of errors in the z-motion, bridging and swing.

deformation, vertical - In the relative orientation of two photographs, the cumulative warpage of the models affecting the vertical datum because of errors introduced by x-tilt and y-tilt.

degaussing (v.) - Reducing the magnetization of a body as nearly to zero as possible.

degaussing range - SEE range, degaussing.

degree - (1) A unit of angle (measure) equal to 1/360-th of the angle through which a line rotates in going from its original position around through a complete circle and back to its original position. Equivalently, the amount of angle contained between two intersecting straight lines which cut off 1/360-th of the circumference of a circle whose center is the point of intersection. (2) A unit of arc equal to 1/360-th of a complete circle. SEE ALSO plane angle.

degree, length of - SEE length of degree.

degree, square - (1) A unit of measure of solid angle corresponding to the area, on a sphere of unit radius, of a quadrangle composed of great arcs 1° on a side. There are $129\ 600/\pi$ (approximately 41 250.87) square degrees on a sphere, and approximately 3 282.64 square degrees in a steradian. (2) A unit of area equal to 1/64 800 of the surface of a unit sphere. This definition and the preceding one do not define the same quantity. A solid angle (figure) having a solid angle (measure) of 1 square degree is not necessarily square. The quadrangle used in the first definition is not rectangular - the internal angles are slightly greater than 90° and the sides bow out.

degree of curve - The number of degrees of angular measure, at the center of a circle, subtended by an arc 100 feet in length. In early surveys of highways and railroads, a 100-foot chord has been used instead of a 100-foot arc in defining degree of curve.

degrees of freedom, number of - SEE number of degrees of freedom.

Delaborne prism - SEE Dove prism.

Delaunay element - One of a set of orbital elements derived by a transformation of the Keplerian elements. The equations of the orbit are sometimes simpler when Delaunay elements are used.

delimitation - The drawing of boundaries on a map.

delineation - (1) The technique of selecting and outlining, on pictorial sources of information or on a map manuscript, those features worthy of being included on a map. (2) A preliminary step in compiling a map.

de Lisle's map projection - A perspective map projection onto a cone, with two standard parallels usually taken as lying equidistant between the central and extreme parallels to be mapped. The parallels of latitude are mapped as equidistantly-spaced arcs of concentric circles. The map projection was introduced by de Lisle in 1745 and is commonly used in atlases.

Dellen's method - SEE Döllen's method.

delta - A flat, fan-shaped region at the mouth of a river and composed of sediment deposited by the river.

delta angle - SEE angle, central.

demise (v.) - (1) Convey or create an estate for years or life. (2) Lease. This is the usual and operative word in leases: have a granted, demised, and to farm let, and by these presents do grant, demise and to farm let.

demise (n.) - (1) A conveyance of an estate to another for years, for life, or at will. Most commonly, for years. (2) A lease. Demise is synonymous with lease or let. Use of the term in a lease imports a convenant for quite enjoyment.

demography - The statistical study of human populations, particularly with reference to size, constitution, density and distribution.

denary - SEE decimal.

densification - SEE control densification; network densification.

densification network - (1) A survey network connected to and contained in a survey network of the same or less accuracy. (2) A survey network connected to and contained in a survey network of the same or less accuracy, and with shorter distances between stations. (3) A survey network connected to and contained in a survey network of the same or greater accuracy, and with shorter distances between stations. A densification network is established to provide control which is more accessible than and at least as accurate as that of the containing network. Distances between stations in the densification network are usually much shorter than distances between stations in the network containing it.

densification survey - A survey done to increase the number of control stations within a given region, the additional stations being of the same or a lower order of accuracy than that already existing in the region.

densitometer - An instrument for measuring the opacity of translucent materials such as photographic negatives, optical filters and liquids. It passes a beam of light through the material and measures the ratio of the intensity of the light emerging from the material to the intensity of the light entering the material. This ratio is the transmissivity; the (optical) density is the logarithm of the reciprocal of the transmissivity.

Density and transmissivity depend not only on the material but also on the spectral constitution of the light, the diameter and direction of the beam, and other factors. If the beam is of very small diameter at its place of entry into the material, the instrument is called a micro-densitometer.

density (physics) - (1) The ratio of the mass of any substance to the volume occupied by the substance. Also called, in physical oceanography, specific gravity and standard volume. Specific gravity is not the same as density, in general, and should not be used as a synonym for density except when quoting an original use. Specific volume is the reciprocal of density. (2) SEE gravity, specific. (3) The number of times an event or quantity occurs within a unit interval.

density (photography) - SEE density, photographic.

density (real estate) - (1) The number of persons per unit area in a specified region. (2) The number of families (or occupants of building-type units) per unit area. (3) The number of building-type units per unit area.

density (statistics) - The quantity given by a frequency function.

density, optical - SEE density, photographic.

density, photographic - A measure of the blackness of a photograph, taken, in the case of a translucent substance such as a photographic negative or transparency, as the logarithm of the reciprocal of the transmittance T: $D = \log(1/T)$ or in the case of an opaque substance, the logarithm of the reciprocal of the reflectance. Also called optical density. Usually referred to simply as the density where the meaning is clear. The density of a photographic image is approximately proportional to the amount of light creating the image. The characteristic curve is a better approximation.

density, potential - The density (1) a sample of water would have if it were brought adiabatically to the surface at atmospheric pressure.

density altitude - The distance above sea level at which the existing density of the atmosphere would be duplicated by the standard atmosphere.

density exposure curve - SEE curve, characteristic.

density function - SEE frequency function.

density layer - (1) A mathematical surface and a function, ρ, defined at each point of the surface, whose values have the dimensions (mass per unit area). The surface is called a single-layer surface if it is considered one-sided. It is called a double-layer surface if it is considered two-sided and if at each point the value $+\rho$ is associated with one side and $-\rho$ with the other. The solution to Poisson's differential equation can be expressed as the sum of the potentials from three distributions of density: a distribution on a single-layer surface, a distribution on a double-layer surface, and a distribution throughout the volume. (2) A mathematical surface and a function ρ, defined at each point of the surface, whose values are (δΦ/δn)/4π (single-layer surface density) or Φ/4π (double-layer surface density). Φ is a function twice-differentiable in the entire space and continuously differentiable on the surface and (δ/δn) denotes the gradient of Φ at the surface.

density slicing - The value of the density (opacity) at each point in an image is replaced by the value of that value of a stepped gray-scale in which the actual value lies. Also called equidensitometry and level slicing. In this way, the infinitely divisible range of actual densities in an image is replace by a range of discrete intervals of densities.

density zoning - Zoning restricting the largest number of houses per unit area that may be built within a particular subdivision.

departure - The orthogonal projection of a line, on the ground, onto an east-west axis of reference. The departure of a line is the difference of the meridional distances or longitudes of the ends of the line. It is east, or positive, and is sometimes called the easting, for a line whose azimuth or bearing is in the northeast or southeast quadrant. It is west, or negative, and is sometimes called the westing, for a line whose azimuth or bearing is in the southwest or northwest quadrant. It is abbreviated as dep. in notes.

departure, total - SEE abscissa (surveying).

depesas (Spanish-American law) - Spaces of ground, in towns, reserved for commons or public pasturage.

depression, angular - (1) The angle from a horizontal plane through a point to a half-line through the point, and taken as positive in the downward direction. Also called angle of depression, depression angle and descending vertical angle. It is the negative of angular elevation. (2) The vertical angle between the true horizon and the photograph perpendicular, measured at the perspective center.

depression, true angular - The angle between a ray from the exposure station through the principal point in the image plane of an oblique camera, and a ray to the boundary of a horizontal plane through the perspective center.

depression angle - SEE depression, angular.

depression angle, true - SEE depression, true angular.

depression contour - A contour inside of which the ground is at a lower elevation than the ground outside.

depth - (1) The vertical distance, in the direction of gravity, from a specified equipotential surface (the geoid, if not specified to be otherwise) to a specified point. (2) The vertical distance, in the direction of gravity, from a specified sea level to a specified point. In particular, the vertical distance from a specified sea level to the bottom of the sea. SEE ALSO chart depth.

depth, charted - The vertical distance, in the direction of gravity, as determined by echo-sounding, from the tidal datum to the sea bottom, using an assumed speed of sound of 800 fathoms (4800 feet) per second and making no corrections for actual speed or actual path of the sound ray.

depth, controlling - The least depth, in the approach or channel leading to a region such as a parort or anchorage, governing the greatest draft of the craft that can enter.

depth, digital - A depth (distance) obtained in digital form as the direct output from an echo sounder.

depth, dynamic - The work needed to move a unit mass from one equipotential surface (usually the geoid) to a lower equipotential surface. The term as used by oceanographers takes account only of work done against gravity. It does not include work done by expansion or contraction, or work against pressure from currents, for example. Also called dynamic height.

depth, graphic - A depth (distance) obtained from a measurement on the graphic record of an echo sounding.

depth, mean-sphere - The uniform depth to which water would cover the Earth if the solid surface were smoothed off and parallel to the geoid.

depth, observed - Depth as measured by sounding equipment, without corrections.

depth, residual - SEE depth anomaly, residual.

depth, standard - A depth, below the water's surface, at which properties of the water should be measured and reported, according to a proposal by the International Association of Physical oceanography in 1936. The proposed standard depths, in meters, are: 0, 10, 20, 30, 50, 75, 100, 150, 200, 300, 400, 500, 600, 1000, 1200, 1500, 2000, 3000, 4000, 5000, 6000, 7000, 8000, 9000, 10 000.

depth, thermometric - The depth, in meters, at which paired, protected and unprotected thermometers attached to a Nansen bottle are reversed. The difference between the corrected readings of the two thermometers represents the effect of the hydrostatic pressure at the depth of reversal. This depth may been be determined by formula or from a graph of depth anomaly.

depth anomaly - (1) The difference between the computed or thermometric depth and the ideal or assumed depth at which thermometers attached to a Nansen bottle or other instrument are reversed. (2) A variation of the depth of the oceanic bottom from that calculated by assuming that the depth increased with the square root of age until it is about 70 million years old. The assumption does not hold for some regions of considerable size. (3) SEE depth anomaly, residual.

depth anomaly, dynamic - The quantity ΔD defined by

$$\Delta D = \int_{P_1}^{P_2} [(1/\rho) - (1/\rho_o)] \, dP,$$

in which P_1 and P_2 are the pressures at two depths, ρ is the density of the water at pressure P and ρ_o is the density of water at salinity 35° /oo, temperature 0° C. and at pressure P. Also called dynamic height anomaly, dynamic height, anomaly of geopotential difference and geopotential anomaly. It is the product of the specific-volume anomaly and the difference in pressure, in decibars (assumed to equal the difference in depth, in meters).

depth anomaly, residual - The quantity $\Delta d = d(t) + S(\rho_c - \rho_w)/(\rho_m - \rho_s)$, in which $d(t)$ is the depth expected at time t, of a portion of the oceanic floor, S is the thickness of the sediment and ρ_s, ρ_w and ρ_m are the average densities of sediment, water and mantle, respectively. Equivalently, the difference between the depth expected for an oceanic bottom of a given age and the observed depth of the basement, corrected for isostatic loading by sediment and depth of sediment. The expected depth is derived from an empirical relationship of depth to age. Also called residual depth and residual elevation.

depth contour - A contour line indicating depth. Also called a depth curve. This definition is inconsistent with the definition of contour and may be erroneous. SEE depth curve.

depth curve -(1) A contour line indicating depth. Also called fathom curve and fathom line. (2) SEE depth contour.

depth factor - The number by which the apparent depth of the water measured stereoscopically is multiplied to give the true depth. This factor is a ratio of the tangent of the angle of incidence to tangent of the angle of refraction.

depth finder - An instrument for determining the depth of the water. In particular, an echo sounder.

depth finder, sonic - SEE echo sounder.

depth number - (1) A number placed on a depth contour-line to denote the depth of the contour below a given surface. (2) A number placed on a chart to indicate a depth found at the indicated point by a sounding.

depth of compensation - The depth (distance) to which a column of matter must extend in order to keep itself in hydrostatic equilibrium with surrounding matter. In the Pratt-Hayford theory of isostasy, the depth of compensation is calculated from the surface downward and is a constant; the density of matter in a column depends on the longitude and latitude but not on the depth. In the Airy-Heiskanen theory of isostasy, the depth of compensation is calculated from the geoid downward and varies with longitude and latitude, but the density of the column is constant. SEE ALSO isostasy; compensation, isostatic.

depth of field - The distance between those points nearest to and farthest from the camera which are imaged with acceptable sharpness.

depth of focus - The distance that the focal plane can be moved forward or backward from the point of exact focus and still give an image of acceptable sharpness. Also called focal range.

depth of isostatic compensation - That depth below sea level at which the condition of isostatic equilibrium is complete.

depth of no motion - The depth, in the oceans, at which there is no horizontal motion of the water. The depth of the layer of no motion.

depth sounder - (1) A device for measuring depth. (2) A mechanical device for measuring depth. SEE ALSO echo sounder.

dereliction - (1) The gradual, natural and more or less permanent recession of water from land so as to change the boundary between land and water. Also called reliction. This is a legal term and has only incidental relation to the physical processes involved. Two varieties of dereliction are usually recognized the recession of the waters of a lake or sea or other body of water, and the imperceptible recession of the water of a stream from one side and consequently its gradual encroachment on the other. (2) A recession of water from the sea or other body of water, leaving the land dry.

description - (1) A written description of the boundaries of a piece of land, presumably sufficient and accurate enough that the boundaries can be found and identified from the description. Four types of description are common in the USA: description by metes and bounds, verbal description by reference to a map, description by reference to natural objects and adjoiners, and the United States Public Land System. If the map to which reference is made is a plat, the description is said to be by platting. In particular, a document listing the metes and bounds of a property, usually prepared by a surveyor and included in a conveyance by the lawyer preparing the conveyance. (2) The formal, published data on each triangulation station, bench mark, etc. The data include sufficient information of the location and type of mark that anyone can go to the immediate locality and identify the mark with certainty. SEE ALSO bounds description; land description.

description, center-line - A method of describing a long and comparatively narrow strip of land by establishing a central line along the strip and defining the boundaries by distances from the central line.

description, legal - (1) A description, recognized by law, which definitely locates property by reference to governmental surveys, coordinate systems or recorded maps. (2) The same as definition (1) but leaving out the word definitely. (3) A description which is sufficient to locate the property without oral testimony.

description, metes-and-bounds - (1) The designation of a parcel of land by stating the courses and distances around it, or by calling for natural features or recorded monuments. Often the area of the parcel is stated as part of the description. Note that in the opinion of some authorities, the monument need not be a recorded monument for the term metes-and-bounds description to apply. (2) The designation of a parcel of land by giving the courses and distances around it. Also called description by metes and bounds. SEE ALSO description (1); metes and bounds.

description, plat - SEE description by platting.

description by aliquot parts (USPLS) - A description as aliquote parts of large parcels, the aliquote parts of which are not less than 10 acres. This method of description is popular in Alaska.

description by metes and bounds - SEE description, metes-and-bounds.

description by platting - (1) Designation of a parcel of land by referring to a plat (map) which has been filed in a proper public office. (2) Designation of a parcel of land by referring to a plat and giving a description. Also called a plat description.

design matrix - SEE observation matrix.

detail - The small items or particulars of information delineated from aerial photographs shown on a map by lines, symbols and lettering which, when considered as a whole, furnish the comprehensive representation of the physical and cultural features. The fewer the details, the more generalized the map.

detail, cultural - SEE culture.

detail, hydrographic - Hydrographic features considered cartographically.

detail, hypsographic - SEE feature, hypsographic.

detail, natural - SEE feature, natural.

detailing - The process of tying topographic details to the control network.

detail point - (1) A ground-control point established by a subsidiary survey (not a survey for establishing geodetic control), placed at or near an important cultural feature or other feature to be included in the final map, and identifiable on the photograph. The number of detail points selected will vary with the nature of the terrain and the amount of detail needed on the map. Ten detail points per photograph are usually sufficient. (2) A selected, identifiable point used, particularly in oblique photographs, to assist in correctly locating features displaced because of elevation.

detection - (1) Discovery of the existence of an object. Also called sensing. (2) Discovery of the existence of an object, without recognition of the object. This meaning is particularly common in photo-interpretation.

Deutsche Industrie-Norm - The German industrial standard for products made by or used in engineering, industry and science. It is also used outside of Germany for many products such as film speed and electrical connectors. Usually written as DIN.

develop (v.) - Use a developer (q.v.).

developable (adj.) - Of a surface, able to be flattened into a flat surface without compressing or stretching.

developer - A selective reducing agent, reducing those silver ions in silver halides in exposed regions of a photograph to metallic silver much more rapidly than it reduces the ions in unexposed parts.

development - The process of subjecting an exposed, light-sensitive surface to chemical agents in order to make visible the latent image formed by the action of light during exposure.

development method - A method of deriving, from distances measured at or computed for the Earth's surface, equivalent distances on the ellipsoid of reference. The distances on the Earth's surface are reduced to the those distances on the geoid which lie on the same normals to the ellipsoid. Angles measured at or computed for the Earth's surface are referred to the geoid without change except for those angles determined at large elevations, when a small correction may be applied. Then, starting from the origin, the distances and angles on the geoid are developed onto the ellipsoid in such a manner that distances and angles remain unchanged throughout the process.

development paper - A sheet of paper having, on its surface, a layer of emulsion containing light-sensitive halides, and used for printing images photographically.

deviation - The difference ε between the actual value x_o of a randomly varying quantity and the value x_p expected from theory: $\varepsilon = x_o - x_p$. Other names are residual, residual error and discrepancy. Deviation is not the same as error.

deviation, average - (1) The average, taken without regard to sign, of the differences between the members of a set of numbers and some fixed number. SEE ALSO average, arithmetic; error, average. (2) The average, taken without regard to sign, of the differences between the members of a set of numbers and the average value of the members.

deviation, magnetic - (1) The angle by which a magnetic compass aboard a craft deviates from magnetic north as a result of instrumental error and the magnetic effects of the craft and its contents. (2) The angle between the magnetic meridian and the axis of a compass card, expressed in degrees east or west to indicate the amount by which the northern end of the compass card is offset from magnetic north.

deviation, mean - SEE error, average.

deviation, mean radial - SEE error, mean radial.

deviation, residual - The magnetic deviation of a compass after compensation or adjustment.

deviation, standard - The positive square root of the sum of the squares of the differences between the N values of a variable and the average value of that variable, divided by the square root of N-1. If the differences are taken from some value other than the average, the resulting quantity is properly called the root-mean-square value and the divisor should be N instead of N -1. In particular, (a) For a set $\{x_n\}$ of discrete numbers, the quantity $+ \sqrt{[\Sigma (x_n - \xi)^2/(N-1)]}$, where the summation is over N random values of the set $\{x_n\}$ whose average value is ξ. Usually denoted by s_x. If to each x_n corresponds a weight w_n (e.g., if each different x_n occurs w_n times), the standard deviation is usually defined as $\sqrt{\{\Sigma w_n(x_n - \xi)^2/(\Sigma N-1)\}}$. (b) For a continuous, random variable x with frequency function p(x), the quantity $+[\int p(x)(x - \xi)^2 \, dx]^{1/2}$, in which ξ is the average value of x over the specified limits of integration. Also called standard error, mean error and dispersion. Those terms have unfortunately acquired many other meanings and should not be used where clarity is needed. (c) Either of the above definitions, with the condition that the random numbers have a Gaussian distribution.

deviation of compass - The deflection of the needle of a magnetic compass by the magnetic field of metallic masses within the craft on which the compass is located. SEE ALSO deviation, magnetic.

deviation of the vertical - SEE deflection of the vertical.

device - (1) Something made to assist in doing a task. (2) A simple or crude instrument.

device, averaging - Any device that takes the average of a number of readings, e.g., a bubble sextant.

device, centering - SEE centering device.

device, charge-coupled - An array of metal-oxide semiconductors or photodiode semiconductors in which the light falling on each semiconductor builds up a charge proportional to the integrated intensity of the light impinging on its surface. The charges are collected in capacitors and then transferred from the array to an amplifier whose output is a sequence of corresponding voltages. The voltages are digitized and the values stored or used immediately for constructing an image.

device, point-marking - A device for identifying and marking points on a diapositive by making a small hole in the emulsion at the point or making a small ring around the point itself. Also called point-transfer device and point marker.

device, point-transfer - A stereoscope containing a device which marks corresponding image-points on a stereoscopic pair of photographs. Also called a point marker and transcriber. SEE ALSO device, point-marking.

device, stilling - Any device or structure placed near a tide or stream gage to reduce the effect of waves and other random motions on the reading of the gage.

devise (v) - Transfer real property by will or last testament.

devise (n.) - The transfer of real property by will or last testament.

devisee - The person to whom real property is given by will.

devisor - The giver of real property by will.

diagonal check - Measurements made across opposite corners of the outside lines of a graticule to check the accuracy of the graticule or to check the scale of the graticule. A diagonal check has meaning only for graticules bounded by rectangles.

diagram, isometric - A diagram simulating the third dimension, in which the scale is correct along three axes. I.e., a diagram drawn so that the object represented appears to have three dimensions. The diagram is usually drawn by first putting in the three axes of a coordinate system, with suitable angles (such as 120° between each) between the axes. The x- and y-coordinates of points on the object are then plotted as usual on two of the axes; the z-coordinates of visible points are plotted using the third axis.

diagram, polar - A diagram drawn in a polar coordinate system.

diagram, trimming-and-mounting - A sketch showing how the prints of the transformed photographs taken by a multiple-lens camera should be changed to get the equivalent of a single photograph made by a single-lens camera. The sketch gives distances referred to the fiducial marks on the photographs and is the result of the calibration of the particular camera used.

diagram on the plane of the celestial meridian - A diagram in which the local celestial meridian appears as a circle with the zenith at the top and the horizon as a horizontal diameter. SEE ALSO time diagram.

diagram on the plane of the equinoctial - SEE time diagram.

diameter - (1) A straight line drawn from one point on a ellipse to the point on the other side and passing through the center of the ellipse. The diameters of an ellipse vary in size. That diameter passing through a focus is longer than any other diameter; that diameter perpendicular to the longest diameter is shorter than any other diameter. If the ellipse is a circle, all diameters are of equal length, but a diameter is longer than any other line drawn from side to side and not passing through the center. (2) A measure of the degree of magnification or enlargement of an object or figure. If a circle has a diameter of 1 unit and it is magnified to a circle with a diameter of 2 units, the magnification is said to be 2 diameters.

diameter (of a circle) - A line segment joining two points of a circle and passing through the center of the circle.

diameter (of a conic) - A line segment joining two points of a conic and bisecting each member of a family of parallel chords.

diameter, angular - The angle subtended at the observer by a diameter of a distant spherical body, the diameter being perpendicular to the line from the observer to the center of the body.

diameter, geocentric - The angular diameter, usually measured in seconds of arc, of a celestial body as the body would appear if viewed from the center of the Earth.

diameter, polar - (1) The length of the polar axis of a spheroid. (2) The distance between the North Pole and South Pole of the Earth.

diameter enlargement - SEE enlargement, x-diameter.

diaphragm - (1) A thin, opaque piece of material with a hole through which light can pass and placed in an optical system to keep unwanted light from reaching the image plane or detector, e.g., to limit the field of view or to absorb internally scattered light. The stop is the theoretical counterpart of the diaphragm. A shutter is a diaphragm whose opening can be opened or closed, usually in a specified interval of time. The mechanism that does the opening and closing is sometimes called the shutter mechanism but is also often considered, together with the diaphragm, to constitute the shutter. (2) The thin disk of glass on which lines forming a reticle are placed. However, the term reticle is often used to denote both the diaphragm and the lines.

diapositive - (1) A positive photograph on a transparent substance, usually plastic or glass. The term generally refers to a transparent positive on a glass plate, used in a stereoscopic plotter, a projector, or a measuring device (comparator). In Europe, it usually refers to a positive-film transparency which would be identified as a slide by an American user. (2) A photographic image external to the aerial camera by which it was formed. (obsolete)

diapositive printer - A printer for making diapositives from photographic negatives. Also called a reduction printer. There are two general types: contact printers and fixed-ratio diapositive printers.

diapositive printer, fixed-ratio - A photographic apparatus for producing diapositives whose scale is in a predetermined ratio to the scale of the photographic negative from which the diapositive is made. Also called a fixed-ratio projection printer. It contains an optical system placed between the photographic negative and the sensitized plate or film which is to become the diapositive, the distances of negative and diapositive being set at nominal values, within narrow limits.

diazo print - A copy made by placing the original in contact with paper sensitized with a diazo dye and illuminating the original. The exposed paper is treated with fumes of ammonia to develop the image, or it may be passed through an aqueous solution of ammonia.

diazo process - A photographic process of reproduction making use of paper impregnated with light-sensitive diazonium compounds. Also called the white-line process. There are two basic varieties: the dry process, in which the exposed sheet is treated with ammonia vapor, and the moist process, in which the exposed sheet is carried over a roller moistened with aqueous ammonia.

difference, ascensional - The difference between right ascension and oblique ascension.

difference, hour-angle - SEE difference, meridian-angle.

difference, meridian-angle - (1) The angle between two meridians. (2) The difference between the hour angle of a celestial body and the hour angle used as argument for entering a table.

difference, meridional - The difference between the meridional parts of any two given parallels of latitude. Also called the meridian difference and meridional difference of latitude. It is found by subtraction if the two parallels are on the same side of the equator and by addition if they are on opposite sides.

difference, tidal - The difference in time or height of a high or low water, as these occur at a reference station and a subordinate station (reference tide-gage and subordinate tide-gage). The difference is applied to the time or height determined at the reference station to obtain the corresponding quantity at the subordinate station. These differences are available in tables of the tides.

difference of elevation - SEE elevation difference.

difference of latitude - SEE latitude difference.

difference of longitude - SEE longitude difference.

diffraction - The bending of waves, electromagnetic or sound, when the waves meet an opaque obstacle. The waves are bent inwards producing vibrations in what would otherwise be a shadowed region. The phenomenon is not readily explained by the particle-theory of electromagnetic radiation.

diffraction grating - (1) A set of precisely spaced, parallel lines which cause incident electromagnetic radiation to be diffracted either upon reflection or upon transmission between the lines. (2) The ruled piece of metal or other material causing diffraction. Diffraction gratings for radio waves usually consist of closely spaced wires held in a frame. Diffraction gratings for radiation at wavelengths from the infrared to ultraviolet are ruled by a graving tool on a plane or curved sheet of metal, or may be produced by photographing onto a glass plate a set of inked lines on paper. At very short wavelengths, the lattices of crystals serve as diffraction gratings.

diffusion transfer - The simultaneous production, by photographic development, of a positive image in an absorbent layer in contact with a (chemically) developed negative image. While the negative image is developing in the exposed layer, the developer dissolves the unexposed silver halide. The dissolved silver halide diffuses into the receiving layer where a positive image is formed as the dissolved salt is reduced by the developing agent in the presence of catalytic nuclei present in the receiving layer. Although the principle was suggested by Lafevre in 1857, it was not until 1939 that a practical application was developed.

digit, least significant - That digit, in a number, which contributes least to the value of the number.

digit, most significant - That digit, in a number, which contributes most to the value of the number.

digit, octal - A digit which is a member of the set of eight digits 0,1,2,3,4,5,6,7 used in the number system having the radix 8.

Digital Line Graph (DLG) - Digital representation of planmetric map features shown especially on topographic quadrangles.

digitize (v.) - Convert a continuous signal or data in analog form to numerical values representing the signal or data.

Digitize Elevation Model (DEM) - Computer representation of terrain features portrayed on a map or aerial photography.

digitizer - A machine which converts analog signals to discontinuous, discrete levels of amplitude and displays or records these levels as numbers. Also called a digitizing system. In particular, in cartography, equipment which converts data derived from a map or photograph in analog form into numerical form. Digitizers are used in measuring devices and in stereoscopic plotting instruments to convert manual and automatic measurements in analog form to the corresponding values in digital form. They are also used in most electronic distance-measuring instruments to display and record the distances digitally.

dilatation - The increase in volume of a continuous material per unit volume of the material. For example, local dilatation of the crust has been found to be a precursor of some earthquakes.

dilatation, cubical - The ration between the change in volume of a small part of body, in passing from an unstrained state to a strained state, and the value of that part in the unstrained state.

dilution of precision, geometric - A formula giving the effect of errors in the measured angles and lengths in a geometric figure on the error in an angle or length calculated from the measured values. Often referred to as GDOP. The name refers to the well-known fact that the error which results in calculating the value of an angle or length in a geometric figure from the values of other angles and lengths depends very much on the values of those other angles and lengths. There are firm rules set for the shapes allowable in planning a triangulation network, based on this fact.

diluvion - The gradual and imperceptible washing-away and resultant loss of soil along a watercourse. The opposite of alluvion. Note that this definition implies that alluvion is a process, not a material.

dimensional stability - Ability of drafting media to withstand variations in humidty and temperature without experiencing shrinkage or stretching.

diopter - A unit of measurement for the magnifying power of a lens, equal to the reciprocal of the focal length in meters.

dioptric (adj.) - Containing only refractive elements (lenses and/or prisms).

dip - (1) The angle which the plane of a stratum or fault makes with a horizontal plane. Also called angle of dip and true dip. (2) The inclination of the geomagnetic field. (3) The first detectable increase in the angular elevation of a celestial body after the body reaches its greatest angular elevation on or near its transit of the meridian. (4) SEE dip of the horizon.

dip, geometrical - The vertical angle, at the eye of an observer, between the horizontal plane there and a straight line tangent to the surface of the Earth. It is larger than the dip of the horizon by the amount of the terrestrial refraction.

dip, magnetic - (1) The angle which a freely suspended, symmetrically magnetized needle, influenced by no transient, artificial, magnetic disturbance, will make with the plane of the horizon. Also called magnetic inclination. The term plane of the horizon is, more properly, a horizontal plane through the center of the needle. (2) The angle between the horizontal and the direction of a line of force of the Earth's magnetic field at any point. Also called dip and magnetic inclination. (3) SEE inclination, magnetic.

dip, true - SEE dip (1).

dip angle - (1) The vertical angle, at the point of observation, between the true horizon and a line of sight to the apparent horizon. The true horizon is the horizontal plane passing through the point of observation. (2) The vertical angle, at an air station, between the true horizon and the apparent horizon, which results from refraction, curvature of the Earth and flight height.

dip circle - (1) An instrument for measuring geomagnetic inclination and consisting of a magnetic needle suspended so that it can rotate in a vertical plane, the amount of rotation being indicated by a graduated vertical circle. Also called a dip needle. (2) An instrument for detecting strong magnetic anomalies by movement of a magnetic needle suspended so that it can rotate in a vertical plane.

dip circle, magnetic - SEE dip circle.

dip needle - (1) The magnetic needle in an instrument used for determining the inclination of the magnetic field, i.e., the needle in a dip circle or inclinometer. Also called, unnecessarily, a magnetic dip needle. (2) SEE dip circle (1).

dip needle, magnetic - SEE dip needle.

dip of the horizon - (1) The angle between a horizontal plane at the observer and a line from the observer to the apparent (i.e., visible) horizon. The apparent horizon lies below the horizontal plane because of the curvature of the Earth, but the amount is affected by upward or downward refraction of light travelling from the horizon to the observer. (2) The vertical angle between the plane of the horizon and a line tangent to the apparent (visible) horizon.

dipole antenna - An antenna consisting of two separate, open-ended conductors in line and a short distance apart. The adjacent ends are connected to the terminals of a radio receiver or radio transmitter. Usually called a dipole for short when no confusion is likely. Dipole antennas are easily made, easily erected, and take up little space if made for the frequencies at which most satellites transmit or receive.

dip pole - SEE pole, magnetic.

Dirac function - (1) The function $\delta(x - x_o)$, which has the value 0 for all values of x other than x_o and has the value 1 at x_o. (2) The function $\delta(x - x_o)$ such that

$$\int_{-\Sigma}^{+\Sigma} \delta(x - x_o) = x_o.$$

direction - (1) In surveying and mapping, the angle between a line or plane and an arbitrarily chosen reference line or plane. At a triangulation station, observed horizontal angles are referred to a common reference line and termed horizontal direction. They are usually collected into a single list of directions, with the direction of 0° placed first and the other directions arranged in order of increase clockwise. (2) A line, real or imaginary, pointing away from some specified point or locality toward another point. Direction has two meanings: that of a numerical value and that of a pointing line. Two lines must be specified for the first definition to be valid; only one line need be specified for the second meaning. (3) An indication of the location of one point with respect to another without involving the distance between the two points. It is usually thought of as a short segment (of a straight line between the two points) having one end at one of the points and having an arrow-like symbol at the other end. Note that a direction is not an angle.

direction (of gravity) - The direction, at a point, in which an infinitesimal body starting from rest would fall if there were no forces other than atmospheric or gravity acting on the body. Equivalently, (a) the downward direction, at the point of suspension of a massive body, if the suspending thread is extremely thin and very short, and the suspended body is a very small, solid sphere suspended at its center and if no other forces than gravity are present. Or, (b) the downward-pointing tangent to the vertical at a point. Also called the direction of the vertical. Gravity is a force, and therefore has both magnitude and direction. Its direction is independent of any coordinate system but the components of the direction are not. The components with respect to two mutually perpendicular planes through the local normal (one plane being that of the local meridian, are called the deflection of the plumb line or the deflection of the vertical in those planes. At a point outside the Earth's surface, the direction of gravity is toward the surface; at a point inside the surface, the direction is away from the surface. In both instances, the direction is approximately toward the center of the Earth. Note that the direction of gravity and the direction of the plumb line are generally taken as the same, while the direction of the vertical is more usually taken contrary to the direction of gravity, i.e., in the direction of the zenith rather than the nadir. SEE ALSO nadir; zenith.

direction (of the vertical) - The direction, at a point, of the downward-pointing tangent to the vertical at that point.

direction, astronomical - A direction referred to astronomical north.

direction, cardinal - One of the astronomical directions on the surface of the Earth: north, east, south, west. The term cardinal without qualification, is sometimes used to indicate any or all of the above directions. The exact meaning is then to be inferred from the text.

direction, great-circle - The angle from a specified great-circle used as reference to the great circle in question.

direction, horizontal - A direction in a horizontal plane.

direction, principal - One of the two orthogonal directions, at any point on an ellipsoid, which remain orthogonal directions on the planar map showing that point. Also called a principal tangent. SEE ALSO curvature, normal.

direction, relative - A horizontal angle expressed with respect to a heading.

direction, rhumb-line - The horizontal direction of a rhumb line, expressed as the angle from a reference direction. Also called a Mercator direction and a rhumb direction.

direction, true - A horizontal direction expressed as an angle from astronomic north.

direction angle - One of three angles specifying the direction of a line with respect to the three axes of a Cartesian coordinate system in 3-space by specifying the angle between that line and each of the three axes. The three angles are not completely independent, but unless the quadrant in which the line lies is known, three angles must be given to avoid ambiguity.

direction cosine - The cosine of a direction angle.

direction finder - An instrument which can determine the direction to a body emitting radiation - generally at radio wavelengths.

direction instrument - A theodolite in which the horizontal circle remains fixed during a series of observations on different targets. Also called a triangulation theodolite. SEE ALSO direction theodolite.

direction method (of adjusting a traverse) - The method of adjusting a survey traverse by using the directions (azimuths) of courses as the measured quantities.

direction method (of adjusting triangulation) - A method of adjusting a triangulation network, using adjustment by conditions, by determining corrections to measured directions. Each angle is considered to be the difference between two directions, for each of which a separate correction is determined. The direction method is used in adjusting triangulation networks composed of overlapping triangles, but for some work, where the network consists of a chain of single triangles, the angle method of adjusting triangulation may be preferred.

direction method (of determining azimuth) - Determining azimuth by measuring the difference in horizontal direction between a circumpolar star and a given mark. Because the direction of the star is changing at each instant, the time of each measurement must also be determined accurately. SEE ALSO Black's method.

direction number - One of a set of any three numbers proportional to the three direction cosines of a line.

direction of tilt - (1) The direction (azimuth) of the principal plane of a photograph. (2) The direction (azimuth) of the principal line on a photograph.

direction reduction to the spheroid - The computation of the angle between a direction at a point P on a reference ellipsoid and a geodesic from P to another point Q on the ellipsoid, from the measurement of the angle, at a point P' a known height above P, between that same direction and a plane through P', P and a point Q' a known height above Q. The computation involves two separate changes to the measured angle - one because there is a deflection of the vertical at P', the other because the normals at P and Q are skew to each other instead of being coplanar. SEE ALSO reduction to the spheroid.

direction system - A set of N unit vectors, radiating from a common point, to which angles are referred in specifying the direction of a point in the space of the vectors.

direction theodolite - A theodolite in which the graduated horizontal circle remains in a fixed position during a set of observations, while the telescope is pointed at a number of targets in succession and the direction of each is read on the circle. Also called a direction instrument. For greater accuracy, and to detect blunders, a number of sets of measurements is taken, with the circle being rotated systematically between sets, so that each direction is measured on a number of different parts of the circle.

directory call - A call which directs to a neighborhood where the location specified precisely by another call is to be found.

discontinuity - (1) Any abrupt change in geological structure or characteristics. (2) A region in which sudden changes in structure or characteristics can be inferred from observed changes in the speed of seismic waves. The principal discontinuities of this kind are the Mohorovicic and the Conrad discontinuities. SEE ALSO Birch discontinuity; Conrad discontinuity; Förtsch discontinuity; Lehmann discontinuity; Mohorovicic discontinuity; Riel discontinuity; Wiechert-Gutenberg discontinuity.

discontinuity, seismic - An abrupt change in the speed of seismic waves with depth. The term is used when no conjecture about the geological cause of the discontinuity is justified.

discrepancy - The difference between two values of the same quantity when only one value of that quantity is expected. In particular, (a) a difference between results of duplicate or comparable measures of a quantity; (b) a difference in computed values of a quantity; or (c) a difference in computed values of a quantity obtained by different processes using data from the same survey. Examples are: the difference in the length of two measures of the same line; or the amount by which the values of the location of the third point of a triangle, as computed from the other two points, may fail to agree when the values have not been corrected for misclosure. Discrepancy is closely associated with but is not identical with misclosure.

discrepancy, accumulated - The sum of the separate discrepancies which occur in the various steps of making a survey or of computing the results of a survey. For example, if two level-lines, run independently over the same set of bench marks, are computed separately, differences between the two sets of elevations will accumulate. This does not mean that the accumulated discrepancy will necessarily increase in magnitude.

disk - (1) A circle and its interior. (2) That region, in the plane, consisting of all points with norm less than (sometimes defined to be less than or equal to) 1. (3) SEE bench mark, permanent.

dispersion - (1) The differential change in phase or direction undergone by radiation of different frequencies on traveling through matter. All matter affects the velocity of radiation passing through it. The size of the effect depends on the frequency of the radiation. With suitable geometric relationship between beam and medium, the waves in a polychromatic beam will be separated and spread out or dispersed according to frequency. A medium having a strongly dispersive effect is called a dispersive medium. Dispersion is effective only over a limited part of the spectrum. E.g., the ionosphere is a dispersive medium for radio waves longer than a few centimeters but not for shorter radio waves or for light. A well-known example of dispersion is the effect of a glass prism on light. A beam of white light entering one side of a triangular glass prism is fanned out into a spectrum of different colors on leaving the prism. SEE ALSO refraction. (2) The occurrence of several (usually many) values of a quantity when only one value is theoretically correct. (3) The standard deviation of a quantity when referring to a characteristic as distinguished from a measurement. For example, one would speak of the dispersion of the heights in a particular terrain but one would speak of the standard deviation of the measurements of those heights. This is a subtle distinction which is sometimes important. (4) Either standard deviation or variance.

dispersion matrix - SEE covariance matrix.

dispersion prism - SEE prism, dispersing.

displacement - (1) The horizontal shift of the plotted location of a topographic feature from its true location, caused by required adherence to prescribed sizes of symbols and widths of lines. (2) Any change, from photograph to photograph, in the relative locations of the images of a point (in object space), which is independent of the characteristics of the cameras. For example, relief displacement, a change in position of the camera with respect to the object, or ordinary changes in atmospheric refraction can cause a change in the location of the image on the focal plane. SEE ALSO relief displacement; tilt displacement.

displacement, radial - SEE principle of radial displacement.

displacement, refractive - Displacement of images radially outward from the photograph nadir because of atmospheric refraction. It is assumed that the refraction is symmetrical about the nadir.

displacement method (of determining tilt) - A method of determining the amount of tilt of an aerial photograph by comparing the displacements of image points for which the vertical and horizontal coordinates of the corresponding ground points are known.

dissipation function - The function F in the equation $d(\delta T/\delta q_k)/dt - \delta T/\delta q_k + \delta F/\delta q_k = Q_k$ for k = 1 to K. (δ denotes partial derivation.) In this equation, q_k is a generalized coordinate of a physical system which has kinetic energy T and which is subject to the general force Q_k.

distance - (1) A number or quantity indicative of the separation or amount of space between two points, without regard to any physical or imagined connection between them. The lack of connection distinguishes distance from length. Length always implies that the two points have a material connection, real or hypothetical, between them. Distance is used when the two points are thought of as being physically separate and unconnected. For example, one writes of the distance from the Earth to the Moon or the distance travelled by an electron in one second in a vacuum. One writes of the length of a yardstick or the length of an aircraft. One refers to the distance between the ends of a base line but to the length of the base line, even tho the line proper has no physical existence. SEE ALSO length. (2) SEE distance, angular (1). SEE ALSO ground distance; skip distance; slant distance.

distance, angular - (1) The angle between two directions. It is numerically equal to the value of the angle between two lines extending in the given directions. Astronomers frequently refer to angular distance merely as distance. (2) The arc, expressed in angular units, of the great circle joining two points on a sphere. (3) The distance between two points, expressed in angular units of a specified frequency. It is equal to the number of wavelengths between the two points, multiplied by 2π if expressed in radians, by 360° if expressed in degrees and 400g if expressed in gons.

distance, astronomical - The distance of a celestial unit expressed in parsecs, light years or astronomical unit.

distance, back focal - (1) The distance, measured along the optical axis, from the vertex of the last surface of a lens system to the plane of best average definition. Also called the back focal length and back focus. This is the quantity used in setting the lens system in an aerial camera. (2) The distance, measured along the optical axis, from the vertex of the last surface of a lens system to the back focal point.

distance, calibrated principal - SEE length, calibrated focal.

distance, conjugate - One of the corresponding distances of object-point and image-point from the nodal points of a lens. The conjugate distances L_o and L_i and the focal length f of the lens are related by the equation $1/f = 1/L_o + 1/L_i$. The total distance from object-point to image-point equals the sum of the two conjugate distances plus or minus (depending on the design of the lens) the distance between nodal points.

distance, double meridian - SEE meridian distance, double.

distance, double parallel - SEE parallel distance, double.

distance, double zenith - SEE angle, double-zenith.

distance, ecliptic polar - The complement of the ecliptic latitude.

distance, electrical - Length measured in terms of the distance travelled by radio waves in unit time.

distance, external - The distance from the intersection of the tangents (vertex) of a circular curve to its midpoint, measured along the line extending between the vertex and its center of radius.

distance, focal - (1) The limiting value f of the ratio of a linear quantity y' in an sharply defined image to the angle ß subtended by the corresponding distance or length in object-space: $f = -\lim(y'/\tan ß) = -\lim(y'/ß)$ as ß goes to 0. Also called focal length. (2) SEE length, focal.

distance, front focal - (1) The distance, measured along the optical axis, from the vertex of the first surface of a lens system to the plane of best average definition. Also called the front focal length and front focus. (2) The distance, measured along the optical axis, from the vertex of the first surface of a lens system to the front focal point.

distance, geodetic - Distance measured along a geodesic. Also called geodesic distance.

distance, great-circle - The length of the shortest arc of the great circle joining two points. It is usually expressed in nautical miles.

distance, hyperfocal - The shortest distance at which an object is acceptably in focus when viewed through an optical system focussed at infinity.

distance, interocular - SEE distance, interpupilary.

distance, interpupilary - The distance between the centers of the pupils of the eyes. It is also called interocular distance or eye base.

distance, ionospheric correction to measured - SEE correction, ionospheric.

distance, lunar - The angle between the line of sight toward the Moon and the line of sight toward another celestial body, as measured by an observer on the Earth. It was formerly used in determining longitude at sea.

distance, mean - The semimajor axis of an elliptical orbit.

distance, meridional - (1) Distance measured along a meridian on an ellipsoid. (2) The ratio, for a given parallel of latitude, of the length of one minute of latitude, at a specific latitude, to the length of one minute of longitude at the equator. Also called a meridional part. This is the definition given in many books on map projections. It does not correspond in any way, however, to the way the term is actually used. A definition corresponding to actual usage is that given for meridional parts. (3) SEE meridian distance.

distance, plus - (1) The distance along a surveyed line from a survey station or the last whole-numbered survey-point to a supplementary point. For example, if a stake is set at 515.56 feet from the initial point of a surveyed line and is does not mark the end of that line, the whole-numbered 500-foot point on the line is station No. 5; the stake is a plus station (No. 5+15.56), and the 15.56 is a plus distance. (2) A fractional part of the length of a course from the beginning of the course to a specified point on the course.

distance, polar - An angle which is the complement of the declination, i.e., 90° minus the declination. Also called co-declination. The arc corresponding to polar distance is one side (celestial body to pole) of the astronomical triangle. It is opposite the zenith.

distance, principal - (1) The perpendicular distance from the internal perspective center of a projector to the plane, in the projector, of the photographic copy (finished negative or print). This distance is equal to the calibrated focal length of the camera that took the original photograph, corrected both for the enlargement or reduction ratio and for the shrinkage or expansion of the film or paper since the photograph was taken. It preserves the same perspective angles at the internal perspective center to points on the photographic copy as existed in the camera taking the picture at the moment of exposure. This is a geometrical property of each particular, finished negative or print. It is also called the effective focal length. (2) The perpendicular distance from the internal perspective center of the projector of a stereoscopic plotter to the plane, in the projector, of that side of the diapositive on which the emulsion lies.

distance, relative - Distance from a specified point which is, usually, in motion.

distance, rhumb-line - Distance along a rhumb line.

distance, tangent - The distance from the point of intersection (vertex) of a curve to its point of tangency or point of curvature.

distance, taped - A distance which has been measured with a surveyor's tape. Also called a taped length if the two points involved are thought of as being connected by a line. Even if the surveyor's tape were perfect, the measured distance would still be in error because a real tape cannot be laid in a perfectly horizontal straight line while measuring. The measured distance is therefore changed by the addition of certain small amounts corrections to give the distance which would have been measured had both tape and process been perfect. SEE ALSO alinement correction; grade correction; sag correction.

distance angle - An angle, in a triangle, opposite a side used as a base in the solution of the triangulation equations, or a side whose length is to be computed. In a chain of single triangles, two sides in each triangle are used as the computation proceeds through the chain: a known side and a side to be determined. The angles opposite these sides are the distance angles.

distance determination - The determination of distance by calculation. The calculation may convert some measured quantity such as time or angle to the corresponding distance, or it may convert a given (measured or calculated) distance in one space to a corresponding distance in another space. SEE ALSO distance measurement, optical; tangential method (of determining distance).

distance measurement - The measurement of a distance between two points. Distance measurement involves the actual measurement of distance; distance determination involves the calculation of distance from a given distance or from some quantity other than distance such as time or angle. The distinction is often not clear and the two terms are then equivalent. Distance measurement is sometimes called telemetry. However, such usage has deprecated because telemetry is used to mean making measurements at a distance.

distance measurement, electromagnetic - (1) Distance measurement using an instrument which (a) measures the difference between the time or phase at which an electromagnetic wave was emitted and the time or phase at which the wave returns from the point to which it was sent, and (b) converts this difference to the corresponding distance, which is displayed to the operator or is recorded. Sometimes referred to, erroneously, as electronic distance- measurement. (2) SEE distance measurement, electro- optical.

distance measurement, electronic - (1) Distance measurement using an instrument containing electronic devices which take part in the measuring process. Contrasted to mechanical distance-measurement as by using a surveyor's tape or by using a stadia rod and tachymeter. The term is unfortunately often used for electromagnetic distance measurement, which is something quite different. (2) SEE distance measurement, electromagnetic.

distance measurement, electro-optical - Electromagnetic distance-measurement using a beam of pulsed or otherwise modulated light; the modulation is done by imposing an electrical field on a suitable medium through which the light passes in the instrument. The term is also often applied to measurements made using infrared radiation instead of light.

Also called, ambiguously, electromagnetic distance measurement.

distance measurement, ionospheric correction to - SEE correction, ionospheric.

distance measurement, optical - Distance measurement by using light. Classical methods use the fact that light travels in a straight line (except for the effect of refraction) by making the light rays the two sides of a isosceles triangle in which the observer is either at the vertex of the triangle and uses the measured angle there together with the known length of the base, or he is at the base and uses the two opposite angles there together with the known length of the base. This is actually distance determination if the observer actually calculates the distance from the measured angle or angles; it is distance measurement if the instrument used does the calculating and displays the corresponding distance. The stadia method and the subtense-bar method are examples. Newer methods do not use the properties of a triangle but measure the distance directly. The instrument measures the difference between the time or phase at which light is emitted and the time or phase at which the light returns from the distant point and converts this quantity to a distance. The term is also applied to methods in which infrared radiation is used instead of light.

distance-measuring equipment - SEE equipment, distance-measuring.

distance-measuring instrument - SEE instrument, distance-measuring.

distance meter, electronic - SEE instrument, electro-optical distance-measuring.

distance modulus - The difference between absolute and apparent magnitudes of a celestial body. E.g., the Sun's distance modulus is -31.57 m.

distance point - SEE point, distant.

distance-ratio method - A method of adjusting the results of a trilateration by assuming that the ratio of distances measured from a station does not change and adjusting the ratios of distances rather than the distances themselves. It is also called the line-ratio method, Robertson's method and relative lateration. It was invented by K.D. Robertson. Its usefulness depends on the fact that the ratio of two distances measured from the same point at times close together is less sensitive to the effects of atmospheric refraction than are the distances themselves.

distance reduction - The process of finding that distance, on a reference ellipsoid or on a particular equipotential surface, corresponding to a distance measured on the ground or, in some instances, measured between points above the ground. SEE ALSO reduction to the ellipsoid; reduction to the geoid; reduction to mean sea level; reduction to the spheroid.

distance reduction to chord length - SEE reduction to chord length.

distance reduction to mean sea level - SEE reduction to mean sea level.

distance reduction to the ellipsoid - SEE reduction to the ellipsoid.

distance reduction to the geoid - SEE reduction to the geoid.

distance reduction to the spheroid - SEE reduction to the spheroid.

distance wedge - A thin prism placed in front of the objective of a telescope and rotatable through an accurately measured angle about a line (horizontal or vertical) transverse to the optical axis. Rotating the prism through a known angle deflects the line of sight through a measurable distance on a graduated staff. For a given angle of rotation, the distance of the observer from the staff is a function of the distance the line of sight is deflected along the staff and can be determined from these two quantities. The principle is used in some tachymeters.

distortion - (1) An optical aberration which causes the scale of the image to change from point to point in the image. It is also called distortional aberration and optical distortion; it is sometimes called lens distortion if caused by a lens system. Photogrammetrists commonly distinguish two components: radial distortion, which is a change of scale along a line radially outward from the optical axis and tangential distortion, which is a change of scale in a direction tangent to a circle about the optical axis. Distortion is usually measured by imaging a rectangular grid or reseau of known dimensions. If the lines in the image appear curved and concave toward the optical axis, the system is said to have barrel distortion. If the lines appear convex as seen from the optical axis, the system is said to have pincushion distortion. Most optical systems show distortions that are combinations of the two kinds. In a good optical system, radial distortion is small and radially symmetric; tangential distortion is extremely small and is irregularly distributed. (2) Any shift in the position of a feature, on a photograph, which alters the perspective characteristics of the photograph. Such distortion is caused by aberrations, differential shrinkage of the emulsion and motion of the film or camera. SEE ALSO barrel distortion; decentering distortion; lens distortion; pincushion distortion; point of zero distortion; scan positional distortion.

distortion, angular - (1) The failure of a lens system to reproduce accurately, in image-space, the angle subtended by two points in object-space. (2) Distortion, in a map projection, because of non-conformality. (3) The change in shape of a small circle on a sphere when it is transformed to a plane by a projection.

distortion, calibrated - Radial distortion calculated so as to minimize the errors in locations of individual image-points with respect to each other.

distortion, characteristic - The radial distortion characteristic of a particular camera.

distortion, differential - Those dimensional changes in length and width of film, paper, etc., which deform photographic or printed images.

distortion, image-motion-compensation - In photography with a panoramic camera, the displacement of the images of points on the ground from their expected locations on the cylindrical surface of the film because the lens system or surface of the film was translated to compensate for motion of the image during exposure.

distortion, linear - The failure of a lens system to reproduce accurately, to scale, all lengths or distances in the object.

distortion, negative - SEE barrel distortion.

distortion, optical - SEE distortion (1).

distortion, panoramic - The displacement of images of object points from their expected positions in perspective, caused by the cylindrical shape of the negative and the scanning action of the lens in a panoramic camera.

distortion, positive - SEE pincushion distortion.

distortion, radial - The change in scale, in an image, with radial distance from the center of the image (optical axis).

distortion, tangential - The change in scale, in an image, in a direction tangent to a circle about the optical axis (i.e., perpendicular to the radial direction).

distortion, tipped panoramic - Displacement of images of ground points from their expected locations on a panoramic photograph because the scanning axis is tipped in the vertical plane through the direction of flight. This distortion is small and adds directly to the distortion caused by the cylindrical projection and by incomplete compensation for image motion and other factors.

distortion caused by image-motion compensation - SEE distortion, image-motion-compensation.

distortion compensation - That correction applied to offset the effect, in a double-projection, direct-viewing stereoscopic plotter, of radial distortion introduced in an original negative by the objective lens of an aerial camera.

distortion curve - A curve representing the linear distortion caused by a lens system. It is plotted with radial distances of image-points from the optical axis as abscissas and with radial displacements from their correct (undistorted) distances as ordinates.

distribution - (1) The pattern created by one or more random variables or by the occurrence or non-occurrence of different kinds of events. A distribution is most accurately represented by a tabulation or plot scatter diagram of the values or events in the order they appeared. It may be represented less accurately but more concisely by a mathematical formula which gives the frequency with which a specified value or event will occur (frequency function) or the total frequency with which all values or events below a specified limit will occur (distribution function). (2) SEE frequency function. (3) SEE distribution function; distribution function, cumulative. (4) SEE frequency. SEE ALSO Poisson distribution; Rayleigh distribution.

distribution, Bernoullian - SEE distribution, binomial.

distribution, binomial - The distribution characteristic of a random variable or event which can assume one of two values and which in N possible events, has the probability p of having the one value and the probability (1 -p) of having the other in each event. Also called Bernoulli's distribution. The usual example of a binomial distribution is the pattern of heads and tails obtained by tossing a (possibly weighted) coin. However, the distribution also occurs in the theory of switching circuits and in other situations where a random event either will or will not occur.

distribution, bivariate - The distribution characteristic of two quantities varying simultaneously and randomly. SEE ALSO frequency function, bivariate.

distribution, Gaussian - SEE Gaussian distribution.

distribution, multivariate - A distribution involving more than one random variable, i.e., the pattern characteristic of a randomly varying vector of more than one component.

distribution, normal - SEE Gaussian distribution.

distribution curve - A line, continuous or broken, drawn so that the area between it and a given straight line is exactly 1, while the area between any segment of the first line and the straight line is proportional to the frequency of the quantity represented by the line. Alternatively, it is the graph of the corresponding frequency function.

distribution density - SEE frequency.

distribution-density function - SEE frequency function.

distribution function - (1) Either a cumulative distribution function or a frequency function. (2) SEE distribution function, cumulative. (3) SEE frequency function.

distribution function, binomial - The cumulative distribution function $\Sigma(n) \theta k (1 - \theta)n-k$, giving the probability that a random sample of size n, with replacement, contains x or fewer objects of a certain type if the sample is taken from a population of size N containing $N\theta$ of that type. Also called the cumulative binomial distribution function, Bernoulli distribution function, etc.

distribution function, cumulative - A function giving the number of times, or proportional number of times, a set of random variables takes on a value lying within a specified interval - conventionally, an interval having no lower limit (i.e., having a lower limit of $-\Sigma$) Alternatively, the total frequency with which a value equal to or less than a specified value of a random variable has occurred. It is also called a distribution function, density function, cumulative frequency function, cumulative probability function and probability function. It is abbreviated as c.d.f. The term distribution function is preferred by many statisticians who consider cumulative distribution function a pleonasm. Unfortunately, distribution function now has several meanings, making the term ambiguous at best and vague at worst. Probability function is also ambiguous. Cumulative distribution function, on the other hand, has only one meaning at present and is therefore to be preferred where clarity is desired. Density function is most often used as a synonym for frequency function and its occurrence with the meaning of cumulative distribution function must be regarded as an aberration.

distribution function, bivariate Gaussian - SEE Gaussian distribution function, bivariate.

distribution function, Gaussian - SEE Gaussian distribution function.

distribution function, normal bivariate - SEE Gaussian distribution function, bivariate.

distribution function, normal cumulative - (1) SEE Gaussian distribution function. (2) SEE Gaussian frequency function.

distribution map - A map showing the geographic occurrence of a specific product, commodity, or formation. The map may show merely that the product, etc., occurs in certain regions, or it may also show how much is produced or occurs there.

district, approach-zone - All that region outward from the end of a runway in which the heights of structures or other hazards to aircraft are restricted. The slopes and dimensions are usually fixed by zoning commissions. The surface defining maximum heights is the floor of the district.

disturbance, gravitational - The difference between a measured value of gravity and the vertical gradient of the disturbing potential.

disturbance, local magnetic - SEE anomaly, local magnetic.

disturbance, magnetic - (1) An irregular, large-amplitude, rapid change of the Earth's magnetic field which occurs at approximately the same time all over the world. Also called a magnetic storm. It is usually associated with the occurrence of solar flares or other strong solar activity. (2) SEE variation, daily magnetic.

ditch - An artificial, open channel or waterway constructed through earth or rock for the purpose of carrying water. A ditch is smaller than a canal, although the line of demarcation between the two is indefinite. A ditch usually has sharper curvature in its alinement, is not constructed to such refinement or uniformity of grade or cross-section, and is seldom lined with impervious material to prevent seepage. SEE ALSO canal.

ditch, diversion - SEE diversion ditch.

divergence - The difference between the numerical values of two measurements over the same section of a level line. Divergence and partial are practically synonymous.

divergence, accumulated - The algebraic sum of the divergence (partials) for the sections of a level line, from the beginning of the line to the end of any section at which the total divergence is to be computed.

divergence of the vertical - SEE deflection of the vertical, gravimetric.

diversion ditch - (1) A ditch constructed to divert water from its natural channel. (2) An open, artificial waterway approximately parallel to the top of a cut backslope for preventing surficial water from flowing over the slopes of a caut or against the foot of an embankment. Its purpose is protection of the slope from erosion.

dividing engine - SEE engine, dividing.

division - (1) The placing of marks on an instrument or device to represent standard values. (2) The marks, on an instrument or device, which represent standard values. The term is applicable especially to a circle because the number of equally-spaced marks placed on the circle determines the size of a unit, rather than the size of a unit determining the spacing and number of marks. The term graduation is almost synonymous with division but is more often applied to the placing of intermediate marks on an instrument or device (tape, thermometer, etc.) by interpolation.

D-layer - The lowest layer of the ionosphere, at a distance of 70 -80 km above the Earth's surface. It exists only during the daytime.

D-logE curve - SEE curve, characteristic.

dock - A marine structure for the mooring of ships, loading and unloading cargo, or getting passengers on and off.

doctrine of appropriation - The doctrine that the water of natural streams belong to the public generally, but an individual may appropriate it for a beneficial use. The first person making such use of the water thereby establishes a prior right to the continued use of the water for that particular use against all others.

dodging - The process of reducing the amount of exposurein certain parts of sensitized material to avoid overexposure in those parts while getting proper exposure in other parts.

DoD standard indexing system - A system, used within the U.S. Department of Defense, for indexing all aerial photographs held by the U.S. Government. The flights made for aerial photography are plotted on sheets of acetate covering 1° quadrangles of the world at a scale of 1:250 000.

Döllen's method - (1) A method for determining time or longitude by observing the passage (transit) of one or more circumpolar stars (time-stars) through the vertical plane through Polaris. (2) The same as in the previous definition, but passage through the meridian is used instead. Also spelled Dellen's method and Doellen's method.

dolphin - A cluster of wooden or steel-pipe piles fastened together and designed to absorb the shocks of impact from docking or mooring ships.

domain - The set of values, of a vector x, for which a function of x is defined. SEE ALSO range.

domain, eminent - (1) The power of a sovereign government, or some person or group authorized by that government to appropriate part or all of the property within a State or the extent of authority of that government. Allied concepts of geodetic importance are easement and accessibility, e.g., the right of a surveyor to enter private property in doing a governmental survey. (2) The right by which a sovereign government, or some person or agent acting in its name and under its authority, may acquire private property for public or quasi-public use upon payment of reasonable compensation and without consent of the owner. (3) The right or power of a government to take private property for public use on making just compensation therefor. SEE ALSO condemnation.

domain, public - In the USA, the territory ceded to the Federal Government by the original thirteen states, together with certain subsequent additions by cession, treaty, and purchase. At its greatest extent, the public domain contained over 1 820 000 000 acres and included the present States of Alaska, Alabama, Arizona, Arkansas, California, Colorado, Florida, Idaho, Illinois, Indiana, Iowa, Kansas, Louisiana, Michigan, Minnesota, Mississippi, Missouri, Montana, Nebraska, Nevada, New Mexico, North Dakota, Ohio, Oklahoma, Oregon, South Dakota, Utah, Washington, Wisconsin and Wyoming. It now contains about 677 800 000 acres (2 743 000 sq. km.), administered by Federal agencies.

Domesday - (1) SEE Domesday Book. (2) Any of various records of authority similar to that of the Domesday Book (q.v.). E.g., the Domesday of St.Pauls, the record of a survey of the estates of that Chapter, made in 1181. (3) Any of various abstracts based upon the Domesday Book. E.g., the Exon Domesday, the Exchequer Domesday, etc.

Domesday Book - The ancient record of the Grand, or Great Inquest or Survey of the lands of England (1085-86) by order of Willliam the Conqueror. So called as of final authority. Also called Domesday and Doomsday Book. It consists of two volumes; a quarto, sometimes called the Little Domesday dealing only with Essex, Norfolk and Suffolk, and a large folio, the first to be published and sometimes called the Great Domesday, dealing with the rest of England. The Domesday Book gives a census-like description of the realm, with the names of the proprietors, the nature, extent, value, liabilities, etc., of their properties.

Dominion Standard Yard - (1) The length of a bronze bar, a copy of the Imperial Standard Yard bar A, at 61.91° F. This standard served from 1874 to 1951. (2) 0.9144 of the International Meter. This is the official redefinition of 1951.

dominium - In the civil and old English law: (1) Ownership. (2) Property in the larger sense, including both the right of property and the right of possession or use. (3) The mere right of property, as distinguished from the possession or usufruct. (4) The right which a lord had in the fee of his tenant. In this sense, the word is very clearly distinguished by Bracton from dominicum.

donation - The voluntary conveyance of private property to public ownership and use, without compensation to the owner.

donation lands - (1) Lands granted from the public domain to an individual as a bounty, gift or donation. Particularly, in early history of Pennsylvania, lands thus granted to soldiers of the Revolutionary War. (2) Lands granted from the public domain to a corporation. E.g., lands so granted to railroad companies as incentives to the construction of railroads.

donee - (1) A recipient. (2) One to whom property is given.

Doodson number - A number, also called an argument number, identifying a particular component of the tides, according to a scheme devised by Doodson. In the equation for the amplitude of the tide (atmospheric, oceanic, or earth) as a function of time, the trigonometric series used contains as argument a linear function $a\tau + bs + ch + dp + eN' + fps$, in which b,c,d,e,f are integers lying between -4 and +4, and a takes on only positive values and 0. τ is local mean lunar time; s is the mean longitude of the Moon's orbit; h is the mean longitude of the Sun; p is the longitude of lunar perigee; N' is the negative of the mean longitude of the Moon's ascending node; and ps is the mean longitude of perihelion. A particular term is therefore identified by the coefficients a,b,c,d,e,f in the argument. The Doodson number is the number aBC.DEF, where B = (b+5), C = (c+5), D = (d+5), E = (e+5), F = (f +5). For example, the M2 component, the semidiurnal tide caused by the Moon, has the Doodson number 255.555; the S2 term has the Doodson number 273.555; and so on. A few tidal components may have values of b or c greater than 4 in absolute value. Various conventions are used to deal with these rare cases.

Doolittle method - The Gaussian method of solving a set of simultaneous linear equations by successively solving for each unknown in terms of the others and then substituting this solution into the remaining equations, as organized by Doolittle for greater ease of computation for easier checking of the computations. It is also called the Gauss-Doolittle method, which may be preferable because Gauss invented the basic method.

Doomsday - SEE Domesday.

Doppel projection - SEE map projection, double.

Doppler, airborne - Airborne equipment using the Doppler effect to give the aircraft's velocity with respect to the ground. SEE ALSO Doppler navigation.

Doppler, differential - The measurement, at two different points, of the difference in frequency of the Doppler shift. Also called frequency-difference-of-arrival method (FDOA).

Doppler count - The number of times the amplitude of the Doppler shift in a received signal passes from a negative to a positive value in a specified length of time. The Doppler count is approximately equal to the change in the distance between source and observer during the specified interval of time.

Doppler effect - (1) The fact that the frequency observed when sound or electromagnetic radiation is received from a source moving with respect to the receiver is higher or lower than the frequency actually emitted by (or reflected from) the source. The effect was first noted in sound waves, by J.C. Doppler in 1842. It was first noted in electromagnetic radiation by astronomers, as a shifting of lines in stellar spectra. SEE ALSO Doppler frequency; Doppler shift. (2) The frequency f_o measured when radiation emitted at frequency f_s is received at a point other than the source if source and/or receiver are moving with respect to the medium in which the radiation propagates. If v is the velocity of the source relative to the medium and c is the velocity of the radiation, then the Doppler frequency f_o is given by $f_o = f_s [1/(1 - v \cdot c/c \cdot c)]$ if the receiver is stationary. If the source is stationary and the receiver is moving, the Doppler frequency is $f_o = f_s [1 + v \cdot c /c \cdot c]$. In the case of electromagnetic radiation, c is the same for all observers. The Doppler frequency is then $f_o = f_s [1 - v \cdot v/c \cdot c]^{1/2} / [1 - v \cdot c/c \cdot c]$, where v is the velocity of the receiver relative to the source (c being the velocity of light in vacuo). The observed (apparent) direction θ' of the source is given by $\cos \theta' = \sin \theta [(1 - v \cdot v/c \cdot c)/(1 + v \cdot c/c \cdot c)]$, in which θ, the actual direction, is given by $\cos \theta = v \cdot c$. Also called Doppler frequency and, erroneously, Doppler shift.

Doppler effect, radial - That part of the Doppler effect which depends on the direction of the relative velocity between the source and the observer. It is the analog, for an electromagnetic wave, of the Doppler effect in sound.

Doppler effect, transverse - That part of the Doppler effect which occurs when the source of radiation is moving in a direction perpendicular to the direction of the radiation received by the observer.

Doppler error - The error which occurs, in determining the radial velocity of a source by using Doppler radar, because of atmospheric refraction. Such an error may result from (a) the false assumption of a constant wave-velocity in a non-homogeneous atmosphere or (b) the refraction of the ray so that the ray's path deviates from the straight line between radar source.

Doppler-Fizeau effect - The Doppler effect applied to a source of light.

Doppler frequency - (1) SEE Doppler effect (2). (2) The change in the measured frequency with time. This usage may have arisen from a misunderstanding of the meaning of the term and is not common. SEE Doppler shift.

Doppler location - Location determined using measurements of the Doppler shift in radio signals emitted by artificial satellites.

Doppler navigation - (1) Navigation using the shift in frequency (Doppler shift) of sound waves reflected from the ocean's bottom to determine the velocity of the vessel. (2) Navigation using the shift in frequency (Doppler shift) of radio waves reflected from the ground to determine the velocity of an aircraft. (3) Navigation using the shift in frequency (Doppler shift) of radio waves from an orbiting radio-transmitter to determine the location of a receiver on a vessel or aircraft. (4) SEE Doppler navigation system.

Doppler navigation system - (1) Any navigation system using the measured shift in the frequency of a received signal to determine the velocity of the receiver relative to the point from which the signal is transmitted or returned, and from this, to determine by dead reckoning or other means, the location of the receiver. The transmitter and receiver may be located together or the receiver may be located at the user and the transmitter at a fixed or moving unit of known location. The signal may consist of sound waves, radio waves, or light or infrared radiation. Acoustic signals are used almost exclusively by vessels. They may originate at beacons placed at known locations on the bottom or they may originate at generators in the vessel's hull, be reflected from the bottom and received at sound detectors in the hull. Signals at radio frequencies, with transmitter and receiver placed together, are used in aircraft. The signals are radiated downward, scattered by the ground, and part of the scattered radiation picked up by a receiver in the aircraft. The receiver-transmitter combination is often employed in sets of three, the waves being directed as beams forward, to the left, and to the right, so that the complete velocity can be determined. Doppler navigation systems operating at radio frequencies and using signals from fixed transmitters are rare, but systems using signals from transmitters carried in artificial satellites are fairly common, at least on board large vessels (the orbits of the satellites are of course sufficiently well known to satisfy the demands of navigation). Such systems are little used on aircraft because the accuracy of the system depends greatly on the accuracy with which the velocity of the vessel or aircraft is known, and this accuracy is not very high for aircraft. (2) In particular, an airborne navigation system which measures speed relative to the ground and drift by means of radio signals emitted from the aircraft and reflected from the ground. Also called a Doppler navigation. The system depends on the difference in frequencies between emitted and received signals caused by the aircraft's motion.

Doppler navigator - A self-contained apparatus transmitting two or more beams of electromagnetic or acoustic energy downward toward a reflecting surface and using the change in returned frequency (caused by the motion of the vehicle) to measure the speed of the vehicle with respect to the reflecting surface.

Doppler positioning - (1) Placing an object or observer at a specified point by using a Doppler navigation system to determine the difference between the actual location and the desired location and minimizing this difference. Also called Doppler tracking. (2) SEE Doppler surveying.

Doppler positioning system - A positioning system consisting of a radio receiver at the points whose coordinates are to be determined, one or more beacons in orbit about the Earth, a number of radio receivers at points whose locations are known, and computing systems for determining the orbits of the beacons and for determining the location of the point of interest. The difference between the frequency of the radio wave as received and the frequency at which the radio wave was transmitted by the beacon is a function of the radial velocity of the beacon with respect to the receiver. Given the ephemeris of the beacon, as determined from data gathered by the receivers at known locations, the coordinates of the receiver at the point of interest can be calculated from the data gathered there and from the ephemeris.

Doppler radar - (1) An instrument determining radial velocity of an object by comparing the frequency of radiation returned from an object with the frequency of radiation transmitted toward the object. E.g., Doppler radar on an aircraft determines the velocity of the aircraft, referred to a coordinate system on the aircraft, by measuring the Doppler shift experienced by microwaves transmitted towards the Earth and scattered back to a receiver at the aircraft. The radio waves are transmitted as suitably oriented beams. At least three beams are required; a fourth beam has recently been added (in radars operating coherently) for a check. The term Doppler radio system is preferable, since the so-called Doppler radar is not, properly speaking, radar. (2) An instrument which determines the radial velocity of an object by comparing the frequency of radiation received from the object with the known frequency of the radiation emitted by it. This type of Doppler radar has been used for determining the trajectories of spacecraft sent to other planets, etc. The term is also applied to equipment and apparatus incorporating an instrument of either of the types described in the definitions. The term Doppler radio system is preferable, since the so-called Doppler radar is not, properly speaking, radar.

Doppler radar, pulsed - Radar capable of determining not only distance to an object but also the object's radial velocity by measuring the Doppler shift in the frequency of the pulses.

Doppler Satellite Datum - A datum defined for use with the TRANSIT system of navigational satellites (also called the Navy Navigation Satellite System).

Doppler Satellite Survey System - A combination of the TRANSIT satellites and a receiver capable of using the signals from the satellites to determine the location of the receiver.

Doppler shift - The difference between the frequency fo of radiation received at a point and the frequency fs of the radiation when it was emitted, when observer and source are moving with respect to each other. If relativistic effects are ignored, the Doppler shift is given by $\Delta f = f_o - f_s = -f_s [(c + v)/(c - v)]^{1/2}$, in which c is the speed of the radiation and v is the speed of the observer with respect to the source (positive for the two points moving apart).

Doppler surveying - Determining the coordinates of points on the Earth, or the angles and/or distances between these points, by measuring the Doppler shift in the radio waves emitted by a satellite whose orbit is known. Also called Doppler tracking. The Doppler shift depends on (a) the frequency of the radio waves; (b) the orbit of the satellite; and (c) the location of the receiver with respect to the orbit.

Doppler tracking - (1) Pointing an instrument at an object by using the Doppler shift in the object's radiation to determine the direction of the object. (2) Using a radio receiver which puts out the difference between the actual frequency of a received signal and a constant frequency close to the of the signal actually emitted. (3) SEE Doppler positioning; Doppler surveying.

dot grid - A positive photographic film showing an array of regularly spaced dots. The grid is used as an overlay for determining areason maps and aerial photographs.

dot map - A distribution map in which dots (usually of uniform size) are used, each dot representing a specific number or amount of the item whose distribution is being shown.

dot screen - A photographic negative covered with equal-sized dots placed at equal intervals in a rectangular array. Also called a flat-tint screen. It is used to print different tones of a single color.

double-sighting - (1) Taking two readings on a transit, once with the telescope in its normal position and once with the telescope inverted, and then taking the average of the two readings. (2) SEE centering, double.

Dove prism - A prism which reverts an image without deviating or displacing the beam of rays; a given angular rotation of the prism about its longitudinal axis causes the image to rotate through twice the angle. Also called a rotating prism and Delaborne prism.

dower - That portion of, or interest in, the real estate of a deceased husband which the law gives to his widow for life. The extent varies with statutory provisions.

draft correction, dynamic - The correction applied to a measured depth to take into account the difference between the draft of the ship when moving and the draft when motionless.

drafting - The art of drawing from given specifications of the object drawn.

drafting, lithographic - SEE tusching.

drafting guide - SEE scribing guide.

drag - (1) A slight movement of the graduated circle of a theodolite, produced by the rotation of the alidade. Also called instrumental drag. It may be caused by excessive friction in the instrument's centers, by excessive spacing in the fit of the centers, or by instability in the instrument's supports. (2) A wire or pipe extended between two craft and weighted so that it lies at a predetermined depth below the surface of the water for a given separation of the craft. Also called a pipe drag or wire drag. The wire or pipe is pulled by the craft along a selected course; objects extending above that depth intercept the wire or pipe and are detected. (3) SEE drag, fluid. (4) SEE air drag. (5) SEE drag, atmospheric. SEE ALSO air drag.

drag, atmospheric - The retarding force exerted by an atmosphere on a body moving through it. For atmospheres similar to the Earth's, the formula is the same as that given in the definition of air drag. However, atmospheres like those of Jupiter, Venus or the Sun may induce other forces which may have greater effects on bodies moving through them.

drag, fluid - The retarding force exerted by a fluid on a body moving through it. Force perpendicular to the drag is called lift. SEE ALSO air drag; drag, atmospheric.

drag, instrumental - SEE drag (2).

drag coefficient - The quantity k in the formula $F = \frac{1}{2} k \rho A v^2$, in which F is the force exerted by a fluid on a moving body, ρ is the density of the fluid, A is the frontal area of the body in the direction of motion and v is the velocity of the body relative to the fluid.

drainage - In cartography, all features associated with water, such as shorelines, rivers, lakes, marshes, etc.

drainage area - SEE watershed.

drainage basin - SEE watershed.

drainage pattern - The pattern, or overall appearance, made by the network of drainage features on a map or chart.

drawdown - The lowering of the recorded level of water, at a tide gage, caused by the presence of streams near the tide gage.

drawing, as-built - An architectural drawing showing the precise details of construction and where equipment and utility lines are located.

drawing, color-separation - One of a set of drawings that contain similar or related features, such as drainage or culture, to be printed in different colors. There are as many drawings as there are colors to the shown on the printed copy.

drift - (1) The slow motion of a body under the influence of external forces. In particular, the gradual lateral movement of a ship or aircraft caused by currents, winds, or other external forces acting laterally. Also, the gradual horizontal movement of large, continent- sized parts of the Earth's crust. This is known as continental drift or, with some modification, plate motion. (2) The slow and secular change in precision or accuracy of an instrument. For example, the slow changes in readings of a gravimeter, caused by changes in the instrument's structure, is called drift. (3) The vertical component of that part of the precession of the rotational axis of a gyroscope which is caused by the Earth's rotation. (4) All clay, silt, sand, gravel and boulders transported by a glacier and deposited by or from the glacier directly or by water running from the glacier. (5) A horizontal opening in or near a mineral deposit and parallel to the course of the vein or long dimension of the deposit.

drift (of a gravimeter) - A gradual change in the characteristics of a gravimeter, as determined by calibration.

drift, anomalistic - The slow change of frequency of a source of periodic signals for example, the change in the frequency emitted by a crystal oscillator. The change may be caused by varying temperature, aging of the crystal or other components of the oscillator, etc., none of which can be predicted or completely controlled.

drift, continental - The hypothesis that the continents of today were at one time integral parts of a single large continent (Pangea) or of a northern continent (Laurasia) separated by the Tethys Sea from a southern continent (Gondwanaland). These integral parts then separated and drifted to their present locations. The original idea is traceable to Antonio Snider (1858); scientific evidence for the hypothesis was provided by Suess (1900), Taylor (1910), and, very substantially, by Wegener (1929) and du Toit (1937). However, the evidence was not considered strong enough for general acceptance until paleomagnetic data on sea-floor spreading and movement of the magnetic pole were produced in the late 1960's. Discovery of paleomagnetic variations showed that the motion of the magnetic pole, as determined from measurements on rocks in one continent, did not correspond to the motion as determined by similar measurements in another continent. The theory of continental drift was further strengthened by the discovery that the floor of the oceans had apparently been spread apart by material coming up from the mantle through the mid-oceanic ridges. The theories of continental drift and plate motions (plate tectonics) are almost identical. The principal difference has been that the former considers the continents themselves as moving, isolated blocks. The latter considers the continents move as integral parts of larger blocks (plates) that are contiguous and move as wholes. The difference has diminished in recent years with the introduction, into the theory of plate motion, of the hypothesis that some or all of the blocks are separated by regions of plastic material which does not move with the blocks.

drift, instrumental - A gradual change, in some property of an instrument, which affects the measurements made with that instrument.

drift, littoral - The material that moves more or less parallel to the shore-line under the influence of oceanic forces.

drift, total - The algebraic sum of the movement of the rotational axis of a gyroscope because of precession caused by the Earth's rotation and precession caused by other forces.

drift angle - The angle, measured in degrees, between the heading of an aircraft or ship and the track made good. Drift angle is designated as being right or left to indicate the direction of measurement.

drift station - (1) That ground station which transmits SHORAN signals and about which an aircraft flies to maintain a circular path. The other ground station is then referred to as the rate station. The term is rarely used. (2) That navigation station, of a pair, from which a vessel doing a hydrographic survey tries to maintain a fixed distance. The second station of the pair is then called the rate station.

driftways - SEE byroad.

drive - A winding, scenic road (trafficway) without noticeable buildings.

drop chaining - Chaining in which the graduated tape is kept horizontal at each step, using a plumb line to correlate the mark on the tape with the mark on the ground. Also written drop-chaining.

drop-out - A halftone negative which has received a modified or supplementary exposure so that the extreme highlights or the white background of the artwork will not produce a printing dot on the negative.

drumlin - An elongated hill or ridge composed of till, usually oval and shaped like half an egg, having glacial origins.

drum plotter - SEE plotter, drum-type.

drum scanner - A scanner in which the image being scanned is placed on a rotating drum which rotates slowly while the scanning beam moves rapidly back and forth parallel to the axis of rotation.

duck - One of the lead weights used to hold a spline in place.

dummy (printing) - (1) A preliminary drawing or layout showing the positions of the illustrations and text as they will appear in the final reproduction. (2) A set of blank pages made up to show the size, shape, and general style of a book, booklet or pamphlet.

dummy pendulum - A pendulum similar in structure to working pendulums but equipped with a thermometer and fastened rigidly in its container so that it cannot swing during observations. The dummy pendulum is subject to the same temperatures as the working pendulums, and is used for determining their temperature when in use.

dump - SEE waste bank.

dumpy level - SEE level, dumpy.

Dupin's indicatrix - The curve obtained by the intersection of a plane with a surface as the plane approaches tangency with the surface. The indicatrix of Dupin characterized the curvature of the surface at the point of tangency.

Durchmusterung - A star catalog whose principal purpose is to identify stars and to give their coordinates with sufficient accuracy that the stars can be located. The stars are usually identified by giving the name of the Durchmusterung in which they occur and the star's number in that catalog. SEE ALSO Bonner Durchmusterung; Cordoba Durchmusterung.

dynamic number - SEE number, dynamic; number, normal dynamic.

dynamostat - An apparatus for applying tension to a surveyor's tape and consisting basically of a lever having two arms, one short and one long, at an oblique angle to one another. The lever is pivoted at the junction of the two arms.

The pivot moves in a slide. A weight is hung from the end of the long arm and one end of the tape is attached to the end of the short arm.

dyne - A force which, acting on a mass of one gram, imparts to that mass an acceleration of one centimeter per second per second. It is the unit of force in the c.g.s. system of units. It has been replaced in the Système International d'Unités (SI) by the newton. SEE ALSO gal.

E

eagre - SEE bore, tidal.

earth - (1) The land surface of the Earth, as distinguished from the waters and the atmosphere. (2) Soil or dirt. (3) SEE Earth.

Earth - The world we inhabit; the third planet from the Sun in the Solar System. Also referred to, in literature, as the terrestrial sphere. It is an approximately spherical body revolving around the Sun in a period of one year at a distance of approximately 150 000 000 km. It rotates at the rate of one complete rotation per day, although the length of the day in atomic seconds varies irregularly and in any event depends on how the axis of rotation is defined. It has one satellite, the Moon; this has mass about 1/81 that of the Earth and is at an average distance of about 384 000 km. The characteristics most important to geodesy are given in the Appendix.

Earth, effective radius of the - SEE radius of the Earth, effective.

Earth, hydrostatic - The Earth considered as a fluid in hydrostatic equilibrium (under gravitational and centrifugal forces).

Earth, model - (1) Any massive body having approximately the geometrical and physical characteristics of the Earth. In particular, a massive body whose shape approximates the geoid and whose gravitational field approximates the Earth's. Also called a normal Earth. (2) A theoretical solid whose bounding surface is an equipotential surface and whose gravity field agrees with that of the actual Earth where observations are possible. The agreement will not be exact for any solid that can be represented mathematically. It will usually be quite good for values of the gravity field averaged over sizable areas. (3) A surface close to the geoid and obtained from the actual surface of the Earth by smoothing (in theory) the topography in some suitable manner. Hunter's Model Earth is of this type. (4) SEE Model Earth.

Earth, Model - SEE Model Earth.

Earth, normal - (1) A body, rotating with the Earth, whose external equipotential surface is an oblate, rotational ellipsoid and whose gravity field on that surface is given by a standard gravity formula. (2) A body which rotates with the same angular velocity as the Earth and has the same axis of rotation, and whose gravitational potential is given by the zero-th and second-degree terms in the expression of that potential as a series of Legendre functions. Also called a model Earth. (3) SEE Model Earth.

Earth, standard - (1) A body defined in such a way as to approximate the Earth's mass, dimensions, and rate of rotation but representable by simple mathematical formulae. The simplest form is a solid ellipsoid of rotation with defined major axis and flattening, whose internal density is distributed in such a way as to make its surface an equi-potential surface of its attraction and the centrifugal force, with its center of mass coinciding that of the Earth. (2) A theoretical solid in which each surface of equal seismic speed of P-waves or S-waves is spherical and encloses the same volume as the corresponding surface of equal speed in the actual Earth. The seismological tables of Jeffreys and Bullen relate to this standard Earth. SEE ALSO Airy's figure of the Earth; radius (of the Earth), effective; radius (of the Earth), gravitational; figure of the Earth; figure of the Earth, hydrostatic; flattening (of the Earth); model (of the Earth; shape of the Earth.

Earth curvature - SEE curvature of the Earth.

Earth ellipsoid - SEE ellipsoid, terrestrial.

Earth ellipsoid, mean - (1) An ellipsoidal body which has the same mass and the same rotational velocity, about the shortest axis, as the Earth and the same value of the coefficient C_2 ($-J_2$) in the representation of the Earth's gravity-potential by a Legendre series. (2) An ellipsoidal body whose surface is at constant potential and whose four defining constants are determined by an adjustment of more than four interrelated constants. E.g., the values for the mass of the Earth, gravity at the equator, flattening, potential on the geoid, equatorial radius, and the ratio of the gravitational force to the potential on the geoid have been determined simultaneously, even though only four of them are independent. (3) That ellipsoid which most closely approximates the geoid.

earth inductor - An instrument used in magnetic surveys for determining the magnetic dip; it works on the principle of a small dynamo whose magnetic field is provided by the Earth. The current generated by the dynamo is sent to a galvanometer which indicates the amount of dip.

Earth Model - (1) A mathematical representation of the density or similar characteristic of the Earth as a function of depth. (2) SEE Model Earth. SEE ALSO model of the Earth.

Earth reference model, geodetic - A rotational ellipsoid of given dimensions and rate of rotation, containing a given mass and on which the gravity potential has a given value.

Earth spheroid - (1) An equipotential surface of a body which has the same mass as the Earth, whose center of mass coincides with that of the Earth, and which rotates with the Earth. It is axially symmetric and is symmetric with respect to a plane through the center of mass and perpendicular to the axis of rotation. (2) The equipotential surface assumed to be the Earth's outer surface and to have the same volume as the geoid.

Earth spherop - (1) The surface obtained by assigning to the spheropotential function a value such that the volume inclosed by the resulting surface is the same as the volume inclosed by the geoid. Because the volume of the Earth is not directly measurable but must be calculated from data on elevations of the Earth's surface above the geoid and on geoidal heights, the Earth spherop is known only approximately. (2) A spherop containing a volume equal to that contained in the geoid.

earth tide - That periodic movement of the lithosphere which is caused by the attraction of the Moon, Sun and other planets. Also called bodily tide or Earth tide. It is completely analogous to the tides in the open oceans but is less than half their amplitude.

earthwork - (1) Any operation involving moving large amounts of earth. (2) In particular, the operations connected with excavations and embankments of earth, in preparing foundations of buildings, in constructing canals, railroads, highways, etc. (3) An embankment or other construction made of earth.

easel - The flat surface, in a rectifier or printer, onto which the image is projected.

easement - The right, privilege, or liberty given to a person or group to use land belonging to another for a specific and definite purpose. A common easement is that giving a company the right to bring electrical transmission lines across private property. Another is the legal establishment of a public trail across private property. It is approximately equivalent to servitude (Canada), but the latter term includes restrictive covenants and profit a prendre.

easement, appurtenant - An easement attached to a parcel of land which passes with that parcel to heirs and assigns of the owner of the land.

easement, augmenting - An easement lying outside the parcel of land being conveyed but of benefit to the parcel.

easement, implied - The privilege, by operation of law, to use the land of another for a particular purpose that arises in connection with a conveyance when it was obvious that continued use before the conveyance was meant to be permanent and reasonably necessary for beneficial enjoyment of the land conveyed or retained.

easement, necessary - An easement providing the only reasonable means by which the dominant tenement can be enjoyed. Also called easement by necessity.

easement, negative - An easement precluding the owner of the land upon which the easement exists from acting on the land in a way one would be entitled to act if no easement existed. For example, a solar easement where building a structure on the servient estate might block sunlight from reaching the dominant easement.

easement, overhead - The right to use the space at a designated distance above the surface of the land e.g., for power lines, aviation and air rights.

easement, prescriptive - A right acquired by an adverse user to use the land of another thru prescriptive rights.

easement, subsurface - The right to use the land at a designated distance below the surface of the land e.g., for pipelines, electric and telephone circuits and cables, storage facilities, etc.

easement, surface - SEE surface easement.

easement by necessity - SEE easement, necessary.

easement by prescription - SEE easement, prescriptive.

easement curve - A curve the radius of which varies to provide a gradual transition between a tangent and a simple curve (or between two simple curves of different radii). SEE ALSO transition curve.

east, magnetic - The direction shown by a magnetic compass to be east.

easterly - A direction within 22.5° of east.

easting - (1) The distance eastward (positive) or westward (negative) of a point from a particular meridian taken as reference. (It is common practice to use positive westings instead of negative eastings. In particular, (2) the distance eastward (positive) or westward (negative) from the central (zero) meridian (the line of zero eastings or the y-axis) on a gridded map. This meridian is frequently replaced by another meridian, called the false meridian, sufficiently far to the west of the central meridian that all points on the map having eastings are then positive with respect to the false meridian. SEE ALSO departure.

easting, false - A constant value added to all eastings, or from which all westings are subtracted, to produce only positive values.

east point - The eastern intersection of the plane through the vertical and perpendicular to the astronomical meridian with the horizon.

ebb current - The movement of a tidal current away from shore or down a tidal river or estuary.

ebb tide - (1) That portion of the oceanic tide which occurs between high water and the following low water. Also called a falling tide. (2) A seaward current caused by that portion of the tide which occurs between high water and the following low water.

Eberhard effect - The increase in density of small regions at the expense of larger, neighboring regions. Also called the adjacency effect, Nachbareffekt, neighborhood effect, etc.

eccentric (n.) - SEE station, eccentric.

eccentric error - SEE centering error.

eccentricity - (1) A quantity e indicating the amount by which a given conic section differs from a circle, and given by the formula $e = (\sqrt{[a - k^2 (b/a)^2]}$, in which a is half the length of the major axis, b half the length of the minor axis and k^2 a constant equal to +1 for an ellipse, 0 for a parabola and -1 for a hyperbola. It is also called the first eccentricity when it must be distinguished from the second eccentricity. (2) The eccentricity of a surface is the eccentricity (definition 1) of a designated intersection of that surface with a plane. In particular, (a) The eccentricity of a rotational ellipsoid (spheroid) is the eccentricity of an ellipse formed by meridional section of the spheroid. (b) An ellipsoid has in general two eccentricities: the meridional eccentricity, which is the eccentricity of the ellipse formed by a section containing the longest and the shortest axes (one of which is the polar axis), and the equatorial eccentricity, which is the eccentricity of the ellipse formed by a section through the center and perpendicular to the polar axis. (c) The eccentricity of a rotational ellipsoid (spheroid), is given by $+ \sqrt{[(a^2 + b^2)/a^2]}$, in which a and b are the lengths of the equatorial and polar semi-axes, respectively. (3) The distance of a point from a center or axis. The term is generally applied either to (a) the distance of a surveying instrument from the point it should be occupying or to (b) the rotation of a part of the instrument about an axis other than the proper axis.

eccentricity, angular - The angle whose sine is the eccentricity. Because the eccentricity always lies between 0 and +1, the angular eccentricity always lies between 0 and $\pi/2$.

eccentricity, equatorial - That eccentricity of an ellipsoid which is the eccentricity of an equatorial section of the ellipsoid.

eccentricity, first - SEE eccentricity (1).

eccentricity, linear - The quantity $\sqrt{(a^2 - b^2)}$, in which a and b are half the lengths of the major and minor diameters, respectively, of a rotational ellipsoid. It is equal to ae, in which e is the eccentricity.

eccentricity, meridional - That eccentricity of an ellipsoid which is the eccentricity of a section by a plane through the longest and shortest axes.

eccentricity, orbital - (1) The eccentricity (1) of an elliptical orbit. (2) The eccentricity (1) of the osculating ellipse or average ellipse of an orbit.

eccentricity, second - The quantity e' defined, for a conic section, as $e' = \sqrt{[(1/k^2)(a/b)^2 - 1]}$, in which a is half the length of the major axis, b is half the length of the minor axis and k^2 is a constant equal to +1 for an ellipse, 0 for a parabola and -1 for a hyperbola. It is related to the eccentricity (first eccentricity) by the equation $1 = (1 - e^2)(1 + e'^2)$.

eccentricity correction - That correction which must be applied to a direction observed by an instrument with either the instrument or signal (target), or both, eccentric, to reduce the observed direction to what it would have been if there had been no eccentricity. Also called eccentric reduction.

eccentricity of alidade - The distance between the center defined by the index points on the alidade and the center defined by the graduated circle. The index points (on vernier or micrometer microscope) are on the alidade, and any eccentricity of alidade combines with eccentricity of circle to form eccentricity of instrument.

eccentricity of an ellipse - SEE eccentricity.

eccentricity of a spheroid - SEE eccentricity.

eccentricity of circle - The distance between the center of figure of a graduated circle and the axis of rotation of the part whose rotation is referred to the circle. Eccentricity of circle is usually expressed in terms of its equivalence in seconds on the circle. It may be made quite small by adjustment. Its effect on an observed direction is eliminated by reading the circle at equally spaced points around its circumference. SEE ALSO eccentricity of instrument.

eccentricity of compass - A non-linear error, in the readings from a compass, resulting from one or more of the following conditions: (a) a straight line through the ends of the magnetic needle fails to pass through the center of rotation of the needle; (b) the center of rotation of the needle does not coincide with the center of figure of the graduated circle; or (c) the line of sight fails to pass through the vertical axis of the instrument.

eccentricity of instrument - The condition resulting from the combination of eccentricity of circle and eccentricity of alidade. The effect of eccentricity of instrument on an observed direction is eliminated by having the verniers or micrometer microscopes with which the circle is read spaced at equal distances around the circle.

eccentricity of vernier - A displacement of the two verniers, on the horizontal circle of a transit or theodolite, so that the line joining their indices does not pass through the axis about which the upper plate rotates.

echo - (1) That part of an emitted signal which is reflected or scattered back to a receiving instrument such as the ear, a fathometer, or radar. (2) The mark made on a recording medium (such as an echogram or as an impulse on magnetic tape) made bo the echo (sense 1 above) from a sonic pulse. (3) The mark displayed, on a cathode-ray tube, as evidence of a received pulse. Also called a blip or echo pip.

echogram - The graphic record produced by an echo sounder and showing, as a function of time, the strength of the echo and the time taken for the echo to return. If the ship's velocity is constant, the echogram is a distorted profile of the bottom or, if the sound penetrates the bottom for a considerable distance, a distorted cross-section of the underlying layers.

echo pip - SEE echo (3).

echo sounder - An instrument for determining the depth of water by measuring the time it takes a sound-pulse to travel from the water's surface to the bottom and back, multiplying by the speed of sound, and dividing by 2. One type consists of an oscillator which generates the pulses, a hydrophone for detecting the echo, a clock for measuring the time elapsed between sending out the pulse and receiving the echo, and a recorder which converts the elapsed time to a corresponding depth and plots that depth on a continuous chart. Another type displays the depth digitally, records the value on magnetic tape as a function of location, and plots the depth on a fair sheet or chart. Early models of echo sounders used hammers or similar devices for generating pulses. Present-day models use submerged loudspeakers or magnetostrictive or piezo-electric oscillators.

echo-sounder calibration - SEE bar check.

echo sounding - Determining the distance from the surface of a body of water to the bottom by measuring the interval of time it takes a sonic pulse to travel from the surface to the bottom and back again. The principal source of error is incomplete correction for refraction; another is schools of fish which reflect the sonic pulses before the pulses reach the bottom. The idea of echo sounding was suggested by D. Arago in 1807. Eels, in a U.S. patent of 1907, described the principles completely. The first echo sounder known to be used successfully was designed by A. Behm a few years later.

Eckert's map projection - One of six map projections devised by M. Eckert for mapping the entire sphere onto the plane in such a way that the poles are represented by straight lines half the length of the equator. Eckert's map projections are usually identified by Roman numerals from I to VI. Numbers I and II are map projections into the interior of a trapezoid; number I is an equidistant map projection; number II is an equal-area map projection. Numbers III and IV represent meridians by elliptical arcs (number III is also known as Ortelius's map projection); number IV is an equal-area map projection. Numbers V and VI represent meridians by sinusoidal arcs; number VI is an equal-area map projection.

eclipse - (1) The partial or total shadowing (obscuration), relative to an observer, of one celestial body by another. The term occultation is used for those eclipses in which some star other than the Sun provides the light, and the light is obscured by the Moon or by some other body in the Solar system. Occultations and eclipses have been used in the determination of longitude and latitude. The principle is the same for either phenomenon. The time at which the Moon's shadow passes a point of known longitude and latitude is measured, as is the time the same part of the shadow passes a second point whose longitude and latitude are to be determined. The difference in times is a function of (a) the differences in longitude and latitude, and (b) the velocity of the Moon's shadow across the Earth's surface. The last of these quantities can be calculated from the data in astronomical ephemerides. Observations of several eclipses give enough data for calculating the longitudinal and latitudinal differences. In the past, the necessary calculation have used the Besselian elements listed in ephemerides and almanacs. (2) The period of time during which one celestial body obscures the light received from another.

eclipse, annular - (1) A solar eclipse in which the Sun's disk is never completely covered but is seen as an annulus or ring at maximal eclipse. It occurs when the apparent disk of the Moon is smaller than that of the Sun. (2) An eclipse in which a thin ring of light appears around the obscuring body.

eclipse, lunar - An eclipse in which the Moon passes through the shadow cast by the Earth. The eclipse may be total (the Moon passing completely into the Earth's umbra), partial (the Moon passing partially into the Earth's umbra at the time of greatest eclipse) or penumbral (the Moon passing only through the Earth's penumbra).

eclipse, solar - An eclipse in which the Earth passes through the shadow cast by the Moon. Also called a solar occultation. It may be total (observer in the Moon's umbra), partial (observer in the Moon's penumbra) or annular (Moon appears surrounded by a ring of light). Although sunlight does not reach the Moon's visible face directly during a solar eclipse, some light does reach it by reflection from the Earth, producing an effect called the ashen Moon.

eclipse year - The interval of time between two successive conjunctions of the Sun with the same node of the Moon's orbit. It averaged 346d 14h 53m 50.7s in 1900 and was increasing by 2.8s per century.

ecliptic - (1) That circle which is the intersection of the celestial sphere with a plane defined by the Sun's bary-center, the barycenter of the Earth-Moon system and the velocity of that barycenter, as determined from the theory of planetary motions. The points in which the ecliptic intersects the celestial equator are the equinoxes; the angle at which the ecliptic intersects the celestial equator is the obliquity of the ecliptic. (2) The apparent path of the center of the Sun as seen from the barycenter of the Earth-Moon system. (3) The average plane of the Earth's orbit around the Sun. SEE ALSO obliquity of the ecliptic; pole of the ecliptic.

ecliptic, true - The actual ecliptic at a specified instant. The term distinguishes this concept from that of a mathematically derived ecliptic such as an average ecliptic. It has the defect that the ecliptic is itself a mathematical concept.

ecliptic coordinate system - SEE coordinate system, ecliptic.

ecliptic latitude - SEE latitude, celestial.

ecliptic longitude - SEE longitude, celestial.

ecliptic meridian - SEE meridian of the ecliptic.

ecliptic node - SEE node.

ecliptic parallel - SEE parallel, ecliptic.

ecliptic pole - SEE pole of the ecliptic.

ecology - The study of the mutual relationships between organisms and their environments.

edge, bleeding - SEE bleed.

edge detection - The detection, in image space, of the image of the edge or line of an object in object space. This is a special kind of line detection (q.v.).

edge enhancement - A process by which the boundaries between the images of distinct objects in object-space and the background are emphasized. SEE ALSO dodging.

edge fog - The darkening of unexposed photographic film by light leaking between the flanges of the spool on which the film is wound.

edge gradient - The sensitometric density of the photographic image of a sharply defined edge (knife edge), as a function of distance from a line parallel to the theoretical position of the image of the edge.

editing - The process of checking a map or chart in its preparatory stages to ensure accuracy, completeness and, in its printed form, legibility and precision. The various stages of editing are also referred to, in jargon, as edits.

edition - A particular issue of a map, chart, or atlas different in time of issue and in some details from other issues.

edition, new - An issue of a previously published map, chart, or atlas differing from previous issues in its factual content, layout, or design.

edition, preliminary - SEE edition, provisional.

edition, provisional (cartography) - A map or chart published for temporary use only, with the provision that it will be superseded later.

EDM corner reflector - A cube-corner reflector used in distance measurement by electro-optical distance measuring.

EDM instrument - SEE instrument, electronic distance-measuring.

effect, electrostatic - The lengthening of a pendulum period by an electrostatic charge on the pendulum. A non-metallic pendulum such as one made of quartz picks up an electrostatic charge. This results in a change in the period, but can be prevented by placing a small amount of radioactive material in the container to ionize the gas and so let the charge leak off.

effect, indirect - (1) The effect, upon gravity, of that massive layer which lies between the geoid and the new position of the corresponding level surface when masses are transferred (hypothetically) from outside the geoid to inside the geoid according to some systematic method. Also called the Bowie effect and Bowie gravity correction. Those terms, however, are best reserved for the effect which results in the Bowie gravity correction; indirect effect is a more general term applicable for any method of transferring masses. The indirect effect was discussed by Stokes, Bruns, Clarke, Helmert, and others, but it was under the direction of Bowie that a practical means was provided for computing its value for a given station. (2) The corresponding effect on gravity potential. (3) The corresponding effect on the geoid. (4) SEE Bowie gravity correction.

effect, luni-solar - The effect of the gravitational attractions of the Moon and the Sun.

effect, magnetic - A lengthening of the period of a pendulum, caused by the action, on the pendulum, of the Earth's magnetic field or of a permanent magnetic field in the instrument. The period of an invar pendulum may be increased by as much as 0.1 microsecond if the magnetic field increases by 0.02 Oersted. This corresponds to 0.05 mgal. In regions of large magnetic anomaly, the error may amount to tens of mgals. If the instrument itself is generating the magnetic field, it should be demagnetized by being placed in a Helmholz coil. The best solution is to make the pendulum of a non-magnetic material such as quartz. Quartz, however, accumulates an electrostatic charge which affects the period.

Egault level - A leveling instrument, no longer in general use, having the spirit level attached to a leveling bar with wyes in which the telescope rests. In the so-called wye level, the spirit level is attached to the telescope and is reversed with it. In Egault levels, the telescope can be reversed in the wyes without disturbing the spirit level.

egress - (1) The act or right of going from a place of real or seeming confinement. (2) A place or means of exit; an outlet. (3) The emergence of a celestial body from eclipse, occultation, or transit.

Einstein convention - SEE summation convention.

Einstein summation convention - SEE summation convention.

Ekman bottle - SEE bottle, water-sampling.

Ekman spiral - In general, a polar diagram showing how rotational velocity in a fluid varies with depth. In particular, a polar diagram showing how the velocity of a wind-driven, oceanic current varies with depth. The Ekman spiral is often represented as a three-dimensional model (a helix); arrows representing speed and direction of the current are attached horizontally to a vertical rod on which the corresponding depths are shown.

elastic (adj.) - A term applied to any substance which offers resistance to forces that tend to deform it and which resumes its original form when the forces cease, provided the latter are within a certain limit called the elastic limit.

elasticity - The ability of a body to return to its original size and shape after deformation. It deals, properly, only with reversible relations between stress and strain.

elasticity correction - SEE tension correction (to tape length).

Electrokinetograph, Geomagnetic - SEE Geomagnetic Electro-kinetograph.

electronic-position-indicator - Distance-measuring equipment operating at 2 MHz and involving the measurement of the time it takes radio pulses to travel between the vessel carrying the indicator and each of two stations on land. Abbreviated as E.P.I. or written as an acronym EPI. It has been used primarily in hydrographic surveying. Distances of over 800 km can be measured. The accuracy is between 60 and 500 meters.

element - Something fundamental, essential, or irreducible which helps make up or define a larger whole. SEE, in particular, element, optical; element, orbital.

element (of a fix) - The specific value of one of the coordinates used to specify a location.

element (of an orbit) - SEE element, orbital.

element, Besselian - SEE Besselian element.

element, Delaunay - SEE Delaunay element.

element, front - That element of an optical system which is closest to the object. SEE ALSO element, optical.

element, geomagnetic - A set of seven quantities that characterize the Earth's magnetic field: declination, horizontal intensity, vertical intensity, total intensity, inclination, strength of the geomagnetic force towards geographic north and strength of the force towards geographic east. Also called a magnetic element.

element, geometrical - (1) A point, line, surface, solid or hyper-solid which exists as an undefined part, postulate or geometrical theory. (2) A point, line, surface or solid, or a combination of these, which is the geometrical counterpart or representation of a physical object.

element, Keplerian - SEE Keplerian element.

element, magnetic - SEE element, geomagnetic.

element, mean - One of the elements, of an orbit adopted as reference, which approximates the actual, perturbed orbit. Mean elements serve as the basis for calculating perturbations.

element, optical - A single, indivisible entity of an optical system, usually a lens. It is a single piece of optical material, such as a lens, mirror or prism, which is designed to reflect, refract or diffract light in a specified way. The material between lenses, mirrors, etc., may itself be an optical element in certain optical systems.

element, orbital - (1) One of the six constants of integration resulting from the solution of the three second-order differential equations describing the motion of a point mass in a gravitational field. These six constants are enough to specify the path of the point mass. Together with the time, they specify the orbit. They are often referred to simply as elements. When the differential equations are written in terms of Cartesian coordinates, three of the elements specify the location of the body at a particular time; the other three elements are the components of the point-mass's velocity at that time. When polar coordinates are used, Keplerian elements are the natural result. (2) SEE Keplerian element.

element, osculating - (1) One of the six orbital elements which specify that elliptical orbit which is tangent to the actual orbit at a particular instant and which the orbiting point-mass would follow if the primary were replaced by a point of the same total mass and all perturbing forces were to vanish at that instant. Alternatively, one of the six orbital elements which specify the instantaneous location and velocity of a point-mass in its actual orbit. The two definitions are equivalent because the location and velocity at an instant are known if the elliptical orbit is known, and vice versa.

element, rear - That element of an optical system which is closest to the image plane or surface. SEE ALSO element, optical.

elevation - (1) The distance of a point above a specified surface of constant gravity-potential, measured along the direction of gravity between the point and the surface. The surface is understood to be the geoid unless some other surface is specified, and the elevation is then the orthometric elevation. Although mean sea level was long specified, that surface is not at constant gravity-potential. It lies above or below the geoid by as much as a meter in some places but its exact shape is not known. When mean sea level is specified, the term should be understood to mean either the geoid or a surface of constant gravity-potential, passing through a specific point at mean sea level and of limited extent about that point. The terms altitude and height are often used uncritically as synonyms for elevation. However altitude more precisely means distance above the Earth's physical surface or above a surface of known and constant atmospheric pressure. Height is commonly used to mean distance vertically between two points on an object (e.g., the base and top of a mountain or two marks on a leveling rod) or in geodesy, the perpendicular distance from a specified, geometric surface such as a reference ellipsoid or reference plane. (In this glossary, elevation is used only with the meaning given for it here; altitude and height are not used as synonyms for it but are used with the meanings just given.) In geodetic formulae, H is usually used for elevations, h for height above the ellipsoid. (2) An approximate or theoretical value of the distance of a point above a surface (usually the geoid) defined in terms of the gravity potential or measured differences of elevation (1). This kind of elevation is usually given a name e.g., Baranov elevation, orthometric elevation, etc. (3) The quantity obtained by adding differences of elevation (1) i.e., the quantity obtained by spirit leveling without correcting the data for deflection of the vertical. (4) A region of limited extent and of nearly uniform elevation. (5) A vertical distance above sea level or other datum. (6) SEE elevation, angular. SEE ALSO Baranov elevation; barometer elevation; field elevation; geopotential elevation; ground elevation, assumed; ground elevation, mean; ground elevation, optimum; Helmert elevation; Niethammer elevation; spot elevation; Vignal elevation.

elevation (of the tide) - SEE height (of the tide).

elevation, adjusted - (1) The elevation resulting from adding an adjustment correction to an orthometric elevation. (2) The elevation resulting from adding both an orthometric correction and an adjustment correction to a preliminary elevation.

elevation, angular - The angle, at a point, between the horizontal plane through that point and a line from the point to a designated object or in a designated direction. It is also called altitude, angular altitude, ascending vertical angle, angle of elevation and elevation. It is the complement of zenith distance. It is positive if taken upward and negative if taken downward from the horizontal.

elevation, barometric - An elevation determined by measuring atmospheric pressure with a barometer. The scale of atmospheric pressures indicated by the barometer can be calibrated to give altitude above some surface at standard pressure. This altitude (barometric altitude) must then be corrected for the elevation of the standard-pressure surface above the geoid. The term is sometimes applied in error to the barometric altitude.

elevation, checked - An elevation determined by two or more independent sets of measurements, or by a closed traverse, in which the results agree within a specified limit.

elevation, critical - The greatest elevation in any group of related and more or less contiguous formations shown on a map.

elevation, dynamic - SEE height, dynamic. Although this quantity is neither an elevation nor a height, it is termed a height in common usage.

elevation, fixed - An elevation which has been obtained either from tidal observations or from a previous adjustment of leveling, and which is held at its adopted value in any subsequent adjustment.

elevation, geometric - The quantity $rH(r+H)$, in which r is the radius of the Earth at 45° 32' 40" latitude and H is the dynamic height in dynamic meters.

elevation, highest - The elevation of the point of greatest elevation represented within the limits of a map. SEE ALSO elevation, critical.

elevation, low-tide - A naturally formed region of land surrounded by and above water at low tide but submerged at high tide. This definition was accepted by the Geneva Convention.

elevation, modified spheroidal - The elevation obtained by integrating the value of gravity between the geoid and the point B on the surface $dHms = \int (g/\tau_B) \, dH$, in which g is the measured value of gravity, τ_B is the theoretical value of gravity at B and H is the measured difference of elevation.

elevation, Molodenski - SEE height, normal.

elevation, normal - SEE height, normal.

elevation, normal orthometric - The value HnoN calculated for the elevation of a point PN according to the formula $HnoN = [\int_{Pmsl}^{PN} \tau_\phi \, dH]/\tau_N$, in which τ_ϕ is a theoretical value of gravity acceleration calculated from a gravity formula for elevation H and latitude ϕ, dH is the corresponding increment in elevation and the integration is taken over the route used for leveling between point Pmsl at mean sea level at some tide gage and PN on the Earth's surface. τ_N is the value of gravity acceleration calculated by the same gravity formula for the latitude of PN and for an elevation midway between the elevation of PN and the elevation of Pmsl. If the gravity formula does not contain a term involving H, τ_N may be calculated by adding a free-air gravity correction to the computed value.

elevation, orthometric - (1) The distance between the geoid and a point, measured along the vertical through the point and taken positive upward from the geoid. Also called orthometric height and normal height. Also referred to simply as elevation. If g is a value of gravity acceleration on the surface of the Earth and g' is the value of the acceleration on the same level surface as g but on the vertical through the point PN, then the orthometric elevation HN of PN relative to a point P_o on the geoid is given by $HN = \int_{P_o}^{PN} (g/g') \, dH$, in which g is measured over a continuous route from P_o to PN. The value g' is not measurable and must be calculated. Niethammer's elevation or some modification of it is the closest approximation to HN. Orthometric elevation differs from elevation in that use of the latter term does not require that the geoid be the level surface of reference. (2) A preliminary elevation to which the orthometric leveling correction has been applied.

elevation, practical - SEE height, practical.

elevation, preliminary - An elevation arrived at in the office after the index, level, rod and temperature corrections have been applied to the observed differences of elevation and new elevations have been calculated.

elevation, quasi-dynamic - A value HqdNh found for the elevation at a point PN using the formula $HqdNh = [\int_{P_o}^{PN} g \, dH] / \tau_\phi$, in which the integral is taken along the route between point P_o on the geoid and point PN on the ground, and τ_ϕ is a theoretical value of gravity acceleration calculated from a gravity formula for some chosen latitude ϕ (which may be the same as the latitude of PN) and for a point halfway between the equipotential surfaces through P_o and PN. When the gravity formula does not contain a term that takes geodetic height into account, a free-air gravity correction is added to the calculated gravity-acceleration to obtain a theoretical value at the mid-point.

elevation, quasigeoidal - Elevation referred to the quasi-geoid.

elevation, relative - The distance of a point above an arbitrary, specified surface of reference, taken along the vertical through the point.

elevation, residual - SEE depth anomaly, residual.

elevation, spheroidal - The value HsN calculated for the elevation of a point PN by dividing a theoretical value WN of the potential function at PN (assuming the value to be 0 on the geoid) by a theoretical, average value gsN of gravity acceleration between PN and the geoid. The theoretical value τ_n of gravity acceleration is calculated from a standard gravity formula $\tau_n = \tau_o [1+ (\beta/2) \cos 2\phi_n - 2Hn/R]$ instead of from measured values. τ_n is the value of acceleration at latitude ϕ_n and elevation Hn; τ_o and ß are constants and R is an appropriate radius of the Earth.

elevation, standard - An adjusted elevation based on the sea-level datum of 1929 (now known as the National Geodetic Vertical Datum of 1929) or some definite epoch. In some localities, this is modified by a regional adjustment, e.g. Pacific Northwest Suplemental Adjustment.

elevation angle - SEE elevation, angular.

elevation anomaly, dynamic - SEE height anomaly, dynamic.

elevation datum - A point or surface used as reference for elevations.

elevation difference - (1) The difference in elevation at two points. Called height difference by some geodesists. (2) The observed difference in the readings on the forward and backward leveling rods in spirit leveling. I.e., the difference in height at two points. This definition may be in very limited use.

elevation limits - SEE tint, hypsometric.

elevation meter - A mechanical or electromechanical device on wheels which measures slope and distance, and automatically and continuously integrates their product into difference of elevation. The version developed by the U.S. Geological Survey is called a Ground Elevation Meter.

elevation tint - SEE tint, hpsometric.

11/10 peg adjustment - SEE peg adjustment.

elinvar - A variety of invar, composed of approximately 34% nickel, 57% iron, 4% chromium and 2 % tungsten. The springs in some spring-type gravimeters are of elinvar.

ellipse - A second-degree, plane curve characterized geometrically as the locus of all points whose distances d_1 and d_2 from two fixed points (called the foci) have a constant sum: $d_1 + d_2$ = constant. It is characterized algebraically by the fact that the discriminant of its second-degree equation in x and y $ax^2 + by^2 + cxy + dx + ey + f = 0$ is greater than zero. The longest straight line joining two points on the ellipse is called the major axis or major diameter; it passes through the two foci. The center of the ellipse is the point midway between the ends of the major axis. The shortest straight line joining two points on the ellipse is called the minor axis or minor diameter; it is perpendicular to the major axis at the center. An ellipse with center at (x_o, y_o), with the major axis parallel to the x-axis and of length 2a, and with a minor axis of length 2b has the equation $(x -x_o)^2/a^2 + (y - y_o)^2/b^2 = 1$. The general form of the equation is obtained by rotating the coordinate system through an arbitrary angle. SEE ALSO error ellipse; flattening.

ellipse, aberrational - The elliptical path, on the celestial sphere, apparently followed yearly by a star because of the combination of the Earth's revolution about the Sun with the finite velocity of light. Also called the parallactic ellipse.

ellipse, great - The ellipse in which a plane through the center of an ellipsoid cuts the ellipsoid. When the ellipsoid is rotationally symmetrical, the great ellipse may be a circle; when the ellipsoid is a sphere, the great ellipse is a great circle.

ellipse, osculating - An ellipse tangent to a curve. SEE ALSO orbit, osculating.

ellipse, probable-error - An ellipse containing exactly 50% of a set of points having a bivariate Gaussian distribution. The equation of the ellipse is $(\xi/\sigma_x)^2 - 2\rho(\xi/\sigma_x)(\eta/\sigma_y) + (\eta/\sigma_y)^2 = (1 - \xi^2)k^2$, in which k^2 is such that $[1 - \exp(-k^2/2)]$ is exactly 0.5. σ_x and σ_y are the standard deviations and ρ is the correlation coefficient. k^2 is approximately 1.1777.

ellipse, parallactic - SEE ellipse, aberrational.

ellipse, rectilinear - The straight line (a degenerate ellipse) formed by letting the eccentricity of an ellipse approach zero while keeping the major axis fixed.

ellipse, standard-error - An ellipse for which the quantity $(\xi/\sigma_x)^2 - 2\rho(\xi/\sigma_x)(\eta/\sigma_y) + (\eta/\sigma_y)^2 = (1 - \rho^2)$; σ_x and σ_y are standard deviations of the coordinates (x,y) of a point and ρ is the correlation coefficient. Alternatively, an ellipse bounded by the lines $\xi = \pm \sigma_x$ and $\eta = \pm \sigma_y$. About 39.4% of the points lie within this ellipse. Also called the Helmert error-ellipse.

ellipse of distortion - SEE Tissot's indicatrix.

ellipse of error - SEE error ellipse.

ellipsoid - Geometrically, a closed surface all planar sections of which are ellipses. Algebraically, the set of points with coordinates (x,y,z) satisfying the equation $(x -x_o)^2 / a^2 + (y - y_o)^2/b^2 + (z -z_o)^2/c^2 = 1$, in which a,b,c are the lengths of the semi-axes of the ellipsoid and x_o, y_o, z_o are the coordinates of the center; the axes of the coordinate system are parallel to the axes of the ellipsoid. The general form of the equation is obtained by rotating the coordinate system to a new position. An ellipsoid is specified by giving three independent dimensions. The most common specification gives the lengths a,b,c of the semi-axes. Almost as common is the specification of the length of the longest semi-axis a and the eccentricities of the equatorial and polar sections. A rotational ellipsoid or spheroid is the ellipsoid obtained by rotating an ellipse about its major or minor diameter. An ellipsoid with different lengths for all three axes is sometimes called a general ellipsoid, heteroaxial ellipsoid, three-axis ellipsoid or triaxial ellipsoid. The term heteroaxial ellipsoid is used in this glossary. SEE ALSO Carte de France ellipsoid; Helmert ellipsoid; Jacobi ellipsoid; Krassovsky ellipsoid 1940; Plessis ellipsoid; reference ellipsoid; spheroid; Struve ellipsoid.

ellipsoid, biaxial - SEE ellipsoid of revolution.

ellipsoid, datum-centered - An ellipsoid which best fits the astro-geodetic network in a particular datum and hence does not necessarily have its center at the Earth's center. More precisely, an ellipsoid whose dimensions and position have been selected to best fit the astronomic and geodetic coordinates of a particular geodetic network.

ellipsoid, Earth-centered - An ellipsoid representing the Earth, having its center at the Earth's center of mass, and having its minor axis coincident with the Earth's axis of rotation.

ellipsoid, equipotential - An ellipsoid on which a potential is defined and constant.

ellipsoid, flattened - (1) A non-spherical ellipsoid. (2) SEE ellipsoid, oblate rotational.

ellipsoid, general - (1) An ellipsoid considered without specifying the relative lengths of its axes. (2) SEE ellipsoid, heteroaxial.

ellipsoid, heteroaxial - An ellipsoid all of whose axes are of unequal lengths. Also called a general ellipsoid, three-axis ellipsoid or triaxial ellipsoid.

Ellipsoid, International - SEE International Ellipsoid.

ellipsoid, level - An rotational ellipsoid on which the gravity potential is constant and on which the value of gravity is given by a standard gravity formula.

ellipsoid, mean-error - An ellipsoid having the value 1 for its quadratic form. Alternatively, an ellipsoid bounded by the planes $\xi = \pm \sigma_x$, $\eta = \pm \sigma_y$, $\zeta = \pm \sigma_z$, in which σ_x, σ_y, σ_z are the standard deviations of the set of variables x,y,z having a Gaussian distribution. The probability that an error lies within the ellipsoid is 0.19874. Also called a standard-error ellipsoid and Helmert mean-error ellipsoid.

ellipsoid, momental - That ellipsoid which has for its three coefficients the moments of inertia, of a rigid body, with respect to the principal axes. Also called the ellipsoid of inertia and the ellipsoid of Poinsot.

ellipsoid, normal - (1) An ellipsoid which is at a specified, constant gravity-potential and which envelopes but rotates with the Earth. (2) An ellipsoid on which gravity is given by a standard gravity formula and which belongs to a normal Earth in the sense of Molodenski. (3) An ellipsoidal body having a given mass, a given rate of rotation, and given dimensions, and subject to the condition that the gravity potential at its surface be a constant.

ellipsoid, oblate - SEE ellipsoid, oblate rotational; spheroid, oblate.

ellipsoid, oblate rotational - A rotational ellipsoid symmetrical about the shortest axis; an ellipsoid formed by rotating an ellipse about its minor axis. Also called, in geometrical geodesy and in physics, an oblate spheroid. Also called, rarely and inappropriately, a flattened ellipsoid, or ambiguously, an oblate ellipsoid.

ellipsoid, probable-error - An ellipsoid such that 50% of the points in a set of points having a Gaussian distribution in three-dimensional space lie within the ellipsoid.

ellipsoid, prolate - SEE spheroid, prolate.

ellipsoid, rotational - SEE ellipsoid of revolution.

ellipsoid, terrestrial - (1) A body which rotates with the same angular velocity as the Earth and has the same axis of rotation, and which has a gravitational potential given by the zero-th and second-degree terms in a representation of the Earth's gravitational potential as a series of Legendre functions. Also called an Earth ellipsoid and normal Earth. (2) An ellipsoidal solid whose volume is equal to the volume of the geoid, whose center of mass and equatorial plane coincide with the Earth's center of mass and equatorial plane, and the sum of whose deflections of the vertical with respect to the geoid are a minimum.

ellipsoid, three-axis - SEE ellipsoid, heteroaxial.

ellipsoid, tri-axial - An ellipsoid having three axes, each a different length. Also called a general ellipsoid, heteroaxial ellipsoid or three- axis ellipsoid.

ellipsoid of error - SEE error ellipsoid.

ellipsoid of inertia - SEE ellipsoid, momental.

ellipsoid of Poinsot - SEE ellipsoid, momental.

ellipsoid of reference - SEE reference ellipsoid; reference spheroid.

ellipsoid of revolution - An ellipsoid formed by rotating an ellipse about either the major axis or the minor axis. Also called a spheroid, ellipsoid of rotation, rotational ellipsoid and biaxial ellipsoid. If the ellipse is rotated about the major axis, the result is called a prolate rotational ellipsoid or, more usually, a prolate spheroid. If it is rotated about the minor axis, the result is called an oblate rotational ellipsoid or, more usually, an oblate spheroid. The Earth deviates in shape from an oblate spheroid by less than 1 part in 500.

ellipsoid of revolution, flattened - SEE spheroid, oblate; ellipsoidof revolution.

ellipsoid of rotation - SEE ellipsoid of revolution.

ellipsoid of rotation, oblate - SEE ellipsoid, oblate rotational.

ellipticity (of an ellipse) - The ratio of the difference in lengths of the major and minor axes of an ellipse to the length of the major axis. Also called flattening.

ellipticity (of a spheroid) - SEE flattening. In the older literature, ellipticity was sometimes used when the shape was being considered mathematically and flattening was used when the rotational ellipsoid was considered to represent the shape of the Earth. SEE ALSO Jeffreys-Radau ellipticity.

ellipticity, dynamical - The quantity $(C - A)/C$, in which A and C are the moments of inertia of a rotationally ellipsoidal solid about a major axis and about the minor axis, respectively. Also called mechanical ellipticity. The Earth's dynamical ellipticity can be determined from measurements of the luni-solar precession. It is affected by other factors such as the ratio of the mass of the Moon to the mass of the Earth.

ellipticity, mechanical - SEE ellipticity, dynamical.

elongation - (1) The angular distance of a celestial body from the Sun, as viewed from the Earth. An elongation of 0° is called conjunction, one of 180° is called opposition and one of 90° is called quadrature. (2) The geocentric angle between a planet and the Sun, measured in the plane of the planet, Earth and Sun. Elongation is measured from 0° to 180° east or west of the Sun. (3) The geocentric angle between a planetary satellite and its primary, measured in the plane of the satellite, planet, and Earth. Elongation is measured from 0° east or west of the planet. (4) That point, in the apparent daily motion of a star about the celestial pole, at which the star's rate of change of azimuth becomes zero; the point at which the star is seen to cease increasing its bearing east or west of the celestial pole and to reverse its direction of motion eastward or westward. An equivalent definition is, that point at which the parallactic angle of star and observer is 90°.

elongation, greatest - The instants when the geocentric angles between Venus or Mercury and the Sun are the greatest.

elongation, planetary - SEE elongation (2).

embankment - An artificial (man-made) rise composed of earth on and above the original surface of the ground. For example, a dike, a highway or a railroad, constructed to grade across a valley or over a plain.

embayment - An indentation of a coast, regardless of width at the entrance or depth of penetration into land.

embedment - The length of a pile from the surface of the ground, or from a cutoff below the ground, to the point of the pile.

emergency run - A reprint, of a map, necessitated by unusual conditions before extensive revisions can be made. Also called a tide-over run (jargon).

emissivity - The amount of energy given off by a body in unit time, relative to the amount per unit time given off by a black-body at the same temperature. It is usually expressed as a positive number between 0 and 1.

emulsion - (1) A colloid in which globules of a liquid are dispersed in another liquid with which it is immiscible. (2) SEE emulsion, photographic.

emulsion, color - SEE color emulsion.

emulsion, infrared - A photographic emulsion whose sensitivity extends into the near-infrared portions of the spectrum.

emulsion, panchromatic - A photographic emulsion having a sensitivity to color similar to that of the human eye, but producing black-and-white pictures only.

emulsion, photographic - A suspension of a light-sensitive salt of silver (especially silver chloride or silver bromide) in a colloidal medium (usually gelatin, which is placed as a coating on rigid or flexible materials such as glass plates, plastic film, or paper to be used in photography. It is usually referred to simply as the emulsion, although it is not a true emulsion. Types in common use are panchromatic (producing black-and-white pictures), color negative (producing pictures in colors complementary to those of the object), color positive (producing pictures in colors the same as those of the object), infrared-color (producing pictures in which infrared radiation from the object is represented by colors in the picture), and infrared black-and-white (producing pictures in which the density of a spot in the picture is proportional to the amount of infrared illumination radiation that has illuminated that spot).

emulsion-to-base (adj.) - SEE exposure, emulsion-to-base.

emulsion-to-emulsion (adj.) - SEE exposure, emulsion-to-emulsion.

encroachment - An intrusion or invasion of an adjoining landowner's real estate by a neighbor through some means or structure, permanent in nature.

encumbrance - An interest or right in real property which diminishes the value of the fee but does not prevent conveyance of the fee by its owner. Mortgages, taxes due and judgments are encumbrances known as liens. Restrictions, easements and reservations are encumbrances, though not liens.

end lap - SEE overlap (photography).

endosphere - The mantle and core of the Earth.

endowment - (1) The act of endowing, or bestowing a dower, fund or permanent provision for support. (2) That which is bestowed or settled on a person or an institution.

end standard of length - A bar or block whose ends or sides are flat and contain a defined or otherwise known length between them. The ends are usually optically flat (i.e., flat to within a fraction of a wavelength of light) and polished to a mirror-like finish so that the length can be measured interferometrically.

energy - The immediate or indirect capacity, or a mathematical or numerical representation of that capacity, for changing the location, position, or shape of a body. The concept of immediate capacity is also called, specifically, mechanical energy, which was the first kind identified (by Galileo in his study of moving bodies). Other kinds of energy such as thermal, chemical, electromagnetic and the neutrino, are concepts introduced to avoid having energy appear in or disappear from apparently closed physical systems. Such concepts allow a universal law of conservation of energy to hold. Energy is conventionally considered to be the sum of two kinds of energy: kinetic energy, which depends on velocity, and potential energy, which depends on location and, perhaps, other factors. Alternatively, it can be considered the sum of kinetic energy, a potential energy which depends only on location, and an energy which depends on other factors such as temperature or state of magnetization. SEE ALSO law of the conservation of energy.

energy, available potential - That portion of the total potential energy which can be converted to kinetic energy in an adiabatically enclosed system.

energy, gravitational potential - (1) The gravitational potential energy at a point P is the amount of work done by (or from) a conservative field in carrying a unit mass to P from infinity. (2) The difference ΔW in gravitational potential energy between points P and Q is the amount of work done, by a conservative field, in carrying a unit mass from P to Q.

energy, kinetic - The capacity of a system for doing work by virtue of the velocities of the various moving masses in the system. The kinetic energy (k.e.) of a body of mass M moving with velocity v is $\frac{1}{2} M v^2$. It is equal to the work that would have to be expended to bring the body to rest. It is additive, in the sense that the k.e. of a system composed of several moving bodies is the sum of the k.e.s of the individual bodies. A distinction is sometimes made between a body's internal k.e. and its external or mechanical k.e. The internal k.e. is the sum of the k.e.'s of the individual particles making up the body, with respect to a coordinate system fixed in the body. The external k.e. is the k.e. appearing as translation and rotation of the body as a whole in some external coordinate system. Unless stated otherwise, kinetic energy means external kinetic energy.

energy, law of the conservation of - SEE law of the conservation of energy.

energy, mechanical - SEE energy, mechanical potential.

energy, mechanical potential - (1) The work done in changing a physical system from one distribution of masses and velocities (state I) to another (state II), if the same amount of work can be recovered by the system's changing from state II to state I. Such a system is called a reversible mechanical system. (2) The work done in changing a physical system from one distribution of masses and velocities (state I) to another (state II), if the work done is a function only of the initial and final configurations and not of the trajectories along which the work was done. In speaking of systems composed of point-masses, the phrase a function only of the initial and final coordinates of the masses is often used. The two definitions are almost equivalent, and both are referred to as definitions of mechanical potential energy or, if no confusion is likely, as definitions simply of potential energy. (3) The difference in mechanical potential energy of a body at two points P and Q is the amount of work done in moving that body from P to Q. It is, for a point-mass, the product of the mass of the body by the negative of the difference in potential at the two points.

energy, potential - (1) The work done in changing a physical system from some standard configuration (state) to its present configuration (state). The implication of potential is that work done on the system remains in the system and can later be taken from the system to do the same amount of work on something else. This assumption is false on the macroscopic scale. Some of the work done on a system is stored or converted to a form which cannot be used by the system to do work. Classical mechanics therefore considers only reversible forms of potential energy and limits the term potential energy to such forms. SEE ALSO energy, mechanical potential. An alternative definition differs from that just given in starting out: The work done by a physical system in changing from. . . The two versions are equivalent except for producing opposite signs for the potential energy. Which version is used depends on the user's viewpoint. (2) The work done by a force-field in moving a body from one point in the field to another. An alternative definition differs from this by substituting the word against for by. The two different kinds of potential energy differ only in their sign. Potential energy differs from potential in that the former depends on the characteristics of both the force field and the body moved. The latter is, or at least is defined only in terms of, the characteristics of the force field.

engine, dividing - A machine, also called a ruling engine, for precisely drawing or incising equally-spaced lines in material. It is used to produce diffraction gratings, the graduated circles on telescopes, transits, and theodolites, and so on. In making graduated circles, only the prototype is usually graduated on the dividing engine. Copies are made by photoetching or some other photographic process.

engine, ruling - SEE engine, dividing.

engineering, human - That part of engineering which deals with the design of devices, machines, and structures in such a way as to make them best fitted to use by humans.

engineering geology - The application of geology to engineering to ensure that the geological factors affecting determination of route or site, location, design, construction, operation and/or maintenance of the work are recognized and are taken adequately into consideration.

engineering map - A map showing information essential for planning an engineering project or development and for estimating its cost. An engineering map is usually a large-scale map of a comparatively small area or of a route. It may be made entirely from the data gathered by a survey made specifically for the purpose or may be complied from information collected from various sources and assembled on a base map.

engineering survey - A survey made to obtain information essential for planning an engineering project and estimating its cost. The information obtained may, in part, be recorded in the form of a map.

engineering surveying - Surveying done in preparation for or as a part of the execution of an engineering project.

engineer's chain - A chain 100 feet long and composed of 100 links each 1 foot long. At the end of every 10th link, a brass tag is fastened notched to show the number of preceding 10 link segments.

engineer's level - A leveling instrument whose telescope and level are connected rigidly to a vertical axle, so that the line of sight must be leveled using the screws in the base (or using the compensator). A contrasting variety is the tilting level.

engineer's leveling instrument - SEE engineer's level.

engineer's scale - A piece of wood or metal having a triangular cross-section and marked, along each face adjacent to an edge, at regular intervals; the marks are numbered so as to indicate the ratio between lengths along the edge and lengths or distances along an object. An engineer's scale therefore provides six different graphic scales, two along each edge on opposing faces.

engineer's transit - (1) An angle-measuring instrument consisting basically of a telescope so mounted that it can be rotated through 360° in a horizontal plane and through 180° in a vertical plane, and capable of measuring the amount of rotation in those planes. A long level-vial is fastened to the longitudinal axis of the telescope. Also called an engineer's transit theodolite. An instrument similar to the engineer's transit but lacking the long level-vial and the ability to measure vertical angles is called a plain transit. SEE ALSO theodolite, repeating. (2) A transit or theodolite usable for measuring both horizontal and vertical angles with an accuracy sufficient for civil engineering and construction engineering but not sufficient for establishing geodetic control. (3) SEE theodolite, repeating.

engineer's transit theodolite - SEE engineer's transit.

English foot - SEE foot.

English map projection - SEE map projection, globular.

engraver (cartography) - SEE scriber.

engraver, rigid-tripod - A scribing tool that has contact with the scribed material at three points, to compensate for the normal imbalance of the pressure exerted by the operator.

engraver subdivider - A scribing device that permits selection of uniformly spaced ticks in subdividing or putting ticks on graticules or maps.

engraving - (1) The cutting of a design into a solid surface. The design may be made by cutting away material so that the design is raised above the rest of the surface, or it may be made by cutting so that the design is in the cut-away portion and lies beneath the rest of the surface. The most common form of engraving is of the second kind, the design being formed by lines cut (graved) into the surface. Such engravings are usually used for printing on paper, the surface being covered with ink and then wiped, so that ink remains only in the cut lines. The finest maps were formerly made from engravings because of the delicate effects that could be produced by an artist. The most common materials for engravings have been soft metals such as copper, and wood (the result being called woodblocks). However, much commercial printing is now done from engravings made by cutting lines through the photographic emulsion on a negative. This kind of engraving is called, in cartographic jargon, scribing. SEE ALSO graver. (2) The result of engraving (1) or the print made from such an engraving.

engraving, electronic - Engraving in which the original is scanned linearly and an engraver is controlled by the output from the scanner.

engraving, negative - SEE negative engraving.

engraving, positive - SEE positive engraving.

enlargement, x-diameter - The ratio (x+1):1 of the length of the diagonal of an enlarged copy to that of the rectangular original.

enlargement (reduction), two-step - Projecting and printing a small image, then copying and projecting it again to the required size. This is often necessary when the size of the copy or the limitations of the copying camera do not allow enlargement (reduction) in a single operation.

enlargement factor - The ratio of the size of an original to the size of the enlarged copy. Also called scale of reproduction.

enlargement/reduction diagram - A chart showing the necessary extensions of lens system and copyboard required for various enlargements and reductions.

enlarger - An apparatus which creates an enlarged image of the picture on a photographic transparency. Some enlargers allow tilting the surface (easel) onto which the image is projected. If the plane of the photographic transparency can also be tilted, the enlarger is usually called a rectifier.

entrance pupil - That image (real or virtual) of the aperture stop formed in object space by all those elements of the optical system on the object-space side of the aperture stop.

entrance window - That image of the field stop formed in object space by all elements (of an optical system) on the object-space side of the field stop.

entropy - (1) The thermodynamic function S defined by $\Delta S = \int (dQ/T)$, in which ΔS is the change in S which occurs when integrating over given limits, Q is the amount of heat added (reversibly) to a system at absolute temperature T. The concept and term were invented by R. Clausius in 1865. The concept has been applied in statistical mechanics and information theory. (2) A quantity proportional to the logarithm of the state of a physical system, as defined by $S = k \ln P + \text{constant}$, in which S is the entropy, k is the Boltzmann constant and P is the statistical probability of the state. (3) The amount of information per elementary unit or symbol of a message. The change in the function $S = -k \Sigma \theta \ln pi$ was shown by Claude Shannon in 1948 to measure the amount of information in any message. The receiver assigns the probability θ to the i-th message out of all possible messages. k is a constant. SEE ALSO method of maximum entropy.

entropy, method of maximum - SEE method of maximum entropy.

entry - A passage for workers, haulage, or ventilation deep below the surface and of a permanent nature (i.e., not in the ore to be removed). SEE ALSO portal.

entry, desert-land - An entry of irrigable and agricultural public lands under the act of March 3, 1877, as amended, which the entryman reclaims, irrigates, and cultivates in part and for which he pays a certain amount.

entry, right-of-survey - The right to enter property temporarily to make surveys and investigations for proposed improvements.

entryman - A person who makes an entry of land under the public-land laws of the USA.

envelope line - A line, in a marginal sea, every point of which is at a distance, from the nearest point of the baseline, equal to the breadth of the marginal sea. Geometrically, it is the locus of the center, of a circle of fixed radius, which moves so that the circumference is always in contact with the baseline. The name is derived from the fact that the line forms a continuous series of intersecting arcs farthest seaward of all the possible arcs that can be drawn from the baseline with the same radius, thus enveloping all arcs that fall short of the seaward arcs. SEE ALSO baseline.

environment - The set of all conditions external to an organism or community which act upon that organism or community to influence its development or its existence. Included in the environment are all forms of life, air, soil and other such material, water, and foreign particles and organisms within them, and light, temperature and wind.

Eötvös balance - SEE torsion balance.

Eötvös correction - (1) The difference δg_E between the value of gravity-acceleration actually measured by a horizontally moving gravimeter and the value that would be measured if the gravimeter were stationary. The correction is the sum of the vertical components of the accelerations produced by the Coriolis effect and the centrifugal acceleration caused by the gravimeter's motion. It is given by $\delta g_E = (R_\phi + h)(v_\phi v_e \sin\phi \sin A_z + v^2) / R_\phi$, in which v is the speed of the gravimeter, v_e its eastward component, v_ϕ the eastward speed of the Earth at geodetic latitude ϕ, A_z the azimuth of the direction in which the gravimeter is moving, R_ϕ the radius of the reference ellipsoid at the location of the gravimeter and h the geodetic height of the gravimeter. (2) The quantity $4.0 v \sin\alpha \cos\phi$.

Eötvös effect - The fictitious, vertical force experienced by a body on the rotating Earth. It is the vertical component of the Coriolis force.

Eötvös pendulum - SEE torsion balance.

Eötvös unit - A unit of measure for rate of change of acceleration with distance i.e., the gradient of acceleration. It has magnitude and dimensions 10^{-9} m/(s²m) or 10^{-9} cm/(s²cm). The quantity is sometimes given as 10^{-9} s^{-2}. This form is dimensionally correct but physically wrong and should not be used. The Eötvös unit is used principally in expressing values of gravity gradient. Gravity gradiometers typically have sensitivities of 0.1 to 1.0 Eötvös units. E.U. and E have been used as symbols for the unit; the former is preferred.

ephemeris - A tabulation of the locations and related data for a celestial body for given epochs (dates) at uniform intervals of time. In particular, a publication containing such data for a number of celestial bodies. The Astronomical Ephemeris (formerly the American Ephemeris and Nautical Almanac) is such a publication. It contains, for specified instants, the coordinates of the principal celestial bodies; the constants specifying the position of the coordinate system used; numbers used in calculating the effects of changes in the location of the observer; and, in general, data on all those phenomena relating to celestial bodies which may be regarded as functions of time and which are of interest to navigation and related arts. The plural of ephemeris is ephemerides. While custom has approved the use of the singular form to designate the publication as a whole, the plural form is used when tabulations for specific bodies are discussed e.g., the ephemerides of Mercury and Venus. When tabulation for one specific body is discussed, the singular form is used e.g., the ephemeris of Mercury.

ephemeris, astrometric - An ephemeris, of a celestial body in the Solar System, in which the tabulated positions are directly comparable to the mean places of stars listed in catalogs for some standard epoch. An astrometric position is obtained by adding to the geometric position, calculated from gravitational theory, the correction for light time and the E-terms of annual aberration.

ephemeris, broadcast - An ephemeris which is broadcast from the satellite to whose orbit it refers.

ephemeris, precise - An ephemeris which contains coordinates and times (and possibly velocities) more precise than those of a broadcast ephemeris. The term has been used, in particular, to refer to the ephemerides of the TRANSIT satellites. Those ephemerides are prepared from data received from several tracking stations over the world and are calculated with great precision using the method of least squares.

ephemeris calendar - A calendar based on the ephemeris second and the ephemeris day, and on an epoch given in ephemeris time.

ephemeris date, Julian - SEE Julian ephemeris date.

ephemeris day - A day containing 84,400 ephemeris seconds.

ephemeris day number, Julian - SEE Julian ephemeris day number.

ephemeris hour angle - An hour angle referred to the ephemeris meridian.

ephemeris longitude - Longitude measured eastward from the ephemeris meridian.

ephemeris meridian - (1) The meridian 15.04l07 T east of the meridian of Greenwich, where T is a number equal to the difference, in seconds of time, between ephemeris time and Universal Time. (2) The meridian where the Greenwich mean astronomical meridian would have been if the Earth had rotated uniformly at the rate implicit in the definition of ephemeris time. It is located at 1.002738 T east of the actual meridian of Greenwich on the Earth's surface.

ephemeris second - The unit of time equal to $1/(31\,556\,925.974\,7)$ part of the tropical year for 1900 January 0 at 12 hours ephemeris time. This definition was adopted in 1960 by the general Conference on Weights and Measures, the previous definition in terms of the rotation of the Earth being abolished. It is theoretically equivalent to the system of time used by Simon Newcomb in his theory of the motions of the Solar System. The ephemeris second was replaced, in 1967, by the atomic second.

ephemeris time - (1) That time or independent variable which is determined by the equations of motion of the planets and moons of the Solar System and by determinations of the coordinates of these bodies. Also written Ephemeris Time. It is about 4 seconds later than Universal Time at 1900 January 0, at 12 hours. (2) Those values which, when substituted for the variable t in the equations of motion of members of the Solar System, cause the predicted coordinates to best agree with those determined from measurements. (3) Time as reckoned from the instant specified by the 10th General Assembly of the International Astronomical Union (1958) as the epoch for Ephemeris time, and using the ephemeris second as adopted by the Committee International des Poids et Mesures (1957) for the fundamental interval of time. Also called dynamical time or ephemeris time. The fundamental unit of time, from which the ephemeris second is derived, is the tropical year. The epoch is specified as that instant, near the beginning of A.D. 1900, when the geometric mean longitude of the Sun was 279° 41' 48.04". Ephemeris time was then 0d 12h January 1900, precisely.

ephemeris transit - The passage of a celestial body or point across the ephemeris meridian.

epipolarity - The relationship existing between two perspective (e.g., aerial photographic) images such that a plane through the object- point and the two image-points also passes through the epipoles.

epipole - One of the two points where the planes of two aerial photographs are cut by the airbase (the extended line joining the two perspective centers). In the case of a pair of truly vertical photographs, the epipoles are infinitely distant from the principal points.

epoch - (1) A particular instant of time from which an event or a series of events is calculated; a starting point in time, to which events are referred. In particular, a date and instant corresponding to the position of a coordinate system, all subsequent positions being referred to the position at that epoch. In this sense, it is equivalent to date (1). Many writers prefer to use date for this meaning, since it denotes an instant unambiguously. However, the two terms have different connotations. Epoch is associated with a point in time to which events are referred; date is associated with a value of time and is associated with only the event that occurred then. Also, date is commonly used for civil date, which is not an instant. (2) The angular lag of the maximum of the observed tide or tidal current behind the corresponding maximum of the same constituent of the theoretical equilibrium-tide. When the theoretical, local equilibrium-tide is meant, the epoch is represented by κ. When the corresponding equilibrium tide at Greenwich is meant, it is called the Greenwich epoch and is represented by the symbol G. (3) A specific period of time.

Epoch, National Tidal Datum - SEE National Tidal Datum Epoch.

epoch, standard - A date and a time which specify the reference system to which celestial coordinates are referred. Coordinates in star catalogs are commonly referred to the mean equinox and equator of the beginning of a Besselian year.

epoch, tidal - The period of time specified for uninterrupted measurement of tides at a tide gage. The U.S. National Ocean Service specifies a period of at least 19 years.

equal-altitude method - Any method of determining azimuth, observing the same celestial body when it first passed upward through a particular circle of constant angular elevation and when it next passed downward through the same circle, or (b) observing several different celestial bodies as they pass through a particular circle of constant angular elevation. Also called an equal angular-elevation method and equal zenithal-distance method.

equal-altitude method (of determining azimuth) - (1) Horizontal angles are measured from a celestial body to the indicated direction at two different times - when the body first reaches a specified angular elevation, and when the same body reaches that same angular elevation again. The average of the two angles is the azimuth of the direction. In determining azimuths in the northern hemisphere, the direction should be southerly from the observer, while if the point is in the southern hemisphere, the direction should be northerly from the observer. If the body is the Sun, a correction must be made for the Sun's change in declination between the morning's and the afternoon's observations. It is also important to remember to make all observations on the same limb. Except in the case of the Sun or Moon, the body's coordinates at the time of observation need not be known. SEE ALSO equal-altitude method, solar. (2) The azimuths at which three or more celestial bodies (usually stars) cross a specified circle of constant angular elevation are measured and corrections δ_A to the observed azimuths are calculated using the equation $\Delta A = \delta_A + \operatorname{cosec} \sigma \cos\theta$

cosec ζ dϕ - cot σ cosec ζ dζ, in which ΔA is the difference between measured and calculated azimuths at the times of crossing, σ is the parallactic angle, θ the calculated hour angle and A the calculated azimuth. δ_A, δ_ϕ and δ_ζ are the corrections to the assumed values of azimuth, latitude and zenith distance ζ, respectively. (3) The angle is measured between the azimuth mark and the vertical plane through the point at which a star crosses that almucantar whose angular elevation is the same as the latitude of the observer. The time of crossing, used for calculating the azimuth of the star, is also observed, and the azimuth of the point of crossing is calculated. Also called the almucantar-crossing method.

equal-altitude method (of determining latitude) - Measurement of the azimuths of three or more stars at the instants these stars pass through a circle of specified, constant angular elevation θ, and calculation of a correction $\Delta\phi$ to the assumed latitude ϕ, according to the equation $\Delta A_o = \delta_A$ - sec θ_o cosec σ_o cos $\tau \Delta\phi$ - sec θ_o cot σ_o $\Delta\theta$, in which ΔA_o is the difference between observed and calculated values of a star's azimuth, δ_A is a correction to readings on the horizontal circle, θ_o is the assumed angular elevation, $\Delta\theta$ is the correction to the angular elevation, σ_o is the calculated parallactic angle and τ is the calculated hour-angle of the star. (There is one such equation for each observation).

equal-altitude method (of determining latitude and time) - Measurement of the times when three or more stars, all at the same angular elevation, pass through the vertical cross-hair of the theodolite at a constant azimuth, and determination of increments to the measured times and assumed latitudes by solving the equations (one for each star) $\Delta t = \Delta t_c$ - cot A$_o$ sec ϕ_o $\Delta\phi$ + cosec A$_o$ sec ϕ_o $\Delta\epsilon$, for the adjustments Δt_c, $\Delta\phi$ and $\Delta\epsilon$. Δt is the difference between computed and measured times, A$_o$ is the computed azimuth for the assumed latitude ϕ_o and angular elevation ϵ. Also called the Gauss method or Gauss's method.

equal-altitude method (of determining longitude) - The times at which each of a pair of celestial bodies reaches the same angular elevation are measured, and the longitude calculated from these data and an ephemeris of the body. The Tsinger method of determining longitude or time is a variant of this method. Also called the equal angular-elevation method and equal zenith distance method. The same method can be used for determining the time if the longitudes are measured.

equal-altitude method (of determining time) - SEE equal-altitude method (of determining longitude).

equal-altitude method, solar (of determining azimuth) - The equal-altitude method (1) of determining azimuth, using the Sun as the observed body. Horizontal angles are measured from a southerly reference-point to the Sun's limbs at an identical vertical angle; if measured to the right limb before noon, it should be measured to the left limb after noon. The same limb should be observed in vertical angles. The average of the two horizontal angles, with small correction for the change in the Sun's declination during the interval from a.m. to p.m. readings, gives a resulting, horizontal angle to the meridian.

equal angular-elevation method - SEE equal-altitude method (of determining azimuth).

equality - The situation that arises when a single point on a route alinement has two values because of joining preliminary and final stationing. Thus, station 123+45.6 ahead equals 123+54.3 back.

equal zenith-distance method - SEE equal-altitude method (of determining azimuth).

equation - (1) A stated equality between two functions f_1 and f_2 of the same quantities: $f_1(x,y,z,....) = f_2(x,y,z,....)$. f_1 is usually a constant; x,y,z, etc., are usually variables. In applied mathematics, it is customary to distinguish two kinds of equations, depending on the nature of f_1: observation equations and condition equations. When f_1 is a measurable or measured quantity, or is the difference between two such quantities, the equation is an observation equation. When f_1 is a quantity fixed by mathematical considerations or by definition (and therefore not dependent on the measurements), the equation is a condition equation. (2) In general, any astronomical quantity which differs in value from what it should be. In particular, the term is applied in particular to (a) a cyclic perturbation of an orbit, and (b) differences between what an observation should be and what it actually is. SEE (for example) equation, personal. SEE ALSO angle equation; azimuth equation; Clairaut's equation; condition equation; correlate equation; error equation; Laplace equation; latitude equation; length equation; longitude equation; observation equation; perpendicular equation; side equation.
latitude, longitude, or time which depends either on (a)

equation, annual - An inequality (perturbation) in the Moon's motion, having an amplitude of -11' 8.93" and a period of one anomalistic year.

equation, conditional - SEE condition equation.

equation, conditioned - SEE condition equation.

equation, geostrophic - An equation which represents a balance between the horizontal difference of pressure, in a fluid, and the Coriolis force per unit area. The equation used to compute the speed of the current is $v = k(D_a - D_b) n/L$, in which D_a and D_b are the dynamic height anomalies at stations a and b, respectively; n is a unit-distance conversion factor, L is the distance between stations a and b, v is the speed of the current and k is defined as $1/(2\omega \sin\phi)$, in which ω is the Earth's rate of rotation and ϕ is the latitude.

equation, hydrostatic - The equation expressing the rate of change of pressure p with geometric height (or depth) h as a function of density ρ and gravity acceleration g, when all Coriolis forces, friction and vertical accelerations, and the curvature of the Earth are considered to have negligible effects. The equation is usually written $p/dh = -rg$.

equation, multiquadric - An equation in rectangular, Cartesian coordinates, giving the quantity z as the sum of an arbitrary number of quadratic functions of x, y and z. The quadratic functions usually have the form $(-x_k)^2 + (y - y_k)^2 +$

$(z - z_k)^2$, in which x_k, y_k, z_k are constants. They have been used in surface-fitting.

equation, normal - (1) One of a set of simultaneous equations equal in number to the number of unknowns. (2) One of the equations resulting when a system of N linear equations in M unknowns (N greater than M) is converted to a system of M linear equations in M unknowns by imposing the condition that the sum of the squares of the residuals obtained from the original set be a minimum. The residuals are obtained by assigning more or less arbitrary values to the unknowns in the first system, evaluating the equations, and subtracting the results from the originally given values of the equations. In matrix notation, the original system can be written as $y = Ax$, in which x is the vector of unknowns. The system of normal equations is then $A^T y = A^T Ax$, in which A^T is the transpose of A. The matrix $A^T A$ is referred to as the normal matrix and is often denoted by N. It is a symmetric, square matrix. In most geodetic problems, it is positive-definite and of rank M.

equation, observational - SEE observation equation.

equation, parametric - One of a set of equations in which the independent variables are each expressed in terms of one or more other independent variables called parameters. For example, instead of studying the equation $F(x,y) = 0$ directly, one may find it easier to express both x and y in terms of a parameter u: $x = g(u); y = h(u)$. The parameter may or may not have a useful geometric or physical interpretation.

equation, personal - The interval of time between an observer's perception of a phenomenon and his reaction. SEE ALSO error, personal.

equation, photogrammetric - The equation of a straight line which joins an identified point in object-space to the corresponding point in the image and passes through the perspective center. A system of six photogrammetric equations referring to points on a single photograph suffices to determine the location and orientation of the photograph if the coordinates of the points in object space and in the image are known.

equation, Schreiber's - SEE Schreiber's method (of adjustment).

equation, thin-lens - (1) The equation $1/d_o + 1/d_i = 1/f$, in which d_o and d_i are the distances of object-point and image-point, respectively, from the center of a thin lens and f is the focal length of the lens. This equation is also called Gauss's equation (for a thin lens), the Gauss lens-equation, and Gauss's lens-law. (2) Any equation valid only for lenses or lens systems whose principal points can be considered coincident. Also called a thin-lens formula. SEE ALSO lens equation; Newton's equations (fora a thin lens).

equation of center - The difference between the true anomaly and the mean anomaly. The difference, in elliptical motion, between the actual location in the orbit and the location the body would have had if motion were uniform.

equation of condition - SEE observation equation.

equation of ephemeris time - The difference: hour angle of true Sun minus hour angle of fictitious mean sun.

equation of error - SEE error equation.

equation of the equinoxes - The right ascension of the mean equinox, referred to the true equator and true equinox. The term nutation in right ascension was used before 1960. Also called apparent sidereal time minus mean sidereal time.

equation of time - (1) The difference in hour angle between mean solar time and apparent solar time. An equivalent definition is the difference between the hour angle of the true Sun and the hour angle of the mean Sun. As the equation of time may be expressed as a correction to either apparent solar time or to mean solar time, its sign must be observed carefully. (2) The difference between mean and apparent time. From the beginning of the year until near the middle of April, mean time is ahead of apparent time, the difference reaching a maximum of about 15 minutes near the middle of February. From the middle of April to the middle of June, mean time is behind apparent time but the difference is less than 5 minutes. From the middle of June to the first part of September, mean time is again ahead of apparent time, with a greatest difference less than 7 minutes. From the first part of September until the latter part of December, mean time is again behind apparent time, the difference reaching a maximum of nearly 17 minutes in the early part of November. The analemma is a graphical display of the equation of time.

equation of Universal time - The difference: apparent solar time minus mean solar time.

equator - (1) The ellipse (or, in particular, the circle) in which a plane through the center of an ellipsoid (or, in particular, a sphere) and perpendicular to the minor axis (or a diameter) intersects the ellipsoid. (2) A closed line, on a celestial body or on an ellipsoid representing such a body, along which a specified quantity has a maximum or minimum. SEE, in particular: equator, astronomic; equator, geodetic; equator, geomagnetic; equator, magnetic. (3) SEE equator, celestial. (4) SEE zone, equatorial. SEE ALSO grid equator.

equator, astronomic - That line, on the surface of the Earth, on which the astronomic latitude at every point is 0°. Also termed the terrestrial equator and frequently written the Equator. Because of the deflection of the vertical, the astronomic equator is not a plane curve. However, the verticals at all points on it are parallel to one and the same plane, the plane of the celestial equator. That is, the zenith at every point on the astronomic equator lies in the celestial equator.

equator, celestial - (1) A great circle, on the celestial sphere, whose plane passes through a point on or in the Earth and is perpendicular to the Earth's axis of rotation. Also called the equinoctial. Sometimes called the equator by astronomers. In astronomy, the exact point in the Earth through which the plane of the celestial equator passes is not usually important because the celestial sphere is, in effect, at infinity. So an astronomer's definition often does not specify that the plane pass through any particular point. (2) The projection, onto the celestial sphere, of the Earth's equator. This definition is invalid if the astronomic equator is meant. It is valid only for an equator defined by a plane through the Earth's center and perpendicular to the axis of rotation.

equator, fictitious - A line, other than the equator, on the reference ellipsoid and to which angles analogous to latitude are referred.

equator, galactic - That great circle of the celestial sphere which constitutes the primary circle of reference for galactic latitude and which passes approximately through the center of the Milky Way. It is inclined at about 62° to the celestial equator.

equator, geodetic - (1) That ellipse, on an ellipsoid representing the Earth, in which a plane through the center of the ellipsoid and perpendicular to the shortest axis cuts the surface. If the ellipsoid is an oblate spheroid (oblate rotational ellipsoid), the ellipse is a circle. (2) That circle, on an oblate rotational ellipsoid, midway between the ends of the rotational axis. The geodetic equator is the line on which geodetic latitude is 0° and from which geodetic latitudes are reckoned, north and south, to 90° at the poles. The plane of the geodetic equator cuts the celestial sphere in a line coinciding with the celestial equator if the axis of the ellipsoid is parallel to the Earth's axis of rotation.

equator, geomagnetic - The terrestrial great circle, on a sphere representing the Earth, everywhere 90° from the geomagnetic poles. SEE ALSO equator, magnetic.

equator, inverse - SEE equator, transverse.

equator, lunar celestial - A great circle, on the celestial circle, in the plane of the Moon's equator i.e., in a plane perpendicular to the Moon's axis of rotation.

equator, magnetic - (1) The imaginary line, on the Earth's surface, at which the magnetic inclination is zero degrees i.e., the magnetic field is horizontal. (2) A line drawn on a map or chart and corresponding to a line on the Earth's surface along which the magnetic inclination (dip) is zero for a specified epoch. Also called an aclinic line and dip equator. SEE ALSO equator, geomagnetic.

equator, oblique - The ellipse which is the intersection of an ellipsoid with a plane through the center of the ellipsoid and perpendicular to a line oblique to the axes of the ellipsoid.

equator, terrestrial - SEE equator, astronomic.

equator, transverse - The ellipse which is the intersection of an ellipsoid with a plane through the center of the ellipsoid and containing the smallest axis of the ellipsoid. Also called an inverse equator.

equator, true - The actual celestial equator at a specified instant. This term distinguishes the actual equator from an equator which is an average or otherwise mathematically defined.

equatorial (adj.) - Having the center of a map at one of the poles.

equatorial (n.) - SEE telescope, equatorial.

equator system - SEE coordinate system, equatorial.

equiangulator - (1) An astrolabe containing a 60° prism for observing celestial bodies at a constant angular elevation of 60°. (2) SEE astrolabe. The term is obsolescent.

equidensitometry - SEE density slicing.

equilibrium - (1) That condition, of a system, in which no change occurs in the system's state as long as the system's surroundings are not altered. The definition is imprecise insofar as the concept of state is imprecise. E.g., a system may be in equilibrium under one definition of state and not in equilibrium under another definition. (2) The condition in which a particle, or the constituent particles of a body, are at rest or in unaccelerated motion, with respect to an inertial reference frame. Some physicists deny the validity of definitions which rely on the concept of elemental particles in describing characteristics of solids or liquids. The above definition can, however, be re-worded so as to apply to continuous matter without introducing elemental particles. Balance is an acceptable synonym for equilibrium in both senses.

equilibrium, geostrophic - The condition, of a moving, inviscid fluid, in which the horizontal force per unit area exactly balances the horizontal pressure at all points of the field in question: $2 \Omega \times V = - \delta_H p \rho$, in which Ω is the angular velocity of the Earth, V the geostrophic velocity of the fluid, p the pressure, ρ the specific gravity and δ_H the horizontal del-operator. In geodetic applications, dynamic equilibrium results when the difference in hydro-static pressure between two points is balanced by the Coriolis force per unit area.

equilibrium, hydrostatic - The condition of a fluid when gravity per unit area and pressure are in equilibrium everywhere. The relationship between distance above an equipotential surface and pressure is given by the equation of hydrostatic equilibrium (the hydrostatic equation). SEE ALSO Jeffrey's theorem.

equilibrium, indifferent - SEE equilibrium, labile.

equilibrium, isostatic - That condition which exists, in the Earth's crust, when the downward pressure of masses above the geoid is exactly balanced by the upward pressure of masses below the geoid, so that the crust neither rises nor sinks (except, of course, for changes produced by erosion, glaciation, etc.). The equilibrium is between the crust and the sub-crustal material and probably not within the crust itself. The existence of isostatic equilibrium is based on the idea that the mass of each unit section of the Earth's crust exerts the same pressure on the sub-crustal material.

equilibrium, labile - That state of a body (in oceanography, a packet of sea water) in which a slight displacement of the body leaves it in mechanical equilibrium with its surroundings, so that it has no tendency either to return to its former position or to increase its displacement. Also called indifferent equilibrium and neutral equilibrium. The state of water which is always in equilibrium with its surroundings at any depth.

equilibrium, neutral - SEE equilibrium, labile; equilibrium, unstable.

equilibrium, stable - The condition, of a system, such that a small, temporary change in state results only in small, bounded, variations about the initial state.

equilibrium, static - Hydrostatic equilibrium assumed to be present over a very small oceanic area.

equilibrium, thermal - The condition in which two bodies in physical contact exist if they have the same (but not necessarily constant) temperature.

equilibrium, thermodynamic - A system is said to be in thermo-dynamic equilibrium when its entropy is a maximum.

equilibrium, unstable - The condition, of a system, such that a small, temporary change in the state causes a gradually increasing departure from the initial state. Also called neutral equilibrium.

equilibrium argument - The theoretical phase of a constituent of the equilibrium tide. It is usually represented as the sum of two angles V and u, in which V is a uniformly and rapidly changing angle involving multiples of the hour angle of the mean Sun, mean longitudes of the Sun and Moon, and mean longitude of the lunar or solar perigee. u is a slowly changing angle depending on the longitude of the ascending node of the Moon's orbit.

equilibrium spheroid - The shape that the Earth would assume if it were entirely covered by a tideless ocean of constant depth. This definition is incomplete. It does not specify that the ocean be also without winds or currents, and does not specify how solid materials such as land formerly above water is to be redistributed.

equilibrium theory (of the tides) - A theory which assumes that the waters covering the face of the Earth respond instantly to the tide-producing attractions of the Moon and Sun, and form a surface which is in equilibrium under the action of those attractions. Friction and inertia are ignored, as is the irregular distribution of the land.

equilibrium tide - (1) A hypothetical tide caused by the tide-producing attractions of the Moon and Sun as called for by the equilibrium theory. Also called gravitational tide. (2) The oceanic tide which would occur, in theory, if the oceans covered the Earth completely and uniformly, and the Moon and Sun had fixed distances from the Earth and fixed geographical longitudes and latitudes. Also called the astronomic tide, astronomical tide and gravitational tide. (3) The oceanic tide which would occur, in theory, if the oceans covered the Earth completely and uniformly and responded instantaneously to the gravitational attraction of the Moon and Sun as the distances, geographical longitudes, and latitudes of these bodies changed. The second and third definitions are almost equivalent, in that the second definition implies a different configuration of the equilibrium tide for each different configuration of Earth, Moon, and Sun, while the third definition requires a theory able to deal with a liquid having mass but not inertia. (4) SEE tide, gravitational.

equinoctial - SEE equator, celestial.

equinox - (1) One of the two points in which the celestial equator intersects the ecliptic. Also called an equinoctial point. The two equinoxes are the vernal equinox and the autumnal equinox. (2) The time at which the Sun passes through one of the two points in which the celestial equator intersects the ecliptic i.e., when the apparent longitude of the Sun is 0° or 180°. SEE ALSO catalog equinox; equinox, dynamical.

equinox, autumnal - (1) That point at which the celestial equator intersects the ecliptic and at which the Sun apparently passes from north to south. (2) That time at which the Sun apparently passes from north to south across the celestial equator. It occurs about 23 September.

equinox, dynamical - The ascending node of the Earth's orbit; i.e., that intersection of the celestial equator with the ecliptic at which the Sun's declination is changing from south to north.

equinox, mean - The average location of the equinox as affected by nutation.

equinox, true - The actual equinox at a specified instant. This term distinguishes the concept from that of an average or other mathematically derived equinox.

equinox, vernal - (1) That point at which the celestial equator intersects the ecliptic and at which the Sun apparently passes from south to north. It is also called the First Point of Aries and First of Aries. It is the point from which right ascension is reckoned along the celestial equator and from which celestial longitude is reckoned along the ecliptic. The true location of the equinox is affected by precession and nutation, while the average location (the mean equinox) is affected by precession but not by nutation. (2) The time at which the Sun apparently passes from south to north across the celestial equator. Also called the March equinox. It occurs about 21 March.

equinox and equator, mean - The celestial reference system determined by ignoring small variations of short period in the motions of the celestial equator and ecliptic. Coordinates in stellar catalogs are normally referred to the average catalog equinox and equator of the beginning of a Besselian year.

equinox and equator, true - The celestial reference system determined by the instantaneous positions of the celestial equator and ecliptic. The motion of this system is caused by the progressive effect of precession and the short-term, periodic variations of nutation.

equinox of date - One (usually the vernal equinox) of the two points on the celestial sphere in which the celestial equator intersects the ecliptic at the time of an observation. Both the celestial equator and the ecliptic rotate in inertial space. The epoch (date) as well as coordinates of the points of intersection must therefore be given in order to specify completely the locations of the equinoxes. The epoch is specified either by giving the Besselian year and fraction or by using the words of date, implying that the epoch is that of the instant of observation.

equinox of date, mean - SEE equinox, mean.

equipment - In geodesy, an instrument, together with all other material needed to obtain results with the instrument. Equipment used for triangulation, for example, includes at least one theodolite, signals, towers, radios, tools, and so on. It does not, by convention, include the personnel. Equipment is a more general term than apparatus and is a less general term than system; the last includes not only the equipment but also the procedures, personnel and so on.

equipment, acoustical distance-measuring - Distance-measuring equipment which measures the difference between the time at which a pulse of sound was emitted and the time at which the echo is received, and converts this to a distance. This type of equipment is used primarily for determining depths, in which case it is known as echo- sounding equipment or for determining underwater distances to vessels such as submarines. It has not found a geodetic application so far, although it has been proposed to use the Sofar channel to determine intercontinental distances.

equipment, distance-measuring - (1) Equipment used for measuring distances between points. Distance-measuring equipment differs from a distance-measuring instrument in consisting of a number of distinct and separable parts, each carrying out a separate job but all being necessary to measure distance. An example would be the combination of engineer's transit and stadia rod, or an electromagnetic distance-measuring instrument and a retro-directive reflector. Another example would be the equipment consisting of a surveyor's tape, tensiometer, marking pins, stakes,. etc. However, careless, inaccurate writing does sometimes confuse distance-measuring equipment with distance-measuring instrument. Distance-measuring equipment is often denoted by DME. It may be purely mechanical in nature, such as base-line apparatus, or electro-magnetic in nature, or may be a combination of electromagnetic and mechanical parts, such as an optical range-finder. (2) SEE radio distance-measuring equipment.

equipment, electro-infrared distance-measuring - Electromagnetic distance-measuring equipment containing an electro-infrared distance-measuring instrument as one component. Some geodesists consider it to be electro-optical distance-measuring equipment.

equipment, electromagnetic distance-measuring - Distance-measuring equipment which includes an electromagnetic distance-measuring instrument as a component. Also called electronic distance-measuring equipment, although this term actually refers merely to distance-measuring equipment which depends contains electronic circuits.

equipment, electronic distance-measuring - Electronic equipment used for measuring distances.

equipment, electro-optical distance-measuring - Distance-measuring equipment which includes an electro-optical distance-measuring instrument as a component. Also called laser radar if a laser is the source of the radiation, or lidar.

equipment, frequency-measuring - Equipment which, by measuring the frequency of electromagnetic radiation from a moving body, can determine the velocity of that body. At present, the most common kind of frequency-measuring equipment measures the Doppler shift in radio waves.

equipment, infrared distance-measuring - Distance-measuring equipment having an infrared distance-measuring instrument as one component. The term appears to be used synonymous with electro-infrared distance-measuring equipment.

equipment, optical distance-measuring - Distance-measuring equipment which contains an optical system essential to the measuring process; in particular, distance-measuring equipment which contains an optical distance-measuring instrument as one component. There are at least two types: that which involves the formation of an image, and that which involves the modulation of light. The first type includes the equipment used in the stadia method of measuring distances. The second includes the Väisälä base line apparatus and electro-optical distance-measuring instruments with reflectors.

equipotential (adj.) - At the same potential everywhere. E.g., an equipotential surface is one on which a potential is defined and is the same at every point; an equipotential body is one at every point of which a potential is defined and is the same.

Equivalence Principle - SEE Principle of Equivalence.

erosion - (1) In riparian law, the gradual and imperceptible washing away of land by a stream or body of water such as a lake or sea. (BLM) The results of erosion of a stream may be distinguished from those of avulsion by the absence of identifiable upland between former channels and new channels. (2) The wearing away of land or structures by running water, glaciers, wind, or waves. Weathering, although sometimes considered a form of erosion, is a distinctly different process; it does not imply that material is removed.

error - (1) The difference between the measured value of a quantity and the theoretical or defined value of that quantity: ε(error) = y (measured) - y (theoretical). (2) The difference between an observed or calculated value of a quantity and the ideal or true value of that quantity. Logically, definitions (1) and (2) are distinctly different. In practice, they are equivalent except for the second definition's allowing calculated values to be used instead of measured values. (3) The difference between an approximate number and the correct number. (4) The difference between the theoretical or defined value of a quantity and the measured value of that quantity: ε (error) = y (theoretical) - y (measured). This definition differs from the first definition only in changing the sign of the error.

error, absolute - The value, taken without regard to sign, of the difference between some value of a quantity and the true value of that quantity.

error, absolute personal - Personal error determined by comparison with values that are correct by assumption or definition.

error, accidental - (1) An error which does not always recur when a measurement is repeated under the same conditions. (2) SEE error, random. The term accidental error has been preferred by astronomers and geodesists; the term random error has been preferred by physicists and statisticians and is gradually replacing accidental error in other sciences.

error, accumulative - SEE error, cumulative.

error, actual - (1) The measured value of a physical quantity minus the true value of that quantity. (2) The sum of all those systematic and random errors which have not been eliminated from the final, adopted, measured value.

error, along-track - That component, of the difference between the location of a satellite's location as computed from theory and the location estimated from observations by the method of least squares, which lies in the direction of the satellite's velocity at a particular instant.

error, apparent - SEE error, residual (2).

error, average - (1) The average, taken without regard to sign, of all the errors in a set. εaverage = Σ_o^n (y_i measured - y_i theoretical) n / N, in which N is the number of values in the set. (2) The average, taken without regard to sign, of the differences between the actual values in a set and the average of those values. εaverage = Σ ($y_n - y_m$) / (N-1), in which y_n is one of the N values in the set and y_m is the average of all the y_n. Average error is sometimes defined as the average, without regard to sign, of the average of the positive errors and the average of the negative errors. Because there will be about the same number of positive errors as negative errors in a well-balanced series of measurements, this method of obtaining the average error will give practically the same result as the other methods. Also called average absolute error, average deviation, mean deviation and mean of the errors. It is frequently called mean error in American usage but this should not be confused with European use of the term for root-mean-square error. SEE ALSO error ellipse; error, ellipsoid; index error; normal law of error; propagation of error; rejection of error; truncation error.

error, average absolute - SEE error, average.

error, chronometric - The time at which a clock indicated that an event occurred, minus the actual time of occurrence. The negative of the chronometric correction.

error, circular - The radius of a circle such that there is a given probability that any point whose location is expressed as a function of two variables will be within the given circle.

error, circular near-certainty - The radius of a circle such that the probability that a point will be within the given circle is about 99.78%.

error, circular probable - The radius of a circle containing 50% of the points of a population of randomly distributed points in the plane. Also called circular error probable. The center of the circle is usually at the barycenter of the population and the coordinates usually have a bivariate Gaussian distribution. Under such conditions, the radius is approximately 1.1774 σ, where σ is the common standard deviation of the two coordinates.

error, circular standard - The radius of a circle such that there is a probability of about 39.35% that a particular point will lie within that circle.

error, clamping - A systematic error, occurring in observations made with a repeating theodolite, caused by strains set up by the clamping devices of the instrument.

error, closing - SEE misclosure.

error, compensating - An error which tends to offset another error and thus obscure or reduce the effect of each. Also called a compensatory error.

error, compensatory - SEE error, compensating.

error, constant - An error which is the same in both magnitude and sign throughout a given series of measurements. A constant error tends to have the same effect upon all the measurements of a series or any part of a series. Therefore it does not usually have any influence on the standard deviation of the series. An example of a constant error is the index error of a precise instrument. Also called bias and cumulative error.

error, cross-track - That component, of the difference between the location of a satellite's location as computed from theory and the location estimated from observations by the method of least squares, which is perpendicular to the direction of the satellite's velocity at a particular instant and perpendicular to the osculating plane at that instant.

error, cumulative - (1) An error which adds to or increases the total error. Also called an accumulative error. The simplest type of cumulative error is the constant error. A systematic error which retains the same sign, but not necessarily the same magnitude, throughout a series of measurements is also a cumulative error. (2) A constant error (q.v.). This definition is inconsistent with that given in (1) and does not cover many errors usually considered cumulative.

error, curved-path - The difference between the length of a ray refracted by the atmosphere and the distance along a straight line between the ends of the ray.

error, eccentric - SEE centering error.

error, erratic - An error caused by an incomplete or deficient component of an instrument e.g., backlash in a gear train.

error, external - A systematic error arising from natural physical conditions outside the observer. Also called theoretical error. Examples of external errors are the effect of atmospheric refraction of spirit leveling, the changes in the length of a surveyor's tape because of thermal expansion, and the effects of atmospheric pressure on barometric altitudes. Such errors may be controlled to some extent by making measurements only when conditions are favorable: for example, by measuring the length of a line with a steel tape at night or in cloudy weather. They may be determined by means of formulas and measurements made specially for the purpose. For example, the length of a metallic tape may be determined by using a formula which involves the coefficient of thermal expansion and the measured temperature of the tape.

error, gross - (1) A very large error. The usual implication is that the error was made by a person. If this is to be made explicit, the error is termed a blunder. (2) An error which is the result of carelessness or a mistake.

error, inherent - SEE error, inherited.

error, inherited - The error, in a quantity, which was present at the beginning of a set of operations on that quantity. Also called inherent error.

error, instrumental - A systematic error arising from imperfections in the instrument used. Instrumental errors may arise from imperfections which are built into an instrument, such as errors in the graduation of a horizontal circle, or they may arise from lack of complete adjustment of some part of the instrument, such as the error of collimation. They can be determined in the laboratory. They may be eliminated from a result by a suitable process of measurement or by applying suitable corrections.

error, irregular - SEE error, accidental.

error, law of propagation of - SEE law of propagation of error.

error, linear - (1) A systematic error which is a linear function of some variable. (2) A random error having a Gaussian frequency function in one variable.

error, mean - (1) SEE error, average. (This is American usage.) (2) SEE error, root-mean-square. (This is European usage.)

error, mean radial - A radius of length $\sigma (\pi/2)^{1/2}$, in which σ is the standard deviation of each of two variables having a Gaussian distribution and specifying the location of a point in a plane. Also called the mean radial deviation.

error, mean-square - A quantity s^2 measuring the variation of a random variable about some standard or accepted value, and given by $s^2 = [\Sigma (x_n - x_{nm})^2] / N$ (n = 1 to N), in which x_n is one of a set $\{x_n\}$ of N random values and x_{nm} the corresponding accepted value out of the set $\{x_{nm}\}$. In many circumstances, all the x_{nm} have the same value, the average value of the set $\{x_n\}$. Also called standard error.

error, natural - An error arising from variations in natural phenomena such as geomagnetism, gravity, humidity, refraction, temperature or wind.

error, near-certainty - The difference (error) between the average value of a random quantity having a Gaussian distribution and the value of that quantity at three standard deviations from the average. (jargon)

error, observed - (1) The average value of a random quantity minus a particular value of a measurement of that quantity. (2) The average value of a set of random values, minus the average value of a subset of that set. The term is inappropriate for either definition. Errors cannot be observed, and the sign of an observed error is the opposite of the sign of an error.

error, octantal - An error in direction or measured bearing, caused usually by the departure of an antenna pattern from an ideal shape, varying sinusoidally throughout the 360° and having four positive and four negative maxima.

error, orthometric - That part of the difference between a measured elevation and the true elevation of point caused by the non-parallelism of the equipotential various level surfaces along the route along which the leveling was done. Alternatively, the negative of the orthometric correction. Level surfaces at different elevations are not exactly parallel. A lake at a high elevation in latitude 45° N would be nearer the geoid at its northern end than it would at its southern end. This would cause an apparent error in a level line which rose from the geoid to the south end of the lake, then followed the lake's surface to the northern end, then dropped down to the geoid again, and finally followed the geoid back to the starting point.

error, parallactic - An error caused by personal or instrumental parallax.

error, periodic - An error whose amplitude and direction vary systematically with time.

error, personal - (1) A systematic error caused by an observer's personal manner of making observations, or by the deviation of his physical characteristics from those of an ideal observer. Also called personal equation. A personal error arising from a personal manner of observing, such as the observer's standing in the same position relative to the end of a surveyor's tape when measuring the length of a line, may be eliminated from a result by having the observer shift his position so that the error is positive for half the observations and negative for the remainder. A personal error arising from an observer's physical characteristics can often be eliminated by calibrating the observer. For example, most observers mark the time of transit of a star shortly before or shortly after the actual transit, and this error is approximately constant for and characteristic of each observer. The amount of lead or lag can be determined by tests. The classic example of this type of error is given by the case of the astronomer Maskelyn's firing of his assistant because his and his assistant's measurements differed consistently. Maskelyn attributed the difference to his assistant's carelessness; the difference probably was caused by personal errors in either or both person's measurements. SEE ALSO equation, personal. (2) An error caused by an observer's personal habits, mental or physical reactions, or inability to perceive dimensional quantities exactly. The error may be accidental or systematic. A systematic personal error is called a personal equation.

error, positional - The amount by which the actual location of an imaged cartographic feature fails to agree with the feature's true position.

error, principal-distance - An error, in a stereoscopic plotting instrument, resulting from improper calibration of the camera, printer used for making the diapositive, or the projector. The error is of little importance in a model of a flat surface, but the effects of the error increase in proportion to the relief in the model.

error, principal-point - A personal error committed by improperly orienting a diapositive in the printer or, later, in the projector of a stereoscopic plotting instrument, or both, and resulting in unequal displacement of the principal points in the instrument and an error in the vertical distances.

error, prismatic - The error caused by lack of parallelism of the two faces of an optical element such as an optical filter.

error, probable - A quantity of such size that the probability that an error larger than that quantity will occur is the same as the probability that an error smaller than that quantity will occur. Probable error $p_{1/2}$ is related to standard deviation σ by the formula $p_{1/2} = 0.6745\, \sigma$ if the errors have a Gaussian distribution. If the errors in a series of measurements be arranged in order of magnitude without regard to sign, and if the series be indefinitely large, the probable error will fit the middle place in that list of errors. Expressed in another way, the probable error of a result is a quantity such that a second determination obtained under the same conditions as the first has an equal chance of being less than or greater than the probable error. The probable error of the result of a series of measurements is a function of the random error attending the individual measurements of the series. A systematic error may often remain in the series with little effect on the size of the probable error. Probable errors are useful in comparing the precision or accuracy of similar measurements and may be used as criteria for prescribing degrees of precision and accuracy to be obtained. Probable error is currently little used in the sciences; standard deviation is used instead. It should therefore be used only if a definite advantage results. In such a case, the error should be identified specifically as probable error.

error, quadrantal - An error in direction or measured bearing, frequently caused by irregularities in the antenna or goniometer, which varies sinusoidally throughout the 360° and has two positive and two negative maxima.

error, quasi-probable - The square of the average of the square root of the absolute value of the true error.

error, random - An error produced by irregular causes whose effects upon individual measurements are governed by no known law connecting them with circumstances and which therefore can never be subjected to computation a priori. Sometimes called an irregular error or, more commonly, an accidental error. The theory of elemental errors assumes that a random error is assumed to be composed of an infinite number of independent, infinitesimal errors, all of equal magnitude, each as likely to be positive as negative. In practice, a random error is composed of an indefinitely large number of finitely small errors, each as apt to be positive as negative. With these assumptions, probabilities can be associated with the results of the method of least squares, and it is to the elimination of random errors only that the method of least squares properly may be applied.

error, regular - SEE error, systematic.

error, relative - The ratio of the difference between an approximate value and the correct value, to the correct value.

error, residual - (1) The value, corrected for known systematic errors, of one measurement in a series, minus the value of the quantity obtained statistically as a representative of the series. Residual errors are sometimes called, simply, errors or residuals. (The latter term is used mostly in referring to actual values in a specific computation). In practice, it is the residual errors which enter into a computation of standard deviation. (2) The difference between a measured value and its computed value after adjustment. (3) SEE residual. A residual error is obtained by explicitly including corrections for systematic errors; a residual is obtained without necessarily including such corrections.

error, resultant - SEE error, true.

error, root-mean-square - (1) A quantity s measuring the deviation of a random variable from some standard or accepted value, and given by $s = \sqrt{[\Sigma (x_n - x_{nm})^2 / N]}$, in which x_n is one of the set $\{x_n\}$ of N random variables and x_{nm} is the corresponding value from the set $\{x_{nm}\}$ of accepted values. Also called standard error. The x_{nm} are often all equal to a single value x_m, the average of the set $\{x_{nm}\}$, $x_m = (\Sigma x_n W_n)/(\Sigma W_n)$, in which W_n is the weight assigned to x_n. Note that the root-mean-square error is not, in general, the same as the standard deviation. (2) The square root of the integral of the square of the difference between a random variable x and some constant value x_c, the square being multiplied by the frequency-function of x.

error, rounding - SEE error, round-off.

error, round-off - An error which occurs, during a computation, when the number of significant figures in a number is decreased. Also called rounding error. The process itself is called rounding off and usually proceeds according to some definite rule.

error, standard - (1) The standard deviation of the errors associated with physical measurements of an unknown quantity, or statistical estimates of an unknown quantity or of a random variable. (2) SEE deviation, standard. (3) SEE error, root-mean-square. (4) SEE error, mean-square.

error, stochastic - SEE error, random.

error, systematic - (1) An error whose algebraic sign and, to some extent, magnitude bear a fixed relation to some condition or set of conditions. In other words, an error which is, in theory at least, predictable and therefore is not a random error. Systematic errors are regular and therefore can be determined a priori. They are usually eliminated from a set of observations before applying the method of least squares to reduce or eliminate random errors. They are classified as theoretical (external) errors, instrumental errors, and personal errors, according to their origin and nature. (2) The difference between the limiting value of the average of a set of measurements of a quantity and the true value of that quantity. Also called bias.

error, theoretical - (1) An error whose value can be predicted from theory. (2) SEE error, external.

error, true - (1) The measured or calculated value of a quantity minus the true value of that quantity. Also called resultant error. (2) An error. In general, the terms true error and error are used interchangeably. However, it is best to limit the meaning of true error to that given above.

error budget - A catalog or listing of the natures and magnitudes of the errors which affect the results of a project. The term is used when the project is a hypothetical or planned one as well as when the project is actual.

error circle - A circle containing a specified percentage of randomly distributed points or having a specified value for the quadratic form $(\xi/\sigma_x)^2 + (\eta/\sigma_y)^2$, in which σ_x and σ_y are the standard deviations of the random variables x and y, respectively. Also called circle of error. It is a special case of the error ellipse.

error ellipse - An ellipse containing a specified percentage of randomly distributed points or having a specified value for the quadratic form $[(\xi/\sigma_x)^2 - 2\rho(\xi/\sigma_x)(\eta/\sigma_y) + (\eta/\sigma_y)^2]/(1-\rho^2)$, in which σ_x and σ_y are the standard deviations of x and y, respectively and ρ is the correlation coefficient. Also called an ellipse of error and contour ellipse. Unless specifically stated otherwise, the distribution is assumed to be bivariate Gaussian. SEE ALSO ellipse, probable-error; ellipse, standard-error.

error ellipsoid - An ellipsoid containing a specified percentage of randomly distributed points in three-dimensional space, or having a specified value of the quadratic form for the ellipsoid.

error ellipsoid, mean - SEE ellipsoid, mean-error.

error ellipsoid, probable - SEE ellipsoid, probable-error.

error equation - An equation stating that the probability of an event is given by the Gaussian frequency function. SEE ALSO frequency function, Gaussian. This term is not in general use.

error function - SEE Gaussian frequency function.

error of closure - SEE misclosure.

error of closure, angular - SEE misclosure, angular.

error of closure, horizontal - SEE misclosure of the horizon.

error of closure, linear - SEE misclosure, linear.

error of closure, relative - SEE misclosure, relative.

error of closure in azimuth - SEE misclosure in azimuth.

error of closure in leveling - SEE misclosure in leveling.

error of closure of horizon - SEE misclosure of the horizon.

error of closure of traverse - SEE misclosure of traverse.

error of closure of triangle - SEE misclosure of triangle.

error of collimation - The angle between the line of sight (line of collimation) of a telescope and the telescope's collimation axis. Also called collimation error. However, if the collimation axis is not horizontal, error of collimation and collimation error are different. When an instrument is perfectly collimated (which is never the situation in practice), the line of sight and the collimation axis coincide and the error of collimation is zero. In practice, adjustment for collimation is continued until the error is so small that it may be considered negligible. In precise work, after collimation is complete, the remaining error is either determined by observation and applied as a correction, or is eliminated from the results by a suitable program of observing. Error of collimation is a systematic error, so in a series of observations it is usually treated as a constant error.

error of commission - That error introduced in a computation by using incorrect data or incorrect theory.

error of misclosure - SEE misclosure.

error of observation - The difference between the measured value of a quantity and a value adopted as representing the ideal or true value of that quantity. Errors of observation are composed of either one or both of two general classes of error: accidental errors and systematic errors. Constant errors are sometimes placed in a third class but are more often included among the systematic errors as they should be. Errors of observation are also classified according to their origin as external errors, instrumental errors, and personal errors. The algebraic sign of the error of observation is determined from the equation: error of observation = measured value - adopted (ideal) value. The term error of observation is dropping out of use and is being replaced by measurement error.

error of omission - An error introduced into a computation by leaving out pertinent data or by omitting terms of an equation.

error of representation - The error committed by assuming that one member of or sample from a statistical population is a valid representative of that population. In particular, in the analysis of data on gravity, the error committed by using the value of gravity at one or more points in a region instead of using the average value of gravity in the region. The term is also used in this way with respect to gravity anomalies.

error of run - The difference between the intended or nominal value of one turn of a micrometer screw and the actual value of one turn, in seconds of arc. Also called the run of the micrometer.

error of survey - SEE misclosure of traverse.

error of the level - SEE C-factor.

error of the mean - SEE error of the mean, standard.

error of the mean, standard - The standard deviation of the averages calculated from a large number of sets of samples. Also called error of the mean. If the standard deviation of all measurements (from all samples) is σ_p and if the number of measurements in a set is N, then the standard error σ_m of the mean is given by $\sigma_m = \sigma_p / \sqrt{N}$.

error probable, circular - SEE error, circular probable.

error propagation - SEE propagation of error.

escape speed - SEE escape velocity.

escape velocity - The lowest radial velocity that one body in the force field (such as a gravitational field) of a second, more massive body must have in order to ensure that its distance from the second body increases monotonically. Also called escape speed and speed of escape. If the field is gravitational, the mass of the second body is M, and the first body starts at a distance r from the center of the second body, the escape velocity is $(2GM/r)^{1/2}$; G is the gravitational constant. Escape velocity is numerically equal to (but is opposite in direction to) the velocity the first body would attain if it were dropped onto the second body from infinity. Escape speed from the Earth's surface is about 11.2 km/s; from the Moon's surface, about 2.4 km/s; and from the Sun's photosphere, about 617.7 km/s.

escheat - Reversion of property to the state when there is no one competent or available to inherit.

escrow - (1) A deed delivered to a disinterested person to be delivered to the grantee upon the fulfillment or performance of some act or condition. (2) Money or something of value deposited with a disinterested person, to be released upon fulfillment of performance of some act or condition, or to be returned to the depositor if these conditions are not met.

esker - A long, narrow ridge system composed of generally sand and gravel which was once the bed of a stream flowing under or in the ice of a glacier.

establishment - The interval of time between upper or lower transit of the Moon and the next high water at a place. It is sometimes called establishment of the port, although that term should be reserved for average values of the interval. SEE ALSO interval, lunitidal.

establishment, common - SEE establishment of the port.

establishment, corrected - The average of all high-water intervals. This is usually 10 to 15 minutes less than the establishment of the port. SEE ALSO interval, mean high water lunitidal.

establishment, vulgar - SEE establishment of the port.

establishment of the port - The average high-water interval, at a particular place, on days of new and full moon. Also called common establishment, high water full and change, and vulgar establishment.

estadal - A measure of land of sixteen square varas, or yards. (Spanish law).

estate - (1) A person's right or interest in property and its use i.e., the entire property of a person. (2) The property and debts of a deceased or bankrupt person.

estate, dominant - An estate, the owners of which are entitled to an easement on another's property. That particular parcel of land that benefits from an easement.

estate, real - Property consisting of land, natural resources on or in the land, and any man-made improvements established thereon. Also called realty. Abbreviated to R.E.

estimate, sufficient - Y is a sufficient estimate for y, the average value of the population, if, for any other function υ of the same random variables, the distribution of the conditional random variable $f(\upsilon,Y)$ does not depend on y and if the average of Y, given y, is 0. Also called a sufficient estimator. y is usually a function of a random vector x and Y a function of samples of x.

estimate, unbiased - Y is an unbiased estimate for y, the average value of the population, if the average value of Y is y. Also called an unbiased estimator. y is usually a function of a random vector x and Y a function of samples of x.

estimator, sufficient - SEE estimate, sufficient.

estimator, unbiased - SEE estimate, unbiased.

estop (v.) - (1) Stop, bar. (2) Impede.

estoppel - The doctrine by which a party is prevented by his own act from claiming a right to the detriment of aother party who was entitled to rely upon the first party's actions and has acted accordingly. The situation can arise in disputes over a boundary if one of the parties tries to make a claim rendered suspect or invalid by his own actions or record.

estuary - (1) An embayment of the coast in which fresh water entering at its head mixes with the relatively saline water of the ocean. (2) The lower reaches and mouth of a river emptying directly into the sea where tidal mixing takes place. Also called a river estuary.

estuary, tidal - An estuary in which tidal action is the principal mixing agent.

etch, deep - A method of making lithographic plates by etching the design slightly into the surface of the plate.

etch slip - A penci-shaped piece of abrasive used in removing unwanted marks from a metallic pressplate.

et con. (law) - And husband.

et seq. (abbr.) - And following.

et ux. (law) - And wife.

E-terms (of aberration of light) - Terms of annual aberration which depend on the eccentricity and longitude of perihelion of the Earth.

et vir. (law) - And husband.

Euclidean space - A space in which a coordinate system can be so chosen that the square %w the distance between two points is given by the sum of the squares of the differences in like coordinates.

Euler angle - SEE Eulerian angle.

Euler equation - Any one of a number of equations discovered by Euler. In particular, Euler's equations of motion or the equation relating the various characteristics of an elliptical orbit.

Euler equations of motion - SEE Euler's equations of motion.

Eulerian angle - One of a set of three angles (ϕ, θ, ψ) specifying the orientation of the three axes (x,y,z) of one rectangular Cartesian coordinate system with respect to the three axes (X, Y, Z) of another. The angles are generated by rotating the first system about the z-axis through the angle ϕ until the x-axis, in its new position, is parallel to the XY plane, then rotating about the new position of the x-axis through the angle θ until the new xy-plane is parallel to the XY plane, and then rotating about the new position of the z-axis through the angle ψ until all three axes of the first system are parallel to the corresponding axes of the second. Also called Euler angle or Euler's angle. The definition given follows astronomical usage and American usage in theoretical mechanics. British usage is to bring the y-axis parallel to the XY plane and then rotate about the y-axis in its new position. The essential features, in any event, are to rotate first about a particular axis, then about the new position of a second axis, and finally about the new position of the first axis. This distinguishes Eulerian angles from Cardan angles, which result by rotation first about one axis, then about the new position of a second axis, and finally about the new position of the third axis. Various symbols are used for Eulerian angles. Those given here are probably the most common. However, if Eulerian angles and geographic coordinates are to be used together, another symbol should be used for the Eulerian angle ϕ - ψ is a common choice.

Eulerian coordinate system - A coordinate system associated with a fluid in such a way that properties of the fluid (e.g., density or velocity) are associated with points in space at each instant of time, without any attempt to identify individual parcels of the fluid from one instant to the next. In other words, a coordinate system which does not move with the fluid. The term is not used to describe events connected with solids. If it were, a Eulerian coordinate system associated with, for example, a projectile fixed on the Earth and would be used for giving the coordinates of points on the trajectory. A Lagrangian coordinate system, on the other hand, would be fixed in the projectile and would be used for giving the changes in the body's density, temperature, and shape at different instants of time. SEE ALSO Lagrangian coordinate system.

Eulerian free period - The period of nutation of a rotating, rigid body not under the influence of external forces. The term is applied, in particular, to the period of nutation a rigid Earth would have if not affected by external forces. It is about 305 mean solar days.

Eulerian nutation - SEE nutation.

Euler-MacLaurin formula - The formula
$$\Sigma_n f_n(x_n) = \int_0^n f(x)dx + \Sigma_m [B_{2m}(f_o(2m-1) - f_n(2m-1)]/(2m)!$$

for the sum of N terms of a series in terms of an integral and the first and last terms of the series; B_{2m} is the 2m-th Bernoulli number.

Euler's angle - SEE Eulerian angle.

Euler's dynamical equations - SEE Euler's equations of motion.

Euler's equation - An equation relating the lengths r_1 and r_2 of two radius vectors in a parabolic orbit to the length of chord and interval of time between them. Denoting the length of the chord joining the ends of the radii by s and the interval of time by $t_2 - t_1$, Euler's equation is $6k(t_2 - t_1) = (r_1 + r_2 + s)^{3/2} \pm (r_1 + r_2 - s)^{3/2}$ in which k^2 is the gravitational constant.

Euler's equations of motion - The equations $M_i = A_i \Omega_i - A_j - A_k) \Omega_j \Omega_k$, in which i, j, k take on the values 1,2,3 cyclically, M_i is a component of the total force applied to a rigid body about an axis in that body, Ω_i is the corresponding angular velocity and A_i is the corresponding moment of inertia. The axes are the principal axes of the body. Also called Euler equations of motion, Euler's dynamical equations or simply, the Euler equations.

Euler's formula - SEE Euler's theorem.

Euler's map projection - A conical, equidistant map-projection with two standard parallels such that the absolute scale errors along the lines representing the central and extreme parallels of latitude are exact.

Euler's spiral - SEE clothoid.

Euler's theorem - The theorem relating the radius of curvature R of a normal section in any azimuth A on the reference ellipsoid to the radius of curvature M in the meridian and radius of curvature N in the prime vertical. Also called Euler's formula. It is usually written as $1/R = (\cos^2 A)/M + (\sin^2 A)/N$. The theorem actually applies to surfaces considerably more complex than an ellipsoid.

eumorphism - The property, of some map projections, of representing regions of specific relative area on the ellipsoid by regions having the same relative area on the map. SEE ALSO map projection, equal-area.

European Datum 1950 - A horizontal-control datum defined by the following coordinates of the origin and the deflection of the vertical on the International Ellipsoid; the origin is at the control point in Helmert Tower, Potsdam:

longitude of origin	13° 03' 58.9283" E.
latitude of origin	52° 22' 51.4456" N.
deflection of the vertical at origin	
in the meridian	3.36"
in the prime vertical	1.78"

The horizontal control in Europe has been re-adjusted by several countries, and the Bomford geoid of 1963 was recommended for use in these re-adjustments. Geoids other than Bomford's have been used. In the original adjustment, a specific origin was not used. Also called Datum Europeén 1950.

European Datum 1979 - A horizontal-control datum having its origin at the point D783t on the northern tower of the Frauenkirche, in MÜnich, with the same coordinates as those of the European Datum 1950, and on the International ellipsoid.

eustasy - (1) Those changes of oceanic mean sea level which are peculiar to the oceans, as opposed to those changes which result from changes in crustal elevation.
(2) The changes in average oceanic level, regardless of the cause (i.e., regardless of the kind of surface to which elevations are referred) and implying vertical movement of the average oceanic surface at a particular point.

evection - The largest periodic perturbation in the Moon's longitude. The term is given, in Brown's theory of the Moon's motion, by $1° 16' 26/4" \sin(\lambda - 2\lambda' + \lambda_p)$, in which λ and λ' are, respectively, the average longitude of Moon and of Sun and λ_p is the Moon's longitude of perigee. The term was known to Hipparchus, although Ptolemy is often credited with its discovery. The period is about 31 4/5 days. The evection is a maximum when the Sun is passing the Moon's line of apsides and a minimum when the Sun is at right angles to it.

Everest spheroid 1830 - A rotational ellipsoid having the dimensions

semi-major axis	20 922 931.80 feet
semi-minor axis	20 853 374.584 feet.
flattening (derived)	1/300.80 .

The above values are in Indian feet, derived from the Indian 10-foot Standard Bar A and are used in the Geodetic Surveys of India and Pakistan. G. Bomford used the ratio 1 Indian foot equals 0.304 798 41 meters to get the value 6 377 276.345 m for the semi-major axis. There is also an Everest spheroid of 1847, but this has not been used in triangulation.

Everest spheroid 1847 - SEE Everest spheroid 1830.

eviction - (1) The act or process of evicting, or state of being evicted. (2) In particular, the recovery of lands, tenements, etc., from another's possession by due course of law or dispossession in virtue of a permanent title. (3) Also, dispossession of a tenant by his landlord.

evidence (law) - (1) That which is legally submitted to a competent tribunal as a means of ascertaining the truth of any alleged matter of fact under investigation before it.
(2) A means of making proof. (3) Medium of proof.

evidence, extrinsic (law) - (1) Evidence that is not contained in the body of an agreement, contract, and the like. (2) Evidence not legitimately before the tribunal in which the determination is made.

evidence, prima facie - Evidence deemed by law to be sufficient to establish a fact if the evidence is not disputed.

evidencing of title - The submission of proof of title to a tract of land, or subdivision, as shown by an abstract of the recorded patent and deeds of transfer, inheritance, court decree or other means of establishing the title by such evidence as may be approved by competent judicial opinions, including court decrees in the event of controversy.

exaggeration, stereoscopic - SEE hyperstereoscopy.

exaggeration, vertical - (1) The factor by which scale in the vertical direction is greater than scale in the horizontal direction, on a model of terrain. Also called exaggeration of relief. (2) The change made in a surface by raising proportionately the apparent height of all points above the base while retaining the same scale in the base.

exaggeration of relief - SEE exaggeration, vertical.

exception (in a deed) - The retaining, in the grantor, some portion of the former estate, which, by the exception, is taken out of or excluded from the grant; whatever is thus excluded remains in him as of his former right of title because it is not granted.

exception, doubled - The exception, from the description of land, from a previous exception. Doubled exceptions should always be avoided. Lot A, except the east 50 feet, except the south 50 feet conveys two meanings: 1) Lot A, except the easterly 50 feet of all of Lot A and except the south 50 feet of Lot A and 2) Lot A, except the east 50 feet of all of Lot A, except that the south 50 feet of the east 50 feet is reserved by the grantor. In a doubled exception, the second exception may refer to the exception or to the lot.

excess, spherical - The amount by which the sum of the three angles of a spherical triangle exceeds 180°. The amount of spherical excess depends on the radius of curvature and on the area of the triangle, and is about 1" for each 195.8 square kilometers on a rotational ellipsoid representing the Earth. In geodesy, in the computation of triangles, the difference between spherical angles and spheroidal angles is generally neglected; spherical angles are used and Legendre's Theorem is applied to the distribution of the spherical excess. That is, approximately one-third of the spherical excess of a given spherical triangle is subtracted for each angle of the triangle.

excess, spheroidal - The amount by which the sum of the three angles of a spheroidal triangle exceeds 180°. In geodesy, spherical angles are used instead of spheroidal angles; the difference between the two kinds of angles is quite small even for the largest triangles and is considered negligible for ordinary triangulation.

excess, triangular - The amount by which the sum of the internal angles of a triangle exceeds of falls short of 180°. Triangular excess is positive for triangles on a sphere or spheroid, zero for triangles on a plane and negative for triangles on a hyperboloid.

excess condemnation - SEE condemnation, excess.

executor - A person named in a will to carry out its provisions.

exit pupil - That image of the aperture-stop formed in image space by all those elements of the optical system on the image-space side of the aperture stop. The ratio of the diameter of the entrance pupil to the diameter of the exit pupil is equal to the magnifying power of the optical system.

exit window - That image of the field stop formed in object space by all elements (of an optical system) on the image-space side of the field stop.

exosphere - That portion of the Earth, extending from about 1600 to 3000 km above the surface, in which many particles move at speeds comparable to that needed for escape from the Earth's gravitation.

expansion (mathematics) - The conversion of a function into a finite or infinite series.

expansion, coefficient of thermal - The relative change (expansion or contraction) in a linear dimension of a body corresponding to a change of 1 in the temperature of the body. The coefficient of expansion, as it is generally called, may be in terms of the Celsius, Fahrenheit or other temperature scale. To a very great extent, its magnitude is peculiar to the material; a steel tape has a coefficient of expansion about 25 times as great as that of a base line, it enters as a correction that is a product of the coefficient of expansion of the apparatus used, the length measured and the difference between the temperature at which the measurement was made and the temperature (standard or calibration temperature) at which the length of the apparatus is known.

ex parte - On one side only; by or for one party in a case.

expectation - SEE average. Also called the expected value. The term should not be confused with the average in the real world.

experience radar prediction - The prediction of size, shape, and relative intensity of returned radar signals and of radar shadow and regions of no returned signal, based primarily on the knowledge and experience of the individual making the prediction rather than on proven formulae, tables or graphs. Also called artwork prediction.

explement (of an angle) - SEE angle, explementary.

exploration, geophysical - The determination of the presence and location of layers of solid rock, the detection and identification of minerals, the classification of solids and underlying strata and so on, by measurements of gravity, the speed of sound in the ground, electrical resistivity of the ground and other physical measurements.

exposure - (1) The amount of electromagnetic radiation per unit area which has fallen on a radiation-sensitive surface in a given interval of time. The term is applied in particular to radiation at wavelengths shorter than radio wavelengths, such as infrared radiation and light. (2) The amount of radiation, of any kind, per unit area which has fallen on a sensitive surface in a given interval of time. (3) The act of exposing a sensitive surface to radiation. SEE ALSO contact exposure.

exposure, emulsion-to-base - A contact exposure in which the base of the copying film is in contact with the emulsion side of the sheet being copied. SEE ALSO exposure, emulsion-to-emulsion.

exposure, emulsion-to-emulsion - A contact exposure in which the emulsion side of the copying film is in contact with the emulsion side of the sheet being copied. SEE ALSO exposure, emulsion-to-base.

exposure, photographic - The interval of time during which a photographic emulsion is exposed to light, times the illuminance (intensity of the light).

exposure camera - A camera used for taking pictures; a camera. The term is used only when a camera proper must be distinguished from a camera used as a projector.

exposure factor - Any factor which must be taken into consideration when calculating the correct time-interval for exposure. Among the important exposure-factors are the relative aperture of the lens system, type of photographic film used, reflecting power of the object to be photographed, season of the year, time of day, color of the light, geographical location, altitude at which photograph is being taken (if an aerial photograph) and atmospheric conditions.

exposure index, aerial - The reciprocal of the length of time of exposure, expressed in meter-candle-seconds, at the point on the toe of the characteristic curve where the slope equals 0.6 gamma when recommended conditions for processing are used.

exposure interval - The interval between exposures.

exposure station - SEE camera station.

exposure time - The length of time a sensitive surface is exposed.

expression, harmonic - A series of harmonic functions used to represent a periodic phenomenon.

expression, topographic - The effect achieved by shaping and spacing contour lines so that topographic features can be interpreted with the greatest ease and fidelity. Good topographic expression is achieved by paying care to the spacing and width of the contour lines, due consideration being given to the scale and contour interval of the map. The contour lines must be adjusted sometimes (without exceeding the specified tolerances) to (a) show features otherwise missed; (b) emphasize significant features of the terrain, or (c) omit minor features relatively unimportant or leading to an incorrect interpretation of the map. Also called configuration of terrain.

expressway - A highway with full or partial control of access and with important crossroads separated in grade from the pavements for through traffic of all types. Expressways are usually divided highways. Some streets may be left as crossings at grade, to be eliminated at a later time. In some instances, the term has been incorrectly applied to named highways not having the characteristics of an expressway. SEE ALSO highway; highway, divided; freeway; road.

expropriation - (1) A voluntary surrender of rights or claims. (2) The act of divesting oneself of that which was previously as one's own, or renouncing it. In this sense, it is the opposite of appropriation.

extension - (1) The establishment of control of the same order and class as the order and class of the control to which it is or will be connected. Establishment of control of lower order and/or class is then called densification. However densification is also used with both meanings. (2) The determination, by photogrammetric methods, of horizontal and/or vertical coordinates of ground points to be used for the absolute orientation of other photogrammetric models. SEE ALSO bridging; cantilever extension. The latter is extension from a region containing ground control into a region without such control. Bridging is extension of phototriangulation from two, non-adjacent regions containing ground control into the region between them which does not contain such control. The term is usually further modified as horizontal or vertical according as the primary purpose is determining horizontal or vertical ground control.

extension, cantilever - (1) The extension and adjustment of a photogrammetric survey from a region containing ground control into a region without ground control. It is in contrast to bridging, which is the extension of a survey between regions containing ground control. (2) Aerotriangulation from a region containing ground control to a region without ground control. (3) The connection, by relative orientation and scaling, of a series of photographs in a strip to obtain coordinates of points in the strip. Also called the cantilever method. The term is often shortened to extension.

extension, horizontal - SEE extension.

extension, vertical - Photogrammetric extension of vertical control. SEE ALSO extension.

extension network, horizontal-control - SEE triangulation network, base-extension.

extension of control - SEE extension.

extension survey - A survey made to add to an existing, partial survey. An extension survey does not, however, complete a survey of boundaries, townships, sections or subdivisions thereof.

extensometer - (1) An apparatus for measuring small changes in distance between two points on the Earth. Also called a creep meter and a strainmeter. The Sassa extensometer, for example, consists of a wire stretched under tension between the two points. A weight hung from the center of the wire moves up or down as the distance between the two points increases or decreases. Another type is the laser interferometer which, operating on the principle of the Fabry-Perot interferometer, measures changes of less than 0.1 nm over a distance of 25 meters. (2) SEE strain gage.

exterior angle - (1) The angle between two adjacent sides of a polygon measured outside the polygon. Also called an external angle. (2) An explementary angle. This is surveying usage and is rare. (3) SEE angle, external. (4) An element of a simple horizontal circular curve measured from the mid point of the curve to the point of intersection.

exterior boundary - The seaward boundary of a marginal or territorial sea.

exterior orientation - SEE orientation, outer.

exterior perspective center - SEE center of perspective; center of perspective, external.

exterior to a curve - (1) Any part of a surface adjacent to a curve and lying toward the convex side of the curve. (2) That part of a surface not included within the circle of which the curve is part of the circumference.

external (adj.)(law) - (1) Apparent; capable of being perceived. (2) Outward. (3) Visible from the outside.

extinction - The decrease in intensity of a ray, as it passes through matter, because of absorption, refraction and scattering.

extinction, atmospheric - Extinction encountered in passing through a planetary or stellar atmosphere. In particular, extinction of stellar radiation on passage through the Earth's atmosphere.

extinguishment - (1) The act of rendering legally nonexistent. (2) The act of destroying or rendering void; nullifying. (3) The act of avoiding, as by payment, setoff, limitation of actions, merger of an interest in a greater one etc. Extinguishment, the act of extinguishing, is distinguished from the mere transfer, passing, or suspension of a right or obligation.

extrapolation - The process of estimating the value of a function at a specified point outside a certain region, from known values of the function at points within that region. Sometimes called prediction or filtering.

extrusion - The extension of detail, in a map, to the region outside the neat line.

eye-and-ear method - A method of measuring the time at which a stellar image crosses a particular line in a reticle, by counting off the seconds or half-seconds to oneself while observing the passage. Counting is started by looking at the chronometer and counting in cadence with its ticks. Attention is then diverted to the telescope but counting is continued. The interval from the tick just before crossing to the moment of crossing is estimated, and the times of the ticks themselves remembered. The time of crossing is calculated from these values.

eye base - SEE distance, interpupillary.

eyott - A small island arising in a river.

eyepiece - A lens assembly which magnifies an image and presents the magnified image to the eye. Also called an ocular. The eyepieces used in geodetic instruments are classified as direct or inverting, according as the observer sees the image right side up or upside down. SEE ALSO shutter eyepiece.

eyepiece micrometer - SEE micrometer, ocular.

F

Fabry-Perot interferometer - (1) A pair of perfectly flat, partly-reflecting surfaces on glass or other transparent substance, placed parallel to and facing each other a short distance apart. In one form, two slightly wedge-shaped plates of glass whose facing surfaces are exactly parallel, flat to 1/30 to 1/100 wavelength of green light, and coated with thin, partially reflecting metallic films are spaced a short distance apart. They are kept parallel by spacers. (2) A pair of parallel surfaces made highly reflecting by applying metallic films to them. Fabry-Perot interferometers working at infrared and radio wavelengths are common.

Fabry-Perot plate - Two precisely parallel, flat, transparent (usually glass) plates, usually a few wavelengths apart, through which light is transmitted so as to cause interference. The plates may be made more reflective by thin, metallic films.

face - (1) That side of a negative or photographic plate which has the emulsion on it. (2) That side of a printing plate which does the actual printing. (3) That side of a leveling rod which has graduations on it or which carries the graduated side of a scale. (4) The end wall of a drift, cross cut or a bed (of deposit).

face left - SEE circle left.

face right - SEE circle right.

factor - (1) A quantity which is an exact divisor of a given quantity. It is called a factor of that quantity. (2) An independent variable, in a function, which is an observable quantity. E.g., the temperature of the air is a factor affecting measurements of distance by electro-optical distance-measuring instruments. SEE ALSO conversion factor.

factor, augmenting - A factor used in making a harmonic analysis of tides or tidal current, because the tabulated hourly heights or speeds used in the summation for any constituent other than S do not in general occur on the exact constituent hours to which they are assigned but at times which may differ from the assigned times by as much as half an hour.

factor, perturbing - SEE force, perturbing.

factor, sea-level - That factor which is applied to measures of distance to give the equivalent distance on a specified datum.

factor, weighting - (1) A number equal to the number of times a particular value occurs among a set of values. Also called the weight. In geodesy, however, the term weight is now applied to the square of the weighting factor. The weighting factor is a particular kind of weight. The larger the weighting factor, the greater the its effect on the average and on the standard deviation. Hence the name. (2) The relative reliability (or worth) of a particular value of a quantity as compared to other values of the same quantity; the greatest weight is assigned the value with the greatest reliability. If one value has a weighting factor of 2 and another value of the same quantity has a weighting factor of 1, the first value is worth twice the second value, and an average value would be obtained by taking a weighted average: twice the first value plus once times the second value, the sum being divided by 3. In most applications, definitions (1) and (2) are equivalent. (3) A number by which both sides of an observation equation are multiplied to affect the solution according to the reliability of the measurement or the number of times that particular value was obtained. It is usually taken to be the reciprocal of the square of the standard deviation. If a set of linear equations is weighted, and if the measurements in the set are not correlated, the weighting factors when arranged as elements in a matrix form a diagonal matrix. If the measurements are correlated, off-diagonal elements in the matrix are not zero.

Fahrenheit temperature scale - A temperature scale in which 32° denotes the freezing point and 212° the boiling point of water at 760 mm barometric pressure. Fahrenheit assigned the value 0 to the freezing point of a saturated solution of salt in water and the value 96 to the normal temperature of man in arriving at the scale used.

fairway - The central, deepest or best navigable channel. This term is used in defining water-boundaries between states and is applied to water boundaries in sounds, bays, straits, gulfs, estuaries and other arms of the sea. It is also applied to bounding lakes and land-locked seas in which there is a channel suitable for deep-water sailing.

fall, free - The motion of a body moving under the influence of gravitational forces only. A body resting on the Earth's surface is not in free fall because it is subject not only to the Earth's gravitational force but also to the supporting force of the surface.

falling - The distance, measured along an established line from its intersection with a random line, to the corner on which the random line was intended to close.

family of surfaces - That set of surfaces obtained, in a three-dimensional coordinate system, by assigning all possible values to the parameters a, b, c,... in the equation $f(q,r,s; a,b,c,...)$, in which q,r,s are the coordinates of points in that system.

fan - A gently sloping, fan-shaped mass of loose, rocklike material or silt which has been worn off or disintegrated from a steep slope and deposited at a place where the slope is less.

fan, alluvial - A fan consisting of silt and other fine material deposited by a stream or river at its mouth or where the slope of the bed abruptly becomes less.

fan camera - An assemblage of three or more cameras positioned at fixed angles relative to each other and providing wide lateral coverage with overlapping images.

Faraday effect - Rotation of the plane of polarization of plane-polarized radiation when the radiation traverses an isotropic, transparent medium placed in a magnetic field having a component in the direction of propagation. Also referred to as Faraday rotation. The Faraday effect changes the direction of polarization of radio signals transmitted through the ionosphere from beacons or transponders in artificial satellites. The type of antenna chosen for transmitter and receiver must therefore take this phenomenon into account.

Faraday rotation - SEE Faraday effect.

farl (law) - One-quarter of (something).

farm - A tract of land devoted to agricultural purposes, managed as a unit by one person, a family or similar group, and including continuous, uncultivated land.

farm crossing - A crossing or path joining parts of a farm that has been intersected by a railroad line.

farmland - (1) Land used specifically for agriculture such as the raising of crops or livestock. It may consist of part or all of a single farm, or it may be made up of many contiguous farms. (2) Land designated, by zoning laws or ordinances, for agricultural use.

fathogram - A graphic record of depth measurements obtained by echo-sounding equipment.

fathom - A measure of depth equal to 6 feet, exactly.

fathom curve - SEE depth curve

fathom line - SEE depth curve.

fault - A fracture or zone of fractures along which there has been displacement of the sides relative to one another and parallel to the fracture. Movement along a fault is detectable by the continuing displacement of features on the surface with respect to one another. It is determined by measuring the displacement between neighboring points on opposite sides of the fault. This may be done by establishing a triangulation network which straddles the fault, and remeasuring the network as often as desirable. If the fault is sufficiently narrow, the movement may be measured by extensometers. SEE ALSO transform fault.

fault, strike-slip - A fault in which the movement is parallel to the fault plane as it intersects the horizon.

fault, transcurrent - An extensive fault in which the movement is parallel to the horizontal direction of the fault and the surface of the fault is steeply inclined.

fault, translational - A fault in which there has been translation and no rotation. The dip of the sides of the fault remains the same.

fault line - The intersection of the surface of a fault with the surface of the earth or with any artificial surface of reference.

Faye anomaly - SEE Faye's gravity anomaly.

Faye correction - SEE gravity correction, free-air.

Faye's gravity anomaly - (1) That condensation gravity anomaly in which the matter is condensed onto the reference surface (geoid or terrestrial ellipsoid). (2) That condensation gravity anomaly in which the matter is condensed onto the surface of the reference surface, and the cylindrical Bouguer gravity anomaly instead of the Bouguer gravity anomaly is subtracted. (3) SEE gravity anomaly, free-air. (4) The difference between free-air gravity anomaly and the topographic gravity correction.

Faye's gravity correction - SEE gravity correction, free-air.

Faye's gravity reduction - (1) A condensation gravity reduction in which the layer of condensed matter is at depth zero i.e., the layer is on the surface of the geoid (or other reference surface being used). (2) SEE gravity reduction, free-air. (3) SEE condensation gravity reduction.

feather (v.) - To thin the edge of a photographic print made by abrading the back surface with sandpaper or emery paper. Also termed to feather-edge. Prints are feathered before assembling them into a mosaic to obtain a smooth surface. When overlapping edges are feathered, shadows and sharp changes in contrast are reduced by dodging.

feather-edging (n.) - (1) The technique of progressively dropping contours to avoid congestion on steep slopes, and tapering the width of the line near the end of the contour to be dropped. Also called feathering. (2) The act of feathering the edges of prints.

feather-edging (v.) - SEE feather. Also spelled feather-edging.

feathering - SEE feather.

feature, areal - A topographic feature such as sand, swamp or vegetation, which extends over a considerable area. It is represented on a published map by a solid or patterned color, by a prepared symbol, or by a delimiting line.

feature, cartographic - The natural objects of culture shown on a map.

feature, cultural - SEE culture.

feature, hydrographic - Features along the shore and the submerged parts of bodies of water. Also called hydrographic detail if being considered for inclusion on a map.

feature, hypsographic - A topographic feature with elevations referred to the geoid.

feature, man-made - SEE culture.

feature, natural - A feature such as a stream, lake, forest, or mountain, which is not the work of man. Also called natural detail.

feature, physiographic - A prominent or conspicuous physiographic landform or a salient part thereof.

feature, topographic - (1) A prominent or conspicuous topographic form, or a salient part thereof. (2) A relief feature generally of third-order, but sometimes of second-order magnitude.

fee - (1) The form of remuneration received by professional people (such as lawyers, engineers, surveyors, appraisers, etc.). It is to be understood that the fee is paid by the person to whom the service was rendered, and not by the employer of the professional person. A person working for and employed by someone who then provides his services to others receives a salary. (2) An estate of inheritance in real property; i.e., an estate in fee. SEE ALSO fee simple.

fee, determinable - (1) An estate which may rely on the happening of a merely possible event. (2) An estate created with special limitations which delimit the duration of estate in land. Also called a base fee or qualified fee.

fee road - A road built on land purchased specifically for that purpose.

fee simple - (1) An absolute fee (sense 2); a fee without limitations to any particular class or heirs or restrictions, but subject to the limitations of eminent domain, escheat, police power and taxation. (2) An inheritable estate.

fee tail - An estate or inheritance limited to some particular class of person to whom it is granted. Estates in fee tail have been abolished in most States, converting them to fee-simple estates.

felt side - That side (the top or smooth side) which is in contact with the felt belt for extracting moisture during manufacture. This is the correct side of the paper for printing on.

fence - A hedge, structure, or partition erected to enclose a piece of land or to divide a piece of land into district portions or to separate two continuous estates.

fence, ancient - The line of a fence around a tract, maintained for at least 30 (probably 50) years, which may be taken as fixing the correct boundaries of the tract as against a later survey.

feria - A day of the week, identified by number e.g., the 4th feria for Wednesday.

Fermat's principle - The path along which electromagnetic radiation travels between any two points will be that for which the time taken for the journey is an extremum. The time need not be a minimum; it can be a maximum.

Fermat's spiral - The spiral given by the equation $r^2 = k\theta$, in which r and θ are polar coordinates and k is a constant.

fermenting-dough theory - SEE isostasy.

Ferrel's law - A statement of the Coriolis effect as applied to the motion of winds.

Ferrero's formula - The approximate sf to the standard deviation of the angles in a triangulation network, according to the formula $sf = [(\Sigma \varepsilon n^2)/3N]^{1/2}$, in which εn is the misclosure, in angle, of triangle n and N is the number of triangles in the network. The formula was developed by General Ferraro as a measure of the accuracy of a network. It is widely used as an estimate of the accuracy when only misclosures of triangles are available.

Ferro meridian - The meridian 17° 37' 45" west of Greenwich. Also called the Hierro Meridian. It was originally chosen to be exactly 20° west of the Paris Meridian. Many European geographers reckoned longitude from this meridian, which was taken as the dividing line between the Eastern and Western Hemispheres. It was established in 1634 by order of Louis XIII as a basis for maps and surveys of France, where it was used until about 1800, and outside of France for a much longer time. The Ferro Meridian does not pass through the nearby island of the same name. It still appears on some maps of Czechoslovakia, Hungary, and Romania.

ferrotype (v.) - To give a gloss to a photographic print by squeezing the print face down, on a polished plate, while wet and allowing to dry.

FG achromatism - That property of a properly designed telescopic lens system which ensures that light at a wavelength of the Fraunhofer F-line at 486.1 nanometers and light at the wavelength of the G-line at 430.8 nanometers are focused at the same point and produce images of the same size. Also called actinic achromatism and achromatism.

fiber optics - That part of optics which concerns itself with the propagation of light along fibers of glass, quartz or other transparent substances. One geodetic application of fiber optics is to the construction of ring gyroscopes capable of measuring the Earth's rate of rotation.

fiber pendulum - SEE pendulum, filar.

fictitious (adj.) - In cartography, pertaining to or measured from an arbitrary line used as reference.

field (adj.) - Referring to work done or equipment used in the field. For example, field party, field data and field computations.

field (n.) - (1) A region (space) and a quantity defined at every point of that region (space). (2) A region at every point of which a vector is defined. (3) A region (space) and a set of vectors defined at every point of that region (space). (4) A region (space) and a force (vector) defined at every point of that region (space). In physics, this is the most common definition of field. The type of field is often identified by prefixing to field the name of the quantity involved for example, gravity field or geomagnetic field. It is sometimes called a force field to distinguish it from a field in which the vectors do not represent forces but, for example, velocities. Gravitational, gravity and magnetic fields are all of particular geodetic importance because the same theory can often be used for all three. The region may or may not contain matter. (5) A region of space at every point of which is defined a scalar or vectorial quantity. The definition is sometimes generalized by specifying a tensorial instead of a scalar of vectorial quantity. SEE ALSO gravity field.

field, angular - SEE angle of field.

field, electric - A field at every point of which the vectorial intensity of the electric force is defined.

field, electromagnetic - A region (space) at every point of which are defined two forces (vectors) - a force capable of accelerating electrically charged particles by an amount proportional to the charge and a force capable of accelerating moving, electrically charged particles by an amount proportional to the charge and the speed. Alternatively, it can be defined as a region in which there exist an electrostatic field and a magnetic field, with the two fields related through Maxwell's equations.

field, electrostatic - A region (space) at every point of which is defined a force (vector) able to cause a charged particle to be accelerated by an amount proportional to the charge on the particle. The electrostatic field is attributable to the presence of an electric charge either within the region or outside of it.

field, geodesic - A one-parameter family of geodesic lines forming a field of tangent vectors over a portion of a surface.

field, geomagnetic - The magnetic field of the Earth. Sometimes referred to, ambiguously, as the magnetic field. The intensity of the geomagnetic field varies from about 0.25 weber/m² in a small region around northern Argentina to over 0.70 weber/cm² near the south magnetic pole. The magnetic force is vertical at the northern and southern magnetic poles; elsewhere, its direction varies considerably. The field is representable, to a first approximation, as a magnetic dipole with a moment of about 1.01×10^{17} weber meters and an angle of about 11° with the Earth's axis of rotation. It can be represented well as one large magnetic dipole and 8 to 10 small magnetic dipoles placed about 1640 km from the Earth's center at various longitudes and latitudes. The geomagnetic field has a slow, secular variation, fairly periodic variations which depend on the locations of the Sun and Moon, and large, random variations which are related to disturbances in the Sun's atmosphere. The field does not extend out to infinity, as was once thought, but is inclosed in a cavity in the ionized gases continuously ejected from the Sun. The field lines deflect the gases from the Earth, forming the cavity; at the same time, the lines of magnetic force are bent backward, away from the Sun, by the stream of ionized solar gases. The cavity and its boundaries are called the magnetosphere. The surface separating the geomagnetic field from the interplanetary magnetic field is called the magnetospheric sheath. These is no sharp separation between the two fields; the region of transition is called the magnetopause.

field, gravitational - A field in which the force at each point is caused by gravitation only.

field, magnetic - (1) A region at every point of which is defined a force (vector) evidenced by the accelerational effect upon a magnetic substance or upon a moving electric charge. A stationary electric charge is surrounded by an electrostatic field i.e., there is a force surrounding the charge which affects other electric charges. If the charge moves, the field moves with it, albeit changed in shape. At the same time, a new field, the magnetic field, is created. This may be considered the relativistic consequence of the electrostatic field i.e., it is the electrostatic field as viewed from a moving point. (2) SEE field intensity, magnetic. (3) SEE field, geomagnetic.

field, main geomagnetic - That part of the geomagnetic field left after daily and short-term variations (of a few days) are removed.

field, regional magnetic - SEE magnetism, regional.

field, residual geomagnetic - The difference between the measured intensity of the geomagnetic field and the intensity of that dipolar magnetic field which best approximates it. Also called a non-dipole magnetic field.

field, stationary - Any natural field of force, such as a gravitational or magnetic field.

field calibration - The checking of instrumental accuracy by a combination of computations in the field and in the office. Adjustments other than normal adjustments by the operator cannot be made during field calibration.

field check - The checking, on the ground, of map compilation.

field classification - Inspection and identification, in the field, of features which a map compiler is unable to delineate: identification and delineation of political boundaries, place names, road class, buildings hidden by trees, and so on. It may be included as part of the control survey and normally is completed before compilation begins. SEE ALSO field inspection.

field classification survey - SEE field classification.

field comparator - A short line whose length has been measured both accurately and precisely and which is used to check the lengths of base-line apparatus or tapes used in surveying. Also called a comparator base.

field completion - SEE survey, field completion.

field contouring - Determining the contours for a topographic map by making a planetable survey on a prepared sheet or by stadia surveying. This operation is generally used for mapping terrain unsuitable for mapping by photogrammetric methods. It is also used in regions of limited extent when the engineering plan requires contour lines at 1-foot (0.3 m) intervals.

field control - SEE ground control.

field correction - An adjustment made to measurements of such quantities as angles or distances to correct for discrepancies in distance or geometry.

field correction copy - A map or tracing made in the field and showing corrections for subsequent reproduction of a map.

field elevation - An elevation taken from the computation, in the field, of a line of levels.

field inspection - The job of comparing what is shown on a set of aerial photographs with what is actually on the ground and thereby supplementing or clarifying what cannot be inferred from the photographs alone. Surveying is not a part of field inspection but, together with the field inspection, constitutes field completion.

field intensity (at a point) - The force exerted at a point, on a body of unit mass, charge, etc. Also called field strength (at a point).

field intensity, horizontal magnetic - The intensity of the horizontal component of the magnetic field in the plane of the magnetic meridian.

field intensity, magnetic - (1) The force exerted by a magnetic field on an imaginary magnetic pole of unit strength placed at any specified point in space. Also called magnetic field, magnetic field strength and magnetic intensity. It is a vector, and its direction is taken as the direction toward which a north magnetic pole would tend to move under the influence of the field.

field intensity, total (magnetic) - The vectorial resultant of the intensity of the horizontal and vertical components of the Earth's magnetic field at a specified point. Also called total magnetic intensity and total intensity.

field intensity, vertical (magnetic) - The intensity (magnitude) of the vertical component of the Earth's magnetic field, reckoned positive if the field is directed downward and negative if the field is directed upward.

field notes - The permanent, detailed record made by a surveyor during a field survey that details procedures, measurements and observations pertinent to that survey.

field of view - (1) That angle which the entrance window, in an optical system, subtends at the center of the entrance pupil. (2) The angular extent of the portion of object-space imaged on photographic material. In photogrammetry, it is the angle subtended at the rear nodal point by some characteristic dimension of the photograph such as a diagonal of the photograph or, less frequently, by a diameter of the largest circle that can be inscribed in the photograph. SEE ALSO angle of view; angle of coverage.

field of view, angular - SEE field of view, true.

field of view, apparent - The projected angle of the field stop, as seen through the eyepiece.

field of view, instantaneous - The smallest solid angle resolvable by a scanner, expressed in radians.

field of view, real - SEE field of view, true.

field of view, true - The angle of view of the outside world, as limited by the field-stop diameter of the eyepiece. Also called angular vield of view and real field of view.

field position - A location calculated while work in the field is in progress, to determine whether or not the measurements are acceptable or to provide a preliminary location for other purposes.

field sheet - (1) The sheet, used in the field, on which is plotted the details of a hydrographic survey during the survey. Also called a boat sheet. (2) A sheet of stable material, generally used on a planetable, on which work is plotted in the field. Also called a planetable sheet.

field standardization (of tape) - Comparison of the length of a surveyor's tape with the length of a standard tape (calibrated tape) to determine the true length of the former.

field stop - That opening, in an optical system, which determines the largest angle for which principal rays from points in image space will pass entirely through the system. The field stop determines the field of view of the optical system; the aperture stop determines the brightness of the image.

field strength - SEE field intensity.

field strength, magnetic - SEE field intensity, magnetic.

field variation, geomagnetic - SEE variation, magnetic.

figure adjustment - SEE adjustment of figure.

figure of the Earth - (1) A mathematical representation of the geoid. (2) Those elements used in defining a mathematical surface which approximate the geoid. (3) The physical surface of the Earth. (4) A general term, of indefinite meaning, indicating some aspect of the geoid or the shape of the Earth for example, mean sea level and its continuation under the continents. (5) SEE shape of the Earth. (6) SEE geoid.

figure of the Earth, hydrostatic - (1) The equipotential surface fitting an Earth completely in hydrostatic equilibrium. (2) The surface of an ellipsoidal solid having the same mass and rate of rotation of the Earth, which is in hydrostatic equilibrium, and on which the gravity potential is constant.

film - (1) A material having the form of a thin sheet or closed surface. The film may be a sheet only few molecules thick, as an oil film on water or an anti-reflection coating on glass, or it may be as thick as the plastic sheets used as a base for photographic emulsions. Or it may be globular like a soap bubble or like the oxide coating on an aluminum bearing. In particular, photographic film. (2) SEE film, photographic. SEE ALSO film, autoscreen.

film, aerial - Photographic film designed specially for use in aerial cameras. It is usually supplied in rolls in many lengths and widths, with various kinds of emulsions.

film, autopositive - A photographic film which gives a positive copy from a positive transparency (or a negative from a negative) by direct processing. Also called direct copy or direct positive.

film, autoscreen - A photographic film embodying a halftone screen which automatically produces a halftone negative from continuous-tone copy.

film, cartographic - Photographic film having sufficient dimensional stability that photographs taken with it can be used in making maps. Cartographic films are usually referred to by their trade names.

film, color-blind - Photographic film sensitive only to wavelengths of about 400 - 500 nanometers (the lower third of the visible spectrum).

film, color-infrared - Photographic film manufactured to record the green, red, and near-infrared components of a scene using dyes such as yellow, magenta and cyan to create color. Also called false-color film.

film, false-color - (1) SEE film, color-infrared. (2) Photographic film which records reflected wavelengths from features but renders them as different colors through development.

film, infrared (photographic) - (1) Photographic film sensitive to infrared radiation. (2) In particular, photographic film sensitive to radiation in the near-infrared part of the spectrum.

film, latent-image - Photographic film exposed but not developed.

film, photographic - A thin, flexible, transparent sheet (film) of stable, plastic material (the base or film base) coated with a light-sensitive material (the emulsion) and used for taking photographs. Commonly referred to as film if it is perfectly clear from the context that photographic film is meant. Unfortunately, film is often used as if it were a complete synonym, with resulting confusion in some places. Photographic film and film are also often applied to the final product after exposure, development and fixing. However, that product is more commonly referred to as a negative or a positive, according to the nature of the final image. Photographic film is distinguished from photographic plate in that the latter has the emulsion coated onto a rigid material, usually glass.

film base - The film used as the support for a layer of light-sensitive emulsion i.e., as the material onto which the emulsion is coated.

film distortion - The distortion of photographic film with changes in humidity or temperature, or from aging, handling or other such causes.

film mosaic - SEE panel base.

film negative - SEE negative.

film plane - The position occupied, in a camera, by the photographic film or photographic plate.

film platen - The device which holds the photographic film in the focal plane during exposure.

film positive - SEE positive, photographic.

film speed - The response or sensitivity of a photographic film to radiation in the near infrared, light, or ultraviolet parts of the spectrum. It is often expressed numerically in one of several systems such as the Hand D, DIN, Scheiner or ASA exposure indices. SEE ALSO aperture, relative; speed, aerial-film.

film speed, aerial - SEE speed, aerial-film.

filter - A device or procedure which separates the input into two sets: a desired set and an undesired set. Only the desired set appears in the output proper; in particular, a device which transmits a desired range of matter or energy while substantially attenuating all other ranges. A mathematical filter operates on a set of numbers or variables to remove those numbers or variables not passing certain criteria; an electric filter operates on electrical currents or voltages to attenuate components having undesired frequencies; an optical filter operates on light to attenuate components having undesired frequencies. SEE ALSO color filter; filter, optical; Kalman filter, vignetting filter.

filter, anti-vignetting - An optical filter whose density decreases radially from the center to compensate for the radially decreasing illumination given by certain kinds of lens systems, particularly wide-angle lens systems. The filter may be a separate piece of glass with a partly transparent coating of varying thickness, or it may be the coating itself deposited directly on one of the elements of the lens system. It does not prevent vignetting; it merely obscures the effect of vignetting by reducing the amount of light reaching non-vignetted portions of the image.

filter, band-pass - A substance or device that transmits only that part of a periodic input which has frequencies or wavelengths lying between certain limits. The limits are called cut-off points, but no filter can completely include or exclude wavelengths or frequencies outside these limits. Filters operating on electromagnetic radiation at optical and near-optical frequencies are usually simple elements, while filters operating at radio frequencies are usually complex circuits. In a small region in the submillimeter part of the electromagnetic spectrum, gratings are often used as filtering elements.

filter, band-stop - A substance or device that transmits only that part of a periodic input which has frequencies or wavelengths lying outside certain limits. SEE ALSO filter; filter, band-pass.

filter, dichroic - (1) A partially reflecting surface produced by placing several thin metallic coatings on the surfaces of glass plates or prisms to produce interference and permit only a narrow band of frequencies to pass through. (2) A colorless optical filter containing uniformly oriented molecules which permit only light of a certain polarization to pass.

filter, digital - (1) An electrical filter which operates on digitized versions on electrical currents or voltages to attenuate unwanted frequencies. Its operation can be represented by a system of difference equations. (2) SEE filter, mathematical.

filter, electric - A circuit which substantially attenuates the amplitudes of those components of a varying current or voltage which are present within a specified range of frequencies. A low-pass filter removes all frequencies lower than the one specified, a high-pass filter removes all frequencies higher than the one specified, and a band-pass filter removes all filters above or below a specified range. A passive filter contains no source of power and is a combination of resistances, capacitances and/or inductances. An active filter contains an amplifier or similar element such as a transistor which differentially amplifies the wanted frequencies or attenuates the unwanted ones.

filter, high-pass - A substance or device that transmits only that part of a periodic input which has frequencies lying above (or wavelengths lying below) a certain limit. SEE ALSO filter; filter, band-pass.

filter, infrared - A device which transmits a specified band or bands of frequencies in the infrared part of the spectrum while substantially attenuating radiation at other frequencies. The various types of optical filters have their counterparts among infrared filters.

filter, low-pass - A substance or device that transmits only that part of a periodic input which has frequencies lying below (or wavelengths lying above) a certain limit. SEE ALSO filter; filter, band-pass.

filter, mathematical - (1) A mathematical function or algorithm which removes, from a finite or infinite set of numbers, those members of the set lying outside prescribed limits. A smoothing function is a mathematical filter. (2) SEE function, smoothing.

filter, non-recursive - A mathematical function or algorithm which selects some numbers from a finite or infinite set of numbers and rejects the rest according to preset criteria, and creates other numbers to represent those rejected. The function or algorithm does not, in making its selection, make any use of the values it creates.

filter, neutral - An optical filter which reduces the intensity of light reaching a light-sensitive surface such as photographic film, without affecting the rendition of the color tones in the original scene.

filter, optical - A device which transmits a specified band or bands of frequencies in the optical (light) portion of the spectrum but substantially attenuates all other frequencies. The term is also applied to a similar device which works on infrared radiation but such a device is better termed an infrared filter. The most common type of optical filter is the absorption filter, which transmits the desired frequencies with little attenuation but absorbs most of the unwanted radiation. It is quite inexpensive but does not cut off the undesired colors sharply. An interference filter reflects the incident light between semi-reflecting surfaces in such a way that light at the undesired frequencies is canceled by interference but light at the desired frequencies is strengthened. Such a filter is considerably more expensive than an absorption filter but can (in fact usually does) produce an extremely narrow band of transmitted light.

filter, polarizing - An optical filter which transmits principally light of a certain polarization and substantially attenuates light polarized in other directions. The more common varieties are made of plastic in which polarization-selective crystals are suspended. The term polarizing filter is misleading. The filter does not polarize the light transmitted; it merely does not transmit well light polarized in directions other than those permitted by the filter's structure. This is why a scene viewed through a polarizing filter appears darker than when seen without; much of the light does not get through.

filter, recursive - A mathematical filter which uses previously computed values of the numbers as well as current values of the given set in selecting the representative function or algorithm.

filter, spatial - (1) A device consisting of a metallic plate with a small hole onto which coherent light is focused. The hole is slightly larger than the undiffracted image, so that only the undisturbed beam with a spherical wavefront free of unwanted modes, reflections, etc., passes through the hole. (2) A device which transmits periodic patterns of certain periods while attenuating those patterns having other periods. Also called a spatial frequency filter.

filter, spatial frequency - SEE filter, spatial.

filter, vignetting - SEE vignetting filter.

filter factor - (1) The effective transmittance of an optical filter. (2) A number indicating the increase in amount of time needed for exposure of a photographic emulsion when using an optical filter, as compared to the amount of time needed when no filter is used.

filtering - (1) The process of separating, from a set of data, the part that does not contain wanted information from the part that does. The part that does not is called noise, error or misinformation; the part that does is called signal or information. (2) The process of obtaining an estimate of the value of a quantity at a specified time t from data available for times up to and including t. When this definition is used, smoothing is considered to mean the same process applied to times before t, while prediction or extrapolation means the same process applied to times after t. These special meanings are most common in control engineering and related subjects like inertial surveying. (3) SEE curve fitting. (4) SEE extrapolation.

filtering, maximum-likelihood - The method of maximum likelihood.

filtering, optical - (1) The removal of light within selected portions of the spectrum by passing the light through colored glass or other colored substance such as gelatin or plastic. (2) Doing mathematical operations such as Fourier or Laplace transforms on data by displaying the data in the form of diffraction or interference patterns and then operating on these patterns by means of suitable optical systems. Also called spatial filtering. (3) Removing, by optical means, certain spatial frequencies from a periodic pattern.

filtering, spatial - SEE filtering, optical.

finder - SEE finder telescope.

finder circle - A small, vertical circular disk with coarse graduations and with an accessory spirit-level of low sensitivity, attached to an astronomical telescope of the type used in geodetic astronomy. It is placed perpendicular to the horizontal axis of the telescope and is used to put the telescope in approximate position for observing. Using a computed angular elevation at the proposed time of observation, the telescope is pointed in the direction of the object to be observed. When the object comes into the field of view, the finder circle will have served its purpose and the observations will be made with the more precise apparatus provided for the purpose. Also called a setting circle.

finder telescope - A small, optical telescope of low power and wide angular field of view, attached to a larger telescope of very restricted field of view, and used for finding the object at which the larger telescope is to be pointed. Also called a finder or finding telescope. The object is located by looking through the finder telescope. When the image has been brought to the center of the field, the object will also be visible in the larger telescope.

finite-element method - A method of obtaining an approximate solution to a problem for which the governing equations and boundary conditions are known, by dividing the region of interest into numerous, interconnected subregions (finite elements) over which simple, approximating functions are used to represent the unknown quantities. The basic idea originated in the early 1900's as a method of solving for strains in bridges and buildings. It was further developed for analyzing the structure of aircraft in the early 1950's. The method, supported by rigorous mathematical theory, has now been applied to problems in heat flow, hydraulics, geodynamics and other disciplines.

firing range - A piece of land set aside for the testing or firing of rifles, artillery and missiles.

First of Aries - SEE First Point of Aries.

First Point of Aries - That point, on the celestial sphere, at which the ecliptic and the celestial equator intersect and at which the Sun appears to cross from south to north (i.e., the Sun's declination changes from negative to positive). Also called the vernal equinox, First of Aries and gamma (Brit.). However, the term vernal equinox is ambiguous and therefore less desirable. Because of precession, the First Point of Aries is no longer in the constellation of Aries but is in Pisces; in 300 years it will be in Aquarius. By definition, the right ascension and declination of the First Point of Aries are zero ($\alpha = 0$, $\beta = 0$.)

First Point of Cancer - SEE summer solstice.

First Point of Capricornus - SEE winter solstice.

First Point of Libra - That point, on the celestial sphere, at which the ecliptic and the celestial equator intersect and at which the Sun appears to cross from north to south. Also called the autumnal equinox.

Fischer Ellipsoid of 1960 - SEE Fischer 1960 spheroid.

Fischer level - A binocular dumpy level, constructed of invar, having the spirit level placed in the tube of the telescope and quite close to the line of collimation. The length of bubble can be adjusted by use of an air chamber. An arrangement of mirrors and prisms brings an image of the bubble to the one eyepiece; the leveling rod is viewed through the other. The telescope is so mounted that its inclination can be adjusted through a small angle without changing its elevation. The observer views the leveling rod with the right eye and at the same time views the bubble with the left eye. By turning a screw, the inclination of the telescope is adjusted until the bubble is seen in the middle of its tube. The leveling rod is read with the bubble in the average position. The Fischer level was designed by E.G. Fischer and adopted by the U.S. Coast and Geodetic Survey in 1899 (the use of invar in its construction is of more recent date.) It is no longer in use, although it is one of the most sensitive and accurate leveling instruments. It has been replaced by types easier to set up and adjust.

Fischer leveling instrument - SEE Fischer level.

Fisher leveling rod - SEE U.S. Coast and Geodetic first-order leveling rod.

Fischer 1960 spheroid - One of the two rotational ellipsoids calculated by I. Fischer. (1) The ellipsoid used for the Mercury datum has the dimensions

    semi-major axis   6 378 166.0 m
    flattening   1/298.3 .

It was used by the United States of America for charts showing Loran lines, etc. (2) The ellipsoid used for the South-Asia datum has the dimensions

    semi-major axis   6 278 155.0 m
    flattening   1/298.3 .

Fischer's polygnomonic map - SEE map, polygnomonic.

fit (n.) - (1) The representation of a set of N points (pairs of numbers) by a smooth function, of a single variable, containing fewer than N defining constants. This is a special case of smoothing and curve-fitting. (2) Representation of a set of N points in space by a smooth surface in that space, the surface being specified by too few constants to allow it to pass through all the points. A fit is also called a smoothing or an adjustment; the process is called fitting, curve-fitting or adjustment. (3) An expression of how well a curve or surface represents a given set of numbers i.e., a word or number used to give an idea of how far the given points lie from the curve or surface representing them. E.g.., to say that a fit is good is to imply that the points lie acceptably close to their geometric representation; to say that the fit is poor is to imply the opposite. The root-mean- square value of the distances is frequently used to indicate fit.

fit (v.) - To represent a given set of points by the corresponding points of a smooth function, curve, or surface; to produce a fit (n.). Also called to smooth or, misleadingly, to adjust. The method of least squares is usually preferred, because it produces the best fit in a statistical sense if the points have a Gaussian distribution. However, curves are often fit by hand and give quite good results without requiring difficult calculations. Fit and smooth have similar meanings but not the same connotations. One fits a curve to a limited number, usually quite small, number of points. One smooths a set of values or set of points, but the number of values or points may be very large or even unlimited.

fitting - SEE curve fitting; fit; surface fitting.

fix (n.) - (1) A location obtained, from astronomical observations, as the intersection of lines of position. The act of obtaining such a location is called obtaining a fix or getting a fix. (2) The location, obtained by surveying and indicated on a map, of a point at which observations were made. (3) The act of obtaining, by surveying, the location of the point at which the observations were made, and then indicating this location on a map. (4) The location, obtained by surveying, of a point at which observations were made. (5) The act of obtaining, by surveying, the location of a point at which observations were made. (6) A location obtained by computation from data obtained at other locations and used as the location of a control point. (7) A relatively accurate location determined without reference to any former location. A pinpoint is a very accurate fix, usually established by passing over or very close to an aid to navigation or a landmark of small extent.

fix (v.) - (1) To make a developed, photographic image permanent by removing the unexposed and still light- sensitive material, so that only the developed image remains. (2) To determine the location of a point from which observations are made, by surveying e.g., by resection.

fix, running - The intersection of two or more lines of position not obtained simultaneously but adjusted to a common time.

fixed (adj.) - A bench mark, point, position, station, etc., for which geodetic coordinates (longitude and latitude) or elevations have been determined by a previous adjustment or by a more precise survey, and are to be held without change in a newer survey or adjustment.

fixing - The process of rendering a developed, photographic image permanent by removing the unaffected, light-sensitive material.

fixing system - SEE locating system.

flagperson - (1) The person, in a surveying party, who carries the flag or rod. (2) A person assigned to warn or control approaching traffic of the presence of surveyors or other work crews along a road or railway.

Flamsteed's map projection - SEE map projection, sinusoidal.

flare - (1) Patches of light or unwanted images which appear on the focal surface of a camera as a result of poor optical conditions within the camera. There are two kinds of flare: mechanical flare, which occurs when light is reflected from some bright spot on the mounting of a lens or on a faulty shade, and optical flare, which occurs when light is reflected as it strikes the surfaces of the individual lenses or prisms. (2) SEE hotspot. (3) A bright source of light, usually chemical in nature, of fairly constant intensity and lasting for some time e.g., several minutes. A flare lasts longer than a flash, which usually lasts only for a small fraction of a second and which is often produced electrically. Both flares and flashes have been used as signals in triangulation.

flare triangulation - A method of triangulation in which observations of a flare at very high altitude are made simultaneously from a number of stations of known and unknown location. The flare is usually carried aloft by a rocket or airplane and then ejected to float downward supported by a parachute. However, it may be towed by an aircraft and ignited at the desired altitude. The flare gives off light continuously. Photographs are taken of it at timed instances at each station, and the vertical and horizontal circles of the instruments are photographed at the same times. Flare triangulation has been used for extending triangulation over lines so long that the ends are not intervisible. For example, rocket-borne flares were used in a stellar triangulation connecting Bermuda to the USA.

flash apparatus - An apparatus auxiliary to the use of a pendulum for determining gravity, which produces a flash of light every time the pendulum passes through a position used as reference; the flash may be recorded on moving photographic paper whose rate of motion is measured, or may be detected by a photoelectric cell and timed electronically.

flash plate - SEE calibration plate.

flash triangulation - A method of triangulation in which observations of a flashing light at very high altitudes are made simultaneously from a number of stations of known and unknown location. The flashes occur at timed instances and can be distinguished by those times, so that observations made on the same flash are simultaneous.

flat - (1) A level surface, without elevation, relief, or prominence; a plain. (2) A level tract along the banks of a river. (U.S.A.) (3) A level tract lying at little depth below the surface of the water or, a level tract alternately covered and left bare by the tide; a shoal, shallow, or strand. (4) An assemblage of photographic negatives or positives in the positions required by the complete picture to be printed. The term is part of the jargon of the graphic arts and is in common use. There are three common types of flat: (a) the goldenrod flat, an assemblage of negatives on goldenrod paper (which serves as a layout, mask, and support for the negatives); (b) the blueline flat, an assemblage of photographic films or stripfilm sections in register with blueline images on glass or on a plastic sheet; and (c) the complementary flat, one of a pair or set of flats, each of which contributes its detail, by successive exposures, to the same printing plate. SEE ALSO key flat; layout.

flat, optical - A metallic or vitreous body having one surface ground and polished to a flatness within a fractional part of a wavelength of light. Also called, ambiguously, an optical plane. The usual material is hard glass or quartz.

flat, tidal - A sandy or muddy, flat, coastal zone or marsh which is covered and uncovered by the rise and fall of the tide. SEE ALSO marsh, tidal; mud flat; salt marsh; slough.

flatness - A measure of the shape of a three-dimensional object. A. Cailleux's formula is $(a+b)/2c$, in which a is the longest dimension of the object, c the shortest and b and a representative, intermediate dimension.

flattening - (1) The ratio of the difference in the lengths 2a and 2b of the major and minor axes, respectively, of an ellipse, to the length of the major axis of the ellipse: $(a - b)/a$. Also called ellipticity. It is usually denoted by f in English and by a (abplattung) in German works on geodesy. (2) The ratio of the difference in the lengths 2a and 2c of an equatorial diameter and of the polar diameter, respectively, of a rotational ellipsoid (spheroid), to the length of an equatorial diameter: $(a - c)/a$. Alternatively, the flattening (1) of the ellipse which was rotated to create the ellipsoid. Note that the flattening is positive if the ellipsoid is an oblate spheroid, negative if the ellipsoid is a prolate spheroid. (3) The flattening (1) of the rotational ellipsoid used to represent the geoid (figure of the Earth) or the shape of the Earth. (4) The quantity $(a - c)/a$, in which 2a is the length of the longest diameter perpendicular to the axis of symmetry of a rotationally symmetric figure and 2c is the length of the axis of symmetry of that body. The original definition assumed that the body in question was the geoid, that the geoid was symmetrical with respect to the equatorial plane, and that a unique flattening could be defined in terms of a and c. (5) (of a spheroid) - The flattening of an ellipse cut from the spheroid (rotational ellipsoid) by a plane containing the ellipsoid's polar axis (axis of rotation). Also called, sometimes, ellipticity or compression. In some literature, flattening refers to the Earth, not to the ellipsoid representing the Earth. The exact meaning of the term, when used in this way, is uncertain. For a rotational ellipsoid with major diameter 2a and minor diameter 2b, the flattening is $(a - b)/a$. It is usually denoted by f. (6) SEE flattening (of the Earth).

flattening (of the Earth) - The flattening $(a - b)/a$ of a rotational ellipsoid having a major axis of diameter 2a and a minor axis of diameter 2b, taken to represent the Earth or the geoid. Also called compression of the Earth and compression of the geoid.

flattening, gravitational - The ratio of the difference between values of gravity, as given by a standard gravity formula, at the pole and at the equator, to the value at the equator. Also, and more properly, called gravity flattening.

flattening, hydrostatic - The flattening of a rotational ellipsoid representing an Earth in hydrostatic equilibrium.

flexure - (1) The curved or bent state of a loaded beam. Alternatively, the change in shape of a body under tension or pressure. SEE ALSO flexure (of the pendulum support). (2) (of a pendulum) The bending a swinging pendulum undergoes because the material of the pendulum is not perfectly rigid. The effect of the bending is to increase the period of oscillation. That is, a flexible pendulum swings less rapidly than a rigid pendulum. The slower oscillation may be regarded as caused either by changes in the moments of the various parts of the pendulum as it bends, or caused by the storage of energy in the bent pendulum. Bending increases the period of oscillation in both positions of the reversible pendulum. In determining relative gravity, the bending is ignored. Flexure is then taken to mean that movement of the pendulum's supports caused by the swinging of the pendulum. SEE ALSO flexure (1); correction, flexural. (3) (of the pendulum's support) The forced movement, caused by the motion of a swinging pendulum, of the structure supporting the pendulum. The horizontal component of the force acting on the knife edge of a swinging pendulum causes the supporting structure to move slightly in unison with the pendulum, thereby affecting the period of the pendulum. The measured period must be corrected for this small effect. SEE ALSO correction, flexural.

flicker comparator - SEE blink comparator.

flicker method (of comparing images) - (1) First one and then the other of a pair of corresponding images (photographic or other) are projected alternately and in rapid succession onto a flat surface such as a table or screen, or into the optical system of a photogrammetric apparatus for simultaneous viewing with both eyes. The viewer sees a flickering in those parts of the composite image where the corresponding images differ in detail. (2) First one picture of a stereoscopic pair is viewed by one eye and then the other picture is viewed by the other eye, the viewings taking place alternately and in rapid succession, as by blinking first of one eye and then the other. The viewer gets the impression of a single, fused picture. A flickering will be noticed in those regions of the composite where the two pictures differ appreciably in detail. The effect is usually obtained without making the viewer blink, by placing the two pictures side by side but separated by a thin partition placed so that each eye of the viewer can see only one picture. The pictures are then illuminated rapidly and alternately.

flight altitude - The vertical distance above sea level of an aircraft in flight. In the past, flight altitude has been used interchangeably with flight height. In aerial navigation however, the altitude of an aircraft is always stated in reference to sea level. In the interest of consistency, flight altitude should also be used only in reference to sea level.

flight block - A set of overlapping strips of photographs, considered as a unit for photogrammetric adjustment. The smallest flight-block consists of at least three, overlapping strips.

flight chart - SEE route chart.

flight height - (1) The altitude of the camera above the average elevation of the ground at the instant an aerial photograph is taken. (2) The altitude of the camera's lens-system above the ground nadir or above the average elevation of the ground shown in the photograph or above a specified datum. Also called flying height; however, that term is more often used for the barometric altitude. The difference between flight altitude and flight height is the average elevation above sea level of the region being photographed. The rear nodal point of the lens system would be a more precise specification than merely lens system in the definition, but the difference is probably negligible in most situations.

flight line - (1) The path taken by an airplane during an aerial survey. (2) The line, drawn on a map, representing the path followed by an airplane while taking aerial photographs. (3) A line drawn on a map to represent the track of an aircraft.

flight map - (1) The map on which are shown the lines along which an aircraft is to fly when aerial photographs are being taken, and/or the locations at which such photographs are to be taken. (2) The map on which have been plotted selected points at which aerial photographs were taken and the line of flight of the aircraft between those points.

flight strip - A succession of overlapping aerial photographs taken along a single course.

float - An object made so that it will float in water in a certain position. Floats are often used to mark regions of particular importance to navigation. They are also used in oceanography to mark points of observation in currents. Such floats are designed to float either on the surfaced or at a predetermined depth below the surface. The first kind are called surface floats; the second kind are called subsurface floats. SEE ALSO rod float.

float, subsurface - SEE float.

float, surface - SEE float.

float gage - An instrument for measuring the height of the tide or of a stream by measuring the height of a float on the water's surface. The float is confined within a vertical pipe or channel open, at the bottom, to the water. In the more primitive float gages, the height of the float is measured directly by reference to a graduated staff fixed to the pipe. In more advanced models, a counter weighted and graduated tape or chain attached to the float passes over a pulley geared to a recording mechanism.

floating - The technique of making minor adjustments of detail on a map to maintain the proper relative positions of symbols.

float well - A vertical pipe or box with a relatively small opening in the bottom, used to inclose the float of a float gage so that the float will not be affected by non-tidal motions of the water. Also called a stilling well.

flood current - The movement of a tidal current toward the shore or up a tidal river or estuary.

flood hydrograph - The hydrograph of flow during a flood.

flood plain - (1) Land, along the course of a stream, which is subject to inundation during periods when the water's height exceeds the normal for being full to the banks. (2) Land which is generally parallel to the stream with approximately level ground, gentle longitudinal slope corresponding to the gradient of the stream, and very flat backslope. (3) Natural terrain frequently consisting of low-lying, timbered land interspersed with swamp, marsh, small lakes, ponds and bayous.

flood tide - (1) That portion of the oceanic tide occurring between low water and the following high water. Also called a rising tide. (2) The landward current caused by that portion of the tide occurring between low water and the following high water.

flow - (1) A general term for the motion of a fluid such as water. (2) The volume or mass of water passing through a specified surface per unit of time.

flow, geostrophic - Motion of a fluid when the horizontal component of the pressure is balanced by the horizontal component of the Coriolis force per unit area. Also called geostrophic current. The state of balance between opposing pressures is referred to as geostrophic equilibrium. Flow in the oceans is never exactly geostrophic, but in many regions it can be treated as geostrophic without significant error.

flow, non-geostrophic - Motion of a fluid when the horizontal component of the pressure does not balance the horizontal component of Coriolis force per unit of area.

flowage - The natural flow or movement of water from an upper estate to a lower estate. This is a servitude which the owner of the lower estate must bear even tho the flowage be not in a natural water course with well-defined banks.

flowage line - A contour or line around a reservoir, pond, lake, or along a stream, corresponding to some definite waterlevel (highest, mean, low, spillway, crest, etc.). The term is usually used in connection with the acquisition of rights to flood lands for storing water.

flowchart - A diagram showing the steps to be taken in solving a problem, doing a computation or carrying out a project. The flowchart for computational purposes was originated by von Neumann and has been accepted as an essential first step in encoding an algorithm for calculation on a computer. The symbols have also been standardized.

flowline - The slope put into the plastic surface of a plastic-relief map, extending from the heights along the neatline to a flat portion beyond the neatline, at an angle no greater than 45°, to avoid forming the plastic sheet at an angle of 90° at the neatline.

flowmeter - An instrument which measures the rate at which water flows past or through it. The term is more general than current meter.

fluid - A gas or a liquid. A state of matter in which only a uniform isotropic pressure can be maintained without infinite distortion.

fluid, opaquing - A liquid having a dense, opaque substance in suspension and painted on a transparent material to render that material opaque. Opaquing fluid is used in the graphic arts to correct errors made in photoengraving, to change parts of a picture, etc.

fluxmeter - SEE magnetometer.

fly-by method (of barometric altimetry) - A method of determining barometric altitude identical to the two-base method, except that the roving barometers are carried by air and read in the aircraft as it passes on a level with the topographic feature whose altitude is wanted. The method is used to determine approximate elevations in regions where the terrain is extremely rugged.

fly leveling - SEE leveling, flying.

flying height - SEE flight height.

flying leveling - SEE leveling, flying.

f-number - SEE aperture, relative.

f-number, effective - SEE aperture, effective.

focal distance - SEE distance, back focal; distance, focal; distance, front focal.

focal length - SEE length, focal; length, calibrated focal; length, effective focal; length, equivalent focal; length, minimal focal.

focal point - SEE point, focal; point, first focal; point, second focal.

focus - (1) The point to which rays of light converge after passing through a lens system. If the incident rays are parallel and close to the optical axis of the system, they come to a focus in the (primary) focal plane. The distance along the axis of the lens system, from the second nodal point to this plane, is called the focal length of the lens system. (2) Any small region in which rays of light from a point source converge after passing through an optical system; the set of all image-points corresponding to one point in object-space. (3) The point to which rays of light converge after passing through an optical system. (4) Any small region in which monochromatic rays of electromagnetic radiation converge after passing through a suitable system. (5) SEE ALSO coudé focus.

focus, back - SEE distance, back focal.

focus, coude - SEE coude' focus.

focus, front - SEE distance, front focal.

focus, principal - SEE focus.

focus, sidereal - The location of that plane which is perpendicular to the principal axis of an optical system and onto which a stellar image is focused. In an ideal optical system, the sidereal focus is called the principal focal plane. A camera or telescope is said to be in sidereal focus when rays incident from a great distance come to a focus in the plane of the photographic plate or of the reticle. The sidereal focus is sometimes called the solar focus.

focus, solar - SEE focus, sidereal.

fod - In the Virgin Islands, a unit of length equal to 1.029 feet.

fog - (1) Droplets of water or, more rarely, ice crystals, suspended in the air in sufficient concentration to reduce visibility appreciably. (2) A veil of silver of very low density covering uniformly all or parts of photographic film or paper. It is usually caused by unwanted radiation reaching the film through multiple reflection by the lens, by aged or impure photographic materials, etc. (3) A darkening of a photographic negative or print by a deposit of silver which does not form part of the image. Fog tends to increase density and decrease contrast. SEE ALSO edge fog.

fog density - The photographic density of an unexposed but processed photographic emulsion.

foot - A unit of length, in the American system of units, now defined as (1) exactly 1/3 of 36/39.37 of the International Prototype Metre, or 1/3 of a yard. The legal and practical definitions of the foot are not necessarily the same. Until 1893, both the American foot and the British foot (also called the English foot) were defined as (2) 1/3 of the length of the British imperial (standard) yard (a brass bar in London, or copies thereof). The American foot, by a law passed by Congress on 28 July 1866, and by the Mendenhall Order of 1893 (U.S. Department of the Treasury, 5 April 1893), was established as the length specified in definition (1). This foot continues in use up to the present, at least for surveying in the U.S.A. On 1 July 1959, by agreement between the U.S. National Bureau of Standards and the corresponding organizations of some other countries, the international yard was defined to be exactly 0.9144 meter long. The foot, by this new definition, is then 0.3048 meter long. It is, however, for scientific use only and does not legally replace the foot of 1893. The foot used in the triangulation of Britain was defined as (3) 1/10 of the length of the 10-foot bar O1 of the Ordnance Survey. One foot of O1 is 0.304 800 756 International meters. The Indian foot used by the U.S.A. and the United Kingdom for computing triangulation of India and neighboring countries is 0.304 798 41 meters long; the foot used at present by the Survey of India is 0.304 799 6 meters long. The legal value, adopted in 1956, of the Indian foot is 0.3048 meters, exactly. SEE ALSO survey foot.

foot, American - SEE foot; survey foot.

foot, British - SEE foot.

foot, English - SEE foot.

foot, French - A unit of length equal exactly to 1/6 toise. The French foot contained exactly 12 lignes and was approximately 0.31 meters long.

foot, Indian - SEE foot.

foot, square - A unit of area, in the English system of measure, equal to the area contained within a square 1 foot on a side. It is not usually used as a measure of land area; the acre is preferred except where, as in large cities, land is very costly. It is approximately equal to 0.092 903 04 square meter. Note that it is not a primary unit of area but is defined in terms of the foot.

foot-candle - A unit of illuminance equal to one lumen of incident light per square foot.

foot-lambert - A unit of luminance equivalent to $1/\pi$ candela per square foot. Equivalently, the luminance of a perfectly flat, diffusing surface radiating one lumen per square foot.

foot plate - (1) The plate or other type of bearing surface on which the screws used for leveling an instrument rest. (2) A metallic plate used to support a leveling rod at a turning point in leveling. Also called a turning plate. In the USA, metallic pins (turning pins) are preferred to foot plates for supporting the leveling rods, except on hard surfaces such as roads.

footprint - The power-density contours, at the ground, of the radiation from a transmitting antenna or the pattern of a receiving antenna carried on an aircraft or artificial satellite. The term has a number of other meanings as well, including, apparently, that of resolution. The term ground coverage is preferable.

foot wall - The wall or rock under a vein or under other steeply inclined formation of minerals.

force - (1) The product $f = ma$ of the mass m and the acceleration a of a moving body. It has the dimensions MLT^{-2}. It can be defined for a stationary body by introducing fictitious and balancing accelerations (i.e., by using Newton's law that to every action there is an equal and opposite reaction). (2) A quantity f measured, for a moving body, by the product ma of the mass m and the acceleration a of the body. For a stationary body, f must be measured by a process which can be traced to kinematics. SEE previous definition. The two definitions are similar in form but differ fundamentally in physical meaning. By the first definition, force is merely a name for the product of mass times acceleration and does not exist otherwise. By the second definition, force is a physical quantity which exists independently of the concepts of mass and acceleration. (3) The cause of change of momentum of a free, movable body. (4) SEE ALSO Coriolis force; force, electrostatic.

force, apparent - A force introduced into a relative coordinate system to satisfy Newton's laws as expressed in that system. E.g., centrifugal force.

force, central - A force which, for purposes of computation or theory, can be considered to be a function only of the distance from a particular point; a force which is always directed toward a fixed point. Gravitational force is considered a central force in celestial mechanics except in those cases where the mass of one (or both) of the bodies involved is not uniformly distributed in spherical layers about a center and the two bodies are so close together that the asymmetric distribution of mass affects the orbits. For example, the Moon is so close to the Earth that the Earth's non-spherical shape appreciably affects the Moon's orbit.

force, centrifugal - The force, in the opposite direction to the normal, with which a body constrained to move along a closed, curved path reacts to the constraint. Centrifugal force acts in the direction away from the center of curvature of the path followed by the body. As a force caused by the rotation of the Earth, it is opposed to gravitation and combines with it to make a resultant force called gravity. SEE ALSO force, centripetal.

force, centripetal - The force, in the direction of the normal, required to constrain a body to move in a closed, curved path. Centripetal force acts in the direction toward the center of curvature of the path followed by the body. It is equal and opposite to centrifugal force. It, and the centrifugal force, are sometimes called fictitious forces or quasi-forces because the observed effects can be explained without introducing such forces.

force, compound centrifugal - SEE Coriolis force.

force, deflecting - SEE Coriolis force.

force, electrostatic - The force which attracts two particles of negligible mass and opposite electrical charges towards each other or repels two particles of negligible mass and the same kind of electrical charges away from each other. Also called a Coulomb force. Denoting the charges by q_1 and q_2 and the vector connecting the particles by r, the force f between the particles is given by $f = k_e^2 \, q_1 \, q_2 \, r/r^3$, in which k_e is a constant and r is the absolute value of r. Electrostatic force is related to magnetic force by Maxwell's equations.

force, fictitious - A quantity introduced into an equation or description to account for the behavior of a body, but which can be dispensed with by using a different coordinate system for the equations or a different viewpoint in the description. The Coriolis force and centrifugal force are typical of such forces.

force, gravitational - The force exerted between two bodies because they have mass. If the bodies have masses m_1 and m_2, respectively, and are of negligible size in comparison to the distance r between them, then the force f acting between the two bodies is $f = - G m_1 m_2 r/r^3$. G is the gravitational constant.

force, magnetic - SEE field intensity, magnetic.

force, perturbing - (1) The difference between the resultant of the actual forces acting on a body and the resultant of an ideal, hypothetical, or assumed set of forces acting on that body. (2) The difference between the resultant of a set of forces assumed to be acting on a body and the resultant of a subset of that set. By either definition, the gradient of the disturbing potential. Also called a perturbing factor.

force, specific - (1) The force acting on a body, divided by the mass of the body. (2) The difference between the inertial acceleration and the gravitational acceleration of a body. This definition is completely inconsistent with the usual definitions of force and acceleration and should not be used.

force, tide-producing - The slight, local difference between the gravitational attraction of two bodies and the centrifugal force which holds them apart. Also called the tractive force. These forces are exactly equal and opposite at the center of gravity of either of the two bodies but, because gravitational attraction is inversely proportional to the square of the distance, the forces vary from point to point on the surface of each body. Therefore, gravitational attraction predominates at those points of the surface nearest the other body, while centrifugal force predominates at those points farthest from the other body.

force, tractive - SEE force, tide-producing.

force field - SEE field (4).

force function - That scalar function of location whose gradient at a point is the force at that point. There is considerable difference in usage in giving signs to potential, force function, and potential energy; the following conventions are fairly common. (a) In abstract fields of force, the potential and force function are identical and are the negative of the potential energy. (b) In Newtonian fields, the potential at P because of a unit element at Q is 1/r; if the elements attract, potential and force function are the same and are the negative of potential energy; if the elements repel, the potential is the negative of the force function and is the same as the potential energy. SEE ALSO potential.

forepoint - The point to which an observation is made in surveying i.e., the point to which a foresight is directed.

foreshore - (1) According to riparian law, that strip of land between the high and low water marks which is alternately covered and uncovered by the flow of the tide. (2) That part of the shore which lies between the mean high water line and the mean low water line and is daily traversed by the water as the tide rises and falls. (3) That part of the shore lying between the crest of the seaward berm (or upper limit of wave action at high tide) and the mean low water mark, and which is ordinarily traversed by the uprush and backrush of the waves as the tides rise and fall. (4) That part of a beach normally exposed to the action of waves. Also called the beach face.

foreshore slope - The slope of the foreshore, taken as the angle, at the highwater mark, between a horizontal plane and a line running seaward from that mark and more or less parallel to the foreshore.

foresight - (1) In a survey made between an initial point and a final point through a succession of intermediate points, a sight made to a succeeding point of the reading or measurement obtained by that sight. The situation existing at the time sights are taken is called a set-up, and the distinction between preceding point and succeeding point, and hence between backsight and foresight, is peculiar to each set-up. Ordinarily, what was the succeeding point in one set-up becomes the preceding point in the next, and a new point becomes the succeeding point. (2) In leveling from an initial point to a final point through a sequence of intermediate points, a sight to or a reading on a leveling rod held on a succeeding point of the sequence. The sight to or reading on the preceding point is called the backsight. If the sequence consists of only two points, it does not matter which sight is called the backsight and which the foresight. Note that in these definitions, the survey must be made through the sequence of points. Sights taken to points not directly in the sequence are called side sights or side shots (except extra foresight). SEE ALSO backsight.

foresight, extra - The reading made at an instrument station in a level-line and on a leveling rod standing on a bench mark or other point not in the continuous level-line. In spirit leveling, there may be one or more extra foresights from a single instrument station or set-up, but there can be only one backsight and one primary foresight from any one instrument station.

forestry map - A map normally showing the size, density and specifies of trees and other vegetation in a forested area.

form, physiographic - SEE landform, physiographic.

form, topographic - SEE landform, topographic.

format - (1) The size, arrangement, and other factors affecting the appearance of visual material. (2) The dimensions of that part of the negative which lies within the field of view in the focal plane of a camera. (jargon) (3) The dimensions of a map. (jargon)

format, bimargin - The format of a map on which the cartographic detail is extended to two edges of the sheet (usually north and east), thus leaving only two margins. SEE ALSO bleed.

form factor, dynamical - The negative of the coefficient of the second-degree, zonal harmonic in the expansion of the Earth's gravitational potential into spherical harmonics, normalized and adjusted to be dimensionless. Its non-normalized value is on the order of 10^{-3}.

form line - A line, usually dashed, drawn to represent the contour of the terrain. Unlike a contour line, a form line is drawn without regard to a vertical datum and without any regular vertical interval. It is sketched in on the basis of visual observations or unverified data.

formula, barometric - SEE altimetry, barometric.

formula, prismoidal - A formula for the volume of a prismoid. A typical formula, used for calculating the volume of a cut or fill shaped like a prismoid, is $V = L(A + 4B + C)/6$, in which V is the volume, A and C are the cross-sectional areas of the ends, B is the cross-sectional area at the middle of the cut or fill, and L is the length of a uniformly tapering part of the cut or fill. The formula is used, for example, in grading the routes of railroads or highways.

formula, tachymetric - One of two formulae commonly used in tachymetry to determine the distance d and difference Δh of height from the measured angular elevation α and the measured interval l on the rod. Common forms are $d = (k_1 + k_2 l) \cos^2\alpha$ and $\Delta h = (k_1 + k_2 l) \sin\alpha \cos\alpha$. Here, k_1 is the addition constant of the telescope and k_2 is a multiplicative constant (as a rule, approximately 100).

formula, thin-lens - SEE equation, thin-lens.

formula for gravity - SEE gravity formula.

formula for theoretical gravity - SEE gravity formula.

Fortin barometer - SEE cistern barometer.

Förtsch discontinuity - A discontinuity supposedly lying between the Conrad and Mohorovicic discontinuities.

Foucault pendulum - A pendulum consisting of a very heavy, spherical mass suspended by a very long thread or wire and set to swinging in a plane. If the site for the pendulum is carefully chosen to be free from microseisms and air currents, the pendulum will keep the plane in which it swings fixed in space while the Earth rotates under it. The plane of oscillation then appears to rotate about the point of lowest descent, with an angular speed $d\theta/dt$ given by $\omega \sin\phi'$, in which ω is the Earth's rate of rotation and ϕ' is the geocentric latitude of the point of suspension.

Fourier analysis - SEE analysis, harmonic (1).

fraction, representative - (1) A fractional scale whose numerator is unity. Often abbreviated to R.F. (2) SEE scale, fractional.

fracture zone - An extensive, linear region (of the sea floor) of unusually irregular relief and having lare seamounts and steep-sided or irregular ridges, troughs or escarpments.

frame - An individual exposure contained in a continuous sequence of photographs.

frame, contact-printing - A device for holding in contact the negative photographic film and the light-sensitive material on which the image is to be placed, during the exposure. The source of light may or may not be a part of the device. If the frame contains a vacuum pump to exhaust air within the frame (to ensure contact between the negative photographic film and the light-sensitive material, it is called a contact vacuum printing frame.

frame camera - A camera in which the entire film format is exposed through a lens system fixed relative to the focal plane and the recording medium. Contrasted with the continuous-strip camera and the panoramic camera.

frame of reference - (1) A set of lines or surfaces from which angles or distances are measured to a point, and which are associated with a physical system. (2) A coordinate system associated with a physical system. (3) Any collection of points and lines made use of to set up a system of coordinates for a space. (4) A coordinate system in which measurements of distance and/or angle can be made and locations of points determined.

frame of reference, Galilean - SEE frame of reference, inertial.

frame of reference, inertial - Properly, any frame of reference which has no translational acceleration and which is not rotating with respect to the distant galaxies. Also called a Galilean frame of reference. Less precisely, any frame of reference which can be considered non-rotating.

framework of control - SEE survey network; control network.

fraud - A misstatement of a material fact, made with intent to deceive or made with a reckless disregard of the truth that actually does deceive.

freehold - (1) An estate of inheritance. (2) An inheritance for life or an estate during the life of a third person.

freeway - A highway with full control of access and with all crossroads separated to grade from the pavements for thru-traffic. SEE ALSO expressway; highway, divided; road.

freezing point - That temperature above which a liquid remains liquid and at which a liquid solidifies. The term is applied, in particular, to the behavior of pure water, which has its freezing point at 0° C (32° F.). Note that the melting point and freezing point may not be the same.

French foot - SEE foot.

French legal metre - The unit of length defined by the Metre des Archives. SEE ALSO meter (unit of length); Metre des Archives.

frequency - (1) The number or proportionate number of times, per unit of time, that an event occurs within a specified unit-interval. In particular, the number of times a periodic phenomenon recurs (i.e., returns to its original form). For example, the frequency of a light wave at a particular point in space is the number of times per second the average intensity takes on the same sequence of values there. Most periodic phenomena are composite that is, they can be represented as the sum of a number of periodic phenomena which vary sinusoidally. (2) The number, or proportionate number, of times a randomly varying vector assumes a value within a specified unit interval in its range. Also called a distribution, distribution density and probability density. In many instances, it is convenient to take the range to be from minus to plus infinity and to set the frequency equal to zero for values outside the original range. A function which gives the frequencies in unit intervals is called the frequency function. Frequency is usually expressed as a percentage of the total number of times. This proportion, multiplied by 100, is commonly interpreted as the probability that the value will lie in the corresponding interval.

frequency, audio - The range of frequencies corresponding to those of normally audible tones. Audio frequencies generally range from 16 Hz to 30 kHz, but the range is often taken to be much less.

frequency, bivariate - SEE frequency function, bivariate.

frequency, characteristic - SEE frequency, natural.

frequency, limiting - A frequency which forms a least upper bound or greatest lower bound of a given band of frequencies.

frequency, natural - The frequency at which a system will oscillate freely in the absence of external forces: for example, after release from a simple displacement from a state of equilibrium. Also called the characteristic frequency. The natural frequency is determined by the dynamical characteristics of the system.

frequency, spatial - (1) The number of times a design is repeated over a given distance. (2) The number of times a design is repeated per unit distance in a specified direction. In both definitions, the design is assumed to be fixed and not vary with time while the spatial frequency is being determined.

frequency, very low - A frequency lying between 3 kHz and 30 kHz. The wavelength of electromagnetic radiation therefore lies between 10 km and 100 km.

frequency band - (1) The range of frequencies needed to represent a particular sound, color, electromagnetic signal, etc., as a Fourier series; alternatively, the range of frequencies present when such a phenomenon is separated into its simplest component frequencies. (2) The radiation (acoustic, electromagnetic, etc.) having frequencies between specified limits. Electromagnetic radiation is classified roughly according to the way the radiation is perceived, as radio waves or radio radiation, infrared radiation, light, ultraviolet radiation, X-rays, and gamma rays. Within each class, and particularly within the class of radio waves, limiting frequencies of internationally accepted values create subclasses SEE ALSO radiation, electromagnetic; radiation, infrared; radio wave; light;

frequency curve, altimetric - A diagram showing the frequency with which groups of elevations occur within a region.

frequency-difference of arrival method - SEE Doppler, differential.

frequency distribution - SEE frequency function.

frequency drift - A slow, nearly monotonic variation (drift) in frequency. For example, a slow rise or fall of the temperature of a crystal oscillator, or aging of the components in the circuit, will cause the oscillator's frequency to change slowly in one direction. Frequency drift cannot, in general, be completely controlled or its amount predicted.

frequency function - (1) A function giving the number, or proportionate number, of times a randomly varying vector assumes a value lying within a unit interval in the vector's range. Also called frequency distribution, distribution, distribution function, differential distribution function, probability distribution function, distribution density, probability density, probability density function and density function. The frequency function usually gives the proportionate number per unit interval, the total number being taken as 1. This number is then the same, numerically, as the probability of the value's occurring within that interval. (2) A function assigning to each possible value of the independent variable a number between 0 and 1 indicating the probability that value will occur. Also called a distribution function. SEE ALSO Poisson frequency function; Rayleigh frequency function.

frequency function, bivariate - A frequency function which is a function of two variables.

frequency function, bivariate Gaussian - SEE Gaussian frequency function, bivariate.

frequency function, cumulative - SEE distribution function, cumulative.

frequency function, Gaussian - SEE Gaussian frequency function.

frequency function, Gaussian multivariate - SEE Gaussian frequency function.

frequency-measuring equipment - SEE equipment, frequency-measuring.

frequency standard - A source, natural or artificial, of events believed to occur at such uniform intervals of time that the frequency of the events can be used as a standard against which other frequencies or intervals of time can be compared. The Earth's rate of rotation with respect to the distant galaxies or fixed stars was at one time considered suitable as a frequency standard. The period of rotation was subdivided into smaller units by clocks. When the rate of rotation was found to be irregular, the rate of revolution of the Earth in 1900.0 was adopted. This standard was replaced in 1968 by the frequency of light emitted by the cesium atom in a specified transition between atomic energy levels. Frequency standards whose accuracy is highest are called primary standards; those of high precision but of lower accuracy are called secondary standards.

frequency standard stability - SEE stability.

Fresnel lens - A lens having a saw-toothed cross-section in the radial direction; one side of each tooth is straight and vertical; the other side slants and has the curvature of the corresponding zone of a convex (convergent) lens. A Fresnel lens has the same optical properties as a regular lens of the same diameter and focal length but is very much lighter because the lens does not increase in thickness from edge to center as does a regular convex or plano-convex lens. Fresnel lenses are therefore used in situations calling for a very large lens of reasonable weight and not requiring the ultimate in faithfulness of the image e.g., as collimating lenses for the lamps in lighthouses. However, they have also been used in cameras, photographic projectors, and telescopes.

frilling - The separation, along the edges of its base, of a photographic emulsion from its base.

fringe - Lines, in a spectrum, formed by interference between waves of electromagnetic radiation. Fringes are usually observed as faint lines on each side of the strong central line produced by light received directly, without interference. Fringes of light can be observed as a series of parallel light and dark lines. If the light is not monochromatic, the fringes have colored edges. The distance between two neighboring fringes depends on the difference in phase of the interfering waves and can be used to determine the difference in distance traveled by the two waves. The Väisälä base line apparatus depends for its accuracy on the counting of fringes. The same principle is used in radio interferometry.

fringe, interferometric - The product of the voltages of the intermediate frequency stages at two separate antennas receiving signals from the same celestial object, after passage through a low-pass filter. The output has a sinusoidal variation because of the Earth's rotation with respect to the object. Each fringe has an amplitude proportional to the intensity of the source and a width proportional to a change of one wavelength between signals at the two antennas.

frog - A section of railroad track placed at the intersection of two rails and designed to provide support for the wheels and passageway for the flanges.

frog angle - The angle formed by the intersecting gage-lines of a frog.

frog heel - SEE heel (of frog).

frog point - SEE point of frog.

frontage - That length of that part of a property which abutts a street or body of water.

frontage road - A local street or road auxiliary to and located on the side of an arterial highway for service to abutting property and adjacent localities, and for control of access.

front and rear - Used to specify the dimensions at each end of a lot or part, as, e.g. 50 feet front and rear and implying the length in front as the length on the street or easement line and the length in the rear as the length on the portion of the boundary opposite the front. Obviously, the term's meaning is uncertain if it is applied to a corner lot with adjacent frontages on the streets. It is also indeterminate if applied to a lot having double frontage on the street; i.e., frontage on each end of the lot.

frost heave - A gradual rising of the surface in cold weather, caused by the freezing and consequent expansion of water in the soil. Any part of the water table near the surface has, in cold climates, the potential for causing frost heave especially in silts and poorly drained soils.

frost line - The depth to which ground becomes frozen. This depth may pertain to a particular winter, be the average for several winters, or be the extreme depth to which frost has penetrated.

full and change - SEE establishment of the port.

full and change, low-water - The average interval of time between the upper or lower transit of the full or new Moon and the next low water.

function - (1) A relationship between variables such that when the values of all but a particular one of the variables are known, the value of that variable is determined. Also called a correspondence, mapping or transformation. By convention, a function also exists if the value determined for the particular variable is not unique but is one of a denumerable set of values e.g., the arcsin function. The particular variable is called the dependent variable; the other variables are called the independent variables. The relationship is often expressed in the form: y is a function of the N quantities x_1, x_2, x_3, ... x_N if, to every allowable value of the quantities in the set $\{x_N\}$ there corresponds one value (or a small, finite set of values) of y. y and the x_N may be vectors. (2) That variable, or set of values of that variable, which is related to another variable or set of values by a function as defined above. Also called a map. E.g., if y is a function of x, then y is also called the map of x (there is some variation in the usage of this term). (3) A quantity which is derived from one or more other quantities by mathematical operations on those quantities. This is a special case of function as defined in (2). (4) For named functions, see under that name; e.g., Walsh function.

function, disturbing - SEE potential, disturbing (2).

function, elliptic - A function w, of a complex variable z, satisfying the following conditions: (a) w is doubly periodic with two finite, primitive periods whose ratio is not a real number; and (b) the only singularities of w in the finite portion of the plane are poles.

function, free - A randomly varying function which is not correlated with the other functions under consideration.

function, geodetic - A spherical harmonic multiplied by the quantity $3GM_m R_o^2 / (4 d_m^3)$, in which G is the gravitational constant, M_m is the mass of the Moon, d_m is the average distance of the Moon from the Earth and R_o is the average radius of the Earth. The term is applied also to the quantity obtained by substituting M_s, the mass of the Sun, for M_m and d_s, the Sun's distance, for d_m.

function, harmonic - (1) Any function which is a solution of Laplace's equation and which has continuous first and second derivatives. (2) A function satisfying Laplace's partial differential equation, i.e., a solution of that equation. Often referred to, when the meaning is clear, as a harmonic. Laplace's equation was first solved using Fourier series, the terms of which (sines and cosines) are harmonically related. The term harmonic function was subsequently applied to all solutions of the equation. Other functions less obviously harmonic than sines and cosines are also solutions. Among them are Bessel functions, Legendre functions, and Lamé functions. The kind of harmonic function most suitable as a solution depends on the coordinate system in which the equation is expressed. Fourier series are best for Cartesian coordinate systems, Bessel functions for cylindrical coordinate systems, Lamé functions for ellipsoidal coordinate systems and so on.

function, line-spread - (1) The density, as a function of distance from the axis of the image, of the photograph of a linear source in object-space. It is the line integral of the point-spread function. The line in object space is a geometric line lying in a plane perpendicular to the optical axis of the optical system. (2) A function giving the illuminance in the image, on a plane perpendicular to the optical axis, of a linear source of light as a function of distance from the line along which the illumination is greatest. The line spread function can be derived from the point spread function by integrating the point spread function in a definite direction.

function, periodic - A function which assumes the same values at equal intervals of the independent variable or variables. E.g., a function f(x) such that f(x+k) = f(x), in which k is a constant called the period of the function, is called a singly-periodic function. The sine and cosine functions are singly-periodic functions. Elliptic functions are doubly-periodic functions.

function, point-spread - (1) The density of the photographic image of a point source in object-space as a function of location in the image - usually of distance from the center of the image. (2) A function giving the illuminance in the image, on a plane perpendicular to the optical axis, of a point-source as a function of distance from the center of the image.

function, sampling - (1) A function which is nearly zero over most of the domain of its argument x, but is finite and not zero within a small interval about some specific value of x. The Dirac function is a sampling function, as is (sinx)/x. (2) The function (sin x)/x. This function has the value 1 at x = 0. It is symmetric about the y-axis and oscillates about the x-axis with an amplitude decreasing hyperbolically with increasing x. Written as $[\sin(x - x_o)] / (x - x_o)$, it has a maximum at x_o. SEE ALSO Dirac function; Walsh function.

function, sectorial harmonic - A spherical harmonic function of only the longitude or an analogous angle, and not of the radial distance or latitude. Also called a sectorial harmonic.

function, smoothing - (1) A function taken to represent a set of N or more points but containing fewer than N defining constants. Also called a mathematical filter or, according to the kind of function used, a smoothing polynomial, smoothing integral, moving average, etc. (2) An algorithm which embodies a function of the type specified in the previous definition i.e., an algorithm which accepts N or more (usually many more) points as input and produces as output corresponding points calculated using a function having fewer than N defining constants.

function, specific dissipation - SEE storage factor.

function, spherical harmonic - A solution, in spherical coordinates, of Laplace's partial differential equation in spherical coordinates. Also called a spherical harmonic.

function, spheropotential - SEE spheropotential function.

function, tesseral harmonic - A spherical harmonic function having only longitude and latitude (or their equivalents, such as right ascension and declination) as independent variables. A surface harmonic function in spherical coordinates. Also called a tesseral harmonic.

function, testing - A real-valued function, of a real variable, which can be differentiated an arbitrary number of times and which is identically zero outside a finite interval. SEE ALSO function, sampling.

function, zonal harmonic - A spherical harmonic function having only latitude (or an analogous quantity) as independent variable. I.e., a spherical harmonic function in which radial distance and longitude are held fixed. Also called a zonal harmonic.

functional - A function whose independent variables are functions.

function space - A set whose elements are functions.

furlong - A length of 1/8 mile or 220 yards.

furrow (hydrography) - A fissure that penetrates into a continental or insular shelf in a direction approximately perpendicular to the shore line.

fusion, stereoscopic - The mental process which combines two perspective views of an object to give an impression of a three-dimensional object.

G

gage - (1) An instrument for measuring the state of, or recording facts about, a phenomenon. Also spelled gauge. (2) The horizontal distance between the inside heads of the two rails of a railroad track. The distance is measured between points 5/8 inches below the top of the rails. The standard gage in the United States is 4 feet 8½ inches.

gage, automatic - SEE gage, self-registering.

gage, self-registering - Any tide gage or stream gage that provides a continuous record of the variation of the level of the tide or stream with time, and which can operate, unattended, for a number of days. Also called an automatic gage. SEE ALSO pressure gage, gas-purged; tide gage, mechanical; tide gage, sonic.

gage height - The elevation of the surface of water referred to some arbitrary datum.

gage line - A line 5/8 inch below the top of the center line of the head of a running rail or the corresponding location of other types of track and taken along that side nearer the center of track.

gain-riding technique - SEE technique, gain-riding.

gal - A unit of acceleration equal to one centimeter per second per second. The milligal, 0.001 gal, is more commonly used in geophysics. However, neither gal nor milligal is an approved unit in the Système International d'Unités.

Galilean telescope - An optical telescope having a positive objective-lens and a negative eyepiece, producing, in combination, a final image which is seen as erect but no real image is formed internally. Also called a terrestrial telescope.

galley proof - A proof made by printing from type on a printer's galley before it is set up in pages.

Gall's map projection - (1) Any one of three map projections: a stereographic map projection, a modified Plate Carrée map projection and an orthographic map projection. The first of these is the best known and is usually the one meant when Gall's map projection is mentioned. (2) SEE Gall's stereographic map projection.

Gall's stereographic map projection - A map projection defined, for mapping the sphere onto the plane, by the equations $x = (½\sqrt{2}) k R \lambda$ and $y = (1 + ½\sqrt{2}) k R \tan(\phi/2)$ in which x and y are the rectangular Cartesian coordinates of that point, in the plane, whose coordinates on the sphere are λ (longitude) and ϕ (latitude). R is the radius of the sphere and k is the scale factor. Sometimes referred to simply as Gall's map projection. This is a modified stereographic map projection and is not conformal. Scale is exact along the lines representing parallels of 45° N and 45° S latitude.

gamma - (1) The tangent τ of the angle which the straight portion of the characteristic curve makes with the axis of abscissas (the axis showing log-exposure). It is a measure of the extent of development and contrast of the photo-graphic material. SEE ALSO curve, characteristic. (2) A unit, commonly denoted by τ, primarily of magnetic flux density and equal to 10^{-5} gauss, but also used as a unit of magnetic field intensity and equal to 10^{-5} oersted (gilberts per centimeter) or to $7.957\,747\,2 \times 10^{-4}$ ampere per meter, approximately. The gamma is used principally for expressing intensity or measures of flux density of the Earth's magnetic field. The field has an intensity varying from about 25 000 gamma on the Tropic of Capricorn to over 70 000 gamma at the South Magnetic Pole. The quantity measured by magnetometers is often magnetic flux density and not magnetic field density, which is the proper quantity for describing the geomagnetic field. The gamma is therefore not a good unit to use in describing the measurements of the field itself. Adoption (1960) of the Système International d'Unités, in which the unit of magnetic field intensity is the ampere per meter, disqualifies the gamma from use. (3) SEE First Point of Aries.

gangway - A main road, underground, for hauling ore, etc.

gap - (1) A lack of photographs, or of stereoscopic pairs of photographs, of a small region between two regions each of which is shown on photographs or on stereoscopic pairs of photographs. The term may be applied either to the region not shown or to the missing photo-graphs of that region. (2) An unintentional lack of the required photographs of a small region in between two properly photographed regions, when such lack results from human error or from the equipment's malfunctioning. Also called a hiatus or holiday. The term is used of aerial photographs in particular. A gap occurs when, for example, the specified overlap between photographs is not obtained. When the missing photographs are of the interior of a region otherwise shown entirely on photographs, the lack is sometimes called a hole. (3) The existence of a space between the description of two pieces of property which supposedly adjoin according to the deeds or other records of surveys of the individual properties. The condition arises when portions of property are described from two directions. When there is a gap and two pieces of property do not meet, the owner of neither piece has title to the surplus because the original grantor did not sell it. A similar condition may exist when descriptions by metes and bounds, intended to indicate that two pieces of property are contiguous, are given from two directions. SEE ALSO overlap. (4) A break or opening in a mountain ridge. (5) The horizontal component measuring the separation of the two parts of a fault, taken parallel to the strike of the strata, with the faulted bed absent from the measured interval.

Garfinkel correction - The quantity H in Garfinkel's single-sight equation.

gauge - SEE gage.

gauss - A magnetic flux density of one field line per square cm of area perpendicular to the direction of the field lines. The gauss is properly defined only for B, the magnetic flux density of a magnetic field. However, it has also been used as a unit of measure for H, the magnetic field intensity. As a unit of magnetic flux density, it is equal to 10^{-4} webers per square meter (tesla). As a unit of magnetic field intensity, it is equal to one oersted, or approximately 79.577 472 amperes per meter. The Earth's magnetic field varies in flux density from about 0.25×10^{-4} webers per square meter to 0.70 webers per square meter and is almost equal, numerically, to the magnetic field intensity. Geophysicists have therefore commonly treated flux density and field intensity as if they were equivalent and used the gamma (10^{-5} gauss) as a unit of magnetic field intensity. Under the rules of the Système International d'Unités, this equivalence does not exist. For this reason and because the term is ambiguous, gauss should be used only in quoting or referring to old measurements made in this unit.

Gauss-Boaga map projection - SEE Mercator map projection, transverse.

Gauss conformal conical map projection - SEE Lambert conformal conical map projection.

Gauss conformal map projection - SEE Gauss-Krüger map projection; Mercator map projection, transverse.

Gauss-Doolittle method - Gauss's method for solving a system of linear equations, as arranged by M. Doolittle in 1878 for routinely solving such systems. Doolittle's arrangement provides an internal check on the accuracy of the calculations and is particularly suitable for hand computing by non-mathematicians.

Gauss elimination - Gauss's method (of elimination).

Gaussian constant - The gravitational constant expressed in units of the Sun's mass, the average distance of the Earth from the Sun and the ephemeris day and defined to be exactly 0.017 202 098 950 000 00 000 $(A.U.)^{3/2} (M_s)^{-1/2} (E.D.)^{-1}$. Also called the Gaussian gravitational constant. SEE ALSO constant, gravitational.

Gaussian coordinate - One of the two coordinates λ_g, ϕ_g defined on the ellipsoid by the two conditions: (a) the first fundamental form for element ds of length shall have the form $ds^2 = E(\lambda g, \phi g)(d\lambda g^2 + d\phi g^2)$, and (b) scale shall be exact along the central meridian. This is equivalent, on the sphere, to requiring that one of the two coordinates be measured along the central meridian to the foot of the perpendicular great circle through the point, and the other coordinate be given by $R \ln \tan \frac{1}{4}(2\xi + \pi)$, where ß is defined as $x_{soldner}/R$, $x_{soldner}$ being the distance measured along a geodesic from the central meridian to the point. SEE ALSO Riemann homogeneous coordinate.

Gaussian curvature - The product of the principal curvatures of a surface in Euclidian space. SEE ALSO curvature of a surface.

Gaussian distribution - A distribution characteristic of a random variable having the following properties: positive and negative values of about the same size are about equal in number, a small absolute value is more frequent than a large absolute value and the arithmetic average is the most probable value. Also called a normal distribution. The errors that occur in most kinds of measurement show a Gaussian or similar distribution. The distribution can be derived as a limiting case of the binomial distribution when the number of events becomes very large.

Gaussian distribution function - (1) The cumulative distribution function $(1 / \sigma \sqrt{2\pi}) \int \exp[-\frac{1}{2}(z - \zeta)^2 / \sigma^2] dz$ giving the probability that a continuous, random variable with average value ζ and standard deviation σ will have a value less than or equal to x. Also called a normal cumulative distribution function, normal distribution function, etc. (2) SEE Gaussian frequency function.

Gaussian distribution function, bivariate - (1) A cumulative distribution function P(x') as given by

$$P(x') \equiv K \int^{x_2} -\Sigma \int^{x_1} -\Sigma f(x) dx$$

in which K is $(\sqrt{\sigma}) / 2\pi$ and the frequency function f(x,y) is $\exp(-\frac{1}{2} Q)$; σ is the bivariate covariance-matrix, Q is the scalar $xT \sigma x$ and x is the vector with components x_1 and x_2. (2) SEE frequency function, Gaussian bivariate.

Gaussian frequency function - The function $f(x) = [1 / (\sigma \sqrt{(2\pi)}] e - [x - x_i]^2 / 2\sigma^2]$, which assigns the proportionate frequency f(x) dx to the occurrence of a value between x and x + dx; x_i is the average value of x and σ is the standard deviation. Also called a normal distribution and normal frequency function. The multi-variate Gaussian frequency function is $f(x) = [\Sigma - 2 / (2\pi) k/2] e [-\frac{1}{2} (x-x_i) T \Sigma^{-2} (x - x_i)]$, in which Σ^2 is the covariance matrix of x.

Gaussian frequency function, bivariate - A frequency function which is a function of two variables and is a Gaussian frequency function of either variable alone. I.e., that case of the multivariate Gaussian frequency function in which the function if a function of only two variables. Also called a normal bivariate frequency function.

Gaussian frequency function, multivariate - SEE Gaussian frequency function.

Gaussian gravitational constant - SEE Gaussian constant; constant, gravitational.

Gaussian lens-equation - SEE equation, thin-lens.

Gaussian mean radius of curvature - SEE Gaussian radius of curvature.

Gaussian radius - SEE Gaussian radius of curvature.

Gaussian radius of curvature - The geometric average of the greatest and least radii of curvature at a point on a surface. Also called the mean radius of curvature, Gaussian radius and the Gaussian mean radius of curvature.

Gauss-Krüger coordinate - (1) One of a pair of coordinates x and y of a point on a rotational ellipsoid: y is measured along the central meridian to the intersection with the parallel of latitude through the point; x is measured from the central meridian, along the parallel of latitude, to the point. Some authors interchange x and y in the definition. (2) One of a pair of coordinates in the coordinate system of the transverse Mercator map projection. (3) A Gaussian coordinate with the magnification factor 1, using Bessel's ellipsoid and a system of meridional sectors 3° wide.

Gauss-Krüger grid - SEE Mercator grid, transverse.

Gauss-Krüger map projection - The transverse Mercator map-projection derived by mapping directly from the rotational ellipsoid onto the plane. Also called the Gauss map projection and the Gauss conformal map projection. Nomenclature is highly inconsistent. Some cartographers limit the term Gauss-Krüger map projection to meaning the transverse Mercator map-projection from a rotational ellipsoid onto a plane as defined above, and limit the term transverse Mercator map-projection to mean projection from a sphere onto the plane. Other usages exist. However, most American cartographers consider the terms Gauss-Krüger map projection and transverse Mercator map-projection to be equivalent. (There is also a Gauss-Krüger stereographic map projection.) The Gauss-Krüger map-projection, being identical to the transverse-Mercator map-projection in all respects except perhaps the manner of its derivation, is conformal and scale is exact along the line representing the central meridian.

Gauss-Laborde map projection - SEE Gauss-Schreiber map projection.

Gauss magnetometer - A magnetometer measuring the horizontal component, H, of the total intensity F of the magnetic field. A magnet suspended horizontally is set to swinging and its period of oscillation is measured. The period is a function of the intensity M x H at the magnet, where M is the magnetic-dipole moment. The suspended magnet is then taken down and placed near a suspended, standard magnet of the same kind. Interaction of the two magnets causes one of them to turn by an amount proportional to M/H. The value of H can then be calculated from the two sets of measurements.

Gauss map projection - SEE Gauss-Krüger map projection.

Gauss method - SEE equal-altitude method (of determining latitude and time).

Gauss's algorithm - SEE Gauss's method.

Gauss-Schreiber map projection - The transverse Mercator map-projection derived by a conformal transformation from the rotational ellipsoid to an auxiliary sphere and thence, by another conformal transformation, to the plane. Also called the Gauss-Laborde map projection and the Schreiber double map-projection. The procedure was devised by Gauss about 1820 but it was first described by Schreiber in 1866. Note the distinction between the Gauss-Krüger map projection and the Gauss-Schreiber map projection. Both are the transverse Mercator map projection.

Gauss's equation - The equation $(1/s) + (1/s') = (1/f)$, in which f is the focal length of a thin lens, s is the distance of an axial point in object space from the lens and s' is the distance of the image from the lens. Also called the thin-lens formula. It applies only to monochromatic, paraxial rays going through a thin lens. However, it is an useful approximation even for thick lenses.

Gauss's method - (1) The method of solving a system of linear equations by successively eliminating one unknown and one equation at a time from the system until only one equation in one unknown remains. The value thus obtained for the last unknown is then substituted into the previous equation containing two unknowns, and so on until all unknowns have been determined. Also called Gauss's algorithm. At each state, a particular equation and a particular unknown in that equation are selected, using various criteria, for elimination. This equation is then multiplied by a constant which makes the coefficient of the selected unknown the same as the coefficient of that unknown in a second equation of the reduced system. The selected equation with its new coefficients is then subtracted from the second equation, eliminating the selected unknown from the second equation. This process is repeated for a third equation, a fourth, and so on until the selected coefficient has been eliminated from the reduced system. The result is a new, reduced system containing one fewer equations and one fewer unknowns. (2) SEE equal-altitude method (of determining latitude and time).

Gauss's variational equations - A set of six first-order differential equations relating the rate of change of each of the six orbital elements to the rectangular components of the disturbing function. The components are usually those of the moving trihedron i.e., the tangent, normal and binormal to the orbit.

gazetteer - An alphabetical list of names of places, giving for each name the kind of place (city, river, mountain, etc.) and the geographic or grid coordinates of the place. Some gazetteers also give the approximate populations of cities, town and countries. However, a gazetteer giving more than the minimal amount of information is usually called a geographical dictionary or geographical encyclopedia.

geländereduktion - A correction to gravity, needed to account for attraction by the terrain (topography). It is closely related to the topographic gravity correction and is sometimes used as a synonym for that correction.

generalization - The gradual loss of resemblance between a symbol on a map and the object represented by that symbol, as the scale of the map becomes smaller. For example, on a large-scale map, the streets of a town may be shown specifically and accurately by lines indicating the sides and widths of the streets, and the size and locations of buildings may be shown by properly placed, filled-in figures. On a medium-scale map of the same region, the symbols representing individual buildings will have disappeared and streets will be represented, if shown at all, by single lines. Only the size and extent of the town as a whole may be shown. On a small-scale map, the town will be represented merely by a dot or circle indicating the area or population of the town.

General Land Office - Formerly, an office of the United States government, a division of the Department of the Interior and originally constituted by Act of Congress in 1812, and having charge of all executive action relating to the public lands, including their surveying, sale, or other disposition and patenting. The General Land Office (GLO) and the U.S. Grazing Service were consolidated into the Bureau of Land Management (BLM), under the Department of the Interior, by the 1946 Reorganization Plan No. 3,403.

generation (n) - A particular, photographic positive or negative produced in one of a succession of stages of reproduction from an original negative. The original negative is called the first generation. The first positive produced from it is called the second generation product; the negative produced from this positive is called the third-generation product and so on. Quality deteriorates with each successive generation.

geocartography - Mapping of the Earth and terrestrial phenomena, in contrast to the mapping of extraterrestrial bodies.

geocentric (adj.) - Referred to the center of the Earth. The Earth has many different centers; which is meant must usually be inferred from the context. The following kinds of center are in common use: (a) the center of an ellipsoid representing the Earth's shape; (b) the center of an ellipsoid representing the geoid; (c) the Earth's center of mass; (d) the point at which the Earth's axis of rotation intersects the plane of the celestial equator.

geodesic (adj.) - Having a length greater than or shorter than the length of any other line in the same space between the same two points. Unless specified otherwise, the shortest length is usually meant. The word has more precise definitions framed to include lines in more general spaces than the ordinary Euclidian ones, but those are of interest primarily to mathematicians.

geodesic (n.) - That line, between two points, which is longer or shorter than any other line between those points. Also called an orthodrome, geodesic line or geodetic line. (The last term should be applied only to a geodesic on a rotational ellipsoid.) In geodesy, geodesic is understood to mean the shortest line, on a specified surface, between two points on that surface. The geodesic between two points on a plane is the segment of the straight line joining the points. The geodesic between two points on a sphere is the shorter arc of the great circle joining the points. The geodesic between two points on an ellipsoid is, in general, not a second-degree curve.

geodesic, spheroidal - A geodesic on a spheroid (rotational ellipsoid).

geodesic coordinate - SEE coordinate, geodesic.

geodesic distance - SEE distance, geodetic.

geodesy - (1) The science concerned with determining the size and shape of the Earth. This is essentially Helmert's definition of 1880. In practice, it involves determining, in some convenient coordinate-system, the coordinates of points on the Earth's surface. Each country originally set up its own coordinate system and developed its own surveying system. Most of these coordinate systems can now be referred to a common coordinate system for geodetic purposes, but many countries still adhere to their own systems for internal use. (2) The investigation of any scientific question connected with the shape and dimen-sions of the Earth. (3) That science which deals with the size and shape of the Earth, the Earth's external grav-itational field, the locations of points above, on, or under the Earth's surface, and, to a limited extent, the internal structure of the Earth. (4) That branch of surveying in which the curvature of the Earth must be taken into account when determining angles and distances. (5) That branch of surveying which is concerned with providing an accurate framework (i.e., set of coordinates, or distances and directions between points) in a country, to which surveys covering smaller regions made at different times and in scattered places can be connected without any fear of discontinuities at the junctions. (6) The art which locates points and determines the Earth's gravity field. Geodesy may be classified as lower geodesy (also called surveying) and higher geodesy (also called geodesy). Lower geodesy contains the techniques and instru-mentation of geodesy; higher geodesy contains the theory. Alternatively, lower geodesy may contain that part of geodesy which does not require a knowledge of the Earth's curvature, while higher geodesy does require such knowledge. Disciplines part of geodesy by the above definitions are often considered by their practitioners to be separate from geodesy e.g., cartography, photogrammetric mapping, and nautical charting. (7) The determination of the gravitational field of the Earth and the study of temporal variations such as Earth tides, polar motion, and rotation of the Earth.

geodesy, astronomic - SEE geodesy, astronomical.

geodesy, astronomical - The determination of longitudes, latitudes, and azimuths by observations of the directions of stars, planets, the Moon or other celestial bodies. Also called astronomic geodesy, astronomical geodesy, celestial geodesy and geodetic astronomy. Coordinates determined in this way are called astronomic coordinates or astro-geodetic coordinates. The most accurate values are obtained by observations on stars other than the Sun. Special techniques involving observations of eclipses, occultations or transits of the planets have been used with some success but have been replaced, for the most part, by methods involving observations on artificial satellites.

geodesy, celestial - (1) That branch of geodesy which uses observations on near celestial bodies (including artificial satellites of the Earth) to determine the size and shape of the Earth. (2) SEE geodesy, astronomical.

geodesy, coastal - That branch of geodesy which deals with the relationships between sea level, water level and elevations.

geodesy, ellipsoidal - The study of methods for solving geodetic problems on a reference ellipsoid. This is a more general term than spheroidal geodesy.

geodesy, four-dimensional - Geodesy which considers the natural changes which occur in the locations of points on the Earth.

geodesy, geometrical - That part of geodesy which does not depend on a knowledge of the Earth's gravity.

geodesy, gravimetric - That part of geodesy which concerns (a) measurement of the magnitude and/or direction of gravity, and (b) determination of the geoid from such measurements. Also called physical geodesy. The direction of gravity at a point is related, for geodetic uses, to the direction of the normal (to the reference spheroid) at that point. The difference between the two directions is called the deflection of the vertical. Determining this difference may be considered part of gravimetric geodesy or of astronomic geodesy.

geodesy, higher - (1) That part of geodesy which gives the theory on which the techniques and instrumentation of geodesy are based. Also called mathematical geodesy and theoretical geodesy. (2) That part of geodesy which explicitly involves the Earth's curvature i.e., which requires that the Earth be represented by a curved surface. Also called advanced surveying, particularly in American textbooks. (3) That part of geodesy which is concerned with determining the size and shape or figure of the Earth.

geodesy, intrinsic - A geodetic theory, introduced by Antonio Marussi, in which the coordinate system is completely defined by the Earth's gravity field and rotation, and geodesy is developed using only that coordinate system. The reference ellipsoid, geoidal heights and similar concepts are therefore not relevant in intrinsic geodesy. The theory was initially formulated using a special algebra devised by Marussi. It was translated by M. Hotine into the language of tensor analysis.

geodesy, marine - That part of geodesy which concerns the shapes of the upper surfaces and bottoms of the oceans and similar bodies of water, and with the locations of points on those surfaces. By convention, marine geodesy concerns itself only with the average positions of the upper surfaces. The instantaneous positions are the concern of navigation and physical oceanography. The oceanic bottoms are also the concern of oceanography and their determination is called hydrographic surveying or hydrography in that science. Other, broader definitions of marine geodesy have been proposed.

geodesy, mathematical - (1) The theoretical part of geodesy, as contrasted to the practical part i.e., surveying. Also called higher geodesy and geodesy. (2) That part of geodesy which concerns the geometric rather than the physical aspects.

geodesy, photogrammetric - Geodesy by photogrammetric methods.

geodesy, physical - SEE geodesy, gravimetric.

geodesy, spherical - That part of geodesy which deals with problems on a sphere used as a reference surface. Alternatively, geodesy which assumes that a sphere is adequate as a reference surface.

geodesy, spheroidal - (1) That part of geodesy which deals with problems on a rotational ellipsoid used as a reference surface. (2) The study of the geometry of the terrestrial ellipsoid and representation of important parts of its surface on the sphere and plane.

geodesy, three-dimensional - Geodesy in which the size and shape of the Earth's surface (or parts thereof) are determined primarily by geometric methods and gravity is not explicitly involved. Geometric satellite geodesy is a form of three-dimensional geodesy. SEE ALSO satellite geodesy.

geodesy, tri-dimensional - SEE geodesy, three-dimensional.

Geodetic Reference System 1967 (GRS 67) - The following set of numbers adopted in 1967 by the International Union of Geodesy and Geophysics:

a 6 378 160 meters
GM 398 603 x 10^9 m^3/s^2
$J_2 (-C_{20})$ 0.001 082 7

a is the length of the semi-major axis of an equipotential, rotational ellipsoid, GM is the product of the gravitational constant G by the mass M of the Earth and C_{20} is the coefficient of the second-degree Legendre function in the representation V of the Earth's gravitational potential V = $(GM/a) \Sigma_n [(a/r)^{n+1} \Sigma_n \{(C_{nm} \cos m\lambda + S_{nm} \sin m\lambda) P_{nm} (\sin \phi)\}$, in which r is geocentric distance, ϕ is geocentric latitude and λ is geodetic longitude.

Geodetic Reference System 1980 (GRS 80) - The following set of numbers adopted in 1979 by the International Union of Geodesy and Geophysics:

a 6 378 137 meters
GM 398 600.6 x 10^9 m^3/s^2
$J_2 (-C_{20})$ -0.001 082 63
ω 729 211.5 x 10^{-11} radians/s .

ω is the angular rate of rotation of the ellipsoidal body; the other symbols have the same meanings as those used for the Geodetic Reference System 1967.

geodynamics - (1) The science dealing with natural changes of the Earth's form or structure. (2) The science dealing with the forces and processes of the interior of the Earth. The dynamics of the atmosphere is usually considered part of meteorology; the dynamics of the oceans is dealt with by oceanography and, in particular, by physical oceanography. The dynamics of the lithosphere has, in the past, been called physical geology or, where only the surface is of interest, geomorphology, physiography and physical geomorphology. The complement to geodynamics is geostatics, the science dealing with balanced states of the Earth.

Geographer's Line - The surveyed line beginning at the point where the boundary of Pennsylvania intersected the Ohio River, and intended to run due west (at an azimuth of 270° from astronomic north). It was surveyed by Thomas Hutchins, a geographer of the USA and was done according to the plan provided for in the Ordinance of 1785.

geographic (adj.) - Related to the Earth or the life on the Earth. The form generally favored is geographical. However, geodesists prefer the form geographic in terms such as geographic coordinates, geographic position, geographic north, etc.

geographical (adj.) - SEE geographic.

geography - (1) The science which concerns itself with the Earth or the life on the Earth. (2) The science which concerns itself with all aspects of the Earth's surface, including its natural and political divisions, the distribution and differentiation of regions, and, often, man in relation to his environment.

geoid - (1) That equipotential surface of the Earth's gravity field which best fits, in a least-squares sense, mean sea level. Also called figure of the Earth and, infrequently, shape of the Earth. In practice, the average position of mean sea level and a corresponding average over the time-varying potential must be used. (2) That equipotential surface of the Earth's gravity field which coincides with the surface the oceans would have if they were motionless and affected only by the Earth's gravity field. Sometimes called ideal sea level by oceanographers. This is the original definition given by Listing (1872). (3) The equipotential surface (of the Earth's gravity field) coinciding with mean sea level in the oceans and extended under the continents. The definition is that of Gauss, who specified, however, mean sea level of the quiet ocean. The definition is probably equivalent to Listings [definition (2) above]. (4) That equipotential surface (of the Earth's gravity field) through a given point near mean sea level, which would exist if only the rotation of the Earth and the Earth's gravitational potential affected the potential. (5) The equipotential surface consisting of those points for which the potential W is given by $W = \int^M (1/s)dM + \frac{1}{2} A \omega r^2$, in which W is given a unique, defined value, dM is an element of the mass M of the Earth, s is the distance from the element dM to a point on the surface being defined, r is the distance of that point from the Earth's axis of rotation and ω is the rate of rotation. The definition was given by Bruns. (6) The surface of mean sea level extended continuously (e.g., through frictionless, narrow channels or tunnels) through the continents. This definition is commonly used where a pictorial concept the geoid is wanted and a rigorous definition is not needed. (7) An equipotential surface which coincides with mean sea level on the open ocean. The definition is inconsistent with the fact that sea level is not an equipotential surface. SEE ALSO Bomford's geoid, 1963; Brillouin's geoid; flattening; satellite geoid.

geoid, astro-geodetic - An approximation to the geoid, obtained by combining measurements of distance, angle or direction, and elevation with astronomical determinations of longitude and latitude.

geoid, compensated - (1) A mathematical surface derived from the geoid by applying values of the deflection of the vertical adjusted for the effects of topography and isostatic compensation. The data for adjustment are obtained from measurements on topographic maps; the effect is then computed in accordance with the assumptions made about the depth and distribution of isostatic compensation. A method for doing this was developed by Hayford, who accepted Pratt's theory of isostasy. (2) SEE co-geoid.

geoid, gravimetric - An approximation to the geoid, as determined from gravimetric data.

geoid, ideal - The mathematical surface which correspond to a hypothetical Earth in the same way that the geoid corresponds to the real Earth.

geoid, isostatic - (1) A mathematical surface derived from the spheroid of reference by applying values of the deflection of the vertical adjusted for the effects of topography and isostatic compensation. The adjusted values of the deflection used in obtaining the isostatic geoid are similar to those used for obtaining the compensated geoid, but are of opposite sign. If the theory and assumptions about the nature of isostasy were correct and exact, and if there were no gravity anomalies, the isostatic geoid would agree with the geoid. (2) A surface lying above (or below) a theoretical, equipotential surface represented by a formula for the geopotential, by the same distance as the geoid lies above the co-geoid.

geoid, marine - An equipotential surface which would coincide with the average level of the oceans if the waters were at rest. This definitions assumes that the average level of water at rest must be an equipotential surface. It is therefore equivalent to defining the marine geoid as an equipotential surface which would coincide with the average level of the oceans if the oceanic surface were an equipotential surface.

geoid, regularized - (1) That equipotential surface which results by removing from the Earth all masses outside the geoid and then placing these masses back inside the geoid in such a way as to give an equipotential surface differing little from the geoid. The new surface is not the geoid but is close to it. (2) SEE co-geoid.

geoid anomaly - A departure from constant geoidal height.

geoid contour - SEE contour, geoidal.

geoid determination - SEE astro-geodetic method; astro-gravimetric method; gravimetric method.

geoid height - SEE height, geoidal.

geoid separation - SEE height, geoidal.

geoid undulation - (1) SEE undulation, geoidal. (2) SEE height, geoidal.

geokinetics - Local and global motion of the Earth or sea, the measurement of such motion, and its effect on and isolation from the functioning of precise instruments and devices using gyroscopes or other such inertia-dependent devices. The word seems to be part of the jargon of the instrument-making industries and not much used outside of that group.

geology - The study of the Earth - the materials of which it is made, the processes that act on these materials, the products formed, and the history of the planet and the life on it since its origin. Geology considers the physical forces which act on the Earth (physical geology or geophysics), the chemistry of the materials of which the Earth is made (geochemistry), the shape of the surface (geomorphology) and the biology of past life as evidenced by fossils. It differs from geography, of which it is a part, in not being concerned with present-day life on the Earth.

geology, engineering - SEE engineering geology.

Geomagnetic Electrokinetograph - An instrument for measuring the electromotive potential (EMF) generated in an oceanic current by the movement of the water across the Earth's lines of magnetic force. The instrument consists of a pair of electrodes immersed in the sea 200 meters apart and joined by cables (ordinarily three times the length of the ship) to a sensitive, Wenner-type potentiometer with recorder attached. The instrument is usually referred to as a GEK. It is used in depths greater than 200 meters (100 fathoms).

geomagnetism - (1) The magnetic phenomena, considered as a whole, exhibited by the Earth. Also called terrestrial magnetism. (2) The study of the magnetic field of the Earth. The magnetic field affects many geodetic instruments such as the magnetic compass, the automatic leveling instrument, and electromagnetic instruments used for measuring distance. There is a theoretical relationship between the intensity of the magnetic field in a particular direction and the gradient of the gravity field in that direction.

geometer - A surveyor. The term is no longer used in English except when it is used to denote a surveyor in ancient times e.g., an Egyptian or Roman surveyor. It may be used if there is a reason to distinguish a land or similar surveyor from other types; the English word surveyor originally meant a steward, a person who kept track of property.

geometronics - The art consisting of mapping, (or cartography) photogrammetry and surveying. The term was suggested in the 1950's by Walter S. Dix in an attempt to find a single term covering both cartography and surveying.

geometry, affine - A geometry having the usual postulates for incidence of points, lines, and planes, with the additional postulate that: if a point p is not on line L, there exists one and only line M such that M contains p and is parallel to L.

geomorphology - That branch of physical geology which deals with the form of the Earth, the general configuration of its surface, and the changes which take place in the evolution of land forms. Formerly called physiography.

geon - (1) An entity consisting of an electromagnetic field held together by the gravitational force arising from the energy, and hence mass, of the field. (2) A mathematical surface coinciding with the physical surface of the lithosphere.

geop - SEE surface, level. J. de Graaf-Hunter used the term (1958) to designate a level surface of the actual Earth, as distinguished from a spherop, which denotes a level surface defined by some standard gravity formula.

geophone - A device for listening to faint sounds transmitted through the earth. The sounds are those produced by explosion or by mechanical devices.

geophysics - (1) The science concerned with the forces which act in and on the Earth and with the changes caused by these forces. (2) The study of the Earth by quantitative physical methods. (3) The physics of the Earth and its environment. Geophysics is a subdivision of geology and commonly is divided into two disciplines: geostatics, which concerns the Earth as a body in equilibrium; and geodynamics, which concerns the Earth as a body changing in response to applied forces. Although special terms have been invented to denote the application of geophysical methods to the Moon and other planetary bodies, the present tendency is to apply geophysics to all such studies.

geopotential - (1) The sum U of the gravitational potential V_g of the Earth at a point P and the rotational potential V_c at the same point: $U = V_g [= G \int_S (\rho/r)dS] + V_c [= \omega^2 p^2/2]$ in which G is the gravitational constant, ρ the density at a point Q in the Earth, dS an element of volume S, r the distance from P to Q, p the distance from P to the axis of rotation and ω the rate of rotation of the Earth. Integration is over the entire volume S of the Earth, including the atmosphere. The gravitational potential V_g is commonly denoted simply by V. In French and German literature, W is commonly used instead of U. This definition is preferred by geodesists. (2) The negative of the sum of the gravitational potential V_g and the rotational potential V_c: $U = -(V_g + V_c)$. This definition is much used by physicists. (3) The potential energy of a unit mass relative to the geoid, numerically equal to the work which would be done in lifting the mass from the geoid to the elevation at which the mass is actually located.

geopotential anomaly - SEE depth anomaly, dynamic.

geopotential elevation - (1) The integral $\int_0^{H'} (g/g_o)dH$ taken along a line between the geoid and a point P of orthometric elevation H', where g(H) is the acceleration of gravity on the geoid and on the same vertical as P. Also called geopotential altitude and geopotential height. (2) The same as the preceding definition but using 980 cm/s² instead of g_o. (3) The quantity K defined by $K = rH/(r+H)$, in which r is the radius (approximately 6 356 766 meters) of the Earth at latitude 45° 32' 40" N and H is the elevation. (4) The work done in lifting a unit mass from the geoid to the point of interest, through a region in which the acceleration of gravity is uniformly 9.80665 m/s². (5) The elevation, of a given point, in units proportional to the geopotential at this elevation.

geopotential height - SEE geopotential elevation.

geopotential meter - The dynamic meter divided by 10 g_o, in which g_o is some standard value of gravity.

geopotential number - (1) The sum, from i = 1 to i = I-1, of the quantities ΔH_i gi obtained in going from a level surface of reference, through a sequence of level surfaces numbered

from i = 1 to i = I-1, to the level surface through the point for which the equipotential number is wanted. Successive surfaces S_{i-1} and S_i are separated by the very small distance ΔH_i, expressed in meter and g_i is the average value of the acceleration of gravity, expressed in kilogals, along the distance ΔH_i. The equipotential number at a point is approximately equal to the elevation of that point, in meter. (2) The quantity given by the integral
$$\int_1^2 g\, dH,$$
in which g is the acceleration of gravity along the vertical joining points 1 and 2, and H is the vertical distance between those two points. (3) The difference between the actual gravity-potential at a point and the gravity-potential on the geoid.

geopotential number, normal - (1) The quantity WN given by the integral $WN = \int_{P_o}^{PN} \tau_\phi\, dH$, in which H is the vertical distance from point P_o on a spherop through the zero-elevation mark on a tide gage, to a spherop through the point PN being considered and τ_ϕ is the value of the acceleration of gravity at latitude ϕ, calculated from a standard gravity formula. H is measured in the direction of the gradient to the spherop. Also called a spheropotential number. (2) SEE spheropotential number.

geopotential surface - (1) A surface on which the gravity potential of the Earth is defined, specified, or known. (2) SEE surface, level.

geopotential topography - The topography of any surface as represented by lines of equal geopotential. Also called absolute geopotential topography. The lines of equal geopotential represent the contours in which the actual surface intersects the surfaces of constant geopotential and are spaced at equal intervals of dynamic height.

geopotential topography, absolute - SEE geopotential topography.

geopotential unit - (1) A quantity expressed in kilogal- meters is said to be expressed in geopotential units. It is equal to the difference in gravitational potential of two points separated by 1 meter when the gravitational field has a strength of 10 meters per second squared and is directed along the line joining the points. (2) The quantity 1 kilogalmeter.

GEOREF - SEE GEOREF grid.

GEOREF coordinate system - An alphanumeric coordinate system adopted by the U.S. Air Force for designating the locations of points on the Earth's surface. The name is derived from The World Geographic Reference System.

GEOREF grid - A military grid enabling users to specify the location of a point in terms of certain letters and numbers. Also called GEOREF. It is used on aeronautical charts.

geosphere - (1) The solid and liquid portions of the Earth. I.e., the lithosphere plus the hydrosphere; all of the Earth except the atmosphere. (2) The solid part of the Earth. Also called the lithosphere. (3) A sphere having its center at the center of mass of the Earth and a radius such that the sphere is below the surface of the Earth at all points. It is understood that the radius mentioned is the largest radius. (4) The totality of all those regions contributing to the structure of a region: the atmosphere, hydrosphere, lithosphere, biosphere and anthroposphere.

geostatics - The science concerned with the static relations between natural forces in the Earth. It is that part of geo- physics which deals with the forces and the effects of those forces in the Earth considered as a body in balance.

ghost - Any faint image, on a negative or plate, which barely prints or fails to print.

gilbert - The magnetomotive force resulting from the passage of a current of 4π abamperes through one turn of a coil.

gimbal - A ring supported externally by a pair of trunnions and having internally another pair of trunnions placed so that the axis of the internal pair is perpendicular to the axis of the external pair. The internal pair supports bearings and an axle to which is fastened the object of interest. If the object is initially at rest, it will remain at rest regardless of any rotation of the framework in which the gimbal is placed, except for a rotation in the plane of the ring and except for the amount of motion transmitted through the trunions and bearings by friction. A second gimbal is often nested within the first in such a way that the object remains at rest regardless of the direction in which the supporting framework rotates. Gimbals are used in geodesy and in navigation to support a magnetic or gyroscopic compass so that the indicator remains horizon- tal. They are also used in inertial surveying systems to allow the gyroscopes to keep fixed orientations.

g-inverse - SEE matrix.

gisement - The angle between the grid meridian and the geodetic meridian. Gisement is sometimes called the declination of grid north and reckoned east and west from geodetic north. The term is used principally, in the USA, with military grids and is not used with the State Plane Coordinate Systems, in which the corresponding angle on the grid used with the transverse Mercator map projection is the conver- gence of the local meridian with the central geodetic meridian. In the grid associated with the Lambert map projection, the angle is known as the mapping angle. It was formerly designated by θ and called the theta angle.

gisement method - SEE bearing method.

glacier - A large mass of ice formed, at least in part, on land by compaction and recrystallization of snow, and creeping slowly downward or outward in all directions under the force of its own weight, and surviving from year to year. The term covers ice sheets and ice shelves which are fed in part by ice formed on land, as well as the small glaciers formed in mountainous valleys. It is sometimes restricted to denote only glaciers occurring in mountainous valleys. Such glaciers are otherwise distinguished as Alpine glaciers, valley glaciers or ice streams. The first systematic mapping of glaciers was done by L. Agassiz about 1840, using a plane table. S. Finsterwalder used tachymetry and triangulation about 1886 for mapping glaciers, and introduced photogrammetric surveying in 1889 for the same purpose.

glare - SEE hotspot.

glare stop - (1) A blackened diaphragm placed at an image of the aperture stop, in an optical system, to block out stray radiation (i.e., prevent unwanted light from outside, or scattered light from inside, the containing structure from reaching the receptor). (2) In general, any blackened diaphragm placed in an optical system to block unwanted light from the receptor.

glass screen, crossline - SEE screen, halftone.

glebe - The land belonging to or yielding revenue to a parish church or ecclesiastical benefice.

Glennie's gravity anomaly - (1) The quantity Δg_G obtained by subtracting, from a Bouguer gravity anomaly Δg_B, Bullard's term B, a topographic gravity correction $\delta g_t'$ for Hayford zones 18 to 1, and an isostatic gravity correction $\delta g_{is}'$ for the same zones: $\Delta g_G = \Delta g_B - B - \delta g_t' - \delta g_{is}'$. Also called a combined gravity anomaly. Bullard's term converts the Bouguer gravity anomaly to a modified Bouguer gravity anomaly, substituting a spherical cap for the Bouguer plate. This takes care of Hayford zones A through O. The last two corrections then account for the other zones. (2) The magnitude of gravity as measured and the magnitude of gravity as calculated, by a gravity formula, for the surface of the terrestrial ellipsoid.

glint - Angular variations in the direction from which radio waves arrive at an antenna, and caused by interference at the object.

Global Positioning System (GPS) - The NAVigation Satellite Timing and Ranging (NAVSTAR) GPS is a passive, satellite-based, navigation system operated and maintained by the Department of Defense (DoD). Its primary mission is to provide passive global positioning/navigation for land-, air- and sea-based strategic and tactical forces. A GPS receiver is simply a range measurement device: distances are measured between the receiver point and the satellites, and the position is determined from the intersections of the range vectors. These distances are determined by a GPS receiver which accurately measures the time it takes a signal to travel from the satellite to the station. This measurement process is similar to that used in conventional pulsing marine navigation systems and in phase comparison electronic distance measurement land surveying equipment.

globe - (1) A spherical body. (2) The Earth or the surface of the Earth. (3) A sphere on which a map is shown. In astronomy and geodesy, globe usually denotes a sphere on which is a map of the Earth (terrestrial globe) or of the heavens (celestial globe). SEE ALSO horizon ring; meridian ring; time dial.

globe, celestial - A sphere on which the locations and brightness of stars and other celestial bodies are shown by symbols. The outlines or boundaries of constellations usually are shown also. The celestial globe commonly has a blue background on which stars are indicated by black or white dots whose sizes are proportional to the stellar magnitudes. Great circles on the globe indicate meridians of right ascension; small circles parallel to the equatorial circle indicate parallels of declination.

globe, generating - (1) The sphere upon which lie the points to be mapped, by perspective projection, onto the plane. The radius of the generating globe bears the same relationship to the radius of the Earth as does a distance on the map to the corresponding distance on the repre-senting ellipsoid (as indicated by the scale of the map). (2) An ellipsoid whose characteristic dimensions bear a specified ratio to the corresponding dimensions of an ellipsoid representing the Earth; the ratio is chosen to be close to the scale at which the Earth is to be mapped.

globe, terrestrial - A sphere on which the features of the Earth's surface are shown, in their relationships to one another, by symbols and reference lines. It is usually made by printing an map or sections (gores) of a map on a flat sheet of paper, cutting the map or sections to the proper form, fitting the pieces to the globe, and cementing them in place. Terrestrial globes are usually between 20 to 100 cm in diameter. Larger terrestrial globes up to several meters in diameter have been made and show the Earth's surface in sculptured relief.

globe gore - SEE gore (2).

gnomon - An object used to cast a shadow of the Sun for determining the time or astronomical north.

Godson's constant - The quantity used, in reduction of distances measured with radio waves, to account for the effect of atmospheric refraction on the measured length.

Godson's method (of reducing measured distances to mean sea level) - A method of reducing distances, measured using radio waves, to mean sea level (or the geoid) by combining the reduction with the making of corrections for index or refraction, curvature of path, slope, chord to arc and location of sea level.

goldenrod paper - A sheet of yellowish-orange paper used for preparing flats from negatives. Also called masking paper. Goldenrod paper serves as a base for drawing the layout and attaching the negatives. When openings are cut in it, the remainder serves as a mask because the yellowish-orange color prevents most of the light used for exposure from getting through.

Golofnin Bay Datum - The horizontal-control datum defined by the following coordinates of the origin and azimuth (clockwise from South) on the Clarke spheroid of 1866; the origin is Golofnin Bay longitude station:
```
longitude of origin                 162° 52' 13.700" W.
latitude of origin                   64° 27' 11.461" N.
azimuth from origin to azimuth mark  180° 00' 42.9" .
```
The above coordinates and azimuth are based on astronomical observations made in 1899; the longitude was determined by the chronometric method.

gon - 1/400 of a circumference. Also called grad (q.v.), new grad and Neugrad. Because the term grad has been used in continental Europe for both the degree (1/360 of a circumference) and for the Neugrad (1/400 of a circumfer-ence), the term gon has been legally adopted in Germany to replace Neugrad. Gon is an accepted angular unit in SI.

gon, square - SEE grad, square.

goniasmometer - A compact surveying instrument for measuring horizontal angles and bearings, consisting of a vertical cylinder divided horizontally into two parts: the lower edge of the upper part is graduated in circular measure and revolves on the lower part, which carries a vernier on its upper edge. A magnetic needle is centered in the upper part, which is also provided with slits or a telescope for sighting. The goniasmometer has been used in France for the same purposes as a surveyor's compass or transit.

goniometer - In general, any one of various instruments used for measuring angles, as in crystallography, anthropology and metrology. In surveying in particular, an instrument such as a theodolite for measuring horizontal angles.

Goode's homalographic map - The interrupted map obtained by using Mollweide's map projection to map regions on the spheroid individually onto the plane in such a way that the maps of the individual regions match along the Equator and the centers of the maps are on five or six different meridians (the exact number depends on the regions, such as land or water, to be emphasized. Also called Goode's homolograhic map, homalographic map and Goode's homalographic map projection. It was first described by J.P. Goode in 1916.

Goode's homolographic map - SEE Goode's homalographic map.

Goode's polar equal-area map - A map produced by applying drawing three meridians so that each passes through one of the three major continental regions (North and South America, Europe and Africa, Asia and Australia) dividing these regions from each other by suitable meridians, and applying Werner's map projection to each lobe thus obtained. New meridians are chosen as desired to keep the central line approximately in the center of the region.

gore - (1) A small, irregular or triangular piece of land usually occurring as a gap between adjoining pieces of land or as a piece common to two such pieces of land. Also called a gap or hiatus, especially when no connotation of shape is wanted. The gore may result from an error in surveying the land, from an error in recording the titles, or from using different, independent origins for the surveys leading to the titles, etc. (2) A lune-shaped part of a map, which may be fitted to the surface of a globe with a negligible amount of distortion. Also called a globe gore. A terrestrial or celestial globe may be constructed by covering a sphere with a complete set of gores each having the form of a lune of such narrow width that it may be mounted on the sphere without appreciable stretching or shrinking. (3) A map of the narrow region between two meridians close together.

gorge - A narrow, deep valley with nearly vertical, rocky walls, inclosed by mountains. It is smaller than a canyon and more steep-sided than a ravine.

Goulier collimator level - A leveling instrument designed by G.M. Goulier to keep the line of sight level by having it pass through a prism mounted on a pendulum having a Cardan suspension. It was a forerunner of the present-day, automatic leveling instrument.

Goulier prism - A pentagonal prism having four dihedral angles of 112.5° between faces and one dihedral angle of 90° between faces.

government corner (USPLS) - SEE corner.

government lot - The fractional part of a section (of a public-land survey) protracted by office procedures from field notes and designated by boundary limits, location and (not always) number on the township plat. A typical U.S. patent description could be: Government Lot 1, Section 2, T 112N, R17W of 5th Principal Meridian.

GPS - SEE Global Positioning System.

Graaf-Hunter gravity reduction - (1) The process of applying, to a measured or calculated value of gravity, a quantity δ_H defined as $\delta_H = 2\pi G \rho (H-H') - 2\tau (H-H')(1/R - 1/R') + \delta g_t$, in which δg_t is the topographic gravity correction appropriate to the representation used for the real Earth; H and H' are the elevations of the representing surface and the terrestrial ellipsoid, respectively; R and R' are the radii of curvature of the corresponding surfaces; ρ is the density of the Earth and τ is the theoretical value of the acceleration of gravity appropriate to the terrestrial ellipsoid used. The representation involved is obtained by averaging elevations over regions 150 km in radius. This gravity reduction does not involve the concept of isostasy. SEE ALSO Model Earth. (2) The quantity δ_H defined in the previous definition.

graben - A block of rock or sediment dropped between two approximately parallel faults, with the length of the block being much greater than the width and parallel to the bounding faults.

grad - (1) The angle subtended, at the center of a circle, by an arc of 1/400-th of the circumference. Also called centesimal grade and Neugrad (German for new grad), although the latter term is now replaced by gon. (2) A degree. This usage was common in continental Europe but is so no longer. It has never been common in America.

grad, square - A unit of measure of solid angle, defined as the solid angle which intercepts, on a sphere of radius 1, an equiangular quadrangle with sides each 1 grad long. Also called the square gon.

gradation - The range of tones, in a picture, from the brightest highlights to the deepest shadows.

grade - (1) The degree of inclination of an inclined surface; the rate at which a sloping surface (a slope) descends or ascends. For example, a 1 to 100 grade is a slope on which the elevation changes by 1 meter per 100 meters of horizontal distance; this would also be called a 1-percent grade. SEE ALSO gradient. (2) A difference in elevation above or below a reference surface. This usage is limited mostly to surveying for construction. (3) The continuously descending bed of a stream just steep enough for the current to flow and transport its load of sediment. (4) An ascending or descending stretch of road.

grade, broken - The change in slope which occurs when the middle point of a surveyor's tape is not in a straight line with the ends. If the support for the middle of the tape is not in a straight line with the supports for the ends, that fact is noted as broken grade at. . ., naming the particular length which contains the broken grade.

grade, centesimal - SEE grad (1).

grade, hydraulic - The elevation, above a level surface used as reference, of the surface of the liquid in a conduit.

grade correction - A correction to a distance measured with surveyor's tape, between two points on a slope, to obtain the equivalent horizontal distance between the vertical lines through the points. Also called slope correction, correction for slope and correction for inclination. Because the vertical lines through the end points are not parallel, the correction should be such that the distance at the average elevation will be obtained. In practice, this is done by considering each measured distance a separate line and correcting it individually. In measuring base lines, the difference in elevation of the ends of the tape is used in computing the correction, rather than the inclination (of the line) expressed as an angle. In effect, this treatment applies the correction to the nominal length of the tape in determining its effective length in measuring the distance.

grade intersection - An intersection at which the two roads are at the same level. SEE ALSO intersection at grade.

grade line - (1) The slope in the longitudinal direction of a roadbed. It is usually expressed in percent, which is the number of units of change in elevation per 100 units of horizontal distance. (2) The highway's profile.

grade stake - A stake upon which a difference of elevation has been marked. The top of the stake may be driven to the desired difference of elevation or, more commonly, a horizontal mark is made with arrow or crow's foot to mark the difference. If the difference is below the reference level, the distance is marked C for cut. If it is above the reference level, the distance up to the reference level is marked F for fill. Grade stakes are frequently placed on an offset line, with the amount of offset and its direction from the line of survey indicated on the side of the stake.

gradient - (1) The gradient of a scalar v, in a particular Cartesian coordinate system, is the vector [$\delta v/\delta x$, $\delta v/\delta y$, $\delta v/\delta z$]. For any other coordinate system, the gradient is defined as the vector obtained by transforming the above components into the desired coordinate system. The gradient is perpendicular, at each point, to the curve or surface of constant v. (2) A grade expressed as the tangent of the angle of inclination. Also called the slope and percent of slope. For example, a grade of 1 meter per 100 meters is a gradient of 0.01. (3) (photography) The slope of the characteristic curve at any point.

gradient, gravitational - (1) The change in the value of gravitational force per unit distance. (2) SEE gravity gradient.

gradient, hydraulic - (1) The slope of the line drawn between two points on the surface of and in the direction of flow of the liquid in a conduit. (2) The rate of change of pressure head. (3) The ratio of the loss in the sum of the pressure head and position head to the distance of flow. In open channels, it is the slope of the water's surface and is frequently considered parallel to the invert. In closed conduits under pressure, it is the slope of the line joining the elevations to which water would rise in pipes freely vented and under atmospheric pressure. A positive slope is usually one which drops in the direction of flow. In dikes, embankments, earth-dams and earth-structures retaining water, it is the line of saturation to the earth.

gradient boundary - A line located midway between the lower level of the flowing water (of a stream) that just reaches the cut-bank and the higher level of it that approaches but does not overtop the cut-bank. This kind of boundary was created by judicial action in the settlement of the boundary between Oklahoma and Texas.

gradient current - A current associated with horizontal differences of pressure in the ocean. In particular, a geostrophic current.

gradienter - An attachment, to an engineer's transit, with which an angle of inclination is measured in terms of the tangent of the angle instead of in degrees and minutes (i.e., instead of in angular measure). A gradienter may be used for determining horizontal distances.

gradient speed - The film speed of a photographic material, determined on the basis of the exposure corresponding to a particular gradient of the characteristic curve.

gradient tint - SEE tint, hypsometric.

grading plan - A plan showing (a) the original contours of the ground and (b) showing, superposed, contours of the highway, subdivision or other embankment or excavation to be completed. The superposed contour lines are shown connected, at the edge of the limits set for construction, to the contour lines of the original contours.

gradiometer - An instrument which measures the rate of change of some physical quantity with distance, in some specified direction or set of directions. The quantity measured is usually specified as part of the name of the instrument, as, e.g., gravity gradiometer.

gradiometer, vertical - A torsion balance designed to be particularly sensitive to vertical differences of gravity.

gradual and imperceptible (adj.) - A term used to describe changes in riparian lands which bring such changes within the scope of the doctrine of accretion and erosion. The test of what is gradual and imperceptible has been held to be that, Though the witnesses may see, from time to time, that progress has been made, they could not perceive it while the progress was going on.

graduation - (1) SEE division. (2) SEE smoothing.

graduation error - An error in the graduation of the scale of an instrument.

grain - A minute particle of silver resulting from the chemical reduction of silver salts when an exposed, light-sensitive photographic emulsion is developed.

graining - The roughing, as by grinding or scouring, of the surface of a metallic plate to be used in a printing press, to increase the area of the surface and to improve the ability of the surface to absorb water.

grant - (1) A thing or property granted; a gift. In particular, (2) A tract of land, a monopoly or the like granted by the government.

grant deed - A deed conveying the fee title of the land described and owned by the grantee. If, at a later date, the grantor acquires a better title to the land conveyed, the grantee immediately acquires (after rights) the better title without formal documents. In some states, by law, the grantor warrants the deed against acts of his own volition.

grantee - A person to whom property is transferred by deed or to whom property rights are granted by a trust instrument or other document.

grantor - A person who transfers property by deed or grants property rights through a trust instrument or other document.

granularity - (1) The amount of clumping-together of silver grains resulting from chemical development. (2) A measure of the average size of clumps of silver resulting from chemical development.

graticule - (1) A network of lines, on a map, representing geodetic meridians and geodetic parallels of latitude. Also called a grid or a planisphere. Equivalently, the set of lines, in a plane, resulting from applying a map projection to lines of constant longitude and constant latitude on an ellipsoid. (2) An imaginary network of meridians and parallels on the surface of the Earth or other celestial body. In some parts of the world, a graticule is also referred to as a grid. However, the term grid is more usually reserved to mean a more or less arbitrary network of regularly spaced lines (usually forming a rectangular Cartesian or, more rarely, polar coordinate system).

graticule, azimuthal equal-area - The graticule resulting by use of the azimuthal equal-area map-projection. Also called a Lambert equal-area polar grid.

graticule, fictitious - A graticule whose lines represent fictitious meridians and fictitious parallels of latitude.

graticule, oblique - A graticule based upon an oblique map projection.

graticule, transverse - (1) A graticule created by a transverse map projection. (2) A graticule representing fictitious longitudes and latitudes and created by a transverse map projection.

graticule convergence - SEE projection convergence.

grating - Any set of parallel lines placed in the path of a beam of electromagnetic waves to diffract the waves and create a diffraction pattern. The two principal kinds of gratings are (a) transmission gratings and (b) reflection (or scattering) gratings. Transmission gratings for optical wavelengths are made by photographing a set of closely-spaced, parallel lines or by removing, in closely-spaced, parallel lines, material from an opaque coating on a transparent base such as glass or plastic. At radio wavelengths, a grating consists of a series of closely-spaced, parallel wires set in a frame. Reflection gratings are made by cutting closely-spaced, parallel grooves into a suitable material such as metal or glass. Both gratings and prisms are used to produce spectra. However, the grating can produce a longer and more detailed spectrum and can act over a range of wavelengths limited principally by the closeness with which the lines can be ruled and the thickness of the lines. A prism's action is limited principally to the optical wavelengths, although special materials do allow the action to be shifted into the infrared or ultraviolet wavelengths. A disadvantage of the grating is that the grating produces a whole sequence of similar spectra, all similar but progressively wider and fainter. This disadvantage is partly overcome by shaping the ruled grooves to divert more radiation into a particular spectrum of the sequence. SEE ALSO objective grating.

grating, objective - SEE objective grating.

Gravatt level - Originally, a dumpy level having the spirit level mounted on top of a short telescope with a large objective lens. Later, the telescope was mounted on wyes. This is said to be the original dumpy level (a distinction also claimed for the Troughton level.)

Gravatt leveling rod - A self-reading leveling-rod on which graduations are marked by rectangles each 0.01 foot high. The rectangles at the tenth-of-a-foot marks are longer; those at the half-tenth marks are identified by dots. The rod is no longer in general use.

gravel - SEE sand.

graver - A tool for cutting through the plastic surface of a printing plate without cutting into the material under it. SEE ALSO scriber.

graver, straight-line - A variation of the rigid-tripod graver, designed so that the scribing point, the vertical vane, and one supporting leg lie on one line. It is used with a straight-edge for drawing long, straight lines.

gravimeter - (1) An instrument for measuring either (a) the force or acceleration of gravity at a point, or (b) difference in the force or acceleration of gravity between two points. Also called a gravity meter. A gravimeter which measures gravity at a point is called an absolute gravimeter; one which measures the difference in gravity at two points is called a relative gravimeter. Gravimeters are also classified as static gravimeters, which essentially measure the force of gravity by balancing that force against the force in an extended spring or twisted thread, and dynamic gravi-meters, which measure the acceleration of gravity by measuring the rate of fall of a small body. (2) SEE gravimeter, spring-type. SEE ALSO drift (of a gravimeter); Haalck gravimeter; sea gravimeter; tare (of a gravimeter).

gravimeter, absolute - A gravimeter designed to measure the value of gravity at a station.

gravimeter, astatic - A gravimeter in which the sensing mass is placed in a position of nearly unstable equilibrium so that a small change in gravity produces a very large change in the position of the mass. Also called an astatized, unstable or unstable-type gravimeter. The Lacoste-Romberg, the North American and the Worden gravimeters are astatic gravimeters.

gravimeter, astatized - SEE gravimeter, astatic.

gravimeter, ballistic - SEE gravity apparatus, ballistic.

gravimeter, bifilar - A weighted, horizontal disk suspended at its center by a helical spring and at the rim by two threads attached at diametrically opposite points. A change in gravity pulls down and rotates the disk from its position at equilibrium. The amount of rotation is a measure of the change in gravity.

gravimeter, dynamic - A gravimeter which determines the acceleration of gravity by measuring the rate of motion of a freely falling body or of a falling body constrained to move along a fixed path (e.g., a pendulum). Until recently, the pendulum was almost the only type of dynamical gravi-meter in use. The rate of oscillation of a pendulum of fixed length was measured, and this converted, after suitable corrections, to the acceleration of a freely falling body, i.e., the acceleration of gravity. The pendulum apparatus has now been replaced by apparatus which measures the rate at which an unconstrained body falls or by the rate at which an elastic thread vibrates under tension. The first kind is used under carefully controlled conditions in laboratories; the second kind is suitable for general use in the field. Also called a dynamic gravity-meter.

gravimeter, free-fall - SEE gravity apparatus, free-fall.

gravimeter, free-motion - SEE gravity apparatus, free-motion.

gravimeter, geodetic - A gravimeter having such a low rate of drift that it can be used anywhere in the world without needing to be recalibrated or the dials on which readings are made, reset.

gravimeter, laser - SEE gravity apparatus, ballistic.

gravimeter, pendulum-type - SEE pendulum apparatus.

gravimeter, relative - A gravimeter designed to measure the difference between the value of gravity at one station and that at another, or the ratio between the values of gravity at the two stations.

gravimeter, spring-type - A gravimeter which measures gravity by the amount of extension in a spring holding a weight in balance.

gravimeter, stable - (1) A gravimeter of such construction it indicates the value of gravity (within the range for which it was designed) without having to be reset after every change of gravity. (2) A gravimeter having a single weight or spring arranged so that the sensitivity is proportional to the square of its period.

gravimeter, stable-type - A gravimeter having a mechanical or optical connection between the test mass and the scale such that a change of position of the mass can be read directly.

gravimeter, static - A gravimeter which balances the force of gravity against a known and adjustable force, such as that of an extended spring or twisted thread.

gravimeter, straight-line - A gravimeter, of the zero-length-spring type, whose testing mass is constrained to move along an upright axis.

gravimeter, tidal - A gravimeter devoted to measuring the changes of gravity caused by tides in the solid Earth. A tidal gravimeter is similar to an ordinary gravimeter but is refined to be hundreds of times more precise.

gravimeter, trifilar - A weighted, horizontal disk suspended by a spiral spring attached at its center and by three threads attached at equal intervals around its rim. The disk is in torsional equilibrium; the weight of the disk tends to pull the disk down and around so as to lower the tension of the threads, but the torsion of the spring keeps the disk in place. Changing gravity turns the disk, which is brought back to its original position by turning the spring through a measured angle.

gravimeter, unstable - SEE gravimeter, astatic.

gravimeter, unstable-type - SEE gravimeter, astatic.

gravimeter, vibrating-string - A gravimeter consisting of a wire which is kept vibrating in an electrostatic or magnetic field by passing a current of fixed, precise frequency through it. The rate of vibration is proportional to the force of gravity and is determined by the weight of the wire and by the frequency of the current. A change in gravity can be detected by the change in electrical frequency necessary to keep the wire vibrating at its initial, fixed frequency. An accuracy of better than 1 mgal can be obtained in a few minutes. There are many varieties of instrument built on this principle. Vibrating-string instruments have been used, for example, for measuring the density of the medium surrounding the wire.

gravimeter, zero-length-spring - A static gravimeter containing a spring wound in such a manner that elongation of the spring under the weight of the attached mass is equal to the distance between the points of attachment of the spring. The spring has, effectively, a length of zero when it is not elongated.

gravimetric method (of determining the geoid) - A method of determining geoidal heights from measurements of the acceleration of gravity. The acceleration must be known, theoretically, over the whole world. Stokes's formula or a modification of it is usually used.

gravimetry - The art and science of measuring gravity.

gravimetry, absolute - Measurement of the force or acceleration of gravity by any method not requiring a knowledge of the value of gravity elsewhere.

gravimetry, airborne - The measurement of gravity using a gravimeter mounted in an aircraft. If the instrument is carried in an airplane, the Eötvös correction is very large. This difficulty has been avoided by making the measurements from a blimp.

gravimetry, geodetic - Worldwide relative measurements of gravitational acceleration, used in geodetic studies of the Earth.

gravitation - (1) The observed tendency of massive bodies to move towards each other unless restrained from moving. (2) The cause (force) behind the observed tendency of massive bodies to move towards each other unless restrained from moving. Also called gravitational force. It is sometimes called gravity. However, geo-physicists use gravity only to denote the sum of the gravitational and centrifugal forces, and this usage should be adhered to by physicists in general. Gravitation is assumed to be inherent in any massive body and is the weakest of the four types of force recognized by physics. If two bodies are far enough apart that they can be considered points, the force acting to bring them together is proportional to the mass of each body and is inversely proportional to the square of the distance between them. The constant of proportionality is denoted by G and is cal-led the gravitational constant. If the two bodies are so close together that their shapes must be considered, the force is given by $f = G \int_{M2} \int_{M1} [(r_2 - r_1)/(r_2 - r_1)^3] dM_1 dM_2$ in which r_2 and r_1 are the vectors from the origin to elements dM_2 and dM_1, respectively and M_2 and M_1 are the respective masses of the two bodies. SEE ALSO constant, gravitational.

gravity - (1) The effect, on any body, of the gravitational attraction by all other bodies and of any centrifugal force which may act upon it. Gravity may give a body weight or it may give the body an acceleration. (2) The force exerted, at a point, by the gravitational attraction of all bodies acting at that point and by the revolution of that point about a given axis. (3) The magnitude of the sum of the gravitational attraction and the centrifugal force acting at a point. Also called apparent gravity and intensity of gravity; it is usually referred to merely as gravity. It is almost the same as weight, and that term is often used with this meaning. (4) The weight imparted to a body by the gravitational attraction of all other bodies and by the action of any centrifugal force which may act upon it. This definition is almost identical to that given as (2) preceding, because weight has the same dimensions as force. However, weight is usually used for the magnitude of the force and is usually assumed to be exerted on a non-accelerating body e.g., a body at rest on the end of a spring or on the surface of the Earth, etc. (5) The acceleration imparted to an unconstrained body by the gravitational attraction of all other bodies and of any centrifugal force which may act upon it. The acceleration is a function of the body's location but not of the mass, density, shape, size, or other properties of the body. The force acting on the body, however, is a function of both location and mass. (6) The magnitude of the acceleration exerted by the sum of the gravitational attractions and the centrifugal force acting on a body. This is the usual meaning, it being understood that the direction is known or immaterial. E.g., a standard gravity formula for the acceleration of gravity gives the magnitude as a function of geodetic latitude. The direction is assumed to be in the inward direction of the normal to the ellipsoid to which the formula refers. (7) Any of the above definitions, with the restriction that the gravitational attraction of only one body, usually the Earth, is considered. Gravity, in this sense, is usually evaluated on the surface of the attracting body (e.g., the Earth) and therefore is the result of the attraction of two masses: the combined mass of the solid Earth and the bodies of water thereon, and of the atmosphere. Besides the direct effect of the atmosphere's attraction, there is its indirect effect on the buoyancy of the attracted body. (8) A unit of acceleration equal to the acceleration resulting from the average or standard force of gravity at the Earth's surface. (9) SEE gravitation. Many geodesists use gravity indifferently for the force and for the acceleration, and also for both the vector and its magnitude. Thus the word may have any one of several different meanings, the particular one meant being inferred from the context where possible. It is best, when using the word, to make clear at the outset which meaning will be used and to introduce each use of a different meaning with the appropriate warning. (In this glossary, gravity is used when a statement is true whether gravity force or gravity acceleration is meant.) If both meanings are to be used frequently, use gravity force and gravity acceleration instead of gravity [or gravity (f) and gravity (a)]. If no indication is given of whether gravity is considered to be a force or an acceleration, any statements about gravity can be assumed to be true for both the force and the acceleration. However, some geodesists are not careful in their usage, therefore this assumption may be incorrect, particularly where formulas are concerned. SEE ALSO Bouguer gravity; Bouguer gravity, expanded; Bowie gravity correction; Cassinis's formula; direction (of gravity); gravity correction; gravity formula; intensity (of gravity); point gravity; Potsdam standard of gravity; Potsdam system of gravity; Vienna system of gravity.

gravity, absolute - The value of gravity force or gravity acceleration found by direct measurement, as by the use of a spring balance or a pendulum.

gravity, apparent - SEE gravity (3); gravity acceleration.

gravity, free-air - (1) The value of gravity obtained by adding, to a hypothetical value of gravity at a point Q on a reference surface, the amount by which gravity would change in going from Q to a point P a vertical distance H_p above Q, if gravity varied between the two points according to the inverse-square law. Also (rarely) called Bouguer gravity. (2) The value of gravity obtained by subtracting, from a measured value of gravity at a point P on the Earth's surface, the amount by which gravity would change in going from P to a point Q a distance H_p vertically below P on the geoid, if gravity varied between the two points according to the inverse-square law. Also (rarely) called Bouguer gravity. In either of the above definitions, the point Q is assumed to lie on the surface of a ball with a radius equal to the average radius R_o of the Earth and on which the value of gravity is the average value g_m of gravity over the surface of the Earth. The rate of change of gravity between P and Q is then $-2(g_m / R_o) (1 - 3H_p / 2R_o + ...)$.

gravity, horizontal - The value of gravity at a point, multiplied by the rate of change of geodial height with distance at that point. The horizontal component of gravity is zero under all circumstances, so the term is misleading.

gravity, mean - The average value of gravity over a specified area or region.

gravity, normal - (1) A value of gravity acceleration calculated from a formula considered standard by a reasonable number of users in particular, such a standard formula giving the acceleration caused by a rotational ellipsoid having a specified major diameter, a specified constant value of gravity potential over its surface, and having the same rotational velocity as the Earth. Also called standard gravity. Unless stated otherwise, normal gravity can be assumed to mean the acceleration cal-culated using the International Gravity Formula, Gravity Formula 1967, or Gravity formula 1980. (2) The value of gravity acceleration calculated by using a particular formula, not necessarily a standard formula, for gravity. Also, the set of all such calculated values. (3) The value of gravity acceleration on an ellipsoid which is at a constant gravity potential. Most gravity formulae, and almost all standard gravity formulae, give the acceleration on such a surface. For such gravity formulae, definitions (1) and (2) are equivalent. (4) The value of gravity acceleration on an ellipsoidal solid representing the Earth. This definition is almost equivalent to any of the above definitions, because most gravity formulae are derived assuming that the Earth can be represented by an ellipsoidal solid. (5) A value of gravity force, between 978 and 982 g-cm/s^2, close to those values which gravity has on the Earth's surface. It may be an average value or merely a rough approximation. This usage is most common in astronomy but is also found in the theory of inertial navigation and inertial surveying. (6) Gravity along the inward (negative) direction of the normal to a specified ellipsoid.

gravity, observed - The value of gravity actually measured at a station.

gravity, reduced - Measured value of gravity reduced to the value it would have on the geoid or some other reference surface, using one of the procedures for gravity reduction.

gravity, regional - That part of a gravity anomaly caused by irregularities in density at much greater depths than those at which the geological formation sought for can occur. SEE ALSO gravity anomaly, regional.

gravity, relative - The value of gravity found by measuring gravity at one place, the instrument being calibrated by comparison with gravity at another place where the value is taken as known or standard. E.g., values of gravity determined using spring-type gravimeters are relative to the value of gravity at a base station such as that at Potsdam or to the value obtained by averaging over measurements at a number of stations.

gravity, residual - That part of gravity, at a point, left after an amount varying systematically over a region containing that point has been subtracted from the total amount.

gravity, specific - The ratio of the density of a substance at a given temperature and pressure to the density of some other substance used as standard, at the same temperature and pressure. Sometimes called density by oceanographers. For liquids and solids, the usual standard is the density of distilled water at 4° C and at standard atmospheric pressure (760 mm of mercury).

gravity, standard - (1) A value of gravity acceleration given by a standard gravity formula. (2) A value of gravity used as the referent when calculating values of some function of gravity. E.g., the rate of precession of a gyroscope can be separated into (a) the rate in a field of standard gravity and (b) the variation of this rate with variations from standard gravity. (3) As adopted by the International Committee on Weights and Measures, a gravity acceleration equal to 980.665 cm/s^2. (4) The gravity acceleration at 45° latitude and mean sea level, equal to 980.616 cm/s^2. (5) A starting value for gravity acceleration as reduced to mean sea level.

gravity, theoretical - A value of gravity calculated for a particular latitude according to an accepted gravity-formula.

gravity, virtual - The force of gravity acting on a portion of atmosphere, reduced by the amount of the centrifugal force caused by motion of that portion relative to the lithosphere.

gravity acceleration - (1) That acceleration caused by gravity alone (i.e., by the combination of gravitational and centrifugal forces) and not by electrostatic or magnetic forces, etc. (2) The magnitude of the acceleration caused by gravity alone and not by electrostatic or magnetic forces. Also called apparent gravity. Gravity acceleration is properly a vector. However, geodesists and geophysicists usually use the term to mean the magnitude of the accel-eration. This usage results from the fact that direction at a point on the Earth's surface is frequently defined using the direction of gravity as the reference. At the equator and close to the surface, the magnitude of gravity acceleration is approximately 978.03

cm/s²; at the poles, it is approx-imately 983.22 cm/s². The quantity 1 cm/s² is called the gal by geophysicists. (3) The ratio of weight to mass at a point in a reasonably uniform gravitational field. Also called the acceleration of gravity.

gravity adjustment - The process of determining theoretical values of gravity at a number of points on the Earth's surface from measurements made at those points. Gravity adjustment is usually done so that the resulting theoretical values are the best approximation, in a statistical sense, to the measured values.

gravity anomaly - (1) The difference Δg between a value g_p of the magnitude of gravity found by measurement at a point P and a value τ_o calculated for the same longitude and latitude using a standard gravity formula: $\Delta g = g_p - \tau_o$. Also called an anomaly of gravity or, simply, an anomaly. Collections of gravity anomalies for general use frequently define gravity anomaly in this way because to do so does not require that any particular theory be adopted or any particular use for the data be specified. (2) The difference Δg between a value g_p of the magnitude of gravity found by measurement at a point P and a value τ_p calculated for the same point using a value τ_o for gravity on a reference surface such as the terrestrial ellipsoid: $\Delta g = g_p - \tau_p = g_p - (\tau_o + \delta g_p)$, in which δg_p is a theoretical correction (sometimes called a gravity reduction) added to the value of gravity on the reference surface. Also called anomaly of gravity, gravity correction, gravity disturbance and modified gravity anomaly. The value on the geoid is usually taken to be the value given by a gravity formula, although that value belongs, properly, to the reference ellipsoid. This kind of gravity anomaly is particularly useful for studies of isostasy and geology, but it is also suitable for studying the geoid. It was much used by the U.S. Coast and Geodetic Survey in its study of these subjects. (3) The difference Δg between a value g_q of the magnitude of gravity calculated for a point Q on the geoid from a value g_p found by measurement at a point P vertically above Q and a value τ_q of gravity at the corresponding point Q' vertically below Q on the terrestrial ellipsoid, as calculated from a gravity formula: $\Delta g = g_q - \tau_q'$. Also called anomaly of gravity or, simply, an anomaly. This kind of gravity anomaly has been found useful in studies of the geoid and in deriving gravity formulae, because it allows values of gravity on the geoid to be compared directly. (4) The same as the preceding definition, except that the point Q' is on the normal to the geoid at Q. (5) SEE gravity anomaly, vertical. A gravity anomaly is usually an intermediate product in the process of comparing measured values of gravity with theoretical values. To be useful, values derived from measurements and values derived from theory must ultimately refer to the same point. The value given by a gravity formula must therefore be corrected to account for, among other things, the distance between the point for which the gravity is calculated and the point at which the gravity was measured. The value which results from applying corrections to the value calculated using a gravity formula is identified by indicating the kind of correction applied: e.g., free-air gravity anomaly, Bouguer gravity anomaly and Prey gravity anomaly. Gravity anomalies are sometimes distinguished as point gravity anomalies and mean gravity anomalies. The word point is, however, redundant and should not be used in referring to a gravity anomaly. NOTE: The term gravity anomaly was originally defined and used in the sense measured value of gravity minus calculated value of gravity. With the passage of time, it and associated terms such as gravity correction have acquired additional meanings such as calculated value minus measured value and calculated value A minus calculated value B, as well as meanings differing from each other only in minor details. Where clarity will not suffer, only one definition is usually given in full; other definitions are then given as variants of the first. An exception is made if an alternative definition is very widely used; in that case, the alternative definition or definitions are given fully also. SEE ALSO Airy-Heiskanen gravity anomaly; Bouguer gravity anomaly; Bouguer gravity anomaly, cylindrical; Bouguer gravity anomaly, expanded; Bouguer gravity anomaly, modified; Bouguer gravity anomaly, refined; Bouguer gravity anomaly, spherical; condensation gravity anomaly; Faye's gravity anomaly; Glennie's gravity anomaly; Pratt-Hayford gravity anomaly; Prey's gravity anomaly; Rudzki gravity anomaly; surface gravity anomaly; Vening Meinesz gravity anomaly, isostatic.

gravity anomaly, combined - SEE Glennie's gravity anomaly.

gravity anomaly, external - The difference between a value of gravity measured at a point on or outside the Earth's surface and the value of gravity calculated for that or for some other specified point.

gravity anomaly, 5-km - The gravity anomaly obtained by adding, to the measured value, the amount by which gravity is increased by removing all water at depths greater than 5 km and replacing it with rock of density 2.60 g/cm³, and replacing rock at depths less than 5 km by water.

gravity anomaly, free-air - (1) The quantity Δg_f obtained by subtracting, from a gravity anomaly Δg, the free-air gravity correction δg_f: $\Delta g_f = \Delta g - \delta g_f$. The free-air gravity correction (which is negative) is the total change in gravity which occurs in moving (without taking into account any mass along the path) from the reference surface to the point P at which gravity was measured: $\delta g_f = (\delta \tau / \delta h) h$, in which h is the geodetic height of P above the reference surface and τ is the theoretical value of gravity on that surface. (2) The same as the preceding definition, except that the free-air gravity correction is calculated using the elevation H_p of P above the reference surface instead of the geodetic height, and the rate of change of measured values g of gravity instead of the theoretical value: $\delta g_f = (\delta g / \delta H_p) H_p$. This definition is used if the acceleration on the terrestrial ellipsoid or geoid) is to be calculated from the measured value of the station for comparison with the value calculated from a standard gravity formula. Free-air gravity anomaly is sometimes called Faye's anomaly and incomplete Bouguer gravity anomaly.

gravity anomaly, isostatic - (1) The difference Δg_{is} between a measured value g of gravity and a theoretical value τ_{is} which is the sum of the following terms: (a) the theoretical value τ on a particular reference surface, (b) the free-air gravity correction δg_f, (c) the complete topographic gravity correction δg_{tc} and (d) the isostatic gravity correction δg_{is}. $\Delta g_{is} = g - \tau_{is} = g - (\tau + \delta g_f + \delta g_{tc} + \delta g_{is})$. The quantity τ may be known, in which case the gravity anomaly is used to compare the measured value of gravity with that derived by adding the gravity anomaly to τ, or it may be unknown, in which case the gravity anomaly is subtracted from g to give the value of τ. The first alternative is used by geophysicists; the second is used by geodesists. The isostatic gravity corrections are calculated using one of a number of theories on how isostasy is brought about within the crust. The correction is then named for the originator of the theory and, when desirable, for the originator of the particular correction. SEE ALSO Airy-Heiskanen gravity anomaly; Pratt-Hayford gravity anomaly; Vening Meinesz gravity isostatic anomaly. (2) The same as the preceding definition, except that only gravity acceleration is considered. (3) The difference between a measured value of gravity and the value calculated by a method involving the principle of isostasy. Also called an isostatic anomaly.

gravity anomaly, mean - The value obtained by averaging gravity anomalies over a specific area such as a quadrangle 1° by 1° or 5' by 5' on a side.

gravity anomaly, mixed - The difference between a value of gravity measured at a point P and the value calculated for gravity at a different point Q. SEE ALSO gravity anomaly, proper.

gravity anomaly, modified - Any quantity derived from a gravity anomaly by subtracting one or more gravity corrections from it.

gravity anomaly, non-classical - The difference between the magnitude g of gravity as measured on the ground and the corresponding value τ of gravity as calculated for a point on a spherop and on the same vertical.

gravity anomaly, observed - SEE gravity anomaly. NOTE: a gravity anomaly cannot be observed ever, under any circumstances.

gravity anomaly, point - SEE gravity.

gravity anomaly, proper - The difference between the magnitude of a measured value of gravity at a point on the earth's surface and a corresponding, theoretical value of gravity at that same point.

gravity anomaly, regional - A gravity anomaly calculated as the average value of gravity anomalies over a specified region.

gravity anomaly, residual - (1) The difference between a gravity anomaly and the average value of the gravity anomalies calculated over a specified region. (2) The quantity $g_f + 2\pi G \Sigma [h_i (\rho_m - \rho_i)]$, in which g_f is the free-air gravity anomaly, G is the gravitational constant, h_i is the thickness of the i-th layer whose density is ρ_i and ρ_m is the density of the mantle. This is a sort of Bouguer gravity anomaly removing the effects of compensated crustal topography and uncompensated bulges in the mantle.

gravity anomaly, root-mean-square - The root-mean-square value Δg_{rms} of gravity anomalies Δg_n over the entire Earth: $\Delta g_{rms} = +\sqrt{[(\Sigma \Delta g_n^2)/N]}$, in which n goes from 1 to N, the total number of gravity anomalies.

gravity anomaly, topographic - (1) The difference between the value of gravity as measured at a point P and the value as calculated from a standard gravity formula, corrected for the effect of matter between the terrestrial ellipsoid and an equipotential surface through P. (2) A Bouguer gravity anomaly from which is subtracted the topographic gravity correction. Also called a Bouguer gravity reduction (q.v.).

gravity anomaly, two-dimensional pseudo-isostatic - The quantity Δg_{2i} obtained by subtracting from the measured value g the sum of the value τ of gravity on the reference surface, a free-air gravity correction δg_f which is zero in oceanic and similar regions, the two-dimensional value δg_{B2} of the line integral of the gravity in oceans filled with rock and the two-dimensional value (negative) δg_{i2} of the effect of antiroot masses compensating for the filled-in oceans.

gravity anomaly, vertical - The difference between the value of gravity at a point and the value obtained by changing the density of the rocks in a specified manner and within a specified region on the Earth. Also called, simply, gravity anomaly.

gravity anomaly representation - Representation of a gravity anomaly either mathematically, by means of a formula, or physically as the result of the gravitational attraction by point masses of suitable size and rotating with the Earth.

gravity apparatus - An apparatus for measuring gravity or gravity acceleration. SEE ALSO Brown's gravity apparatus.

gravity apparatus, ballistic - An apparatus for determining gravity acceleration by measuring the amount of time it takes an object to rise or fall (or both) between marks a known distance apart vertically. Also called a ballistic gravimeter. There are two varieties: the free-fall gravity apparatus, in which the object merely falls, and the free-motion gravity apparatus, in which the object is propelled upward and both rises and falls. There appears to be no significant difference between the accuracies obtainable using either type, although at present, gravity acceleration has been determined to more significant figures using the free-motion gravity apparatus.

gravity apparatus, free-fall - An apparatus for determining gravity acceleration by dropping an object so that it falls vertically in a vacuum and measuring the amount of time it takes for the body to fall through a known distance. If the distance the body falls is determined by shining a vertical beam of coherent radiation onto the object and counting changes in the number of wavelengths from the source to the object, the apparatus may be called a laser gravimeter or laser-type gravity apparatus.

gravity apparatus, free-motion - An apparatus for determining gravity apparatus by measuring the length of time it takes a body thrown upwards to rise, and then fall, between horizontal, parallel planes a known distance apart. Also called a symmetrical free-fall gravity apparatus or, if a beam of coherent radiation is used to measure the rise and fall, a laser free-motion gravity apparatus or laser free-motion gravimeter.

gravity apparatus, laser-type - A gravity apparatus in which the distance moved by the object is determined by measuring the change in the number of wavelengths in the path of a beam of coherent radiation shining on the object.

gravity apparatus, symmetrical free-fall - SEE gravity apparatus, free-motion.

gravity change, absolute - A change in gravity at a point fixed with respect to a geocentric coordinate system.

gravity change, relative - A change in gravity measured with an instrument fixed to the Earth's surface.

gravity corer - Any type of corer which penetrates the bottom solely because of its weight.

gravity correction - (1) A quantity, usually denoted by δg with a subscript (e.g., δg_f, δg_{Bs}), which is added to a theoretical value of the gravity to obtain a better approximation to the measured value. Equivalently, the quantity (g measured - g computed). g computed is usually denoted by τ, often with a subscript indicating the kind of formula or theory used in deriving it. (2) A quantity, usually denoted by δg with a subscript (e.g., δg_f, δg_{Bs}), which is added to a measured value to obtain a theoretical value of the gravity. The gravity corrections created by the above two definitions have the same absolute value and opposite signs. That created by the first definition has been used extensively for geophysical problems such as determining the state of isostatic compensation, etc. That created by the second definition has been extensively used in determinations of the geoid. Neither definitions is more correct than the other, and which is used in a particular writing must usually be determined from the context. In this glossary, conventions of sign, etc., will follow the first of these definitions, which is consistent with the way correction is commonly used. Data on gravity are ordinarily stored as and used as gravity anomalies - the difference between gravity as measured and gravity as calculated. The calculated value is usually arrived at by adding, to an initial value derived from a formula for gravity on a particular ellipsoidal solid (the terrestrial ellipsoid), small increments also called gravity corrections, the sum of which constitutes the (total) gravity correction defined above. By the adopted convention, these small gravity corrections are to be subtracted from the gravity anomaly which is the difference between the measured value and the initially computed value. Under the second definition, they would be subtracted from the initial gravity anomaly to obtain the modified gravity anomaly. A variation in usage occurs because some corrections increase the calculated value of gravity while others decrease it. Some writers therefore treat gravity corrections as absolute values and add or subtract these as appropriate instead of treating gravity corrections as algebraic quantities. As a result, formulae will be found in which some gravity corrections are added and some subtracted. In such instances, the context must be carefully studied to determine just how each correction is defined. (3) A correction applied to the nominal length of a surveyor's tape to compensate for change in length caused by using the tape under a different gravity from that under which the tape was calibrated. The sag correction for tape length is not affected by a change in gravity; the tension correction is. It is changed by the amount $(1 + \Delta g/g)$, in which g is the value of gravity at the place where the tape was calibrated and Δg is the (algebraic) increase in gravity on going to the place where the tape is used. SEE ALSO Airy-Heiskanen gravity correction; Bouguer gravity correction; Bouguer gravity correction, cylindrical; Bouguer gravity correction, spherical; Browne's correction; condensation gravity correction; Helmert condensation gravity correction; Honkasalo's gravity correction; latitude gravity correction; Pratt-Hayford gravity correction; Rudzki gravity correction.

gravity correction, combined - The sum of the Bouguer gravity correction and the topographic gravity correction.

gravity correction, complete topographic - A gravity correction (δg_{tc}) added to the theoretical value of gravity to account for the presence of matter outside the level surface used as reference. The complete topographic gravity correction may be calculated as the sum of the Bouguer gravity correction δg_B and the topographic gravity correction δg_t, or may be calculated from directly from the topography. If the first method is used, the gravity correction is also called the combined gravity correction.

gravity correction, cross-coupling - A correction made to values of gravity measured using a spring-type gravimeter on a moving vehicle (usually a ship), to remove the errors caused when horizontal and vertical components of acceleration have equal periods and are out of phase. Corrections are calculated from the results of measure-ments on two horizontal accelerometers having their axes mutually perpendicular. They are small for low sea-states and can be calculated with adequate accuracy for moderate sea-states.

gravity correction, Eötvös - SEE Eötvös correction.

gravity correction, free-air - (1) The gravity correction $\delta g_f = \Sigma (h^n/n!)(\delta^n\tau/\delta h^n)$ giving the change in gravity found in going from a point Q on the reference surface to the point P on the normal through Q. h is the distance QP and τ is the value of gravity calculated for Q. The first two terms are, explicitly, $-2\tau h/(R+h)$ and $+3 \tau h^2/(R+h)^2$. R is the radius of the Earth at Q. If P is at an elevation of less than 2 km, terms of degree two and higher are not needed. δg_f is then approximately $-\tau h/R$, or -0.3086 h mgal (or newtons/kg, etc.). The free-air gravity correction is sometimes undesirably referred to as Faye's gravity correction. (2) The same as the preceding definition, but with g, the actual value of gravity at P, used instead of τ and H, the elevation of P, used instead of h. This definition has been used when two theoretical values of gravity acceleration were being compared. A theoretical value gravity acceleration was calculated for Q from a measured value at P and compared with a value calculated for Q from a gravity formula.

gravity correction, isostatic - A gravity correction δg_i added to the theoretical value of gravity to correct for the (assumed) presence of additional mass below the geoid balancing excess mass above the geoid. The isostatic gravity correction is also called an isostatic gravity reduction. It is an ad hoc correction invented to make predicted values of gravity agree better with measured values. There are many different kinds of isostatic gravity correction. Each kind results from a particular assumption on how density varies within the crust and upper mantle. All the assumptions postulate that the crust and part of the upper mantle float on an underlying material of greater density. They differ in the depths to which the floating materials extend, the density of that floating material under the geoid, and in other particulars. Seismological regimes in the crust are only roughly related, if at all, to the regimes involved in explaining the isostatic gravity corrections.

gravity correction, orographic - SEE gravity correction, topographic.

gravity correction, tidal - (1) A quantity added to a theoretical value of gravity or subtracted from a measured value to account for the gravitational attraction of the Sun and Moon on the gravimeter or other instrument. (2) A correction applied to measurements of gravity to remove the effect of earth tides on the measured values. It is commonly included in the drift-correction and may be determined by a series of observations at a fixed base station. (3) A correction applied to the measured period of a pendulum to take into account the effect of earth tides. If the Earth were rigid, the Moon and Sun at the zenith would decrease the magnitude of gravity acceleration by about 0.16 and 0.08 mgal respectively, relative to the values when the Moon and Sun are on the horizon. Because of the yielding of the crust and sea, the effect is a little greater than it would otherwise be. The amount is not entirely predictable and is usually ignored. (4) SEE Honkasalo's gravity correction.

gravity correction, topographic - (1) The difference δg_t, at a point P, between the value of gravity caused by the attraction of the actual amount of matter outside the reference surface and the attraction of the matter in the Bouguer plate through P. (2) The difference δg_t, at a point P, between the value of gravity caused by the attraction of matter contained between a level surface through P and the geoid and the value of gravity caused by the attraction of the Bouguer plate through P. (3) The difference δg_t, at a point P, between the value of gravity caused by the attraction of matter contained between a level surface through P and the geoid, and the value of gravity caused by the attraction of some simple body such as a disk, a sphere or infinite plate through P. Also called a terrain gravity correction, terrain correction, topographic correction, Geländereduktion, orographic gravity correction, etc. (4) The sum of the Bouguer gravity correction, the free-air gravity correction and the topographic gravity correction in the sense of the definition (1) preceding.

gravity determination - SEE gravimetry.

gravity disturbance - (1) The difference between the actual or measured value of gravity at a point on the Earth and the value predicted by theory. Also called a gravity anomaly. In particular, this difference is when the theoretical value is calculated by a standard gravity formula. (2) The difference between the actual or measured value of the gravity vector at a point and the value predicted by theory. (3) SEE disturbance, gravitational.

gravity field - (1) Gravity considered as a whole, rather than as a force or acceleration at a particular point e.g., the Earth's gravity field. (2) A field having gravity as the function of location i.e., a space and a mathematical function giving gravity as a function defined over the points of that spaced.

gravity field, normal - A mathematically-defined function which gives values of gravity close to those of the actual gravity-field and having, usually, at least one equipotential surface which is mathematically simple (e.g., a rotational ellipsoid).

gravity field, standard - The gravity field defined by a standard gravity.

gravity flattening - SEE flattening, gravitational.

gravity formula - A formula giving the value of gravity, at a point, as a function of the coordinates of that point. Most gravity formulae give the gravity acceleration as a product of (a) the value of acceleration at mean sea level or the geoid at some point, and (b) a power series in the sine of the latitude or a Fourier series having latitude as the argument. Usually, no more than two or three terms are present. Some gravity formulae also include terms for the variation of gravity with longitude, elevation, or both. For example, the gravity formula, used by the Meteorological Office since 1 January 1955 (for applying corrections to barometric readings) is $\tau = 980.616 \, (1 \, -0.0026373 \, \cos2\phi + 0.0000059 \, \cos^2 2\phi) - 0.00009406 \, H$, in which τ is the gravity acceleration in cm/s² at latitude ϕ and elevation H in feet. Most gravity formulae are approximations for the acceleration of gravity on or above an ellipsoid which is at a fixed potential. SEE ALSO Helmert's 1901 gravity formula; Helmert's 1915 gravity formula; longitude term (in a gravity formula).

gravity formula, general - A formula representing, as fully as possible, the Earth's gravity field.

gravity formula, International - SEE International Gravity Formula.

gravity formula, normal - (1) A gravity formula giving the value of gravity on or above a body whose surface is mathematically defined and is a level surface with given gravity potential, and which has a rate of rotation equal to that of the Earth. Also called a standard gravity formula. The term is practically equivalent to gravity formula. It has the connotation that the normal gravity formula represents the gravity field of a body (normal Earth, model Earth, etc.) which is a mathematically tractable representation of the Earth. A body with surface a rotational ellipsoid is commonly chosen. Differences between values derived from the actual gravity field and values given by the normal gravity formula are then called gravity perturbations. (2) SEE gravity formula, standard.

gravity formula, standard - A formula giving the value of gravity as a function of geodetic or geocentric latitude (and possibly also of geodetic or geocentric longitude and/or height), and considered standard by a large number of geodesists. Commonly called a normal gravity formula if the formula applies to the gravity field of a rotational ellipsoid on which the gravity potential has one specified value and which rotates as the same rate as the Earth. Standard gravity formulae are used because differences between actual gravity and theoretical gravity are more easily handled than are the actual values themselves and because users of gravity data can more conveniently compare results if a common reference formula is used. The term normal has entered American usage through European geodesy, in which normal has the meaning standard. Four gravity formulae: the Helmert-1901 gravity formula, the International Gravity Formula, Gravity Formula 1967 and gravity Formula 1980, have commonly been considered standard gravity formulae by geodesists. These usually have one of three forms:

$$\tau = \tau_o (1 + \beta_1 \sin^2\phi + \beta_2 \sin^4\phi),$$
$$\tau = \tau_o (1 + \beta_1 \sin^2\phi + \beta_2 \sin^2 2\phi) \text{ or}$$
$$\tau = \tau_o (1 + b_1 \sin^2\phi)(1 - b_2 \sin^2\phi)^{-1/2}$$

The quantities τ_o, β_1, β_2, b_1 and b_2 are constants and ϕ is the geodetic latitude.

gravity formula, theoretical - A gravity formula expressing gravity, on a surface of reference, as a function of geographic location, with the assumption that the potential function has a constant value on the surface.

Gravity Formula 1967 - One of the four gravity formulas

$$\tau = 978.031\,85\,\tau_e (1 + 0.005\,278\,895 \sin^2\phi + 0.000\,023\,462 \sin 4\phi),$$

$$\tau = 978.031\,85\,\tau_e (1 + 0.005\,3024 \sin^2\phi - 0.000\,059 \sin^2 2\phi),$$

$$\tau = \tau_e (1 + k \sin^2\phi) / \sqrt{(1 - e^2 \sin^2\phi)} \text{ and}$$

$$\tau = G_1 [\sqrt{(1 + e'^2 \cos^2\phi)}] + G_2/\sqrt{(1 + e'^2 \cos^2)}$$

$$G_1 = -281.256\,3939 \text{ cm/s}^2$$

$$G_2 = 1264.473\,1218 \text{ cm/s}^2.$$

in which τ_e has the value 978.031 845 58 cm/s², e^2 has the value 0.006 694 605 328 56, k = 0.001 931 663 383 21. The other coefficients were calculated from the geodetic constants in the set Geodetic Reference System 1967. e'^2 is defined as $e^2/(1-e^2)$.

Gravity Formula 1980 - The gravity formula $\tau = \tau_e (1 + 0.005\,279\,0414 \sin^2\phi + 0.000\,023\,2718 \sin^4\phi + 0.000\,000\,1262 \sin^6\phi + 0.000\,000\,0007 \sin^8\phi)$, in which τ_e is 9.780 326 7715 m/s² and ϕ is the geodetic latitude. The numerical coefficients are derived from the constants in the set Geodetic Reference System 1980.

gravity gradient - (1) The change in the value of gravity per unit distance. It is commonly resolved into two horizontal and one vertical component. (2) SEE gradient, gravitational.

gravity gradiometer - An instrument for measuring the rate of change of gravity with a vertical or horizontal change of location. The torsion balance is at present the most sensitive and widely used type of gravity gradiometer. It relies on the static balancing of gravity gradient by torsion on a wire. However, instruments depending on dynamical principles and involving rapidly revolving masses have been invented. Such instruments are sturdier than the torsion balance, require less time to make a measurement, and can be moved rapidly about while in operation.

gravity gradiometer, moving-base - An instrument which, while in motion (as on a ship or aircraft), can measure the rate at which gravity changes with change of location. The torsion balance, another gravity gradiometer, must remain in place while making a measurement. At present there are two types of moving-base gravity gradiometer. One uses rotating masses whose dynamic weights are determined by stress gages. The other uses pairs of accelerometers contained in superconducting circuits.

gravity gradiometer, spaceborne - An instrument carried on a spacecraft and capable of measuring the gradient (spatial rate of change) of gravitational force or acceleration. Note that a spaceborne gravity gradiometer measures gravitational gradient, not gravity gradient. However, the basic design may be the same in both kinds of instruments.

gravity harmonic - One of the terms in the expansion of a gravity field onto a series of spherical harmonics. Also called a gravitational harmonic.

gravity instrument - An instrument for measuring gravity or for determining the difference in gravity between two points.

gravity intensity - SEE intensity (of gravity).

gravity measurement - (1) SEE gravimetry. (2) A measured value of gravity.

gravity meter - SEE gravimeter.

gravity meter, dynamic - SEE gravimeter, dynamic.

gravity network - (1) A diagram consisting of points connected by lines, in which the points represent gravity stations (points on the Earth at which gravity was measured) and the lines represent the sequence in which the gravity stations were occupied indicate the gravity stations between which comparisons of gravity were made.

gravity potential - (1) The potential created, at a point rotating with the Earth, by the combined action of gravitational and centrifugal forces. I.e., a potential such that its directional derivative is gravity. The term is also applied to the potential created on any celestial body, at a point rotating with that body, by the combined actions of the body's gravitational and centrifugal forces. In particular, the potential attributable to the Earth's gravitational attraction and centrifugal force. Gravity potential is equivalent to the work done in bringing a body of unit mass from infinity to a point at the Earth's surface and then giving the body the angular velocity of that point. (2) The gravity potential, as defined in (1), per unit mass of the body creating the potential. (3) The quantity whose directional derivative is the negative of the acceleration of gravity. Also called the geopotential or, if only the Earth's gravity potential is being discussed, potential. (4) The work done in moving a body of unit mass from one point on a rotating, massive body to another point on or rotating with the same body. It is understood that dissipative forces are not present.

gravity reduction - (1) The process of subtracting, from measured values of gravity, various gravity corrections to give a theoretical value for gravity at some other point (generally the geoid or some other reference surface) and for a particular (assumed) distribution of mass in the Earth. (2) The result of subtracting, from a measured value of gravity, various gravity corrections to give a theoretical value for gravity at some other point such as the geoid, for a particular (assumed) distribution of mass in the Earth. (3) The sum of the gravity corrections subtracted from a measured value of gravity (or added to a theoretical value) to give a theoretical value of gravity at some other point such as the geoid (or the Earth's surface), for a particular (assumed) distribution of mass in the Earth. (4) The value, derived from theory, to be added to the value of gravity calculated for the reference ellipsoid (in particular, calculated using an International Gravity Formula) to obtain the corresponding value at the point of measurement. Gravity reduction has been used with other meanings than those given here. There is probably no one meaning in common use by a majority of geodesists or geophysicists, and some geodesists use the term with several different meanings, depending on context. In the following definitions of terms based on gravity reduction, the form of the first definition will be used; this accords best with the commonly accepted meaning of reduction as well as with the most common usage of gravity reduction. The other common meanings (i.e. as process or result) can be obtained in the obvious manner. SEE ALSO Airy-Heiskanen gravity reduction; Bouguer gravity reduction; Bouguer gravity reduction, extended; Bouguer gravity reduction, incomplete; Brillouin's gravity reduction; Bruns gravity reduction; condensation gravity reduction; Faye's gravity reduction; Graaf-Hunter gravity reduction; Helmert gravity reduction; Pratt-Hayford gravity reduction; Prey's gravity reduction; Rudzki gravity reduction; Vening Meinesz gravity reduction.

gravity reduction, free-air - (1) The process of applying the free-air gravity correction to a measured or calculated value of gravity. (2) The free-air gravity correction. (3) The process of adding, to a measured or calculated value of gravity, the sum of the free-air gravity correction δg_f and the topo-graphic gravity correction δg_t. (4) The sum mentioned in the previous definition.

gravity reduction, indirect - (1) The quantity to be added to that used in the free-air gravity reduction to take account of the fact that by removing matter between the geoid and the real surface, the position of the geoid is changed (lowered) so that masses now appear outside the geoid.

gravity reduction, isostatic - (1) The process of applying, to a measured or computed value of gravity, the sum δ_i of the following gravity corrections: (a) an isostatic gravity correction δg_i; (b) a free-air gravity correction δg_f; (c) a Bouguer gravity correction δg_B or a modified Bouguer gravity correction δg_{Bm}; and (d) a topographic gravity correction δg_t or a topographic gravity correction modified to correspond to the modified Bouguer gravity correction. The gravity correction is carried out, theoretically, in the above order. Practically, the isostatic and topographic gravity corrections, or the isostatic, topographic and Bouguer gravity corrections are computed more or less simultaneously, often using gravity tables for this purpose. Also called an isostatic reduction. The isostatic gravity corrections most commonly used are the Pratt-Hayford, Airy-Heiskanen and Vening Meinesz gravity reductions. The U.S. Coast and Geodetic Survey used the Pratt-Hayford gravity reduction as embodied in the Hayford-Bowie and Hayford-Bullard methods. (2) The quantity δ_i defined in the previous definition. (3) Changing the measured value of gravity by (a) removing the topography (masses outside the geoid); (b) removing the effects of isostatic compensation; and (c) doing a free-air reduction to the geoid.

gravity reduction, non-isostatic - A gravity reduction not involving the theory of isostasy.

gravity reduction, refined - The Bouguer gravity reduction plus a correction for the difference between the Bouguer plate and the actual topography.

gravity reduction, simple - SEE Bouguer gravity reduction.

gravity reduction, topographic - A gravity reduction applying the free-air gravity correction δg_f, the Bouguer gravity correction δg_B and the topographic gravity correction δg_t. Also called the Oeländer gravity reduction.

gravity reference station - A station at which the value of gravity is known and with respect to which the differences in gravity at other stations are determined. The absolute value of gravity may or may not be known at the reference station.

gravity representation - The representation of the values of gravity by a formula (such as a gravity formula) or as the result of a hypothetical distribution of masses rotating with the Earth (such as a set of point-masses or massive, simple geometric bodies).

gravity standard - SEE Potsdam standard of gravity; International Gravity Standardization Net 1971.

Gravity Standardization Net 1971, International - SEE International Gravity Standardization Net 1971.

Gravity Standardization System, International - SEE International Gravity Standardization Net 1971.

gravity station - A station at which measurements are made to determine the value of gravity.

gravity station, absolute - SEE station, absolute-gravity.

gravity system - A set of values of gravity (usually, gravity acceleration) referred to the value of gravity, at one or more places, defined or taken as standard. Three gravity systems have been adopted internationally. These are the Vienna gravity system (1900 to 1909), the Potsdam gravity system (1909-1971), and the International Gravity Standardization Net 1971.

Gravity System 1971, International - SEE International Gravity Standardization Net 1971.

gravity table - (1) A mathematical table giving the gravitational effects of simple bodies of various sizes, and used for calculating gravity corrections or deflection-of-the vertical corrections. (2) A tabulation of measured values of gravity at various points. SEE ALSO Bowie's gravity table; Cassinis's gravity table.

gravity table, fundamental - A table giving the deformation of the geoid and the effect of this deformation on gravity, computed for masses of unit density extending to various distances above and below the surface of the geoid. Fundamental gravity tables serve as the basis for preparing special tables using particular assumptions about density, isostasy, and so on. Several such fundamental tables have been prepared, each designed for a particular effect of such assumptions. (a) The tables of Cassinis, Dore and Ballarin (1937) determine the direct effect, on gravity, of masses of unit density extending to various distances above and below the geoid. This direct effect is known as the Hayford effect; it neglects the differences of elevation between the reference ellipsoid and the geoid. (b) The tables of Lambert and Darling (1938) determine the indirect effect of masses of unit density extending to various distances above and below the geoid. This indirect effect is known as the Bowie effect; it takes into account the difference of elevation between the reference ellipsoid and the geoid. (c) The tables of Darling (1949) give the horizontal effect (deflection of the vertical) of masses of unit density extending to various distances above and below the geoid.

gravity tide - Those variations, in the value of gravity at a point, which are caused (a) by the attraction of the Moon, the Sun, and other planets, and (b) by the effects of those attractions on the land and seas.

gravity variometer - SEE gravity gradiometer.

gravity yard - A (railroad) yard in which cars are moved (sorted) by locomotive, with material aid from gravity.

gravure - A method of printing in which the part of the plate covered by the image is recessed below the level of the printing cylinder. Also called photogravure.

Gray code - A sequence of binary numbers arranged so that each number differs from the preceding one in only one digit (bit). Also called a reflection code. The graduated circles in some electronic distance-measuring instruments and theodolites are marked by black and white sectors or by magnetic and non-magnetic sectors. etc., to correspond to the numbers of a Gray code.

gray scale - (1) A strip of film or a glass plate whose transparency diminishes from one end of the strip or plate to the other. (2) SEE step wedge.

gray scale, continuous-tone - A scale of shades from white to black or from transparent to opaque, each shade blending imperceptibly into the next without visible texture or dots. Also called a continuous wedge. Compare with step wedge and gray scale.

gray scale, equal-contrast - A gray scale in which the transparency diminishes by discrete amounts from one end of the scale to the other, the amount being such that the perceived contrast is the same between all pairs of adjacent steps.

grazing lease - A lease that attributes the use of public lands outside of grazing districts (Taylor Grazing Act) for the grazing of livestock for a specific period of time.

Great Lakes Datum 1955, International - SEE International Great Lakes Datum 1955.

Greenwich apparent time - Local apparent time at the Greenwich Meridian.

Greenwich apparent sidereal time - The hour angle between the meridian of the Mean Observatory and the true vernal equinox of date.

Greenwich argument - The equilibrium argument computed for the meridian of Greenwich.

Greenwich Civil Time - Mean solar time for the Greenwich meridian, counted from midnight. Abbreviated to G.C.T. Also called Universal Time. Greenwich Civil Time was used in the American Ephemeris and Nautical Almanac for 1925 and thereafter. Universal Time is no longer used as a synonym for Greenwich Civil Time in ephemerides; Ephemeris time is used instead. However, Greenwich Civil Time is still used in Britain for certain purposes.

Greenwich grid - A grid used in polar regions and based on the Greenwich meridian.

Greenwich hour angle - The angle, measured westwards from the celestial meridian at Greenwich, from the plane of that celestial meridian to the plane of the hour circle passing through the celestial body observed. Greenwich hour-angle is usually abbreviated to G.H.A. or GHA. It may also be defined as the arc of the celestial equator, or the angle at the celestial pole, between the upper branch of the celestial meridian at Greenwich and the hour circle or a point on the celestial sphere, measured westward from the celestial meridian at Greenwich through 24 hours or 360°. It is the local hour angle at the Greenwich meridian.

Greenwich interval - An interval of time based on the Moon's transit of the celestial meridian at Greenwich. This is not the same as a local interval based on the Moon's transit of the local celestial meridian.

Greenwich lunar time - The arc of the celestial equator between the lower branch of the Greenwich celestial meridian and the hour circle of the Moon, measured westward from the lower branch of the Greenwich celestial meridian through 24 hours. Greenwich lunar time is equivalent to local lunar time at the Greenwich meridian and to the Greenwich hour angle of the Moon, expressed in units of time, plus 12 hours.

Greenwich Mean Astronomical Meridian - A meridian determined by the Bureau International de l'Heure as the meridian to which Universal Time 2 is referred. Also called the Mean Observatory and meridian of the Mean Observatory. It is determined from observations at over 40 time-keeping stations. (time services).

Greenwich mean astronomic meridian plane - A plane parallel to an average position of the Earth's axis of rotation and at an angle to the plane of the Greenwich Meridian by an amount defined by the Bureau International de l'Heure (B.I.H.) in such a way that time reckoned from this plane agrees with the Universal-Time 2 determined by the B.I.H. The plane is close to that of the Greenwich Meridian.

Greenwich mean time - The right ascension of the fictitious mean Sun. This definition was changed, in Britain, in 1925. (2) The Greenwich hour-angle of the vernal equinox of date, plus 12 hours. Also called, by the military, z-time and zulu time. This is equivalent to Universal Time 0. The definition was retained for some time but eventually abandoned. SEE ALSO Universal Time 0.

Greenwich mean sidereal time - (1) The hour angle between the Greenwich meridian and the mean equinox of date. Also called Newcomb sidereal time if emphasis is on the equation used. SEE ALSO Newcomb sidereal time. (2) The hour angle between the meridian of the Mean Observatory and the mean vernal equinox of date.

Greenwich Meridian - The astronomic meridian through the center of the Airy transit instrument of the Greenwich Observatory, Greenwich, England. By international agreement in 1884, the Greenwich meridian was adopted as the meridian from which all longitudes, worldwide, would be calculated. Observations to maintain the Greenwich meridian as reference are no longer made at the Airy transit or, for that matter, at the Greenwich Observatory. Instead, they are made (1986) at Herstmanceaux Observatory, England, and the observations made there are reduced, by calculation, to the equivalent values at the Airy transit. SEE ALSO meridian, prime. A meridian close to that of Greenwich but used as a reference for determining time internationally is maintained by the Bureau International de l'Heure. Although the definitions of the Greenwich meridian and of the meridian used as reference for times are valid, neither provides a fixed reference for either longitude or time. The meridian of Greenwich changes its position with respect to meridians at other observatories because of geological changes in the region surrounding the Greenwich Observatory. The meridian provided by the Bureau International de l'Heure is affected by the errors in time signals from the stations monitored by the Bureau.

Greenwich sidereal date - The whole and fractional number of sidereal days (determine by the equinox of date, that have elapsed at Greenwich since the beginning of the sidereal day that was in progress at Julian Date 0.0 i.e., at noon, 1 January 4713 B.C. For example, 1980 January 0.5 U.T. has the Greenwich sidereal date 2 450 931.77.

Greenwich sidereal day number - The integral number of sidereal days (determined by the equinox of date) that have elapsed at Greenwich since the beginning of the sidereal day that was in progress on 1 January 4713 B.C. (Julian day number 0.0). It is the integral part of the Greenwich sidereal date.

Greenwich sidereal time - The arc of the celestial equator, or the angle at the celestial pole, between the upper branch of the Greenwich celestial meridian and the hour circle of the vernal equinox, measured westward from the upper branch of the Greenwich celestial meridian through 24 hours. It is equivalent to local sidereal time at the Greenwich meridian, and to the Greenwich hour-angle of the vernal equinox expressed in units of time.

Greenwich time - Time based on the Greenwich meridian as reference. In distinction from time based on a local meridian or on the meridian of a time zone.

Gregorian calendar - A calendar in which numbered years not divisible by 4, or divisible by 100 but not divisible by 400, contain 365 mean solar days, while numbered years (called leap years) divisible by 400, or divisible by 4 but not by 100, contain 366 mean solar days. Thus, 1985 is not a leap year but 1984 is; 1600 and 2000 are leap years, but 1700, 1800, and 1900 are not. Therefore, only 97 years out of every 400 are longer by 1 day, and the average length of a year in the Gregorian calendar is exactly 365.2425 mean solar days. This average year is longer than the tropical year by about 26 s. The Gregorian calendar was adopted by all Roman Catholic countries in 1582 October 4/15. Great Britain and the American colonies adopted it in 1752 September 3/14. The USSR adopted it for civil purposes in 1918 February 1/14.

Gregorian telescope - A telescope consisting of a paraboloidal primary mirror and an ellipsoidal secondary mirror place beyond the focus of the primary mirror. There is a real, internal image and the final image, projected through a hole in the primary, is erect.

griblet - A map printed in a subdued monochromatic hue such as gray-blue and used as a base for overprints.

grid - (1) A pattern composed of two distinct families of lines such that, except at one or two singular points on the surface, each line from one family intersects each line from the other family and does so at no more than two points. The lines in each family are commonly numbered. A point on the surface may be located, using the grid, by giving the numbers of the pair of intersecting lines closest to the point and using that pair as the axes of a suitable coordinate system having the point of intersection as origin. A grid on a flat surface is most commonly composed of two families of straight lines intersecting at right angles and is called a plane rectangular grid. A less common grid is composed of concentric circles intersected by straight lines radiating from the circles' center. A common grid on a sphere is composed of one family of circles parallel to a given great circle and of another family of great circles perpendicular to the given great circle (and therefore intersecting at two diametrically opposite points on the sphere.) A grid is sometimes named for the map projection used for producing the map to which the grid is applied e.g., the Lambert grid. A grid differs from a coordinate system in being composed of only a finite number of lines. A coordinate system is, in theory, composed of an infinite number of lines, with the lines in each family corresponding to the numbers of the real-number system. A grid differs from a graticule in not necessarily representing lines of longitude and latitude. (2) A grid as defined in (1), except that only those portions of the lines close to the points of intersection are shown. The resulting pattern is a two-dimensional array of small crosses. It is commonly called a reseau in photo-grammetry. (3) Any systematic pattern composed of lines superposed on a map and lettered and/or numbered in such a way that the location of any feature on the map can be precisely defined. (4) SEE grid, plane rectangular. (5) A graticule (q.v.). This is bad usage. All graticules are grids, but not all grids (in the proper sense) are graticules. (6) SEE coordinate system. As explained under the first definition here, a grid and a coordinate system are not the same thing. SEE ALSO atlas grid; Australian Map Grid; GEOREF grid; Greenwich grid; Lambert grid; National Grid; origin, false; plane rectangular grid; State grid; Universal Polar Stereographic Grid; Universal Transverse Mercator Grid.

grid, alphanumeric - SEE atlas grid.

grid, arbitrary - Any reference system developed for use where no grid is available or practicable, or where military security is desired for the reference.

grid, Canadian - The pattern of lines resulting from the perspective projection of a rectangular grid on one plane onto another plane oblique to the first. Also called a Canadian grid. Perspective grids are used in photo-grammetry to transfer, by visual interpolation, detail from an oblique photograph to a map. The appropriate perspective grid is laid over the oblique photograph. SEE ALSO perspective grid method.

grid, circular - A grid consisting of a family of concentric circles and a family of straight lines radiating from the common center of the circles. Also called a polar grid. The straight lines are usually spaced a constant angular distance apart. The circles are not necessarily a constant distance apart.

grid, geographic - A system of meridians and parallels used to locate points on the Earth's surface. The proper name for such a system is geographic coordinate system.

grid, major - That military grid which is designated as the official grid for a region and is depicted on a map by full lines.

grid, military - A grid, quadrillage or system of rectangles associated with rectangular coordinates, referred to a single origin, and superposed on and extending over the whole of a map. A military grid is intended to make it easier to obtain coordinates of points on a map.

grid, Möbius - SEE Möbius grid.

grid, National - SEE National Grid.

grid, overlapping - A grid extending beyond its normal limits on a map so as to cover maps of regions governed by a different grid. Normally, large-scale maps of regions that lie within approximately 25 km of the dividing line between two grids will bear an overlapping grid indicated by short lines (ticks) drawn outward from the neatline of the map.

grid, parallactic - A plane rectangular grid drawn or engraved on some transparent material, usually glass, and placed either over the photographs of a stereoscopic pair or in the optical system of a stereoscope, in order to provide a floating grid rather than (or in addition to) a floating mark.

grid, perspective - SEE grid, Canadian.

grid, point-designation - An arbitrary, rectangular grid placed on an aerial photograph.

grid, polar - (1) A grid composed of a family of concentric circles and a family of straight lines radiating from the common center of the circles. Also called a circular grid. (2) A plane rectangular grid having the x-axis and y-axis alined respectively with the $0°$-$180°$ and $90°$ E-$90°$ W meridional lines of a graticule, the origin being at the point representing the pole. It is used for aerial navigation in the polar regions. When plotted using a transverse-Mercator map projection of the polar regions, it represents a graticule of transverse meridians and parallels whose poles are at the equator and the $0°$-$180°$ meridian.

grid, rectangular - (1) A grid composed of two families of straight lines, the lines of one family intersecting the lines of the other family at right angles. (2) A grid composed of two families of straight lines, the lines of each family being equidistant and parallel and intersecting the lines of the other family at right angles. Also called a reseau. If the spacing is the same in the two families, the grid is sometimes called a square grid or quadrillage. (3) SEE coordinate system, plane rectangular.

grid, secondary - A grid, in regions now covered by the Universal Transverse Mercator grid or Universal Polar Stereographic grid, considered obsolete but indicated also on the map. Indicating a secondary grid is a temporary procedure adopted to provide a common grid on maps that are of the same general region but at different scales.

grid, square - SEE grid, rectangular.

grid amplitude - Amplitude relative to grid east or west. SEE ALSO amplitude.

grid azimuth - The angle, in the plane of projection, measured clockwise from north, between grid north and any other line. In the State Plane Coordinate System established by the U.S. Coast and Geodetic Survey, grid azimuths were reckoned from south ($0°$) clock-wise through $360°$. To reduce geodetic azimuth to grid azimuth, a quantity $[y_2 - y_1)(2x' + x_2')]/(6\ r_o \sin 1")$ should be subtracted from the geodetic azimuth if the line is more than 7 km long and if full accuracy is wanted. (x_i', y_i are the coordinates of the points whose azimuth is wanted, and the denominator is a constant for the coordinate system of a particular State and is given in the tables of plane rectangular coordinates for that State). While essentially a map-related quantity, a grid azimuth may, by mathematical processes, be transformed into a survey-related or ground-related quantity. SEE ALSO convergence; gisement; correction, arc-to-chord.

grid bearing - The bearing, on a map having a grid, from a north-south line of the grid to a line drawn on the grid.

grid convergence - (1) The angle, at a point on a map, between that line representing the meridian through that point and the meridional line (through the same point) of a graticule on that map. (2) The angle between a meridional line of a graticule and the central meridional line of that graticule. Also called meridional convergence. SEE ALSO gisement.

grid coordinate - A coordinate in a grid coordinate system. In particular, in geodesy, a coordinate in a plane, rectangular, Cartesian coordinate system. A grid coordinate, in this sense, commonly means a distance measured from one of the axes of the system, and parallel to the other axis, to the point in question. In surveying, a constant amount is usually added to each measured distance to avoid negative numbers. Equivalently, the origin is assigned large positive coordinates. Grid coordinates can be transformed into geodetic coordinates and vice versa by simple arithmetic operations if tables (usually readily available) are used. The ordinary computations of surveying can then easily be carried out for regions of large extent using the methods of plane surveying that are otherwise usable only for regions of small extent.

grid coordinate system - (1) A set of two families of lines, on a surface, such that through every point on the surface passes at least one line from each family and through all but a finite number of points passes no more than one line from each family. The grid coordinates of a point are the values of the parameters defining the lies through that point. (2) In particular, a plane rectangular, Cartesian coordinate system. Grid coordinate systems are often distinguished as being military or civilian, according to the use to which they are put. Civilian grid coordinate systems, such as State plane coordinate systems, are invariably rectangular and are used primarily by land surveyors because of the simple mathematics involved in their use. Military grid coordinate systems may be rectangular Cartesian coordinate systems or polar coordinate systems and are used primarily for aerial navigation or for plotting and planning fire by artillery. SEE ALSO State plane coordinate system. (3) A coordinate system in which one coordinate (elevation or geodetic height) is given by distance from a horizontal plane.

grid declination - The angle between the direction of grid north and geodetic north. Also called gisement, grid convergence or convergence. It is commonly denoted by τ. The terms convergence or grid convergence are preferred.

grid equator - A line, on a grid, passing through the origin and perpendicular to the line representing the prime meridian.

grid factor - A combination of elevation and scale factors used to convert from distance on the ground to distance on the grid, or vice versa, in the State Plane Coordinate System or State Coordinate System.

grid inverse - The computation of length and azimuth from coordinates on a grid.

gridiron layout - SEE gridiron pattern.

gridiron pattern - The pattern of rectangular blocks found in most American cities. Also called a gridiron layout.

grid length - The distance between two points as obtained by computation from the plane rectangular coordinates of the points. In the State Plane Coordinate Systems, a grid length differs from a geodetic length by the amount of a correction based on the scale factor for the given line.

grid magnetic angle - SEE grivation.

grid meridian - (1) That line, in a grid on a map, parallel to the line representing the central meridian or y-axis. (2) That line, in a rectangular, Cartesian coordinate system applied to a map, parallel to the line representing the y-axis or central meridian.

grid meridian, prime - That line, in a grid on a map, which represents the central meridian.

grid method - SEE perspective-grid method.

grid method, perspective - SEE perspective-grid method.

grid navigation - Navigation in which the location is referred to lines on an arbitrary grid, as contrasted with navigation in which locations are referred to meridians and parallels of latitude. Sometimes referred to as the grid system of navigation or G system. A chart used for grid navigation may contain a set of straight lines (grid meridians) parallel to the 0° and 180° meridians. The lines are used like true meridians on a chart on the Mercator map projection. Great circles and rhumb lines are identical with respect to the grid meridians. Thus, some navigational problems are much simpler when north on the grid is used instead of north on an ordinary chart.

grid north - That direction, along the south-north lines of a grid, which is designated as north i.e., which is approx-imately the direction of north on the map to which the grid is applied. Grid north coincides with astronomic north only along that line which corresponds to the meridian of origin.

grid number - The number identifying a particular line of a grid.

grid origin - That point, usually near the center of a grid, where the line representing a parallel of latitude intersects the line representing a meridian.

grid parallel - A line parallel to that line, on a grid, which represents the equator.

grid plate - A glass plate on which is etched an accurately ruled grid. Also called a reseau. It is sometimes used as a focal-plane plate to provide a means of determining distortion in the film. It is also used for calibrating plotting instruments.

grid reference - The coordinates of a point on a map, expressed in terms of the letters and numbers of the grid or in terms of numerical coordinates only. Conventionally, easting is given before northing.

grid reference system - A grid and a system of referring to points on the grid such that locations can be referred to or directions and distances between points on the grid calculated accurately and consistently. Approximately equivalent to a grid system (q.v.).

grid reference system, military - A grid reference system using a square grid or standard scale, with its origin placed on a graticule so that locations can be referred to or directions and distances between points on the grid calculated accurately and consistently. Also written as Military Grid Reference System.

grid rhumb line - A line making the same oblique angle with all those lines of a grid which represent meridians.

grid system - A collection of grids and schemes for labeling the lines of the grids. SEE ALSO Universal Polar Stereographic Grid System; Universal Transverse Mercator Grid System.

grid system, British - A collection of rectangular grid systems devised or adopted by the British for use on military maps. There is no related global plan for the many grids, belts, and zones making up the British grid system. It is being replaced by the Universal Transverse Mercator (UTM) grid system.

grid system, tangent-plane - A grid system in a plane tangent to the ellipsoid and having its origin at the point of tangency. The origin is usually given the coordinates 10000 E and 10000 N, or some similar values, to keep all coordinates positive. This system never extends for any great distance.

grid tick - One of a pair of short lines drawn at the opposite edges of a map and perpendicular to the edges, indicating that the line of a grid extends between these marks. In other words, grid ticks are those parts of a grid left after most of each line between the edges of the map has been removed. They are usually used when another system of lines such as a graticule is to be shown in full on the map.

grid variation - SEE grivation.

grid zone - An arbitrary division of the Earth's surface into parts given designations without reference to longitude and latitude.

Grinten map projection - SEE van der Grinten map projection.

grivation - The direction of the magnetic meridian, given as an angle measured clockwise through 360° from the corresponding meridian on the grid. Also called a grid magnetic angle. It is the difference between grid north and magnetic north.

groin - A structure, generally consisting of tight, sheetlike piling of creosoted timber, steel or concrete, braced with wales and with round, very long piles, and extending out from the shore in such a way as to prevent of check erosion of the shore by currents or waves.

ground camera - SEE camera, terrestrial.

ground control - Control established by control surveys on the ground, as distinguished from control established by photogrammetry using aerial photographs.

ground control point - A point, established by surveying, on the ground and used to determine scale, orientation, etc., for an aerial photograph or set of photographs. Also called a ground-control point.

ground coordinate - SEE space rectangular coordinate.

ground coverage - That area, on the ground, which is illuminated by a transmitter or visible to a detector carried on an aircraft or satellite. The term is usually understood to refer to a particular instant of time, the terms swath or, more often total ground coverage being used to denote the area illuminated or detected over a specific interval of time.

ground coverage, instantaneous - The ground coverage at a particular instant of time. Also called footprint.

ground distance - The distance, along a great circle, between two points on the ground. Contrasted with slant range, the distance along a straight line between the points. Also called ground range. This definition is defective; a great circle cannot be drawn on the ground. A workable definition can be obtained by substituting geodesic for great circle.

ground elevation, assumed - The elevation presumed to prevail in the locality shown on a particular photograph or group of photographs. It is used particularly to denote the elevation assumed to prevail in the vicinity of a critical point such as a peak or other feature having abrupt local relief.

ground elevation, mean - The average elevation above sea level of the terrain included in a set of aerial photographs.
Ground Elevation Meter - SEE elevation meter.

ground gained forward - The net increase, in distance or area per photograph, in the direction of flight for a specified amount of overlap. Commonly abbreviated as GGF. It is used to compute the number of exposures in a strip of aerial photographs.

ground gained sideways - The net lateral increase, in distance or area per flight, for a specified amount of sidelap. Commonly abbreviated as GGS. It is used to compute the number of flight lines needed to completely photograph a given region.

ground nadir - That point on the ground which is vertically underneath the perspective center of an airborne camera's lens system. Also called ground plumb point and nadir, although the latter term is ambiguous.

ground parallel - That line in which the plane of a photograph intersects the plane of reference on the ground. Also called the ground trace. It is a particular case of the axis of homology (q.v.). SEE ALSO map parallel.

ground photogrammetry - SEE photogrammetry, terrestrial.

ground photograph - SEE photograph, terrestrial.

ground plane - The horizontal plane passing through the ground nadir of a camera station.

ground plumb point - SEE ground nadir.

ground pyramid - A geometric figure (pyramid) whose base is the triangle formed from three control-points on the ground and whose apex is the perspective center of the camera at the instant the photograph showing the control points was taken.

ground range - SEE ground distance.

ground resolution - The size, in length or area, of the smallest pattern or region, on the ground, distinguishable on an image. SEE ALSO pixel.

ground speed - The speed at which an aircraft moves along its track with respect to the ground. It is the magnitude of the resultant of the heading and air speed of an aircraft and the direction and speed of the wind i.e., of the aircraft's velocity with respect to the air and the velocity of the wind with respect to the ground. SEE ALSO air speed.

ground station - A monumented survey station at which is placed equipment used in obtaining locations of an aircraft taking photographs.

ground survey - A survey made on the ground and not by an aerial survey. A ground survey provides coordinates of points on the ground, distances between points on the ground, or angles between points on the ground, or any combination of these quantities. It is used to provide ground control to which photogrammetrically determined coordinates can be adjusted. It may or may not use photographs.

ground swing - (1) A condition, in measuring distances by transmission of radio waves between the points of interest, in which some of the radiation is reflected from the ground or water's surface between the points and interferes with that radiation which travels directly between the points, causing the measuring instrument to give an erroneous value for the distance. (2) The variation of a distance, measured with an electromagnetic distance-measuring instrument at radio frequencies, from the average value of the distance as the wavelength of the measuring frequency is varied. The condition exists because some of the radiation is reflected from the ground or other objects along the way from the instrument to the target and back. This condition is called multipath radiation. Radiation arriving at the instrument by other than the direct path from instrument to target and back interferes with the direct-traveling radiation and causes the instrument to give erroneous or unreadable measurements. By varying the wavelength of the modulating waves, the error can be made to vary cyclically and be averaged out to a great extent. (2) The erroneous measurements themselves.

ground trace - SEE ground parallel.

ground truth - The value or values of any quantity as determined by accurate measurement on the Earth's surface (as contrasted with values determined by some other method of unknown accuracy). (jargon) Also called reference data or ground checked data.

ground water - All water in the ground and lying below the surface of the ground, particularly that part lying in the zone of saturation. Also, and preferably, spelled ground-water.

ground water level - The upper surface of the water in the ground.

group velocity - The speed of the envelope of a group of waves of different frequencies. The energy in a wave travels at the speed with which the envelope travels. Group velocity s_g is related to phase velocity s_p by $s_g = s_p - \lambda (ds_p/d\lambda)$, in which λ is the wavelength of a particular component. For waves on the surface of deep waters, the group velocity is equal to one-half the speed of individual waves in the group; for waves in shallow waters, it is the same as their speed. SEE ALSO phase velocity.

growth, axial - Extension of a growing city outward along the main routes of transportation. The pattern created by axial growth is usually star-shaped.

Guam map projection - An approximation to the azimuthal equidistant map projection from the rotational ellipsoid to the plane and producing an oblique aspect of the graticule, as used by the U.S. National Geodetic Survey for maps of Guam.

guarantee of title - SEE title, insured.

guarantee title - SEE title, insured.

guard stake - A stake driven into the ground near a marker (hub), usually sloped so that the top of the stake is over the marker. The guard stake protects, and its marking identify, the marker.

guidance, map-matching - Navigation (guidance) by map-matching.

guide - SEE scribing guide; guide, color-registration.

guide, color-registration - A display, on a lithographic copy of a chart, which shows accurately the amount and direction of misregistration between the graticule and certain significant overprint. SEE ALSO scribing guide.

guide meridian (USPLS) - A line projected northward along an astronomical meridian, from a point established on the base line or on a standard parallel, and on which township, section, and quarter-section corners are established. Guide meridians are usually spaced at intervals of 24 miles east or west of the principal meridian.

guide meridian, auxiliary (USPLS) - A new guide-meridian established for purposes of control where the original guide-meridians were placed farther apart than 24 miles. Auxiliary guide-meridians may be needed to limit errors of old surveys or to control new surveys. They are established by surveying in the same way as are regular guide-meridians. They may be assigned a local name, such as Grass Valley Guide Meridian or Twelfth Auxiliary Guide Meridian West.

guide telescope - A telescope attached to another and used for keeping that other telescope directed at a particular point in the heavens while observations are being made. Guide telescopes are most often used when the telescope to which it is attached is being used for photographing or spectro-scopy and cannot be seen through. A guide telescope must have about the same magnifying power as the telescope it guides or the aiming will not be accurate enough. In particular, a refracting telescope attached to a telescope used for photographing.

Gulf Coast Low-Water Datum - The tidal datum used as chart datum for the coastal waters of the Gulf Coast of the United States of America and defined as mean lower low water when the tide is of the mixed type and as mean low water when the tides are diurnal. The datum was abandoned by the National Ocean Survey in 1980 and replaced by mean lower low water for the region involved.

Gunter's chain - A measuring device, once used in land surveying, composed of 100 metallic links fastened together with rings; the total length of the chain is 66 feet. Also called a four-pole chain. Invented by the English astronomer Edmund Gunter about 1620, it is the basis for the chain and link, units of length used in surveying the public lands of the USA. The original form of the chain has been replaced by metallic tapes or ribbons graduated in links and chains, but the new forms are still called chains.

Gutenberg discontinuity - SEE Weichert-Gutenberg discontinuity.

Gutenberg-Weichert discontinuity - SEE Weichert-Gutenberg discontinuity.

guyot - (1) A flat-topped seamount having a crest lying at least 100 fathoms (183 meters) below the surface of the sea. (2) A seamount rising more than 500 fathoms (approximately 100 meters) from the oceanic floor and having a comparatively smooth, flat top with minor irregularities. The guyot is also called a tablemount.

Guyou's map projection - A conformal map projection of a hemisphere into a square, with the poles at the centers of opposite sides of the square. The lines representing meridians and parallels of altitude are doubly-periodic, elliptic integrals of the first kind. Also called a doubly-periodic map-projection. The entire sphere is mapped into a rectangle with sides in the ratio 2:1.

gyroazimuth theodolite - SEE gyrotheodolite.

gyrocompass - SEE compass, gyroscopic.

gyrocompassing - The process of aligning a gyroscopic compass in azimuth (north) without any external equipment.

gyro-meridian indicating instrument - SEE gyrotheodolite.

gyroscope - Any device using a spinning mass to establish or maintain a fixed direction. A spinning mass resists any attempt to change the direction in which the axis of rotation points. A torque applied to the mass and perpendicularly to the axis of rotation causes the axis of rotation to precess. Gyroscopes can therefore be used to find North (as in gyroscopic compasses and gyroscopic theodolites) or to maintain the orientation of a structure (as in inertial navigation systems.) The basic law governing the motion of a free gyroscope is $dI\omega / dt = T / I\omega$, in which I^* is the vector giving the angular velocity of the spinning mass, T is the torque applied perpendicularly to axis of rotation, ω is the rate of rotation, I is the magnitude of the angular momentum and t is the time. The name gyroscope was applied to the device by Foucault in 1852.

gyroscope, pendulous - A gyroscope whose axis of rotation is constrained, by a suitable weight, to remain horizontal.

gyroscope, ring-laser - A gyroscope based on the principle that if two mutually coherent rays of light are sent in opposite directions in a circular path which is rotating in one direction, the one ray will be shifted in frequency with respect to the other by an amount proportional to the rate of rotation (the Sagnac effect). Also called a ring laser or laser gyroscope. At present, there are two forms of the ring-laser gyroscope. One form uses mirrors or prisms to direct the rays; the other sends the rays in opposite directions through optical fibers.

gyroscopic effect - The tendency of the axis of rotation of an object rotating at high speed to maintain its initial direction in space without modification.

gyrotheodolite - A theodolite having a gyroscopic compass built into it so that angles can be measured with respect to that North indicated by the compass. Also called a gyro-azimuth theodolite, gyroscopic meridian-indicating instrument, gyro-meridian indicating instrument and gyroscopic theodolite. The instrument is particularly useful in mines and other places (such as the far north) where direction is difficult to establish by other means.

H

Haalck gravimeter - Essentially, a barometer in which the end ordinarily open to the air is sealed off, inclosing a suitable amount of air. The height of the mercury column in the tube then depends on the weight of the mercury and the pressure of the contained air. The pressure of the contained air depends mostly on the air's temperature. An accuracy of 1 mgal can be attained only if the height of the mercury and the temperature of the air are measured with extremely high accuracy.

habendum clause - SEE clause, habendum.

hachure - Short lines running in the direction of greatest slope, the thickness and spacing of the lines indicating the amount of relief. The lines are short, heavy and close together for steep slopes; they are longer, lighter and more widely spaced for gentle slopes. Artistically done, hachure is effective in showing relief, although actual elevations are not indicated. SEE ALSO hachuring; Lehman system.

hachuring - A method of showing, on a map, the slopes of the ground by sketching lines of varying thickness and spacing in the direction in which the ground slopes. The pattern produced on the map is called hachure. Hachuring was introduced in the middle years of the 17th century. Hachuring assumes that the light falls vertically from above. SEE ALSO hachure; Lehman system.

hachure line - SEE hachure.

hack (USPLS) - A horizontal, V-shaped notch cut into the trunk of a tree at about breast height.

hack chronometer - A chronometer observed visually for meeting the schedule of observations, making routine settings of instruments, etc., but not usually used as an accurate source of time or source of recorded time.

hack watch - A watch with a device for stopping the motion of the balance, so that the time may be set precisely.

hairline - The finest line normally produced in printing, approximately 0.08 to 0.1 mm thick.

halation - The spreading of a photographic image beyond its proper boundaries because of, in particular, reflection from that side of the film or plate opposite to that on which the emulsion is coated. Halation is particularly noticeable in photographs of bright objects against a darker background.

half-line - A (straight) line having one point, called the end point, such that the line extends indefinitely from that point in one direction only.

half-mark - SEE index mark.

half-model - The stereoscopic model formed by viewing the overlap of two adjacent right-hand or left-hand exposures of convergent photographs.

half-plane - The set of all points lying on one side only of a given (straight) line. Alternatively, the set of all points such that if any two points are connected by a line, that line either lies entirely on a given straight line, called the edge of the half-plane, or has no point in common with the given line.

half section - SEE section, half.

half-space - The set of all points lying on one side only of a given plane. Alternatively, the set of all points such that if any two points are connected by a line, that line either lies entirely on a given plane, called a face of the half-space, or has no point in common with that plane.

halftone process - SEE process, halftone.

halftone screen - SEE screen, halftone.

Hammer template - A transparent, plastic sheet for laying over a map, having a family of concentric circles and a family of corresponding radii drawn on it according to a scheme devised by S. Hammer, so that the angles between radii and the spacings between circles simplify the calculation of topographic gravity-corrections for geophysical exploration. The template and its accompanying tables were improved by J. Bible (1962) and by V. Douglas and S. Drahl in 1972, to allow calculation to 0.1 mgal. The tables have been further simplified by presentation in the same form as the luni-solar gravity correction.

Hammer-Aitoff map projection - An equal-area map projection transforming the meridional form of the azimuthal equal-area map projection onto a plane tilted 60° about a line tangent to the equator, with the values of the lines representing meridians doubled to range over the entire surface being mapped. Also called Hammer's equal-area map-projection and, improperly, Aitoff's equal-area map-projection. An oblique version with origin on the Greenwich meridian at 45° N is called the Nordic map projection. The asymmetric, oblique version with origin at 10° W, 45° N is called Bomford's map projection and, improperly, an oblique Aitoff map projection. The equations for mapping from the sphere to the plane are:

$$x = 2\sqrt{2} k R (\sin\tfrac{1}{2}\Delta\lambda \cos\phi)/D$$
$$y = \sqrt{2} k R (\sin\phi)/D$$
$$D = (1 + \cos\phi \cos\tfrac{1}{2}\Delta\lambda)^{1/2},$$

in which x and y are the rectangular Cartesian coordinates of that point, in the plane, corresponding to the point, on the sphere, at longitude $\lambda_o + \Delta\lambda$ and latitude ϕ. λ_o is the longitude of the central meridian, R is the radius of the sphere and k is the scale factor. The resulting graticule is bounded by an ellipse in which the length of the line representing the equator (major axis) is double the length of the line representing the central meridian (minor axis). Tables for constructing the map projection can be prepared arithmetically from the tables for the azimuthal equal-area map projection. The map projection devised by Aitoff is not an equal-area map projection. Hammer, in 1892, substituted the azimuthal equal-area map projection for the azimuthal equidistant map projection used by Aitoff, thereby achieving a true equal-area map projection.

hand compass - SEE compass, hand-held.

hand compass, liquid - SEE compass, liquid-damped, hand-held.

H & D curve - SEE curve, characteristic.

handle, clamping - A clamp designed to grip the flat ribbon of a surveyor's tape without kinking the tape and to fit easily in the hand when measuring less than full lengths of tape.

hand lead - (1) A light (3 -7 kg) lead attached usually to a line not more than 20 fathoms (about 40 meters) long. (2) A light (3 - 6 kg) sounding lead attached to a line usually less than or equal to 25 fathoms (about 46 meters). (3) The lead-line consisting of the lead and line referred to in the preceding definition.

hand level - A small leveling instrument designed for holding in the hand while observing. The spirit level is so mounted that the observer can see the bubble at the same time that he/she is observing an object through the telescope. The bubble is viewed by means of a prism or mirror in the telescope. The instrument is used in reconnoitering for surveys, in taking cross-sections, and in other work where great precision or accuracy is not required. The Abney level and Locke's level are well-known varieties. SEE ALSO Abney hand level; Locke hand level.

hand proof - A proof made from a lithographic plate on a press operated entirely manually.

hand signal - Motions of the hands and arms made by the observer to direct his rodperson or chainperson.

hand templet - A templet made by tracing the radials from a photograph onto a transparent sheet. Hand templets are laid out and adjusted by hand to form the radial aerotriangulation.

hand-templet aerotriangulation - SEED hand-templet method.

hand-templet method - Each aerial photograph is photographically reduced onto a transparent templet on which lines have been drawn from the principal point radially through each picture point. The templets are assembled and the intersection of all radials common to each identified picture point represents the actual location, on a map, of that point. Also called hand-templet aerotriangulation, hand-templet plot and hand-templet triangulation.

hand-templet plot - SEE hand-templet method.

hand-templet triangulation - SEE hand-templet method.

Hansen's formula - A formula relating the difference between a star's right ascension α and that star's observed time T of transit across the meridian to the correction (clock correction) ΔT to the observed time, the astronomical latitude Φ of the observer, the star's declination δ and the azimuth constant a, the level constant b and the collimation constant c. It is usually written as $\alpha = T + \Delta T + b \sec \Phi + n (\tan \delta - \tan \Phi) + c \sec \Phi$, in which $n = b \sin \Phi - a \cos \Phi$.

Hansen's method - A method of solving Hansen's problem in resection which occurs when only two points of known location are available. An auxiliary point is established and angles measured from that point and from the unknown point to the two given points.

Hansen's problem - The problem of determining the (unknown) location of an occupied point by measuring angles at that point between an auxiliary point and two points whose locations are known or by measuring directions to the two points of known location. SEE ALSO problem, two-point.

harbor - A stretch of water partly enclosed and so protected from storms as to allow vessels to load, unload, refuel, repair, etc., in safety.

harbor chart - A nautical chart intended for anchorage and navigation in harbors and smaller waterways.

harbor line - That line beyond which wharves and other structures may not be extended.

harmonic - (1) A sinusoidally varying quantity having a frequency equal to an integral multiple of the frequency of a periodic quantity to which it is related. (2) SEE function, harmonic.

harmonic, gravitational - (1) One of the terms in the expansion of a gravitational field into a series of spherical harmonics functions. (2) SEE gravity harmonic.

harmonic, sectorial - SEE function, sectorial harmonic.

harmonic, spherical - SEE function, spherical harmonic.

harmonic, tesseral - SEE function, tesseral harmonic.

harmonic, zonal - SEE function, zonal harmonic.

harmonic coefficient - SEE coefficient, harmonic.

harmonic component - SEE component, harmonic.

harmonic expression - SEE expression, harmonic.

harmonic function - SEE function, harmonic.

Hassler base line apparatus - An optical base-line apparatus consisting of four rectangular iron bars mounted end to end in a wooden box. Each bar was 1 meter long, the combined length of the apparatus being 8 meters. Also called a Hassler base-line measuring apparatus.

Hassler base-line measuring apparatus - SEE Hassler base line apparatus.

hat, cocked - SEE triangle of error.

hatching - (1) Drawing or engraving fine, parallel or crossed lines on a picture to indicate shading; i.e., to give the visual impression of solidity. SEE ALSO hachure; shading. (2) The representation of shading by evenly space, parallel lines.

hatching, cross - SEE cross-hatching.

Hatt's map projection - SEE map projection, azimuthal equidistant.

Hause method - A method of carrying direction from a line of known azimuth on the surface, or at one level, to a horizontal line of unknown azimuth in another level by measuring directions and distances at the surface (or in the first level), using the line of known azimuth, to two plumb lines in a vertical shaft extending to the second level in question, and measuring directions to the two plumb lines from two stations a known distance apart in the second level. Also called the quadrilateral method or the Weiss method. SEE ALSO Weissbach's method.

Hayford-Bowie gravity table - One of a set of tables prepared by J.F. Hayford and W. Bowie (1912) giving combined topographic and isostatic gravity corrections for a depth of compensation of 113.7 km. Also called a Hayford gravity table.

Hayford-Bowie method of gravity reduction - A method devised by Hayford and Bowie for calculating the complete topographic and isostatic gravity corrections in the Pratt-Hayford gravity reduction. The complete topographic gravity correction δg_{tc} accounts for all matter outside the level-surface of reference having an effect at the point of measurement. It is calculated by laying a polar grid (the Hayford template) on a topographic map, with the origin at the point in question, reading off the average elevation within each quadrangle of the grid, and calculating the corresponding complete topographic gravity reduction from a formula or table. To this term is added the Pratt-Hayford gravity correction δg_{PH}.

Hayford-Bowie method of isostatic reduction of gravity - SEE Hayford-Bowie method of gravity reduction.

Hayford-Bullard method of gravity reduction - A modification of the Hayford-Bowie method of gravity reduction. The topographic effect of an infinite slab of density 2.67 g/cm3 and a thickness equal to the elevation of the point at which gravity was measured is first computed; this is then corrected for curvature of the geoid, the effect of topography, and finally for variation of the actual topography from the slab. The first step in the computation gives the ordinary Bouguer gravity correction for topography; the second step takes account of the departure of the actual topography from a plate (or smooth cap, if this is used instead of a slab) whose thickness is equal to the elevation of the point. Applying the correction for curvature in the lettered zones (of the Hayford template) reduces the effect of the flat slab to that of a plateau or cap extending to the outer limits of those zones and curved to fit the reference surface. Topography is taken into account by applying a correction for deviations of the topography from the surface of the plateau or cap. This method was devised by E.C. Bullard as a substitute for the Hayford-Bowie method of gravity reduction in the lettered zones of the Hayford template.

Hayford-Bullard method of isostatic reduction of gravity - SEE Hayford-Bullard method of gravity reduction.

Hayford deflection-of-the-vertical template - A Hayford template having the spacings of circles and radii so proportioned (with respect to azimuth and scale of the accompanying map) that elevations and depths within each compartment formed by adjacent arcs and radii can be averaged easily and the effect of the mass therein upon a plumb line at the station (center of the circles) be, under various hypotheses, calculated. The template was devised by J.F. Hayford for use in connection with studies of the figure of the Earth and isostasy, and has been used since by many geophysicists for the same purpose.

Hayford effect - The direct effect, on gravity, of masses of unit density extending to various distances above and below sea level (the geoid). The Hayford effect does not take into account differences of elevation (geoidal heights) between the reference ellipsoid (terrestrial ellipsoid) and the geoid.

Hayford gravity anomaly - SEE Pratt-Hayford gravity anomaly.

Hayford gravity template - A Hayford template having the spacings of circles so proportioned (with respect to the scale of the accompanying map) that elevations and depths within each compartment formed by adjacent arcs can be averaged easily. The Hayford gravity-template takes no account of the azimuths of the radii; all compartments bounded by a given pair of circles are the same size and shape. It is used for calculating topographic gravity- corrections.

Hayford-Pratt gravity anomaly - SEE Pratt-Hayford gravity anomaly.

Hayford spheroid - (1) A rotational ellipsoid, derived by Hayford in 1909, having the dimensions
 semi-major axis 6 378 388 m ± 18 m (probable error)
 flattening 1/297.0 ± 0.5 (probable error)
The ellipsoid was derived using the theory of isostasy to determine the geoid, to which the spheroid was fitted. Hayford's value for the length of the semi-major axis and the flattening were adopted, but without the probable errors, by the International Association of Geodesy at its general Congress in Madrid in 1924, as the defining values of the International Ellipsoid of Reference. However, because the Hayford spheroid was derived using only measurements made in the USA, it did not fit the geoid elsewhere as well as was thought at the time. (2) SEE International Ellipsoid.

Hayford template - A transparent, plastic sheet for laying over a map, having a family of concentric circles and a family of corresponding radii drawn on it according to a scheme devised by J.F. Hayford, so that the angles between radii and the spacings between circles simplify the calculation of deflections of the vertical or topographic gravity-corrections. A given template can be used only on maps of the scale and on the map projection for which it was drawn.

Hayford zone - A ring-shaped region centered at a point P on a level ellipsoid of rotation, having inner and outer radii specified according to a scheme devised by Hayford for calculating the effects of topographic masses on gravity. The first 15 zones from the center outward are labeled A through O, consecutively. The next 18 zones are labeled 18 through 1, consecutively. Zone A consists of the entire region within a circle 2 m in radius; zone O has an inner radius of 99 km and an outer radius of 166.7 km (1° 29' 58"). Zone 1 extends from 150° 56' to 180°. Zones C through O are frequently subdivided each into two zones differentiated by subscripts 1 and 2: e.g., zone C_1 and zone C_2.

head, leveling - SEE quadribrach; tribrach.

headframe - That structure, at the top of a shaft, which houses hoisting equipment.

heading - (1) The azimuth of the longitudinal axis of a vehicle; the straight-ahead direction. (2) (mining) A horizontal passage or drift of a tunnel. (3) (mining) The end of a drift or galley.

heading sensitivity - The change in reading of an inertial surveying system with change in direction of motion of the system. Ideally, an inertial surveying system should indicate the same distance traveled per unit distance actually traveled, regardless of the direction of travel. In fact, some systems indicate a different distance when the direction of travel is north-south than when the direction of travel is east-west.

headland - (1) A land mass having a considerable elevation. (2) The apex of a salient of the coast, the point of greatest extension of a portion of the land into the water, or a point on the shore at which there is an appreciable change in the direction of the general trend of the coast.

headland-to-headland line - SEE line, headland-to-headland.

head tapeman - SEE tapeman.

headwall - (1) The vertical or sloping structure constructed to prevent an embankment or other earthen material from encroaching over the ends of a culvert. (2) The steep, wall-like cliff at the back of a cirque.

heath - (1) A tract of wasteland. (2) An open, level region clothed with a characteristic vegetation consisting principally of underbrush of the genus Erica. (Brit.)

heave (mining) - The horizontal distance between the parts of a vein that have been separated by a fault, measured along the strike of the fault. SEE ALSO throw (mining).

heave correction - A quantity added to the depth obtained by echo sounding to compensate for the vertical displacement (because of waves) of the sounding vessel above or below the average surface of the water.

Heaviside layer - An ionospheric layer now generally called the E-layer. Also called the Kennelly-Heaviside layer. SEE ionosphere.

hectare - Approximately 2.471 acres; 100 ares; 10 000 square meters.

hedgerow - A row of shrubs or trees, planted for enclosure or separation of fields.

heel (of a frog) - That part of a frog which extends from the point of the frog to the part farthest from the switch. Also called frog heel.

height (n.) - (1) The vertical distance between the top and bottom of an object; e.g., the height of a building or the height of a man. By analogy, one speaks of the height of a mountain when one thinks of the mountain as an object with top and bottom. However, for historical reasons connected with the use of barometers for measuring heights, one speaks of a point on the object as having a certain altitude; e.g., The peak is at an altitude of 3 000 meters. (2) The perpendicular distance between the horizontal lines or planes passing through the top and bottom of an object. (3) The perpendicular distance between a point and a reference surface; e.g., the height of an aircraft. In the case of an aircraft, for example, the reference surface is the surface of the Earth or a plane fitted to that surface. However, the term height may, when applied to an aircraft, be used for altitude in the sense of vertical distance above a surface of constant barometric pressure. Such use should be avoided. In geodesy, the term is often used for geodetic height, the reference surface being an ellipsoid. (4) The vertical distance between a point and a level surface of reference, taken to be positive upwards i.e., the elevation of the point with respect to that surface. There is no general agreement on the use of height in this sense. Some authors use height as a synonym for elevation while others restrict the meaning of height to the first two given above. (5) The distance of a point on the water's surface above a reference surface through a specified point or defined statistically (e.g., mean water level). The distances involved are commonly so small and their determination so inaccurate that no effort is usually made to specify whether the distance is to be measured along the vertical or the perpendicular. Most oceanographers use the term height in preference to elevation for distances measured on a tidal gage, etc. SEE ALSO Baranov elevation; spot height.

height (of instrument) - (1) In spirit leveling, the vertical distance, above the adopted surface of reference, of the intersection of the line of sight with the axis of rotation of a leveling instrument. (2) In stadia surveying, the height of the center of the telescope (horizontal axis) of a transit or telescopic alidade above the ground or station mark. (3) In trigonometrical leveling, the height of the center of the theodolite (horizontal axis) above the ground or station mark.

height (of target) - The height above the reference point that the rod or pole is set on.

height (of the tide) - The vertical distance, at any instant, between sea level and chart datum.

height (v.) - To measure height (n.) above the ground.

height, absolute geoidal - Geoidal height above an ellipsoid of reference having its center at the Earth's center of mass.

height, approximate - SEE height, practical.

height, astrogeodetic - Geoidal height referred to an astrogeodetically oriented ellipsoid.

height, auxiliary - SEE height, normal.

height, barometric - SEE altitude, barometric.

height, dynamic - (1) That vertical distance between the point P_n on the Earth's surface and the point P_o on the geoid which is given by the function $\int (g/\tau_{45}) dH$ (the integral being taken from P_o to P_n), in which g is the value of gravity acceleration along the path of integration, τ_{45} is the value of gravity at 45° latitude, as calculated from a standard gravity formula and dH is the increment of vertical distance along the path of integration. Also called dynamic number. (2) The same as the preceding definition except that the value of τ is calculated for an arbitrarily selected latitude. (3) The same as the first definition except that the value of τ is calculated for the latitude of the point P_n. (4) The same as the first definition except that τ_{45} is the value of gravity acceleration calculated for (mean) sea level at 45° latitude. This definition assumes that the surface of mean sea level is symmetrical with respect to the equator and with respect to the Earth's axis of rotation. (5) The integral $\int g \, dH$ of gravity acceleration with respect to vertical distance H between a point P_o on the geoid and a point on the level surface through P_n, the point for which the dynamic height is wanted (the integral being taken along the path from P_o to P_n). Also called geodynamic height and dynamic elevation. The standard unit for this quantity is the dynamic meter (also called the geodynamic meter), defined as 10 m²/s². It corresponds to a distance about 2% longer than the meter but is of course not itself a distance. SEE ALSO depth, dynamic; depth anomaly, dynamic. (6) A number proportional to, or related to, the difference in gravity potential between two points. (7) The amount of work done when a particle of water of unit mass is moved vertically from one level to another. The dimensions are those of potential energy per unit mass. One of the levels is usually a reference level. An equivalent definition uses the concept of gravity potential instead of work but is practically equivalent to that given here. E.g., the International Great Lakes Datum assigns a value of zero to the gravity potential at mean water level at Pointe-au-Pierre. The quantities defined by (5), (6) and (7) are not heights, nor are they elevations. Dynamic heights according to definitions (5), (6) and (7) are practically equivalent to dynamic number, differing only in small and possibly unintended ways. SEE ALSO number, dynamic. (8) The geopotential number at a point divided by a convenient, constant average value of gravity. (9) SEE depth, dynamic.

height, ellipsoidal - The sum of the elevation of a point and the geodetic height of the geoid above the reference ellipsoid at the corresponding point on the geoid. Ellipsoidal height is neither a height nor an elevation, but the quantity is sometimes convenient as a substitute for the actual vertical distance above a terrestrial ellipsoid.

height, geodetic - The perpendicular distance from a specified ellipsoid (reference ellipsoid) to the point of interest. Also called geometric height. Some British textbooks call it spheroidal height. Note that geodesists distinguish sharply between geodetic height, which is measured along the perpendicular to the reference ellipsoid, and elevation, which is measured along the vertical to the reference surface. Conventionally, h is used for geodetic heights, H for elevations.

height, geodynamic - SEE height, dynamic; depth, dynamic.

height, geoidal - (1) The perpendicular distance from the reference ellipsoid to the geoid. Also called geoidal separation and, erroneously, geoidal undulation. (An undulation is a wave and has no single distance above the ellipsoid.) Geoidal undulation is properly used to mean the varying distance of the geoid from the reference spheroid. Also called geoid height, geoid separation and undulation of the geoid. (2) Distance above the geoid, measured, indifferently, along the normal to a reference ellipsoid or along the vertical to a terrestrial ellipsoid. This is a British usage and has no corresponding term in American usage. SEE ALSO elevation; elevation, orthometric.

height, geometric - (1) The perpendicular distance from a given surface to a specified point. (2) A height determined geometrically as contrasted to a height determined dynamically. (3) SEE height, geodetic.

height, geopotential - SEE geopotential elevation.

height, gravimetric geoidal - Geoidal heights determined gravimetrically.

height, metric - A height (i.e., distance), as distinguished from quantities which are not distances but which are, by convention, called heights (e.g., dynamic height or thermal height).

height, Molodenski - SEE height, normal.

height, normal - (1) The quantity h_{NM} introduced by Molodenski (Molodensky) in 1945 as a substitute of the orthometric elevation and defined by the implicit equation $-\int g \, dH = U(L_N, \Phi_N, h_{NM}) - U(L_N, \Phi_N, 0)$. The integral is determined by measuring elevation differences dH and gravity-acceleration g along a route from point P_o on the geoid to the point P_N of interest, i.e., the integral is taken along a path from P_o to P_N. The quantity $U(L_N, \Phi_N, h_{NM})$ is a theoretical value of the gravity potential which agrees with the gravity potential on the geoid in those regions where there is no mass above the geoid; it is given elsewhere by the zeroth- and second-degree harmonic functions in the expression of the gravity potential as a series of Legendre functions. L_N and Φ_N are the astronomical longitude and latitude of P_N. The normal height can be thought of as the distance, measured along the curved plumb-line, of P_N above a theoretical surface, the quasi-geoid. The normal height is also referred to as the auxiliary height, normal height of Molodenski, Molodenski height and Molodenski elevation. The term has also been used for the Vignal height. (2) The distance from that spherop which has the same value of gravity potential as the geoid to that spherop which has the same value of gravity potential as the point P_N of interest. This is properly referred to as Hirvonen's normal height. It was introduced by R.A. Hirvonen about 1960 as a substitute for orthometric elevation. Other things being equal, Hirvonen's normal height is numerically equal to Molodenski's normal height but has been defined in terms of two theoretical surfaces rather than one. (3) SEE elevation, orthometric.

height, normal dynamic - The quantity hNnd calculated from the formula $hNnd = [U(L(N), \Phi(N), hN) - U(L(N), \Phi(N), hmsl)]/\tau_{45}$, in which $U(L(N), \Phi(N), hN)$ is a specified spheropotential function, τ_{45} is the value of gravity acceleration calculated from $(\delta U/\delta h)$ at 45° latitude for h = 0, and hmsl is that value of h in U which causes the corresponding equipotential surface to go through the mean-sea-level mark at a specified tide-gage. Note: Since, in general, mean sea level and the geoid do not coincide, the values of the spheropotential function on the geoid and at mean sea level will be different.

height, normal orthometric - SEE elevation, normal orthometric.

height, orthometric - (1) SEE elevation, orthometric. (2) Any one of several different quantities similar to the orthometric elevation, e.g., normal orthometric elevation.

height, practical - Any elevation intended for general use and published. Also called approximate height or approximate elevation. The term practical height was introduced to designate elevations derived by formulas known to be and intended to be approximate. Baranov's, Ledersteger's, Ramsayer's and Vignal's elevations are usually considered to be practical heights.

height, quasi-dynamic - The quantity hQ calculated using the function $[\int g(H) dH] / \tau_{f\phi}$, in which g(H) is the measured value of gravity acceleration at elevation H, $\tau_{f\phi}$ is the acceleration τ_o calculated from a standard gravity formula for the latitude ϕ and corrected by half the free-air gravity correction δg_f. Integration is from point P_o on the geoid to the point P_n vertically above it on the surface. ($\tau_{f\phi} = \tau_\phi + \delta g_f/2$). Physically, it is intended to be the distance, taken along the vertical, between the geoid at latitude ϕ and the geop through the point P_n whose elevation is wanted (free-air gravity is calculated at the mid-point between the two surfaces). It is therefore the same as dynamic height except that free-air gravity is used instead of standard gravity.

height, quasi-geoid - The difference between the geodetic height of a point on the Earth's surface above the reference ellipsoid, and the normal height of that point.

height, relative geoidal - (1) At a given point, the difference between the geoidal height there and the geoidal height at another point used as reference. (2) The geoidal height above an ellipsoid of reference having its center at some point other than the Earth's center of mass.

height, sea-level - The elevation of sea level as measured from the shore.

height, spheroidal - SEE height, geodetic.

height above mean sea level - (1) The elevation of a point, calculated from the data of a leveling survey, the elevations of certain tide-gage bench-marks being given specified values. This concept dates from a period when mean sea level and the geoid were treated as equivalent terms. (2) The distance, measured vertically, of a point above the geoid. I.e., the elevation of a point. This definition applied in those instances, such as in the legends on maps, where heights are stated as distances above mean sea level but are actually elevations above the geoid.

height anomaly - (1) The distance from that level ellipsoid used with a standard gravity formula to the quasi-geoid. This is the definition given by Molodenski, who invented the term. A different definition of the term should be distinguished in some way, such as by attaching the name of the author of the new definition. (2) The difference (h - H) between the geodetic height h of a point and the elevation H of that point. (3) The difference (h - H*) between the geodetic height h of a point and the normal height H* of Molodenski or Hirvonen for that same point. (4) The difference in height at two points. (5) SEE elevation difference.

height anomaly, dynamic - The excess of the actual difference of geopotential, between two given surfaces of constant pressure, over the difference of geopotential in a homogeneous column of water of salinity 35 °/oo and temperature 0° C. It is usually considered identical to dynamic depth anomaly. Also called anomaly of potential difference and dynamic depth anomaly.

Height Datum, Australian - SEE Australian Height Datum.

height displacement - SEE relief displacement.

height finder - A stereoscopic instrument similar to a range finder but constructed to give heights rather than slant range (distance along the line of sight).

heighting - SEE leveling.

heighting, barometric - The determination of altitudes by using a barometric altimeter. More usually called barometric altimetry.

heighting, trigonometric - SEE leveling, trigonometric.

height pole - A pole divided at heights above top of floor or top of window sills to top of next floor etc., and used to give the heights of various points in a wall of masonry.

heirs and assigns - Ordinarily, words of limitation and not of purchase. At common law, the words were essential to conveyance granting title in fee simple. They are unnecessary for that or any other purpose under statute law when used in wills or deeds.

heliacal (adj.) - Near the direction of the Sun, first becoming visible just before sunrise or just after sunset, and then setting. Said of a star whose direction is near that of the Sun.

heliocentric (adj.) - With reference to, or pertaining to, the center of the Sun.

heliotrope - A device composed of one or more flat mirrors mounted and arranged so that a beam of sunlight reflected by the mirrors can be aimed in any desired direction. The heliostat was invented by Gauss. Placed at one survey station, a heliostat is used to direct a beam of sunlight toward another, distant survey station where the light can be observed through a theodolite. It provides an excellent target to which horizontal directions can be measured; such targets have been observed at distances approaching 300 kilometers. There are several types of heliostat, known by various names. However, these devices differ only in details of mechanical arrangement and construction.

heliport - A zone devoted to the landing and taking-off or helicopters. The heliport may also provide fuel, a place to eat, etc.

helix - A curve, in three dimensions, having the property that its orthogonal projection onto any plane perpendicular to a given straight line (the axis of the helix) is a spiral. Alternatively, a curve, in three dimensions, such that $r = f(\theta)$ and $z = g(\theta)$, in which r, θ and z are cylindrical coordinates and r and z are monotonic functions of θ. Coiled springs (of the kind used in spring balances) and the threads of screws are shaped like helices.

helix, spherical - SEE rhumb line.

Helmert blocking - The process of carrying out the Helmert blocking method.

Helmert blocking method - A method of network adjustment in which the network is broken up into a hierarchy of smaller and smaller networks. The smallest members of the hierarchy are adjusted first; these adjusted members are then combined by means of common points into next-larger members, and these are then adjusted, and so on. The process continues until the entire network has been adjusted. The final result is the same as that which would result from a simultaneous least-squares adjustment of the entire network but needs a much smaller computer for the calculation. Helmert blocking is a special form of solving a system of simultaneous linear equations by partitioning the normal matrix, the observations and unknowns being partitioned on a geographical basis. It is therefore particularly applicable to the solving the large, sparse systems of equations typical of large geodetic networks. It was used for the re-adjustment (1979) of the European triangulation and (1983-1985) for re-adjusting the North American triangulation.

Helmert condensation gravity correction - A quantity δg_{Hc} added to the value of gravity calculated from a gravity formula (or subtracted from a gravity anomaly) to take into account the attraction of topographic masses condensed into a layer on or below the reference surface. Masses outside the reference surface are condensed into a layer on or inside that surface to prevent, as far as possible, the drastic change in gravity and the geoid which result when matter is simply removed from between the surface and a level surface through the point where gravity was measured.

Helmert deflection of the vertical - SEE deflection of the vertical (1).

Helmert elevation - The value H^H_N found for the elevation of a point P_N by dividing the geopotential number W_N by an approximation g^H_N to the average value of the acceleration of gravity along the vertical between P_N and the geoid: $H^H_N = W_N / g^H_N$. W_N is calculated from the measured differences ΔH_n of elevation between points P_{n-1} and P_n along a route from point P_o on the geoid to P_N, and the measured or otherwise determined values gn of the acceleration of gravity at these points P_n: $W_N = \Sigma g_n \Delta H_n$ (n = 1 to N). The value g^H_N is calculated from the measured value g_N of the acceleration at P_N, the Bouguer gravity correction δg_B, and the free-air gravity correction δg_f: $g^H_N = g_N - \delta g_B - \delta g_f / 2$. This is approximately the value of the acceleration at a point on the vertical through P_N and halfway between P_N and the geoid. The Helmert elevation is also called the Helmert height.

Helmert ellipsoid - A rotational ellipsoid characterized by the following constants:
    length of semi-major axis   6 378 200 meters
    flattening   1/298.3 .
Also called the Helmert spheroid.

Helmert error - SEE ellipse, standard-error.

Helmert error ellipse - SEE ellipse, standard-error.

Helmert gravity reduction - The condensation gravity reduction in which the masses above the geoid are first transferred inside the geoid and as close to it as possible. The best method is to coat the masses onto the geoid (giving the ideele störende Schicht). The process is then sometimes called the Faye gravity reduction.

Helmert height - SEE Helmert elevation.

Helmert spheroid - SEE Helmert ellipsoid.

Helmert's 1901 gravity formula - A development by Helmert (1901) of the formula for gravity acceleration, using the measurements of gravity available at the time but not using any pre-assigned value of the Earth's flattening. Referred to the Potsdam value of gravity, Helmert's gravity formula of 1901 is $\tau_o = 978.030 (1 + 0.005302 \sin^2\phi - 0.000007 \sin^2 2\phi)$. While the International Gravity Formula is based on a pre-assigned value of the Earth's flattening, the above formula is not but was used in determining a value for the flattening, namely, 1/298.2. The coefficients 978.030 (τ_e) and 0.005302 (β_1) were determined from the measurements. Attempts made to determine the coefficient of $\sin^2 2\phi$ in the same way were abandoned because of the large probable errors that resulted, and the value 0.000007 was based on Darwin's and

Weichert's theories of the figure assumed, at equilibrium, by a rotating, liquid body of the size of the Earth. Such a body, unless homogeneous, will not be exactly ellipsoidal in shape but will, in middle latitudes, lie below an ellipsoid (rotational) having the same lengths of major and minor axes. The amount of this depression will depend upon the law used to express the variation of density with respect to distance from the Earth's center. Darwin and Weichert assumed different laws but both found a depression of 3 meters at latitude 45°. This depression corresponds to the difference between 0.0000059 and 0.000007, the corresponding coefficients in the International and the Helmert gravity formulae.

Helmert's 1915 gravity formula - A development by Helmert of the formula for gravity acceleration, based on the assumption of a triaxial ellipsoid and including a term containing the longitude: $\tau_o = 978.052\,[1 + 0.005285\sin^2\phi - 0.000007\sin^2 2\phi + 0.000018\cos^2\phi\cos^2 2(\lambda + 17°)]$, in which ϕ is the latitude and λ is the longitude. The greatest equatorial radius is 115 meters longer than the average equatorial radius and 230 meters longer than the shortest. The longest diameter lies in the meridian plane 163° east of Greenwich. The average flattening of a meridian is 1/296.7. SEE ALSO longitude term (in a gravity formula).

Helmert's projective method - The projective method of reduction to the reference ellipsoid, in which the point P" on the ellipsoid corresponding to the point P at the Earth's surface is the footprint, on the ellipsoid, of the normal through P to the ellipsoid. Also called Helmert's method.

Helmert's spheroid - The surface defined by a term representing centrifugal potential, and by the zero-th, second-, and fourth-degree terms in a representation of the Earth's gravity potential by a series of Legendre functions.

Helmert Tower - That building, at the Geodetic Institute in Potsdam, at which definitive measurements of gravity have been made and which for many years was the reference point for gravity networks.

Helmert transformation - A set of linear formulas giving the small changes in azimuth, latitude, and longitude which take place in a geodetic network when the dimensions and orientation of the rotational ellipsoid carrying the network are changed.

hereditament - Every sort of inheritable property, such as real, personal and corporeal, and incorporate property such as right-of-way.

hereditaments, incorporeal - Intangible rights arising out of real or personal property. E.g., the right to light and air.

Hermetian matrix - A matrix such that the elements of the transposed matrix are equal to the complex conjugates of the corresponding elements of the original matrix.

herpolhode - The path described on the invariable plane by the pole of the instantaneous axis of rotation of a rigid body. SEE ALSO polhode.

Herschel - A unit of brightness in monochromatic radiation of frequency v, being the brightness of an object at a distance of 1 parsec and emitting as much energy in the decade of frequency centered at v as the Sun does at all wavelengths (frequencies). The unit was proposed in 1982 by M. Disney and W. Sparks. It is numerically equal to 3.198×10^{-5} ergs per sq. cm per second per decade of frequency about v. Sirius has a brightness of 5.9 Herschels at about 106 GHz (i.e., in visible light).

hertz - One cycle per second. The symbol required by the S.I. is Hz. For example, 16 Hz means 16 cycles per second. The term cycles per second (symbol c/s) is still in use but is not part of the S.I. It was replaced officially by hertz because of the common, misleading practice of writing cycles when cycles per second was meant.

Heuvelink test - A test for the accuracy of graduations on a graduated circle (e.g., a horizontal circle) by measuring repeatedly the angle between two points (e.g. collimators) subtending a suitable angle (about 40°); a number rounds on each face are taken for each angular increment from 0° to 360°.

hiatus - (1) In land surveying, a region between two surveys of record which are described by the records as having one or more common boundary lines with no space between them. The title to the hiatus, except possibly for adverse possession, would appear to remain where it was before the surveys were placed on record. Judicial opinion and court decree may be required to clarify the record. SEE ALSO gore. (2) (photogrammetry) SEE gap.

hide - An obsolete, English unit of area, commonly used in Domesday Book and old English charters. Its size varied with the nature of the land and with the period of use. The normal hide of the Domesday Book is 120 acres.

Hierro meridian - SEE Ferro Meridian.

highland ice - A comparatively thin but continuous covering of ice, conforming generally to the irregularities of the land upon which it rests.

highlight - (1) That portion of an object from which the greatest amounts of light are reflected to an observer.
(2) The densest parts of a photographic negative and the lightest parts of a print or transparency. (3) The maximal brightness of the picture on a cathoderay tube.

high seas - SEE sea, high.

high water full and change - SEE establishment of the port.

highway - A way open to the public at large for travel and transportation, without restriction except as such is incident to secure to the general public the largest benefit and enjoyment.

highway, divided - A highway with separated roadways for traffic in opposite directions. SEE ALSO expressway; freeway; highway; road.

highway, limited-access - A highway that can be entered or left only at a few, widely-separated places (usually many tens of kilometers).

highway by use - SEE road, prescriptive.

highway by user - SEE road, prescriptive.

highway map - (1) A map showing the planimetric configurations of a highway and its connections, at grade or by controlled access, to other highways. (2) A map showing the details of curvature, roadside and cross-drainage, right-of-way boundaries and delineations regarding adjacent occupancy and land uses. The amount and kind of detail shown in a highway map depends entirely on the scale of the map and the purpose of the map, whether merely for indicating traffic routes or for depicting details of construction.

hill - (1) A well-defined, naturally elevated region of land smaller than a mountain. (2) A rounded elevation of land of comparatively medium height (between 100 to 500 feet [30 to 150 meters] above surrounding, flat land.

hill, abyssal - A well-defined, naturally elevated region (a mound) of the sea bottom rising 600 to 900 meters above the bottom and several kilometers wide.

hillock - (1) A small hill. (2) A small hill (20 to 200 feet above flat land).

hill plane - A plane containing three ground marks which are also control points. A hill plane may be, but rarely is, a horizontal plane.

hill shading - That kind of shading which assumes that the ground depicted is illuminated and throws shadows, the shading being used to simulate the visual effect of such shadows. Also called hillwork. The illumination may be either vertical or oblique. Oblique shading is usually introduced on representations of slopes facing south and east, that is, it is assumed that the region is illuminated from the northwest and that the shading then shows the shadows thrown by this illumination. The source of illumination is conventionally placed at a point in space corresponding to point beyond the top left-hand corner of the map.

Hill variable - One of the six quantities r^*, G, H, r, u, Ω used in the theory of orbits: r is length of the radius vector (the line from the origin to the satellite), u the angle from the equatorial plane to the radius vector and Ω the angle from a reference line in the equatorial plane to the projection of the radius vector onto that plane. r^* is the rate of change of r, G is $r^2(du/dt)$ and H is G cos i, i being the inclination of the orbital plane to the equatorial plane.

hillwork - SEE hill shading.

hinterland - (1) The zone containing the flanks of the beach and the region inland from the coastline to a distance of 5 miles (8 km). (2) The region lying behind the coastal zone.

Hiran (high-precision Shoran) - An improved version of Shoran, emitting a sharper pulse with more carefully controlled amplitude. Phase is also measured more accurately in the receiver. Also written HIRAN and hiran. Other characteristics of Hiran are much the same as those of Shoran. Hiran was developed principally for geodetic use after the spectacular success of Shoran. The standard deviation of a measurement by Hiran is approximately $0.36 \sqrt{K}$ meters, in which K is the measured distance in kilometers. Among the major geodetic projects completed using Hiran were the connection of Crete to Africa and Rhodes in 1953, and the connection of North American Datum 1927 to European Datum from Canada to Norway and Scotland by way of Greenland and Iceland in 1953-1956.

Hiran station - A base station (3) operating as part of Hiran.

Hiran trilateration - Trilateration in which the distances between stations have been determined by using Hiran.

Hirvonen's normal height - SEE height, normal (2).

histogram - A graphical representation of a frequency function, in which the frequency of occurrence is indicated by the height of a rectangle whose base is proportional to the interval within which the events occur with the indicated frequency.

history overlay - A specially prepared, matte, plastic sheet which shows the sources of all soundings used in a bathymetric compilation.

hodograph - The locus described by one end of a moving vector when the other end remains fixed. SEE ALSO Ekman spiral.

holiday - In hydrography, an unintentionally unsurveyed part of a region which was to have been completely surveyed. A holiday is particularly likely to occur at the junction between two surveys. It is similar to a gore or hiatus in land surveying or aerial photographs.

Hollerith card - A punched card having eighty vertical columns each containing the numbers from 1 to 12, the instruction or data being determined by which numbers are punched out.

hologram - A photograph of the pattern created by interference between two beams of mutually coherent radiation, one of which is diffracted by an object before being combined with the second beam for interference. Also called, ambiguously, a holograph. Usually only one original beam of coherent radiation is involved. This beam is split into two parts, one of which travels directly to the camera. The other part is diverted to illuminate an object; the diffracted radiation is then directed towards the camera, combining with the first beam to form the interference pattern.

hologrammetry - (1) The techniques associated with making holograms or with reproducing images from holograms. (2) The science concerned with measurement of holograms. Also called holography.

holograph - (1) A document which is entirely in the handwriting of its author. A holograph usually bears the signature of its writer. (2) SEE hologram.

holography - SEE hologrammetry.

Holweck-Lejay gravimeter - SEE Lejay pendulum.

Holweck-Lejay pendulum-apparatus - SEE Lejay pendulum; Lejay-Holweck pendulum-apparatus.

homalosine map - SEE map, homolosine.

homestead - (1) The home, land and surrounding buildings where the family dwells. (2) A tract of land owned and occupied as the family home.

homestead entry - An entry initiated under the homestead laws and providing for the entrance of patents to entrymen who settle upon and improve agricultural land.

homestead entry, additional - A homestead entry which is made by an individual for public lands additional to those he has already acquired under the homestead laws, the total area covered by his original homestead and additional homestead-entries not exceeding the maximal area allowed for the class of homestead entry involved.

homestead law - A statute protecting a homestead from forced sale by creditors.

homolographic (adj.) - Equal-area.

homology - A geometric relationship between two figures such that through a particular point can be drawn straight lines which pass through corresponding points on the two figures and such that the relationship leaves invariant every point of a given plane and every plane through a particular point in the given plane. Homology is used extensively in the design of rectifiers. SEE ALSO axis of homology.

Honkasalo's gravity correction - The quantity $\delta g_{Honkasalo} = -k(1 - 3\sin^2\phi)$ added to a theoretical value of gravity (or subtracted from a measured value) to correct for the effect of a permanent Earth tide; this produces an increase of gravity with latitude ϕ. The value 0.037 for k gives $\delta g_{Honkasalo}$ in milligals.

hook gage - A hook on a long, graduated shank, having a needle-sharp point, and movable vertically with respect to a base. The hook is lowered to just below the surface of the water and then slowly raised until the point just touches the surface. The zero of the scale can be determined by leveling.

Hooke's coupling - SEE Cardan coupling.

horizon - (1) That line which separates the region visible to an observer (real or hypothetical) from the region not visible to him, under specified conditions. Also called the local horizon. The usual condition is that of a human observer on or close to the Earth's surface and with visibility limited principally by the Earth's curvature. However, a camera or other detector is commonly specified, placed some distance above the Earth's surface; observation may be in the infrared or radio parts of the spectrum. (2) SEE horizon, apparent. (3) SEE horizon, celestial. (4) SEE horizon, geometric. (5) SEE plane of the horizon. SEE ALSO closing the horizon; dip of the horizon; plane of the horizon; radio horizon.

horizon, apparent - That irregular line, on the Earth's surface, which bounds the regions within which points are visible to an observer or detector. (Points outside that same line therefore are not visible from the point of observation.) Also called the horizon, geographic horizon, local horizon, topocentric horizon, visible horizon and visual horizon. It is the line where the visible surface of the Earth appears to meet the sky. When the apparent horizon is formed by the surface of a body of water, it is sometimes used as a reference in observing vertical angles. SEE dip of the horizon. The apparent horizon is farther from the observer than the geometric horizon except where a temperature inversion causes the rays of light (or other radiation) to bend upwards instead of downwards as is normal. Its location depends not only on the topography and on the point of observation but also on the refractivity of the air.

horizon, artificial - A device consisting of a flat, reflecting surface adjustable to coincide with the plane of the horizon. The first practical form for use at sea was invented by A. Becher in 1838. It consisted of a small pendulum with its bob suspended in a cistern of oil. A form used on land consists of a dish or trough filled with mercury (or an amalgam of tin and mercury), the upper surface of which is free and horizontal. This form, modified by overlaying the mercury with a transparent, viscous fluid, has been used in making precise astronomical observations at sea and a similar form has been incorporated in some self-leveling leveling instruments. Another form is a plane mirror of glass or other material fitted with spirit levels and leveling screws so that the mirror's surface can be adjusted into the plane of the horizon. The artificial horizon used for aerial navigation is usually kept parallel to the plane of the horizon by a gyroscopic pendulum. In observing a celestial object such as a star or the Sun, the angle is measured between the body as seen directly (with transit or theodolite) or by reflection (in the horizon glass of a sextant) and its image as seen reflected in the artificial horizon. This is a vertical angle and is double the angular elevation of the body.

horizon, astronomic - SEE horizon, celestial; plane, horizontal.

horizon, astronomical - SEE horizon, celestial; plane, horizontal.

horizon, celestial - The great circle, on the celestial sphere, 90° away from the zenith. Also called the astronomic horizon, astronomical horizon, geometrical horizon and rational horizon. Among navigators and astronomers, it is more commonly referred to simply as the horizon. It is also called the true horizon. It defines the plane of the horizon or can be defined from the plane of the horizon. The definition the great circle in which a plane perpendicular to the vertical at the observer intersects the celestial sphere is practically equivalent.

horizon, false - A line resembling the apparent horizon but lying just above or below it.

horizon, geocentric - A plane through the center of the Earth and parallel to the apparent horizon. Note that the kind of center is not specified by the definition, so there is more than one geocentric horizon.

horizon, geodetic - The plane perpendicular to the normal at a point on the reference ellipsoid.

horizon, geographic - SEE horizon, apparent.

horizon, geoidal - That circle, on the celestial sphere, formed by the intersection of the celestial sphere with the plane tangent to the geoid at the plumb line through the point of observation.

horizon, geometric - (1) That line, on the Earth's surface, at which straight lines from the point of observation are tangent to the Earth's surface. Also called geometrical horizon and true horizon, although the latter term is also used for the plane of the horizon, apparent horizon, and celestial horizon. The term surface in this definition must be interpreted loosely; the surface proper may be hidden by buildings, trees, etc. The geometric horizon is usually closer to the observer than the apparent horizon; the latter is defined by the concave paths of rays from the surface to the observer. (2) That curve, on an ellipsoid representing the Earth, at which lines through the assumed point of observation are tangent to the ellipsoid. Points within the geometric horizon are visible from the point of observation; points outside the geometric horizon are not. (3) The intersection (line) of the celestial sphere with a plane tangent to the Earth's surface and passing through the eye of the observer. Earth's surface actually means, here, a sphere or ellipsoid representing the Earth's surface; tangent to the Earth's surface means parallel to a plane tangent to the representing sphere or ellipsoid. (4) SEE horizon, celestial. (5) A plane through the center of the Earth and parallel to the plane of the apparent horizon (a plane fitted in some suitable manner to the apparent horizon).

horizon, geometrical - SEE horizon, geometric.

horizon, local - SEE horizon; horizon, apparent.

horizon, rational - SEE horizon, celestial.

horizon, sensible - The great circle formed on the celestial sphere by intersection, with the celestial sphere, of a plane perpendicular to the vertical through any point such as the eye of the observer. This definition is practically the same as that of celestial horizon.

horizon, topocentric - SEE horizon, apparent.

horizon, true - SEE horizon, geometric.

horizon, visible - SEE horizon, apparent.

horizon, visual - SEE horizon, apparent.

horizon camera - An aerial camera designed for photographing the horizon at the same time as another camera photographs the ground. The pictures taken by the horizon camera show how far the other camera is off the vertical.

horizon closure - SEE misclosure of the horizon.

horizon map projection, stereographic - A stereographic map projection the center of projection at some parallel of latitude other than the Equator.

horizon photograph - A photograph of the horizon, taken simultaneously with a vertical photograph to obtain an indication of the tilt of the vertical camera at the instant of exposure.

horizon plane - SEE plane of the horizon. Note that this is not the same as a horizontal plane.

horizon-plane method - A graphical method of preparing a map from oblique photographs by determining the location of the horizon trace in the photograph, the principal point and the perspective center, and using these to determine the direction and length of the horizontal ray.

horizon prism - A prism inserted in the optical system of an instrument such as the bubble sextant, to permit the horizon to be seen.

horizon ring - A graduated ring fitted to a globe so that the plane of the ring contains the center of the globe and is adjustable to be parallel to the horizontal plane for any point on the globe. The horizon ring is used for measuring angles around the horizon at a given point.

horizon sweep - The series of measurements, in a clockwise sequence from the farthest, known, visible object, of angles or directions to objects such as tanks, spires or signals, for identification of the objects and subsequent use of their angles or directions. Similar to a round of observations.

horizon system of coordinates - SEE coordinate system, horizontal.

horizontalizing (a stereoscopic model) - SEE leveling.

horizontal-stadia method - Measuring distances by holding the stadia rod horizontal and the stadia hairs of the telescope vertical during observations.

horizontal taping and plumbing - SEE taping and plumbing, horizontal.

horizon trace - An imaginary line, in the plane of a photograph, representing the image of the celestial horizon; it corresponds to the intersection of the plane of a photograph and the horizontal plane containing the perspective center or rear nodal point of the lens.

Horrebow-Talcott level - A very sensitive spirit level placed on the axis of a transit for precise measurement of zenith angles. Also called a Talcott level and Horrebow level attachment.

Horrebow-Talcott method (of determining longitude) - A precise method of determining astronomic latitude by measuring the difference in the meridional distances of two stars of known declination, one north and the other south of the zenith by approximately the same amount and at approximately the same right ascensions. The observations are made with a zenith telescope or with an astronomical transit (such as the meridian telescope and the broken-tube telescope) which can be converted to serve as a zenith telescope. The two stars must have approximately the same zenith distances at the meridian and their culminations must occur within a few minutes of each other. The astronomic latitude of the point of observation will be one-half the sum of the declinations of the two stars, plus or minus one-half the difference of their zenith distances. This method is also known as the zenith-telescope method of determining latitude and as the Talcott method. While Peter Horrebow (not Harrebow) published an account of his method in 1732, in Kobenhavn, the publication was buried in obscurity and there is good reason to believe that Captain Andrew Talcott had no knowledge of Horrebow's work when he announced his own discovery of the method in 1834.

horst - A block of crust, formed by faulting and generally elongate, raised relative to the blocks on either side without much tilting or faulting.

Hotine's rectified skew orthomorphic map projection - An approximation to the oblique Mercator map-projection, based on the use of the aposphere as the surface onto which the rotational ellipsoid is first mapped, before finally being mapped onto the plane. This map projection is precisely conformal and the scale along the chosen central line is nearly enough constant over a restricted region for the map projection to be geodetically useful.

hotspot - (1) In aerial photography, that portion of an image at which light has been received from the Sun directly or by reflection rather than by scattering from the scene. Also called flare, glare, hot spot, shadow-point and no-shadow area. It is overexposed (washed out) with respect to the rest of the image. (2) A region, in the Earth's mantle, so much hotter than its surroundings that it causes volcanic activity in the crust immediately above it. In the theory of continental drift (plate theory), hotspots are responsible for chains of volcanic islands such as the Hawaiian Islands.

hour - An interval of time variously defined as equal to 3600 seconds, 60 minutes, or 1/24th of a day. SEE ALSO constituent hour.

hour, cotidal - The average interval of time between the Moon's transit over the Greenwich meridian and the time of the following high water at a place. The interval may be expressed either in solar or in lunar time. When it is expressed in solar time, it is the same as the Greenwich high-water interval. When it is expressed in lunar time, it is equal to the Greenwich high-water interval multiplied by the factor 0.966.

hour, sidereal - A period containing 3600 sidereal seconds or 1/24th of a sidereal day.

hour angle - The angle, measured westwards, from the plane of the observer's celestial meridian to the plane of the hour circle passing through the celestial body observed. Equivalently, the angular distance on the celestial sphere, measured westward along the celestial equator, from the meridian to the hour circle passing through a celestial object. The hour angle is reckoned from the meridian (0 hours or 0°) westwards through 24 hours (360°). SEE ALSO ephemeris hour angle; Greenwich hour angle.

hour angle, local - The angle, measured westwards, from the plane of the local celestial meridian to the plane of the hour circle passing through the celestial body being observed. It may also be defined as the arc of the celestial equator, or the angle at the celestial pole, measured westwards from the upper branch of the local celestial meridian to the hour circle of a point on the celestial sphere.

hour angle, sidereal - The angle, expressed in degrees and measured westwards, from the plane of the hour circle passing through the vernal equinox to the plane of the hour circle passing through a point on the celestial sphere. It may also be defined as the angular distance on the celestial sphere, expressed in degrees and measured westward along the celestial equator, from the vernal equinox to the hour circle passing through the celestial body being observed. It is equal to 360° minus the right ascension in degrees.

hour-angle method (of determining azimuth) - Any method which involves measuring the horizontal angle between a celestial body and the point whose azimuth is to be determined, and adding this to the azimuth of the celestial body as determined by calculation or from an ephemeris. Among the celestial bodies used for this purpose are Polaris (α Ursa Minoris), σ Octantis and various other readily observable stars.

hour-angle of Polaris method (of determining azimuth) - SEE Polaris hour-angle method.

hour-angle system of coordinates - SEE coordinate system, hour-angle.

hour circle - A great circle, on the celestial sphere, passing through the celestial poles (and therefore perpendicular to the celestial equator). Equivalently, a great circle, on the celestial sphere, lying in a plane perpendicular to the plane of the celestial equator. Also called a circle of declination. That hour circle containing the zenith at a given point is identical with the celestial meridian there.

hovercraft - SEE vehicle, ground-effect.

hoverscope - An optical device used by the pilot of a helicopter to assist him in hovering precisely over a marked point on the ground. The device consists of an auto-collimator mounted vertically on a silicone-damped pendulum. The image of the mark, on the autocollimator's reticle, is superposed on an image of the ground beneath.

hub - A temporary marker, usually a wooden stake with a tack or small nail on top to indicate the exact point of reference for angular and linear measurements, driven flush with the ground at a station.

huebras - In Spanish law, a measure of land equal to as much as a yoke of oxen can plow in one day.

humidity - (1) The amount or proportion of moisture in a gas; the degree of wetness of a gas. (2) The amount or proportion of moisture in the atmosphere; the degree of wetness of the atmosphere.

humidity, absolute - (1) The mass of water present as vapor in a given volume of air. (2) The weight of water present as vapor in a given volume of air. Absolute humidity is usually expressed as grams per cubic meter.

humidity, relative - The ratio of the amount (mass or weight) of water present in a given volume of air, at a given temperature, to the greatest amount possible at that temperature.

humidity, specific - The amount (mass or weight) of water vapor present per unit amount (mass or weight) of the moist air.

hump yard - A (railroad) yard in which cars are moved (sorted) by pushing them over a summit, beyond which they move by gravity.

Hunter's gravity reduction - SEE Graaf-Hunter gravity reduction.

Hurter and Driffield curve - SEE curve, characteristic.

Huygens' principle - The statement that every point on the instantaneous position of a wave front (equal phase) may itself be regarded as a source of spherical waves. The position of the wavefront a moment later is then determined as the envelope of all the waves emitted from points on the preceding wavefront, and this procedure of constructing successive wavefronts can be continued as long as desired. The principle is extremely useful in explaining effects due to refraction, diffraction, and scattering, for acoustic radiation as well as for electromagnetic radiation. The principle can be justified theoretically but its extensive use is based on its practically applications.

hydrograph - A graph of the discharge, speed, available power, or other property of a stream or conduit with respect to time. The most common type of hydrograph, the observed hydrograph, represents the readings on stream gages, plotted at the time of observation. Other types of hydrograph which are derived from the observations include the distribution graph and the unit hydrograph.

hydrographer - (1) A person who makes surveys of bodies of water. (2) In some sections of the United States of America, a person who measures the flow of water in streams or conduits.

hydrography - (1) The science dealing with the physical features (geometry and dynamics) of the surface waters of the Earth. Hydrography differs from physical oceanography in that it deals with rivers and lakes as well as with oceans and seas. It differs from hydrology in dealing with only surficial waters and not with underground waters. (2) The science dealing with the geometric features of the surface waters of the Earth. In this sense, the term is practically synonymous with charting. (3) The science dealing with the surface waters of the Earth. This appears to be the term's original meaning. However, present custom is to use the term oceanography instead, if marine waters are being considered and hydrology if rivers, lakes, etc., are being considered also. (4) That branch of physical oceanography dealing with the measurement and definition of the configuration of the bottoms and adjacent land-regions of oceans, lakes, rivers, harbors and other forms of water on the Earth. (5) The science and art of measuring the oceans, seas, rivers, and other waters, with their marginal lands, inclusive of all the fundamental elements which have to be known for the safe navigation of such regions, and the publication of such information in a form suitable for the use of navigators. This definition resembles definition (2) but is much less general in that it excludes all subjects not concerned with navigation or concerned with navigation but not published. It does include measurement of marginal land, however. (6) That science which deals with the measurement and description of the physical features of the oceans, seas, lakes, rivers and their adjoining coastal regions, with particular attention to the use for navigational purposes.

hydrology - That science dealing with the properties, laws, and phenomena of water; of its physical, chemical and physiological relations; of its distribution throughout the habitable Earth; and of the effect of this distribution on human lives and interests. The trend at present is to have hydrology deal only with the occurrence of water on and under the land, and to put the parallel science of water in seas and oceans under oceanography.

hydrophone - (1) The receiver for the echoes produced in echo sounding. (2) A device for listening to or detecting sound in water.

hydrosphere - The free waters of the Earth. The term may or may not be taken to include ground water. Water in the atmosphere is considered part of the hydrosphere, according to this definition which is in common use. However, many scientists consider atmospheric water to be part of the atmosphere.

hygrometer - An instrument for measuring the relative humidity of the air.

hygrometric (adj.) - Relating to the relative humidity or comparative amount of moisture in the atmosphere. Because the atmosphere penetrates bodies in varying degrees depending on the substances of which they are composed, the amount of moisture in the air will affect the shapes and dimensions of certain instruments and equipment used in surveying and mapping. Therefore, materials to be used in making leveling rods, plane-table sheets, maps, etc., must be selected which are not sensitive to hygrometric conditions.

hygrometry - The science of measuring the humidity (relative water content) of air and other gases.

hyperstereoscopy - Stereoscopic viewing in which the scale (usually the vertical scale) along the line of sight is exaggerated in comparison with the scale perpendicular to the line of sight. Also called appearance ratio, stereoscopic exaggeration, exaggerated stereo and relief stretching.

hypo - Sodium hyposulfite, a common fixing agent used in developing and printing photographs. The term is jargon but is in common use.

hypsograph - A circular instrument of the slide-rule type, used to calculate elevations from vertical angles and horizontal distances. It is commonly used in plane-table surveys.

hypsography - (1) The science or art of describing elevations or heights of land surfaces with reference to a specified surface (usually the geoid). (2) The topography (relief) or a region referred to a specific datum. (3) Those parts of a map which represent relief.

hypsometer - (1) An instrument used in determining altitudes of points on the Earth's surface in relation to sea level by determining atmospheric pressure through observation of the boiling point of water at each point. Explorers have made much use of this method for determining the heights or altitudes of mountaintops. (2) A differential barometric altimeter. I.e., a barometric altimeter indicating changes in altitude with respect to a surface of constant pressure. (3) An instrument used in forestry for determining heights of trees. (4) A person who measures elevations or the land, or who constructs maps or models to show elevations of land. This use is rare.

hypsometric (adj.) - Relating to elevation above a reference surface, usually the geoid. In a limited sense, hypsometric applies to those altitudes determined with a hypsometer.

hypsometry - (1) The art of determining, by any method, elevations of the Earth's surface with respect to the geoid. Determining altitudes by means of either mercurial or aneroid barometers is sometimes called barometric hypsometry or, more commonly, barometric altimetry. (2) Determining distances of the physical surface above some specified reference surface.

hypsometry, barometric - The determination of elevations using barometers. The quantities determined directly by barometric methods are altitudes, not elevations. However, they may sometimes be converted, by adding suitable corrections, to elevations.

I

ICAO atmosphere - The standard atmosphere adopted by the International Civil Aviation Organization (ICAO). Also called the ICAO Standard Atmosphere. Its main features are a pressure, at mean sea level, of 1013.25 millibar and a lapse rate of 6.5 ° C/km from surface to tropopause (situated at 11 km). It was extended to 32 km by international agreement in 1964.

ICAO standard atmosphere - SEE ICAO atmosphere.

ice chart - A chart showing the prevalence of ice, usually with reference to navigable waterways.

ice point - The temperature at which a mixture of pure, air-saturated water, and ice can exist in equilibrium under a pressure of one standard atmosphere.

ice pole - The center of the more consolidated portion of the arctic ice-pack.

ice records - Records of the occurrence of ice. Ice records of importance to engineers are usually obtainable from the following sources: reports by local engineers; statements by local residents; examination of structures damaged by ice; photographs of icy conditions; records of the U.S. Weather Bureau; records of the U.S. Geological Survey of flow of streams.

iconogrammetry - SEE ikonogrammetry.

iconometry - (1) The process of making maps from photographs. (2) The process of constructing a plan and elevation of an object from a perspective view of that object.

identification, control-station - SEE photoidentification.

identification point - SEE identification post.

identification post - A post of wood or other suitable material, appropriately marked and inscribed and placed near a survey station in aid in recovering and identifying the station. Also called an identification point. Identification posts also serve as guards, calling attention to the stations which they mark and thus protecting the stations against accidental destruction.

identity matrix - SEE unit matrix.

ikonogrammetry - The science of inferring the physical dimensions of objects from measurements made on images of the objects. Also written iconogrammetry. Ikono-grammetry includes photogrammetry, which involves measurements on photographs; radargrammetry, which involves measurements on images created on a cathode-ray tube by radio waves; and X-ray photogrammetry. However, photogrammetry is often used instead of the more proper term ikonogrammetry. The definitions given for various kinds of photography can be transformed to the corresponding kinds of ikonogrammetry by substituting image for photograph and electromagnetic radiation or particles for light, where appropriate.

illuminance - The rate at which luminous energy is incident upon a unit of area. Equivalently, the luminous flux incident per unit area. Formerly called illumination. Note that the term applies only to that part of the electromagnetic spectrum which is involved in vision; the corresponding term for radiation in general is irradiance.

illumination - SEE illuminance.

image - (1) A pattern of electromagnetic radiation such that there is approximately a one-to-one correspondence between each point of the pattern and each point of the radiation source. This definition is more general than those given in textbooks on optics because apparatus used for detection often uses radiation at wavelengths outside the optical portion of the spectrum (e.g., at radio or X-ray wavelengths) but otherwise works on principles similar to those on which apparatus working with light is based. The word image is usually preceded by a modifier to indicate the part of the spectrum involved if this is not the optical part or if several parts of the spectrum are involved: e.g., radio image, infrared image, optical image, X-ray image. However, the terms radio image and X-ray image are also applied to optical images formed from images originally at the other wave-lengths. E.g., in radargrammetry, radio images as such exist only at the antennas. These images are converted electronically to optical images on the face of a cathode-ray-tube and these in turn are converted to photographic images. But all three kinds are often referred to as radio images. An observer associates six directions with any particular scene: from down to up, from left to right, from front to back and the three opposites. These directions can also be associated in the same way with an image of that scene if the image gives a visual impression similar to that of the scene itself. For example, down-to-up in the scene will correspond to down-to-up in the image. The image is then said to have a normal aspect. However, one or more pairs of opposite directions may be reversed in the image from the corresponding directions in the scene. For example, a down-to-up direction in the scene may appear as an up-to-down direction in the image, and conversely. The different kinds of reversal possible are called aspects. The simple aspects are the inverted image, the reversed image and the mirror image. These simple aspects combine to produce more complicated aspects. Also called ambiguously and therefore undesirable, imagery. (2) A 2-dimensional representation of an object, such as to give a visual impression similar to that of the original object. The word image is frequently prefaced by a modifier indicating the nature of the substance forming the image e.g., photographic image and retinal image. SEE ALSO mirror image.

image, achromatic - An image transmitting or reflecting light without breaking the light up into component hues e.g., a black-and-white image.

image, blind - SEE blueline.

image, congruent - An image which can, in concept, be superposed on the object exactly with only a change in scale being needed.

image, conjugate - SEE image, corresponding.

image, continuous-tone - An image which contains tones varying gradually from black to white. The source of this definition calls it a continuous tone but this is probably an error and is certainly not in accord with general usage.

image, corresponding - (1) In general, the image or images of a specified object. The image (or images) is said to correspond to or represent the object. (2) In ikono-grammetry, particularly photogrammetry, one of a set of corresponding images (q.v.).

image, diffraction-limited - An image as perfect as diffraction by the aperture stop will allow. The term implies that the total error introduced by optical aberrations is less than the error caused by diffraction. SEE ALSO system, diffraction-limited, optical.

image, enantiomorphic - An image which can, with a change of scale and with reversal of one pair of opposite directions, be superposed, in concept, on the object. SEE ALSO image.

image, erect - (1) An image such that down-to-up motion in the scene causes up-to-down motion in the image. (2) An image appearing upright or in the same relative position as the object.

image, false-color - An image in which other colors are substituted for the natural colors. SEE ALSO film, color-infrared.

image, homologous - (1) One of a set of images of a single object and appearing on each of two or more overlapping photographs having different camera stations. (2) One of a set of images of a single point, on an object, which appear on each of the two or more overlapping photographs having different perspective centers. SEE ALSO images, corresponding.

image, inverted - (1) An image which has been turned upside down, with respect to the original, so that top and bottom are interchanged but right and left sides are not. (2) An image which has been turned upside down, with respect to the original, so that top and bottom are interchanged. Although the first definition is the one usually given, the second definition is the one commonly used. (3) An image such that motion from down-to-up in the scene causes an up-to-down motion in the image. An optical system consisting of objective lens and eyepiece such as is found in astronomical telescopes and some surveyor's telescopes produces an inverted and reversed image.

image, latent - The invisible image produced in radiation-sensitive materials and consisting of those molecules which have absorbed one or more quanta of radiation and are therefore in an excited condition but have not yet changed their chemical nature. When the material is subjected to a process called development, the excited molecules change chemically, agglutinate with other molecules, and create a visible image.

image, mirrored - SEE mirror image.

image, optical - (1) An image formed by light. The apparatus forming an optical image is called an optical system. That part of a light ray not acted upon by the optical system is said to lie in object space; the part that has been acted on by the optical system is said to lie in image space. There are two kinds of optical images: real (optical) images and virtual (optical) images. If rays from a point in object space converge to a point-like volume in image space, the image is said to be real. If the volume is a single point, that point is called the stigmatic image of the corresponding point in object space. If the rays from a point in object space diverge in image space but asymptotes to the rays converge to a point-like volume, the image is said to be virtual. Both real and virtual images can be seen, but real images can be recorded on photographic film while virtual images can not. (2) A visible record, such as a photograph, of an image formed by light.

image, perverted - SEE mirror image.

image, photographic - The image on a photograph. The terms photographic image and photograph are practically synonymous and there seems to be not reason to use the former instead of the latter.

image, pseudoscopic - An image in which the shading is such that the normal impression of hollows and elevations (relief, in aerial photography) is reversed, with high points on the object appearing to be low and low points appearing to be high. This may occur when the image is viewed with shadows pointing away from you.

image, real - The set of points in image space formed by the intersection of rays from corresponding points in object space. The power radiated from each point on the object is concentrated by the optical system to a small region, the real image, in image space. This power can be detected and recorded by a suitable material such as photographic emulsion (if the image is formed by light).

image, reversed - An image such that motion from left-to-right in the scene causes motion from right-to-left in the image. Also called a reverted image. Note that this is not the same as a mirror image.

image, reverted - SEE mirror image.

image, stereo - SEE image, stereoscopic.

image, stereoscopic - The impression that a three-dimensional object is being viewed when one of a pair of overlapping images of the object is viewed with the left eye only and the other one of the pair is viewed with the right eye only, this viewing taking place simultaneously or in rapid succession by each eye. Also called a stereo image.

image, stigmatic - An image in which each point corresponds to a separate point of the object, and each point of the object corresponds to just one point in the image.

image, virtual - If the rays from each point on the object diverge in image space, a virtual image is formed of each such point at the intersection of the asymptotes to these diverging rays, and the virtual image of the object is the collection of the virtual images of the individual points.

image analysis - The science of inferring characteristics of an object by studying images of that object. The term includes image interpretation and ikonography, as well as the more specialized subjects of photogrammetry, photo-interpretation and radargrammetry.

image compression - The representation of an image by a set of numbers and the subsequent reduction of this set to a smaller set that can still be converted back to a satisfactory copy of the original image. Image compression is used primarily to reduce the bandwidth needed to transmit images by radio or wire.

image direction - The orientation of the image on a photographic positive or negative relative to the position of the emulsion.

image-displacement method (of determining tilt) - The determination of tilt by comparing the holes in transparent templets used in a radial adjustment with the locations, on the photographs, of the corresponding points.

image distortion - Distortion of an image. Among the possible causes are distortion by the lens system, differential shrinkage of the film or paper, and motion of the camera while the picture was being taken.

image enhancement - Manipulation of the point-dependent density of an image to bring out or emphasize certain features of the image.

image modulation - The quantity $M(\nu)$ expressing the fidelity with which a periodic pattern (frequency ν) can be reproduced in a photograph, and given by $M(\nu) = (E_{max} - E_{min}) / (E_{max} + E_{min})$, in which E_{max} (E_{min}) is the largest (smallest) value of exposure converted (using the D-logE curve of the emulsion) from the measured density of the photographic image.

image motion - (1) The smearing or blurring of an image on an aerial photograph because of the motion of the camera relative to the ground during exposure. (2) The motion of an image on the focal surface of a camera because of the motion of the camera with respect to the scene or object being imaged. SEE ALSO compensation, image-motion; compensator, image-motion.

image patch - That set of points, in an image, which is the image of a single point on the object.

image plane - (1) The plane in which an image is assumed to lie. (2) That plane, in a camera, which provides the best average focus over the area of the photographic plate or film.

image point - (1) The image of a definite point on the object. Because no optical system is perfect, a single point in object space gives rise to wave fronts which do not converge again to a single point but, at their smallest, still occupy an appreciable volume. The region in which this small volume is intersected by a focal surface is called an image patch. The image point can be a point in a virtual image. It is then (ideally) the point which is the common center of the diverging spherical waves in image space. (2) The image, on a photograph, corresponding to a definite point on the object.

image point, conjugate - SEE image point, corresponding. The latter term is preferred in order to avoid confusing corresponding image point with conjugate point.

image point, corresponding - One of the images, on two or more photographs, of the same point in object space.

image points, corresponding - The images, on two or more photographs, of the same point in image space.

image ray - SEE ray, perspective (1).

image ray, corresponding - One of the rays connecting each of a set of corresponding image points with its particular perspective center.

image registration - (1) The process of matching two pictures to determine the locations of the elements of one picture with respect to the corresponding elements of the other. (2) The process of superposing one picture on another so that corresponding elements have the same positions.

imagery - (1) The process of creating images, particularly images composed of electromagnetic radiation. (2) The visual representation of energy recorded by cameras or other similar instruments. These two definitions are probably equivalent. (3) SEE image. (4) SEE photograph.

images, corresponding - Points or lines which occur on the overlapping portions of two photographs having different perspective centers and are images of the same point or line in the object.

image smoothing - Replacement of the density and/or color of each point in an image by a linear combination of those characteristics with the like characteristics of that point's neighbors.

image space - SEE image, optical.

imaging system - Any device or apparatus producing an image of an object. The term is used to include devices or apparatus not photographic in nature but which create images from radiation at wavelengths longer or shorter than those of light. It also includes apparatus producing sonic images. In most cases, however, the final product of an imaging system is a visual or photographic image.

imbalance - In leveling, the difference in distances from the leveling instrument to the two leveling rods (or two locations of one leveling rod).

Imbitube Datum 1959 - The vertical-control datum of Brazil, based on mean sea level at the tide gage at Imbitube.

Imperial Standard Yard - (1) A brass bar inscribed with a 3-foot scale and adopted by Act of Parliament, 1 January 1826, as embodying the legal definition of the yard. The bar was a copy of an earlier bar made by the Royal Society. It was destroyed by fire in 1834. (2) The distance between marks on two golden plugs in a bronze bar at 62° F and designated legally to be the Imperial Standard Yard. This standard was constructed to replace the one destroyed by the fire of 1834. It is kept at the Board of Trade in London (1951) and, by the Weights and Measures Act of 1872, defines the British Imperial Yard. A comparison with the International Meter in 1894 gave 1 meter = (39.370113/36) yard. Later measurements in 1927 and 1934 gave 47 and 38, respectively, in the last two places of the numerator, but an evaluation in 1945 concluded that there was no evidence for any change in the Yard in the previous 40 years.

imposition - The positioning and assembling of photographic negatives or positives onto those positions, on a flat, in which they will be printed.

imprecision - (1) A measure, such as a standard deviation, of the variation of a quantity from a constant value. Imprecision and precision are frequently used as synonyms. However, they are also frequently used as antonyms. (2) The randomly varying component of the total measure of deviation from a constant value. The other component is then called the systematic error and the total measure is called the uncertainty.

impression - The inked image put on a sheet in a printing press.

impulse - The product of a force and the interval of time during which that force acts.

inch - (1) A unit of length defined to be 1/36 of a yard and equal, since 1866 in the USA, to exactly 1/39.37 meter. This equivalence was established by an Act of Congress, 28 July 1866, but was put into practical effect only after the USA received copies of the International Prototype Meter. With changing definitions of the meter, the definition of the yard and inch change accordingly. SEE ALSO meter (length); yard (length). (2) A unit of length defined for scientific purposes, and by an agreement made in 1959 among the U.S. Bureau of Standards and similar organizations of some other countries, to be 1/36 of the international Yard (2.54 cm exactly).

inch, circular - An area equal to the area of a circle 1 inch in diameter.

inclination - (1) The vertical angle between an horizontal plane and a freely suspended magnetic needle. Also called dip. The magnetic needle gives, practically, the direction of the magnetic field at a point, so the inclination is the vertical component of the field's direction. (2) The angle between two planes or their poles.

inclination (of an orbit) - SEE inclination, orbital.

inclination, magnetic - The vertical angle, at a specified point, between a horizontal plane through that point and the direction of a line of force of the Earth's magnetic field. Also called dip and magnetic dip. This definition and that of inclination (1) are almost synonymous and may be taken to be so for most purposes. However, the first definition gives an angle which may not be exactly the direction of the line of force.

inclination, orbital - The dihedral angle, positive measured northwards, from a reference plane to the plane of an orbit. Northwards means counterclockwise when looking inwards from a point on the orbit to the attracting center. Inclination is usually one of the six elements that completely define an orbit. The plane of the celestial equator is customarily used as the reference plane when specifying the orbit of an artificial satellite, although the plane of the ecliptic has also been used. Also called, simply, inclination. Note that inclination of the orbit and inclination of the plane of the orbit are not, speaking precisely, the same thing. The former is an angle between the tangents to two curves; the latter, an angle between two planes.

inclination of the horizontal axis - The vertical angle between the horizontal axis of an astronomical or surveying instrument and the plane of the horizon. Inclination of the horizontal axis is measured with a striding level or a hanging level. It causes the line of collimation of a theodolite or transit to describe an inclined plane instead of a vertical plane when the telescope is rotated about the horizontal axis of the instrument. This produces an error in a measured horizontal angle and requires that a correction for inclination of the horizontal axis be applied.

inclinometer - A small instrument for measuring the inclination of a surface to a horizontal plane. It consists of a graduated arc held firmly at one end by a rigid arm and traversed by a rotating arm, carrying a level, pivoted on the fixed arm. In use, the fixed arm is placed on the surface whose inclination is to be measured, the graduated arc is held vertically, and the movable arm is rotated until the level indicates that this arm is horizontal. The inclination is then read on the arc.

indefeasible (adj.) - Not able to be defeated, revoked, or made void. The term is usually applied to an estate or right which cannot be defeated.

indemnity land - Land granted to a railroad in lieu of place lands when the place lands had already been granted elsewhere for other purposes.

indemnity selection - SEE lien selection.

indenture - A formal, written instrument made between two or more persons. The term is derived from the ancient practice of indenting or cutting the deed in a waving line.

independence, stochastic - A variable U is said to be stochastically independent of a variable V if the probability of U occurring is the same whether V occurs or not.

independent-model method of adjusting aerotriangulation - A method of adjusting aerotriangulation, in which the x,y,z coordinates of points in each model created from stereoscopic pairs a strip or block of overlapping photographs are measured independently, using a stereoscopic plotting instrument, in the model's own coordinate system. The sets of coordinates are then combined and fitted to ground control and each other by removing computationally the inconsistencies between coordinates from different pairs. The coordinates are recomputed a second or third time, using the improved values each time, until the discrepancies are negligible. Also called the semi-analytical method and the independent-pairs of pictures method (of adjusting aerotriangulation). It was proposed by M. Zeller in 1934.

index, absolute refractive - The ratio of the speed of light in a vacuum to the speed of monochromatic light in the substance of interest. Also called the absolute index of refraction. It is commonly called, the refractive index or index of refraction. However, those two terms are also sometimes applied without warning to the ratio found when the ray emerges into air instead of a vacuum. That ratio should be called a relative refractive index. Equivalently, the ratio of the sine of the angle of incidence to the sine of the angle of refraction, at the point where a monochromatic ray leaves the substance and enters a vacuum. The index of refraction is slightly different for light of different wavelengths. This variation is known as dispersion and, for any given substance, the amount of the deviation is a measure of the dispersive power of that substance. Chromatic aberration is a result of such deviation. It is overcome over a moderate range of wavelengths by using a lens system composed of optical elements having different dispersive powers. (2) SEE index, relative refractive.

index, atmospheric refractive - The absolute refractive index of an atmosphere; in particular, of the Earth's atmosphere.

index, control-document - Photomicrographic copies of control documents, mounted incards arranged by State, meridian and township.

index, diagrammatic - An index depicting the successive arrangement and relationship of one map-sheet to another, or of that part of one aerial photograph which has stereo-scopic coverage with respect to another. The diagrammatic index is usually schematic and is only approximate in scale. It may contain the bounds or only the marked centers of the map sheets or photographs.

index, historical - (1) The narrative part of the new-status records. (2) A summary of and index to all essential actions that affect or have affected the title to or use of lands and resources, public domain.

index, modified refractive - The refractive index of the atmosphere, modified mathematically so that when its gradient is applied to an equation for the path taken by energy over a (hypothetically) flat Earth, it yields a path substantially equivalent to that followed by the energy over the true, curved Earth. A formula frequently used for the modified index n' is n' = (n + h/R), in which n is the actual refractive index, h is the difference between the elevations of emitter and receiver and R is an average radius of the Earth.

index, perspective - The distance, on an aerial photograph, from the isocenter to the true horizon.

index, refractive - (1) SEE index, absolute refractive. (2) SEE index, relative refractive.

index, relative refractive - The ratio of the speed of light in one substance to the speed of light in another. Also called a relative index of refraction or, when the meaning is clear, refractive index or index of refraction. The last two terms are often used when the substance into which the light emerges is air. However, such usage should be avoided unless clear warning is that this particular definition is being used. Equivalent definitions are: the ratio of the (absolute) refractive indices of two substances; and the ratio of the sine of the angle of incidence to the sine of the angle of refraction, at the point where a ray leaves the one substance and enters the other. This last definition is valid because Snell's law of refraction states that the ratio depends on the substances involved and not on the angles of incidence and refraction.

index chart - A chart in outline form, showing the limits and identifying designations of navigational charts, volumes of sailing directions, etc.

index contour - SEE index contour line.

index contour line - A contour line made thicker to distinguish it from the contour lines on either side of it. All primary contour lines are multiples of the interval. Depending on the size of the contour interval, index contour lines are drawn as every fourth or fifth contour line, and usually with the corresponding elevation shown in a gap in the line. Sometimes incorrectly called an index contour.

index correction - (1) A correction applied to the reading from the scale on any graduated device used for measurement, to compensate for a constant error such as would be caused by displacement of the scale from its proper position. (2) The correction applied to an measured difference of elevation to eliminate the error introduced into the measurements when the zero of the graduations on one or both leveling rods does not coincide exactly with the actual, physical foot or bottom surface of the rod. Also called index leveling correction.

index error - (1) A systematic error caused by misplacement of an index mark or zero mark on an instrument having a scale or vernier, so that the instrument gives a non-zero reading when it should give a reading of zero. This is a constant error. (2) The distance upwards (or downwards) from the foot of a leveling rod (the lowest horizontal surface) to the nominal origin (theoretical zero) of the scale.

indexing system, DoD standard - SEE DoD standard indexing system.

index leveling correction - SEE index correction (2).

index map - A map showing where to find collections of related data such as other maps, aerial photographs, statistical data or descriptions of terrain. Also called a key map. Index maps may be used to show references to the origin, place of deposit or filing or other identifying characteristics of such data.

index map, photocontrol - SEE photocontrol index map.

index map, photographic - An index map showing the area covered by each photograph in a set of photographs of a region. Three kinds of photographic-index maps are in common use. (a) A photographic copy of the assemblage of photographs, with each photograph in its proper position relative to the others. The copy is usually made at a scale 1/8 to 1/10 that of the original photographs. (b) A map, planimetric or topographic, on which are drawn the outlines of the area covered by each photograph of the set. (c) A transparent overlay, keyed to a base map, on which are drawn the outlines of the area covered by each photograph. Photographic-index maps of the first kind are usually used only when all the photographs in the set were taken at approximately the same scale and along flight lines. Also called a photo index, photo index map, photo plot and plot map.

index mark - A real mark, such as a cross or dot, lying in the plane or the object-space of a photograph and used alone as a reference in certain types of monocular devices, or as one of a pair forming a floating mark in certain types of stereoscopes. In stereoscopic plotters having pairs of index marks, each mark is called a half-mark. SEE ALSO mark, floating.

index of precision - SEE precision (3).

index of refraction - SEE index, absolute refractive.

index of refraction, absolute - SEE index, absolute refractive.

index of refraction, atmospheric - SEE index, atmospheric refractive.

index of refraction, relative - SEE index, relative refractive.

index prism - A prism, in a sextant, capable of being rotated to any angle, within established limits, corresponding to angular elevation. It is the counterpart, in the bubble or pendulum sextant, of the index mirror of a marine sextant.

Index To Names - An alphabetical list of geographic names keyed to a map series or to maps covering a specific country, giving the designations of features, geographic and grid coordinates, and sheet number for each name appearing in the series. It is almost the same as a gazetteer, but has been established by international agreement.

Indian Datum - A quasi-geodetic datum based on the Everest spheroid, with the origin at triangulation station Kalianpur:
longitude of origin 77° 39' 17.57" E
latitude of origin 24° 07' 11.26" N
deflection of the vertical*
  in the meridian -0.29"
  in the prime vertical +2.89"
geoidal height 0 meters.
* Deflection of the vertical in the prime vertical is defined to be zero, except for adjustment of Laplace azimuths, when it is given the value shown. Deflection of the vertical in the meridian is as stated. These constants are the ones used in the adjustment of 1938 and are the ones usually meant when Indian Datum is specified. However, the datum has had a long and complicated history (from 1802), and coordinates determined before 1938 should be verified as to the actual datum used.

Indian foot - SEE foot.

Indian spring low water - The approximate, average water level determined from all lower low waters at spring tides.

Indian tidal plane - A surface used as a reference for height of the tide at a number of ports in India, and used for a time in Puget Sound and for all Alaskan waters except the mouth of the Yukon River. Also called a harmonic tidal plane, harmonic tide plane and Indian tide plane. Indian tidal plane corresponds closely to a surface 2 feet below lower low water.

indicator, terrain-clearance - SEE altimeter, absolute.

indicatrix - SEE Dupin's indicatrix; Tissot's indicatrix.

indicatrix of Dupin - SEE Dupin's indicatrix.

indicatrix of Tissot - SEE Tissot's indicatrix.

Industrie-Norm, Deutsche - SEE Deutsche Industrie-Norm.

inequality - (1) In astronomy, the angular deviation of a satellite from the location it would have if it were moving in an elliptical orbit. The term is most commonly applied to irregularities in the Moon's orbit but is also used in describing the orbits of the planets and their satellites. (2) In hydrography, a systematic departure from the average value of a tidal quantity. SEE ALSO age of diurnal inequality; age of parallax inequality; age of phase inequality; parallax inequality; phase inequality.

inequality, annual - A seasonal variation in water level or current, more or less periodic, caused by seasonal changes in air pressure, direction of the wind and other meteorological factors.

inequality, diurnal - (1) The difference in elevation between the two high waters or the two low waters of each day. (2) The difference in speed between the two tidal flood currents or the two tidal ebb currents of each day. The difference changes with the declination of the Moon and, to a lesser extent, with the declination of the Sun. In general, the inequality increases with an increasing declination, either north or south, and decreases as the declination decreases. The National Ocean Service specifies that the periods of observation for the diurnal inequalities be certain 19-year long intervals called National Tide Datum Epochs.

inequality, high-water - The difference between the elevations of the two high tides during a tidal day.

inequality, low-water - The difference between the elevations of successive low tides.

inequality, lunar - A deviation from elliptical motion of the Moon, caused by attraction of the moon by the Earth, Sun and other bodies in the Solar System.

inequality, lunar declinational - An inequality caused by the Moon's moving from the celestial equator into the northern and southern hemispheres, and having a period of one-half a tropical month - 13.66 days. Besides the lunar declinational inequality, there is also a solar declinational inequality, which depends on the Sun's declination.

inequality, lunar parallactic - SEE parallax inequality.

inequality, mean diurnal high-(low-)water - One-half the average difference between the two high (low) waters of each day observed over an extended period. It is obtained by subtracting the average of all high (the lower low) waters from the average of the higher high (all low) waters. The definitions given here for mean diurnal high-water or low- water inequalities are applicable only to either semi-diurnal or mixed tides.

inequality, mean diurnal low-water - SEE inequality, mean diurnal high (low) water.

inequality, parallactic - An inequality, in the longitude of the Moon, caused by variations in the Sun's pull on the Moon as the Earth's distance from the Sun varies throughout the year. Also called the parallax inequality. The argument is the difference between solar and lunar longitudes, the amplitude is -124.8" and the period is one mean synodic month (29.530589 days or 29d 12h 44m).

inequality, semi-monthly - An inequality having a period on one-half of a month.

inequality, solar parallactic - The variation in range of tide caused by the varying distance of the Earth from the Sun, as indicated by changes in the solar parallax.

inequality, solar declinational - SEE inequality, lunar declinational.

inequality, tropic - The average difference between the two high waters or two low waters of the day at the time of tropic tides. The first is called tropic high water inequality; the second is called tropic low water inequality.

inequality, tropic high (low) water - SEE inequality, tropic.

inequality, variational - A variation, in the Moon's angular rate of motion, caused mostly by the tangential component of the Sun's attraction.

inertia - That quality of a body which resists any change in the body's velocity, i.e., resists acceleration. SEE ALSO moment of inertia; product of inertia.

inertia, rotational - SEE moment of inertia.

inertial measuring unit - SEE measuring unit, inertial.

in fee - Ownership in land. SEE ALSO fee; fee simple.

information - That part of a sequence of symbols (message) which (a) contains what was put in by the sender and (b) was not predictable by the receiver. This is actually a definition of new information; what is already known to the receiver is not information.

information content - A measure of the amount of information contained in a string of symbols (message). If a message consists of a sequence of symbols x_i (i goes from 1 to some number I), and if the a priori probability that x_i will occur in the sequence is p_i, then the information content of x_i, i.e., the contribution of x_i to the information in the sequence is commonly defined to be $\ln(1/p_i)$.

information content, average - The sum, over i (i goes from 1 to some number I) of $p_i \ln(1/p_i)$, in which p_i is the a priori probability that the symbol x_i will occur in the sequence $\{x_i\}$ of I symbols. Also called the entropy of the sequence (message).

information system, geographic - Any computerized information-system designed for analyzing information related primarily to a non-cadastral aspect of the land.

infrared - SEE radiation, infrared; radiation, near-infrared.

infrared, near - SEE radiation, near-infrared.

ingress (v.) - Go in; enter.

ink, opaque - An opaque ink used to paint out defects or other parts of a film. Also called opaque (jargon) or opaquing ink.

ink, opaquing - SEE ink, opaque.

inlet - A narrow body of water extending into the land from a larger body of water. Also called an arm or tongue of the larger body. A long, narrow inlet with gradually decreasing depth inward is called a ria.

inlet, tidal - A short, narrow body of water with channels at ends diverging at 90° to 180°, connecting a bay, lagoon or similar body of water with a large parent body of water (usually an ocean), and wetted wholly or in part by the tides.

inset - A map separate from a larger map but positioned within the neat line of that map. Three forms are recognized: (a) an inset depicting a region geographically outside that depicted on the larger map but included for convenience of publication and usually at the same scale (e.g., insets of Alaska and the Hawaiian Islands on a map of the USA); (b) a portion of the map at an enlarged scale (e.g., an inset showing Chicago on a map of Illinois); and (c) a map, at a smaller scale, of the region within which that depicted by the larger map lies (e.g., inserts showing designations of maps adjoining the one on which the inset appears).

inset map - A map positioned within the borders of a larger map but otherwise separate from and usually on a different scale than the larger map.

inshore - That zone, of variable width, lying between the shoreface and the seaward limit of the breaker zone.

instability, bench-mark - The (principally) vertical movement of some bench marks because of local changes in the nature of the material surrounding the bench mark or movement of that material. The cause of the instability is usually within about 15 meters of the surface; it may, however, be considerably more deep-rooted.

instrument - (1) In general, a tool. (2) In geodesy and allied sciences, a tool used for measuring and usually of some complexity. A simple measuring tool is often referred to as a measuring device; a measuring tool of some complexity but largely mechanical in nature is referred to as a measuring apparatus. The instruments used in geodesy are sometimes referred to as measuring instruments. They must often be used with other devices which do not measure. E.g., a leveling instrument is usually mounted on a tripod and used with one or two leveling rods; a theodolite is mounted on a tripod and used with one or more targets or signals; etc. The combination of instrument and auxiliaries is called measuring equipment. (3) (legal) SEE instrument, legal. SEE ALSO angle instrument; direction instrument.

instrument, absolute - An instrument which measures a physical quantity in absolute units by means of simple physical measurements.

instrument, alt-azimuth - (1) An instrument on an alt-azimuth mounting, and equipped with both horizontal and vertical circles for the simultaneous measurement of horizontal and vertical directions or angles. Also called an astronomical theodolite, alt-azimuth and universal instrument. The alt-azimuth instrument derives its name from the combination of altitude and azimuth. Many theodolites and transits are alt-azimuth instruments, but the term is little used for surveying and mapping instruments. Its use may properly be restricted to designating an instrument of the theodolite type constructed for use in astronomical work and sometimes called a universal instrument. (2) An instrument on an alt-azimuth mounting; i.e., an instrument rotatable in angular elevation and azimuth. Theodolites and transits are alt- azimuth instruments; leveling instruments, not being freely rotatable in angular elevation, are not.

instrument, analog - An instrument which represents the quantity to be measured by some other, continuous quantity that can be made to vary proportionately and which is then measured instead. SEE ALSO computing machine, analog.

instrument, circumzenithal - SEE astrolabe, circumzenithal.

instrument, distance-measuring - An instrument used, either by itself or as part of an equipment, for measuring distances. Also called a telemeter. Distance-measuring instruments are classified according as they are entirely mechanical or contain electronic devices. The latter kind are further classified according as they do or do not use electro-magnetic radiation for the measurement and, if they do, according to the part of the spectrum used.

instrument, electromagnetic distance-measuring - A distance-measuring instrument which measures the difference between the time or phase at which electro-magnetic waves are emitted and the time or phase at which the waves are received. Also called an electronic telemeter and electronic distance-measuring instrument. According to the part of the spectrum in which the radiation lies, the instrument (electro-magnetic DMI) is called a radio DMI, electro-infrared DMI or electro-optical DMI.

instrument, electronic distance-measuring - A distance-measuring instrument which uses electronic devices in making the measurement. Also called an electronic telemeter or EDM instrument. Most such instruments (electronic DMI) also send out and/or receive electromagnetic radiation as an integral part of the measuring process. They are then known as electromagnetic distance-measuring instruments, in general or have special names according to the part of the spectrum utilized. However, some instruments use the electronic parts only for converting measurements to digital form or for timing, e.g., acoustical pulses, and do not use electromagnetic radiation for measuring distance.

instrument, electro-optical distance-measuring - An electro-magnetic distance-measuring instrument operating at optical wavelengths. Also called an electro-optical rangefinder, phototachymeter, laser radar and lidar. Laser radar, however, always denotes an electro-optical DMI in which a laser is the source of radiation, and it usually denotes an electro-optical DMI so large that it cannot be easily carried about by one or two people, i.e., is either fixed in place or is mounted on a truck. The term is sometimes used to denote instruments that modulate infrared radiation instead of light.

instrument, gyroscopic meridian-indicating - SEE gyrotheodolite.

instrument, infrared distance-measuring instrument - A distance-measuring instrument operating at infrared wavelengths.

instrument, laser-interferometer distance-measuring - An optical distance-measuring instrument for measuring distances and the difference in phases between radiation sent by a laser from one point to another and the radiation of standard phase within the instrument. Such instruments are used primarily in laboratories and tool-makers' shops for measuring lengths or short distances. Some have been used for establishing short but very accurate base lines.

instrument, legal - A document or other writing made and executed as the expression of some act, contract, or legal proceeding. Usually referred to, in legal matters, simply as instrument. However, where the word is used in two or more senses (as in the present glossary), the term legal instrument is preferred.

instrument, mechanical distance-measuring - A distance-measuring instrument entirely or almost entirely mechanical in nature. Examples are the odometer and the surveyor's tape. The former is a self-sufficient instrument; the latter must usually be used with additional devices such as pins, stakes, scales, etc. If the instrument is at all complex, or if it must be prepared or put together in the field, it is referred to as apparatus, e.g., base line apparatus.

instrument, optical distance-measuring - A distance-measuring instrument whose functioning depends on an optical system. Also called an automatic optical range finder. There are two types: that in which the optical system forms an image of the object to which distance is to be measured; and that in which the object's distance is determined by measuring the difference in time or phase between light waves emitted from the instrument and the time or phase of the waves returned from the object. The first type is generally called a range finder or tachymeter. The second type includes lidars (radar-type instruments using light instead of radio waves), Väisälä interfero-meters and similar instruments.

instrument, optical-projection - One of a class of (optical) projectors that throw images of photographic prints or other opaque material onto a map, map manuscript, or other surface for drawing. It is often used for transferring details from near-vertical photographs or similar material.

instrument, repeating - SEE theodolite, repeating.

instrument, restitutive - SEE plotting instrument, stereoscopic.

instrument, single-model - A stereoscopic plotting instrument capable of projecting only one stereoscopic model at a time. Such a plotting instrument is intended for compilation only and depends on supplementary techniques to do stereo-triangulation.

instrument, spirit-level - (1) An instrument consisting of a spirit level or having a spirit-level attached. (2) A leveling instrument carrying a spirit level. (3) An instrument for testing spirit-levels. Also called a level trier.

instrument, stereoscopic - (1) A device using a stereoscopic pair of images to produce a visual effect of depth or solidity. Also called a stereoscope. (2) An instrument using a stereoscopic pair of images to produce a visual effect of depth or solidity and producing numerical data and/or topographic maps showing depths or heights. Called a stereoscopic plotter or stereoscopic plotting instrument if the principal use is actually drawing a map of the object. (3) An instrument using a stereoscopic pair of images to produce numerical data from which a map of the imaged object can be drawn. Usually called an analytical plotter.

instrument, universal - SEE instrument, alt-azimuth.

instrument approach chart - SEE chart, instrumental-approach.

instrument error - SEE calibration error.

instrument parallax - SEE parallax, instrumental.

instrument phototriangulation - SEE stereotriangulation.

instrument station - That point over which a leveling instrument is placed for taking a backsight, a foresight, and such extra foresights as may be necessary from that point. Except in rare instances, the instrument stations in a level line are not marked points. SEE ALSO set-up.

intaglio - A design cut below the level of the surface of a material.

intensity (of gravity) - The magnitude of gravity, as distinct from the direction of gravity. The intensity of gravity is measured by spring balances such as spring-type gravimeters. It is what geophysicists and geodesists usually mean by gravity.

intensity, ambient luminous - The average luminous intensity in the vicinity of a specified object or region.

intensity, horizontal (magnetic) - The intensity of the horizontal component of the Earth's magnetic field in the plane of the magnetic meridian.

intensity, magnetic - SEE field intensity, magnetic.

intensity, total (magnetic) - SEE field intensity, total (magnetic).

intensity, vertical (magnetic) - SEE field-intensity, vertical magnetic.

interchange - An intersection designed for easy and safe transfer from one road to another, usually with the roads being placed at different levels at the intersection.

interest, proprietary - An interest growing out of ownership, as distinguished from a governmental interest which would not necessarily imply ownership.

interest, undivided - Property held jointly by two or more persons by the same title, whether their rights be equal or unequal in quantity or value.

interferometer - A device combining two different parts of the same wave front into a single wave, causing the two parts to alternately reinforce and cancel each other. Most interfero-meters operate by splitting a single beam of radiation into two beams, one of these two beams going directly to a receiver while the other is lead through a longer path or through a different medium and thence to the receiver where the two beams are reunited. Interferometers working with sound waves have been constructed. However, sonic equipment of the type used in seismology and hydrography is not interferometric because the sound waves arrive at separate receivers where they are converted to other forms such as tracings or numerical data before being combined. Interferometers used in surveying work in the radio, infrared, or optical portions of the electromagnetic spectrum. Those operating with light are used principally for measuring short to moderately large distances, for measuring short baselines, or for calibrating leveling rods or other standards of length. Those operating with infrared radiation are used principally for measuring short or moderately large distances. Those operating at radio wavelengths may be used for measuring distances of any size but are particularly useful in measuring large distances (.e.g., long-baseline interferometry). However, the so-called very-long-baseline interferometers are not, properly speaking, interferometers; the separate wave fronts are not combined in a single receiver but are converted first to recordings or numerical data. SEE ALSO Fabry-Perot interferometer; Michelson interferometer; radio interferometer.

interferometer, connected-element - A radio interferometer having the antennas connected directly (usually by cables) to the receiver.

interferometer, optical - A device combining two different parts of a light wave in such a way as to cause interference between the separate parts. The type most used is the Michelson interferometer, in which a single beam of light is split into two parts which are sent over different paths and are then recombined. The Mach-Zender interferometer is a variant of the Michelson interferometer: the medium whose properties are to be examined is inserted in one of the two paths in a Michelson interferometer. Optical interferometers of the Michelson and Fabry-Perot types are used in geodesy for measuring short distances precisely, for measuring the rate of fall of freely-falling prisms, and for measuring flexure in the supports of pendulum apparatus.

interferometer, stellar - SEE Michelson interferometer.

interferometer, very-long-baseline - A pair of radio telescopes (i.e., a pair of antennas each attached to its own receiver and recorder), separated by a great distance (usually 1000 km or more) and recording, simultaneously, the voltage variations in radio waves received from the same distant source. Also called a VLBI. The recordings from the separate radio telescopes are later compared to determine the difference in phase of the radiation received from the source. From this information, the direction and structure of the source, and the distance and direction between telescopes can be determined. The ability to determine the location of one telescope relative to the other to within a few centimeters makes the instrument important to geodesy.

interferometry - The study of the interference occurring when two separate wavefronts (i.e., beams) of radiation are combined. Interferometry involving electromagnetic radiation is more important to geodesy than that involving sound waves; that involving sound waves is of sole interest in hydrography and seismology.

interferometry, very-long-baseline - The obtaining and use of data from very-long-baseline interferometers. Often written as VLBI.

interior angle - An angle between adjacent sides of a closed figure, and lying on the inside of the figure. The three angles within a triangle are interior angles. SEE ALSO polygon.

interior orientation - SEE orientation, inner.

interior perspective center - SEE center of perspective, internal.

interior water - SEE water, inland.

International Atomic Time - Time established by the Bureau International de l'Heure in accordance with the definition of the second in the Système International d'Unités. Shortened to TAI. The epoch is 1958 January 1 00h UT2. The time was defined by the General Congress on Weights and Measures in 1971.

International Date Line - SEE date line, international.

International Ellipsoid - A rotational ellipsoid adopted by the International Association of Geodesy (IAG) in 1924 for international use. It is characterized by the following constants:
length of semi-major axis (a)   6 378 388 meters
flattening (f)   1/297 .
The International Ellipsoid is sometimes erroneously called the Hayford spheroid or Hayford ellipsoid. It is numerically the same as the Hayford spheroid of 1909, which was derived from measurements in the USA, but differs from that spheroid in that the length of the semi-major axis and the flattening are defined; the corresponding values for the Hayford spheroid were calculated. In 1967, the IAG adopted a new ellipsoid, specifying a new value for a but not for f; instead, it specified a value for the coefficient C_2 in the representation of the geopotential by a series of Legendre functions.

International Ellipsoid of Reference - SEE International Ellipsoid.

International Gravity Formula - The formula $\tau = 978.049 (1 + 0.0052884 \sin^2\phi - 0.0000059 \sin^2 2\phi)$, in which ϕ is the geodetic latitude. This is a development of the formula for gravity, based on the representation of the Earth by a body shaped like a rotational ellipsoid having the dimensions of the International Ellipsoid of reference (Madrid, 1924) and rotating about its minor axis one in a sidereal day, and on the assumptions that (a) the surface of the ellipsoid is a level surface and that (b) the value of gravity acceleration at the equator is 978.049 gals. The International Gravity Formula was adopted by the International Association of Geodesy at its meeting in Stockholm in 1930. The purpose was not, primarily, to represent the gravity measurements then available, although the value 978.049 was based on those measurements, but rather it was intended to put the determination of the figure of the Earth from gravity data on the same basis as the determination of the figure of the Earth from deflections of the vertical. The formula is based on the Potsdam value for gravity. It has been replaced, for general used, by Gravity Formula 1967 and by later formulae, but is still used for some purposes.

International Gravity Standardization Net 1971 - A set of adjusted values for the acceleration of gravity at various places throughout the world, adopted by the International Association of Geodesy in 1971 to replace the Potsdam gravity-system as an international system. Generally written as IGSN 71. Also called the International Gravity System 1971 and International Gravity Standardization System. It is based on 10 measurements with ballistic gravity-apparatus, 1 200 measurements with pendulum apparatus, and 24 000 measurements with spring-type gravimeters. 34 of the gravity stations are first-order. The precision is 0.1 mgal.

International Gravity Standardization System - SEE International Gravity Standardization Net 1971.

International Gravity System 1971 - SEE International Gravity Standardization Net 1971.

International Great Lakes Datum 1955 - A vertical-control datum with zero at mean sea level at Pointe au Pere, Quebec, as determined from readings over the period 1941-1956. It is used primarily for hydraulic studies and for the definition of chart datums in the Great Lakes and connecting waterways. Elevations on this datum are based on leveling from Point au Pere to Lake Ontario and along all connecting waterways, to Lake Superior. Elevations on the lakes are derived by water-level transfer based on the assumption that the mean water level in each lake, for the months of June, July, August and September during the period 1941-1956, constituted an equipotential surface. Elevations are stated in units of dynamic number as defined by Bowie and Avers, 1914.

International Hydrographic Bureau - (1) The organization founded in 1921 in Monaco for getting international cooperation on matters affecting hydrography and, in particular, for standardizing charts and oceanic surveys. (2) Since 22 February, 1970, the headquarters of the International Hydrographic Organization. The offices remain those of the previous International Hydrographic Bureau in Monaco.

International Hydrographic Organization - Since 22 February 1970, the organization succeeding to the International Hydrographic Bureau and carrying on the work of that Bureau, with some additional objectives such as development of the sciences of hydrography and descriptive oceanography. The headquarters of the organization is called the International Hydrographic Bureau and is located in Monaco.

International Latitude Service - The organization established in 1900 by the International Astronomical Union and the International Association of Geodesy to be responsible for determining polar motion by measuring the astronomical latitudes of the three to five observatories it uses. Usually abbreviated to I.L.S. The observatories in operation at present, and their nominal astronomic coordinates, are:
Carloforte, Italy   8° 18' 44" E   39° 08' 08.941" N
Kitab, USSR   66° 52' 51" E   39° 08' 01.850" N
Misuzawa, Japan   141° 07' 51" E   39° 08' 03.602" N.
Since its founding in 1900, the Service has maintained uninterrupted service (although with a varying number of observatories) up to the present time. All observatories are located as close as practicable to the parallel of 39° 08' 10" N. They use the same kind of instrument (zenith telescope) and the same method of observation (Horrebow-Talcott). The headquarters have changed, by convention, from observatory to observatory at 6-year intervals or multiples thereof. In 1962 the headquarters of the Service was incorporated in the headquarters of the International Polar Motion Service. However, the set of observatories is still referred to as the I.L.S. stations.

International Low Water - A surface of reference below mean sea level by half the range between mean lower low water and mean higher high water multiplied by 1.5.

International Map of the World - A series of topographic maps started in 1913 with the objective of mapping the world at a uniform scale and on a common map projection, but with each participating country mapping its own region. A scale of 1:1 000 000 and the Lambert conformal conic map projection were adopted, together with specifications on contour intervals, coloring, symbols for boundaries, etc. Large regions, particularly in eastern Europe and Asia, remain unmapped as parts of the original project. However, because various organizations have undertaken to map the world at this scale and to adhere to most of the specifications of the original project, the mapping is virtually complete. SEE ALSO map projection, International Map of the World.

International Map of the World map projection - SEE map projection, International Map of the World.

International Polar Motion Service - The organization established in 1962 to take over the work of the International Latitude Service (I.L.S.) and to include, with the work of I.L.S.'s observatories, data from a large number of other observatories also engaged in determining latitude. A great variety of instruments are used. The principal types are the visual zenith telescope, the photographic zenith telescope and the Danjon astrolabe.

International Prototype Meter - SEE International Prototype Metre.

International Prototype Metre - A bar of 90% platinum and 10% iridium, of approximately X-shaped cross-section, produced under the authority and direction of an International Conference held in Paris in 1875, and accepted as defining the meter by the representatives of 17 governments. Also called the Prototype Meter. It is a line standard and was derived from the Metre des Archives, an end-standard. The International Prototype Meter and two other, similar bars of the same origin are maintained in the International Bureau of Weights and Measures at Severes, France. They were used in calibrating the representative copies supplied to the governments supporting the Convention of 1875. As a signatory of that convention, the United States of America received two such representative copies, one of which is known as the National Prototype Metre. These meters are in the custody of the National Bureau of Standards.

International System of Units - SEE Système International d'Unités.

International Temperature Scale - A temperature scale, introduced in 1927 and revised in 1948 (symbolized as ITS-48) based on six defined points (conditions):

	1948	1968
boiling point of oxygen	-182.970°	-182.962°
triple-point of water	0°	+0.01°
boiling point of water	+100°	+100°
freezing point of zinc		+419.58
boiling point of sulfur	+444.600°	
freezing point of silver	+960.8°	+961.93°
freezing point of gold	+1063.0°	+1064.43°

In 1960, the temperature scale of 1948 was revised by incorporating a new definition of the degree Kelvin (that which results by defining the thermodynamic temperature of the triple-point of water as exactly 273.16° K). The new temperature scale was called the International Practical Temperature Scale (symbolized as IPTS-48). The IPTS-48 has been superseded by the IPTS-68 adopted in 1967, which adds 5 fixed points at the lower end of the IPTS-48, down to -259.34° (the triple-point of equilibrium hydrogen).

International Yard - A unit of length equal to 0.9144 meter. The International Yard was adopted in 1959 by the U.S. Bureau of Standards and certain other national standards laboratories as a unit of length for scientific purposes only.

interpolation - (1) Determination of a value intermediate between given values, from some known or assumed rate or system of change of values between the given ones. Equivalently, determination of the value of a function $y = f(x)$ for some arbitrarily specified value of x, given values (y_1, x_1) and (y_2, x_2) such that the arbitrary value of x lies between the two given values. (2) The process or technique of obtaining an estimate of the value of a quantity at a specified time t, given that data useful for obtaining the estimate are available for times both before and after t. This definition seems to be part of the jargon of control engineering, in which it is called smoothing.

interpolation, double - Determination of the value of a function z of two independent variables x and y for arbitrarily specified values of x and y, given the values of z at given, specified values of x and y between which the arbitrary values lie. The function is often given in tabular form, with the values of the independent variables indexing the rows and columns of the table. SEE ALSO interpolation, spatial.

interpolation, spatial - Determination of the value of a function $z = f(x,y)$ of two variables neither of which is time, for arbitrary values of these variables, given the value of the function for a number of values (usually evenly spaced) of the two independent variables. SEE ALSO interpolation, double; Kriging; surface fitting.

interpretation (photographic) - SEE interpretation, photographic.

interpretation, photographic - (1) The determination of the nature of objects whose images appear on a photograph, and the description of those objects. Also called photointerpretation. The more general term is image interpretation; this covers situations such as those in which the images appear on the screens of cathode-ray tubes. The term remote sensing is sometimes used erroneously for photographic interpretation. Photogrammetry and photographic interpretation are closely related. However, photogrammetry deals with the geometry of the objects; photographic interpretation deals with the purpose, use, origin or identity of the objects.

interpreter - (1) A person who interprets. (2) A person who does photographic interpretation. (3) A person who infers geological structure from geophysical data.

intersection - (1) Determination of the location, on a surface, of a point as the intersection of two lines drawn with specified directions from two other points having known locations. SEE ALSO resection. (2) Determination of the location, on a surface, of a point at the intersection of lines having directions determined photogrammetrically. (3) Determination of the location, on a surface, of a point at the intersection of two or more lines drawn with specified directions from two or more points also on the surface and having known locations. Because the directions on which the intersection is based were usually determine by direct observation, the point located will be misplaced because of errors in the observations. Directions from three or more points are therefore usually observed. The directed lines will generally not intersect at a single point. The most likely location of the point in question is then determined either geometrically or by computation. (4) A place where two or more roads cross or meet each other.

intersection, grade - SEE grade intersection.

intersection at grade - Intersection at the same level (grade).

intersection station - An object or point (unoccupied station) whose horizontal coordinates are determined by measurements from other survey stations, no measurements being made at the object or point itself. If the object is observed from only two stations, the location is called a no-check position, because there is no proof that the measurements were free from blunders. Intersection stations either are at places which it would be difficult to occupy with an instrument, or are survey signals whose locations can be determined with sufficient accuracy without having to occupy the station.

interval - The time, space or angle between two times, points or directions. In particular, in astronomy, the difference in time between two events such as the passage of a celestial body through two arcs on the celestial sphere. SEE ALSO Greenwich interval; rise interval, mean.

interval, equatorial - One of the angles, expressed in units of time, between the various lines of the reticle of an astronomical transit and the average position (central axis) of those lines. The equatorial interval for a given line of a reticle is equal to the length of time required for the image of an equatorial star (declination 0°) to travel from the line in question to the central axis of the reticle, or from the central axis to the line in question, the instrument being adjusted in the meridian. In determining time with an astronomical transit, equatorial intervals are used to reduce incomplete measurements to an average value.

interval, higher high-(higher low-) water - The interval of time between the transit (upper or lower) of the Moon over the local or Greenwich meridian and the next higher high (higher low) water. This term is used when there is considerable diurnal inequality.

interval, higher low-water - SEE interval, higher high-(higher low-) water.

interval, high-water - Either a high-water lunitidal interval or a mean high-water lunitidal interval. SEE ALSO interval, lunitidal.

interval, high-water lunitidal - The interval of time between the upper or lower transit of the Moon and the next high water at a place.

interval, local - An interval of time based on the Moon's transit of the local celestial meridian.

interval, low-water - SEE interval, lunitidal.

interval, lower high-(lower low-) water - The interval of time between the upper or lower transit of the Moon over the local meridian or Greenwich meridian and the next lower high (lower low) water. This term is used when there is considerable diurnal inequality.

interval, lower low-water - SEE interval, lower high (lower low) water.

interval, low-water lunitidal - SEE interval, lunitidal.

interval, lunar - The interval of time between the transit of the Moon over the Greenwich meridian and over a local meridian. The lunar interval equals the difference between the Greenwich and local intervals of a phase of the tide or tidal current. The average value of this interval, in hours, is -0.069 λ, where λ is the local longitude in degrees.

interval, lunitidal - The interval of time between the Moon's upper or lower transit over the local or Greenwich meridian and the following high or low water. Where there is considerable diurnal inequality in the tide, separate intervals may be obtained for the higher high waters, the lower high waters, the higher low waters, and the lower low water. These are designated, respectively, as higher high water interval, lower high water interval, higher low water interval and lower low water interval. In such cases, and also where the tide is diurnal, one must distinguish between the upper and lower transits of the Moon with respect to the Moon's declination. Intervals referred to the Moon's upper transit when the Moon's declination is north or to the Moon's lower transit when the Moon's declination is south are marked a. Intervals referred to the Moon's lower transit when the Moon's declination is north or to the upper transit when the Moon's declination is south are marked b.

interval, mean high-water (low-water) lunitidal - The average of all lunitidal intervals from transit to following high (low) water for all phases of the Moon. Shortened to high water (low water) interval. Also called establishment. The mean high-water lunitidal interval is also called a corrected establishment. The interval is described as local or Greenwich according as reference is to transit of the Moon over the local or over the Greenwich meridian. When not otherwise specified, the local meridian is implied.

interval, meridional - The value of the distance between meridians, at the scale of the chart.

interval, tropic - A lunitidal interval pertaining to either the higher high water or the lower low water at the time of the tropic tides.

interval, tropic higher high-(higher low-) water - The lunitidal interval pertaining to the mean higher high (higher low) waters at the time of tropic tides.

interval, tropic higher low-water - SEE interval, tropic higher high-(higher low-) water.

intervalometer - A device for automatically opening and closing the shutter of a camera at selected intervals of time. In aerial photography, the length of the interval is selected to give the desired amount of overlap between successive photographs when the focal length of the camera, and the velocity and amplitude of the airplane is known.

intervisibility test - Any one of several tests for determining whether two stations in a survey net can see each other.

Invar - An alloy of iron containing about 64% iron, 36% nickel, and small amounts of chromium to increase hardness, manganese to facilitate drawing, and carbon to raise the elastic limit, and having a very low coefficient of thermal expansion (about 1/25 that of steel). Also spelled invar. It was invented by C. Guillaume of the International Bureau of Weights and Measures (Paris). It is used wherever a metal does not change its dimensions appreciably with temp-erature as desired. For this reason, it or a similar alloy has replaced steel in surveyor's tapes and wires, and particularly in those tapes and wires used for measuring geodetic baselines. It is also used for the scale of some leveling rods, in first-order leveling instruments, and in pendulums.

invariant, adiabatic - A physical quantity which, to a certain degree of approximation, remains unchanged during a slow spatial or temporal change in a system.

Invar leveling rod - A leveling rod having the graduations marked on a strip of invar fastened to one side of the rod. Also called an Invar-band leveling rod. The strip is set into the face of the rod, firmly attached to the bottom of the rod (rod shoe), and held in place by guides and a spring at the top of the rod for constant tension. Invar leveling rods are used in first-order leveling.

Invar pendulum - A pendulum made of Invar. In particular, a type of quarter-meter pendulum made of Invar and used by the U.S. Coast and Geodetic Survey beginning in 1920, when it replaced the bronze pendulum for the Survey's measurements of gravity. An Invar pendulum is subject to magnetization but has a coefficient of thermal expansion about 1/15 that of the bronze pendulum it replaced.

Invar scale - A measuring bar made of Invar. One side of the bar is usually graduated in units of the metric system of units and the other in the English system of units.

Invar tape - A surveyor's tape made of Invar.

Invar wire - A surveyor's wire made of Invar.

inventory survey - A survey made to collect and correlate engineering data of a particular type or types over a given region. An inventory survey may be recorded on a map.

inverse - (1) SEE inverse, geodetic. (2) SEE matrix, inverse.

inverse, conditional - SEE matrix, generalized inverse.

inverse, generalized - SEE matrix, generalized inverse.

inverse, geodetic - (1) The computation of distance and direction between two points on a rotational ellipsoid when the coordinates of both points are known. Also called, simply, an inverse. (2) The same as the previous definition but with rotational removed. Unless specifically stated otherwise, rotational ellipsoid is meant.

inverse, geographic - SEE computation, inverse.

inverse computation - SEE computation, inverse.

inversion method of gravity reduction - SEE Rudzki gravity reduction.

inversor - (1) A mechanism used to maintain proper geometric relationship (correct conjugate distances and collinearity of planes of photographic negative and easel) in autofocusing optical devices such as projectors and rectifiers. (2) A mechanical device for ensuring that Newton's law, $xx' = f^2$, is satisfied in a camera; x is the distance of the object, x' the distance of the image and f the focal length. SEE ALSO cam inversor; Carpentier inversor; Peaucellier inversor; Peaucellier-Carpentier inversor; Pythagorean band inversor.

inversor, Pythagorean - SEE inversor, right-angle.

inversor, Pythagorean right-angle - SEE inversor, right-angle.

inversor, right-angle - An inversor consisting of a right-angled lever pivoted at the apex of the angle to a fixed point in the plane of the perspective center of the lens system, and of two rods, one connected rigidly at one end to the plane of the easel and at the other end by a sliding sleeve to the longer arm of the lever. The other rod is connected rigidly at one end to the plane of the photographic negative and at the other end by a sliding sleeve to the shorter arm of the lever.

Also called a Pythagorean inversor and Pythagorean right-angle inversor.

invert - The floor, bottom or lowest part of the internal cross-section of a conduit or pipe.

ionosphere - That part of the atmosphere containing a sufficient number of ions and electrons to cause appreciable dispersion of radio waves of frequency less than 1 GHz. The lower boundary of the ionosphere is at about 50 km (the upper boundary of the stratosphere) and the upper boundary (to the extent that one can be defined) is at about 2000 km. Four layers can be distinguished in the ionosphere. The lowest, the D-layer, has an electron density of about 1000 electrons per cc at maximum; it appears only during the day and disappears at night. The maximum occurs at about 80 km. The E-layer extends from about 85 km to 140 km, contains about 100 000 electrons per cc at periods of least solar activity and about 150 000 electrons per cc at times of greatest solar activity. The F-layer extends from about 140 km outwards. The structure of the F-layer is extremely complex (electron density varies from about 500 000 to over 2 000 000 particles per cc). Two regions can easily be distinguished: the F1 region extends from about 140 km to about 200 km; the F2 region extends from 200 km on out. The ionization is caused principally by solar radiation. The D-layer appears to be caused by hard x-rays, the E-layer by soft x-rays, and the F-region by ultraviolet radiation and light.

iris diaphragm - A continuously variable, approximately circular opening, in a lens system, which regulates the amount of light passing through the system. Also called an iris stop or stop.

iris stop - SEE iris diaphragm.

irradiation - An optical effect - bright objects viewed against a dark background appear to be larger than they really are. This effect has a physical basis and is encountered in photographs of celestial objects as well as when viewing the objects directly. It should not be confused with the Moon paradox - an extended object such as the Moon, for example, appears larger when seen on the horizon than when seen high in the sky.

irrevocability - (1) The quality not being able to be recalled or revoked. (2) The quality (in law) of being unalterable.

isarithm - SEE isopleth.

isarithme - SEE isopleth.

island - (1) A body of land extending above, and completely surrounded by, water at mean high water. This definition is applied extensively in surveying and mapping in the USA and is fully supported by federal-court decisions about the ownership of land. In foreign affairs, however, international negotiations are affected by various different conceptions of what constitutes the boundary line between the land and the water. (2) A naturally formed body of land, surrounded by water, which is above water at high tide. This definition is according to the Geneva Convention. (3) A swamp entirely surrounded by open water. (4) Dry land surrounded by a swamp and higher than the surrounding swamp. The vegetation on the land is usually different from that of the swamp proper. (5) Land of one character entirely surrounded by land of an entirely different character. (6) Land, rocky or otherwise, covered or not covered with water, connected or unconnected with a continent, over which it is impossible to navigate. (7) SEE traffic island. SEE ALSO barrier island.

island arc - A chain of islands describing a segment of an arc, generally convex outward from the continental region, and part of the major, global, active orogenic belt.

isanomal - A line connecting points of equal variation from a normal value.

isentropic (adj.) - Of equal or constant entropy with respect to either space or time.

isobar - (1) A line on which all points are at the same pressure. (2) A line (isobaric line), on a chart or map, representing a line, in the Earth, on which all points are at the same pressure. The isobar is most commonly drawn for points in the atmosphere or below the surface of water in oceans, seas, lakes, etc. However, it may be drawn for points in the crust as well.

isobath - SEE depth contour.

isocenter - (1) That unique point common to the plane of a photograph, the principal plane, and the plane of a hypothetical photograph assumed truly vertical and taken from the same camera station and having the same principal distance. (2) The point, on a photograph, at which the principal aline and the isometric parallel intersect. (3) The point at which the bisector of the angle between the plumb line and the photograph perpendicular intersects the plane of the photograph. The isocenter is significant because it is the center from which tilt-caused displacements of images radiate.

isocenter plot - SEE aerotriangulation, isocenter-radial.

isocenter triangulation - SEE aerotriangulation, radial.

isochrone - A line drawn on a chart or map and connecting all points that correspond to points on the Earth at which a particular event or value occurred simultaneously.

isoclinal - SEE line, isoclinic.

isodiff - One of a set of lines, on a map or chart, connecting points of equal correction or difference in datum. The isodiff is especially useful in readjusting surveys from one datum to another. Isodiffs connecting points of equal corrections in latitude are called isolats; isodiffs connecting points of equal corrections in longitude are called isolongs.

isogonal - SEE line, isogonic.

isogram - SEE isopleth (3).

isogriv - A line, on a map, joining points of equal angular difference between grid north and magnetic north.

isogriv chart - A chart on which lines are drawn connecting points having the same angular difference in direction between grid north and magnetic north.

isohypse, absolute - A line along which both the pressure and the height above mean sea level are constant.

isolat - SEE isodiff.

isoline - (1) A line representing the intersection of the plane of a vertical photograph with the plane of an overlapping, oblique photograph. If the vertical photograph were free of tilt, the isoline would be the isometric parallel of the oblique photograph. (2) SEE isopleth (1).

isolith - SEE isopach.

isolong - SEE isodiff.

isopach - A line along which the thickness of a layer is constant. Also called an isolith when referring to a geological layer.

isoparametric method(of mapping one ellipsoid on another) - Geodetic coordinates on the first ellipsoid are converted to rectangular coordinates using some suitable map projection (e.g., transverse Mercator or State Plane coordinates). These are then converted back into geodetic coordinates using the constants for the second ellipsoid. Distortions introduced by this method are proportional to the distortions of the map projection used for calculating the coordinates.

isopleth - (1) A line along which a particular quantity is constant or is assumed to be constant e.g., an isobar, isopycnic or isotherm. Also called an isarithm or isarithme or isoline. The isopleth was devised by L. Lalanne in 1843 as a modification of von Humboldt's isotherm. (2) A line along which values are constant or are assumed to be constant but the values correspond to quantities which do not exist at any particular point, e.g., the number of persons per square kilometer. (3) The line, on a map, representing an isopleth on the Earth. Also called an isogram. A contour line is an isopleth but a contour is not.

isopleth mapping - SEE mapping, isoplethic.

isopor - A line, on magnetic charts, showing points of equal annual change. Also called a magnetic isoporic line.

isopycnic - (1) A line along which density is constant. (2) The line, on a map, representing a line of constant density on or in the Earth. Also called an isopycnic line. A surface on which density is constant is called an isopycnic surface.

isoradial - A radial (line) drawn from the isocenter of a photograph.

isostasy - A condition of approximate equilibrium, in the outer part of the Earth, such that the gravitational effect of masses extending above the surface of the geoid is counterbalanced by a deficiency of density in the material beneath these masses, while the effect of a deficiency of density in oceanic waters is counterbalanced by an excess of density in the material under the waters. The basic principle of isostasy is that the masses of prismatic columns extending to some constant depth below the surface of the geoid are proportional to the areas of their sections at the geoid, regardless of the elevations of the Earth's surface above the columns. The depth to which these hypothetical columns extend is known as the depth of isostatic compensation and is somewhere between 95 and 15 km (60 and 70 statute miles). The area of the section, at the geoid, of a column for which isostatic compensation is ordinarily complete has not been determined; it may be uniform for all parts of the Earth or may vary with the character of the relief in the same continental regions. While the existence of isostasy is generally accepted as a proven principle, there are several theories as to the relative distribution of the matter producing this condition of equilibrium. The two principal theories are those of Airy and of Pratt. Pratt's theory, announced by him in 1855, postulated that the continents and islands project above the average elevation of the solid surface of the Earth because there is material of less density below them: the higher the surface, the less the density below. Under this theory, complete equilibrium exists at some uniform depth below the geoid and should be at the same depth in oceanic regions as in land regions. It is the theory used in the investigations of isostasy by Hayford and Bowie of the U.S. Coast and Geodetic Survey and, in the form used by them, is sometimes referred to as the Pratt-Hayford theory of isostasy. Airy's theory, announced by him in 1855, postulated that continents and islands are supported hydrostatically on highly plastic or liquid material, with roots or downward projections penetrating the inner material of the Earth just as icebergs extend downward into the water. The greater the elevation, the deeper the penetration. This has also been called the Roots of Mountains theory and was used in the investigations of W. Heiskanen for determining the geoid. In the form used by him, it is sometimes referred to as the Airy-Heiskanen theory of isostasy. The fundamental difference between the theories of Pratt and Airy is that Pratt postulated uniform depth with varying density while Airy postulated uniform density with varying depth. The term isostasy, meaning equal pressure, was first proposed by C. Dutton in a paper presented on 27 April 1889 and published later in Bulletin No. II of the Philosophical Society of Washington. There is also the theory that isostasy as such does not exist; the Earth's crust is never in a state of equilibrium. SEE ALSO Airy's theory of isostasy; adjustment, isostatic; compensation, regional; correction, isostatic; equilibrium, isostatic; gravity anomaly, isostatic; system, isostatic.

isostasy, regional - Isostasy existing over an extended region because of isostatic compensation over that region.

isostere - A line, on a map, representing a line along which atmospheric density is constant.

isotherm - (1) A line along which the temperature is constant. A surface on which the temperature is constant is called an isothermal surface. (2) The line, on a map, representing a line along which the temperature is constant. Also called an isotherm line. It was invented by A. von Humboldt in 1817.

Italian method - A method of resection in which the direction to the point of observation is determined by finding an auxiliary point from two of the known points using the known and measured data, and drawing the wanted direction through this point and the third known point. The method is probably the same as Collins' method. SEE Collin's method (of resection).

J

Jacobi ellipsoid - One of a series of ellipsoids describing the shape of a homogeneous, rotating, fluid body in hydrostatic equilibrium. The ellipsoids are stable at low rates of rotation; above a certain rate, Jacobi ellipsoids merge into a series of rotational ellipsoids called MacLaurin spheroids.

Jacob's staff - A single staff or pole on which a surveyor's compass or other instrument is mounted. A Jacob's staff can be used instead of a tripod. It is fitted with a ball-and-socket joint at its upper end so that the instrument can be adjusted to a level position. A metallic spur is fitted to the foot so that the staff can be pressed firmly into the ground. Many of the early land-surveys in the USA were made with surveyor's compasses mounted on Jacob's staffs.

Jäderin base line apparatus - A base-line apparatus consisting of (1) an invar wire 24 m long and 1.6 -1.7 mm in diameter, with a short, graduated, invar bar at each end; (2) two special supports at each end for the wire; and (3) precisely positionable reference marks with respect to which the distances were measured. The original apparatus (1880) consisted of steel wires; a later version added a brass wire parallel to the steel wire to allow the effect of temperature on the measurements to be determined. SEE ALSO Jäderin wires.

Jäderin's method - A method of measuring distances by using surveyor's measuring tape or wire freely suspended between two marks or supports about 25 meters apart and kept under a constant tension. The method does not involve a support midway between the two marks as do other methods that use much longer tape or wire. It was introduced by E. Jäderin about 1880 and was used in most European countries.

Jäderin wire - (1) The composite wire, composed of separate steel and brass wires, used in the Jäderin base-line apparatus. It was invented by Jäderin in 1885. It is usually 24 m long and 1.63 mm in diameter, with a short, subdivided scale (reglette) at each end. (2) SEE Jäderin base line apparatus.

Jäderin wires - An apparatus consisting of separate steel and brass wires extended under constant tension over reference marks on a line to be measured. SEE ALSO Jäderin base line apparatus.

James's map projection - SEE map projection, perspective. Also called Sir Henry James' map-projection.

Jansky - A unit of density of radiation, equal to 10^{-26} W/(m²Hz). The unit was adopted in 1973 by the International Astronomical Union. It is not part of the SI but is still used in radio astronomy and related field.

Jeans spheroid - The spheroidal shape assumed by a rotating, homogeneous mass distorted by the tides raised by a second body toward which it is moving with constant velocity.

Jeffreys-Radau ellipticity - The quantity $(5m/2) [1 + 25(1 - 3C/2Ma^2)^2/4]$ characterizing the shape of a body of mass M in hydrostatic equilibrium and rotating at the rate ω. The equatorial diameter is 2a, the moment of inertia about the least axis is C and m is the quantity $\omega^2 a^3/GM$, in which G is the gravitational constant.

Jeffreys' theorem - Hydrostatic equilibrium in a fluid body is impossible if the density varies on level surfaces.

jerk - The rate of change of acceleration.

jig transit - A precise transit without vertical or horizontal circles or compass, but with an optical micrometer on the telescope. Also called a jig collimator. A jig transit usually contains a lamp to provide a collimated beam of light.

Johnson tripod - A tripod designed particularly for supporting a plane table, having a ball-and-socket joint clamping the plane table in a suitable position. After about 1887, the Johnson tripod replaced an earlier support which had folding wooden legs fastened to the plane table.

joint, universal - SEE Cardan coupling.

Jolly's method - A method of adjusting a traverse using the method of least squares, with the following modifications from the straightforward method of least squares: (a) the angular misclosure is distributed equally among the angles, with the adjusted angles being used to obtain the bearings of the courses; (b) the coordinates are referred to the barycenter of the traverse; and (c) the weights assigned to distances and angles are constant throughout the traverse. SEE ALSO method of least squares.

Jordan arc - Any curve which is a one-to-one and continuous map of a straight line. Alternatively, a continuous curve without multiple points. Also called a simple arc or a simple curve.

judicate (v.) - Judge.

Julian calendar - A calendar in which those numbered years not divisible by 4 contain 365 mean solar days while those numbered years divisible by 4 contain 366 mean solar days. It was established in 45 B.C. by order of Julius Caesar, on the advice of the astronomer Sosigenes and was based on the assumption that the true length of the tropical year was exactly 365¼ mean solar days. It provided that the year begin on 1 January - a provision that was not universally followed even where other provisions of the order were. The calendar became effective for the year 44 B.C., which became known as the Year of Confusion because, in order to bring the calendar into harmony with the seasons, two additional months were intercalated between November and December, giving it 14 months with a total of 445 days. Until 8 A.D., days were intercalated every three years instead of four. Augustus ordered a return to the correct procedure in 8 A.D. The Julian calendar was superseded, in most nations, by the Gregorian calendar in 1582. SEE ALSO Gregorian calendar.

Julian century - A period of 36,525 days.

Julian date - An instant identified by giving the Julian day number of the Julian day preceding the instant and the fractional part of the day from the preceding noon (12h) U.T. to the instant. Unless otherwise specified, each Julian day is a mean solar day.

Julian date, modified - A date obtained by subtracting 2 400 000.5 from the Julian Date. Also called, incorrectly, Modified Julian Day.

Julian day - A day beginning at Greenwich noon and ending 24 hours later. SEE ALSO Julian day number.

Julian day, modified - SEE Julian day-number, modified.

Julian day-number - The integral number of days that have elapsed since Greenwich mean noon (1200 U.T.) on 1 January 4713 B.C. The day starting at this epoch is therefore Julian day number (J.D.) + 0; the day preceding it is J.D. - 0 (this is on the Julian proleptic calendar). Julian day number 2 415 020 corresponds to 12h on 0 January 1900. The Julian day number for 1950.0 is 2 433 282.423.

Julian day-number, modified - A Julian day number from which a specified constant has been subtracted. The constant can be assumed to be 2 400 000.5 unless otherwise stated.

Julian ephemeris date - (1) A Julian date calculated in days containing 86 400 ephemeris second. (2) A Julian date with the day beginning at 12h Ephemeris Time instead of at 12h Universal Time.

Julian ephemeris day number - A Julian day number calculated using days 86 400 ephemeris-seconds long.

Julian proleptic calendar - A calendar employing the rules of the Julian calendar but extended and applied to times preceding the introduction (44 B.C.) of the Julian calendar.

Julian year - An interval of 365.25 mean solar days, exactly.

junction - In leveling, the place where two or more level-lines are connected together.

junction bench mark - A bench mark selected as the common meeting point for lines of levels or links of levels.

junction closure - SEE junction misclosure.

junction detail - A sketch or diagram showing the details of what leveling has been done at a junction.

junction figure - A configuration, in a triangulation network, in which three or more triangulation arcs meet or two or more arcs intersect.

junction misclosure - The amount by which a new survey into a junction fails to give the previously determined location or elevation for that junction. Also called junction closure.

junior (adj.) - (1) Of more recent date. (2) In particular, of a mortgage, lien or the like, inferior or subordinate as to right of preference.

jurisdiction - The extent to which a body of law extends. It is an acknowledged principle of law that the title and disposition of real property is subject exclusively to the laws of the country where the real property is situated and that those laws alone prescribe the mode by which a title to the property can pass from one person to another.

juxtaposition - A placing side by side. In surveying, the term refers often to the placing of figures which, in a description, have differing units: e.g., thence easterly along the north line of Lot 21, 211 feet can easily be interpreted to mean 21,211 feet instead of Lot 21 a distance of 211 feet. Inserting a phrase such as thence easterly along the north line of Lot 21, a distance of 211 feet is better.

K

Kaestner's method - A method of solving the problem of determining, in a plane, the coordinates of a point A, given the coordinates of three points B, C, D and the measured angles BAC and CAD. The angles that CD and CB make with the x-axis are calculated first, from which the angles CDA and CBA are calculated. The distances CD and CB are next calculated and the coordinates of A then follow easily.

Kalman filter - Sequential analysis applied to problems of mechanics or electrical engineering. In particular, the method of least squares applied to data as they arrive singly or in sets, i.e., sequentially, rather than to all the data at once.

kappa - In photogrammetry, an angle of rotation about the z-axis. Also called swing and usually denoted by the Greek letter K. It is commonly used to denote the angle of rotation of an aerial camera about a vertical axis, the angle of rotation of a photograph about the z-axis of the coordinate system in a stereoscopic instrument, etc. Other definitions of kappa exist, but these are peculiar to particular organizations.

karst - A plateau composed largely of limestone, marked by numerous sinks (deep holes) on the surface and extensive caves below. Bench marks emplaced in a karst must be considered unstable.

Kater pendulum - (1) Any rigid pendulum of a type devised by Kater for comparing the acceleration of gravity at different points on the Earth. (2) SEE pendulum, reversible.

Kavraisky's constant - The constant devised by Kavraisky for determining the latitudes at which to place the standard parallels of a conic map projection, using criteria depending on the size and shape of the region being mapped.

kelp-shore - The land between the high-water mark and the low-water mark.

Kelvin scale - See Kelvin temperature scale.

Kelvin temperature scale - The temperature scale having the degree Kelvin (° K) as its unit and assigning the value 273.16 ° K to the triple-point of water. Also called the Kelvin scale. Until 1954, the degree Kelvin was equal to the degree Centigrade, defined as the 1/100-th part of the interval between the freezing point of water and the boiling point of water under standard pressure. In 1954, the unit was defined by the General Conference on Weights and Measures as 1/273.16 of the thermodynamic temperature of the triple-point of water. The zero point is then at -273.16° K from the triple-point. This must be the same as absolute zero, because thermodynamic temperature is specified. On this scale, the difference between two temperatures T_1 and T_2 is proportional to the amount of heat converted into mechanical work by a Carnot engine operating between the isotherms and adiabats running through T_1 and T_2.

Kennelly-Heaviside layer - An old name for a heavily-ionized layer of the Earth's atmosphere, identified with the E-layer.

Kepler ellipse - SEE Keplerian ellipse.

Keplerian element - One of the following six orbital elements specifying an orbit:
- a   half-diameter of the osculating ellipse
- e   eccentricity of the osculating ellipse
- i   angle (inclination) from a reference plane through the center of attraction to the plane of the osculating ellipse
- ν   angle from a reference line in the reference plane to the intersection (line of nodes) of that plane with the plane of the osculating ellipse
- ω   angle from the line of nodes to the major diameter of the osculating ellipse
- $σ_o$ angle from the radius vector to the body at time t_o to the radius vector to the body at pericenter.

The symbol Ω is sometimes used instead of ν; t_o, the time at which the body is at a specified point in the orbit, or M_o, the mean anomaly at time t_o, are often used instead of $σ_o$. The constants a, e, $σ_o$ characterize the osculating ellipse and the location of the body on that ellipse at a particular time. The constants ν, i and ω are the Eulerian angles specifying the orientation of the ellipse with respect to the reference plane and reference line. The center of attraction is at one of the foci of the ellipse and lies on the reference plane and reference line. However, an ellipse other than the osculating ellipse, an ellipse in average position, for example, may be used to represent the orbit. Also called orbital elements if an exact name is not wanted.

Keplerian ellipse - The ellipse defined by specified values of the Keplerian elements of an orbit at a specific time.

Keplerian orbit - (1) The orbit taken by a body obeying Kepler's laws of motion. By extension, any body following an elliptical orbit. Sometimes called a normal orbit, two-body orbit or unperturbed orbit. (2) A solution of the equation $d^2x/dt^2 = -k^2x/x^3$, in which x denotes the radius vector of a point mass, t the time and k^2 a specific constant. Also called a normal orbit or an unperturbed orbit.

Kepler's equation - The equation $M = E - e \sin E$, relating the mean anomaly M of a satellite to the eccentric anomaly E and the eccentricity e of the orbit. The equation is usually solved for E by an iterative method or by expressing E as an infinite series in M. Similar equations for motion in parabolic and hyperbolic orbits are sometimes also called Kepler's equations.

Kepler's laws - The three laws governing the motions of the planets and enunciated by J. Kepler in 1609 and 1619. (a) Each planet moves in an elliptical path with the Sun at one focus of the ellipse. (b) The line from the Sun to any planet sweeps out equal areas of space in equal lengths of time. (c) The squares of the sidereal periods of the several planets are proportional to the cubes of their average distances from the Sun. Also called Kepler's planetary laws. The first and second laws refer to the motion of an individual planet; the third law expresses a relationship between the motions of different planets. The laws are not strictly true, for they assume that the movement of one planet does not affect the movement of any other, that planetary movements do not change the location of the Sun, and that the Sun has no translational motion of its own.

Kepler's planetary laws - SEE Kepler's laws.

Kerr cell - A small, transparent container holding a transparent medium of large dielectric constant (such as nitrobenzene) and a pair of metallic plates to which is applied a large voltage producing a strong electric field across the medium. The cell is usually placed between a pair of prisms transmitting only plane- polarized light. The assemblage is used in one of two ways. (a) The prisms are so oriented that their planes of polarization are parallel. Light transmitted by the one is then normally transmitted by the other. With no difference of voltage between the plates, the electric field is zero and light can pass through the system. When voltage is applied to the plates, the resulting electric field rotates the plane of polarization of light passing through the Kerr cell and no light passes through. (b) The prisms are so oriented that their planes of polarization are not parallel. No light normally passes through, and applying voltage to the plates then rotates the plane of polarization of the light passing the first prism until it does pass through the second prism. The Kerr cell is used as a shutter, e.g., to generate pulses of light or to time the intervals between pulses. Some distance-measuring instruments use Kerr cells for timing pulses or determining phase difference.

key, photographic-interpretation - A selection of images or descriptions of objects, used by photo-interpreters as an aid to identifying what is shown on a photograph. Also called a photo-interpretation key. The term is often used in the more general sense of a key to the identification of images, including photographic. There are two general types of keys: dichotomous and descriptive.

key, photo-interpretation - SEE key, photographic-interpretation.

key flat - The principal layout used as a guide for positioning on other flats. Also called a layout guide.

key map - SEE index map.

K-factor - The ratio of base length to altitude in aerial photography and photogrammetry. Also called the base-height ratio (q.v.).

kill (Dutch) - (1) A channel or bed of a river. (2) A river or stream. This Dutch term is found used in this sense in descriptions of land in old conveyances.

kilogram - The mass of a platinum-iridium body kept at the International Bureau of Weights and Measures near Paris and accepted internationally as a standard of mass. In the original metric system, the centimeter-gram-second system, the gram was the unit of mass and defined to be the mass of 1 cubic centimeter of water at its maximum density. This definition was later changed to make the gram equal to 1/1000 of the mass of the body called the kilogram. For practical reasons, the kilogram is designated as the unit of mass in the SI.

Kimura term - A minute (0.03") variation, of yearly period, found by Kimura in 1902 in the variation of latitude. Also called the z-term and Kimura's z-term. The reason for the variation is not known. It has been attributed to seasonal variations of the personal equation of the observers, to seasonal refractive effects, and to the effect of a liquid core on the Earth's rotation.

kine-theodolite - SEE cine-theodolite.

kiss plate - A press-plate used to make an addition or correction to a previously printed sheet. Also called a touch plate.

Klimsch-Variomat method (of compilation) - A photographic process by which material for large-scale maps is selectively reproduced so as to retain only indexed contours. The indexed contours are then used to indicate relief on medium-scaled maps. The method is no longer much used.

knife edge - That edge, of a small triangular prismatic, which acts as the pivot for the oscillation of a pendulum. SEE ALSO correction, knife-edge.

knoll - A small, rounded hill rising less than 500 fathoms (about 1000 meters) from the floor of the ocean and of limited extent across the summit. The term is sometimes confused with guyot.

knot - A unit of speed defined (1978) as 1 international nautical mile per hour. The knot was defined previously as 1 nautical mile per hour, but this led to confusion because of the difference between the American and British nautical miles, which differ by 1.184 m. The knot is equal to 1.852 km per hour or 0.5148 meters per second.

Koppe's formula - The formula $mh = \pm (a + b \tan \beta)$ for the standard deviation mh of the error in representing elevations, on a map, as a function of β, the angle giving the slope of the terrain. a and b are constants depending on the error of measurement and on the scale of the photographs from which the map was prepared.

Köster's prism - A prism formed by cementing together two identical prisms having 30°-60°-90° triangles for bases so that a composite prism having equilateral triangles for bases is formed. A semi-reflecting coating is applied to one of the faces opposite the 60° angle before cementing. Also called a Köster's double-image prism. This prism is an element in many optical interferometers and has been used for testing mirrors, flats and other optical elements.

Köster's double-image prism - SEE Köster's prism.

krakovian - A rectangular array of numerical or algebraic quantities (elements) obeying the following algebraic rules.

(a) The sum of two arrays having the same number of rows and the same number of columns is the array whose element in row m and column n is the sum of the elements in row m and column n of the summands. (b) The product of an array by a constant is the array whose element in row m and column n is the constant times the element in row m and column n of the original array. (c) The product of two arrays having the same number of rows is the array such that the element in row m and column n is the sum of the products of each element of column m of the one array by the corresponding element in column n of the second array. The word krakovian is also written as Krakovian, Cracovian and cracovian. The krakovian was invented by T. Banachiewics to lessen the chance of blunders of the kind made in hand calculations involving matrices. Blunders made when doing column-by-column multiplication are less frequent than similar blunders made when doing row-by- column multiplication. However the introduction of large, programmable calculators has eliminated the computational need for krakovians and such arrays are therefore principally of theoretical interest.

Krakovian method - One of two methods devised by T. Banachiewics for solving a system of simultaneous linear equations; the first method corresponds closely to Gauss's method, the second to Cholesky's method.

Krassovski ellipsoid 1940 - An ellipsoid characterized by the following constants:
length of semi-major axis	6 378 245.000 meters
flattening in the meridian	1/298.3
flattening in the equator	1/130 086

The semi-minor axis and the other axis therefore have lengths of, respectively, 6 356 863.019 meters and 6 378 033 meters.

Krassovski spheroid - A rotational ellipsoid used for horizontal control in the USSR, and having the dimensions
semi-major axis	6 378 245 m
flattening	1/298.3.

The Krassovski spheroid is derived from the Krassovski ellipsoid, which has axes of three different lengths.

Kriging - (1) A method of curve fitting which represents the curve as a linear function of the given values, subject to the conditions that (a) the average difference between the fitted curve and the given values shall be zero, and (b) the co-variance matrix of the difference shall be a minimum. (2) The same as the definition (1), with surface fitting and surface substituted for curve fitting and curve, respectively. Two kinds of Kriging are distinguished: simple Kriging, in which the covariance function is constant over the region to which the curve or surface is being fitted, and Universal Kriging, in which this condition is not imposed.

Kripniyuk-Kwiklokchun Datum - The horizontal-control datum defined by the following coordinates of the origin and azimuths (clockwise from South) on the Clarke spheroid of 1866; origin is at Kripniyuk astronomical station:
longitude of origin	165° 19' 27.12" W
latitude of origin	62° 20' 06.58" N
azimuth from origin to station Tent	180° 00' 00.0"
azimuth from origin to station Camp	0° 00' 00.0" .

The above coordinates are based on weighted means of astronomical determinations made in 1898; the longitudes were determined by the chronometric method.

Kronstadt Datum - The zero of the tide gage at Kronstadt (Baltic Sea), used as the zero for leveling networks in the USSR and some neighboring countries.

Krüger's method of adjustment - A method developed for adjusting triangulation, but applicable to solving most systems of linear equations, in which the set of equations is divided into two groups according to certain criteria. One group is solved first, the results applied to the second group, and the process repeated until the solution is considered satisfactory. Also called Krüger's two-group method. It and Schreiber's method of adjustment are mathematically equivalent. They differ only in the sequence of procedures involved.

Kukkamäki's formula - (1) The formula $t = a + bzc$ for the temperature t at a difference z of elevation between the ground and the point in question. a, b and c are constants determined by observations over a long period. The values of a, b and c depend on the climate and on the date. The formula has been used in computing atmospheric refraction as a function of temperature, particularly for computing refractive corrections to elevations. (2) The formula $\Delta H = [(dn/dt)(t_2 - t_1) s / (H_2 c - H_1 c)(H_2 - H_1)^2] \times \{[H_o c (H_2 - H_1) - (H_2 c + 1 - H_1 c + 1)/(c+1)]\}$ for the correction ΔH to a difference of elevation determined by spirit leveling. dn/dt is the rate of change of refractive index with temperature; H_2 and H_1 are the readings on the forward and backward leveling rods, respectively; t_2 and t_1 are the corresponding temperatures; H_o is the height of the instrument above the ground, s is the length of the line of sight and c is the constant in Kukkamäki's formula for temperature.

Kukkamäki's method - A method of determining the C-factor, requiring that the leveling instrument be half way between the two leveling rods for the first set of measurements and that it be on the same line, extended beyond the second leveling rod, for the second set of measurements.

L

laboratory standard - SEE standard, secondary.

Laborde map projection - A map projection from the rotational ellipsoid to the plane, approximately equivalent to an oblique Mercator map-projection but modified so that, within the region mapped, linear distortion is minimal. There are several versions of the Laborde projection. The one used by the U.S. Lake Survey proceeds as follows. (a) The rotational ellipsoid (spheroid) is mapped conformally onto a sphere having the average curvature of the ellipsoid at the proposed origin and tangent to the ellipsoid at that origin. The region on the ellipsoid is then mapped conformally onto the sphere. Differences in longitude and isometric latitude are kept proportional between the ellipsoid and the sphere. (b) The sphere is mapped conformally onto a cylinder tangent along the central meridian, using a transverse Mercator map-projection. (c) The cylinder is mapped onto a plane and the map on the plane is distorted (but kept conformal) to reduce error in scale along an axis oblique to the meridional line (error in scale along the line representing the meridian is thereby increased). The x- and y-axes at the origin are still parallel to and perpendicular to, respectively, the meridian. (d) By a change of scale, the error is made zero along two oblique lines instead of along one. The linear distortion in an oblique direction is thereby minimized. This map projection was first described in 1926 as applied to the mapping of Madagascar, where Laborde served as chief of the Service Geographique. It has also been used for flight charts of the United States Air Force.

lace - One pole (16.5 feet) of land as used in Cornwall.

ladder grid number - A number associated with a line of a grid and identifying that part of the line within the neatline.

lag coefficient - SEE rise time.

lag covariance - SEE covariance.

lagoon - (1) An elongated body of water lying parallel to the coastline and separated from the sea by islands which act as a barrier. (2) The central body of water in an atoll. (3) A tidal lake filled with water from the sea.

Lagrange's variational equations - A set of six first-order, partial differential equations giving the rate of change, with respect to time t, of each of the six orbital elements as a function of the partial derivatives of the perturbing force R with respect to the orbital elements. The Keplerian elements a, e, i, Ω, ω and M_o are usually used; one form of the equation is as follows:

$$da/dt = (2/ma)(\delta R/\delta M_o)$$
$$de/dt = [(1-e^2)/(na^2e)][(\delta R/\delta M_o) - (\delta R/\delta \omega)]$$
$$di/dt = [\cos i\, (\delta R/\delta \omega) - (\delta R/\delta \Omega)]/[na^2(1-e^2)\, \sin i]$$
$$d\Omega/dt = (\delta R/\delta i)/[na^2(1-e^2)\, \sin i]$$
$$d\omega/dt = [((1-e^2)(\delta R/\delta e) - \cot i\, (\delta R/\delta i)]/[na^2(1-e^2)]$$
$$dM_o/dt = n - [(1-e^2)(\delta R/\delta M_o) - 2ae\, (\delta R/\delta a)]/(na^2e).$$

All the elements are those of the osculating (instantaneous elliptical) orbit. a is the length of the semi-major axis, e is the eccentricity and i is the inclination of the orbital plane. Ω is the angle from a reference line in the reference plane to the intersection of that plane with the orbital plane, ω is the angle from the line of intersection to the line or apsides, and Mo is the angle from a specified radius vector in the orbital plane to the radius vector of the satellite. n is the average rate of change of mean anomaly. The constants of the actual orbit are then the elements at a specified time (usually zero). The elements in the variational equations are six additional functions of the time and the perturbing force R. Lagrange's equations are used as the basis for a variety of similar equations sometimes also called Lagrange's variational equations. A particularly elegant form, closely related to Hamilton's canonical equations, is obtained by defining the following elements in terms of the Keplerian elements, the gravitational constant G_a and the mass M_b.

$$(G_a M_b)^{1/2} = L \qquad \int n\, dt = l$$
$$L(1-e^2) = G \qquad \omega = g$$
$$G \cos i = H \qquad \Omega = h$$

Then
$$dL/dt = \delta R/\delta l \qquad dl/dt = -\delta R/\delta L + n$$
$$dG/dt = \delta R/\delta g \qquad dg/dt = \delta R/\delta G$$
$$dH/dt = \delta R/\delta h \qquad dh/dt = \delta R/\delta H.$$

These are also called Delaunay's canonical equations because they contain Delaunay's variables as used by him in his theory of the Moon's motion.

Lagrangian coordinate system - A coordinate system in which points are identified by assigning them coordinates which do not vary with time. Also called a material coordinate system.

Lagrangian multiplier - An unknown quantity introduced into a set of linear equations containing both observation equations and condition equations to make the number of unknowns equal to the number of equations. Also called an undetermined multiplier or correlate. The term correlate, however, is used principally for Lagrangian multipliers used in the method of correlates (a variant of the method of least squares).

La Hire's map projection - SEE map projection, perspective.

lake - (1) A large body of still water situated in a depression in the ground without direct communication with the sea. (2) A region of deep water completely surrounded by land.

Lallemand's formula - The formula $\Delta^2 H = -2a\, \Delta H\, \sin 2\phi\, \Delta\phi$ for converting measured differences ΔH of elevation to true differences $\Delta H + \Delta^2 H$ in a section of leveling; ϕ is the average latitude of the section of leveling, $\Delta\phi$ is the difference in latitude of the ends of the section and a is a constant.

Lallemand's formulas - A set of three formulas derived by L. Lallemand for the growth of error with distance in spirit leveling and adopted as standard by the International Geodetic Association in 1912. The formulas were used by the Association in defining leveling of high precision and by various countries in evaluating their leveling networks. The second geodetic leveling of England and Wales (1912-1921) was evaluated using Lallemand's formulas.

lambert - A unit of luminance (brightness), equal to the brightness of a perfectly diffuse flat surface radiating one lumen per square centimeter.

Lambert azimuthal equal-area map projection - SEE map projection, azimuthal equal-area.

Lambert azimuthal equal-area map projection, polar - SEE map projection, azimuthal equal-area.

Lambert azimuthal meridional equal-area map projection - SEE map projection, azimuthal equal-area.

Lambert azimuthal polar map projection - SEE map projection, azimuthal equal-area.

Lambert bearing - A bearing as measured on a map drawn on the Lambert conformal conic map projection. This approximates a great-circle bearing.

Lambert central equivalent map projection upon the plane of the meridian - SEE map projection, azimuthal equal-area.

Lambert conformal chart - A chart on the Lambert conformal map projection.

Lambert conformal conical map projection - A conformal map projection from the hemisphere or half-rotational ellipsoid to the plane, in which all meridians are represented by straight lines converging to a common point outside the limits of the map, and the parallels of latitude are represented by arcs of circles having this point for a center. Meridians and parallels intersect at right angles. The map projection may represent one parallel of latitude by an arc along which the scale is held exact, or there may be two such parallels of latitude, both represented by lines along which the scale is exact. Also called the Lambert conformal conic map-projection and the Lambert conformal map-projection. The Lambert conformal conical map-projection is given, for the transformation from hemisphere to plane, by the equations

$r = k R \phi_1 [\tan\frac{1}{4}(\pi - 2\phi) / \tan\frac{1}{4}(\pi - 2\phi_1)]^{n} / n$
$\theta = n \omega \lambda$
$n = (\ln \cos \phi_1 - \ln \cos \phi_2) / [\ln \tan\frac{1}{4}(\pi - 2\phi_1) - \ln \tan\frac{1}{4}(\pi - 2\phi_2)]$.

r and θ are the polar coordinates of that point, in the plane, which corresponds to the point at longitude $\lambda_o + \Delta \lambda$ and latitude ϕ on the hemisphere; λ_o is the longitude of the central meridian. ϕ_1 and ϕ_2 are the latitudes of two parallels for which the scale, on the corresponding lines of the graticule, should be exact. R is the radius of the hemisphere and k is the scale factor. This particular form is called the Lambert conformal conical map-projection with two standard parallels. Lambert's conformal conical map-projection with one standard parallel is a special case of the above and results from the above equations by letting ϕ_2 approach ϕ_1 so that $n = \sin\phi_1$. At any point on the map, the scale is the same in every direction. It changes along the lines representing meridians and is constant along each line representing a parallel of latitude. Where two standard parallels are used, the scale between the representing lines is too small; beyond them, it is too large. The Lambert conformal conical map projection with two standard parallels is used in the State Coordinate System for zones of limited extent northward and southward and indefinite extent eastward and westward. In that coordinate system, the standard parallels are placed at distances of approximately one-sixth the width (north to south) of the map from its upper and lower limits. It is also used for the World Aeronautical Charts of the zone between 80° N and 80° S. It was introduced in 1772.

Lambert's conformal conical map projection, modified - Lambert's conformal conical map projection modified for use in polar regions. Also called Ney's map projection.

Lambert's conformal conic map projection - SEE Lambert conformal conical map projection.

Lambert conformal map projection - SEE Lambert conformal conical map projection.

Lambert-Euler equation - SEE Euler's equation; Lambert's equation.

Lambert equal-area meridional map projection - SEE map projection, azimuthal equal-area.

Lambert equal-area polar graticule - SEE graticule, azimuthal equal-area.

Lambert equal-area polar grid - SEE graticule, azimuthal equal-area.

Lambert grid - (1) A rectangular grid superposed on a map prepared using Lambert's conformal conical map projection. One zero-line coincides with the line of the graticule representing the central meridian; the zero-line of the other family is placed below the line (of the graticule) representing the most southerly latitude of the region involved. (2) A rectangular grid superposed on a map prepared using Lambert's conformal conical map projection. In particular, an informal designation for a State plane coordinate system or State coordinate system with two standard parallels. (3) An informal designation for a coordinate system based on Lambert's conformal conic map projection.

Lambertian source - A planar source of electromagnetic radiation for which the radiance does not vary as a function of the angle from which the source is viewed.

Lambertian surface - A surface which has the appearance of a Lambertian source.

Lambert's conformal cylindrical map projection - SEE Mercator map projection, transverse.

Lambert's conical equal-area map projection - A map projection defined, for the sphere, by the equations

$r = 2 k R \sin\frac{1}{4}(\pi - 2\phi)$
$\theta = n \Delta \lambda$,

in which r and θ are the polar coordinates of that point, in the plane, which corresponds to the point, on the sphere, with longitude $\lambda_o + \Delta \lambda$ and latitude ϕ; λ_o is the longitude of the central meridian. R is the radius of the sphere and k is the scale factor. n is a constant called the constant of the cone; if the degree of average latitude be required to preserve its true ratio to the degree of longitude, $n = \sec^2 \frac{1}{2}\phi$. If, at a latitude of 45°, the scale along the line representing the parallel of latitude is made equal to the scale along a line representing a meridian, the entire hemisphere is included between two radii inclosing and angle of 307° 16'. For this particular map projection, n is approximately equal to 1.171. Also called Lambert's isospheric stenoteric map-projection.

Lambert's cylindrical equal-area map projection - SEE map projection, cylindrical equal-area.

Lambert's cylindrical orthomorphic map projection - SEE Mercator map projection, transverse.

Lambert's equal-area meridional map projection - SEE map projection, azimuthal equal-area.

Lambert's equation - An equation relating the lengths r_1 and r_2 of two radii vectors to two points in an elliptical orbit, to the length of the chord joining those two points, and to the interval of time $(t_2 - t_1)$ between the two points. Denoting the length of the semi-major axis by a, the equation is

$$k(t_2 - t_1) = a^{3/2} [2\pi n + (n_1 - \sin^n{_1}) \pm (n_2 - \sin^n{_2})]$$ or
$$k(t_2 - t_1) = a^{3/2} [2\pi(n+1) - (n_1 - \sin^n{_1}) \pm (n_2 - \sin^n{_2})]$$

according as the focus of the ellipse is located outside or inside the figure consisting of the arc and the chord. k^2 is the gravitational constant, n is the fraction of a revolution completed in going from the first point to the second (i.e. between times t_1 and t_2). n_1 and n_2 are defined by the equations $\sin^2(\frac{1}{2} n_1) = (r_1 + r_2 + s)/4a$ and $\sin^2(\frac{1}{2} n_2) = (r_1 + r_2 - s)/4a$ and by the conditions $\pi \geq n_1 \geq 0$ and $\pi \geq n_2 \geq 0$. s is the sum of the sides of the triangle formed by the chord and the two radii vectors.

Lambert's theorem - The time required for a body to move between two points in its orbit depends only on the length of the chord between the two points and on the length of the perimeter of the triangle formed by the chord and the two radii to the points.

Lambert "Twin" tower - A tower consisting of a 9"-diameter vertical pipe supporting the surveying instrument, and a vertical, triangular prism 12" on a side which surrounds the pipe and supports an observing platform at the top. The Lambert "Twin" tower is used primarily for establishing horizontal control. The pipe is kept vertical by cables connecting internally to the prism. The prism is kept vertical by guying cables connecting it to anchors in the ground. If the tower is more that 15 feet high, its rigidity is increased by external bracing wires. Both pipe and prism are made in 15-foot long modules, allowing towers to be built in 15-foot increments to a height of 60 feet.

Lambert zenithal equal-area map projection - SEE map projection, azimuthal equal-area.

Lamé coordinate - A coordinate in an ellipsoidal coordinate system. SEE coordinate, ellipsoidal.

Lame' elastic constant - SEE Lamé's constant.

Lame' elastic parameters - SEE Lamé's constant.

Lamé function - A solution, in ellipsoidal coordinates, of Laplace's equation. Lamé functions are particularly useful for representing the gravitational potential of ellipsoidal bodies.

Lamé's constant - One of the two constants λ and μ in the formula $\sigma = \lambda / [2(\lambda + \mu)]$ for Poisson's ratio σ. Also called a Lame' elastic constant or Lame' elastic parameter. λ is also referred to as Lamé's lambda.

Lamé's lambda - SEE Lamé's constant.

laminate (n.) - A composite material composed of a thin sheet of one kind of material sandwiched between and bonded to two sheets of another kind of material usually a clear plastic.

laminate (v.) - To sandwich a thin sheet of one kind of material between two thin sheets of another kind (usually clear plastic) and bond the three sheets together. Bonding is usually done by applying heat and pressure to the sandwich. In map-making, the intermediate sheet is a map or other graphic; the covering sheets are a polyethylene-polyester plastic. An adhesive is not used.

lamp, signal - SEE lamp, signalling.

lamp, signaling - (1) A lamp used for sending signals by systematically opening and closing the lamp's shutter.
(2) SEE lamp signal.

lamp signal - A lamp placed at a triangulation station and sighted on by observers at other stations. Also called a signal lamp and signalling lamp. Signalling lamp, however, is best used to denote a lamp used principally to send signals. SEE ALSO target.

lamp person - SEE light keeper.

land - Real property and real estate.

land (law) - The surface of the Earth and all that lies under it, down to the center of the Earth, and all that lies above it to the Sky, including all natural things thereon such as trees, crops, or water, plus the minerals below the surface and the air rights above it.

land, acquired - Federal land obtained by purchase, condemnation, or gift under laws other than the public-land laws.

land, appropriated public - Original public-domain lands which are covered by an entry, patent, certification, or other evidence of land disposal, for certain purposes: e.g., public lands which are within a reservation, which contain improvements constructed with the aid of Federal funds, or which are covered by certain classes of leases.

land, fast - Land inshore of the inner edge of a marsh; usually at or above the level of mean high water.

land, filled - A region where the ground has been raised by depositing or dumping dirt, gravel or rock.

land, overflowed - Land covered by unnavigable waters, but not including land between high and low waters of navigable streams or bodies of water, or land covered and uncovered by the ordinary daily ebb and flow of normal tides of navigable waters. The intent of this definition is to include only such land as is normally not covered with water.

land, public - In the United States of America, that portion of the public domain whose title is still vested in the Federal Government.

land, public-domain - SEE domain, public.

land, registered - SEE abstract of title; certificate of title; Torrens Registration System.

land, restricted - Land, the alienation of which is subject to restrictions imposed by Congress.

land, riparian - (1) Strictly, lands bordering on a river. (2) Loosely, the shore of the sea or other tidal water, or of a lake or other considerable body of water not having the character of a watercourse. SEE ALSO littoral; boundary, riparian.

land, saline - Land having salt deposits. Lands having salt springs were granted to 14 States; others received none. In early times, salt was a necessary commodity.

land, submerged - Land covered by water at any stage of the tides. It differs from tideland in that the latter is attached to the mainland or an island and covered and uncovered by the tide, with a high-water line as upper boundary. It differs from overflowed land in that the latter is usually not covered by the tide at any stage of the tide.

land boundary - A line of demarcation between adjoining parcels of land that are politically or legally distinct. The parcels of land may be of the same or of different ownership, but are distinguished at some time in the history of their ownership by separate legal descriptions. A land boundary may be marked on the ground by material monuments placed primarily for this purpose by fences, hedges, ditches, roads and other structures along the line or may be specified by astronomically described points and lines; or by coordinates on a survey system whose position on the ground is witnesses by material monuments which are established without reference to the boundary line. Various other methods are also in use. Although the land boundary is defined to be a line and is determined as such by the surveyor, it is usually considered by law to be a vertical surface extending upwards and downwards from the line.

land claim, private - (1) A claim to a tract of land, based on the assertion that title thereto was granted to the claimant or his predecessors in interest by a foreign government (before the territory in which it is situated was acquired by the USA). (2) A tract of land to which claim is made as based on the assertion as stated in the preceding definition.

land court - A tribunal established to administer legislative statutes relating to land boundaries and titles. Also written Land Court.

Land Court - SEE land court.

Land Department - In legal literature, the generalized term adopted to denote the Secretary of the Interior, The Commissioner of the General Land Office, and, currently, the Director of the Bureau of Land Management, and their predecessors, together with subordinate officials, when acting in their capacities as administrators of the public-land laws.

land description - The exact location of a parcel of land stated in terms of lot, block, and tract; metes and bound; or section, township and range.

land disposal - A transaction that leads to the transfer of title to public lands from the Federal Government.

land effect - SEE refraction, coastal.

landfill, sanitary - A portion of land on which refuse (solid or semi-solid waste) is dumped and then covered with earth. Now carefully designed to meet environmental standards.

landform - (1) A shape into which the Earth's surface has been sculptured by natural forces. (2) One of the many features which, taken together, make up the surface of the Earth. The term includes all such broad features as plains, plateaus, and mountains, as well as such minor features as hills, valleys, slopes, canyons, arroyos, alluvial fans, terraces and glacial deposits. Most of these features result from erosion and resistance to erosion. The term also includes all features resulting from sedimentation and from movements within the crust. A landform can be portrayed by various means and various inferences can be made as to its probable origin, cause or history. Whether it is called a topographic landform or a physiographic landform depends upon how it is being used. However, the term landform is used extensively in place of and with the meaning of the more restrictive term physiographic form. It is also written as two words, land form, which is preferred by some writers. Note that the definition applies not merely to features on land but also to features under water, e.g., submarine canyons.

landform, physiographic - A landform considered with regard to its origin, cause, or history. Also called a physiographic form. SEE ALSO landform.

landform, topographic - A landform considered without regard to its origin, cause or history. Also called a topographic form.

land information system (LIS) - Any computerized information-system designed for storing, using, and retrieving information associated with or related to a cadaster.

land league - A league used as a unit of length on land.

land line - A real or imaginary line drawn on the ground. For example, the line connecting two monuments on a boundary is, in general, an imaginary land-line. A boundary marked by a line drawn on rocks, or marked by a fence, would be a real land-line. SEE ALSO adjustment, land-line.

landmark - (1) Any fixed object (monument or other material mark) used to identify the location of a boundary of a piece of land. (2) Any prominent object, on land, usable for determining a location or direction.

land mass simulator plate - SEE plate, land-mass-simulator.

land mile - SEE mile, statute.

Land Office - A US Government office in which the entries upon and sales of public land are registered, and other business respecting the public lands is transacted.

land registration - SEE Torrens Registration System.

lands, public-domain - SEE domain, public.

LANDSAT - One of a series of satellites, so called, designed to transmit images of portions of the Earth's surface to stations on the ground and launched by the National Aeronautics and Space Administration. The first LANDSAT was launched in 1972 and went into orbit at a height of 919 km. Each image covered an area about 185 km on a side, with a nominal scale of 1:1 000 000. LANDSAT was launched in 1981 into an orbit at a height of 705 km and was programmed to provide images on request.

landscape map - A topographic map made to a relatively large scale and showing all detail consistent with the scale. Such maps are required by architects and landscape gardeners for planning buildings to fit the natural features and for landscaping parks, playgrounds, private estates, etc. They generally represent only small areas and scales vary from 1 inch equals 20 feet (1:240) to 1 foot equals 50 feet (1:50), depending on the amount of detail wanted.

land survey - (1) A survey, or the results of a survey, giving at least the boundaries of a particular piece of land. Also called a property survey or boundary survey. The term cadastral survey is also sometimes used to denote a land survey. However, in the USA that term should be used only for surveys of the nation's public lands. A land survey often gives other information such as the locations of structures, easements, water courses and total area is often included. SEE ALSO survey; surveying, cadastral. (2) SEE land surveying.

land survey, rectangular - SEE centuratio; survey, rectangular.

land surveying - (1) The process of determining boundaries and areas of tracts of land. Rarely called land survey. The term cadastral survey is also used to denote a land survey, but in the U.S. Bureau of Land Management, that term is used only to denote surveys of the public lands of the USA. Boundaries are usually determined by proven claims to ownership of the land within specified boundaries, commencing with the earliest owners, and descending through successive ownerships and partitions. Land surveying includes re-establishing original boundaries and the establishing of such new boundaries as may be required in partitioning the land. (2) The art and science of reestablishing cadastral surveys and the boundaries of land, based on documents of record and historical evidence; planning, designing and establishing boundaries of property; and certifying surveys as required by statute or local ordinance. Land surveying can include associated services such as mapping, constructing layout surveys and analyzing and using the data acquired by surveying.

land surveyor - A person whose profession is the determination, by surveying, of boundaries and areas of tracts of land and often registered by the states in which practicing.

land use - The use to which land is put.

land use control ordinance - SEE ordinance, land-use control.

land warrant - A certificate from the Land office, authorizing a person to assume ownership of public land.

lane - (1) That region, on the ground, within which the phase of a fixed-frequency, continuous wave from a radio transmitter changes by 1 cycle as compared to the phase of a standard originally in phase with the transmitter. The boundaries of a lane are usually taken at whole-wavelength intervals from a starting point where the phase is assumed to be zero. (2) That region, on the ground, within which the sum or difference of the distances from two fixed radio transmitters changes by a whole number (usually 1) of wavelengths of the fixed-frequency, continuous wave transmitted. If the sum is used, the boundaries of the lane are ellipses; if the difference is used, the boundaries of lanes are hyperbolae. (3) A narrow trafficway between trees or buildings on each side.

lap - SEE overlap.

lap, forward - SEE overlap, forward.

lap, side - SEE overlap, side.

Laplace azimuth - A geodetic azimuth computed from an astronomic azimuth by means of the Laplace equation. SEE ALSO Laplace's equation.

Laplace-azimuth mark - SEE mark, Laplace-azimuth.

Laplace condition - The condition imposed by Laplace's equation (1) on a geodetic azimuth or longitude determined at a point where the astronomical azimuth or longitude has also been determined. SEE ALSO Laplace's equation (1).

Laplace control - Geodetic control established for the conversion of astronomic azimuths to geodetic azimuths by determining the deflection of the vertical in the prime vertical.

Laplace correction - The quantity added to an astronomic azimuth to get a geodetic azimuth.

Laplace equation - SEE Laplace's equation.

Laplace point - A point at which the difference between astronomic and geodetic latitude is determined.

Laplace's equation - (1) The equation expressing the relationship between astronomic and geodetic azimuths in terms of astronomic and geodetic longitudes and geodetic latitude. A usual form is $\alpha_A - \alpha_G = + (\lambda_A - \lambda_G) \sin \phi_G$, in which α_A and α_G are astronomic azimuth and geodetic azimuth, respectively; λ_A and λ_G are astronomic longitude and geodetic longitude, respectively; and ϕ_G is the geodetic latitude. The signs of these quantities depend on convention. As written here, latitudes north and longitudes east are considered positive. Also called the Laplace equation and Laplace's formula. The Laplace condition, implied by Laplace's equation, arises because, in the plane of the prime vertical, a component of the deflection of the vertical will give a difference between astronomic and geodetic longitudes and between astronomic and geodetic azimuths. Conversely, the differences between astronomic and geodetic values of the azimuth and of the longitude may both be used to determine the component in the plane of the prime vertical. Because longitudes can be calculated with considerable accuracy when extending a triangulation network but azimuths can not, the value of the deflection of the vertical is usually calculated from the component in longitude. The value of the geodetic azimuth thus obtained is called a Laplace azimuth. The deflection of the vertical at the initial station in a triangulation network will affect any geodetic azimuths determined subsequently. Therefore, in that readjustment of the triangulation network which introduced the North American datum of 1927, geodetic azimuths were controlled by Laplace azimuths distributed throughout the network. The equation is named for Pierre Simon Laplace (1749-1827), who derived a general form for it when comparing actual figure of the Earth with a theoretical figure. (2) The partial differential equation $\nabla^2 V = 0$ or $\nabla^2 V' = 0$ for the gravitational potential V or gravitational-potential function V' or, if the coordinate system is rotating with rate ω, $\nabla^2 W = 2\omega^2 \rho$ for gravity potential W or $\nabla^2 W' = 2\omega^2$ for gravity-potential function W', in which ρ is the density. Also called Laplace's partial differential equation. Because the Earth and the geoid are nearly spherical, Laplace's equation used as a condition on V is conveniently written in (geocentric) spherical coordinates λ, ϕ' and r, the geodetic longitude, the geocentric latitude and the radius. $\delta^2 V/\delta r^2 + (2/r)(\delta V/\delta r) + [\delta\{\cos\phi'(\delta V/\delta\phi')/\delta\phi'\}/(r^2 \cos\phi')] + (\delta^2 V/\phi'\delta\lambda^2)/(r^2 \cos^2\phi') = 0$. The equation applies to the gravity potential function if W is substituted for V and 2ω is substituted for 0. (3) The equation $\nabla^2 V = 0$, where V is any function having second derivatives with respect to the coordinates. Also called Laplace's partial differential equation.

Laplace's formula - SEE Laplace's equation (1).

Laplace's partial differential equation - SEE Laplace's equation (2) and Laplace's equation (3).

Laplace station - A triangulation station or traverse station at which astronomic azimuth and longitude are determined.

Laplacian plane - For a system of satellites, the fixed plane relative to which the vectorial sum of the perturbing forces on the system has no orthogonal component. (2) SEE plane, invariable.

lapse rate - The rate of change of some atmospheric characteristic with altitude. In particular, the negative of the rate of change of temperature with altitude. Normally, the lapse rate is positive: i.e., temperature decreases in going from the Earth's surface up to the tropopause. However, a negative lapse rate is common within a few meters of the surface because the air is heated by contact with the ground.

Larmor precession - The rotation of the rotational axis (spin axis) of a particle with magnetic moment m and angular momentum J about an axis of fixed direction when a magnetic field of intensity H is applied. The frequency of the precession is mH/J.

laser - Any device capable of generating a narrow beam of coherent, nearly single-frequency radiation in the infrared and optical parts through long ultraviolet part of the spectrum by exciting a medium electrically or with light in such a way that the wave so generated is amplified as it moves through the medium. Also called an optical maser (the analog, in the radio part of the spectrum, of a laser is called a maser). The name arose as an acronym for light amplification by stimulated emission of radiation. Radiation at wavelengths from sum-millimeter to 0.1 nm has been generated, and some lasers have been constructed capable of emitting at more than one wavelength. The idea was proposed by Schawlow and Townes in 1958 as a reasonable extension of the maser, which had already been developed for amplifying radio waves (and applied to construction of an atomic clock). The first laser used ruby as the medium and was demonstrated by Maiman in 1960. It emitted a beam of red light. The ruby laser is still one of the most powerful lasers for its size and is used in many geodetic instruments. Lasers in general are used in geodesy primarily to generate a narrow beam of modulated infrared radiation or light for measuring distances, or for generating an unmodulated beam of coherent radiation usable for interferometric measurements of lengths. Lasers are also used for alinement of points and, in some leveling instruments, to establish a real line of sight.

laser, self-leveling - A laser rotatable about the standing axis (vertical axis) of a mounting which carries two levels connected electrically to a servo-mechanism. In response to signals from the levels, the servo-mechanism brings the standing axis into a vertical position, so that the laser's beam is horizontal. SEE ALSO laser leveling instrument.

laser altimeter - An instrument which determines altitude by measuring the length of time needed for a pulse of laser-generated light to travel from the instrument to the ground below and back again, and multiplying half this by the speed of light in the atmosphere. Also called a lidar altimeter. For the distance measured by the altimeter to be the altitude, the following conditions must be met. (a) The path followed by the pulse must be close to the line of the vertical through the instrument. (b) The area covered by the pulse must either be so narrow that high ground not directly under the instrument does not send an early, misleading echo back to the instrument, or so wide that significant variations in elevation over the region covered by the pulse are averaged out. The beam of a laser altimeter is so narrow that its direction must be closely controlled or indicated by a gyroscopic system or similar system.

laser gravimeter - SEE gravity apparatus, ballistic.

laser level - SEE laser leveling instrument.

laser leveling instrument - A laser mounted on pair of trunnions in such a way that when the axis (standing axis) determined by the trunnions is brought to a vertical position, the laser can be rotated about that axis so that its beam describes a horizontal plane. Also called a laser-type leveling instrument and a laser-type level. The laser's mounting also carries a pair of levels which (a) detect when the standing axis is not vertical and (b) control a servomechanism which brings the standing axis vertical. In use, the beam is pointed at a leveling rod to which a photodetector is attached. The photodetector can be moved up or down the rod until an indicator shows that the center of the photodetector is on, the axis of the beam. The scale on the leveling rod is then read by the rodman.

laser leveling rod - A leveling rod having as target a photodetector in a mounting which can be slid up or down on the rod. Also referred to as a laser-type leveling-rod. It is used together with a laser-type leveling-instrument to indicate differences of elevation above or below the plane described by rotation of the laser beam about the standing axis of the laser-type leveling instrument. SEE ALSO laser leveling instrument.

laser plummet - An optical device projecting a coherent, monochromatic beam of light or infrared radiation vertically. Equivalently, a laser mounted so that its beam is directed vertically. It is used primarily for aligning points vertically.

laser radar - A distance-measuring instrument or equipment which determines distance from measurements of the length of time it takes a pulse of light or infrared radiation to travel from the instrument to the object and back, using a laser as the source of radiation. SEE ALSO instrument, electro-optical distance measuring; lidar.

laser ranging - Measuring distances using pulses produced by a laser.

Laser Terrain Profile Recorder - SEE Airborne Profile Recorder.

lateration - SEE distance-ratio method.

lateration, relative - SEE distance-ratio method.

latitude - (1) The angle from a specified reference-plane to a line of specified type through the point of interest. The definition is so general that many different kinds of angle can be included under it by specifying different positions for the reference plane and for the apex of the angle. The kinds most often used in geodesy have the apex of the angle at the center of the Earth or at a point representing the center, and the reference plane is placed perpendicular to the Earth's axis of rotation or to a line representing that axis. Geocentric latitude and geodetic latitude are of this kind. Astronomic latitude uses the same kind of reference plane but the apex of the angle is at the Earth's surface. In all three cases, the reference plane is called the equatorial plane and the latitude is considered positive if described in the northward direction. These three kinds are used in specifying the location of a point on the Earth. The same terms may be used to designate the corresponding angles for points on the Moon or other planets, although some astronomers prefer using more specific terms such as aerocentric latitude for the latitude of a point on Mars. The term is also applied to similar angles used in specifying the directions of celestial bodies such as members of the Solar System and stars. Such latitudes are then distinguished as celestial latitude, galactic latitude and so on. (Such latitudes, although peculiar to astronomy, are not commonly referred to as astronomic latitudes.) In particular, for a point on a sphere representing the Earth, the latitude is the angle from the reference plane (equatorial plane) to a line through the point on the surface and perpendicular to the surface. For a point on an ellipsoid representing the Earth, the latitude is the angle from the reference plane (equatorial plane) through the center and perpendicular to the longest axis, to the line through the point on the surface and perpendicular to the surface. (2) An angle represented as a function of latitude as defined in (1) and used to give a simpler appearance to formulas or equations involving that latitude. Examples are the authalic latitude, isomeric latitude and parametric latitude. Such latitudes may not be easily defined geometrically. (3) The length of a course multiplied by the cosine of the bearing. (4) The perpendicular distance of the middle of a line from the east-west reference-line. (5) SEE latitude, difference of (2). (6) The range of values of brightness that a photographic emulsion can record. SEE ALSO latitude difference; variation of latitude.

latitude, assumed - A value of latitude chosen for convenience or because the correct value is not known.

latitude, astronomic - (1) The angle, at a point, from the plane of the celestial equator to the vertical at that point. (2) The angle from the equatorial plane to the vertical through the point in question. (3) The smaller of the two angles between the plane of the horizon and the Earth's axis of rotation. This angle is usually referred to as the latitude if the meaning is clear. Astronomic latitude is determined directly from observations on celestial bodies; the amount of the deflection of the vertical (station error) is not subtracted. In the United States of America, this may be as much as 25". The term is applied only to locations on the Earth and is reckoned from the celestial equator (0°) northward to +90° and southward to -90°. (5) SEE latitude, geodetic.

latitude, authalic - The authalic latitude ß is related to the geodetic latitude ϕ (of a point on a rotational ellipsoid with semi-major axis of length a and eccentricity e) by the equation $a_A{}^2 \sin\beta = a^2 (1 - e^2) \{[(½ \sin\phi)/(1 - e^2 \sin^2\phi)] + (1/4e) \ln[(1 + e \sin\phi)/(1 - e \sin\phi)]\}$, in which a_A is the radius of the sphere whose area equals the area of the rotational ellipsoid. Also called equal-area latitude and equivalent latitude. It is used in the theory of equal-area map projections.

latitude, auxiliary - That angle, corresponding to geodetic latitude, which results when a rotational ellipsoid is represented by a sphere. The three common forms are authalic latitude, equidistant latitude and isomeric latitude.

latitude, celestial - The arc of a great circle perpendicular to the ecliptic and included between the ecliptic and the point whose latitude is to be given. Also called ecliptic latitude. Celestial latitude is not the same as astronomic latitude and does not enter into the usual problems of surveying.

latitude, conformal - The quantity χ defined by the equation $\{\tan[(2\chi + \pi)/4] = \tan[(2\phi + \pi)/4][(1 - e\sin\phi)/(1 + e\sin\phi)]e/2\}$ in which ϕ is the geodetic latitude and e is the eccentricity of the rotational ellipsoid. Also called the isometric latitude. It is used in mapping the rotational ellipsoid conformally onto a sphere. By transforming geodetic latitudes on the ellipsoid into isometric latitudes on a sphere, a conformal map projection can be obtained using formulas of spherical geometry only.

latitude, eccentric - SEE latitude, parametric.

latitude, ecliptic - SEE latitude, celestial.

latitude, equal-area - SEE latitude, authalic.

latitude, equidistant - SEE latitude, rectifying.

latitude, equivalent - SEE latitude, authalic.

latitude, fictitious - (1) The angle between the normal to a point on the ellipsoid and a reference plane other than the plane of the equator. (2) That angle, in a graticule, which corresponds to the angle specified in the preceding definition. The line in which the reference plane intersects the ellipsoid is called the fictitious equator. The angle is taken to be positive upwards from the reference plane.

latitude, foot-point - The latitude indicated, on a map, by dropping a perpendicular from a point onto the y (north- south) axis. If the lines representing parallels of latitude, on the graticule, are straight lines, then the foot-point latitude and geodetic latitude are the same. If the lines are curved, then the two latitudes are usually different.

latitude, galactic - SEE coordinate, galactic.

latitude, geocentric - (1) The angle between a line from the center of an ellipsoid to the point whose latitude is wanted, and the equatorial plane of that ellipsoid. The center of the ellipsoid is placed at a specified center (mass, geometric, etc.) of the Earth or other celestial body. The equatorial plane passes through that same center and is perpendicular to a specified axis of the ellipsoid. The axis is usually (but not always) specified to be parallel to a rotational axis of the Earth. (2) The angle formed with the major axis of the ellipse, in a meridional section of the rotational ellipsoid, by the radius vector from the center of the ellipse to the given point. Geocentric latitude is used in some astronomical and geodetic calculations. In astronomy, geocentric latitude is also called reduced latitude, a term sometimes applied in geodesy to parametric latitude.

latitude, geodetic - The angle between the normal to the rotational ellipsoid and the equatorial plane of that ellipsoid. Also called topographical latitude and, erroneously, astronomic latitude. Geodetic latitudes are reckoned from the equatorial plane but in the horizontal-control survey of the United States of America they have been calculated from the latitude of station Meades Ranch as prescribed in the North American datum of 1927. In recording a geodetic location, the geodetic datum on which it is based must also be stated. A geodetic latitude differs from the corresponding astronomic latitude by the amount of the meridional component of the local deflection of the vertical. In the USA, this may amount to more than 25". Note that geodetic latitude is defined only for rotational ellipsoids. It can be defined in exactly the same manner for ellipsoids in general by leaving out the word rotational in the above definition.

latitude, geographic - A general term applying alike to astronomic latitude, geocentric latitude and geodetic latitude. Also called terrestrial latitude.

latitude, geomagnetic - The angle from the geomagnetic equator, described northward or southward through 90° and labeled N or S to indicate the direction in which it is described. Geomagnetic latitude should not be confused with magnetic latitude.

latitude, geometric - SEE latitude, parametric.

latitude, high - A polar or sub-polar region of the Earth.

latitude, inverse - SEE latitude, transverse.

latitude, isometric - (1) The quantity ln $\{\tan[(2\phi + \pi)/4][(1 - e\sin\phi)/(1 + e\sin\phi)]e/2\}$, in which ϕ is the geodetic latitude and e is the eccentricity of the rotational ellipsoid. It is the natural logarithm of $\tan[(2\chi + \pi)/4]$, where χ is the conformal latitude. (2) SEE latitude, conformal.

latitude, low - A tropical or sub-tropical region of the Earth.

latitude, magnetic - The angle, at any point on the Earth's surface, whose tangent is one-half the tangent of the magnetic dip at that point.

latitude, middle - One-half of the arithmetical sum of the latitudes of two places on the same side of the Equator. Also called mid-latitude. It is labeled N or S to indicate whether it is taken north or south of the Equator.

latitude, oblique - Latitude measured from an oblique equator.

latitude, parametric - (1) The angle, at the center of a sphere tangent to the rotational ellipsoid along the geodetic equator, between the equatorial plane and the radius to that point on the sphere where a straight line through the point of interest and perpendicular to the equatorial plane hits the sphere. Also called reduced latitude. However, when the term reduced latitude is used in astronomy, geocentric latitude is meant. Parametric latitude is an auxiliary quantity used in some geodetic problems. (2) The angle θ in the two parametric equations defining an ellipse in the plane: $x = a\cos\theta$ and $y = b\sin\theta$, in which 2a is the length of the major axis and 2b the length of the minor axis of the ellipse, and x and y are the rectangular Cartesian coordinates of the ellipse. Also called geometric latitude.

latitude, rectifying - Given a sphere of radius such that a meridional circle on it has the same length as a meridional ellipse on the rotational ellipsoid, the rectifying latitude is a latitude such that all lengths along a meridian from the equator to that latitude are exactly equal to the same lengths on the rotational ellipsoid. Also called an equidistant latitude.

latitude, reduced - (1) SEE latitude, geocentric. (2) SEE latitude, parametric.

latitude, selenographic - The angle, at the origin of the selenographic coordinate system, from the equatorial plane to a line from the origin to the point of interest, described in a positive direction towards lunar north, i.e., toward Mare Crisium.

latitude, terrestrial - SEE latitude, geographic.

latitude, topographical - SEE latitude, geodetic.

latitude, total - SEE ordinate.

latitude, transverse - Angular distance from a plane perpendicular to the equatorial plane and designated the transverse equator. Also called an inverse latitude.

latitude band - SEE zone (3).

latitude correction - (1) A correction to the angular distance traveled in latitude in one course of a traverse. (2) SEE latitude gravity correction.

latitude determination - SEE equal-altitude method; Polaris-observation method; secant method; Sterneck's method.

latitude difference - (1) In plane surveying, the difference, in a plane tangent to the ellipsoid representing the Earth locally, of the perpendicular distances of the two ends of a line from an east-west axis of reference in that plane. Also called latitude of the line and difference of latitude. Alternatively, the length of the projection of that line on the north-south axis in the plane. By custom, this term has been shortened to latitude or lat. It is north, or positive (and sometimes termed the northing), for a line whose azimuth or bearing is in the northwest or northeast quadrant; it is south, or negative (and sometimes termed the southing), for a line whose azimuth or bearing is in the southeast or southwest quadrant. (2) The shorter of the two segments into which a meridian is divided by the parallels of latitude through two places, expressed in angular measure.

latitude equation - A condition equation expressing the relationship between the fixed latitudes of two points connected by traverse or triangulation. When traverse or triangulation connects two points whose latitudes have been fixed by direct observation or by previous surveys, a latitude equation is used to make the latitude of either point, as computed through the survey from the other point, agree with its latitude as previously fixed.

latitude factor - The change in latitude, along an astronomical line of position, per 1' change in longitude.

latitude gravity correction - The increment or decrement applied to measured values of gravity to refer them to the value at an arbitrarily chosen latitude.

latitude level - A sensitive spirit level attached in such a way to the telescope of an instrument used for observing astronomic latitude that when the telescope is clamped in position, the level indicates, in a vertical plane, variations in the direction of the line of collimation. A latitude level is more sensitive than the level attached to the finder circle of an astronomical telescope.

latitude of the line - SEE latitude difference (1).

lava delta - A delta-like body of lava formed where a lava flow enters the sea. A coast consisting of such deltas formed by recent lava flows has a convex shoreline and is called a lava-flow coast.

law - That body of rules by which a society governs itself, i.e., which each member of that society obeys either voluntarily or under pressure from the other members. SEE ALSO admiralty law; case law.

law, adjective - Rules of procedure and methods for maintaining or enforcing rights.

law, black-letter - Legal principles derived from statutes and cases recognized by the majority of judges in the country.

law, civil - The law or laws that determine private rights, as distinguished from criminal law.

law, common - The body of judicial decisions developed in England based on immemorial usage. It is unwritten law, as opposed to statute, or written, law. The common law of England forms the foundation for the system of law in the USA.

law, riparian - That branch of the law dealing with the rights in land bordering on a river, lake or sea.

law, unwritten - That body of law not promulgated and recorded, but nevertheless observed and administered in the courts of the country. It has no certain repository, but is collected from reports of the decisions of the courts and from the treatises of learned men.

law of boundaries - That collection of common-law statutes and rulings which influence decisions in determining boundaries where conflicting evidence exists.

law of constant relative proportions - Regardless of the concentration of total dissolved solids in sea water, the ratios between the concentrations of the more abundant substances are virtually constant throughout the world's oceans.

law of error, normal - The Gaussian frequency function as it applies to errors.

law of propagation of error - The standard deviation of the sum of two or more quantities is equal to the square root of the sum of squares of their standard deviations. Equivalently, the variance of the sum of two or more quantities is equal to the sum of the variances and covariances of those quantities.

law of reflection - The angle between the reflected ray (normal to the wave front) and the normal to the reflecting surface is the same as the angle between this normal and the incident ray, provided the reflected wave is of the same type (travels with the same speed) as the incident wave.

law of refraction - SEE Snell's law (of refraction).

law of the conservation of energy - The statement that, in any closed physical system, energy cannot be created or destroyed. The law's validity depends on the definition of energy rather than on physical reality. Whenever a situation has been encountered in which the law appeared to be violated, a new form of energy has been postulated or defined to account for the discrepancy. For example, a pendulum in motion has kinetic energy because of this motion. Energy is completely stored as potential energy in the pendulum at the extremes of its arc and is gradually converted to kinetic energy during the following swing. However, the motion dies down gradually so that the sum of kinetic and potential energies becomes zero. To keep the law valid, the existence of heat is postulated. The vanished kinetic and potential energies are supposed to have been transferred to the surrounding gas as turbulent motion and heating of the gas. Heat is defined to be a form of energy. It manifests itself as a rise in the temperature of the surrounding gas and of the supports.

law of the sea - Those internationally agreed-upon laws which specify those boundaries, on the oceans, outside of which national laws do not apply and which define the rights and responsibilities of those who use the oceans outside such boundaries. Also called the International Law of the Sea.

lay (v.) - To assemble a photo-mosaic.

laydown - SEE mosaic.

layer, reversing - The dense, lower region of the Sun's atmosphere in immediate contact with the photosphere.

layer of no motion - A layer, at some depth in the ocean, in which the water is assumed to be at rest. The existence of a layer of no motion implies that the isobaric surfaces within that layer are level and hence that such surfaces can be used as reference surfaces when calculating absolute- gradient currents. Such a layer may, however, not exist.

layout - (1) Material to be printed arranged to satisfy the limitations on size and shape set by the printing process, the size of the paper to be used, etc. Also called lay (jargon). (2) A preliminary drawing or outline of the intended arrangement of the material to be printed, and used for positioning the negatives or positives when making up a flat.

layout engineer - SEE construction surveyor.

layout work - The work done by a construction surveyor for the construction.

lead - (1) The weight attached to the end of a long, graduated cord or wire (the lead-line) to pull the end of the wire down to the bottom when depth is being measured and to keep the wire more or less vertical during the measurement. Also called a sounding lead. (2) SEE lead line. SEE ALSO hand lead.

lead, deep-sea - A heavy (about 10 to 40 kg) sounding-lead usually having a line 200 meters (100 fathoms) or more in length.

lead, sounding - SEE lead (1).

lead line - A long, graduated chain, stranded cord, or wire having a leaden weight (lead) on one end and used for measuring depths of water. Also called a sounding lead and sounding line. The lead-line is commonly used when measuring depths by hand in water less than 50 meters deep. Lead-lines used for hydrography in deep water are usually 200 meters (100 fathoms) long. Lines of hemp or cotton are in common use by surveyors but because such lines shrink and stretch easily, many surveyors prefer to use a chain. Some prefer a steel chain with brazed links; this is very light and strong. In hydrographic surveying, the standard is a braided cotton rope with a phosphor-bronze wire at the center and is usually marked to feet and fathoms with attached cords, leather strips, and colored buntings, and usually with a 3-kg to 6 kg-weight (the lead or sounding lead) attached to one end. It may be used when measuring depths of up to 100 fathoms when the surveying party is not equipped with electronic depth-measuring equipment, or for checking such equipment.

league - (1) A unit of distance varying in size, in different countries and at different times, from about 2.4 to 4.6 miles but is estimated, in English-speaking countries, to be about 3 miles. If the miles are statute miles, the unit is called a land league; if they are nautical miles, the unit is called a marine league. (2) A unit of area - a square league (5 760 acres for the English land-league; about 4 439 acres or 1 796 hectares for the old Spanish land-league used in surveys in California). (3) A stone placed under a public road to mark a distance of one league.

league, marine - SEE league (1).

leap-frog method (of barometric altimetry) - A method for quickly obtaining altitudes of points along a route between two points of known altitude, by using four barometric altimeters in pairs. One pair of altimeters remains at the starting point while the other pair is advanced to the first point at which altitude is to be determined. At this point, the altimeters and weather instruments are read and the values recorded. The pair at the starting point are now moved to the second point at which altitude is to be determined and the procedure repeated. The pairs are advanced alternately, the one past the other, until the last point of known altitude (and this may be the starting point) is reached.

leap second - (1) The adjustment made to Universal Time Coordinated (UTC) to compensate for the approximately

1-second increase UTC gains over UT1 or UT2 each year. Normally, UTC is decreased by exactly 1 second (the leap second at 24h on the last day of December and/or June). (2) A second added to or subtracted from Coordinated Universal Time to keep the time within 0.7" of Universal Time 1.

leap year - A year, in the Gregorian calendar, having 366 days, the extra day (over 365) being added to the month of February and designated 29 February. Also called an embolismic year.

lease - A written document by which the possession of land and/or a building is given by the owner to another person for a specified period of time and for the speriod specified.

leasehold - (1) A property held under the tenure of lease. (2) A property consisting of the right of use and co-occupancy of real property by virtue of lease agreement. (3) The right of a lessee to use and enjoy real estate for a stated term and upon certain conditions, such as the payment of rent.

least squares - SEE method of least squares.

least-squares, collocation method of - SEE collocation method (of least squares).

least-squares method - SEE method of least squares.

least-squares method (of adjusting a traverse) - (1) The adjustment of the values of either the measured angles or the measured distances in a traverse by determining the corrections so that the sum of the squares of the corrections is a minimum. (2) SEE Jolly's method.

ledge - (1) A ridge or reef, especially one under water near the shore. (2) (mining) A mass of rock that constitutes a valuable deposit of mineral. (3) Lode. (4) Vein.

leeard - When referring to the movement of sand at a shore, the direction toward which the prevailing littoral drift moves.

leeway - That component of the speed of a waterborne craft which is along the transverse axis and caused by the wind.

left, vertical-circle - SEE circle left.

leg - SEE course.

legend - That information printed on a map or chart to explain the symbols used, the sources for the map, the date of the map, the scales of the map, etc. The title of a map or chart was formerly considered part of the legend, but this usage is obsolete.

Legendre equation - The second-order, ordinary differential equation $(1 - z^2)(d^2y/dz^2) - 2z(dy/dz) + [n(n+1) - m^2/(1 - z^2)]y = 0$ in which m and n are constants.

Legendre function - Any function satisfying the Legendre equation. Because the Legendre equation is of second order, two independent functions can satisfy the equation. These two are called Legendre functions of the first kind and Legendre functions of the second kind. When m is 0, n is a non-negative integer and z is a real variable between -1 and +1, Legendre functions of the first kind can be expressed as polynomials, called Legendre polynomials, in z. Legendre functions are important in geodesy because they are solutions of Laplace's equation when that equation is expressed in spherical coordinates. When m is zero, Legendre functions can be derived as the coefficients in the expression of the generating function as a series: $1 / [1 - 2\rho z + \rho^2]^{1/2} = \Sigma P_n(z) \rho^n$.

Legendre function, associated - Any function satisfying the Legendre equation for general, integral values of m.

Legendre polynomial - A function $P_n(z)$ satisfying the Legendre equation, n being a non-negative integer, m zero and the solution finite at the boundary values. Given that $P_0(z) = 1$ and $P_1(z) = z$, the other polynomials can be found by using the recursion formula $(n+1) P_{n+1}(z) - (2n+1)z P_n(z) + n P_{n-1}(z) = 0$. SEE ALSO Legendre equation; Legendre function; Legendre function, associated.

Legendre series - A series whose terms are Legendre functions or Legendre polynomials.

Legendre's theorem - The angles A, B and C of a spherical triangle whose sides are a, b and c (supposed very small in comparison to the radius of the sphere) are equal to the corresponding angles of a plane triangle whose sides are a, b and c each increased by one-third the spherical excess of the triangle. Using Legendre's theorem greatly simplifies the treatment of spherical triangles by permitting their solution as plane triangles. This theorem is named for Adrien Marie Legendre, French mathematician, who demonstrated it.

Lehman discontinuity - A discontinuity, at a depth of about 220 km, at which there is an increase of 3.5 -4.5 % in seismic speeds.

Lehman's method of hachuring - The light is supposed to fall vertically onto the terrain, the amount of illumination received by each slope varying in proportion to its divergence from the plane of the horizon. Rays falling on a plane at an angle of 45° are reflected horizontally and are represented by black on the scale of shades, while rays falling on horizontal planes are reflected vertically and represented by white. Intermediate slopes are divided into nine groups, each comprising a 5° range of slopes. The proportion of white to black varies according to the degree of slope. The first use of this method was in a map drawn by Lehman in 1798.

Lehman's method of intersection - SEE triangle-of-error method.

Lehman's system - The scheme or system according to which the proper shade of gray is matched, by Lehman's method, to the slope of the ground depicted.

Lejay pendulum - An inverted pendulum supported by a leaf spring, so that the period of oscillation depends on the elasticity of the spring as well as on the mass, length, and moment of inertia (about the point of suspension) of the pendulum. Also called a Holweck-Lejay pendulum. The period T of the pendulum is given by the formula $T = \pi [I/(k M g L)]^{1/2}$, in which k is a constant expressing the elasticity of the band, M is the mass of the pendulum, L is the length of the pendulum, I is the moment of inertia of the pendulum about the point of support and g is the value of the gravity acceleration. SEE ALSO Lejay-Holweck pendulum-apparatus.

Lejay-Holweck pendulum apparatus - An inverted pendulum contained in an evacuated glass tube, with a period of 6 to 7 seconds. Also called the Holweck-Lejay pendulum apparatus. Forty to sixty minutes of observation gives results with a precision better than 1 mgal.

length - (1) A number or quantity indicating how far apart two physically connected points (i.e., points on the same body) are. (The connection may be imagined instead of actually existing.) For example, one speaks of the length of a river or of a metallic bar. One also speaks of the length of a geometric line, although the line has no physical existence. Compare this to distance, which denotes a similar quantity but without the requirement that the two points be physically connected. Length and distance are not synonyms. The values of length and distance may be the same for a given pair of points or may be quite different. For example, the length of a surveyor's tape is the same whether the tape is stretched out or is rolled up, but the distance between two points on the tape depends on how the tape is placed. (2) A number or quantity indicating how far apart two points are in space or time, e.g., the length of a line or the length of a (timelike) interval. SEE ALSO end standard of length; line standard of length; standard of length; unit of length.

length (of a degree) - The length of one degree, measured along a meridian of longitude, on an ellipsoid representing the Earth. The length of a degree varies somewhat with the latitude. Lengths near the poles are longer and lengths near the equator are shorter. The length also depends on the dimensions of the chosen ellipsoid.

length (of a wave) - SEE wavelength.

length, back focal - SEE distance, back focal.

length, calibrated focal - (1) An adjusted value of the equivalent focal length, calculated so as to distribute the effect of distortion more evenly over the entire field of view. (2) The distance, along the optical axis, from the internal perspective center to the image-plane, with the internal perspective center chosen to balance the positive and negative areas of distortion for some specified field of view. Also called the calibrated principal distance (Brit.) and the camera constant.

length, concluded - The length of a side not measured but introduced into a trilateration network initially containing a non-triangular figure, in order to make the network consist entirely of triangles.

length, effective focal - The distance from the second (or first) principal point to the second (or first) focal point.

length, electrical - (1) The quantity $2\pi/\beta$, in which β is the shift in phase per unit length, of a monochromatic wave. (2) The quantity $\sqrt{k} L$, in which k is the dielectric constant and L is the physical length of the conductor.

length, equivalent focal - (1) The distance, measured along the optical axis, from the rear nodal point of a lens system to the plane of best average definition. (2) The distance, measured along the optical axis, from the rear nodal point to that point near which the definition on the corresponding plane is best (plane of best axial definition). This second meaning is rarely used in photogrammetry. (3) The focal length of a thin lens forming images most nearly duplicating those of a given compound lens, thick lens or lens system.

length, focal - (1) The distance, measured along the optical axis, from the rear nodal point of the lens to the plane of critical focus of a object at infinity. Also called focal distance. (2) SEE distance, focal.

length, minimal focal - The least focal length giving an acceptable amount of chromatic aberration.

length, standard (of surveyor's tape) - The length of a surveyor's tape when the tape is supported throughout its length under a specified, standard tension.

length, taped - A length which has been measured with surveyor's tape. Also called a taped distance. However, lengths are very seldom measured in surveying; it is almost always the distance between two points which is measured. However, these distances are usually reduced to the equivalent lengths of the geodesic or normal section between the corresponding points on a reference ellipsoid, so the term taped length is often applied to the measured distance. SEE ALSO distance, taped.

length correction - The difference between the nominal length of surveyor's tape and the effective length of that tape as determined by calibration. The calibrated (standard) length is usually expressed by a number of whole units (the nominal length) plus or minus a small amount which is the length-correction defined above (calibrated minus nominal). Length correction and calibration correction do not mean the same.

length equation - A condition equation expressing the relationship between the fixed lengths of two lines connected by triangulation. When a section of triangulation connects two lines whose lengths are fixed by direct measurement or by previous triangulation, a length equation is used to make the length of either line, as computed through the adjusted survey from the other line, agree with its length as previously fixed.

length-ratio method - SEE distance-ratio method.

Lenker direct-reading leveling rod - SEE Lenker leveling rod.

Lenker leveling rod - A direct-reading leveling-rod having a continuous, graduated face adjustable to give direct readings of elevations. Also called the Lenker direct-reading leveling rod.

Lenoir level - A leveling instrument in which the telescope passes through steel blocks, one near each end of the telescope. The upper and lower faces of the blocks are flat and closely parallel. The lower faces rest upon a brass circle and the upper faces support a spirit level which is reversed in leveling the instrument. The Lenoir level is no longer in general use.

lens - (1) An object of transparent material, bounded by curved surfaces (usually spherical and radially symmetrical), redirecting electromagnetic radiation to form an image of the radiation's source. (Similar objects operating on non-electromagnetic radiation such as sound and electrons are also called lenses.) Lenses for light are commonly made of glass or quartz; lenses for infrared radiation are made of silica, fused halides, etc.; lenses for radio waves have been made from gases inclosed in plastic bags, etc. (2) SEE lens system. Properly speaking, a lens is one continuous substance. However, a combination of two or more such bodies in direct contact or separated by a film is also often called a lens. Such a combination, when used as part of a larger assembly, is properly referred to as a component or, if used alone, as a compound lens (also called a doublet, triplet, etc.). Any combination or assemblage of lenses assembled as a unit is also often referred to as a lens but is better called a lens system. The fixed combination of a lens system and the mechanical structure holding the components rigidly together is sometimes called the lens barrel or in aerial photography, the lens cone. SEE ALSO combination lens; Fresnel lens; lens system; Luneburg lens; objective lens; process lens.

lens, achromatic - A lens system whose chromatic aberration is corrected for two different wavelengths. Preferably called an achromatic lens system. SEE lens system, achromatic.

lens, anallactic - A convergent lens placed between the objective lens system and the diaphragm of a telescope to bring the apex of the triangle having the stadia rod as its base, into the center of the telescope and so permit distance to be measured directly.

lens, anamorphic - A lens or lens system producing an image having different scales in different directions. The most common type of anamorphic lens is a lens shaped like a half-cylinder, so that it has circular curvature in one direction and zero curvature in the orthogonal direction.

lens, anastigmatic - A compound lens imaging a point in object space into a point in image space. Also called a stigmatic lens and preferably, an anastigmatic lens-system.

lens, aplanatic - (1) A lens system transmitting light without introducing spherical aberration into the image and producing uniform magnification for all angles within the angular aperture. (2) A lens system free of both spherical aberration and coma. Preferably called an aplanatic lens system. SEE lens system, aplanatic.

lens, apochromatic - A lens system corrected for spherical and chromatic aberration. Preferably called an apochromatic lens system. A single lens cannot correct for either aberration. SEE lens system, apochromatic.

lens, aspheric - SEE lens, aspherical.

lens, aspherical - A lens whose bounding surfaces are not spherical or flat. Also called an aspheric lens. Some aspherical lenses are made by figuring the lens's surface in zones, each zone having a different radius of curvature.

lens, astigmatic - A lens imaging a point in object-space into a small area on the focal surface in image-space. A lens which does not image points into points.

lens, bloomed - SEE lens, coated.

lens, coated - (1) A lens coated with a transparent film of such thickness and refractive index as to reduce the amount of light reflected from the surface of the lens. Also called a bloomed lens (jargon). An uncoated lens reflects about 4% of the light incident on the surface facing the air. (2) A lens coated to selectively absorb light in a particular band of wavelengths (color) so as to reduce the amount of light of that color from passing through the lens.

lens, collimating - A lens or lens system converting a bundle of convergent or divergent rays into a bundle of parallel rays.

lens, compensating - A lens or lens system introduced into an optical system to correct for radial distortion.

lens, compound - A combination of two or more lenses in contact or separated only by thin films, and acting as a unit. A compound lens consisting of two lenses is called a doublet; a compound lens consisting of three lenses is called a triplet.

lens, convergent - A lens or lens system bringing originally parallel or divergent rays together to a focus. The term is sometimes restricted to cover the case where the rays originate from a point, i.e., a lens creating a real image of a point-source.

lens, convertible - SEE lens system, convertible.

lens, cylindrical - A lens one of whose surfaces is cylindrical, i.e., has a constant, finite radius of curvature in one direction and an infinite radius of curvature in the orthogonal direction.

lens, decentered - A lens or lens system which is part of an optical system but whose optical axis is parallel to but not on the optical axis of the rest of the system. A decentered lens has the effect of a similar, centered lens plus a weak prism. The effect is therefore sometimes called prism aberration.

lens, divergent - A lens or lens system causing originally parallel or convergent rays to diverge and thus appear to emanate from some point other than the actual source.

lens, fish-eye - A lens or lens system providing a field of view of 180°. Also called, erroneously, a spherical lens. The simplest type of fish-eye lens is a hemispherical lens such as can be made by cutting a homogeneous, spherical ball of glass in half along an equator. Such a lens produces a distorted and chromatic image. Lens systems are made which considerably reduce such aberrations. SEE ALSO Luneberg lens.

lens, narrow-angle - (1) A lens system having a field of view not greater than 60°. (2) A lens system having a focal length approximately two or more times the greatest dimension of the image.

lens, negative - A lens or lens system causing rays of light to diverge and produce a virtual image.

lens, normal-angle - (1) A lens system having a field of view between 60° and 75° wide. (2) A lens system having a focal length approximately equal to the greatest dimension of the image.

lens, objective - SEE objective lens.

lens, positive - A lens or lens system causing rays of light to converge to a focus and thereby produce a real image. Also called a converging lens or convex lens.

lens, spherical - (1) A lens having the shape of a sphere. The Luneburg lens is a particular kind of spherical lens. The fish-eye lens, being a hemisphere, is not a spherical lens. (2) A lens all of whose surfaces are sectors of spheres. Most lenses are of this type.

lens, super-wide-angle - (1) A lens system having a field of view greater than 100°. (2) A lens system whose focal length is appreciably less than half the greatest dimension of the image.

lens, telephoto - A lens system consisting of a positive element as the first element and a negative element as the last element; the focal length of the system is greater than the distance from the first surface to the focal plane. A telephoto lens is used in relatively compact, long-focus cameras.

lens, thick - A lens sufficiently thick that the thickness must be considered in calculating the progress of rays through the lens; all distances are then measured from the nodal point instead of from the center of the lens.

lens, thin - A lens so thin that the thickness of the lens can be ignored in calculating the progress of rays through the lens and all distances can be measured from the center of the lens. Alternatively, a lens whose thickness is negligible in comparison to the focal length. All real lenses have some thickness, but many can be considered thin lenses if an approximate calculation is sufficient.

lens, varifocal - SEE lens system, varifocal.

lens, wide-angle - (1) A lens system having a field of view of between 75° and 100°. (2) A lens system having a focal length approximately half the greatest dimension of the image.

lens, zoom - SEE lens system, varifocal.

lens axis - SEE axis, optical.

lens calibration - SEE calibration (of a lens-system).

lens component - (1) A combination of two or more lenses without intervening spaces. (2) SEE lens element.

lens cone - (1) In an aerial camera, the mechanical structure holding the components of the lens system rigidly in placed with respect to each other. Also called a camera cone. (2) SEE lens cone assembly.

lens cone assembly - In an aerial camera, the lens system together with the mechanical structure holding the components rigidly in place with respect to each other. Also called a lens cone.

lens distortion - Optical distortion caused by or inherent in a lens system. SEE ALSO distortion (1).

lens element - A single, isolated lens in a lens system. Also, but undesirable, called a lens component.

lens equation - The equation relating the distance from a point in object-space to the front nodal point of an optical system, the distance from the corresponding point in image space to the rear nodal point and the focal length of the system. Also called the lens law. SEE ALSO equation, thin-lens; Newton's equation (for a thin lens).

lens-equation, Gaussian - SEE equation, thin-lens.

lens-equation, Newtonian - SEE Newton's equation (of the lens).

lens law - SEE lens equation.

lens stereoscope - SEE stereoscope, lens-type.

lens system - Any combination of lenses assembled to act as a unit for a specific purpose. Also called a lens assembly and in jargon, a lens. The individual lenses in the system are called elements of the system if they are separate and continuous. Combinations of lenses in direct contact are called components. The foremost member or front member consists of all elements and components in front of the aperture stop; the back or rear member consists of all elements and components in back of the aperture stop. The optical axis of a lens system consists of the joined principal axes of all the elements of the system. SEE ALSO calibration (of a lens system); focus, sidereal; point, nodal; process lens.

lens system, achromatic - (1) A lens system whose chromatic aberration is corrected at two different wave-lengths. Also called an achromatic lens. Alternatively, a lens system capable of bringing rays at two different wavelengths from a single point in object-space to a focus at a single point in image-space. A lens system is made achromatic by making the elements of glass (or other suitable material) of different refractive indices. The indices are proportioned so that the chromatic aberration in one element is compensated by that in the other. Achromatic lens systems are usually also made aplanatic, i.e., free of spherical aberration. (2) A lens system capable of transmitting light without separating it into component colors. No real lens system is achromatic in this sense.

lens system, anastigmatic - SEE lens, anastigmatic.

lens system, aplanatic - (1) A lens system capable of transmitting light without introducing spherical aberration and of producing uniform magnification for all angles within the angular aperture. (2) A lens system satisfying the equation (Abbe's sine-condition) $ny\sin\theta = n'y'\sin u'$ = constant, in which n is the refractive index in object space, y is the distance of a point P from the optical axis and u is the angle made with the optical axis by a ray through the foot-point of P on the optical axis. n', y' and u' are the corresponding quantities in image-space.

lens system, apochromatic - A lens system corrected for spherical and chromatic aberration.

lens system, convertible - A lens system containing two or more elements that can be used individually or in combination.

lens system, diffraction-limited - A lens system in which the quality of the image is limited by diffraction and not by aberrations.

lens system, narrow-angle - (1) A lens system having a field of view not greater than 60°. (2) A lens system having a focal length approximately twice the greatest dimension of the image.

lens system, normal-angle - (1) A lens system having a field of view containing between 60° and 75°. (2) A lens system having a focal length approximately equal to the greatest dimension of the image.

lens system, objective - SEE objective lens.

lens system, pancratic - SEE lens system, varifocal.

lens system, process - SEE process lens.

lens system, super-wide-angle - (1) A lens system having a field of view greater than 100°. (2) A lens system whose focal length is appreciably less than half the greatest dimension of the image.

lens system, varifocal - A lens system whose effective focal length can be varied by moving one or more components along the optical axis. Also called a pancratic lens system, zoom lens system, zoom lens and varifocal lens. The lens system is so designed that the location of the image plane is fairly constant as the effective focal length is varied.

lens system, wide-angle - (1) A lens system having a field of view of between 75° and 100°. (2) A lens system whose focal length is approximately half the greatest dimension of the image.

lens-system calibration - SEE calibration (of a lens system).

lessee - A person who possesses the right to use or occupy a property under lease agreement.

lessor - A person who holds title to and conveys the right to use and occupy property under lease agreement.

letterpress printing - Printing in which the designs are raised above the level of the rest of the material holding the design. It is contrasted with intaglio, in which the design is cut below the level of the rest of the material.

Levallois' modification (of Clairaut's equation) - The equation $(a-b)/a + (\tau_p - \tau_e)/\tau_e = (\omega^2 b/\tau_e) e'V_2/(V_2 - V_1)$, in which τ_p and τ_e are the theoretical values of gravity at the poles and at the equator, respectively, of an equi-potential, rotational ellipsoid of equatorial diameter 2a and polar diameter 2b, rotating at the rate ω. e' is $\sqrt{[(a^2-b^2)/b^2]}$ and the quantities V_1 and V_2 are, respectively, $\frac{1}{2}(\arctan e' - e e')/(a e 3)$ and $(e' - \arctan e')/(b e'3)$. Also called the Clairaut-Levallois equation.

level (adj.) - (1) Locally horizontal, i.e., horizontal at a particular point. E.g., an instrument is level if a flat surface used in it as a reference is tangent to the equipotential surface there as shown by spirit levels. The term level line may mean a line horizontal at a particular point. It may also have another meaning in which level is not used as an adjective in the present sense. (2) Everywhere perpendicular to the direction of gravity. E.g., a level surface is a surface everywhere perpendicular to the direction of gravity. An equipotential surface (of gravity) is therefore a level surface in this sense.

level (n.) - (1) Any device sensitive to the direction of gravity and used to indicate directions perpendicular to that of gravity at a point. (2) By extension, any instrument incorporating such a device and used for the same purpose. In particular, a leveling instrument. (3) The location of a level surface. E.g., mean sea level, higher level. Level used in this sense is similar in meaning to elevation. However, elevation as used in surveying means a number; level merely gives a sense of place without necessarily involving a numerical value. (4) A horizontal deposit along which the mining of an ore is performed. The term includes all drifts, cross cuts and other horizontal workings on any one such deposit or level. (5) A horizontal passage or drift driven from a shaft to a deposit or vein. The levels are customarily spaced at regular intervals in depth and are either numbered consecutively from the surface or are designated by their actual depth below the top of the shaft. (6) SEE spirit level. SEE ALSO Abney level; axis (5); base level; Chezy level; Egault level; engineer's level; Fischer level; Goulier collimator level; Gravatt level; hand level; Horrebow-Talcott level; Lenoir level; Liesegang level; mercury level; Ramsden level; rod level; spirit level; spirit level, chambered; Stampfer level; striding level; Troughton level; U.S. Geological Survey level; water level; wye level.

level (v.) - (1) To make a systematic series of measurements of differences of elevation; to do leveling. (2) To make a line or surface horizontal at a particular point. E.g., to level an instrument. (3) To bring a piece of ground to a common level, as by removing earth from heights and using the earth to fill hollows.

level, automatic - SEE leveling instrument, automatic.

level, base-line - A level-line run along a base line to determine and establish the elevations of the stations along and at the ends of the base line.

level, bull's-eye - SEE level, circular.

level, chambered spirit - SEE spirit level, chambered.

level, circular - A spirit level having the inner surface of its upper part ground to a spherical shape. Also called a universal level, bull's-eye level and spherical level. The bubble has a circular shape and the graduations on the glass above the bubble are concentric circles. This form of spirit level replaces two spirit levels of toroidal shape and is used where great precision is not required, as in plumbing a leveling rod or setting an instrument approximately upright.

level, cylindrical - A spirit level approximately cylindrical in shape.

level, dumpy - A leveling instrument in which the telescope is permanently attached to the base carrying the spirit levels, either rigidly or by a hinge about which the telescope can be rotated by means of a micrometer screw. The dumpy level takes its name from the dumpy appearance of the early form of this instrument, invented in 1845 by W. Gravatt. The telescope was short and had a large objective lens. The dumpy level is characterized by the way in which the telescope is mounted however, and not by the shape of the instrument. It constitutes one of the two principal classes to which most leveling instruments used in surveying belong. The other class is represented by the wye (Y-) level. The Fischer level formerly used by the U.S.Coast and Geodetic Survey on first-order and second-order leveling is a dumpy level, as are most automatic leveling instruments.

level, first-order - SEE leveling instrument.

level, ground-water - SEE ground water level.

level, half-tide - SEE water level, half-tide.

level, hanging - A spirit level so mounted that, when in use, its tube is lower than its points of support. A hanging level literally hangs from that part of the surveying or astronomical instrument whose position it indicates with respect to a horizontal plane. It is used in place of a striding level to indicate the inclination of the horizontal axis of the broken-telescope transit. (2) A spirit level fitted with a hooked, vertical wire or extension at each end. In use, the hanging level is hung by the hooks from a cord or wire strung between two points at nearly the same elevation.

level, military - SEE level, prismatic; leveling instrument, military.

level, precise - SEE leveling instrument, precise.

level, prism - SEE level, prismatic.

level, prismatic - Also called a military, semi-precise, or prism level.

level, reversible - SEE leveling instrument, reversible.

Level, Revised Local - SEE Revised Local Level.

level, self-adjusting - SEE leveling instrument, automatic.

level, self-aligning - SEE leveling instrument, self-leveling.

level, self-leveling - SEE leveling instrument, self-leveling.

level, semi-precise - SEE level, prismatic.

level, spherical - SEE level, circular.

level, spirit - SEE spirit level. **level, still-water** - (1) The height which the surface of a body of water would assume in the absence of wind-waves. (2) The surface which a body of water would have in the absence of wind-waves. Still-water level is not the same as the level of water in the absence of wind; swell might still be present. It is definitely not the same as mean sea level, which is affected by the presence of currents, differences in salinity, and other inhomogeneities.

level, stride - SEE striding level. (jargon)

level, striding - SEE striding level.

level, Talcott - SEE Horrebow-Talcott level.

level, tilting - SEE leveling instrument, tilting.

level, toric - A level vial shaped internally like a segment of a torus. This is the usual form for the vials of spirit levels used in precise work.

level, universal - SEE level, circular.

level axis - SEE axis (5).

level circuit - SEE level line, duplicate.

level circuit, double-rodded - SEE level circuit, simultaneous.

level circuit, multiple - SEE level line, multiple.

level circuit, simultaneous - A leveling network consisting of a single loop connecting two end points, in which two separate turning points are used for calculating the height of the instrument at each successive setup. Also called a double-rodded level circuit. Note that this is not the same as a simultaneous level-line.

level collimation correction - SEE level correction.

level collimation correction factor - SEE collimation correction factor; C-factor.

level collimation error - (1) The small angle by which the line of sight (determined by the horizontal cross-line in the reticle of a leveling instrument) departs from the actual level surface through the instrument when the instrument is considered leveled. (The definitions of level collimation error (1), level constant (1) and level error differ only slightly in their wording. The meanings, however, may be significantly different if the definitions are taken as written. The framers of the definitions possibly intended the terms to refer to the same concept. Their intentions are unfortunately unknown. It is therefore seems best to accept the definitions as given instead of guessing at the intentions and possibly causing further confusion.) (2) SEE C-factor.

level collimation factor - The collimation factor as used with a leveling instrument.

level constant - (1) The amount by which the actual line of sight through a leveling instrument departs (when the bubble is centered in the vial) from the truly horizontal line through the center of the instrument. SEE ALSO level collimation error. (2) SEE C-factor.

level control - A series of bench marks or other points of known elevation, established throughout a project.

level correction - (1) That correction which is applied to an observed difference of elevation to correct for the error introduced because the line or sight through the leveling instrument is not absolutely horizontal when the bubble is centered in its vial. Also called the level collimation correction and collimation correction. It may be calculated by multiplying the C-factor by the product of the stadia constant and the difference in distances to the two leveling rods. (2) The quantity added to a measured difference of elevation to correct for the error introduced when the line of sight through the telescope of the leveling instrument is not perfectly horizontal even though the bubble is centered in its vial. Also referred to as the level leveling correction. (The definitions (1) and (2) of level correction and of leveling correction (1) differ only slightly in wording. The meanings of the definitions are substantially different. It is possible that the framers of the definitions intended them to refer to a common concept. In absence of knowledge of the framers' actual intentions, it seems best to accept the definitions as they were given.)

level error - (1) The angle between the horizontal axis of a telescope's mounting and a horizontal plane through the mounting. SEE ALSO level factor. (2) The angle from a horizontal plane through the standing axis of a leveling instrument to the line of sight of the instrument. It is the sum of the collimation error (the amount by which the line of sight is not perpendicular to the standing axis of the instrument) and the amount by which the standing axis is not exactly vertical. SEE ALSO level collimation error. (3) The difference between the angular elevation of a celestial body as measured from the apparent horizon, and the true angular elevation measured from a plane through the center of the Earth and parallel to the plane of the apparent horizon.

level factor - The quantity, usually denoted by b, in Bessel's formula or Mayer's formula and equal to the angle between the horizontal axis of a telescope's mounting and the horizontal plane through the mounting. Also called b-factor or level error.

leveling - (1) The art or process of finding vertical distances, called elevations, from a selected level surface to points on the Earth's surface or of finding differences of elevation. Also called heighting (Brit.) The usual procedure is to find the elevations as the sum of incremental vertical displacements of a graduated rod (differential leveling) or by measuring vertical angles (trigonometrical leveling). Unless some other method is specified, leveling is understood to mean differential leveling. The term is also applied by some surveyors to barometric altimetry and topographic photogrammetry. It is also occasionally applied to the astronomic determination of geoidal heights, in the form astrogeodetic leveling or astronomic leveling. This usage is becoming rare. The level surface specified in the definition is, ideally, the geoid and is so specified by many geodesists. Because the geoid is not directly accessible, the surface actually used is one assumed to be close to the geoid (e.g., mean sea level.) It is not necessarily a level surface. (2) The art or process of finding the elevation of mean sea level above the geoid. This is more properly called oceanographic leveling. (3) The process of relating a stereoscopic model to the plane to which vertical distances (heights) are referred. (4) SEE leveling network. (5) SEE leveling, differential. SEE ALSO compensator leveling; pipeline leveling; profile leveling; spirit leveling; stadia leveling; water leveling.

leveling (a model) - In the absolute orientation of a stereoscopic model, bringing the datum of the model parallel to a plane of reference. The plane of reference is usually the tabletop of the stereoscopic plotting instrument.

leveling, astro-geodetic - SEE geoid determination, astro-geodetic method (of geoid determination).

leveling, astrographic - Determination of geoidal heights by using values of gravity along the line of levels to determine the deflections of the vertical along that line. The values are often obtained from maps on which the contour lines indicate contours along which the gravity is constant.

leveling, astro-gravimetric - A process of determining difference of geodetic height between two points P and Q by determining the difference of elevation between the two points by differential leveling; determining the deflection of the vertical at a few well-chosen, secondary points along the route between P and Q; and determining the deflection at points intermediate between the secondary points by gravimetric methods.

leveling, astronomic - SEE astro-geodetic method (of determining the geoid).

leveling, barometric - SEE altimetry, barometric.

leveling, differential - (1) Determining the difference in elevation between two points by holding a graduated rod (leveling rod) vertically at each point and reading off the number, on each rod, where a horizontal line of sight intersects that rod. The difference in readings is approximately the difference in elevation between the two points. Also called geometric leveling and spirit leveling. The horizontal line of sight is obtained by sighting through a telescope rotatable about a vertical axis. The leveling rods are usually placed within 100 meters of each other horizontally and on points not more than 2 to 3 meters apart vertically, with the telescope (leveling instrument) approximately midway between them. The difference in readings must be corrected, if high accuracy is wanted, for errors in the graduated rods, for the effect of refraction on the line of sight, and for the effect of the Earth's curvature. If the points are more than 2 to 3 meters apart, the horizontal distance is broken up into shorter intervals within which the differences of elevation can be measured, and the total difference in elevation taken as the sum of the observed smaller differences. If the two points are farther apart than about 100 meters horizontally, the horizontal distance is broken up into equal, shorter distances that fall within the allowed distance apart and the leveling rods are set up successively on the end points of these intervals, with the leveling instrument between them, and the differences summed. The differences observed by differential leveling are not properly summable because the leveling rods are held vertical at their respective points and hence are not quite parallel to each other. The effect of this lack of parallelism is called the orthometric error. The true elevation is called the orthometric elevation. (2) Determining differences in elevation between points some distance apart. (3) The process of establishing bench marks.

leveling, direct - (1) Determining differences of elevation by measuring a continuous series of short, horizontal lines the vertical distances from which, to marks underneath, are determined by direct observations on graduated rods with a leveling instrument. (2) SEE spirit leveling.

leveling, double-rodded - Leveling carried out with a single observer and a single leveling instrument but with two sets of leveling rods. Foresights are taken on each of two rods at two separate turning points and backsights are taken on two other rods at two other, separate turning points, all sights being taken from the same location of the leveling instrument.

leveling, double-run - Leveling done by proceeding from a starting point to a final point and then returning in the opposite direction. Also called a double-run survey. Usually the same general route is taken on returning as on going out, and points occupied in the morning on the forward part of the journey are occupied in the afternoon on the returning part, and vice versa. This and other standardized procedures are taken to cancel or identify systematic errors as well as to reduce random errors.

leveling, double-simultaneous - (1) Leveling done with leveling rods showing two different scales, both on the same face or one on each face, the one scale being read immediately after the other. (2) SEE leveling, double-rodded. A blunder can usually be detected by comparing the difference of elevation as computed from the two different scales. Statistically, the procedure gives the same effect as would using two foresights and two backsights. It will not detect other kinds of random or systemic errors.

leveling, first-order - (1) Leveling resulting in first-order vertical control. SEE classification (of leveling); classification (of leveling networks). (2) Leveling by the procedures specified for first-order leveling.

leveling, first-order class III - Leveling done according to the 1976 specifications, slightly modified, of the Federal Geodetic Control Committee, for first-order class I leveling, but run in one direction only (single-run leveling).

leveling, fly - SEE leveling, flying.

leveling, flying - (1) Leveling done at the close of a working day to check the results of leveling an extended line run in one direction only; longer sights and fewer set-ups are used, as the purpose is to detect blunders. (2) Leveling done with the ordinary equipment but by a procedure yielding results with a distinctly lower order of accuracy. Sometimes referred to as fourth-order leveling or fly leveling. Greater tolerances are allowed in balancing the lines of sight, the misclosures, and so on. Flying leveling may also be done with alidade and planetable, etc. Misclosures may be as much as 20 cm per sq km.

leveling, fourth-order - SEE leveling, flying.

leveling, geodetic - (1) Leveling of a high order of accuracy, usually extended over large areas, to furnish accurate vertical control for surveying and mapping. Formerly called geodesic leveling. (2) The same as the preceding definition but with the proviso that systematic errors in the leveling are reduced to tolerable limits by correcting the measurements while reducing the data. (3) The process of applying, during the reduction of data from leveling, corrections for systematic errors in the measurements. This usage is rare and in any event is not recommended. (4) The same as the first definition, but the leveling is required to be done with a leveling instrument having a precise spirit-level. (5) Differential leveling done to establish first-order control. (Brit.) (6) Leveling done for geodetic purposes or in accordance with the specifications of the U.S. National Geodetic Survey.

leveling, geometric - SEE spirit leveling.

leveling, geostrophic - The art or process of determining the difference in elevation between two points at mean sea level by measuring the speed of the current between known points and assuming that the water is in geostrophic equilibrium, i.e., that the hydrostatic pressure is balanced by the Coriolis force per unit area. Originally called quasi-nivellement by A. de Cour, Danish geodesist, in 1913. Also called quasileveling. The basic equation is $v_2 - v_1 = (P_2 - P_1) / (k_g \rho L)$, in which v_2 and v_1 are the speeds of the current at stations 2 and 1; P_2 and P_1 are the magnitudes of the corresponding pressures there; k_g is the geostrophic parameter $2\omega \sin\phi$; ρ is the density of the water; and L is the distance between the two stations. ω is the Earth's angular rate of rotation, $7.292\ 116 \times 10^5$ radians per second and ϕ is the geodetic latitude.

leveling, gravimetric - The process of determining differences of elevation by measuring differences of gravity at various points.

leveling, hydrodynamic - The process of determining the difference in elevation between two points at mean sea level by making use of the fact that a difference in elevation causes a difference in pressure, and a difference in pressure will cause water to flow. Unless hindered by a barrier, the water will flow, in response to the combined effects of pressure and Coriolis force, in a direction at right angles to the line joining the two points. By measuring the velocity of flow, the difference in pressure, and hence the difference in elevation, can be calculated. This is called geostrophic leveling.

leveling, hydrostatic - (1) Determining the difference in elevation between two points by connecting the points with a tube open at both ends and filled almost to the ends with a suitable liquid. The liquid at the end near the lower of the two points is brought to the level of the point there. The vertical distance of the point at the other end from the level of the liquid there is the difference in elevation between the two points. The ends of the tube may be graduated so that the distances at each end of the points from the liquid's surface can be measured without having to raise or lower the entire tube. The term is applied, in particular, to pipeline leveling. (2) Determining the difference in elevation between two points at sea level by measuring the salinity, temperature and pressure of the water and assuming that the water is in hydrostatic equilibrium above some identifiable, isobaric surface which coincides with an equipotential surface. Also called steric leveling. The basic equations are those relating the density ρ to the salinity S, temperature T and pressure P, $\rho = \rho(S, T, P)$ and the hydrostatic equation $dP/dZ = g\rho$, in which g is the acceleration of gravity and z is the depth.

leveling, indirect - (1) Determination, by methods other than differential leveling, of differences of elevation, for example, from (a) measured vertical angles and horizontal distances (trigonometric leveling); (b) altitudes derived from measurements of atmospheric pressure using a barometer (barometric altimetry); or (c) altitudes derived from measurements of the boiling point of water or other suitable liquid, using a hypsometer (thermometric altimetry). (2) SEE leveling, trigonometric.

leveling, motorized - Leveling in which the leveling rods and leveling instruments are carried on individual vehicles during a project. Leveling rods are fastened by linkages to the vehicles carrying them so that they can easily be swung into position on the ground; the leveling instrument may be supported in a similar manner or may be mounted firmly on and used from the vehicle.

leveling, oceanic - Determination of the elevation of mean sea level when atmospheric as well as hydrostatic pressure is taken into account. G. Dietrich originated this usage in 1936.

leveling, oceanographic - Determining the vertical distance between mean sea level and the geoid (i.e., determining the elevation of mean sea level) by calculating the distance from measurements of density, velocity, and/or other characteristics of the water. The two principal kinds of oceanographic leveling are hydrostatic leveling, in which elevations are determined from measurements of salinity, temperature and pressure; and hydrodynamic leveling, in which the velocity of currents is measured.

leveling, precise - (1) A category of leveling in some classifications of leveling (e.g., the classification adopted by the U.S. Coast and Geodetic Survey in 1925) roughly equivalent to the present category: first-order class II leveling. (2) Leveling done with careful and frequent adjustment and calibration of the instruments used. The term is here used to distinguish the process from geodetic leveling, in which systematic errors are reduced to tolerable limits by applying corrections to the measurements during reduction of the data. (3) Leveling that is precise. (4) Leveling done by the National Geodetic Survey.

leveling, reciprocal - Determining the difference of elevation between two points by taking foresights to leveling rods on both points from one set-up and backsights to both leveling rods from a second set-up. The method is used principally when it is not feasible to level between the two points by the ordinary method of differential leveling, as when the points are separated by a wide river or a steep valley. The method eliminates errors caused by the curvature or the Earth or by lack of horizontality of the line of sight but does not usually eliminate all of the error caused by refraction.

leveling, reverse - Determining an elevation by determining the distance from a point above the Earth's surface (e.g., from an airplane or satellite) to the desired point on the surface.

leveling, second-order - Leveling resulting in a second-order leveling network or second-order vertical control, or leveling by procedures and instruments satisfying the specifications for second-order leveling. The U.S. National Geodetic Survey has two categories of second-order leveling: second-order class I and second-order class II. They differ in the linear density of the points between which differences of elevation are determined and the precision of the measurements. The results obtained by leveling in these categories are expected to have the accuracies associated with the categories of the same name established for vertical control.

leveling, single-run - Leveling done by proceeding from the starting point to the final point of a level line and not leveling back to the starting point. Also called a single-run survey. Single-run leveling is faster than double-run leveling but is not, usually, as accurate.

leveling, special-order - (1) A category of leveling, in Canadian practice, based on the requirement that the discrepancy between forward and backward runnings of a section shall not be greater than $3\sqrt{K}$, where K is the distance, in kilometers, between bench marks measured along the route used for leveling. (2) Leveling in which $3\text{ mm }\sqrt{K}$ is the greatest difference permitted between independent levelings.

leveling, spirit - SEE spirit leveling.

leveling, steric - SEE leveling, hydrostatic (2).

leveling, thermometric - SEE altimetry, thermometric.

leveling, third-order - (1) Leveling resulting in third-order vertical control. Third-order leveling results in differences of elevation having a precision indicated, roughly, by the requirement that the standard deviation of the difference in elevation between two points shall not be greater, in absolute value, than $2.0\sqrt{K}$ mm, in which K is the distance, in km, between the points as measured along a level-line. (2) Leveling done by the procedures specified for third-order leveling. The specifications published in 1984 by the U.S. Federal Geodetic Control Commission specify not only the procedures and equipment to be used, but also the linear density of the elevations.

leveling, three-wire - Leveling in which a leveling instrument containing a reticle on which there are three horizontal lines is used. The scale on the leveling rod is read at each of the three lines and the average is used for the final result.

leveling, trig - SEE leveling, trigonometric.

leveling, trigonometric - Determining differences of elevation by measuring the vertical angles between points. The horizontal distances between the points must be known or the straight-line distances between them measured and converted to horizontal distances. The differences of elevation are then calculated using a trigonometrical formula. Also called trig leveling, vertical angulation (USA) and indirect leveling. This last term is usually used in a more general sense.

leveling adjustment - (1) Determining corrections to measured elevations of, or differences of elevation between, points in a leveling network in such a way that the resulting elevations or differences are the best obtainable under the given conditions and from the measurements used. Adjustment removes those inconsistencies that result from accumulation of small random and systematic errors. However, the process of correcting measured elevations or differences for refraction, curvature of the earth, leveling-rod errors, etc., is not usually considered part of leveling adjustment. (2) The process of changing measured differences of orthometric elevation or to orthometric elevations to make the elevations of all bench marks consistent and independent of the misclosures of the circuits.

leveling base - SEE quadribrach; tribrach.

leveling circuit - SEE level line, duplicate.

leveling classification - SEE classification (of leveling); classification (of leveling networks).

leveling correction - (1) The amount by which the line of sight of a leveling instrument must be turned to make the line horizontal. SEE ALSO level correction. (2) A correction to an elevation, or difference of elevation, obtained by leveling. Leveling corrections are commonly applied to correct systematic errors whose magnitudes are known or can be calculated. SEE ALSO adjustment correction; correction, orthometric; index correction; level correction; rod correction; temperature correction.

leveling correction, dynamic - The quantity added to the measured or calculated elevation of a point to obtain its dynamic height.

leveling correction, orthometric - SEE correction, orthometric.

leveling correction, temperature - SEE temperature correction.

leveling datum - An equipotential surface to which measurements of elevation are referred. The surface is usually the geoid, but need not be. SEE ALSO datum, vertical control.

leveling equipment - The equipment used in leveling and consisting of a leveling instrument, leveling rods, turning pins, and other devices used particularly for leveling. Leveling equipment by itself does not constitute a leveling system.

leveling error, orthometric - SEE error, orthometric.

leveling error of closure - SEE misclosure in leveling.

leveling head - SEE quadribrach; tribrach.

leveling instrument - An instrument consisting of a telescope and a mechanism for making the line of sight horizontal, and intended solely or primarily for establishing a horizontal line of sight. Also called a level. A leveling instrument can be moved freely about a more or less vertical axis (the standing axis) but motion vertically is restricted just as much as is needed for adjusting the line of sight to a horizontal position. Note that a leveling instrument is not an instrument in the sense of making measurements itself. Measurements are actually made on the leveling rods; the leveling instrument merely provides the horizontal line of sight. SEE ALSO compensator leveling instrument; construction leveling instrument; laser leveling instrument; pendulum leveling instrument; prism leveling instrument.

leveling instrument, automatic - (1) A leveling instrument whose line of sight is kept horizontal automatically even when the standing axis, about which the telescope rotates, is not exactly vertical. Also called an automatic level, self-leveling leveling instrument, self-leveling level, self-leveling leveling instrument and self-adjusting level. Leveling instruments using a pendulous component in the optical system are also called compensator leveling instruments or (improperly) collimating leveling instruments. Most automatic leveling instruments do use such a pendulous component. Some, however, do not; for example, some use a small pool of mercury to provide the level surface. (2) A leveling instrument all or most of whose leveling functions are carried out without intervention by the operator.

leveling instrument, dumpy - SEE level, dumpy.

leveling instrument, first-order - A leveling instrument designed for use in first-order leveling. Sometimes called a geodetic level or geodetic leveling instrument. A first-order leveling-instrument is generally characterized by special design and by superior workmanship in construction. A general but not always exact distinction between first-order leveling-instruments and other leveling instruments is the following: a first-order leveling-instrument is carefully adjusted, and any remaining instrumental errors are determined by observation and are applied as corrections to the measured differences of elevation; other leveling instruments are carefully adjusted, but thereafter no account is taken of errors remaining in the instrument.

leveling instrument, geodetic - (1) A leveling instrument suitable for establishing vertical control. (2) SEE leveling instrument, first-order.

leveling instrument, mercury-pool - A leveling instrument in which the line of sight is kept horizontal by the reflecting surface of a pool of mercury acting as an element in the telescope's optical system. This is a type of automatic leveling-instrument. Like types having pendulous elements, the mercury-pool type must first make the line of sight nearly horizontal by using a circular level. The instrument then automatically compensates for the remaining amount of non-horizontality.

leveling instrument, military - A dumpy level made more compact and sturdy, for use by the U.S. military.

leveling instrument, precise - A leveling instrument designed for use where highly precise or accurate results are required. Also called, ungrammatically, a precision leveling instrument.

leveling instrument, precision - SEE leveling instrument, precise.

leveling instrument, reversible - A leveling instrument whose telescope can be rotated through 180° about a horizontal axis.

leveling instrument, self-adjusting - SEE leveling instrument, automatic; leveling instrument, self-leveling.

leveling instrument, self-leveling - A leveling instrument in which the line of sight is made level without the operator's doing more than first making the line of sight approximately level (to within a degree or so). Also called a self-aligning level, self-leveling level, automatic level, automatic leveling instrument, etc. SEE ALSO leveling instrument, automatic.

leveling instrument, tilting - A leveling instrument in which the line of sight is brought into its final, level position by rotating the telescope on its trunnions.

leveling instrument, Troughton - SEE Troughton level.

leveling instrument, U.S. Geological Survey - SEE U.S. Geological Survey level.

leveling instrument, wye - SEE wye level.

leveling instrument, zenithal - An optical device mounted in such a manner that line of sight can be directed vertically. The usual form introduces, between the ocular lens and the reticle of a telescope, a prism deflecting the optical axis upwards through 90°. The device is otherwise leveled in the same way as an ordinary leveling-instrument.

leveling line - A unit of field work consisting of bench marks and temporary bench marks connected by chains of differential leveling called runnings. Note that this is not the same as a level line.

leveling misclosure - SEE misclosure in leveling.

leveling net - SEE leveling network.

leveling network - A collection of level lines connected together to form a network of loops or circuits extending over a region. Also called a leveling net, vertical-control survey network, vertical network and level network. The last two terms are ambiguous.

leveling network, fundamental - SEE level net, fundamental.

leveling network classification - SEE classification (of leveling networks).

leveling of high precision - A category of leveling adopted by the International Geodetic Association in 1912 and redefined by the International Association of Geodesy in 1936 and 1948. (a) The original definition was every line, set of lines, or network which is run twice in opposite directions, on different dates as far as possible and whose errors, accidental and systematic, when computed by the formulas hereinafter given do not exceed ± 1 mm per km for the probable accidental error, or ± 1.5 mm per km for the mean accidental error; ± 0.2 mm per km for the probable systematic error or ± 0.3 mm per km for the mean systematic error. The formulas referred to are those of Lallemand. (b) The definition of 1936 was: the total probable error e, calculated by means of the preceding formulas, does not exceed 2 mm per km. The formulas referred to are those of Vignal. (c) The definition of 1948 was: t shall not be greater than 2 mm, where t is the probable total limiting error per kilometer, computed from the above formulas. The formulas referred to are Vignals formulas slightly modified.

leveling party - The group of people directly involved in leveling in the field. Also called an observing team, observing crew, observing unit or leveling unit.

leveling rod - A straight rod or bar designed to be held vertically and graduated so that the vertical distance between a horizontal line of sight aimed at the rod and the ground (or foot of the rod) is indicated by the reading on the rod. Also called a levelling rod, level rod and levelling staff. A leveling rod is usually made of wood and has a flat face that either is itself graduated in terms of some linear unit and fractions thereof, or has attached a metallic strip so graduated. The zero of the graduations is at the lower end of the rod. A leveling rod having the graduations on a strip of invar is generally called an invar leveling rod. On some rods the numerical values of the graduations are intended to be read by the observer at the leveling instrument; such leveling rods are called speaking leveling rods or self-reading leveling rods.

Another type, called a target leveling-rod, carries a disk (the target) which is moved up or down on the leveling rod according to directions from the observer; when the disk is bisected by the line of sight, the numerical value of the corresponding graduation is read by the rodman. The word leveling is commonly omitted: e.g., Boston rod, invar rod or Philadelphia rod. SEE ALSO Barlow leveling rod; Boston leveling rod; Chicago leveling rod; Gravatt leveling rod; invar leveling rod; laser leveling rod; Lenker leveling rod; Molitor precise leveling rod; New York leveling rod; Pemberton leveling rod; Philadelphia leveling rod; Stephenson leveling rod; target leveling rod; U.S.Coast and Geodetic Survey leveling rod; U.S. Coast and Geodetic Survey first-order leveling rod; U.S. Engineer precise leveling rod; U.S. Geological Survey precise leveling rod; yard leveling rod.

leveling rod, double-target - Any leveling rod having a target and having graduations on two opposite faces.

leveling rod, foot-meter - A leveling rod marked in feet and tenths of feet on one side and in meters and fractions of a meter on the other.

leveling rod, Invar - SEE Invar leveling rod.

leveling rod, Invar-band - SEE Invar leveling rod.

leveling rod, laser-type - SEE laser leveling rod.

leveling rod, long - A leveling rod extendable to permit readings of as much as 13 feet. Also called a high rod and long rod. The Philadelphia leveling rod is a long leveling rod.

leveling rod, precise - SEE U.S. Coast and Geodetic Survey first-order leveling rod.

leveling rod, self-reading - A leveling rod with graduations designed so that they can be read by the observer. Also called a speaking rod and speaking leveling rod. The reading may be taken on the place on the rod where line of sight through the central, horizontal line on the reticle of the leveling instruments intersects the leveling rod, or it may be calculated as the average of the places where the lines of sight through the upper and lower horizontal lines on the reticle intersect the rod. The leveling rods used in modern geodetic leveling are of this type.

leveling rod, short - A leveling rod permitting only readings of 7 feet or less. Also called a short rod.

leveling rod, single-target - Any target leveling-rod having graduations on one face only.

leveling rod, speaking - SEE leveling-rod, self-reading.

leveling rod, target - SEE target leveling rod.

leveling-rod constant - SEE constant, leveling-rod.

leveling-rod stop - SEE stop, leveling-rod.

leveling-rod unit - SEE rod unit.

leveling system - The complete assemblage of men, equipment and techniques used for a particular project in leveling. A leveling system may include only what is used for the actual measurement (i.e., the field work) or, in some interpretations, may include everything and everyone involved in producing the elevations. In either case, the term is usually used only for complex or expensive assemblages. For example, the term motorized leveling system denotes the assemblage of personnel, leveling instruments, leveling rods, vehicles, computers and special instruments and techniques involved in motorized leveling.

leveling unit - SEE leveling party (1).

level leveling correction - SEE level correction. Also called level correction, C-correction or collimation correction.

level line - (1) The set of points, presented in the order of their measurement on the ground, between which differences of elevation have been determined by differential leveling. (2) The set of measured differences of elevation, presented in the order of their measurement, as determined by differential leveling. It is customary to refer, loosely, to either the set of points or the set of measurements as a level line, so the context must be interpreted to find out which meaning is intended. However, the term line of levels has been defined as applying particularly to the set of differences of elevation. It would therefore be good practice to use level line to refer only to the set of points, not to the set of differences. SEE ALSO spur level line.

level line, double-run - A level line in which the measurements were made from starting point to end point (the end point may be the same as the starting point so that the level line forms a loop) and then back again in the opposite direction. SEE ALSO leveling, double-run.

level line, duplicate - A level line composed of two single level-lines run over the same route but in opposite directions and using different turning points for each. Also called a level circuit or leveling circuit. Some definitions require that the network be established by spirit leveling.

level line, multiple - Two or more single level-lines run between the same terminal points but along different routes. Also called a multiple level circuit.

level line, simultaneous - A level line composed of two single lines run over the same route and in the same direction, but using different turning points for each. Also called a double- rodded line and a simultaneous double line. It is not the same as a simultaneous level-circuit.

level line, single - A level line consisting of a single set of points between which differences of elevation were determined. The points are ordered in the order of their measurement, from the starting point to the end point, and the starting and end points are different.

level line, single-run - A level line established by single-run leveling.

level line, spur - SEE spur level line.

levelperson - That person, in a leveling party, who observes at the leveling instrument.

level net - SEE leveling network.

level net, fundamental - The various lines of first-order leveling considered as a whole.

level network - (1) A network lying on a level surface or on a flat surface of limited horizontal extent. (2) SEE leveling network.

level of no motion - The depth at which it is assumed that isobaric surfaces in a layer of water coincide with level surfaces. At a level of no motion, there are no horizontal components of pressure gradient and, therefore, no horizontal motions of the water.

leveling specifications - SEE specifications (for leveling).

level rod - SEE leveling rod.

level slicing - SEE density slicing.

level spheroid - (1) A rotational ellipsoid representing the Earth and on whose surface the gravity potential is constant. (2) Any surface approximating the geoid, and on which the gravity potential is constant. In particular, such a surface which is nearly spherical.

level surface - SEE surface, level.

level trier - A device used for measuring the angular value of the divisions of a spirit level. One form of level trier consists of a beam mounted on a stable base. One end of the beam is hinged to the base; the other end can be moved vertically by means of a graduated screw, the angular amount of movement being measured with the screw. The spirit level is mounted on the beam and the angle through which the beam moves while the bubble of the spirit level travels the length of a division is measured with the screw.

level unit - SEE leveling party (2).

level vial - The transparent, closed container which is part of a spirit level. The liquid part of the spirit level is not a part of the level vial.

Levi-Civita variable - The quantity $[\varepsilon/(1 + k)] ek$, in which ε is the eccentricity of an elliptical orbit and k^2 is $(1 - \varepsilon^2)$.

libration - (1) An oscillatory, rotatory motion. (2) That relative motion between an observer and a body which makes the body seem to rotate slightly back and forth about an average position. There are two kinds of libration: geometric and physical. Physical libration is rotation with respect to the line joining the centers of mass of the body being observed and the body on which the observer is located and is independent of the location of the observer. Geometric libration is rotation with respect to the line joining the center of mass of the observed body to the observer and depends on the location of the observer. (3) In particular, a variation in the Moon's motions allowing more than half the Moon's surface to be seen from the Earth.

libration, dynamic - SEE libration, physical.

libration, geometric - Libration resulting from changes in the location of the observer with respect to the body. For example, part of the geometric libration of the Moon occurs because the observer is rotating with the Earth and so sees the Moon first of to one side and then off to the other a half-day later, seeing each time a little farther around the edge of the Moon than when he sees the Moon on the meridian. An additional amount of geometric libration is caused by the ellipticity of the Moon's orbit and by the inclination of the Moon's axis of rotation to the plane of the Moon's orbit. The Moon's geometric libration is about 8° in longitude and about 7° in latitude.

libration, physical - Libration resulting from the actual rotation of the body observed, about an average position. Also called dynamic libration. For example, because the Moon has a small bulge on the side nearest the Earth, the Earth's gravitational attraction on this bulge causes a physical libration of the Moon. The Moon's physical libration is about 0.03° in longitude and about 0.04° in latitude.

license, cartographic - The freedom to adjust, add, or omit features on the map, within certain limits, to attain the best appearance of the map while still adhering to specifications.

license, topographic - The freedom to adjust, add or omit contour lines, within allowable limits, to attain the best appearance of the map. Topographic license must not be construed as permitting contours to be adjusted by amounts that significantly impair their accuracy.

lidar - (1) An instrument capable of measuring distance and direction to an object by emitting timed pulses of light in a measured direction and converting to the equivalent distance the measured interval of time between when a pulse was emitted and when its echo was received. Also called laser radar. The term was first used in meteorology for an instrument designed to measure distances and directions to clouds. It predates the introduction of lasers. (2) SEE equipment, electro-optical distance-measuring.

lidar altimeter - SEE laser altimeter.

lien - A charge against property whereby the property is made security for the payment of a debt.

lien selection - A selection in exchange for which the applicant relinquished his rights or title to other lands which he for some reason cannot or does not wish to acquire or hold. Also called indemnity selection.

Liesegang level - The oldest form (about 1785) of a modern type of leveling instrument. The telescope had two objective lenses, two reticles, and an interchangeable eyepiece.

life estate - An estate in property, the duration of which is confined to the life of one or more persons or is contingent upon certain happenings. SEE ALSO life interest; life tenant.

life interest - The estate which a person has in a property that will endure only during his/her own or someone else's lifetime. SEE ALSO life estate; life tenant.

life tenant - A person who owns an estate in real property for his/her own life or for another person's life or for an indefinite period limited by a lifetime. SEE ALSO life estate; life interest.

lift - SEE drag.

light - That form of electromagnetic radiation detectable by the normal, human eye and perceived by normal, human vision. The eye can detect electromagnetic radiation having wavelengths between 0.4 and 0.8 micrometer. (Most eyes are unable to detect radiation longer than about 0.75 micrometer.) The range is bounded on the longer-wavelength side by the infrared part of the spectrum and on the shorter-wavelength side by the ultraviolet part. An exceptional human eye may see slightly into the infrared or the ultraviolet. Most commercial black-and-white photographic film is sensitive to about the same range, extending somewhat farther into the ultraviolet and somewhat less into the infrared parts. So called infrared film is not sensitive much beyond 0.9 micrometer. SEE ALSO speed (of light); velocity (of light).

light, actinic - Light capable of causing changes in a sensitive emulsion such as that used for photographic film.

light, coherent - Monochromatic light retaining, at a point, the same relationship between the phases of its components as was present between the phases at the two or more sources of the light.

light, infrared - SEE radiation, infrared.

light beam - (1) A stream of photons traveling within a cone having a very small apex angle. (2) Light whose intensity in a direction transverse to the direction of travel falls off extremely rapidly beyond a certain distance from a line drawn in the direction of travel. Alternatively, a bundle of light-pencils whose central axes all have approximately the same direction. Also called a beam of light.

light keeper - That member of a surveying team who has the task of attending and keeping in working condition the lamps used as beacons in surveying. Also called the lamp man or lightkeeper. He puts the lamps in place at a station, turns them on, aims their beams at the stations occupied by the observer and turns them off and removes them when the work is completed.

light level - (1) SEE intensity, luminous. (2) SEE illuminance.

light level, ambient - SEE intensity, ambient luminous.

light pencil - Light originating at, or direct at, a single point. Alternatively, a bundle of light rays originating at or directed towards a single point.

light ray - (1) A line drawn perpendicular to the sequence of surfaces defined by a particular phase of a light wave as it moves through space. (2) The path followed by a photon. Alternatively, the geometrical concept of a single element of light propagated along a geodesic and of infinitesimal cross-section. Also called a ray of light.

light time - The time needed for light to travel between two specified points. The term is used in astronomy to denote, in particular, the time it takes light to travel from a celestial body such as a star to the observer.

light year - The distance traveled by light or other electromagnetic radiation in a vacuum, in one year. Also written light-year.

ligne - An obsolete French unit of length, equal to 1/144 of a French foot. A ligne is about 1/443.296 meter. It is important geodetically because it was used in every survey of geodetic importance in the 17th and 18th centuries, and even well into the 19th century. E.g., Bouguer's investigation of the deflection of the vertical in Peru gave measurements in lignes.

limb - The edge of the apparent disk of a celestial body. That half of the limb having the lesser angular elevation above the horizon is called the lower limb; that half having the greater angular elevation is called the upper limb.

limb, lower - That half of the edge of the apparent disk, of a celestial body, which has the lesser angular elevation above the horizon.

limb, upper - That half of the edge of the apparent disk, of a celestial object, which has the greater angular elevation above the horizon.

limit, proportional - The greatest stress a material is capable of withstanding without any deviation from proportionality of stress to strain (Hooke's law).

limitation - (1) The limiting or marking out of an estate in property. E.g., the granting by deed of real estate without words of limitation will convey only a life estate. (2) The creation, by deed or devise, of a lesser estate or estates out of a fee. E.g., the gift of a life estate to A, remainder for life to B, remainder to C. (3) A certain period, limited by statute, after which the claimant shall not enforce his claims by suit.

limits, geographic - (1) Those lines, on a map, which have definite longitudes and latitudes and which bound the region shown on a chart. On a map whose graticule consists of straight lines, the geographic limits would be the edges of the map. SEE ALSO line, neat.

line - (1) A geometrical concept usually undefined and having meaning only in relation to the axioms and postulates using it. Intuitively, it is usually taken to mean a connected sequence of points such that if A and B are any two points in the sequence, then there is a third point C also in the line and such that the three points occur in the sequence ACB. A line is usually assumed to extend indefinitely in both directions from any given point on it unless specifically stated otherwise. A line on which there are given two (and only two) points such that the line extends in only one direction at each point is sometimes called a segment, segment of a line or line segment. (In this glossary, line is used indifferently for a line of indefinite extent and for a segment.) There are geometries based on a finite number of points. These have no application to surveying. (2) SEE line, straight. (3) In physics, the sequence of points in space occupied by a moving point. (4) SEE line, straight. (5) In geodesy, the term is used with both of the above meanings and, in addition, as a term whose exact meaning may depend on the context. Unless the context specifies otherwise, line is commonly taken to mean a straight line. (6) In the mechanical arts, any mark representing or approximating the geometric or physical concept of a line. In cartography, line may be combined with another word to distinguish the other word's physical aspect from its geometric aspect. For example, a contour is a set of points having a certain common elevation on the ground; a contour line is the set of points drawn on a map and representing the set of points on the ground. SEE ALSO center line; cut line; envelope line; flight line; lorhumb line; lubber line; make line; match line; property line; reference line; refraction line; range line; rhumb line; ridge line; sheet line.

line (on an ellipsoid) - SEE line, geodetic; line, straight.

line, aclinic - SEE equator, magnetic.

line, agonic - A line, on the Earth's surface, on which the magnetic declination is zero. Equivalently, it is the locus of all points, on the Earth's surface, at which magnetic north and astronomic north coincide. It is a particular case of an isogonic line.

line, branch - A stretch of railroad tracks carrying a significantly smaller amount of traffic than the tracks from which it diverges. A stretch of railroad of secondary importance to the main line.

line, broken - A continuous line consisting of a finite number of connected segments of straight lines in the plane.

line, cardinal (USPLS) - A line running north-south or east-west.

line, co-range - A line representing a contour along which the range of the tide is the same at all points.

line, co-tidal - (1) A line representing a contour joining points at which high water (or low water) occurs simultaneously. The times may be expressed as differences from times at a standard port or as intervals after the time of the Moon's transit of the meridian. (2) A line, on a chart, representing a line on the Earth passing through all points where high water (or low water) occurs simultaneously. The lines show the average lapse of time, usually in lunar hours, between the Moon's transit over a reference meridian (usually Greenwich) and the occurrence of high water (or low water) for any point lying on the line.

line, cutoff - SEE cutoff line.

line, dead-reckoning - A line, drawn on a chart, representing a ship's course as determined by dead-reckoning.

line, directed - A line (1) so defined that it is distinguished from another line having the same points by the order in which the points occur, i.e., the line containing the points ABC in that order is defined to be different from the line containing the same points but in the order CBA.

line, double-rodded - (1) A line of levels resulting from leveling in which two sets of turning points are used to give independent sets of measurements between bench marks. (2) SEE level line, simultaneous.

line, established - A surveyed line.

line, geodesic - SEE geodesic (n.); line, geodetic.

line, geodetic - A geodesic on a rotational ellipsoid. Sometimes called a geodesic line, a geodetic line is a line of double curvature and usually lies between the two normal-section lines determined by two points on the ellipsoid. If the two terminal points are in nearly the same latitude, the geodetic line may cross one of the normal-section lines. Except along the equator and along the meridian, the geodetic line is not a plane curve and a single line of sight cannot be directed from one terminal point to the other and still follow the geodetic line. However, for conventional triangulation, the lengths and directions of geodetic lines differ inappreciably from those of corresponding pairs of normal-section lines. The distinction between a geodesic and a geodetic line is a fine one and is usually unnecessary.

line, great-circle (USPLS) - In land surveying, the line of intersection of the surface of the Earth and the plane of a great circle on the celestial sphere. The definition does not define a line uniquely unless the plane is also required to pass through a specified point in or on the Earth. Because the Earth is not a sphere, any such line is only an approximation to a great circle and, because of the deflection of the vertical, will not, as surveyed, lie in a single plane.

line, half - SEE half-line.

line, headland-to-headland - The line joining the termini at the outer headlands of a coastal indentation determined to inclose inland waters by the semicircular rule or on historic

grounds. A headland-to-headland line marks the seaward limit of inland waters.

line, high-water - (1) In tidal water, that line which is the intersection of the plane of mean high water with the shore. (2) The line, along the shore, to which the water normally reaches at high water. (3) The line formed by the intersection of the land with the surface of the water at high water. The high-water line varies daily with changing lunar phases and with meteorological conditions. (4) SEE mark, high-water. Use of high-water line for high-water mark should be discouraged.

line, isoclinic - A line drawn on a chart or map and joining points at which the values of equal magnetic dip at corresponding points on the Earth are equal. Also called an isoclinal.

line, isodynamic - A line connecting points of equal magnitude of force.

line, isogonic - (1) A line drawn on a chart or map and connecting all points representing points on the earth having equal magnetic declination at a given time. Also called an isogonal. SEE ALSO line, agonic. (2) A line, on the ground, along which the magnetic declination is constant.

line, isopycnic - SEE isopycnic.

line, level - (1) A line tangent, at some specified point, to an equipotential surface of the Earth's gravity field. (2) A line on a level surface. (3) SEE level line.

line, local lunar - That arc of the celestial equator, or that angle at the celestial pole, lying between the lower branch of the local celestial meridian and the hour circle of the Moon, taken positive westward from the lower branch of the local celestial meridian through 24 hours.

line, lorhumb - SEE lorhumb line.

line, lower low-water - A line or mark indicating lower low water.

line, low-water - (1) The line defined by the boundary of a body of water at its lowest elevation (stage). Because there may be appreciable variation with time in the elevation of the low-water surface at a designated locality, an average value of the elevation of low water should be used in determining the location of the low-water line on the ground. (2) In tidal waters, the intersection of the plane of mean low water with the shore. In this definition, it may be assumed that plane of mean low water means a horizontal plane through the average location of the water's surface on a tide gage during low water. (3) The line formed by the intersection of the land with the water's surface at low water. As low water may vary considerably with time at a designated locality, an average value of the height at low water should be used in determining the position, on land, of the low-water line. SEE ALSO water, mean high (low). (4) In tidal water, the line formed by the intersection of the land with the surface of mean low water. (5) The intersection of the plane of low water with the shore.

line, magnetic isoporic - SEE isopor.

line, meander - SEE meander line.

line, mean higher high-water - (1) The line formed by the intersection of the land with the water's surface at the height of mean higher high water. (2) That line, on a map, which represents the intersection of the land with the water's surface at the height of mean higher high water.

line, mean high-water - (1) The line formed by the intersection of the land with a surface which is at the height of mean high water. SEE ALSO shore line. (2) That line, on a map, which represents the intersection of the land with the water's surface at mean high water. (3) That line, in tidal water, formed by the intersection of the sloped surface of the land with the water's surface at the height of mean high water which is the arithmetic average of the water's heights measured for a series of points over a specific 19-year Metonic cycle (235 lunations) also known as the National Tidal Datum Epoch and as held by the U.S. Supreme Court in the Borax Company vs. City of Los Angeles (296 U.S.10, 1935). The mean high water line may or may not coincide with lines of ownership or lines of vegetation.

line, mean lower low-water - (1) The line formed by the intersection of the land with the water's surface at the height of mean lower low water. (2) That line, on a map, which represents the intersection of the land with the water's surface at the height of mean lower low water.

line, median - SEE median line.

line, meridian - SEE meridian line.

line, neat - A line bounding or intended to bound the detail of a map, i.e. bounding the map proper. Also called an inner neat line to distinguish it from the border drawn outside of the neat line and separated by a space from the map proper. Also written neatline. SEE ALSO sheet, line.

line, non-riparian meander - SEE meander line, non-riparian.

line, normal-section - (1) That line, on an ellipsoid, which is the intersection of the ellipsoid with a plane perpendicular to it at a given point. A normal-section line is in general elliptical. It is not, in general, a geodesic. (2) A line connecting two points on an ellipsoid and lying in the plane through the perpendicular at one of the points. This definition differs from the preceding in that the line it defines has definite end-points. In general, the normal-section line through one of the points will be different from the normal-section line through the other point. However, there is only one shortest line (geodesic) through the two points.

line, offset - A supplementary line close to and approximately parallel to a main line to which it is referred by measured offsets. Where the line for which measurements are desired is in such a position that it is difficult to make measurements along it, the required measurements are obtained by establishing an offset line in a convenient location and measuring offsets from it to salient points on the other line. The desired measurements are then made on the offset line and corrected, if necessary, for the distance between the two lines.

line, pecked - A cartographic symbol consisting of a line broken at regular intervals.

line, plumb - SEE plumb line.

line, pointing - SEE line of collimation.

line, principal - The intersection of the plane of a photograph with that plane which contains both the vertical through the rear nodal point and the photograph perpendicular at that point. Also called a principal meridian. It is the line through the principal point and the nadir point.

line, quarter - In public-land surveys made on a township, range, and section basis, one of two lines joining opposite quarter-section corners, and the original survey of a section of land is divided into two parts by the two lines.

line, radial - A radius of a circular curve, drawn to a designated point on the curve. If the line is extended beyond the convex side of the curve, it is called a prolongation of the radial line.

line, random - A line run, as nearly as circumstances permit, from one survey station toward another survey station out of sight from the first survey station. A correction to the initial azimuth of the random line can be calculated from the misclosure (the amount by which the second station is missed). Offsets from the random line, to establish points on the line between the stations, can also be calculated from the misclosure.

line, right - A straight line between corners (of a Public Land Survey). The term is obsolete.

line, set-back - (1) A line outside the right-of-way along a highway, established by public authority, on the highway side of which the erection of buildings or other permanent changes is controlled. (2) A line established by law, restrictions in deeds or custom and fixing the least distance of the external face of the building, walls, and any other construction from a street or highway right-of-way. SEE ALSO set-back.

line, side - As applied to a strip of land such as a street or a right-of-way, the boundaries of that strip. It does not apply to the ends of a strip.

line, simultaneous double - SEE level line, simultaneous.

line, sounding - SEE lead line.

line, spur - SEE spur level line.

line, standard (USPLS) - Any base line, standard parallel, principal meridian, or guide meridian. The term is not in common use.

line, straight - (1) A geodesic in Euclidean space. (2) A line having the following properties: (a) there is one and only one straight line connecting any two given points; (b) a straight line and a point not on that straight line determine, by intersection and connection, a set, called a plane of lines and points such that any straight line in the plane intersects all other lines in the plane; (c) a plane and a point not in that plane determine, by intersection and connection, a set, called a space, of points, lines and planes such that every plane in the space intersects all other planes and the intersections are straight lines. Usually referred to simply as a line unless the meaning would not be clear. However, the term curve is usually used to denote a non-straight line. (3) The term as used in geodesy and in law does not have a precise meaning. C. Schott gave (1901) eight different possible meanings for the term as applied to a line between two points A and B on a rotational ellipsoid: (a) the normal-section line determined by a plane through B and the perpendicular at A; (b) the line from A to B such that the line of sight from one point in the line to the succeeding point lies in the plane through the perpendicular at the preceding point and through the succeeding point; (c) the line from A to B such that the line of sight to B, at any point of the line, lies in the plane through B and through the perpendicular at that point; (d), (e), (f) the lines defined by interchanging A and B in the preceding definitions; (g) the line of alinement (curve of alinement) between A and B; and (h) the geodesic between A and B. Definitions (a) and (d) cover the definitions of the other six lines. Lines (b), (e) and (g) approach the geodesic as the distance between successive points is decreased.

line, stray - An ungraduated portion of line connected with the current pole used in measuring current. The stray line is usually about 100 feet long and permits the pole to acquire the speed of the current at some distance from the disturbed waters in the immediate vicinity of the observing vessel before the speed of the current is read from the graduated portion of the line.

line, surveyed - (1) An imaginary line, on the Earth's surface, whose direction or length has been determined by surveying. (2) A line on the geoid or ellipsoid whose direction or length has been determined from measurements on the corresponding line on the Earth's surface.

line, tangent - The limiting position of a straight line cutting a curve in two points, as the second point approaches the first along the curve. Also called a tangent.

line, true (USPLS) - (1) A straight line of constant bearing between corners. Formerly called a right line. (2) The direct, forward bearing from one monument to the next, as distinguished from a random line.

line, vanishing - That straight line, on an image formed by perspective projection, upon which lie all the vanishing points of all systems of parallel lines on a plane.

lineament - A linear image of a linear, geological feature.

line copy - (1) Any document (copy) suitable for reproduction without using a (halftone) screen. (2) Copy composed of lines as distinguished from continuous-tone copy.

line-crossing method (of trilateration) - Trilateration in which the distances between points in the network are determined by measuring the distances from an airborne distance-measuring apparatus to each end of every line of the network. The aircraft flies back and forth across every line and the sum of the distances to each end is calculated or plotted. That point in the path at which the sum of the distances is least is assumed to lie in a vertical plane through the line. The sum of distances is then reduced to the length of the corresponding line on the reference ellipsoid. SEE ALSO Shoran-line crossing; Sodano azimuth survey.

line detection - The detection, in image space, of the image of a line in, object space.

line map - SEE map, planimetric.

line printer - A typewriter-like machine in which an entire line of characters is printed, by impact, at the same time.

linen tape - A surveyor's tape ten meters long and made of linen. A linen tape is extremely pliable, easily carried and able to be rolled up into a small volume. It is used by land surveyors but is less accurate than tapes made of steel or invar. It is also much less expensive.

line of alinement - SEE curve of alinement.

line of allocation - A line delimited through the high seas or an unexplored region to allocate lands without conveying sovereignty over the high seas. The line of allocation may be unofficial (as drawn by a mapmaker) or official. The simplest way to specify a line of allocation is by giving the meridians and parallels which the line must follow, without using diagonal lines.

line of apsides - (1) The major axis of an elliptical orbit, extended indefinitely in both directions. (2) The line joining those two points, in an orbit, at which the distance of the moving body from the center of attraction is an extremum. SEE ALSO apsis. Also called an apse line.

line of collimation - The line passing through the second nodal point of the objective lens-system of a telescope and the center of the reticle. It is variously called the line of sight, sight line, pointing line and aiming line of the instrument incorporating the telescope. In the telescope of a transit, the center of the reticle may be defined by the intersection of cross-hairs, by the mid-point of a fixed, vertical wire, or by the mid-point of a micrometer wire in its average position. In the telescope of a leveling instrument, the center of the reticle may be the mid-point of a fixed, horizontal wire. In lens systems, a line of collimation and the line of sight in the usual sense need not be the same.

line of constant scale - Any line, on a photograph, parallel to the true horizon or to the isometric parallel. Also called a line of equal scale.

line of equal scale - SEE line of constant scale.

line of force - A line the directed tangent to which at any point has the direction of the force acting at that point. A line indicating the direction in which a force acts.

line of leveling - SEE level line; line of levels.

line of levels - (1) A continuous sequence of measured differences of elevation. Also called a level line. The individual differences may be single measurements, as in the case of single-run leveling, or may be averages of repeated measurements as in the case of double-run leveling. (2) SEE level line.

line of levels, spur - SEE spur level line.

line of nodes - (1) The straight line connecting the two points in which the plane of the ecliptic intersects the orbit of a planet, asteroid, or comet. (2) The straight line joining the two points in which the orbit of a satellite is intersected by the equatorial plane of the primary body. Equivalently, the intersection of the plane of an osculating orbit with the reference plane.

line of position - A line drawn on a chart to indicate the set of possible locations of the point from which observations were. Also called a position line. Often abbreviated to L.O.P. or LOP. Line of position is a more general term than Sumner line.

line of position, advanced - A line of position which has been moved forward parallel to itself and along the line indicating the course, to obtain a line of position for a later time. A retired line of position is the opposite of an advanced line of position.

line of position, celestial - A line of position determined from measurements on a celestial body.

line of position, electronic - A line of position obtained using electronic instruments.

line of position, hyperbolic - A line of position determined by measuring the difference in distance to two fixed points. Such a line is a segment of a hyperbola. Lines of position determined using Loran are an example.

line of position, retired - A line of position which has been moved backward parallel to itself and along the line indicating the course, to obtain a line of position for an earlier time.

line of sight - (1) A straight line drawn between two reference points in an optical system and extended indefinitely in a forward direction (towards object-space). In telescopes used in surveying instruments, one of these points is usually a mark on a reticle, while the other is the rear nodal point of the objective lens-system. The line of sight need not coincide with the optical axis of the telescope. When it should but does not, the angle from the optical axis to the actual line of sight is called the collimation error or internal collimation error. Collimation error also, however, designates the divergence between the line of sight and the collimation axis. (2) The straight line between two points. This line is in the direction of a great circle but does not follow the curvature of the Earth. (3) SEE line of collimation. The line of sight and the line of collimation are identical in lens systems but not, in general, for catoptric or catadioptric optical systems.

line of soundings - (1) A line along which a series of soundings has been made by a moving vessel, usually at regular intervals. (2) The series of soundings made, usually at regular intervals, by a moving vessel.

line of tide, nodal - In a region having tides, the line about which the tide oscillates and along which there is little or no rise and fall of the tide.

line rod - SEE range rod.

lines, antiparallel - A pair of parallel, directed lines having opposite directions.

line scanner - A device which views only a very small spot in a scene at any one time, but which moves from spot to spot in a straight line from side to side of the scene; when it reaches a side, it moves the view down and proceeds in the opposite direction parallel to the previous line. This movement starts at one corner of the scene and continues unit the entire scene has been viewed piecemeal. A television camera, for example, is usually a line scanner. Line scanners usually work in one of two ways: (a) the scene is viewed through an optical system with a very narrow field of view; or (b) the scene is projected in its entirety onto the screen of a special type of cathode-ray tube and a narrow beam of electrons creates a spot which is then viewed. Some scanners move from spot to spot in a spiral fashion. Such scanners are not considered line-scanners. The most important geodetic application of line scanners is to the measurement of photographs. SEE ALSO scanner, linear.

line scanner, infrared - A line scanner which views the scene in the infrared radiation emitted by the scene. Infrared line-scanners are often mounted in aircraft and used for reconnaissance. Such a scanner consists of a rotating mirror projecting the incoming beam onto a detector sensitive only to radiation in a narrow band between 1 and 14 micro-meters. The optics is such that the instantaneous field of view is very small, but the mirror scans the scene rapidly from side to side.

line-spread function - SEE function, line-spread.

line standard of length - A bar or block whose sides are flat and having on one side two marks containing a defined or otherwise known length between them.

line symbol - A distinctive kind of line used either to represent a feature whose width is very small in proportion to its length, or a boundary between regions to which particular characters or values are attributed. An isogram is a line symbol.

line tree (USPLS) - (1) A tree intersected by a surveyed line on which legal corners are established and are reported in the field notes of the survey. Also called a sight tree or a fore-and-aft tree. (2) A tree intersected by a surveyed line, reported in the field notes of the survey, and marked in some manner on each of the sides facing the line. Also called a station tree, sight tree and fore-and-aft tree, but line tree is preferred. A tree whose trunk lies entirely upon one side of the boundary belongs exclusively to the owner of the land on that side. A tree whose trunk stands on the land of two or more adjoining owners belongs to the owners in common.

line work - A design which is to be reproduced and which consists only of completely black and/or completely white regions.

lining pole - SEE range rod.

lining rod - SEE range rod.

link - (1) One one-hundredth of a chain. It is equivalent to 7.92 inches. SEE ALSO Gunter's chain. (2) A level line, part of a level line, or a combination of level lines or parts of level-lines which, taken as a whole, make a continuous piece of leveling directly from one junction bench-mark to another junction bench-mark without passing through or over any other junction bench-marks.

Linnik's rule - The rule that a particular measurement of value x, out of N measurements, is to be rejected if the ratio $(x - x_m)/s$ is greater than the ratio computed, for N and a pre-assigned probability P, using Student's distribution. s is the standard deviation of the sample of size N with average x_m.

liquid hand-compass - A hand-held magnetic compass in which the needle is immersed in a liquid that damps the needle's movement.

lis pendens - A notice of the pendency of an action involving real estate recorded in the registry of deeds. He who purchases during the pendency of the suit is bound by the decree that may be made against the person from whom he derives his title.

list - SEE x-tilt.

list of directions - A list of the objects observed at a triangulation station, together with the horizontal directions in terms of circular arc from the direction of one of the objects assigned the direction zero. The directions listed are usually the average of a number of readings on the horizontal circle or of sets of measured directions.

lithography - A method of printing in which the design to be printed is placed in a greasy and ink-receptive form on the surface of the plate that is to make the prints, while the rest of the plate is dampened with water to make it ink-repellant. Sometimes erroneously called photolithography. In the original process, a special variety of limestone (hence the name of the process) was used for the plate. At present, specially prepared plates of plastic or of metal such as zinc are used. There are four major methods in use for reproducing the image: (a) drawing the image directly on the plate; (b) transferring the image by direct contract from transfer paper on which the image has been drawn in special ink or transferred from another plate; (c) printing by direct contact with a sensitized plate from a drawing on translucent paper (VanDyke process); and (d) making a drawing, photographing this, and printing from the negative (helio-zincography) or from a positive (photo-etching) by direct contact upon a sensitized plate. SEE ALSO photolithography.

lithography, flat-bed - Lithography in which the printing plate is fixed on a flat, horizontal support called the bed.

lithography, screenless - Photolithography in which the image is transferred from a continuous-tone, positive transparency to a positive-working, highly sensitive resin coated on a grained and anodized sheet of aluminum, without using a half-tone screen as intermediary.

lithosphere - (1) The rocklike part of the Earth, as distinguished from the gaseous part (the atmosphere) and the watery part (the hydrosphere). The lithosphere is usually considered to be made up from a core and two concentric shells, the regions being distinguished from each other by the kinds and speeds of seismic waves that can travel in them. The crust, the outer, rocklike shell, extends from the surface of the Earth to a depth of 10 km to 60 km. The mantle, the inner shell, extends from the bottom of the crust to a depth of about 2900 km and is apparently quite plastic or nearly liquid. The core extends from the bottom of the mantle to the Earth's center at a depth of about 6400 km. (2) The crust of the Earth. (3) That part of the crust and the upper mantle in which the speed of seismic waves is high and the rocks posses considerable rigidity. This is the layer taking part, supposedly, as a unit in continental drift. When lithosphere is used with this meaning, one or more layers are commonly introduced between this layer and the mantle: the asthenosphere, a region of considerable plasticity; the rheosphere; and the metasphere. There is as yet no commonly agreed-upon, exact definitions of these terms. However, lithosphere is probably equivalent, roughly, to crust.

litigate (v.) - (1) Make the subject of a lawsuit. (2) Contest in law. (3) Prosecute or defend by pleading, evidence, and debate in a court.

litigation - (1) A contest in a court of law to enforce a right. (2) A judicial controversy, a suit at law. (3) Civil actions.

littoral (adj.) - Of or pertaining to a shore; in particular, of or pertaining to a seashore.

littoral (n.) - (1) A coastal region. (2) That zone of the sea floor lying between tide levels. SEE ALSO seaboard.

littoral current - SEE current, littoral.

littoral drift - SEE drift, littoral.

littoral zone - SEE zone, littoral.

livery of seisin - A method of transferring land, in which the receiving person is taken out to the land and, either on it or in sight of it, a ritual of transfer was gone through, with the receiver being handed something from the land to symbolize the transfer. Livery of seisin was one of the early methods of transferring land; that is, the buyer was seized of the land. A written instrument was not sufficient to transfer a land title without actual delivery. Because it was physically impossible to take the land to the buyer, the buyer was taken on the land and there, in the presence of friends and neighbors as witnesses, the delivery was made. This was quite a ceremony, with the chief characters repeating a required form. The delivery was symbolized by the seller giving the buyer a twig from a tree, a handful of soil, or the doorknob. Transfers in this manner were liver in deed, but if the ceremony was performed in sight of the land and not on it, it was constructive delivery and sufficient in law. This method of transfer was used rather extensively in some foreign countries but was used very little in the USA.

load number - A set of numbers h', k', l' defined by the ratios u_i/u_a, u_i/u_a and v_i/v_a, respectively. u_i denotes vertical displacement of the Earth's surface under a load, u_a is the vertical displacement of a level surface caused by displacement of water, u_i is vertical displacement because of the indirect effect, v_i is the horizontal displacement of the surface under the load and v_a is the horizontal displacement caused by displacement of water.

lobe - (1) A conical region, in space, containing a local maximum of the power radiated by an antenna and bounded by lines of local minimum. The power is usually expressed as a fraction of the total power radiated by the antenna (intensity may be used instead of power.) If the local maximum is also the maximum of the entire field, the lobe is called the main lobe; otherwise, it is called a side lobe. (2) The same as the previous definition, with received by in place of radiated by. The two definitions are, in theory, completely equivalent because they indicate the same pattern. (3) The region between directions of two adjacent minima in the interference pattern of an antenna.

lobe, main - That lobe, of an antenna pattern, into which more power is concentrated than into any other lobe.

lobe, minor - Any lobe, of an antenna pattern, containing less power than the main lobe. Also called a side lobe.

localization - The process of estimating the location and motion of a body.

locating (v.) - (1) Placing an object at a specified point. (2) Finding the point at which a specified object is located.

locating a point by resection - Locating a point by drawing a line of known direction but of indefinite length from one known point, occupying a station on that line, and determining the direction of a line through a second, known point. This line, the resection line, will cut the first line at the location of the point in question.

locating system - A system used for determining the location of an object such as a vehicle, vessel, or aircraft. Also called a positioning device, positioning system, fixing system and location system.

location - (1) The numerical or other identification of a point or object, sufficiently precise that the point or object can be found from the identification. Also called position. Location and position are allied terms and are frequently used as synonyms. In precise usage, however, location has the specific meaning given above and can be applied to both points and objects. Position has the meaning location and orientation or, sometimes, of orientation alone; it can be used for objects but not for points. For example, the location of a point on a plane can be specified by a pair of numbers (coordinates) such as the distances of the point from two fixed and given points. The location of a point in space can be specified by a set of three numbers (coordinates) such as the distance of the point from a fixed, given point and the two angles defining direction of the point with respect to two given lines. Location may be specified in other terms than numerical coordinates. It may be specified, as is done in land surveying, as being at the intersection of two named lines or by identifying it with some prominent and known feature (e.g., at the intersection of the center-lines of Viers Mill Road and Old Georgetown Road; or at the summit of Pike's Peak.) (2) The coordinates defining the location of a point on the geoid or on a rotational ellipsoid representing the geoid. Also called a position. SEE ALSO paper location.

location, geographic - SEE coordinate, geographic.

location monument - A monument established to supplement a monument of the U.S. Public Land Survey. Formerly called a mineral monument. Location monuments are established in connection with the official patent survey of mining claims in regions where subdivision surveys have not been made or monumentation is inadequate and where there is no corner of the Public Land Survey within 2 miles of the claim.

location survey - The establishment, on the ground, of points and lines in locations previously determined by computation or graphical methods. The plans for an engineering project (road, canal, etc.) are prepared in the office from survey data obtained in the field. These plans form a paper location and are the basis for the location survey.

location system - SEE locating system.

location system, active - A navigation system in which a satellite signals the craft and the craft responds.

loch - (1) A lake. (Scot.) Also called a lough. (2) A bay or inlet so inclosed as to be almost a lake. (Scot.) Also called a lough. (3) SEE lough.

Locke hand level - A simple metallic tube containing (a) a prism; (b) a sliding eyepiece in which is mounted half of a convex lens dividing the field of view into a right half and a left half. In the top of the tube and just above the prism is a hole through which can be seen an adjustable cross-wire and the bubble of a level.

Locke's level - SEE Locke hand level.

lode claim - A mining claim embracing public lands that contain minerals occurring in a lode or vein.

Loernermark base line - A base line 576 m long at Loernermark, Holland, established using the Väisälä baseline apparatus.

lofting (v.) - Laying out the plans for a structure on a large, level surface and assembling the parts of the structure according to the plan (and often on the plan). Surveying procedures have been used frequently to ensure the proper alinement of parts of the structure. This kind of surveying is usually classified as engineering surveying.

log line - A graduated line used to measure the speed of a vessel through the water or to measure the speed of the current from a vessel at anchor.

long-arc method - Determining the coordinates of an observer from observations on a satellite over a portion of the orbit so long that the orbit must be represented by as faithful a set of equations as possible. A long-arc method is usually considered to require more than a quarter of one complete revolution of the satellite. Anything less than this is treated by the short-arc method.

longitude - A coordinate indicating linear or angular distance from a north-south line used as a reference on a surface, or from a north-south plane used as a reference in space. SEE ALSO ephemeris longitude.

longitude, astronomic - The angle from the plane of an initial, arbitrarily chosen celestial meridian to the plane of the celestial meridian at the point whose astronomic longitude is wanted. Astronomic longitude results directly from observations on celestial bodies. It differs from geodetic longitude by the amount of the deflection of the vertical in the prime vertical times the secant of the latitude. It is measured by the angle, at the celestial pole, between the tangents to the local and the initial meridians, or by the arc intercepted on the celestial equator by those meridians. At an international convention held in Washington, D.C. in 1884, the Meridian of Greenwich was adopted as the initial meridian (prime meridian) for all longitudes, and is now generally so used. A value of longitude is sometimes accompanied by a statement of the method used in its determination e.g., wireless longitude. This may also indicate its precision. Astronomic longitude may be determined by the following methods, among others: (a) timing the occurrence of celestial events such as eclipses; (b) comparing, by clocks, local and Greenwich time at a station; (c) comparing time determined locally with Greenwich time transmitted by radio or telegraph to the station; (d) measuring the Moon's angular elevation or its culmination. SEE ALSO longitude determination.

longitude, assumed - A value of longitude chosen for convenience or because the correct value is not known.

longitude, celestial - The arc of the ecliptic intercepted between the vernal equinox and the foot of a great circle perpendicular to the ecliptic and passing through the object whose celestial longitude is to be determined. Also called ecliptic longitude. Celestial longitude is measured along the ecliptic from west to east. It is not the same as astronomic longitude and does not enter into the usual problems of surveying.

longitude, ecliptic - SEE longitude, celestial.

longitude, fictitious - (1) The dihedral angle between two planes perpendicular to a common plane of reference through the center of an ellipsoid, the one plane serving as a reference and the other containing the normal at the point whose longitude is desired. The angle is positive taken clockwise from the plane of reference. (2) The difference between the values assigned two lines representing, on a graticule, meridians on an ellipsoid; the one line is taken as reference and the other is the one representing the meridian through the point in question. (3) The arc, on a great circle denoted the fictitious equator, between two great circles passing through the poles of the fictitious equator, one of the great circles serving as a meridian of reference (fictitious prime meridian) and the other as a local meridian (fictitious meridian). A fictitious longitude may be called a grid longitude, oblique longitude or transverse longitude according as the fictitious equator is a line on a grid or is a great circle oblique or transverse to the actual equator.

longitude, galactic - SEE coordinate, galactic.

longitude, geocentric - The dihedral angle between two planes passing through the least axis of the ellipsoid. The angle is taken counterclockwise, as viewed from the north, from one plane used as reference to the other plane associated in a specified manner with the point in question. Also referred to as longitude if the meaning is clear. The planes used in the definition are usually selected in one of two ways. (a) The reference plane is required to contain a specified point on the ellipsoid; the other plane contains the point of interest. (b) The reference plane is parallel to the normal at a specified point on the ellipsoid; the other plane is parallel to the normal at the point of interest. If the ellipsoid is rotationally symmetric, these two ways are equivalent, and geocentric longitude becomes the same as geodetic longitude.

longitude, geodetic - The dihedral angle between two planes parallel to the least axis of an ellipsoid. The angle is taken counterclockwise, as viewed from the north, from one plane used as reference to the plane containing the normal at the point of interest. Also referred to a longitude if the meaning is clear. The reference plane is usually specified in one of three ways. (a) It contains the normal at some specified point on the ellipsoid. (b) It contains the least axis of the ellipsoid and is parallel to the normal at the specified point. (c) It contains the least axis and the specified point. Properly, only (a) should be used; (b) and (c) are more appropriately used for geocentric longitudes. However, all three ways are identical if the ellipsoid is rotationally symmetrical. A geodetic longitude on the rotational ellipsoid can be measured by the angle, at one of the poles of the ellipsoid, between local and initial meridians, or by the arc intercepted by these meridians on the equator. A geodetic longitude differs from the corresponding astronomic longitude by the amount of the component, in the prime vertical, of the local deflection of the vertical divided by the cosine of the latitude. In the USA, this may be as much as 25". In recording a geodetic longitude, the datum to which it is referred must be stated. In the USA, before the introduction of the North American Datum of 1983, geodetic longitudes were numbered from the Meridian of Greenwich but were computed from the meridian of station Meades Ranch as prescribed by the North American Datum of 1927.

longitude, geographic - A general term including astronomic, geocentric, and geodetic longitudes.

longitude, inverse - SEE longitude, transverse.

longitude, oblique - The angle between an oblique meridian used as reference and any given oblique meridian.

longitude, selenographic - The angle from the reference plane through the Moon's average axis of rotation to the plane through the same axis and the point of interest, taken as positive when measured from east to west (i.e, toward Mare Crisium).

longitude, terrestrial - The angle between the plane of the reference meridian and the plane of the meridian of a point on the Earth.

longitude, transverse - The angle between a transverse meridian used as reference and any given transverse meridian. Also called an inverse longitude.

longitude determination - The determination of astronomic longitude, unless some other kind of longitude is specified. Unlike geodetic longitudes, astronomic longitudes can be determined by measuring the difference between the time when a star or other suitable celestial body passes through the local meridian and the time when that same body passes through the meridian of Greenwich (or other meridian chosen as reference.) A number of methods have been invented for determining astronomic longitude either alone or together with astronomic latitude or azimuth. The problem of determining longitude is inseparable from the problem of determining time with respect to Greenwich, and the same methods often serve both purposes. The methods may be classified according to the kind of equipment used: the chronometric method, the radio method, and the telegraphic method. Or they may be classified according to the celestial body observed: lunar, solar or stellar. Or by the manner in which the observations are made, such as at equal angular elevation or at culmination. SEE ALSO Bessel's formula; chronometric method; Döllen's method; equal-altitude method of determining longitude; lunar culmination method; Mayer's formula; radio method; solar-observation methods (of determining longitude); telegraphic method; Tsinger's method.

longitude difference - The smaller angle at the pole, or the shorter arc of a parallel, between the meridians of two places, expressed in angular measure. Also called departure (q.v.).

longitude equation - A condition equation which expresses the relationship between the fixed longitudes of two points connected by traverse or by triangulation. When traverse or triangulation connects two points whose longitudes have been fixed by direct observation or by previous surveys, a longitude equation is used to make the longitude of either point, as computed through the survey from the other point, agree with its longitude as previously fixed.

longitude factor - The change in longitude, along a celestial position-line, per 1' change in latitude.

longitude of pericenter - An orbital element specifying the orientation of an orbit. It is a broken angle consisting of the angle, in the plane of the ecliptic, from the vernal equinox to the ascending node, plus the angle, in the plane of the orbit, from the ascending node to the pericenter.

longitude signal - A signal made at a particular time, observable at different stations, and used in comparing local times at those stations to determine the differences of longitude. SEE ALSO longitude determination.

longitude term (in a gravity formula) - That term, in a formula for the value of gravity, explicitly containing the geodetic longitude λ and geodetic latitude ϕ as arguments and implying that the geoid is represented by an ellipsoid having three unequal axes. With this additional term (longitude term), the gravity formula would be written as $\tau_o = \tau_e [(1 + \beta_1 \sin^2\phi + \beta_2 \sin^2 2\phi + \beta_3 \cos^2\phi \cos 2(\lambda - \lambda_o)]$, in which τ_e is the average value of gravity at the equator and β_3, like β_1 and β_2, is a dimensionless number. As always the gravity formula implies a surface of definite shape; when the surface is an ellipsoid having three axes of different lengths, the shortest axis is the polar axis, corresponding to the Earth's rotational axis, while the two longer axes lie in the equatorial plane. The longest axis lies in the plane at longitudes λ_o and $\lambda_o \pm 180°$. The two equatorial axes differ in length by $4a\beta_3$, in which a is the length of the longer semi-major axis. The idea of representing the Earth by an ellipsoidal body having axes of three unequal lengths is by no means new. There was, however, no continuing interest in the idea until Helmert and his assistant, Berroth, each published in 1915 his calculations on which were based gravity formulae containing a longitude term of the type given above. Since 1915, various discussions of the accumulating amount of data on gravity have tended to confirm the belief that the geoid is closely approximated by an ellipsoid having a size and character like that proposed by Helmert. Studies of the deflection of the vertical have likewise tended to confirm that belief. But such a surface is not likely to be used as a reference surface for geodetic networks, maps or deflections of the vertical. The mathematical complications are two great. Also, it is convenient to use the same reference surface for both gravity and deflections of the vertical and therefore to omit the longitude term from the gravity formula. Besides Helmert's formula, a number of other gravity formulae have contained a longitude term. The values are given, in the table below, for the constants in some of these formulae.

		τ_e	β_1	β_2	β_3	λ_o
Heiskanen	(1938)	978.052	5297	5.9	29	25.2°
Zhongolovich	(1952)	978.057	5268	5.9	16	6°
Heiskanen & Uotila	(1967)	978.052	5291	5.9	11	6°

(The values of β_1, β_2 and β_3 were multiplied by 10^6.) A significant variation of gravity with longitude has been verified by analysis of the orbits of artificial satellites. Such analysis also shows that gravity varies significantly with the third power of the sine of latitude. When such variations must be taken into account, a general gravity formula is more useful than a standard (normal) gravity formula.

look angle - (1) The direction in which an antenna is pointed when transmitting to or receiving from a particular element of area. (2) The direction in which an observer must look or an instrument be pointed in order to see something. The term is frequently used in special ephemerides to denote the direction in which an artificial satellite will be observable from a particular spot.

loop - A pattern of measurements in the field, such that the final measurement is made at the same place as the first measurement.

loop bar - A bar formed by the junction of the ends of two spits on the mainland side of an offshore island being eroded by waves.

loop closure - SEE misclosure (of a loop).

loop error of closure - SEE misclosure (of a loop).

loop misclosure - SEE misclosure (of a loop).

loop traverse - SEE traverse, closed.

Lorac - A system of four radio transmitters operating at four different frequencies between 1.650 and 1.700 MHz, forming two families of hyperbolas (iso-phase lines) along each of which the difference of phase of the signals from a pair of stations is a constant. A receiver's location can be determined by measuring the differences of phase in signals from the two pairs of stations and plotting the location at the intersection of a pair of hyperbolas. A plotting accuracy of 1 meter is attainable. The accuracy with which a location can be determined depends on many factors such as the location of the receiver within the pattern of iso-phase lines.

Loran (LOng RAnge Navigation) - A hyperbolic navigation-system for measuring differences between distances from a radio receiver to three or more fixed transmitters of known geographic location. One type of Loran operates in a narrow band about 2 MHz; another in a narrow band about 100 kHz. The receiver's location is found to lie on a particular hyperbola-like curve on the ellipsoid by measuring the difference in times of arrival of pulses sent synchronously from a single pair of transmitters. This curve is one of the two loci of all points whose distances from the two stations differ by the linear equivalent of the time-interval measured. By measuring the time-interval using a different pair of stations, a second hyperbola-like curve is determined. The receiver is therefore located at the intersection of these two curves. Note that each difference actually determines a pair of hyperbolas (i.e., a complete hyperbola of two branches.) However, one branch is usually too far distant from the approximately known location of the receiver to need consideration. Furthermore, it is usually on land.

Loran, standard - SEE Loran-A.

Loran-A - A hyperbolic navigation-system of the Loran type, operating at about 1.8 to 2 MHz (carrier) and emitting 40 or 50-μs pulses. The receiver matches the envelopes, not the phases, of the received pulses. Also called standard Loran. Loran-A is not as accurate as other navigating systems which compare phase, and such systems (e.g., OMEGA and Loran-C) have made Loran-A obsolete. At a distance of 400 km from the master station, a good intersection would have an accuracy of about 2 km. Loran-A is useful out to a few hundred kilometers from the stations, although locations may be obtained at a distance of 1400 km (750 nautical miles) in the daytime and 2500 km (1400 nautical miles at night). Accuracy is very nearly independent of weather. The precision of a location is comparable to that obtained by high-grade celestial observations with a sextant. Loran-A has been used for navigation at great distances from land and also for hydrographic surveying when more precise method were not available. It is planned to discontinue Loran-A stations after Loran-C becomes operational in the regions covered by Loran-A.

Loran-B - An unsuccessful version of Loran-A, in which both envelope and phase of the pulses were matched.

Loran-C - A hyperbolic navigation-system of the Loran type operating at 100 kHz. The system is similar to Loran-A but has greater range and higher accuracy. It matches the phase of the signals instead of their envelopes, uses a different coding of the signals, incorporates a cesium clock to give more accurate control of the timing, and puts more power into the signals, which are in Golay code. Locations can be determined at distances up to 1 500 km over land and 3 000 km over water, but less accuracy is obtainable at greater distances. At a distance of 600 km from the master station, a good intersection will have an accuracy of about 150 meters. At 2 000 km, over water, an accuracy of about 500 meters may be obtained. However, an error of one cycle in determining phase would produce an error of about 8 km. Predictable accuracy is better than 400 meters within the region covered, and the precision is better than 18-90 meters.

Loran chain - (1) A combination of three or more Loran-A stations forming two or more pairs of stations for navigation. (2) A combination of a master station and two or more slave stations, all of the Loran-C type, forming two or more pairs of stations for navigation.

Loran chart - A chart on which are shown the ground-wave lines of position and sky-wave corrections for Loran.

Loran-D - A modification of Loran-C in which the baseline is considerably shortened.

Loran line - A line along which the difference in time of arrival between Loran signals from a master station and a slave station is constant.

Loran station - A base station (3) which is an operating part of Loran.

Lorgna's map projection - SEE map projection, azimuthal equal-area.

LOROP photography - SEE photography, long-range oblique.

lorhumb line - A line along which the rates of change of the values of two intersecting families of hyperbolae are constant.

lot - (1) A subdivision of a section of land not described as an aliquot part of the section but designated by number, e.g., LOT 2. (2) A parcel of land, generally a subdivision of a city, town or village, represented and identified by a recorded plat. (3) An individual parcel of land intended to be conveyed, in its entirety, to a prospective buyer; also, that same parcel after it has been purchased.

lot, common - (1) Land held in common, as by a community. The term is now often used to mean a park, as Boston Common. (2) The right (arising either from a grant or contract, or from prescription or operation of a statue) of taking a profit in the land of another, in common either with the owner or with other persons.

lot line - (1) That line shown on the map as creating the lot. A lot line is permanent and does not change when openings are made to the street. (2) A line marking or otherwise indicating the boundary of a lot. SEE ALSO zero lot line. Note: Not always the same as the ownership line or boundary.

lough - (1) A region of shallow water, usually not more that 5 feet deep, more or less, surrounded by land. (Australian) (2) SEE loch (1). (3) SEE loch (2).

louvre shutter - A shutter consisting of a number of thin, parallel, and overlapping metallic strips pivoting simultaneously and individually, like the slats of a venetian blind, about their longitudinal axes to form either an opaque metallic screen or a group of slots separated by very thin strips of metal. A louvre shutter is usually located just in front of or just behind the lens system.

Lovar - An alloy having a coefficient of expansion between that of steel and that of Invar but costing considerably less than Invar.

Lovar tape - A surveyor's tape made of the alloy lovar. Lovar is less expensive than invar but has a larger coefficient of thermal expansion than invar, but less than steal.

Love number - (1) One of the two numbers designated by h and k, where h is the ratio of height of the Earth tide to height of the corresponding, static oceanic tide and k is the ratio of the additional, gravitational potential produced by an Earth tide to the gravitational potential producing the Earth tide. The numbers are defined for points on the Earth's surface. The numbers were introduced by Love in 1909. h is approximately 0.6; k is approximately 0.3. (2) One of the four numbers designated by h, k, l and f, where h and k have the same meaning as in the previous definition, l is the Shide number, and f is the ratio between the cubical expansion and the height of the corresponding static tide.

Love wave - (1) A transverse wave propagated along the boundary between two elastic layers, both of which have elasticity. (2) A seismic wave on the Earth's surface, in which the particles vibrate transversely in a horizontal plane.

Lowry's map projection - SEE map projection, perspective.

loxodrome - SEE rhumb line.

loxodrome, fictitious - SEE rhumb line, fictitious.

loxodrome curve - SEE rhumb line.

loxodrome spiral - SEE rhumb line.

lubber line - A pair of marks or a line, on the housing of a compass, indicating the fore and aft directions on the vehicle (e.g., ship) on which the compass is mounted. Also called a lubber's line and lubber's point. The angle between the lubber line and north shown by the compass indicates approximately the direction in which the vehicle is pointed.

lubber's line - SEE lubber line.

Ludovici system (of units) - A system of units in which the fundamental quantities are: the constant of universal gravitation, the charge on the electron, the magnetic constant and the electric constant. Length, mass and time are derived quantities.

lumen - The luminous flux through 1 steradian from a uniform point of one candela.

luminance - The amount of luminous energy per unit time moving, at a given point, in a given direction per unit solid angle and per unit projected area perpendicular to the specified direction. Formerly called luminosity, and often still so called by astronomers. The corresponding term for electromagnetic radiation in general is radiance.

lunar-culmination method (of determining longitude) - The right ascension of the Moon is determined at a place of known astronomic longitude and at the place whose astronomic longitude is wanted, at the time of the Moon's culmination (the passage of the Moon's center through the celestial meridian) by comparing the Moon's coordinates with the coordinates of several stars nearby. The difference in longitude is then the difference in right ascension divided by the Moon's hourly rate of motion in right ascension.

lunar equation - A factor used to reduce observations of celestial bodies to the barycenter of the Earth-Moon system.

lunar methods of determining longitude - SEE lunar culmination method; lunar-observation method.

lunar-observation method (of determining longitude) - Any method in which the Moon's right ascension is determined at two points, one of which is at a known astronomic longitude and the other is at the longitude to be determined. The difference in longitude is determined from the difference in right ascension and the Moon's hourly rate of motion in right ascension. The Moon may also be observed at equal angular elevations, or at any angular elevation. Longitude may be determined by measuring the angle between the Moon and the Sun, or between the Moon and the zodiacal stars. Occultations may be used for determining longitudes; these usually give information about latitude also.

lunation - SEE month, synodical.

lune - A part of a spherical surface, bounded by halves of great circles and including the bounding semicircles.

Luneburg bipolar coordinate system - A coordinate system, in space, consisting of a reference plane and two points in that plane. The bipolar coordinates of the point of interest are the dihedral angle between the reference plane and the plane through the three points, and the two angles between the line connecting the two points of reference and the two lines from those points to the point of interest.

Luneburg lens - A spherical lens in which the index of refraction n varies with the distance from the center according to the formula $n = rm-1/(1 + r2m)$, in which m is some suitable constant equal to or greater than 1. When m is 1, the lens is a spherical lens. The general Luneburg lens has been used principally in microwave antennas for radar. SEE ALSO lens, fish-eye.

lunicentric (adj.) - SEE selenocentric.

lux - A unit of illuminance, equal to 1 lumen per square meter.

Luzon Datum - The horizontal-control datum in the Philippine Islands which is defined by the following coordinates of the origin and by the azimuth (clockwise from South) on the Clarke spheroid of 1866; the origin is at triangulation station Balanacan:

longitude of origin	121° 52' 03.00" E.
latitude of origin	13° 33' 41.00" N.
azimuth from origin to station Balanacan	9° 12' 37.00"

Luzon Datum was adopted in 1911 and was derived from observations on Luzon. It has been extended to all parts of the Philippine Islands except a few that are remotely situated. Note that the longitude and latitude are defined quantities.

M

MacCullagh's formula - The approximation $V \approx -(GM/r)[1 + (A + B + C - 3I)/(2Mr^2)]$ to the gravitational potential V at a distance r from the center of mass of a body whose mass is M, whose moments of inertia about the principal axes are A, B and C, and whose moment of inertia about the line from the center of mass to the point in question is I. G is the gravitational constant.

machine, calculating - SEE calculator (1).

machine, computing - Any machine, mechanical or electrical, that carries out mathematical calculations according to a predetermined set of algorithms. Also commonly called a computer when such use is not likely to be ambiguous.

machine, forming - The equipment for making relief maps of plastic by applying heat and vacuum to plastic placed over a mold representing the terrain of the region mapped. Also called a map-forming machine.

machine, map-forming - SEE machine, forming.

machine, sounding - SEE sounding machine.

machine coordinate - One of the two coordinates of a point, on a stereoscopic pair of images, as determined by measuring with a stereoscopic plotting instrument.

machine language - That code with which a computer works directly. Machine language is usually numerical in nature.

MacLaurin ellipsoid - SEE MacLaurin spheroid.

Maclaurin spheroid - One of a series of stable, rotational ellipsoids formed by the surface of a homogeneous, rotating liquid body. Another series consists of heteraxial ellipsoids (Jacobi ellipsoids). The two series merge when the product of density by rate of rotation assumes a certain value; the Maclaurin spheroids are stable at higher values of the product, the Jacobi ellipsoids at lower values. Also called, erroneously, a MacLaurin ellipsoid.

magazine - A container for photographic plates or rolled photographic film, attached to the body of a camera in such a way that plates or film can be moved directly from the magazine into the camera's chamber and into the focal plane of the camera. Small magazines are also called cassettes or, if they are cylindrical, cartridges. Magazines containing photographic plates are also called plate holders. Magazines attached to aerial cameras have a mechanism for automatically advancing the photographic material between exposures.

magnet - SEE magnetism.

magnetic anomaly, local - Also called local attraction.

magnetism - (1) That quality of a body and the region surrounding that body which causes other bodies made of ferromagnetic material to be pulled towards that body, or which causes electrical currents to be generated in bodies moving through the field, or which causes charged particles moving in the field to be deflected from a straight line. The attracting body is called a magnet. The surrounding region is called the "magnetic field" of the magnet. The same theory can explain all three phenomena as different aspects of the same underlying cause. Just as a current is induced in a conductor moving through a magnetic field, so a current moving in a conductor creates a magnetic field. The magnetic field of a magnet can therefore be attributed to the motion of the orbital electrons within the magnet. (2) The ability to attract ferromagnetic material such as iron and steel. SEE ALSO field, magnetic.

magnetism, blue - The magnetism displayed by the south-seeking end of a freely suspended magnet. This is the magnetism of the Earth's northern magnetic pole.

magnetism, red - The magnetism displayed by the south-seeking end of a freely suspended magnet. This is the magnetism of the Earth's southern magnetic pole.

magnetism, regional - That observed part of the magnetic field which is attributed to the Earth's magnetic field or to effects which are too deep, too broad, or too high to be possible expressions of geologic structure or other features of interest. Better called a regional magnetic field.

magnetism, terrestrial - SEE geomagnetism.

magnetometer - An instrument used for determining the intensity of a magnetic field or a component thereof. In particular, in geomagnetism, an instrument used for measuring the intensity of the Earth's magnetic field. The types in common use today are: the dip circle and field balance (mechanical); fluxgate (saturable-core reactor); proton-precession magnetometer; and instruments measuring the frequency of atomic or molecular resonance in a magnetic field. SEE ALSO Gauss magnetometer; sine magnetometer.

magnetometer, absolute - A magnetometer which measures the intensity of a magnetic field without reference to measurements by other magnetometers.

magnetometer, flux-gate - A magnetometer determining the intensity of the Earth's magnetic field by measuring the field's effect on a saturable-core reactor or on a pair of oppositely wound, saturable-core reactors. The reactor is driven through saturation by a low-frequency (about 1000 Hz) oscillator and the second harmonic or the sum of all even harmonics of the generated EMF of the reactor is measured. Such a magnetometer has a sensitivity of about 10^{-5} oersted. Fluxgate magnetometers are simple, inexpensive, and sturdy and have been much used in aeromagnetic surveys.

magnetometer, nuclear - (1) A magnetometer determining the intensity of a magnetic field by measuring the effect of that field on some property of atomic nuclei. Typical nuclear magnetometers are the proton-precession magnetometer and the spin-precession magnetometer (also called an Overhauser magnetometer). These magnetometers are sensitive to 10^{-6} oersted or less. (2) SEE magnetometer, proton-precession.

magnetometer, nuclear-precession - SEE magnetometer, proton-precession.

magnetometer, optical-precession - A magnetometer determining the intensity of a magnetic field by measuring the frequency (Larmor precession frequency) with which atomic electrons in the field precess. The electrons in an atom move in orbits about the nucleus. As moving charges, they experience a force, in a magnetic field, proportional to the intensity of the field and in a direction perpendicular to the plane of the orbit. The orbit therefore precesses about the nucleus at a rate proportional to the intensity of the field. The frequency is measured by passing light of the proper frequency through a gas and measuring the frequency (equal to the Larmor frequency) at which the light is strongly absorbed when modulated by a weak magnetic field applied to the gas.

magnetometer, optical-pumping - A magnetometer using a gas (helium or alkali metal) to absorb energy from a beam of infrared radiation. The absorption is a maximum when an FM oscillator is tuned to the resonant frequency of the gas. The resonant frequency is proportional to the intensity of the Earth's magnetic field at the point of measurement.

magnetometer, proton-precession - A magnetometer determining the intensity of a magnetic field by measuring the effect of the field on the spin of hydrogen nuclei (protons). Also called a proton magnetometer or nuclear-precession magnetometer, although the latter term properly applies also to magnetometers other than the proton-precession magnetometer. A proton is a spinning, positive charge. As a moving charge, it is acted on by a magnetic field with a force proportional to the intensity of the field. An ambient magnetic field therefore forces the proton's axis of spin to precess. The rate of precession is proportional to the strength of the impressed field. When the impressed field is removed, the electromotive force generated by the precession can be detected and its period measured. Also called a proton magnetometer.

magnetometer, semiconducting quantum-interference - SEE SQUID.

magnetometer survey - A survey in which the Earth's magnetic field or a portion thereof is measured using a magnetometer.

magnetopause - The outermost portion of the Earth's ionosphere, at a distance of 8 to 14 Earth-radii on the Sunward side, where the Earth's magnetic field ends and the Earth's ionosphere merges into the Sun's.

magnification - (1) A ratio between the distance between two points on an image and the distance, projected orthogonally onto the plane of the image, between the corresponding points on the object. (2) A ratio between the angular size of an image as seen by the unaided eye and the angular size as seen with an optical instrument. If the distance or size is expressed in linear units such as millimeters, the ratio is called linear magnification. If it is expressed in angular units such as radians, it is called angular magnification.

magnification, angular - SEE magnification (2).

magnification, lateral - Magnification in a direction transverse to the optical axis.

magnification, linear - SEE magnification (1).

magnification, longitudinal - Magnification in the direction of the optical axis.

magnification, unidimensional - Transformation of one rectangle into another of different proportions.

magnitude - (1) The positive square-root of the sum of the squares of the components of a vector. (2) SEE magnitude, stellar.

magnitude, absolute - The apparent brightness a celestial object would have, relative to a standard, if the object were at a distance of 10 parsecs. Designating the absolute magnitude by M, the apparent magnitude by m, and the distance, in parsecs, by d, the formula for absolute magnitude is $M = m + 5 - 5 \log d$. The absolute brightness of the Sun is +4.87; its apparent magnitude is -26.7.

magnitude, apparent - The number m characterizing a celestial object of brightness b_m, according to the formula $m = \text{constant} - 2.5 \log b_m$, in which the value of the constant depends on the assigned magnitude n and apparent brightness b_n of another object taken as standard: $\text{constant} = n - 2.5 \log b_n$, so that $m - n = -2.5 \log (b_m - b_n)$. This defines the scale of apparent magnitudes except for the location of the zero point. The zero point of the scale of visual magnitudes is, by convention, found by assigning the magnitude +1 to the average brightness of the two stars Aldebaran and Altair. On this scale, the Sun has an apparent magnitude of -26.7. A star of first (visual) magnitude provides illumination of 2.43 lumens per square meter or 3.1×10^{-6} ergs/cm²-sec. A star of sixth magnitude (just barely visible) sends 3000 photos per second into the eye (an aperture of 5.5 mm). A star of 21st magnitude (near the limit of telescopic visibility) sends approximately 0.01 photos per second into an area of 1 cm².

magnitude, bolometric - The magnitude of an object calculated from the object's total radiation over the entire spectrum. The term originated from the practice of using bolometers for the measurements.

magnitude, photoelectric - The magnitude, absolute or apparent, of an object, as determined by measuring the radiation with a photoelectric photometer. The photoelectric photometers most commonly used are much more sensitive to the blue end of the spectrum than the red end, so that the photoelectric magnitudes of red objects are larger than their visual magnitudes, while the photoelectric magnitudes of blue objects are less than their visual magnitudes.

magnitude, photographic - The magnitude, absolute or apparent, of an object determined from measurements made on a photograph of the object. Because photographic material is sensitive to different wavelengths than the human eye is, photographic magnitudes are different from visual magnitudes. The difference (the color index) between the two is a rough indication of the color of the object.

magnitude, photovisual - The magnitude of an object, determined by measuring the photographic densities on a photograph taken using emulsions filters with combined sensitivities equivalent to the sensitivity of the human eye. Hence, photovisual magnitudes are practically equivalent to visual magnitudes.

magnitude, stellar - A measure of the brightness of a celestial object relative to the brightness of a standard, i.e., to a brightness of specified value. Usually referred to simply as magnitude if the meaning is clear. The concept was first introduced by Hipparchus in 150 B.C. He designated the 15 brightest stars as stars of the first magnitude and stars just visible to the naked eye as stars of the sixth magnitude. The idea was extended by Ptolemy (150 B.C.) and given its modern form by Pogson in 1856. To keep Hipparchus's scheme as intact as possible, John Herschel, about 1830, made a difference of 5 magnitudes equivalent to a ratio of exactly 100 in brightness. In other words, two stars differing in magnitude by 5 differ in brightness by a ratio of exactly 100 to 1. Any particular object will have several magnitudes, depending upon what portion of the object's spectrum is used for viewing. Unless specifically stated otherwise, the object is assumed to be viewed in the optical part (light) of the spectrum and apparent magnitude is ordinarily meant. If several parts of the spectrum are being considered, the term visual magnitude is used for the magnitude of the object as seen in light. (Note that although the term is stellar magnitude, it is applied also to non-stellar objects.)

magnitude, visual - (1) The astronomical magnitude of an object as judged by the human eye, but taking into account the sensitivity of the eye and the color of the object. (2) The magnitude of a celestial object as estimated by the eye.

magnitude correction, bolometric - SEE correction, bolometric.

magnitude equation - The nearly constant amount by which a particular observer's timing of the transit of faint stars is later than his timing of the passage of bright stars.

make line - An accurately scaled line indicating the size to which original material (copy) is to be enlarged or reduced. Also called make size.

makeready - All the preparations needed to be done to a lithographic plate on the press so that it will print correctly (correct registration and color). Also written make-ready. It may also include the associated preparation of the printing press.

make size - SEE make line.

makeup - The arrangement of components used in printing, to compose a page or other printed form.

Malmros prism - SEE Thompson prism.

manhole - A short, vertical shaft allowing access to an underground passage such as a sewer or tunnel carrying telephone or power lines. The opening at ground level is usually covered by a cast iron or steel plate.

manometer - A gage for measuring the pressure of a gas. A common form is a U-shaped tube containing mercury and closed at one end only; the other end is open and exposed to the gas whose pressure is to be measured. Another form is a U-shaped tube open at both ends; one end is connected to a gas at a known pressure and the other is connected to the gas whose pressure is to be measured.

mantle - That portion of the lithosphere extending downward from the bottom of the crust (the Mohorovicic discontinuity) to about 2 900 km. At the upper surface of the mantle, the speed of longitudinal seismic waves jumps from about 7 km/s to over 8 km/s; at the lower boundary, only longitudinal seismic waves can travel on down into the core. Close to the upper boundary of the mantle, about 100 - 200 km below the Earth's surface, is a zone in which seismic waves travel at speeds lower than those occurring at greater depths. This zone in the upper mantle has been called the asthenosphere, rheosphere and upper mantle. However, the term asthenosphere has also been applied to the mantle as a whole.

manuscript - The original drawing of a map as compiled or constructed from various data such as ground surveys and photographs. SEE ALSO compilation manuscript.

manuscript map - The original drawing of a map as compiled or constructed from various data such as ground surveys or photographs.

map - (1) A conventional representation, usually on a plane surface and at an established scale, of the physical features (natural, artificial, or both) of a part or whole of the Earth's surface by means of signs and symbols and with the means of orientation indicated. (2) A conventional representation, usually on a plane surface and at an established scale, of the physical features (natural, artificial or both) of a part or whole of the celestial sphere by means of signs and symbols and with the means of orientation indicated. Because of the widespread use of such maps for navigation, a map of the celestial sphere is usually referred to as a star chart. A map may emphasize, generalize, or omit the representation of certain features to satisfy specific requirements. The type of information a map is designed primarily to convey is frequently indicated by an adjective, to distinguish it from maps of other types. The map projection used for the map should always be indicated. The oldest known map is a Babylonian map in baked clay. The oldest known printed map is a woodcut, appearing in a Chinese encyclopedia of about 1155 A.D., depicting western China. (3) SEE plan. (4) SEE function. (5) SEE plat. SEE ALSO base map; butterfly map; contour map; distribution map; dot map; engineering map; flight map; Goode's homalographic map; index map; inset map; landscape map; outline map; plastic relief map; radar map; reconnaissance map; relief map; sketch map; State map.

map, administrative - A map having on it graphical information regarding administrative matters such as supply, medical facilities or boundaries.

map, aerial - SEE chart, aeronautical.

map, aerial-navigation - SEE chart, aeronautical.

map, anaglyphic - A map printed in two complementary colors (such as red and blue-green) in such a way that, when viewed through spectacles fitted with filters of corresponding colors, the map appears to have depth, i.e., shows the relief in three dimensions.

map, arbitrary - A map made by a title company, assessor, or other person or organization for its own convenience in locating property in a region in which all the descriptions are by metes and bounds. On the map, the individual pieces of land (lots) are given arbitrary numbers. The deeds and other documents affecting these pieces of land are placed in what is called an arbitrary account.

map, assessment - SEE assessment map.

map, bathymetric - A map showing the contours of the bottom of a body of water. It is a topographic map of underwater regions. It differs from a bathymetric chart in containing more topographic information and little or no navigational information.

map, cadastral - A map showing the boundaries of subdivisions of land, usually with the bearings and lengths thereof and the areas of individual tracts, for purposes of describing and recording ownership. A cadastral map may also show culture, drainage, and other features affecting the value and use of the land. Also called a property map.

map, chorographic - (1) Any map representing large regions, countries, or continents on a small scale. The term has been almost entirely replaced by small-scale map. Atlases and wall maps are chorographic maps. (2) A map at a larger scale than those used for maps of the entire world but at a smaller scale than those used for metropolitan regions, counties, etc. Also called a chorographic-scale map or, more commonly, an intermediate-scale map.

map, chorographic-scale - SEE map, chorographic (2).

map, choropleth - SEE map, choroplethic.

map, choroplethic - A map on which the frequency with which a certain quantity occurs is shown by patches of constant density or color, each patch covering a region within which the quantity has a particular, constant value or has a particular, average value, or is consistently greater than or less than some particular, constant value. Also called a choropleth map. A topographic map on which elevations are indicated by hypsometric tints is a choroplethic map. Each patch of a particular color indicates that in the corresponding region of the Earth, elevations lie within corresponding, particular limits. Density of population, rate of heat flow through the crust, etc., are also shown as choroplethic maps. In general, any frequency that can be shown on a choroplethic map can be shown on an isoplethic map.

map, compiled - Any map containing information collected from various sources other than surveys made for the particular purpose of constructing the map. Most small-scale maps of large areas are compiled maps.

map, composite - A map containing information of two or more general types. A composite map is usually a compiled map, bringing together for comparison data originally shown on separate maps. For example, a map showing air-routes and roads would be a composite map, particularly if the air-routes and roads had originally been shown on separate maps.

map, controlled - A map based on precise horizontal and vertical ground-control.

map, dasymetric - A map in which color or shading is applied to regions which are homogeneous, within specified limits, and in which the color or shading need not be restricted by statistical or administrative boundaries.

map, demographic - A map showing primarily political or social data such as political divisions, populations or occupations.

map, digital - A set of numbers which completely represent a map and which can be used, without further data, to prepare a complete map. In particular, a set of numbers usable by a computer-controlled plotter for producing a map.

map, domestic - A map of a region within the limits of the USA. The term applies, more generally, to a map of a region within the limits of that country whose maps are under discussion.

map, dynamic - (1) A map showing events involving motion, action, or change. A dynamic map involves time. The term may be applied to maps depicting the flow of traffic, migration, military movements, progress in an engineering project, historical geography and so on. Various symbols, such as flow lines and arrows, are used to show movement. A dynamic map may also be composed of two or more static maps, showing comparable data at stated but different times or dates. (2) A map on which changes take place. A map showing continental drift by having the images of continents actually move about would be a dynamic map. A dynamic map may be created by photographing a sequence of maps, each map showing a different stage in the sequence of events, and then displaying the photographs as a moving picture; or by creating a digital map, displaying the re-created map on the screen of a cathode ray tube, and using a computer to make the necessary changes from instant to instant.

map, equal-area - A map such that the area within each closed curve in the original figure is the same as the area within the corresponding closed curve on the map. This does not imply that each closed curve in the original figure preserves its shape also when mapped. In fact, the shape will usually be considerably distorted in the map.

map, fire-control - A map made specifically for planning and use in controlling the firing of artillery. Also called an artillery map.

map, flood-control - A map prepared for studying and planning the control of floods in regions subject to floods.

map, fluorescent - A map printed using fluorescent inks or on fluorescent paper. The map can be read in darkness under ultraviolet radiation.

map, general - A map showing a variety of geographical features (e.g., coastlines, political boundaries, transportation lines) and used for planning, location, reference, etc. A general map is contrasted with a thematic map and with a special-purpose map.

map, general-purpose - A map that provides a large variety of information and satisfies the needs of many different kinds of users.

map, geoidal - A map showing, by contour lines, the height of the geoid above or below a specified ellipsoid of reference.

map, geological - A map showing the structure and mineralogical composition of the Earth's crust.

map, geomorphic - SEE map, geomorphological.

map, geomorphologic - SEE map, geomorphological.

map, geomorphological - A map showing, by appropriate symbols, the appearance, dimensions, slopes and ages of features in the region mapped. Also called a geomorphologic map and a geomorphic map. Topographic maps form the basis for geomorphological maps. A considerable amount of information on the geological interpretation of the topography and on the structural geology of the region must be added to the topographic map.

map, gravimetric - A map on which isolines represent lines on the ground or other surface along which the gravitational acceleration is constant.

map, gravity-anomaly - (1) A map showing the locations and magnitudes of gravity anomalies. (2) A map on which contour lines represents lines on the Earth along which gravity anomalies are constant.

map, hemispherical - A map of one-half of the Earth's surface (or of one-half of a sphere) and bounded by the Equator or by another great circle. The Earth is usually considered to be divided either at the Equator, into the Northern and Southern Hemispheres, or along some meridian (continued around the globe) between Europe and America, into Eastern and Western Hemispheres. Cartographers usually divide the Earth along the meridians of longitude passing through 20° West and 160° East. The Americas are considered to be in the Western Hemisphere.

map, homalosine - SEE map, homolosine.

map, homolographic - SEE Goode's homalographic map.

map, homolosine - The interrupted map obtained by using five or six different central meridians (the number depends on whether land or oceanic regions are to be emphasized) for mapping the ellipsoid onto the plane in such a way that the maps match along the equator and the center of each map is on a separate meridian. The sinusoidal map projection is used for the zone between 40° N and 40° S, approximately; Mollweide's map projection is used for the regions outside this zone. Also called a homalosine map and homolosine map projection. A homolosine map is similar to Goode's homalographic map except that it uses the sinusoidal map projection for the zone between about 40° N and 40° S. The map is widely used in atlases and for thematic maps because the interruptions can be made to occur in regions of little interest.

map, hydrographic - SEE chart, nautical.

map, hypsographic - A topographic map on which elevations are shown. Sometimes called, erroneously, a hypsometric map. Called a hypsographic chart when it has the features of a chart.

map, hypsometric - A map showing relief by any convention, such as by tints, hachures, contours or shading. Also called a relief map. Hypsographic maps and topographic maps are special kinds of hypsometric maps. It is bad practice to use hypsometric as a synonym for hypsographic.

map, intermediate-scale - A map at a scale smaller than that commonly used for maps of counties or metropolitan regions but larger than that used for maps of continents or the world. The term may be considered roughly equivalent to medium-scale map. A map at a scale between 1:50 000 and 1:500 000 would usually be considered an intermediate-scale map.

map, interrupted - A map produced by fitting together maps of different regions not all mapped by the same kind of map projection, or not all mapped using the same constant in a particular kind of map projection. Also called a recentered map projection and interrupted map projection. For example, the homolosine map projection is produced by using Mollweide's map projection for one portion of the map and the sinusoidal map projection for other portions. The gnomonic map on a cube is produced using only the gnomonic map projection but using six different locations of the tangent points. Interrupted maps were produced as early as 1507 by Waldseemüller and by Rumold Mercator.

map, isarithmic - SEE map, isoplethic.

map, isogonic - A map showing lines along which magnetic declination was constant on a particular date. Lines of equal annual change in declination are generally shown also. The date for which the map is valid must be shown on the map. If the map is intended for use in navigation, it is termed an isogonic chart.

map, isometric - A map on which lines are drawn true to scale. It is not possible to draw an isometric map of features on a sphere or any part thereof. But if the scale of the map is large enough and the area of the depicted part is small enough, error in scale can be made insignificant.

map, isoplethic - A map carrying lines joining points having the same numerical values. Also called an isarithmic map. The terms isoplethic and isarithmic appear to be synonyms, and there does not seem to be any good reason for using the second, less common term. Examples of isoplethic maps are: topographic maps having contour lines; population-density maps outlining regions along whose boundaries the density of population is constant; and thermal-flow maps on which lines join points at which the rate at which heat is flowing out is the same.

map, land-classification - A map showing the different types of soil, rock, etc. Each type is usually indicated by a characteristic shading or color.

map, large-scale - Commonly, a map at a scale of 1:50 000 or larger. Also called a topographic-scale map. However, the criterion for large-scale maps is by no means standard, e.g., the value 1:75 000 is often used as a criterion.

map, magnetic - SEE chart, magnetic.

map, marine - SEE chart, nautical.

map, mean - SEE chart, mean.

map, medium-scale - Commonly, a map at a scale between 1:50 000 and 1:500 000, inclusive. These criteria are by no means universally accepted. SEE ALSO map, intermediate-scale.

map, minimum-error - A map such that the sums of squares of errors in scale in mutually perpendicular directions and integrated throughout the map, is minimal for a given type of map projection.

map, military - A map designed particularly for military use.

map, morphographic - A small-scale map showing physiographic features by means of standardized, pictorial symbols based on the appearance such features would have if viewed obliquely from the air. This definition is very similar to that of physiographic pictorial map, and it is quite possible that the terms were originally intended to denote the same concept.

map, native - A map of a foreign region produced by the government or private agency of the foreign country.

map, nautical - SEE chart, nautical.

map, official - The map outlining present and future rights-of-way for streets or other public land, and adopted by the legislative body of a city or other governmental unit.

map, photocontrol index - SEE photocontrol index map.

map, photogrammetric - A topographic map produced photogrammetrically from aerial photographs and geodetic control.

map, photo-revised - A topographic or planimetric map that has been revised by photogrammetric methods.

map, physical - A map representing the surface of the land or the floor of the oceans. In addition to representing relief, physical maps show some natural phenomena such as oceanic currents, swamps, sands and deserts. Man-made structures or vegetation are not shown.

map, physiographic pictorial - A map with relief depicted by systematic application of a standardized set of conventional, pictorial symbols based on the simplified appearance of the physical features they represent, as viewed obliquely from the air at an angular elevation of about 45°. Compare with the definition of morphographic map.

map, planimetric - A map showing only the horizontal positions of the features represented. Also called a line map. Unlike a topographic map, it does not show relief in measurable form. The natural features usually shown on a planimetric map include rivers, lakes and seas; mountains, valleys and plains; forests, prairies, marshes and deserts. Man-made features include cities, farms, transportation routes and public-utility facilities; and political and private boundary lines.

map, planimetric-base - A planimetric map prepared photogrammetrically for later inclusion of contour lines drawn in the field.

map, plastic relief - SEE plastic relief map.

map, polygnomonic - An interrupted map made by mapping the sphere onto an inclosing polyhedron, using the gnomonic map projection for each face. Fisher's polygnomonic map is a mapping of the sphere onto a circumscribed icosahedron.

map, polyhedral - SEE map, polyhedric.

map, polyhedric - The map of an ellipsoid onto an inclosing polyhedron. Also called a polyhedral map or polyhedric map projection. Polyhedric maps were first produced systematically by the Kgl. Prüssische Topographische Bureau and is therefore sometimes called the Prussian polyhedral projection. It was intended to project the rotational ellipsoid orthogonally onto each face, but a conical map projection was adopted instead because this would allow adjacent sheets to be matched without gaps.

map, provisional - Any material used as a map in the absence of a map that was issued officially.

map, recentered - SEE map, interrupted.

map, right-of-way - A plan of a highway improvement showing its relationship to adjacent property, the parcels or portions thereof needed for the highway and other pertinent information.

map, shaded-relief - A map on which elevations and depressions are made to appear three-dimensional by the used of graded tones of shadow. Usually, the shadows are shaded as though the features depicted were illuminated from the northwest. A shaded-relief map may also combine contour lines or hachures with the shading.

map, small-scale - Commonly, a map at a scale equal to or smaller than 1:500 000. This criterion is by no means universally used.

map, special-purpose - Any map designed for a special purpose rather than for general use. Usually the information on a special-purpose map is emphasized by omitting or subordinating other information of a general character. A word or phrase, such as geological, road or traffic-flow, for example, is usually used to describe the type of information a special-purpose map is designed to present.

map, standard-accuracy - A map complying with the U.S. National Map Accuracy Standards.

map, standard-content - A map representing natural and artificial features according to current standards and specifications.

map, static - A map portraying information available at a single date or time. Most maps are static maps. Static maps presenting comparable information valid at different dates may be combined into a single map called a dynamic map.

map, strategic - A military map of medium or smaller scale, designed for strategic use.

map, tactical - A military map of large-scale, designed for tactical use.

map, tactile - A map on which certain symbols are represented by raised portions of the map so that they can be located and identified by touch. Also called a tactual map. Geographic names and other information are given in braille.

map, tactual - SEE map, tactile.

map, thematic - A map whose principal purpose is depicting information other than that about the Earth's physical surface. Also called a topical map. Among the kinds of information shown on thematic maps are density of population, kind or amount of crops grown, types of soil, amount of rainfall and administrative subdivisions.

map, topical - SEE map, thematic.

map, topographic - (1) A map showing the horizontal and vertical locations of the natural and man-made features represented. Also called a relief map. A topographic map is distinguished from a planimetric map by the presence, in the former, of symbols showing relief in measurable form. A topographic map usually shows the same features as a planimetric map but uses numbered contour lines or comparable symbols to show mountains, valleys and plains. In the case of hydrographic charts, it uses symbols and numbers to show depths in bodies of water. A topographic map differs from a hypsographic map in that, on the latter, vertical distances are shown with respect to the geoid, while on the former, vertical distances may be referred to any suitable and specified surface. (2) A map whose principal purpose is to portray and identify the natural and artificial features on the Earth's surface as faithfully as possible within the limitations imposed by scale.

map, traffic-circulation - A map showing routes and the measures used for regulating traffic on them. It indicates the roads to be used by certain classes of traffic, location of traffic-control stations and the directions in which traffic may move. Also called a circulation map.

map, trapezoidal - A map characterized by equidistant, straight lines representing parallels of latitude and convergent, straight lines representing meridians. Also called a trapezoidal map projection. The trapezoidal map is the earliest form of polyhedric map, having been used on star charts as early as 1426. It was used for maps of the world by Nicolaus Germanus in 1466.

map, uncontrolled - A map based at least in part on data from an original survey but based on a number of fixed points insufficient to maintain accuracy of scale and location consistently. This is not the same as a sketch map.

map, urban - (1) A map of an urban region. (2) SEE city map.

map accuracy - The accuracy with which a map represents the region it is intended to depict. Three types of error commonly occur on maps: errors of representation, which occur because conventional signs must be used to represent natural or man-made features such as forests, buildings and cities; errors of identification, which occur because a nonexistent feature is shown or is mis-identified; and errors of position, which occur when an object is shown in the wrong position. Errors of position are commonly classified into two types: errors of horizontal location and errors of elevation. A third type, often neglected, is errors of orientation. SEE ALSO Koppe's formula; U.S. National Map Accuracy Standards.

map accuracy, circular - The radius of that circle within which a specified percent of the errors in horizontal location occur on a map.

map accuracy, relative - (1) The accuracy with which details on a map are located with respect to each other, as compared to their correct relative locations, i.e., the relative locations of the corresponding details on the ground. (2) The accuracy with which details on a map are located with respect to the graticule on the map, regardless of any error in the graticule or the datum defining the graticule.

map accuracy, vertical - The accuracy or error with which the elevations shown on a map correspond to the actual elevations of the corresponding points on the ground.

map accuracy specifications - Specifications as to the accuracy that a specified map or set of maps must have. E.g., the U.S. National Map Accuracy Standards are specifications for the accuracy of maps prepared for agencies of the U.S. government, and also used by other organizations and cartographers.

map accuracy standards - SEE U.S. National Map Accuracy Standards.

map adjustment - Any change in the horizontal position of a map to make the map fit certain control points, or to make it fit a specific grid plotted on the graticule at the scale at which the map was compiled.

map angle - SEE gisement.

map chart - A map showing both land and sea, on which the symbols characteristic of a chart are used to represent the oceanic regions. Also called a combat chart.

map compilation - Compilation for producing a map.

map data - Specific cartographic information plotted in relation to base data.

map deformation - SEE map distortion.

map distortion - Alteration in the shape of a figure on the ellipsoid, caused by mapping the figure onto a plane. Also called map deformation.

map edit - SEE map editing.

map editing - The preparation of an additional drawing containing all new and additional lettering, data and symbols, deletion of obsolete information from previous such drawings, and review of the overall content and style of a map before printing. Also called map edit (jargon).

map face - That part of a sheet, on which a map is printed, which is inclosed by the neat line around the map.

map grid - A grid superposed on a map to provide a coordinate system more convenient than that provided by the graticule. The most common grid consists of two mutually perpendicular families of equidistantly-spaced lines. The spacing may be the same for both families (square map grid) or different (rectangular map grid). However, grids consisting of families of circles and intersecting straight lines (polar grids) are also common, while grids consisting of intersecting hyperbolae are commonly used with charts. SEE ALSO graticule.

Map Grid, Australian - SEE Australian Map Grid.

Map Grid System, Australian - SEE Australian Map Grid.

map legend - SEE legend.

map matching - The comparison of a map of a region with the appearance of that region as it is being observed by a scanning device aboard a craft above that region. The technique is used for guiding aircraft, rockets, spacecraft, ships and submarines (the two latter type of craft using sonic scanners).

map matching, stellar - SEE matching, star-chart; navigation by star-chart matching.

map nadir - That point on a map corresponding to the point on the ground vertically underneath the perspective center of the camera's lens system. Also called the nadir, although that term is ambiguous.

Map of the United States, Topographic - SEE Topographic Map of the United States.

Map of the World, International - SEE International Map of the World.

map parallel - The intersection of the plane of the photograph with the plane of the map. This is a particular case of the axis of homology (q.v.). SEE ALSO ground parallel.

mapping - (1) The process of designing or making maps. The term is used by some cartographers to mean the making of planimetric or topographic maps only. It is used by some geographers and engineers to mean the gathering of geographical data of any sort. Neither usage is widespread and neither is particularly suitable. Both lead to difficulties. (2) A function or functional relationship. The term, used in this sense, is popular because of its pictorial associations, but is not suitable as a general synonym for function. SEE function (1).

mapping, aerial - The making of planimetric or topographic maps from aerial photographs. Also called aerocartography.

mapping, choroplethic - Depicting a quantity, on a map, by means of choropleths.

mapping, coastal - (1) Mapping of a coastal region. (2) Mapping of either the landward portion alone of a coast; the landward portion together with that extending to the low-water line, non-floating aids to navigation and landmarks; or the topography of the seaward portion out to the greatest depth that can be determined from aerial photographs.

mapping, isarithmic - SEE mapping, isoplethic.

mapping, isometric - A function preserving distances between certain points. Also called a length-preserving mapping. Given certain points (i.e., the coordinates of certain points in a specific coordinate system), an isometric mapping on these points produces a new set of coordinates in another coordinate system, but the distances between the points in the new coordinate system are the same as the distances between these points in the first coordinate system. Note, however, that the word distances does not necessarily mean distances in straight lines.

mapping, isoplethic - Depicting a quantity by means of isopleths on a map. Also called isarithmic mapping and isopleth mapping.

mapping, length-preserving - SEE mapping, isometric.

mapping, photogrammetric - Mapping which uses photographs as the primary source of data and relies on photogrammetry to prepare maps from these data.

mapping camera - A camera designed particularly for taking photographs to be used in mapping. The camera contains mechanisms to indicate and maintain the internal orientation of the photographs accurately enough for mapping. The camera may be an aerial mapping camera or a terrestrial mapping camera. Also called a surveying camera.

mapping photography - Photography for purposes of mapping. Also called cartographic photography.

map point - A supplemental-control point having a horizontal location that can be obtained by measuring the coordinates from a map or chart on which the point can be identified.

map projection - (1) A pair of functions relating coordinates of points on a specified surface (usually a rotational ellipsoid) to coordinates of corresponding points on a plane. Equivalently, a relation between a coordinate system on a specified surface and a coordinate system on a plane. A map projection is therefore a pair of functions establishing a one-to-one correspondence (except possibly at a small number of exceptional points) between points on the specified surface and points on the plane. The specified surface is usually a sphere or a more general rotational ellipsoid. It is rarely a heteraxial ellipsoid and very rarely a more complicated surface. (2) Any systematic method of representing the whole or a part of the curved surface of the Earth upon another, usually flat, surface. (3) SEE graticule. SEE ALSO Adams' map projection; Adams' rhombic conformal map projection; Airy's map projection; Aitoff's map projection; Albers map projection; August's map projection; Behrmann's map projection; Bonne map projection; Breusing's map projection; Briesemeister's map projection; Cassini's map projection; Chebychev's map projection; Clarke's map projection; de Lisle's map projection; Eckert's map projection; Euler's map projection; Gall's map projection; Gall's stereographic map projection; Gauss-Krüger map projection; Gauss-Schreiber map projection; Guam map projection; Guyou's map projection; Hammer-Aitoff map projection; Hotine's rectified skew orthomorphic map projection; Laborde map projection; Lambert conformal conical map projection; Lambert's conical equal-area map projection; master map projection; Mercator map projection; Mercator map projection, decumenal; Mercator map projection, oblique; Mercator map projection, space-oblique; Mercator map projection, transverse; Mercator Map Projection, Universal Transverse; Miller's cylindrical map projection; Miller's modified oblique stereographic map projection; Mollweide's map projection; Nell's map projection; Nicolosi's map projection; Parent's map projection; plane perspective map projection; Plate Carrée map projection; Roussilhe's map projection; Schreiber's double map projection; Soldner map projection; Universal Polar Stereographic Map Projection; van der Grinten's map projection; Werner's map projection; Wiechel's map projection.

map projection, American polyconic - SEE map projection, polyconic.

map projection, annular equal-area - SEE map projection, azimuthal equal-area.

map projection, aphylactic - A map projection which is not a conformal, equal-area, or equidistant map projection. Also called an arbitrary map projection. The term aphylatic was introduced by Tissot in 1881.

map projection, approximately-equidistant - A map projection in which the particular scale in a direction radial from the point of zero distortion or perpendicular to the line of zero distortion remains constant throughout the map but is not equal to the principal scale. If such a map projection is to be useful, the departure of this particular scale from unity should be small.

map projection, arbitrary - SEE map projection, aphylactic.

map projection, authalic - SEE map projection, equal-area.

map projection, autogonal - SEE map projection, conformal. This term is seldom used.

map projection, azimuthal - A map projection producing a graticule on which the azimuths or directions of all lines radiating from a central point or pole are the same as the azimuths or directions of the corresponding lines on the ellipsoid. Also called a zenithal map projection or central map projection. As a class, azimuthal map projections include a number of special, named map-projections; these are described under their individual names.

map projection, azimuthal equal-area - (1) An azimuthal map projection mapping a hemisphere onto a plane in such a manner that the directions of points from a particular point in the plane are the same as the directions between corresponding points on the hemisphere, and such that there is a constant ratio between the area of a region on the plane and the area of the corresponding region on the hemisphere. There are three important, special cases. (a) The center of the graticule corresponds to one of the poles of the sphere. The equations are $r = 2 k R \sin \frac{1}{4}(\pi - 2\phi)$ and $\theta = \Delta\lambda$ in which r and θ are polar coordinates of a point in the plane; $\lambda_o + \Delta\lambda$ and ϕ are longitude and latitude, respectively, of the corresponding point on the sphere; λ_o is the longitude of the central meridian; R is the radius of the sphere; and k is a scale factor. This version is also called Lambert's central equivalent map projection (m.p), Lambert's equal-area m.p.; the polar azimuthal equal-area m.p.; polar Lambert azimuthal equal-area m.p.; Lambert azimuthal polar equal-area m.p.; Lorgna's m.p., etc. (b) The center of the graticule corresponds to a point on the equator. The equations are $x = 2 k R \sin A_z \sin \frac{1}{2} *\theta$ and $y = 2 k R \cos A_z \sin \frac{1}{2} *\theta$. A_z is the azimuth from the selected point on the equator to another point; *θ is the angle between the two points; and x and y are rectangular Cartesian coordinates. This map projection has also been called the meridional azimuthal equal-area m.p., transverse azimuthal equal-area m.p.; Lambert azimuthal meridional equal-area m.p, Lambert equal-area meridional m.p. "Lambert central equivalent m.p. upon the plane of the meridian, etc. (c) The oblique case, in which the center of the graticule is neither at a pole nor on the equator. The equations are $x = k R (\cos \phi_1 \sin \phi - \sin \phi_1 \cos \phi \cos \Delta\lambda) / D$ and $y = k R (\cos \phi \sin \Delta\lambda) / D$ and $D = \sqrt{[(1 + \sin\phi_1 \sin \phi + \cos \phi_1 \cos \phi \cos \Delta\lambda) / 2]}$, the symbols having the same meanings as above. (d) Also called the Lambert zenithal equal-area m.p. (2) A map projection mapping a hemisphere onto a plane in such a manner that the directions of points with respect to a particular circle in the plane are the same as the directions with respect to the similar circle on the sphere, and such that there is a constant ratio between the area of a region on the plane and the area of the corresponding region on the hemisphere. Also called the annular equivalent projection (Driencourt and Laborde 1932). This definition is the same as the preceding one, except that the word point there is replaced by the word circle here. This version is not strictly azimuthal but it is equal-area.

map projection, azimuthal equidistant - An azimuthal map projection from the rotational ellipsoid to the plane; it is given, for the special case of the sphere and the polar aspect, by the equations $r = k (½π ± \phi)$ and $θ = \Delta\lambda$, in which r and θ are the polar coordinates of that point, in the plane, which corresponds to a point, on the sphere, at longitude $\lambda_o + \Delta\lambda$ and latitude ϕ; λ_o is the longitude of the central meridian. k is the scale factor. Also called Postel's map projection. This map projection is neither equal-area nor conformal. Straight lines radiating from the pole or center of projection represent great circles in their true azimuths from the center, and lengths along these lines are of exact scale. Equations for the transformation from the rotational ellipsoid have been derived. A particular approximation to the oblique aspect of the graticule is produced by a map projection called Hatt's map projection; it has been used in Greece, since 1952, for topographic and cadastral maps.

map projection, bipolar oblique conformal conical - SEE map projection, bipolar oblique conical conformal.

map projection, bipolar oblique conical conformal - A conformal map projection mapping the sphere onto two cones with their axes at a 104° angle. This map projection was developed by O.M. Miller and W.A. Briesemeister for a map of the Americas and is an adaptation of the Lambert conformal conic map projection. It has been used by the United States Geological Survey for some geological maps but is being replaced, for this purpose, by the transverse Mercator map projection. Also called the bipolar oblique conformal conical map projection.

map projection, central - SEE map projection, azimuthal; map projection, gnomonic.

map projection, combined - A map projection derived from two or more map projections, or from two or more different aspects of the same map projection, by expressing the coordinates in the plane as some average of the corresponding coordinates given by the original map projections. Breusing's map projection and the eumorphic map projection are examples.

map projection, conformal - A map projection producing a map with the property that, at any point, the angle between two arbitrary lines through a point on the map is the same as the angle between the corresponding lines on the mapped surface at the corresponding point. Alternatively, a map projection such that the scale at any point on the map is the same in all directions. Also called an autogonal map projection or orthomorphic map projection. Among the more important conformal map projections are the Mercator map projection (m.p.), the stereographic m.p., the transverse Mercator m.p., and the Lambert conformal conical m.p. The latter two are used in the State Coordinate Systems and State Plane Coordinate Systems.

map projection, conformal conical - A conformal map projection mapping a region on the rotational ellipsoid onto a region on a cone in such a way that at each point, the ratio of scales in two orthogonal directions is unity. Also called a conical conformal map projection and a conical orthomorphic map projection. The most common such map projections are Lambert's conformal conical map projections in the polar aspect, but others, such as those of Kavraisky and Vitkowsky, also have some use. SEE ALSO Lambert conformal conical map projection.

map projection, conic - SEE map projection, conical.

map projection, conical - A map projection mapping the rotational ellipsoid onto a tangent or secant cone. Also called a conic map projection. The map projection has the fundamental property that the principal scale is preserved along the line representing the arc of a small circle or along the two lines representing the arcs of two small circles. There are several methods of mapping from the rotational ellipsoid to the cone. Most of them are analytical in character and can not be constructed by simple, graphic methods. (A perspective map projection from the center of the rotational ellipsoid could be represented graphically.) Conical map projections may be considered to include cylindrical map projections (the apex of the cone is then at an infinite distance from the sphere or ellipsoid), and projections onto a tangent or secant plane (the apex of the cone is then at the center of the plane.) They may project onto a single cone tangent to the ellipsoid or, conceptually, secant to the ellipsoid along two parallels; or they may project onto a series of tangent cones, all having the same axis which passes through the center of the ellipsoid but with apices at constantly increasing distances from the ellipsoid. While a cone or cones may be used in descriptive illustrations of conical map projections, care must be taken not to consider such projections as geometrical concepts, because most are not.

map projection, conical conformal - SEE map projection, conformal conical.

map projection, conical equal-area - SEE Albers map projection; Lambert conical equal-area map projection.

map projection, conical equidistant - An equidistant map projection onto the cone, so that scale is exact along all lines radiating from the apex of the cone (the lines representing meridians) or along all the arcs of concentric circles (representing parallels of latitude). Also called a simple conical map projection. SEE ALSO Euler's map projection.

map projection, conical orthomorphic - SEE map projection, conformal conical.

map projection, conventional - Any map projection which is not azimuthal, cylindrical or conical.

map projection, cordiform - A map projection producing a graticule bounded by a heart-shaped curve symmetrical about the central meridian, and representing parallels of latitude by arcs centered on the point representing a pole.

map projection, cylindrical - (1) A map projection having the fundamental property that the scale is specified and constant along some line which represents the arc of a great circle. (2) A map projection which first projects the geographic meridians and parallels of latitude onto a cylinder either tangent or secant to the surface of a rotational ellipsoid, and then develops the cylinder into a plane. There are several ways of projecting from the rotational spheroid to the cylinder. As with conical map projections, some of these are purely analytical in character and do not have graphic counterparts. For example, the Mercator map projection is a cylindrical map projection easily described algebraically but not graphically. Other methods may be defined geometrically and the projection carried out graphically. For example, the cylindrical, equal-spaced map projection can easily be done graphically. Both types of map projection are called cylindrical map projections. The three major types of cylindrical map-projection are: the Plate Carrée, or simple cylindrical map projection; the equal-area cylindrical map-projection; and the conformal cylindrical map-projection. There are transverse and oblique versions of each of these. SEE ALSO Behrmann's map projection.

map projection, cylindrical equal-area - An equal-area map-projection mapping a rotational spheroid onto a cylinder in such a way that there is a constant ratio between the area of a region on the ellipsoid and the area of the corresponding region on the cylinder. When the surface is mapped onto a cylinder tangent at the equator, the map projection used is often referred to as Lambert's cylindrical equal-area map-projection. Among other cylindrical, equal-area map projections are the modified Plate Carrée map-projection (the Plate Carrée itself is not an equal-area map projection), and transverse or oblique aspects of it.

map projection, cylindrical equally-spaced - (1) A map projection, from a sphere to a cylinder, mapping meridians into equally-spaced straight lines and parallels of latitude into equally spaced, straight lines perpendicular to the lines representing meridians. (2) A map projection similar to that defined above, but mapping the rotational ellipsoid onto a cylinder. The two principal forms are the Plate Carrée map projection, in which spacing is the same for the two families of lines, and the modified (equi-rectangular) Plate Carrée map projection, which uses different spacings for the two families. (3) A map projection similar to that defined in (1) or (2), but with the axis of the cylinder inclined at an angle to the polar axis of the ellipsoid.

map projection, cylindrical equidistant - SEE Plate Carrée map projection.

map projection, cylindrical perspective - SEE map projection, perspective.

map projection, doppelprojektion - SEE map projection, double.

map projection, double - (1) A map projection produced by applying two map projections, one after the other. (2) A conformal mapping from the rotational ellipsoid to the sphere, followed by a second conformal mapping from the sphere to the plane. Also called Doppelprojektion and doppelprojektion map projection. In particular, that version of the transverse Mercator map projection from ellipsoid to plane produced by first mapping the ellipsoid onto a sphere and then mapping the sphere onto a plane. SEE ALSO Gauss-Schreiber map projection.

map projection, doubly azimuthal - An azimuthal map projection producing a map on which there are two points such that the azimuths of lines through these points are the same as the azimuths of the corresponding lines on the original surface. SEE ALSO map projection, two-point azimuthal.

map projection, doubly-periodic - SEE Guyou's map projection.

map projection, elliptical - A map projection of the whole sphere into the region within an ellipse.

map projection, equal-area - A map projection preserving a constant ratio between the area of a region on the surface being mapped and the area of the corresponding region on the plane. Also called an authalic map projection, orthembadic map projection, equivalent map projection and homalographic map projection. The area inclosed by any given boundary on the sphere would bear the same ratio to the area inclosed by the line into which that boundary maps on the plane as the area of the mapped portion of the sphere would bear to the area of the map. A definite area (such as a square centimeter) on an equal-area map will represent a constant area on the mapped surface regardless of the location of the boundaries of that area on the map. An approximately equal-area map projection was used by Ptolemy for a map of the world, and the type was used for a number of maps by the Arabs in the 9th century. The principal equal-area map projections (m.p.) are Albers' m.p., the azimuthal equal-area m.p., the Bonne m.p., Lambert's conical equal-area m.p., Mollweide's m.p., the Hammer-Aitoff m.p. and the sinusoidal m.p.

map projection, equally-spaced - SEE Plate Carrée map projection.

map projection, equatorial - (1) SEE map projection, polar. (2) SEE map projection, meridional.

map projection, equatorial cylindrical orthomorphic - SEE Mercator map projection.

map projection, equidistance - SEE map projection, equidistant.

map projection, equidistant - (1) A map projection, from the ellipsoid to the plane, keeping the scale exact along all lines radiating from the point at which distortion is zero, or along all lines perpendicular to the line along which distortion is zero. (2) A map projection producing a graticule on which lines representing parallels of latitude are spaced at equal distances along the lines representing meridians, on which lines representing meridians are equally spaced along lines representing parallels of latitude. Also called an equidistance map projection. This definition and the preceding one are practically equivalent. (3) A map projection such that scale is maintained along lines joining certain pairs of points.

map projection, equidistant cylindrical - SEE Plate Carrée map projection, modified.

map projection, equidistant polar - SEE map projection, azimuthal equidistant.

map projection, equidistant polyconic - A map projection such that on lines representing meridians, equal meridional distances are intercepted everywhere between the same lines representing parallels of latitude. This is not strictly a polyconic map projection.

map projection, equi-rectangular - SEE Plate Carrée map projection.

map projection, equivalent - SEE map projection, equal-area.

map projection, eumorphic - An equal-area map projection for which the rectangular Cartesian coordinate y of a point is the arithmetic average of the point's rectangular coordinate ym in the Mollweide map projection and ys in the sinusoidal map projection, while the coordinate x is so determined that the equal-area condition holds. Also called Boggs' map projection and Boggs' eumorphic map projection.

map projection, external perspective - An azimuthal map projection based on a perspective projection in which the center of perspective lies outside the closed surface (e.g., ellipsoid) being mapped.

map projection, geometric - SEE map projection, perspective.

map projection, globular - (1) A map projection, other than one producing the transverse aspect of the azimuthal map projection, of a hemisphere onto a disk. The earliest globular map projection appears to have been derived by G. Fournier about 1640 (an earlier one by R. Bacon in 1265 is reported by J.P. Snyder). Nell's modified globular map projection produces a graticule in which the rectangular Cartesian coordinates of a point are the arithmetic average of those produced by Nicolosi's map projection and those produced by the transverse aspect of the stereographic map projection. (2) SEE Nicolosi's map projection.

map projection, gnomonic - (1) A perspective map projection from the sphere onto a tangent plane, with the point of projection at the center of the sphere. Also called a central map projection. The form having the plane tangent at the equator has been called the meridian central (map) projection. The projection is given by the equations

$$x = k R (\sin \Delta\lambda \cos \phi) / D$$
$$y = k R (\cos \phi_o \sin \phi - \sin \phi_o \cos \phi \cos \Delta\lambda) / D$$
$$D = \sin \phi_o \sin \phi + \cos \phi_o \cos \phi \cos \Delta\lambda ,$$

in which x and y are the rectangular Cartesian coordinates of that point, in the plane, corresponding to the point, on the sphere, at longitude $\lambda_o + \Delta\lambda$ and at latitude ϕ, where λ_o is the longitude of the central meridian and ϕ_o is the latitude of the point at which the plane is tangent. R is the radius of the sphere and k is the scale factor. The gnomonic map projection is neither conformal nor equal-area. It is the only map projection representing great circles on the sphere by straight lines. It may be the oldest map projection; it appears to have been used by Thales (B.C. 640 -548) for charts of the heavens. (2) The same as the preceding definition, but mapping an ellipsoid onto a tangent plane, with the point of projection being at the center of the ellipsoid.

map projection, homalographic - (1) SEE map projection, equal-area. (2) SEE Mollweide's map projection. (3) SEE Goode's homalographic map.

map projection, homalosine - SEE map, homolosine.

map projection, homolographic - SEE map projection, equal-area.

map projection, homolosine - SEE map, homolosine.

map projection, International Map of the World - (1) For maps prepared before 1962, a modified polyconic map projection with two standard meridians along which the scale is held exact. The geographic meridians are represented by straight lines connecting points on the top and bottom parallels of latitude. The lines representing the parallels of latitude are non-concentric arcs of circles. Up to latitude 60°, each sheet of the series depicts a region between 4° of latitude and 6° of longitude. Between latitudes 60° and 76°, the sheets are double width, so each covers 4° of latitude and 12° of longitude. From 76° to 84°, the sheets are 24° wide. The scale of the map is 1:1 000 000; for this reason, it is also called the millionth-scale map of the world. The map projection was devised by Lallemand about 1909. (2) In 1962, this map projection was replaced, for the regions between 84° N and 80° S, by Lambert's conformal conical map projection with two standard parallels. At the poles, the polar stereographic map projection was used.

map projection, inverse cylindrical orthomorphic - SEE Mercator map projection, transverse.

map projection, isometric - A map projection mapping any particular set of geodesics on one surface into a set of lines of equal length on the other surface. Also called a length-preserving map-projection. All isometric map projections are conformal map projections, but the converse is not true.

map projection, isoperimetric - A map projection, from the ellipsoid to the plane, such that exact scale is maintained along that closed curve on the plane which corresponds to a given, closed curve on the ellipsoid.

map projection, meridional - A map projection producing a graticule having its center on the line representing the equator.

map projection, meridional central - SEE map projection, gnomonic.

map projection, meridional orthographic - SEE map projection, orthographic.

map projection, minimum-error - A map projection producing a map such that the sums of squares of errors in scale in particular, mutually perpendicular directions and integrated throughout the map, is minimal for a given type of map projection.

map projection, modified - (1) A map projection based on some common, known map projection which has been changed by relocating the point or line or zero distortion and by specifying the scale along different lines. (2) A map projection differing slightly from the named map projection on which it is based. (3) A graticule having the oblique aspect of the named map projection and obtained by calculation.

map projection, modified polyconic - (1) Any of various map projections obtained by altering the polyconic map projection. (2) A map projection obtained from the polyconic map projection by so altering the scale along the central meridian that scale is exact along two standard meridians, one on either side of and equidistant from the central meridian. Before 1962, a modified polyconic map projection was used for the International Map of the World; the scale was made exact along straight lines representing meridians located, usually, 2° on either side of the central meridian. A slightly different kind of modified polyconic map projections is also used for some State maps formerly published by the U.S. Geological Survey. A third type is used for maps of extremely northern Canada. (3) A rectangular polyconic map projection.

map projection, non-perspective - Any map projection not able to be constructed by linear projection (i.e., by perspective) from a single point. The Mercator and the azimuthal equal-area map projections, for example, are non-perspective map projections.

map projection, non-perspective azimuthal - A map projection not based on imaginary lines of sight from a single center of perspective. Azimuthal equal-area and azimuthal equidistant map projections are of this type.

map projection, normal-aspect - SEE aspect, normal.

map projection, oblique - SEE aspect.

map projection, oblique cylindrical orthomorphic - SEE Mercator map projection, oblique.

map projection, ordinary polyconic - SEE map projection, polyconic (2).

map projection, orthembadic - SEE map projection, equal-area.

map projection, ortho-apsidal - The orthographic map projection of some surface other than the sphere or rotational ellipsoid in general.

map projection, orthodromic - SEE map projection, two-point azimuthal.

map projection, orthogonal - SEE map projection, orthographic.

map projection, orthographic - A map projection from the ellipsoid to the plane by placing the plane in a suitable position with respect to the ellipsoid and then identifying each point on the ellipsoid with that point in the plane which lies on the perpendicular from the point on the ellipsoid. Also called an orthogonal map projection. The transformation from the sphere to the plane is given by the equations $x = k R (\cos \phi_1 \sin \phi - \sin \phi_1 \cos \phi \cos (\Delta\lambda)$ and $y = k R \cos \phi \sin \Delta\lambda$, in which x and y are the rectangular, Cartesian coordinates of that point, in the plane, which corresponds to the point, on the sphere, at longitude $\lambda_0 + \Delta\lambda$ and at latitude ϕ; λ_0 is the longitude of the central meridian. ϕ_1 is the latitude of the point represented by the center of the graticule. R is the radius of the sphere and k is the scale factor. In general, the meridians are represented by ellipses having their centers coincident at the origin; the parallels of latitude are represented by ellipses having their centers on the x-axis. The orthographic map-projection is used, for maps of the Earth, in three forms: (a) a polar orthographic map projection, in which the plane of the map is perpendicular to the Earth's axis of rotation; the parallels of latitude are then mapped into full circles true to scale and the meridians are mapped into straight lines; (b) the meridional orthographic map-projection, in which the plane of the map is parallel to the plane of some selected meridian; the parallels of latitude and the central meridian then map into straight lines, the outer meridian maps into a full circle, and the other meridians map into arcs of ellipses; and (c) an oblique orthographic map projection, in which the plane is perpendicular to some radius neither in the equatorial plane nor on the axis of rotation. (This is the most common form.)

map projection, orthomorphic - SEE map projection, conformal. American usage favors conformal; British usage favors orthomorphic.

map projection, parabolic - An equal-area map projection, invented by J.E.E. Craster (1929), representing parallels of latitude by straight lines parallel to the line representing the equator and the meridians by parabolas with vertices at the line representing the equator.

map projection, perspective - A map projection from an ellipsoid to a developable surface, such that corresponding points on the two surfaces lie on a common line through a single, specified point (the perspective center). A perspective map-projection is equivalent, geometrically, to the process of drawing straight lines radially from the selected point (called the perspective center or center of projection) through points on the ellipsoid, to the developable surface. Of the various developable surfaces, only the plane, the cylinder and the cone have been used extensively. Each of these surfaces characterizes an important family of map projections using that surface. (a) The simplest family is that in which the ellipsoid is mapped onto a plane. The plane is usually tangent to the ellipsoid at the center of the region to be mapped. If the ellipsoid is a sphere, the center of projection is usually some selected point on that diameter which passes through the point of tangency. If the center of projection is at the center of the sphere, the gnomonic map projection results; if it is at the opposite end of the diameter from the point of tangency, the stereographic map projection results; and if it is at an infinite distance along that diameter from the point of tangency, the orthographic map projection results. Special names have been given to perspective map projections mapping the sphere onto a plane tangent at a pole, according to the distance of the center of projection from the center of the sphere. Denoting that distance by D and denoting the radius of the sphere by R, the principal named varieties are: Clarke's twilight map-projection ($D = 1.4 R$); James's map-projection ($D = 1.367R$); La Hire's map-projection ($D = R + \frac{1}{2} R \sqrt{2}$); Lowry's map-projection ($D = 1.69 R$); the Parent-II map-projection ($D = R \sqrt{3}$); and the Parent-III map-projection ($D = 2.105 R$). Airy's map-projection is often considered part of the family of perspective map-projections because the principle Airy used in deriving it was also used in deriving some truly perspective map-projections. However, it is not, itself, a perspective map-projection. (b) Perspective map-projections using the cylinder as the developable surface are next up in the order of complexity. Although any ellipsoid can in principle be mapped onto a cylinder by a perspective map-projection, this is rarely done. The projection is almost always from a sphere to the cylinder. The cylinder is usually tangent to the sphere and the center of projection is at the center of the sphere. If the cylinder is tangent at the equator, meridians map into equally spaced, parallel, straight lines; the parallels of latitude map into parallel, straight lines; and the two families of straight lines are perpendicular to each other. The lines representing parallels of latitude are not spaced evenly; if the center of projection is at the center of the sphere, the distance between them increases as the tangent of the latitude and the line representing the pole is at an infinite distance from the line representing the equator. (This map projection must not be confused with the Mercator map-projection which it resembles.) (c) Perspective map-projections using the cone as the developable surface are the most complex and, like the perspective map-projection onto a cylinder, is almost always done using a sphere as the original surface. The cone is tangent or secant to the sphere and the center of projection is almost always placed on the axis of the cone - usually at the center of the sphere.

map projection, polar - (1) A map projection producing a graticule with center representing one of the poles of the ellipsoid. Also called an equatorial map-projection, although this latter term is also used for a meridional map-projection. (2) The polar aspect of a graticule. SEE aspect (of a map).

map projection, polar equal-area - SEE Goode's polar equal-area map.

map projection, polar orthographic - SEE map projection, orthographic.

map projection, polar stereographic - SEE map projection, stereographic.

map projection, polyconic - (1) A map projection, from the rotational ellipsoid to the plane, producing a graticule in which the parallels of latitude are represented by non-concentric circular (or higher-degree) arcs with centers on the central meridian. The most common version is that which keeps scale exact along lines representing the central meridian and parallels of altitude. It is usually the one meant by the term polyconic map projection. SEE definition (2), following. (2) A map projection, from the rotational ellipsoid to the plane, producing a graticule in which the parallels of latitude are represented by non-concentric, circular arcs with centers on the central meridian, and keeping the scale exact along the central meridian and parallels of latitude. Also called the ordinary polyconic map-projection and simple polyconic map-projection. The polyconic map-projection is defined algebraically, for mapping a rotational ellipsoid on a plane, by the equations $x = k N \cot \phi \sin (\Delta\lambda \sin \phi)$ and $y = 2 k N \cot \phi \sin^2 \frac{1}{2}(\Delta\lambda \sin \phi) + ks$, in which x and y are the rectangular, Cartesian coordinates, in the plane, of that point which corresponds to the point, on the rotational ellipsoid, which is at longitude $\lambda_o + \Delta\lambda$ and at latitude ϕ; λ_o is the longitude of the central meridian. s is the length of the arc between latitude ϕ and the equator, N is the radius of curvature in the prime vertical at the point on the ellipsoid and k is the scale factor. The polyconic map-projection was devised by F.R. Hassler, organizer and first superintendent of the U.S. Coast Survey. For this reason, it is sometimes called the American polyconic map-projection or American ordinary polyconic map-projection. The central meridian maps into a straight line in the plane. The parallels of latitude map into arcs of circles whose centers lie on the line representing the central meridian and whose radii are determined by the lengths of the elements of cones tangent along the parallels. All meridians except the central one are curved. This map projection is neither conformal nor equal-area, but it has been much used for maps of small regions because it can be constructed easily. It was used, until the 1950's, for maps in the series of the Topographic Map of the United States (U.S. Geological Survey) and, in a modified form, for maps of larger regions. The rectangular polyconic map-projection resembles the ordinary polyconic map-projection defined above but scale is exact only along one line representing a parallel of latitude. Distances along circular arcs representing other parallels of latitude are scaled so that the arcs intersect lines representing meridians at right angles. This map also is neither conformal nor equal-area. The modified polyconic map-projection, also called the polyconic map-projection with two standard meridians, was devised by Lallemand and is characterized by the fact that scale is exact along two lines representing meridians. There are also polyconic map-projections having the equal-area property, e.g., the National Geographic Society's equal-area polyconic map-projection.

map projection, polygnomonic - SEE map, polygnomonic.

map projection, polyhedral - SEE map, polyhedric.

map projection, polyhedric - SEE map, polyhedric.

map projection, pseudo-azimuthal - A class of map projections mapping, in their normal aspect, parallels of latitude into concentric, circular arcs and mapping meridians into curved lines converging at the pole with their true, angular value. The best-known pseudo-azimuthal map- projections is Wiechel's map-projection (1879).

map projection, pseudo-conical - A class of map projections mapping one standard parallel and the central meridian into lines along which principal scale is preserved. In the normal aspect of the graticule, parallels of latitude are represented by concentric circular arcs and meridians by convergent curves. The Bonne map-projection is the most common.

map projection, pseudo-cylindrical - A class of map projections mapping the equator and the central meridian into lines along which principal scale is preserved. In the normal aspect of the graticule, parallels of latitude are represented by a family of parallel, straight lines and the meridians are represented by a family of convergent curves. The sinusoidal map-projection is one of the best known of this class; Eckert's-III map-projection (Ortelius's map-projection) is another.

map projection, quincuncial - A conformal map projection of the sphere into the interior of a square, using elliptical functions, with one pole at the center of the square and the other pole at the four corners. Also called Peirce's map-projection.

map projection, recentered - SEE map, interrupted.

map projection, rectangular - SEE Plate Carrée map projection.

map projection, rectangular polyconic - SEE map projection, polyconic.

map projection, retro-azimuthal - A map projection mapping the rotational ellipsoid into the plane in such a way that the azimuth or direction from any point on the ellipsoid to a particular point on the ellipsoid remains unchanged when taken in the corresponding sense between the corresponding points on the map. This is the converse of the azimuthal map-projection, which preserves azimuths from the point specified to an arbitrary point. The outstanding example of the retro-azimuthal map-projection is that called the Mecca map-projection, in which Mecca is the point specified. It was first described by J.I. Craig in 1909.

map projection, secant conic - A conic map-projection in which the cone is secant to the spheroid.

map projection, simple conic - A conical map projection thought of as the projection of the sphere or ellipsoid onto a tangent cone which is then spread out to form a plane. Also called tangent conic map projection.

map projection, simple conical - SEE map projection, conical equidistant.

map projection, simple cylindrical - SEE Plate Carrée map projection.

map projection, simple polyconic - SEE map projection, polyconic.

map projection, sinusoidal - An equal-area map projection mapping all parallels of latitude into truly-spaced, parallel, straight lines along which exact scale is preserved. The equator is used as the standard parallel of latitude. The sinusoidal map-projection is also known as Mercator's equal-area map-projection and the Sanson-Flamsteed map-projection. It was used in the Mercator-Hondius atlases as early as 1606 and may for this reason be called the Mercator-Sanson map-projection. It is improperly referred to as Flamsteed's map-projection, modified Flamsteed's map-projection and Sanson's map-projection. It is a variant of the Bonne map-projection in that the equator is used as standard parallel of latitude, and a refinement of the Plate Carrée map-projection in that the x-coordinates are corrected for convergence of the meridians. It is defined for the mapping of the sphere onto the plane by the formulas $x = k R_1 \Delta\lambda \cos \phi$ and $y = k R_2 \phi$, in which x and y are the rectangular, Cartesian coordinates of that point, in the plane, which corresponds to the point, on the sphere, at longitude $\lambda_0 + \Delta\lambda$ and latitude ϕ; λ_0 is the longitude of the central meridian. R_1 and R_2 are the radii of curvature along the parallel of latitude and along the meridian, respectively, and k is the scale factor. The corresponding formulae for transformation from the rotational ellipsoid to the plane are obtained by substituting N, the radius of curvature in the prime vertical, for R_1 and s, the length of arc corresponding to ϕ, for $R_2 \phi$.

map projection, skewed - Any map projection mapping the meridians into lines not pointing approximately north and south at the equator, with respect to the neat lines of the map. Alternatively, any map projection mapping the equator into a line meeting the neat lines at an angle other than a right angle.

map projection, skew orthomorphic - SEE Mercator map projection, oblique.

map projection, stereographic - (1) An azimuthal map-projection mapping a hemisphere onto a plane by projecting from a point on the sphere onto the plane tangent to the sphere at the opposite end of the diameter through the center of projection. It is given by the formulas

$x = k R (\cos \phi_1 \sin \phi - \sin \phi_1 \cos \phi \cos \Delta\lambda) / D$
$y = k R (\cos \phi \sin \Delta\lambda) / D$
$D = \frac{1}{2}(1 + \sin \phi_1 \sin\phi + \cos \phi_1 \cos \phi \cos\Delta\lambda)$,

in which x and y are the rectangular, Cartesian coordinates of that point, in the plane, corresponding to the point, on the sphere, at longitude $\lambda_0 + \Delta\lambda$ and latitude ϕ; λ_0 is the longitude of the central meridian. ϕ_1 is the latitude of that point, on the sphere, which corresponds to the center of the graticule. R is the radius of the sphere and k is the scale factor. When the center of projection is at one of the poles of the sphere, the projection is called a polar stereographic map projection; when the center is on the equator, the projection is called a meridional stereographic map projection; and when the center is on some other, selected parallel of latitude, the projection is called an oblique stereographic map-projection or a horizon stereographic map-projection. However, the terminology is extensive and inconsistent, and many other names exist. The stereographic map-projection is conformal and is the only azimuthal map-projection having that quality. It is one of the most widely known of all map projections and has been much used for maps of an entire hemisphere. It dates back to the days of ancient Greece, having been used by Hipparchus (160 -125 B.C.) (2) A conformal map projection mapping the upper or lower half of a rotational ellipsoid onto the plane by mapping the half-surface from the pole not on that surface to the plane tangent to the surface at the other pole. It is given by the functions $x = \rho \cos \Delta\lambda$ and $y = \rho \sin \Delta\lambda$, in which x and y are the rectangular, Cartesian coordinates of that point, in the plane, corresponding to the point, on the rotational ellipsoid, at longitude $\lambda_o + \Delta\lambda$ and at latitude ϕ; λ_o is the longitude of the central meridian. ρ is a function of ϕ, the length a of the semi-major axis of the ellipsoid, and the eccentricity e: $\rho = 2k_o (a^2/b) \tan¼ (\pi - 2\phi)[[(1 - e) (1- e \sin \phi)] / [(1 + e)(1 + e \sin \phi)] e / 2$. b is the length of the semi-minor axis and k_o is the scale factor at the pole. It is not a true perspective. It is commonly used instead of the transverse Mercator map-projection for maps of the polar regions because it gives a good representation of disk-like regions whereas the transverse Mercator map-projection is better for strip-like regions.

map projection, stereographic meridional - SEE map projection, stereographic (1).

map projection, tangent conic - SEE map projection, simple conic.

map projection, transverse - The traverse aspect of a particular map projection.

map projection, transverse cylindrical orthomorphic - A transverse aspect of the Mercator map-projection.

map projection, transverse polyconic - A polyconic map projection obtained by substituting, for the central meridian, a great circle perpendicular to that meridian. This is done to provide an axis along which will lie the centers of the circular arcs representing lines of tangency of cones with the surface of the sphere. This is a complicated map-projection and the graticule is difficult to construct. However, it is useful for mapping, with comparatively little distortion, a narrow region of large extent eastward and westward.

map projection, trimetric - Map projection mapping the three great arcs of a spherical triangle on the sphere as straight lines on the plane and laying off, from the vertices of the planar triangle, the true distances, to scale, of points on the sphere. To each point on the sphere then corresponds a small triangle in the plane. The center of each small triangle is taken as the representation of the corresponding point on the sphere. This map projection, developed by W. Chamberlin, has been used for a number of maps published by the National Geographic Society.

map projection, tronconic - SEE map projection, truncated conical.

map projection, truncated conical - A map projection which represents the pole as the arc of a circle of finite length.

map projection, two-point azimuthal - A map projection from the sphere to the plane, developed from the gnomonic map projection by projecting the graticule from that projection on to a plane tilted at a suitable angle, expanding or contracting the graticule so that the coordinates in the chosen directions are changed in a constant ratio. Also called the doubly- azimuthal map-projection, Maurer's map projection, Maurer's orthodromic map projection and the orthodromic map projection. It was invented independently by H. Maurer and C. Close as a variant of the gnomonic map-projection. There are two points in the graticule at which azimuths to all other points are correct. All great circles are represented by straight lines.

map projection, two-point equidistant - A map projection with the property that the two distances of any point on the map from two fixed points are true to scale. It was first described by C. Close in 1921.

Map Projection, Universal Polar Stereographic - SEE Universal Polar Stereographic Map Projection.

map projection, zenithal - (1) A map projection from the ellipsoid to the plane such that all points at the same distance from a specified distance on the ellipsoid are mapped onto a circle on the plane. (2) SEE map projection, azimuthal.

map-projection aspect - SEE aspect (of a map projection).

map-projection classification - SEE classification.

map projection with oval isolines, TsNIIGAik - SEE TsNiiGAik map projection with oval isolines.

map revision - (1) The process of making changes to an existing map to improve its accuracy. (2) Bringing a map into agreement with the current local conditions and the latest requirements as to their content and design. Three grades of map revision are recognized: limited (also called provisional), standard and complete. Each grade involves more corrections than the grade preceding it. SEE ALSO photo-revision; planetable revision.

map revision, complete - A map revision making all corrections to the map that are possible using existing surveying data, aerial photography, and other sources. Cartographic specifications current at the time of revision are used.

map revision, limited - A map revision correcting the map for a few, specified changes in the region depicted. Also called a provisional map-revision. It is usually made only when major changes are necessary and there is no time to make a standard map-revision.

map revision, provisional - SEE map revision, limited.

map revision, standard - A map revision correcting the map for all known (mappable) changes that have occurred in the region since the last previous revision of the map. No changes are made in distances or elevations of unchanged points, and the cartographic specifications used for the previous edition are adhered to.

MAPSAT - A design proposed by A.P. Colvocoresses and the U.S. Geological Survey for a satellite and auxiliary equipment to map the Earth.

map scale - SEE scale (of a map).

map scale, bar - SEE scale (of a map).

map scale, equivalent - SEE scale, verbal.

map scale, fractional - SEE scale, fractional.

map scale, graphic - SEE scale (of a map); scale, graphic.

map scale, verbal - SEE scale, equivalent.

map scale number - SEE scale number.

map series - (1) A set of maps produced at a uniform scale, in a common style, and planned to represent a specified region. (2) A set of maps, usually of the same scale and conforming to the same specifications, having some unifying characteristic and identified collectively by the organization producing it.

map sheet - A map, usually one of a map series. The term is used when the map as a printed product, rather than the map as a representation of a region, is meant.

map substitute - Any graphic material such as aerial photographs, photomaps, etc., produced or used as a substitute for a map. Map substitutes are usually produced when maps cannot be produced in time.

map test - Any method used for testing or checking the accuracy of a topographic map. Accuracy can be tested, e.g., by running traverses and level-lines across selected regions of a map sheet and comparing the horizontal coordinates of features determined from the map with the coordinates of the same features determined by traverse, and the elevations of features as given on the map with the elevations of those features as determined by leveling.

March equinox - SEE equinox, vernal.

Marek's method of resection - A method of determining simultaneously the locations of two unknown points by resection from four known points.

mareogram - SEE marigram.

mareograph - SEE marigraph.

mareograph spherop - A surface defined by assigning a particular value to the spheropotential function and requiring that the surface pass through a point at mean sea level at a specified tide-gage.

margin - (1) Properly, the line or edge along which the surface of a body of water meets the land. (2) In law, as applied to the boundary of a piece of land bounded by a stream, the center of the stream forming a boundary of the land. (3) That part of a map lying outside the border. Also called a surround. (Brit.)

margin, continental - A zone separating the emergent continent from the deep-sea bottom. It usually comprises the continental shelf, the continental slope, and the continental rise.

marigram - A graphic record of the rise and fall of the tide. Also called a mareogram. The record is usually represented in a rectangular, Cartesian coordinate system in which time is the abscissa and the elevation of the tide is the ordinate.

marigraph - (1) An instrument for measuring and recording tides. Also called a mareograph. (2) A place where tides are measured.

marina - A section of waterfront and adjacent water at which both local and transient pleasure-boats are accommodated, usually with other facilities for fueling boats, feeding passengers, etc.

Marinus map projection - SEE Plate Carrée map projection.

mark - (1) A dot or the intersection of a pair of crossed lines or other physical point corresponding to a point in a survey; the physical point to which distances, elevations, heights or other coordinates refer. (2) SEE marker. (3) SEE monument. (4) SEE water mark. SEE ALSO contact mark; index mark; reference mark; witness mark.

mark, astronomic-azimuth - SEE mark, Laplace-azimuth.

mark, collimating - A mark, on the stage (easel) of a reducing printer or projector, to which the position of a diapositive or photographic negative is referred. SEE ALSO mark, fiducial.

mark, fiducial - (1) The image of a fiducial marker in photogrammetry. (2) A point or line used as a reference or origin in surveying. (3) SEE marker, fiducial.

mark, floating - A mark which, while actually part of the optical system of a stereoscope, appears to occupy a position in the three-dimensional space perceived by stereoscopic fusion of a pair of photographs, and used as a reference in examining or measuring the stereoscopic model. The mark may be formed, in a binocular system, (a) by one real mark lying in the image-space of the system; (b) by two real marks lying in that image-space; (c) by two real marks lying in the plane of the photographs themselves; or (d) by two virtual marks lying in the planes of the images.

mark, geodetic-azimuth - SEE azimuth mark, geodetic.

mark, half - SEE index mark.

mark, high-water - (1) The place, on the bank or shore, where the usual and long-continued presence and action of water has impressed on the bed of the stream a character distinct from that of the banks with respect to vegetation and the nature of the soil. Also called, undesirably, the high-water line. The high-water mark is the boundary line between the bed and the bank of a stream. The average high-water line usually determines the boundary of the land of the proprietor having riparian rights. (2) An established mark, on a structure or natural object, which indicates the greatest observed elevation to which the tide has reached. (3) A line or mark left upon tidal flats, beach or objects on the shore and indicating the height or elevation to which high water has intruded. The mark may be a line of oil or scum on objects, or a more or less continuous deposit of fine shells or debris on the foreshore or berm. It is physical evidence of the general height reached by the waves at recent high waters. It should not be confused with the mean high water line or mean higher high water line. (4) The line which the water impresses on the soil by covering it for periods sufficient to deprive it of vegetation characteristic of uplands. (5) An established reference mark, on a structure or natural object, which indicates the greatest observed height of the tide.

mark, Laplace-azimuth - A mark whose astronomic azimuth from a station which is also a Laplace station is known.

mark, low-water - A physical indication of a former, persistent, low level of water. While most low-water marks are destroyed by the water's action, some indicators of low water do persist. Most of these are left by animals which live in the inter-tidal region.

mark, permanent - A readily identifiable, relatively permanent, recoverable marker designating precisely the location of a survey point.

mark, underground - SEE marker, underground.

marker - (1) An object identifying the location of a station. (2) The object, such as a metallic disk, on which a dot or pair of crossed lines is incised or stamped to correspond to a point in a survey. Also called a survey marker. (3) SEE survey marker.

marker, fiducial - One of a set of (usually 4 or 8) small objects rigidly fastened to the interior of a camera's body near the focal plane, in such a manner as to be photographed along with the scene when taking a picture. Also called a fiducial mark, although this term is more properly used for the mark made on the film. The two lines joining images of fiducial markers on diametrically opposite sides of the picture intersect at the principal point of the photograph; the orientation of the photograph is frequently referred to them. A fiducial marker is usually a small, triangular, metallic tab projecting from the body or film holder and fastened directly in front of the film. Such tabs image on the film during exposure. If the camera is used for taking pictures under conditions in which the light from the scene is not bright enough to cast sharp shadows of the markers, an optical fiber leading light from a small lamp in the camera to a spot at the edge of the film may be used as a fiducial marker.

marker, geodetic - SEE survey marker.

marker, permanent - SEE mark, permanent.

marker, reference - SEE reference mark.

marker, underground - A survey marker set vertically below the center of a marker on the surface and separated from it so as to preserve the station in case of an accident to the marker on the surface.

market price - SEE market value.

market value - (1) As defined by the courts, the highest price estimated in terms of money which a property would bring if exposed for sale in the open market, allowing a reasonable time to find a purchaser who buys with a knowledge of all the uses to which the property is adapted and for which it is capable of being used. (2) The price at which a willing seller would sell and a willing buyer would buy, neither being under abnormal pressure. (3) The price expectable if a reasonable time is allowed to find a purchaser and if both seller and buyer are fully informed. The essential difference between market price and market value lies in the premises of intelligence, knowledge, and willingness, all of which are contemplated in setting the market value but not in the market price. Stated differently, at any given moment of time the value connotes what a property is actually worth and market price what it may be sold for. Note, however, that the concept of actually worth is itself defined by the definition of market value and has no meaning apart from it.

marksetter - A person whose principal task is in placing survey markers.

Marsden chart - A numbered grid made by dividing the graticule of a Mercator map projection between 90° north and 90° south (or between 80° north and 80° south) into quadrangles (Marsden squares), each extending 10° in latitude and 10° in longitude. The quadrangles are numbered systematically to indicate their location in the graticule. (The quadrangles may be divided further into quarters or into one-degree quadrangles numbered from 00 to 99). The Marsden chart commonly shows, in each quadrangle, a number indicating some average characteristic of the ocean or atmosphere in the region covered by that quadrangle. For example, a Marsden chart of sea state shows, in each quadrangle, the average or other characteristic height of the waves in the corresponding region.

Marsden square - One of the rectangles on a Marsden chart, containing data on sea state, etc.

marsh - (1) A region of low-lying, wet ground subject to frequent or regular flooding, or ordinarily covered by shallow water. The vegetation of a marsh consists chiefly of reeds, grasses and grass-like plants. SEE ALSO swamp. (2) SEE salt marsh, coastal. SEE ALSO salt marsh; salt marsh, coastal.

marsh, coastal salt - SEE salt marsh, coastal.

marsh, salt - SEE salt marsh.

marsh, tidal - A marsh whose surface is covered and uncovered by the flow of tidal waters.

Marussi's coordinate system - A coordinate system based on a terrestrial ellipsoid and using geodetic longitude and latitude for two of the coordinates of a point, but using as the third coordinate the potential of the level surface through that point.

Marussi tensor - The tensor whose elements M_{ij} are the second derivatives, in a rectangular, Cartesian coordinate-system rotating with the Earth, of the gravitational potential of the Earth.

mascon - A hypothetical concentration of mass, used to explain a large gravity anomaly. The term is a contraction of mass concentration. It was first introduced as a name for the masses causing variations observed in the trajectory of a spacecraft moving around the Moon. The concept itself, however, has long been common in geophysics.

maser - A device using the monochromatic emission from a narrow band in the spectrum of a suitable molecule (e.g., ammonia or methane) or atom to control the frequency of a circuit resonant at the frequency of radio waves. The principal geodetic application has been to the generation of precise frequencies for clocks. The earliest masers used in this way depended on ammonia for controlling the frequency. At present, the most accurate control is obtained by using masers depending on rubidium, cesium or hydrogen gas.

mask - (1) A sheet of thin, black paper, metal, or plastic used to obtain white margins on a photograph by covering part of the photographic emulsion during exposure. (2) Any material used to cover part of an emulsion and thus keep that part from recording an image.

mask, burn-out - A sheet (carrier or carrier sheet) of transparent material to which are affixed pieces (masks) of opaque material. It is placed over an unexposed emulsion and the combination exposed to uniform illumination. Those portions not covered by the masks are completely exposed (burnt out).

mass - A measure of the amount of matter contained in or constituting a body, as evidences by that body's resistance to a change of velocity. Also called, erroneously, weight. Weight is a measure of the force exerted on a body and changes with the force exerted; it is zero in the absence of a force external to that body. E.g., the mass of an astronaut is the same whether he is on Earth or in outer space. His weight, however, is near zero in interplanetary space if the spacecraft is not accelerating. SEE ALSO test mass.

mass, proof - SEE test mass.

mass, topographic - Solid or liquid matter between the geoid and the lower boundary of the atmosphere. The topographic masses are all those parts of the hydrosphere and lithosphere lying outside the geoid.

mass diagram - A cumulative graph of the amount of earth moved from one place to another.

master map projection - An originally computed and constructed graticule from which copies are made and which serves as the master for making copies of all graticules contained within the same pair of standard parallels. This term is used principally for graticules on the transverse Mercator map-projection. Also called a master projection.

master model - SEE model, original.

master print - A photograph representative of the region photographed for making a mosaic and used as a guide during the developing process to ensure the tonal match of subsequent prints.

master projection - SEE master map projection.

master station - That base station (3), in a group of base stations, which controls transmission or reception at the other base stations. If pulses are transmitted, the master station controls the timing between the pulses sent out by the base stations. SEE ALSO slave station; LORAN.

master title plat - (1) The plat on which is shown survey data necessary to identify and describe lands in the public domain, and on which is indicated those sections which currently limit or restrict the use or availability of public lands and resources. (2) A composite of the plats, of a township, on which is shown the ownership and legal status of the land.

matching (cartography) - The comparing, adjusting, and correcting of details on the overlapping part of a new map to agree with the same details on the overlapped part of an existing map.

matching (navigation) - SEE map-matching; matching, star-chart; navigation by star-chart matching.

matching, star-chart - Comparing a star chart with the appearance of the sky as it is being observed by a scanning device aboard an vehicle on the surface, in the air, or in outer space. The technique is used for guiding aircraft, rockets, spacecraft, ships and sometimes, land-based vehicles.

match line - A line normally drawn with a grease pencil on a photograph inward from a torn edge of the print, serving as a guide to registration when inserting the print into a mosaic. Usually, several match lines are drawn on each photograph of a mosaic.

match strip - SEE tie strip.

matrix - A rectangular array of numbers or algebraic quantities (called elements) identified by the numbered rows and columns in which they appear, and added to or multiplied by a similar rectangular array according to the following rules. (a) The sum of two arrays A and B each having N-rows and K-columns is the array C having N rows and K columns; the element in row n and column k of C is the sum of the elements in the n-th row and k-th column of the arrays A and B. (b) The product C = AB of the array A having N rows and K columns times the array B with K rows and M columns is the array C having N rows and M columns, the element in row n and column m of C being the sum over k of the products of the elements in row n, column k of A times the element in row k, column m of B. The symbols for arrays satisfying the above rules for addition and multiplication can be added (or subtracted) and multiplied together as if they were simple algebraic quantities. If the reciprocal (inverse) of a matrix B is defined as a matrix B^{-1} satisfying the equation $B^{-1} B = 1$, then the quotient C of A divided by B can be defined as the product of A by B^{-1}. This definition requires that the dimensions of B^{-1} satisfy the requirements for its multiplication by A and that the reciprocal exist. (2) The more or less homogeneous material, in rock, containing large, imbedded grains or crystals of mineral. SEE ALSO observation matrix; rank (of a matrix); rotation matrix; Stieltjes matrix; trace (of a matrix); unit matrix; unit matrix, extended; weight matrix.

matrix, augmented - The matrix consisting of the coefficients of the unknowns, in a system of simultaneous linear equations, together with the constant terms.

matrix, conditional inverse - SEE matrix, generalized inverse.

matrix, covariance - SEE covariance matrix.

matrix, generalized inverse - (1) G is a generalized inverse (reciprocal) matrix of the matrix B if B G B = B. Also called a generalized inverse, rectangular inverse matrix, g-inverse and conditional inverse matrix. There are numerous other definitions of generalized inverse matrix, as well as special varieties such as the pseudo-inverse matrix. (2) SEE matrix, pseudo-inverse.

matrix, Hermetian - SEE Hermetian matrix.

matrix, inverse - The matrix B^{-1} related to a matrix B by the equation $B^{-1} B = I$, in which I is a square matrix having all elements in the n-th row and n-th column equal to 1 and having all other elements equal to zero. Also called an inverse. The inverse matrix exists, in general, only for square matrices and then only for those which do not have their determinant (the determinant of the array) equal to zero. Inverse matrices of rectangular matrices can be made to exist by making special definitions for them. Such inverse matrices are called generalized inverse matrices, g-inverses or pseudo-inverse matrices.

matrix, irreducible - A matrix, having N rows and N columns, such that it cannot be put into the form
$$A_{r,r} \quad B_{r,n-r}$$
$$0 \quad C_{n-r,n-r}$$
by any interchange of rows or interchange of columns.

matrix, normal - The matrix composed of the coefficients of the independent variables in a set of linear, simultaneous equations having exactly as many equations as independent variables. The normal matrix occurring when adjusting a set of differences of elevation is an irreducible Stieltjes matrix.

matrix, positive-definite - A matrix A such that for every real vector x the scalar quantity $x^t A x$ is a positive real number.

matrix, pseudo-inverse - A matrix G satisfying the four conditions MGM = M, GMG = G, $(MG)^t$ = MG and $(GM)^t$ = GM, in which M is the matrix of which G is the pseudo-inverse. Also called the generalized inverse matrix.

matrix, rectangular - A matrix having N rows and M columns, with N different from M.

matrix, rectangular inverse - SEE matrix, generalized inverse.

matrix, rotational - A square matrix which leaves unchanged the length of any vector it multiplies. Alternatively, a matrix whose determinant is equal to 1, and such that the adjoint is equal to the reciprocal. Also called a rotation matrix. The rotational matrix of order 2 is often defined as
$$\begin{matrix} \cos\theta & \sin\theta \\ -\sin\theta & \cos\theta \end{matrix}$$
Rotational matrices of third order can be constructed using the 2nd-order rotational matrix as a building block.

matrix, transposed - A matrix A^T derived from a matrix A by interchanging the rows and columns of A. Also called a transpose matrix, matrix transpose or transpose.

matrix, unitary - A Hermetian matrix which is equal to its inverse matrix.

matrix, variance-covariance - A matrix whose diagonal elements are variances and whose off-diagonal elements are covariances.

matrix algebra - An algebra in which the variables represent matrices instead of scalars; the set of rules for using matrices.

matte print - SEE print, matte.

Maurer's map projection - SEE map projection, two-point azimuthal.

Maurer's orthodromic map projection - SEE map projection, two-point azimuthal.

maximum entropy - SEE method of maximum entropy.

maximum thermometer - A thermometer which automatically registers the highest temperature to which it has been exposed since its last setting.

Mayer potential - A function U(x,dx/dt,t) having a force f as its gradient, according to the formula f = $*\Delta_1$ U - (d/dt) ($*\Delta_2$ U), in which $*\Delta_1$ denotes the gradient with respect to x and $*\Delta_2$ denotes the gradient with respect to dx/dt.

Mayer's formula - The formula $\alpha = T_{obs} \Delta T + a \sin(\phi' - \delta) \sec \delta + b \cos(\phi' - \delta) + c^* \sec \delta$, in which α is the right ascension of the star observed, T_{obs} is the time at which the star was observed to pass through the meridian, ΔT is the clock correction, ϕ' is the latitude of the observer and δ is the declination of the star. a, b, and c* are constants characteristic of the instrument used: a is the azimuth error, b the level error and c* the collimation error. This formula applies to observations made at the upper culmination of the star. For use on observations made at lower culmination, the sign of δ is changed. Bessel's formula and Mayer's formula are equally suitable for computing when, as is almost always the case, the errors in azimuth, leveling and collimation are small. When these errors are large, Bessel's formula should be used.

M crust - That part of the Earth above the Mohorovicic discontinuity.

M-discontinuity - SEE Mohorovicic discontinuity.

Meades Ranch - SEE North American Datum 1927.

mean - SEE average; error of the mean, standard.

mean, arithmetic - SEE average, arithmetic.

mean, consecutive - SEE average, moving.

mean, geographic - That point, on a map, whose coordinates are the average of the coordinates of a mapped distribution of points.

mean, geometric - The N-th root of the product of the N numbers in a set. Also called a geometric average.

mean, harmonic - The reciprocal of the sum of the reciprocals of the numbers in a set. Also called a harmonic average.

mean, overlapping - SEE average, moving.

mean, running - SEE average, moving.

mean, standard error of the - SEE error of the mean, standard.

mean, weighted - SEE average, weighted.

meander (n.) - A turn or wandering, as of a stream or the edge of a lake.

meander (v.) - To survey, by traversing, the margin of a body of water such as a stream or lake.

meander corner - (1) A corner established at the intersection of a township, range, or section line with the banks of a navigable stream or a body of water around which a meander line has been established. (2) A corner established at the intersection of a standard township or section line with the meander line near the banks of a navigable stream or a meanderable body of water.

meander corner, auxiliary - A corner, established at a suitable point on the meander line of a lake lying entirely within a quarter section or on the meander line of an island falling entirely within a section, which is found to be too small to subdivide. A line is run connecting the corner to a regular corner on the section's boundary line.

meander corner, special (USPLS) - A corner established at (a) the intersection of a surveyed subdivision-of-section line and a meander line; or (b) the intersection of a computed center line or a section and a meander line. In the latter case, the center line of the section is calculated and surveyed on a theoretical bearing to an intersection with the meander line of a lake (over 50 acres in area) which is located entirely within a section.

meander factor - The length of the geodesic between two points, on an ellipsoid representing the Earth, divided by the length of the trace, on that same ellipsoid, of the path actually taken in traveling between the two points.

meander line (USPLS) - (1) A traverse of the margin of a body of water. (2) The line followed in meandering. Meander lines are not surveyed as boundary lines but are surveyed to determine the sinuosities of the bank or shore line and the quantity of land not under water. Practice of the U.S. Bureau of Land Management is to meander only permanent, natural bodies of water. (3) A metes-and-bounds traverse made approximately along the mean high-water line of a permanent, natural body of water. By following the sinuosities of the bank or shoreline, the meander line allows one to compute the area of land remaining after the area covered by water has been segregated. A meander line (2) differs from lines established by other kinds of metes-and-bounds surveys in that it ordinarily does not determine or fix boundaries.

meander line, non-riparian - A meander line (2) established by judicial opinion as a meander line even tho the meander did not follow the actual bank or shore. A non-riparian meander line exists as the result of a judicial opinion that a gross error was committed in recording the notes relating to the meander of a bank or shore line, the opinion being that the meander line as recorded did not follow the actual bank or shore as it existed at the time of survey but that, instead, a large tract, situated between the actual bank or shore and the location of the meander line as recorded, was omitted. Ordinarily, the opinion annuls the usual doctrine that the bank or shore line here should be construed as the boundary of the survey, and contrarily gives that status to the meander line as recorded.

Mean Observatory - A fictitious observatory situated on the Greenwich Mean Astronomical Meridian. I.e., a fictitious, ideal observatory situated so that times determined by it are exactly the same as the times established by the Bureau International de l'Heure as the average of times determined at many real observatories.

mean of the errors - SEE error, average.

mean sea level - SEE sea level, mean.

measurement - (1) The act or deliberately sensing an event or thing, noting the circumstances, and assigning a numerical value to the event or thing. Equivalently, an observation together with the act of associating a numerical value with the observation. SEE ALSO observation. (2) The numerical value associated with the observation. The terms measurement and observation are often used as synonyms, particularly in astronomy and surveying. In surveying, reading is often used as a synonym for measurement in the sense of a numerical value. SEE ALSO measurement process; method of measurement.

measurement, direct - Measurement. Also called, ambiguously, direct observation.

measurement, double-proportionate (USPLS) - A procedure for restoring a lost corner of four townships or four internal sections by using the rule that the measured distance from a known corner to the restored corner shall have the same ratio to the distance given in the original record as the measured distance between known corners has to the distance given in the original record.

measurement, indirect - (1) A computed value derived from measurements. An indirect measurement is not a measurement and the term is ambiguous. The term is used with another meaning in the U.S. Bureau of Land Management. (2) A distance determined by computation from measured distances and angles. (3) SEE measurement, intermediary.

measurement, intermediary - A measurement made of a quantity which is a function of the quantities whose values are wanted. Also called an indirect measurement.

measurement, mark-to-mark - A measurement from the marker identifying one control point on the ground to the marker identifying another control point on the ground. The definition given is the original version. It will fit actual practice better if mark is substituted for marker in it.

measurement, proportionate - A measurement applying an even distribution of a determined excess or deficiency of measurement, ascertained by retracing and established line, to provide concordant relations between all parts.

measurement, single-proportionate - (1) A procedure for restoring a lost corner by reference to two known corners (both east and west or north and south of the lost corner) and an alinement in the transverse direction, by using the rule that the measured distance from a known corner to the restored corner shall have the same ratio to the corresponding, recorded distance (to the original corner) as the total, measured distance between known corners has to the distance given in the original record, and that the alinement be preserved. (2) A method of proportioning measurements, in the restoration of a lost corner whose location is determined with respect to alinement in one direction, by distributing the excess or deficiency in such a way that the amount given to each interval shall have the same proportion to the whole difference as the recorded length of the interval bears to the whole distance.

measurement process - The realization of a method of measurement.

measuring engine - An instrument for measuring distances between points on a photograph or other flat surface. It consists either of a microscope moved, by accurately measured amounts, with respect to the surface on which measurements are made, or of a carriage moving that surface by accurately measured amounts under the microscope. Also called a comparator or coordinatograph. The term comparator is common in photogrammetry; the term coordinatograph in cartography. The movable member (microscope or carriage) is usually connected mechanically to two very precisely-machined, mutually perpendicular screws in such a manner that each screw can move that member independently in one of two orthogonal directions. The distance the microscope or carriage is moved by either screw is measured by dials showing the equivalent, in distance, of the number of rotations and fraction of a rotation each screw has made. The distances measured by a perfectly constructed measuring engine are equivalent to rectangular, Cartesian coordinates. An actual measuring engine is equivalent to a Cartesian coordinate system in which the axes are not mutually perpendicular and the scale along each axis is divided into intervals differing slightly in length from each other. Every measuring engine must therefore be calibrated by making measurements with it on a grid or other pattern having accurately-known divisions. Some measuring engines have only one screw. These engines use the screw for moving the microscope; the object being measured is fastened to a platform that can be rotated through 90°. The object is therefore first measured with the platform in one position and then with the platform rotated through 90°. If the engine is of very high quality, the platform is sometimes graduated in angular units around the circumference so that polar coordinates can be determined.

measuring engine, stereoscopic - A measuring engine having binocular optics creating a stereoscopic image from stereoscopic pair of photographs and allowing the operator to determine coordinates in a three-dimensional, rectangular, Cartesian coordinate system.

measuring rod - One of a pair of similar (usually graduated) wooden rods 3 to 5 meters long, each having a flat, metallic ferrule at one end and ending in a knife-edge at the other. Measuring rods were formerly used in measuring distance by laying the rods down, one after the other, knife-edge against flat end, from the starting point to just past the end point.

measuring unit, inertial - That part of an inertial-surveying system which measures acceleration by using accelerometers. The data from the inertial measuring unit go to a computer in which the coordinates of the surveying system in a suitable coordinate system are computed. The computer also sends commands, based on these data, to the inertial measuring unit and, possibly, to the navigator.

Mecca map projection - SEE map projection, retro-azimuthal.

mechanical (n.) - A complete piece of copy for reproduction, consisting of an accurate assemblage of paste-ups of text, display matter, line drawings, and illustrations. (jargon)

mechanical-templet method (of aerotriangulation) - Radial aerotriangulation using slotted, spider-type, or other form of mechanical templet for graphical adjustment. Also called a mechanical-templet plot.

mechanics, celestial - (1) That part of positional astronomy concerned with the mathematical theory used for describing or predicting the motions of celestial bodies. Equivalently, it is dynamics limited to the consideration of the motions of celestial bodies. (2) That part of mechanics which treats of the motions of two or more bodies in space under the influence of their mutual gravitational attraction. Also called orbit theory, dynamical astronomy or astrodynamics. Orbit theory, however, covers a broader subject than merely the motion of celestial bodies; it also covers, for example, the motion of charged particles in an electromagnetic field. Dynamical astronomy and astrodynamics are fully equivalent to celestial mechanics in meaning but have more popular appeal.

median - (1) The central number, in a set of numbers arranged in order of size, if the set contains an odd number of numbers; it is the average of the two central numbers if the set contains an even number of numbers. (2) A longitudinal strip of road separating opposing streams of traffic. A median may be raised above the level of the road or be flush with it; it may consist of reinforced rails or concrete barriers.

median, geographic - That point, on a map, whose coordinates are the medians of the coordinates of mapped points.

median line - (1) A line, in a body of water, every point of which is equidistant from the nearest points on opposite banks or other points of reference. (2) The line lying at the same distance from two non-parallel lines. (3) A line every point of which is equidistant from the two nearest points on the baselines from which the breadth of the territorial sea of each of two coastal nations is measured. The term was adopted at the Geneva Conference on the Law of the Sea in 1958.

medimarimeter - An instrument for determining mean sea level.

mediterranean - A large body of salt water, or of inland seas surrounded by land, which may have one or more narrow openings to the ocean or to another sea.

Mediterranean - That large body of water lying between and effectively separating Europe from Africa, and connecting to the Atlantic Ocean through the Straits of Gibrater. Also called the Mediterranean Sea. The name comes from Latin and means, essentially, at the center of the world.

Mediterranean Sea - SEE Mediterranean.

megagram - SEE tonne.

megalopolis - An extensive, densely populated, metropolitan region consisting of a number of large urban regions. SEE ALSO area, metropolitan.

Meinesz formula for deflection of the vertical - SEE Vening Meinesz formula for deflection of the vertical.

meizoseismal (adj.) - Having the greatest destructive, seismic effects. A line or curve representing points at which destruction was greatest in the region around an earthquake's perimeter is called a meizoseismal curve.

member (of an optical system) - All the components either ahead of the aperture stop (front member) or behind the aperture stop (rear member).

member, front - SEE member (of an optical system).

member, optical - SEE member (of an optical system).

member, rear - SEE member (of an optical system).

memorial (USPLS) - A durable article deposited in the ground at the location of a corner to perpetuate that location should the monument there be removed or destroyed. The memorial is usually deposited at the base of the monument and may consist of anything durable such as glass or stoneware, a marked stone, charred stake or quantity of charcoal.

memory, random-access - That part of computer's memory into which numbers or numerically-coded information can be placed and later retrieved independently of the order in which the numbers were placed there. Usually referred to as RAM.

memory, read-only - That part of a computer's memory into which numbers or numerically-coded information has been placed and from which they can be later retrieved, but into which no further information can be placed. Usually referred to as ROM.

Mendenhall Act - An order by the office of Weights and Measures officially establishing the definition of the yard as 3600/3937 of the meter, and approved by the Secretary of the Treasury on 5 April 1893. This order applied specifically to weights and measures used officially by the U.S. Government and by the separate States.

Mendenhall pendulum - An invariable pendulum, one-quarter meter long, with a period of vibration of one-half second, and consisting of a lenticular bob on a thin stem, swung in near vacuum in an airtight case. The pendulum was designed in 1890 by Mendenhall and assistants of the U.S. Coast and Geodetic Survey.

mensuration - (1) The act, art, or process of measuring. (2) That part of applied mathematics dealing with the determination of height, length, area or volume.

menu - A list of choices. (jargon)

Mercator bearing - The bearing, of a straight line plotted on a chart on a Mercator map projection, as taken from the chart.

It is the angle from a line representing the meridian to the straight line.

Mercator chart - A chart drawn on the Mercator map projection. Also called an equatorial, cylindrical, orthomorphic chart.

Mercator chart, inverse - SEE Mercator chart, transverse.

Mercator chart, oblique - A Mercator chart having an oblique aspect.

Mercator chart, transverse - A chart made using the transverse Mercator map projection. Also called an inverse cylindrical orthomorphic chart and an inverse Mercator chart.

Mercator coordinate - One of the two coordinates λ_m, ϕ_m of a point on a sphere of radius R and at longitude $\lambda_o + \Delta\lambda$ and latitude ϕ, as defined by the equations $\lambda_m = \Delta\lambda$ or $R\Delta\lambda$; $\phi_m = \ln \tan\frac{1}{4}(\pi+2\phi)$ or $R [\ln \tan\frac{1}{4}(\pi+2\phi)]$. Mercator coordinates can be defined in a similar manner for a point on a rotational ellipsoid.

Mercator coordinate, Universal Transverse - One of the two coordinates of a point in the Universal Transverse Mercator Grid system. The particular grid (zone) in which the point lies must be designated. The designation may be considered an additional coordinate.

Mercator direction - SEE direction, rhumb-line.

Mercator equal-area map projection - SEE map projection, sinusoidal.

Mercator grid, transverse - (1) A plane rectangular grid placed on maps drawn on the transverse Mercator map projection. This was originally a military grid. (2) An informal designation for a State plane coordinate system or State coordinate system based on a transverse Mercator map projection.

Mercator Grid, Universal Transverse - SEE Universal Transverse Mercator Grid.

Mercator Grid System, Universal Transverse - SEE Universal Transverse Mercator Grid System (UTM).

Mercator map projection - A conformal map projection from the rotational ellipsoid to the plane, as defined by the equations $x = k a \Delta\lambda$ and $y = k a \ln \{\tan \frac{1}{4} (\pi+2\phi) [(1 - \epsilon \sin \phi)/(1 + \epsilon \sin \phi)]^{\epsilon/2}\}$, in which x and y are the rectangular Cartesian coordinates of that point, in the plane, which corresponds to a point, on the rotational ellipsoid, with longitude $\lambda_o + \Delta\lambda$ and latitude ϕ; λ_o is the longitude of the central meridian. a is the length of the semi-major axis of the ellipsoid and ϵ is the eccentricity. k is the scale factor. The equator is represented by a straight line (usually true to scale; the two lines representing parallels of latitude equidistant from the equator are sometimes made true to scale instead). The meridians are represented by parallel straight lines perpendicular to the line representing the equator. The parallels of latitude are represented by a second family of straight lines; they are parallel to the line representing the equator. The projection is made conformal by increasing the spacing between lines representing parallels of latitude as the distance from the equator increases, to conform with the expanding along the lines representing parallels of latitude. The Mercator map projection is considered one of the most valuable of all map projections. Its most valuable feature is that a line of constant azimuth (bearing) on a sphere is represented on the map by a straight line. Although Mercator is commonly credited with introducing the map projection in 1569 and Wright is credited with giving a theoretical description in 1599, the projection was used by Etzlaub in 1511. Also called an equatorial, cylindrical, orthomorphic map-projection and a Mercator projection. SEE ALSO Mercator map projection, transverse.

Mercator map projection, decumenal - That special case of the oblique Mercator map-projection, for the mapping of a sphere onto a cylinder, in which the cylinder's axis passes through the point at longitude $180^\circ + \lambda_o$ and latitude $90^\circ - \phi_o$; λ_o and ϕ_o are the coordinates of the center of the region to be mapped.

Mercator map projection, inverse - SEE Mercator map projection, transverse.

Mercator map projection, oblique - (1) A conformal map projection mapping the sphere so that a particular great circle which is neither a meridian nor the equator is mapped into a line along which the scale is constant (true to scale). Also called the transverse Mercator map projection (American) and skew Mercator map-projection. (2) A map projection mapping a rotational ellipsoid in such a manner that the map has approximately the characteristics of a map produced according to the preceding definition. The rotational ellipsoid cannot be mapped on the plane so as to preserve both conformality and constant scale along the oblique line. Most practical applications retain conformality but allow the scale along the oblique line to vary slightly. Such variations are named after the originator; for example, the Rosenmund map projection and the Hotine oblique Mercator map-projection. The oblique Mercator map-projection has been used principally for maps of Malaya, Borneo and Switzerland, and for the State Coordinate System in Alaska. It has also been used for maps made from pictures taken by cameras on various satellites.

Mercator map projection, skew - SEE Mercator map projection, oblique.

Mercator map projection, space-oblique - A variant of the oblique Mercator map-projection used for maps made from pictures taken by LANDSAT satellites. The central line follows the undulating ground-track (foot-point) of the satellite rather than an ellipse or great circle. The initial concept was due to A. Colvocoresses.

Mercator map projection, transverse - (1) A conformal map projection from the rotational ellipsoid to the plane in such a way that the central meridian, or a pair of lines virtually parallel to the central meridian, is mapped into a straight line and exactly to scale. It is essentially the same as a Mercator map-projection preceded by a rotation of the ellipsoid through $90°$ about a major axis. Equivalently, it is the same as a Mercator map-projection calculated for a cylinder with axis in the equatorial plane. The map projection is derived either as a direct mapping from rotational ellipsoid to plane (and is then called, by some cartographers, the Gauss-Krüger map-projection) or as the product of a mapping from ellipsoid to sphere followed by a mapping from sphere to plane (and then called, by some cartographers, the Gauss-Schreiber map-projection or Gauss-Laborde map-projection. It is also called Lambert's cylindrical orthomorphic map projection, Lambert's conformal cylindrical map projection and the Gauss conformal map projection. Except for the central meridian, no meridians are mapped into straight lines and except for the equator, no parallels of latitude are mapped into straight lines. The theory for mapping the sphere was developed by J.H. Lambert (1772) and an analytical theory for the rotational ellipsoid was given by Gauss (1825). The map projection was extensively discussed by L. Krüger in two publications (1912 and 1919) and was adopted, under the name Gauss-Krüger map-projection by Germany in 1927 for its maps. (2) Either the transverse Mercator map projection as defined above or an oblique Mercator map projection. (American) Also called an inverse cylindrical orthomorphic map projection.

Mercator Map Projection, Universal Transverse - SEE Universal Transverse Mercator Map Projection.

Mercator projection - SEE Mercator map-projection.

Mercator-Sanson map projection - SEE map projection, sinusoidal.

Mercator track - A rhumb line on a map constructed using a Mercator map projection. SEE ALSO rhumb line.

mercury barometer - SEE barometer, mercurial.

Mercury Datum 1960 - A geodetic datum based on the Mercury ellipsoid 1960 and located with respect to a number of major datums of the world by a large, given set of constants. It was used principally by the USA in computing orbits and trajectories from observations made at tracking stations on that datum, in particular, those tracking stations used in Project Mercury. It has also been used as the datum for locating OMEGA, LORAN-C and LORAN-A stations.

Mercury Datum 1968, Modified - SEE Modified Mercury Datum 1968.

mercury level - A level consisting of a pool of mercury in a container attachable to a transit in such a manner that the surface of the mercury is visible and can be used for indicating a level surface. In one form, called a mercury leveler, the surface is viewed through a telescope (autocollimator) attached to the instrument to be leveled. When the cross-hairs of this telescope appear to coincide with the reflection of the cross-hairs from the surface of the mercury, the line or sight is level.

mercury leveler - SEE mercury level.

Mercury spheroid 1960 - A rotational ellipsoid having the dimensions

    semi-major axis   6 378 166 m
    flattening   1/298.3 .

Also called the Fischer 1960 spheroid. The Mercury 1960 spheroid was used principally by the USA for calculating orbits and trajectories of rockets and satellites.

mere - (1) A monument marking a point on a boundary. (2) A wall marking a boundary. (3) SEE boundary.

mere stone - A monument marking a point on a boundary and made of stone. Equivalently, but less precisely, a mere made of stone.

meridian - (1) A line running north and south and from which differences of longitude (or departures) and azimuths are reckoned. (2) A curve all points of which have the same longitude. (3) The intersection of a plane containing the optical axis of an optical system with a plane perpendicular to that axis. (4) SEE meridian plane. SEE ALSO ephemeris meridian; Ferro Meridian; grid meridian, prime; guide meridian; Greenwich Meridian; Greenwich Mean Astronomical Meridian; grid meridian; Paris Meridian; Pulkova meridian; time meridian; Washington Meridian; zero meridian.

meridian, assumed - A meridian adopted as a matter of convenience.

meridian, astronomic - (1) A line, on the surface of the Earth, having the same astronomic longitude at all points. Also called a terrestrial meridian and true meridian. Because the deflection of the vertical is not the same at all points on the Earth, an astronomic meridian is an irregular line and does not lie in a plane. The astronomic meridian and the line whose astronomic azimuth at every point is due north (0°) or due south (180°) are not necessarily coincident, although in land surveying the term astronomic meridian is sometimes used for a line running north and south and having its starting point on a prescribed astronomic meridian in the usual sense. (2) A plane parallel to the Earth's axis of rotation and passing through the vertical at a given point. Also called the plane of the astronomic meridian. (3) That great circle, on the celestial sphere, lying on the plane parallel to the Earth's axis of rotation and passing through the vertical at a given point. Also called a celestial meridian. Equivalently, the great circle passing through the celestial poles and through the zenith of a point on the Earth. For planetary observations, a meridian is defined to be half that great circle. (4) The curve, on the celestial sphere, defined by the zeniths of all points with the same astronomic longitude.

meridian, celestial - That hour-circle which contains the zenith, or that vertical circle containing the celestial pole. The two definitions are equivalent. The plane of the celestial meridian is parallel to the Earth's axis of rotation but, because of the deflection of the vertical, usually does not contain it. The intersection of the plane of the celestial meridian with the plane of the horizon is the meridian line used in plane surveying. It should not be confused with a meridian in an ecliptic (celestial) coordinate system. Note that the term celestial meridian does not denote a meridian of the ecliptic.

meridian, central - (1) The line of constant longitude at the center of a graticule. The central meridian is used as a base for constructing the other lines of the graticule. (2) The meridian used as y-axis in computing tables for a State Plane

Coordinate system. The central meridian of the coordinate system usually passes close to the geometric center of the region or zone for which the tables are computed but, to avoid using negative values, is given a large positive value which must be added to all x-coordinates. (3) That line, on a graticule, which represents a meridian and which is an axis of symmetry for the geometric properties of the graticule.

meridian, ecliptic - SEE meridian of the ecliptic.

meridian, fictitious - The line on an ellipsoid, or the corresponding line of a graticule, in which a suitably defined plane perpendicular to the plane of the fictitious equator intersects the ellipsoid. If the chosen plane is also a reference plane, the resulting fictitious meridian is called the prime fictitious meridian.

meridian, geodetic - The line in which a plane through the shortest axis of an ellipsoid intersects the ellipsoid. Equivalently, a line, on the ellipsoid, having the same geodetic longitude at every point. Also called, erroneously, a geographic meridian. Geodetic meridians on a heteraxial ellipsoid or on a rotational ellipsoid are ellipses; geodetic meridians on a sphere are great circles.

meridian, geographic - (1) A general term for either an astronomic or a geodetic meridian. (2) SEE meridian, geodetic.

meridian, geomagnetic - A line, in a geomagnetic coordinate system, having an approximately north and south direction at the equator. Not to be confused with a magnetic meridian.

meridian, inverse - SEE meridian, transverse.

meridian, local - The meridian through any particular place or observer, serving as the reference for local time. Also called a reference meridian.

meridian, magnetic - (1) The vertical plane in which a freely suspended, symmetrically magnetized needle, influenced by no transient, artificial, magnetic disturbance, will come to rest. (2) The curve, on the Earth's surface at all points of which the vertical plane described in the preceding definition is tangent to the curve. (3) The direction, at any point, of the horizontal component of the Earth's magnetic field.

meridian, normal - That meridian, lying west of Greenwich, at which the beginning of the Besselian year coincides with the beginning of the Julian year. The Besselian year begins at the same time all over the world. For example, the Besselian year 1966.0 began on 1966 January 0.799d Greenwich mean sidereal time. At this instant, the Julian date was 1966 January 0.00d on the meridian at 72.36 degrees West.

meridian, oblique - A great circle perpendicular to an oblique equator. An oblique meridian used as referent for longitude or time is called the prime oblique meridian. SEE ALSO meridian, fictitious. This definition applies only to meridians on a sphere. It must be modified if it is to apply to ellipsoids in general.

meridian, photographic - SEE photograph meridian.

meridian, primary - SEE meridian, prime.

meridian, prime - Any meridian from which the longitudes of other meridians are reckoned. Also called a null meridian, primary meridian or zero meridian. At the International Meridian Conference held in Washington, D.C. in 1884, representatives of 22 governments approved adoption of the Greenwich Meridian as the primary meridian for the Earth (one representative objected and two representatives abstained from voting.) Note that the Greenwich Meridian is a primary meridian but a primary meridian is not necessarily the Greenwich Meridian. Although this meridian was accepted internationally, various countries still include their older, national meridians on some maps. Among these meridians are Amsterdam Meridian (4° 53' 01" E), Athens Meridian (23° 42' 59" E), Berne Meridian (7° 26' 25" E), Djakarta Meridian (106° 48' 28" E) and Istambul Meridian (28° 58' 50" E).

meridian, prime fictitious - A fictitious meridian used as reference.

meridian, prime inverse - SEE meridian, prime transverse.

meridian, prime oblique - An oblique meridian chosen as the referent from which longitudes and time are calculated.

meridian, prime transverse - A transverse meridian chosen as the referent from which longitudes and time are calculated. Also called a prime inverse meridian.

meridian, principal (USPLS) - (1) A line extending north and south along the astronomical meridian passing through the initial point and along which township, section and quarter-section corners are established. Also called a basis meridian in some very early surveys. A principal meridian is the line from which the survey of the boundaries of the township along the parallels is started. Additional meridians, known as guide meridians, are established along astronomical meridians, usually at intervals of 24 miles east or west of the principal meridian. There are 35 principal meridians in the USA. The first few are numbered, e.g., First Principal Meridian, Second Principal Meridian. The rest are named after localities, e.g., Salt Lake Meridian. (2) SEE line, principal.

meridian, standard - (1) That meridian on which mean solar time is the same as the standard time throughout a particular zone. Throughout most of the world, the standard meridians are those whose longitudes are exactly divisible by 15. (2) That line, on a graticule, representing a meridian and along which the scale of representation is as stated.

meridian, terrestrial - SEE meridian, astronomic. For particular use, the term astronomic meridian is preferred over terrestrial meridian. For general use, the preferred term is geographic meridian, which applies alike to astronomic and geodetic meridians.

meridian, transverse - On the sphere, a great circle perpendicular to a great circle (the transverse equator) passing through the poles of the sphere. The analogous figure on a rotational ellipsoid would be an ellipse whose major diameter is perpendicular to the minor axis of the ellipsoid and perpendicular to a specified plane (the transverse equatorial plane) through the minor axis. Also called an inverse meridian.

meridian, true - SEE meridian, astronomic.

meridian altitude - SEE altitude, meridional.

meridian angle - The arc at the celestial equator, or the angle at the celestial pole, between the upper branch of the local celestial meridian and the hour angle of the object of interest, measured eastward or westward from the local celestial meridian through 180° and labeled to indicate the direction in which the angle is taken.

meridian circle - A telescope mounted on a horizontal axis transverse to the optical axis; the ends of the horizontal axis are carried in bearings fixed so that the line of sight lies permanently in the local meridian. Graduated circles on the axis of rotation allow the angle of rotation in the vertical plane to be measured very precisely. Also called a transit circle.

meridian difference - SEE difference, meridional.

meridian distance - (1) The average distance of the ends of a line from the meridian. (2) The distance of the middle of a course from the meridian. (3) The perpendicular distance, in a horizontal plane, of a point from a meridian of reference. Also called meridional distance. The difference of the meridian distances of the ends of a line is called the departure of the line. (4) The hour angle of a celestial body when close to but not exactly on the celestial meridian.

meridian distance, double (DMD) - The sum of the meridian distances of the ends of a surveyed line or course. In practice, assuming a closed traverse with the meridian of reference passing through the initial station, the double meridian-distance of the first course is equal to the difference of the meridian distances of the ends of that course; the double meridian-distance of each succeeding course is equal to the double meridian-distance of the preceding course plus the difference of the meridian distances of the ends of the current course. If the double meridian-distance of each course is multiplied by the length of the arc between the parallels of latitude bounding that course, areas double those between the courses and the meridian are obtained; from these areas, the area of the tract being surveyed can be calculated. By a strictly analogous procedure, double parallel-distances can be calculated, using the latitudinal differences (in linear measure) of the successive courses instead of meridian differences and differences of meridian difference, and using a line running east and west as the parallel of reference.

meridian extension - That part, of a line representing a meridian on a graticule, which extends above the top construction line of the graticule.

meridian line - The line in which the plane of the celestial meridian intersects the plane of the horizon. The astronomical azimuth of a meridian line is 0° or 180°.

meridian of the ecliptic - A circle, on the celestial sphere, containing all points of the same celestial longitude.

meridian of the Mean Observatory - SEE Mean Observatory.

meridian plane - (1) A plane containing the least axis of an ellipsoid representing the Earth; the plane is usually taken to be parallel to the Earth's axis of rotation. (2) A plane containing the normal at a point on an ellipsoid representing the Earth, and parallel to the least axis of the ellipsoid. (3) A plane, containing a meridian on the celestial sphere. (4) SEE plane of the astronomic meridian. SEE ALSO Greenwich mean astronomic meridian plane.

meridian plane, astronomic - SEE plane of the astronomic meridian.

meridian plane, astronomical - SEE plane of the astronomic meridian.

meridian plane, geodetic - SEE plane of the geodetic meridian.

meridian plane, mean astronomic - SEE plane of the mean astronomic meridian.

meridian ring - A graduated ring fitted to a globe in such a manner that its plane contains the poles of the globe and so that it can be adjusted into the plane of any given great circle representing a meridian. A meridian ring is used for measuring angles along a line representing a meridian.

meridian seeker - SEE telescope, meridian-seeking.

meridian telescope - (1) A portable optical telescope so designed that it can be used as an astronomical transit or be quickly converted for use as a zenith telescope. This type of telescope is also known as the Davidson meridian instrument, after G. Davidson of the U.S. Coast Survey, who designed it in 1858. (2) An optical telescope mounted permanently on a horizontal axle transverse to the optical axis and to the plane of the meridian; the ends are carried in bearings fixed permanently so that the telescope's line of sight lies in the local meridian. SEE ALSO meridian circle.

meridian transit - The movement of a celestial body across a celestial meridian. Also called meridian passage.

meridional (adj.) - Having the center of the graticule, on a map, on that line which represents the equator. SEE ALSO part, meridional; table of meridional parts.

meridional parts table - SEE table of meridional parts.

mesa - (1) A nearly level region standing distinctly above the surrounding country, bounded by steeply sloping sides, and capped by layers of erosion-resistant rock. (2) A very broad, flat-topped, usually isolated hill or mountain of moderate height bounded on at least one side by a steep cliff or slope and created by erosion of the surrounding land. A mesa is similar to a butte but has a more extensive summit.

mesne conveyance - A title, in the chain of title, located somewhere between the original conveyance and the present title-holder; not the original or present title but an intermediate conveyance.

mesosphere - (1) The region, in the Earth, extending from about 300 km below the surface to 700 km below the surface, as indicated by data obtained from reflected seismic waves. (2) A region in the atmosphere, extending from about 55 km to 80 km above the surface, in which the temperature decreases with height at an average rate of about 4° - 5° per kilometer.

metacartography - Portrayal of spatial properties on maps considered in competition with other devices such as photographs, pictures, graphs, language and mathematics.

meter (instrument) - (1) A device measuring something and indicating the value of the measurement. Meter and instrument have much the same meaning in geodesy. However, instrument may have other meanings than that given.

meter (unit of length) - A unit of length originally (1791) intended to be 1 part in 10 000 000 of the distance from Equator to Pole, measured along the meridian passing through Dunkirk and a point close to Barcelona. As measurements of length became more accurate, the definition, was changed from time to time. The successive definitions are as follows. (1) A length equal exactly to 443.44 lignes of the Toise de Peru, an iron standard of length made by La Condamine in 1735. This was the metre provisoire recommended by the French Academy in 1793. It was based on various measurements of meridional arcs and related to the arc between Dunkirk and Barcelona (then still being measured.) It was the Academy's best estimate of 1:10 000 000 of that quadrant of the meridional arc passing through Dunkirk and a point close to Barcelona (as proposed by the Academy in 1791). The recommendation was adopted by the National Convention that same year and realized in a line-standard of brass. (2) A length equal to exactly 443.296 lignes of the toise of Peru (3 pieds, 11 lignes, 296 millemes). The new meter, based on the newly completed arc from Dunkirk to Barcelona, was decreed by the French Government on 10 December 1799 to be the national standard of length, was designated the metre vrai et definitif (true and definitive meter) and the old meter abolished. It was realized in four end-standards of platinum, of which one, the Metre des Archives, was deposited in the State Archives. (3) The distance between centers of two marks on a platinum-iridium bar designated by the First General Conference on Weights and Measures (1875) as the Prototype Meter, and located at Breteuil. It was specified that the bar be at 0° C, under 1 atmosphere pressure, and that it be resting horizontally and symmetrically on rollers 571 mm apart and at least 10 mm in diameter. Formulae were derived for the distance under other conditions of temperature, pressure and orientation. This definition was in effect from 1875 to 1960. In 1889, 30 copies of the bar were completed and distributed to members of the International Conference on Weights and Measures. The USA received copies numbered 21 and 27. Copy number 27 was adopted as the primary standard, copy 21 as a secondary or working standard. (4) A length equal exactly to 1 650 763.73 times the wavelength of the orange line (2 $p10^{-5}$ d5) emitted by Kr3086 in an Englehard hot-cathode discharge tube inclosed in a cryostat at the triple-point of nitrogen (63 K). This definition was adopted in 1960 at the 11th Annual Conference on Weights and Measures. In 1927, the value 643.846 96 x 10^{-9} meter was adopted for the wavelength of the red line of cadmium as a secondary standard of length. (5) The length of the path traveled by light in vacuum during a time interval of 1/ 299 792 458 of one second. This definition was adopted by the General Conference on Weights and Measures in October 1983. It was derived from the definition of the speed of light as 299 792 458 meters per second. The standard unit of time, the second, was left unchanged, so that the meter became a derived unit, not a fundamental unit as it was formerly. The spelling metre is that recommended by the International Conference on Weights and Measures and is the one used in official translations of the International Standards Organization's Recommendations. The spelling commonly used in the USA is meter. SEE ALSO Committee Meter; French legal metre; geopotential meter; International Prototype Meter; Metre des Archives; National Prototype Meter.

meter, acoustic ocean-current - An instrument which determines the speed of a current by transmitting acoustic pulses in opposite directions parallel to the direction of flow of the current and measuring the difference in between the times of travel in the two directions.

meter, dynamic - The standard unit of dynamic height, equal to 10 square meters per second. Also called the dynamic metre and geodynamic meter.

meter, geodynamic - SEE meter, dynamic.

meter, international - (1) A unit of length defined by international agreement to have a length of 1 metre (1 meter). In particular, when written International Metre. (2) The unit of length defined by the International Prototype Metre (Meter). The term is best not used with this meaning because the meters defined in 1960 and 1983 are also international meters.

meter, pygmy - SEE pygmy current-meter.

meter, square - An area equal to the area inclosed within a square one meter on a side.

meter-candela-second - An exposure (of photographic film) for 1 second at a distance of 1 meter to a light-source of 1 candela.

metering, remote - SEE telemetering.

mete - Boundary. The term is archaic and is now found chiefly in the plural, in the term metes and bounds (q.v.).

metes and bounds - (1) Those characteristics of a piece of land which are used in defining its boundary. Metes and bounds include measurements as well as monuments, etc. A metes and bounds description of a piece of land is a complete description of the boundary, with the distances (metes) and directions given sequentially around the perimeter (bounds) of the piece. (2) The boundary lines or limits of a piece of land.

metes and bounds, true - A description containing full information on the vicinity of the piece of land, a call for each tie and monument determining the boundaries, all references to adjoining lands by name and record and a full recital of distances and directions of all courses on the external boundary of the piece of land.

metes-and-bounds description - SEE description, metes-and-bounds.

method - For terms beginning with this word, see under the name of the specific method, e.g., for the method of equal altitudes or the equal-altitude method, see equal-altitude method.

method of bearings - SEE bearing method (of adjusting triangulation).

method of correlates - The method of least squares applied to a combination of observation equations and condition equations by introducing additional unknowns (correlates) and solving first for the correlates and then, using this solution, solving for the original unknowns in the observation equations.

method of equal angular elevation - SEE equal-altitude method.

method of equal zenith-distance - SEE equal-altitude method.

method of general perturbations - SEE method of perturbations, general.

method of horizontal taping and plumbing - SEE taping and plumbing, horizontal.

method of least squares - (1) Any method of solving a system of simultaneous equations in which values of the dependent variable are given and the requirement is laid on the solution that the sum of the squares of the differences between the given values and those calculated using the solved-for values of the independent variables shall be less than the corresponding sum calculated using any other values of the independent variables. Also called the least squares method, least squares and minimum squares. (2) The same as the previous definition, but each equation is multiplied by an arbitrary constant (called the weight). Also called weighted least-squares method or the method of weighted least-squares. The method of least squares applied to a system in which there are fewer equations than there are unknowns (values of the independent variables) does not give a set of unique values for the unknowns. Applied to a system in which the number of equations is equal to the number of unknowns, it gives the same answer as any other mathematically rigorous method of solving. Applied to a system in which there are more equations than there are unknowns, it usually gives a set of unique values, whereas the methods of ordinary algebra would declare that no solution existed. It can be shown that, under certain very liberal conditions, the values found by the least-squares method are the most probable values of the unknowns, given the probabilities of the given (observed) values of the dependent variables. This solution is therefore also called the best solution, although the term best has a subjective connotation. Credit for the discovery of the method of least squares is sometimes given to Lagrange, who in 1770 published a description. His method was lost or forgotten for some time. Credit for the discovery is more often given to Gauss, who used it in 1795 but did not publish his results until 1809, while it had been demonstrated by Adrain in 1808. Laplace demonstrated it in 1810 and in 1812, and its theory was compared with practice by Bessel in 1818. The method's usefulness depends on the methods to provide the most probable set of values for the unknowns.

method of lunar observations - Any method for determining longitude, latitude, azimuth and/or time by using lunar observations. SEE ALSO lunar-observation method of determining longitude; lunar-culmination method; occultation.

method of maximum entropy - A method of finding particular values of a set of randomly variable unknowns by requiring that the entropy - $\Sigma \, p_i \log p_i$ of the set of probabilities p_i associated with the unknowns x_i be a maximum. The method has been applied mostly to determining the frequencies present in signals that can be represented as a mixture of frequencies, i.e., in spectral analysis.

method of measurement - The method used in making measurements of a specific kind, as defined collectively by specifying the apparatus and auxiliary equipment used, the operations performed and their sequence and the conditions under which they are performed.

method of meridional zenith distances - SEE Sterneck's method.

method of perturbations - A method of solving the partial differential equations governing the motion of a body moving under various forces (particularly gravitational forces) by first solving the equations for a simple field and then solving the equations for the effects of the differences between the actual field and the simple field.

method of perturbations, general - A method of perturbations in which the solution is obtained as an analytical solution for the deviations from a standard (usually Keplerian) orbit as a function of time. Also called the method of general perturbations.

method of perturbations, special - A method of perturbations employing numerical integration with respect to time. Also called the method of special perturbations.

method of quadrilaterals - SEE quadrilateration (1).

method of repetition (for determining azimuth) - SEE repetition method.

method of rounds - A procedure for making angular measurements, in which the theodolite is pointed successively, in order of their azimuths, at each of the targets involved, starting with a particular target (called the initial target) and the reading of the horizontal circle recorded for each pointing, up to and including the pointing at the initial target again. The succession of pointings and the ordered set of readings is called a round.

method of solar observations - Any method for determining longitude, latitude, azimuth and/or time by using lunar observations. SEE ALSO solar-observation method of determining longitude.

method of special perturbations - SEE method of perturbations, special.

method of steepest descent - (1) A method of finding complete asymptotic expansions for integrals of the form $I(x) = \int \{g(z) \exp[x f(z)]\} dz$ as x approaches infinity. Integration is carried out along a line of steepest descent. (2) A method of solving a system of linear equations by starting off with an initial guess at the value of the unknown vector, and the making successive corrections by taking the gradient of the residual sum of squares and taking the correction in the direction of steepest negative gradient. The method has been widely used in the method of least squares.

Metonic cycle - A period of 235 lunations (about 19 years). Devised by the Athenian astronomer Meton (c. 450 B.C.) for obtaining a period in which new and full moon would recur on the same day of the year. Taking the Julian year of 365.25 days and the synodic month as 29.530 588 days, the 19-year period is 6939.75 days long, as compared with the 235 lunations of 6939.69 days, a difference of only 0.06 days.

metre - SEE meter (unit of length). SEE ALSO Committee Meter; French legal metre; geopotential meter; international meter; International Prototype Meter; Metre des Archives; National Prototype Meter.

metre, dynamic - SEE meter, dynamic.

Metre des Archives - A bar defining the length of the French legal meter. It is an end-standard made of platinum and was used in determining a length for the International Prototype Meter. It is now preserved as a museum piece in the International Bureau of Weights and Measures at Sevres, France.

metrication - The process of converting from some system of units other than the c.g.s. or SI system of units to the c.g.s. or SI system of units.

metric system - SEE system, metric.

Metrogon lens - The trade name for a lens system.

Meusnier's theorem - At any given point (x,y) on a surface, the curvature κ of every curve, in the plane and through that point, whose osculating plane makes an angle α with the plane of the normal section through the same tangent is $\kappa_n / \cos \alpha$, in which κ_n is the curvature of the normal section through the tangent at the point.

Michelson interferometer - One of two types of interferometer invented by A.A Michelson for two different purposes. (1) A device in which a beam of light is split into two beams by a semi-transparent, optically flat surface, the two beams so created being sent along different paths and then redirected and recombined at the same plate to produce interference. This type is incorporated, for example, in the Väisälä base-line apparatus. Versions of it are used for calibrating surveyors' tapes. (2) A device in which light from a very distance object is picked up by two flat mirrors placed at the ends of a very long beam fastened to the front end of a telescope. The two resulting beams, one from each mirror, are combined in the telescope to produce interference. Also called Michelson's stellar interferometer. The principle is essentially the same as that used in short-baseline interferometers working at radio wavelengths.

microclimate - The climate in the layer (about 2 meters thick) of air next to the ground. The microclimate undergoes large variations in temperature and humidity.

microdensitometer - A densitometer measuring the density in a very small region of a photograph or similar object. Microdensitometers usually focus the light from a lamp into a very small, fixed spot on the stage holding the object; the transmitted light is picked up by a microscope and sent to a detector. The stage is shifted under the microscope so that eventually the entire region of interest has been investigated. Some microdensitometers, however, use a rapidly moving, very small spot of light to scan the object.

microfeature - A feature of relief, drainage, and other landforms, identifiable on photographs but too small to appear on maps.

micro-geodesy - Geodetic principles or techniques applied within small regions, typically regions one kilometer or less in greatest extent.

MicrOmega - SEE Omega, differential.

micrometer - (1) In general, any instrument for measuring small distances or angles very accurately. (2) In astronomy and geodesy, a device, for attachment to a telescope or microscope, consisting of a mark moved across the field of view by a screw connected to a graduated drum and vernier. If the mark is a hairlike filament, the micrometer is called a filar micrometer. (3) SEE micrometre. SEE ALSO run of the micrometer; transit micrometer.

micrometer, automatic - SEE micrometer, impersonal.

micrometer, filar - A device, to be attached to a telescope or microscope, consisting of a hairlike filament (hair, segment of spider-web or wire) connected to a screw so that as the screw is turned, the filament is translated across the field of view and in the focal plane of the telescope or microscope. The screw has a finely-cut thread and the amount of rotation is indicated by graduations on the head of the screw or by other means. The screws used in micrometers for geodetic instruments are graduated in angular measure (seconds, on American instruments). Turning the screw moves the filament a distance corresponding to the angle, at the instrument, between the two positions of the filament. Any small angle can be measured by noting the whole and fractional number of turns of the screw made in moving the filament from one position to the other and multiplying this by the angular value of one turn.

micrometer, impersonal - A filar micrometer using a motor to move the filament at a constant but controllable rate and sending signals to a recorder to indicate the position of the moving filament. Also called an automatic micrometer.

micrometer, ocular - A filar micrometer placed so that its moving filament moves in the principal focal-plane of a telescope. Also called an eyepiece micrometer. The ocular micrometer is used in surveying and in astronomical observations for making precise measurements of small angles between lines to objects viewed with the telescope. It is mounted in a frame perpendicular to the longitudinal axis of the telescope and can be rotated to measure an angle in any plane containing that axis. It can be adjusted to measure angles in the plane containing the longitudinal and horizontal axes of the telescope, as in determining time using the transit micrometer (a form of ocular micrometer). It can be adjusted to measure angles in a plane containing the longitudinal axis and a vertical line, as in determining latitude using the zenith telescope.

micrometer, optical - (1) A device consisting of a prism or lens placed in the path of light entering a telescope and rotatable, by means of a graduated linkage, about a horizontal axis perpendicular to the optical axis of the telescope axis. Also called an optical-mechanical compensator. The device is usually placed in front of the objective of a telescope, but may be placed immediately after it. The parallel-plate optical micrometer is the form usually found in leveling instruments. (2) A telescope containing such a device.

micrometer, parallel-plate optical - SEE micrometer, parallel-plate.

micrometer, parallel-plate - (1) A telescope or microscope having, in front of the objective lens, a small, square or rectangular piece of glass with flat and parallel faces. The piece can be rotated, by a precisely graduated linkage, about a horizontal axis perpendicular to the optical axis of the telescope. Also called a parallel-plate optical micrometer. As the piece of glass is rotated, light rays passing through are displaced laterally, causing the image to be displaced with respect to the reticle within the telescope. Very small displacements can be measured in this way. The device is frequently a part of or attached to a level. (2) The device, in the parallel-plate micrometer defined above, consisting of the rotating optical element and the graduated linkage alone. This usage is usual for leveling instruments.

micrometer correction - SEE contact correction; correction for run of micrometer.

micrometer method (of determining azimuth) - Determination of the astronomic azimuth of a line by measuring, with an ocular micrometer attached to a theodolite or transit is used to measure the horizontal angle between vertical plane through a star and a vertical plane through a marker on the ground (close to the vertical plane through the star). That angle is applied to the azimuth of the star computed for the epoch of the observation to give the azimuth of the marker. At elongation, the apparent motion, in azimuth, of a closely circumpolar star like Polaris is very small for an appreciable length of time, and a series of observations can be made by the micrometer method without re-orienting the instrument. A correction for inclination of the horizontal axis, depending on the angular elevations of the star and of the marker, is applied to the observed angle, and additional corrections are applied for curvature of the apparent path of the star the field of view, for variation of latitude, and for diurnal aberration.

micrometer microscope - A filar micrometer so placed that its moving filament moves in the focal plane of a microscope. Readings made on a graduated circle by using a microscope micrometer are highly precise and accurate. The micrometer is adjusted so that an even number of turns will carry the filament exactly the distance between adjacent marks on the circle. Any intermediate point on the circle can be read by interpolation. If the adjustment is imperfect, the amount by which an even number of turns of the screw fails to carry the line exactly the distance between marks is treated as a systematic error known, preferably, as the run of micrometer. The term correction for run of micrometer is also acceptable.

micrometer screw, reversing point of a - SEE reversing point (of micrometer screw).

micrometer transit - A transit to which is fastened a micrometer for following the motion of a star through the field of view and determining the instant at which a star passes through a given vertical circle. The micrometer has one or more movable threads which lie in the focal plane of the telescope. The threads are moved in the direction of azimuth by a single screw with a graduated head. The star is followed by rotating the screw and recording an observation at each of the main graduations. Local sidereal time can be determined if the telescope is set up to move in the meridional plane or to determine latitude if set up to rotate in the plane of the prime vertical.

micrometre - A unit of length defined to be exactly 0.000 001 meter. The spelling micrometer is more common in the USA, as is the equivalent term micron (q.v.).

micron - SEE micrometre. Although this term should, according to the rules of the SI, be called a micrometre, it is still widely used in photogrammetry and other sciences. It was already in use by 1891 and was used by the British Board of Trade as early as 1896.

microphotometer - SEE photometer.

microtriangulation - Triangulation in which the sides of triangles are from a few meters to a few hundreds of meters long.

microtrilateration - Trilateration in which the courses are from a few meters to a few hundreds of meters long.

microwave (adj.) - (1) Having a wavelength less than a few meters but longer than a millimeter. (2) Having a wavelength between 30 cm and 1 mm. There is no general agreement on the exact distinction between microwave radiation and radiation of longer or shorter wavelengths.

middletone - (1) In the halftone process, any neutral tone intermediate between the highlights and shadows of an original and the resulting reproduction. (2) Any tone, in a reproduction, which lies between highlight and shadow.

mid-latitude - (1) That line which, on a graticule produced by the Mercator map projection, represents a parallel of latitude and is at the same scale as that indicated for the projection. The line may not lie on the map itself, i.e., may be at a latitude outside the confines of the map. (2) A region lying between the sub-tropical and sub-polar regions of the Earth. (3) SEE latitude, middle.

Mie scattering - The scattering of electromagnetic radiation by particles whose characteristic size is of the same order of magnitude as the wavelength of the radiation.

mil - (1) The angle subtended by 1/6400 of the circumference of a circle. The mil is approximately 1/1000 radian or 0.05265°. It is used principally by artillerists. (2) An angle having a tangent of 0.001. (3) A length of 1/1000 of an inch. Also called a mille-inch. (3) A distance of 4.68 miles (Virgin Islands).

mil, circular - The area of a circle having a diameter of 0.001 inch. This is approximately equal to 7.85×10^{-7} square inch.

mil, true - An angle subtending $1/(2000 \pi)$ of a circumference. This is exactly 1/1000 radian.

mile - A unit of distance, variously defined and usually preceded by an adjective indicating which of the various definitions is meant, e.g., nautical mile and statute mile. The word mile is derived from the Latin mille for one thousand, and mille passum meant one thousand paces of about 5 feet each. The Roman mile was therefore about 5 000 feet long, a value that underwent many changes as the mile came into use among the other western nations. In general usage, mile means the statute mile of 5 280 feet. The nautical mile is almost never referred to simply as a mile unless the meaning is obvious from the context. SEE ALSO radar mile.

mile, aeronautical - SEE mile, international nautical.

mile, geographical - (1) The length of one minute of arc, measured along the equator of a rotational ellipsoid representing the Earth. Its value is determined by the dimensions of the ellipsoid chosen. (2) SEE mile, international nautical. SEE ALSO mile, nautical.

mile, international nautical - The nautical mile of exactly 1 852 meters. Also called the aeronautical mile, air mile and geographical mile. The definition was proposed in 1929 by the International Hydrographic Bureau because the many different kinds of nautical mile then in use was causing confusion. Since then, the unit was adopted for maritime use by most maritime nations. It was adopted by the U.S. Department of Commerce and the U.S. Department of Defense on 1 July 1954. Until 1970 the unit used by the United Kingdom was the nautical mile of 6 080 feet or 1 853.18 meters.

mile, nautical - One of a number of units of distance all related to the length of an arc, on a sphere, subtending an angle of one minute. Also called a sea mile, geographical mile and geographic mile. The nautical mile may be taken as equal to the length of a minute of arc along the equator or a minute of latitude anywhere on a map. It is used principally for expressing distances over water. However, it is the unit of length used for defining the knot (a speed of one nautical mile per hour), and as such is involved in aerial navigation as well as in marine navigation. The nautical mile of the USA is defined to be a distance equal to the length of one-sixtieth of a degree of a great circle on a sphere having an area equal to the area of a spheroid (rotational ellipsoid) representing the Earth. Its value, calculated for the Clarke spheroid of 1866, is 1 853.248 meters (6 080.2 feet). The international nautical mile is 1 852 meters (6 076.1 feet).

mile, square - The area of a flat surface bounded by a square 1 mile on a side. This unit is now used almost entirely by the USA. In using it, care must be taken to specify whether the nautical mile or the statute mile is meant. A square (statute) mile is equal to approximately 2.5899985 square kilometers or 258.9985 hectares; it is equal exactly to 640 acres or 27 878 400 square feet.

mile, statute - A unit of length defined to be exactly 5 280 feet (approximately 1 609.347 meters). It is used primarily in stating distances on land in the USA. In marine regions, the (international) nautical mile is used. Also called a land mile.

mileage chart - A chart showing distances between various points.

Military Grid Reference System - SEE grid reference system, military.

mille-inch - SEE mil (3).

Miller's cylindrical map projection - A map projection mapping the rotational ellipsoid onto the plane according to the formula, for the sphere, $x = k R (\lambda - \lambda_o)$ and $y = 4 k R [\ln \tan¼(\pi + m\phi)]/n$, in which x and y are the rectangular Cartesian coordinates, in the plane, of that point which, on the sphere, has longitude λ and latitude ϕ; λ_o is the longitude of the central meridian on the map. R is the radius of the sphere and k is the scale factor. m and n are constants equal, in this instance, to 1.6 and 5, respectively. This map projection is neither conformal nor equal-area. Meridians and parallels are represented as straight lines, however. The projection was proposed by O.M. Miller in 1942 as a map projection giving a reasonable compromise between areal distortion and angular distortion. It was used by the United States Geological Survey for the National Atlas of the United States (1970). Values for m and n other than those given above have been used by some cartographers.

Miller's modified oblique stereographic map projection - A stereographic map projection from the sphere to the plane, as defined by the formulas

$$x = KX[1 + Q(3Y^2 - X^2)/12]$$
$$y = KY[1 + Q(3X^2 - Y^2)/12]$$
$$X = 2\tan(\sigma/2)\sin\theta \qquad Y = 2\tan(\sigma/2)\cos\theta$$
$$\sigma = \arccos[\sin\phi\sin\phi_o + \cos\phi\cos\phi_o\cos(\lambda - \lambda_o)]$$
$$\Theta = [\sin\phi\sin(\lambda - \lambda_o)] / \sin\sigma$$

in which x and y are the rectangular, Cartesian coordinates of that point, in the plane, which has, on the sphere, the corresponding longitude λ and latitude ϕ, respectively. λ_o and ϕ_o are the longitude and latitude of that point at which the plane used in an oblique stereographic map projection is tangent to the sphere. K and Q are constants. The constants K and Q were chosen by O.M. Miller so that Europe and Africa could be shown on the map within an oval on which the scale was constant. The map projection is conformal.

millies - SEE tonne.

milligal - An acceleration of 0.001 cm/sec².

millimap - A map on which there are 1000 dots representing the distribution of some quantity, each dot representing 1/000 of the total amount and being placed on the map according to the evidence on the geographical occurrence of the quantity.

Mills cross - SEE antenna, Mills-cross.

mineral lands - Public lands that have been classified as containing or being known to contain valuable minerals.

mineral rights - The rights or title to all, or to certain specified, minerals in a given tract.

mineral survey - A survey made by U.S. mineral survyeors to mark the legal boundaries of mineral deposits or ore-bearing formations in the public domain, where the boundaries are to be determined by lines other than those of the normal subdivision of the public lands.

miners inch - The discharge from an orifice 1-inch square under a definitely specified head (pressure). The value of a miners inch has been fixed by statute in various Sates as follows: Arizona, California, Montana and Oregon: 40 miners inches are the equivalent of 1 cubic foot per second; Idaho, Nebraska, Nevada, New Mexico, North Dakota, South Dakota and Utah: 50 miners inches are the equivalent of 1 cubic foot per second; Colorado: 38.440 miners inches are accepted as equivalent to 1 cubic foot per second; British Columbia: 35.7 miners inches are equivalent to 1 cubic foot per second; in some parts of California: 40 miners inches to 1 cubic foot per second is used, whereas in southern California: 50 miners inches to 1 cubic foot per second is used regardless of the legal definition.

mine survey - A survey to determine the locations and dimensions of underground passages of a mine and of the natural and artificial features, underground and on the surface, relating to the mine. The data include both horizontal and vertical coordinates, lengths, and directions and inclinations of tunnels; topographic and geological characteristics of the particular locality; ownership of the land and the mine; etc.

minimum focal length - SEE length, minimal focal.

minimum pendulum - A pendulum whose form is such that a change of temperature causes minimal change in the length of the pendulum.

minimum squares - SEE method of least squares.

minimum thermometer - A thermometer which automatically registers the lowest temperature to which it has been exposed since its last setting.

mining transit - An engineer's transit designed particularly for surveying in mines. A mining transit is usually provided with an auxiliary telescope or other means of taking very sharply inclined (or even vertical) sights.

minus sight - SEE foresight.

minute - (1) An angle exactly 1/60 of a degree or containing exactly 60 seconds. Sometimes called minute of arc. That term, however, is almost always unnecessary and should not be used unless confusion with the minute of time is likely. The symbol for minute (of arc) is ('). In astronomy, it is always placed immediately after the last integral digit of the number, i.e., above the decimal point if there is one; in other sciences, it is usually placed immediately after the last digit of the number. E.g., 35.'599 in astronomy and 35.599' in surveying. (2) An interval of time containing exactly 60 seconds of time or 1/60 of an hour. The minute is not a fundamental unit and therefore it has always been defined in terms of the second, the hour, or the day (which contains 1440 minutes.) The symbol for minute of time is (m). In astronomy, it is placed immediately after the last integral digit of the number, i.e., above the decimal point if there is one; in other sciences, it is placed immediately after the last digit of the number. E.g., 35.^{m}599 in astronomy and 35.599 m in surveying.

minute of standard length - The length of 1 minute of longitude on the Equator, on a spheroid representing the Earth.

mirror, front-surface - A mirror reflecting light or other radiation mostly from the surface closer to the source. The word mirror as used by astronomers and other scientists almost always means a front-surface mirror. Front-surface mirrors are sometimes made entirely of a substance which is a good reflector, e.g., silver. However, they are more usually made by evaporating a metallic such as silver or aluminum onto a suitably shaped material such as glass or quartz. To prevent the metal from tarnishing, silicon monoxide is usually evaporated onto the coated surface, forming a very thin layer of transparent silicon dioxide.

mirror, rear-surface - A mirror reflecting light or other radiation mostly from the surface farther from the source. In non-technical writing, mirror means a rear-surface mirror 7unless otherwise unless otherwise stated. A highly reflecting, metallic coating is placed (usually by chemical deposition) on one surface of a transparent material like glass or quartz. A tough coating, usually black, is then painted over the metal to protect it from corrosion and wear. For use in some optical systems, rear-surface mirrors are made by evaporating a metallic coating onto the rear surface.

mirror image - An image such that back-to-front motion in the scene causes front-to-back motion in the image. Also called a mirrored image or perverted image. The image that would be seen if the scene were viewed in a mirror. Note that right and left are not, contrary to a popular misconception, interchanged by the image. If the scene were to be rotated so as to face in the same direction as the image, it would be found that right and left sides in the scene were indeed reversed with respect to the image. But this is because the back-to front directions were reversed, not because right and left were reversed. A single mirror produces a mirror image; a pair of mirrors produces a normal aspect. Also called a mirrored image, perverted image or reverted image.

mirror stereoscope - SEE stereoscope, mirror-type.

misclosure - (1) In general, the amount by which a value (of a quantity) obtained by surveying fails to agree with a value (of the same quantity) determined, e.g., by an earlier survey, an arbitrarily assigned starting value, or from theory. Also called closure, closing error and error of closure. (2) In leveling, the amount by which the two values for the elevation of the same bench mark, derived by different surveys, by the same survey made along two different routes, or by independent measurements, fail to exactly equal each other. The misclosure may occur in a line of leveling beginning and ending on different bench marks whose elevations are held fixed, or beginning and ending on the same bench mark. (3) (traversing) The amount by which a value for a component of the location (of a traverse station) obtained by computation fails to agree with another value for the component as determined by a set of measurements over the same or a different route. Also called closure, error in closure, traverse error in closure, closure in position, total closure and misclosure in position. The traverse may run between two stations whose locations are held fixed or it may begin and end at the same station. In either case, there are two values for the location of the final station: one known before the traverse was computed and the other obtained by computation using the measurements made on the traverse. The difference between these is the misclosure. It may be resolved into misclosure in latitude, misclosure in longitude (departure), or both. Although the definition calls for the difference (computed value minus measured value), in practice the difference may be taken either way and corrections applied in an ad hoc manner. SEE ALSO junction misclosure; misclosure in azimuth.

misclosure (of a loop) - The amount by which a value of a quantity measured at the starting point of a survey fails to be exactly equal value found by measurement when the survey returns, by a loop, to the starting point. The definition applies, actually, only to surveys in which the value obtained at each successive point of the loop is a function of the measurement obtained at that point and the value obtained at the preceding points. Surveying in a loop does not protect against systematic errors in measurements of distance or against making blunders in starting location or azimuth. Also called loop error of closure.

misclosure, angular - The amount by which the actual sum of a series of angles fails to equal the theoretical value of that sum. SEE ALSO misclosure of the horizon; misclosure of triangle.

misclosure, linear - The distance between the beginning and final points of a closed traverse, as calculated from the measurements.

misclosure, relative - The value obtained by dividing the misclosure of a traverse by the length of the traverse. Relative misclosure is commonly expressed as a fraction whose numerator is 1, e.g., 1/1320.

misclosure in azimuth - The amount by which two values of the azimuth of a line, derived by different surveys or along different routes, differ. Usually, one value is derived by computations using the measurements made during the survey (traverse, triangulation, or trilateration); the other is an adjusted or fixed value determined by an earlier or more precise survey or by independent, astronomical observations.

misclosure in departure (traversing) - The amount by which the sum of the directed excursions east and west of a line running north and south fail to add up to the directed distance of the final station; in particular and usually, the amount by which they fail to add up to zero upon return to the original station.

misclosure in latitude (traversing) - The amount by which the sum of the directed excursions north and south of a line running east and west fail to add up to the directed distance of the final station; in particular and usually, the amount by which they fail to add up to zero upon return to the original station.

misclosure in leveling - The amount by which two values of the elevation of the same bench mark, derived by different surveys or along different routes or by independent measurements, fail to be exactly equal to each other. Also called error of closure (in leveling), leveling error of closure and leveling misclosure.

misclosure of the horizon - The amount by which the sum of a series of horizontal angles measured between adjacent lines in a complete circuit of the horizon fails to equal exactly 360°. SEE ALSO closing the horizon.

misclosure of traverse - The amount by which a value of the location of a traverse station, as obtained by computation through a traverse, fails to be exactly equal to another value at the same station as determined from a different set of measurements or routes of survey. Also called closure of traverse, error or survey, horizontal closure error and traverse error of closure.

misclosure of triangle - The amount by which the sum of three measured angles of a triangle fails to equal exactly 180° plus the spherical excess of the triangle. SEE ALSO angle equation.

mistake - SEE blunder; error. In surveying, mistake and error are frequently given different meanings.

M.K.S. system of units - A system of units, devised by G. Giorgi, in which the fundamental units are: the kilogram (for mass), the meter (for length), the second (for time) and an electromagnetic unit, commonly the coulomb (for electrical charge). In an M.K.S. system, the volt and the ampere are derived units. The M.K.S. system is the forerunner and prototype of the Systém International d'Unitès (SI).

Möbius grid - The projected image of a rectangular grid.

Möbius transformation - SEE transformation, linear fractional.

mock-up - SEE style sheet.

mode - That number, in a set of numbers, which occurs most frequently. If two or more numbers occur with equal frequency, the mode does not exist.

mode, orbital - A method of determining the location of a station from observations made on a satellite, and depending on a knowledge of the satellite's orbit for determining the satellite's location at the instant of observation.

mode, simultaneous - That method (of observing an artificial satellite) in which measurements are made from four or more stations simultaneously; the locations of at least three of the stations with respect to each other must known.

model - (1) A copy in three dimensions of an object, usually on a smaller scale but occasionally on a larger scale. Models are used extensively in engineering; tests are made on the model instead of on the original. The results of such tests are then made to apply to the original by dimensional analogy. Models are also used in other sciences such as physical oceanography and crustal physics. They are little used in surveying or geodesy except as representations of terrain and usually in regions of small area. (2) In particular, in photogrammetry, the visual or optical image (called a stereoscopic model) produced by combining the images from two photographs of the same object from different angles or by creating a hologram. (3) A simplified, theoretical, not necessarily mathematical representation of a real thing or event and acting like the original in situations of interest. (4) A mathematical function or representation.

model, flat - Any stereoscopic model capable of being positioned so that the z-axis is parallel to the direction in which heights are measured in the model. Differences of z-coordinates are then the same as differences of height in the stereoscopic model and are proportional to differences of height in the original. The model is then said to be level. A model not capable of being made level is then said to be warped.

model, gross - The entire overlap of a stereoscopic pair of photographs. SEE ALSO model, neat.

model, half - SEE half-model.

model, neat - That portion of the overlap of a pair of photographs which is actually used in a photogrammetric mapping project. Usually, the neat model is nearly rectangular with a width equal to the air base and length equal to the distance (width) between flights.

model, original - The developed, original model of terrain which bears, in miniature, the same spatial relationships as the actual ground it represents. Also called a master model.

model, perspective spatial - That optical reconstruction of a region produced by viewing a stereoscopic pair of photographs through a stereoscope and showing depth.

model, plastic relief - SEE plastic relief map.

model, spatial - SEE model, stereoscopic.

model, stereoscopic - (1) An image, either actual or visual, formed by the intersecting, homologous rays from the corresponding points in the separate images of a stereoscopic pair of images. Also called stereomodel and spatial model. (2) The points of an image formed by the intersecting, homologous rays from the separate images of a stereoscopic pair, or the points which would exist if such an image were formed. (3) By extension, those portions of a stereoscopic pair of images which are used in creating a stereoscopic image as previously defined. The term stereoscopic image should not be used in this sense because doing so makes for ambiguity and is, in any event, unnecessary.

model, topographic - SEE relief model.

model, warped - A stereoscopic model so distorted that distances measured parallel to the z-axis are not everywhere parallel to lines along which heights in the model should be measured. That is, the model cannot be positioned in the measuring device so that the directions corresponding to that of heights in the model are parallel to the z-axis of the measuring device. Some points in the model will be given erroneous heights and the heights calculated for corresponding points on the ground will be wrong. Also called a warped stereoscopic model.

model, warped stereoscopic - SEE model, warped.

model base - The line, or the length of the line, at the scale of the stereoscopic model, joining the perspective centers as reproduced by the stereoscopic instrument.

model coordinate - One of the three space coordinates of any point imaged on an overlapping pair of photographs, which specify its location with respect to the air base. The model coordinates correspond, with respect to the location of the origin and the directions of the axes, to a spherical coordinate system in which an air base is the polar axis. One such system, suggested by Fourcade, can be defined as follows: origin - the left-hand air station; x-axis - the line of the air base to the right; z-axis - the line perpendicular to the x-axis in the basal plane containing the principal point of the left-hand photograph; y-axis - the line forming a right-handed, rectangular Cartesian coordinate system with the other two axes. (The ground is considered to be in the negative direction.)

model datum - That imaginary surface which is thought of as present in a stereoscopic model and to which heights or elevations are referred. The term is often modified to designate the type of photography used, e.g., convergent model datum and transverse model datum.

model datum, convergent - SEE model datum.

model datum, transverse - SEE model datum.

Model Earth - (1) A fictitious Earth obtained by smoothing the terrain of the real Earth by the following procedure. The elevation H_{me} of the fictitious topography over a region of radius b is the weighted, average value of the elevations H of the actual topography of the same region, with the condition that no slope, between adjacent regions, of the fictitious topography be greater than 0.01. The weighing factor and radius are not specified in the definition but a value for b of not more than 100 miles (150 km) was suggested by the originator of the concept (de Graaf Hunter). Also written as model Earth. (2) SEE Earth, model.

model of the Earth - A mathematical representation of a particular characteristic of the Earth, such as the Earth's density or gravity field.

model scale - SEE scale (of a stereoscopic model).

mode of reduction, orbital - A method of solving for the location of a survey station by first deriving the orbit of a satellite from observations made at stations elsewhere and then using observations on the satellite from the survey station, together with the known orbit, to derive the location of the station. The survey-station's location is found in the coordinate system used for the locations of the stations whose data were used in determining the orbit. Two variants of this method are in common use: the short-arc method and the long-arc method. In the short-arc method, the stations used for determining the orbit observe the satellite shortly before or after the survey station. A very short part of the actual orbit is therefore used, and this can usually be approximated by a curve much simpler than the actual orbit. In the long-arc method, the stations determining the orbit observe long before and long after the survey station. The orbital mode is distinguished from the simultaneous mode in that the latter requires that observations be made simultaneously by the survey station and by the stations whose coordinates are known. No orbit need be determined in this method.

Modified Mercury Datum 1968 - A datum derived in 1968 as an improvement on the Mercury 1960 Datum. A rotational ellipsoid (Mercury spheroid 1968) of the same flattening but with semi-major axis 6 378 150 meters long was used. The other defining constants (those relating coordinates in the datum to coordinates in other national or continental datums) were also changed.

modulation - The addition of harmonics to a basic frequency. In physics, the basic frequency of a periodic phenomenon is called the carrier frequency and the radiation corresponding to this frequency is called the carrier wave. The added frequencies are called the modulating frequencies or harmonic frequencies. In geodesy, electromagnetic phenomena are of primary interest and the modulation is usually amplitude or phase.

modulation error - In electromagnetic distance-measuring instruments, the difference between the modulating frequencies actually obtained from an oscillator and the frequencies required for a correct measurement of distance.

modulation transfer function - The magnitude of the optical transfer function or the magnitude of the Fourier transform of the line-spread function. Commonly denoted by MTF. A common method for determining the MTF is to plot contrast, $(I_{max} - I_{min}) / (I_{max} + I_{min})$, in which Imax is the greatest and Imin the least luminance of sets of equally-spaced lines or the greatest and least illumination of the images of those lines, taking the ratio of the contrast in the object and the image, and plotting this ratio against the spacing of the lines. SEE ALSO transfer function, optical.

modulus of precision - The quantity $1/\sigma$, in which σ is the standard deviation of measurements.

Moho - SEE Mohorovicic discontinuity.

Moho discontinuity - SEE Mohorovicic discontinuity.

Mohorovicic discontinuity - A region, lying at depths of 5-10 km under oceanic basins to more than 60 km under mountain ranges, across which the speed of seismic P-waves jumps from about 6.7 km/s to about 8.1 km/s. Also called the M-discontinuity, moho and Moho discontinuity. It has been assumed that this discontinuity marks the boundary between crust and mantle. However, there is still not an adequate explanation of its relationship to isostatic compensation or to continental drift - two other hypotheses on the structure of the crust and mantle.

Moiré fringe - One of the lines in a Moiré pattern. The term is generally used in the plural as a synonym for Moiré pattern.

Moiré pattern - (1) The pattern formed by transmitting light through two separate, overlapping families of parallel lines. Also called Moiré fringes. As one family of parallel lines is translated or rotated with respect to the other, the pattern shifts or otherwise changes. Because a small change in relative position of the two families can cause a large apparent change in the pattern, the relative movement of two flat surfaces can be measured by affixing to each surface another flat surface on which is ruled a family of closely- spaced parallel lines, the ruled surfaces being slightly rotated with respect to each other. Light is passed through the two surfaces in such a way as to create a Moire' pattern and the changes in this pattern are measured to give the corresponding changes in the positions of the two surfaces of primary interest. This principle has been used extensively in the design of measuring engines. (2) In particular, in printing, the pattern resulting from the overlapping or overprinting of halftones or tints whose screens are at angles large enough to make the pattern conspicuous.

mol - SEE mole (1).

mold, positive - The cast made from a negative mold when making a relief map.

mole - (1) M units of mass of a substance whose molecular weight is M. Also called a mol. Unless otherwise specified, one mole means M grams of mass. (2) The amount of substance containing as many elementary entities as there are carbon atoms in 0.012 kilograms of carbon-12. The elementary entities must be specified; they may be atoms, molecules, ions, electrons or other particles, or may be specified groups of such particles. This definition is that of the General Conference on Weights and Measures (1971). However, applying this definition will be difficult. (3) A breakwater which is at the same time a wharf alongside which vessels can lie for loading and discharging.

Molitor precise leveling rod - A self-reading leveling-rod of T-shaped cross-section, with graduations marked by triangles and rectangles. The smallest division is 2 millimeters. A thermometer and circular level are attached. This type is no longer used.

Mollweide homolographic map projection - SEE Mollweide's map projection.

Mollweide's map projection - (1) An equal-area map-projection mapping the sphere onto the plane and defined by the equations $x = 2\sqrt{2} k (R/\pi)(\lambda - \lambda_o) \cos t$ and $y = \sqrt{2} k R \sin t$, in which x and y are the rectangular, Cartesian coordinates, in the plane, of that point which corresponds to the point, on the sphere, whose longitude is λ and whose latitude is ϕ. t is a parameter defined by the implicit equation $2t + \sin 2t = \pi \sin \phi$. λ_o is the longitude of the central meridian. R is the radius of the sphere and k is the scale factor. Also called Babinet's map projection and Mollweide's homalographic map projection. Mollweide's map projection maps the equator and other parallels of latitude into straight lines and maps the meridians other than the central meridian into elliptical arcs. The central meridian is represented by a straight line, and the meridian 90° from the center is shown as a full circle. When the map projection is used to show the entire surface of the Earth, the map is bounded by an ellipse whose major axis, representing 360° of longitude, is double the length of its minor axis, representing 180° of latitude. When it is used to represent a hemisphere, the boundary of the map will be a full circle. When the projection is applied so that the major axis of the inclosing ellipse is the line representing the meridians 30° West and 150° East, it is called the Atlantis map projection. This was introduced about 1951 and used in many atlases after 1955. (2) The same as the preceding definition, but mapping a rotational ellipse onto the plane.

Molodensky correction - (1) The amount, obtained using Molodensky's formulas, to be added to the length of a geodesic between two specified points on the geoid to obtain the length of the geodesic between the two corresponding points on that ellipsoid taken to represent the geoid. In older methods of calculating coordinates in triangulation networks, the length of the measured base line on the ground was reduced to its equivalent length on the geoid, but this length was then assumed to be the length of the corresponding geodesic on the ellipsoid. The Molodensky correction eliminates this error. (2) The amount $(h - h_c)$ to be added to the assumed height h_c of the geoid above the ellipsoid to obtain the actual height h, as given by Molodensky's formulas.

Molodensky height - SEE height, normal.

Molodensky's formula - One of the two formulas $h = hc + R A \sin \theta + B \cos \theta$ and $\zeta = \zeta_c + A \cos \theta - (B/R \sin x \theta)$ in which h is the geoidal height and ζ is the deflection of the vertical at a point, in a chain of triangulation; h_c and ζ_c are the corresponding, assumed geoidal height and deflection; Θ is the angular distance along the chain; R is the radius, at the midpoint of the chain, of the rotational ellipsoid used; A and B are constants given by $A = \zeta_o - \zeta_{co} - (1/R) \int h_c \cos \theta \, d\theta$ and $B = h_o - h_\infty + \int h_c \sin \theta \, d\theta$, in which the subscript o denotes the value at $\theta = 0$. Also called the formula for reduction to the spheroid. The quantity $(h - h_c)$ is called the Molodensky correction.

Molodensky's modification (of Clairaut's equation) - The equation $(a-b)/a + (\tau_p - \tau_e)/\tau_e = (\Omega^2 b/\tau_e)(1 + 53f/35 - 209 f^2/35 + \ldots)$, in which τ_p and τ_e are the theoretical values of gravity at the poles and at the equator, respectively, of an equipotential, rotational ellipsoid of equatorial diameter 2a and polar diameter 2b, rotating at the rate Ω. f is the flattening (a - b)/a. Also called the Clairaut-Molodensky equation.

moment - The product of the magnitude of a force by the distance of the line of action of that force from a specified point or from a specified line not coplanar with the force. Also called a torque. The moment is said to be with respect to the specified point or the specified line.

moment, magnetic - The product of the distance between two magnetic poles by the average strength of the poles' magnetic fields.

moment of inertia - The sum of the products formed by multiplying the mass (or its equivalent) of each element of a body by the square of that element's distance from a specified line. Also called rotational inertia. There are two mutually perpendicular lines through a point such that (a) the moment of inertia about one of the lines is less than the moment of inertia about any other line through that point, and (b) the moment of inertia about the other one is greater than the moment of inertia about any other line through that point. If a third line is drawn through that point and perpendicular to the other two, the three lines are called the principal axes of inertia of the body or simply, the principle axes.

moment of inertia, axial - The sum of the component masses of a rotating body times the square of the distance of each component mass from the axis of rotation.

momentum - (1) The product of mass and velocity. This is a vector and has the same direction as the velocity. (2) The product of mass and speed.

momentum, absolute - The sum of the (vectorial) momentum of a particle relative to the Earth and the momentum of the particle caused by the Earth's rotation.

momentum, angular - A vector characteristic of a rotating or revolving mass, equal in magnitude to $mrv \cos \theta$, in which m is the mass of the body, r the distance from the origin to the body and v the speed of the body. θ is the angle between the line from the origin to the body and the axis about which the body is rotating or revolving. This definition applies, properly, only to bodies which are point-masses. For a real body, the above quantity must be integrated over all elemental masses making up the body.

momentum, linear - A vector characteristic of a body moving in a straight line and equal in magnitude to the product of the mass by the velocity. Also referred to as momentum when linear momentum is clearly meant. If the body is moving along a curve or otherwise rotating while moving, linear momentum should be used for the product of mass by instantaneous velocity and total momentum for the sum of the linear and angular momenta.

monochromator - An optical system isolating a very narrow portion of the spectrum for viewing or recording, and allowing selection of the portion isolated.

monocomparator - A measuring engine used in photogrammetry for measuring distances between points, or determining plane coordinates of individual points, on one photograph at a time. Also called a monoscopic comparator. Although a monocomparator may have a binocular microscope for viewing single photographs, it cannot be used as a stereocomparator, which measures distances on a stereoscopic model formed from two photographs viewed simultaneously.

monopulse - A technique for determining the direction of an object by comparing the radio signal received, either by scattering from the object or from a transmitter in it, at two different antennas simultaneously.

monopulse radar - Radar incorporating monopulse for determining direction.

Monte Carlo method - A method for solving statistically, mathematical formulas not easily solvable by analytical methods or numerical analysis. A frequency function simply related to the formula in question is found, a large number of random values of the variables involved are generated, their empirical frequency functions are determined, and these are used in a numerical evaluation of the formula.

month - (1) The period of time between two successive passages of the Moon across a specified great-circle on the celestial sphere. Alternatively, the period of revolution of the Moon about the Earth. The great circle may be chosen to pass through the celestial poles (sidereal month or tropical month) or through the poles of the ecliptic (synodical month), or may be the celestial equator (nodical month). Other choices can be made, but the length of the month almost always lies between 27 and 30 days. (2) A period of time approximating that between two successive passages of the Moon across a specified great circle on the celestial sphere, but containing an integral number of days; the exact number is fixed by law or custom. SEE ALSO calendar month.

month, anomalistic - The interval of time between two successive passages of the Moon through the perigee of its orbit. The anomalistic month contains approximately 27.55455 mean solar days.

month, calendarial - SEE calendar month.

month, draconic - SEE month, nodical.

month, lunar - SEE month, synodical.

month, nodical - The interval of time between two successive passages of the Moon through the same node in the orbit. Also called a draconic month. The nodical month contains, on the average, 27.21222 mean solar days.

month, ordinary - SEE month, synodical.

month, sidereal - The interval of time between two successive passages of the Moon through that great circle which contains the celestial poles and passes through a specific star. The sidereal month can be measured by the length of time it takes the Moon to revolve from a given celestial longitude back to the same longitude, reckoned from a fixed equinox. It contains, on the average, 27.321661 mean solar days. Because the orbit is perturbed, the actual length of time varies some 7 hours. The difference between the lengths of the sidereal and tropical months is caused by the precession of the equinoxes.

month, synodic - SEE month, synodical.

month, synodical - The interval of time between two successive conjunctions (new moons) or oppositions (full moons) of the Moon. Also called a lunation, lunar month, ordinary month and synodic month. The synodical month contains, on the average, 29.530589 mean solar days. It is what is meant when a lunar month is specified.

month, tropical - The interval of time between two successive passes of the Moon through the same equinox. The tropical month can be measured by the length of time it takes the Moon to revolve from a given celestial longitude back to the same longitude, reckoned from an equinox affected by precession. The tropical month contains, on the average, 27.321528 mean solar days. Because of perturbations of the Moon's orbit, the actual length varies by about 7 hours. The difference between the lengths of the sidereal and tropical months is caused by the precession of the equinoxes.

monument (USPLS) - (1) In general, any material object or collection of objects indicating the location, on the ground, of a survey station or corner. Also called a beacon (Brit.). The term monument has been used for the mark at the station or corner and for the marker at the station or corner, as well as for all marks or markers associated with the station or corner. However, it is best used only for the entire object consisting of the mark, the marker on which the mark is placed, and the structure holding the marker in place. In land surveying, the term is also used for objects such as roads, ditches or fences forming a boundary for the land. It is advisable, therefore, in writing about survey stations, to use the designations, station mark, center mark, reference mark and witness mark (or corresponding terms for the markers or monuments), for the separate marks (or markers or monuments). In land surveying, unless monument is used in a general sense, use the terms corner, corner mark, reference mark and witness mark (or marker). Note that a witness mark has more authority in land surveying than in control surveying. (2) A structure marking the location of a corner or point determined by surveying. Monument and corner are not synonymous, although often used in land surveying as if they were. SEE ALSO boundary monument; location monument; record monument.

monument, artificial - (1) A monument made by a person, as distinct from a natural monument (a natural object used, in place, as a monument). (2) A relatively permanent object used to identify the location of a survey station or corner. Objects used include abutments, stones, concrete piers and railroad rails.

monument, initial - The physical structure that marks the location of an initial point in the rectangular system of surveys.

monument, judicial - A monument, set by order of a court, at a location determined by that court for a boundary corner.

monument, natural - A permanent object which is the work of nature, such as a stream, river, lake, pond, tree, ledge or outcropping of rock, and which is considered as a monument.

monument, physical - An existing feature, such as a stone, stake, tree, hill, ocean, river or lake, considered as a monument.

moon - A satellite of a planet. The word is not capitalized when used in this sense, e.g., a moon of Mars or an artificial moon.

Moon - The Earth's only natural satellite. Almost always referred to as the Moon. The only exception occurs when, in literature, the Moon is considered a person, e.g., ugly old Moon. The Moon has been used, in geodesy, for determining long distances between points not connectable by ordinary triangulation or traverse, e.g., using the Moon's occultations of stars, using the dual-rate moon-camera, or using laser-type distance-measuring equipment in conjunction with retro-directive reflectors placed on the Moon. The principal characteristics of the Moon are given in the Appendix.

Moon camera, dual-rate - A camera, invented by W. Markowitz, which allows both the Moon and the stars in the background to be photographed simultaneously by tracking the stars at the sidereal rate while keeping the Moon's image stationary with respect to the stars. The camera is attached to a telescope with an equatorial mounting, so that the stars' images are kept stationary on the camera's focal plane. A small glass disk is interposed between the Moon and its image. The disk is rotated about a diameter at such a rate that refraction counteracts the apparent motion of the Moon among the stars, causing the Moon's image to remain stationary on the focal plane. The camera was first used to determine the Moon's orbit more precisely as an aid to determining ephemeris time. It was later used geodetically to determine the coordinates of the camera. Also called the Markowitz Moon camera.

moonrise (moonset) - (1) That position of the Moon, at rising (setting), in which the apparent upper limb of the Moon is on the astronomical horizon. (2) That position of the Moon, at rising (setting) in which the true zenith distance, referred to the center of the Earth, of the central point of the disk is 90° 34'+ a-p; 34' is the conventional value of horizontal refraction, a is the length of the Moon's semidiameter and p is the horizontal parallax.

moonset - SEE moonrise (moonset).

moraine - A mound, ridge, or other distinct accumulation of assorted, stratified dirt, pebbles rock, etc., deposited chiefly by a glacier.

moraine, lateral - A low, ridge-like moraine carried on or deposited at or near the side of a glacier.

moraine, terminal - That moraine, extending in an arc across a glacial valley, which marks the farthest advance or greatest extent of a glacier.

morass - A tract of land frequently covered by shallow water and growths of water weed.

more or less (adj. phrase) - A phrase used in legal descriptions of property to indicate that the actual distance, direction, or area may not be exactly that given in the description. The phrase is used when it would be impractical to determine the exact value and when the exact value is not important. The words in their ordinary use are taken as words of caution, denoting some uncertainty in the mind of the person using them and a desire not to misrepresent. When used in connection with quantity and distance, more or less are words of safety and precaution, intended merely to cover some slight or unimportant inaccuracy. When 125 feet more or less to the point of beginning is used in a deed, the more or less indicates that the 125 feet is an informative term, whereas to the point of beginning is the controlling term. A phrase such as about 12 acres is indefinite and should be avoided because the word about is very broad in meaning.

morgen - A land measure of various sizes, used now or formerly in various Germanic countries, and meaning originally the amount of land one team or man could plow or mow in one morning. The old Dutch morgen was equal to about 8 500 square meters.

Moritz's modification (of Clairaut's equation) - The equation
$(a-b)/a + (\tau_p - \tau_e)/\tau_e = (\omega^2 b/\tau_e)(1 + e'q'/2q)$, in which τ_p and τ_e are the theoretical values of gravity at the poles and at the equator, respectively, of an equipotential, rotational ellipsoid of equatorial diameter 2a and polar diameter 2b, rotating at the rate ω. e' is the second eccentricity, $\sqrt{[(a^2 - b^2)/b^2]}$ and the quantities q and q' are, respectively, $\frac{1}{2}[1 + 3/e'^2]\arctan e' - 3/e']$ and $3(1 + 1/e'^2)([1 - \arctan e')/e'] - 1$. Also called the Clairaut-Moritz equation.

Morse's method (of determining tilt) - A method of determining the tilt of an aerial photograph by using three or more ground points whose horizontal and vertical coordinates are known, and another point of known horizontal coordinates whose image is near the center of the photograph.

mortgage - (1) A conveyance of property that is contingent upon a failure of specific performance such as the payment of a debt. (2) The (legal) instrument making such a conveyance.

mortgage deed - A deed by way of mortgage, which has the effect of the mortgage on the property conveyed and imposes a lien on the granted states.

mortgagee - A person who advances the funds for a loan secured by a mortgage.

mortgagor - A person who gives a mortgage as security for a loan.

mosaic - (1) A picture or pattern created by assembling many small, individual pictures or patterns. (2) SEE mosaic, photographic. (3) SEE mosaic, aerial. Also called a laydown (jargon). SEE ALSO mosaic, orthophoto; strip mosaic.

mosaic, aerial - An composite picture made up of overlapping, aerial photographs assembled and matched, usually by tearing or cutting away parts of the overlap, to form a continuous, photographic picture of a portion of the Earth's surface. Also called a mosaic, photomosaic or photographic mosaic. Distortions in the aerial mosaic may be ignored or they may be controlled by adjusting them to measurements on the ground.

mosaic, controlled - An aerial mosaic oriented with distances between images of horizontal ground-control points true to scale. A controlled mosaic is usually assembled from rectified photographs. Note that all distances are not true to scale.

mosaic, map-controlled - An aerial mosaic constructed by using topographic maps to provide orientation and control. Both controlled and semi-controlled aerial mosaics (particularly the latter) may be map-controlled mosaics.

mosaic, orthophoto - An aerial mosaic made up of orthophotos at a uniform scale.

mosaic, photographic - An assemblage of photographs, each of which shows part of a region or object, put together in such a way that each point in the individual photographs appears once and only once in the assemblage, and variation of scale from part to part of the assemblage is minimized. Also called a mosaic or photomosaic. A photographic mosaic is assembled by trimming, warping and fitting together the individual photographs. If the photographs were taken at different distances from the object, the photographs must be enlarged or reduced to a common scale. They may be assembled without any control over the distortion, or the assemblage may be controlled by adjusting distortions to measurements on the object.

mosaic, scale-ratio - An assemblage of aerial photographs brought to a common scale by printing each photograph to a scale obtained by measuring distances on a map and making corresponding distances on the mosaic true to scale.

mosaic, semi-controlled - An aerial mosaic composed of photographs mutually oriented by reference to some coordinate system other than that determined by ground control.

mosaic, uncontrolled - An aerial mosaic in which the photographs have not been adjusted either in orientation or scale by reference to ground control. The only adjustment is by matching detail point by point on adjoining photographs. Sometimes called a staple mosaic if used for general locations of photo images.

mosaick (v.) - To assemble and fit together separate, individual photographs so that the assemblage is a continuous, consistent picture of the entire region or object depicted by the individual photographs.

mosaicking (n.) - The process of constructing a mosaic from photographs. SEE ALSO breakaway-strip method; center-to-center method.

Mother Hubbard clause - A clause, written into a conveyance of title, often of the form all abutting strips of land owned by the grantor, whether owned of record or by virtue of the Statue of Limitations. In most States, this allows the tacking on of adverse or possession rights. If only the written title is conveyed, an interruption of possession occurs and the time necessary for possession must start again.

motion, absolute - (1) Motion relative to a point fixed on the Earth or to an apparently fixed celestial point. (2) The motion, of an object, described by its measurement in a coordinate system preferred over all other coordinate systems.

motion, absolute proper - (1) Proper motion in a well-defined coordinate system based on the geometric relationship between Earth and Sun. I.e., proper motion in an inertial coordinate system or in a coordinate system (such as that associated with a meridian circle) not based on the coordinates of other stars. (2) The changes in location of a star caused by the star's motion with respect to an inertial coordinate system with origin at the barycenter of the Solar system, and not caused by motion of the coordinate system itself. Also called proper motion.

motion, actual - (1) Motion relative to the Earth. (American) (2) SEE motion, proper (Brit.); motion, apparent.

motion, apparent - Motion relative to a specified or implied reference which may itself be in motion. Also called actual motion. The expression usually refers to movement of celestial bodies as observed from the Earth. SEE ALSO motion, relative.

motion, Chandlerian - SEE Chandlerian motion.

motion, compound harmonic - (1) Motion which can be described by the function $y = a \cos(\omega_1 t - \theta_1) + b \sin(\omega_2 t - \theta_2)$, in which a, b, ω_1, ω_1, θ_1 and θ_2 are constants and t represents time. Also called harmonic motion. Alternatively, motion such that its projections on each of two orthogonal axes are harmonic motions. (2) Motion as defined above, with the proviso that ω_1 and ω_2 have different values.

motion, crustal - The secular motion of parts of the Earth's crust with respect to other parts or with respect to the mantle. Crustal motion should be clearly distinguished for motion within the crust, which is motion originating above the base of the crust. Crustal motion is not relevant to the stability of bench marks. Motion within the crust is relevant and must be minimized.

motion, direct - (1) Motion, in the solar system, counter-clockwise in the orbit as seen from the north pole of the ecliptic. (2) Motion, of an object observed on the celestial sphere, from east to west and resulting from the relative motion of the object and the Earth. Also called progressive motion.

motion, diurnal - The apparent, daily motion of a celestial body.

motion, harmonic - Any motion describable by a function of the form $a \cos(\omega t - \theta)$ or $b \sin(\omega t - \theta)$, in which a, b, θ and ω are constants and t is the time. Also called simple harmonic motion. Alternatively, the motion of the projection, onto a diameter of a circle, of a point moving with uniform motion along the circumference of that circle.

motion, lost - Such motion, in a mechanism, as is not transmitted to connected or related parts because parts of the mechanism fit loosely.

motion, lower - Rotation of the horizontal circle of a repeating theodolite.

motion, peculiar - (1) The motion of a celestial object relative to a group of objects of the same kind. (2) That part of proper motion which is caused by a star's own motion as distinguished from the observer's motion.

motion, polar - (1) The irregularly varying motion of the Earth's pole of rotation with respect to the Earth's crust. (2) The movement of the Earth's instantaneous axis or rotation with respect to the axis of figure. Also called variation of latitude, variation of the pole and, in recent works, wobble, wander, spin perturbation and grooving. Sometimes erroneously called Chandlerian motion; Chandlerian motion is the largest component of polar motion, but the term properly applies only to the component having a period of about 1.2 years. Polar motion and variation of latitude are not the same. Polar motion is accompanied by a variation in astronomic latitude, but astronomic latitude may vary without an accompanying motion of the pole. Polar motion was first definitely detected by Chandler in the late 1800's. The International Latitude Service was created in 1899 to determine polar motion by its effect on latitude and is still active. The International Polar Motion Service and the Bureau International de l'Heure are also responsible for determining polar motion. The Horrebow-Talcott method is still used for determining polar motion (variation of latitude), but it is being replaced for that purpose by radio interferometry and from the orbits of artificial satellites. The direction of the Earth's axis of rotation is nearly invariable in inertial space. The Earth's axis of figure (that axis, within the Earth, about which the moment of inertia is a maximum) does not coincide with the axis of rotation. Extended to the celestial sphere, the poles of the axis of figure describe a path about each pole of the axis of rotation - a path which is known to be irregular although an annual periodicity and a periodicity of 14 months have been detected as components. The greatest variation of the angle between the axis of figure and the axis of rotation is between 0.1" and 0.3". Polar motion affects the determination of astronomic latitude, longitude, and azimuth. Compensation for the motion is important only in first-order surveying.

motion, progressive - Motion, in an orbit, in the usual direction of celestial bodies within a given system. Direct motion is a specific type of progressive motion.

motion, proper - (1) The projection, onto the celestial sphere, of the motion of a star or similar body relative to the solar system thus, the transverse component, with respect to a point in the Solar System, of the motion of a star or similar object. (2) That part of the apparent, angular motion of a star with respect to the observer, not caused by the movement of the coordinate system, i.e., by precession and nutation or revolution of the Earth. In star catalogs, coordinates are usually given for a specific epoch without taking into account proper motion of the star during the period between the observation and the epoch. To determine coordinates at a different epoch, account must be taken of (a) movement of the coordinate system since that epoch and (b) the proper motion since the time of observation. (3) The actual motion of a craft, as contrasted to apparent or relative motion. (Brit.) (4) SEE motion, absolute proper.

motion, radial - Motion toward or away from a specified point.

motion, reduced proper - Proper motion, as measured in seconds of arc per year, reduced to absolute proper motion in kilometers per second of time. Reduced proper motion can be calculated only if the distance or parallax of the star is known.

motion, relative - Apparent motion, relative motion, and relative movement have the same denotation: all denote the changing position of two bodies with respect to each other. Apparent motion, however, is the motion of one of the bodies as it appears to or is measured by someone at the other body; relative motion usually connotes merely the measurements, without regard to where the measurer is placed; relative movement is usually used to refer to changes of position of different parts of the same body. SEE ALSO motion, apparent.

motion, relative proper - Proper motion determined by comparing a star's location at different times with the locations of other stars at those times.

motion, restricted proper - The angular motion of a star relative to other stars.

motion, retrograde - (1) Motion of a body, in its orbit in a solar system, in a sense contrary to that in which a majority of the other bodies in the same solar system revolve. (2) The apparent motion of a celestial body, in its orbit in the Solar System, in a westward direction among the stars. Also called retrogression.

motion, secular polar - A constant change in the angle between the axis of rotation and the axis of figure. Secular polar motion was discovered by B. Wanach in 1916 from an analysis of the data on variation of latitude between 1900 and 1915. He found a motion of 0.003" per year along the 55° West meridian. Markowitz in 1960, from an analysis of more data found a value of 0.0032" per year along the 60° W meridian.

motion, simple harmonic - SEE motion, harmonic.

motion, upper - Rotation of the upper plate of a repeating transit or theodolite.

Motion Service, International Polar - SEE International Polar Motion Service.

mound - A round elevation of land under about 20 feet in height.

mounds and pits - SEE pits and mounds.

mount - (1) A part of a mountain less high than the mountain itself but higher than much of the rest. (2) SEE mounting.

mount, azimuth - A mounting, for an aerial camera, consisting of a fixed mounting to which is added a ring and motor which rotates the camera about its optical axis.

mount, fixed - A mounting, for an aerial camera, consisting of a simple metallic frame to which the camera is attached in such a position that the camera points vertically downwards (or in a desired oblique direction) when the aircraft is in level flight.

mountain - A region of high land or rock rising more than 500 feet above the relatively flat, surrounding terrain.

mountain barometer - (1) A barometer constructed for use in barometric altimetry. It is a general term, little used, for a barometer of any type which is constructed for safe transportation under difficult conditions in the field. Such an instrument may be used in surveying. SEE altimeter; barometer, aneroid. (2) SEE siphon barometer.

mountain range - A connected chain of mountains. I.e., a number of mountains connected together and extending more or less in one direction. Also called, when the meaning is clear, a range.

mounting - (1) A mechanical structure serving to connect a detachable instrument to a supporting base. Also called a mount. The term is applied, in geodesy, to the device to which a telescope or a distance-measuring instrument is attached and which is itself attached to a tripod. Mountings to which cameras are attached are usually attached either to the fuselage of an airplane or, as in the case of cameras used for terrestrial photogrammetry, to a tripod. If the mounting is attached to a vehicle or airplane, shock absorbers are usually interposed between the mounting and the camera as well as between the mounting and the moving structure. Mountings for telescopes are classified by the kind of motion allowed the telescope (alt-azimuth; right-ascension/declination; triaxial) or by the particular design (equatorial, yoke, Springfield, English, etc.) (2) The mechanical structure holding the elements of an optical component or system together. Also called a barrel or lens cone. (3) The process of placing a set of photographs in proper relationship and fastening the assemblage to a flat surface.

mounting, altazimuth - SEE mounting, alt-azimuth.

mounting, alt-azimuth - A mounting permitting the device or instrument attached to it to be rotated either in azimuth or in angular elevation. Also written altazimuth mounting. A theodolite incorporates an alt-azimuth mounting for the telescope. The mounting has graduated circles which indicate the amount of rotation in either direction. Some astronomical telescopes are supported by alt-azimuth mountings. In that case, the telescope's motion is controlled by eye and hand if the telescope is small or automatically by a computer if the telescope is large.

mounting, dry - A method for cementing a photographic print to a base by means of a thin tissue of thermoplastic material. The tissue is placed between the print and the base, and sufficient heat applied to melt the tissue.

mounting, English - An equatorial mounting in which the two bearings for the polar axis are situated one so far above the declination axis and the other so far below it that the telescope rotates in a meridian between them. The polar axis may be thought of as splitting in half lengthwise and spreading the halves apart to accommodate the telescope and declination axis between them.

mounting, equatorial - A mounting permitting the instrument to be rotated about a fixed axis (right-ascension axis or polar axis) parallel to the Earth's axis of rotation and/or about an axis (declination axis) perpendicular to the polar axis.

mounting, German - An equatorial mounting in which the two bearings for the polar axis are both below the declination axis.

mounting, inertial - Any instrumental support whose orientation is held fixed with respect to a non-rotating coordinate system. The support should in theory be one which is not rotating with respect to the totality of mass in the universe or with respect to the very distant galaxies. In practice, the requirement is usually weakened to mean fixed with respect to bodies known to be rotating. The usual way of fixing the support's orientation is to attach at least two gyroscopes with non-parallel axes of rotation to the support.

mounting, right-ascension/declination - A mounting allowing an instrument to be rotated in right ascension and/or declination. Also called a polar mounting. A right-ascension-/declination mounting is almost always an equatorial mounting.

mounting, stabilized - A mounting whose orientation is kept constant in spite of small changes in the orientation of the structure to which the mounting is fastened. Also called a stabilized platform.

mounting, triaxial - A mounting allowing the instrument to be rotated independently about any of three axes (usually perpendicular to each other).

mounting, x-y - A mounting allowing the instrument to be rotated about a fixed horizontal axis and/or an axis perpendicular thereto. Note that the x-y mounting differs from, the alt-azimuth mounting in that in the latter type, the vertical axis is fixed.

movement, relative - SEE motion, relative.

movement (of a projector), rotational - One of a set of systematic, rotational adjustments of a projector in a photogrammetric device. The projector is usually supported in an assembly of nested gimbals (Cardan suspension), or its equivalent, allowing rotation about any of three mutually perpendicular axes (called x-axis, y-axis and z-axis). Rotation of the projector within the inner gimbal is about the z-axis and is called swing. The inner gimbal rotates about the x-axis (secondary axis); this rotation is called x-tilt. The outer gimbal rotates about the y-axis (primary axis); this rotation is called y-tilt. (The same names are applied to rotations about these axes when the projected is supported by some mechanism other than nested gimbals.)

movement, translational - The systematic movement of the projectors, in a stereoscopic plotter, in the direction of the line of flight of the aircraft which took the pictures.

mud flat - A tidal flat composed of mud.

multilateration - Measurement, simultaneously, of distances to or from fixed transmitters or receivers on the ground to receivers or transmitters, respectively, on a vehicle.

multilateration, hyperbolic - Measurement, simultaneously, of differences of distances between a number of receiving or transmitting stations and a transmitter or receiver, respectively, on a vehicle.

multiplex - A stereoscopic plotter consisting of two or more projectors that can be translated or rotated independently, each projecting an image in either red or blue-green light. The images being projected are on small transparencies (called diapositives) reduced from the original, photographic negatives. When the projectors are spaced an appropriate distance apart and the projected images (i.e., the images formed by projection onto a suitable surface) are viewed through spectacles having a red filter in one frame and a blue-green filter in the other, a stereoscopic image (called the stereoscopic model) is seen. Measurements made on it would, in an ideal situation, be directly proportional to corresponding measurements on the original object. The pair of differently colored, projected images is called an anaglyph.
multiplex control - Photogrammetric control for the multiplex.

multiplex triangulation - Stereotriangulation using the multiplex.

multiplier, undetermined - SEE Lagrangian multiplier.

muniment - (1) The evidences or writing whereby one is enabled to defend the title to an estate or maintain a claim to rights or privileges. (2) Specifically, title, deeds and papers.

muniment room - A room devoted to the keeping of muniment.

Murphy's law - If anything can go wrong, it will.

Murphy's method (of adjusting a traverse) - Adjusting the angles in a traverse using the method of least squares, subject to the conditions that (a) the geodetic coordinates of the initial and final points are known, and (b) the azimuths at the initial and final points are also known. The angular changes are weighted inversely as the squares of the lengths of the corresponding courses.

N

nadir - (1) That point on the celestial sphere which is diametrically opposite the zenith. Equivalently, that point at which the direction of the plumb line extended below the horizontal plane meets the celestial sphere. Also called the nadir point. (2) SEE ground nadir. (3) SEE map nadir. (4) SEE photograph nadir. SEE ALSO ground nadir; map nadir.

nadir point - SEE nadir; photograph nadir.

nadir radial - (1) A radial drawn from the nadir point as radial center. (2) SEE isoradial.

name, contrary - A name which denotes the opposite of that possessed by something else, e.g., declination has the contrary name to latitude if the one is north and the other is south. SEE ALSO name, same.

name, geographic - The proper name or expression by which a particular geographic feature such as a river, lake, town or other natural or artificial object is known.

name, same - A name which is the same as that possessed by something else, e.g., declination has the same name as latitude if both are north or both are south. SEE ALSO name, contrary.

name index - A listing of records of title by the names of the grantor and grantee, mortgagor and mortgagee, etc.

Nansen bottle - A container used for taking samples of sea water. It consists of a brass cylinder fitted with a plug-valve at each end and capable of being reversed in its position on the cable to which it is attached when lowered. The two valves are closed simultaneously and the bottle reversed by a weight (messenger) dispatched along the supporting cable from the surface. Also called a Nansen water-bottle. Newer versions are lined with plastic to prevent internal corrosion and consequent faulty chemical analyses. A pair of reversing thermometers is usually attached to the bottle to obtain the temperature of the water at the depth at which the sample was taken.

Nansen water-bottle - SEE Nansen bottle.

NAP Datum - SEE Normaal Amsterdams Peil.

National Geodetic Vertical Datum - (1) The vertical-control datum used (1980 and later) by the National Geodetic Survey for vertical control. (2) In the form National Geodetic Vertical Datum of 1929, a synonym for Sea-level Datum of 1929. This term was officially adopted by the National Geodetic Survey on 17 May 1976.

National Geodetic Vertical Datum of 1929 - The name, after 10 May 1973, of Sea-level Datum of 1929.

National Grid - The grid used on the British Ordnance Survey's maps of the United Kingdom. The origin is south-west of the Scilly Isles, and all maps are drawn so that the margins coincide with lines of the grid. Each 100-m square of the grid is identified by a pair of letters. The first letter identifies the 500-meter square within which the 100-meter square lies; the second letter identifies the 100-meter square itself. The primary axes of the grid are at 2° West and 49° North; the false origin is transferred 400 km west and 100 km north. Also called The National Grid.

National Map Accuracy Standards - SEE U.S. National Map Accuracy Standards.

National Prototype Meter - Copy number 27, given to the USA in 1899, of the International Prototype Meter.

National Tidal Datum Control Network - Those tide stations of the U.S. National Ocean Survey which provide the basic tidal datums for coastal boundaries and for chart datums of the USA. Geodetic leveling between stations of the Network is not required.

National Tidal Datum Convention of 1980 - The convention adopted by the USA effective 28 November 1980 and making the following changes to the previously established rules governing tidal datums. (a) One uniform, continuous tidal datum was established for all tidal waters of the USA (including Commonwealths, territories and U.S. Trust Territories under jurisdiction of the USA). (b) It provided a tidal datum independent of computations based on the type of tide. (c) It lowered chart datum from mean low water to mean lower low water along the Atlantic coast of the United States of America. (d) It changed the National Tidal Datum Epoch from the period 1941 through 1959 to the period 1960 to 1978. (e) It changed the name Gulf Coast Low Water Datum to mean lower low water. (f) It introduced the tidal datum of mean higher high water in regions of predominantly diurnal tides. (g) It lowered mean high water in regions of predominantly diurnal tides.

National Tidal Datum Epoch - A period of 19 years adopted by the National Ocean Survey as the period over which observations of tides are to be taken and reduced to average values for tidal datums. The epoch is designated by giving the year the period began and the year it ended, e.g., National Tidal Datum Epoch of 1941 through 1959.

Nautical Almanac, The - An annual publication of the U.S. Naval Observatory and the Royal Greenwich Observatory, listing the Greenwich hour-angle and declination of various celestial bodies to a precision of 0.1' at hourly intervals, times of sunrise, sunset, moonrise, moonset and other astronomical information useful to navigators.

Navier-Stokes equation - The equation of motion for a viscous fluid: it may be written $dv/dt = -(1/\rho) \Delta p + F + \kappa * \Delta^2 v$

+ ½ *Δ (*Δ.v), in which p is the pressure, ρ the density, F the total external force, v the fluid's velocity and κ the kinematic viscosity. For an incompressible fluid, the term in *Δ.v vanishes and the viscosity then plays a role analogous to that of temperature in thermal conduction and to that of density in simple diffusion. Analytical solutions of the Navier-Stokes equation have been obtained only in a small number of special cases. More general solutions are obtained by numerical methods.

naviface - (1) The surface separating air from the sea. (2) The surface of the sea.

navigable (adj.) - Able to be navigated by a vessel. The question of whether a stream or body of water is navigable is a question of fact with respect to the character of the stream or lake in its natural and ordinary condition as it was at the date of Statehood. At such time, the channel should have been a passage for commerce in the customary modes of trade and travel on water. Legal navigability includes the navigability of all coastal waters and tidal rivers, bays, bayous, lakes and their connecting tidal passages.

navigation - The art of making a vehicle, vessel or aircraft follow a prescribed course or making it reach a prescribed destination. The allied art of determining where the vehicle, etc., actually is, as contrasted to where it should be, is called locating, locationing, fixing the location, obtaining a location, obtaining a fix, obtaining a position and positioning. (Positioning is also used to denote the process of placing a vehicle in a specified location or position.) The equipment and techniques used for navigation, considered as a unit, are called a navigation system. A particular set of equipment and techniques used for locating or positioning is called a positioning system. The same system is usually used for both navigation and locating. If, however, locating is done for surveying, different systems may be used, with the positioning system being the more precise. SEE ALSO arc navigation; grid navigation; radio navigation; survey navigation.

navigation, acoustic - Navigation by means of sound waves, whether the sound is within the audible range or not. Also called sonic navigation. Although the term applies to navigation done by listening to fog horns and other beacons above water, it is more commonly applied to navigation done using beacons underwater.

navigation, aerial - Navigation of aircraft. Also called airborne navigation.

navigation, airborne - SEE navigation, aerial.

navigation, analytic inertial - Inertial navigation in which the outputs of accelerometers that have gyroscopically- maintained orientations are converted to geodetic coordinates by automatic computers.

navigation, astronomical - Navigation depending on locations determined from observations on celestial bodies. Also called celestial navigation. The observations are usually measurements of angular elevations of several stars during the night and angular elevations of the Sun during the day.

navigation, bathymetric - The art of navigating in oceanic regions by determining the vessel's location with respect to known locations of geological features on the bottom. This method is particularly useful to submarines; it is valueless to most surface vessels.

navigation, celestial - SEE navigation, astronomical.

navigation, Doppler - SEE Doppler navigation.

navigation, hyperbolic - Navigation in which locations are determined as differences of distance from three or more fixed points on land. Equivalently, navigation using a hyperbolic navigation system. A hyperbolic navigation system determines a difference in distance (measured in wavelengths) of a mobile unit from three or more stationary, fixed units. The locus of points all of which have the same difference of distance from two of these units is a hyperbola. Taking the fixed units in two pairs (one unit common to both, if there are only three units), two intersecting hyperbolas are determined. The mobile unit is located at one of the two points where the two hyperbolas intersect.

navigation, inertial - (1) Dead reckoning by double integration of the output of accelerometers whose orientations are fixed in inertial space, i.e., fixed with respect to the distant galaxies. (2) The process of measuring a craft's velocity, orientation (in some applications), and displacement from a known starting point by measuring the accelerations acting on the craft in known directions, by means of devices such as accelerometers and gyroscopes.

navigation, rho-theta - Navigation in which locations are determined by measuring distances and directions to a station of known location.

navigation, short-distance - Navigation depending on equipment having a range longer than that used for approaches but shorter than that used for long-range navigation. There are no generally accepted demarcations between the three types.

navigation by star-chart matching - Navigation of a vehicle or craft by star-chart matching (q.v.).

navigation chart, long-range - SEE chart, long-range navigation.

navigation satellite - An artificial satellite (of the Earth) constructed and put into orbit as a beacon for use in navigation.

navigation station - A device, apparatus, or equipment more or less permanently emplaced at a known location to assist in navigation.

navigation system - A set of equipment and techniques by which an aircraft, vessel or vehicle can determine its location at suitable intervals and thereby follow a planned course or reach a desired destination. Also called a positioning device and a positioning system. However, navigation system and positioning system are not, in general, synonymous. A

navigation system differs from a positioning system in that the locations provided by the latter are used primarily for placing the user at a desired location or position and are not used, except incidentally, for navigation. The locations provided by positioning systems are usually more precise or more accurate than those provided by a navigation system.
SEE ALSO satellite navigation system; survey navigation system.

navigation system, Doppler - SEE Doppler navigation system.

navigation system, global - A navigation system usable anywhere in the world. Examples are: celestial, Omega, TRANSIT and NAVSTAR.

navigation system, hybrid - A navigation system consisting of two different and independent navigation systems combined in such a manner that the combination gives better results than either system separately would give. A combination of a Doppler and an inertial navigation-system, or a combination of Loran-C and a Doppler navigation- system, are examples.

navigation system, hyperbolic - A navigation system using the measured differences in distance (measured in wavelengths) of a mobile unit from three or more fixed units to determine the location of the mobile unit. Only the fractional part of one wavelength is actually measured. Most hyperbolic navigation-systems keep count of the changes in difference by one whole wavelength, so that one the entire distance is known, the system continues to indicate the total difference, regardless of the movement of the unit. To get a total difference to start with, or to recover one if the system stops functioning for a while, some systems use auxiliary signals at much longer wavelengths also; this allows the smaller difference to be accurately located within the one determined using the longer wavelengths.

navigation system, inertial - (1) A navigation system in which gyroscopes or accelerometers are used to provide a coordinate system having a fixed orientation with respect to the distant galaxies and accelerometers to provide (by double integration) changes in location within that coordinate system. The gyroscopic compass is a particularly simple form of an inertial navigation system. A more complicated form giving not only orientation but location, and used for navigating ships, is called SINS (Ships Inertial Navigation System). (2) A navigation system not dependent on artificial sources of electromagnetic radiation for signals. This definition is not particularly useful; it includes navigation systems such as the lead line, the sextant and so on.

navigation system, rho-rho - A navigation system determining the distance of a mobile unit from two or more fixed stations, by measuring the time needed for a signal to travel between the mobile unit and each of the fixed stations. One measured distance locates the mobile unit on a circle with its center at a known station and a known radius equal; a second measured distance locates the mobile unit on a second circle with known center and radius. If the unit has not moved between the two measurements, it is located at one of the two intersections of the two circles. If it has moved, an estimated velocity and location are used to correct for the lapse in time.

navigation system, rho-theta - A navigation system determining the distance of a mobile unit from a single, fixed station and the direction of that unit from a pair of fixed stations (one of which may be at the same point as the station used in determining distance). The mobile unit determines its distance from the single station by measuring the time it takes a signal to travel between it and that station. This locates the unit somewhere on a circle of known center and radius. At the same time, it determines the difference in its distances from the two fixed stations, locating the station somewhere on a hyperbola of known position. However, because the two stations of the pair are very close together, this difference can be converted into an approximate direction to the mid-point of the line joining the pair. The combination of distance and direction gives the location of the mobile unit.

navigation system, short-range - SEE Shoran.

navigation system, theta-theta - A navigation system deriving location by determining the direction two or from two stations.

NAVSTAR - NAVigation Satellite Timing and Ranging. SEE Global Positioning System (GPS).

Navy Navigation Satellite System (NNNS) - A set of 5 satellites in polar orbit, all at approximately 1000 km altitude; a set of four tracking stations, all in the USA; and a computation center in California. Also called the TRANSIT System or TRANSIT (q.v.).

neap - SEE neap tide.

neap high water, mean - A tidal datum without a precise meaning, adopted in some instances by U.S. courts as defining ordinary (average) high water mark.

neap range, mean - The average, semidiurnal difference in the height of the water at the time of quadrature. Mean neap range is smaller than the mean range where the tide is either semidiurnal or mixed, and is of no practical significance where the tide is diurnal.

neap rise - The height of mean high water neaps above the chart datum.

neaps, high-(low-)water - (1) A high (low) water recorded during quadrature. SEE ALSO neap tide. (2) The average elevation of all high (low) waters of neap tides recorded during quadrature over an established period, usually 19 years, or computed over an equivalent period. Also called neap high-(low-)water.

neaps, low-water - SEE neaps, high-(low-)water.

neaps, mean high-(low-)water - (1) The average height of high (low) water recorded during quadrature over an established period, usually 19 years, or over a computed, equivalent period. (2) The average height of the high (low) waters of neap tides. Also called high water neaps and mean high water.

neaps, mean low-water - SEE neaps, mean high-(low-)water. Also called low-water neaps and low-water quadrature.

neap tide - A tide of decreased range occurring semi-monthly as the result of the Moon's being in quadrature. The neap range (N_p) of the tide is the average, semidiurnal range occurring at the time of neap tide. It is smaller than the mean range where the tide is either semidiurnal or mixed, and is of no practical significance where the tide is diurnal. The average height of the high waters of the neap tides is called neap high water or high water neaps; the average height of the corresponding low waters is called neap low water or low water neaps.

nearshore - That part of a beach between the shoreline and the line at which the waves break.

neat line - SEE line, neat.

negative - (1) A photograph which is dark where the original scene was light and light where the original scene was dark, etc. Also called a photographic negative, i.e., a photograph in which the scale of shading from light to dark of the original scene is reversed, progressing instead from dark to light. (2) A photograph whose colors are the complements of those in the original scene.

negative, open-window - A negative having open parts (i.e., having small parts cut out) and used as a mask where screens, rulings or tints are to be printed in the uncovered parts.

negative, original - That photographic negative which was developed from the photographic film which was in a camera's magazine at the instant of exposure.

negative, photographic - SEE negative.

negative correction - A correction made to a photographic negative.

negative engraving - (1) Making a negative manually by removing portions of a coating applied to a glass plate. (2) Making corrections and additions to wet-plate negatives.

negative proving - Preparing a color proof photomechanically from the photographic negatives. Also called the neg proof process (jargon).

negative scribing - SEE scribing.

negotiate (v.) - (1) Transact business, i.e., treat with another respecting a purchase or sale. (2) Hold intercourse. (3) Bargain or trade. (4) Conduct communications or conferences. (5) Conclude by bargain, treaty or agreement. (6) Discuss or arrange sale or bargain.

negotiation - In the process of constructing, enlarging, or improving a highway, the process by which property is sought to be acquired for the highway, the process proceeding through discussion, conference, and final agreement upon the terms of a voluntary transfer of such property.

neighborhood - An urban or suburban residential or commercial locality having considerable homogeneity as to housing, tenancy, income and characteristics of the people in it. Neighborhoods are often outlined by physical barriers such as railroad tracks, streams, commercial or industrial buildings, type or age of buildings, hills or ravines, or by lines created by the construction of subdivisions, differences in zoning ordinances, or restrictions in deeds.

neighborhood effect - SEE Eberhard effect.

neighborhood unit - A neighborhood created according to a plan and whose characteristics are determined by that plan.

Nell's map projection - One of two map projections devised by Nell: a modification of the globular map projection, and an equal-area map projection combining the Bonne map projection and Lambert's conical equal-area map projection.

Nell's modified globular map projection - SEE map projection, globular.

Nenonen camera - An aerial camera which, in addition to photographing the ground, also photographs the horizon and the differential barometer (statoscope).

net - SEE base net; network. The term net has been used extensively in the older literature of surveying but is being replaced gradually by network.

net, vertical-control - SEE network, leveling.

Netherlands Normal - The tidal datum for the Netherlands, referred to the standard tide-gage at Amsterdam (Normaal Amsterdams Peil). This tidal datum has been used by many European countries as a common datum for leveling.

network - A set consisting of (a) stations whose geometric relationships (distances, directions, coordinates, etc.) have been determined and which are related in such a way that removal of one station from the set will affect the relationships between the other stations; and (b) lines connecting the stations in such a way as to show this interdependence. The lines represent adjusted distances, directions, or angles, or the sequence in which measurements were carried out. For example, a triangulation network consist of stations whose horizontal coordinates with respect to each other have been determined, and lines representing distances, directions or sides of angles between the stations. A level network consists of a network of level lines. A gravity network consists of gravity stations occupied in sequence, together with lines connecting them in the sequence in which they were occupied. Networks contain only closed circuits. All lines connect at least two different points, and each point lies on at least two different lines. This follows from the condition that stations' relationships are interdependent. SEE ALSO base network; control network; densification network; gravity network; level network.

network, astrogeodetic - A geodetic network consisting of stations whose horizontal coordinates have been determined and at some of which the astronomic azimuth and, possibly, also latitude or longitude have been determined.

network, extension - SEE triangulation network, base-extension.

network, floating - SEE network, free; network, independent.

network, free - (1) A network in which the geometric relationships between the stations and the numbers associated with each station are not related to stations (or their associated numbers) not in the network. (2) SEE network, free geodetic.

network, free geodetic - (1) A geodetic network whose position with respect to a recognized datum is not specified. Distances between points of the network, and angles between lines of the network, are known but coordinates or directions are not. (2) A geodetic network in which there are no fixed lengths, distances, directions (with respect to an external line) or coordinates.

network, geodetic - A network whose stations are related to each other by geodetic surveys or are gravity stations.

network, independent - A survey network in a datum defined for that network only and not in the datum selected for the other geodetic control of a nation. Also called a floating network. The terms floating datum and independent datum have also been used but are misleading.

network, leveling - A network in which the observations consist of elevation differences between vertical bench marks and turning points. Also called a vertical-control network or vertical-control net. SEE ALSO survey network.

Network, National Tidal Datum Control - SEE National Tidal Datum Control Network.

network, vertical-control - SEE network, leveling.

network adjustment - The process of changing the values associated with the stations of a network, and/or the geometric relationships (distances, angles, etc.) between the stations according to some principle or rule, in order to obtain better values or geometric relationships.

network adjustment, free - Network adjustment in which the number of quantities to be changed is greater than the number of equations relating these quantities to each other or to given quantities.

network adjustment, Helmert blocking method of - SEE Helmert blocking method.

network densification - The process of adding control points and observations to a network in a region already containing control points.

Newcomb sidereal time - The quantity T given by T = 6h 38m 45.836s + 8640184.542s t + 0.0929 t^2, in which t is the time in units of 36525 mean solar days (100 Julian years) counted from 31 December 1899, 12h U.T. This equation, derived by S. Newcomb, is in general use for calculating ephemerides. SEE ALSO time, Greenwich mean sidereal.

New England Datum - A datum identical with the horizontal-control datum for which the name United States Standard Datum was adopted in 1901; that name was later changed to North American Datum. Before 1901, this datum was defined by the following coordinates of the origin and azimuth (clockwise from South) in the Clarke spheroid of 1866; the origin was triangulation station Principio:

longitude of origin	76° 00' 16.407" W
latitude of origin	39° 35' 36.692" N
azimuth from origin to Turkey Point	1° 34' 36.413"

Newlyn Datum - A vertical-control datum defined by mean sea level at Newlyn, England. SEE ALSO Ordnance Newlyn Third Geodetic Leveling Datum.

Newton - The force needed to give a mass of 1 kilogram an acceleration of 1 meter per second^2. This is a derived unit in the SI. It is equal to 105 dynes.

Newton-Cotes formula - A general formula for numerical integration of a function f(x) between limits x_o and $x_o + N\Delta x$:

$$\int_{x_o}^{x_o+N\Delta x} f(x)dx \approx \Sigma_n a_n y_n \ (n = 0 \text{ to } N)$$

in which the y_n are given values of f(x) at equal intervals of x from x_o to $x_o + N\Delta x$ and the a_n are constants given by

$$a_n = (-1)^{N-1} \Delta x / [n! (N-n)!] \{\int_o^N [z(z-1)(z-2)...(z-n)/(z-n)!] \, dz$$

The trapezoidal rule, Simpson's rule, Boole's rule and Weddel's rule are simple, special cases of the Newton-Cotes formula for N equal to 1, 2, 4 and 6.

O

object, celestial - (1) A celestial phenomenon appearing to be a body but actually of unknown or uncertain nature. (2) SEE body, celestial.

objective - The lens or assemblage of lenses focusing light from the object to form the first real image of the object. Also called an objective lens or objective lens-system. In optical instruments such as telescopes and microscopes, the image formed by the objective is magnified by a second lens-system called the eyepiece. In cameras, the image formed by the objective is the final image. Objectives in astronomical telescopes usually consist of only one or two elements (separate lenses). Objectives in telescopes used for surveying and in cameras usually contain several elements.

objective grating - A grating placed in front of the objective lens of a telescope to produce stellar spectra. The image of each star in the field is then a short spectrum instead of a dot. The grating must be coarse to keep the spectra short and to reduce overlapping between neighboring images. Objective prisms are also used to produce stellar spectra.

objective lens - That part of a telescope or microscope which receives light directly from the object and forms the first image. In a camera, the image formed by the objective lens is the final image. Also called an objective.

objective prism - A thin, wedge-shaped prism placed in front of the objective lens of a telescope to disperse starlight into spectra on the photograph.

object observed, eccentric - An eccentric signal to which observations are made. SEE ALSO signal, eccentric.

object space - The space of all points from which light or other electromagnetic radiation can enter an optical or other system forming images. The term is often limited to denote that part of space from which rays can enter and pass entirely through the image-forming system. SEE ALSO image, optical.

oblate (adj.) - Flattened, as opposed to prolate (lengthened). In mathematics and physics, it is used mostly to denote a spheroid (rotational ellipsoid) having the minor axis shorter than the other two axes, as contrasted to a prolate spheroid, which has its minor axis longer than the other two axes. It is used in the same way in geodesy except in gravimetric geodesy, where rotational ellipsoid is used instead of spheroid.

oblateness - (1) An eccentricity (of a rotational ellipsoid) which is a real number. A rotational ellipsoid whose eccentricity is an imaginary number is a prolate, rotational ellipsoid; the absolute value of the eccentricity is called the prolateness. (2) The eccentricity of a rotational ellipsoid. (3) SEE flattening.

oblatum - An oblate, rotational ellipsoid, i.e., an oblate spheroid.

oblique (n.) - An aerial photograph taken at an angle to the vertical. Preferably called an oblique aerial photograph (q.v.).

oblique, gridded - An oblique aerial photograph on which is printed a grid to assist in identifying a particular region or point within the photograph. Gridded obliques are used chiefly for spotting targets for artillery.

oblique plotting instrument - A plotting instrument for plotting from oblique photographs.

obliquity - (1) In general, the angle between a satellite's axis of revolution and its axis of rotation. (2) In particular, the angle between the plane of the Earth's equator and the plane of the ecliptic, and usually referred to as the obliquity of the ecliptic. SEE ALSO ecliptic.

obliquity factor - A factor, in the expression for a constituent tide or tidal current, involving the angle of inclination of the plane of the Moon's orbit to the plane of the Earth's equator.

obliquity of the ecliptic - (1) The acute angle ε between the ecliptic and the celestial equator. (2) The acute angle between the pole of the ecliptic and the mean north celestial pole. It is given as a function of time by $\varepsilon = 23° \ 27' \ 08.26" = 0.4684" \ T - 0.000006" \ T^2$, in which T is the time in tropical years since 1900.0. The obliquity of the ecliptic is therefore about 23.45° and is decreasing at the rate of about ½" per year.

observable (n.) - (1) Any thing or any event able to be seen or otherwise detected or sensed, and noted. (2) Any measurable thing or event. (3) A quantity present as an unknown in an equation and whose value is to be found. This usage is misleading and appears to be used as jargon by some statisticians.

observability condition - A set of linear equations between two vectors x and y, $y = Ax + b$, which is solvable for x in terms of y, the matrix A and the constant b. This usage is misleading and, like observable (3) appears to be used as jargon by some statisticians.

observation - (1) The act of deliberately sensing an event or thing and noting the circumstances. An observation is distinguished from a sensing by being deliberate and involving the noting (as by making a record) of the circumstances. It differs from a measurement in that a number, the measurement, need not be given to the thing or event observed. That is, a measurement is an observation and the act of assigning a numerical value to the observation. (2) The notes or records of the circumstances accompanying the sensing of a thing or event. (3) SEE measurement.

observation, celestial - (1) The observation of events in the heavens, i.e., of celestial phenomena. This does not imply making measurements. (2) Measurement of the altitude or azimuth, or both, of a celestial body. (3) The data obtained from measurements made of the location or position of a celestial body.

observation, conditioned - A measurement subject to the condition that a function of it and other quantities have some specified value. More precisely, a measurable quantity whose values are subject to the condition that some function of the quantity and other quantities have some specified value when the measured value is inserted. For example, the measured value of one angle in a plane triangle is a conditioned observation if the sum of the three angles in the triangle must be 180°.

observation, direct - (1) An observation in which the event or thing itself is observed. It is the antithesis of an indirect observation, in which an image or other surrogate for the event or thing is actually observed. (2) SEE measurement, direct.

observation, equal-altitude - Observation of celestial bodies at a fixed, angular elevation (as by an astrolabe), the observations being taken in more or less uniformly-spaced directions around the horizon. Also called an equal zenith- distance observation and equal angular-elevation observation. The purpose is to obtain a number of lines of position by a method which reduces the effects of vertical refraction.

observation, ex-meridional - (1) An observation of a celestial body near the celestial meridian of the observer. (2) The measured angular elevation of a celestial body near the celestial meridian of the observer.

observation, indirect - (1) An observation made on a particular effect or aspect of an event or thing, rather than on the event or thing itself. (2) SEE measurement, indirect.

observation, intermediary - SEE measurement, intermediary.

observation, lunar - A measurement of the Moon's angular elevation, right ascension, time of crossing a particular circle on the celestial sphere, etc. SEE ALSO method of lunar observations.

observation, reciprocal - One of two measurements made as a pair to reduce the size of some systematic error in the individual measurements. In particular, one or a pair of measurements taken forward and backward at the ends of a line. For example, in trigonometric leveling, the angle measured from the horizontal plane of the observer to the top of some feature is usually paired with the angle (the reciprocal observation) measured from the horizontal plane at the top of the feature to the point from which the first angle was measured. The term should be applied only to measurements made simultaneously or within a short time of each other, because only then will conditions be such as to allow systematic errors to cancel each other.

observation, solar - In astronomical geodesy, a measurement of the Sun's angular elevation, right ascension, time of crossing a particular plane, etc. SEE ALSO method of solar-observations.

observation equation - Any equation containing an observed or measured value or a correction to a measured value, or functions or differences of these values. Also called an observational equation. Formerly called an equation of condition; this should not be confused with condition equation. The measured value (or its correction) may occur as an implicit function of the unknowns or, as is more usual in geodetic problems, may appear alone on one side of the equation and the unknowns on the other side.

observation matrix - A matrix A whose elements are the coefficients of the unknowns x in a set of linear equations of the form $y = Ax$, in which y is the vector of measured values. Also called an adjustment matrix, condition matrix, constraint matrix and design matrix. The terms condition matrix and constraint matrix more usually have quite a different meaning from that of observation matrix.

observations, simultaneous - Observations or measurements made simultaneously from several different points.

observation star catalog - SEE star catalog, independent.

Observatory, Mean - SEE Mean Observatory.

observing tower - A tower constructed to permit sighting from an observing instrument to a distant point. The tower may be of wood, steel, or other strong material. It may be of triangular or rectangular cross-section. If very precise measurements are to be made from it, it usually consists of two separate structures: an inner structure supporting only the instrument and an outer structure supporting the observer and auxiliary equipment.

occultation - (1) The disappearance (obscuration) of one celestial body behind another of larger apparent size. (2) In particular, the disappearance of a star or planet because of the Moon has passed between it and the observer. The star or planet is said to be occulted. (3) The disappearance of a satellite behind the disk of its primary. If the primary source of illumination of a reflecting body is cut off by the occultation, the phenomenon is called an eclipse. The occultation of the Sun by the Moon is a solar eclipse. The differences between the three similar terms eclipse, occultation and transit are exemplified by the changes in the appearance of the moons of Jupiter. An eclipse occurs when a moon passes into the planet's shadow; an occultation occurs when the moon passes directly behind the planet, so that it could not be seen even if it were illuminated; and a transit occurs when the satellite passes between the observer and the planet, i.e., in front of the planet instead of behind it. Occultations of stars by the Moon have been timed as a means of determining the distances between the observing stations.

occultation, solar - SEE eclipse, solar.

occultation surveying - Surveying done by timing the instants at which selected stars are occulted by the Moon. At the instant when the star is occulted, the observer lies on an almost straight line joining his location, the star and a point on the Moon's edge. Other points of observation are chosen so that the occultation of the same star occurs at the same point on the Moon's edge but at different times. The intervals between occultations are known functions of the Moon's apparent velocity and of the distances between points of observation. If several sets of observations are made from the same unknown point in conjunction with simultaneous observations from known points, the location of the unknown point with respect to the known points can be determined.

occupancy - The act of taking possession of a thing having no owner, thus acquiring a title by occupation.

occupancy, open - A residential, rental property, the occupancy of which is not restricted by race, color or religion or any other illegal form of discrimination. The term usually refers to residential developments that have been constructed with the aid of government fonds.

occupy (v.) - Be at a station and make observations from or at that station.

ocean - (1) Properly, one of three large bodies of water: the Atlantic Ocean, the Pacific Ocean and the Indian Ocean. Also called a sea. A fourth ocean, called the Antarctic Ocean, is distinguished by some oceanographers but is considered by others to be merely parts of the other three oceans. There are no easily and universally recognized boundaries between the four oceans, because they do form a continuous body of water. (2) The single body of water formed by the union of the Antarctic, Atlantic, Pacific and Indian Oceans. The continuity of the three last-mentioned bodies of water is often recognized by referring to them as a single entity, the ocean or the world ocean. A number of smaller but still large bodies of water adjacent to and connected to the oceans are distinguished from the oceans proper and are called seas, for example the Mediterranean Sea and the Black Sea, the Persian Gulf and the Gulf of Mexico. The boundaries between the oceans and seas are often indistinct and more a matter of arbitrary definition or convention than of physical difference.

ocean, effective - A hypothetical layer of water, on the Earth, having a thickness such that the ratio of its area to its volume is equal to the ratio between the surficial area of all the world's oceans to the volume of all those oceans.

ocean, standard - An ocean whose temperature is 0° C and whose salinity is 35 o/oo.

ocean bottom - SEE sea bottom.

oceanography - The study of the oceans and seas in all their aspects - their extent, their dimensions, their composition, their movements, the plant and animal life in them, etc. Also called oceanology. Study of the crust under the oceans and seas is by some considered part of oceanography, by some a part of geology. It is commonly referred to as marine geology. Oceanography is otherwise divided into two quite different disciplines: physical oceanography, which is concerned with the waters of the oceans and seas; and biological oceanography or marine biology, which is concerned with the life living in the oceans and seas.

oceanology - (1) The study of oceans and seas and their relation to the rest of the world and the life on it. Oceanology in this sense is a more comprehensive term than oceanography; it includes such topics as navigation, the technology used in exploiting the oceans, etc. (2) SEE oceanography.

ocean shoreline - (1) The narrow strip or zone embracing that part of the land which is affected by waves both above and below the surface of a lake, sea or ocean. The term does not apply to tidal flats or marshes which are overflowed by the tides, but essentially to strips of land which has an appreciable slope towards the water. It is questionable whether the above definition can be used or not. That part of a wave which is underwater can affect the bottom for a considerable distance seaward from the line marking the uttermost extent to which the land appears periodically above the water. (2) The line along which the land meets the surface of a lake, sea, or ocean. (3) The intersection of a specified marine surface with the shore or beach. The line delineating the ocean shoreline on nautical charts of the U.S. National Ocean Survey approximates the line of mean high water. (4) The line along which the surface of the land meets the surface of a lake, sea, or ocean. (5) That part of the land which is affected by the action of waves. The term does not apply to tidal flats or marshes which the tides overflow, but essentially to strips where the land has an appreciable slope outward to the water.

octant - (1) A one-eighth part of a circle. (2) An instrument similar to a sextant but having an arc of only 45°.

odograph - An instrument containing a rotating part that moves or turns by an amount proportional to the actual distance traveled, a compass giving a reference direction with respect to the distance which is resolved into orthogonal components and an integrator for determining the components of the distance. SEE ALSO odometer.

odometer - A counter attached to a wheel of a vehicle to count the number of turns made by the wheel in traveling over the ground. Most odometers indicate the distance traveled by the vehicle, i.e., they multiply the number of turns by the circumference of the wheel.

Oeländer gravity reduction - SEE gravity reduction, topographic.

oersted - The strength of a magnetic field existing at a point where a unit magnetic pole in vacuo experiences a force of one dyne. Also called, before 1930, the gauss. The oersted is a unit of magnetic-field strength in the c.g.s. system and is expressed in abamperes per centimeter or amperes per meter. The strength represented by one oersted is small; the strength of the Earth's magnetic field is of the order of 0.2 to 0.3 oersteds.

offer, jurisdictional - A written notice to an owner of an intent to purchase certain lands or interests therein for a stipulated amount of consideration. This is a prerequisite to condemnation.

office computation - A calculation based on measurements made in the field and including all calculations relating to the reduction of survey notes to graphic form for any type of survey or for the continuation of work in the field.

offset (USPLS) - (1) A short line perpendicular to a surveyed line and measured to a line or point for which data are desired, thus locating the second line or point with respect to the surveyed line. Offsets may be measured from a surveyed line or lines to the edges of an irregular body of water or to any irregular line whose position is wanted. (2) A short segment of a surveyed or other line, at the start of which the original line makes an abrupt change of direction and at the end of which it returns to its original direction. (3) A perpendicular distance measured from a great-circle line to a parallel of latitude, to locate a section corner on that parallel. SEE ALSO secant method; tangent method. (4) The stake or mark placed at a known and specified distance and angle (or direction) from a point, to preserve the location of that point when it is anticipated that the original point might be destroyed or disturbed. (5) SEE lithography. (6) An intentional difference between a realized value and the nominal value of a quantity. SEE ALSO relief offset, dry.

offset, meridional - A small distance applied to the length of a meridian to create the curves of the top and bottom latitudes of a graticule.

offset, normalized - The difference between the realized value of a quantity and the nominal value of that quantity, divided by the nominal value.

offset line - A supplementary line close to a roughly parallel to a main line, to which it is referred by measured offsets. Where the line to which it is desired to measure is difficult to get to, the required data are obtained by surveying an offset line in a convenient location and measuring offsets from it to salient points on the other line.

offset lithography - Lithography in which the inked image on the plate is first printed onto a rubber mat or blanket, from which the inked impression is then transferred (offset) onto the sheet of paper.

offset press - A printing press that contains an extra, rubber-covered cylinder upon which the image is first printed and then transferred from this cylinder to the paper.

offset process - A printing process in which the design, instead of being transferred directly from the plate carrying the design to the paper, is transferred first to the surface of a rotating, rubber cylinder and then to the paper.

offshore - (1) The zone extending from a specified shoreline seaward for an indefinite distance. Also called the offshore zone. On the Atlantic and Gulf coasts of the USA, the shoreline specified is the line of mean low water; of the Pacific coast, it is the line of mean lower low water. (2) The comparatively flat region, of variable width, extending from the outer margin of the rather steeply sloping shore face to the edge of the continental shelf.

offshore zone - SEE offshore (1).

Old Hawaiian Datum - (1) The horizontal-control datum which is defined by the following coordinates of the origin and by the azimuth (clockwise from South) on the Clarke spheroid of 1866; the origin is at triangulation station Oahu West Base:
longitude of origin   157° 50' 55.79" W
latitude of origin   21° 18' 13.89" N
azimuth from origin to Oahu East Base   291° 29' 36.0"

The datum is based on an astronomic longitude obtained from observations of lunar culminations, star occultations, etc., and on an adjusted latitude derived from a number of astronomic latitudes in various parts of the islands. (2) A datum having the same constants as the Old Hawaiian Datum (1) proper except that the International Ellipsoid is used instead of the Clarke spheroid of 1886. This definition contradicts the first and the term Old Hawaiian Datum so used should be qualified to make its meaning clear, e.g., Old Hawaiian Datum (International Ellipsoid).

Omega - A hyperbolic navigation-system operating at frequencies between 10 and 14 kHz. The phase difference in the signal received from two stations is determined by the receiver. The receiver's location is known well enough, a priori, that the hyperbola determined by the measured phase difference can be drawn. The phase difference between signals from a second pair of stations is likewise measured, and the corresponding, second hyperbola drawn. The receiver is therefore located at one of the two intersections of these hyperbolas. In practice, only small hyperbolic segments in the vicinity of the estimated position are drawn. Omega is useful for navigation to about 9 000 km from the transmitting stations. The root-mean-square-error of a location is between 1 and 3 km; the lower values occur during daylight, when the ionosphere is lower and denser. By 1978 there were enough stations to make navigation by this method possible anywhere at sea.

Omega, differential - A form of Omega allowing one to obtain the distance of a radio receiver from a station on shore with about 100 meters error at distances of up to 200 km. Also called MicroOmega. The method requires that another radio station of known location also be available.

one-projector method - SEE one-swing method.

one-swing method - The method, used in getting relative orientation of a pair of vertical photographs on a stereoscopic plotter with two projectors, for removing parallax in y by keeping one projector of the pair in a fixed position and making all adjustments by moving the second projector with respect to the first. Also called the one- projector method, y-swing method and single-swing projector method.

opacity - The resistance of a material to the transmission of light through it. The measure of opacity is the reciprocal of the transmissivity.

opaque (v.) - Paint a transparent sheet or plate with opaque paint, to prevent the corresponding part of an underlying, photosensitive material from being exposed to light. SEE ink, opaque.

opaqueing - The hiding of incorrectly scribed features, scratches, and other undesirable marks by painting over the lines with an opaque liquid using a fine brush.

operating system - The set of instructions that a computing machine uses to perform its basic functions.

opposition - That configuration taken by the Sun, Earth, and another planet, in which the geocentric (celestial) longitude of the planet differs by 180° from that of the Sun.

optical system - SEE system, optical.

optics, folded - SEE system, folded optical.

optimum ground elevation - That elevation of an assumed, horizontal surface, in the region shown on a set of aerial photographs, which would be projected at the best distance in a stereographic plotter.

option - An agreement that permits one to buy or sell property or an interest therein within a stipulated time, in accordance with the terms of the agreement.

orbit - (1) The path followed by the center of mass of a body moving in response to a central force (.e.g.,gravitation, electrical, etc.) forces, together with the times at which the points in the path were occupied. The terms orbit, trajectory and path denote, in common usage, similar but different concepts and should never be treated as synonyms. They all denote the set of point occupied by a moving, point-like body. Path, however, denotes only this. Trajectory and orbit denote the set of points and the times at which these points were occupied. A trajectory, however, extends only between two fixed points - a starting point and an ending point; an orbit is considered to extend indefinitely forward and backward in time. In algebraic terms, a path is a solution (of the equations of motion) from which the time has been eliminated; a trajectory is a solution obtained by integrating between fixed points or definite times; an orbit is a solution obtained by integrating between indefinite limits. (2) The function giving the location, as a function of time, of a body moving in response to natural forces such as gravitation. Among the forces, other than gravitation, which may affect the motion are air drag, radiation pressure, and magnetic or electrical fields. (3) The path followed by the center of mass of a body moving in response to the force of gravitation and other natural forces. (4) The path followed by a celestial body about some point used as reference. (5) The path taken by a particle of water participating in the motion of waves. The orbit in deep water is nearly circular; in shallow water, it is highly elliptical. In general, the orbit is slightly open in the direction in which the wave is moving, causing a net transport of mass in that direction. SEE ALSO element, orbital; perturbation, luni-solar; reference orbit.

orbit, central-force - The orbit described by a body of negligible mass acted upon by a single force located at a point.

orbit, Earth-synchronous - A circular orbit about the Earth, having its direction the same as the Earth's direction of rotation and having a period of 24 hours. Also called a geosynchronous orbit (q.v.) or a synchronous orbit. The radius is approximately 42 200 km.

orbit, equatorial - An orbit whose plane coincides, or almost coincides, with the plane of the equator.

orbit, geostationary - An orbit in which the satellite has a period exactly equal to the Earth's period of rotation, moves in the same direction, and lies in the equatorial plane, i.e., a geosynchronous, equatorial orbit. So called because (except for slight variations caused by non-circularity of the orbit and by gravitational perturbations) the satellite appears to be stationary in the sky. Also called an Earth-stationary orbit or synchronous orbit; that term, however, is ambiguous.

orbit, geosynchronous - An orbit in which the satellite has a period exactly equal to the Earth's period of rotation and in the same direction. If the orbit is circular and inclined to the equator, the satellite appears to describe a figure 8 once every 24 hours. If the orbit is equatorial, the satellite appears to be stationary in the sky. Because no orbit is perfectly circular, the satellite's motion will vary slightly about the average, ideal location.

orbit, intermediate - (1) A fictitious orbit that a satellite moving in the field of a non-central force would follow if a central force were substituted for the real force at a given instant and the direction of motion were to be the same at that instant. SEE ALSO orbit, osculating. (2) Any fictitious orbit used as an intermediary and approximation in calculating an actual orbit.

orbit, normal - SEE Keplerian orbit.

orbit, osculating - That orbit a satellite would follow, under the influence of several forces, if all forces other than the gravitational attraction of the primary were suddenly removed (set equal to zero). The body would then follow an elliptical orbit about the primary. This orbit would be tangent to the actual orbit at the point where the perturbing forces were removed. Because most satellites follow orbits which are approximately Keplerian for short intervals of time, the actual orbit is often conveniently expressed in terms of an osculating orbit whose elements (size, shape and orientation) are changing slowly.

orbit, perturbed - An orbit differing from another orbit considered standard or ideal for that same body. The perturbations are the differences between the actual orbit and the standard or ideal orbit. An osculating orbit at some suitable time is commonly selected as the standard. A Keplerian orbit chosen so that the average values of the perturbations are small is also often used.

orbit, polar - An orbit having an inclination of about 90° to the equator of the primary and therefore passing near the poles of the primary.

orbit, stationary - An orbit in which the satellite revolves about the primary at the same angular rate as the primary rotates on its axis. From the primary, the satellite appears to be stationary over a point on the primary if the orbit is in the equatorial plane. Otherwise, the satellite will appear to oscillate back and forth across the equator, along a fixed meridian. SEE ALSO satellite, synchronous.

orbit, Sun-synchronous - (1) A nearly polar orbit about the Earth, with a radius such that the satellite passes twice daily, at local solar time, over all places on the Earth having the same latitude. (2) An orbit with a radius and inclination such that the orbit maintains its initial orientation relative to the Sun. Sun-synchronous orbits are retrograde, having inclinations between 95.7° and 180° and altitudes up to 5980 km. E.g., a noon-midnight orbit can be selected to allow good photographic conditions for about one-half of every revolution, or a twilight orbit can be selected so that the satellite is never in shadow (allowing solar power to be used continuously). The relation between the average radius r of the orbit and the inclination i of the orbit is given, approximately, by $9.97(R/r)^{3.5} \cos i = -0.9856$, in which R is the average radius of the Earth.

orbit, synchronous - SEE orbit, Earth-stationary; orbit, geostationary; orbit, geosynchronous.

orbit, two-body - SEE Keplerian orbit.

orbit, unperturbed - SEE Keplerian orbit.

order (of accuracy) - A ranking, according to accuracy, of one measurement or survey with respect to other measurements or surveys. Accuracy is greatest in the first order, less in the second order, and so on. Each order of accuracy is expressed in numerical terms, or such terms are to be understood as being present.

ordinance - A municipal law, comparable to a statute on State and Federal levels.

ordinance, land-use control - An ordinance that combines into one law the zoning regulations, subdivision regulations, and possibly other land-use controls.

ordinance, planned community-development - An ordinance providing for and regulating development within a planned community.

ordinance, zoning - SEE zoning.

ordinary (adj.) - When used with such terms as low water or high water, the term is considered, in law, to be equivalent to average or mean.

ordinate - The distance, in a plane, Cartesian coordinate system, from the axis of abscissas, along a line parallel to the axis of ordinates, to the point in question. Also called total latitude.

ordinate, middle - The distance from the mid-point of a chord to the mid-point of the corresponding circular arc.

ordinate axis - SEE axis of ordinates.

Ordnance Datum - The datum or series of datums established on the mainland and adjacent islands of the British Isles as the point of origin for the leveling network (vertical control) on land.

Ordnance (Liverpool) Datum - The point of origin of an early leveling network used in Great Britain and now obsolete. It has been used in determining changes of sea level along the coast.

Ordnance Newlyn Third Geodetic Leveling Datum - The vertical-control datum used in Great Britain for the Third Geodetic Leveling. It is based on mean sea level at Newlyn, England, during a specific period.

ore pass - A vertical or diagonal passage between levels to permit ore to be dropped from the one level to the other.

orientation - (1) The rotation or set of rotations needed to make the axes of one rectangular Cartesian coordinate system parallel to the axes of another. Equivalently, the set of angles made by the axes of one such coordinate system with the axes of the other. (2) The set of angles made by a set of non-coplanar axes in a body with respect to another, defined set of non-coplanar axes. Also called the attitude or the position of the body. Attitude is more commonly used in engineering than in surveying or geodesy. (3) The set of angles and coordinates defining the relative location and orientation between an image and the corresponding object. This definition appears to be peculiar to photogrammetry in making orientation practically synonymous with position. Outside of photogrammetry, the geometric relationship between an object and its surroundings is expressed by three related concepts: location, which is where the object is placed, orientation, which is how it is pointed there and position, which is the object's location and orientation. (4) The act or process of rotating one coordinate system into coincidence with or parallelism to another. In particular, in photogrammetry, the process of bringing the coordinate system in an image into proper relationship to a coordinate system in the object. In surveying, it is the act of rotating an instrument and/or parts of it until certain marks in the instrument are aline with corresponding points of the compass. A map is said to be in orientation or oriented when straight lines joining points on the map are parallel to the corresponding lines on the ground. A plane table is in orientation when lines connecting points on the sheet fastened to the plane table are parallel to the lines connecting the corresponding points on the ground. A surveyor's transit is in orientation if the horizontal circle reads 0° when the line of sight is parallel to the direction it had at an earlier (initial) position of the instrument or to a standard line used as reference. If the line of reference is the meridian, the circle will show azimuths referred to that meridian. A photograph is in orientation when lines drawn between points on the photograph are parallel to lines drawn between corresponding points on the ground, or when images on the photograph appear in the same direction from the point of observation as do corresponding symbols on a map.

orientation (of an astro-geodetic datum) - SEE datum orientation, astrogeodetic.

orientation, absolute - (1) The process of fixing the scale, location, and orientation, with respect to a coordinate system in object space, of a stereoscopic model which has been formed using a pair of photographs in correct relative orientation. (2) The orientation of a photograph (or the camera taking the photograph) with respect to the object being photographed.

orientation, analytical - (1) Those computational steps required to determine the location and orientation of a camera's coordinate system in the coordinate system used for object space. (2) Those computational steps required to determine tilt, direction of the principal line, flight height, preparation of templets at the scale used for rectification, distances and angles, and preparing aerial photographs for rectification. The resulting values are set on the circles and scales of a rectifier or transforming printer. (3) Adjustment of a rectifier by setting, on the rectifier's controls, the precise values needed for orientation; these values having been calculated beforehand.

orientation, basal - The determination of the locations of both ends of an air base with respect to a system of coordinates on the ground. Six quantities - the three coordinates of each end of the air base or their equivalents - are required. In practice, these quantities may also be expressed as (a) the coordinates of one end of the air base and the three differences between these and the coordinates of the other end, or (b) the coordinates of one end of the air base, the length of the air base, and the direction (two angles) of the air base.

orientation, empirical - (1) Those quantities - the magnification, swing, tilt of easel, x-displacement and y-displacement - used to adjust a rectifier in correctly re-duplicating, in the projected image, the exact conditions existing in the photographic negative at the time of its exposure. (2) Adjustment of a rectifier by matching certain elements of the projected image with corresponding, controlling elements on a map or template. Unlike analytical orientation of a rectifier, empirical orientation does not require that the tilt, direction of principal line, flight height, etc., be calculated exactly. SEE ALSO point-matching method (of rectification).

orientation, exterior - SEE orientation, outer.

orientation, inner - (1) Determination of the calibrated focal length and the coordinates, in the focal plane, of the principal point of the camera, and the numbers giving the radial and tangential distortion. Also called interior orientation. (2) Determination of the quantities defining the coordinate system of a photograph with respect to the coordinate system of the camera. This includes the position (location and orientation) of a photograph with respect to the coordinate system of the camera or with respect to the coordinate system of the photogrammetric plotter in which the photograph is used, and the variation of scale along the axes of the photograph's coordinate system. Because the fiducial marks in the camera are reproduced on the photograph, the orientation of the photograph's coordinate system is assumed to be the same as that of the camera or to be able to be made the same as that of the projector.

orientation, interior - SEE orientation, inner.

orientation, outer - (1) Determining, analytically or in a photogrammetric instrument, the location of the camera station and the orientation of the camera at the instant of exposure. Also called exterior orientation. Outer orientation, in a stereoscopic plotter, is of two kinds: relative and absolute. (2) The set of quantities fixing the location and orientation of the camera at the instant the photograph was taken. Such a set consists of three coordinates (location) and three angles (orientation). Unless the angles are very small, the order of taking the angles makes a difference in the final orientation of the camera.

orientation, preliminary - The rough orientation of a stereoscopic model before doing the relative orientation.

orientation, relative - (1) The reconstruction (mathematically or in a photogrammetric instrument) of the same geometric relationships, except for scale, between a pair of photographs that existed when the photographs were taken. In photogrammetric plotters, this is achieved by a systematic sequence of rotational or translational movements of the projectors. It is sometimes referred to as the clearing of the y-parallax. (2) The placing of two photographs in the same relative position, except for scale, that they occupied when they were expressed. If the two photographs have enough pairs of image-points corresponding to the same object-points (at least 5), the one photograph can be oriented relatively to the other.

orientation, total - The process of determining the exterior and interior orientation of a camera while solving at the same time for the coordinates of the object photographed.

orientation inset - SEE inset.

orientation point - A point selected, in a region common to a vertical photograph and the corresponding oblique photograph, to help establish the relationship between the vertical and the oblique photograph.

origin (of a coordinate system) - (1) That point, in a coordinate system, which has defined coordinates and not coordinates determined by measurement. This point is usually given the coordinates (0,0) in a coordinate system in the plane and (0,0,0) in a coordinate system in space. In surveying, when a plane rectangular coordinate system is used, the coordinates of the origin are often given large, positive values. These values are called false easting (or false westing) and false northing (or false southing). (2) The point to which the coordinates (0,0,0,....0) are assigned, regardless of the location of that point with respect to the axes of the coordinate system. (3) The point from which coordinates of other points in the coordinate system are reckoned.

origin (of a datum) - A point whose coordinates are defined and are part of a datum (usually a geodetic datum). The origin is usually a survey station.

Origin, Conventional International - SEE Conventional International Origin.

origin, false - An arbitrary point to the south and west of a grid overlaying a map, from which the lines of the grid are numbered and from which distances are taken as positive eastward and northward on the grid.

origin of coordinates - SEE origin (of a coordinate system).

Ortelius's map projection - SEE Eckert's map projection.

orthochromatic (adj.) - (1) Of, pertaining to, or producing shades of black and white corresponding to the natural shades or gradations of color. (2) (of a photographic emulsion) Sensitive to blue and green light but not to red light.

orthodrome - A geodesic on an ellipsoid; in particular, a great circle. The term is obsolete.

orthoembadic (adj.) - Equal-area. The term was proposed by Lenox-Conyngham in 1944.

orthoembadism - The property, of a map projection, of preserving areas of figures.

orthogonal method (of depicting relief) - The method of representing relief, on a map, so as to give an impression of seeing a three-dimensional model illuminated obliquely, by plotting the intersections of a series of parallel, equally spaced, inclined planes and conventional contours. Through these intersections, isograms are drawn to represent perpendicular distances above or below a hypothetical surface of reference intersecting the geoid (assumed flat) at a constant and predetermined angle. The isograms are sometimes called inclined contours. The method was first proposed by Kitiro in 1932.

orthographic-relief method - The creation, on a map, of the impression that the scene is obliquely illuminated, by plotting the intersections of a series of parallel, equally-spaced, inclined planes with conventional contour lines. Through these intersections, isograms are drawn representing the perpendicular distances above or below a hypothetical plane which intersects the plane of mean sea level at a constant and predetermined angle. These isograms are sometimes described as inclined contours. The method was first proposed by Kitiro in 1932.

orthomorphism - SEE conformality.

orthophotograph - A photograph prepared from a perspective photograph by removing those displacements of points caused by tilt, relief and central projection (perspective). The removal of the relief is often in zones and not specific. Sometimes called an orthophotomap.

orthophotography - The art of making orthophotographs.

orthophotomap - (1) A map made by assembling a number of orthophotographs into a single, composite picture. Also called an orthophotomosaic. A grid is usually added. The orthophotomap may be further improved, cartographically, by photographically emphasizing edges in the picture, by adding color or symbols or by a combination of these methods. (2) SEE orthophotograph.

orthophotomosaic - SEE mosaic, orthophoto.

orthophotoquad - An orthophotograph or mosaic of orthophotographs having the size of a standard quadrangle to a scale of 1:24000, with little or no cartographic work done on it.

orthophotoscope - A device used in conjunction with a double-projection, analytical type of stereoscopic plotter to produce orthophotographs from ordinary photographs.

orthopictomap - A pictomap based on an orthophotmap.

orthostereoscopy - A condition affecting the stereoscopic viewing of images by making horizontal and vertical distances appear to be at the same scale.

oscillation - In general, a repeated movement, in equal intervals of time, from one point to another and back again. Also called a double oscillation or a complete oscillation in some texts; a vibration is then called a simple oscillation. In particular, the double motion, one in each direction, of a pendulum. Because dissipative forces affect most motions, real bodies oscillate between points which gradually approach each other, i.e., the oscillations die down. If the mechanism is very carefully constructed and if dissipative forces like friction and air drag are made small, the oscillations can continue undisturbed for long periods. The pendulums of clocks are kept oscillating by the periodic addition of a small amount of energy just sufficient to overcome the energy dissipated.

oscillation (of a pendulum) - (1) The total movement of a pendulum from the time it passes in a specified direction through a specified position until it again passes in the same direction through the same position. (2) The total movement of a pendulum between its passing from one extreme position to its next passage in the same direction to an extreme position.

oscillation, complete - SEE oscillation.

oscillation, double - SEE oscillation.

oscillation, forced - An oscillation caused by the continued application of force.

oscillation, free - Oscillation not affected by a dissipative force.

oscillation, simple - SEE vibration.

osculation - The condition in which one curve is tangent to another at a point. The term is frequently used in celestial mechanics to denote the tangency of a particular ellipse to the actual path of a satellite.

ouady - SEE wadi.

out - A distance equivalent to 5 chains. The term was used in some field notes and deeds in the early 1800's in America. This was later called a "tally".

outboundary - (1) A township line or range line forming part of the perimeter of a surveyed region. (2) The controlling lines around the perimeter of an independent resurvey.

outcrop - The exposed portion of a stratum of rock or of a vein at the surface of the Earth. In describing a survey station, the exposed part of a large boulder is sometimes mistaken for an outcrop.

outkeeper - On old styles of surveyors' compasses, a scale numbered 1 to 16, with a pointer turned by a milled knob to keep track of the outs.

outlier - (1) Any value of a randomly varying quantity lying outside certain arbitrary limits. (2) Any value, of a randomly varying quantity, larger in absolute value than expected and therefore suspected to be a blunder. (3) A measurement which does not fit the other measurements of the same quantity, and no reason can be found for the discrepancy. SEE ALSO rejection of a measurement.

outline map - A map showing just enough geographic information to permit the correct placement of additional details. Outline maps correspond to what were at one time known as base maps; they usually show only coast lines, principal rivers, major civil boundaries and large cities. As much space as possible is left for placing the particular details or data desired. An outline map presents less detail than what is now known as a base map.

outwash - Sand and gravel deposited by a stream of melted ice in front of or on the margin of an active glacier.

outwash plain - An alluvial plain formed by streams originating from the melting ice of a glacier. Also called a sandr.

overflow - (1) Any device or structure over which or through which any excess of water or sewage beyond the capacity of the conduit or container is allowed to flow or go to waste. (2) The excess of water which overflows the ordinary extent of the water, such as the banks of a stream, the spillway crest or the ordinary level of a container. (3) The act of covering or inundating with water or other fluid.

overhang - The additional exposures made beyond the boundary of a region to be photographed. It consists, usually, of two exposures at the ends of each strip, to ensure complete, stereoscopic coverage.

Overhauser magnetometer - SEE magnetometer, nuclear.

overlap - (1) Those parts of two separate photographs which show the same part of the object or scene. Also called stereoscopic overlap. (2) A measure of the area, in two photographs, devoted to showing the same part of the object or scene. It is customarily expressed as a percentage of the total amount of area of one photograph. For example, 60% overlap means that 60% of the area of one photographs shows the same part of the object or scene as does the other photograph. SEE ALSO overlap, forward; overlap, side. (3) That region, on a map or chart, showing the same geographical region as an adjoining map or chart. (4) A region of land common to two titles; a piece of land lying within the boundaries claimed by two distinct titles. SEE overlap (land surveying).

overlap (land surveying) - The condition that exists when a survey carried out according to the descriptions of two separate pieces of property indicates that the properties overlap, while the descriptors had intended that the properties be contiguous. The condition occurs when descriptions by metes and bounds of the two parcels are described from two directions. For example: beginning at the northwest corner of the Southeast Quarter of said Section, thence southerly along the west line of said Southeast Quarter a distance of 1300 feet; thence easterly and beginning at the southwest corner of the Southeast Quarter of said Section; thence northerly along the west line of said Southeast Quarter a distance of 1340 feet; thence easterly will lead to an overlap.

overlap (photography) - Those parts of two aerial photographs, taken from adjacent air-stations, showing the same region on the ground. Also called lap. SEE ALSO overlap, forward; overlap, side.

overlap, forward - The overlap between two successive photographs in the line of flight. Also called forward lap and end lap.

overlap, side - The overlap between photographs taken on adjacent lines of flight. Also called side lap.

overlap, stereoscopic - SEE overlap (1).

overlay - A collection of symbols on a transparent sheet which is to be laid over another collection of symbols so that the two collections appear as one. For example, maps showing original land-grants or patents may be prepared on transparent tracing-cloth so that they can be correlated with maps showing present ownership. Or an overlay may be used in compiling a manuscript of a map; such an overlay is described by a name telling what kind of symbols appear on it or how it is to be used, e.g., lettering overlay.

overprint - An additional printing, usually in a distinctive color, onto a map which is complete in itself. Overprints may be used to reproduce specialized information on a general map (e.g., aeronautical information on a topographic map) or to incorporate revised data without having to change existing plates.

oversheet - A sheet accompanying the manuscript of a map and carrying only such information as names of features, descriptive labels, classification of symbols for roads and drainage and explanatory notes.

overtide - A tidal constituent or tidal-current constituent whose speed is an exact multiple of the speed of one of the fundamental constituents. The presence of overtides is usually attributed to conditions in shallow water.

owner, abutting - An owner or occupant whose property adjoins or touches other land.

owner, riparian - One who owns land bordering on the bank of a river. Usage has broadened the meaning of the term to include ownership of land along the sea or other tidal water, but strictly speaking, the proper designation for such situations is littoral.

ownership - (1) The state, relation or fact of being an owner. (2) A lawful claim or title. (3) Property. (4) Proprietorship. (5) Dominium. All ownership is by purchase or descent.

ownership plat - A plat (at an approximate scale of thirty chains to one inch; 1980 feet to one inch) of the new-status records, on which is shown survey data necessary to identify and describe public-domain lands and on which is indicated those actions and transactions which currently limit or restrict the use of public lands and resources.

P

pace - (1) The distance from where the heel of one foot strikes the ground, when walking normally on level ground, to where the same heel next strikes the ground. Also called a step and a stride. It is about 5 feet for the average man; it would be somewhat less for the average woman. On gentle slopes, the distance is less going up the slope and more going down the slope; the average of the two is about equal to the distance on level ground. The Roman mile was 1000 paces (milia pasuum) and was the basis for the mile used by other nations. SEE ALSO pace, Roman. (2) A distance of 60 inches. Also called the geometric pace. (3) SEE step.

pace, geometric - SEE pace (2).

pace, military - SEE step.

pace, regulation - SEE step.

pace, Roman - Five Roman feet or 58.1 inches, English measure.

pacing - The process of determining the distance between two points on the ground by counting the number of paces taken in walking from the one point to the other in a straight line.

pair, overlapping - Two aerial photographs taken at different places (exposure stations) in such a manner that a portion of one photograph shows the same terrain as is shown on a portion of the other photograph. The term covers the general case and does not imply that the photographs were taken for stereoscopic examination or form a stereoscopic pair.

pair, stereoscopic - A pair of images showing the same object from two different points of view in such a way that, when one eye sees a part of one image and the other eye sees the corresponding part of the other image, a three-dimensional effect is produced visually.

palm - A unit of measure equal to 3 to 4 inches in the United States and to 3.94 inches in the Netherlands.

Panama-Colon Datum - The horizontal-control datum which is defined by the following coordinates of the origin and by the azimuth (clockwise from South) to triangulation station Salud on the Clarke spheroid of 1866; the origin is at triangulation station Balboa Hill:

longitude of origin	79° 43' 50.313" W
latitude of origin	9° 04' 57.637" N
azimuth from origin to Salud	185° 02' 39.54"

panchromatic (photography) (adj.) - Sensitive to light of all colors.

panel - (1) A flat piece of cloth, wood, or other material placed, either alone or with other pieces of material of the same kind, on the ground at a control point to identify and locate that point in an aerial photograph. Dyed or painted cloth or colored plastic sheet is most commonly used. Panels are usually displayed in a symmetrical pattern about the control point, with the control point at the center of symmetry. (2) SEE panel base.

panel base - The complete assembly of pieces of film positives onto a grid or graticule used as a base for compilation. Also called film mosaic or panel.

paneling (n.) - An assemblage of panels identifying a point.

paneling (v.) - (1) Placing panels at a ground point to make it easier to find and identify the station on aerial photographs. (2) Cutting a photograph, or a map containing some distortion, into several pieces and then cementing the pieces onto a gridded, stable material in such a way that the error is distributed in small amounts throughout the region depicted.

pantograph - A linkage made of wooden or metallic bars and shaped like a parallelogram pinned at the corners; two opposite sides of the parallelogram are extended slightly beyond the corners and in opposite directions. The pantograph is used for copying pictures or diagrams at magnified or reduced scales. It also occurs as a component of other mechanisms.

pantograph, two-dimensional - A machine used in making relief models and permitting the three-dimensional terrain base to be cut using a flat contour-map as a guide.

pantograph, variable-ratio - SEE pantograph.

paper, autopositive - Photographic paper which gives a positive copy from a positive transparency or a negative copy from a negative transparency by direct processing. Also called direct copy paper or direct positive paper.

paper, masking - SEE goldenrod paper.

paper, photographic - A sheet of paper having on its surface a layer of barium oxide to provide a reflecting surface and, over this, a layer of emulsion containing light-sensitive halides.

paper location - The plans, for an engineering project, prepared from data obtained from record data and not based on current field survey.

paper-strip method (of rectification) - A graphical method of rectifying a photograph point by point, using the invariance of the cross ratio in a geometric projection from ranges of points in the original to corresponding points in the rectified version.

parabola - A plane curve defined (a) geometrically as the locus of all points for which the distance from a fixed line (called the directrix) is equal to the distance from a fixed point (called the focus), and defined (b) algebraically as the set of points whose coordinates satisfy the equation $y = ax^2$, in which a is a constant. The two definitions are equivalent. The parabola's principle use in surveying is as a transition curve.

parallax - (1) The difference in direction of an object as seen from two different points. (2) The apparent displacement of the location of a body, with respect to a reference point or framework, caused by a shift in the point of observation. (3) The angle, at an object, subtended by the line joining two designated points. This should not be confused with parallactic angle. The parallax of one point in space with respect to a second point used as reference is the angle of convergence of the lines from two observation stations to the point of reference, minus the angle of convergence of the lines from the same two observation stations to the first point. (On a pair of photographs of the points in space taken from the two observation stations, parallax is customarily measured by distances on the photographs rather than by angles.) (4) The angle, at one point, between lines to two other, distinct points such as the two ends of a baseline. This definition is approximately equivalent to the preceding definition (3). It is numerically equivalent to the first definition but differs in the placement of the observer. Parallax enters into many problems of astronomy, surveying and mapping. It is used in many ways, so wherever clarity demands a more precise term, a defining adjective is attached. SEE parallax, annual; parallax, diurnal; parallax, instrumental; parallax, optical; parallax, personal; parallax, secular. Parallax is also a basic term in photogrammetry, where it is identified as linear parallax, stereoscopic parallax, etc.

parallax, absolute - SEE parallax, absolute stereoscopic.

parallax, absolute stereoscopic - (1) The algebraic difference of the distances of two images of a point from the respective photograph nadirs, measured in a horizontal plane and parallel to the air base. (2) The same as the preceding definition except that the distances are measured in the plane of the photograph and in the direction of flight. The term is often shortened to parallax. Other names are absolute parallax, horizontal parallax, linear parallax, stereoscopic parallax and x-parallax. Usually, absolute stereoscopic parallax is used, to denote similar measurements made when the ideal conditions of truly vertical photographs are not attained, as, for example, when measuring parallax on unrectified aerial photographs.

parallax, angular - The angle subtended, at the object viewed, by the observer's eyes. Also called angle of convergence and undesirably, parallactic angle.

parallax, annual - (1) The difference between the direction from the center of the Earth to a celestial object and the direction from the center of the Sun to that object. More precisely, it is half the difference between the direction from the Earth's barycenter at one specified point in its orbit, and the direction from the diametrically opposite point in the orbit. Also called heliocentric parallax and stellar parallax. (2) The angle subtended, at a celestial object, by the average or a specified radius of the Earth's orbit. The difference in length of radii at different times of the year will cause the parallax of even the closest star to vary by less than 0.01%, an amount well below the resolving power of astronomical instruments at present. (3) The angle subtended, at a celestial object, by the average or a specified radius of the Earth's orbit, the radius forming one side of a right triangle of which the line from the center of the Sun to the celestial object is the hypotenuse. The annual parallax defined in this way may have a yearly variation of as much as 0.03%. The coordinates of stars and similar objects are usually given in ephemerides without any correction for annual parallax having been applied to the measurements.

parallax, daily - SEE parallax, diurnal.

parallax, diurnal - (1) The difference between the direction from a point on the surface of the Earth to a celestial object, and the direction from the center of the Earth to the same object. (2) The difference between the direction from a point on the surface of a rotational ellipsoid representing the Earth to a celestial object, and the direction from the center of that ellipsoid to the same object. (3) The angle subtended, at a celestial object, by the line from a specified point on the Earth (or on a specified rotational ellipsoid) to the center of the Earth (or to the center of that ellipsoid). Diurnal parallax is also called daily parallax and geocentric parallax. It is equal to the horizontal parallax when the celestial object is on a horizontal plane through the observer. It is equal to the equatorial horizontal parallax when the observer is, at the same time, on the equator. The coordinates given in ephemerides for celestial bodies within the Solar System have usually been derived from observations by applying a correction for diurnal parallax. Such a correction for the parallax of a body outside the Solar System is insignificant.

parallax, equatorial horizontal - (1) The greatest angle subtended, at a celestial body, by an equatorial radius of the Earth or of a rotational ellipsoid representing the Earth.
(2) The angle subtended, at a celestial body, by a line from a hypothetical observer to the center of the Earth or a rotational ellipsoid representing the Earth, when the observer is at the equator and the celestial body is on the observer's horizon. If a denotes the equatorial radius of the ellipsoid (definition 1) or the distance of the observer from the center of the Earth or the ellipsoid (definition 2), and if r is the geocentric distance of the body, then the equatorial horizontal parallax is $\arcsin(a/r)$.

parallax, false - The apparent vertical displacement of an object from its true location when viewed stereoscopically, because the object itself as well as the point of observation has moved.

parallax, geocentric - SEE parallax, diurnal.

parallax, heliocentric - SEE parallax, annual.

parallax, horizontal - (1) The angle subtended, at a celestial object, by a radius of the Earth (or of an ellipsoid representing the Earth), the radius forming one side of a right triangle and the line from the center of the Earth or ellipsoid forming the hypotenuse. (2) The angle subtended, at a celestial object, by a line from the center of the Earth to a point on the surface, when the object is on the horizon of that point. (3) SEE parallax, absolute stereoscopic.

parallax, instrumental - (1) A change in the apparent position of an object, with respect to the reference mark(s) on an instrument, caused by imperfect adjustment of the instrument or by a change in the position of the observer or both. Also called instrument parallax, parallax error and error of parallax. When a telescope is so poorly focused that the image of the object does not lie in the plane of the reticle (cross-hairs), a movement of the eye transverse to the line of sight will cause an apparent movement of the image with respect to the marks on the reticle (or to the cross-hairs). This is a common form of instrumental parallax, and the term optical parallax has been applied to it. Instrumental parallax may also result if an observer stands in the wrong position with respect to the fiducial marks on an instrument, e.g., when reading a vernier or marking the end of a tape. This kind of instrumental parallax has been called personal parallax. It can usually be avoided by keeping the observer's line of sight perpendicular to the scale on which the readings are marked, e.g., in reading the scale on liquid-in-glass thermometers. (2) SEE y-parallax.

parallax, linear - SEE parallax, absolute stereoscopic.

parallax, lunar - The geocentric parallax of the Moon.

parallax, mean equatorial horizontal - The equatorial horizontal parallax of a celestial body at the average geocentric distance of that body. This quantity differs significantly from the equatorial horizontal parallax only for bodies within the Solar System.

parallax, optical - SEE parallax, instrumental.

parallax, personal - SEE parallax, instrumental.

parallax, residual - Small amounts of parallax that may remain in a stereoscopic model after finishing relative orientation.

parallax, secular - That apparent, secular change in direction of distant stars which is caused by movement of the Solar System with respect to the rest of the Galaxy.

parallax, solar - (1) The angle subtended at the Sun by a line from the center of the Earth to an observer, on the equator, who has the Sun on his horizon. Solar parallax is the basis for the determination of the astronomical unit (the length of the semi-major axis of the Earth's orbit), and is often used, in astronomical calculations, instead of the astronomical unit for expressing distances to stars. The solar parallax is 8.794 148". (2) The quantity 8.794 148" adopted in 1977 by the International Astronomical Union as a standard value for the solar parallax.

parallax, spectroscopic - The parallax of a star determined from those differences in stellar spectra which depend upon the pressure in the star's atmosphere. Atmospheric pressure can be used to determine, from theory, the star's luminance, the absolute magnitude and finally the distance.

parallax, stellar - (1) The angle subtended at a star by the radius of the Earth's orbit (or 1 astronomical unit). Stellar parallax can be measured to about 0.01" optically but is accurate to only about 0.5". (2) SEE parallax, annual.

parallax, stereoscopic - (1) The apparent shift in direction of a point when the point is viewed first by one eye and then by the other. It is very similar to angular parallax. (2) SEE parallax, absolute stereoscopic.

parallax, vertical - SEE y-parallax.

parallax, x- (1) SEE x-parallax. (2) SEE parallax, absolute stereoscopic.

parallax, y- SEE parallax, instrumental; y-parallax.

parallax age - SEE age of parallax inequality.

parallax bar - A bar carrying two co-planar glass plates, each engraved with a dot, cross or other mark, and each movable along the bar, which is graduated. Also called a stereometer although that term properly applies to a larger class of instruments than does parallax bar. The parallax bar is used to determine the height of an object by measuring the absolute stereoscopic parallax between corresponding image points on a pair of overlapping photographs (a stereoscopic pair). One plate can be slid along the bar by large amounts and is used for coarse adjustment of the distance between the pair; the other can be moved slightly but precisely and is used for fine adjustment of the distance.

parallax difference - The difference in the absolute stereoscopic parallaxes of two points imaged on a pair of photographs. It is customarily used in determining the differences in heights of objects.

parallax error - An error introduced into a measurement by sighting along the wrong mark on the scale with the object or lack of correct focus of the observing optics. Also called error of parallax. A common error of this type is made in reading the scale of a liquid-in-glass thermometer.

parallax in altitude - Diurnal parallax at any altitude.

parallax inequality - (1) The variation in the range of the tide or in the speed of a tidal current, caused by the varying distance of the Moon from the Earth. The range of tides and speed of tidal currents tends to increase as the Moon approaches perigee and to decrease as it approaches apogee, the complete cycle being the anomalistic month. There is a similar but relatively unimportant inequality due to the Sun; this cycle is the anomalistic year. (2) SEE inequality, parallactic.

parallel (adj.) - Separated everywhere by the same distance. The term is used, in particular, in respect of lines and surfaces.

parallel (n.) - (1) A line having the same distance everywhere from another line or a surface. Preferably called a parallel line. (2) SEE parallel, astronomic. (3) SEE parallel, geodetic. (4) SEE parallel, geographic. The term parallel alone should be used only where no doubt is possible as to what kind of parallel is meant. SEE ALSO ecliptic parallel; ground parallel; map parallel.

parallel, astronomic - A line, on the surface of the Earth, having the same astronomic latitude at every point. Also called simply, a parallel. Because the deflection of the vertical is not the same at all points on the Earth, an astronomic parallel is an irregular line not lying in a plane. The astronomic parallel of 0° astronomic latitude is called the astronomic equator.

parallel, auxiliary standard (USPLS) - A new standard parallel established, when required, for control where the original standard parallels or correction lines were placed at intervals of 30 or 36 miles. Auxiliary standard parallels are used in extending old surveys and for controlling new surveys. Such a line may be given a local name such as Fifth Auxiliary Standard Parallel North or Cedar Creek Correction Line.

parallel, celestial - (1) A circle, on the celestial sphere, all points of which have the same declination. Also called a parallel of declination.

parallel, ecliptic - A circle, on the celestial sphere, all points of which have the same celestial latitude.

parallel, fictitious - A circle or line parallel to a fictitious equator and connecting all points having the same fictitious latitude. A fictitious parallel may be further specified as a transverse, oblique or grid parallel, depending on the type.

parallel, geodetic - A line, on an ellipsoid, every point of which is at the same geodetic latitude. It is often referred to as a parallel of latitude, although that term is better reserved for the geographic latitude. It may also be referred to simply as a parallel. A geodetic parallel on a heteraxial ellipsoid is a circle only at the equator (latitude zero). At other latitudes, the parallel is not a planar curve.

parallel, geographic - A line, on the Earth or on a representation, every point of which is at the same (astronomic or geodetic) latitude. Also called a parallel of latitude or simply, a parallel (when the intent is clear). The term is applicable alike to an astronomic parallel or to a geodetic parallel.

parallel, inverse - SEE parallel, transverse.

parallel, isometric - A line, on a photograph, which passes through the isocenter and is the image of a horizontal line perpendicular to the principal plane of the photograph.

parallel, oblique - A line or circle parallel to an oblique equator.

parallel, principal - That line, on a photograph, which is the image of a straight line in object space, passes through the principal point and is perpendicular to the principal plane.

parallel, standard - (1) A geodetic parallel used as a control line in calculating a map projection. In graticules made by conic map projections, a standard parallel usually represents a line of tangency or of intersection of a cone with the surface of the ellipsoid used. Such graticules are not exact. (2) An auxiliary governing line established along the astronomic parallel, starting at a selected township-corner on a principal meridian, and on which standard township, section, and quarter-section corners are established. Also called a correction line. Standard parallels are spaced, usually, at intervals of 24 miles from the base line. They are established to limit the convergence of range lines from the south. (3) That parallel of latitude which, in a conical or pseudo-conical map-projection of normal aspect, is transformed into a line along which the principal scale is preserved.

parallel, transverse - A circle or line parallel to a transverse equator, connecting all lines of equal transverse latitude. Also called an inverse parallel. SEE ALSO parallel, fictitious.

parallel distance, double - The sum of the distances of the ends of a surveyed line from a specified parallel. If the line consists of separate segments or courses, the double parallel-distance is the sum of the double parallel-distances of the segments. It is therefore the north-south component of the last course of a line plus twice the sum of the north-south components of all preceding courses. SEE ALSO meridian distance, double.

parallel of altitude - SEE almucantar.

parallel of declination - A small circle on the celestial sphere and parallel to the celestial equator. SEE ALSO parallel, celestial.

parallel of latitude - (1) A circle, on the celestial sphere, parallel to the plane of the ecliptic and connecting points of equal celestial latitude. Also called an altitude circle. (2) SEE parallel, geographic.

parallelogram - A figure having four sides, opposite sides being parallel to each other. Opposite sides are also equal in length and opposite angles are equal in magnitude. The line joining opposite vertices (the points in which sides join) are called diagonals. The area of a parallelogram is equal to the product of the length of one side by the perpendicular distance to the opposite side.

parallelogram of Zeiss - The parallelogram formed, in an opto-mechanical, stereoscopic plotter, by (a) the line joining the lower pivots of the two space bars, (b) the line joining the lower and upper pivots of one of the space bars and (c), (d) the two corresponding lines parallel to these two and making up the parallelogram. Also called the Zeiss parallelogram. The space bars are the rigid bars passing through the pivots representing projection-centers; they represent rays from the two photographs to the corresponding points in the photogrammetric (stereoscopic) model. The parallelogram of Zeiss is used, in some plotting instruments, to overcome the fact that the distance between photographs, and therefore between the upper pivots, is too large to allow the lower ends of the space bars to come together. I.e, because the pair of rays defining a point in the model do not intersect physically in the model, they are force by the optics to intersect with a line of fixed length which remains parallel to the x-axis of the instrument. SEE ALSO base in; base out.

parameter - (1) A quantity appearing as a constant in an equation containing variable quantities, but whose value can be assigned arbitrarily. If the equation represents a line or a surface, then to each value of the parameter corresponds a different line or surface. For example in the equation y = ax + b for a straight line, a and b are parameters. To each value of a corresponds a different straight line through the point (0, b), so that the parameter a corresponds to a pencil of lines through (0, b). To each value of b corresponds a different straight line parallel to a line of slope a through the origin. The Keplerian elements are parameters in the equation of an elliptic orbit; each different value of one of the elements identifies a different orbit (but not necessarily a different path). This fact is used in the method called the variation of parameters method for solving the equations of motion of a satellite. A standard orbit is established by assigning specific, constant values to the elements, and the changes of the orbit with time are realized by changing the values of the elements (parameters). (2) An independent variable in terms of which the coordinates (x,y) of points on a line or one of two independent variables in terms of which the coordinates (x,y,z) of points on a surface are given. Also (and perhaps preferably) called a parametric variable. In general, one of M independent variables in terms of which the N coordinates (N>M) of points of an M-dimensional surface in N-space are given. E.g, the equation of a circle in the plane can be written as $x^2 + y^2 = r^2$ or as $x = a \cos \theta$; $y = a \sin \theta$. θ here is the parameter. (3) A quantity which has only one value in a particular case, but may be assigned another value in another case. Also called an arbitrary constant. (4) Any quantity, in a problem, which is not an independent variable. (5) Any quantity. Definitions (4) and (5) are widely used at present. They have no particular utility other than allowing one to use an impressive-looking word for the more common terms variable, quantity and number. (6) Any numerical constant derived from a set of numbers or from a distribution. Specifically, it is an arbitrary constant in a cumulative distribution function or in a frequency function. The term is used primarily by statisticians. (7) Any quantity or element occurring in meteorology or oceanography. (8) Anything. Definitions (6), (7) and (8) are much like definitions (4) and (5) and are best not used.

parameter, geodetic - The quantity $a^3 b \omega/GM$, in which G is the gravitational constant, M the mass of the Earth, a the equatorial radius and b of an ellipsoid representing the Earth, and ω the angular rate of rotation.

parameter, geostrophic - The quantity $2 \omega \sin \phi$, in which ω is the rate of rotation of the Earth ($7.292\ 116 \times 10^{-3}$ radians per second) and ϕ is the geodetic latitude. Oceanographers usually denote the quantity by f or by kg.

parameter, latent - A quantity on which the results of an adjustment depend, but which does not appear as an unknown in the observation or condition equations. The concept appears to be equivalent to that called a hidden variable in quantum mechanics.

parameterization - The representation of physical effects, in a dynamic physical system, in terms of admittedly over-simplified equations containing arbitrary constants, rather than representing them by equations in which the constants have physical meaning and are theoretical consequences of the system's dynamics. Use of such techniques is dictated by lack of adequate knowledge of the physics of the system and often is a matter of mathematical convenience. The validity of the representation must be judged by the faithfulness of the representation to measurements on the system.

parametric method (of adjustment) - Adjustment of a network by expressing the changes in coordinates as functions of changes in the measurements (observations).

parcel - (1) A single piece of land described in a single description in a deed or as one of a number of lots on a plat, separately owned either publicly or privately and capable of being conveyed separately. (2) A piece of land not describable (identifiable) by lot number.

parcel, cadastral - A continuous region (or volume of land) in which unique, homogeneous interests or rights are recognized. The parcel includes superjacend and subjacent rights in addition to surficial rights.

parcenter - SEE tenant in common.

Parent's map projection - Any one of three azimuthal and perspective map-projections introduced by Parent in 1704. SEE ALSO map projection, perspective.

parent tract - The original tract from which a parcel was taken.

Paris meridian - The meridian through the axis of the south front of the central pavilion of the Observatory of Paris. The Paris Meridian was used as the primary meridian by many countries before the Greenwich Meridian was adopted internationally for that purpose. It is 2° 20' 13.95" east of Greenwich.

parish - A division of land in Louisiana corresponding to a county in other States. It is an outgrowth of the original districts set up by the church.

park, industrial - A locale zoned for industrial use, containing sites for many separate industries, and created and managed as a unit. It usually provides common services to the users.

Parkhurst theodolite - A direction theodolite designed by D. Parkhurst and having a non-binding center, ball-bearing clamping ring, illumination through the central axis and discontinuous conical bearings. The Parkhurst theodolite was formerly used on first-order and second-order control- surveys in the USA.

parkway - A highway giving the appearance, to a driver, of being in a park.

parol (adj.) - Executed or made by word of mouth, or by a writing not under seal.

parol (n.) - (1) A pleading. (2) The pleadings by either or both parties in an action.

parol agreement - An oral agreement between owners, establishing the boundaries dividing adjacent land.

parol contract - A contract made orally or in writing but not under seal.

parol evidence - (1) Oral evidence; evidence that is given by word of mouth. This is the ordinary kind of evidence given by witnesses in court. (2) In a particular sense, and with reference to contracts, deeds, wills and other writings, extraneous evidence or evidence aliunde.

parsec - The distance at which the angle (annual parallax) subtended by the average radius of the Earth's orbit is 1". Equivalently, the distance at which one astronomical unit subtends an angle of 1". It is approximately 3.26 light years or 30.9×10^{12} kilometers.

part, aliquote - A number which is an even divisor of a given number is said to be an aliquote part of that number.

part, meridional - (1) The length of the arc of a meridian between the equator and a parallel of latitude, expressed in units of one minute of longitude at the equator, these lengths being measured on corresponding lines of a graticule drawn using the Mercator map projection from a given rotational ellipsoid. In other words, it is the ratio of the length of line representing, on a graticule drawn using the Mercator map projection from a given rotational ellipsoid, an arc of the meridian from the equator to the given parallel of latitude, to the length of the line representing one minute of arc on the equator. Also called a meridional distance, although the definition usually given for that term does not correspond to actual usage. (2) SEE distance, meridional.

partial - SEE divergence.

parties, third - All parties who are not parties to the contract, agreement, or instrument of writing by which their interest in the thing conveyed is sought to be affected. It is difficult to give a very definite idea of the concept of third parties, for sometimes those persons who are not parties to the contract, but who represent the rights of the original parties, as executors, are not to be considered third parties.

partition - (1) The dividing of lands held by joint tenants or tenants in common, into distinct portions, so that the tenants may hold them in severalty. (2) In a technical sense, any division of real or personal property between co-owners or co-proprietors. Partition does not create or convey a new or additional title or interest but merely severs the unity of possession.

party - In surveying, a group of people working together, in the field, to do a survey. Also called a unit. E.g., a party for a traverse would consist of at least the observer and two tapemen. If many records are to be kept, a recorder would be added. A flagman might be added if the lines of sight are long. A driver who brings the personnel and equipment to and from the site of the survey may or may not be considered a member of the party. (2) A group of subgroups (units) consisting of people who are working together in the field to do a survey; all the subgroups are working simultaneously on the same project or in the same region.

party wall - A wall dividing two adjoining properties, usually (but not always) having half its thickness on each property, and in which each of the owners of the adjoining properties has rights of enjoyment. The parties may own it and the ground on which it stands as tenants in common (the usual common-law party-wall, or they may each own a part, as a half, in severalty and have easements in the other half, or one may own it and the other have easements on it.

pascal - The pressure exerted by a force of 1 Newton on an area of 1 square meter. It is a unit of pressure and is equal to 10 dynes per square centimeter or to 105 bars.

Pascal's law - Pressure exerted at any point upon a confined liquid is transmitted undiminished in all directions.

pass - (1) One complete set of measurements on a specific photograph. (2) The period of time, or the path traversed by a satellite in that period, during which a ground station received signals from the satellite. (3) One revolution of a satellite. This use of the term is inexplicable. (4) A narrow channel connecting two bodies of water. (5) The inlet through a barrier-reef atoll or sand bar. (6) A navigable channel at a river's mouth.

passage - A narrow, navigable pass or channel between two landmasses or shoals.

passometer - (1) A pocket-sized instrument registering the number of steps or paces taken by the pedestrian carrying it. Formerly called a pedometer. The passometer is housed in a case resembling that of a watch and is attached in an upright position to the body or to a leg. The distance walked is obtained by multiplying the number of steps or paces taken by the length of the wearer's step or pace.
(2) SEE pedometer.

pass point - (1) A point whose horizontal or vertical location is determined by photogrammetric methods and which is intended for use (in the manner of a supplemental control point) in the absolute orientation of other photographs. Also called a photogrammetric point. (2) A point used in determining the relative orientation of two adjacent photographs in a strip of photographs. (3) SEE elevation, supplemental.

pass point, horizontal - SEE position, supplementary.

pass point, vertical - A pass point whose elevation has been determined photogrammetrically and is intended for use in orienting other photographs. Also called a supplemental elevation.

patent - The title deed by which a government, either state or federal, conveys its lands.

patent ambiguity - SEE ambiguity, patent.

patentee - One to whom a grant is made, or a privilege secured by a patent.

path - SEE path, optical; path, orbital.

path, optical - The line followed by a ray of light through an optical system. Alternatively, one of the set of lines perpendicular to the wavefronts of light passing through an optical system.

path, orbital - (1) The set of all points occupied by the center of mass of a satellite during its motion about the center of attraction. Usually referred to simply as the path of the satellite. It differs from the orbit in that time does not enter explicitly into its definition. It differs from the trajectory of a body in that a trajectory has definite beginning and end points; the beginning and end of a path can be left undefined. (2) The projection of an orbit onto the surface of the primary body. In particular, the projection of an orbit lying in some plane other than one perpendicular to the axis of rotation of the primary, i.e., lying in some plane other than the equatorial plane.

path, Shoran-wave - The path taken by the signal from a mobile (airborne or shipborne) Shoran station to the ground station.

pattern, representative - (1) An accurate portrayal of the surface of the Earth in the region being mapped. (2) A portrayal of the most prominent of a dense group of similar features.

PC-1000 camera - A camera made from a long-focal-length aerial camera of the U.S. Air Force and adapted to photographing artificial satellites. It was used considerably in the 1950's.

peak - A comparatively sharp apex of a mountain.

Peaucellier inversor - An inversor consisting of four pivots connected by four rods to form an equilateral rhombus having one diagonal vertical and the other horizontal. The two pivots on the horizontal diagonal are connected by still another pair of rods, shorter than the others and pivoted at a point on the vertical diagonal. The lowest pivot is a fixed distance above the easel, which does not move, and the upper most pivot is the same, fixed distance (the focal length of the lens system) below the plane of the photographic negative, which does move. The other pivot on the vertical axis is in the movable plane of the lens system. Also called a scissors inversor.

Peaucellier-Carpentier inversor - A modified Carpentier inversor coupled to the linkage of a Peaucellier inversor.

Pechan prism - SEE Thompson prism.

pedometer - A pocket-sized instrument registering the distance in linear units traversed by the pedestrian carrying it. The term was formerly applied to an instrument registering the number of steps or paces taken, but to which the term passometer is now applied. The pedometer, like the passometer, is housed in a case resembling that of a watch and is carried in an upright position attached to the body or to a leg. It registers the linear distance traveled in kilometers or other units. The number of steps or paces is transformed mechanically into linear units. This transformation depends on the length of the step or pace, and this may be different for different persons. The pedometer can be adjusted to the length of step of the person carrying it.

peepsight - A device used for sighting and consisting of standards with vertical slits. SEE ALSO peepsight compass.

peepsight alidade - An alidade consisting of a peepsight mounted on a straightedge in such a manner that the edge of the straightedge is parallel to the vertical plane in which the line of sight rotates.

peepsight compass - A compass having sights formed by standards with slits for sighting through, rather than a telescope.

peg adjustment - Adjustment of the line of sight of a leveling instrument of the dumpy type to make the line parallel to the axis of the spirit level by placing leveling rods on two stable marks (pegs) placed the same distance apart, on the ground, as the usual distance used for sighting between leveling rods, and sighting on the two rods first from a point midway between them and then from a point much closer to the one than to the other. Also called an 11/10 peg adjustment. A wye-level can be adjusted by peg adjustment but is more readily adjusted by reversing the telescope in its wyes. A special form of peg adjustment is used in determining the collimation error in first-order leveling instruments of the dumpy type. SEE ALSO two-peg method (of adjustment); leveling instrument, peg adjustment of.

peg adjustment, 11/10 - SEE peg adjustment.

peg test - A test for the magnitude of the collimation error, carried out by leveling around a small loop which exaggerates the difference between backsight and foresight distances and produces a misclosure proportional to the collimation error of the leveling instrument.

Peil, Normaal Amsterdams - SEE Normaal Amsterdams Peil.

Peirce's criterion - The numerical statement of the limits in Peirce's rule for rejecting observations. It is preferred to Chauvenet's criterion when more than one unknown is involved.

Peirce's map projection - SEE map projection, quincuncial.

Peirce's rule - The rule that an observations should be rejected when the probability of the set of errors obtained by keeping it is less than the probability of the set of errors obtained by rejecting it, multiplied by the probability of making exactly one abnormal observation.

pel - SEE pixel.

Pellenin term - A term added to the Vening Meinesz equations (for the deflection of the vertical) to correct for the effect of condensing topographic masses onto the reference surface.

Pemberton leveling rod - A self-reading leveling-rod on which the graduations are marked by rows of alternately circular and diamond-shaped dots running diagonally across the rod. This leveling rod is no longer in general use.

pencil - A set of straight lines radiating from a point.

pencil (of light) - SEE light pencil.

pendulum - (1) In general, a body so suspended that it can swing freely back and forth under the influence of gravity. A pendulum used to determine gravity-acceleration is usually a simple bar of metal or quartz. Older types had a wedge-shaped prism (knife edge) mounted near one end of the bar. This acted as a bearing, resting on a flat surface of agate mounted on the supports and carrying the weight of the bar when the bar was swinging. Newer types (after 1898 in the U.S.C.& G.S.) have the wedge mounted on the supports and the flat surface mounted on the bar. When a pendulum used for measuring gravity acceleration is not in use, it is kept, together with its support, in a strong, sealed case called the receiver. SEE ALSO pendulum, simple; pendulum, compound. (2) SEE pendulum apparatus. SEE ALSO bronze pendulum; dummy pendulum; fiber pendulum; flexure (of a pendulum); Foucault pendulum; Holweck-Lejay pendulum; invar pendulum; Kater pendulum; Lejay pendulum; Mendenhall pendulum; minimum pendulum; oscillation (of a pendulum); quartz pendulum; receiver (for a pendulum); Schuler pendulum; Sterneck pendulum; swing (of a pendulum).

pendulum, bronze - SEE bronze pendulum.

pendulum, compound - Any pendulum not having its entire mass concentrated at one point. Any real pendulum is a compound pendulum. It can be considered to be made up of a large number of material particles, each at a different distance from the center of suspension and each constituting the bob of a simple pendulum. The period of oscillation of the compound pendulum is the resultant of the periods of the simple pendulums composing it.

pendulum, dummy - SEE dummy pendulum.

pendulum, filar - A pendulum in which the mass is hung at the end of a long, thin, light wire. Also called a fiber pendulum or a Besselian pendulum. However, the latter term is usually reserved for filar pendulums of the approximate lengths of those used by Bessel, or for the Besselian pendulum-apparatus. A filar pendulum comes lose to being a simple pendulum. Unfortunately, the wire's elasticity introduces other problems which have so far been more severe than the problems associated with using a compound pendulum. Nevertheless, the advantages are great enough that the type was much used in early work on gravity and that research is still being done with it. (E.g., a filar pendulum 200 meters long has been used in Finland.)

pendulum, free-swinging - A pendulum having an initial momentum imparted to it by mechanical or other means and moving afterwards wholly under the influence of gravity. In measuring gravity, the initial momentum can be imparted by drawing the pendulum slightly out of plumb and then releasing it.

pendulum, half-second - A pendulum taking approximately half a second to swing through the lowest point of its path twice in the same direction. A half-second pendulum is about 25 cm long.

pendulum, horizontal - A horizontal rod with a heavy mass at one end and suspended in such a manner that the mass can move freely in a horizontal plane. The most common form of horizontal pendulum has the rod resting at one end in a socket in a rigid, vertical post and connected near the other end (the one carrying the mass) by a thread to the top of the post, so that thread, rod, and the part of the post from the top down to the socket form a right triangle. When the post moves because of seismic waves, the mass tends to remain in the same place because of inertia. It therefore appears to oscillate horizontally, although it is actually the post which is moving. In the Zöllner form of horizontal pendulum, the end opposite that carrying the mass does not rest on the post but is connected by a thread to the bottom of the post, thus keeping the rod horizontal.

pendulum, ideal - A simple pendulum subjected to no forces other than gravity and the tension in the connecting rod.

pendulum, idle - A working pendulum placed in the receiver well before use, so that it will assume the same temperature as the dummy pendulum.

pendulum, Invar - SEE Invar pendulum.

pendulum, invariable - A pendulum so designed and supported that it can be used in only one position. The centers of suspension and of oscillation of an invariable pendulum are not interchangeable. Pendulums of this type (constructed by Kater, Peirce and Mendenhall, among others) can be used only for relative measurements of gravity acceleration.

pendulum, inverted - A pendulum in which the point of support is below the center of mass. In this position, the inverted pendulum is in unstable equilibrium and is extremely sensitive to changes in the location of the point of support or to changes in the value of gravity. The Lejay pendulum and the Lejay-Holweck pendulum-apparatus used for gravi-metry have the pendulum supported by a short, flat, vertical band of elastic metal, one end of the band being fastened to the support, the other end being fastened to the lower end of the rod carrying the mass.

pendulum, mathematical - A hypothetical pendulum whose motion is described exactly by mathematical formulas.

pendulum, minimum - SEE minimum pendulum.

pendulum, physical - An actual, real pendulum, as contrasted to a hypothetical, non-existent pendulum. The motion of a physical pendulum cannot be described exactly by mathematical formulas.

pendulum, quarter-meter - A pendulum, about 25 cm long, having a period of 1 second. It is often called a half-second pendulum.

pendulum, quartz - SEE quartz pendulum.

pendulum, relative - A pendulum designed specifically for determining differences between the values of gravity-acceleration at two stations. The term contrasts this type of pendulum with the type designed to measure the total value of gravity-acceleration at a single station.

pendulum, reversible - A pendulum so designed and supported that it can be used with either end up. Also called a reversion pendulum. The centers of suspension and of a reversible pendulum are interchangeable. Pendulums of this type were constructed by Kater, Peirce and Repsold. They have been used for absolute measurements of the intensity of gravity at base stations.

pendulum, reversion - SEE pendulum, reversible.

pendulum, simple - A hypothetical pendulum consisting of a point-mass attached to one end of a massless, rigid rod of zero cross-section. The other end of the rod pivots without friction on a horizontal axle of zero radius, so that the point mass is free to oscillate in a vertical plane about the horizontal axis. A real pendulum differs from a simple pendulum in many ways, of which the following are particularly prominent. First, its mass is not concentrated at a point but is distributed over an extended region. Second, the rod is flexible rather than rigid and has a finite cross-section, so that its motion is resisted by the medium in which the pendulum swings. Third, the pivot on which a real pendulum swings is affected by friction. Fourth, the pivot is not an ideal line; the pendulum and pivot are in contact along a surface, and the shape of this surface changes with the position of the pendulum. Lastly, the support for the pivot also flexes and sways. Each of these differences of the real pendulum from a simple pendulum requires that a correction be made to the equation governing the motion of a simple pendulum.

pendulum, wire - SEE pendulum, filar.

pendulum, working - A pendulum whose period of oscillation is measured to determine the intensity of gravity.

pendulum alidade - An alidade consisting of a telescope mounted on a straightedge and having a pendulum instead of a spirit level for establishing the direction of the horizontal line or reference for vertical angles.

pendulum apparatus - An instrument consisting of one or more pendulums, with accessories, designed for determining the value of gravity. Also called a pendulum. However, that term should be reserved to denote the actual pendulum. The usual accessories are: a mechanism for lifting the pendulum from its support, releasing it into motion and replacing it on its support; a device for recording the period of oscillation or the positions of the pendulums as a function of time; a clock; thermometers; and a barometer. As many as four pendulums have been included in a single pendulum-apparatus to compensate for flexure of the supports or for the motion of the apparatus as a whole, or simply to provide several independent measurements which can be averaged to get a more precise value for gravity. The pendulum apparatus is kept in a strong, sealed container (the receiver) when not in use and being moved. SEE ALSO apparatus, two-pendulum; Brown pendulum-apparatus; Cambridge pendulum-apparatus; Lejay-Holweck pendulum-apparatus; Sterneck pendulum-apparatus; Vening Meinesz pendulum-apparatus.

pendulum astrolabe - An astrolabe whose distinctive feature is a mirror fixed to the top of a pendulum suspended in such a way as to form an artificial horizon for the line of sight. The telescope is placed so that observations are made at a constant angular elevation. In one form, the instrument consists of a V-shaped casting carrying the objective and eyepiece lenses at the tops of the V (placed vertically). The mirror, which rests on top of the pendulum and forms the artificial horizon, is located at the bottom of the V. The pendulum is suspended so that it is free to swing in either of two planes at right angles to each other (e.g., in north-south and east-west planes). The pendulum's motion is highly damped so that the mirror comes to rest quickly and remains steady under normal conditions.

pendulum clock - A clock in which the intervals of time are determined by the period of a pendulum. Energy is fed to the pendulum by a mechanism or electromagnet called the escapement. To a good first approximation, the period is independent of the mass of the pendulum and depends only on the length of the pendulum and on the force of gravity at the clock. Pendulum clocks were formerly the most accurate timepieces known, and national timekeeping services and observatories used such designs as the Rieffler, Schuler and Short clocks until well into the 1950's.

pendulum day - The length of time needed for the plane of an ideal Foucault pendulum to rotate through 2π radians. The plane rotates through 2π radians in $2\pi/(2\omega \sin \phi)$ days, in which ω is the rate of rotation of the Earth in radians per day and ϕ is the latitude. SEE ALSO day, half-pendulum.

pendulum gravity apparatus - SEE pendulum apparatus.

pendulum hour - The period T, in sidereal hours, given, for a point at geocentric latitude ϕ', by $T = 1/\sin \phi'$.

pendulum level - SEE pendulum leveling instrument.

pendulum leveling instrument - A leveling instrument in which the line of sight is kept horizontal by a prism or mirror mounted on a pendulum. Also called a compensator leveling instrument. This is the earliest type of automatic leveling instrument.

pendulum period - (1) The interval of time required for a pendulum to pass twice through the lowest point of the path, once in each direction. (2) The interval of time required for a pendulum to pass from its extreme position at one end of its path to the extreme position at the other end. Also called the period of one vibration, period of vibration and period of a pendulum. The correction to a pendulum period is half the correction to a period. For corrections to the pendulum period, SEE under the names of individual corrections: arc correction; buoyancy correction; damping correction; pressure correction; temperature correction; temperature-correction, dynamic; correction, aerodynamic; correction, air-drag; correction, clock-rate; correction, flexural; correction, knife-edge.

pendulum support - SEE flexure (of a pendulum's support).

pentaprism - SEE prism, pentagonal; prism, right-angle.

penumbra - That portion of a shadow in which light from an extended source is partially but not completely cut off by the intervening body. I.e., the region of partial shadow surrounding the umbra, the part to which no light penetrates.

peonia - (1) In Spanish law, a portion of land which was formerly given to a simple soldier on the conquest of a country. (2) A quantity of land of different size in different provinces of Spain, deriving from the quantity of land given a simple soldier on the conquest of a country. (3) In the Spanish possessions in America, a quantity of land 50 feet long in front and 100 feet deep.

percent of enlargement - The factor, expressed in percent, by which an original is to be enlarged. (length on copy) = (corresponding length of original) x (1 + factor).

percent of reduction - The factor, expressed as percent, by which an original is to be reduced. (length on copy) = (corresponding length of original) x (1 - factor).

percent of slope - SEE gradient.

perch - A measure of length, now obsolete, equal, by statute in Great Britain and the USA, to 16.5 feet exactly. The perch was used extensively in the early surveys of the public lands and is equivalent, in length, to a rod or pole. Its length does, however, vary locally in different countries.

peri - A prefix meaning the point closest to. It is usually placed before the name of the body or point the point is closest to. For example, perigee is that point, of an orbit, closest to the Earth's center of mass; perihelion is that point, of an orbit, closest to the Sun. Note that if a satellite has appreciable mass compared to the mass of the primary body, the center of mass of the two-body system is meant, e.g., lunar perigee refers to that point, in the Moon's orbit, closest to the center of mass of the Earth and Moon together.

periapsis - The point, in the orbit of a body of negligible mass, at which that body is closest to the attracting body. SEE ALSO perifocus. (2) SEE perifocus.

pericenter - (1) One of the two points, in an elliptical orbit, at which the orbiting body is closest to the center of the ellipse. (2) SEE perifocus.

pericynthion - SEE perilune.

perifocus - The point, in the osculating orbit of one body of a pair of massive bodies, at which that body is closest to the center of mass of the pair. Also called periapsis and pericenter. Perifocus is a more general term than periapsis, because periapsis applies properly only when the mass of the second body is concentrated at a point and is very much greater than the mass of the first component. Pericenter should be used only in the sense of closest to the center.

perigee - The point, in the orbit of a satellite of the Earth, at which the satellite (or its center of mass) is closest to the Earth's center of mass.

perihelion - The point, in the orbit of a body moving in an orbit about the Sun, is closest to the Sun's center of mass. The time at which the Earth passes through its perihelion is the starting time for the anomalistic year.

period - (1) The interval of time between two events in a sequence of events all alike and recurring at equal intervals of time. (2) SEE period of oscillation; period of vibration.

period (of a pendulum) - SEE creep (of period); effect, electrostatic; effect, magnetic.

period (of a satellite) - SEE period, orbital.

period, anomalistic - The interval of time between two successive passages of a body through perigee or, in general, through perifocus. Also called the perigee-to-perigee period.

period, nodical - The length of time required for an orbiting body to move from one node in its orbit back to the same node.

period, orbital - The time needed for a satellite to make successive passages through a specified point in its orbit. Also called simply, the period (of the satellite). The period depends on how the point is specified and will usually not be constant.

period, perigee-to-perigee - SEE period, anomalistic.

period, sidereal - (1) The time taken by a satellite to complete one revolution about its primary, as referred to a fixed star. (2) The time required for a satellite of the Earth to make one complete revolution about the Earth, as referred to the same geocentric right-ascension.

period, synodic - The average interval of time between two consecutive conjunctions of a planet or the Moon with the Sun, or a satellite with its primary, as seen from the Earth.

periodicity (of the tide) - The period or combination of periods during which the tide goes through the cycle of its most pronounced variations: semi-diurnal, mixed but mostly semidiurnal, mixed but mostly diurnal, or diurnal, according as the value of the ratio $(K_1 + O_1) / (M_2 + S_2)$ is less than 0.25, between 0.25 and 1.5, between 1.5 and 3.0 or greater than 3.0, respectively. K_1 and O_1 are, respectively, the magnitudes of the two most pronounced diurnal tides, while M_2 and S_2 are the magnitudes of the two largest semi-diurnal tides. Also called the type of tide. A simpler classification combines the two types of mixed periodicity into a single class, mixed.

period of oscillation - The interval of time between two successive passes, in the same direction, of a pendulum through the lowest point in its path. Also called simply, period if the movement of a pendulum only is being considered. It is the interval of time required for a complete movement from one extreme position to another and back again (one oscillation). It is affected by the amount and kind of gas surrounding the pendulum, by the temperature, by the Earth's magnetic field, and by many other factors. The term must not be confused with pendulum period, which is the interval of time required for the pendulum to pass twice through the lowest point, once in each direction (one swing; one vibration). Because a real pendulum swings in an arc whose amplitude is steadily decreasing, the period required for one swing is slightly more than half the period required for the oscillation containing that swing. SEE ALSO period of vibration.

period of vibration - (1) The interval of time needed for any body which regularly moves through two points to move from the one point to the other. (2) The interval of time needed for any body which moves regularly through a fixed point to move twice, once in each direction, through that point. These two definitions are not equivalent; their numerical values may be significantly different. It is recommended that the period of oscillation be used instead or, if this is undesirable, that the precise definition used for period of vibration be given before the term is used. (3) SEE pendulum period.

periscope, solar - An optical device attached to an aerial camera and allowing the ground and the Sun to be photographed simultaneously. The camera's orientation can be determined in this way without relying on ground control to furnish the orientation. The solar periscope was invented by E. Santoni.

permafrost - (1) Naturally occurring earth whose temperature is below 0° winter and summer, irrespective of the state of any moisture present. (2) Permanently frozen subsoil.

permafrost table - A more or less irregular surface representing the upper limit of permafrost.

permit - A written license or permission given by a person or persons having the authority to give the permit.

perpendicular (adj.) - Making an angle of 90° with a straight line or plane. Also called normal.

perpendicular (n.) - (1) A straight line intersecting another straight line at right angles. Also called a normal. The term is made to cover the case where the line intersected is not straight by changing the definition slightly, as follows: A straight line intersecting the tangent to another line at right angles. (2) A straight line intersecting a plane at right angles to all lines lying in the plane and passing through the point of intersection. Also called a normal. The term is made to cover the case where the surface intersected is not a plane by changing the definitions slightly, as follows: A straight line intersecting the tangent plane to a surface at right angles. The terms perpendicular and normal are sometimes distinguished by applying perpendicular only to those situations where the line or surface intersected is a straight line or plane and using normal for situations where the line or surface is curved. In Euclidean space, the perpendicular from a point to a line or surface is also, locally, the shortest line from that point to the line or surface.

perpendicular equation - A condition equation requiring that the algebraic sum of the projections of the separate lines of a traverse upon perpendiculars to a fixed line (with which the traverse forms a closed figure) shall be zero. The perpendicular equation and the azimuth equation allow, when used together, the misclosure to be removed from a traverse which forms a loop with some fixed line such as a line of adjusted triangulation, by determining corrections to the measured angles of the traverse. The projections of the lines of the traverse upon lines which are perpendicular to a fixed line correspond to departures of those lines when the fixed line is considered the meridian of reference.

perpetuity - A state of continuing without limit. The term is usually applied to an amount which accrues periodically for an indefinite number of installments, i.e., without known or expected limitations.

personality - All articles or property other than real estate.

perspective (adj.) - Related by lying in pairs on straight lines through a given point. Two sets of points are said to be in perspective from a given point if corresponding points in the two sets lie on a straight line passing through the given point.

perspective (n.) - (1) A projection relating points on two surfaces by drawing straight lines from a given point through points on the one surface to the points of intersection with the second surface. Also called a perspective projection and perspective transformation. The given point is called the perspective center or center of perspective. Perspective and projection are not synonyms; a perspective is a particular type of projection. (2) The set of points, on a surface, consisting of the intersections with that surface of straight lines from a given point through points on another surface.

perspective, central - A perspective in which both surfaces are at finite distances from the perspective center.

perspective, convergent - That configuration, in an optical system consisting of a mirror followed by a lens, in which the chief rays passing through the center of aperture of the lens converge to a point in image space. This type of optical system is used for taking photographs (of a model) having more or less arbitrary viewpoints, i.e., from more or less arbitrary perspectives.

perspective, inverse central - That configuration, in an optical system consisting of a mirror followed by a lens, in which the chief rays passing through the center of aperture of the lens diverge in image space (i.e., create a virtual image).

This type of optical system is used for taking photographs (of a model) having more or less arbitrary viewpoints, i.e., from more or less arbitrary perspectives.

perspective, isometric - A perspective in which the two surfaces intersect but distances parallel to the coordinate axes in the one surface are equal to the corresponding distances in the other surface. Also called an isometric projection and parallel perspective. The second term should be avoided in technical writing, however, because it is also, and better, used for the orthogonal perspective.

perspective, orthogonal - A perspective in which both surfaces are at an infinite distance from the perspective center and all lines from the perspective center are perpendicular to one of the surfaces. Also called a projection and orthogonal projection.

perspective, parallel - That configuration, in an optical system consisting of a mirror followed by a lens, in which the chief rays passing through the center of aperture of the emerge parallel in image space. This type of optical system is used for taking photographs (of a model) having more or less arbitrary viewpoints, i.e., from more or less arbitrary perspectives. SEE ALSO perspective, isometric; perspective, orthogonal.

perspective axis - SEE axis of homology.

perspective center - SEE center, perspective.

perspective center, interior - SEE perspective center.

perspective chart - SEE chart, perspective.

perspective grid - SEE grid, perspective.

perspective-grid method - The plotting of detail from an oblique photograph onto a map by superposing a perspective grid on the photograph and transferring, by visual interpolation, from a location referred to the perspective grid to the corresponding location in the grid of the map. Equivalently, rectification by superposing a perspective grid on an oblique photograph, and plotting the locations given by that grid's coordinate system onto a separate sheet on which there is a rectangular grid. Also called the grid method.

perspective map-projection - SEE map projection, perspective.

perspective map-projection onto a tangent cylinder - SEE map projection, perspective.

perspective plane - SEE plane, perspective.

perspective projection - SEE perspective (n.).

perspective ray - SEE ray, perspective.

perspectivity - The correspondence existing between the points, lines or planes of two figures if all lines connecting corresponding points in the two figures intersect at a common point (which may be at infinity). SEE ALSO projectivity.

perturbation - (1) A change in the normal order of things because of external factors or forces. (2) SEE perturbation, orbital.

perturbation, absolute - SEE perturbation, general.

perturbation, general - An orbital perturbation calculated analytically, by successive approximations or otherwise, to obtain solutions in analytical form, yielding the perturbation at any given time. Also called an absolute perturbation.

perturbation, gravitational - An orbital perturbation caused by gravitational forces. Among the causes are tides, non-central sources of gravitational force and relativistic changes of mass, time and distance.

perturbation, long-period - A periodic perturbation of an orbit, having a period longer than that of the orbit.

perturbation, luni-solar - (1) A perturbation of the orbit of a body in the Solar System, caused by the attraction of the Moon and the Sun. (2) A perturbation of the orbit of a satellite of the Earth, caused by the attractions of the Sun and Moon. The most important perturbations are secular variations in the mean anomaly, in the right ascension of the ascending node, and in the argument of perigee. All other orbital elements except the length of the semi-major axis undergo changes having a long period.

perturbation, orbital - (1) The deviation of an orbit from some orbit taken as standard, i.e., a perturbation of an orbit. In particular, the deviation from a Keplerian orbit. It is usually referred to simply as a perturbation if the meaning is clear. In celestial mechanics, the forces causing deviations from a Keplerian orbit are called perturbing forces or disturbing forces. The principal perturbations of an orbit of a satellite of the Earth are caused by the non-central part of the Earth's gravitational field, the gravitational attractions of the Sun and Moon, atmospheric drag, pressure of solar radiation and the Earth's magnetic field. The principal perturbations of the Moon's orbit are caused by the gravitational attraction of the other planets and of the Sun. The principal perturbations of the Earth's orbit are caused by the gravitational attraction of the other planets and of the Moon. Perturbations are usually calculated as quantities to be added to the coordinates of the standard orbit to obtain the precise coordinates. (2) A force causing deviations from a Keplerian or other standard orbit. This usage is best avoided, because it makes the term ambiguous.

perturbation, periodic - An orbital perturbation increasing and decreasing periodically. Perturbations of many different periods may be present at the same time. SEE ALSO perturbation, long-period; perturbation, short-period.

perturbation, radiation-pressure - An orbital perturbation caused by the pressure of electromagnetic radiation on a satellite. The principal cause of radiation-pressure perturbations is solar, electromagnetic radiation pressing on the satellite. It may come either directly from the Sun or may arrive indirectly by reflection and scattering from the Earth. Particulate radiation (the solar wind) from the Sun also exerts pressure on satellites but by convention is not considered to be involved in radiation-pressure perturbation.

perturbation, secular - An orbital perturbation increasing monotonically with time. It is in contrast to a periodic perturbation, which increases and decreases with time. A periodic perturbation of very long period may be indistinguishable from a secular perturbation.

perturbation, short-period - An orbital perturbation having a period less than the period of the unperturbed orbit.

perturbation, special - An orbital perturbation obtained by integrating the appropriate differential equations numerically with respect to time, yielding the perturbation at times between the starting time and the upper limit.

perturbation, terrestrial - Any orbital perturbation caused by the non-centrality of the Earth's gravitational field.

Peter's formulas - A pair of formulas for the root-mean-square error or probable error of a single measurement and the root mean square error or probable error of the average. They are r.m.s. of single measurement = $1.25 \Sigma \delta_x / \sqrt{[n(n-1)]}$ and r.m.s. average = $1.25 \Sigma \delta_x /[n\sqrt{(n-1)}]$; the factor 1.25 is replace by 0.8453 if the probable error is wanted. δ_x is the absolute value of a deviation and n is the number of deviations.

Petzval condition - (1) The condition an anastigmatic lens-system must satisfy in order to produce a planar image. It is expressed by the equation $\Sigma [1/(f_i n_i)] = 0$, in which f_i is the focal length and n_i the refractive index of the i-th lens. (2) The condition that the Petzval curvature will vanish if an optical system is to be free of curvature of field.

Petzval curvature - The quantity $\Sigma [1/(f_i n_i)]$, in which f_i and n_i are, respectively, the focal length and refractive index of the i-th lens in a lens system.

Petzval surface - The surface, in the image-space of a lens system, whose curvature is the Petzval curvature.

Pevtsov's method (of determining latitude) - The hour angles h_1 and h_2 of a pair of stars at corresponding declination δ_1 and δ_2 are measured when the stars are at the same angular elevation. The astronomic latitude Φ is then given by $\tan \Phi = -(\cos h_2 \cos \delta_2 - \cos h_1 \cos \delta_1) / (\sin \delta_2 - \sin \delta_1)$. Correction is made for the inclination of the telescope as shown by the Horrebow level of the telescope. To reduce the effects of refraction and other sources of error, pairs close to the meridian (within 10° to 40°) are selected when possible.

P-factor - Any one of the many factors determining the greatest altitude at which photographs can be taken without violating the specifications for the map to be compiled from the photographs. Among these factors are the configuration and albedo of the terrain to be photographed, the transparency of the atmosphere and the capabilities of the photographic and photogrammetric systems to be used. The term is usually used in a collective sense to denote the set of all contributing factors.

phase - (1) A particular stage in the history of a phenomenon going through a cyclic sequence of stages. Examples are the phases of the Moon, the phase of an alternating current and the phase of a variable star. (2) The visible aspect of an object. The exact meaning varies somewhat, according to the application. In particular, the following special meanings are common. (3) In surveying, the appearance of a signal which presents regions of varying brightness to the observer, e.g. a round pole, illuminated from the side; a square pole, of which the observer sees two sides, one more strongly illuminated than the other. Phase causes the telescope to be pointed in a direction slightly different from that of the vertical axis of the pole or rod. The error is of the same character and requires the same treatment as an error caused by observing an eccentric object. Phase may be closely associated with asymmetry of object (target), but the two terms are not synonymous. (4) The ratio of the illuminated area of the apparent disk of a celestial body to the area of the entire apparent disk taken as a circle. For the Moon, phases (total, partial, penumbral, etc.) are designated according to specific configurations of the Sun, Earth and Moon. For eclipses, designations provide general descriptions of the phenomena (solar eclipse, lunar eclipse, annular eclipse, etc.)

phase, lunar - One of the times, designated New Moon, First Quarter, Full Moon and Last Quarter, at which the excess of the apparent celestial longitude of the Moon over that of the sun is 0°, 90°, 180° and 270° respectively.

phase age - SEE age of phase inequality.

phase angle - The angle, at the center of a partially illuminated body, between the source of light and the observer.

phase inequality - A variation in the tide or tidal currents associated with changes in the phase of the Moon. At new moon and full moon (springs), the tide-producing forces of the Sun and Moon act in conjunction, causing greater than average tides (spring tides) and speed of tidal currents. At first and last quarters of the Moon (neaps), the tide-producing forces oppose each other, causing smaller than average tide (neap tide) and tidal currents.

phase of target - Uneven illumination of a target, causing an error in pointing the telescope on a surveying instrument.

phase space - The set of all sets of six coordinates specifying the location and momentum of a body. Equivalently, a six-dimensional space with a coordinate system having three axes along which components of distance are measured and three axes along which components of momentum are measured.

phase transfer function - The phase of the optical transfer function.

phase velocity - (1) The distance per unit time that a particular amplitude of vibration in a wave appears to travel perpendicularly to the wave front. (2) If the amplitude h at a point of a wave is given, as a function of the vector r and the time t by the formula $h(r,t) = {}^*R_e \{a(r) \exp[i(gr - wt)]\}$, in which ${}^*R_e$ means the real value of, a is a function of r and g and w are constants, then the phase velocity is $w/(\delta g/\delta n)$: the ratio of w to the rate of change of g in the direction n. The speed w with which each of the co-phase surfaces advances is approximately equal, for short wavelengths of electromagnetic radiation, to $c/\sqrt{(\epsilon\mu)}$, in which the refractive index given by $\sqrt{(\epsilon\mu)}$ is less than 1. SEE ALSO velocity, group.

Philadelphia leveling rod - A two-piece leveling rod having a target but with graduations so styled that the rod may also be used as a self-reading leveling rod. Also called a Philadelphia rod. If a length greater than 7 feet is needed, the target is clamped at 7 feet and raised by extending the rod. When the target is used, the rod is read by vernier to 0.001 foot. When the rod is used as a self-reading leveling rod, the rod is read to 0.005 foot. In practice, the target is seldom used.

phosphor - Any substance emitting electromagnetic radiation for any reason other than being heated. By convention, only substances which radiate above radio frequencies are considered phosphors: i.e., substances radiating in the red, optical, and higher wavelengths. The only class of phosphors important to surveying and the allied arts is that of substances used as coatings on the inner surfaces of cathode-ray tubes for displaying pictures.

photo - SEE photograph.

photoalidade - A telescopic alidade, together with a frame for holding a photographic plate and a straight-edge, all mounted on a tripod. The photoalidade is similar in principle to the plane-table alidade. It allows lines indicating horizontal directions to be plotted and vertical or depression angles to selected features appearing on aerial and terrestrial photographs to be measured. It is suited to the mapping of regions of high relief to be mapped with contour intervals of 15 or 20 meters, where trees or other obstacles hinder observation from the ground. It is used for making maps from oblique aerial photographs.

photoangulator - A mechanical linkage for plotting true horizontal directions from oblique aerial photographs.

photo base - The air base or length of the air base as represented on a photograph or pair of photographs. SEE ALSO base length, photographic.

photobathymetry - Determining depths to the bottom of a body of water from aerial photographs, using photogrammetric methods.

photocompose (v.) - Impose one or more images by a process of step-and-repeat exposing to predetermined positions on a pressplate or photographic negative by means of a machine built for that purpose.

photo-contour process - SEE process, photo-contour.

photo control - SEE control, photogrammetric.

photocontrol base - SEE control base.

photocontrol diagram - Any selected base map or photographic index on which ground control and ground points proposed for use in control networks or as pass points are depicted and identified. SEE ALSO photocontrol index map.

photocontrol index map - Any base map or photo index on which ground control and ground points identified on photographs are shown and identified. SEE ALSO photocontrol diagram.

photocontrol point - Any station, in a network of horizontal or vertical control, identified on a photograph and used for correlating the data obtained from that photograph.

photo coordinate - SEE photograph coordinate.

photo coverage - SEE coverage, aerial.

photodeliniation - SEE delineation.

photoengraving - The process of photographically transferring an image to the surface of a metallic plate or other material and selectively etching the material so that the printing surface is in relief. The term is sometimes used, erroneously, for photogravure, which is a process resulting in a plate printing from the portion not in relief.

photogoniometer - An instrument for measuring angles from the true center of perspective to points on a photograph.

photogrammetry - (1) The science or art of deducing the physical dimensions of objects from measurements on photographs. Also called rarely and erroneously, metrical photography. A modifier is customarily used with the term to indicate the type of radiation causing the photographic image when this is not light, e.g., X-ray photogrammetry, infrared photogrammetry and acoustic photogrammetry. The principal application of photogrammetry is to the mapping of the Earth's surface, but it is also used for mapping the surface of other bodies in the Solar System, for recording the geometry of architectural and archaeological objects, for making anthropometric measurements, and in many other sciences and technologies. (2) SEE ikonogrammetry. SEE ALSO satellite photogrammetry.

photogrammetry, aerial - Photogrammetry in which photographs taken of the Earth from aircraft or satellites are used.

photogrammetry, analytical - Photogrammetry in which the size and shape of an object are determined mathematically from measurements made directly on the images, rather than mechanically from measurements on the stereoscopic model.

photogrammetry, close-range - Photogrammetry using photographs of objects at distances of a few millimeters to a few hundred meters from the camera. Also called non-topographic photogrammetry. However, that term is better used with its literal meaning because it is used to cover applications in which the object is outside the limits set for close-range photogrammetry. Among the applications are the metric recording and analysis of structures in general, of monuments and buildings in particular, the measurement amounts of materials in bulk, and X-ray tomography.

photogrammetry, geodetic - Photogrammetry applied to geodesy. The term usually denotes aerotriangulation, aerial mapping, and terrestrial photogrammetry. However, it can also cover the photogrammetric practices used for deriving geodetic information from photographs of artificial satellites.

photogrammetry, GPS - Aerial photogrammetry performed with a GPS receiver in the aircraft for the purpose of flight line navigation or control reduction.

photogrammetry, metrical - SEE photogrammetry.

photogrammetry, monoscopic - Photogrammetry using only a single photograph, as contrasted to photogrammetry using a stereoscopic pair. Also called perspective photogrammetry.

photogrammetry, non-topographic - Photogrammetry applied to purposes other than mapping. Non-topographic photogrammetry should not be confused with close-range photogrammetry, although the two terms are sometimes used as if they were synonyms.

photogrammetry, perspective - SEE photogrammetry, monoscopic.

photogrammetry, planimetric - Photogrammetry applied to determining the horizontal coordinates of points on the Earth' surface or on the surface of the Moon or other celestial body. I.e., photogrammetry applied to determining the coordinates in a coordinate system on the surface of the body, so that each point on the surface has only two coordinates.

photogrammetry, semi-analytical - Measurement of the coordinates of a stereoscopic model on an analog-type of stereoscopic plotter, and the transformation of these coordinates to coordinates of the original object by computation. Also called semi-analytical triangulation.

photogrammetry, stereo - SEE photogrammetry, stereoscopic.

photogrammetry, stereoscopic - Photogrammetry using stereoscopic equipment and methods. Also called stereo photogrammetry and stereophotogrammetry.

photogrammetry, terrestrial - Photogrammetry using photographs taken at the Earth's surface and of the Earth's surface. Also called phototopography and ground photography; the latter term is not preferred.

photogrammetry, topographic - (1) Photogrammetry applied to determining the geometry of surfaces, as distinguished from photogrammetry applied to determining the geometry of figures on surfaces. I.e., photogrammetry applied to determining the coordinates of the surface in a three-dimensional coordinate system. (2) In particular, photogrammetry applied to mapping the surface of the Earth, or other celestial body. One of the coordinates is always height, either geodetic height or height in a rectangular, Cartesian coordinate system. However, this coordinate is usually converted to elevation when the photogrammetric data are used for making a map.

photograph - (1) A picture or pattern created by the action of electromagnetic radiation or particles of high energy on an emulsion containing a radiation-sensitive substance such as silver bromide. The term should, properly, be applied only to pictures or patterns created by the action of visible and near-infrared wavelengths. However, custom and the lack of suitable alternatives sanction the more general definition. (2) A picture or pattern made by exposing a light-sensitive emulsion, using a camera. Also called a photo in the vernacular. The photograph may reproduce the light and dark regions as they appeared in the object, in which event the photograph is called a positive photograph or positive, or it may invert the shades by having dark regions where the object appeared light and light regions where the object appeared dark, in which event the photograph is called a negative photograph or negative. (Note: Some photogrammetrists use photography or imagery for photographs. This peculiar misuse should be discontinued.) (3) The print made photographically from the negative or positive photograph. SEE ALSO horizon photograph; wing photograph.

photograph, aerial - A photograph taken from an aircraft.

photograph, composite - An assemblage of the separate aerial photographs taken simultaneously by the several lens-systems of a multiple-lens camera into the equivalent of a single photograph taken by a camera with a single, wide-angle lens-system.

photograph, equivalent vertical - A hypothetical photograph, theoretically taken by a camera having the optical axis vertical and taken at the same camera station as the actual photograph.

photograph, high-oblique - An oblique photograph showing the horizon.

photograph, horizontal - A photograph taken with the camera's axis horizontal.

photograph, horizontally-controlled - (1) A photograph showing one or more horizontal-control points. (2) A photograph obtained by horizontally-controlled photography.

photograph, lateral-oblique - A photograph taken with the camera pointed obliquely downwards and as nearly perpendicular to the line of flight as possible.

photograph, lorop - A photograph taken with a long-focal-length camera (over 250 cm) having a narrow field of view.

photograph, low-oblique - An oblique photograph which does not show the horizon.

photograph, multiple-lens - A photograph made with a multiple-lens camera.

photograph, negative - A photographic image having the grading of tones or colors on it reversed from the grading of tones or colors in the original scene. Also called a negative. Where the original scene appeared light, the negative photograph appears dark, etc.

photograph, oblique - An aerial photograph taken with the optical axis of the camera deliberately pointed away from the vertical. Vertical photographs are usually taken with the optical axis of the camera kept within 3° of the vertical. Oblique photographs could therefore be defined as photographs taken with the optical axis more than 3° from the vertical.

photograph, panoramic - A photograph that by itself encompasses a panorama of the scene, as distinguished from a set of photographs which, pieced together, encompass a panorama.

photograph, perspective - A photograph related to the object photographed by a perspective (projection).

photograph, positive - A photographic image having approximately the same grading of tones or the same colors as the original scene. Also called a positive.

photograph, supplemental - An aerial or terrestrial photograph not used directly for mapping but used to add to or improve details in maps or charts.

photograph, terrestrial - A photograph taken from the ground and taken of a part of the Earth's surface. Also called a ground photograph.

photograph, transverse - An oblique photograph taken in a direction transverse to the direction of flight.

photograph, tricamera - One of a set of three photographs obtained by simultaneous exposure from three aerial cameras placed in such a way that the overlap is constant.

photograph, vertical - An aerial photograph taken with the optical axis of the camera as nearly vertical as possible, so that the resultant photograph lies approximately in a horizontal plane. The optical axis is usually kept within 5° of the vertical.

photograph center - The center of a photograph as indicated by the images of the fiducial marks of the camera. Also called the center of photograph and center of the photograph. In a perfectly adjusted camera, the photograph center and the principal point coincide.

photograph coordinate - One of two coordinates of a point on a photograph.

photograph coordinates - A coordinate system defining the coordinates of points on a photograph. Note that this use of the term is probably peculiar to photogrammetry. The origin is usually at the principal point, but may be at the nadir point, isocenter, a fiducial mark, etc. The axes are usually either the lines through the fiducial marks or the principal line and a photograph parallel.

photograph coordinate system - A Cartesian coordinate system having its origin at the intersection of the lines joining opposite fiducial marks, the x-axis and y-axis coincide with the lines joining the fiducial marks and the z-axis is perpendicular to the x- and y-axes and positive in the direction of the camera's perspective center. Also called a photographic coordinate system. If the lines joining opposite fiducial marks are perpendicular, the photograph coordinate system is a rectangular Cartesian coordinate system. SEE ALSO photograph coordinates.

photograph meridian - The image, on a photograph, of any horizontal line in object-space parallel to the principal plane. Also called a photographic meridian. Because all such lines meet at the point at infinity, the image of their point of intersection is at the intersection of the principal line and the horizon trace, and all photographic meridians pass through that point.

photograph nadir - That point at which a vertical line through the perspective center of the camera's lens system intersects the plane of the photograph. Also called nadir and nadir point, although those terms are ambiguous. Also called the photograph plumb-point.

photograph parallel - The image, on a photograph, of any horizontal line perpendicular to the principal plane. All photograph parallels are perpendicular to the principal line.

photograph perpendicular - The perpendicular from the internal perspective center to the plane of the photograph.

photograph plane - The flat surface, in a camera, on which the photographic plate or photographic film is held. It is not exactly the primary focal plane of the lens system, but is a surface placed so as to secure the best balance of sharp focus on all parts of the photographic plate or film. Also called an image plane. However, the true image-plane is that plane on which an image is formed. It need not be the same as the topograph plane.

photograph plumb-point - SEE photograph nadir.

photograph pyramid - A geometric figure (pyramid) whose base is the triangle formed by three points on a photograph and whose apex is the perspective center of the photograph. Also called a photogrammetric pyramic and photo pyramid.

photographs, convergent - A pair of photographs taken with two cameras so directed that their optical axes converge.

photographs, high-oblique convergent - Convergent photographs so taken that both photographs show the horizon.

photographs, homologous - Two or more overlapping photographs having different camera stations, i.e., taken from different points.

photographs, low-oblique convergent - Convergent photographs so taken that neither photograph shows the horizon.

photography - (1) The art or process of producing images on sensitized material through the action of light. The analogous art or process using radiation other than light is called ikonography or iconography. (2) The science concerned with the chemical and optical processes involved in taking photographs and transforming them in various ways. (3) SEE NOTE UNDER photograph. SEE ALSO process photography; strip photography; twin low-oblique photography.

photography, aerial - The art or process of taking photographs from an aircraft. Also called aerophotography.

photography, aerial cartographic - SEE mapping photography.

photography, analytical - Photography done either by motion-picture camera or other type of camera to determine whether a particular phenomenon does or does not occur.

photography, cartographic - SEE mapping photography.

photography, continuous-strip - The process of photographing a moving scene (such as the ground as seen from an aircraft) on a long, narrow strip of film. The resulting picture resembles a mosaic made by assembling a sequence of photographs taken in separate frames.

photography, control-point - Electronically-controlled aerial photography carried out on four flight-lines flown in a cloverleaf pattern from the four cardinal directions and with the flight lines intersecting over an identifiable mark or secondary-control point.

photography, convergent - Oblique photography, usually done with a pair of cameras acting as a unit, so that one photograph is taken looking forward along the line of flight and the succeeding, overlapping photographs taken looking back along the line of flight. Convergent photography can provide photographs with a much larger base-height ratio, but it is much more difficult to carry out than vertical photography.

photography, convergent low-oblique - SEE photography, low-oblique convergent.

photography, convergent high-oblique - SEE photography, high-oblique convergent.

photography, cross-flight - Aerial photography in which photographs are taken in sequences forming strips having overlap between exposures and having a direction at right angles to that of other, existing strips of photographs.

photography, divergent low-oblique - SEE photography, low-oblique divergent.

photography, high-oblique - Aerial photography in which the optical axis of the camera (or cameras) is pointed so that the apparent horizon lies within the field of view of the camera.

photography, high-oblique convergent - Convergent photography done so that the cameras' fields of view include the horizon.

photography, horizontally-controlled - Aerial photography in which distances between the taking camera and each of two or more points whose geodetic coordinates are known are determined simultaneously with the taking of the pictures. The distances have usually been determined using Shoran, Hiran or Shiran.

photography, indirect - Photography in which the camera takes a picture of the image cast upon a screen by a cathode-ray tube, as in radar, television, etc. or other light-emitting display.

photography, inertial-reference - Photography taken at the same time that the location and orientation of the camera are obtained from an inertial-navigation system and recorded. The method has been used in taking aerial photographs.

photography, LOROP - SEE photography, long-range oblique.

photography, long-range oblique - Photography using a long-focal-length (greater than 100 inches) camera with a narrow lens. Also called LOROP photography. The term is also used to refer to the photographs themselves, but this practice is undesirable.

photography, low-oblique convergent - (1) Convergent photography done so that the horizon does not lie within the cameras' fields of view. (2) Twin low-oblique photography using a pair of cameras whose optical axes point downwards and towards each other along the line of flight. This kind of photography gives a greater base-height ratio.

photography, low-oblique divergent - Twin low-oblique photography done using cameras whose optical axes point downwards and away from each other crosswise of the line of flight. This kind of photography gives increased angular coverage.

photography, low-oblique transverse - SEE photography, transverse low-oblique.

photography, mapping - SEE mapping photography.

photography, metric - Photography done so precisely as give results that can be used to give precise information on the locations and positions of the objects photographed.

photography, multiband - (1) Photography producing more than one photograph of a given region, each photograph being taken in a different part of the spectrum. (2) A process of recording images of a given region, each image being shown in a different part of the spectrum.

photography, near-infrared - Photography done using radiation in the 0.8 to 1.5 µ part of the spectrum.

photography, oblique - Aerial photography in which one or more cameras are set deliberately with their optical axes at an angle to the vertical.

photography, shoran-controlled - Photography in which the location of the camera is determined by measuring distances from Shoran equipment aboard the photographing aircraft to each of two fixed Shoran stations while photographs are being taken.

photography, split - SEE photography, split-vertical.

photography, split-vertical - Photography done using a split camera.

photography, terrain-profile - Aerial photography in which data on the height, with respect to some pre-established datum, of the terrain along the ground vertically below the aircraft are obtained at the same time as the photographs are taken. The terrain-profile recorder is commonly used as the measuring instrument. The recorder actually measures the distance of the aircraft above the ground. The location of the aircraft with respect to some datum is determined by other means, and the height of the terrain is then obtained by combining the two sets of information.

photography, terrestrial - Photography from the Earth's surface. In particular, in photogrammetry, photography on and of the Earth's surface.

photography, transverse low-oblique - Aerial photography with a pair of cameras placed so that the optical axes point far enough below the horizon that the horizon will not appear on the photograph, and are perpendicular to the longitudinal axis of the aircraft, i.e., are presumably perpendicular to the line of flight.

photography, tricamera - Aerial photography done with three cameras arranged at fixed angles to one another in such a way that adjacent photographs, taken simultaneously, overlap. Usually, the cameras are arranged so that the central camera takes vertical photographs and the other two cameras take high-oblique photographs. The assemblage is often referred to as a trimetrogon camera assembly because Metrogon lens-systems were widely used in early photography with three cameras. The process is then called trimetrogon photography. This type of photography has been supplanted, for most purposes, by photography using single cameras having extremely wide fields of view.

photography, trimetrogon - Photography using an assemblage of three cameras: one pointing vertically downwards and two mounted to point obliquely downwards and as nearly perpendicular to the line of flight as possible. The oblique photographs, taken at an angle of 60° from the vertical, together with the vertical, produce a picture continuous from horizon to horizon. Exposures are ordinarily made at intervals providing pictures overlapping along the line of flight. SEE ALSO photography, tricamera.

photography, vertical - Aerial photography done with the camera pointed as nearly vertically downwards as practicable.

photogravure - The process of transferring an image photographically to the surface of a metallic plate or other material and selectively etching the material so that printing is done by the regions not in relief. Sometimes erroneously called photoengraving.

photoidentification - (1) The detection, identification and marking of survey stations on aerial photographs. Also called control-station identification. Positive identification and location is required if survey data are to be used to control photogrammetric compilation of maps. (2) The process of showing the location, on a photograph, of ground points in relation to surrounding detail and identifying on the ground the points marked and described on the photograph.

photo index - SEE index map, photographic.

photo-interpretation - SEE interpretation, photographic.

photo-interpretation key - SEE key, photographic-interpretation.

photolithography - (1) Lithography using photography to transfer the image to the intermediate surface and then treating the gelatin of the exposed and developed image to produce regions which will not absorb the ink. (2) SEE lithography.

photomap - (1) An assemblage of one or more photographs fitted to ground control and carrying symbols making the assemblage usable as a map with local small scale variations. (2) A photomosaic showing a specified region of land and containing marginal information, descriptive data and a reference grid or graticule.

photomapping - The process of making maps from various types of photographs, using information from surveys and/or other maps.

photomechanical (adj.) - Pertaining to or designating any reproduction process using a combination of photographic and mechanical procedures.

photometer - An instrument for measuring the luminous intensity, luminous flux density or illumination.

photomosaic - SEE mosaic.

photomultiplier - A vacuum tube having one surface capable of emitting one or more electrons when illuminated, and a sequence of other surfaces which emit several electrons for each incident electron. The electrons emitted by the light-sensitive surface are guided by an electrostatic field to a second surface which then emits many more electrons, and the process is carried through several more such surfaces until a cascade of electrons is formed. These form a current which is amplified and measured.

photo plot - SEE index map, photographic.

photopolygonometry - Aerotriangulation in which the length of the air base is derived from the altitude at which the photographs were taken, as measured by an altimeter (typically, a radar altimeter). Otherwise similar to radial aerotriangulation.

photo pyramid - SEE photograph pyramid.

photorevision - The process of revising maps using aerial photographs and other available sources to show planimetric changes since the date of the latest existing map.

photo scale - SEE scale.

photosphere - That intensely bright portion of the Sun which is visible to the naked eye.

phototachymeter - SEE distance-measuring instrument, electro-optical.

phototheodolite - (1) A surveying instrument combining a surveying camera and a theodolite in such a way that the direction of the camera's optical axis is the same as or can be determined from the direction of the theodolite's line of sight. Also called a camera transit. (3) A surveying instrument in which a camera replaces or is affixed to the telescope of a theodolite. In older models, the horizontal and vertical circles had to be read by the observer. In newer models, the circles can be photographed at the instant the camera photographs the object. The phototheodolite differs from the cine-theodolite principally in that the cine-theodolite takes photographs through the telescope with a moving-picture camera and usually tracks the moving object being photographed.

phototopography - That method of surveying in which the detail is plotted entirely from photographs taken at suitable stations on the ground. SEE ALSO photogrammetry, terrestrial.

phototriangulation - (1) That process whereby the measurements of angles and/or distances on overlapping photographs are related to determine (the coordinates of) horizontal and/or vertical control, using photogrammetric principles and techniques. Also called photogrammetric triangulation or vaguely, triangulation. Usually this process involves using aerial photographs and is called aerotri-angulation or aerial triangulation. Less often, terrestrial photographs are used, and the process is called terrestrial phototriangulation. (2) That process whereby the measurements of angles and/or distances on overlapping photographs are related to determine (the coordinates of) horizontal control. The basic principles of phototriangulation were established in 1759 by T. Lambert of Switzerland. He considered perspective geometry only. R. Sturm and G. Hauk, in 1883, related photographic images to projective geometry. (3) The method employed in the process as defined in (1).

phototriangulation, analog - Phototriangulation using a stereoscopic plotter or other analog-type device.

phototriangulation, analytical - Phototriangulation in which the coordinates of the control points are obtained purely by computation. When done using aerial photographs, the process is referred to as analytical triangulation or analytical aerotriangulation.

phototriangulation, mechanical - The mechanical determination of coordinates of points on an object, using stereoscopic pairs of photographs of the object. The term should be usable whether the photographs are stereoscopic pairs or not, but common usage is as given.

phototriangulation, radial - SEE aerotriangulation, radial.

phototriangulation, terrestrial - Phototriangulation using photographs taken from points on the ground. In its original form, the method was developed from plane-table surveying. The photographs were set up vertically on a drafting table in their proper directions from a point taken as the camera station, and then viewed through a special telescope similar to that used in plane-table surveying and using a similar procedure for drawing the map.

phototriangulation adjustment - Determination of corrections to assumed or approximate coordinates of objects, using phototriangulation.

phototriangulation adjustment, radial - An analytical or mechanical phototriangulation adjustment in which photographs are fitted together in such a way that radials are deformed as little as possible.

phototrig - A method of determining heights from vertical angles measured on the ground and from horizontal distances measured on a map constructed by stereocompilation.

phototrig traverse - SEE phototrig; traverse, phototrig.

phototypesetter - Equipment for lettered material on photographic positives that can be incorporated directly with other material that is to be reproduced and printed. The equipment consists of two independent units: an electric typewriter that produces typewritten copy and a perforated tape; and a photographic unit to which the tape is fed at any convenient time and which then produces a right-reading film suitable for stick-up work.

physiography - SEE geomorphology.

picket - SEE range pole.

pictochrome process - The process of producing a pictomap by decomposing a photomosaic into three differently-colored pictures (by photographing a photomosaic through three different filters), masking out unwanted details, combining the result and placing on the composite names and other required information.

pictoline process - The process of producing, from a photograph in which the tones are continuous, a map-like picture in which the tones are sharply separated, giving distinct edges to details.

pictomap - A reproduction, by photographing a black-and-white photomosaic through various colored filters, such that the new picture is in cartographically meaningful colors and appropriate cartographic symbols have been added. SEE ALSO pictochrome process.

pictotone process - Photolithography in which film for reproduction and transfer to printing plates is produced for printing monochromatic photomaps and pictomaps. The process produces a randomly granular effect which visibly sharpens the definition of features, separates tones and in many instances, is superior to the halftone process.

picture control point - SEE control point.

picture plane - A plane upon which is projected a set of lines or rays from an object to form an image or picture. In photogrammetry, the photograph is the picture plane.

picture point - A point, on a map, corresponding to an identifiable point in a picture.

pied - One-sixth of a toise.

pie graph - A circle whose interior is divided into sectors by radii; the proportionate size of each sector indicates the proportion the quantity indicated by that sector bears to the whole of which it is a component.

pier - (1) A structure, usually of greater length than width, built of timber, stone, concrete or other material, having a platform, and projecting from the shore into navigable waters so that vessels can be moored alongside for loading and unloading cargo, or for storage of cargo during the transfer. (2) A vertical support for the adjacent ends of two spans of a bridge. (3) A wharf consisting of a superstructure of timber, steel or concrete supported by a series of pile bents.

pierhead line - SEE bulkhead line.

pile - SEE test pile.

pillbox level - SEE level, circular.

pilotage chart - An aeronautical chart at a scale of 1:500,000 used for planning before a flight and for navigation over short distances using dead reckoning and visual piloting.

pilotage chart, aeronautical - An aeronautical chart designed specifically for aerial navigation. The term is used to distinguish it from aeronautical charts used for military planning, etc.

pilot chart - A chart issued quarterly to show meteorological, oceanographic, and hydrographic data in a specified part of the ocean for each of the three months in the quarter. It is for use with conventional charts. Timely articles of professional interest to the seafarer are printed on the back. Pilot charts cover all the oceans of the world.

pilot sheet - A sample of a new map, made for inspection before printing all the required maps, to disclose errors or problems which may occur in compilation, drafting and reproduction.

pin - (1) SEE turning pin. (2) SEE chaining pin.

pin, turning-point - SEE turning pin.

pincushion distortion - Optical distortion in which magnification in the image increases radially (toward the edges, away from the optical axis). Also called positive distortion.

pinhole - A tiny, clear spot on photographic negatives, caused by dust, air bubbles or undissolved chemicals.

pinhole camera - A camera whose optical system consists simply of a small hole through which the light is admitted. A pinhole camera has extremely high geometric fidelity. However, its resolution is limited by diffraction. The very small aperture needed to ensure geometrical fidelity also cuts down the amount of light admitted, so that either very long exposures must be used or only very bright objects photographed. It has been used for architectural photogrammetry, solar photography and other projects where very high accuracy is important. X-ray photographs are taken with pinhole cameras.

pinnacle - A columnar rock rising from the depths and constituting a menace to navigation on the surface.

pinpoint - A very accurately determined location, usually established by passing over or very close to an aid to navigation or a landmark of small extent.

pipe drag - SEE drag (3).

pipeline leveling - Establishing a common level at two points by running an open-ended pipe or tube nearly filled with water from one point to the other. Also called hydrostatic leveling. However, pipeline leveling is more often used to denote hydrostatic leveling in which the liquid is water and great accuracy is not needed. The pipe is bent upward at each end; the water's surface at each end is then at the same level, i.e., lie on the same level surface. Knowing the elevation of the water surface at one end, the elevation of the points at each end can be determined by measuring the vertical distances of the points from the water's surfaces.

pipe locator - An electronic device for locating buried metallic pipes. It is also useful for locating buried, metallic monuments, etc.

pit - SEE test pit.

pit, borrow - A pit made to provide earth for embankments, fills, etc.

pitch - (1) The angle, measured in a vertical plane, between the longitudinal axis of a vehicle, vessel or aircraft and a horizontal plane. (2) The angle between the longitudinal (roll) axis of a rotated body and the plane through the original, unrotated positions of the longitudinal and transverse (pitch) axes of that body. In particular, in aerial navigation, the angle of rotation of an aircraft about that horizontal axis which is perpendicular to its longitudinal axis. (3) The rotational motion of a vehicle, vessel or aircraft about that horizontal axis which is perpendicular to the longitudinal axis. Pitch is positive if it causes a vessel's or aircraft's front end to rise. (4) The angle through which a camera is rotated about the transverse axis of the aircraft in which it is mounted. (5) The angle through which the coordinate system of a photograph is rotated about the y-axis of the photograph's coordinate system or about the Y-axis of the ground coordinate-system. Also called tip, longitudinal tilt or y-tilt.

Pitot tube - A tube, used for measuring the pressure in a moving fluid, bent into the form of an L. Both ends are open. The base of the L lies in the stream parallel to the direction of flow, with the open end pointed upstream. The upright projects above the surface of the stream. The height of the fluid in the upright is proportional to the pressure of the fluid upon the open end in the stream. The pressure is usually proportional to the speed of the fluid, so that the Pitot tube is also used to determine speed either of the fluid or of the body on which the Pitot tube is mounted. The device was first described by H. Pitot (1732), who used it for determining the speed of water at different depths in a stream.

pits and mounds - The name of a system of witnessing corner stakes of a public-land survey in a prairie.

pivot - A spindle or pin supporting a movable part of an instrument in such a way that the part is free to turn. The pivot of a compass is usually a fixed point, at the end of a pin, on which the needle hangs by a single, jewelled socket.

pivot error - That error in longitude, time, declination or latitude which is caused by deviation of the trunnions of a telescope from a perfectly cylindrical shape.

pivot inequality - Any difference in the diameters, or any irregularities in the form, of the pivots of the horizontal axis of a surveying or astronomical instrument (telescope). A pivot inequality causes a pivot error. Formerly, corrections for pivot inequality were made to astronomical measurements taken for geodetic surveys. Such corrections are not needed for measurements made with precise, modern instruments.

pixel - (1) A geometric element of surface, resulting from subdividing an image into identically-shaped figures such that each figure gives information about the location, intensity and perhaps color of the source, but no smaller subdivision gives more information. The concept of pixel (the term is a corruption of pictorial element) is valid only for images composed of discrete patches. It is not valid for continuous images. The term is not precisely equivalent to resolution even when resolution is expressed in terms of area, because resolution can be defined for images which are not composed of discrete patches. The size of a pixel is determined principally by the size of the smallest, individual, radiation-sensitive element in the device creating the image. For example, in the human eye, the pixel is the size of the region occupied, on the retina, by a cone. On a photograph, it is the size of the smallest clump of silver. (2) That geometric region, on the object, which corresponds to the smallest information-containing geometric element in the image.

Pizzetti's projective method - The projective method of reduction to the reference ellipsoid, by proceeding in two steps: first, the vertical at the point P on the surface is followed downward to the geoid to give the footpoint P' there; then the normal to the ellipsoid is taken through P' and the footpoint of the vertical is the desired map P" of P. Also called Pizzetti's method.

place - (1) The angular coordinates of a celestial body. (2) The location of a celestial body on the celestial sphere.

place, apparent - Those angular coordinates, referred to the true equinox and equator of date, a celestial body would have if it were observed from the center of the Earth. Also called apparent position. Apparent place is calculated from the topocentric place with the same equinox and equator by correcting for atmospheric refraction, diurnal aberration and geocentric parallax. If greater precision is wanted, corrections may also be made for the variation of latitude. Apparent place is calculated from true place by transferring the origin of the coordinate system from the center of the Solar System to the center of the Earth and adding the effects of aberration - stellar aberration, if only distant bodies are involved; planetary aberration if bodies inside the Solar System are involved. It is calculated from the observed place by transferring the origin from the point at which the measurements were made to the center of the Earth, i.e., by taking into account the effects of daily aberration and geocentric parallax. SEE ALSO place, true.

place, heliocentric - The angular coordinates of a celestial body referred to a coordinate system having its origin at the center of the Sun.

place, mean - The angular coordinates of a celestial body in a heliocentric coordinate system, referred to mean equator and equinox of a standard epoch. A mean place is determined by removing from the directly measured angular coordinates the effects of refraction, geocentric and stellar parallax, and stellar aberration and by referring the coordinates to the mean equinox and equator of a standard epoch (e.g., of date). Alternatively, it may be considered the result of removing from the apparent place the annual aberration and the annual parallax. In compiling star catalogs, it has been the practice not to remove the secular part of stellar aberration or the elliptic part of annual aberration.

place, normal - A single direction or location which is the weighted average of a number of observed directions or locations.

place, observed - The angular coordinates, in a topocentric coordinate system, of a celestial body, determined by direct measurements from the Earth's surface and corrected only for errors in the instrument used and errors in the method of measurement (errors of collimation, levelling error, index error, dip of the horizon and so on).

place, topocentric - The angular coordinates of a celestial body referred to a coordinate system having its origin at the observer.

place, true - The angular coordinates of a celestial body in a heliocentric coordinate system, referred to the true equinox and true equator at a standard epoch (e.g., of date). True place is calculated from apparent place by removing the effects of nutation. It is calculated from measurements in a topocentric coordinate system by (a) removing the effects of refraction, diurnal aberration and geocentric parallax; (b) removing the effects of annual aberration and annual parallax (distance of the center of the Earth from the center of the Sun); and (c) removing the effects of precession and nutation.

place, true mean - The angular coordinates, referred to the mean equinox at the beginning of the year, corresponding to the direction of the celestial body with respect to the center of the Earth. True mean place is calculated from the mean place by removing the effects of diurnal aberration and diurnal parallax and by antedating the time of observation by the length of time required for light to travel from the body to the observer.

place lands - Lands granted to a railroad and located on each side of the railroad. The location of the lands became fixed upon adoption of a center line. Indemnity lands were selected in lieu of place lands which had already been granted for other purposes.

place name - SEE toponym.

placer - (1) A place where gold is obtained by washing out the metal from its admixture with sand or gravel. (2) An alluvial or glacial deposited, as of sand or gravel, containing particles of gold or other valuable mineral. In mining law of the United States of America, mineral deposits (not veins in place) are treated as placers, so far as locating, holding and patenting are concerned. Various minerals, besides metallic ores, have been held to fall under this provision, but not coal, oil or salt.

plain - An extensive, treeless or almost treeless expanse of nearly flat land.

plain, alluvial - (1) A plain formed by the deposition of alluvium, usually adjacent to a river which overflows periodically. (2) A level region bordering a river and formed by the deposition of material (usually sand and silt) which the river has carried in suspension.

plaintiff - (1) A person who brings an action (at law). (2) The party who complains or sues in a personal action and is so named on the record.

plan - (1) A method or scheme of action, procedure or arrangement, project, outline of program or schedule. (2) A draft or form, properly drawn on a plane as a map, especially a view from above or the representation of a horizontal section. I.e., a graphical and symbolic representation, usually on a flat surface and at a fixed scale, of the geometric relationships between the parts of a structure. (3) An orthographic projection onto a horizontal plane. (4) One of a number of planes conceived as perpendicular to the line of vision and interposed between the eye and the pictured objects. SEE ALSO grading plan; plat.

plan, grading - SEE grading plan.

plane - (1a) Geometrically, a surface such that a straight line connecting any two points of the surface lies entirely on the surface and contains at least three distinct points of that surface. (b) Algebraically, the surface defined by all values of x, y, z satisfying the equation $ax + by + cz = d$, in which a, b, c, d are constants. (2) A small region on a level surface, as, for example, in the term plane surveying. (3) SEE plane, tidal. SEE ALSO ground plane; hill plane; image plane; reference plane.

plane, basal - SEE plane, epipolar.

plane, epipolar - Any plane containing the epipoles; therefore, any plane containing the airbase. Also called a basal plane.

plane, equatorial - (1) Of a sphere, the plane determined by a specified great circle or, if a diameter of the sphere is specified instead, the plane through the center of the sphere and perpendicular to the specified diameter. (2) Of a rotational ellipsoid, the plane through the center of the ellipsoid and containing the two axes of equal length. On an oblate rotational ellipsoid, the equatorial plane corresponds to the plane center and perpendicular to the shortest axis; on a prolate rotational ellipsoid, it is the plane through the center and perpendicular to the longest axis. (3) Of an ellipsoid with three axes of unequal lengths (a heteraxial ellipsoid), the plane through a specified pair of axes. In geodesy, this is usually the plane through the center and perpendicular to the shortest axis.

plane, first (second) principal - That one of the two principal planes designated as the first (second) principal plane. It is usually the one closest to (farthest from) the source of the illumination.

plane, focal - (1) A plane perpendicular to the optical axis of an optical system and passing through a focal point on that axis. (2) That plane, perpendicular to the optical axis of a lens system, onto which the images of points in object-space are projected. (3) The locus of all points in image space which are the images of points on a plane in object space. The focal plane is a plane only for ideal imaging systems. Most optical systems project points on a plane in object space onto a more or less curved surface, the focal surface. (4) The plane of best average definition. The plane so defined need not pass through a focal point but could be merely the plane onto which the image is projected. For this reason, a focal plane through a focal point is sometimes called a principal focal plane. However, that term is preferably used for a plane through the second focal point.

plane, fundamental - SEE Besselian elements.

plane, galactic - A plane placed, with respect to the stars of the Galaxy, so that half the stars are on one side of the plane and half on the other.

plane, geodetic-meridian - SEE plane of the geodetic meridian.

plane, harmonic tidal - SEE plane, Indian tidal.

plane, horizontal - (1) A plane perpendicular to the direction of gravity at a given point. Equivalently, a plane tangent to a level surface at a given point. In general, a horizontal plane is perpendicular to the direction of gravity only at the point at which it is defined. The term is not synonymous with plane of the horizon; the plane of the horizon is defined by the horizon, not by the direction of gravity. (2) A plane parallel to a plane tangent to the geoid at a given point. (3) A plane perpendicular to the direction of gravity at a given point and within which or one which angles and distances are measured. For any planimetric survey, the direction of gravity at all points of the survey is assumed to be constant and all horizontal planes therein are parallel.

plane, invariable - The plane through the center of mass of the Solar System and perpendicular to the angular-momentum vector of the Solar System.

plane, meridional - (1) A plane through the optical axis of an optical system. (2) SEE meridian plane.

plane, nodal - A plane perpendicular to the optical axis at a nodal point.

plane, optical - SEE flat, optical.

plane, orbital - The plane of an orbit; the plane in which an orbit lies.

plane, perspective - Any plane containing the perspective center. Because the intersection of two planes is a straight line, the intersection of a perspective plane with a horizontal plane will always appear as a straight line in the photograph. Any straight line in object space will therefore also appear as a straight line in the photograph.

plane, prime vertical - That plane, at a given point, which is perpendicular to the astronomic or geodetic meridian through that point and which contains there the normal to the ellipsoid of reference.

plane, principal - (1) That portion of the principal surface which lies sufficiently near the optical axis that it can be considered planar.

plane, principal focal - The focal plane through the point at which entering, monochromatic rays parallel to and near the optical axis come to a focus. Also called the primary focal plane. The term should not be used to mean a focal plane through a focal point.

plane, tangent - (1) A plane touching a curved surface at a point in such a way that all straight lines through that point and in the plane are tangent to curves in the surface. A plane tangent to an ellipsoid at any point is perpendicular to the normal (to the ellipsoid) at that point. A small region on the ellipsoid can be projected conveniently onto a plane tangent near the center of the region. A tangent plane may contain more than one point of the surface, e.g., a plane tangent to a cylinder is tangent along the whole length of a straight line on the cylinder.

plane, tidal - (1) Mean sea level in the immediate vicinity of a tide gage. SEE ALSO datum, tidal. (2) A reference surface with respect to which the height of the tide is measured in a particular region. Also called a tide plane.

plane, vertical - (1) Any plane passing through a point on the Earth and containing the zenith (and nadir) of that point. A vertical plane containing the zenith must also contain the nadir. The planes of the celestial meridian and of the prime vertical circle are vertical planes. (2) A plane perpendicular to a horizontal plane and within which angles and directions are measured. (3) A plane containing a plumb line. This is probably an empty definition, because a plumb line is in general not planar.

plane angle - (1) The figure formed by two intersecting, straight half-lines terminating at their point of intersection. The two lines are called the sides of the plane angle; the point of intersection is called the vertex. (2) The figure formed by rotating a straight line about one of its terminal points. The initial position of the line is called the initial side and the final position the terminal side. Rotation is, by convention, considered positive if done in the counter-clockwise direction. An angle formed by rotation is therefore a directed angle. It can be represented by a vector whose length represents the size of the angle and whose direction is positive or negative according as the rotation was counter-clockwise or clockwise. (3) A number (ratio) expressing the rate at which the sides of a plane angle (f) diverge. The ratio is obtained by drawing an arc of a circle from the initial side of the angle (f) to the terminal side, using the vertex as center, with an arbitrary radius and dividing the length of the arc by the radius of the circle. A ratio of exactly one is called a radian (symbol r). An angle whose two sides coincide has a size of either 0 radians or 2π radians; the angle between a line and the perpendicular to it is $\pi/2$ radians. For example, 1.57085 93268r. The number obtained by multiplying the size of an angle in radians by $360/2\pi$ is called a sexagesimal measure of the angle in degrees. One degree contains 60 minutes of arc and one minute of arc contains 60 seconds of arc, so that the number of minutes less than a degree in the angle is obtained by multiplying the fractional part of the angle in degrees by 60 to get minutes and fractional parts of a minute and multiplying the fractional part of a minute by 60 to get the number of seconds and fractional parts of a second. The symbols ° , ', " denote degrees, minutes and seconds respectively. For example, 57° 17' 44.866". The number obtained by multiplying by 400 instead of by 360 is called a centesimal measure of the angle in gons. The fractional part may be expressed in centesimal sub-units by a process similar to that given for obtaining minutes and seconds, but multiplying in each step by 100 to get centigons and centicentigons, respectively. For example, 63g 66cg 19.77ccg. The number obtained by multiplying by $24/2\pi$ is called the hour-angle measure of the angle. It is used almost solely by astronomers and is derived from the astronomical method of measuring time. Sub-units in minutes of time and seconds of time are obtained by multiplying the fractional part by 60, exactly as for minutes of arc and seconds of arc. However, there are 15 minutes of arc in one second of time and 15 seconds of arc in one second of time. For example, 3h 49m 10.997s. Note that radian, sexagesimal and centesimal measures are nondimensional, whereas hour-angle measure has the dimension of time.

plane coordinate - On of a pair of numbers identifying a point in a plane.

plane coordinate, assumed - One of the coordinates used in an assumed plane coordinate system.

plane coordinate system - A coordinate system in the plane.

plane coordinate system, assumed - A plane coordinate system set up at the convenience of the surveyor to be used locally. The axes are usually defined so that all points of the survey are in the first quadrant. The y-axis is usually directed northward.

plane coordinate system, State - SEE State plane coordinate system; State coordinate system.

plane curve - (1) A curved line in a plane. (2) SEE plane elliptic arc.

plane elliptic arc - A segment of an ellipse. Also called a plane curve. This and the term itself are probably the results of ignorance and not jargon.

plane of incidence - That plane which is defined by the ray incident on a surface and the normal to the surface at the point of incidence. The plane of incidence also includes the reflected or refracted ray, so it can be defined by any two of the three lines.

plane of the apparent horizon - A plane fitted in some suitable and defined manner to the apparent horizon.

plane of the astronomic meridian - A plane parallel to the Earth's axis of rotation and containing the vertical at the observer. Also called meridian plane and plane of the meridian.

plane of the geodetic meridian - That plane which contains the least (polar) diameter of an ellipsoid representing the Earth and the perpendicular to that ellipsoid at a given point.

plane of the horizon - (1) A plane through the celestial horizon. Also called a horizon plane and misleadingly, a horizontal plane. A horizontal plane is defined to be perpendicular to a specified vertical and to pass through a specific point on that vertical. Hence it in turn defines a unique great circle on the celestial sphere. The plane of the horizon, on the other hand, is defined to pass through a particular great circle on the celestial sphere; its intersection with the corresponding vertical is not defined thereby. (2) A plane passing through the average position of the apparent horizon. (3) A plane fitted in some suitable and defined manner to a defined horizon.

plane of the mean astronomic meridian - (1) A plane parallel to the average position of the Earth's axis of rotation and containing the average position of the vertical at the observer. (2) A plane parallel to the average position of the Earth's axis of rotation and containing the vertical at that point on the geoid vertically below the observer.

plane perspective - A perspective satisfying the conditions that (1) corresponding points are collinear with the center of perspective, and (b) corresponding lines intersect on the axis of homology. Figures projectively related in space also satisfy these conditions.

plane perspective map projection - A perspective map-projection from an ellipsoid onto a plane.

plane polarization - The act or process of modifying electro-magnetic radiation in such a way that the electro-magnetic field has the same value (intensity and phase) at all points in a plane perpendicular to the direction of propagation of the radiation. A more picturesque description would be that all the waves in the beam go up and down in planes parallel to each other and to the direction of propagation.

plane polygon - (1) A polygon lying entirely in one plane. (2) A portion of a plane bounded by straight lines. This definition and the preceding one are approximately equivalent. (3) The geometric figure formed by choosing a sequence of points $P_1, P_2, ... P_N$ on a plane, joining each point with the next in sequence by a straight line having the two points as end points, and joining the last point, P_N, to the first point, P_1, by a similar straight line. (4) A planar figure consisting of N straight lines and N points such that each line is included between just two points and each point lies on just two lines. Definitions (3) and (4) are equivalent for all practical purposes. The figure defined either way is also called a simple polygon. Each line (called a side) is a segment terminating at the two points (called vertices) on which it lies. A simple polygon divides the plane into two parts - a part (called the interior) internal to the polygon and a part (called the exterior) external to the polygon. The two parts are distinguished by the characteristic that any line through an internal point either coincides with a side of the polygon or intersects the polygon in exactly two points. If the polygon is a convex polygon, lines can be drawn through points in the exterior without intersecting the polygon. If the polygon is not convex, there is at least one point in the exterior through which lines can be drawn not intersecting the polygon. All points which can be connected to this point by segments not intersecting the polygon then constitute the exterior of the polygon. Angles measured in the interior, between sides, are called internal angles or interior angles; angles measured in the exterior, between sides, are called external angles or exterior angles. The sum of all internal angles is $\pi(N-2)$.

plane rectangular coordinate - One of a pair of coordinates in a rectangular coordinate system in the plane, i.e., in a plane rectangular coordinate system (1). In geodesy, plane rectangular coordinates are usually calculated from data given in the form of distances and directions (bearing or azimuth) between points. The methods used are those of plane geometry and trigonometry. The location of a point on the Earth may be given by plane rectangular coordinates on a plane tangent to the representing ellipsoid or by plane Cartesian coordinates on a cone or cylinder tangent or secant to that ellipsoid.

plane rectangular grid - The pattern formed by two families of straight lines in a plane, each line of one family intersecting each line of the other family at right angles. A rectangular grid in the plane. Also called simply, a rectangular grid.

plane survey - A survey done using a plane as the surface to which horizontal and vertical measurements are referred. SEE ALSO plane surveying.

plane surveying - That branch of surveying in which the surface chosen to represent the Earth's surface locally is a plane. In plane surveying, the Earth's curvature is ignored and the formulas of plane geometry and plane trigonometry are used in the computations. In general, plane surveying is used in surveying regions and boundaries whose greatest dimension is less than 150 km or where the accuracy required is so low that corrections for the effect of curvature would be negligible in comparison to the errors of measurement. For regions of small extent, accurate results can be obtained with the methods of plane surveying, but the accuracy and precision of such results will decrease as the region increases in size.

planet - A celestial body revolving around a sun in a nearly circular orbit. This definition excludes comets, which have highly elliptical orbits. However, it may also exclude some of the asteroids.

planet, giant - SEE planet, major.

planet, inferior - A planet closer, on the average, to the Sun than is the Earth: the planets Mercury and Venus.

planet, inner - One of the four planets closest to the Sun: Mercury, Venus, Earth and Mars. SEE ALSO planet, major.

planet, major - One of the four largest planets: Jupiter, Saturn, Uranus and Neptune. Also called a giant planet. SEE ALSO planet, inner.

planet, minor - (1) A planet having a mass equal to or less than that of the Earth. In the Solar System, this would be the planets: Mercury, Venus, Mars and Pluto. (2) SEE asteroid.

planet, navigational - One of the four planets commonly used for celestial navigation: Venus, Mars, Jupiter and Saturn.

planet, outer - Those planets which lie farther from the Sun than Mars: Jupiter, Saturn, Uranus, Neptune and Pluto.

planet, principal - The larger bodies revolving about the Sun in nearly circular orbits. The known principal planets, in the order of their distances from the Sun: Mercury, Venus, Earth, Mars, Jupiter, Saturn, Uranus, Neptune and Pluto.

planet, superior - A planet which is farther, on the average, from the Sun than is the Earth.

planet, terrestrial - A planet approximating the Earth in size and physical structure: Mercury, Venus, Mars and Pluto.

planetable - A drawing board mounted on a tripod, together with a straight-edge having a sighting device such as a telescope attached for pointing the straight-edge at the object observed. The planetable allows an observed direction to be plotted directly on a sheet fastened to the drawing board. The straight-edge is frequently graduated so that the distance to the object can be laid off to scale also. The term planetable is usually applied to the assemblage of drawing board, tripod and alidade (straight-edge plus sighting device) but may also be applied to the drawing board alone or to the combination of drawing board and tripod. SEE ALSO traverse planetable.

planetable revision - (1) The revision of a map by making a survey with a planetable. (2) Revision done directly on maps, or on transparent sheets overlying maps, fastened to a planetable and carried out by the usually procedures for surveying using a planetable.

planetable sheet - SEE field sheet (2).

planetable surveying - Surveying in which a line parallel to the line of sight from the observer to observed object (target) is drawn on a horizontal sheet of paper or plastic while the measurement is being made, using a straight-edge fixed so as to be parallel to the line of sight. The usual procedure is to fasten a sheet of paper to a board (the plane table) mounted on a tripod. A dot on the sheet indicates the observer's location. A telescope free to rotate in a vertical plane is attached by a vertical pedestal to a straight-edge which can be moved anywhere on the board. The straight-edge is positioned to pass through the point representing the observer, the telescope is pointed at the target and a short line drawn to pass through the presumed location of the point representing the target. This procedure is repeated for other targets and other points of observation, so that the desired locations of targets are determined as the intersections of lines on the sheet.

planetable surveying, topographic - Plane-table surveying in which vertical angles are also measured, using a telescope to which an angular scale is attached. Also called topographic surveying; that term, however, is ambiguous.

planetable traverse - A survey traverse made using a plane-table to determine angles and directions. A surveyor's tape is usually used to measure distances, although stadia may be used (with the proper type of alidade).

plane triangle - SEE triangle. All triangles are plane (i.e., planar) triangles.

planimeter - An instrument for measuring the area of a planar figure mechanically, by tracing the perimeter of the figure. Alternatively, an instrument which serves to determine the value of a definite integral mechanically by tracing the line which represents the function to be integrated. The curve need not be smooth but it must be smooth. The planimeter consists, in its simplest form, of a rod and the tracing arm, on one end of which is mounted the stylus which follows the perimeter. The other end of this arm may be constrained to motion along a straight line or it may be made to move in a circular path by being joined by a hinge to one end of another rod, the polar arm, whose other end is fix (to a pivot). The first form is called a linear planimeter; the second form is called a polar planimeter. Other common forms are the rolling planimeter and the suspended planimeter. The one most often used for calculating areas of regions on maps is the polar planimeter.

planimeter, linear - SEE planimeter.

planimeter, polar - SEE planimeter.

planimetry - (1) The theory and practice of measuring areas of figures on a plane. (2) The determination of horizontal distances, angles and areas from measurements on a map. (3) The determination of horizontal distances, angles or areas by photogrammetric methods. (4) Horizontal measurements. (5) The information contained, in a map, on horizontal distances and angles. (6) Those parts of a map which represent everything except relief, that is, those parts which represents the works of man and natural features such as woods and water.

planisphere - A representation, on a plane, of a family of circles (parallels, meridians, etc.) on a sphere. One particularly well-known form of planisphere is a device carrying a map obtained by polar projection of the celestial sphere and having movable parts adjustable to show the visible aspect of the heavens at any given time. It is used to select and identify stars for various purposes. The planisphere is the basis of the planispheric astrolabe. Many map projections produce graticules which are planispheres. SEE ALSO Aitoff's planisphere.

planisphere, Hammer's - SEE Hammer's map projection.

planning, land-use - Planning for those sort-term and long-term uses of land which will best serve the general welfare, together with formulating ways and means for achieving such uses.

planning commission - (1) A committee composed of citizens appointed by the chief executive or legislative body of a city, village or other governmental body, whose purpose is to review plans for subdivision submitted to them for approval and to advise upon, approve or otherwise act upon such applications. (2) A staff of professional planners hired by one or more governmental organizations to develop comprehensive plans, review plans for subdivision, study matters having to do with planning and advise the governmental organization on such matters.

plan-position-indicator - A display, on a cathode ray tube, produced by having the beam begin its sweep from the center of the screen and proceed in the direction corresponding to that of the main lobe of an antenna, and to a distance proportional to the time interval between the emission of a pulse from the antenna and the reception of the same pulse. Written as the acronym PPI or as the abbreviation P.P.I. Also written as plan position indicator. SEE ALSO radar, plan-position-indicator.

plastic (adj.) - The property, of a material, of remaining rigid under a sufficiently small stress, breaking if the stress is beyond a certain limit, yet distorting without fracture if the stress lies between those limits. SEE ALSO elastic (adj.).

plastic relief map - A relief map made by forming a plastic material into the desired shape.

plat - A diagram (map), drawn to scale, showing all essential data pertaining to the boundaries and subdivisions of a tract of land, as determined by survey or protraction. Also called plan, plot and protraction. A plat should show all data required for a complete and accurate description of the land which it delineates, including the bearings (or azimuths) and lengths of the boundaries of each subdivision. This is, in fact, the distinguishing characteristic of a plat as compared to other types of maps. A plat may constitute a legal description of the land and be used in lieu of a written description. (2) A map showing, accurately, the boundaries and dimensions of land and containing the essential data of a cadastral survey. (3) The map or drawing representing the particular region included in a survey, such as a township, private land claim or mineral claim and the lines surveyed, established, retraced or resurveyed; showing the direction and length of each line; the relation to the adjoining official surveys; the boundaries, descriptions and area of each parcel of land subdivided; and as nearly as may be practicable, a representation of the relief and improvements within the limits of the survey. A public-land survey in this sense does not obtain complete official or legal status until the field notes and the plat have been approved by the proper supervising officer and accepted by the Director of the Bureau of Land Management. SEE ALSO abstractor plat; base plat; master title plat; ownership plat; status plat; subdivision plat; survey plat; use plat.

plat, recorded - SEE subdivision plat.

plat, right-of-way - A plan of a highway improvement showing the old and new highways and the right of way (or interest in lands) to be acquired.

plat, supplemental (USPLS) - A plat prepared entirely from office records and designed to show a revised subdivision of one or more sections without change in the boundaries of the sections and without other modification of the subsisting record. Supplemental plats are required where the subsisting plat fails to provide units suitable for administration or disposal, or where a modification of its showing is necessary. It is also required to show the segregation of alienated land from public land, where the former is included in an irregular survey or a patented mineral or other private claim made subsequent to the plat of the subsisting survey, or where segregation of the claims was overlooked at the time of its approval.

plat book - A public record showing the location, size and name of owner of various recorded plats in the municipality or county.

plat description - SEE description by platting.

plate - (1) All states of an engraved map reproduced from the same engraved printing plate. (Brit.) (2) All detail to appear on a map which will be reproduced from a single printing plate, e.g., the blue plate or the contour plate. SEE ALSO color plate. (3) SEE plate, photographic. (4) SEE plate, printing. (5) SEE plate, lithospheric. SEE ALSO calibration plate; color plate; combination plate; compensation plate; foot plate; pressure plate; printing plate; process plate; surface plate.

plate, color-separation - A printing plate prepared from a drawing or scribed plate showing only one color and used for reproducing that part of the map to be printed in that one color.

plate, compensating - SEE compensation plate.

plate, correcting - An optical element introduced into an optical system to correct for small distortions in the objective lens-system of a camera, stereoscopic plotter or rectifier.

plate, etched zinc - SEE zinc plate, etched.

plate, ferrotype - A sheet of thin, enameled or chromium-plated iron or stainless steel used in making glossy prints.

plate, focal-plane - A glass plate set in a camera in such a position that the surface away from the lens system coincides with the focal plane. Also called a contact glass or contact plate. When the photographic film is mechanically pressed into contact with the focal-plane plate, the emulsion on the film is in the focal plane.

plate, land-mass-simulator - One of the plates used in a land-mass simulator. Also called a factored transparency.

plate, lithospheric - One of the large, continent-sized portions of the crust assumed to move as a whole over the mantle. Also called simply, a plate.

plate, parallel - SEE plate, plane-parallel.

plate, photographic - A thick, flat, rigid piece of material, usually made of glass, supporting a light-sensitive emulsion used for taking photographs. Commonly referred to as a plate when the meaning is clear. (2) SEE plate, press.

plate, plane-parallel - A plate, usually of glass, whose two planar surfaces are optically flat and parallel. Also called a parallel plate, although that term is ambiguous. Plane-parallel plates are used in astrometric telescopes, in optical micrometers and as the base for photographic emulsions when the highest precision is needed.

plate, scribed - SEE sheet, scribed.

plateau - An elevated, level expanse of land.

plateau, continental - A large, elevated part of the crust, coinciding approximately with a continent.

Plate Carrée map projection - (1) The pair of functions $x = k R \Delta\lambda$ and $y = k R \phi$ mapping the sphere onto the plane. x and y are the rectangular, Cartesian coordinates of that point, in the plane, corresponding to the point, on the sphere, at longitude $\lambda_o + \Delta\lambda$ and at latitude ϕ; λ_o is the longitude of the central meridian. R is the radius of the sphere and k is the scale factor. Also called the simple cylindrical map projection, cylindrical equidistant map projection, equally-spaced map projection and equi-rectangular map projection. This is the simplest and, after the gnomonic, the oldest of map projections, having been used by Anaximander (c. 550 B.C.) The term modified Plate Carrée map-projection is used for a map projection such that the spacing between lines representing parallels of latitude is constant but different from that between lines representing meridians. This is also called the equally-spaced map projection, equi-rectangular map projection, rectangular map-projection and simple rectangular map-projection. The most notable example of it is its use by Marinus of Tyre (about 100 B.C.), who used it for a map of regions in the eastern Mediterranean. x is given by $k R \Delta\lambda \cos \phi_o$, in which ϕ_o is some selected latitude. Scale is exact along the lines representing two non-equatorial parallels of latitude.

Plate Carrée map projection, modified - SEE Plate Carrée map projection.

plate constant - One of a set of constants in the pair of equations relating the standard coordinates of a celestial body to the rectangular, Cartesian coordinates of the photographic image of that body. SEE ALSO Turner's plate constant.

plate coordinate - One of a set of two coordinates, in a rectangular Cartesian coordinate system, giving the location of a stellar image on a photographic plate. SEE ALSO Turner's method.

platen - (1) That part of the tracing table, in a projection-type of stereographic plotting instrument, onto which the image is projected and which has on it the reference mark (floating mark). (2) That part of an orthophotoscope onto which the image is projected and which contains the slit. (3) SEE back, locating; vacuum back.

plate reduction - The process of deriving astronomical coordinates of a celestial object whose images appear on one or more photographs, by measuring the plane coordinates of that image and of images of stars (of known astronomical coordinates) on the same photographs.

plate tectonics - That part of geology which describes large-scale changes, such as folding and faulting, in the crust to the movement of large, continent-sized segments (called plates) of the crust, and which explains such changes in terms of the relative movements of the plates and the consequent pressure of plate upon plate.

platform - (1) A flat, level surface devoted to supporting certain kinds of activity. (2) A mounting, container, vehicle, ship, aircraft or other structure devoted to some particular use or activity. (3) A vehicle holding a sensor.

platform, aerial - The support for an aerial camera at a camera station.

platform, inertial - (1) A mounting, for an instrument, whose orientation with respect to a non-rotating coordinate-system is held fixed. There is no precise definition of a non-rotating coordinate-system in the physical sense. Such a system has been defined as one that is not rotating with respect to the average positions of the distant galaxies. However, the definition is frequently weakened by applying the term to any coordinate system with respect to which other coordinate systems are considered to be rotating. Orientation is usually fixed by coupling the support to two or more gyroscopes whose axes are mutually perpendicular. (2) A support, for an instrument, whose orientation with respect to a non-rotating coordinate-system is known.

platform, offshore - A structure constructed on pilings driven into the continental shelf or elsewhere where water is sufficiently shallow, and from which drilling is done or oceanographic observations are made. An example of this kind of platform is the Texas tower, a fixed tower constructed specifically for drilling into the continental shelf.

platform, stable - SEE mounting, stabilized.

platform, stabilized - SEE mounting, stabilized.

platform tide - SEE stand of tide.

Platonic year - SEE year, great.

plat statute - A state law adopted by a State's legislature to govern the process of subdividing land.

platting - The process of preparing a plat of a subdivision according to a State's statutes governing such subdivision. SEE ALSO description by platting.

Plessis ellipsoid - A rotational ellipsoid characterized by the following constants:
  length of semi-major axis (a)   6 376 985 meters
  length of semi-minor axis (b)   6 356 323 meters
  (calculated) flattening (f)   1/308.64
and used as part of a datum on which the triangulation of France was calculated. Also called the Carte de France ellipsoid.

plot (n.) - (1) A drawing composed of lines or points whose locations in the drawing have been determined from their coordinates. (2) A small parcel of land. (3) SEE plat. (4) SEE aerotriangulation, graphic radial.

plot (v.) - (1) To make a drawing by placing points on a sheet according to their coordinates. (2) To place data from a survey upon a map. In the past, no clearly defined difference existed between plat and plot. It has been suggested that a difference be established by limiting the term plat to the graphical representation of a survey (i.e., the map), and plot (in the geodetic sense) to the cartographic operations involved in constructing a map.

plot, direct-radial - SEE radial method, direct.

plot, hand-templet - SEE hand-templet method.

plot, mechanical-templet - SEE mechanical-templet method.

plot, minor-control - SEE triangulation, radial.

plot, nadir-point - SEE aerotriangulation, nadir-point.

plot, radial - SEE aerotriangulation, radial.

plot, slotted-templet - SEE aerotriangulation, radial; slotted-templet method.

plot, spider-templet - SEE aerotriangulation, radial; spider-templet method.

plot, topographic - (1) The representation, by means of contour lines, of the ground relief in a region, as shown in a stereoscopic model. (2) SEE compilation.

plottage damages - Damages to a parcel of land by reason of acquisition of such portion of it as to impair the vaile of the remainder because of the diminished size or resulting character of such remainder, in terms of highest or best use.

plotter - (1) An device capable of converting data in the form of coordinates into a graphical form as points, lines or surfaces. Alternatively, a device capable of preparing drawings from data given in the form of coordinates. Most plotters draw lines on a sheet according to instructions given by a computer or as forced by mechanical or optical linkages to a mark moving in a stereoscopic image. Some plotters create the drawing by focusing a directed beam of light onto a photographic emulsion. (2) A person who plots. (3) SEE plotting instrument.

plotter, analytical - A photogrammetric plotting instrument consisting of a stereoscopic measuring device connected to a computer which controls a plotter, the combination generating a map of the region viewed and measured. Also called an analytical plotting instrument.

plotter, double-projection direct-viewing - SEE plotting instrument, double-projection direct-viewing.

plotter, drum-type - A plotter in which the sheet on which the drawing is made is wrapped around a cylinder. One coordinate of a point is its distance from one edge of the cylinder; the other coordinate is the angle through which the drum must be rotated from a reference position. Also called a drum plotter.

plotter, flat-bed - A plotter in which the sheet on which the drawing is made is held against a rigid, flat plate. The coordinates of a point are the distances of that point from the axes of a rectangular, Cartesian coordinate system defined by a pair of mutually perpendicular, long screws which move a carriage on which the drafting tool is carried. The plotter consists of a flat, hard, rectangular plate mounted on a heavy, rigid stand. A beam moves over the surface at right angles to a guide fixed along one edge of the plate. A carriage holding pen, pencil, scriber or other drafting tool moves along the beam. Both motions are given by very accurately machined screws. The distance moved is determined in some plotters by handwheels turning the screws; it is determined in others by computers controlling electric motors connected to the screws. The sheet of paper or other material is held flat against the plate by air pressure or electrostatically (or by clamps in less expensive machines).

plotter, orthographic - SEE plotting instrument, orthographic.

plotter, paper-print - SEE plotting instrument, paper-print.

plotter, photogrammetric - SEE plotting instrument, photogrammetric.

plotter, planimetric - SEE plotting instrument, planimetric.

plotter, radial - SEE plotting instrument, radial.

plotter, radial-line - SEE plotter, radial.

plotter, rectangular-coordinate - SEE coordinatograph.

plotter, rectoblique - SEE angulator.

plotter, single-model stereoscopic - SEE plotting instrument, single-model stereoscopic.

plotter, stereoblique - SEE plotting instrument, stereoblique.

plotter, stereoscopic - SEE plotting instrument, stereoscopic.

plotting instrument - An instrument combining a stereoscopic measuring device with either a plotter (for converting the measurements into a drawing (map) or a recorder (for recording the measurements in a form suitable for plotting later). Also called a plotting machine. A plotting instrument is also, and perhaps more frequently, called a plotter. However, it is useful to retain the nomenclature distinguishing between the plotter, which merely draws, and the plotting instrument, which both measures and (directly or indirectly) draws. Most plotting instruments are designed for use in photogrammetry. They are classified as either analog plotting instruments or as analytic plotting instruments (analytic plotters), according as the connection of the measuring device to the plotter is direct, through mechanical and optical linkages, or indirect through a computer.

plotting instrument, analytical - SEE plotter, analytical.

plotting instrument, automatic stereoscopic - A stereoscopic plotting instrument which does not require a human operator to view the photographs or stereoscopic images during the measuring and plotting.

plotting instrument, double-projection direct-viewing - A stereoscopic plotting instrument consisting of a pair of projectors places to project light through photographic transparencies onto a flat surface. The relative positions of the two projectors are adjustable. One photographic transparency (diapositive) of a stereoscopic pair is placed in each projector and the images projected onto a flat surface. Each projector emits light of a different color (red or blue-green, usually) or light of a different polarization. The projectors are adjusted until they occupy the same relative positions, except for scale, as did the cameras when the pictures were taken. The projected images, when viewed through filters of the proper colors or polarizations, produce a stereoscopic model on which measurements can be made directly.

plotting instrument, Multiplex - SEE Multiplex.

plotting instrument, oblique - SEE oblique plotting instrument.

plotting instrument, orthographic - A stereoscopic plotting instrument designed to draw maps or portions of maps from aerial photographs, correcting for tilt of the photographs and simultaneously providing an orthographic projection at the required scale. Also called an orthographic plotter.

plotting instrument, paper-print - A plotting instrument designed to draw maps or portions of maps from photographic photographs on paper (paper prints). Also called a paper-print plotter.

plotting instrument, photogrammetric - A plotting instrument designed for photogrammetric use. Also called a photogrammetric plotter.

plotting instrument, planimetric - A plotting instrument designed for drawing maps from aerial photographs but showing no information on heights or elevations. Alternatively, a plotting instrument designed for plotting planimetric detail from aerial photographs. Also called a planimetric plotter.

plotting instrument, radial - A stereoscopic plotting instrument in which an overlapping pair of photographs is viewed stereoscopically and the planimetric details in the part common to the pair are drawn on a map or base sheet through a mechanical linkage embodying the radial-line principle. Also called a radial-line plotter.

plotting instrument, single-model stereoscopic - A stereoscopic plotting instrument able to accommodate only one stereoscopic model at a time. The term is applied primarily to double-projection, direct-viewing plotting instruments.

plotting instrument, stereoblique - A stereoscopic plotting instrument consisting of two photoangulators linked under a stereoscope and connected by mechanical linkage to a plotter.

plotting instrument, stereoscopic - A plotting instrument allowing measurements to be made on a stereoscopic model and the transfer of these measurements by mechanical or optical linkages to the plotter. Also called a restitutive instrument, stereoscopic plotter, stereoplotter and stereo plotter. Most modern stereoscopic plotting instruments not only direct the motion of the plotter but also display and/or record the measurements numerically.

plotting machine - SEE plotter; plotting instrument, stereoscopic.

plotting table - The flat surface, in a stereoscopic plotting instrument, on which is placed the sheet on which a map is to be drawn. In particular, the flat surface on which the tracing table used by Multiplex moves about.

Plücker coordinate - One of an ordered set of (N+1) coordinates defining a M-dimensional space (e.g., point, line, M+1 plane) in a N-dimensional space. In terms of the rectangular, Cartesian coordinates of the M-dimensional space, a Plücker coordinate π (ijk...n) is the determinant whose rows consist of the M+1 coordinates i,j,k,...n, in N-dimensional space, of the M+1 points defining the M-dimensional space. The set of Plücker coordinates of a space may be multiplied by an arbitrary constant without affecting the space. Plücker coordinates are useful when one wants to treat points, lines and planes as interchangeable concepts, e.g., in working with projective transformations.

plumb (adj.) - Conforming to the direction of a line attached to a plumb bob; vertical.

plumb (n.) - SEE plumb bob.

plumb (v.) - (1) To adjust or test using a plumb line. (2) To cause to be vertical or perpendicular.

plumb bob - A small mass or weight of lead or other heavy material (commonly brass and conical) attached to a line and used to indicate the direction of gravity (i.e., the vertical) over short vertical distances. Also called a plumb and plummet.

plumb line - (1) A length of string (or wire) to one end of which a small, pointed weight of heavy material (usually brass in a conical shape) is attached. When the weight (the plumb bob) is hanging suspended by the string, the string indicates the direction of gravity at the tip of the weight. More exactly, it indicates an average direction of gravity between the point of suspension and the weight. (2) The average direction indicated by a length of string (or wire) to one end of which is attached a heavy weight, when the weight is allowed to hang freely and the other end of the string is fixed. The direction is customarily assumed to be downward from the point of suspension. However, this convention is often ignored and the context should always be studied to determine the exact definition. (3) A line, the tangent to which at each point indicates the direction (or opposite direction) of gravity at that point. The tangent is also called the vertical at the point when it is in the direction in which the gravity is decreasing. SEE ALSO vertical. (4) SEE direction of gravity. SEE ALSO vertical. (5) SEE vertical.

plumb line deflection - SEE deflection of the plumb line; deflection of the vertical.

plumb point (of a photograph) - SEE photograph nadir.

plummet - SEE plumb bob; laser plummet.

plummet, optical - A small, telescope having a 90° bend in its optical axis and attached to an instrument in such a way that the line of sight proceeds horizontally from the eyepiece to a point on the vertical axis of the instrument and from that point vertically downwards in a properly adjusted and levelled instrument. In use, the observer, looking into the plummet, brings a point on the instrument vertically above a specified point (usually a geodetic or other mark) below it. An optical plummet is not affected by wind and is therefore superior to the plumb line in this respect. Most modern theodolites have an optical plummet built into the base of the instrument, so that the upright section of the optical axis of the plummet coincides with the vertical axis of the theodolite. The eyepiece is usually located near the base. SEE ALSO collimator, vertical.

plunge (v.) - To reverse or transit the direction of a telescope by rotating it about its horizontal axis.

Pockels cell - A material placed between a pair of conducting plates and having the property that a plane-polarized wave incident on it becomes elliptically polarized when a voltage is applied across the plates.

pocket compass - SEE compass, prismatic.

pocket transit - A small compass having circular and vial levels, folding sights and a pendulum-type clinometer.

Poincaré-Prey gravity gradient - The quantity -0.0848 mGal/m.

Poincaré-Prey gravity reduction - SEE Prey's gravity reduction.

point - (1) In surveying, a geometric point or a pointlike physical mark whose location on the Earth has been determined by surveying. Also called a fix or a station. (2) In law, an extremity of a line. SEE ALSO angle point; annex point; control point; detail point; ground control point; Laplace point; orientation point; pass point; pass point, vertical; picture point; tie point; turning point; wingpoint; witness point.

point (of a switch) - SEE point of switch.

point, amphidromic - (1) A point at which the range of the tides is zero and from which the co-tidal lines radiate, progressing through all phases of the tidal cycle. Also called a nodal point. (2) A point where the vertical amplitude of a certain tidal constituent vanishes. Also called an amphidrome.

point, anallactic - That point, on the optical axis of a telescope, for which a proportionality exists, for a constant size of image, between the distance of an object and its size. Also called anallatic center and anallatic point. For telescopes of usual form, the anallactic point and the focal point coincide.

point, antisolar - That point, on the celestial sphere, which is 180° from the center of the Sun.

point, aplanatic - One of a pair of points, one in object space and one in image space, for which there is no spherical aberration. Alternatively, one of a pair of points, on the optical axis of an optical system, such that rays radiating from one of the points appear to converge to the other point.

point, astrogravimetric - A point whose astronomic longitude and latitude have been determined and corrected to geodetic coordinates by subtracting amount of the deflection of the vertical.

point, back focal - SEE point, first focal.

point, cardinal - (1) One of the four directions (North, East, South and West) indicated on a compass. (2) One of the four directions: North, East, South and West. Sometimes called a cardinal (jargon). (3) One of a set of points used in tracing the paths of rays through an optical system. The cardinal points usually used are the principal points, the nodal points and the focal points. These points are very helpful in designing a lens system and in predicting the performance of an ideal optical system.

point, classical - One of six symmetrically placed points, in a stereoscopic model, at which y-parallax is removed during relative orientation of the photographs. Four of the points are near the corners of the model; the other two are near the x-axis and the boundaries of the model.

point, central - SEE point, principal (3).

point, critical - A peak or high ground with abrupt local relief and requiring investigation to avoid "hidden ground" in aerial photographs.

point, curve-spiral - The point of tangency common to a circular curve and a spiral where the circular curve ends and the spiral begins. Also called C.S. and curve to spiral.

point, distant - A point similar to a tie point but appearing only on the outward-facing, oblique photographs used on the perimeter of a compilation manuscript. Also called a distance point.

point, emergent nodal - SEE point, second nodal.

point, equinoctial - SEE equinox.

point, first focal - That focal point to which rays entering parallel to and near the optical axis of an optical system converge or appear to converge. Also called the back focal point and rear focal point. By convention, object-space is placed to the left and image-space to the right in tracing a ray through an optical system, so that the first focal point is defined by rays entering the optical system from the left.

point, first nodal - That nodal point at which rays enter the optical system. Also called the front nodal point and the incident nodal point. SEE point, front nodal.

point, first (second) principal - That point in which the first (second) principal plane intersects the optical axis. The first (second) principal plane is usually that closest to the object (image).

point, focal - That point to which monochromatic rays entering an optical system parallel to and near the optical axis converge, or appear to converge after leaving the system. Also called a focus point. There are actually two focal points: the back focal point and the front focal point; which is found depends on which of the two ends of the optical system is chosen for the rays' entrance. However, the one chosen is usually that through which rays enter when the instrument is used normally. E.g., it would be that end of a telescope which contains the objective lens.

point, front focal - SEE point, second focal.

point, front nodal - SEE nodal point, first.

point, horizontal-control - A control point whose longitude and latitude, but not necessarily its elevation or height, have been defined or determined. Also called a horizontal-control station.

point, incident nodal - SEE nodal point, first.

point, incident nodal - SEE point, first nodal.

point, initial (USPLS) - The point from which starts the survey of the principal meridian and base line controlling the survey of the public lands (of the USA) within a given region.

point, intercardinal - Any one of the four directions midway between the cardinal points: northeast, southeast, southwest, northwest.

point, intersected - SEE station, intersected.

point, mid - SEE point, principal (3).

point, middle - That point, on a circular curve, which is at the same distance from both ends of the curve. Abbreviated as M.P.

point, nadir - SEE nadir; photograph nadir.

point, near - That object which is closest to the camera and still seen with acceptable sharpness when the camera is focused for a given distance.

point, nodal - (1) One of a pair of points, in an optical system, so placed that a ray oblique to the optical axis and directed at one of them appears to emerge from the other point parallel to its original direction. (2) One of a pair of points, in a lens system, such that, when all distances D_o to objects and all distances D_i to the corresponding images are measured, these distances satisfy the formula $1/f = 1/D_o + 1/D_i$, in which f is the focal distance. Equivalently, one of the two points, on the optical axis of a lens system, at which a ray emerging from the second point is parallel to the ray incident at the first. Also called a node. The first nodal point is also called the front nodal point, incident nodal point or nodal point of incidence; the second point is called the rear nodal point, emergent nodal point or nodal point of emergence. (3) SEE amphidromic point.

point, peripheral - A point, in the vicinity of a control point, which is intended for use with the control point but is not itself a control point. Reference marks and azimuth marks are examples.

point, photogrammetric - SEE point, pass (1).

point, photogrammetric control - SEE control-point, photogrammetric.

point, plus - An intermediate point, on a course of a traverse, located by a plus distance from the beginning of the course.

point, principal - (1) The point in which a principal surface intersects the optical axis of the optical system. (2) The foot of the perpendicular from the interior perspective center to the plane of the photograph. (3) That point at which lines through corresponding fiducial marks on a photograph intersect, or the point taken as the average location of such points of intersection. This is merely an approximation to the principal point as defined in (2), and is better termed the mid- point or central point. (4) That point at which the optical axis of the lens system of a projector intersects the plane of the image being projected.

point, rear nodal - SEE point, second nodal.

point, resected - That point, in resection, from which the angles are measured and whose location is to be determined.

point, reversing - SEE reversing point (of a micrometer screw).

point, second focal - That focal point from which rays leaving parallel to and near the optical axis of an optical system diverge or appear to diverge. Also called the front focal point. By convention, object-space is placed on the left and image space on the right when tracing a ray through an optical system, so that the second focal point is defined by rays leaving the optical system to the right.

point, second nodal - That nodal point at which rays emerge from the optical system. Also called the emergent nodal point and rear nodal point.

point, singular - SEE coordinate system.

point, solsitial - SEE solstice.

point, south - The southern intersection of the celestial meridian with the horizon.

point, spiral-curve - The point of tangency common to a spiral and a circular curve, where the spiral ends and circular curvature begins. Abbreviated to S.C.

point, spiral-tangent - The point where a spiral ends and a tangent, straight line begins. Abbreviated to S.T.

point, stationary - The position, of a planet, at which the rate of change of the apparent right ascension is momentarily zero. This phenomenon results from the relative motion of the planet and the Earth, rather than from the inherent heliocentric motion of the planet in its orbit.

point, subastral - SEE point, substellar.

point, sublunar - That point, on the Earth, at which the Moon is at the zenith at a specified time. SEE ALSO point, sub-satellite.

point, sub-satellite - (1) The point which is the foot of a perpendicular from a satellite (or the center of the satellite) to an ellipsoid representing the Earth. (2) A point vertically below a satellite and on the Earth's surface.

point, sub-solar - That point, on the Earth, at which the Sun is at the zenith.

point, supplemental control - SEE control point, supplemental.

point, tangent-spiral - The point where a straight line ends and a tangent spiral begins. Abbreviated to T.S.

point, triple (of water) - The temperature at which or state in which ice and air-saturated water are in equilibrium under a pressure of 101 325.0 pascals (1 standard atmosphere). Also called the ice point. The water should have the isotopic composition of oceanic water. The temperature of the mixture is, by definition, 273.16 ° K or 0.0100 ° C.

point, turning - SEE turning point.

point, vanishing - The point, in image-space, toward which the images of a family of parallel lines in object-space converge.

point, vertical-control - SEE control point, vertical.

point anomaly - SEE gravity anomaly, point.

point base - A manuscript which contains radial centers, picture points, pass points, control points and tie points from the photographs used in radial aerotriangulation.

Pointers - The second-magnitude stars alpha and beta Ursa Majoris (in the Big Dipper). Usually called the Pointers. A line drawn through the Pointers points to Polaris. They are in the outer side of the bowl of the Big Dipper, away from the handle.

point gravity - The value of gravity at a specific point.

pointing (v.) - (1) Placing the reticle or index mark of a precise measuring instrument, such as a measuring engine, so that it appears to rest on the geometric center, or center of density, of an image whose coordinates are to be determined. (2) Moving the tracing table of a stereoscopic plotting instrument to specific points in the image during orientation of a stereoscopic model. (3) Placing the line of sight of a telescope or other device so that it hits a desired spot.

pointing accuracy - The exactness, in surveying or photogrammetry, with which the line of sight can be directed toward a target or a floating mark.

pointing error - The error committed in placing the floating mark of a stereoscopic plotting instrument on a sharply-defined point of the model.

pointing the instrument - Turning the surveying instrument to a position where the crosshairs or the central mark on the reticle are accurately aligned with the target.

point mass - A geometric point to which is assigned a mass. Point masses do not exist in nature; a point mass would have infinite density. A black hole is a close approximation to a point mass.

point-mass anomaly - SEE representation of gravity anomaly, point-mass.

point-mass anomaly model - SEE representation of gravity anomaly, point-mass.

point-matching method (of orienting a rectifier) - SEE point-matching method (of rectification).

point-matching method (of rectification) - A method of removing the effects of tilt in a photograph by manually matching the image points projected using an autofocus rectifier with those image points plotted in their correct horizontal locations on a film used as templet. Also called the point-matching method of orienting a rectifier.

Point of Aries, First - SEE First Point of Aries.

point of autocollimation, principal - That point, in the focal plane (or surface on which photographic film is placed), which coincides with its image formed by reflection in a plane mirror in object space, the mirror being accurately parallel to the focal plane.

point of beginning - In a description of a survey by metes-and-bounds, the first point mentioned on the boundary of the property being described. After passing through the successive courses, the description returns to the point of beginning. When descriptions start at a reference point not contiguous to the property being described, the starting point is called the point of commencement, and the description starts from the point of commencement and proceeds to the point of beginning.

Point of Beginning - That point, on the western boundary of the State of Pennsylvania at the northern bank of the Ohio River, which is the point of beginning for the survey of the public lands of the USA. Also called The Point of Beginning. The point was marked, on 20 August 1785, by a stake.

point of certainty - In a simple problem of intersection from two points, the point where the two intersecting rays cross, if the location of the point is confirmed by the intersection there of a third (check) ray.

point of commencement - In a description of a survey by metes and bounds, the point with which the description starts if that point is not part of the boundary. Also called a point of commencing. Note that the point of commencement cannot be the same as the point of beginning.

point of commencing - SEE point of commencement.

point of compound curvature - The point at which a circular curve of one radius is tangent to a circular curve of a different radius, both curves lying on the same side of their common tangent. Abbreviated as P.C.C.

point of curvature - The point, on a line, at which a tangent ends and a circular curve begins. Also called the point of curve (P.C.) and beginning of curve (B.C.). It is the point at which a straight line, in a survey, changes to a circular curve. SEE ALSO point of tangency.

point of curve - SEE point of curvature.

point of cusp - The point at which two curves are tangent, the curves proceeding in opposite directions from that point. I.e., if one proceeds along one curve towards the point, such motion ends at the point and must proceed in the opposite direction along the other curve.

point of emergence, nodal - SEE point, nodal.

point of frog - That part of a frog lying between the extensions of the gage lines from their intersection toward the part farthest from the switch (the heel end). Also called frog point.

point of incidence, nodal - SEE point, nodal.

point of intersection - The point at which the two tangents at the extremities of a circular arc intersect. Also called the vertex of curve. Abbreviated to P.I.

Point of Libra, First - First Point of Libra.

point of observation - A point at which one or more observations are made. More precisely, a point at which one or more observations are considered to be made.

point of perspective - SEE center, perspective.

point of reverse curvature - The point where two circular curves are tangent, the two curves lying on opposite sides of their common tangent. Abbreviated to P.R.C.

point of suspension - That point, in a pendulum, at which the moving is suspended, i.e., that point about which the pendulum swings. A real pendulum is supported along a surface, not at a point. However, the surface is usually so narrow that it can be considered a line. Because the pendulum is symmetric with respect to a plane perpendicular to that line, the real pendulum can be replaced, in theory, by a planar figure which is supported at a point in it.

point of switch - That point at which one rail of a railroad track separates from another and diverges to divert a train from one track to another. Often referred to, simply, as point.

point of symmetry - That point, in the focal plane of a lens system, about which all distortions are symmetrical. If the lens system in a camera were perfectly made and mounted, the point of symmetry would coincide with the principal point.

point of tangency - The point where a circular curve ends and a tangent begins. Abbreviated to P.T. The point of tangency and point of curvature are both points of tangency to a curve, their different designations being determined by the direction of progress along the line. The point of curvature is reached first.

point of tangent - SEE point of tangency.

point of the compass - One of 32 equally-spaced marks, on a compass, indicating directions. The points are 11.25° apart. The four points 90° apart and labeled N(orth), E(ast), S(outh) and W(est) are called cardinal points.

point of vertical curve (PVC) - The point at which a line changes from a line of uniform slope to a vertical curve.

point of vertical tangent (PVT) - The point at which a curve changes from a vertical curve to a line of uniform slope.

point of zero distortion - One of those points, on a graticule, at which the principal scale is preserved.

point positioning - (1) The process of using a single receiver to receive signals from TRANSIT satellites, and using the resulting data to determine the location of a specific point (such as the phase center) on the receiver's antenna. SEE ALSO translocation. (2) In general, any method by which the location of a point is determined from data obtained at that point only.

point symbol - A symbol used to indicate that a particular phenomenon occurs at, or a particular value may be attributed to, a region represented by a point on a map.

Poisson distribution - A distribution characteristic of a random variable showing a binomial distribution such that, as the number n of events becomes very large, the probability p of a single event's being successful approaches zero in such a way that np remains constant. The distribution is characteristic of certain random variables that occur frequently in physics and other sciences, e.g., in the theory of nuclear disintegration, and in the theory of traffic control. Some errors occurring in surveying also have such a distribution.

Poisson frequency function - The function $f(x) = (e - x_i x_i x) / x!$ which gives the frequency f(x) of x discrete occurrences of an event in a very large number N of trials, if the frequency a of the events occurring in one trial is very small and $Na = x_i$.

Poisson ratio - The ratio of the transverse (contracting) strain to the longitudinal (stretching) strain when a rod is stretched by forces applied to its ends and parallel to the rod's axis. Also called Poisson's ratio. The Poisson ratio σ is related to Lame's constant λ and μ by the equation $\sigma = \lambda / [2(\lambda + \mu)]$.

Poisson's equation - (1) The second-order, partial-differential equation usually written as $(\delta^2 V/\delta x^2) + (\delta^2 V/\delta y^2) + (\delta^2 V/\delta z^2) = -4\pi Q(x,y,z)$, in which x,y,z are rectangular, Cartesian coordinates, V is a scalar function of the coordinates and Q(x,y,Z) is a scalar function of location. In theories of the potential, Q is usually the density ρ of mass or charge at x,y,z. (2) Any equation into which the above equation can be transformed by changing the coordinate system. In vectorial notation, Poisson's equation is then $*\Delta^2 V = -4\pi Q(q_1, q_2, q_3)$ or $*\Delta^2 V = -4\pi \rho$. Poisson's equation is a generalization of Laplace's equation. Its solution gives the potential V determined from the distribution of mass (or electrical charge, etc.). In geodesy, V is usually the gravitational potential of a body. If the coordinate system is rotating with respect to a stationary coordinate system, the right-hand side of the equation is augmented by terms representing the effects of centrifugal force, Coriolis force, etc.

Poisson's formula - The formula $V = (1/G) (\Delta B/\Delta \sigma) (\delta U/\delta i)$ for the anomalous magnetic potential V in terms of gravitational anomaly U, the anomalous magnetization B of the source and the direction i of that magnetization and the anomalous density σ of the source. G is the gravitational constant.

Poisson's ratio - SEE Poisson ratio.

Poisson term - A term, in the representation of an orbit by a Fourier series, which is the product of a periodic term and the time.

polar (adj.) - In cartography, having the center of the map at a point representing a pole.

Polaris - The second-magnitude star α Ursa Minor in the constellation Ursa Minor (the Little Dipper). Also called the Pole star and North Star, because of its proximity to the north pole of the celestial sphere. Polaris is well situated for the determination of astronomical azimuth and the direction of the celestial meridian. It is at the extreme outer end of the handle of the Little Dipper. SEE ALSO Pointers.

Polaris correction - A quantity to be added to the measured angular elevation of Polaris to obtain the latitude.

Polaris hour-angle method (of determining azimuth) - A method of determining azimuth by the astronomic-direction method or by the micrometer method, using Polaris as the star observed.

Polaris method (of determining azimuth) - One of several closely related methods of determining azimuths by measuring horizontal angles from the direction of Polaris and either applying corrections for the distance of Polaris from the celestial pole or suitably averaging several angles measured at different directions of Polaris symmetric about the pole. SEE ALSO Polaris hour-angle method (of determining azimuth).

Polaris method (of determining latitude) - SEE Polaris-observation method (of determining azimuth).

Polaris-observation method (of determining azimuth) - A method of determining latitude by measuring the angular elevation of Polaris (α Ursa Minor) at a specific time. At the moment that Polaris attains its greatest (or least) angular elevation as seen by the observer, its angular elevation is equal to the latitude of the observer plus (or minus) its declination. The latitude can be determined by observing Polaris at a time other than that of culmination if the hour angle of Polaris at that time is also determined or known.

polarization - The state, or the process of bringing about that state, of a varying electromagnetic field in which the electric and magnetic vectors at each point rotate in a plane perpendicular to the direction in which the field appears to be moving. SEE ALSO plane polarization.

polarization filter - SEE filter, polarizing.

Polastrodial - A mechanical device, developed in the U.S. Geological Survey, for determining graphically the angle between Polaris and the observer's meridian at any hour angle anywhere in the Northern Hemisphere.

pole - (1) One of the two points in which an axis of rotation intersects the figure generated by the complete rotation of a closed, planar curve about that axis. In particular, one of the following: (a) Either of the two points in which the minor axis of an oblate rotational ellipsoid intersects the ellipsoid; (b) Either of two points equidistant from a given great circle on the sphere; (c) Either of the two points (North Pole and South Pole) equidistant from the equator. An equivalent, and perhaps better definition, is: either of the two points in which the Earth's axis of rotation intersects the Earth's surface; (d) SEE pole, celestial. (2) A unit of length, in English measure, legally equal to 5½ yards. Also called a rod or perch. The pole is no longer used but was common in England and in the original thirteen English colonies in descriptions in deeds. In the nineteenth century, surveyors of the public lands were required to use chains 2 poles long in surveying. These were later replaced by four pole chains. (3) The origin of a polar coordinate system. (4) Any point, on a surface, designated as a pole. (5) Any point around which something centers, e.g., the ice pole. (6) SEE range pole. (7) SEE pole, magnetic. SEE ALSO pole of the ecliptic; ice pole; North Pole; north geomagnetic pole; sounding pole; South Pole.

pole, average terrestrial - The average location, on the Earth, of one of the points in which the Earth's axis of rotation intersects the surface, averaged over a specified period of time. SEE ALSO Conventional International Origin.

pole, celestial - (1) Either of the two points at which the Earth's axis of rotation, extended, intersects the celestial sphere. (2) Either of the two points in which a body's axis or rotation intersects the celestial sphere. Equivalently, either of two points equidistant from a celestial equator.

pole, depressed - That one of the celestial poles which is below the observer's horizon.

pole, ecliptic - SEE pole of the ecliptic.

pole, elevated - That one of the celestial poles which is above the observer's horizon.

pole, fictitious - One of the two points 90° from a fictitious equator. It may be called an oblique pole or a transverse pole, depending on the type of fictitious equator used.

pole, galactic - Either of the two points 90° from the galactic equator.

pole, geographic - SEE pole, geographical.

pole, geographical - Either of the two points on the Earth at which the Earth's rotational axis intersects the surface. Also called the geographic pole and terrestrial pole. SEE ALSO pole, average terrestrial; Conventional International Origin.

pole, geomagnetic - Either of two points marking the intersection of the Earth's surface with the extended axis of a hypothetical, powerful bar-magnet assumed to be located at the center of the Earth and having a field approximating the actual magnetic field of the Earth. The geomagnetic pole should not be confused with the magnetic pole, which relates to the actual magnetic field of the Earth. The geomagnetic poles were located (1966) approximately at 78.5° N, 69.0° W and at 78.5° S, 111.0° E. The intensity of the field is about 65 µT near the poles and 25µT near the equator.

pole, lining - SEE range rod.

pole, magnetic - Either of the two spots on the Earth's surface at which a magnetic needle, free to rotate in a vertical plane, points vertically. Also called a dip pole. The poles are known as the North Magnetic Pole (at approximately 73° 25' West, 92° 20' North) and the South Magnetic Pole (at approximately 148° East, 70° South). Lines of equal magnetic declination converge on the poles and lines of equal magnetic inclination encircle them. The magnetic poles should not be confused with the geomagnetic poles.

pole, north - SEE North Pole.

pole, oblique - Either of the two points 90° from an oblique equator.

pole, ranging - SEE range rod.

pole, sounding - SEE sounding pole.

pole, south - SEE South Pole.

pole, south geographical - SEE South Pole.

pole, south geomagnetic - That geomagnetic pole which lies in the southern hemisphere. It is 180° from the north geomagnetic pole.

pole, terrestrial - SEE pole, geographical.

pole, transverse - A point 90° from a transverse equator.

pole of figure - One of the points in which the axis of figure of a celestial body intersects the celestial sphere. SEE ALSO axis (2).

pole of rotation - One of the points in which the axis of rotation of a celestial body intersects the celestial sphere. SEE ALSO axis (3).

pole of the ecliptic - Either of the two points, on the celestial sphere, 90° from the plane of the ecliptic.

Pole star - SEE Polaris.

pole tide - A tide caused by the Chandlerian motion of the Earth, having a period of 433 - 437 days and an amplitude of about 6 mm in the oceans.

polhode - The line in which the cone traced out by the angular-velocity vector of a rotating, rigid body not under external torque intersects the momenta ellipsoid (the ellipsoid of Poinsot). SEE ALSO herpolhode.

polhody - A graph of the movement of the instantaneous axis of rotation of a rigid body as a function of time. SEE herpolhode.

polygon - (1) A geometric figure consisting of N intersecting lines on a surface and their points of intersection. (2) The geometric figure formed by choosing a sequence of points $P_1, P_2, \ldots P_N$ on a surface, joining each point with the next in sequence by a straight line having the two points as end points, and joining the last point, P_N, to the first point, P_1, by a similar straight line. (3) The geometric figure formed as in the preceding definition (2), but with the proviso that no two lines intersect except at points of the chosen sequence. The lines are called the sides of the polygon; the points are called the vertices of the polygon. (4) SEE plane polygon.

polygon, concave - A plane polygon one or more internal angles of which are greater than 180°. Also called a re-entrant polygon.

polygon, convex - A plane polygon none of whose angles is greater than 180°. Alternatively, a plane polygon lying entirely on one side of any line containing a side of the polygon. This form of the definition can easily be extended to cover non-planar convex polygons.

polygon, curvilinear - A closed figure, on a surface, all of whose sides are geodesics.

polygon, directed - A polygon each of whose sides has a direction associated with it. Also called a traverse. Equivalently, a polygon whose vertices are numbered and always taken in a specified order.

polygon, re-entrant - SEE polygon, concave.

polygon, regular - A plane polygon having all sides and all internal angles equal.

polygon, spherical - A closed figure, on the sphere, composed of connected arcs of (usually) great circles intersecting only at the ends of the arcs.

polygonization - SEE traversing.

polyhedron - A convex, closed surface consisting of 4 or more planes intersecting in straight lines which intersect in points. The planes are called faces of the polyhedron, the straight lines are called edges, and the points are called vertices.

polyhedron, regular - A polyhedron all of whose faces form congruent regular polygons.

pond, great - In Maine and Massachusetts, a natural pond (lake) having an area of more than 10 acres.

Porro-Koppe principle - The principle, used in some photogrammetric plotting instruments, that the effects of distortion in a lens system can be minimized by using lens systems in the plotting instrument which have the same distortion as the lens systems in the camera used for taking the original photographs. The plotting instrument, which may have one projector or two, then reverses, in effect, the optical and geometric relationships existing when the photograph was taken and creates, according to the law of reversibility of light, a distortion-free image of the scene photographed. The focal plane replaces the original scene photographed. It is imaged at infinity by parallel rays emerging from the lens system. The chief ray assumes its correct direction, and the cone of rays is identical to that which had its vertex at the first nodal point of the camera's lens-system at the instant of taking the photograph. The parallel rays can be observed by means of a telescope focused at infinity and rotatable about the first nodal point of the lens system.

Porro prism - A prism consisting of two right-angled prisms cemented together so that the prism inverts an image and deviates the axis of the beam through 180°. If light enters through the diagonal side, the prism is sometimes called a single Porro prism.

Porro prism, single - SEE Porro prism.

port - A harbor which has wharves (or piers) and such buildings and amenities as are needed for berthing vessels, storing cargo, etc.

port, standard - SEE reference station.

portal - (1) A more or less horizontal entrance to a mine. (2) An underground passage used as a road for hauling or ventilation. Also called an entry.

Port Clarence Datum - The horizontal-control datum defined by the following coordinates of the origin and azimuth (clockwise from South) to the azimuth mark on the Clarke spheroid of 1866; the origin is at Port Clarence astronomical station:

longitude of origin	166° 50' 08.065" W
latitude of origin	65° 16' 40.18" N
azimuth from origin to azimuth mark	0° 06' 17.0"

The longitude is an assumed value, the astronomically determined value being 166° 50' 45.60"; the latitude was determined astronomically in 1900. The azimuth was derived from observations of time.

portolan - SEE portolan chart.

portolan chart - A chart of the sea and adjacent coastlines which is not drawn on a specific map projection and whose outlines are drawn over a network of rhumb lines corresponding to the directions of the winds and which show approximate compass bearings and estimated distances. Portolan charts date from the 13th to 17th centuries. The term originally described written sailing instructions for sailors, sometimes accompanied by charts. In the late 19th century it was mistakenly applied to the charts themselves, and in particular to the earlier charts of the 13th to 15th centuries.

port plan - A large-scale map of a port and its immediate neighborhood, showing piers, railheads, repair shops, pilots' office, customhouse and other features not involved in navigation.

position - (1) A numerical or other description of the location and orientation of an object. Position is usually given as the coordinates of some point in the object, together with three angles between three lines (axes) in the object with respect to the three axes of the coordinate system. It may be given, particularly in the description of a piece of land, as a verbal description of the lengths and directions of the boundaries of the land or as a statement of what other pieces of land adjoin it. (2) A prescribed setting or reading of the horizontal circle of a theodolite, when the line of sight is directed at the initial station of a sequence of stations to which observations are to be made. The term indicates a definite position of the horizontal circle. For example, in first-order triangulation using a two-micrometer direction theodolite: in Position Number 1, the circle is set to read 0° 00' 40" when the initial station is observed; in Position Number 2, the circle is set to read 11° 01' 50" when the initial station is observed; and so on. (3) The place occupied by or the location of a point on the surface of the Earth. (4) The direction, in celestial coordinates, of a celestial object. (5) SEE location (1). (6) SEE location (2). (7) SEE orientation. SEE ALSO field position.

position, absolute - A location (set of geodetic coordinates) given in a geodetic datum which specifies an oblate rotational ellipsoid having its center at the Earth's center of mass and having its minor axis coincident with the average position of the Earth's axis of rotation.

position, adjusted - Adjusted values of the location (coordinates) of a point on the Earth. In the adjustment of a horizontal-control survey, discrepancies arising from errors in the measurements are removed to give adjusted coordinates in an adopted coordinate-system (geodetic datum or plane coordinate-system). The coordinates obtained by the adjustment are called adjusted coordinates or adjusted positions, and when used as control for other surveys are referred to as fixed coordinates or fixed positions. In the adjustment of data from a vertical-control survey, the values obtained are called adjusted elevations or, when used to control other surveys, fixed elevations.

position, apparent - (1) The direction, in celestial coordinates, from which the radiation from a celestial body appears to come. The apparent position (also called apparent direction) differs from the direction of the straight line between observer and body because of the aberration of light, atmospheric refraction of the radiation and aberrations in the optical or other system used for observing. (2) SEE place, apparent.

position, astrographic - SEE position, astrometric.

position, astrometric - (1) The celestial coordinates obtained by adding planetary aberration to the coordinates given in an ephemeris of the geometrically true coordinates, and then subtracting stellar aberration from which the aberration attributable to the eccentricity and longitude of perihelion of the Earth's orbit has been omitted. (2) The celestial coordinates of a celestial body or spacecraft, corrected for stellar aberration but not for planetary aberration. Also called an astrographic position. Astrometric positions are used in determining the celestial coordinates of an object from measurements, on a photograph, of the objects location with respect to neighboring stars.

position, astronomic - The celestial coordinates of a celestial body, corrected for annual and diurnal aberration but not for planetary aberration.

position, astronomical - (1) A point, on the Earth, whose coordinates have been determined from observations on celestial bodies. Also called a celestial position. The expression is commonly used in connection with determining locations of land with great accuracy for surveying. (2) A point, on the Earth, whose location is given in terms of astronomical longitude and latitude.

position, celestial - SEE position, astronomical.

position, convergent - The placing of a pair of convergent cameras, or of single, split camera, so that the plane containing the cameras' axes is parallel to the line of flight.

position, dead-reckoning - A location determined by dead reckoning.

position, fixed - Adjusted coordinates of a point used to control other surveys.

position, geocentric - SEE coordinate, geocentric.

position, geodetic - (1) The location of a point on the Earth's surface, expressed in terms of geodetic latitude and geodetic longitude. Use of the term geodetic position implies that a particular geodetic datum was adopted. A complete record of a geodetic position must state the datum adopted. (2) A point, on the Earth, whose geodetic longitude and latitude are known.

position, geographic - The location of a point on the Earth's surface, expressed in terms of longitude and latitude, either geodetic or astronomical. Also called a geographical position. SEE ALSO coordinate, geographic.

position, geographical - (1) That point, on the Earth or on a reference ellipsoid, at which a specified celestial body is directly overhead. If the body is the Sun or some other star, the geographical position is called the sub-solar point or the sub-stellar point, accordingly. (2) SEE position, geographic.

position, mean - SEE place, mean.

position, no-check - An intersection station observed from only two stations.

position, preliminary - A location derived from measurements and selected for use in forming condition equations for longitude and latitude.

position, relative - The location of a point with respect to the locations of other points. The relative positions of two points whose locations are given in the same coordinate system are given by the differences of their coordinates or by a direction and distance from the one point to the other.

position, supplemental - A point whose horizontal coordinates have been determined photogrammetrically and are intended for use in orienting other photographs.

position, transverse - The position of a split camera placed so that the plane containing the cameras' axes is perpendicular to the line of flight.

position, true - SEE place, true.

position angle - The angle, measured eastward on the celestial sphere, from a line connecting a specified point with the north celestial pole to a line connecting the specified point with the point whose position angle is wanted. For example, the position angle of one of the stars in a binary system is measured by placing the origin of a polar micrometer on the center of the other star and rotating the movable hair-line from an initial line through the north celestial pole to a final line through the first star.

position computation, direct - SEE computation, direct.

position computation, inverse - SEE computation, inverse.

position error - (1) An error in position. (2) An error in location. (3) That error, in a measurement, caused by an error in either the location or orientation of the instrument.

position finding system - A locating system. SEE locating system; positioning system.

position finding technique - Any method of determining the location of a craft, i.e., of determining the craft's location.

position fix - A determination of the location of a point, usually for navigation.

position fixing - Locating, i.e., determining coordinates. SEE locating.

position fixing system - A locating system. SEE locating system; positioning system.

position indicator, electronic - SEE electronic-position-indicator.

positioning (adj.) - (1) Dealing with the placing of an object in a desired location and/or orientation. (2) Dealing with the determining of an object's location.

positioning (n.) - (1) The process of placing an object into a desired position. (2) The process of determining the position of an object. (3) The process of determining the location of an object.

positioning (v.) - (1) Placing an object in a required location and/or orientation. (2) Determining the location and orientation of an object. (3) SEE locating. SEE ALSO point positioning.

positioning, acoustic - The method, process or technique of determining the location of a vessel by using the difference in times at which acoustic signals from sources of known location are received. The most common and accurate method involves sending sonic pulses from a vessel to transponders on the bottom and measuring the time needed for the pulses to travel from the emitter to the transponder and back. The time needed is multiplied by the speed of sound in water to get the corresponding distances. The vessel's location is found by resection. The error in location is strongly influenced not only by the errors in the locations of the transponders but also by errors in the assumed paths and speeds of the pulses. A less common method uses the pulses for mapping the bottom in the vicinity of the vessel. The approximate location of the vessel must be known. The map is then compared with an accurate map of the bottom and vessel's location adjusted until the two maps agree.

positioning, astronomic - The determination of location by observations on celestial bodies such as stars, the Sun or the Moon. The term is commonly used for only locations on the Earth.

positioning, global - The determination of location in a coordinate system extending throughout the world.

positioning, hypervisual - A method of navigating a vessel along a hyperbolic line of position while measuring, with a sextant, the angle between two objects (such as hydrographic signals on the shore). The objects must straddle (lie on opposite sides of) the hyperbolic line.

positioning camera - A camera mounted coaxially with the radar in the Airborne Profile Recorder and used to photograph the region irradiated by the radar. The photographs are used to relate the data from the radar with the terrain.

positioning device - SEE locating system; navigation system; radio positioning-device.

positioning-device, acoustic - That part of an acoustic positioning-system which is located at the point of unknown location.

positioning-device, electronic - That part of an electronic positioning-system which is located at the point of unknown location.

positioning-device, hyperbolic - That part of a hyperbolic positioning-system which is located at the point of unknown location.

positioning-device, range-range - That part of a range-range positioning-system which is located at the point of unknown location.

positioning-device, rho-theta - That part of a rho-theta positioning system which is located at the point of unknown location.

positioning system - (1) In general, a system (equipment, procedures and personnel) used for placing an object (instrument, vehicle, etc.) in a specified location or orientation. Positioning systems are used, for example, in placing an exploratory ship over a predetermined location when cores are to be obtained. (2) A system used for determining the position of an object such as a vehicle, vessel or aircraft. (3) A system used for determining the location of a vehicle, vessel or aircraft. Also called a navigation system. However, the terms navigation system and positioning system (in the sense of the third definition) are not synonymous in general. A navigating system is used for navigating and determines locations as part of the process. A positioning system, on the other hand, determines position or location only and may or may not be usable for navigation. Also, the locations determined by a positioning system are usually more accurate than those determined by a navigation system. Positioning systems are classified in the same way as are navigation systems. That is, they are classified according to the nature of the equipment used (e.g., radio, inertial, optical or acoustic) or according to the kind of coordinate system which is inherent in the positioning system (e.g., hyperbolic, range-range or range-angle). (4) SEE navigation system. SEE ALSO Doppler positioning-system; radio positioning system; satellite positioning-system.

positioning-system, acoustic - (1) In general, any positioning system making use of sonic pulses. (2) A positioning system which determines the location of a vessel or aircraft by measuring the time required for sonic pulses to travel to and/or from that craft to two or more points of known location. The most common type of acoustic positioning- system consists of a sonic-pulse generator on the vessel, transponders on the bottom and receiving and measuring equipment on the vessel. (3) SEE positioning-device, acoustic. SEE ALSO positioning, acoustic.

positioning-system, circular - SEE positioning-system, range-range.

positioning-system, electronic - (1) A positioning system depending on electronic components for its functioning. (2) A positioning system using radio waves to determine distance or angle. (3) SEE positioning-device, electronic.

positioning-system, global - (1) A navigation system which can be used anywhere in the world for navigation. (2) A surveying system which can be used to determine locations anywhere in the world on a single geodetic datum. (3) A locating system, navigation system or positioning system having world-wide coverage and providing coordinates in one coordinate system.

positioning-system, hyperbolic - (1) A positioning system in which the observer measures the difference in time of reception of signals from two stations whose coordinates are known. The difference in time is converted to a difference in distance. The locus of all points lying at a fixed difference in distance from two points is two branches of a hyperbola. A third station operates in conjunction with the other two to provide the observer with a second pair of stations which can be used to determine another pair of hyperbolas. The observer is located at one of the intersections of the hyperbolas. (2) SEE positioning-device, hyperbolic.

positioning-system, inertial - SEE navigation-system, inertial; surveying-system, inertial.

positioning-system, range-range - (1) A positioning system for determining the location of a vehicle, vessel or aircraft by measuring, as nearly simultaneously as possible, the time needed for a signal to travel from the vehicle, etc., to each of two points of known location and back again. Also called a rho-rho positioning-system and circular positioning-system. (2) SEE positioning-device, range-range.

positioning-system, rho-theta - A positioning system containing at least two radio stations of fixed location and one mobile radio-station. The moving station determined distance to one of the fixed stations and direction to the other (usually, the angle between the line joining it to one station and the line joining the two fixed location). SEE ALSO navigation-system, rho-theta.

position line - SEE line of position.

position-line method (of determining position) - SEE position determination, position-line method of.

position location - Determination of a location. This term appears to be synonymous with locating and is completely unnecessary.

position problem - SEE position problem, direct.

position problem, direct - The problem of calculating the location of a station (point). SEE ALSO position computation, direct.

position problem, inverse - The inverse to the problem of calculating the location (position) of a station. SEE position computation, inverse.

position reference system, acoustic - A locating system which provides continuous information on a vessel's location with respect to a set of three hydrophones of known location on the bottom of the sea, by transmitting an ultrasonic signal to the three hydrophones and measuring the length of time it takes the signal to be returned by each hydrophone.

position vector - SEE radius vector.

positive (n.) - (1) A diagram drawn in ink on a sheet. (2) A document in which linework or detail is in a darker color than the background (or is opaque) and which is right reading when viewed from the emulsion or image side. (3) SEE positive, photographic. SEE ALSO checking positive.

positive, cutting - SEE cutting positive.

positive, direct - SEE film, autopositive; paper, autopositive.

positive, photographic - A photographic image having approximately the same grading of tones or the same colors as the original subject. Also called film positive, positive or a positive photograph. Alternatively, a photograph having approximately the same rendition of light and shade as the original scene. SEE ALSO diapositive; transparency.

positive, reversal - A document in which linework or detail is in a darker color than the background (or is opaque) and which is laterally reversed when viewed from the emulsion or image side.

positive scribing - Scribing using a white-coated surface which, when backed by black paper after scribing, becomes the equivalent of a positive drawing.

possession, adverse - A claim, usually based on actual occupancy, to possession of a piece of land, in opposition to any other persons' claims. These requirements vary with each state.

possession, exclusive - Adverse possession (of a piece of land) such that the adverse possessor must show an exclusive dominion over the land and an appropriation of it to his own use and benefit.

possession, hostile - A situation in which the possessor (of a piece of land) claims to hold possession in the character of an owner, and therefore denies all validity to claims set up by any and all other persons.

possession, notorious - (1) Possession so conspicuous that it is generally known and talked of by the public or the people in the neighborhood. (2) Possession or the character or holding, having in its nature such elements of notoriety that the owner may be presumed to have notice of it and of its extent.

possession, open - Of real property, possession held without concealment or attempt at secrecy, or without being covered up in the name of a third person, or otherwise attempted to be withdrawn from sight, but in such a manner that any person interested can ascertain who is actually in possession by proper observation and inquiry.

post - SEE corner post; identification post.

Postel's map projection - SEE map projection, azimuthal equidistant.

Potenot-Snellius problem - SEE three-point problem.

Potenot's problem - SEE three-point problem.

potential - A function $V(x)$ of location x in a force-field $f(x)$, which is the line integral of f over a path QP from a reference point Q to point P at x: $V(x) = \int_Q^P f\, ds$ under the condition that $V(x)$ be independent of the path taken in going from Q to P; ds is an element of length of the path. Although potential and potential energy are similar, they are not the same. Potential is defined in terms only of the force-field, i.e., in terms of the attracting force, although that force may implicitly involve the attracted body; potential energy involves, explicitly, both the attracting force and the attracted body. The force field in this definition is said to be conservative. Potentials of non-conservative force-fields have been defined but are rarely met with in geodesy. For most force-fields, f is zero at infinity; Q is then conveniently placed at infinity also. (2) A function $V(x)$ which is a function of location only and whose gradient at a point P at x in a force field is the negative of the force at P: $*\Delta V(x) = -f(x)$. (3) Any scalar function of location, in a vectorial field, whose gradient at a point is the negative of the vector at that point. This definition is a generalization of the second definition and is particularly useful in describing velocity-field. (4) SEE energy, potential. SEE ALSO gravity potential; Mayer potential; velocity potential.

potential, centrifugal - SEE potential, rotational.

potential, complex - An analytic function whose real part is the velocity potential and whose imaginary part is the stream function.

potential, disturbing - (1) The difference between the actual value of the potential at a point and the value predicted by theory or taken for reference. Also called the potential disturbance or, particularly in orbital theory, the perturbing potential. SEE ALSO potential, perturbing. (2) In the theory of the geopotential, the difference between the actual value of the gravity potential at a point and the value calculated from a standard formula for the gravity potential. (3) The difference between the total gravitational potential of a body and the potential resulting from a spherical distribution of mass. Also called a disturbing function.

potential, gravitational - (1) The potential attributable to the presence of a gravitational field (i.e., the field generated by a massive body). From the definition of a field and from Newton's law of gravitation, it follows that the potential $V(x)$, at a point P at x, attributable to the gravitational attraction of a body of total mass M is $V(x) = \int_M dm/r$ in which the integral is taken over the entire mass M and r is the distance between P and an element dm of mass of the body. (2) A scalar function W of location is a gravitational potential if its gradient $*\Delta W$ is a gravitational force (or, in special applications, an acceleration). (3) SEE energy, gravitational potential. The three definitions are equivalent in a conservative field, and ΔW is then independent of the path taken between P and Q.

potential, luni-solar - The sum of those potentials from which the tide-raising forces of the Moon and Sun are derived.

potential, perturbing - Any potential causing an orbiting body to deviate from a simpler orbit or orbit used as reference. Equivalently, any potential affecting the motion of a body but not included in the formulas used in calculating the orbit. Also called a disturbing potential or potential disturbance. However, disturbing potential is commonly used in a more general sense than perturbing potential. A Keplerian orbit is usually used as reference, so the perturbing potential would be the total potential minus the Newtonian potential. E.g., the potentials of the Sun's and Moon's gravitational attractions would be considered perturbing potentials for any orbit taking account only of the Earth's gravitational potential. If the potential of a homogeneous, oblate, rotational ellipsoid of known mass is adopted as a standard for computing the orbit of a satellite, then any potential additional to this would be considered a perturbing potential.

potential, rotational - The quantity V_c at a point P revolving at rate ω about an axis at a distance r from that axis, as given by the formula $V_c = \frac{1}{2} \omega^2 r^2$. Also called the centrifugal potential.

potential, tidal - A potential whose gradient is the force causing a tide.

potential disturbance - SEE potential, disturbing.

potential energy - SEE energy, potential.

potential function - The quantity: potential energy per unit mass. This is probably the original meaning of the term. However, some geodesists confuse the term with potential energy.

Pothenot-Snellius problem - SEE three-point problem.

Pothenot's problem - SEE three-point problem.

Potsdam standard of gravity - An assigned value for the acceleration of gravity at the gravity pier at Potsdam. The value g = 981.247 gals ± 0.003 gals determined by Kühnen and Furtwangler between 1898 and 1905, using reversible pendulums, was the average of measurements made at two piers in the northeast corner of the pendulum room of the Geodetic Institute in Potsdam. The Potsdam standard was therefore defined to be 981.274 00 gals at the point midway between the two piers. The coordinates of this point are
λ =+13° 04.06' (approximately),
ϕ =+52° 22.86' (approximately),
H=86.24 m (exactly) above Netherlands Normal (N.N.)
From this value of gravity acceleration and from the coordinates of the midpoint, the following values are derived:
    g (on geoid)   981.301 gal,
    τ (from Helmert's formula of 1901)   981.277 gal,
    τ (from Helmert's formula of 1915)   981.289 gal.
A large number of other measurements throughout the world, using pendulum apparatus, after 1906 showed that the value defined above was probably higher than the actual value by about 14 mgal. Therefore, in 1967, the International Association of Geodesy defined the Potsdam standard of gravity to be exactly 981.260 gals (cm/s²). This standard has, however, been replaced for practical purposes in many parts of the world by the value of gravity acceleration in the International Gravity Standardization Net 1971 (IGSN 71).

Potsdam system of gravity - The set of values of gravity acceleration determined by reference to the Potsdam standard of gravity. Most determinations of gravity are derived from measurements made with spring-type gravimeters, which are calibrated with respect to a known value at some base. Such calibrations were referred to the Potsdam standard of gravity from about 1909 to about 1945, replacing the Vienna system used from about 1900 to 1909. In the latter part of the 1940s, a need was felt for a gravity system which did not depend critically on measurements at a single point. A program was therefore begun to establish a world-wide system of mutually dependent gravity stations. In 1971, the International Association of Geodesy (I.A.G.) adopted the new system, called the International Gravity Standardization Net 1971 (ISGN 71). This has largely replaced the Potsdam system. The Potsdam system had a precision of 3 mgal but an accuracy of about 14 mgal. The error was corrected at the 1967 meeting of the I.A.G.

pouce - One-twelfth of a pied. An obsolete French unit of measure.

power (of a telescope) - The magnification afforded by a telescope when focused for an object at infinite distance.

power (optics) - SEE power, magnifying.

power, covering - The ability of a lens system to give a sharply defined image to the edges of sensitized material it is designed to cover at the largest possible aperture. SEE ALSO angle of view.

power, magnifying (optics) - The magnification afforded by an optical system, expressed in terms of the ratio between the angle subtended by the object and the angle subtended by the image. Usually called power when the intent is clear.

power spectrum - The square of the amplitude of the complex coefficients in the Fourier series representing a given periodic function. Thus, if the function f(t) is periodic with period T, the coefficients F(n) of its Fourier series are (1/T) ∫ f(t) e(-n ω t) dt, in which ω is 2 π/T, and the power spectrum of f(t) is the set |F(n)|², where | | denotes the absolute value. Here, n takes integral values and the spectrum is discrete. The term power spectrum originated in electrical theory, where F(n) can have the dimensions of power. However, the term power spectrum is now applied to many other kinds of functions and the power spectrum frequently has no relation to power whatsoever.

Practical Salinity Scale 1978 - A set of equations relating the salinity of sea water to the conductivity of sea water and the conductivity of standard samples of sea water, and recommended in 1978 by the International Association on Physical Oceanography for use by oceanographers.

Pratt-Hayford gravity anomaly - An isostatic gravity anomaly Δg_{PH} obtained by subtracting from the magnitude g of gravity as measured on the ground the following corresponding quantities: (a) the theoretical value of the magnitude τ of gravity on a reference surface; (b) the free-air gravity correction δg_f; (c) the complete topographic gravity correction δg_{PH}; and (d) the Pratt-Hayford gravity correction δg_{PH}. Also called the Hayford gravity anomaly, the Hayford anomaly and the Pratt-Hayford isostatic gravity anomaly.

Pratt-Hayford gravity correction - An isostatic gravity correction δg_{PH} added to the value τ of gravity given by a gravity formula (or subtracted from a gravity anomaly or measured gravity) to correct the theoretical value according to the isostatic theory of Pratt and Hayford. The theory of Pratt and Hayford assumes that the crust extends everywhere to the same depth but that the density of matter between the geoid and that depth is a function of the elevation of the land. The formula is $\Delta \rho = H \rho / T$, in which ρ is the density of matter outside the reference surface, H the elevation, T the distance to the bottom of the crust and $\Delta \rho$ is the difference between the density of crust under the reference surface and the density of crust above that surface. The crust is divided into vertical prisms having cross sections of the same shape as the quadrangles in the Hayford template. The value of the Pratt-Hayford gravity correction is then found using the formula $\delta g_{PH} = (2 \pi G \rho h / D) [(r_1 + D^2)^{1/2} - (r_2 + D^2)^{1/2} - r_1 + r_2)]$, in which G is the gravitational constant, ρ the density of the matter in the quadrangular prism, D the assumed depth of compensation and r_1 and r_2 are the inner and outer radii, respectively, of the zone in which the quadrangular prism lies. h is the distance of the point (for which the correction is calculated) above the base of the prism. For prisms lying between 29 and 167 km, approximately, Hayford prepared a table giving the small corrections needed to allow the formula to be used.

Pratt-Hayford gravity reduction - (1) The process of applying, to a computed or measured value of gravity, the sum of the following gravity corrections: (a) the Pratt-Hayford gravity correction δg_{PH}, (b) the free-air gravity correction δg_f and (c) the complete topographic gravity correction δg_{tc}. In the past, the Pratt-Hayford gravity correction and the complete topographic gravity correction have commonly been applied simultaneously by using special tables for distances out to about 167 km from the point at which the reduction is to be applied. Such tables are now obsolescent; the needed values are calculated directly on computers. (2) The sum δ_{PH} of the gravity corrections listed in the previous definition: $\delta_{PH} = \delta g_{PH} + \delta g_f + \delta g_{tc}$.

Pratt-Hayford isostatic gravity anomaly - SEE gravity anomaly, Pratt-Hayford.

Pratt's hypothesis - The hypothesis that the pressure of matter above some equipotential surface of reference is balanced by the pressure of matter of suitably varying density beneath that surface and extending down to a constant depth (the depth of compensation). Also called the fermenting-dough theory. SEE ALSO isostasy.

Pratt's theory of isostasy - SEE isostasy.

pre-amplifier - An amplifier having a very low noise-figure and hence not lowering the ratio of signal to noise. Pre-amplifiers are used, for example, in radio interferometers to amplify the signal close to the antenna, so that noise entering the system between the antenna and the receiver will be less obtrusive.

precession - (1) In general, the motion of the instantaneous axis of rotation of a body about a line whose direction is fixed in space. It is customary to limit the term's meaning to secular motion (or motion of very long period) and to denote by nutation the smaller, periodic excursions of the axis of rotation from its average position. (2) SEE precession, general. SEE ALSO constant of precession; constant of general precession; Larmor precession; precession in declination; precession in right ascension.

precession, annual - SEE precession, general.

precession, apparent - The apparent change in the direction of the rotational axis of a body, caused by the rotation of the Earth. Also called apparent wander and wander. The term is applied, in particular, to the precession of the rotational axis of a gyroscope.

precession, general - (1) The very-long-period motion of the instantaneous axis of rotation of the Earth about a line whose direction is fixed in space. Also called annual precession and where the meaning is clear, precession. General precession is observed as a westward motion of the equinox along the ecliptic at a rate of about 50" per year. The rate, as given by Newcomb in 1905, is 5025.64" + 2.22" T, in which T is the time in tropical centuries since 1900.0. The value adopted by the International Astronomical Union in 1976 is 5029.0966" at epoch 2000. The value of the coefficient of T was not specified. General precession causes the difference between the lengths of the sidereal and tropical years. (2) The motion of the equinoxes westward along the ecliptic at the rate of about 50.3" per year. Also called the precession of the equinoxes. (This definition is practically equivalent to the preceding one).

precession, geodetic - A relativistic motion of the equinox along the ecliptic. It is similar to the general precession but is in the opposite sense and amounts to 1.915" per century.

precession, gyroscopic - The oscillation of the axis of a north-seeking gyroscope about the meridian.

precession, induced - SEE precession, real.

precession, luni-solar - That part of the general precession (motion of the equinoxes along the ecliptic) attributed to the pull of the Moon and Sun on the Earth's equatorial bulge. The luni-solar precession is a westward motion.

precession, planetary - That part of the general precession (motion of the equinoxes along the ecliptic) attributed to the pull of the other planets on the Earth's equatorial bulge. The planetary precession is an eastward motion. It amounts to about 0.13" per year.

precession, real - Precession resulting from torque applied to the rotating body. Also called induced precession.

precession in declination - That component of general precession which is in the direction of the celestial meridian. It amounts to about 20.0" per year.

precession in right ascension - That component of general precession which is parallel to the celestial equator. It amounts to about 46.1" per year.

precession of the equinoxes - SEE precession, general.

precipitation - (1) Rain, ice, snow, sleet, hail and other forms of water condensed from atmospheric water vapor and falling or fallen to the Earth's surface. (2) The total, measurable amount of water received directly from clouds as rain, snow and/or hail falling within a specific area over a specific period of time. Precipitation is usually expressed as the thickness of a layer spread over a level surface, covering the specified area, and having the same total volume as the precipitation.

precipitation, annual - Precipitation expressed as the thickness of an equivalent layer spread over a level surface, covering the specified area, and having the same total volume as the amount of precipitation over a one-year period.

precipitation, daily - Precipitation expressed as the thickness of an equivalent layer spread over a level surface, covering the specified area, and having the same total volume as the amount of precipitation over a one-day period.

precipitation, hydrologic - (1) Water falling in liquid or solid form out of the atmosphere, usually onto land or water. The term is probably synonymous with precipitation. However, the latter term is sometimes used to denote any form of natural, particulate matter falling from the sky. (2) The amount of water that has fallen from the atmosphere, measured as a liquid.

precipitation, monthly - Precipitation expressed as the thickness of an equivalent layer spread over a level surface, covering the specified area, and having the same total volume as the amount of precipitation over a one-month period.

precise - A method, procedure or instrument capable of yielding results showing very little variation with repetition. Also called precision. There is no excuse for using precision as a synonym for precise. The distinction between the two is well known, clear cut and observed among those concerned with writing well.

precision (adj.) - SEE precise.

precision (n.) - (1) In statistics, a measure of the tendency of a set of random numbers to cluster about a number determined by the set. The usual measure is either the standard deviation (with respect to the average), or the reciprocal of this quantity. Precision is not the same as accuracy; the latter is a measure of the tendency of a set of numbers to cluster about some number determined not by the set but specified in some more or less arbitrary manner independent of the set. (2) In physics, in metrology and in the art of measuring in general, a measure of the quality of the method by which measurements are made. In this sense, precision differs from accuracy in that the latter relates the quality of the results of the measurements, not the quality of the method used. Precision applies not only to the fidelity with which required operations are preformed, but by custom has been applied to methods and instruments employed in obtaining results of a high order of precision. Precision is exemplified by the number of decimal places to which a computation is carried out and a result stated. In a general way, the accuracy of a result should determine the precision with which it is expressed. Two different kinds of precision are sometimes distinguished: repeatability and reproducibility. Repeatability is then a measure of the amount by which measurements made by one instrument differ from each other. Reproducibility is a measure of the amount by which measurements made by different instruments differ from each other. (3) The quantity $1/(\sigma \sqrt{2})$, in which σ is the standard deviation of the random variable involved. Also called a precision index or index of precision. SEE ALSO modulus of precision.

precision, index of - SEE precision (3).

precision altimeter - SEE altimeter, precise; precise.

precision camera - SEE camera, precise.

precision index - SEE precision (3).

precision level - SEE leveling instrument, precise.

prediction - In statistics, the process of obtaining an estimate of the value of a quantity at a specified time t from data available up to but not including the time.

preliminary (adj.) - Adopted for temporary use with the understanding that it will be changed later. In adjusting triangulation, the term preliminary is applied to triangles and geographic coordinates derived from selected observations.

pre-press proof - A proof made directly from screen-separation film and the resulting sheets registered to get the full effect.

prescription - The process of obtaining title or right by long possession.

press proof - (1) A proof taken from among the first copies made on the press and used to check the correctness of the printing. Also called a press pull. (2) The same as the preceding definition, but restricted to proofs made by lithography.

press plate - SEE printing plate.

press pull - SEE press proof.

pressure - (1) The force, per unit area of a surface, exerted perpendicularly to that surface. (2) A type of stress characterized by its uniformity in all directions and measured by the force exerted per unit area. (3) SEE stress. (4) SEE pressure, atmospheric.

pressure, atmospheric - The pressure exerted by the atmosphere on a level surface at a specified location.

pressure, mean-sea-level - Atmospheric pressure at mean sea level (at a specified location). Abbreviated to M.S.L. pressure.

pressure, solar-radiation - The pressure exerted by electromagnetic radiation from the sun. The Sun also emits particulate radiation, but this is not taken into account when calculating the pressure of solar radiation on artificial satellites. SEE ALSO perturbation, radiation-pressure.

pressure, standard - (1) A pressure of one standard atmosphere. (2) An atmospheric pressure of 1000 millibar. Other values may be used as standard for special purposes.

pressure altimeter - SEE altimeter, barometric.

pressure altitude - The altitude or elevation corresponding to a given value of atmospheric pressure according to the ICAO standard atmosphere. It is the indicated altitude above the surface at which atmospheric pressure is 1013.2 millibar.

pressure altitude, rectified - SEE altitude, apparent.

pressure back - A locating back which holds the photographic film by applying pressure to the front surface of the film.

pressure correction - he quantity added to the period of a pendulum to account for the buoyant effect of the surrounding gas on the pendulum, the resistance of the gas to the pendulum's motion and the increase in the moving mass by adsorption of gas on the pendulum. The gas reduces the effective mass of the pendulum by buoying it up, thereby increasing the period. It increases the effective mass by adhering, by adsorption, in a thin layer to the pendulum, thereby decreasing the period. It drags against the moving pendulum, acting as if the effective mass of the pendulum were reduced, and thereby again increasing the period. A formula commonly used for the correction Δt in seconds when only the buoyancy and drag are to be taken into account is $\Delta t = k_1 s + k_2 \sqrt{s}$, in which $s = P/[760(1 + 0.003665 T)]$, P is the pressure (fully corrected for temperature and humidity) in mm and T is the temperature in °C. k_1 and k_2 are empirical constants.

pressure gage - A tide gage placed at the bottom of a body of water whose changes in depth are to be determined, and recording changes in the height of the tide by differences in pressure caused by the rise and fall of the tide.

pressure gage, gas-purged - An instrument indicating the pressure in a liquid by measuring the equivalent pressure needed to eject a continuous stream of bubbles from an orifice placed beneath the surface and connected to a tank of gas. Also called a bubbler gage. It is now widely used in determining changes in water level, because the pressure is proportional to the height of the water's surface above the orifice.

pressure plate - A flat plate pressing the film, in a camera, into contact with the focal plane. SEE ALSO plate, focal-plane.

pressure scale height - SEE scale height.

Prey reduction - SEE Prey gravity reduction.

Prey's gravity anomaly - (1) The quantity Δg_P obtained by subtracting, from a gravity anomaly Δg, the following gravity corrections: the free-air gravity correction δg_f; the Bouguer gravity correction δg_B; the topographic gravity correction δg_t; a gravity correction δg_d for the variation of the density of actual masses from the average density used in the previous gravity corrections; gravity corrections $\delta g_t'$ and $\delta g_d'$ which result by restoring the removed masses. $\Delta g_P = \Delta g - (\delta g_f + \delta g_B + \delta g_t + \delta g_d + \delta g_t' + \delta g_d')$. For practical reasons, δg_d and $\delta g_d'$ are usually set equal to zero, while δg_t and $\delta g_t'$ are assumed to be equal. Prey's gravity anomaly is then $\Delta g_P = \Delta g - (\delta g_f + \delta g_B + 2\delta g_t)$. (2) The difference between the value of the magnitude of gravity on the geoid and the corresponding value of gravity on a terrestrial ellipsoid as calculated using a gravity formula. This definition is practically the same as (1) when the value of gravity on the geoid is calculated from a value measured at a gravity station and when the effects of geoidal heights are ignored. (3) Prey's gravity anomaly (1) minus an isostatic gravity correction.

Prey's gravity reduction - (1) The process of applying, to a calculated or measured value of gravity, the sum of the following gravity corrections: (a) the free-air gravity correction δg_f, the Bouguer gravity correction δg_B, topographic gravity correction δg_t and the density gravity correction δg_d required for calculating the value of gravity on the reference surface from the value of gravity on the physical surface; and (b) the Bouguer gravity correction $\delta g_B'$, the topographic gravity correction $\delta g_t'$ and the density gravity correction $\delta g_d'$ required for restoring the masses removed in step (a). Isostatically compensating masses may be taken into account also. Also called the Poincaré-Prey gravity reduction. The mathematical procedure is equivalent, when calculating gravity from a theoretical value at a point on the reference surface, to first removing the topographic masses between the reference surface and the physical surface (and possibly taking into account isostatic compensation), moving the point from the reference surface to the physical surface and then restoring the masses that had been removed (and if necessary subtracting the effects of the compensating masses). The effect of the differing densities in the first and second steps may be accounted for by adding a gravity correction δg_e as the difference between the nominal value (usually 2.67 g/cc) used for density during the rest of the reduction and the actual value of the density of the topographic masses. Removing the topographic masses from above the point of calculation and then replacing them below the point in its new location adds, in each case, to the calculated value of gravity. The change brought about by Prey's gravity reduction is therefore about twice as large as that brought about by a Bouguer gravity reduction. (2) The sum of the gravity corrections listed in the previous definition.

Pribilof Islands Datum - SEE St.George Island Datum; St.Paul Island Datum.

primary - SEE body, primary.

prime vertical - SEE circle, prime vertical; vertical, prime.

Princeton Standard Test for Levels - A procedure for determining the combined errors of rod setting and bubble (or pendulum-compensator) settings which contribute jointly to the rod-reading error by making readings from a leveling instrument set up at various points between two leveling rods. The two leveling rods at twice the typical distance specified for sighting from instrument to rod in a survey. Eight equally-spaced, intermediate stations are set on line between the two rods and measurements made at each station just as for prices leveling. Either 16 or 32 sets of elevation differences (two or four from each station) are measured, and the results reduced by the method of least squares to give the required errors.

Princeton test - SEE Princeton Standard Test for Levels.

principal-plane method - A graphical method of determining horizontal and vertical angles from an oblique photograph by determining first the perspective center of the photograph, and then using the intersection of the principal plane with the photograph as an additional element in the process. Also called the Crone method and the Survey-of-India method.

principal-point method (of phototriangulation) - SEE aerotriangulation, radial.

Principle of Equivalence - The principle, or axiom, that at every point in an arbitrary gravitational field in four- dimensional space (space and time), it is possible to choose an inertial coordinate system valid in the neighborhood of the point and in which the laws of nature take the same form as in unaccelerated Cartesian coordinate-systems in the absence of gravitation. This principle is at the basis of the theory of relativity.

principle of proportioning excess or deficiency - The principle, governed by several rules, that the amount by which a measurement is over (excess) or under (deficiency) what is required must be allocated to several components of that measurement. The rules differ according to what was measured, custom, etc. For example, if deeds or plans state that the frontage of 10 lots in a city block totals 300 meters, but measurement shows that the frontage is 303.33 meters, then if the principle applies, the excess, 3.33 meters, must be divided up between the 10 lots in some manner.

principle of radial displacement - The photogrammetric principle that if the nadir point N' on the photograph is vertically above the point N on the ground, and if point P' on the photograph corresponds to point P on the ground, then the distance of P' from N' depends not only on the horizontal distance of P from N and on the angle to the vertical at which the photograph was taken, but also on the relative heights of P and N.

principle of the semi-circular rule for bays - The principle or postulate that a bay whose area is equal to that contained within a semi-circle, the diameter of which is a line joining the headlands, is on the borderline between an open and a closed bay.

print - A photographic copy of an original photograph, made by projection or by contact printing from a photographic negative or transparency. SEE ALSO contact print; diazo print; projection print; ratio print.

print, blackline - A copy made on a transparent sheet by placing the original in contact with sensitized paper, illuminating the original and developing the image on the exposed paper. Blackline prints can be made from blueprints by making the blueprint on a transparent, sensitized sheet and then making a blueprint from this. However, a blackline print can better be made directly, by using specially sensitized paper.

print, composite - SEE composite.

print, contact - SEE contact print.

print, glossy - A print made on a photographic paper having a shiny surface. Referred to, in photographer's jargon, as a glossy.

print, master - SEE master print.

print, matte - A print made on a photographic paper having a dull (matte) finish. A matte print is more suitable for annotating with pen or pencil than is a glossy print.

print, rectified - A print made by removing the displacements caused by tilt of the original photograph and bringing the image to the desired scale.

print, semi-matte - (1) A photographic print on a non-glossy paper having a surface only faintly lustrous. (2) A non-glossy, photographic paper having surface only faintly lustrous.

print, transformed - A photographic print made by projection in a transforming printer.

printer - SEE contact printer; diapositive contact printer; diapositive printer; diapositive printer, fixed-ratio; projection printer.

printer, ink-jet - A machine that prints text character by character by spitting drops of ink from a reservoir between electrostatic plates that deflect the stream of droplets to form the proper characters on the paper.

printer, rectifying - SEE rectifier; printer, transforming.

printer, transforming - A projection printer having fixed geometric relationships between the original image, the lens system and the final image and designed specifically for transforming photographs, by projection, from one plane to another. Also called a rectifying printer. In particular, such an apparatus designed specifically for transforming photographs taken by the oblique components of a pair of coupled cameras, a multiple-lens camera or a panoramic camera.

printer, universal transforming - A projection printer, designed specially for making diapositives on glass plates, in which the effect of a known distortion in the camera is eliminated, compensated for, or, in some cases, introduced. The diapositives may be enlarged, reproduced at the original scale or reduced.

printing - SEE contact printing.

printing, anastatic - A method of reproducing either drawn or printed images by moistening the image with dilute acid and pressing it against a zinc plate. The acid etches the zinc wherever it is in contact with unprinted portions. The plate is then washed and inked and can be used for printing.

printing, dyeline - The process of making a diazo print.

printing, lithographic - SEE lithography.

printing, offset - A process whereby the image is transferred from an inked plate on a rotating roller to a rubber mat on another roller, and from the mat to the paper, instead of directly from the plate to the paper. Printing of different products is scheduled by color to avoid having to change the ink on the press frequently. For example, about 15 different maps would be printed in one color before the rollers were washed, the ink changed and new plates installed.

printing plate - A thin metallic, plastic or paper sheet carrying the image which is to be printed and whose surface is treated to make only that part carrying the image receptive to ink. Also called a press plate or pressplate. Usually called a plate if the meaning is clear.

printing paper - A matted or felted sheet composed of minute, vegetable fibers formed by sedimentation on a screen from an aqueous suspension and specially treated to make it suitable for printing on. Paper used for printing maps must not shrink or expand appreciably with changes in humidity and must give clear sharp impressions. The paper should resist deterioration as it ages. If the maps are to be used in the field, the paper should also retain its strength when wet and be able to last in spite of rough handling.

prism - (1) A geometric surface consisting of two congruent polygons in parallel planes, the parallelograms formed by joining corresponding vertices of the two polygons with straight lines and the interiors (planar surfaces) of the polygons and parallelograms. (2) The geometric solid consisting of the surface defined in (1) and the interior of that surface. The two polygonal surfaces are called the bases of the prism; the other surfaces are called the sides or faces. In geodesy, prisms are used mostly as elements of volume of a body whose gravitational attraction is calculated as the integral over the gravitation attractions of the masses inclosed by the elemental volumes and as approximations in estimating the volumes or irregular cuts and fills. (3) Any transparent solid, usually having the shape of a prism as defined above, and used for deviating a ray of light by refraction or internal reflection. (4) Any transparent solid used for deviating rays by internal reflection, regardless of its shape. There are three common uses for prisms: the dispersion of polychromatic light into its monochromatic components, reflection of light and polarization of light. In geodesy, prisms are used to deviate light, by internal reflection, through 90° or 180° ; as part of a chain of optical elements carrying an image from one part of an instrument to another; and to invert or revert images so that the object appears in its normal position. SEE ALSO Abbe prism; Amici prism; Bauernfeind four-sided prism; Dove prism; Goulier prism; horizon prism; index prism; Köster's prism; Köster's double-image prism; Nicol prism; objective prism; Porro prism; Roelof's prism; Thompson prism; Wollaston prism.

prism, deflecting - (1) SEE prism, reflecting. (2) SEE prism, deviating.

prism, deviating - A prism which changes the direction of a light ray by refraction. Also called a deflecting prism and refracting prism. However, the first term is more often used for a prism which changes the direction of light rays by internal reflection. The second term is used when the prism's principal use is for breaking light up into a spectrum, not for merely changing the direction of a ray.

prism, dispersing - A prism which produces a spectrum by changing the directions of the chromatic components of an incident ray by amounts depending on their wavelengths. Also called a dispersion prism.

prism, double-pentagonal - SEE square, optical.

prism, pentagonal - A prism having five sides (faces), i.e., having pentagonal bases. Also called a pentaprism. It commonly occurs in two forms: as a Goulier prism, having four dihedral angles of the same size and one dihedral angle of a different size between faces; or as a Wollaston prism, having three dihedral angles of one size and two dihedral angle of another. SEE ALSO square, optical.

prism, reflecting - A prism capable of changing the direction of light by reflection from the internal side of its faces. Also called a deflecting prism, although that term is also in the sense of a reflecting prism mounted so that it can be rotated about an axis perpendicular to the bases.

prism, refracting - A prism which deviates a ray of light by refraction only. More precisely, a prism whose principal use is the spreading of light, by refraction, into a spectrum. It is often called a deviating prism if its principal use is merely changing the direction of a ray. The angle through which the ray is bent is a function of the wavelength of the light of which the ray is composed. A refracting prism therefore spreads a ray of white light into a spectrum.

prism, retrodirective - Also called a retroreflecting prism and retro-reflector prism.

prism, retroreflecting - SEE prism, retrodirective.

prism, retro-reflector - A prismatic corner-cube reflector.

prism, rhomboidal - A prism having a rhomboidal cross-section and displacing a ray of light laterally only. Also called a rhomboid prism.

prism, right-angle - (1) A prism turning a beam of incident light through 90°. Depending on the orientation of the prism with respect to the source, a right-angle prism may invert or revert the image. The pentagonal prism is the most common form used in surveying. (2) A prism whose bases are isosceles right-triangles.

prism, roof-angle - A prism having two long, narrow faces (the roof) including a dihedral angle of 90° or less, to revert an image. Also called a roof prism. The Amici prism is the best-known form.

prism, solar - SEE Roelof's prism.

prism, three-sided - A prism having three sides (faces). The type used in surveying for establishing right angles has right-angled, isosceles triangles as bases and is called the Bauernfeind prism or the Bauernfeind three-sided prism (to distinguish it from the four-sided prism introduced into surveying by Bauernfeind).

prism, triple - SEE reflector, cube-corner.

prismatoid - (1) A polyhedron of two bases in parallel planes and having the other faces triangular or trapezoidal with one side common with one base and another side common with the other. (2) A polyhedron formed by connecting vertices of two polygons in parallel planes in such a manner as to form triangles or trapezoids.

prism level - (1) SEE prism leveling instrument. (2) SEE level, U.S. Geological Survey.

prism leveling instrument - A type of dumpy level in which the bubble of the level can be viewed from the eyepiece end by means of an attached prism, and the scale on the leveling rod can be read at the same time.

prismoid - (1) A solid bounded by planes, two of which are parallel. (2) A solid bounded by surfaces which were initially plane, two of the planes having been parallel, but were then warped. (3) A surface or solid having bases which are parallel and composed of similar but not necessarily congruent figures and sides which are quadrilaterals with two parallel sides each. (4) A polyhedron the faces of which are formed by two polygons in parallel planes and triangles or trapezoids formed by lines joining the vertices of the two polygons.

prism stereoscope - SEE stereoscope, prism-type.

probability - A positive number between 0 and 1, inclusive, indicating the ratio between the number of times an event occurs (or is expected to occur) and the total number N of times that event and all alternatives will occur, as N is increased without limit. There are many different definitions of probability. That given here was used by von Mises in developing his theory of probability and is sometimes called an objective definition. While the term can be defined precisely in mathematics, such a definition will usually not agree with what happens in the real world.

probability, a posteriori - SEE probability, empirical.

probability, a priori - Probability based on the number of times an event would be expected to occur in N opportunities for it to occur, as the number N is increased without limit. A priori probability is often calculated by enumerating all the possible events that might occur, counting the number of times, in this enumeration, that the selected event is found, and dividing the latter number by the former.

probability, empirical - A probability based on the number of times an event did actually occur out of a known, total number of times it might have occurred, i.e., it or one of the possible alternative events did occur. Also called an a posteriori probability.

probability, prior - The probability of a particular event's occurring before the event has actually occurred.

probability density - The probability, expressed in percent, that a randomly varying vector will assume a value lying within a specified interval of unit length. SEE ALSO frequency.

probability density function - SEE frequency function.

probability function - SEE distribution function, cumulative.

probability function, cumulative - SEE distribution function, cumulative.

probate (v.) - To prove, at a court, the terms and validity of the last will and testament of a deceased person.

probe (n.) - (1) In surveying, a device used to search for monuments. (2) In civil engineering, a rod or tube used to determine the properties of the ground in or on which a structure is to be erected.

probe (v.) - Drive a rod into the ground to determine the hardness and other properties of the ground.

probing - The technique or process of pushing or driving a steel rod down through the soil as far as it will penetrate to estimate the resistance to penetration at various depths. Also called sounding. Because samples cannot be obtained by this method, no information about the physical properties of or moisture in the soil is obtained. Changes in strata are noted only when there is a definite change in resistance to penetration, as in passing from silt to gravel or from gravel to clay. Sounding is most useful in soft materials to determine the depth of a solid stratum.

problem, boundary-value - (1) Any problem in which a partial differential equation is to be solved for the dependent functions, values of the function and/or its derivatives being given on the boundary of the region over which the function is defined. (2) (geodesy) The problem of finding the shape of the geoid within a specified region, given the value of gravity over that region and given the boundary of the region.

Because there is only one geoid, the specified region is, properly, the entire surface of the earth. However, the value of gravity is not known over the entire surface of the Earth. Therefore the problem has been solved either approximately only, by assuming values of gravity where they are not known, or by solving exactly over smaller regions within which the value is known exactly. The problem is not a boundary value-problem in the mathematical sense.

problem, inverse position - SEE position problem, inverse.

problem, one-body - The problem of determining the location of a body as a function of time, when the field of force in which the body moves and the initial location and velocity of the body are known. Determining the orbit of an artificial satellite resolves itself into a one-body problem because the mass of the satellite is too small to contribute significantly to the field. The motion of the primary is therefore not affected by the motion of the satellite.

problem, three-body - The problem of determining the orbits of three mutually-attracting (or repelling) bodies, given the laws of attraction or repulsion and the locations and velocities of the three bodies at some specified instant. Unlike the two-body problem, the three-body problem has no solution attainable in a finite number of steps, except in a few, simple cases. There is no general solution for the general case. Most solutions are arrived at by numerical integration. The problem occurs in geodesy principally in determining the orbit of an artificial satellite of the Earth; this is a very degenerate case of the problem, because the motion of such a satellite does not appreciably affect the motion of either the Earth or the Moon.

problem, three-point - The problem of determining the location of a point, given the locations of three other points on the same surface and the angles, at the point in question, between geodesics to the points of known location. Also called Pothenot's problem or the Pothenot-Snellius problem. Pothenot is sometimes spelled Potenot in translations from the Russian. The procedure for solving the problem is called resection. If there are more than three points of known location and the corresponding angles given, the same name is given to the problem and to the procedure for solving the problem. In plane-table surveying, the problem is solved graphically by trial and error. It can also be solved graphically by using a three-arm protractor. The edge of each arm is made to pass through one of the known points. The sought-for point is then at the center of the protractor. The problem can be solved analytically by using trigonometric formulae. SEE ALSO resection; Cassini's method; Collins' method; Kaestner's method; Marek's method; triangle of error; two-point problem.

problem, two-body - (1) The problem, in mechanics, of determining the coordinates of the centers of mass of two bodies as a function of time, given the law of attraction or repulsion between the bodies and the initial locations and velocities of the bodies. The usual procedure is to integrate the differential equations of motion of the two bodies and determine, from the initial conditions, the constants of integration. In geodesy, the problem appears principally as that of determining the orbit of a body of negligible mass moving in the gravitational field of the Earth. (2) In particular, the problem of determining the coordinates of two point-masses as a function of time, given the law of attraction or repulsion between the bodies and the initial locations and velocities of the bodies.

problem, two-point - The problem of determining the location of a point occupied by an observer when the locations of two other points, inaccessible but observable, are known. Also called Hansen's problem. The method involves measuring the angle between the two inaccessible points, moving the point of observation to a nearby point and measuring angles from there also. It is quite time-consuming. SEE ALSO Marek's method.

problem of inverse position - SEE position problem, inverse.

process, anastatic - SEE printing, anastatic.

process, color-proof - A photomechanical printing process which makes it possible to combine negative separations by successive exposures to produce a composite color proof on a plastic sheet. The method is usually referred to by the trade name for the materials used. SEE ALSO color proof.

process, halftone - Photoengraving in which a mesh (screen) is interposed between the original image or object and the light-sensitive negative, the image produced by the camera being thereby broken up into black and white dots too small, usually, to be seen by the unaided eye. The dots blend, visually, to give the same impression of light and shade as the original image or object. The picture to be reproduced is projected onto a negative immediately in front of which is the mesh (halftone screen). After exposure and development, the dots in the negative are proportional in size to the relative intensity of the light reaching them in the projected image.

process, one-step cyclic - SEE Seidel's method.

process, open-window - A method of preparing color-separation negatives or positives by peeling an opaque layer from its transparent base in those regions where a particular color is desired. The process is normally used for preparing maps of large regions covered by vegetation or open water.

process, photo-contour - A process by which contour lines are put on aerial photographs by drawing the contour lines with a conventional stereoscopic plotter, rectifying the photographs to remove the effect of tilt, removing the effects of relief displacement by a zone printer or similar device and then superposing the contour lines on the photographs.

process, photomechanical - Any method or process not involving a lens system but making use of light-sensitive materials for reproduction by contact.

process, stochastic - A time series in which the time-dependent variables are random variables.

process, polyvinyl-alcohol - A photographic process using a water-soluble, synthetic resin, polyvinyl alcohol, which, when sensitized, forms the basis for making photo-mechanical prints on glass, plastic or other material. Abbreviated to P.V.A. The finished print is usually referred to as a P.V.A. positive or P.V.A. negative.

process, strip-masking - A method of producing a negative or positive mask by stripping or peeling off a thin, opaque, film of plastic from a translucent base.

process, three-color - A process of producing pictures or diagrams in several colors by printing with three complementary colors, other tints being obtained by combinations of two or three of those in solid colors, half-tones or rulings.

process, whiteline - SEE diazo process.

process camera - A camera designed particularly for photographing an original document in such a way that the image can be used to prepare a printing surface.

processing unit, central - That part of a computing machine in which instructions are carried out and calculations done according to a prearranged algorithm.

process lens - (1) A lens system designed specifically for photographically copying, enlarging or projecting images, nearly free from aberrations and usually of small aperture and of symmetrical construction. Also called a process lens system. (2) A type of lens system usually having low aperture (f/10, approximately), narrow field of view, long focal length, symmetrical structure and limited range of magnification (0.5x to 2x) and used in large copying cameras.

process lens system - SEE process lens.

process photography - The preparation of line photographs and halftone photographs for subsequent use in the preparation of plates for printing.

process plate - One of a set of printing plates, each producing a separate color, combined to produce other colors and shades. SEE ALSO plate, color; plate, combination; plate, press.

product of inertia - The integral $\int x_i x_j \, dm$, over the mass m of a body, with respect to the Cartesian coordinate system having axes x_1, x_2, x_3.

profile - A representation of the intersection between a vertical plane and a portion of the Earth's surface or the surface of an underlying stratum, or both. Alternatively, a representation of the intersection of a developable surface with the Earth's surface. SEE ALSO check profile.

profile grade - The elevation or gradient of the intersection of a vertical plane with the top surface of the proposed wearing surface of a road, usually along the longitudinal center-line of the roadbed. Whether elevation or gradient is meant must be determined from the context of the term.

profile leveling - Determining elevations at closely-spaced points along a line, in order to determine the profile of the ground along that line. In the United States of America, the points are marked by stakes 25 feet, 50 feet or 100 feet apart. The points at 100-foot intervals, starting from the beginning of the line, are called full stations; all other points are called plus stations. (2) Plotting changes in differences of elevation determined by comparing the results of two surveys done at different times.

program - (1) Any planned set of actions. (2) In the theory and use of automatic computing machinery, any set of instructions (algorithm) telling the computing machine what to do. Also spelled programme.

programme - SEE program (2).

programming - Putting together a set of instructions for a computer to follow.

programming, linear - A procedure for finding the best solution to a problem by expressing the problem as a set of linear inequalities in the variables, and then converting the inequalities to equations by introducing additional variables. The procedure has been used, for example, in planning surveys.

progress sketch - A map or sketch showing the progress made in a survey. In traversing and in triangulation, each point established, each line observed over and each base line measured is shown on the progress sketch. In leveling, the progress sketch shows the route followed and the towns passed through, but not necessarily the locations of the bench marks.

project - A definite undertaking, usually with definite limits and with specified standards.

projection (n.) - (1) A set of functions, or the corresponding geometric constructions, relating points on one surface to points on another surface in such a manner that to every point on the first surface corresponds exactly one point on the second surface. A projection differs from a map projection in that the latter deals only with situations in which one surface is an ellipsoid and the other a developable surface. (2) The extension of a line beyond the points determining its character and position. (3) The transfer of a sequence of surveyed lines to a single, theoretical line by lines perpendicular to the theoretical line. In traversing, a sequence of short, measured lines may be projected onto a single, long line, connecting two main stations of the traverse; the long line is then treated as a measured line of the traverse.

projection, arbitrary - SEE map projection, aphylactic.

projection, axonometric - A drawing which shows an object's position inclined with respect to the planes of projection. Also called an isometric projection.

projection, conformal double - A method, developed by O. Schreiber, for mapping the rotational ellipsoid onto the plane by first mapping conformally onto a sphere and then mapping conformally from the sphere onto the plane. Also called simply, double projection.

projection, conic - SEE projection, perspective.

projection, double - (1) Finding the point Q, on the reference ellipsoid, corresponding to a point P on the ground, by first finding a point Q' on the geoid corresponding to P and then finding the point Q corresponding to Q'. Two methods have been used. The point Q' has been taken as the point vertically below P on the geoid and Q has then been taken as the orthogonal projection of Q' onto the referenced ellipsoid. This method was invented by Pizetti and Meinesz. The other method is to first find Q' as before and then project the geoid isometrically onto the reference ellipsoid. Q is then identified with the projection of Q'. This method was devised by A. Marussi. SEE ALSO projection, conformal double. (2) SEE map projection, double.

projection, globular - SEE map projection, globular.

projection, isometric - (1) A mapping from a region on one ellipsoid, A, to a corresponding region on another ellipsoid, B, by first using a map projection to map from A to the plane, and then using an inverse mapping to go from the plane to B. (2) SEE projection, axonometric.

projection, orthographic - (1) The projection defined by a set of parallel, straight lines through corresponding points on the two surfaces. Equivalently, a perspective projection in which the perspective center is placed at infinity. The orthographic projection is regularly used in mechanical drawing. When so used, the surface in question is mapped onto each of three mutually perpendicular planes defining one quadrant. The Y-Z plane and the Z-X plane are then rotated about their intersections with the X-Y plane to show all three maps in one plane. (2) SEE map projection, orthographic.

projection, perspective - (1) The projection defined by a set of straight lines passing through corresponding points on the two surfaces and through a single point common to the set. Also called a conic projection if one of the surfaces is a plane. The lines form a sheaf. The common point is called the perspective center or center of perspective. One of the two surfaces is usually a plane or a developable surface. Unless stated otherwise, the perspective center is understood to be within a finite distance from the surface onto which points are mapped. (2) SEE map projection, perspective.

projection, polyconic - SEE map projection, polyconic.

projection, recentered - SEE map, interrupted.

projection convergence - The difference, on a graticule, between corresponding lines made by a specified kind of curve with the lines representing two meridians. More properly called graticule convergence.

projection cross - A small cross, on a map, indicating points at which lines through two pairs of ticks indicating lines of the graticule would intersect within the map.

Projection de la Carte de France - SEE Bonne map projection.

Projection du Depot de la Guerre - SEE Bonne map projection.

projection distance - The distance from the external node of a projector's lens system to the plane onto which the image is to be projected.

projection print - A print made by projecting the image from the photographic negative or transparency through a lens system onto the sensitized surface.

projection printer - A photographic apparatus which enlarges or reduces the scale of the image, on a positive or negative transparency, by projecting the image onto a sensitized surface movable toward or away from the projecting lens system. This term is applied also to various kinds of photographic apparatus which are not printers - rectifiers and photographic enlargers, for example.

projection printer, fixed-ratio - SEE diapositive printer, fixed-ratio.

projection station - That position which a projector has, in a stereoscopic plotting instrument, when absolute orientation has been finished. Also called a projector station. The projector station reproduces the geometric conditions existing at the corresponding camera station at the instant the photograph was taken.

projection stereoplotter, double-optical - SEE stereoplotter, double-projection, direct-viewing.

projection table - A table for determining the relationship between grid coordinates and graticule coordinates.

projection tick - One of a pair of very short lines (ticks) drawn perpendicularly to the neat lines of a map and representing the ends of a line of the graticule when the graticule itself, or that particular line, is not shown on the map. Projection ticks are used when a grid is present on the map, to avoid the confusion that might occur if both grid and graticule were shown.

projective method (of reduction to the ellipsoid) - A method of mapping points found by surveying at the Earth's surface to corresponding points on the ellipsoid, by projecting perpendicularly onto the reference-ellipsoid. There are two variations: Helmert's projective method which does the projecting in one step and Pizzetti's method, which does it in two steps. SEE ALSO Helmert's projective method; Pizzetti's projective method.

projectivity - The relationship between two figures, one of which can be derived from the other by a chain of perspectivities.

projector - An optical device projecting the image of a photograph, map or other graphic onto a surface where it can be viewed or photographed. It usually consists of a lamp for illuminating the image to be projected and an optical system for projecting the image. It does not include either the surface onto which the image is projected or, usually, the structure holding the device in a suitable position with respect to that surface. However, in photography and photogrammetry, the term is applied, in particular, together with the stage (the flat surface onto which the image is projected) and a structure allowing the relative positions of lamp, optical system and stage to be adjusted precisely but, once set, kept fixed. The entire apparatus: lamp, optical system and stage, together with a mechanism for adjusting and fixing their relative positions, are then also called a projector or projection printer. The stage is also called a copying table or copy board if the primary purpose of the apparatus is the copying of the image being projected.

projector, reflecting - A projector using mirrors as part of the optical system. In particular, an optical device projecting, by means of mirrors and lenses, the image of an aerial photograph onto a map.

projector station - SEE projection station.

project staff - A staff gage designed particularly for determining differences of height of the water's surface during a short period and for a specific project. The staff need not be elaborate, e.g., a 2" x 4" timber with a folding rule attached may be adequate.

prolong (v.) - To lengthen in extent or range.

prolongation - (1) The extension of a directed line beyond its end-point and in the same direction. (a) Applied to a segment in a line composed of segments of straight lines (as in a traverse), the prolongation is the extension of that segment closest to the last-noted (recited) intersection, geodetic marker or monument. (b) Applied to a directed, curved line, the prolongation is understood to be the extension, beyond the end point, of the tangent to the curve. However, prolongation in this application is sometimes used as if it meant continuation. A straight line is prolonged; a curve is continued. (c) Applied to a radial line drawn to circular arc, the prolongation is the extension of the radial line beyond its intersection with the arc. (2) A radial line drawn to a circular arc and extended beyond the curve.

proof - A printed copy made for inspection, trial or as an intermediate stage in printing a final product. SEE ALSO acetate proof; color proof; galley proof; pre-press proof; press proof.

proof, final - A detailed statement, by an entryman and his witnesses, supporting to prove that he has fully complied with the public-land laws relating to his entry.

proof, progressive - One of a series of proofs of color prints showing the individually separated color-printings and their progressive combinations as each color is superposed on the result of the previous printing.

proofing (v.) - Printing copies to be used for editing, proof-reading, approval of text, etc., before printing the final version.

proof mass - SEE test mass.

proof-reading - The process of reading and making corrections to a printed copy of a document before the final version is printed. The verbal form is reading the proof or reading the proofs, not proofreading. After proofreading, the annotated copy is sent to the printer, who uses it for making final corrections to the printing plates before printing the final version.

propagation, standard - The movement of radio waves over a smooth, spherical model of the Earth, of uniform dielectric constant and conductivity, and under conditions of standard refraction in the atmosphere, i.e., in an atmosphere in which the index of refraction decreases uniformly with altitude at a rate of 12 N-units per 1000 feet.

propagation of error - The effect of error in a quantity on a function of that quantity. Also called error propagation. SEE ALSO law of propagation of error.

property - (1) That which belongs exclusively to one person or legal entity. (2) Everything, corporeal or incorporeal, tangible or intangible, to which one person or groups of persons has ownership or the exclusive right to possess or use. (3) Everything having an exchangeable value or which goes to make up wealth or estate. SEE ALSO property, personal.

property, abutting - The buttings or boundings of lands, showing to what other lands they belong or on which they abut.

property, immovable - Real property and some chattels affixed to the realty.

property, personal - Property consisting of movable items, that is, items not permanently affixed to and a part of real estate. In deciding whether or not a thing is personal property or real estate, there must be considered, usually, (a) the manner in which the thing is annexed; (b) the intention of the person who annexed it (that is, did he intend to leave it permanently or to remove it sometime); (c) the purposes for which the place where the thing was located was used. Generally, but with exceptions, things remain personal property if they can be removed without serious injury either to the real estate or to the item itself.SEE ALSO property.

property, real - (1) The interests, benefits and rights inherent in the ownership of the physical real-estate. (2) The bundle of rights with which the ownership of real estate endowed. In some States, this term, as defined by statute, is synonymous with real estate.

property line - The line dividing two parcels of land or dividing a parcel of land from the street.

property map - SEE map, cadastral.

property right - The right of an owner in the property owned; a right growing out of the ownership of land.

property right, absolute - The uncontrolled dominion, of the person in whom it inheres, over the external objects of

property he owns, for all times and for all purposes. Also called an absolute right.

property right, qualified - The right, of the person in whom it inheres, to the object for certain purposes or under certain circumstances only. Also called a qualified right.

property survey - SEE land survey.

proportioning excess of deficiency (principle of) - SEE principle of proportioning excess or deficiency.

prorate (v.) - To divide or distribute proportionally.

proration - (1) A method of distributing an excess or deficiency of land, as discovered by a survey, between parties having equal rights or proportionate rights to the excess or deficiency. (2) A method of calibrating a surveyor's tape used in a recent survey against the tape used in the original survey, by dividing up (distributing) the excess or deficiency of the recently-used tape proportionately along the tape.

proton magnetometer - SEE magnetometer, proton- precession.

Prototype - An international or national standard of length or mass in the metric system of units. E.g., Prototype meter and Prototype of mass. The term is no longer used in this sense.

Prototype Meter - SEE International Prototype Meter.

Prototype Meter, International - SEE International Prototype Meter.

Prototype Meter, National - SEE National Prototype Meter.

protraction - (1) The process of continuing a curve or prolonging a straight line. (2) That straight line or curve which is created by continuation or prolongation. (3) The process of making a plat. (4) SEE plat. (5) In the surveying of land, the subdividing of land by drawing or extending lines on maps or plats of the region being subdivided. The lines are drawn before surveying and marking the boundaries of the subdivision and are therefore indicated by dashed, straight lines.

protractor - A plate marked, usually along the perimeter, with units of circular measure; the marks radiate from a marked point, called the center. A protractor is used in laying out, on a flat or curved surface, an angle of desired magnitude or in determining the magnitude of the angle between two lines on a plane. They are made in many sizes and forms to suit the purpose and convenience of the user. SEE ALSO protractor, coordinate; protractor, three-arm. SEE ALSO coordinate protractor.

protractor, three-arm - A circular protractor equipped with three arms whose edges, extended, pass through the center of the circle. The central arm is fixed and reads 0° on the graduated circle. The other arms are movable, and their positions on the circle are read using verniers. Also called a station pointer. The two movable arms can be set to the two angles measured between three fixed points (signals) of known location. The three arms are made to pass through the locations of the three points as shown on a map. The observer's location is then at that point, on the map, under the center of the protractor. This is a graphical solution of the three-point problem.

proving - SEE ALSO negative proving.

proving, lithographic - Using a lithographic printing plate to prepare a proof (usually on paper). SEE ALSO proving, negative.

proving, photomechanical - Preparing a proof on plastic or other material by a photomechanical process.

pseudo-inverse - SEE matrix, pseudo-inverse.

pseudoscopy - The sensing, when viewing the image of an object, of relief contrary in direction to the relief of the object itself, e.g., hills are seen as valleys, valleys as hills, etc. Also called reversal of relief. Pseudoscopy is most commonly seen when an image is illuminated from the wrong direction with respect to the viewer. When it occurs in viewing a stereoscopic pair of images, it may be caused by placing the images in the inverted order, so that the left eye views what the right eye should be seeing, and conversely. This is also referred to as pseudoscopic stereo, false stereo, inverted stereo, pseudo-stereoscopy and reverse stereo.

pseudo-stereoscopy - SEE pseudoscopy.

psychrometer - An assemblage of two ventilated thermometers, the bulb of one of the thermometers being kept moist by a bit of damp fabric. The difference of temperature indicated by the two thermometers is an indication of the relative humidity of the air. The thermometers are usually attached to a chain or pivoted arm so that they can be ventilated by being swung about rapidly. More elaborate psychrometers have motor or hand-driven fans for ventilation.

publication - In governmental affairs, the act of making known, to the general public, a law, regulation, notice or other official proclamation, by printing the proclamation in an official journal or its equivalent.

pueblo (Spanish law) - (1) People. (2) All the inhabitants of any country or place, without distinction. (3) Town, township or municipality. (4) A small settlement or gathering of people, a steady community. The term in its original signification means people or population, but is used in the sense of the English word town. It has the indefiniteness of that term and, like it, is sometimes applied to a mere collection of individuals residing at a particular place, a settlement or village, as well as to a regularly organized community.

Puerto Rico Datum - The horizontal-control datum which is defined by the following coordinates of the origin and by the azimuth (clockwise from South) to triangulation station Ponce Southwest base, on the Clarke spheroid of 1866; the origin is at triangulation station Cardona island Lighthouse:
longitude of origin   66° 38' 07.520" W

latitude of origin 17° 57' 31.400" N
azimuth from origin to Ponce
Southwest Base 128° 36' 26.2"

Adopted in 1901 or soon thereafter, Puerto Rico Datum is derived from observations on the island of Puerto Rico and the Virgin islands.

Puissant's formula - The formula developed by L. Puissant (1842) for the geodetic coordinates λ_2, ϕ_2 of a point on a rotational ellipsoid, given the coordinates λ_1, ϕ_1 of another point and the distance L and azimuth A from the given point to the point of interest, according to the formulas $\phi_2 = \phi_1 + LB \cos A - L^2C \sin^2 A - (LB \cos A) L^2D \sin^2 A - (\Delta\phi)^2 E$ and $\lambda_2 = \lambda_1 + \arcsin[\sin(L/v_2) \sin A \sec \phi_2]$, in which B,C,D,E are functions of ϕ_1, ϕ_2 and the radii of curvature at the given point; $\Delta\phi$ is the sum of the preceding three terms; and v_2 is the radius of curvature in the prime vertical. Puissant's formula has been much used by the U.S. Coast and Geodetic Survey. It is good to 7 figures up to a distance about 80 km, beyond which the errors increase rapidly.

Pulkova Datum 1932 - The geodetic datum based on the Bessel ellipsoid and having as origin the center point of the Round Hall at Pulkova Observatory:

longitude of origin 30° 19' 42.09" E.
latitude of origin 59° 46' 18.55" W.
azimuth from origin to Signal "A" 317° 02' 50.62"

The same origin, corrected for the deflection (of the vertical) resulting from an astrogeodetic adjustment and computed on the Krasovsky ellipsoid is known as Pulkova Datum 1942.

Pulkova Datum 1942 - A geodetic datum based on the Krasovsky ellipsoid,

semi-major axis 6 378 245 meters,
flattening 1/298.3;

and with the same origin at Pulkova Observatory as Pulkova 1932 Datum:

longitude of origin 30° 19' 42.09" E
latitude of origin 59° 46' 18.55" N
deflection of the vertical
  in the meridian 0.16" S
  in the prime vertical 1.78" N
  azimuth to Bugry 121° 41' 38.19"
  geoidal height at origin 0 meters

It is officially called the 1942 Pulkova System of Survey Coordinates.

Pulkova meridian - The meridian passing through the meridian circle at Pulkova Observatory. It has the longitude 30° 19' 39" east of Greenwich. It was used on Russian maps until 1920.

pull-up - SEE selection overlay.

pulsar - A star periodically emitting pulses at radio wavelengths. The period ranges from 0.03 seconds to over 3 seconds. Pulsars are believed to be rotating, magnetic neutron-stars. They have been used as sources of radio signals for radio interferometry.

pulse, coherent - In navigation systems operating at radio frequencies, pulses in which the phase of the radio-frequency waves within the pulse is retained for measurement.

punch, register-hole - A punch for punching holes in the margins of neighboring charts, maps, overlapping photographs, etc., so that when corresponding holes are lined up, as by inserting studs into the holes, the edges of the charts, etc., will match. A similar device has been used in punching holes for templets used in various mechanical forms of photogrammetric adjustment.

pupil - In an optical system, an opening, or the image of an opening, which limits amount of light that can pass through the system or a designated part of the system. SEE ALSO entrance pupil; exit pupil.

pushbroom method - A method of scanning the ground from aircraft or artificial satellite, in which the scanner senses in one instantaneous view an entire line, on the ground in the direction transverse to that in which the aircraft or satellite is moving. I.e., a linear set of points on the ground is mapped (imaged) instantaneously onto a line of discrete detectors in the scanner.

pushbroom scanner - A scanner which views the ground along a line transverse to the direction in which the scanner itself is moving. The term is applied, in particular, to a scanner in which there are many small detectors each viewing a different small spot on the ground, but all the detectors are placed in a single row transverse to the direction in which the scanner as a whole is carried.

pygmy current meter - A small current-meter intended for use in measuring low speeds in shallow water and consisting of small cups rotating about an axis connected to a counter. Also called a pygmy meter.

pygmy meter - SEE pygmy current meter.

pyramid - A closed surface with flat faces one of which (the base) is a polygon (and its interior), and the others are the triangles (and their interiors) each of which is formed by one side of the polygon, a point (the vertex) not in the same plane as the base, and the two lines joining the vertex to the two ends of the side. SEE ALSO photograph pyramid.

pyramid, photogrammetric - The double pyramid formed by the combination, at their vertices, of a ground pyramid with the corresponding photograph pyramid. SEE ALSO photograph pyramid.

Pythagorean band inversor - An inversor in which the movements of the planes of lens system's perspective center and photographic negative are regulated by bands connecting these planes and the plane of the easel through a system of fixed and movable pulleys. SEE ALSO inversor.

PZS triangle - A triangle, on the celestial sphere, having one vertex (P) at the pole, one point (S) at the star in question, and the third point (Z) at the zenith of the observer.

Q

Q - SEE storage factor.

Q-factor - SEE storage factor.

quad - (1) A region, on an ellipsoid representing the Earth, bounded by two meridional arcs and two geodetic parallels all of the same angular extent. For example, a 1° x 1° quad is a region bounded by two meridional arcs 1° apart and by two geodetic parallels 1° apart. (2) SEE quadrangle (3).

quadrangle - (1) A figure, in the plane, consisting of four specified points and the lines or line segments on which they lie. The quadrangle and the quadrilateral may define the same shape. They differ in that the quadrangle is defined by four specified points, the quadrilateral by four specified lines or line-segments. (2) In particular, on a sphere representing the Earth, a geometric figure having two sides which lie along meridians and two other sides which lie either along great circles perpendicular to the first two or along geodetic parallels between the two meridians. (3) A map or plot of a rectangular or nearly rectangular region, usually bounded by given meridians of longitude and parallels of latitude. Sometimes called a quad (jargon) or quadrangular map. In the USA, the standard map is the quadrangle of 7.5' of longitude and latitude. Other sizes are integral multiples of the standard.

quadrangle, complete - A figure, in the plane, consisting of four specified points and the six lines they determine.

quadrangle, simple - A figure, in the plane, consisting of four specified non-collinear points and the four line-segments joining these point, each point being the end point of exactly two line-segments.

quadrangle map - SEE quadrangle (3).

quadrangle report - A brief history of the mapping of a quadrangle. The report accompanies the mapping material through each phase of production and is filed with that material.

quadrant - (1) A region, on the plane, bounded by two mutually perpendicular half-lines meeting at a point. Two full lines intersecting at right angles divide the plane into four parts; each of the four parts is a quadrant. The quadrants are customarily numbered in Roman numerals counter-clockwise about the point of intersection, the number I being given to the quadrant in the upper right. (2) A sector of a circle, having an arc of 90°. I.e., the smaller of the two regions into which the circle and its interior are divided by a pair of radii drawn at right-angles to each other. (3) A surveying or astronomical instrument composed of a graduated arc encompassing about 90°, and a sighting device (alidade) pivoting about the center of the circle of which the arc is a segment. The quadrant can be considered a mechanical form of the sector. Some quadrants combine both surveying and astronomical functions, having two arcs, one horizontal and the other vertical. (4) A region, on the Earth's surface, included with a quadrangle ¼° on a side. (6) A map of a region, on the Earth's surface, included within a quadrangle ¼° on a side.

quadrature - (1) In general, the relationship between two periodic quantities whose relative phase differs by a quarter of a cycle. (2) An astronomical configuration in which two celestial bodies have apparent celestial longitudes differing by 90° as viewed from a third body. (3) In particular, that astronomical configuration in which lines drawn from the center of the Earth to the centers of the two principal tide-producing bodies, the Sun and the Moon, are nearly at right angles. The Moon is then in quadrature in its first or last quarter. (4) The process of solving, or the solution of, the differential equation $dn y/dx n = f(x)$ with boundary conditions, the process or the solution being given as a multiple definite integral. (5) The process of solving for the length of the side of a square having the same area as the area inclosed within a given, closed curve. This is the original meaning of the term. It can be shown to be equivalent to definition (3) when n is 1, or can be made equivalent to (3) by changing square to hypercurve and area to volume, etc.

quadrature, high-(low-)water - (1) The average high (low) water interval when the Moon is at quadrature. (2) SEE neaps, mean high (low) water.

quadrature, low-water - SEE quadrature, high-(low-)water.

quadribrach - A four-armed base of a surveying instrument which carries the foot-screws used in leveling the instrument. Also called a leveling base or leveling head. The quadribrach is unsuited for establishing control of third-order or higher because three screws are sufficient for leveling an instrument. Adding a fourth screw constrains the motion and may introduce strains into the base of the instrument. This in effect may affect the direction in which the instrument is pointed during observation.

quadrilateral - (1) A figure, in the plane, consisting of four specified, non-concurrent lines or line segments and the points in which they intersect. (2) A figure, on a surface, consisting of four specified line-segments or arcs connected at their end-points, each end-point connecting two and only two line-segments or arcs. This definition allows two different shapes: a loop-shaped figure and an hourglass-shaped figure. Unless specifically stated otherwise, the term denotes a loop-shaped figure (also called a simple quadrilateral). In geodesy, the surface is usually an ellipsoid or a plane. The lines are usually, but not necessarily, geodesics. They may or may not be ordered, i.e., be numbered in sequence. If they are ordered, the points have the corresponding order. The lines may be directed, i.e., each line may have one direction associated with it. (3) SEE quadrilateral, simple. (4) SEE quadrilateral, braced.

quadrilateral, braced - The set of six lines, on a surface, connecting four points, no three of which are collinear, in pairs. Also called a doubly-braced quadrilateral or simply, a quadrilateral. The lines are usually geodesics. The four lines forming the perimeter of the quadrilateral are called the sides; the other two lines are called the diagonals. In particular, the set of six straight lines, on a plane, connecting four points, no two of which are collinear, in pairs. SEE ALSO quadrilateral, complete; quadrilateral, singly-braced.

quadrilateral, centered - The figure obtained from a simple quadrilateral by adding a fifth point in the interior and drawing lines from that point to the other four points.

quadrilateral, complete - A figure, in the plane, consisting of four specified straight lines and the six points in which they intersect.

quadrilateral, double - A figure prepared from two simple quadrilaterals having a common side by connecting each of the six points to each of the other five points. This figure has been used in trilateration where great distances between points can be measured.

quadrilateral, doubly-braced - SEE quadrilateral, braced.

quadrilateral, simple - A figure, in the plane, consisting of four specified line-segments and four points, each line-segment included between just two points and each point lying on just two line-segments. Also called a simple quadrilateral.

quadrilateral, singly-braced - A geometrical figure consisting of four points, the four non-intersecting lines joining these points and a fifth line joining two opposite points. Equivalently, a braced quadrilateral with one of the two diagonals removed. A singly-braced quadrilateral would be treated, in adjustment, as a pair of triangles with a common side.

quadrilateral method - SEE Hause method.

quadrilateration - (1) The method or procedure of making a survey in which a network consisting of simple or braced quadrilaterals is established. The method is also called the method of quadrilaterals. (2) A network composed of simple or braced quadrilaterals. (3) A method of determining the coordinates of a point by measuring distances simultaneously, from that point and from three points whose coordinates are known, to a satellite. Distances must be measured to at least three different locations of the satellite.

quadrillage - (1) A grid composed of two families of equally-spaced, parallel, straight lines, the one family intersecting the other at right angles so as to divide the plane into squares. (2) The same as the previous definition, but with squares replaced by rectangles.

quadripod - A structure consisting of four straight beams joined rigidly together to form the edges of a four-sided pyramid, with a vertical beam extending up through the vertex of the structure. The quadripod has been used in triangulation as a support for a flag or marked disk.

quality control - A set of standards for the quality of a product, a set of procedures for taking samples of the product and a set of statistical procedures for analyzing the results of the sampling and for evaluating the quality of the product based on the analysis.

quantity - (1) Any arithmetic, algebraic or analytic expression which is concerned with value rather than with relations between such expressions, i.e., whatever can be added, subtracted, multiplied or divided. Also called a mathematical quantity to distinguish it from quantity as defined in (2) following. For example, 5, x, (x + y) are or can be quantities; y < 2 is not. A parameter is a quantity having a very specific property in mathematics; quantities are not parameters. SEE ALSO quantity, algebraic; quantity, numerical; parameter. (2) Anything which, by its nature or by definition, has a number or numbers associated with it. Also called a physical quantity to distinguish it from quantity as defined in (1) above. SEE ALSO quantity, physical. For example, length, speed, mass, pressure and temperature are quantities. The associated number is called the value of that quantity. A quantity may have only one value (e.g., the speed of light in a vacuum), several values (e.g., the velocity of light in a vacuum has a value indicating its magnitude and one or more values indicating its direction) or an infinite number of possible values (e.g., the temperature of a cooling body). The term parameter is sometimes erroneously used as a synonym for quantity in this sense.

quantity, algebraic - A mathematical quantity to which a value can be assigned more or less arbitrarily or according to defined rules.

quantity, numerical - A number considered as a quantity.

quantity, physical - The product of a real number and a (physical) unit. A unit here means a quantity having the value 1 and having specific dimensions in some system of units. E.g., the quantity 5 kilograms is the real number 5 and the unit 1 kilogram.

quantity surveyor - A surveyor specializing in determining volumes and quantities of material.

quantization - (1) Basically, assigning a numerical value to a physical quantity. In simple language, measurement. A piece of land may be classified as large, medium, or small in extent. If large areas are assigned the value 3, medium-sized areas the value 2 and small areas the value 1, then the concept of piece of land has undergone quantization.

quantum theory - (1) The theory that electromagnetic radiation interacts with matter in discrete packets (quanta) of energy, and not as continuous waves. Many phenomena, however, can best be explained by assuming that electromagnetic radiation exists both as continuous waves and as quanta. (2) The theory dealing with solutions to Schroedingr's equations (or Heisenberg's matrices).

quarter line - SEE line, quarter.

quarter-section - SEE section, quarter.

quartz horizontal magnetometer - A magnetometer measuring the intensity and direction of the horizontal component H of the Earth's magnetic field. It consists of a short, magnetic bar suspended at its middle by a long, quartz fiber attached at its upper end to the instrument's casing. The torque exerted on the magnet by the magnetic field is balanced by controlling the torque exerted by the quartz fiber.

quartz pendulum - A pendulum made of quartz, usually of fused quartz. The coefficient of thermal expansion of quartz is only one-fourth that of invar.

quasar - A celestial object resembling a star when observed visually or photographed, but having an extremely high redshift. Objects within the Galaxy and identified as stars have redshifts less than 0.02. The lowest redshift found for a quasar is 0.06. Many objects known to be galaxies have very high redshifts, but they do not resemble stars. The magnitude of an object's redshift is associated, through Hubble's constant, with the object's distance from the Galaxy. A quasar may be a star at the same distance as a very distant galaxy, in which case its absolute magnitude must be incredibly great, or it may be a galaxy so far away as to appear starlike under any magnification but so condensed as to give the same brightness. At present, quasars are believed to be very dense concentrations of stellar material at galactic centers. They are of geodetic interest because of their usefulness in radio interferometry and similar applications. The term was derived from the name quasi-stellar or QSO.

quasi-geoid - The surface whose distance from the normal ellipsoid (in the sense of Molodensky) is the same as the distance to the Earth's surface from that level surface which corresponds to the normal potential (in the sense of Molodensky). The quasi-geoid coincides with the geoid in the still ocean; the difference in height between geoid and quasi-geoid will never exceed 2 m even in mountainous regions with heights up to 4000 m.

quasi-leveling - SEE leveling, geostrophic.

quasi-nivellement - SEE leveling, geostrophic. The term was invented by D. la Cour in 1913.

quay - A wharf consisting of a bulkhead filled with rock, sand, etc. and suitably finished on top. SEE ALSO wharf.

quiet title - A court action to obtain a clear title, remove cloud of title and establish ownership.

quintant - An instrument resembling a sextant but having a graduated arc covering 72° (one-fifth of a circle), i.e., covering a range of 144°.

quitclaim deed - SEE deed, quitclaim.

R

radar (RAdio Detection And Range) - An instrument for detecting the presence of an object and determining its location, by means of electromagnetic radiation emitted or reflected from that object. Direction is usually also obtained, either from a graduated disk and a rotating antenna, or by means of a movable antenna-field pattern. The term is applied properly only to instruments measuring distances at radio frequencies. It is sometimes applied also to instruments measuring only angles or frequencies and to instruments using infrared radiation or light. A better term for an instrument operating at infrared or optical frequencies is lidar or electro-optical distance-measuring instrument (EDMI). The usual radar consists of a transmitter sending a narrow beam of pulses in a specific direction, a receiver which amplifies the returned signal, a circuit which determines the difference in phase or time between the emitted and returned signal and calculates from this the distance of the object. Large radar is usually mounted on an alt-azimuth mounting in which angular motion in azimuth and angular elevation is measured by resolvers, allowing the direction of the object to be determined. Some radars measure the Doppler shift in the returned signal, allowing the objects radial velocity to be determined as well. It has become common to refer to instruments determining radial velocity only as radar. SEE ALSO lidar. SEE ALSO angle radar, side-looking; Doppler radar; Doppler radar, pulsed; laser radar; monopulse radar.

radar, imaging - Radar displaying the returned signal in such a form (as on the screen of a cathode-ray tube) as to show the shape of the reflecting object. The display is often photographed when a record is wanted for later analysis.

radar, optical - SEE distance-measuring instrument, optical; lidar.

radar, panchromatic - Radar operating over a wide, continuous band of frequencies.

radar, plan-position-indicator - Radar having a rotating antenna with a narrow but high main lobe which scans around all or part of a complete circle and showing the horizontal locations of objects within the scanned region on a plan-position-indicator (a display, on the screen of a cathode-ray-tube, which indicates objects as spots in a polar coordinate-system).

radar, polypanchromatic - Radar operating over several wide, continuous bands of frequencies, each band centered about one of a small number of separate frequencies.

radar, satellite-surveillance - Radar designed specifically for determining distance and/or direction to artificial satellites.

radar, side-looking - Radar designed so that its beam is pointed, either by scanning or by shaping, in a direction perpendicular to the longitudinal axis of the aircraft or spacecraft carrying the radar. The returned radiation comes from a long, narrow strip of ground approximately perpendicular to the craft's line of flight. Also called side-looking airborne radar and SLAR, although radar of this type has been carried on spacecraft as well as on aircraft. The ground is mapped as a sequence of adjacent, overlapping strips.

radar, side-looking airborne - Side-looking radar mounted in an aircraft.

radar, side-looking angle - SEE angle radar, side-looking.

radar, synthetic-aperture - Radar in which a moving antenna sequentially occupies all those positions which would have been occupied by a larger, stationary antenna (a virtual antenna). The signals received by the moving antenna are combined to produce a signal equivalent to that which would have been received by the virtual antenna. The angular resolution of the actual antenna is much less than the angular resolution of a much larger antenna. The amplitude and phase of the signal received at the actual antenna are stored for each position which the antenna occupies as it moves. These quantities are subsequently combined to yield a signal giving the angular resolution which would have been obtained had a larger antenna been used instead.

radar altimeter - An altimeter which emits pulses at radio frequencies, measures the time it takes a pulse to travel from the altimeter to the ground below and back and calculates the distance to the ground from this, using the speed of radio waves in the atmosphere. The pulses are emitted in a broad beam, so that the radiation returning first can be expected to come from the closest point of the surface (assumed to be that part perpendicularly below the alti-meter). Radar altimeters are used primarily as a necessary aid to aerial navigation or to provide supplemental information in aerotriangulation or aerial photography. They have been mounted in artificial satellites as well as in aircraft.

radar altimetry - (1) Altimetry done using a radar altimeter; determination of the distance of a radar altimeter above the physical surface of the Earth or other celestial body. Distances above the Earth are usually determined by radar altimeters mounted in aircraft and are used primarily as such, for navigation. Distances determined by radar altimeters mounted in satellites are usually converted to geodetic heights of the surface. (2) Determining geodetic heights of points on the Earth's surface (or surface of some other celestial body) from measurements made by a radar altimeter mounted in an aircraft or satellite. Such determination requires that the location of the altimeter be known in the same geodetic datum.

radar altitude - The altitude indicated by an airborne radar. Also called absolute altitude.

radar astronomy - Astronomy carried out at radio wavelengths using pulses generated by radar and reflected from or scattered by celestial bodies. I.e., radio astronomy using radar.

radar chart - A chart intended primarily for use with radar, or a chart suitable for such use.

radar equation - An equation giving the (electric) power received at the antenna of a radar as a function of the maximal power transmitted (peak power), the gain of the antenna, the wavelength used, the equivalent area of the reflecting object and the distance of the object. Other factors may also be included, but those enumerated are the most common.

radargrammetry - A method of surveying, using radar, in which pulses returned from a region scanned by the beam of an airborne radar are combined to form, on the cathode-ray screen, a picture of the region. The picture is photographed and measurements made on the picture to determine, by suitable photogrammetric methods, horizontal and vertical distances between points on the ground. Radargrammetry is the analog, in the radio part of the spectrum, of photogrammetry as done in the optical part of the spectrum. It is usually considered part of photogrammetry.

radar imagery - The entire process of producing images by use of radar-type equipment. SEE ALSO imagery.

radar map - A map produced by applying techniques associated with radar, for example, by photographing the display on the cathode-ray tube of an airborne radar-altimeter.

radar map matching - Matching a radar map with a printed map or photo map.

radar mile - An interval of time equal to 10.75 microseconds. This is the length of time it takes a radio signal to travel two miles (i.e., to an object one mile away and back).

radar photography - Photography dealing with the taking of photographs of the display on the viewing screen of a radar. Radar photography is photogrammetrically important when the radar is carried in an aircraft and scans the ground underneath. SEE ALSO radargrammetry.

radar recording camera - A camera mounted with the lens pointing at the screen of a cathode-ray tube in radar so that a photograph is taken of the image on the screen.

radar reflector - Any device made to reflect radio waves or pulsate the frequencies used by radar.

radar shadow - A shadow thrown by an obstacle to radio waves from radar, i.e., a region in which radio waves emitted by radar are either totally absent or considerably reduced in intensity, because of the obstacle intervening between the radar and the region.

radar surveying - Determining the distance between two radar beacons on the ground, by measuring distances to the individual beacons from a radar carried on an aircraft. For example, HIRAN and SHORAN.

radar target - Any object reflecting or scattering sufficient energy at frequencies used by radar to produce a signal on a radar's screen.

radar telescope - A radar transmitter, antenna, receiver and associated equipment used for radar astronomy. Also called a radio telescope.

radial (adj.) - Directed toward or away from a designated point; directed along a radius of a circle. E.g., radial velocity is the velocity of an point in a direction toward or away from a designated point such as an observer. Direction away from is usually considered positive, so a positive radial velocity would be a velocity away from the designated point.

radial (n.) - (1) A line or direction from the radial center of a photograph to a point on the photograph. Unless otherwise specified, the radial center is assumed to be the principal point of the photograph. SEE ALSO nadir radial; isoradial. (2) SEE line, radial.

radial, principal-point - A radial drawn from the principal point as radial center. Unless some other point is designated as the radial center, the principal point is assumed.

radial assumption - The assumption that angles between lines drawn from a particular point (the radial center) on a photograph to other points on the photograph are equal to the angles between corresponding lines drawn on the ground. Also called the radial-line assumption. This assumption is usually not correct; the conditions that it be correct - perpendicularity of the camera's optical axis to the horizontal plane, lack of distortion in the photograph, absence of relief on the ground, etc. - are usually not satisfied. In an aerial photograph taken from well off the vertical and taken of ground having considerable relief, neither the nadir point nor the isocenter is the theoretically correct radial center. The photographic nadir point should be used as the radial center if relief is the major contributor to distortion, and the isocenter should be used if tilt is the major contributor. SEE ALSO assumption, principal-point.

radial center - That point, on a photograph, from which lines (radials) are drawn, or distances measured, to other points on the photography. Also called the center of radiation and center point. The principal point, the nadir point, the isocenter or a point (the substitute center) which is close to one of these and easily identifiable is chosen to be the radial center.

radial-line method - SEE aerotriangulation, radial.

radial method, direct - Graphic radial aerotriangulation done by tracing the directions from successive radial centers directly onto a transparent sheet rather than by laying out the triangulation with templets. Also called direct radial plot.

radial plot - SEE triangulation, radial.

radial plotter - SEE plotter, radial.

radial triangulation - SEE triangulation, radial.

radian - A unit of angular measure equal to the angle subtended at the center of a circle by an arc whose length equals the length of the radius of the circle. Alternatively, the angle subtended at the center of a circle of radius 1 by an arc of length 1. The circumference therefore encompasses 2π radians (360°), and one radian is approximately equal to 57.295 779 513° or to 206 264.806 247". The milliradian, a frequently used unit, is approximately equal to 3.5' and 0.1 milliradian is approximately equal to 20".

radiation - (1) The movement of energy away from a source. (2) A method of surveying in which a theodolite is placed at any convenient point from which the observer can see all points whose locations are wanted. Directions and distances are then measured from the point of observation to the other points. (3) Elementary particles such as electrons or atomic nuclei moving through space.

radiation, electromagnetic - That form of energy which moves through space or matter, interacting with matter to give rise to electric currents or voltages and magnetic fields. Also called electromagnetic energy and radiation.

radiation, gamma - Electromagnetic radiation having wavelengths shorter than about 10^{-12} m. The spectrum of gamma radiation overlaps slightly the region called hard x-rays.

radiation, infrared - That part of the electromagnetic spectrum containing radiation at wavelengths from about 0.8 μm to 300 μm. Also referred to as infrared, infra-red or infrared light.

radiation, near-infrared - Infrared radiation having wavelengths between 0.8 μm and 4 μm.

radiation, ultraviolet - Electromagnetic radiation at wavelengths shorter than about 400 micrometers (4000 Angstroms).

radiation pattern - A surface showing the amount of power radiated into or detected from different directions, relative to the total amount of power radiated or received. Also called the antenna pattern. The radiation pattern for power radiated from an antenna is the same as that for power received by the antenna. That is, the proportion in which an antenna emits power in different directions is the same as the proportion in which the sensitivity of the antenna varies for power coming in from those directions.

radiation pressure - The pressure exerted upon an object by radiation incident upon that object. In particular, the pressure exerted by electromagnetic radiation from the Sun. Pressure exerted by the solar wind is not usually called radiation pressure.

radio - A device capable of emitting and/or receiving signals in the form of radio waves; i.e. a device capable of converting signals put into it in some suitable form into corresponding radio waves modulated by the signals or detecting and converting radio waves modulated by signals into corresponding signals in a form suitable for use, e.g., as speech, Morse code or pictures. Radio is understood to mean the device for detecting and converting radio signals unless stated otherwise; if confusion is possible, the term radio receiver should be used, with radio transmitter or transmitter being used for the form generating modulated radio waves. The term radio is also used for the form combining both radio receiver and radio transmitter.

radio altimeter - An altimeter emitting radio waves in a downward direction and comparing the time or phase of the scattered or reflected return from the ground with the time or phase at which the radio waves were emitted, to determine a distance from the altimeter to the ground. The two most common types of radio altimeter are the FM altimeter, which emits frequency-modulated radio waves continuously and compares phase of returned waves with phase of emitted waves and the radar altimeter, which emits pulses whose times of emission are compared with times of reception.

radio astronomy - The part of astronomy in which observations are made at radio wavelengths, i.e., at wavelengths longer than 200 micrometers. It is sometimes taken to include only astronomy by observations on radiation originating at the object being studied, as distinct from astronomy by observations on radio waves generated on Earth and reflected or scattered by the object. The latter kind is then called radar astronomy. Properly, radio astronomy covers both kinds. Radar astronomy, in any event, limits itself to observing bodies within the Solar System, since at present we cannot generate radio waves powerful enough to be detected after scattering from bodies outside our System.

radio beacon - A radio transmitter which emits a distinctive or characteristic signal used for navigation or surveying.

radio distance measurement - The measurement of distance by using radio waves as yardsticks.

radio distance-measuring equipment - Electromagnetic distance-measuring equipment which determines the distance between two points by measuring the difference in time or phase between radio waves emitted by an instrument at one point and the time or phase of the waves received at the other, or by measuring the same difference with the same instrument acting as both transmitter and receiver, the waves traveling from the first point to the second and back again. It is usually abbreviated to radio DME. It is also referred to in some places simply as distance-measuring equipment; this usage, however, is probably restricted to radio technicians. There are two major types: equipment which sends out and receives pulses and measures the difference in time of transmittal and time of reception; and equipment which sends out continuous radio waves and measures the difference in phase between emitted and received waves. Pulsed radar is the outstanding example of the first type; radio DME used by surveyors is usually of the second type. Equipment in which pulses are emitted from two or more points on the ground and the receiver is located on an aircraft has been used successfully for determining long distances. SEE ALSO radar; radio interferometer.

radio distance-measuring instrument - An electromagnetic distance-measuring instrument using the radio-wave portion of the spectrum (i.e., from about 0.3 cm and longer). Also called a radar; however, that term is more usually applied to radio distance-measuring instruments too large to be easily carried about by one or two people. Radio DMI and electro-optical DMI operate on approximately the same physical principles, but radio DMI can be used successfully in foggy or rainy weather, whereas electro-optical DMI works well only in fairly clear weather.

radio frequency - The number of cycles per second gone through at a specific point by monochromatic electro-magnetic radiation in the radio portion of the spectrum, i.e., the portion containing radiation at wavelengths longer than 1 mm. It is customary to group (classify) radio frequencies and radio wavelengths into bands, each band containing radio waves with frequencies or wavelengths lying between sharply defined limits. One internationally recognized scheme of classification places each band between the limits $3 \times 10n$ and $3 \times 10^{n-1}$ cycles per second in frequency or between $10m$ and $10m^{-1}$ meters in wavelength. n is an integer equal to or less than 11; m is an integer equal to or greater than -3. (The factor 3 was chosen so that the wavelength corresponding to a specified limiting frequency is equal to 300 000 000 m/s (the approximate speed of radio waves in a vacuum) divided by the frequency.) Each band has a name: extremely high frequencies (EHF, 300 Gc/s to 30 Gc/s); super-high frequencies (SHF, 30 Gc/s to 3 Gc/s); ultra-high frequencies (UHF, 3 Gc/s to 300 Mc/s); very-high frequencies (VHF, 300 Mc/s to 30 Mc/s); high frequencies (HF, 30 Mc/s to 3 Mc/s); medium frequencies (MF, 3 Mc/s to 300 kc/s); low frequencies (300 kc/s to 30 kc/s); and very-low frequencies (VLF, 30 kc/s and lower). Bands arranged according to wavelength take the same names as the corresponding bands arranged according to frequency. Radar was developed in great secrecy during World War II. To preserve this secrecy, the Allies gave arbitrary, literal designations to certain bands of frequencies when research was being done. These designations are still common in American literature even where radar is not being discussed. There is, however, no general agreement even among Americans on what the precise limits of the designated ranges are, and there is considerable variation in the published literature and in practice. A common classification is as follows: UHF (0.3 to 3 Gc/s); L (1 to 2 Gc/s); S (2 to 4 Gc/s); C (4 to 8 Gc/s); X (8 to 12.5 Gc/s); Ku (12.5 to 18 Gc/s); K (18 to 26.5 Gc/s); Ka (26.5 to 40 Gc/s); millimeter (above 40 Gc/s). No classification of this kind can be trusted to apply everywhere, and some periodicals now refuse to accept designations of this kind.

radio horizon - A line at which direct rays from a radio antenna become tangent to the Earth's surface.

radio interferometer - A device combining two separate but coherent radio waves in such a way that the waves alternately reinforce and cancel each other, creating either a standing wave or a wave of lower and a wave of higher frequency. One type of radio interferometer is of the Fabry-Perot type and is used for measuring distances. One part of the radio wave is sent directly to the interferometer (called, in the case of radio waves, a Fabry-Perot resonator); the other part travels to the end of the line being measured and is sent to the resonator upon its return. Radio interferometers used in astronomy and in some geodetic work intercept the radio waves from a distant object at two different antennas. The voltages generated at the antennas are conducted by cable to a single amplifier and mixer (the interferometer). So-called very-long-baseline interferometers do not combine the radio waves at a single receiver but record the voltage variations on magnetic tape either digitally or in analog form. These data are then combined and analyzed. The results are equivalent to those obtained by analyzing the data from a true interferometer.

radiometer - An instrument for measuring the intensity of electromagnetic radiation in some band of wavelengths in any part of the spectrum. The term usually is modified to indicate the band investigated, e.g., infrared radiometer or microwave radiometer. The term does not denote an instrument operating only in the radio band.

radio method (of determining longitude) - Determining astronomic longitude by comparing local sidereal time with the time broadcast by a radio station whose exact astronomic longitude is known and whose approximate distance from the unknown station is known. Also called the wireless method of determining longitude. The approximate distance between stations must be known in order to correct the observed time of reception for the time it took the signal to travel between stations. Because radio signals travel at the speed of light, even a large error in distance causes only a small error in the correction. However, if the stations are very far apart, the radio wave may zigzag back and forth between ionosphere and ground one or more times before reaching its destination. This will add substantially to the total time taken for the radio wave to go between stations and must be taken into account. The astronomic longitude is the longitude of the known station plus the angular equivalent of the difference in sidereal times at the two stations.

radio navigation - Navigation in which location or velocity is inferred from measurements on radio waves. The term is usually applied only to one of the following methods of navigation: (a) measuring direction or distance to two or more radio transmitters; (b) measuring differences of distance to two or more pairs of radio transmitters; (c) measuring the Doppler shift in the frequency of a signal from an orbiting beacon or beacons. It may be used also of navigation depending on measuring the Doppler shift in radio waves emitted from the moving unit and returned by scattering from the ground and converting this to a velocity. But it usually does not refer that part of aerial navigation involved in launching or landing an aircraft.

radio positioning-device - That part of a radio positioning-system which is located at the point of unknown location.

radio positioning system - (1) A positioning-system in which the travel-time or phase shift of radio waves is measured to determine distances or angles. Also called an electronic positioning-system, although that term is properly used for positioning systems which use electronic components and therefore includes positioning systems using infrared radiation, etc. (2) SEE positioning-device, radio.

radio receiver - A device which takes the variations in voltage or current induced in an antenna, amplifies them, can convert them to a form immediately usable for generating intelligible signals such as sound or code or to a form suitable for printing or display. Commonly referred to as a radio and receiver. The radio receiver is the intermediary between the antenna on which the radio waves impinge and the final sound or symbol generator such as loudspeaker or printer.

radio refraction - Refraction of radio waves by the atmosphere.

radiosonde - A group of instruments sent aloft by balloon, rocket or other craft to gather meteorological data and send the data back by radio. SEE ALSO sonde.

radiosonde balloon - A sounding balloon carrying a radiosonde.

radio source - In astronomy, any celestial object such as the Sun, Jupiter, certain stars, nebulae, galaxies and gaseous clouds, emitting detectable radio waves. Radio sources were originally called radio stars because stars were believed to be the sources for the radio waves detected.

radio star - (1) A star emitting radio waves in an amount detected by radio telescopes. (2) SEE radio source.

radio telescope - (1) A radio receiver used for receiving, amplifying, heterodyning, detecting and recording radio waves from natural sources of radio radiation. A radio telescope consists of an antenna which collects power from incoming waves (in the form of varying voltages and currents), an amplifier for amplifying these voltages, a detecting or other circuit for modifying the variations to a usable form and a recording device such as a tape recorder for recording the output. Because the radiation to be detected is extremely weak, the antennas are usually very large in order to collect as much power as possible. Most radio telescopes employ antennas similar in theory to reflecting optical systems (catoptric optical systems), e.g., of the Cassegrainian type. Radio telescopes of small or medium size are usually directed by pointing the antenna itself in the desired direction. Those of very large size are directed by complex circuitry which points the main lobe and not the antenna itself. SEE ALSO antenna; radio interferometer. (2) SEE telescope, radar.

radio transmitter - A device which takes the variations in voltage or current generated by a microphone, printer or similar device and converts them to amplified and modulated AC voltage or current suitable for radiation from an antenna. Commonly referred to as a transmitter.

radio waves - Electromagnetic waves (radiation) having wavelengths longer than 1 mm or, in some classifications, longer than 0.3 mm. Radiation shorter than 0.3 mm is generally considered to be infrared radiation. The region between 1 mm and 0.3 mm may be considered to be either radio waves or infrared radiation.

radio waves, classification of - SEE classification (of radio waves); radio frequency.

radius - (1) A straight line drawn between the center of a circle and a point on the circle. (2) A straight line drawn between the center of a figure (closed surface) and a point on that figure. (3) The distance, along a straight line, between the center of a figure and a point on the figure. A radius is not considered a directed line or signed distance unless specifically stated to be so, even tho it may be defined as being drawn or measured from the center to a point on the circle or figure. (4) The length of a straight line drawn from the center of a figure to a point on the figure. The radius of a circle and the radius of an n-dimensional sphere have a constant length, i.e., all radii of the figure have the same length. In celestial mechanics, the vector from the focus to the body is often referred to as the radius vector, but this is not the same as the radius of the orbit.

radius (of the Earth) - (1) A line drawn from the center of an ellipsoid representing the Earth to the ellipsoid. (2) The length of the line defined in (1). (3) A line drawn from a specified center of the Earth to the surface of the Earth. (4) The length of the line defined in (3).

radius (of the Earth), effective - A fictitious value of the Earth's radius which, when used in a formula for the distance from a point to the geometric horizon, gives the same answer as would using the true value of the radius in a formula for the distance to the apparent horizon. If the Earth were spherical and without air, tangents drawn to the surface from a point above it would define a circle (the geometric horizon) outside of which no points on the sphere would be visible from the chosen point but all points inside it would be. If an atmosphere were added, points outside the circle would become visible because the line of sight would curve and extend to points beyond the geometric horizon and as far as a circle (the apparent horizon) of greater radius. The distance to the apparent horizon can be calculated using the same formula as is used for the distance to the geometric horizon if, in that formula, a larger value, the effective radius, is used for the radius of the Earth. The effective radius, for points of observation within the troposphere, is greater than the true radius and increases with wavelength of the radiation at which the observations are made. It is particularly useful in determining, without long calculation, whether two points a considerable distance apart will be intervisible or able to receive each others signals. For radio waves at ultra-high frequencies, the effective radius is about 4/3 the true radius.

radius, gravitational - (1) The radius r_g of a circular orbit, given by the formula $r_g = [GMT^2/(4\pi^2)]1/3$, in which G is the gravitational constant, M the mass of the primary body and T the period of the orbit. In particular, the radius r_o given by the formula when M is the mass of the Earth and T is the period (Schuler period) of revolution of a satellite about the Earth. (2) The radius of a circular orbit lying just above the surface of a spherical primary body. Equivalently, the length of a Schuler pendulum. (3) The length GM/c^2, in which G is the gravitational constant, M is the mass of the body and c is the speed of electromagnetic radiation in a vacuum. The gravational radius of the Earth is 5 mm; the gravitational radius of the Sun is 1.47 km.

radius, hydraulic - The ratio of the cross-sectional area of a stream or a pipe to its wetted perimeter.

radius, polar - One-half the polar diameter.

radius of curvature - The radius of the osculating circle in a normal section at a point on a surface. At every point of a surface there is a plane (a normal plane) which contains the normal at that point and that osculating circle whose radius is greater than the radius of the osculating circle in any other normal plane at that point. There is another normal plane at that point which contains an osculating circle whose radius is less than the radius of the osculating circle in any other normal plane at the point. These two planes are called the principal planes at the point; the reciprocals of the radii are called the principal curvatures of the surface at that point. Half the sum of the principal curvatures is called the mean curvature of the surface at the point, while the product is called the Gaussian curvature or simply, the curvature of the surface there.

radius of curvature, mean - (1) Half the sum of the greatest and least radii of curvature at a point on a surface. (2) SEE radius of curvature, Gaussian.

radius of curvature, normal - That radius of curvature (at a point on a surface) in a plane through the normal there and such that the length of the radius of curvature is a maximum. SEE ALSO radius of curvature.

radius of curvature, principal - (1) One of the two radii of curvature at a point on a surface that have the greatest and smallest lengths there. (2) That radius of curvature (at a point on a surface) in a plane through the normal there and such that the radius of curvature is a minimum. SEE ALSO radius of curvature.

radius of curvature in the meridian - The radius of curvature (of an ellipsoid representing the Earth) in a plane through a specified normal and the polar axis (i.e., in the plane of the meridian).

radius of curvature in the prime vertical - The radius of curvature (of an ellipsoid representing the Earth) in a plane through a specified normal and perpendicular to the meridian through that normal.

radius of gyration - The distance from a specified line to a point such that, if the entire mass of the body were concentrated at that point, the moment of inertia of the concentrated mass about the specified line would be the same as the moment of inertia of the body about that line. The point at which the mass is imagined to be concentrated is called the center of gyration.

radius vector - (1) The directed line or distance from the origin of a polar or spherical coordinate system to a specified point. (2) A line drawn from the pericenter of an orbit to a point on that orbit. Also called a position vector. Note that the radius vector so defined is not a directed line.

radius vector, geocentric - The vector from the center of the Earth, or from an ellipsoid representing the Earth, to a specified point. Note that this is not the same as a geocentric radius; a geocentric radius may be merely a geometric figure or it may have a length associated with it. A radius vector has both length and direction as part of its definition.

raise - A vertical or inclined passage driven upwards in ore from a level. SEE ALSO wintze.

Ramsden level - A wye-type leveling instrument invented by Ramsden at about the same time as the Chazy level was invented and differing from that instrument only in minor details. The Ramsden level was a brilliant success and, with the Chazy level, played a great part in making leveling instruments depending on spirit levels instead of pendulums practical.

rancho (Spanish) - (1) A small collection of men or their dwellings. (2) A hamlet. (3) As used in Mexico, and in the Spanish law formerly prevailing in California, a ranch or large tract of land suitable for grazing purposes, where horses or cattle are raised. It is distinguished from hacienda, a cultivated farm or plantation.

random (adj.) - Non-systematic; unpredictable by nature.

range - (1) In surveying, in general, two points lying in a straight line with the point of observation, or the straight line itself. Boundaries crossing water and corners of boundaries in regions covered by water where permanent markers cannot be placed are sometimes defined by intersections of lines through ranges or by one such line and a distance from a marker, the lines being marked by permanent monuments on land. (2) The line defined by a straight-sided object such as the side of a building or a fence, and the point found by extending that line to its intersection with a surveyed line. The point thus determined is said to be in range with the side of the building, with the fence, etc. Also called a range tie. (3) Two suitably located objects on shore used to keep a vessel moving along a straight line defined by points on the two objects, or the straight line itself. (4) Specially placed structures or other objects delimiting channels which are to be followed by vessels so as to be clear of dangers. Such structures are often permanently emplaced and suitable lighted and are given identifying names such as Honolulu Channel Front Light and Honolulu Channel Rear Light. (5) Distance in a single direction or along a great circle. (6) The distance between a target and a distance-measuring device such as radar. (7) The greatest distance at which an object can be detected. (8) The greatest distance to which a missile can be sent effectively. (9) The difference in elevation reached by a tide between successive high and low water. (10) The set of values taken by a function f(x) for that set (the domain) of values of x for which x is defined. (11) The least and greatest values taken by a varying quantity. Also called the amplitude. (12) In the U.S. Rectangular System, any series of contiguous townships, or of sections within a township, situated north and south of each other. (USPLS) Ranges of townships are numbered consecutively east and west from a principal meridian. Thus, range 3 east indicates the third column of townships to the east from a principal meridian. The word range is used with the appropriate designation of a township to indicate the township's location. Thus, township 14 north, range 3 east indicates that township which is the 14th township north of the base line and the 3rd township east of the principal meridian. (13) Before uniform adoption of the U.S. Rectangular system, any one of various arrangements of townships or similar parcels of land situated approximately in a cardinal direction. (14) SEE mountain range. (15) SEE range, firing. SEE ALSO firing range; neap range, mean; spring range, mean.

range (of the tide) - SEE range, mean (2).

range, apogean - The average minimal difference of elevations of tides immediately following the time when the Moon is farthest from the Earth. It is usually about 20% less than the average difference in elevation of the tide.

range, critical - (1) A pair of values, of a quantity, between which an occurrence of the quantity indicates a significant change in the phenomenon or material corresponding to that quantity. (2) A pair of values, of a quantity, between which there is an uncertainty in how the values should be understood.

range, degaussing - A locality and facility for determining the magnetic characteristics (signatures) of ships and other marine craft. Such data are used to determine the current settings required in degaussing coils and other corrective actions.

range, diurnal - SEE range, great diurnal.

range, focal - SEE focus, depth of.

range, geographic - The greatest distance at which an object or light can be seen when limited only by the curvature of the Earth and the elevations of the object and observer.

range, great diurnal - The difference in elevation or height between mean higher high water and mean lower low water. The term may also be used in its contracted form, diurnal range.

range, great tropic - The difference in elevation or height between tropic higher high water and tropic lower low water. The term may also be used in its contracted form, tropic range.

range, mean - (1) The average difference in the extreme values of a quantity, as, e.g., the mean range of the tide. (2) The difference in elevation between mean high water and mean low water.

range, mean tropic - The average difference between the great tropic range and the small tropic range. The mean tropic range is insignificant only where the tide is either semidiurnal or mixed.

range, slant - SEE distance, slant.

range, small diurnal - The difference in elevation or height between mean lower high water and mean higher low water. The term is applicable only where the tide is either semi-diurnal or mixed.

range, small tropic - The difference in elevation or height between tropic lower high water and tropic higher low water. The term is applicable only where the tide is either semi-diurnal or mixed.

range, visual - The greatest distance at which a source of light is visible, taking into account both the geographic condition and the detectability of the light. The geographic condition is the greatest distance at which the curvature of the Earth permits a source of light at a given height to be seen from a given height of the eye. The detectability is determined from the nominal distance at which the source would just be detectable and from the existing atmospheric conditions such as fog.

range finder - SEE rangefinder.

rangefinder - (1) An instrument for determining the distance from a single point of observation to other points at which no instruments are placed. Also written range finder. The term rangefinder used without any qualifier usually denotes an optical rangefinder. SEE ALSO tachymeter. (2) SEE rangefinder, optical. (3) SEE distance-measuring instrument, electro-optical.

range finder, automatic optical - SEE distance-measuring instrument, optical.

rangefinder, electro-optical - SEE distance-measuring instrument, electro-optical.

rangefinder, optical - A rangefinder which determines distance to a point by measuring (visually) the difference in direction of the point from opposite ends of a very short base line contained within the instrument. Also called a telemeter. The term rangefinder used alone usually denotes an optical rangefinder. The optical rangefinder allows distances to inaccessible points to be determined. The precision of the optical and mechanical parts of the instrument is very great, but because the lines of sight intersect at a very small angle at the point of interest, the distances obtained are not highly precise. SEE ALSO distance-measuring instrument; tachymeter.

range-in - SEE ranging-in.

range line - (1) An external boundary of a township, extending in a north-south direction. (USPLS) (2) The line determined by two objects and on which an observer places himself. Also called a range. Alternatively, the line determined by a range (1).

range of tide - SEE range (9).

range of tide, mean - SEE range, mean (2).

range of visibility - The greatest distance to which one can usefully see. SEE ALSO range, visual.

range pile - Any pile serving as a guide for hydrographic surveying.

range pole - SEE range rod.

range rate - (1) The rate of change of measured distance. (2) Radial velocity measured by radar. (3) SEE velocity, radial.

range rod - A simple rod, round or octagonal in section, 6 to 8 feet long, 1 inch or less in diameter, fitted with a sharp-pointed, shoe of steel and usually painted alternately in red and white bands at 1-foot intervals. Also called a lining pole, line rod, lining rod, picket, range pole, ranging pole, sight rod, etc. A range rod may be made of wood or metal. It is used to line up a point of a survey or to show the observer at the theodolite the location of a point on the ground.

range signal - A buoy, rod, flag or similar object used to mark and identify points when taking soundings during a hydrographic survey.

range tie - SEE range (2).

ranging (v.) - Measuring distance.

ranging-in - A process of trial and error whereby a surveyor's instrument is placed on an established line, the instrument being moved to one side or the other of the line joining the two points until the line of sight passes through both points. SEE ALSO centering, double; wiggling in.

rank (of a matrix) - The largest number R such that at least one R-th order determinant formed from the matrix by deleting rows and/or columns is different from zero.

Rankine temperature scale - A temperature scale in which the unit is that of the Fahrenheit temperature scale, the ice-point is 491.69° and the boiling point of water, at standard temperature and pressure, is 671.69°. The Rankine temperature scale is used almost solely by engineers in the USA.

raster - The regular pattern of lines traced by a moving beam such as the beam of a cathode-ray tube (CRT) to provide substantially uniform coverage of a region on the screen. The meaning of the term is sometimes extended to denote a similar pattern on other devices than a CRT. The most common raster is one in which the dot created by the beam moves from left to right across the screen, varying in color and intensity as it does so to create one line of a picture. When the dot reaches the edge of the screen, the beam shuts off (the dot vanishes), moves back to the left-hand edge of the screen and down the width of one line, then turns back on (the dot reappears) and the process is repeated until the bottom of the screen is reached, when the beam returns to the upper left-hand corner of the screen. This type is used, with variations, in most television receivers and is called a horizontal raster. Other rasters may be spiral or circular, etc. These are used principally by radars.

rate, angular - SEE speed, angular.

rate, annual - SEE change, annual magnetic.

rate of change, annual - SEE change, annual magnetic.

rate station - That station, in a pair of fixed base stations, which is not used as the drift station.

ratio, anharmonic - SEE ratio, cross.

ratio, altitude-contour - SEE C-factor.

ratio, base-altitude - SEE ratio, base-height.

ratio, base-height - The ratio of the length of an air base to the average altitude at which a stereoscopic pair of photographs was taken. This ratio is also referred to as the K-factor, base-altitude ratio or B/H. Base-altitude ratio is the most appropriate term.

ratio, cross - The ratio $[(z_1 - z_4)(z_3 - z_2)]/[(z_1 - z_2)(z_3 - z_4)]$ of four complex numbers z_1, z_2, z_3, z_4. Also called the anharmonic ratio of these numbers. For any transformation $w = (az + b)/(cz + d)$, in which a, b, c, d are constants, the cross ratio of the z's equals the cross-ratio of the w's. The cross ratio is real if and only if the corresponding points in the complex plane lie on a straight line or circle. It is used in the theory of photogrammetric rectification because it is invariant under the induced transformations.

ratio, signal-to-noise - The ratio of a measure of that part of an event which is considered to be signal to a measure of that part which is considered to be non-signal, i.e., noise. The measure most often used is power, but energy or amplitude are also used.

ratiograph - SEE templet ratiograph.

ratiometer - An instrument determining ratios between scales, from which, through formulas, a rectifying printer can be set to make a rectified print to the scale of a photomosaic. Also called a ratiograph.

rationalization method (of relative orientation) - A method of orienting successive stereoscopic models which takes into consideration the limiting factors of the equipment being used, the nature and variations of tilt and crab at successive camera-stations and which provides approximate adjustments of the projectors so that the final adjustment can be reached in a shorter length of time.

ratio print - A print whose scale has been changed, by projection, from that of the original photographic negative or transparency.

ray - (1) A line perpendicular to the sequence of surfaces (wave fronts) identifying a propagating wave. Alternatively, a line drawn to represent the actual or ideal path of a photon. (2) SEE half-line.

ray (of light) - SEE light ray.

ray, chief - A ray directed toward the center of the entrance pupil of an optical system. Also called a principal ray.

ray, epipolar - (1) That line, on the plane of a photograph, joining the epipole to the image of any point. (2) The intersection of an epipolar plane with the plane of the photograph.

ray, incident - That part of a ray which is followed by the radiation before being refracted at a surface. Equivalently, a ray indicating the direction in which energy is propagating before being reflected or refracted.

ray, meridional - A ray coplanar with the optical axis.

ray, paraxial - A light ray so close to the optical axis of an optical system that Gauss's equations can be used without significant error.

ray, perspective - (1) A straight line joining a point in object space or image space to a perspective center. Also called an image ray. (2) The broken line consisting of the line joining a point in object space to the front nodal point of an optical system, the line joining the front nodal point to the rear nodal point and the line joining the rear nodal point to the corresponding image-point.

ray, principal - (1) A ray directed toward the first principal point of an optical system. (2) A ray directed toward the optical center of an optical system. (3) SEE ray, chief.

ray, refracted - A ray indicating the direction into which energy is propagated after passing from a substance with one refractive index into a substance with a different refractive index.

Rayleigh distribution - The distribution of the magnitude (modulus) of a randomly varying vector.

Rayleigh frequency function - The function $f(r) = (r / \sigma^2) e^{(-r^2 / 2\sigma^2)}$, in which r is the modulus of a vector of two randomly varying components and σ is the standard deviation. f(r) gives the frequency of the magnitude r of the vector.

Rayleigh scattering - The scattering of electromagnetic radiation by particles whose characteristic size is considerably smaller than the wavelength of the radiation. For example, scattering of light by molecules is a form of Rayleigh scattering and is responsible for the blue color of

the sky. Radar signals undergo Rayleigh scattering by minute particles in the air.

Rayleigh's criterion - The criterion for two pointlike sources of radiation to be visible: the images of the two sources shall not be so close together that the first dark ring in the image of one source lies on the center of the image of the other source. SEE ALSO resolution; resolving power.

rays, homologous - The two perspective rays corresponding to a pair of homologous image-points.

ray tracing - Calculation of the path of a light ray through an optical system.

reach - (1) An extended region of water or land. (2) A straight part of a stream or river. (3) A level stretch, as between locks in a canal. (4) An arm of the sea extending from the land.

reactor, saturable - An induction coil enclosing a metallic core capable of being magnetized to saturation by the current in the coil. Among other applications is that of regulating pulses in a fluxgate magnetometer (q.v.).

reading (n.) - (1) The number obtained by noting and/or recording that number which an instrument indicates is a result of the measurement. (2) The act of noting and/or recording that number which an instrument indicates is a measured value or a result of the measurement. Instruments which have graduated scales and indicate the result of a measurement by a movable mark placed on the scale usually do not indicate a specific number. I.e., the movable mark usually comes to rest at a point, on the scale, between two graduations. The correct reading is then determined according to rules set down or agreed on for that type of measurement. It may be taken as the number of the graduation closest to the mark, a number estimated from the location of the mark relative to the two graduations on either side of it, etc.

reading (v.) - The observing, on an instrument, of that value which the instrument indicates is a measured value. SEE ALSO reading (n.).

realty - SEE estate, real.

réaumer scale - SEE Réaumer temperature scale.

Réaumer temperature scale - A temperature scale in which the value 0 is given to the freezing point of water and 80 to the boiling point of water at 760 mm barometric pressure. Also called the Réaumer scale. This temperature scale was invented in 1730 by Réaumer in studying the thermal expansion of a mixture of alcohol and water. In some of the earliest geodetic work in Europe, the scale was used in stating the temperature of apparatus used in measuring base lines (base apparatus).

rebar - A section of metal rod used for reinforcing concrete. Rebars are often used by surveyors as monuments or parts of monuments.

rebound, glacial - SEE rebound, isostatic.

rebound, isostatic - The rising of the ground in a large region after the glacier covering the ground has retreated. Also called glacial rebound and post-glacial rebound. The idea, first proposed by F. Jamison in 1865, is that the crust is depressed under the load imposed by a glacier, but recovers its former position when the load is removed. In the 1890's, observations around the region of the Great Lakes and Baltic Sea gave evidence of such rebound. The process is still going on in these regions and can be detected by precise leveling.

rebound, post-glacial - SEE rebound, isostatic. This term is unneeded.

recast (v.) - Change the graticule of a map from one horizontal datum to another by appropriately changing the longitudes and latitudes of the lines of the graticule.

receiver - (1) A device for converting electromagnetic radiation into perceptible signals. The term is applied mostly to devices, called radio receivers, for converting radio-waves into audible or visual signals, but it is often applied to instruments operating in other parts of the spectrum. (2) (for a pendulum) A heavy box of cast metal, within which the pendulum is suspended and some auxiliary equipment placed when measuring gravity-acceleration. (3) SEE radio receiver.

recession (of water) - In law, the gradual, natural and more or less permanent lowering of the elevation of a lake's surface, or the complete disappearance of that surface, when referred to what was once regarded as the normal (average) elevation, as at the date of an established survey.

recital - (1) Repetition of the words of another, as by reading or from memory. (2) The formal statement, or setting forth, of some related matter of fact in any deed or writing, as to explain the reasons for a transaction, to evidence the existence of facts, or, in pleading, to introduce a positive allegation.

reckoning, dead - The calculation of a vessel's or aircraft's location, given the initial location and measurements of the vessel's or aircraft's velocity and drift, without further observations. Also written dead-reckoning. The term originated in the practice of writing in the log, after a location determined by dead reckoning, ded. for deduced.

reclamation homestead entry - A homestead entry, not exceeding 160 acres, initiated under the act of 17 June 1902, which provides for the homesteading of public lands within reclamation projects.

recompilation - The process of producing an essentially new map replacing a previously published map. Recompilation usually involves making significant changes to the placement of details on the map, revising elevations or improving the planimetric or navigational data.

reconnaissance - (1) The gathering of information by personal inspection, as a preliminary to further action or more intensive study. SEE ALSO reconnaissance, aerial; reconnaissance, aerial photographic. (2) In surveying, an investigation of a region where a survey is to be made, to determine the best locations for observing stations, geodetic markers, etc. The reconnaissance usually includes getting permission to enter and use land, etc.

reconnaissance, aerial - The gathering of information about a region by inspection of the region from an aircraft.

reconnaissance, aerial photographic - Aerial reconnaissance by inspecting photographs taken from the aircraft.

reconnaissance map - A map incorporating both the information obtained in a reconnaissance survey and the data obtained from other sources. A reconnaissance map differs from a map based on an exploratory survey in that it contains more detail. The detail is selected to serve either special or general purposes. SEE ALSO reconnaissance sketch.

reconnaissance sketch - A drawing resembling a reconnaissance map but lacking sufficient information to make it usable as a map of the region depicted.

reconnaissance survey - (1) A preliminary survey, usually done rapidly and at relatively low cost. Also called a windshield survey in the jargon of some surveyors. The information obtained is recorded, to some extent, in the form of a map or sketch. (2) A survey of a region to determine generalities, overall relationships and feasibility, and identify controls for a project. Also called a reconnoissance survey; the other form is preferred. (3) SEE survey, preliminary (2).

record monument - (1) An adjoining tract of land (an adjoiner), such as a street or particular parcel of land, called for in a deed. Frequently, the boundary line of the adjoiner is referred to as the record monument; actually, the entire property, rather than the line, is the record monument. (2) An object or feature, marked or unmarked, used as reference (called for) in a title or deed.

records, tidal - The principal sources are: 1) annual tables of the tides, tidal currents and specific reports of the U.S. Coast and Geodetic Survey; 2) records of the nearest tidegage of the National Ocean Survey; 3) the Beach Erosion Board; 4) records from the nearest station of the United States Coast Guard; 5) records from the nearest station of the United States Weather Bureau; 6) local offices of the Corps of Engineers; 7) local reports, records and statements of individuals.

recover (v.) - Get back or regain something lost, abandoned or lost track of. The term is sometimes used, erroneously and unjustifiably, as if it meant solve.

recovery - In general, the process of getting back or regaining something which has been lost, abandoned or not kept track of for a long time. The term has a particular meaning in geodesy. SEE recovery (of a station).

recovery (of a station) - The finding and identification of the geodetic marker of a previously established survey-station, and the making of a record attesting to the fact that the marker was found, identified and proven to be authentic, and in its original location. Also called recovery of station and station recovery. Recovery is tested by checking the measurements of distance and azimuth (or bearing) from the station to a reference mark. Witness marks are aids to the recovery but afford only secondary evidence of the location of the recovered station. Exact recovery sometimes requires that the original, surveyed distances and/or directions between the station and at least two adjacent stations of the same survey and class be tested and found unchanged.

rectagraver - A scribing tool that rests on the surface being cut, and only the cutter arm moves to scribe each symbol.

rectangle - A closed figure having four angles of 90° each and four sides. The rectangle is a special form of a parallelogram and a more general form than the square. Opposite sides of a rectangle are of equal length.

rectification - The process of producing, from a tilted or oblique photograph, a photograph from which displacement due to tilt has been removed. Also called transformation. A photograph may be rectified optically, graphically or mathematically. The optical method is the most common. The photograph is given the same orientation with respect to an unexposed photographic film or plate as the ground plane had to the photograph when the picture was taken. The image on the photograph is then projected onto the emulsion to reverse, in effect, the original process by which the photograph was taken. Graphical rectification involves plotting selected points from the photograph onto another sheet, using a graphical method of correcting for displacement because of tilt. Neither method compensates for radial displacement caused by relief. Mathematical rectification uses a computer as intermediary between the original photograph and the rectified image, correcting not only for non-verticality of the photograph but also, in suitable cases, for relief-caused displacement. SEE ALSO computed-data method; grid method; paper-strip method; point-matching method; rectification, graphical; rectification, mathematical; rectification, optical.

rectification, analytical - SEE rectification, mathematical.

rectification, differential - The process of removing the effects of tilt, relief and other distortions from photography by considering the picture as being made up of many small portions and correcting each portion independently. SEE orthophotograph.

rectification, graphical - Rectification in which selected points on the original photograph are replotted onto another sheet using a special grid, special scales or other graphical methods to converting a non-vertical photograph to an equivalent vertical photograph. SEE ALSO grid method (of rectification); paper-strip method (of rectification).

rectification, mathematical - Rectification in which a computer is used to couple a light-beam scanning the original photograph to a modulated light-beam scanning an unexposed photographic film or plate in such a way that tilt-caused distortions in the original photograph are removed in the copy created by scanning.

rectification, multiple-stage - Rectification of a photograph by a series of projective transformations. A photograph can be rectified in several stages using simple equipment, whereas complicated and expensive equipment is usually needed if a high-oblique photograph is to be rectified in a single step.

rectification, optical - Rectification of a photograph by projecting an image of that photograph onto an unexposed photographic film or plate held at a suitable angle with respect to the original. SEE ALSO rectification.

rectification, optical-mechanical - Rectification done by a combination of optical systems and mechanical linkages.

rectification, photogrammetric - The process of establishing the projective relationship between the inclined plane of a tilted photograph and the horizontal plane of a map.

rectification system, universal analog photographic - An electronic rectifier permitting detail to be transferred rapidly from trimetrogon, panoramic or other kinds of photographs, to its proper, rectified position on a sheet, and consisting of a light beam scanning the original photograph, a computer, a controller and an x-y plotter.

rectifier - (1) A projection printer designed specifically to allow the geometric relationship between an aerial photograph, center of projection and projected image (copy) to be varied in such a way that the effect of tilt can be eliminated from the aerial photograph. Also called a rectifying camera, transformer and transforming camera. There are two basic types: that in which the optical axis of the rectifier's lens-system is the common reference or basic direction of the device and that in which the line between the principal point of the photograph and the rectifier's lens-system is the common reference. (2) A device designed particularly for making, from a non-vertical aerial photograph, a copy in which the effects of the photograph's not being taken vertically have been removed. SEE ALSO orientation, analytical; orientation, empirical; point-matching method.

rectifier, autofocus - A precise, vertical rectifier connecting lens system, original photographic negative, and enlargement by a linkage which automatically keeps the projected image in sharp focus. The linkage is usually driven by a motor. SEE ALSO inversor.

rectifier, automatic - Any rectifier in which a mechanism ensures that the lens law and the Scheimpflug condition are automatically satisfied. Such a mechanism, called an inversor, provides a mechanical solution for the linear and angular elements of rectification.

rectifier, nontilting-lens - A rectifier in which the lens system can move only in the direction of its fixed axis.

rectifier, nontilting-negative-plane - A rectifier in which the carrier of the photographic negative can not be tilted, i.e., must remain horizontal.

rectifier, optical - A rectifier in which the transformation between original photograph and final image is done by an optical system recreating the geometry of the situation under which the photograph was taken.

rectifier, scanning - A rectifier coupling a beam scanning the original photograph to a beam scanning the image plane in such a way that effects (other than radial displacement caused by relief) of tilt in the original photograph are not present in the image.

rectifier, tilting-lens - A rectifier in which the principal point is fixed on the axis of swing of the lens system and cannot be displaced.

rectoplanigraph - A device, used for preparing planimetric maps, in which vertical photographs are mounted vertically.

redshift - The ratio $(\lambda - \lambda_o) / \lambda_o$, in which λ is the measured wavelength of an identified line in the spectrum of a moving source and λ_o is the wavelength the same line would have if the source were stationary. It is commonly denoted by z. The wavelength of the line from a stationary source is determined by making the substance responsible for the line emit that radiation in the laboratory. The redshift in the radiation from a celestial object indicates the radial velocity of that object with respect to the Earth (actually, with respect to the observer). A positive value of z indicates that the source is receding from the observer; a negative value indicates that the source is approaching.

reduce (v.) - (1) Make smaller. (2) Operate on (data) to simplify them, correct them, or obtain a solution to a problem involving those data.

reduction - (1) The process or act of making the size or scale of an object smaller. SEE ALSO reduction, photographic. (2) The process of calculating theoretical (corrected) values from observational data. (3) A number subtracted from a measured value of gravity to get the corresponding value on the geoid or similar surface. Alternatively, a number added to a measured value of gravity to get the corresponding value, etc. Whether the number is added or subtracted depend's on the user's conventions as to signs of the quantities involved; there is no general agreement on the conventions. In gravimetry, a reduction is the sum of individual corrections. SEE ALSO gravity reduction. (4) The process of adding a reduction to (or subtracting a reduction from) a measured value of gravity to get the corresponding value on the geoid or similar surface. SEE ALSO gravity reduction. (5) The process of computing angles or distances on a reference surface such as an ellipsoid or the geoid from measured angles or distances. SEE ALSO development method (of reduction).

reduction, eccentric - SEE eccentric reduction; reduction to center.

reduction, free-air - SEE gravity reduction, free-air.

reduction, graphical - Reduction of the scale of an image by superposing a network of lines on the original and redrawing this network to a similar pattern of lines at the smaller scale.

reduction, isostatic - SEE gravity reduction, isostatic.

reduction, lunar - A correction to the Moon's orbit, calculated from the difference between the true longitude of the Moon and its tabulated longitude.

reduction, mechanical - Reduction of the scale of an image by a mechanical device such as a pantograph.

reduction, optical - Reduction of the scale of an image by projecting the image through a lens system onto a flat surface over which a new drawing can be made.

reduction, photographic - Reduction of the scale of an image by projecting the image onto a photographic emulsion at a scale smaller than the original. Also called reduction when the meaning is clear.

reduction, tachymetric - The calculation of horizontal distances from data obtained by tachymetry.

reduction, two-step - SEE enlargement (reduction), two-step.

reduction by development - SEE development method of reduction.

reduction by projection - SEE projection method of reduction.

reduction factor - SEE scale of reduction.

reduction of direction, eccentric - SEE eccentric reduction.

reduction printer - SEE diapositive printer.

reduction to bench-mark level - The introduction of a correction for the height of the line of sight, at an observing station, above the bench mark.

reduction to center - (1) The amount which must be applied to a direction observed at an eccentric station or to an eccentric signal, to obtain what the direction would have been had there been no such eccentricity. (2) One of the values used in finding the equation of time.

reduction to chord length - Calculation of the length of that straight line (chord), between two points on a reference ellipsoid, which corresponds to the length of a measured line between the two corresponding points on the ground, or to the length of the corresponding line on the geoid or other level surface.

reduction to mean sea level - (1) Reduction of a measured angle or distance to the corresponding value the angle or distance would have had if it had been measured at mean sea level. Also called reduction to sea level. This definition is usable only when the angle or distance has been measured close to a location at which mean sea level has been measured also. Otherwise, and usually even when this condition is satisfied, the term means reduction to the surface defined by the vertical datum in the region or to the geoid. (2) (of a distance) The process of finding, from a distance measured on or above the Earth's surface, a corresponding length on an equipotential surface passing through mean sea level at some point. The process is the same as that used in finding a corresponding length on the geoid but is applicable only in a limited region about the point at which mean sea level has been determined. The term is sometimes applied, erroneously, to the reduction of a distance to the geoid. SEE ALSO reduction (of a distance) to the geoid; reduction to the geoid.

reduction to sea level - (1) Calculation, from a measured distance, of the equivalent distance length on the surface to which elevations are referred locally. (2) SEE reduction to mean sea level (1).

reduction to the ellipsoid - (1) Calculation of that angle or length on the reference ellipsoid which corresponds to (a) a measured angle or distance, or to (b) an angle or length on the geoid which has been derived from a measured angle or distance. Also called "reduction to the spheroid when, as is usually the case, the ellipsoid is a rotational ellipsoid. (2) (of an angle) Calculation of an angle on the ellipsoid from the corresponding angle on the geoid or on the Earth's surface, both angles having their vertices on the same normal. (3) (of a distance) (a) The process of finding, from a distance measured on or above the Earth's surface or from a distance determined on the geoid, a corresponding length on the reference ellipsoid, the end points having the same longitudes and latitudes on each surface. Two common methods are the projection method and the development method. In the projection method, the required length is the length of that geodesic, on the ellipsoid, which connects the foot points of the normals through the end points of the given distance. In the development method, the required length has the same value as the given distance and makes the same angle with some given line (e.g., has the same azimuth). Before 1950, the position of the ellipsoid with respect to the Earth's surface was not well known except in the immediate vicinity of the origin. It was therefore common practice to first reduce the distance to the geoid or to mean sea level and to assume that the distance from geoid to ellipsoid was negligible and did not affect the length. When the ellipsoid is a rotational ellipsoid (spheroid), the process is traditionally referred to as distance reduction to the spheroid or as simply, reduction to the spheroid. SEE ALSO base line reduction to the ellipsoid. (b) The amount subtracted from or added to the horizontal equivalent of a measured distance, at average elevation, to obtain the length of the corresponding line on the reference ellipsoid. Also called reduction (of a length) to the ellipsoid and distance reduction to the ellipsoid. The height used will differ (approximately) by the amount of the geoidal height from the elevation used in reduction to mean sea level. (4) (of a length) The amount added to or subtracted from the equivalent, on a horizontal surface at average elevation, of a measured length to obtain the corresponding length on the reference ellipsoid. Also called length reduction to the ellipsoid. The term is used primarily for the reduction of lengths of base lines and other short distances for which the concept of length is reasonable.

reduction to the geoid - (1) The calculation of an angle or the length of a line on the geoid from a measured angle, length or distance. If the reduction is not to the geoid but to a surface having the elevation of mean sea level at one or more points, the process is often called reduction to the geoid, even though this is a misnomer. It is also called more properly, reduction to mean sea level or reduction to sea level. (2) (of a distance) The process of finding, from a distance measured on or above the Earth's surface, the length of a corresponding line on the geoid. Also called erroneously, reduction to mean sea level. In one method, the length on the geoid is the length of the line joining the foot-points of the verticals through the end points of the given distance. In another method, the foot-points of the normals to the reference ellipsoid are used instead. The second method must be used if, as is usually the case, the corresponding length on the ellipsoid is eventually to be determined. However, it requires that the deflection of the vertical be known at both ends of the line and that these deflections be able to be converted to the corresponding angles on the geoid. SEE ALSO base line reduction to the geoid.

reduction to the meridian - The process of applying a correction to an angular elevation measured when a body is near the celestial meridian of the observer, to find the angular elevation when the body is on the meridian. The angular elevation at the time of such a measurement is called an ex-meridian altitude or ex-meridional altitude.

reduction to the spheroid - (1) Reduction to the ellipsoid in that common case in which the ellipsoid is a spheroid (rotational ellipsoid). (2) (of an angle) Calculation of an angle on the rotational ellipsoid from the corresponding angle on the geoid or on the Earth's surface, both angles having their vertices on the same normal. Also called angle reduction to the ellipsoid, although that term is more general. The value of the derived angle will depend not only on the value of the original angle but also on the directions of the sides of the original angle. Therefore it is often best to reduce the direction of each side independently. SEE ALSO direction reduction. (3) (of a distance) Reduction of an distance to the ellipsoid in the case that the ellipsoid is a spheroid (rotational ellipsoid). (4) (of a length) Reduction of a length to the ellipsoid in the case that the ellipsoid is a spheroid (a rotational ellipsoid).

reef - A submarine elevation of rock or coral forming a menace to navigation on the surface. It may or may not be uncovered at low tide and may or may not be connected to the shore.

reference antenna - (1) That antenna, in a set of antennas engaged in radio interferometry, used as a referent in establishing an epoch for the time of arrival of signals at all the antennas. The term is used, in particular, in very-long-baseline interferometry. (2) An antenna used as a standard to which the gain of other antennas is referred.

reference datum - In general, any datum, plane, or other surface used as a reference or base from which other quantities can be measured.

reference direction - A direction from which the angles to other directions are calculated or measured.

reference ellipsoid - (1) An ellipsoid of specified dimensions, fixed in a specified position in the Earth, and having an associated coordinate system (e.g., longitude, latitude and distance from the surface). Coordinates given in this system are said to be with respect to the reference ellipsoid. The constants specifying the ellipsoid's dimensions and position constitute a geodetic datum. Reference ellipsoids are most commonly rotational ellipsoids (i.e., have two axes of equal length). (2) An ellipsoid on which the gravity potential is constant and to which the gravity potentials at other points are referred. (3) An ellipsoid on which the gravity potential is constant and which approximates the geoid in size and position. SEE ALSO ellipsoid, terrestrial.

reference frame - SEE frame of reference; coordinate system.

reference frame, absolute - SEE coordinate system, absolute.

reference grid - SEE grid.

reference level - SEE datum plane.

reference line - Any line usable as a reference or base for measuring other quantities. Also called a datum line.

reference mark (USPLS) - A survey marker of permanent character, close to a survey station to which the marker is related by an accurately measured distance and azimuth (or bearing). The connection between a survey station and its reference marks should be sufficiently precise and accurate to permit the station to be re-established on the ground should its markers be destroyed, or to permit the reference mark to be used in place of the survey station in extending surveys. Reference marks are used to identify locations of boundary corners situated in places (as in water) where permanent marks cannot be placed. SEE ALSO corner, witness.

reference meridian - SEE meridian, local.

reference monument (USPLS) - An iron post or rock cap used where the place at which a corner monument is to be located is such that, for practical purposes, a permanent monument cannot be established or, if one is established, a full complement of bearing trees or bearing objects is not obtainable.

reference orbit - A hypothetical orbit to which the actual or other orbits are referred by subtracting from the actual coordinates of the satellite the coordinates in the reference orbit for the same instant.

reference plane - A plane from which angles or distances are measured as standard practice or by definition.

reference point - A point used as an origin from which measurements are taken or to which measurements are referred. Also called a datum point.

reference signal - The signal to which other signals are compared, to determine differences in time, phase, frequency or other values or quantities. In particular, in telemetry, the signal to which the times, phase, frequencies or other characteristics of modulated waves carrying data are compared.

reference spheroid - (1) A rotational ellipsoid to which or one which the coordinates of points on the Earth are referred. Also called a spheroid of reference and reference ellipsoid. (2) An approximately spherical surface, defined theoretically, whose dimensions closely approach those of the geoid. (3) A rotational ellipsoid of the approximate size and shape of the Earth, to which the coordinates of points on the Earth's surface are referred. Also called a spheroid of reference and ellipsoid of reference. The last term, however, is ambiguous. The ellipsoid is positioned in a suitable and prescribed manner with respect to the Earth's surface, a reference meridian (zero meridian) is drawn on the ellipsoid, and lines are drawn perpendicularly from points on the Earth's surface to the ellipsoid. The coordinates of a point are then the geodetic longitude and latitude of the foot of the perpendicular on the ellipsoid and the length of the perpendicular.

reference standard - SEE standard, reference.

reference station - (1) A place where tidal or tidal-current constants have been determined from measurements, and whose tides or tidal currents are used as a standard to which measurements made at the same time at a subordinate station can be compared. (2) A place where tidal or tidal-current constants have been determined and from whose tides or tidal currents the corresponding quantities at other locations can be predicted by means of differences or factors. Also called a standard port or standard station.

Reference System 1967, Geodetic - SEE Geodetic Reference System 1967.

Reference System 1980, Geodetic - SEE Geodetic Reference System 1980.

reference table - The flat surface to which the coordinates of a stereoscopic model are referred and on which the paper used for drawing a map is placed. The reference table is usually the same as the plotting table or the supporting table.

referencing - The process of measuring and recording the horizontal distances and directions (azimuths or bearings) from a survey station to nearby landmarks, reference marks and other marks which can be used in recovery of the station.

reflectance - That fraction of the radiant flux incident upon a surface which is reflected.

reflection - (1) An abrupt change in the direction in which radiation (sonic, electromagnetic or particulate) is traveling when the radiation hits a suitable surface. The change consists of the change in sign of one or more of the angles specifying the direction of travel. SEE ALSO law of reflection. (2) The image of an object seen after reflection of light from the object.

reflection, diffuse - Reflection in which the energy of the radiation leaves the reflecting surface in many directions, regardless of the direction of the radiation hitting the surface. Diffuse reflection is a characteristic of irregular or rough (matte) surfaces made up of innumerable small surfaces each making an abrupt change of direction with its neighbors. Each small element of the surface reflects the radiation in a different direction, although not necessarily with the same intensity.

reflection, mixed - Reflection which is partly diffuse and partly specular.

reflection, regular - SEE reflection, specular.

reflection, specular - Reflection in which the direction of the radiation after reflection, the direction before reflection and the perpendicular to the surface at the point of reflection all lie in the same plane, and the angle at which the radiation leaves the surface is equal in magnitude to the angle at which it hits the surface. Also called reflection and regular reflection. Specular reflection is a characteristic of highly reflecting, smooth surfaces such as those of mirrors, and is therefore also called mirror-reflectance.

reflection, total internal - Reflection occurring when radiation traveling in a medium having the refractive index n_i hits the surface of a medium with refractive index n_r, at an angle (with respect to the normal) greater than arctan (n_r/n_i).

reflection angle - SEE angle of reflection.

reflection code - SEE Gray code.

reflector - (1) A surface, or combination of surfaces, capable of reversing at least one component of the direction of travel of a substantial portion of the radiation hitting the surface, i.e., capable of causing reflection. All surfaces reflect some of the incident radiation. Some physicists therefore prefer to consider all surfaces to be reflectors, merely classifying a surface as a poor reflector if only a small amount of energy is reflected or as a good reflector if most of the energy is reflected. General usage, however, is that a reflector is a good reflector. (2) A piece of material having a smooth, shaped surface capable of specular reflection. Commonly called a mirror if made to reflect light. (3) SEE telescope, reflecting.

reflector, corner-cube - SEE reflector, cube-corner.

reflector, cube-corner - An arrangement of three flat, reflecting surfaces to form a tetrahedron having 90° dihedral angles between adjoining faces at the apex. Also called a corner- cube reflector or rarely, a corner reflector or triple prism. The device can be made from a cube by slicing the cube along a plane through three corners having no edges in common. Its most common use is for completely reversing the direction in which a beam of light or other electromagnetic radiation is traveling, i.e., turning the direction of propagation through 180°. There are two common forms. One, used for radiation at all wavelengths shorter than X-ray wavelengths, consists of three flat reflectors fastened together so as to be mutually perpendicular. The other, useful only for radiation at wavelengths from the ultraviolet to the infrared, consists of a solid prism having three mutually perpendicular faces and a fourth face making equal angles with the other three. Radiation incident on the fourth face then undergoes total internal reflection from the other three faces.

reflector, dihedral - SEE corner reflector.

reflector, parabolic - (1) A reflector whose reflecting surface has a parabolic cross-section in any plane perpendicular to a particular straight line. The surface can be generated by translating a parabola in a direction perpendicular to the plane of the parabola. Parabolic reflectors of light are little used in surveying. Similar reflectors of radio waves have been used for focusing radio waves onto dipole antennas or for collimating radiation from such antennas. (2) SEE reflector, paraboloidal.

reflector, paraboloidal - A reflector whose reflecting surface has a parabolic cross-section in any plane through a particular straight line. Also called erroneously, a parabolic reflector. The surface can be generated by rotating a parabola about its axis of symmetry. At optical wavelengths, a paraboloidal mirror is usually a thin, metallic coating on a thick, figured disk of quartz or special glass of low thermal coefficient of expansion. At radio wavelengths, it usually consists of a figured metallic sheet or metallic mesh supported by suitably shaped girders. A dipole or horn antenna is placed at the focus.

reflector, retrodirective - A reflector capable of turning the direction of incident radiation through 180°, so that the radiation proceeds, after reflection, in the exactly opposite direction from that in which it came. Also called a retroreflector. Cube-corner reflectors are the most common type of retrodirective reflector for electromagnetic and sonic radiation.

reflector, spherical - A reflector whose reflecting surface has a circular cross-section in any plane through a particular straight line.

reflector constant - The amount that a measurement of distance must be reduced, when using cube-corner reflectors working by internal reflection in a prism, because the speed of light is less in the material of the prism than it is in air, plus the difference in distance between the equivalent plane of reflection in the prism and the point, in the prism's mounting, vertically above the survey point on the ground.

refraction - (1) The bending of sonic or electromagnetic rays by the substance through which the rays pass. The amount and direction of bending are determined by a characteristic, of the substance, called the refractive index. (Relativistic bending by a gravitational field is not refraction.) (2) (of light) The change or amount of change in the direction in which a ray of light is traveling, when the ray passes from a region in which the speed of light has one value to a region where the speed has a different value. Refraction of light by the Earth's atmosphere is called atmospheric refraction; two kinds are considered: astronomical refraction, which is the refraction undergone by light coming to the Earth's surface from outside the atmosphere; and terrestrial refraction, which is the refraction undergone by light in traveling from one point on the surface to another. (3) (of radio waves) The change or amount of change in the direction in which of a ray of radiation at radio wavelengths is traveling, when the ray passes from a region in which the speed of the radiation has one value to a region in which the speed has a different value. Also called electronic refraction; this is probably dialect. SEE ALSO angle of refraction; coefficient of refraction; radio refraction.

refraction, astronomic - SEE refraction, astronomical.

refraction, astronomical - (1) Refraction, by the Earth's atmosphere, of light from a source outside the atmosphere. Also called astronomic refraction. Light from a star or planet passes through the entire depth of the atmosphere before reaching the surface of the Earth. Refraction causes the ray to follow a curved path concave toward the surface. (2) The angle between the direction of a light ray where the ray enters the Earth's atmosphere and the direction at the point of observation. Also called celestial refraction. The angle is greatest when the source is near the horizon and decreases to a minimum near the zenith. It reaches zero when the ray is perpendicular to the surface separating layers of different densities. It depends on the elevation of the observer and on atmospheric conditions. SEE ALSO refraction of light.

refraction, atmospheric - The bending of electromagnetic radiation by the atmosphere. Most of the bending of light and infrared radiation is done in the troposphere. The stratosphere and ionosphere have little effect because they are too rarified. Radio waves longer than about 30 cm are bent more by the ionosphere than by the troposphere. Shorter radio waves are bent less by the ionosphere and more by the troposphere. SEE ALSO refraction, terrestrial.

refraction, celestial - SEE refraction, astronomical.

refraction, coastal - The apparent change in the direction in which a radio wave is traveling when the wave crosses a coastline obliquely. Also called the land effect.

refraction, differential - The difference, caused by refraction, in true direction between two bodies lying in the same line of sight. I.e., the differing amounts by which atmospheric refraction affects the apparent direction of two sources of radiation because the two sources are at different directions or different distances from the observer. SEE ALSO refraction, parallactic.

refraction, dynamic - The change in refraction caused by a body's changing angular elevation.

refraction, electromagnetic - The bending of rays of electromagnetic radiation on passing through a substance, as distinguished from the bending of sound rays, etc. Also called electronic refraction; this usage is rare and is probably dialect.

refraction, electronic - SEE refraction, electromagnetic; refraction (of radio waves).

refraction, geodetic - Refraction affecting geodetic operations such as leveling, traversing, triangulation and trilateration.

refraction, horizontal - (1) The bending of light rays in an approximately horizontal direction by the atmosphere. In particular, such bending when it occurs as light travels from its source to the observer's instrument. Also called lateral refraction. Horizontal refraction is caused by variations in the density of the atmosphere along the path taken by the light. Under some conditions, horizontal refraction can be of such magnitude as to seriously affect measurements of horizontal angles. A careful reconnaissance is required, especially for surveys in cities, to prevent inclusion of lines significantly affected by horizontal refraction. SEE ALSO refraction, terrestrial. (2) The amount of bending, in an approximately horizontal direction, undergone by a light ray in passing through the atmosphere. (3) The numerical value (about 35') of the angle of refraction when a celestial body is on the horizon at rising or setting. Use of this definition leads to confusion and should be dropped.

refraction, ionospheric - Refraction of electromagnetic radiation by the ionosphere. The ionosphere affects chiefly radio waves. It has little effect on measurements of distance made using infrared radiation or light.

refraction, longitudinal - The change in speed undergone by light or infrared radiation when passing through air of varying density. Longitudinal refraction is more important than is geometrical (i.e., horizontal or lateral) when measuring distances with an electro-optical distance-measuring instrument.

refraction, lateral - SEE refraction, horizontal.

refraction, mean - (1) The value of astronomical refraction when barometric pressure is 30 inches of mercury and the temperature is 50° Fahrenheit. (2) The amount of refraction, in a vertical plane, for average conditions of barometric pressure and temperature.

refraction, parallactic - The difference between the actual amount of refraction undergone by the line of sight to a body in the atmosphere, and the astronomical refraction for a body in the same line of sight. Also called satellite refraction. The astronomical refraction can be calculated from the angle, at the observer, between the line of sight and the zenith. The actual amount of refraction depends on the distance of the body in the atmosphere. SEE ALSO refraction, differential.

refraction, photogrammetric - The refraction of light by the Earth's atmosphere, as it affects aerial photographs or images recorded by devices on spacecraft.

refraction, stratospheric - Refraction of electromagnetic radiation by the stratosphere. Approximately 1/3 of the total amount of refraction undergone by infrared radiation, light and short (3 cm or less) radio waves occurs in the stratosphere.

refraction, terrestrial - Refraction, by the Earth's atmosphere, of electromagnetic radiation from a terrestrial source. The term is understood by geodesists to apply to the refraction of light unless otherwise stated. In geodetic applications, the path of light from a terrestrial source is usually almost horizontal. It passes through only the lower layers of the atmosphere. The light is therefore refracted all along its path. Because the atmospheric layers are not exactly symmetrical, in form and density, with respect to the observer, the path of a light ray in the atmosphere is a complex, non-planar curve. The curvature of the path is greatest, however, close to a vertical plane, where the magnitude of the refraction is important for measurements referred to the zenith or to the plane of the horizon. Its effect is usually to make the apparent angular elevation of an object greater than the true angular elevation, though this effect may be reversed under special atmospheric conditions. In triangulation, a station which is normally just below the apparent horizon may, be cause of refraction under certain atmospheric conditions, become visible for a short period of time.

refraction, tropospheric - The refraction of electromagnetic radiation by the atmosphere in the tropospheric region (approximately that part lying below 11 km). Tropospheric refraction has a significant effect on light, less on radio waves in millimeter range of wavelengths, and little on radio waves of long wavelength. The refractive index is independent of frequency up to about 15 GHz.

refraction, vertical - That component of refraction, in the Earth's atmosphere, which occurs in a vertical plane. Two kinds of vertical refraction affect surveying: that caused by the curvature of the atmospheric layers of different densities (important principally in aerial photogrammetry), and that caused by temperature gradients near the ground (important principally in spirit leveling and trigonometrical leveling).

refraction angle - SEE angle of refraction.

refraction constant - The constant k in the formula k tan ζ for the difference between the angle between the true direction of a star and the zenith, and the angle ζ between the apparent direction of the star and the zenith. SEE ALSO refraction, coefficient of.

refraction displacement - Displacement of images radially outward from the photograph nadir because of atmospheric refraction. The refraction is assumed to be symmetrical about the direction of the nadir.

refraction error, astronomical - SEE refraction, astronomical.

refraction line - A line of sight to a survey signal which is visible only because of the bending of the light rays by refraction. Refraction lines often extend over bodies of water. The observed signal normally lies below the apparent horizon, but it becomes visible during the day at the time of greatest refraction, when it appears to rise out of the water. It disappears downward, apparently into the water, as refraction decreases.

refractive index - SEE index of refraction; index of refraction, atmospheric.

refractivity - (1) That property of a substance which bends the path of light or other electromagnetic radiation passing through. (2) The quantity (n-1), in which n is the absolute refractive index.

refractivity, potential - That value of refractivity which a parcel of air would attain if brought adiabatically to a standard pressure of 1000 mbar.

refractor - SEE telescope, refracting.

region, amphidromic - The region surrounding an amphidromic point.

region, inter-tidal - The region, on a shore, lying between the lowest and highest levels reached by the tides.

register (n.) - (1) The position of one component of a composite image in relation to the positions of the other components. The components are said to be in register. (2) The amount by which two successive components of a printed image, as printed by successive impressions, may differ in position from their required positions. SEE ALSO register, hairline.

register (v.) - Bring two images together so that the two are in some required position with respect to each other. The process is called registration.

register, hairline - A difference of position, in register, so small that fine lines 0.08 to 0.01 mm thick are superposed in the printed version.

registered (adj.) - Entered or recorded in some official register or record or list.

registering (or recording) a deed - A deed is registered or recorded to give constructive notice of conveyance to purchasers and creditors. A deed may be valid as between the grantor and grantee but will fail to give constructive notice to others if not registered or recorded.

register mark - A mark such as a small cross, circle or other pattern made, along with others of the same kind, on original material before reproduction to make it easier to register and to indicate the relative positions of successive impressions. Also called a corner mark, tick, corner tick, register tick or registration tick.

register stud - A small peg placed through the holes punched in sheets of maps, photographs, etc., for putting adjacent sheets into matching positions.

register tick - SEE register mark.

registration - (1) The process of matching details on an overlay with the details of a map so that the details are in their respectively correct places. Equivalently, bringing an overlay into register with a map. (2) The process of matching details originally placed on separate surfaces so that when combined in different colors they will produce an accurate map. (3) SEE register (n.)

registration (of land) - SEE abstract of title; certificate of title; Torrens title system.

reglette - The graduated scale fastened to the end of a surveyor's wire. In particular, the form used on wires designed for measuring base lines.

regression - (1) A backward motion. The antonym is progression. (2) A function for the systematic component of a random variation in terms of one or more independent, non-random variables. The term is much used in the biological and social sciences but is little used in engineering or the physical sciences, where the terms function and relationship are preferred.

regression, linear - A linear function.

regression, non-linear - A non-linear function.

regression of the nodes - (1) The motion of the line in which the plane of an orbit intersects a reference plane. (2) The motion of the pair of points in which the line of intersection of the plane of an orbit with a reference plane intersects the celestial sphere.

regularization - (1) Calculation of those changes to the measured value of gravity which must be made to convert it to the value it would have on the equipotential surface resulting from transferring matter outside the geoid to inside the geoid in a particular manner. The new surface is called the regularized geoid, although it is not the geoid.
(2) Modification of the observation equations, in an adjustment, to allow solving the system of equations even though it may be poorly conditioned.

regulation pace - SEE step.

reiteration theodolite - SEE theodolite, repeating.

rejection of a measurement - Removing a measurement from a set of measurements and using only the reduced set in further computations. Also called rejection of an observation and rejection of an error. The measurement removed is called an outlier if the principal reason for its removal is its greatest difference from the other measurements or from the average or expected value. Several rules have been proposed for determining if a measurement should be removed or not. The first rule appears to have been given by Galileo, who recommended that a measurement be rejected if it deviated from the majority. The most familiar rule is Chauvenet's rule. Another, frequently used by engineers, is to reject a measurement if the residual (difference between the measured and the average value) is greater than a pre-selected number of times the standard deviation of the set of values. The number is usually 2 or 3. Still another rule frequently used is to reject the measurement if the probability of that measurement's occurring is less than 10% or 5%, assuming that the measured values have a Gaussian distribution. A serious objection to removing a measurement because it failed a test like those mentioned is that such removal changes the nature of the distribution and invalidates the assumptions on which the rule was based. Some statisticians attempt to compensate for removal of measurements from a set having a Gaussian distribution by also removing some measurements which passed the test. Scientists usually do not reject a measurement unless the reason for the discrepancy can be found and identified as definitely an error.

rejection of an observation - SEE rejection of a measurement.

rejection of error - SEE rejection of a measurement.

relativity - (1) The principle that all descriptions of the universe in terms of the laws of physics must be equivalent, regardless of the location of the observer in the universe. (2) SEE relativity theory. SEE ALSO Principle of Equivalence.

relativity theory - (1) Any theory developing the physical consequences of relativity. (2) One of two theories proposed by A. Einstein: special relativity and general relativity. (3) A theory based on the postulate that the same laws of electrodynamics and optics will be valid for all frames of reference for which the equations of mechanics hold good.

relativity theory, general - A relativity theory proposed by A. Einstein, based on the same postulates as the special relativity-theory, but generalized to relate measurements by observers in accelerated systems to each other. One of the basic postulates is that there is no detectable difference between phenomena in an unaccelerated system and the same phenomena in a system accelerated by gravitation. One consequence of this postulate is that gravitation is equivalent to a curvature of space.

relativity theory, special - A relativity theory proposed by A. Einstein, based on the two postulates: (a) physical laws are the same in all unaccelerated coordinate systems; and (b) the speed of light in a vacuum is the same for all observers. Special relativity-theory deals only with unaccelerated systems. This is one reason why paradoxes such as the twin paradox cannot be resolved by resort only to special relativity-theory.

relaxation method - An empirical method solving systems of simultaneous linear equations, or linear ordinary or partial differential equations, by starting with an initial guess at the answer and successively correcting the guess according to discrepancies between calculated and expected values at points in the domain of the independent variable vector. It may be considered a variant on the scheme of computation used in Seidel's method. Two forms of the relaxation method are recognized: the over-relaxation method, in which corrections are deliberately too large, and the under- relaxation method, in which corrections are deliberately too small.

relaxation time - SEE rise time.

relay - A device transmitting an amplified or restored version of a received signal.

release - The relinquishment, concession or giving up of a right, claim or privilege, by the person in whom it exists or to whom it accrues, to the person against whom it might have been demanded or enforced.

relevement - An iterative process of resection, in which one selects an approximate location for a point as close to the unknown point as possible; then one calculates by how much and in what direction the initial guess was in error, using the triangle of error as a guide. The French word relevement means, simply, resection. The American word, although derived from the French, denotes a particular kind of resection.

reliability - The probability that an device will work satisfactorily for a specified length of time. A formula often used is reliability = exp(-kt), in which t is the length of time the device is to be used and k is the average rate at which the device will deteriorate.

reliction - (1) The gradual recession of water, resulting in an uncovering of land once submerged. Also called dereliction. Reliction does not cover seasonal fluctuations in levels of the water. SEE ALSO accretion. (2) Land once submerged but now left uncovered by a gradual recession of the water which had covered it.

relief - (1) The actual, physical surface of a region, i.e., of a part of the Earth. (2) The amount of deviation of the actual, physical surface of a region from some surface used as reference. The usual surfaces to which relief is referred are the geoid, the rotational ellipsoid (spheroid), the sphere or the plane. In photogrammetric geodesy, man-made objects are usually considered part of the actual, physical surface when calculating the amount of deviation; dense vegetation may also be so considered. Otherwise, only the natural surface of the region is considered and relief is then commonly called topography. (3) The deviation of a surface from some surface used as reference. Outside of geodesy, this concept is more commonly called the topography of a surface, e.g., the topography of a diffusing surface. (4) SEE relief, relative. SEE ALSO orthogonal method (of depicting relief).

relief, absolute - Elevation above mean sea level.

relief, illuminated - The representation, on a map, of relief in such a way as to give the appearance of a three-dimensional object lighted from one or more directions.

relief, relative - The difference between elevations of high and low points in a region. Sometimes called, simply, relief.

relief, shaded - The depiction of relief, on a map, by darkening or lightening the grayness on a slope to indicate the degree of slope, as if shadows were cast by the relief. The illumination is generally assumed to come from the northwest.

relief, submarine - (1) The shape of that part of the Earth's surface which lies beneath the oceans. (2) The elevations of that part of the Earth's surface which lies beneath the oceans. (3) The depiction, on a chart, map or relief-map, of that part of the Earth's surface which lies beneath the oceans.

relief contour - SEE relief contour method.

relief displacement - Displacement of an image-point radially from the principal point when the object-point is moved parallel to the optical axis. Equivalently, the difference in radial distance of two image points which correspond to object points that are the same distance from the optical axis but at different distances from the camera. Also called image displacement caused by relief. In particular, the radial displacement of an image-point toward or away from the nadir according as the corresponding ground point is below or above the ground nadir. Also called height displacement.

relief distortion - SEE relief displacement.

relief map - (1) A map whose surface is shaped to represent the topography in a region. Also called a terrain model or relief model. The most common is the plastic relief-map. This is made by printing an ordinary topographic map on a plastic sheet. The sheet is then placed on a plaster mold carved to represent the topography. Heat and pressure are applied to fix the plastic sheet permanently into the shape of the mold. Another kind, less common and more costly but showing more detail, and having greater accuracy, is the solid relief-map, made by carving the topography, etc., into a suitable substance such as plaster, and then painting or drawing further detail on the model. (2) SEE map, hypsometric. (3) SEE map, topographic.

relief map, shaded - SEE map, shaded-relief.

relief model - (1) A model of the whole or part of the Earth's surface. (2) A representation, to scale in three dimensions, of a section of the crust of the Earth or of another celestial body. A relief model made primarily to display features on the Earth's surface is sometimes called a topographic model. (3) SEE map, relief.

relief offset, dry - A plate printing by relief but prepared for use on an offset press to eliminate the need for dampeners.

relief representation - Any technique used to depict the configuration of the Earth's (or other celestial body's) surface on a map (e.g. by contour lines, hill shading or layer tinting).

relief shading - The technique used in making shaded relief.

relief stretching - SEE hyperstereoscopy.

relinquishment - The transfer of title to right-of-way, encompassing developed or undeveloped land from the current holder of title to right-of-way to another governmental authority.

remainder - (1) That portion of a larger parcel remaining in the fee owner after a partial taking in condemnation. (2) An estate in property created simultaneously with other estates by a single grant and consisting of the rights and interest continent upon, and remaining after, termination of the other estates.

remnant - A remainder so small or irregular that it usually has little or no economic value to the owner.

remnant rule - The width of lot (frontage of which is not specified on the plat specifying frontage of all other lots in the same block) is the length of the block, minus the total width of the other lots.

remote sensing - SEE sensing, remote.

repeatability - (1) A measure of how close the values in a set of measurements are to a particular value. The particular value need not be a member of the set. For example, the difference between the largest and smallest values in the set, or the number of times the most frequent value is repeated, may be used as the repeatability. In particular, the square root of 1/(N-1) times the sum of the squares of the differences between the values and the average of the values in the set. I.e., the standard deviation with respect to the average of the values in the set. However, this latter quantity is more often called the precision (q.v.). (2) A measure if the variation in accuracy of a surveying instrument when tests are made over the line(s) at different times of the year, with different operators, and with different but equivalent instruments, always using the same procedures. (3) A measure of the amount by which measurements all made by a single instrument differ from each other. SEE ALSO reproducibility.

repetition (surveying) - SEE theodolite, repeating.

repetition method (of determining azimuth) - The method of determining astronomic azimuth by accumulating, on the horizontal circle of a repeating theodolite, the sum of a series of measures of the horizontal angle between a selected star and a suitable mark, and applying the average of such measures to the azimuth of the star computed for the average epoch of the measurements. A correction for inclination of the horizontal axis, the magnitude depending on the angular elevations of the star and of the mark, is applied to the measured angle, and corrections for curvature of the apparent path of the sat, variation of latitude and diurnal aberration also enter into the computations. The method is, theoretically, very precise and accurate. In practice, it is not as satisfactory as the direction method.

repetition of angle - A procedure for determining horizontal angles with a transit or theodolite allowing the horizontal circle to be coupled to and uncoupled from the telescope's motion. The telescope is pointed at the initial point and the direction read. The telescope is next aimed at the second point, the horizontal circle is clamped to the telescope, and telescope and circle are swung around together to point back at the initial point. The horizontal circle is then unclamped, the telescope pointed again and the second point, and this procedure repeated the desired number of times. The final reading on the circle is the sum, minus the initial reading, of the individual measurements, divided by the number of measurements.

representation (of a gravity anomaly), point-mass - SEE gravity anomaly representation.

representation (of gravity), point-mass - SEE gravity representation.

reproducibility - (1) That characteristic of an operation which allows the operation to be repeated again and again without significant differences in operation or the results. (2) A measure of the amount by which measurements made by different instruments differ from each other, the conditions of measurement being the same. SEE ALSO precision; repeatability. (3) The standard deviation of the measurements produced by a set of independent instruments of the same design.

reproduction - (1) The summation of all the processes involved in printing copies from an original drawing, pattern, picture, etc. The principal processes are the diazo process, lithography, engraving, etching and printing. (2) A printed copy of an original drawing, pattern, picture, etc., made by a diazo process, lithography or other similar process.

reproduction ratio - SEE scale of reproduction.

Repsold base line apparatus - An optical base-line apparatus (1) consisting of a steel bar, approximately 4 meters long, whose precise length at any temperature was known and whose temperature was determined by means of a metallic thermometer composed of the steel measuring bar and similar bar of zinc, the two being fastened together at their middle points. COMPARE Bessel's base-line apparatus.

reseau - (1) A rectangular grid on glass, placed against the photographic film in a camera in such a manner that the grid is photographed together with the scene. Also called a grid plate. (2) A rectangular array formed by holes in the platen against which photographic film in a camera is pressed during exposure; the holes are illuminated from the back of the platen. Changes in the film after exposure distort the image of the reseau as well as the image of the scene. The effects of such distortion can be made small by restoring the image of the reseau to its correct shape and size, the image of the scene being thereby restored also. (3) SEE grid, rectangular.

resection - (1) Any procedure for determining the location of a point by extending lines of known direction from that point to two other points whose locations are known. Equivalently, a procedure of determining the location by extending lines making known angles with a third line whose position and length are known. The procedure may be done on the ground using lines of sight as the lines extended, or may be done geometrically by plotting the points and drawing the lines to scale on a sheet, or may be done algebraically by using the coordinates, angles, etc., of the points and lines involved. SEE ALSO intersection. (2) Any procedure for determining the location of a point, on a horizontal surface or plane, by using angles measured at that point between points whose locations are known. The problem, called the two-point problem, of determining the location of a point by measuring the angle, at that point, between two known points is indeterminate, in general. However, given some kinds of additional information, the problem can be solved. Among the methods devised are those of Pothenot and Hansen. While many procedures have been devised for solving the three-point problem, i.e., determining the location of one unknown point from known angles, at that point, between three points of known location, only three are frequently used: those of Cassini, Collins and Kaestner. Cassini's method introduces two auxiliary points; Collins's method uses only one auxiliary point and reduces the problem to one involving two intersections; Kaestner's method does not involve any auxiliary points. The problem is indeterminate if the unknown point lies on the same circle as the three points of known location. The solution is most precise if the unknown point is at the center of the circle determined by the other three points. The methods of Collins and Cassini are useful when the geometry is poor. The solution may be obtained by mechanical means (as is done in hydrographic surveying, with a three-arm protractor), graphically (by drawing angles and lines to scale on a sheet, as in planetable surveying), or algebraically by using the known values for coordinates, distances and angles. The solution is often called a fix. If only three points are used, and the problem is one in surveying, resection is sometimes called trilinear surveying. SEE ALSO Cassini's method; Collins's method; Hansen's method; Italian method; Kaestner's method; Marek's method; problem, three-point; Runge's method. (3) SEE resection, photogrammetric. (4) SEE resection line. SEE ALSO locating a point by resection.

resection, analytical three-point - SEE aerotriangulation, radial.

resection, Italian - SEE Italian method.

resection, photogrammetric - Determination of the location or height of a camera, or of the photograph taken by that camera, with respect to the coordinate system external to the camera. Often called simply, resection if the meaning is clear.

resection, three-point - Determining the location of a point from measurements of angles, at that point, between three points of known location.

resection line - The line drawn from an unknown but occupied station through the plotted location of a known station, using the known direction to the known station. Also called a resection.

resection station - A station located by resection.

reservation - A provision, in a deed, by which the grantor of the deed withholds or takes back some interest in the property conveyed by the deed. A reservation creates some right or privilege for the grantor in the land described as granted. It is distinguished from exception, which is similar, in that the latter definitely refuses to grant a certain interest. Reservations are used in preference to exceptions in creating separate estates in easements, rights of way and similar matters.

reserve (v.) - To take back or withhold, in a deed, something otherwise granted by the deed; to make a reservation in a deed.

reservoir - A region of artificially stored water, formed by a dam on a watercourse and under the control of a duly appointed authority.

resettability - The unavoidable deviation between measurements produced by an instrument, when specified parameters are independently adjusted under stated conditions of use.

residual - (1) The difference: (observed value) - (computed value). The negative value of this difference is often called the residual, espeically in European usage. SEE ALSO error, residual. (2) Any quantity remaining after some other quantity has been subtracted from it.

resolution - (1) In general, a measure of the finest detail distinguishable in an object or phenomenon. (2) In particular and commonly, a measure of the finest detail distinguishable in an image. Also called spatial resolution. Resolution usually varies from point to point of an image, so an average value (area-weighted average resolution, or AWAR) is often used for the resolution of the entire image. (3) A measure of the shortest distance over which differences of gravity or of magnetic intensity can be distinguished. In particular, the length of arc, along a great circle of a sphere representing the Earth, corresponding to the term of highest degree present in a representation, by spherical harmonics, of the gravity potential or magnetic potential. (4) The reciprocal of the width of the beam from a unidirectional antenna, measured in degrees. (5) The process of separating a vector into its components. (6) SEE resolving power (3). Resolution and resolving power are often used as if they were synonymous. However, it is better to use resolution only when referring to details of the object and resolving power only when referring to the capability of the instrument for observing fine detail. SEE ALSO ground resolution.

resolution, area-weighted average - (1) The average resolution over the entire area of an image, as obtained by weighting the resolution, in each region where the resolution is uniform, by the area of the region, summing such products over the entire image, and dividing by the entire area of the image. (2) The average resolution over the field of view of an optical instrument, computed as a weighted average of the measured resolution in each small region of the field, the weighting factors being the areas of those regions. The term is usually used only in the form of its acronym, AWAR. SEE ALSO resolution.

resolution, spatial - SEE resolution (2).

resolution chart - A sheet carrying a pattern consisting, usually, of groups of parallel, black lines on a white background. Within each group the lines are of equal width are spaced at equal intervals. The thickness and spacing increases systematically from group to group, and the resolution of a groups is expressed as the number of lines per millimeter. Resolution charts are used to test the resolving power of optical instruments. SEE ALSO test chart.

resolution in bearing - The smallest detectable separation of two objects, or two parts of the same object, observed at the same distance and elevation, expressed as the horizontal angle between such objects.

resolution in distance - The smallest detectable difference between the distances of two objects, or of two parts of the same object, in the same line of sight. Also called resolution in range. Note that the requirement that the objects be on the same line of sight cannot be satisfied exactly because the one object would then be hidden behind the other.

resolution in elevation - The smallest detectable separation between two objects, or between two parts of the same object, at the same distance and same bearing, expressed as a vertical angle.

resolution in range - SEE resolution in distance.

resolution limit - The least distance, between two gravitating or magnetic bodies, at which the two bodies cease to be separately detectable.

resolving power - (1) A measure of the ability of an optical instrument to distinguish fine detail, as measured by the angular distance, θ, at the center of perspective, between two point-sources of equal brightness when the images of these points are so close together that the brightness at a point between the two images is half the greatest intensity in the two images. If photographic images are used instead of point sources of light, this definition still applies if photographic density is substituted for brightness. (2) A measure of the ability of an optical device to produce perceptibly-separate images of objects close together. (3) In photography, the closest spacing, expressed as the number of lines per millimeter or per inch, of equally-spaced, equally-wide black lines on a white background, allowing individual lines to be distinguished. The term is a misnomer; it should be resolution. However, it is now firmly fixed in the jargon of the photographic trade. The quantity defined is not a precise measure of resolution, since it depends for its value on the method used for distinguishing between lines.

responder - A device which indicates the reception of an electromagnetic signal. SEE ALSO transponder.

restitution - (1) Determination of the correct, (mapped) position of an image when the original image of the object appears distorted or displaced on an aerial photograph. Restitution corrects for distortion introduced both by tilt of the photograph and by relief displacement. (2) Reconstruction of the geometry of an object from the geometry of its stereoscopic image. The term appears to be synonymous with stereophotogrammetry.

restitution camera - The projector in a stereoscopic plotting instrument. The term is used to distinguish the projector from the camera used for taking the picture originally. The same camera (sometimes called the exposure camera) that was used for taking the original picture is often used as the projector, hence the distinctive name. However, the term is not needed and disagrees with the definition of camera. Its use is discouraged.

restitution instrument - SEE plotting instrument, stereoscopic.

restoration - (1) The replacing of a survey monument which has disappeared or been destroyed, or the substantial renewal of an existing survey monument by a satisfactory equivalent. In particular, the replacement of one or more lost corners or obliterated monuments in an approved manner, including the substantial renewal of one or more monuments, as required for a survey. (2) The recovery of one or more lines or corner locations or both, of a previously approved survey.

restriction - A provision limiting the free use of a piece of land.

resurvey - A retracing, on the ground, of the lines of an earlier survey when all recovered points of the earlier survey are held fixed and used as control. If too few points of the earlier survey have been recovered to satisfy the requirements of the resurvey for control, a new survey may be made. A resurvey is related directly to an original survey, although there may have been several previous resurveys. The terms original survey, resurvey and new survey have different meanings in land surveying.

resurvey, dependent - (1) A resurvey made to restore a missing corner, based on the original, recorded conditions. The dependent resurvey is made, first by identifying existing corners and other recognized and acceptable control points of the original survey, and second, by restoring the required corners by proportionate measurement in agreement with the original survey. This type of survey is used where there is fair agreement between the conditions on the ground and the records of the original survey. Titles, areas and descriptions should remain unchanged. (2) A retracement and re-establishment of the lines of an original survey in their true, original locations according to the best available evidence of the locations of the original corners. It includes the restoration of lost corners in accordance with the procedures set by the Bureau of Land management.

resurvey, independent - A resurvey which does not depend on the records of the original survey but is intended to supersede them and establish new boundaries and subdivisions. Independent surveys are made in regions usually (but not necessarily) having both private and public lands present in the tract to be resurveyed and where the evidence, on the ground, of the original survey has been lost or the descriptions of the earlier survey cannot be reconciled.

resurvey survey - SEE resurvey.

retard (v.) - Delay. The term is sometimes used as the equivalent of retire (meaning, to move back). This usage is not considered appropriate.

retardation - (1) The amount of time by which corresponding phases of the tide grow later day by day. It averages approximately 50 minutes. SEE ALSO retardation, lunar.
(2) The amount of delay in time or phase introduced by the resistivity of the surface over which a radio wave used in navigation is passing.

retardation, lunar - The approximately 50 minutes needed for a particular meridian on the Earth to catch up with the Moon one day (24 hours) after the meridian first passes the Moon. It occurs because the Moon revolves about the Earth in the same direction as the Earth rotates. Note that lunar retardation and tidal retardation (SEE retardation (1)) are different things although they relate to the same phenomenon.

retardation, tidal - SEE retardation (1).

reticle - A planar assemblage of filaments (threads, etched lines, etc.) placed perpendicularly to the optical axis at the principal focus of an optical instrument. Also called a reticule. That word, however, also means a purse. SEE ALSO graticule.

reticle ring - The ring across which the filaments of a reticle are stretched or, if the reticle is of glass, the ring supporting the disk on which the pattern is etched or cut.

reticule - SEE reticle.

retire (v.) - Move back, as to move a line of position back, parallel to itself, along a course line to get a line of position at an earlier time. The antonym is advance.

retouching (adj.) - The process of correcting the image on an photographic negative or positive, plate or copy by hand, as by using a pen, pencil or brush.

retracement - A survey made to verify the directions and lengths of lines, and to identify monuments and markers of an earlier, established survey. Monuments on recovered corners are repaired, but monuments at lost corners are not restored and lines through timber are not re-blazed.

retracing - The process of making a retracement.

retroreflector - SEE reflector, retrodirective.

retrogression - SEE motion, retrograde.

reversing in altitude and azimuth - SEE centering, double.

reversing point (of a micrometer screw) - That setting, of the head of the screw in a micrometer placed on a spirit level, at which the bubble remains in the center of the vial when the instrument is level and is rotated about its axis.

reversion - (1) The principle or method of leveling a spirit level by rotating the spirit level about the standing axis and bringing the bubble back half-way to the center by turning the leveling screws. The process is repeated until the bubble does not change its position. (2) SEE image, reverted. (3) In law, the returning of an estate to the grantor or his heirs after a particular estate is ended.

reversion, double - SEE centering, double.

revert (v.) - Interchange the right and left sides of an image without changing the relative positions of the top and bottom. Equivalently, produce a mirror image.

Revised Local Level - The name given to a vertical datum established at a tide gage and defined to the nearest 0.1 m below the elevation of the bench mark connected directly to the tide gage. The appropriate year of adoption is appended.
Revised Local Reference - The name given to an elevation defined arbitrarily as being at a specified, integral number of decimeters below the elevation of the bench mark to which the reference surface for tidal elevations is referred. The integer is chosen so that sea level averaged over one year shall have a value between 70 and 71 decimeters. Abbreviated to RLR. Values of mean sea level published by the Permanent Service for Mean Sea Level since 1968 have been referred to Revised Local References.

revision - (1) The process of revising a map or publication to show the current topographic status of a region. Typically, revision of a map will not entail significant changes in the horizontal or vertical coordinates of features, but some improvement is made in planimetric data. Normally, publications are revised, not recompiled. (2) A map or publication which has undergone revision as defined above in (1). SEE ALSO planetable revision.

revision, photo - SEE photorevision.

revolution - (1) A body's turning motion about an external point or axis. SEE ALSO rotation. (2) The process of turning a body through 360° about an external point or axis. (3) The process of turning a geometrical figure about an external point or axis. We say that the Earth rotates on its axis and revolves about the Sun, or that an ellipse is rotated about a minor axis to form a spheroid or is revolved about an (external) axis to form a torus. However, usage has approved ellipsoid of revolution.

revolver - The pair of horizontal angles between three points, as observed at any point on the circle defined by the three points. (Brit.) Also called a swinger. This is the situation in which such angles do not establish a unique point.

revulsion - SEE avulsion (1).

rhomboid - A four-sided, planar figure having opposite sides equal and containing one pair of oblique angles. This is the general version of the parallelogram. The rectangle is a special case of the parallelogram and is not a rhomboid.

rhomboid prism - SEE prism, rhomboidal.

rhombus - A four-sided, planar figure having all sides of the same length. Opposite sides are parallel and opposite angles are equal. The rhombus is a special case of the parallelogram, and the square is a special case of the rhombus.

rhumb - (1) A point of a mariner's compass. (2) SEE rhumb line.

rhumb bearing - SEE bearing, rhumb-line.

rhumb direction - SEE direction, rhumb-line.

rhumb distance - SEE distance, rhumb-line.

rhumb line - A line crossing successive meridians at a constant angle. Also called a rhumb, loxodrome, loxodromic curve, loxodrome spiral, spiral helix, equiangular spiral and Mercator track. Alternatively, a curve spiraling in toward the poles in a constant direction with North. Parallels and meridians can be considered special cases of the rhumb line. The Mercator map projection is the only map projection mapping rhumb lines into straight lines. SEE ALSO grid rhumb-line.

rhumb line, fictitious - A line making the same oblique angle with all fictitious meridians. It may be called a transverse rhumb-line, oblique rhumb line or grid rhumb-line, according to the type of fictitious meridian used.

rhumb-line, inverse - SEE rhumb-line, transverse.

rhumb-line, oblique - (1) A line making the same, oblique angle with all fictitious meridians of a graticule produced by an oblique Mercator map projection. (2) Any rhumb line, real or fictitious, making an oblique angle with meridians. The term is used in this sense to denote a rhumb line which is not a parallel or meridian.

rhumb line, transverse - A line making the same oblique angle with all fictitious meridians in a graticule produced by a transverse Mercator map projection. Also called an inverse rhumb-line.

ria - SEE inlet.

Rice's template - One of a set of templates devised by D. Rice for calculating deflections of the vertical from the formulae of Vening Meinesz.

ridge, mid-oceanic - A submarine range of mountains extending continuously through the Atlantic, Indian, Antarctic and Pacific Oceans, with an average height above the adjacent bottom of 1-3 km and a width of more than 1500 km. That portion which extends into the Pacific Ocean is also called a rise or oceanic rise.

ridge line - (1) A graphic representation of a major ridge in a region. Ridge lines are used particularly on maps or regions in which radar stations are located, to show those places where the beam is blocked, partially or totally, at low angular elevations. (2) A line through points of greatest elevation along a ridge.

Riel discontinuity - A region, between 25 and 35 km deep in Canada, within which the speed of seismic P-waves changes from 6.5 km/s to 7.2 km/s. It may be a continuation of the Conrad discontinuity in Europe.

Riemann homogeneous coordinate - One of the two coordinates u,v defined on a surface by the two conditions: the first fundamental form for the element ds of length shall have the form $ds^2 = E(u^2+v^2)(du^2+ dv^2)$ and $E(0)=1$. Riemann homogeneous coordinates are obtained by a conformal transformation from the plane to the surface in question.

Riemann sphere - A sphere having the complex plane mapped onto it by a stereographic map projection.

right - A claim or title to, or interest in anything whatsoever, which is enforceable by law. The type of right involved depends, among other things, on the thing to which claim or title is made and on the situation of the person making the claim. Often, each type involves several distinct rights; for this reason, rights is sometimes used instead of right even when only one right is involved and right should, properly, be used. Such rights include abutter's, access, air, avigation, flowage, littoral, reversionary, riparian, squatter's, subsurface, surface and water rights. SEE ALSO access right; air rights; mineral rights; slope rights; squatter's rights; timber rights.

right, absolute - SEE property-right, absolute.

right, qualified - SEE property-right, qualified.

right, reversionary - The right to repossess and resume the full and sole use and ownership of real property which temporarily has been alienated by lease, easement or otherwise. According to the terms of the controlling instrument, a reversionary right becomes effective at a stated time or under certain conditions, such as the termination of a leasehold, abandonment of a right of way or at the end of the estimated economic life of improvements.

right, riparian - A right incident to land bordering on navigable water. [Florida Statute 217.09 (1) (1957)]. The exact definition of riparian right varies from State to State. It originated in the common law, which allowed each riparian owner to require the waters of a stream to reach his land undiminished in quantity and unaffected in quality except for minor domestic uses. It has been abrogated in a number of western States and greatly modified in others, and in general, at present, allows each riparian owner to make a reasonable use of the water upon his riparian land, the extent of such use depending on the reasonable needs and requirements of other riparian owners and the quantity of water available. SEE ALSO rights, riparian.

right, vertical-circle - SEE circle right.

right ascension - SEE ascension, right.

right of access - SEE access right.

right of entry - The right acquired to enter onto private property for a specific reason or purpose.

right of survey entry - SEE entry, right-of-survey.

right-of-way - Any strip or piece of land, including surface, overhead or underground, granted by fee or easement, for construction and maintenance according to designated use. E.g., for such uses as drainage and irrigation canals and ditches; electricpower lines; telegraph and telephone lines; gas, oil, water and other pipelines; highways and other roadways, including right of portage; sewers; flowage or impoundment of surface water; and tunnels.

right-reading (adj.) - Appearing correctly oriented when viewed through the transparent sheet forming the base. Said of a photographic image on a transparent base. Also called normal-reading or natural reading.

rights, aquatic - Rights for fishing or navigation in a specific region.

rights, junior - SEE rights, senior.

rights, reversionary - SEE right, reversionary.

rights, riparian - The legal right which assures, to the owner of land abutting on a stream or other natural body of water, the use of such water and the rights to the banks, bed and travel on the water. Also called water rights. SEE ALSO right, reversionary.

rights, senior - The rights in a parcel of land, or several parcels, created in sequence with a lapse of time between them. A person conveying part of his land to another (senior) cannot, at a later date, convey the same land to another (junior). A buyer (senior) has the right to all land called for in a deed; the seller (junior) owns the remainder.

rigidity, residual - That property of material which enables it to maintain its form under stresses which act for a long time but are within the elastic limit of the material.

riparian (adj.) - Pertaining to anything connected with or adjacent to the banks of a stream or other body of water. SEE ALSO owner, riparian; rights, riparian.

rise (v.) - (1) Cross the visible horizon while ascending. (2) Become detectable by a radio receiver while ascending. Equivalently, cross the radio horizon while ascending.

rise, continental - A gently sloped bottom, in the ocean, with a generally smooth surface and rising toward the foot of the continental slope.

rise, mean - The height of mean high water above the reference level of chart datum.

rise, oceanic - SEE ridge, mid-ocean.

rise interval, mean - (1) The interval of time, in hours and minutes, between the transit of the Moon and the elevation of the tide measured above chart datum. (2) The average interval between a transit of the Moon and the middle of the period during which the tide is rising. The mean rise interval may be referred either to the local or to the Greenwich meridian.

rise of tide, mean - SEE rise, mean.

rise time - (1) The length of time it takes the output of a system to change from a specified, small fraction (usually 5% or 10%) of its steady-state amplitude to a specified, large fraction (usually 90% or 95%). Also called the time constant, lag coefficient and relaxation time of the system. (2) The length of time required for a pulse to increase from a reference level (usually zero) to its greatest amplitude. Also called time of rising. (3) The published time at which a satellite's signal can be received by a suitable radio receiver.

Ritchey-Chretien telescope - A reflecting telescope of the Cassegrainian type, whose two mirrors have been figured to give an image, at the Cassegrainian focus, free from spherical aberration and coma.

Rittenhouse compass - A large, tripod-mounted magnetic compass equipped with two collinear arms on each of which vertical brackets with peepholes are fastened. This compass was used for surveying public lands in the USA from about 1800 until after 1836.

river - (1) A natural stream of greater volume than a creek or rivulet, having a source in fresh water, flowing in general in one direction toward a sea, lake or other river, in a more or less permanent bed or channel, with a current which may be either continuous in one direction or affected by the ebb and flow of the tidal current. The low of fresh water is controlled by the topographic difference in the elevation of the water at the source and the elevation of the receiving body of water. Where the flow is under the influence of tidal currents, the average flow of saline water up river over a long period of time must be zero. (2) A large, natural, flowing body of water emptying into an ocean, lake or other body of water and usually fed along its course by converging tributaries. (3) A large body of flowing water confined to a channel. (4) A constantly flowing watercourse so named from its source to its outfall. A river is usually considered to be larger than a stream.

river bed - SEE bed of a stream.

river bore - A bore in a river.

river crossing - SEE water crossing.

river estuary - SEE estuary (2).

river gage - SEE stream gage.

river level - The elevation of the surface of a river above a reference surface. Also called the water level of the river.

river level, mean - The average elevation of the surface of a river at any point for all stages of the tide observed over a 19-year period. It is usually determined from hourly readings of the height of the surface at a graduated staff or other device. Unusual variations of the level caused by abnormally large runoff or discharge are excluded from the computation.

river shoreline - That line which is washed by the water wherever it covers the bed of the river within its banks.
(2) The line lying along the bank at the average level attained by the waters of river when they reach and wash the bank without overflowing it.

river terrace - A flat, level or nearly level tract of land bordering a river, bounded on at least one side by a definite, steep slope rising upward from it and on the other side by a downward slope.

road - (1) An open place or public passage for vehicles, persons and animals. A road is generally understood to be a public passage for vehicles, etc., outside of an urban district, as distinct from a street, which is a public passage for vehicles, etc., in an urban district. Road and highway have the same meaning. The road includes not only the roadway but such things as ditches and culverts. (2) A track for travel or for conveying goods, etc., forming. A paved trafficway leading from town to town. (3) A paved trafficway leading from town to town.

road, prescriptive - An adverse holding of a road under a color of right, e.g., a road that has been used and enjoyed by the public for a long and continuous period, and which use has ripened into a right, which is ordinarily an easement. Also called a highway by use or highway by user. Depending on the State, particular requirements, which are usually the same as those for adverse possession, must be met.

roadbed - The graded portion of a road, usually considered as the part between the intersections of top and side slopes, upon which the base course, surface course, shoulders and median are constructed.

road map - A map, usually at medium or small scale, showing the roads in a region.

roadway - (1) That part of a highway, including shoulders, made for use by vehicles. A divided highway has two or more roadways. (2) In the specifications for construction of a highway, that part of the highway lying within the limits of construction.

Robertson's method - SEE distance-ratio method.

Roche's law - The formula $D = D_o (1 - 0.764\, r^2)$ for the variation of density D from the center of the Earth to the outside. D_o is the density at the center of the Earth and r is the average radius of a layer.

rock platform, high-water - A coastal terrace a little below the level of high tide and coinciding with the ground-water table resulting from the erosion, by waves, of rocks which are softer or more decomposed above the ground-water table than below.

rod - (1) A unit of length, in the English system, legally equal to 16½ or 5½ yards or to 198 inches or 25 links. Also called a perch or pole. These units were commonly used in deeds written in the time of the 13 American colonies, to describe boundaries of parcels of land. The rod is equivalent to 5.029 210 meters, approximately. It was established, as a unit of length, by an ordinance of Edward I, in 1303, as equal exactly to 5½ ulnae, the ulna being defined by the length of a standard iron bar which was the prototype of the yard. A rod 16 feet in length was used in some surveys in colonial New England, and one of 18 feet is common in some parts of the USA. There are other definitions, depending on the locale. (2) A unit of area equal to the area of a square one rod on a side. (3) Any slender bar, as of wood or metal. Specifically, a bar or staff for measuring. (4) SEE leveling rod. (5) SEE range rod. (6) SEE stadia rod. SEE ALSO leveling rod; lining rod; stadia rod; target rod; wading rod.

rod, foot-meter - SEE stadia-rod, foot-meter.

rod, high - SEE leveling rod, long.

rod, leveling - SEE leveling rod.

rod, long - SEE leveling rod, long.

rod, measuring - SEE measuring rod.

rod, short - SEE leveling rod, short.

rod, speaking - SEE leveling rod.

rod constant - SEE leveling-rod constant.

rod correction - The quantity added to a measured difference of elevation to correct for the error introduced when the intervals on the leveling rods are not actually of the lengths indicated by the numbers on the graduations. Also called a leveling-rod correction and rod leveling-correction. It is traditionally applied as a correction to the scale.

rod excess - The difference between the actual and the numerical distance between extreme marks on a leveling rod, divided by the nominal length of the rod.

rod float - A small, cylindrical tube of any material, closed at the bottom and weighted with shot so that it floats in an upright position with part of the tube projecting above the water. Rod floats used for determining the speeds of currents are usually weighted so that about 5 to 15 cm of the tube projecting.

rod level - A circular level or a pair of mutually perpendicular toric levels mounted so that it can be placed against a leveling rod or stadia rod in such a way as to aid in making the rod vertical.

rod leveling correction - SEE rod correction.

rod mark - A rod assembled from sections of stainless-steel rod and driven into the ground to a depth below any expected layer of disturbance; the top of the rod is about 10 cm below the surface of the ground; it is used to mark the site of a bench mark.

rod person - A person who is responsible for moving a leveling rod from point to point during leveling, for ensuring that the rod is held perpendicularly while it is being observed, and for other associated duties such as moving the target if the leveling rod has one.

rod stop - SEE leveling-rod stop.

rod sum - The algebraic sum of plus and minus sights in a given level line.

rod unit - The smallest interval into which the scale of a leveling rod is divided.

Roelof's prism - A prism made to be attached to the front of the objective lens-system of a theodolite's telescope and permitting precise pointing at the Sun's center by creating four overlapping images of the Sun. The images overlap in such a way that all four images intersect at a common point on the optical axis when the telescope is pointed at the center of the Sun. Also called Roelof's solar prism and solar prism.

Roelof's solar prism - SEE Roelof's prism.

roll - (1) The angular motion of a vessel or aircraft about its longitudinal axis. (2) The angular deviation of a plane through the longitudinal and transverse axes of a vessel or aircraft from a horizontal plane. Roll is often specified to be positive clockwise when viewed in the direction of forward motion along the longitudinal axis. (3) Angular deviation of the focal plane of an aerial camera from a horizontal plane. (4) Angular deviation of the plane of an aerial photograph about the x-axis of the photograph's coordinate-system or about the X-axis in object space. Roll is commonly denoted by ω.

Romanization - (1) The process of recording in Roman letters or numbers either the sounds of a language or the non-Roman, graphic symbols of a written language. Also written romanization. (2) A part, of a language, which has undergone the process of Romanization.

rood - (1) A measure of area usually equal: in England and Scotland, to one-fourth of an acre, or 40 square rods; in the Union of South Africa, to 17.07 square yards or 14.28 square meters. (2) A measure of length or distance varying, locally, from 5½ to 8 yards.

roof - SEE wall, hanging.

roof prism - SEE prism, roof-angle.

roots-of-mountains theory - SEE Airy's theory of isostasy.

rose - A set of lines radiating from a point on a map and marked to indicate velocity or some other quantity having direction. SEE ALSO compass rose; current rose; wind rose.

Rosenmund map projection - SEE Mercator map projection, oblique.

rotation - (1) A turning motion of a body about an internal axis. For example, the Earth rotates on its axis; the alidade of a theodolite rotates about a vertical axis. Also called, erroneously, spin by some geophysicists. SEE ALSO revolution. (2) The angle through which a point is revolved or a body is rotated. Three different kinds of rotation are common in photogrammetry. The first two refer to the orientation of a photograph with respect to the camera and the orientation of the camera with respect to the object. The third kind refers to the orientation of the photograph, when placed in a measuring engine, with respect to the horizontal axes of the engine. SEE ALSO orientation.

rotation, anti-cyclonic - SEE rotation, clockwise.

rotation, clockwise - A direction of rotation which, when viewed looking towards the rotated object from a point on the axis of rotation, is the same as the direction in which the hands of a clock rotate. Clockwise rotation is also called anti-cyclonic rotation (in the Northern Hemisphere) or rotation cum sole. Any sequence of rotations about different axes can be combined into a single rotation about one axis.

rotation, counter-clockwise - A direction of rotation which, viewed looking towards the rotated object from a point on the axis of rotation, is opposite to the direction in which the hands of a clock rotate. Also called cyclonic rotation (in the Northern Hemisphere) or rotation contra sole.

rotation, cyclonic - SEE rotation, counter-clockwise.

rotation contra sole - SEE rotation, counter-clockwise.

rotation cum sole - SEE rotation, clockwise.

rotation matrix - A matrix such that it leaves unchanged the length of any vector it multiplies.

round - SEE method of rounds.

roundoff - The changing, by elimination or substitution of different digits, of one or more of the digits in a number, starting at the least significant digit, so that the resulting number contains fewer significant digits than did the original. The corresponding verb is to round off. A number may be rounded off to eliminate digits which imply a greater precision than is warranted. For example, if a length is measured using a scale graduated only to millimeters, a value of 37.43 mm for the length would certainly be rounded off at least to 37.4 and possibly to 37. Or if 27.1 is divided by 3.0 and both numbers contain the largest number of significant figures possible, a quotient of 9.3333 implies a completely wrong precision of the original numbers. A number may be rounded off by the machine doing the calculating. A mechanical or electronic calculator operates on numbers having a limited number of digits, the exact number being fixed partly by the operator but largely by the machine's structure. Numbers containing more digits will be rounded off by the machine. Truncation is a particular kind of roundoff: digits are simply dropped from the number without any other change being made.

roundoff error - The difference between a number before roundoff and after.

Roussilhe's map projection - A conformal map-projection from the rotational ellipsoid to the plane, producing a graticule symmetric with respect to the line (axial line) representing the central meridian, and having the ordinate y of a point on the axial line given (by analogy with the stereographic map-projection) by the formula $y = 2 k R_o \tan[(Y - Y_o) / 2R_o]$, in which R_o is the average radius of curvature at the point P, on the ellipsoid, corresponding to the origin of the coordinate system in the plane; Y is the length of the arc of the meridian from the equator to a parallel of latitude with latitude ϕ; Y_o is the length of the arc of the meridian from the equator to the parallel of latitude through P. Also called the Russel map projection, although this name may be the result of incorrect translation from the Russian.

route chart - (1) A chart showing routes between various places, usually with distances shown also. (2) An aeronautical chart covering the route between specific terminals, and usually of such scale as to include the entire route on a single sheet. Also called a flight chart.

route survey - (1) A survey made of a region (a) to determine the feasibility of constructing a road, railroad, canal, pipe line, transmission line or similar structure in that region, and (b) to compare routes and select the most feasible one for the road, etc. A complete route-survey will provide sufficient data for indicating the feasible alinement, grades and cross-sections, and proposed right-of-way lines. A route survey does not provide such data as can be gotten from a preliminary survey. Rather, it provides data pertaining to general location, possibilities, feasibility and probable costs of right-of-way, construction, use and maintenance. (2) A survey made to provide all information needed for the planning and construction of a road, railroad, canal, pipe line, transmission line, etc. It comprises all reconnaissance, preliminary and location surveys, and the surveys made during construction.

route surveying - Surveying done incident to locating, designing and constructing a railroad, road, canal, pipe line, transmission line or other similar structures extending in a long, narrow strip across country. Route surveying is done in three steps: a reconnaissance survey consisting of a general study and an investigation of all land affected by the route; a preliminary survey, to further investigate tentative locations selected by the reconnaissance, prepare a topographic map and profile for each of the locations, and establish a tentative center line along the route finally selected; and the location survey, in which the final center line is laid out and leveling done to determine profiles.

rubidium clock - A atomic clock in which the basic interval of time is determined by reference to a characteristic frequency of the rubidium atom. Like the cesium clock, the rubidium clock determines the basic interval by counting a definite number of cycles of radiation absorbed by the atoms at a specified frequency. The rubidium clock has a better short-term stability than the cesium clock but a poorer long-term stability.

Rudoe's formula - A formula developed by Rudoe for calculating either: the coordinates of a point on the rotational ellipsoid, given the coordinates of another point and the distance and azimuth of the first of these points from the second (the direct formula); or the distance and direction between two points, given the coordinates of the two points (inverse formula). The direct and inverse formulae are essentially the same, differing only in the way they are used. Rudoe's formula is accurate but laborious to use.

Rudzki gravity anomaly - The quantity Δg_R obtained by subtracting from a gravity anomaly Δg the free-air gravity correction δg_f and the Rudzki gravity correction δg_R. Subtracting the Rudzki gravity correction is equivalent to moving masses from inside the geoid to outside the geoid in such a way that the potential on the geoid does not change. Because an element of mass inside the geoid has a slightly greater effect on the potential has a slightly greater effect on the potential than has an element of mass at the corresponding distance outside, the space between geoid and an equipotential surface through the gravity station cannot be filled with matter from inside without changing the potential. An additional gravity correction is sometimes subtracted to take this into account.

Rudzki gravity correction - (1) The change in the value of gravity, at a point P, caused by converting mass outside the geoid to an equivalent mass inside the geoid in such a way (by inversion with respect to a sphere) that the shape of the geoid is not altered thereby. The geoid, in the neighborhood of a particular point P outside the geoid, is assumed to be spherical with radius R. Denoting an element of mass at P by dm and the elevation of P by h, then the element dm is replaced by an element dm' at a depth h' below the geoid, with $dm' = R/(R+h)\, dm$ and $h' = R[1 - R/(R+h)]$. This process is to be carried out for all masses above the geoid. The change in gravity at any point P is then the difference between the contribution of extra-geoidal masses to gravity there, i.e., the topographic gravity correction and the contribution of the masses after transfer to within the geoid. The Rudzki gravity correction δg_R is therefore the integral of the vertical component (at P) of the gravity contributed by each element of mass dm' as given above. (2) The same as the above definition, except that the correction is calculated for a point vertically below P on the geoid.

Rudzki gravity reduction - A gravity reduction applying, to a calculated or measured value of gravity, the following gravity corrections: (a) the free-air gravity correction δg_f; (b) the Rudzki gravity correction δg_R; and (c) a topographic gravity correction δg_{tR} corresponding to the plate, sphere or cylinder assumed to correspond to the masses between the point of measurement and the point of calculation. Also called the inversion method of gravity reduction and Rudzki's inversion method of gravity reduction. The reduction transfers masses between the reference surface and the physical surface in such a way that the geoid's position remains unchanged.

Rudzki inversion - SEE Rudzki gravity anomaly; Rudzki gravity reduction; Rudzki gravity correction; Rudzki inversion method.

Rudzki inversion method - A method, developed by Rudzki in 1911, of converting, in theory, masses outside the geoid to equivalent masses inside the geoid in such a way that the shape of the geoid is not changed, although the gravity field outside the geoid is changed. The geoid, in a region around a selected point, is replaced there by a sphere of suitable radius R. An element of mass dm_1 outside the sphere and at a distance r_1 from the center is replaced by an element of mass dm_2 on the same radius but inside the sphere at a distance r_2 from the center. The element dm_2 and the distance r_2 are related to the element dm_1 and the distance r_1 by the equations $dm_2 = R\, dm_1 / r_1$ and $r_2 = R^2 / r_1$. Also called Rudzki's method and the inversion method.

Rudzki's method - SEE Rudzki inversion method; Rudzki gravity reduction.

rule, distance-prorate - The rule that the angles in a survey traverse must be held to their measured value during an adjustment, while the corrections to distances between stations in the traverse are prorated. The bearing is referred to a convenient meridian and the corrections to distances are prorated through a trigonometric process.

rule, semi-circular - A geometric method, using a semicircle, for determining when an indentation of a coast should be regarded as part of the inland waters of a country and when it should be regarded as part of the marginal sea. The method was first proposed in 1930 by the U.S. Coast and Geodetic Survey. SEE ALSO principle of the semi-circular rule for bays.

rule, ten-mile - The rule which limits inland waters at coastal indentations to a distance of 10 miles between headlands. For indentations 10 miles or less wide at the entrance, a headland-to-headland line would mark the limits; for indentations wider than 10 miles, the limit would be a line drawn across the bay in the part nearest the entrance at the first point where the width does not exceed 10 miles.

rule, trapezoidal - That rule for determining the area under a curve by replacing the curve by a sequence of short chords; calculating the area of each trapezoid formed by a chord, the ordinates of the end points and the projection of the chord onto the axis of abscissas; and using the sum of the individual areas as the desired approximation to the area under the curve. The formula is extremely simple but is, in most cases, considerably less accurate than Simpson's rule.

rule of thalweg - The rule holding that, where a navigable river separates two nations, the middle of the main channel, rather than the geographical middle of the river, is the boundary between them. Also called the thalweg doctrine. The rule of thalweg has also been applied to other boundaries where the boundary is described as being the middle (or center) of the main channel of a navigable river.

rule of the tidemark - The rule that where a coastline is relatively straight, or where only slight curvatures exist, the baseline follows the sinuosities of the coast as defined by a tidal plane.

run - (1) In a series of measurements or observations, the repeated occurrence of measurements or observations of the same type. (2) The number of impressions made, using a printing press, for a given sheet (map). (3) SEE run of the micrometer. (4) SEE running.

run, tide-over - SEE emergency run.

running - (1) A sequence of differences of elevation, measured set-up by set-up in one direction along a section of a level line. Also called a run. A running results in a determination of the elevation between the bench marks or other points, either temporary or permanent, at the ends of the section. (2) The set of differences of elevation determined by leveling in one direction between the two ends of a level line. Equivalently, the line of levels for leveling in one direction. (3) The chain of small differences of elevation measured between the ends of a section. (4) The surveying done in going from beginning to end of a section of a level line. (5) The survey of a traverse, or a segment of a traverse, done in one direction only.

run of the micrometer - The difference, in seconds of arc, between the nominal value of one complete turn of the screw in a micrometer and the actual value determined by measuring, with the micrometer, the distance between two adjacent graduation-marks on an accurate scale such as a graduated circle. This quantity is sometimes called the error of run, but run of the micrometer (or run of micrometer) is more generally used and preferred, although it is an error. The error is kept quite small by adjusting the micrometer, and its effect on measurements is minimized by suitable methods of observing. A correction (called the correction for run of micrometer) was applied, in the past, to readings of the horizontal circle or vertical circle of a theodolite when such theodolites had micrometer-equipped microscopes for making such readings. However, the correction was later made unnecessary by requiring that the instrument be adjusted before using for observations and by adopting a different method of observing.

Runge's method - Resection using angles to three known points. The point P is determined from angles at P' between points A, B, C by calculating the triangle BC'P' and BA'P', in which BC' = 1/BC, BA' = 1/BA, angle BC'P' is equal to measured angle BPC and angle BA'P' is equal to measured angle BPA. P' is therefore found by intersection from A' and C', and BP is found from BP' by calculation.

running with the land - A phrase describing a covenant for something related to land which inures to the benefit of subsequent owners in the chain of title.

runway - A straight road, within a landing strip, normally used by airplanes for landing and taking off.

Russel map projection - SEE Roussilhe's map projection.

S

sac - An indentation in the contour lines of equal depth showing submarine relief. A sac is the submarine analog of a gulf on the surface. (Brit.)

saddle - (1) A low point on a ridge or between two seamounts. (2) A smooth, curved and portable structure used for supporting a surveyor's tape at intermediate points.

safelight - A lamp giving light of a color and intensity not affecting photographic materials in a darkroom but still sufficient to let the worker see what he/she is doing.

sag - The vertical distance between the lowest point of a surveyor's tape or wire suspended between two points and the straight line joining the two points. The sag in a surveyor's tape is approximately $W^2L^3/24T^2$ in which W is the weight of that part of the tape between the supports, L is the length of that part and T is the tension on the tape. SEE ALSO sag correction.

sag correction - The difference between the effective length of a tape (or part of a tape) when supported continuously throughout its length, and when supported at a limited number of intermediate points. Also called a catenary correction (to taped length). Tapes used for measuring base lines are usually supported at 3 or 5 points and hang in curves (catenaries) between adjacent supports. A tape may also be supported throughout, as on a rail, or at 4 points. Correction for sag is not required when the same method is used to support the tape as was used in calibrating the tape, otherwise, a correction is required. The formula used by the U.S. Coast and Geodetic Survey for the correction ΔL to the taped length L, if the tape was lying flat when calibrated but was supported at N+2 points in use is $\Delta L = -L(W/NT)^2/24$, in which W is the total weight of the tape, T the tension and N the number of sections into which the tape was divided by equally spaced, frictionless supports. If the sag d is used instead of the tension, the formula is $\Delta L = -8d^2N/(3L)$. If the tape is calibrated while supported at N+2 points but was lying flat when used, the same formulas apply but with opposite signs. SEE ALSO catenary; sag.

Sagnac effect - SEE gyroscope, ring-laser.

sailing chart - A small-scale chart used for offshore sailing between distant coastal ports. It shows offshore soundings and the most important lights, outer buoys and natural landmarks that are visible at considerable distances. A sailing chart is at the smallest scale used for planning, determining location at sea and plotting the dead-reckoning location. It is usually 1: 600 000 or smaller. The navigator plots on it his location when he is out of sight of land or is approaching the coast from the open ocean.

St. George Island Datum - The horizontal-control datum defined by the following coordinates of the origin and by the azimuth (clockwise from South) to a meridian mark, on the Clarke spheroid of 1866; the origin is at St.George Island astronomic station:

longitude of origin	169° 32' 36.000" W
latitude of origin	56° 36' 11.31" N
azimuth from origin to meridian mark	0° 00' 06"

The longitude is based on longitudes of astronomic stations which were determined by the chronometric method, a weighted average being obtained using triangulation connecting the two islands. The latitude and azimuth are based on independent astronomical determinations.

St. Hilaire's method - A method of determining a line of position from the measurement of the angular elevation of a celestial body by using an assumed location, the difference between the measured and computed angular elevations, and the azimuth. The angular elevation and azimuth of a celestial body are calculated for a given instant of time and for a location where the vessel is supposed to be. The difference (called the intercept between the angular elevation as measured with a sextant and the angular elevation as calculated is then determined. A line is drawn through the point, corresponding in direction to the assumed azimuth. The intercept is laid off along this line. The desired line of position is drawn through the new point on this line and perpendicular to it. Also called the St. Hilaire method. It is an adaptation of Sumner's method and simplifies the usual method of calculating a location from Sumner lines. It is particularly effective and simple near the geographic poles, and is the method commonly used by modern navigators. It was invented in 1874 by Marcq de St. Hilaire, a French naval officer.

St. Michael Datum - The horizontal-control datum defined by the following coordinates of the origin and by the azimuth (clockwise from South) of the azimuth mark, on the Clarke spheroid of 1866; the origin is at St. Michael astronomic station:

longitude of origin	162° 01' 06.000" W
latitude of origin	63° 28' 41.51" N
azimuth of azimuth mark	359° 59' 55.6"

The above coordinates depend on astronomical observations made in 1891, the longitude having been determined by lunar culmination methods and by stellar occultations.

St. Paul Island Datum - The horizontal-control datum defined by the following coordinates of the origin and by the azimuth (clockwise from South) of the azimuth mark, on the Clarke spheroid of 1866; the origin is at St.Paul Island astronomic station:

longitude of origin	170° 16' 24.000" W
latitude of origin	57° 07' 16.86" N
azimuth from origin to azimuth mark	179° 59' 12"

The longitude is based on longitudes of the St.Paul Island and St.George island astronomic stations, which were determined by the chronometric method, a weighted average being obtained by triangulation connecting the two islands. The latitude and azimuth are based on independent astronomical observations.

salinity - The total amount of dissolved, solid matter, in grams, contained in one kilogram of sea water when all the carbonate has been converted to oxide, the bromine and iodine to chlorine, and all organic matter has been completely oxidized. (Some definitions omit the word dissolved, but the word is present implicitly.) This definition, based on the original methods of determining salinity, is still accepted as the theoretical definition, but it is no longer used directly in present-day methods. Salinity is now usually determined by titration or, even more commonly, by measuring the conductivity of the water. It is commonly denoted by S and is given in parts per thousand (o/oo). It was at one time determined by titration with silver nitrate, which relates the salinity to the concentration of chlorine in the water. Hence a relation, first set by Knudsen (1901) between salinity and chlorinity, Cl, was established: $S = 1.8050\ Cl + 0.030$, all concentrations being given in parts per thousand. Since 1966, with the publication of the International Oceanographic Tables, the definition $S = 1.80655\ Cl$ has been used. If the conductivity of the sea water is used, salinity is defined by $S = -0.08996 + 28.29720R + 12.80832\ R^2 - 10.67869\ R^3 + 5.98624\ R^4 - 1.32311\ R^5$, in which R is the ratio of the conductivity, at 15° C and 1 standard atmosphere, of the sample of sea water to the conductivity of sea water having a salinity of exactly 35o/oo.

salinity scale - An equation or set of equations relating the salinity of sea water to its conductivity and the conductivity of standard samples of sea water.

Salinity Scale 1978, Practical - SEE Practical Salinity Scale 1978.

salinometer - An instrument for determining the salinity of a liquid. In its most common form, it consists of a hydrometer graduated to indicate the percentage of salt in the liquid. In a more modern form used for determining the salinity of sea water, it is an ohmmeter with a scale or circuit which converts the measured conductivity of the water to salinity.

salt marsh - (1) A flat, poorly-drained, coastal swamp (marsh) flooded by most high tides. (2) A region of low-lying, wet ground containing a high proportion of salt or alkali. Salt marshes of this type are generally found in arid regions. SEE ALSO salt march, coastal; swamp.

salt marsh, coastal - A marsh located along or near the coast and with a surface below the elevation of mean high water. Sometimes referred to as a salt mars", although not all salt marshes are coastal. In cadastral surveying, it is usually called a marsh.

sample function - SEE function, sampling.

sand - Loose material consisting of small but easily distinguishable, separate grains between 0.0625 and 2.000 mm in diameter. If the grains are smaller, the material is called silt. If they are larger, they are called gravel and if considerably larger, shingle (Brit.)

sandr - SEE outwash plain.

Sanson-Flamsteed map projection - SEE map projection, sinusoidal.

Sanson map projection - SEE map projection, sinusoidal.

Saros - A period of 223 synodic months, corresponding to approximately 18.03 Julian years (6585.32 days) or about 19 passages (6585.78 days) of the Sun through a particular node of the Moon's orbit. The difference of 0.46 day causes an eclipse repeated after one Saros to fall about 120° west of the place where the eclipse occurred on the previous Saros.

satellite - Any body revolving about another body without significantly moving that body. The body about which the satellite revolves is called the primary body or primary. All systems of two or more attracting and revolving bodies revolve, actually, about the center of mass of the system as a whole. However, one body is usually so much more massive than the others that its motion is negligible when considering the motions of the other bodies, the satellites. For example, the planets, meteorites and comets in the Solar System are satellites of the Sun; the Moon is a satellite of the Earth, and the moons of Mars are satellites of Mars. SEE ALSO communications satellite.

satellite, active - A satellite observable by radiation generated by the satellite itself. Only a few natural satellites in the Solar System are active satellites - Jupiter, which radiates appreciably at radio frequencies, is an example. Many artificial satellites are active satellites; they are able to transmit signals they themselves generate or to receive, amplify and re-transmit signals received from elsewhere. Satellites whose power for signals derives from solar cells are considered active satellites. SEE ALSO satellite, passive.

satellite, artificial - A man-made satellite, such as SPUTNIK I, as distinguished from a satellite such as, the Moon or the Earth, made by natural forces. The most important and numerous artificial satellites are those which have been put into orbit about the Earth. A few artificial satellites have been put into orbit about the Moon and Sun. There are a vast number of artificial satellites of the Earth which exist as the debris incident to the launching of satellites, e.g., spent rockets and fragments of rockets exploding prematurely. Spacecraft which are deliberately placed in orbit for a short time only or whose trajectories include only a short segment of an orbit are usually called space probes.

satellite, drag-free - An artificial satellite containing small jet- or rocket type motors governed by a computer that uses these motors to periodically compensate for the effects of drag on the satellite and so keep the satellite in the orbit it would follow if there were no drag.

satellite, equatorial - A satellite whose orbit lies in or near the plane of the equator of the primary body, e.g., a satellite whose orbit lies in the plane of the Earth's equator.

satellite, fixed - SEE satellite, synchronous.

satellite, geodetic - (1) A satellite designed specifically for, but not necessarily solely for, use in geodetic surveying. (2) Any satellite usable for geodetic surveying.

satellite, geostationary - (1) A geosynchronous satellite in an equatorial orbit. Also called ambiguously, a synchronous satellite. SEE ALSO orbit, geostationary. (2) SEE satellite, geosynchronous. (3) SEE satellite, synchronous.

satellite, geosynchronous - (1) A synchronous satellite of the Earth. Also called a geostationary satellite, 24-hour satellite, twenty-four-hour satellite and fixed satellite. Also called ambiguously, a synchronous satellite. The orbit of a geosynchronous satellite has a largest radius of about 42240 km. (2) A satellite in a geosynchronous orbit.

satellite, natural - Any body revolving in a more or less permanent orbit about another body, and not made by man. The Moon is, as far as we know, the only natural satellite of the Earth. The Earth is a natural satellite of the Sun. The Moon can be considered a natural satellite of the Sun, although this concept is not particularly useful.

satellite, passive - (1) In general, any satellite not emitting signals the power for which is provided by the satellite. In particular, any artificial satellite which does not radiate energy generated within the satellite. (2) Any geodetic satellite not emitting signals the power for which is generated within the satellite.

satellite, polar - A satellite whose orbit lies in or near a plane through the rotational axis of the primary body, e.g., a satellite whose orbit carries it over the polar regions.

satellite, selenoid - SEE satellite, synodic.

satellite, Sun-synchronous - A satellite whose orbit is such that local solar time at points on the ground directly under the satellite remains constant.

satellite, synchronous - (1) A satellite whose period is the same as that period of rotation of the primary body about which it is in orbit, and which revolves in the same direction as the primary body rotates. Also called a fixed satellite, geostationary satellite, geosynchronous satellite and 24-hour satellite, although the last three terms should properly be applied only to synchronous satellites of the Earth. (2) SEE satellite, geostationary.

satellite, synodic - A satellite of the Earth, moving with the same angular velocity as the Moon. Also called a selenoid satellite.

satellite, twenty-four-hour - SEE satellite, geosynchronous.

satellite, 24-hour - SEE satellite, geosynchronous.

satellite altimeter - An altimeter mounted in a satellite. Also called a satellite-borne altimeter. All satellite altimeters operate either at radio, infrared or optical wavelengths. Barometric altimeters cannot be used at the altitudes occupied by satellites.

satellite altimetry - (1) A technique or procedure for determining the height or vertical distance of a satellite above the surface of the Earth, Moon or other celestial body using an altimeter mounted in the satellite. (2) A technique of procedure for determining elevations or geodetic heights from measurements made by an altimeter mounted in a satellite.

satellite camera - (1) A camera designed for photographing artificial satellites. (2) A camera designed for taking pictures from an artificial satellite.

satellite city - A planned city, in the path along which a nearby, larger city is expected to grow, and intended to stop uncoordinated spread (city sprawl) of the larger city into the suburbs and to supplement and aid expansion of the larger city.

satellite geodesy - Geodetic surveying in which satellites of the Earth are used as objects of observation. Also called satellite surveying. Two kinds of satellite geodesy are usually distinguished, according to the way in which the observations are made or used: geometric satellite-geodesy and dynamic satellite-geodesy. The satellite (or satellites) may be natural or artificial. However, satellite geodesy involving the Moon is usually considered part of astronomical geodesy. Mapping of the Earth (or the Moon) by photography from artificial satellites is considered part of photogrammetric geodesy rather than of satellite geodesy, while mapping of the Moon or other bodies in the Solar System by photographing from the Earth is considered part of astronomy.

satellite geodesy, dynamic - Satellite geodesy making use of the satellite's orbit to determine the satellite's location at the instant it was observed. Dynamic satellite geodesy treats the observations as data in the problem of (a) determining an orbit from observations at a single point or several points whose locations are known, (b) determining the location of one or more points of observation from observations on a known orbit or (c) determining both orbit and points of observation. The problem is solved by using celestial mechanics, although when both simultaneous and non-simultaneous observations are plentiful, both geometric and orbital theories may be combined. SEE ALSO long-arc method; satellite geodesy, geometric; short-arc method.

satellite geodesy, geometric - Satellite geodesy making little or no use of the satellite's orbit, only the geometric relationship between observation stations and satellite at the instant of observation being important. Also called satellite triangulation and stellar triangulation. Geometric satellite geodesy treats the observations as data in a geometric problem and solves the problem by purely geometric methods (except for the introduction of corrections for refraction, etc.) Because satellites move rapidly, observations must be made nearly simultaneously from several points to make the geometric assumptions valid. SEE ALSO satellite geodesy, dynamic; simultaneous method (of satellite geodesy).

satellite geodesy, simultaneous method of - SEE simultaneous method (of satellite geodesy).

satellite geodesy, simultaneous mode of - SEE simultaneous method (of satellite geodesy).

satellite geoid - An approximation to the geoid, derived by analyzing orbits of Earth satellites.

satellite navigation system - A navigation system having beacons or transponders placed on satellites rather than at fixed points on land. The navigational principles embodied in satellite navigation systems are the same as those embodied in navigation systems using beacons or transponders at the Earth's surface. The only difference is that the location of the beacon or transponder is constantly changing and must be available to the user from other sources. SEE ALSO global positioning system; positioning system, satellite; TRANSIT.

satellite photogrammetry - Photogrammetry using photographs taken from satellites or spacecraft. This should not be confused with photogrammetry using photographs of satellites.

satellite photography - Photography from an artificial satellite.

satellite positioning-system - (1) A positioning system containing a radio receiver or a combination of radio receiver and radio transmitter, at the point whose location is to be determined, one or more beacons or transponders in orbit about the Earth and a computing system for calculating the location of the point of interest. The beacons are considered points of known location. The radio receiver (or receiver/transmitter) may measure times of travel of radio pulses, directions to the beacons or the Doppler shift in the frequency of the radio waves emitted by the beacon. (2) The satellite positioning-system as defined above, with the addition of similar radio receivers or combinations of transmitter and receiver, together with computers, for determining the orbits of the beacons and predicting the locations of the beacons. (3) SEE system, satellite- positioning.

satellite station - SEE station, eccentric.

satellite surveying - The process of determining the locations of points on the Earth's surface by using artificial satellites as points to which or from which measurements are made. SEE ALSO satellite geodesy.

satellite trail - The streak made by the image of a satellite on the photograph taken by a camera that does not follow the motion of the satellite.

satellite triangulation - Any method of determining the coordinates of points on the Earth by measuring directions from those points to one or more artificial satellites. Also called rarely and erroneously, satellite geodesy. The most common method is to photograph the satellites against a stellar background, the stars then furnishing the directions. Although directions could be (and have been) determined by measuring the angular elevation and azimuth of the satellites as given by graduated circles on the observing instrument, such a method is considerably less accurate than the photographic. The term usually implies that observations are made simultaneously, or nearly so, from two or more points on the ground.

Savonius current meter - A current meter composed of two semi-cylindrical vanes disposed about a vertical axis to form an S-shaped rotor and connected to a counter. Also called a Savonius-rotor current-meter. The meter is sensitive to low rates of flow and is responsive to a wide range of horizontal speeds.

scaffold - A temporary, elevated platform erected to support craftsmen, their tools and materials while they are working.

scalar - A quantity having magnitude only. For example, a quantity each of whose values is a single number. A quantity having both magnitude and direction is called a vector.

scale (n.) - (1) The ratio of two numbers, one of which is the length of a characteristic dimension of some object, and the other is the length of the corresponding dimension in a representation (map or model) of that object. A scale is referred to as the scale of the representation, not the scale of the original. It is commonly expressed as a simple fraction a/b, where a and b are integers, or as a proportion a:b, and usually with b adjusted so as to make a equal to 1. It is sometimes written as a decimal fraction. SEE ALSO scale, map; scale of a map. (2) A set of marks placed at uniform intervals along a straight line and numbered, either wholly or in part, to indicate the distance of each mark from the first mark on the line. SEE ALSO scale, graphical; engineer's scale; temperature scale.

scale (of a map) - The ratio of a specified distance, or the average ratio of specified distances, on a map to the corresponding distance or distances on the ellipsoid used in making the map (i.e., used in the map projection). The terms fractional scale, graphic scale, map scale, scale, representative fraction and scale of a map are often used without any particular distinction between their meanings. The corresponding distance on the ground, rather than on the ellipsoid, is sometimes specified. This is not what is actually meant, being neither practicable nor true to practice. The scale of a map may vary from point to point. The variation is small on maps depicting a small region, large on maps depicting large regions. Maps are commonly classified by their average, principal or predominant scales into three groups: large-scale, medium-scale or small-scale maps. There is no general agreement on the boundaries between these groups. A fairly common practice is to classify maps with scales not smaller that 1/50000 as large scale maps; maps with scales between 1/50000 and 1/500000 as medium- scale maps; and maps with scales smaller than 1/500000 as small-scale maps. The U.S. Geological Survey classifies maps with scales of 1/20000 to 1/25000 as medium-scale maps; maps at scales of 1/50000 to 1/100000 as intermediate-scale maps; and maps at scales of 1/250000 or less as small-scale maps. The three major official British mapping agencies also use three groups in classifying maps by scale, but each agency uses different criteria. SEE ALSO bar scale; map scale, bar; compilation scale.

scale (of a stereoscopic model) - The ratio between a distance measured on a stereoscopic model and the corresponding distance on the actual object (e.g., the ground).

scale (v.) - (1) Measure lengths or distances between points on an object. Sometimes written as scale off. The term is not in common use; it seems to be part of the jargon of draftsmen, cartographers, navigators and others who frequently use or make graphic scales. (2) Make a model of an object. To scale down is to make the model smaller than the original; to scale up is to make the model larger than the original. When a map is prepared from photographs, it is usually scaled down from the photographs. (3) Change the scale of a stereoscopic model so as to bring the size of the model into agreement with a plot of the horizontal control.

scale, equivalent - SEE scale, verbal.

scale, fractional - The ratio any small distance on a map bears to the corresponding distance on the ellipsoid used in producing the map. Also called representative fraction, linear scale, map scale, natural scale, numerical scale and scale of the map. It should not be confused with principal scale, which is a single number characterizing the scale of the map as a whole. SEE ALSO scale (of a map).

scale, graphic - A line with graduations numbered and designated so that the length of an interval on the line is to the difference between the corresponding numbers as the distance between two points on the map is to the distance between the corresponding points on the ellipsoid used in making the map. Also called a bar scale, bar map-scale, graphic map-scale, graphical scale, linear scale and map scale. The graphic scale is usually shown outside the border-line of the map. However, on maps showing a considerable amount of ocean, the graphic scale is sometimes shown in the oceanic part of the map. SEE ALSO scale (of a map). Also called a bar map scale, bar scale and graphic map scale.

scale, graphical - SEE map scale; scale, graphic; scale (of a map).

scale, gray - SEE gray scale.

scale, linear - SEE scale, fractional; scale, graphic.

scale, natural - (1) True size, as it exists in nature, without magnification or reduction. Some tables for map projections give distances on the map at natural scale to permit the map to be drawn conveniently at any scale. (2) The scale of a map expressed as a fraction or ratio and not giving linear units of measure. This usage is discouraged. SEE ALSO scale, fractional.

scale, nominal - SEE scale, principal.

scale, particular - The ratio between an infinitesimal linear distance in any direction at any point on a map, and the corresponding distance on the ellipsoid used in making the map. Also called relative scale.

scale, principal - The scale of a reduced sphere representing that sphere used to represent the Earth, as defined by the fractional relation of their respective radii. Also called a nominal scale. The definition assumes that the body mapped (the Earth, for example) is represented by a sphere of approximately the same size, and that this is then mapped onto a much smaller sphere which is mapped directly onto the plane.

scale, relative - SEE scale, particular.

scale, verbal - The ratio or proportion between a small distance on a map and the corresponding distance on an ellipsoid representing the Earth, expressed verbally as an equivalence: e.g., 1 centimeter (on the map) equals 25 kilometers (on the ground). Sometimes called an equivalent scale or a verbal map-scale. The expression also appears in abbreviated forms: e.g., inch to the mile and mile to the inch (the former is preferred).

scale checking - (1) Determining the scale of an aerial photograph or, more precisely, that altitude above sea level wich best fits the photograph. (2) In the compilation of maps using a stereoscopic plotter, determining the scale of a vertical photograph for points at a specific height and the subsequent derivation of distance and direction.

scale error - A systematic error, in a surveyed distance, proportional to the distance.

scale factor - (1) A number by which a distance obtained from a map by computation or measurement is multiplied to obtain the actual distance on the datum of the map. (2) A number by which a length or distance calculated in the State Plane Coordinate System is multiplied to obtain the length of the corresponding geodesic, or conversely. Lengths of distances calculated on the State Plane Coordinate System are affected by the distortion introduced by the map projection. (3) The number by which the reading on the scale of a gravimeter must be multiplied to get the corresponding value of gravity.

scale factor for length - The quantity GM/W_o, in which G is the gravitational constant, M the mass of the Earth and W_o the gravity potential on the geoid.

scale height - (1) The quantity kT/gm, in which g is the gravity acceleration, m is the average molecular mass of the atmosphere below the height where the temperature is T and k is the Boltzman constant, 1.380662×10^{-23} Joules/° Kelvin. Also called height of the homogeneous atmosphere and to distinguish it from density scale height derived from it by using the equation of state of the atmosphere, pressure scale height. It is not a true height and should not be used as if it were. (2) The height at which a given characteristic of the atmosphere falls to 1/e of its value at the surface.

scale indicator - A device carrying a scale graduated logarithmically, for determining the natural scale of a map from the divisions marked on the graphic scales, or from the intervals of latitude on the map. The device is convenient to use and gives the answer rapidly.

scale number - The denominator of a fractional map-scale. E.g., if the fractional map-scale is 1/50000, then the scale number is 50000.

scale of reduction - Also called the reduction factor.

scale of reproduction - The ratio of a representative or average length in the original to the corresponding length in the copy. Also called reproduction ratio, enlargement factor or reduction factor.

scale point - A point, on a photograph, at which the scale is known. A scale point is considered an infinitesimal segment of a line whose scale can be determined.

scale-point method (of determining tilt) - A method of determining the amount of tilt in an aerial photograph by using the scale at three points (scale points) in the photograph for which the elevations of the corresponding ground-points are known. There are various scale-point methods; they vary principally in details of procedure. Anderson's scale-point method is a particularly well-known variant.

scaling - (1) The process of determining the scale of something, such as of a photograph or map. (2) A step, in the absolute orientation of a stereoscopic model, in which the scale of the model is adjusted. Also called scaling the model. (3) The process of changing the scale, in a stereoscopic model, to bring the size of the model into agreement with a plot of horizontal control.

scaling, cartometric - The accurate measurement of geographic or grid coordinates on a map, by means of a scale.

scaling the model - SEE scaling (2).

scan (n.) - The picture or pattern produced by scanning. (jargon)

scan (v.) - (1) Direct a beam of radiation successively over all points of a given region. (2) Select, by means of a small, rotating mirror or other device, the radiation emitted by small elemental areas of a region.

scanner - A device for examining a given region by the systematic examination, in a systematic pattern, of small consecutive elements of the region. There are two types of scanners: (a) the type which directs a beam of radiant energy successively over all consecutive elements of the region; and (b) the type which uses radiation from the region itself but accepts, at any instant, only radiation coming from a definite direction (i.e., from a single element of the region). A scanner of the first type correlates the energy in the beam, after it has been scattered or reflected by the region of interest, with the energy in the original beam and with the direction of that beam. The scanner of the second type uses a mirror, pierced disk or other device to select a portion of the scene; it compares the energy received with that of a standard or reference. In either case, the region is examined element by element in a continuous, systematic manner until the entire region has been examined. The pattern is usually a succession of linear sweeps from side to side of the region, but it may be spiral or radial. SEE ALSO drum scanner; scanner, pushbroom.

scanner, cartographic - A scanner which examines a map or diagram line by line, recording the type of each element (light or dark) and the rectangular coordinates of that element.

scanner, flying-spot - A scanner in which the image of a dot on the face of a cathode-ray tube is focussed onto the object being scanned (usually a photographic negative) and the transmitted or reflect light focussed onto a detector. As the dot moves across the face of the cathode-ray tube, its image moves across the object and is modulated by the density or reflectivity of the object.

scanner, infrared - A scanner operating in the infrared part of the electromagnetic spectrum. SEE ALSO line scanner, infrared.

scanner, infrared linear - SEE line scanner, infra-red.

scanner, linear - A scanner which examines a region by linear sweeps across the region. Also called a line scanner. The scanner may start each sweep just below the end of the preceding sweep, moving back in the reverse direction to end up even with the beginning sweep; or it may move back swiftly to an element just below the first element in the preceding sweep and then move forward, parallel to the path of and in the same direction as the preceding sweep. Scanners mounted in airplanes and satellites frequently provide only the side-to-side motion. The movement forward to a new line is provided by the forward motion of the vehicle itself. Such a scanner is also called a pushbroom scanner.

scanner, mechanical - A scanner in which the image is scanned by viewing through holes in a revolving disk or other purely mechanical arrangement.

scanner, multi-spectral - A scanner separating the returned energy into several different bands of the spectrum and treating each band separately.

scanner, opto-mechanical - A scanner using a rotating mirror to deflect the outgoing beam to each of the elemental areas of a region, or to select a small cone of radiation from successive elemental areas of the region.

scanner, solid-state - A scanner in which the image is focused on an array of semi-conductors acting as detectors. The image is sampled but switching electronically between detectors in the array. The linear scanner is an example.

scanning - The process of viewing the elements of a picture sequentially rather than viewing the whole picture at once.

scanning, opto-mechanical - Scanning done by a combination of mechanical and optical devices. The term is used to distinguish this type of scanning from that done by a beam of electrons or other electronically controlled device.

scan positional distortion - In photography using a panoramic camera, the displacement of the image of ground points from their expected locations on the cylindrical surface of the film because the camera was being moved forward while the lens system scanned.

scatter, forward - SEE scattering.

scattering - The random redistribution of the direction in which radiation incident on a small area is propagated, the radiation going from a single direction into a cone of directions. Scattering occurs either because the surface is composed of many small, randomly oriented, reflecting facets or because the surface first absorbs and then re-emits the radiation. Opaque surfaces scatter radiation only into the hemisphere from which the radiation comes; that part sent back in the direction of the source is called the backscatter. Translucent materials redistribute the radiation into both hemispheres; the portion scattered into the hemisphere opposite to that containing the incident radiation is called the forward scatter. SEE ALSO Mie scattering; Rayleigh scattering.

scatterometer - An instrument for measuring the amount of energy scattered from a particular part of a surface or region. The source of the energy may be the instrument itself, e.g., radar-type scatterometers direct radio pulses at the region and measure the amount of radio energy returned. Or the source may be completely independent of the instrument.

scatterometry - The measurement of the amount of energy scattered from a specific region. In particular, the use of radar as a scatterometer. Sea state, albedo, composition of the scattering surface and other characteristics have been investigated by scatterometry.

Scheimpflug condition - The requirement that object-plane, image-plane and plane of the lens system (the plane through the center of perspective and perpendicular to the optical axis) intersect in the same line (be collinear) for any direct-focusing projector to focus sharply. A rectifier is focused sharply when the Scheimpflug condition is satisfied and when the distances from photographic negative to lens and from lens to easel satisfy Gauss's equation. These conditions are fulfilled automatically by devices containing inversors.

schlieren - Small, irregular packets or layers of air or water whose temperatures and densities differ slightly from those of their surroundings. Schlieren can be made visible in the laboratory by interferometric techniques. They are responsible for scintillation.

Schmidt prism - SEE Thompson prism.

Schmidt telescope - A reflecting telescope which has a primary mirror with a spherical figure and has a correcting plate across the aperture to correct for coma. The Schmidt telescope is often called a Schmidt camera, because it is used almost solely for photographing. It has a large field of view for its focal length typically, 50° for a focal length of 0.6 m. The focal surface is curved, so the film is cut in the form of a Maltese cross and this is then fitted onto the focal surface. Variations of the Schmidt telescope have been used for photographing meteors (Harvard super-Schmidt camera) and artificial satellites (Baker-Nunn camera and Hewitt camera).

Schott base line apparatus - A contact and compensating base-line apparatus consisting of three parallel bars, the middle bar being of zinc and the outer bars of steel. One end of each steel bar was free; the other end was fastened to an end of the zinc bar, a different end for each steel bar. The lengths of the bars were so proportioned with respect to their coefficients of thermal expansion that a constant distance was kept between the free ends of the steel bars.

Schott 1900 spheroid - SEE Schott's osculating spheroid.

Schott's osculating spheroid - A rotational ellipsoid developed by C.A. Schott in 1902, as best fitting the region traversed by the Eastern Oblique Arc. The ellipsoid's dimensions are

   semi-major axis   6 378 157 m ± 90 m
   flattening   1/(304.5 ± 1.9)

The Eastern Oblique arc is a chain of triangulation extending from Calais, Maine, to New Orleans, Louisiana. Also called Schott 1900 spheroid.

Schreiber's double map projection - A set of formulas devised by Schreiber in 1876 for producing the Gauss-Krüger map-projection by first transforming from the rotational ellipsoid to the sphere and then from the sphere to the plane. Also called the Gauss-Schreiber map-projection and Schreiber's conformal double-projection map-projection. SEE ALSO Gauss-Schreiber map projection.

Schreiber's equation - SEE Schreiber's method (of adjustment).

Schreiber's method (of adjustment) - A method of eliminating a particular unknown from a set of equations by adding to the set a fictitious observation-equation in which each coefficient is equal to the sum of all corresponding coefficients in the set multiplied by (-1/n), where n is the total number of observed directions in that set. More generally, the normal equation of the variable is appended with negative weight. (This equation is sometimes called a Schreiber's equation.] The principle underlying the method is called Schreiber's rule. The method has lost most of its usefulness since calculating machines replaced computation by hand.

Schreiber's method (of measuring angles) - A systematic procedure for measuring angles so that the angles between each station and every other station visible from that station are observed independently. Also called Schreiber's method (of observing angles), Schrieber's method (of observation) and the method of all combinations. Each angle is measured once clockwise and once counterclockwise or, depending on the particular variation used, each angle and its supplement are measured. A modification of this method involves measuring the explement of each angle on alternate arcs. Schreiber's method was first used by P. Hansen in 1871 but had been considered earlier by Gauss and by C. Gerling. Schreiber's contributions were presenting the method in a standardized form and applying it to a large triangulation network (Prussia's first-order triangulation, 1864-1874).

Schreiber's rule - SEE Schreiber's method (of adjustment).

Schuler pendulum - A hypothetical, simple pendulum having its point of support at the Earth's surface, its other end (the massive one) at the Earth's center of mass and a length equal to the radius of the Earth. A Schuler pendulum with a length of 6 378 km has a period of about 84.4 minutes. The term is also applied to a similar pendulum having its point of suspension on a vessel or aircraft. The length is then equal to the distance from the center of the Earth to the point of suspension.

Schuler period - A period of oscillation of about 84.4 minutes or, more exactly, $2\pi \sqrt{(R/g)}$, in which R is the radius of the Earth and g is the acceleration of gravity.

Schuler-tuned (adj.) - Insensitive to applied accelerations because of having its natural period of oscillation set to about 84.4 minutes. Said of a gyroscopic device in a vehicle, aircraft or vessel. M. Schuler determined that if the gyroscope was to remain unaffected by the horizontal motions of the craft, the device should have a natural period of oscillation of $2\pi \sqrt{(R/g)}$, in which R is the radius of the Earth and g is the gravity acceleration. This is about 84.4 minutes.

Schuler tuning - The designing or adjustment of a gyroscopic device so that its period of oscillation will be about 84.4 minutes.

Schupman telescope - An optical telescope whose optical system consists of a double-concave lens backed by a mirror, so that light entering the system is refracted, reflected and again refracted. The Schupman telescope is used in some surveying instruments.

Schwendener's method - A method of calibrating an electronic distance-measuring instrument, or of measuring a base line with such an instrument, by dividing a base line into six sections and measuring the lengths of all combinations of these sections. The sections are laid out so that the 21 different combinations cover the half-wavelengths of the modulation on the instrument's carrier-wave and the lengths vary from 20 m to about 500 m or 1000 m.

scintillation - In general, rapid variations in the apparent location, brightness or color of a distant, luminous object seen through the atmosphere. Also called shimmer, but that term should properly be used only for variations in apparent location, not for scintillation in general. If the object lies outside the Earth's atmosphere, as is the case with stars, the phenomenon is called terrestrial scintillation. Almost all scintillation is caused by anomalous refraction occurring in rather small parcels or strata of air called schlieren. Winds transporting such schlieren across the observer's line of sight produce the irregular fluctuations characteristic of scintillations. Also called twinkling. SEE ALSO seeing. (2) SEE scintillation, astronomical. SEE ALSO amplitude scintillation.

scintillation, angular - SEE seeing.

scintillation, astronomical - (1) The irregular changes in brightness of an image, caused by inhomogeneities of refractive index in the air between an extra-terrestrial object and its image. Also called glitter, twinkle and scintillation, although this last name also has a more general meaning. Astronomical scintillation should not be used as a name for seeing (shimmer). Stellar images scintillate so rapidly that the variations can be averaged well enough, when photographing, by exposing a photographic emulsion for several seconds. (2) Scintillation in the image of an extra-terrestrial object.

scintillation, chromatic - The changing color of a star, caused by varying refractive index in the Earth's atmosphere.

scintillation, terrestrial - Scintillation in the image of an object inside the Earth's atmosphere.

scissors inversor - SEE Peaucellier inversor.

screen - A sheet of transparent film, glass or plastic carrying a grid or other regularly repeated pattern which can be used in conjunction with a mask, either photographically or photo-mechanically, to reproduce small regions in the pattern. In other words, a transparent sheet carrying a pattern and used to modify the appearance of a picture seen or photographed through the sheet.

screen, beaded - A surface covered with minute glass beads. Each bead acts as a retro-reflector, directing a high proportion of incident light back through a narrow angle.

screen, biangle - A photographic negative containing a composite of two screens each having a rectangular array of dots, the rectangles being placed at a 30° angle. The biangle screen is used to print colors for those features, in a map, with thin lines.

screen, flat-tint - SEE dot screen.

screen, halftone - A screen carrying a pattern of fine, opaque, crossed lines. Halftone screens of glass consist of a pattern of fine, engraved, opaque, parallel lines on the inside surfaces of each of two plates of glass cemented together so that the lies cross at right angles. The screen is placed immediately in front of the photographic negative onto which the image is to be projected. The resulting screen is used for printing halftone screens on photographic positives, giving the appearance of being continuous in tone. SEE ALSO dot screen.

screen angle - The angle which the rows of dots in a halftone screen make with the vertical when right-reading. The angle is measured clockwise, with 0° at the top.

screening - The application of dots or lines to a region on a map, to provide distinctive tones or tints.

scribe (v.) - Cut lines in an opaque coating to (a) make a design to be printed, or (b) remove marks in an opaque coating for such a purpose. The tool used in scribing is called a scriber.

scribe coat - SEE scribe coating.

scribe coating - An opaque coating, on a translucent or transparent sheet, through which lines are cut so that light can later pass through to a photographic emulsion. Also called a scribing coat and scribe coat.

scriber - A tool having a shaped point and used for cutting through the emulsion of a photographic material, for engraving the surface of transparent sheets of plastic, etc. I.e., a tool used for scribing. Also called a scribing instrument.

scribing (n.) - The process of cutting lines into a prepared coating, to make or remove marks or designs, as one step in preparing a map or other design. Scribing is an efficient way of making pictures for reproduction and is taking the place of the traditional method of drawing. It first came into general use in the early 1950's. Before that time, drawing in ink on plastic sheets had been the usual method.

scribing, negative - SEE negative scribing.

scribing, positive - SEE positive scribing.

scribing coat - SEE scribe coating.

scribing guide - A surface bearing the image of a map to be traced by drafting or scribing for reproduction. Also called color-separation guide, drafting guide and when no other interpretation is posible, guide.

scribing instrument - SEE scriber.

scribing point - A needle or blade ground and sharpened to prescribed 4 dimensions and used in a scriber for scribing. Scribing points come in various diameters or cross-sections to suit the kind of material being worked on.

scrivener - (1) A professional or public writer; a scribe. (2) A notary. (3) A person whose occupation is to draw contracts or to prepare writings.

sea - (1) One of a number of bodies of water of considerable extent, but smaller than oceans, which resemble oceans in their physical and chemical characteristics but are sufficiently inclosed by land that they are geographically distinguishable for oceans, although connected to them. Two types of sea are distinguished, the type depending on the extent of inclosure: mediterranean seas, which are rather obviously almost entirely inclosed; and marginal seas, which are much more open. (2) The character of the surface of a body of water, particularly the height, length (or period) and direction of travel of waves generated locally. (3) SEE ocean. SEE ALSO shelf sea.

sea, adjacent - SEE sea, marginal.

sea, epicontinental - A shallow sea occupying wide portions of a continental shelf or lying in the interior of a continent.

sea, high - (1) That water (the open sea), beyond and adjacent to the marginal sea, which is subject to the exclusive jurisdiction of no one nation. Also called the high seas. A territorial sea is subject to the exclusive jurisdiction of only one nation, while the high sea is basically international territory. Nevertheless, littoral nations frequently exercise limited jurisdiction over portions of the high sea adjacent to their coasts for enforcing customs and other regulations. The Geneva Convention On The High Seas defines high seas as "all parts of the sea that are not included in the territorial sea or in the internal waters of a state." (2) The ocean or sea in general. Also called the high seas.

sea, marginal - (1) Those waters bordering a nation over which the nation has exclusive jurisdiction except for the right of innocent passage of foreign vessels. Also called territorial sea, territorial waters, adjacent sea, marine belt and maritime belt. (2) A partially inclosed sea adjacent to, widely open to and connected to the oceans at the waters' surface but bounded below by submarine ridges. For example, the Yellow Sea and Hudson Bay. (3) SEE waters, territorial.

sea, mediterranean - A sea almost entirely inclosed by land. The Mediterranean Sea is the best known example, but the Caribbean Sea, with or without inclusion of the Gulf of Mexico, is another and is sometimes called the American Mediterranean Sea.

sea, territorial - SEE sea, marginal.

seaboard - A broad region of land bordering the sea. Also called littoral. The terms seaboard, coast and littoral have nearly the same meanings. Seaboard is a general term used somewhat loosely to denote a rather extensive region bordering the sea. Coast is the relatively narrow strip of land in immediate contact with the sea. Littoral is used when other parts of a region are also being discussed.

sea bottom - The interface between the lithosphere and the hydrosphere. Also called ocean bottom and oceanic bottom or, if the meaning is clear, bottom. A definition more in accord with usage would be that part of the lithosphere which is in contact with marine waters or that part of marine waters which is in contact with the lithosphere.

sea boundary - SEE boundary, marine.

sea floor - SEE bottom, deep-sea.

sea floor spreading - SEE spreading, sea-floor.

sea gravimeter - A gravimeter designed specifically to operate on board a ship at sea. A sea gravimeter is of more sturdy construction than other types and is usually mounted on a gyroscopically stabilized mounting.

sea level - (1) In general, the surface of the sea or ocean used as a reference surface from which elevations are measured. Also called physical sea-level if sea level is to be explicitly distinguished from mean sea level or some other calculated quantity. (2) The elevation of the surface of the sea, ocean or a portion of either. (3) The height of the surface of the sea at any time. (4) SEE sea level, mean. Use of sea level for mean sea level is confusing and undesirable.

sea level, derived mean - The average of measured heights of water level at a particular place over a specified period, the height being referred the elevation of a specified bench mark.
sea level, geodetic - SEE sea level, mean.

sea level, ideal - SEE geoid.

sea level, instantaneous - The free surface of the sea at a particular instant.

sea level, mean - (1) The average location of the interface between ocean and atmosphere, over a period of time long enough that all random and periodic variations of short duration average to zero. The U.S. National Ocean Service has set 19 years as the period suitable for measuring mean sea-level (in the sense given in (2) below) at tide gages. If an average over a shorter period is meant, this period should be stated (see definition (3) following). (2) The arithmetic average of heights of the water's surface observed hourly over a specific cycle of 19 years. Also called geodetic sea-level. This quantity is not suitable for most oceanographic or geodetic work, because such a mean sea-level exists only at places where measurements have actually been made. Definition (1) must be adopted if the usually definition of the geoid as an average equipotential-surface best fitting mean sea-level is interpreted to mean mean sea-level everywhere rather than only at tide gages. (3) The arithmetic average of heights of the water's surface observed hourly (or at other short intervals of time). If this definition is used, the period over which observations are made should be specified, e.g., monthly mean sea-level or annual mean sea-level. (The average referred to in all three definitions is an average over time, not over area.) The first definition defines a single surface extending over the entire globe. This kind of mean sea level must be clearly distinguished from that defined, as in (2), at a specific place or places only. The term mean water level is sometimes used instead of mean sea level if definition (2) is intended. (3) The same as definition (2) but leaving out the requirement that the heights be observed.

sea level, mean instantaneous - The average elevation of the water's surface at a specified instant, over a specified region.

sea level, physical - SEE sea level.

sea level, relative - The difference between sea level (at corresponding parts of the cycle) at different points.

sea level, steric - (1) Sea level determined by hydrostatic leveling, i.e., by considering the differences in the density of sea water at different points in the ocean. (2) The quantity expressing the excess elevation of an actual column of sea water above a specified isobaric surface, over the elevation of a similar column of standard sea water.

sea-level datum - SEE datum.

Sea-level Datum of 1929 - A vertical-control datum established for vertical control in the USA by the general adjustment of 1929. Mean sea level was held fixed at the sites of 26 tide gages (21 in the USA and 5 in Canada). The datum is defined by the observed heights of mean sea level at these 26 tide gages and by the set of elevations of all bench marks resulting from the adjustment. A total of 106 724 km of leveling was involved, constituting 246 closed circuits and 25 circuits at sea level. The datum is not mean sea level, the geoid or any other equipotential surface. It was therefore renamed, in 1973, the National Geodetic Vertical Datum of 1929.

sea-level factor - SEE factor, sea-level.

sea mile - (1) The length of one minute of arc measured along the meridian of a specified rotational ellipsoid representing the Earth, at the latitude of a specified location. The length of a sea mile varies both with the latitude and with the rotational ellipsoid selected. (2) SEE mile, nautical.

seamount - A more or less isolated region of the sea floor which is nearly circular or elliptical in cross section and rises at least l km above the sea floor. The definition applies to seamounts having pointed tops as well as to those having flat tops (the latter are called guyots).

seamount, flat-topped - SEE guyot.

seas, high - SEE sea, high.

sea surveying - (1) Any surveying done using techniques or methods adapted particularly for use on or in the sea. (2) Surveying done on or in the sea.

seawall - A self-contained, thick wall kept in place by its own weight and intended to maintain a stable shore-line. Usually constructed of rock laid so as to have a trapezoidal cross-section.

sea water - The common water of the open oceans. Also spelled seawater and sea-water. The properties of sea water have been tabulated as functions of salinity, temperature and pressure, and have also been expressed as equations of state. SEE ALSO water, normal.

secant (adj.) - Cutting at two or more points. A straight line is secant to a curve or a surface if it cuts that line or surface in at least two distinct points, does not lie on the line or surface, and is not tangent at either point. It is tangent to a line or surface if it has only one point in common with that line or surface (within a neighborhood of that point). One surface is secant to another if the two surfaces have a line (straight or curved) in common and if, after a small change in the parameters of one of the surfaces, the two surfaces still have a line in common. E.g., a plane is secant to a sphere if the two surfaces have a circle in common. It is tangent to the sphere if they have only one point in common. A cone is secant to a sphere if they have two circles in common. A cone tangent to a sphere will have a circle in common with the sphere, but this circle will disappear if the angle at the apex of the cone is increased.

secant (n.) - (1) A line cutting a geometric curve or surface at two or more points. (2) The ratio of the hypotenuse of a right triangle to the base of the triangle. More precisely, that function of an angle, in a right triangle, which is the length of the hypotenuse divided by the length of the side adjacent to the angle.

secant method (USPLS) - A method of determining the parallel of latitude for the survey of a base line or standard parallel by offsets from a great circle line which cuts the parallel at the first-mile and fifth-mile corners of the township's boundary. The secant method is a modification of the tangent method. The lengths of offsets made from the projected great-circle line to the parallel are minimal. At the first-mile and fifth-mile stations on the secant, the offsets are zero; between these stations, the offsets are measured to the south; before and after those stations, the offsets are measured to the north.

second - (1) A unit of time, variously defined, but always defined so to be approximately 1/86400 of the ordinary day from one noon to the next. Until 1960, the second in general use was the mean solar second, defined in terms of the rate of rotation of the Earth with respect to the Sun. This rate had been known for a long time to be irregular. In 1960, therefore, the mean solar second was replaced, as the unit of time, by the ephemeris second, approximately equal to the mean solar second but defined in terms of the period of revolution of the Earth about the Sun. This interval was found to be difficult to determine accurately and rapidly. It was therefore replaced, in 1967, by the atomic second, defined in terms of the time needed for a specific number of vibrations in a particular color in light emitted by the cesium atom. (2) A unit of angle defined to be exactly 1/60 of an angular minute. Also called second of arc and arc-second when confusion with a second of time is possible. There are 3600 seconds in one degree and 216 000 seconds in the angle subtended by a full circle. By convention, one second of time is taken as equivalent to 15 seconds of arc when converting angular measures of the Earth's rotation to measures in terms of time. SEE ALSO ephemeris second; leap second; minute.

second, atomic - By a resolution adopted by the General Conference on Weights and Measures in 1967, the ephemeris second was replaced as a unit time in the metric system by the atomic second, defined as the duration of 9192631 770 periods of the radiation corresponding to the transition, unperturbed by external fields, from the hyperfine level (F=4, MF=0) to the hyperfine level (F=3, MF=0) of the ground state of the cesium-133 atom. This is the same as the Système Internationale (SI) second. By this definition, each cesium clock becomes a primary standard for time, and each cesium clock would be a separate time standard. To resolve this difficulty, the General Conference on Weights and Measures decided, in 1972, that the second shall be that second found by the Bureau International de l'Heure as the average given by a number of cesium clocks whose times are monitored by the Bureau.

second, mean sidereal - The fraction 1/86400 of the time needed for one complete rotation of the Earth on its axis, with respect to the distant nebulae.

second, mean solar - The fraction 1/86400 of a mean solar day. The mean solar second was accepted internationally as a fundamental unit of time until 1960, when it was replaced by the ephemeris second as a fundamental unit.

second, SI - SEE second, atomic.

secondary - SEE circle, secondary great.

second of arc - SEE second (2).

second-order work - SEE survey, second-order.

section - (1) That portion of a level line which is recorded and abstracted as a unit and constitutes a self-consistent, self-sufficient set of measurements of differences of elevation. A section begins and ends on a bench mark, either temporary or permanent, whether it is part of the main level-line or is for a spur from a level-line. In the case of a spur to a point whose elevation is determined by taking an additional foresight, the section must begin on a temporary or permanent bench-mark and end at the point on which the leveling rod was held when the additional foresight was taken. (2) A segment of a level line consisting of two neighboring markers connected by leveling in one direction. (3) (USPLS) The unit into which a township is subdivided, normally a quadrangle one mile square with boundaries conforming to meridians and parallels with established limits and containing 640 acres as near as may be. A township is normally divided into 36 sections by lines 1 mile apart as measured from the southern and eastern boundaries. Any excess or deficiency created by the measurements is placed in the northern tier (row) or western range (column) of sections. The sections within a township are numbered consecutively commencing with number 1 in the northeastern section and progressing west and east alternately within each tier, to number 36 in the southeastern section.

section, fractional - (1) (USPLS) A section containing appreciably less than 640 acres, usually because it has been invaded by a segregated body of water or by other land which cannot be properly surveyed or disposed of as part of that section. (2) A section which in its original form contained one or more subdivisions of less than 40 acres because of irregular external boundaries or because a meandered body of water encroached upon it, or by the extension into it of boundaries of other land which could not be properly surveyed or disposed of as an aliquot part of that section. Sections are also frequently turned into fractional sections in closing the surveys on the northern and western boundaries of the township, because deficiencies in measurement caused by errors of survey or convergence of meridians are placed in the first half-mile closing against these boundaries.

section, geodetic - The process of determining the location of a point on a surface, given the locations of certain other points and certain angles in the geometric figure formed by the known and unknown points. The term has been used to designate the processes of intersection and resection.

section, geoidal - A profile of the geoid with respect to the ellipsoid of reference.

section, half (USPLS) - One-half of a section, formed by dividing a section into two parts by a line connecting two opposite quarter-section corners, and containing, as nearly as possible, 320 acres. The half section is a unit used in identifying public lands, e.g., the east half, section 10 is the legal identification of that portion of section 10 of a given township lying east of the north-south central line of the section. SEE ALSO section, quarter.

section, Italian - SEE Collins' method; Italian method.

section, normal - (1) The curve in which a plane through the normal at a specified point of a surface intersects that surface. The normal section of an ellipsoid is an ellipse. If the ellipsoid is rotationally symmetrical, the normal section in a plane perpendicular to the axis of symmetry is a circle. If the ellipsoid is spherical, all normal sections are circles. In general, the normal section specified at one point and required to pass through a second point on the surface differs from the normal section specified at the second point and required to pass through the first point. However, on the sphere, these two normal sections coincide. The geodesic between the two points lies between the two normal sections. (2) The intersection of a solid with a plane through a normal to the surface of the solid. This definition is more common in civil engineering and geology than in geodesy.

section, one-sixteenth - SEE section, quarter-quarter.

section, quarter (USPLS) - One-fourth of a section, formed by dividing a section into four parts by lines connecting the opposite quarter corners of a section, and containing, as nearly as possible, 160 acres. The quarter section is a unit used for identifying public lands, e.g., the northeast quarter, section 10 is the legal identification (description) of that portion of section 10, of a given township, lying east of the north-south, central line and north of the east-west central line of that section.

section, quarter-quarter (USPLS) - One-sixteenth of a section, formed by dividing a quarter section into four parts by lines connecting the midpoints of opposite sides, and containing 40 acres, as near as may be. Sometimes called erroneously, a one-sixteenth section. The quarter-quarter section is a unit used in identifying public lands, e.g., the northeast quarter of the northeast quarter, section 10 is the legal identification of that portion of section 10, of a given township, lying east of the north-south central line and north of the east-west central line of the northeast quarter of that section.

section method - Adjacent pictures are combined to form spatial models from which the coordinates (in the model's system) of points can be obtained.

sector - (1) A geometrical figure bounded by a circular arc and the two radii to the ends of that arc. (2) An instrument composed of a graduated arc to which is fastened a sighting device (alidade) so that the angle between two directions of sighting can be measured. SEE ALSO track sector; zenith sector.

seeing - (1) The disturbing effects produced by atmospheric in homogeneities upon the quality of the image of a luminous body. SEE ALSO scintillation. (2) SEE shimmer.

segmental method - A method for determining the legal status of a bay by using the segment of a circle as the borderline case. The area between the curve of the coast and its chord is equal to the area of a segment of the circle whose center is on the perpendicular to the chord at its middle and which is at a distance from the chord equal to one-half the length of the chord, and which has a radius equal to the distance separating this point from one end of the curve.

seiche - A free oscillation of the water in a closed, oblong basin of variable width and depth.

Seidel aberration - Any one of five different optical aberrations which can prevent an optical system from imaging a point or straight line in object-space as a point or straight line in image-space. SEE ALSO aberration, optical.

Seidel's method - A method of iteratively solving a system of linear equations by successive approximation from a set of assumed values for the unknowns, in which the k-th approximation for the i-th unknown is calculated from the already-calculated k approximations for the i-1 components. Also called the one-step cyclic process of solving a system of linear equations.

selenocentric (adj.) - Referring to the center of the Moon. SEE ALSO coordinate system, selenocentric.

selenotrope - An device similar to the heliotrope but adapted to act as a signal for surveying, by reflecting moonlight rather than sun-light. The selenotrope differs from the heliotrope only in the greater size of the mirror used. It is operated in exactly the same way. In tests made in 1883 and 1887, selenotropes furnished satisfactory lights on which to observe at distances up to 70 miles. The device is no longer in general use.

self-attraction - (1) The attraction of a body by itself. I.e., the attraction of each attracting element of a body by all the other attracting elements of that body. (2) The attraction exerted by a particular body, attraction by other bodies being neglected.

semi-analytical method (of aerotriangulation) - The measurement of the x,y,z coordinates of points in a stereoscopic model on an analog-type of stereoscopic instrument, and the transformation from those coordinates to geodetic coordinates, or coordinates on a map, by calculation.

semi-circular rule (for a bay) - SEE rule, semi-circular.

semi-diameter - (1) One-half the length of a line passing through the center of a closed figure and terminated by the figure. (2) Half the angle, at the observer, subtended by the visible disk of a celestial body. (3) The angle, at an observer, subtended by the equatorial radius of the Sun, Moon or a planet. Also written semidiameter. In making an observation of the Sun or the Moon, the angle should be measured to one edge (or limb) of the body, and the measured angle should be reduced to the center by adding or subtracting the apparent semi-diameter. This quantity may be found in almanacs. For the Sun, it is approximately 16' but varies about 15" either way. For the Moon, its average value is nearly the same, but its actual value is more variable.

semi-tangent - SEE sub-tangent.

sensibility - SEE sensitivity.

sensing, remote - This term is of recent origin. It was invented to designate various particular techniques developed for use with artificial satellites. General definitions encompassing all the special definitions which cover only particular techniques have been proposed. Three common forms are: Sensing an object without touching the object; detecting or inferring the properties of an object without touching the object; detecting, sensing and/or inferring the properties of an object which is far from the detector or sensor. There are, in addition, all the original definitions of more limited scope. These run through all the various combinations possible by substituting ground, Earth's surface or planetary surface for object; adding the proviso by use of radio waves or by use of electromagnetic radiation after sensing or inferring and/or modifying properties to geographical properties. They also include other variations too numerous to list. (4) The response of an device or organism to stimuli from a source far from the device or organism. (5) The response of a device or organism to stimuli from a source far from the device or organism, together with the theory and procedures by which inferences are drawn about the properties of the source. Remote sensing is a composite term and usually implies detection rather than merely sensing.

sensitivity - (1) The change in the reading of an instrument corresponding to a unit change in the property being measured. (2) The change in a system per unit change in some characteristic of that system. For example, one refers to the sensitivity of an orbit to small, real or theoretical changes in the attraction by the primary body.

sensitometer - The instrument used for exposing light-sensitive materials for measurement of the response of that material to light or other forms of radiant energy. Alternatively, a device designed to produce the data from which a characteristic curve can be drawn. A sensitometer for measuring response to light consists of: a standard source of light and a means of regulating the amount of exposure.

sensitometry - (1) The science of measuring the response of a light-sensitive material to the action of light or other forms of radiation.

sensor - A device or organism which responds perceptible to a stimulus.

sensor, remote - A device which responds perceptibly to a stimulus without being in contact with the source of the stimulus; in particular, to a stimulus whose source is far from the device. The term has acquired a variety of other, sometimes inconsistent, meanings.

separation, geoidal - SEE height, geoidal.

separation, longitudinal - Difference in time.

sepia process - SEE Van Dyke process.

sequence - (1) A set of numbers (or other entities) such that every number except one (called the first number in the sequence) is associated with one and only one other number which is said to come before it; and every number except possibly one (called the last number in the sequence) is associated with one and only one other number which is said to follow it. Equivalently, a set whose elements are arranged in the same order as the whole numbers in the order 1,2,3,.... A sequence may be thought of as a set of numbers arranged in a straight line or along a curve, so that each number along the line is next to two other numbers unless the line is infinite at one or both ends or the line is a closed curve. (2) (of tides) The order in which the four tides of a day (two high, two low) occur, with special reference to whether the higher high water immediately precedes or follows the lower low water.

sequence, pseudo-random - A sequence, determined by some defined arithmetic procedure, which is satisfactorily random for a given purpose.

sequester (n.) - In the civil law, a person with whom two or more contending parties deposited the subject-matter of the controversy.

sequestration - The making of a deposit, whereby a neutral depository agrees to hold property in litigation and to restore it to the party to whom it is adjudged to belong.

servitude - SEE easement.

set-back - (1) The horizontal distance from the fiducial mark on the front end of a tape, or on that part of the tape which is in use at the time, back to the point on the tape or monument to which the particular measurement is being made. A set-back is usually very small when it occurs in measuring a base line where stakes supporting the tape are put in place before measuring begins. If the distance between the stakes is too small, the tape will over-run the stakes, and set-backs must be measured. If portable supports such as bucks (saddles) are used, there will seldom be need for measuring set-backs. Set-backs are negative corrections to measured distances. The U.S. National Geodetic Survey avoids the use of set-backs. SEE ALSO set-up. (2) The distance that zoning regulations require between the front surface of a building and the front property-line. (3) The height at which, according to regulations, the upper floors of a building must be set back from the face of a lower part of the building. In very tall buildings there may be more than one set-back.

setting, relative - The dihedral angle between the two planes passing through the principal point of the opposite oblique photographs, the principal point of the vertical photograph and the common exposure station, in a set of photographs from a trimetrogon camera.

setting circle - SEE finder circle.

setting-up - The process of centering a surveying instrument over a marker and leveling the instrument.

set-up - (1) In general, the situation in which a surveying instrument is in position at a point from which observations are made. E.g., measurements made at the last set-up. (2) The actual physical process of placing a leveling instrument over an instrument station. (3) The surveying instrument itself when in position at a point from which observations are made. This last definition may be empty; in any event, its usage with this precise meaning seems to be rare. The terms instrument station and set-up, especially in leveling, are often used interchangeably. However, if any distinction is made, set-up is considered to be the instrument when mounted or set up over the instrument station or point on the ground which is on the axis of rotation of the instrument. (4) The horizontal distance from the fiducial mark on the front end of a tape, or on that end of the tape which is in use at the time, measured forward to the point, on the mark or monument, to which the particular distance is being measured. A set-up is usually very small when measuring a base line with stakes for supporting the tape put in place before measuring begins. If the distance between stakes is too great, the tape will not reach from stake to stake, and set-ups must be measured. If portable supports such as bucks are used, there will seldom be need for measuring set-ups. Set-ups are positive corrections to taped distances. SEE ALSO set-back. (5) The positions given those parts of a stereoscopic plotting instrument which are adjustable but remain fixed while the instrument is in use.

severance - The division of the provisions, rights, liabilities or the like arising under or in something. Specifically, destruction of the unity of interest in a joint estate, the detachment of fixtures from realty or of crops, fruits, timber, minerals, etc., from the soil or separation of a portion of the realty from the whole, as by acquisition for right of way.

severance damages - That loss in value of the remainder of a parcel of land which results from an acquisition. Sometimes called indirect damages.

sewage, domestic - That waste-water which is collected from dwellings, commercial buildings and institutions of a community.

sewer - (1) A pipe or conduit, generally closed but normally not full when flowing, for carrying sewage and other liquid wastes. (2) A conduit through which sewage, storm-water or other waste-waters flow. SEE ALSO trunk sewer.

sewer, building - SEE building sewer.

sewer, combined - A sewer carrying both sewage and the drainage of storm-water.

sewer, private - A privately-owned sewer which is used by one or more properties.

sewer, public - A common sewer controlled by public authority.

sewer, sanitary - (1) A sewer carrying sewage but excluding storm, surface and ground water. (2) A sewer carrying primarily waste-water, but not storm-water or ground-water.

sewer, storm - SEE storm-sewer.

sewer invert - The inner surface of the bottom of sewer pipe. SEE ALSO pipe invert.

sewer slope - The inclination of the invert of a sewer, expressed as the percentage of length, as a decimal, or as 1 foot fall in a given length in feet (or as 1 meter fall in a given length in meters).

sextant - A hand-held surveying instrument for measuring the angle, at the observer, between a celestial object and the horizon, or between two objects. It consists of a graduated arc on which the angle is measured, an alidade for sighting at the object, and a mirror placed so that the observer can see both the object and the horizon at the same time. Also called a hydrographic sextant. The sextant's design is based on the fact that if a ray of light undergoes two successive reflections in the same plane, the angle between the first and second directions of the ray will be twice the angle between the flat mirrors. The graduated arc usually covers one-sixth of a full circle, giving a range of 120°. However, the name "sextant" is often given to instruments constructed on the same principle, even though the arc may be longer or shorter than 60°. Thus, the octant, with a range of 90°, the quintant, with a range of 144° and a quadrant, with a range of 180°, have all been referred to as sextants. Each degree of division on the arc represents 2° of angle between the first and last directions of the ray of light, and is so marked; the 60° division is marked 120°. The last reflection of the ray is made parallel to the direction of the second object. Hence, the angle between the two objects, one seen by double reflection and the other viewed directly, is indicated on the sextant. The instrument is used in navigation for measuring angular elevations of celestial bodies; in hydrographic surveying for measuring horizontal angles, at a point in a moving boat, between objects on shore; and wherever an unstable support makes it impossible to use a theodolite or transit. The sextant has been in use since about 1730. SEE ALSO box sextant; bubble sextant; surveying sextant.

sextant, double - A sextant designed to help an observer rapidly determine his location with respect to three points of known location by measuring, simultaneously, the left and right horizontal angles between the three points.

sextant, hydrographic - SEE sextant.

sextant, marine - A sextant designed primarily for marine navigation.

sextant altitude - SEE altitude, observed.

sextant chart - A chart with curves enabling the three-point problem to be solved graphically rather than with a three-arm protractor.

shade (v.) - Draw fine, parallel or crossed lines on a map to give a visual impression of a third dimension.

shade error - That error, in a measurement made looking through a telescope, caused by refraction in the darkened, transparent glass placed in the line of sight to reduce the intensity of the light reaching the eye.

shade glass - A darkened, transparent piece of glass placed in an optical system to reduce the intensity of the light reaching the eye.

shading - A pattern of fine parallel or crossed lines drawn on a map to give a visual impression of a third dimension. SEE ALSO hill shading.

shading, oblique - Hill shading representing the visual effect of light rays coming in obliquely to illuminate a region.

shadow - (1) Darkness in a region, caused by an obstruction between the source of light and the region. The darkest part of a shadow, that in which light is completely cut off, is called the umbra; a lighter portion surrounding the umbra and in which the light is only partly cut off, is called the penumbra. (2) The partial or total blocking, from a region, of a specific kind of radiation because there is an obstruction between the source of the radiation and the region. A modifier is often added to denote the kind of radiation illuminating the rest of the surface, e.g., radio shadow and X-ray shadow. SEE ALSO radar shadow.

shadow, geometric - The region outlined by drawing straight lines parallel to the direction of approaching rays of radiation and through the extreme edges of a blocking structure or other obstacle to the passage of the rays. The geometric shadow differs from the region actually in shadow (the umbra) because it does not show the effects of refraction and diffraction by the obstacle. The umbra is always smaller than the geometric shadow.

shaft - A vertical or inclined passage extending downward from the surface or from some underground point as a principal passage through which machines, workers and/or ore pass.

shape of the Earth - (1) A geometric figure which coincides, on land, with the surface of the lithosphere and, on water, with a suitably averaged surface of the hydrosphere. Also called the figure of the Earth. Because surfaces of large bodies of water are disturbed by wind waves, seiches and other variations of short duration, some geodesists consider that the shape of the Earth is determined only by the outer surface of the lithosphere and the shape over water is undetermined. (2) SEE figure of the Earth. (3) SEE geoid.

sharpness - A subjective attribute of an image generally indicating how distinctly particular shapes in the image stand out from their surroundings. Acutance is the numerical equivalent or correlate of sharpness.

sheep commons - In colonial days, a parcel of land set aside for the use of the public for grazing of sheep or other animals.

sheering - The process of transforming a rectangle into a parallelogram.

sheet - (1) A body usually a solid, whose dimensions in two directions are very much greater than in the third direction. I.e., a thin, usually solid, body. SEE ALSO bar (2). (2) The material upon which the details of a survey are drawn. SEE ALSO sheet, base; sheet, field. (3) A single map either a complete map in one sheet or a map which is part of a series. SEE ALSO base sheet; boat sheet; carrier sheet; circle sheet; field sheet; pilot sheet.

sheet, control-data - SEE card, control-data.

sheet, fair - SEE sheet, smooth.

sheet, position-plotting - An otherwise blank sheet carrying only a graticule (usually on the Mercator map projection) and a compass rose, so that it can be used as a chart for any longitude.

sheet, smooth - A sheet on which field control and hydrographic data such as soundings, depth curves and regions surveyed with a wire drag are finally plotted or drawn before being used in making a final chart. Also called a smooth chart (USA) or fair chart or fair sheet (Brit.).

sheet line - (1) The outermost border on a sheet carrying a map. (2) A line encompassing the map, on a sheet, and setting a limit to the part of the Earth's surface shown by the map. A sheet line by this second definition may be the same as a neat line.

shelf, continental - (1) That region of the sea bottom extending outward from the shore line, with an average slope of less than 1:100, to a line at which the gradient begins to exceed 1:40 (the continental slope). This is a geological definition. (2) Because the lines specified are difficult to determine, oceanographers have commonly used the practical criterion that the shelf is that part of the sea bottom lying at a depth of less than 200 m and having a slope, outward from the shoreline, of less than 1:1000. (3) That part of the sea bottom extending out from the beach (i.e., that part which is at some times free of overlying water because of tides) to a depth of about 200 m and with a slope of about 1° to 1.5°. (4) That part of the sea bottom extending from land out to a depth of 200 m. This was the definition used in the International Law of the Sea until 1973. (5) Either the entire continental margin, consisting of shelf, slope and rise; or the oceanic bottom (sea bottom) within the 200 nautical mile economic zone, whichever is greater. This definition was adopted at the United Nations Conference on the Law of the Sea (UNCLOS III) in 1973. SEE ALSO shelf, outer continental.

shelf, insular - That part of an island or archipelago which extends from the shore outward to the line where the bottom begins to slope rapidly down to the oceanic depths.

shelf, outer continental - That part of the continental shelf which is seaward of States' boundaries as defined in the Submerged Lands Act (43 U.S.C.A. sect. 1301 et seq.).

shelf sea - A shallow marginal sea less than about 300 meters deep. For example, Hudson Bay is a shelf sea.

shell (of the Earth) - That part of the Earth which is of interest to geographers, extending from the upper boundary of the mantle to the atmosphere.

Shida number - The ratio between a horizontal tidal displacement at the Earth's surface and the horizontal tidal displacement in the corresponding static, oceanic tide. The Shida number was introduced by T. Shida in 1912. A typical value corresponding to the effect of the second-degree harmonic in a series-representation of the tidal potential is 0.8.

shield - A screen or other object substantially reducing the effect of electric or magnetic fields upon radiation-sensitive objects (such as animals, electrical devices or circuits) on the other side.

shift, lateral - The amount by which the greatest value of a gravity anomaly or magnetic anomaly is shifted with a change in the mass or magnetic strength of the anomaly.

shimmer - That apparently irregular motion of the image of an object which is caused by moving irregularities in the refractive index of the atmosphere through which the light from the object passes. Also called scintillation, angular scintillation, dancing and seeing. Astronomers prefer the term seeing to shimmer for this phenomenon, and distinguish it sharply from scintillation, which is used to denote variations in the brightness of the image. Shimmer can cause the apparent location of an object to change by several seconds over a period of a few minutes of time.

shingle - Material consisting of stones considerably larger than those making up gravel. (Brit.)

Ship's Inertial Navigation System - An inertial navigation system designed specifically for use on board a ship. Also, and more commonly, denoted by its acronym SINS. It is often combined with some other navigation system such as a Loran-C or Doppler navigation system.

Shiran - A continuous-wave, electronic distance-measuring system developed from Hiran by translating the frequency to about 3 GHz (S-band). Also writtem SHIRAN. The frequency of 3.312 GHz of the carrier from the ground stations is modulated at 4 frequencies between 664 kHz and 161 Hz to allow the distance to be measured in units of 4 different wavelengths, giving a finest resolution of 1 meter. The standard deviation of distance measurements is about $0.23 \sqrt{K}$ meters, in which K is the distance in kilometers. Shiran has been replaced, for most purposes, by systems making use of artificial satellites.

shoal (hydrographic) - A submarine feature over which the water is shallow, which is detached from the shore, which may be composed of any material other than rock or coral, and which is a menace to navigation on the surface.

shoe - A metallic cone fastened to the free end of a leg of a tripod and designed to be easily driven into the ground. The shoe often has a spur (also called a cleat or foot pad) projecting from it so that the surveyor, by placing his weight on the spur, can drive the pointed end of the shoe firmly into the ground. The spur also helps keep the shoe from sinking further into the ground while observations are being made.

shoot (v.) - (1) Make an astronomical observation, particularly for determining the observer's location. (2) Take a photograph of copy, such as a map, using a camera designed for such work. (jargon)

shoot (n.) - SEE chute.

shooting, double - SEE burn, double.

shop calibration - An adjustment or calibration made to an instrument in a shop having a limited amount of equipment specially suited to the job.

Shoran (short-range navigation system) - A system containing electronic distance-measuring equipment consisting of (a) a radio-transmitter/radio-receiver unit mounted in an aircraft or vessel; (b) two radio-receiver/ radio-transmitter units, in fixed locations on the ground, which receive radio pulses from the mobile unit and re-transmit them to the mobile unit; and (c) a unit which takes the difference between time of transmission and time of reception of each pulse and converts this to a distance. The equipment operates at frequencies between 200 and 300 Mhz. Also written SHORAN. In using Shoran for surveying, the mobile unit was moved back and forth across the line joining the two fixed units, the smallest sum of the distances from the two units found, and the length of the corresponding chord between the two stations calculated. The system indicates distance to the nearest 0.001 mile. The standard deviation of distances found using Shoran is estimated to be $0.56 \sqrt{K}$ meters, in which K is the distance in kilometers. The average distance between fixed units, in a typical project, was about 350 km. Shoran was first developed as a navigating system for bombing aircraft and later adapted to use in photographic reconnaissance and mapping. Its use for geodetic surveying was first tested near Denver in 1945. It was used from 1949 to 1957 to establish the Canadian trilateration network; a relative error of 1:56000 (in distance) was achieved. It is no longer being used for geodetic surveying.

Shoran control - The control of aerial photography by using Shoran to determine the distance of exposure stations from two Shoran stations on the ground.

Shoran line-crossing - SEE Shoran line-crossing method.

Shoran line-crossing method - A method of determining distance between two points by flying across the line joining the two points while using Shoran to determine the distance to each point. SEE ALSO line-crossing method.

Shoran range - The greatest possible distance permissible between the mobile unit of Shoran and the closer of the two units having fixed locations, if the system is to be operable. Shoran range is determined by the elevation at which the mobile unit is flown or the height at which it is mounted (if on a vessel), by the elevation of the fixed unit, the configuration of the intervening land, the distance itself (because of the curvature of the Earth, the time and season when the survey is conducted, etc.

Shoran reduction - The process of converting the sum of the distances measured using Shoran to an equivalent distance on the ellipsoid of reference.

Shoran triangulation - SEE Shoran trilateration.

Shoran trilateration - The process of creating a geodetic network by using the line-crossing method and Shoran to determine distances between points in the network. Also called Shoran triangulation.

shore - (1) Land which is covered and uncovered by the rise and fall of the normal tide. (2) The strip of land in immediate contact with the sea and lying between the high-water line and the low-water line. In its strictest sense, the term applies only to land along tidal waters. (3) Of a sea, that part of the land extending inward from the waters as far as the highest waves reach in winter. This definition is according to civil law. By common law, that region covered and uncovered by the flux and reflux of the sea at ordinary (i.e., average) tides. It is the land between ordinary high and low water marks, the land over which the daily tides ebb and flow. SEE ALSO shoreline, ocean. (4) Of a stream, the region between the bank of a stream and its low-water line. SEE ALSO shore, river. (5) SEE beach.

shoreface - (1) That narrow zone, seaward from the shoreline at low water, which is permanently covered by water and over which the sands and gravel actively move with the action of the waves. (2) The same as the previous definition, but enlarged in scope to apply to lakes as well as seas.

shore line - SEE shoreline.

shoreline - (1) The bounding line between a body of water and the land. In particular, the bounding line between the water and the line marking the extent of high water or mean high water. (2) The intersection of the surface of the land with the surface of a body of water. Also written shore line. The shoreline shown on hydrographic charts represents the line of contact between the land and a water surface having a specified elevation. In regions affected by tidal fluctuations, this line of contact is usually the line of mean high water. In confined, coastal waters of diminished tidal range, the mean water level may be used instead. (3) SEE river shoreline. SEE ALSO ocean shoreline; river shoreline.

shoreline, apparent - The outer edge of marine vegetation (marsh, mangrove, cypress) delineated on aerial photographs where the actual shoreline is obscured.

short-arc method - The method of determining the coordinates of a station of unknown location by using observations of a satellite from that station and from stations of known location lying so close to the unknown station that a short, simple curve can be fitted to the observations, and perturbations of the satellite's orbit during the period of observation can be ignored. Also called the short-arc reduction method.

short-arc reduction method - SEE short-arc method.

Shortt clock - A pendulum clock, formerly much used in timekeeping services and observatories for highest accuracy, which is distinguished by its use of a slaved pendulum to do the mechanical work such as turning the hands of the clock, opening and closing contacts, and giving the impulses to the master pendulum, thus leaving the master pendulum free of most outside influences. The stability is about 0.004 seconds per day (about 1 part in 10^7).

shot - A sighting or measurement from one point to another. (jargon) SEE ALSO side shot.

shoulder - That part of a roadway between the edge of the metalled wearing course (layer) and the top of the foreslope of a ditch or embankment.

shrinkage, differential - The difference between the amount, per unit length, a sheet contracts along the grain of the material, minus the amount, per unit length, the sheet contracts across the grain. The term is frequently applied to differential shrinkage of photographic film, photographic paper and maps.

shutter - That mechanical part of a camera which, when set in motion, permits radiation to reach the sensitized surface of the film or plate for a length of time determined by the operator. SEE ALSO butterfly shutter; louvre shutter.

shutter, between-the-lens - A shutter located between the elements of a camera's lens-system. There are several varieties: the butterfly shutter and the rotating disk shutter are common.

shutter, capping - A shutter mounted in front of a camera's lens-system.

shutter, focal-plane - A shutter located just in front of the focal plane of a camera. The usual form is that of a curtain containing a slit which is pulled across the focal plane to make the exposure.

shutter, Kerr-cell - A shutter consisting of a Kerr cell and a pair of plates capable of establishing a polarizing field. Light is passed into a polarizer which allows only light polarized in a particular plane to pass through. This plane-polarized light is then passed through the Kerr cell, which leaves the plane of polarization alone or rotates it through a predetermined angle, according to the state of the polarizing plates. The light passes through the second polarizer or is stopped there, according to the direction of the plane of polarization.

shutter eyepiece - An eyepiece containing a shutter which is opened and closed at precise intervals by a timepiece, and used for timing the motion of a star's image across the lines of a reticle. The shutter eyepiece was invented by J. de Graaf-Hunter (1932) for use on a transit of theodolite. Tests showed it to be at least as impersonal as the moving-wire micrometer.

shutter mechanism - The mechanism which opens and closes the shutter in a camera. The term shutter is often considered to mean the combination of shutter mechanism and diaphragm.

side equation - A condition equation which expresses the relationship between the various sides in a geometric figure as they may be derived by computation from one another. A side equation is used to make the computed length of a triangle's side the same for all routes through the triangulation by which it is derived.

side lap - SEE overlap, side.

side lobe - SEE lobe, minor.

side overlap - SEE overlap, side.

siderostat - A device consisting of one or more mirrors, one of which is rotated by a motor, for directing light from a selected portion of the sky into a telescope whose optical axis has a fixed direction.

side shot - A sighting or measurement from a survey station to locate a point which is not to be used as a base for extending the survey. A side shot is usually made to determine the location of some object shown on the map.

side test - In triangulation of a quadrilateral or similar figure where lengths can be calculated two different ways, the ratio of the difference between the two calculated results to the length of the line.

siding - Railroad track auxiliary to the main track and used to permit trains on the main track to pass.

sight - (1) The act of looking through an alidade at a distant mark. (2) The act of looking through an alidade (such as a telescope) at a distant mark and recording the corresponding reading. Also called a shot. (jargon) (3) The reading taken as a result of looking through an alidade at a distant mark.

sight, closing - The observation or measurement made to the last point of a survey when that point is also the first point of the survey or is a point whose location has been established previously.

sight, plus - SEE backsight.

sighting - Changing the orientation of an optical system until the line of sight is directed to a specified object or a specified point.

sighting, direct - SEE circle left.

sighting, double - SEE double-sighting; centering, double.

sighting, reversed - SEE circle right.

sighting compass - A magnetic compass with a device attached for sighting an object and determining its direction with respect to the observer. The device usually consists merely of a pair of slotted pieces of metal placed upright on the rim of the compass on opposite sides of the pivot. The uprights are usually hinged so that they can be folded under or over the compass. SEE ALSO compass, prismatic; Rittenhouse compass.

sight rod - SEE range rod.

sights, reciprocal - Observations or measurements made nearly simultaneously from each of two points, toward a point at or near the other point. Reciprocal sights are often made, in spirit leveling, when the level line is to be carried across a wide river or steep valley.

sigma - The density of sea water of a given salinity, temperature and pressure at the surface.

signal - (1) That component, of an event, which is information or carries information. Noise is that component which does not convey information. Which part is signal and which is noise does not depend on the event itself but on the purpose for which the event was observed. For example, to a surveyor determining latitude by measuring the azimuth or zenith distance of a star, the direction which would be observed if the atmosphere were not present is signal and the changes caused by refraction in this direction are noise. To a physicist using the same star to determine refraction, the actual direction of the star is irrelevant, and the changes caused by refraction are the signal. In effect, a signal is any desired component of a transmitted or received message, while noise is the accompanying, undesired component. (2) A natural or artificial object toward which a line of sight is directed in making a measurement. Note that signal and target are not synonyms. The target is that point on the signal at which the line of sight is aimed, while the signal is the entire object. However, the target on a leveling rod is usually understood to be the entire, round object which is moved up and down the rod. SEE ALSO signal, triangulation; signal, survey. (3) Any transmitted, electrical impulse. SEE ALSO lamp signal; longitude signal; reference signal; survey signal; triangulation signal.

signal, aerial - An object suspended from a balloon and marked in such a manner as to be suitable as a survey signal. The balloon is positioned above the mark on the ground so that the mark is vertically under the aerial signal, which may be as much as 60 meters above the mark.

signal, eccentric - A survey signal not in the same vertical line as the station which it represents. A calculated correction is applied to a measured angle to convert the angle to what it would be if the survey signal were in the same vertical line as the station which it represents. The eccentric signal is called the eccentric object observed. SEE ALSO center, reduction to.

signal, geodetic - A signal designed specifically for use in geodetic surveying.

signal, hydrographic - Any object, whether natural or artificial, observed when measuring angles to determine the location of a vessel engaged in sounding.

signal, topographic - A survey signal whose location has been established by surveying, but whose primary purpose is to serve in a topographic survey.

signal lamp - SEE lamp signal.

signal-to-noise ratio - SEE ratio, signal-to noise.

signal tower - A tower made specifically to support a signal, lamp or other object to which observations are made.

signature - (1) Those characteristics of the image of an object which permit that object to be identified on an aerial photograph. (2) Any characteristic or set of characteristics by which a material may be recognized.

signature, spectral - Those lines and bands in the spectrum of electro-magnetic radiation received from a material (by reflection or transmission) that are characteristic of the material.

silt - (1) Water-borne sediment. The term's coverage is usually confined to fine earth, sand or mud, but is sometimes broadened to include all material carried, including both suspended material and bed load. (2) Loose, fine sediment such as that carried by water. The individual particles of silt are finer than particles of sand, being between 1/256 and 1/16 mm in diameter. (3) Deposits of water-borne material, as in a reservoir, delta or on overflowed lands.

silver, atomic-weight - SEE Atomgewichtesilber.

Simpson's formula - SEE Simpson's rule.

Simpson's 1/3 rule - SEE Simpson's rule.

Simpson's rule - The formula $I = \Sigma \Delta x [y(n) + 4y(n+1) + y(n+2)] / 3$, in which I is the approximate value of the integral of $y(x)$ from x_o to x_{N+2}, Δx is the constant $(x_{N+2} - x_o) / (N+2)$, and $y(n)$ is $y(n\Delta x)$, for n from 0 to N. Also called Simpson's formula and Simpson's 1/3 rule. The error is $-\Delta x^5 (d^6y / dy^6)/90$, evaluated at some value of x between x_o and x_{N+2}.

Simpson's three-eighths rule - The formula $I = \Sigma 3\Delta x [y(n) + 3y(n+1) + 3y(n+2) + y(n+3)]/8$, in which I is the approximate value of the integral of $y(x)$ from x_o to x_{N+3}, Δx is the constant $(x_{N+3} - x_o)/(N+3)$, and $y(n)$ is $y(n\Delta x)$, for n from 0 to N.

simulator, land-mass - A device in which topographic and radar-reflection data are stored on glass plates and used to create images. The plates are scanned by a flying-spot scanner and the densities of the images are read by two photo-multiplier tubes. The two planar dimensions of the two images are the x- and y-dimensions of the topographic and reflectance topographic and reflectance data, respectively. The densities of the reflectance images are stored as the intrinsic strength of radar-target reflectance. The images are identical in the x and y values but are separated in one dimension by the optical spacing of the dual read-out elements. The plates are called land mass simulator placates or factored transparencies.

simultaneous method (of satellite geodesy) - A method of determining the coordinates of a point on the ground by measuring directions and/or distances to a satellite simultaneously from that point and from points whose coordinates are known. The orbit of the satellite is then irrelevant. The term also covers the case where the measurements are not exactly simultaneous but are made so nearly at the same time that a simple correction can be made for the error.

sine magnetometer - A magnetometer for measuring the horizontal component H of the Earth's magnetic field and consisting of a magnetic bar suspended by a fiber inside a Helmholz coil wound on a marble cylinder. The amount the bar is deflected from its position at equilibrium is a function of H, of the amount of current passing through the Helmholz coil, and of the angle through which the coil must be moved to preserve the bar's alinement with the cylinder.

single-base method (of barometric altimetry) - A method of barometric altimetry in which two barometers are used. One barometer (designated the base barometer) is left at a central point of known elevation; the other barometer (called the roving barometer) is moved to those points whose elevations are to be determined. At each point occupied by the roving barometer, the pressure, weather conditions and time of observation are recorded. The same quantities are measured and recorded once every five minutes, or other suitable interval, at the central point (the base). Data are later reduced to elevations.

single-projector method - SEE one-swing method.

SINS - SEE Ship's Inertial Navigation System.

siphon barometer - A mercurial barometer consisting of a column of mercury in a glass tube which is so bent as to have two vertical branches, one about one-fourth the length of the other (i.e., bent into a J-shape). The end of the longer branch is closed and the air in it is displaced by mercury, but the shorter branch is left open and the mercury thereby subjected to atmospheric pressure. The difference of the heights of the mercury in the two branches is a measure of the atmospheric pressure.

Sir Henry James's map projection - SEE map projection, perspective.

SI second - SEE second, atomic.

size (v.) - (1) Coat with any of various glutinous liquids used for filling the pores in the surface of paper, fiber or porous board. (2) Calculate the values of the quantities needed to make a photograph, to the desired scale, of a map. (jargon)

sketch map - A map drawn freehand and greatly simplified, preserving essential spatial relationships but not truly preserving scale or orientation.

sketchmaster, universal - A sketchmaster capable of handling vertical or oblique photographs.

skip distance - The distance, measured along the Earth's surface, between the point at which a radio signal transmitted along the ground ceases to be detectable by a radio receiver of normal sensitivity, and the point at which the signal is just detectable after refraction by the ionosphere.

slant distance - The distance between two points not at the same elevation. Also called slant range, slope or slope distance.

slant range - SEE slant distance.

SLAR - SEE radar, side-looking.

slave station - A radio station whose transmissions and/or receptions are governed by another radio station (the master station). Also called remote station, remote (jargon) or slave (jargon).

slope - (1) A stretch of ground lying at a slant. (2) SEE gradient. (3) SEE grade. (4) SEE slope angle.

slope (of a drain) - SEE drain slope.

slope (of a foreshore) - SEE foreshore slope.

slope (of a sewer) - SEE sewer slope.

slope, apparent - The gradient, of a piece of ground, which appears distorted or exaggerated vertically in aerial photographs viewed through a stereoscope.

slope, continental - A declivity seaward from the edge of a continental shelf to a greater depth.

slope angle - The angle between a slope and a horizontal plane. Also called vaguely slope.

slope chaining - Measuring distance along a slope with a graduated tape which is suspended between two points. The measured distance is reduced to horizontal distance by calculation, the slope of the tape or the difference in elevation of the ends of the tape being known.

slope correction - (1) The correction applied to a sounding, obtained using an echo sounder, which is in error because the first echo was returned from a point upslope from its recorded location. (2) SEE grade correction (taped length).

slope rights - The right to extend, adjacent to roads, fill and cut beyond the side lines of the easements for the road as dedicated.

slope stake - A stake set on the line where a finished side of an excavation (cut) or embankment meets the original surface of the ground.

slope taping - Taping during which the tape or chain is held as the slope of the ground requires, the slope of the tape is measured, and the horizontal distance is calculated. Also called slope chaining.

slotted-templet method (of aerotriangulation) - A technique for doing radial aerotriangulation using thin, slotted plates to simulate the photographs. Each plate has a small, central hole in which is placed a pivot representing the principal point. Slots are cut in the templet in the directions of radial lines on the photograph and studs fitting these slots closely enough to allow sliding without wobble are inserted into the slots and serve to connect the templets representing adjoining photographs. When all the templets have been assembled, the locations of the studs identify the corresponding corrected locations of points on the photographs. Also called a slotted-templet plot.

slough - (1) A small marsh. (2) A large wetland such as, for example, a swamp. (3) A backwater region on a river or stream. (4) The region of a cutoff of a river.

smoothing - (1) The process or technique of representing a function whose values vary irregularly, by a function approximating the first but whose values are more regular. Also called filtering or graduation; curve fitting, if the observed values are a function of a single variable. Smoothing may be graphical or arithmetic in nature. In the graphic method, the function is plotted at a number of points and a smooth curve is drawn by eye or with some mechanical assistance so that it passes through or close to as many points as possible. In the arithmetic method, an analytic function of suitable form is chosen and the values of the given function are used to set up a system of equations in which the constants of the analytic function are the unknowns. (2) The process of representing a function which is exactly represented by N terms of a series, by fewer than N terms of another series. Also called filtering. The shorter series is usually obtained by dropping, from the original series, all terms of degree or frequency greater than a selected number less than N. (3) The process or technique of using a function having fewer than N adjustable parameters to represent a set of N points. (4) SEE interpolation. SEE ALSO curve fitting; Vondrak's method.

Snellius-Pothenot method of resection - SEE Pothenot-Snellius method.

Snellius's problem - SEE problem, three-point.

Snell's law (of refraction) - The sine of the angle of incidence divided by the sine of the angle of refraction equals a constant (called the refractive index when one of the substances is air). In algebraic terms, a ray passing from a substance of refractive index n_i to a substance of refractive index n_r undergoes a change $\Delta\theta$ in direction, given by $\Delta\theta = \arcsin[(n_i \sin \theta_i)/(n_r \sin \theta_r)] - \theta_i$, in which θ_i and θ_r are the angles of incidence and of refraction.

snow survey - A set of measurements of the depth and density of snow.

Sodano azimuth survey - A method of determining the direction of a line joining two points, not intervisible, by flying an airplane-carried beacon back and forth across the line joining the two stations and tracking the beacon simultaneously with recording theodolites at both stations.

Sodano's fourth method - A non-iterative method of calculating the length of a geodesic between two points on an ellipsoid, starting with an approximate length calculated on a sphere. The length is then corrected using a number of special quantities which are functions of the eccentricity and the length of the normal.

soil - (1) Finely divided material composed of disintegrated rock mixed with organic matter. (2) The loose material, at the surface of the Earth, in which plants grow. There is actually no single, generally accepted definition of soil, and definitions usually depend on the use made of the material.

soil map - A map showing, for a particular region or the whole world, what kinds of soil occur in that region.

soil rebound - The rising of some kinds of soil after the soil has been depressed by an overlying weight. The phenomenon is important in precise leveling. Soil is compressed when the feet of the support for an instrument are pressed into it. The soil expands when the pressure is removed. The amount of subsequent rise may vary from approximately zero in wet soil or gravel to almost 3 mm in dry soil. The effect on leveling can be reduced by using two leveling rods and observing first on one rod at odd-numbered stations and first on the other rod at even-numbered stations.

soil survey - A survey made (a) to determine the nature of the soils in a region, the soil profile, the density and moisture present, and (b) to classify the soils and determine their boundaries. The procedure used is in three parts: the soil profile is determined; samples are selected for determining the physical properties of the soils in the profile; and the profile is mapped.

soil surveying - Surveying to classify soil by types and to delineate their boundaries.

solarization - A reversal of the usual gradation of gray in the (usually very dense) image obtained on the normal development of photographic emulsions after a very intense or long exposure. A still greater exposure seems to restore the normal gradation.

solar-observation method - Any method of determining azimuth, latitude or longitude measuring the angle between the Sun and some reference plane, or determining the time at which the Sun passes through a specific plane. Special precautions must be taken in making such measurements because (a) the observer's eyes must be protected from direct sunlight, and (b) precise observations must be made on the limb and not on the center.

solar-observation method (of determining azimuth) - SEE azimuth determination from the Sun's angular elevation.

solar-observation method (of determining longitude) - A method of determining astronomic longitude by measuring the differences between the times at which the Sun crosses the meridian of Greenwich and the time at which it crosses the local meridian, and converting this to a difference in longitude. The time at which the Sun crosses the meridian of Greenwich is tabulated in most ephemerides. The time at which it crosses the local meridian can be determined by direct measurement or can be calculated from measured time and angular elevation at which the Sun has prescribed azimuth, and converting this to local hour angle if the latitude of the observer is known.

Soldner coordinate - (1) One of the two coordinates in a Cassini- Soldner coordinate system. Also called a Cassini-Soldner coordinate. On the sphere, x is measured along the central meridian to the foot of the great circle perpendicular thereto and through the point in question; y is measured along that great circle from the central meridian to the point. On a rotational ellipsoid, the same procedures apply, except that geodesic is substituted for great circle. Some writers interchange the roles of x and y. (2) One of the two coordinates in a Soldner coordinate system. SEE ALSO Soldner map projection (2). (3) A coordinate in a rectangular spherical coordinate system. Also called a Cassini coordinate. (4) One of two coordinates λ_s, ϕ_s of a point defined, on an ellipsoid, by the condition that λ_s is the distance from the central meridian to the point, measured along a geodesic through the point and perpendicular to the central meridian; ϕ_s is the distance from the equator to the point where the above-mentioned geodesic intersects the meridian.

Soldner coordinate system - A coordinate system, on the sphere, consisting of two great circles (called the x-axis and y-axis) intersecting at right angles. One intersection is chosen as the origin. The x-coordinate of a point is the distance, measured along a small circle parallel to the x-axis, from the y-axis to the point. The y-coordinate is the distance from the origin to the intersection of the aforementioned small circle with the y-axis. One of the great circles is usually postulated to lie in the plane of a geodetic meridian and the other to lie in the equatorial plane. (2) SEE Cassini- Soldner coordinate system.

Soldner map projection - (1) A map projection from the rotational ellipsoid to the plane such that the x-coordinate of a point P', in the plane, is the same as the distance, to scale, of the corresponding point P, on the ellipsoid, from the central meridian, measured along a parallel of latitude; the y-coordinate of P' is the distance, to the same scale, of a parallel of latitude through P from the equator, measured along the central meridian. This is not the same as Cassini's map projection, which is more common. It was, however, widely used in Germany, France and Russia before Gauss-Krüger coordinates were introduced. (2) SEE Cassini's map projection.

Sollins's tables - A set of tables prepared by A.D. Sollins for calculating the values of deflections of the vertical from gravity anomalies.

solstice - (1) One of two points, on the ecliptic, at which the apparent longitude of the Sun is 90° or 270°. (2) One of the two times at which the apparent longitude of the Sun is either 90° or 270°. Also called the solstitial point. SEE ALSO solstice, summer; solstice, winter; equinox.

solve (v.) - Find a set of numbers which, when substituted for literal quantities in an equation or inequality, satisfy the conditions imposed by that equation or inequality. Reduce is sometimes used for solve, but it is more often used in the sense of to simplify. Recover has been used erroneously as a synonym for solve.

Somigliana's equation - The equation: $\tau = (a \tau_a \cos^2\phi + b \tau_b \sin^2 \phi) / \sqrt{(a^2 \cos^2\phi + b^2 \sin^2\phi)}$, for the gravity τ on a level ellipsoid of rotation, in which a and b are the lengths of the semi-major and semi-minor axes, respectively; τ_a and τ_b are the values of gravity on the equator and poles, respectively, of the ellipsoid; and ϕ is the geodetic latitude at which τ is calculated. Also called Somigliana's formula.

sonar - An apparatus which detects the presence of or determines the distance or direction to an object underwater, by receiving and interpreting sound emitted by or reflected by that object. An acronym for Sound NAvigation and Ranging. The term is applied principally to apparatus which itself generates the sound. The object then merely reflects or scatters the sound and the apparatus receives the echo. However, some sonar consists of a receiver and associated devices only, the sound being generated by the object. At present, that type of sonar (called ASDIC) is used principally by naval vessels for detecting the presence of submarines. However, it has also been used for navigation in which sound-emitting beacons are placed at known locations on the sea bottom.

sonar, side-looking - Sonar emitting timed acoustic pulses in a fan-shaped beam whose axis is lightly below the horizontal and perpendicular to the direction in which the craft is moving. The receiver detects, amplifies and measures the time of travel of the pulses, and these times are then converted to distance and direction. Also called side-scan sonar. The results are recorded as depths, either graphically or digitally. The pulse-generators and receivers are usually mounted in a streamlined cylinder towed at a fixed depth behind the vessel. Two sets of transducers are usually present, radiating in opposite directions sideways from the craft so that the bottom on both sides of the towed cylinders is charted, from directly underneath to as far as 500 meters to each side.

sonar, side-scan - SEE sonar, side-looking.

sonde - An instrument carried aloft by aircraft, rocket or balloon and measuring such characteristics of the atmosphere as pressure, density, humidity and/or temperature. Called a radiosonde if the data are sent to the ground by radio. Sondes may also be used to determined chemical composition or to bring back samples of the air. Balloon-borne sondes equipped with radio and radar are used to determine the velocity or speed of winds. Some sondes are carried aloft inactive and then started functioning when dropped to return to the ground by parachute.

Sonne camera - SEE camera, continuous-strip.

sound - (1) Movement of the atomic and/or molecular particles of a substance in such a manner as to cause waves of alternate compression and expansion within the substance. Regarded as an environment for the sound, the substance is often referred to as the medium in which the sound travels or is propagated. The frequency of the waves may vary considerably from below 1 cycle per second to upwards of 1 megacycle per second, depending on the nature of the source and of the medium. SEE ALSO speed (of sound). (2) That sensation which the brain recognizes as sound and which is transmitted to the brain by nerves connecting the ear to the brain.

sounding - (1) A measured depth in water, usually a measurement of the distance of the bottom below the vessel from which the measurement is made. (2) The process of measuring depths in water. Also referred to as taking soundings. In very shallow water, depths are usually measured by mechanical devices, as by using a sounding pole. In deeper water, a lead-line or an echo sounder may be used. SEE ALSO sounding, echo. (3) A depth of water referred to the datum (usually a tidal datum) shown in the legend of a hydrographic chart. (4) A point, on a map, where the determined depth of the sea, or other body of water, is shown. (5) SEE probing. SEE ALSO echo sounding.

sounding, hydrographic - SEE sounding.

sounding, off - Any region where the depth cannot be measured by a lead-line and generally considered to be seaward of the 100-fathom (200-meter) line.

sounding, on - Any region where the depth can be measured using a lead-line, and generally considered to be landward of the 100-fathom (200-meter) line.

sounding balloon - A free, unmanned balloon carrying meteorological instruments and used for making observations well above the Earth's surface. Sounding balloons are frequently used to determine the temperature and humidity of the air between widely separated points when the distance between the points is determined using light, infrared radiation or radio waves (i.e., using electromagnetic distance-measuring instruments). The term is sometimes used to denote only a radiosonde balloon.

sounding datum - The level (equipotential surface) to which soundings are reduced in the course of a hydrographic survey. This is the datum used for the completed, final tracing. Ideally, it should be the same as the datum used for the final chart.

sounding equipment - Equipment used for measuring depths by sounding. The sounding apparatus may be mechanical, such as a lead-line; acoustical, such as an echo sounder; or optical, such as an airborne laser.

sounding lead - SEE lead (1); lead-line.

sounding line - SEE lead-line.

sounding machine - A machine consisting essentially of a reel of wire to one end of which is attached a weight. The machine carries a device for recording the depth. A crank or motor is provided for reeling in the wire.

sounding pole - A round, wooden pole 5 meters (15 feet) long, used for determining depths of shallow water. It is graduated metrically or in feet and half-feet from the center towards both ends; the graduations are numbered consecutively from the ends towards the center.

sounding sextant - SEE surveying sextant.

sounding wire - A wire used with a sounding machine to determine depths.

South African foot - A unit of length approximately equal to the English foot and having the ratio 0.304 797 265 to the length of the International Prototype Meter. Also called the South African geodetic foot. It should not be confused with the Cape foot used for earlier surveys.

South African geodetic foot - SEE South African foot.

South American Datum 1969 - The geodetic datum based on the South American ellipsoid 1969 with:

semi-major axis	6 378 160 meters
flattening	1/298.25;

and with origin at triangulation station CHUA (Brazil):

	geodetic	astronomic
longitude of origin	48° 06' 04.0639"W	48° 06' 07.80"W
latitude of origin	19° 45' 41.6527"S	9° 44' 41.34"S

azimuth from origin to Uberaba
(clockwise from North) 91° 30' 05.42"
geoidal height 0 meters

South American Provisional Datum 1956 - The geodetic datum based on the International Ellipsoid and with origin at triangulation station LA CANOA:

longitude of origin	296° 08' 25.12" E
latitude of origin	8° 34' 17.17" N.
geoidal height	0.0 meters. (T.G.)

Also called Provisional South American Datum 1956.

South American Spheroid 1969 - A rotational ellipsoid used in the South American datum 1969 and having the dimensions

semi-major axis	6 378 160 m
flattening	1/298.25

southeasterly - A direction within 22.5° of Southeast.

southerly - A direction within 22.5° of South.

South Frigid Zone - SEE Antarctic Circle.

southing - SEE difference of latitude.

South Polar Circle - SEE Antarctic Circle.

South Pole - The geographical pole in the southern hemisphere, at 90° South. Also called the south geographical pole.

southwesterly - A direction within 22.5° of Southwest.

space, inertial - (1) A coordinate system defined by reference to the fixed stars. (2) A coordinate system not undergoing an acceleration with respect to the distant galaxies.

space, translunar - That part of the Universe included (1) within a zone extending 30° on either side of the celestial equator or (2) within zone extending 5° on either side of the ecliptic and lying outside of a sphere containing the Moon's orbit. The term is really quite vague and has no astronomic significance. It is used primarily in discussing projects for spacecraft.

space coordinate - (1) One of the three coordinates necessary and sufficient to locate a point in three-dimensional object-space, as distinguished from the coordinates of a point in image-space. (2) SEE space rectangular coordinate.

space coordinate, rectangular - One of a set of three distances measured from a point to each of three mutually perpendicular planes taken as reference. The three lines of intersection of the three planes are called axes; their common point of intersection is the origin.

space coordinate system - A three-dimensional, rectangular, Cartesian coordinate-system in which the x- and y-axes lie in a plane tangent to the ground at a particular point and the z-coordinate points upward.

spacecraft - Any artificial structure moving inside the Earth's atmosphere and maneuverable. The ability to maneuver distinguishes a spacecraft from a satellite.

space polar coordinate - One of three coordinates of a point in a space polar coordinate system. One of the coordinates represents a distance from a fixed point (pole or center); another represents the angle between a reference line (polar axis) and a line (radius vector) from the pole to the point; and the third represents the angle from a reference plane containing the polar axis to a plane containing the polar axis and the point in question. SEE ALSO coordinate, polar.

space polar coordinate system - A coordinate system consisting of two mutually perpendicular reference lines (called the polar axis and the line of zero longitude), and a plane through the line of zero longitude and perpendicular to the polar axis. SEE ALSO coordinate, space polar; coordinate system, spherical.

space rectangular coordinate - (1) SEE space coordinate, rectangular. (2) In photogrammetry, the x and y coordinates which define the horizontal coordinates of a point on the ground in a rectangular Cartesian coordinate system and the z coordinate, which is the elevation of the point. Also called a ground coordinate.

Space Rectangular Coordinate System, Universal - SEE Universal Space Rectangular Coordinate System.

spad - A nail used in surveying, having a hook for suspending a plumb line from the roof of a mine or tunnel.

span - (1) A unit of length equal to 6 inches. (2) SEE length of span.

Sparrow's criterion - The images of two point-like sources of radiation can not be distinguished if the combined image shows no minimum between the centers of the two images.

species (of constituent) - A classification of tidal constituents according to the period of the constituent. The principal species are: semidiurnal, diurnal and long-period constituents. SEE ALSO tide, types of.

specification - In geodesy, one of a set consisting of such elements of a survey as are considered essential if the survey is to meet certain stated standards of accuracy or precision. Among the elements considered by the U.S. National Geodetic Survey in its specifications are the number of observations to be made, the length of time and sequence of observations, the quality of the instruments used, the greatest and/or least distance permitted between observer and observed object, the shapes of the geometric figures created by the survey and the spacing of control points.

specifications (for a traverse) - Specifications of certain procedures and criteria considered essential to obtaining, in a survey traverse, horizontal control of a desired and specified category. SEE ALSO classification (of control).

specifications (for leveling) - Those procedures and criteria leveling which are considered essential for obtaining elevations of a specified precision or accuracy.

specifications (for triangulation) - The specifications for certain procedures and criteria considered essential for obtaining, by triangulation, horizontal control of a designated order and class. The procedures and criteria considered essential by the National Geodetic Survey have been published in two parts: a part giving the specifications and criteria, and a part expanding on these and explaining them.

specifications (for trilateration) - Specifications of certain procedures, instruments and criteria considered essential for obtaining, by trilateration, geodetic control of a certain order and class.

spectrophotometer - An instrument for measuring transmittance, reflectance or relative emittance in a specific part of the spectrum.

spectroradiophotometer - An instrument for measuring the spectral distribution of radiant energy.

spectrum, continuous - A part of the electromagnetic spectrum in which the wavelengths or frequencies present can be placed in one-to-one correspondence with the real numbers between two limits.

spectrum, electromagnetic - The entire range of electromagnetic radiation, arranged in order of increasing wavelength, or decreasing frequency, extending from gamma rays to the longest radio waves. Light occupies a very small section of the center of the spectrum.

spectrum analysis - Determining the distribution of frequencies present in the spectrum of a signal. This should not be confused with spectral analysis, which is determining what frequencies are present in a signal.

spectrum level - The level of that part of a signal contained within a band one cycle per second, centered at a particular frequency.

speed (mechanics) - The rate of change of the distance of a moving point or body from a fixed point, the distance being measured along the path taken by the moving point or body. Also sometimes called ambiguously velocity. Velocity is speed in a specified direction.

speed (oceanography) - The rate of change of phase of a particular constituent (harmonic component) of the tide or tidal current.

speed (of a lens) - SEE aperture, relative.

speed (of light) - The speed of light or other electromagnetic radiation in a vacuum and far from matter. Also called ambiguously the velocity of light. The term applies to the speed with which an actual disturbance moves, i.e., to the speed of the individual photons. It does not apply to the speed with which the phase of the waves may change. It is usually denoted by the letter c. The speed depends on the substance through which the light is traveling. In a vacuum, the speed is a maximum; it does not depend upon the speed with which the source, the observer, or both are moving, and is not exceeded by any other attainable speed. It is therefore a universal constant. The value adopted for c by the International Association of Geodesy in 1960 was 299 792.4 km/s. The value adopted in 1976 by the International Astronomical Union was 299 792.458 km/s. The value adopted by the International Conference on Weights and Measures in 1983 was 299 792.458 km/s. That same conference made the speed of light a fundamental unit and defined the meter as the length of the path traveled by light in a vacuum during an interval of 1/(299 792 458) second.

speed (of photographic film) - SEE speed (photography); speed, aerial-film.

speed (of sound) - The distance travelled by a particular maximum or minimum in a wave of sound in a unit of time. The speed of sound varies according to the substance in which the sound occurs. It is about: 343 meters/second in air at standard temperature and pressure, 1280 meters/second in hydrogen at 0° C and standard pressure, 1510 meters/second in sea water at 17° C and 3950 meters/second in granite.

speed (photography) - A measure of the sensitivity of photographic emulsion to light. Roughly, the lower the speed, the longer the emulsion must be exposed to light of a given intensity to produce an image of given darkness (density). Speed is often expressed numerically, according to one of several systems such as, H&D, DIN, ASA and Scheiner.

speed, aerial-film - A photographic speed, for aerial film, which is equal to $3/(2H)$, in which H is the amount of exposure in micro-lux-seconds at that point on the characteristic curve where the density is 0.3 above base- plus-fog density on black-and-white film. Aerial-film speed replaces the aerial exposure index, now obsolete.

speed, angular - The amount of rotation or revolution per unit time, without regard to the direction of rotation, of a point or body. Also called angular rate and erroneously angular velocity. Angular velocity is angular speed and the direction of the binormal to the path at the point or body.

speed, linear - The speed of a point or body moving in a straight line.

speed of escape - SEE escape velocity.

sphere - (1) Geometrically, the set of all points equidistant from a given point. (2) Algebraically, the set {x} of coordinates of points x in N-dimensional space, satisfying the equation $r^2 = \Sigma x_n^2$, in which r is a given constant (the radius of the sphere) and x_n is the n-th component of a point x; the summation is over n from 1 to N. Sphere is sometimes used for the solid consisting of the sphere and its interior. The preferable term for such an object is ball. (2) A unit of solid angle equal to the solid angle subtended, at a point, by an entire sphere having that point as its center. The practical unit derived from it is the centisphere, equal to 0.01 sphere or, approximately, 0.1257 steradians. The unit is used principally in calculating illumination.

sphere, authalic - A sphere whose area is equal to that of a specified rotational ellipsoid. If the length of the major axis and the eccentricity of the ellipsoid are 2a and e, respectively, then the diameter d of the authalic sphere is given by $d^2 = a^2 (1-e^2)\{1/(1-e^2) + (1/2e) \ln[(1+e)/(1-e)]\}$.

sphere, celestial - An imaginary sphere of arbitrary radius and specified center, upon which locations of celestial bodies are shown as points. The points are determined by projecting the centers of the celestial bodies onto the sphere by lines drawn to the center of the sphere. As circumstances require, the celestial sphere may be centered at the observer, at the Earth's center, or at any other location. For observations on bodies within the limits of the Solar System, surveyors place the center of the celestial sphere at the center of the Earth. For bodies outside the Solar System, the diurnal and annual parallax are negligible in surveying, and the center of the celestial sphere may be taken at the point of observation. A number of points and great circles on the celestial sphere are used as referents for the establishment of coordinate systems. The most important points are the celestial and ecliptical poles, the equinoxes and the solstices. The important great circles are the meridians, the equator and the ecliptic. SEE ALSO sphere, parallel.

sphere, conformal - A sphere on which the element ds of length is given by $ds^2 = R^2 (d\chi^2 + \cos^2 d\lambda^2)$ in which R is the radius of the sphere, λ is the longitude and χ the conformal latitude of a point, on the rotational ellipsoid, corresponding to the latitude ϕ. The defining equation can also be written as $R = (ds\ N \cos \phi) / (d\sigma \cos \chi)$, in which N is the radius of curvature of the rotational ellipsoid in the prime vertical and $d\sigma$ is the element of length on the ellipsoid corresponding to ds on the sphere. SEE ALSO latitude, conformal (1).

sphere, oblique - The celestial sphere as it appears to an observer between the Equator and a pole. Celestial bodies appear to rise and set obliquely to the horizon.

sphere, osculating - That sphere which passes, in the limiting case, through four points of a curve as the four points approach a fixed one of the four points. Also called the sphere of curvature.

sphere, parallel - The celestial sphere as it appears to an observer at the pole, where celestial bodies appear to move parallel to the horizon.

sphere, reduced - A sphere obtained from a sphere representing the Earth and on which a map has been drawn, by reducing the length of the radius.

sphere, right - The celestial sphere as it appears to an observer on the Equator, where celestial bodies appear to rise vertically above the horizon.

sphere, terrestrial - SEE Earth.

sphere depth, mean - SEE depth, mean-sphere.

sphere of curvature - SEE sphere, osculating.

sphere of mean radius of curvature - A sphere whose radius is equal, at a specified point of an ellipsoid, to the square root of the product of the radii of curvature, at that point, in the meridian and in the prime vertical.

sphericity - The shape of an object expressed in terms of its greatest dimension a, intermediate dimension b and least dimension c, given by the cube root of (b/a)(c/a).

spheroid - (1) A surface generated by rotating an ellipse about its minor axis or about its major axis. In the first case, the spheroid is an oblate spheroid; in the second case it is a prolate spheroid. Equivalently, the spheroid may be defined as a two-dimensional surface having the equation $(x^2/a^2) + (y^2/a^2) + (z^2/c^2) = 1$, in which a and c are constants. If c < a, the spheroid is oblate; if c > a, the spheroid is prolate; if c = a, the spheroid is a sphere. Also called an ellipsoid of revolution and rotational ellipsoid. (2) A spheroid as defined in (1), but having dimensions so chosen as to approximate the Earth as a whole, or so as to make a small portion coincide as nearly as possible with the corresponding portion of the geoid belonging to a particular region. (3) Any surface differing but little from a sphere. (4) A surface derived, by a standard formula for the Earth's gravity potential, by giving the potential a specific value. Such a surface is intended to be an approximation to the geoid. In gravimetric geodesy, the term spheroid may be used in either of the two senses (3) and (4). In geodesy other than gravi-metric, the term is used principally as defined in (2), with occasional use of definition (1). In mathematics and sciences other than geodesy, the term is used with the sense given in (1). (5) A surface close to the geoid or approximating the geoid. (6) An equipotential surface on which the gravity potential is the same as the gravity potential on the geoid, and the value of the gradient of the potential is the same (or is the negative of) the value given by a standard gravity formula. Alternatively, that spherop which has the same constant value for the gravity potential as has the geoid. (7) A geometric figure describing the size and shape of the Earth, the departures from true sphericity being determined from measurements of the Earth's surface. (8) An ellipsoid of revolution with its minor axis parallel to the axis of rotation and having dimensions so chosen as to approximate the Earth as a whole, or so chosen as to make a small portion coincide as nearly as possible with the corresponding portion of the geoid belonging to a particular region. SEE ALSO Airy spheroid 1830; Australian National spheroid; Bessel spheroid 1841; Bruns' spheroid; Clarke spheroid 1858; Clarke spheroid 1866; Clarke spheroid 1880; Clarke spheroid 1880, modified; ellipticity (of a spheroid); Everest spheroid 1830; Everest spheroid 1847; Fischer spheroid 1960; flattening (of a spheroid); Hayford spheroid; Helmert's spheroid; Jeans spheroid; Krassovski spheroid; level spheroid; Maclaurin spheroid; Mercury spheroid 1960; reduction to the ellipsoid; reduction to the spheroid; reference spheroid; Schott's osculating spheroid; South American spheroid 1969.

spheroid, gravimetric - SEE spheroid (5); spheroid (6); spheroid (7).

spheroid, International - SEE International Ellipsoid.

spheroid, level - SEE level spheroid.

spheroid, oblate - A rotational ellipsoid whose axis of rotation is shorter than the other two axes.

spheroid, prolate - A rotational ellipsoid whose axis of rotation is longer than the other two axes.

spheroid, rotational - A body, of approximately the size and mass of the Earth, having rotational symmetry and having a gravitational field defined by some of the lower-degree and order terms in the Legendre series for the Earth's gravity field.

spheroid junction - An accentuated line, on a chart of map, separating two or more major grids or graticules which are on different spheroids.

spheroid of reference - SEE reference spheroid.

spheroid of revolution - SEE spheroid (1).

spherop - (1) An equipotential surface on which the gradient of the potential is a standard gravity field of the earth. Also called a spheropotential surface. The so-called normal gravity field is usually chosen. SEE ALSO spheropotential. (2) The surface defined by assigning a constant value to the gravity potential in some series expansion of the Earth's gravity potential, using only the first few terms of that series. Also called a spheropotential surface. (3) The surface defined by the equation $W = (GM/r)[1 + (C-A)(1 - 3\sin^2\phi)/2] + (\omega r^2 \cos^2\phi)/2$, in which W is a constant, A and C are the Earth's moments of inertia about the principal axis and the minor axis, respectively, G is the gravitational constant, M is the mass of the Earth and r and ϕ are spherical (geocentric) coordinates. SEE ALSO Earth spherop; mareograph spherop.

spheropotential - The potential U calculated from the spheropotential function. Also called normal potential, potential of the normal Earth, etc. This should not be confused with the spherop, which is a surface.

spheropotential function - (1) The function $U = (GM/r)[1 + (C-A)(1 - 3\sin^2\phi')/(2Mr^2)] + (\omega^2 r^2 \cos^2\phi')/2$, in which M is the total mass of the Earth, C and A are moments of inertia about the axis of rotation and about an axis perpendicular to that axis, ω is the rate of rotation of the Earth and G is the gravitational constant. The point at which the potential is to be calculated is at a distance r from the origin of the coordinate system and at an angle $(90° - \phi')$ to the axis of rotation. (2) The function $U = (GM/R) \Sigma [(R/r)n A_n P_n(\phi')] + (\omega^2 r^2 \cos^2\phi')/2$, in which G is the gravitational constant, M is the mass of the Earth, R is the radius of a sphere inclosing all the mass, P_n is an n-th degree Legendre function, the constants A_n are related to n-th degree moments of inertia of the Earth, r is the distance of the point of interest from the center of the sphere, ω is the Earth's rate of rotation and ϕ' is the geocentric latitude of the point. The series is truncated at a suitable, usually low, degree.

spheropotential number - The difference between the spheropotential on a spherop through a point at mean sea level and the spheropotential on a spherop through the point of observation. Also called a normal geopotential number.

spheropotential surface - SEE spherop (2).

spider templet - A mechanical device consisting of long, slotted arms (usually of steel) connected at one end by a clamping stud sitting in the slots. Each arm contains another closely-fitting stud which can be slid along the slot. Also called a mechanical-arm templet and radial-arm templet. Each arm represents a radial and the central stud represents the principal point, nadir point, etc. A spider templet is more expensive than the other kinds of templet, but it can be disassembled and the parts used again. Other kinds can be used only once.

spider-templet method (of aerotriangulation) - A method of radial aerotriangulation in which slotted, metallic arms representing radials are attached to a central stud representing the principal point. The spider-templets representing points and radials on the various photographs are assembled by means of movable studs in the slotted arms. Also called a spider-templet plot.

spin - A rapid rotation. Some geophysicists use the term to denote the rotation of the Earth.

spinor - (1) A vector whose elements are complex numbers. Just as a vector having real elements can be represented as a straight line leading from one point in the real plane to another, so a spinor can be represented as a line leading from one point in the complex plane to another. A spinor is often represented as a 2 x 2 matrix whose elements are real numbers. It may also be represented as a 4-dimensional vector. Spinors are very useful for describing spin in quantum mechanics and rotation in celestial mechanics.

spiral - Any planar curve whose radius of curvature increases or decreases monotonically with distance along the curve. Alternatively, the locus of a point moving so that its distance from a fixed point is constantly increasing or constantly decreasing. This definition is, however, satisfied by a point moving directly toward or directly away from the fixed point, i.e., describing a straight line. Also called a spiral curve, particularly in surveying. The spiral most often used in surveying is the clothoid (Cornu spiral). SEE ALSO Archimedes' spiral; clothoid; Ekman spiral; Fermat's spiral.

spiral, equiangular - (1) SEE spiral, logarithmic. (2) SEE rhumb line.

spiral, hyperbolic - The spiral given by the equation $1/r = k\theta$ in which r and θ are polar coordinates and k is a constant.

spiral, logarithmic - The spiral given by the equation $r = \exp(k\theta)$, in which r and θ are polar coordinates and k is a constant. Also called an equiangular spiral. An alternative form of the equation is $\ln r = k\theta$. The curve cuts all radii at the same angle ϕ: $\phi = \text{arccot } k$.

spiral, loxodromic - SEE rhumb line.

spiral, sinusoidal - The spiral defined by the equation $r^m = a^m \cos(m\theta)$, in which r and θ are polar coordinates and a and m are constants, m being an integer. The logarithmic spiral, the straight line and the circle are special cases of the sinusoidal spiral.

spiral curve - SEE spiral.

spiral curve point - SEE point, spiral-curve.

spiral tangent point - SEE point, spiral-to-tangent.

spirit level - (1) A small, closed container of transparent material (usually glass) having the inner surface of its upper part curved (toric or spherical). The container is nearly filled with a liquid of low viscosity (alcohol or ether), with enough free space for a bubble of air or other gas at the top of the container. The outer surface of its upper part carries an index mark or graduations. Also called a level, although that term normally includes many other types than the spirit level. Most surveying requires that the line of sight lie in a horizontal plane or be referred to such a plane. The spirit level is used to establish such a plane. For highly precise work, a spirit level of great sensitivity is used to determine deviations from the horizontal plane, and corresponding corrections are then applied to the observations. Two types of spirit levels are in common use. One type, the circular level, is shaped like a round pillbox. The top is a glass disk whose underside is curved like part of a sphere. The bubble underneath this cover has therefore a circular shape when viewed from above. The other type is shaped like a tube closed at both ends and bent into a circular arc. It has a very large radius of curvature lengthwise and a small radius of curvature crosswise (like a segment of a doughnut). The bubble has a roughly rectangular shape when viewed from above. This is they type usually meant when the term spirit level is used. The tubular spirit level is used on surveying and astronomical instruments to refer observations precisely to a horizontal plane or to the zenith. When such a spirit level is in adjustment, a tangent at the uppermost point of the bubble defines a horizontal line in the plane of the longitudinal axis of the tube (vial); the accuracy and precision depends on the quality of the workmanship and the sensitivity of the spirit level. The sensitivity (sometimes called the sensibility of a spirit level) depends upon the radius of curvature of the longitudinal section: the longer the radius, the more the tube must be tilted to cause a mark to move from one end of the bubble to the other and hence the more sensitive the spirit level. Sensitivity is expressed by giving the length of a division between graduations on the tube and the equivalent angular value of this division at the center of curvature. For example, the sensitivity of the striding level of meridian telescope No. 9 (U.S. Coast and Geodetic Survey) was expressed as 1 division = 2 mm = 1.884" at 12° C. A usual value for the sensitivity of the long level on a surveyor's transit is 1 division = 2 mm = 30". (2) A leveling instrument incorporating a spirit level as defined in (1) above. (3) SEE level.

spirit level, chambered - A spirit level with a partition near one end partially cutting off a small reservoir of air, so arranged that the length of the bubble can be regulated.

spirit-level axis - SEE axis (of spirit level).

spirit leveling - (1) Leveling done with a leveling instrument depending on spirit levels for making the line of sight horizontal. The term is now sometimes used, erroneously, to denote leveling by using leveling instruments not containing a spirit level, e.g., pendulum leveling-instruments. Such leveling is best referred to simply as leveling. Also called leveling, direct leveling and geometric leveling. The first two terms have more general meanings when used properly; the third term is ambiguous (it could be taken to mean trigonometric leveling). A spirit level is attached to a telescope in such a way that the axis of the level and the line of collimation of the telescope can be made parallel and the level adjusted so that its axis is horizontal. The difference of readings on leveling rods held vertically at two different points is, after a few corrections have been made, the difference in elevation of the points. If the elevation of one of the points is known, the elevation of the other can be calculated as the sum of the elevation of the first and the difference of elevation. By making a progressive sequence of measurements of differences of elevation, the elevations of a sequence of points (bench marks) can be determined. The exact procedure used in spirit leveling depends on the degree of precision wanted and on the conditions under which the leveling is done. Highly precise spirit-leveling requires that observations be repeated many times, that the temperature of the air and of the leveling rods be determined, and so on; spirit leveling done where great precision is not needed (e.g., construction of roads) requires little or no repetition of observations, etc. Similarly, the instruments and leveling rods used in spirit leveling are usually designed to provide specific levels of accuracy and to be used under particular conditions. (2) Leveling done with a leveling instrument, regardless of whether a spirit level is part of the instrument, i.e., differential leveling. Also called geometric leveling.

spirit-level wind - SEE wind (of a spirit level).

spite fence - A fence maintained for the purpose of annoying the owner of adjoining land.

spline - A flexible, curved ruler which can be adjusted so that it can guide a pencil or pen in drawing a curve fitting a series of points. The spline is held in place by leaden weights.

spoil bank - SEE waste bank.

spot 0 - SEE blaze.

spot, hot - SEE hotspot.

spot elevation - The elevation of a point of particular significance and not usually part of a leveling network. A spot elevation is typically determined for the top of a hill or mountain, for an intersection of roads, for a mountain pass and for a plateau. It is commonly printed on a map next to a dot or cross indicating the point to which it corresponds.

spot elevation, checked - An elevation established in the field by spirit leveling, trigonometric leveling or barometric leveling in a closed circuit, or by any other method such that the accuracy of the elevation is proven.

spot elevation, unchecked - An elevation determined in the field but not checked.

spot height - (1) The elevation of a point on the ground, given next to the corresponding, isolated (i.e., not part of a contour line) point on a map. (2) An isolated (i.e., not part of a contour line) point on a map, representing a point on the ground whose elevation is known. This latter definition is probably empty, since explanations following such a definition nevertheless use the first definition. (3) A point, on a map, showing the location of an indicated elevation. The ground point for which a spot height is given is seldom marked on the ground.

spot zoning - A condition in which there is nonconforming use permitted in a locality zoned for a specific purpose.

spread function - SEE line spread function; point spread function.

spreading, sea-floor - The hypothesis that material from the mantle flows very slowly upward through the ridges on the oceanic bottoms, pushing the existing crust aside and, as it solidifies, forming new crust. The hypothesis derives from the observation that, in the vicinity of oceanic ridges, the bottom is magnetized in zones lying more or less parallel to the ridges but of different ages, with age increasing with distance from the ridge. It was first suggested in 1944 by A. Holmes.

spring balance - A balance consisting of a vertical spring attached at one end to a fixed support and with provision, at the other, free end for attaching or holding the body to be weighed. A scale, fastened alongside the spring, is graduated in units of weight. A body placed on the free end extends the spring by an amount proportional to the weight of the body (Hooke's law) and the weight is read on the scale. The spring balance is an accessory (called a tensiometer) used in some methods of taping to apply a known or predetermined amount of tension to the tape. Although other methods of taping use weights to apply the tension, less heavy equipment is needed if a spring balance is used. Note that the spring balance measures weight, which is a force and not a mass. The weight of a given mass therefore varies from place to place on the Earth. The mass is found by dividing the weight by the acceleration of gravity at the place.

spring high water - SEE springs, mean high-water.

spring low water - SEE springs, mean low-water.

spring range - SEE spring range, mean.

spring range, mean - The average semidiurnal range of tide at the time of syzygy, i.e., at the time of spring tides. Also called spring range. It is most conveniently calculated from the harmonic constants. Mean spring range is greater than the mean range of tide where the tide is either semidiurnal or mixed, and is of no practical significance where the tide is diurnal.

spring rise - SEE spring rise, mean.

spring rise, mean - The height or elevation of mean high water springs above the basic surface of reference (chart datum). Also called spring rise.

springs - Those tides which occur twice a month, when the Sun and Moon are lined up with the Earth (have the same right ascension). Also called spring tides. The terms spring and springs have nothing to do with the seasons; they are synonyms for jump and jumps, respectively. The high tide is higher at springs than it is on other days and rises more rapidly.

springs, high-water - SEE springs, mean high-water.

springs, low-water - SEE springs, mean low-water.

springs, lowest low-water - A surface of reference approximating the mean lowest low water during syzygy (spring tides).

springs, low-water equinoctial - Low-water springs near the times of the equinoxes. Also called equinoctial springs low-water. Expressed in terms of the harmonic constants, it is an elevation depressed below mean sea level by an amount equal to the sum of the amplitudes of the constants M_2, S_2 and K_2.

springs, mean higher high-(lower low-)water - The average height of all higher high (lower low) waters recorded during syzygy over a long period such as 19 years, or derived by computation for an equivalent period.

springs, mean high-water - (1) The average height of all high waters recorded during syzygy over a long period such as 19 years, or a computed equivalent period. Also called high water springs. The high-water counterpart of mean low-water springs. (2) The average height of the high waters of the spring tides. Also called spring high water.

springs, mean lower low-water - SEE springs, mean higher high- (lower low-)water.

springs, mean low-water - (1) The average of the heights of low water occurring at the time of the spring tides, observed over an extended period. Also called spring low water and low-water springs. It is usually derived by taking a height depressed below the half-tide level by an amount equal to one-half the spring range of tide, with necessary corrections used to reduce the result to an average value. This quantity is used, to a considerable extent, for hydrographic work outside the USA. It is the level of reference for the Pacific approaches to the Panama Canal. (2) The average of the heights of low water occurring at the time of spring tides observed during the National Tidal Datum Epoch.

springs datum, low-water - An approximation to mean low-water springs, used locally as a tidal datum. It is often retained for an extended period even though a better determination of mean low-water springs may be available.

spring tidal current - A tidal current of increased speed occurring semimonthly as the result of the Moon's being new or full.

spring tide - An oceanic tide of increased range occurring semi- monthly, when the Moon is aligned, in right ascension, with the Earth and the Sun.

spring tide, perigean - The oceanic tide which occurs when the Moon and Sun line up with the Earth (i.e., have the same right ascension) and the Moon is closest to the Earth. Perigean spring tides are extremely high.

spur level line - (1) A level line connected to a leveling network but neither part of the principal leveling network nor itself part of a loop. (2) A level line run as a branch from the main level-line either to determine the elevations of bench marks not conveniently reached by the main level-line or to connect to tidal bench marks or other previously established bench marks. It is also used to obtain checks on old leveling either at the beginning or end of a level line or at intermediate junctions along the new level line. Also called spur line of levels. A branching level-line run from a bench mark on a spur line is called a double-spur line or double- spur. Similarly, a branching level-line run from a bench mark on a double-spur line is called a triple-spur line and so on.

spur line of levels - SEE spur level line.

spur traverse - Any short traverse that branches off from the established traverse to reach some vantage point or location. Also called a stub traverse.

square - (1) A planar figure having four equal sides and four equal angles. The square is a particular case of the parallelogram. All angles are right angles. The area within the square is the square of the length of one side. The lines joining opposite corners of the square are called the diagonals. (2) An open space, in a city or town, into which two or more streets open and without a pronounced circular shape. If the shape is approximately circular, the space is called a circle. (3) SEE block.

square, optical - A small, hand-held device deflecting a beam of light through 90°. The optical square is used to set up right angles in the field. One form consists of two flat mirrors placed at an angle of 45° to each other. In use, the device is held so that one object is sighted directly and another object is placed so that its twice-reflected image appears directly in line with the first object. The lines from the point of observation to the two objects will then meet at a right angle. Another form consists of a single, flat mirror so placed that it makes an angle of 45° with the line of sight when in use. One object is sighted directly, and the other is placed so that its reflected image is also seen on the line of sight. Still other forms are prisms: the pentagonal and double-pentagonal prisms. By using the double- pentagonal prism, the observer can place himself on the line between two visible objects and can project a 90° angle ahead from the line. SEE ALSO sextant, box.

squatter - A person who settles on another's, especially public, land without legal authority.

squatter's rights - Those rights, to occupancy of land, created by virtue of land and undisturbed use, but without legal title or arrangement. It is in the nature of right at common law.

SQUID - A magnetometer consisting, basically, of a superconducting loop interrupted by one (in the radio-frequency SQUID) or two (in the direct-current SQUID) Josephson junctions and biased by a fixed current. Any change in voltage across the loop by an applied magnetic flux is amplified and converted to a current through a coil coupled to the SQUID to produce an equal and opposite flux. SQUID is an acronym for semiconducting quantum-interference magnetometer. The device has been applied to, among other things, the detection of gravity waves and the determination of gravitational gradient.

stability - (1) The tendency, or a measure of the tendency, to remain in a specific state or condition when subjected to influences trying to change that state or condition. Alternatively, the ability, or a measure of the ability, to return to a specified state or condition if disturbed. (2) (of a clock) The ability of a clock to keep marking equal intervals of time. Three categories of stability are in use: drift - a steady lengthening or shortening of the basic interval of time over a long period, long term - the root-mean-square deviation of the basic interval from its average value, after drift has been removed, over a long period (generally in excess of 10 seconds) and short term - defined in the same way as long-term stability except that the averaging period is between 100 microseconds and 10 seconds. The categories are also defined in slightly different ways because the usefulness of the definitions depends somewhat on the stability of the clock being considered. For example, the definition given for short-term stability would be useless if applied to a portable mechanical clock.

stability, long-term - The stability of a system as measured by the average of the system's deviations from a specific state or condition over a long period of time. The term refers, in electronic engineering, to the stability of an oscillating circuit such as a frequency standard. In such cases, the period may be taken to be anywhere from a few hours to a year or more. The term is also used in reference to the stability of planetary orbits, stellar systems, etc., but the period of time referred to here is usually on the order of millennia or millions of years.

stability, short-term - The stability of a system as measured by the average of the system's deviations from a specific state or condition over a short period of time. As used in reference to an oscillating electrical circuit, it refers to the deviations of the frequency of oscillations from its average value over a period of from 100 µs to 10 s.

stabilization, gyroscopic - (1) The technique whereby the orientation of an aircraft, vehicle, or vessel is kept constant by using gyroscopes. (2) The technique of keeping the orientation of a camera constant by using gyroscopes.

Stab-Werner map projection - SEE Werner's map projection.

stade - SEE stadium.

stadia - (1) SEE stadia method; stadia surveying. (2) SEE stadia rod. (3) SEE stadium.

stadia, foot-meter - SEE stadia-rod, foot-meter.

stadia, horizontal - SEE stadiarod method, horizontal.

stadia arc - A graduated, vertical arc on which graduations correspond to those vertical angles for which the difference in height is a simple multiple of the stadia-rod interval. Also called a stadia circle. The graduations may be on the vertical circle of the surveying instrument or on an auxiliary arc. The original design of the stadia arc was invented by W. Beaman of the U.S. Geological Survey. SEE ALSO Beaman stadia arc.

stadia board - A stadia rod designed particularly for use in stadia surveying and usually having markings of unusual design which have a clarity easily read at long distances. SEE ALSO stadia rod.

stadia characteristics - Those characteristics of a surveying instrument which affect the instrument's use for stadia surveying.

stadia circle - SEE Beaman stadia arc; stadia arc.

stadia constant - (1) The sum of the focal length of a telescope and the distance from the vertical axis of the instrument on which the telescope is mounted to the center of the objective lens-system. Also called the anallactic constant and anallatic constant. It is not a constant for internally-focusing telescopes. (2) SEE stadia factor.

stadia diagram - A diagram which is drawn to the scale of the survey being done and provides a means for rapidly reducing readings on the stadia rod. It is usually drawn on cross-section paper.

stadia factor - (1) The constant which multiplies the stadia interval to obtain the distance from the leveling rod or stadia rod to the front focal point of the telescope of a surveying instrument (tachymeter). Also called a stadia constant. (2) The constant which multiplies the sum of the stadia intervals for all observations of a running to convert that sum to the distance included in the running. Also called a stadia constant and stadia ratio. However, the first of those terms is best reserved for the distance from the vertical axis of a surveying instrument to the front focal point of the telescope rotating about that axis. The factor is usually such as to provide the distance in kilometers. If feet, yards or rods are used in the survey, the resulting distances will be in feet and thousands of feet or yards and thousands of yards. (3) The ratio of the focal length of a telescope to that length, on a graduated rod which is seen included between the upper and lower horizontal lines on the reticle.

stadia instrument, self-reading - A surveying instrument in which the customary parallel, horizontal lines on the reticle are replaced by lines whose spacings vary as the line of sight is raised or lowered, the spacings being derived from the trigonometric formulas for stadia reduction. In one variation, the lines are curved. In another, both lines are straight, but one is fixed while the other moves up or down with the line of sight.

stadia intercept - SEE stadia interval.

stadia interval - The length of that portion, of a graduated rod, seen included between the upper and lower horizontal lines on the reticle of a telescope. Also called the stadia intercept.

stadia leveling - (1) Determining a sequence of consecutive differences of elevation by using a transit and stadia rod. (2) Determining a sequence of consecutive differences of elevation by using transit and stadia rod, planetable and stadia rod, etc., to determine horizontal distances and vertical angles, and then calculating the differences of elevation. Equivalently, using stadia surveying to do trigonometric leveling. Also called stadia trigonometric leveling.

stadia method - A method or technique used in measuring distance, wherein the observer reads the interval subtended on a graduated rod between two lines or marks on the reticle of a telescope, the distance to the rod being proportional to the interval read. Also called a stadia. SEE ALSO stadia-reduction method.

stadia method, horizontal - SEE stadia-rod method, horizontal.

stadia ratio - SEE stadia factor.

stadia reduction - The calculation of elevation and horizontal distance from measurements made using a stadia rod. Stadia measurements actually give heights rather than elevations, but the difference is either negligible or can be reduced by computation. SEE ALSO stadia-reduction method; stadia survey.

stadia-reduction method - Any one of a number of methods used to obtain horizontal distance and height or elevation from measurements made using a stadia rod. Measurements made with the line of sight inclined to the horizontal are more frequent than measurements made with the line of sight horizontal. Each measurement made with the line of sight inclined is reduced to give the horizontal distance and difference in elevation. Trigonometric formulas involving the vertical angle and the stadia interval are used to make this reduction. Surveying instruments equipped with a stadia arc have multipliers on the vertical circle for reading both horizontal and vertical distances and eliminate the need to measure vertical angles.

stadia rod - A graduated rod used with a surveying instrument of special design to determine the distance from the instrument to the stadia rod by reading the interval, on the stadia rod, included between two lines on the reticle of the instrument's telescope. Also called a stadia. Either the length of rod subtended by the distance between two fixed lines on the reticle is read, or the distance between adjustable lines on the reticle is read when the distance is adjusted to cover a certain definite interval on the stadia rod. SEE ALSO stadia board.

stadia rod, foot-meter - A stadia rod graduated in feet on one side and in meters on the other. The foot-meter stadia-rod is sometimes used in putting in supplemental control. Measurements are made using both sides. The two values obtained serve to check the accuracy of the survey.

stadia-rod method, horizontal - A method of measuring distances, in which the stadia rod is held in a horizontal position and the stadia hairs of the telescope are vertical during observations. Also called horizontal stadia and the horizontal stadia method.

stadia slide rule - A slide rule having, in addition to the ordinary scales of numbers, two scales especially constructed for reducing the measurements of stadia surveying. One scale gives values of $\log \cos^2 a$ and the other gives values of $\log \frac{1}{2}\sin 2a$ for different values of a. A 10-inch stadia slide-rule gives results sufficiently accurate for all ordinary purposes.

stadia survey - A survey in which (a) distances are determined either from readings of the interval intercepted on a graduated rod by lines of sight through lines a known distance apart on the reticle of a telescope, or from the angle subtended between lines of sight to the ends of a definite, fixed interval on the rod; and (b) angular elevations to the rod or to a point on the rod are measured. Simple trigonometric formulas may be used to determine the elevation (height) and horizontal distance from the readings, or the instrument itself may be so constructed as to make the determination automatically. SEE ALSO stadia instrument, self-reading; stadia-reduction method.

stadia surveying - Surveying in which angles are measured with a transit or theodolite and distances are measured by sighting on a stadia rod. The reticle of the transit or theodolite carries a pair of parallel lines. These intersect the image of the stadia rod. The difference between the readings at the intersections determines the distance of the stadia rod from the instrument. Two methods of stadia surveying are common: horizontal-stadia surveying (also called stadia-subtense-bar surveying), in which the stadia rod is held horizontally and the lines on the reticle are vertical during a measurement, and vertical-stadia surveying (also called stadia-intercept surveying), in which the stadia rod is held vertically and the lines on the reticle are horizontal.

stadia traverse - A survey traverse in which distances are determined using a stadia rod. A stadia traverse is suited to regions of moderate relief with an adequate network of roads. If done carefully, such a traverse can establish elevations accurate enough for compiling maps with any contour standard interval.

stadia trigonometric leveling - SEE stadia leveling.

stadimeter - An instrument for determining the distance to an object of known height by measuring the angle subtended, at the observer, by the object. The instrument is graduated directly in units of distance. SEE ALSO range finder; tachymeter.

stadium - An ancient measure of length based on the length of the field in an architectural structure, also called a stadium in which races and other contests were held. Also called a stade. The Greek stadium (measure) was 600 Greek feet long, but the length of the Greek foot varied according to the length of the stadium (structure) involved. The Olympic stadium (measure) was equal to 192.3 meters; the Roman stadium (measure) was equal to 625 Roman feet or 184.8 meters.

staff gage - The simplest form of tide gage or stream gage, consisting of a graduated rod securely fastened to a pole or other suitable support. The staff gage is so designed that part of the staff will be below lowest low water when placed in position and the rest of the staff, above water, will be in position for direct observation from the shore or other vantage point. SEE ALSO tide gage.

stake - SEE guard stake; slope stake.

Stampfer level - A type of leveling instrument having the telescope so mounted that it can be rotated by a known amount about a horizontal axis by using a striding level and an adjusting screw. Also called a Stampfer leveling instrument and a Vienna level. The mechanical principle involved was quite similar to that of the present-day, first-order leveling instrument. The type was used by the U.S. Coast and Geodetic Survey from 1877-1899, when it was replaced by the Fischer level.

Stampfer leveling instrument - SEE Stampfer level.

stand - SEE stand of the tide (26).

stand, high-(low-)water - (1) The condition existing, at high (low) tide, when there is no change in the water level. (2) A prolonged period of negligible vertical movement near high (low) water. This is a regular feature of the tides in certain localities, while in other places stands are caused by meteorological conditions.

stand, low-water - SEE stand, high-(low-)water.

standard - (1) An object, force or other physical entity which, under specified conditions, defines, represents or records the magnitude of a unit of measurement. For example, the meter bar at Sevres was at one time a standard of length. It has on it two marks, the distance between these marks being defined to have the value 1 meter. All distances expressed in the metric system were given as a multiple or sub-multiple of this distance. The standard of length is now the wavelength of a particular color of light. SEE ALSO laboratory standard; standard of length; meter; second; gram; kilogram; standard of mass; standard of time. (2) A procedure agreed upon, within a particular industry or profession, as one to be followed in producing a particular product or result. (3) SEE standard, industrial. (4) SEE tolerance, standard. (5) SEE standard, primary.

standard, absolute - A body designated as a standard by assigning to it a mass of one unit.

standard, commercial - A standard satisfying the requirements of local certifiers of weights and measures, drafting, mechanical work, etc.

standard, industrial - A set of characteristics, such as dimensions or color, which must, by agreement within a profession or industry, be possessed by a product of that profession or industry. Also called a standard if it is obvious that industrial standard is meant. An industrial standard, called a Deutsches Industrie-Norm in Germany, is established to make sure that products are more or less interchangeable in use, or to promote efficiency in manufacturing. Standards are set, for example, for the size, weight and color of common sizes of paper; the dimensions and material of screws, nuts and bolts; the sizes and shapes of structural steel and the sizes of lumber.

standard, primary - (1) A standard which is correct by definition, and to which measuring devices are referred for calibration. (2) An international or national standard of the highest possible accuracy and precision. A primary standard is often referred to as the standard.

standard, secondary - (1) A standard which has been compared directly to a primary standard. (2) A standard which has been compared directly with a national standard. (3) A measuring device which has been calibrated by direct comparison of its scale or measurements with the scale or measurements of a primary standard, and which has been calibrated with sufficient accuracy to permit its use for calibrating other, similar measuring devices. Also called a laboratory standard.

standard, working - A measuring device which has been calibrated using a secondary standard and which is used for calibrating other measuring devices which will not themselves be used for calibration. A working standard is usually the last and least accurate of the measuring devices in the chain: primary standard, secondary standard and working standard.

Standard (California) Astronomical Datum 1885 - The horizontal-control datum which is defined by the following coordinates of the origin and by the azimuth (clockwise from South) of triangulation station Mt. Diablo, on the Clarke spheroid of 1866; the origin is at triangulation station Mt. Helena:

longitude of origin 122° 38' 01.410" W
latitude of origin 38° 40' 04.260" N
azimuth from origin to Mt. Diablo 324° 01' 31.04"

The name is taken from the so-called old registers of the Division of Geodesy. It is identical with the Yolo base datum used in the report on the California-Nevada boundary line.

standardization - (1) The establishment of standards and the enforcement of their use. In the USA, compliance with standardization is largely voluntary except for those organizations whose principal products are for military use. Russia, Germany and some other countries, however, have established standards governing most products manufactured there. These standards are also widely used outside those countries. (2) SEE calibration. SEE ALSO field standardization.

standard of accuracy, geodetic - One of a set of numbers used for classifying geodetic control or geodetic surveys into categories according to the minumum accuracy or precision of the control or the survey. SEE ALSO classification.

standard of length - A physical representation of a unit of length approved by authority or otherwise commonly agreed on. A standard of length is not independent of temperature, pressure and other physical conditions around it. It is an exact embodiment of the unit it represents only under definite, prescribed conditions. SEE ALSO meter; yard. Standards of length (s.o.l.) are classified according to the accuracy and precision with which they represent the given unit. These are prescribed by their intended use and have been designated by the U.S.C.& G.S. as follows: (a) primary standard - an international or national s.o.l.; (b) secondary standard - a s.o.l. which has been compared directly with the primary standard; (c) reference standard - a s.o.l. suitable for use in constructing precise instruments used in scientific investigations; (d) working standard - a s.o.l. suitable for all ordinary, precise work as in college laboratories, the general manufacture of precise instruments or for the use of State and city sealers of weights and measurers; (e) commercial standard - a s.o.l. satisfying the requirements of local sealers of weights and measures, drafting, machine work, etc. Standards of length are also classified according to their design: (a) line standard - on which the separation (or a specified fraction of the separation) between lines or marks on the side of a bar is defined as the unit of length; (b) end standard - on which the separation (or a specified fraction of the separation) between the ends of a bar is defined as the unit of length; (c) wavelength standard - a specified multiple of some specified wavelength of electromagnetic radiation is defined as the unit of length.

standard of length, end - SEE end standard of length.

standard of length, line - SEE line standard of length.

standard of surveys - SEE classification, control-survey.

standard temperature and pressure - A phrase used in physics and some allied sciences to indicate a temperature of 0° C and a pressure of one standard atmosphere. Abbreviated to S.T.P. or (S.T.P.). Other sciences, such as chemistry, may use other values.

stand of the tide - An interval, at high or low water, when there is no sensible change in the height of the tide. Sometimes called a platform tide. The water level is stationary at high and low water for only an instant, but the change in level near these times is so slow that it is not usually perceptible. In general, the length of the interval will depend upon the range of the tide, being longer for a small range than for a large range. But where there is a tendency for a double tide, the stand may be several hours even with a large range of the tide. There is some confusion in the literature between the use of stand to denote a condition of the tide and the use to denote the length of time that condition lasts. The context must be examined carefully to determine which meaning is intended.

stand of tide - SEE stand of the tide.

standpoint - The point from which or at which, in surveying, an observation is made.

star - The Sun, or any continuous, self-luminous, celestial body like the Sun. Stars are approximately spherical, gaseous bodies obtaining their energy by atomic fission and fusion. Assemblages of stars are called clusters, associations or galaxies, according to their size and shape.

star, circumpolar - (1) A star whose apparent path in the sky does not pass below the horizon of an observer. The latitude of the observer and the declination of the star determine whether the star is circumpolar or not. (2) A star whose declination is equal to or greater than (90° - latitude of observer).

star, equatorial - A star whose declination is close to 0°. Because equatorial stars move more rapidly in azimuth or angular elevation than do stars with greater declinations, they are preferred for observations to determine time and longitude.

star, fixed - Any celestial object so far away that its proper motion is too small to be determined. In particular, the distant galaxies. The term originated at a time when the stars were thought to be immovable and fixed forever in position.

star, fundamental - A star whose coordinates are determined with greatest accuracy and independently of other stars.

star catalog - (1) A compilation of stars and their coordinates. Also called a stellar catalog or star catalogue. Stellar catalogs were made by the Greeks at least as long ago as the time of Plato. The oldest existing stellar catalog is contained in the Almagest of Ptolemy issued about 137 A.D., which is a re-issue of a catalog of Hipparchus, no longer extant. Tycho Brahe's catalog of 1580 marks the start of the modern era of stellar catalogs. At present, stellar catalogs generally give, for each star, an identifying number or name, the photovisual apparent magnitude, the angular coordinates (right ascension and declination), the epoch to which the coordinates refer and the proper motion. There are different types of stellar catalogs, the type depending on the nature of the observations used, the nature or source of the coordinates, the use for which the catalog is intended, etc. (2) A star catalog, in the sense of the preceding definition, except that the coordinates have been fully reduced from the measured values. Star catalogs, such as annual star catalogs, of unreduced coordinates (i.e., of measured coordinates) are not star catalogs in this sense. SEE ALSO Astrographic Catalog; Bonner Durchmusterung; Boss star catalog; compilation star catalog; zone star catalog.

star catalog, absolute - (1) A star catalog containing an extensive list of coordinates referred to a coordinate system associated directly with the observing instrument. E.g., a star catalog containing right ascensions determined using a meridian circle would be an absolute star-catalog. (2) A star catalog in which the coordinates of the stars are in a coordinate system not depending on previously known coordinates of stars, e.g., they were found by reference to a coordinate system based on observations of the asteroids, other planets, etc.

star catalog, annual - A list of measured, not fully reduced, stellar coordinates.

star catalog, astrographic - SEE Astrographic Catalog.

star catalog, fundamental - A star catalog, compiled from a number of sources, in which the coordinates have been adjusted (by analyzing differences between coordinates given in the sources, to give as precise and accurate coordinates as possible. For the geodesist, the most important fundamental star catalogs are those designated as FK3, FK4 and FK5. These give the average right ascensions and declinations for 1535 stars for the epoch and equinox of 1950.0, and the annually-published star-catalog Apparent Places of Fundamental Stars giving coordinates derived from those in the FK3 and its successors.

star catalog, independent - A star catalog giving coordinates which were measured and reduced for that particular catalog and are not based on previously published and fully-reduced coordinates. Also called an observation star catalog.

star catalog, relative - A star catalog giving the coordinates in a coordinate system defined by previously determined coordinates of certain stars, so that coordinates in such a catalog are with respect to the coordinates of these reference stars.

star chart - A diagram showing the directions of stars in such a manner that the stars shown can be found or identified.

star factor - One of the coefficients in Bessel's equation for the difference between observed and calculated times of transit, as used in astronomic geodesy. I.e., a factor accounting for error in azimuth, inclination of axis, or error in collimation of the telescope.

star list - A star catalog containing only sufficient information (principally coordinates) about each star to enable the star to be identified. Star lists are often used in preparing for astronomic observations in the field.

star number, Besselian - SEE Besselian day number.

star-occultation method - SEE occultation surveying.

star place, fundamental - The apparent right ascension and declination of comparison stars obtained by astronomical observatories and published annually. Fundamental star places are published in fundamental-star catalogs.

star trail - The photographic image left by a star when the axis of the camera points in a direction fixed with respect to the Earth, i.e., is not kept pointed at a fixed point in the sky.

state (cartography) - An issue, of a map, which differs in some way from all other issues of the same map. The term has a particular application in the historical study of engraved maps and its use should be confined, in this application, to such issues of a map which were made from the same printing plate.

State, public-land - A State or Territory, of the USA, created out of the public domain. A list of public-land States is given under domain, public.

State base map - A base map of one of the States of the USA. State base maps are suitable for overprinting of information for special purposes. The term is frequently used to designate State maps issued by agencies of the Federal government and used by governmental and private agencies as bases on which special information is placed. Federally issued maps are not copyrighted.

state boundary - (1) A boundary between two states. (2) In particular, a boundary between two states of the USA. Such a boundary may be changed by agreement of the legislatures of the states concerned, but the agreement must be approved by Congress. Congress cannot change a state boundary without the consent of the state, nor can two states by mutual consent change their common boundary without the consent of Congress. Such consent need not be granted by a special act but may be inferred from subsequent legislation. See, however, the creation of West Virginia - a change not concurred in by the State of Virginia. Congress has several times given its consent in advance for adjoining states to fix an otherwise indefinite riparian boundary between them. A boundary between a state and a territory is fixed by joint action of Congress and the state.

State coordinate - A coordinate in a State coordinate system. Also called a State plane coordinate.

State coordinate system - One of the plane rectangular coordinate systems, one for each State in the Union, established by the U.S. Coast and Geodetic Survey for use in specifying locations of geodetic stations in terms of plane rectangular Cartesian coordinates. Also called a state plane coordinate system. However, this latter term is best reserved for the State coordinate system as modified by the National Geodetic Survey and derived from North American Datum 1983. SEE State plane coordinate system. Most States are individually mapped by a conformal map projection in one or more zones, over each of which is placed a rectangular grid. The relationship between the rectangular grid and the graticule generated by the map projection is established mathematically. Zones of limited extent east-west and indefinite extent north-south are mapped by a transverse Mercator map projection. Zones of indefinite extent east-west and limited extent north-south are mapped by the Lambert conformal conic map projection with two standard parallels. Zone One of Alaska is mapped by the oblique Mercator map projection. For a zone having a width of 250 km, the greatest departure from exact scale (scale error) is 1 part in 10,000. Only coordinates adjusted to the North American Datum of 1927 may be properly transformed into locations in a plane coordinate system. All such geodetic locations determined by the U.S. National Geodetic Survey (NGS) or its predecessor are transformed into plane rectangular coordinates on the proper grid, and are distributed by NGS together with the geodetic coordinates.

State grid - A plane rectangular grid placed on a map to show the 10 000 foot intervals in a State Plane Coordinate System. The grid is usually indicated by short lines along the neatlines of the map.

State map - (1) A map of one of the States of the USA. (2) A map produced by an agency of a State of the USA.

State plane coordinate - (1) One of a pair of coordinates in one of the plane rectangular coordinate systems known as State coordinate systems. Also written state plane coordinate. (2) One of a pair of coordinates in one of the plane rectangular coordinate systems known as the State plane rectangular coordinate systems. Every State in the USA has its own State plane rectangular coordinate system or State coordinate system. When necessary, the coordinates in the coordinate system of a particular state are referred to by the name of the State, for example, Georgia plane coordinates. State plane coordinates are used extensively for calculating and recording the results of land surveys.

State plane coordinates datum - The surface onto which each point of interest is transferred mathematically from the corresponding point on the ellipsoid representing the Earth, to give its state plane or state plane-rectangular coordinates.

State plane coordinate system - (1) A plane-rectangular coordinate system identical to the corresponding State coordinate system except for the following differences. (a) The grid is constructed using North American Datum 1983. (b) Distances from the origin are expressed in meters and fractions thereof. (c) The coordinates previously assigned to the origin are the equivalent, in meters, of the former coordinates in feet unless a State elects to have different coordinates assigned, in which event NGS will consider changing them. (d) The Lambert conformal conic map projection with two standard parallels is used in Michigan, rather than the transverse Mercator map projection. (e) The greatest error in computed coordinates do not exceed 0.1 mm for a point within the boundaries of a zone. (2) SEE State coordinate system.

State system of plane coordinates - SEE State plane-coordinate system.

state vector - (1) A vector, in six-dimensional space, drawn from the origin to a point whose coordinates are the three coordinates giving the location of a point and the three components of the velocity of the point, e.g., in Cartesian coordinates, a vector with components x, y, z, dx/dt, dy/dt and dz/dt. Alternatively, a state vector may give the components of a mass-point's momentum instead of the components of the velocity. (2) A vector whose components describe the state of a physical system at a particular instant. The first definition is a particularization of the second.

station - (1) The physical location or site at which, from which, or to which observations have been made. (2) A point representing the physical location or site at which, from which, or to which observations have been made. (3) Any point whose location is given by the total distance of that point from the starting point. (4) SEE station, taping. (5) SEE survey station. (6) SEE station, oceanographic. The principal types of geodetically important station are the air station, gravity station, magnetic station, survey station (including bench marks and triangulation stations) and tide gage or tidal station. Such stations are also classified according to their order of importance: base station, principal station, supplementary station, etc. In particular, in photogrammetry, the term is also used to denote a point (also called common station) at which a photograph has been taken. The point is usually at the camera's center of perspective. If the photograph was taken from an aircraft, the station is referred to as an air station. SEE ALSO A-station; B-station; base station; base line terminal station; camera station; control station; drift station; gravity station; ground station; Hiran station; instrument station; intersection station; Laplace station; Loran station; master station; navigation station; projection station; rate station; reference station; resection station; slave station; taping station; tracking station; traverse station; triangulation station.

station, absolute-gravity - A station at which the value of absolute gravity has been measured.

station, astronomic - A point whose location on the Earth has been determined from observations on celestial bodies. An astronomic station is usually a point whose astronomic longitude and latitude are known.

station, auxiliary - Any station connected to the main-scheme network and dependent on it for the accuracy of its location.

station, common - SEE station (3).

station, eccentric - (1) A point over which a surveying instrument is centered and at which measurements are made, and which is not in the same vertical line with the station which it represents and to which the measurements will be reduced before being combined with measurements at other stations. Sometimes called an eccentric (jargon) or satellite station. In general, an eccentric station is established and occupied when it is impracticable to occupy the intended station or when points to which observations must be made are not visible from the intended station. Because the data resulting from the occupation of an eccentric station are reduced to what they would have been had the measurements been made at the intended station (which should be marked with a permanent geodetic marker), the marker at an eccentric station is usually of a temporary character. SEE ALSO center, reduction to. (2) A station, located above or below the instrument used for measuring, whose elevation difference is measured and used to calculate the elevation of the station.

station, fixed - A base station (3) which is kept at a fixed place of known location.

station, full - SEE profile leveling.

station, horizontal-control - A station whose longitude and latitude or plane rectangular coordinates have been accurately determined and which is used for geodetic control. Also called a horizontal-control point.

station, in-and-out - A recoverable but unoccupied station incorporated into a traverse by recording a fictitious deflection angle of 180° to reverse the azimuth of the course leading into it, so that the next station coincides with the preceding station and the in-and-out station is used for the backsight in continuing the traverse. In computations, an in-and-out station is treated as an ordinary survey station in the traverse.

station, intersected - Also called intersected point.

station, magnetic - A monumented station at which a series of measurements of the Earth's magnetic field have been made. A magnetic station usually consists of a bronze marker, set in stone or concrete, on which the longitude and latitude of the station and the magnetic intensity at the marker are shown.

station, main-scheme - SEE station, principal.

station, master - SEE master station.

station, mobile - A base station (3) which is moved from place to place.

station, occupied - A station, in a particular survey, at or from which observations have been made that can be used in that survey.

station, oceanographic - (1) The place at which a vessel makes an observation or measurement. Also called, simply, a station if the meaning is clear. An oceanographic station is usually identified by the name of the vessel involved, the longitude and latitude of the station and a code identifying the set of observations or measurements, usually by date. The code may be numerical or literal, e.g., BROWN BEAR M-1035 180° E, 0° N, 1977 Nov. 12, 1200 GCT. (2) Oceanographic observations taken at a specific location, from a vessel which is lying to or anchored.

station, plus - An intermediate point, on a surveyed line, which is not at an integral number of tape-lengths from the initial points. SEE ALSO plus distance.

station, primary - SEE station, principal.

station, principal - A survey station used in the extension of a survey network or part of the original network to which the new network is connected. Also called a main-scheme station and primary station. A principal station serves primarily for the continued extension of a survey. Its location must therefore be determined with higher precision and accuracy than if it were used only to control local surveys or to help establish supplementary stations.

station, recovered - A survey station which, upon being revisited, is identified as being authentic and proved to be occupying its original site. SEE ALSO recovery (of a station).

station, remote - SEE slave station.

station, satellite-triangulation - A station equipped with satellite-observing equipment (distance-measuring, direction-measuring, Doppler-shift-meassuring, etc.) and working together with a set of one or more other stations of the same type to make simultaneous observations of artificial satellites, with the purpose of determining the geodetic coordinates of one or more stations of the set.

station, secondary - A station established in conjunction with a triangulation network consisting of principal stations, but observed with less precision, to increase the density of control. A secondary station does not usually contribute to the strength of the network and is not included in the adjustment of that network.

station, stream-gaging - A point, on a stream, at which measurements of discharge of velocity are made periodically, and at which daily or continuous measurements are made and records kept of the height of the water's surface above a given datum.

station, subordinate - (1) One of the places for which predictions of tides or tidal currents are determined by applying a correction to the predictions of a reference station. (2) A tidal station or tidal-current station at which a short series of measurements has been made and the measurements reduced by comparison with measurements made simultaneously at a reference station.

station, subsidiary - A station established to overcome some local obstacle to the progress of a survey, and not to establish location of the station. The term is usually applied to A-stations of a traverse. Subsidiary stations are usually temporary in character and without permanent markers. If the station also supplies control for a local survey, such a station may have a permanent marker. It is then a supplementary station.

station, supplemental - A station established solely to provide supplemental vertical control. Also called a vertical-angle station. It is not usually given a permanent marker (some are merely identified on the photographs) and accuracy does not have to be of the same order as that of the horizontal control to which it is tied.

station, supplementary - A survey station established either to increase the number of control stations in a given region, or to mark a desired location where it is impracticable or unnecessary to establish a principal station. A supplementary station is permanently marked. It is established with a precision and accuracy somewhat lower than those required for principal stations because extensive surveys are not made from them. SEE ALSO station, subsidiary.

station, total - SEE tachymeter. (brochure, Lietz)

station, unoccupied - A station, in a particular survey, at or from which observations that can be used in the survey have not been made.

station, vertical-angle - SEE station, supplemental.

station adjustment - SEE triangulation adjustment, local.

station datum orientation, single astronomic - SEE datum orientation, single astronomic-station.

station elevation - (1) The actual or assumed elevation of a marker from which other elevations are measured or to which the elevations of other points are referred. (2) In the U.S Weather Bureau, the elevation above mean sea level adopted for a station as the basis to which all measurements of air pressure at the station are referred.

station error - SEE deflection of the vertical.

station mark - SEE mark.

station marker, permanent - SEE marker, permanent.

station marker, temporary - SEE marker, temporary.

station pointer - SEE protractor, three-arm.

station position, geocentric - The coordinates of a station in a geocentric coordinate system.

statoscope - A sensitive barometer used, in taking aerial photographs, for measuring small differences in altitude between successive air stations. Also called a differential altimeter.

status of public land - The information with resepct to any particular parcel or tract of public land: its legal description; whether surveyed or unsurveyed; the non-federal rights or privileges, if any, which attach to it or its resources; whether

classified as mineral land; withdrawals or special laws. If any, which may influence the operation of the public-land laws so far as use or disposal of the land is concerned.

status plat - A plat on which has been drawn and noted such information as is necessary to determine Federal ownership, disposition, use or availability of public lands and resources.

statute law - Laws passed by proper, legislative bodies. A statute usually repepals all earlier laws and incativates conflicting common law.

statute mile - SEE mile.

statute of limitations - A statute assigning a certain limit after which specified rights cannot be enforced by action.
steel tape - A surveyor's tape made of steel.

steepest-descent method - SEE method of steepest descent.

stellar (adj.) - Starlike or pertaining to a star or of a star. However, the term stellar magnitude is used for the magnitude of any celestial object.

stellar-occultations method - SEE occultation surveying.

step - (1) The distance, in the normal way of walking on level ground, between the point at which the heel of the left foot first touches the ground and the succeeding point at which the heel of the right foot first touches the ground. Also called a pace, although that term is preferably used for the distance equal to two steps. Ordinarily, the step is estimated at 2½ feet, but in stepping off a distance, the step is extended to 3 feet or to 3.3 feet (1/5 rod). (2) A distance of 30 inches. Also called a pace, military pace and regulation pace.

step cast - The negative or positive reproduction of the stepped-terrain base of a relief model.

Stephenson leveling rod - A self-reading leveling-rod having graduations forming a diagonal scale and having horizontal lines through the tenth-of-a-foot marks. This rod is no longer in general use.

step tablet - SEE step wedge.

step wedge - A gray scale in which the gradation of tones occurs in a sequence of steps going from black through gray to white. Also called a step tablet and gray scale. However, gray scale is best reserved for the scale of tones in general. The step wedge is often used to determine the density in a photograph.

steradian - The solid angle subtended, at the center of a sphere of radius r, by a bounded region, on the surface of the sphere, having an area r^2. The total area of a sphere is therefore 4π steradians, or the area of 4 circles each of radius r.

stere - A unit of volume equal to one cubic meter.

stereo - A word used in jargon as a synonym for any term containing stereo, e.g., stereoscopic perception, stereoscope, stereoscopy, stereoscopic plotting instrument, etc.

stereo, exaggerated - SEE hyperstereoscopy.

stereo, false - SEE pseudoscopy.

stereo, inverted - SEE pseudoscopy.

stereo, pseudoscopic - SEE pseudoscopy.

stereo, reverse - SEE pseudoscopy.

stereocomparagraph - A simple stereoscopic plotting instrument in which each photograph is viewed separately by reflection from a pair of mirrors. The mirrors are mounted on a base having attached a parallax bar, a device for drawing and a parallel-motion mechanism. This simple and portable instrument was invented by B. Talley in 1935.
stereocomparator - (1) A stereoscopic measdevice for measuring the effects of parallax on a pair of photographs. (2) A photogrammetric instrument which allows the stereoscopic parallax of corresponding points in a pair or photographs to be measured. (3) A photogrammetric instrument which allows two photographs to be viewed stereoscopically and which allows the stereoscopic parallax between corresponding points on the photographs to be measured. (4) A photogrammetric instrument which allows two photographs to be viewed stereoscopically and which allows the coordinates of corresponding points, or the coordinates of one point and the parallax of its conjugate, to be measured. (5) A measuring device having an optical system enabling the operator to view a pair of photographs stereoscopically and to make the measurements on the stereoscopic image. SEE ALSO measuring engine, stereoscopic; comparator, stereoscopic.

stereocompilation - (1) Compilation done using a stereoscopic plotting instrument. (2) The data obtained by compilation done using a stereoscopic plotting instrument.

stereocompilation photography - Photography dealing with the taking of photographs to be used in stereoscopic plotting instruments for making maps.

stereogram - A stereoscopic pair of photographs or drawings correctly oriented and mounted or projected for stereoscopic viewing.

stereograph - (1) A stereometer with an attached pencil used to plot topographic detail from a properly oriented stereogram. (2) A stereoscopic plotting instrument, designed by F. Drobyshev, in which the two projectors have a fixed separation in the Y-direction, while the separation in the X-direction is zero.

stereo image - SEE image, stereoscopic.

stereomate - (1) One of the pictures of a stereoscopic pair. (2) An orthophoto modified by the introduction of artificial horizontal parallaxes.

stereometer - A stereoscope containing a micrometer by means of which the separation of two index marks can be changed to measure the effects, in the stereoscopic image, of difference of parallax. Also called a parallax bar.

stereomodel - SEE model, stereoscopic.

stereo-orthophoto technique - The conversion of a pair of stereoscopic photographs to a pair of orthophotos, and the replacement of one of the orthophotos by a modification in which horizontal parallaxes have been introduced artificially.

stereo pair - SEE pair, stereoscopic.

stereo pair, independent - SEE independent-model method.

stereophotogrammetry - Determination of the geometric characteristics of an object from measurements on images of the object as seen from different points. Also called photogrammetry where it is clear that stereophotogrammetry is meant. In particular, determination from measurements on photographs taken from two different points in such a way that the same region is shown on both photographs.

stereoplanigraph - A stereoscopic plotting instrument of high precision and able to accomodate most types of stereoscopic photographs, including terrestrial photographs.

stereoplotter - SEE plotting instrument, stereoscopic.

stereoplotter, automatic - SEE plotting instrument, automatic stereoscopic.

stereoplotter, double-projection direct-viewing - A stereoplotter that projects the images of two correctly-oriented, overlapping, aerial photographs onto a reference surface so that the resulting images can be viewed directly without additional use of the stereoplotter. Also called a double-optical projection stereoplotter.

stereoplotting - Photogrammetry done using a stereoscopic plotting instrument.

stereoplotting instrument - SEE plotting instrument, stereoscopic.

stereoplotting instrument, optical-mechanical - SEE plotting instrument, optical-mechanical.

stereoscope - An optical device which gives the user the impression that he is seeing a three dimensional object, when he is actually viewing two two-dimensional images (such as photographs or diagrams). A stereoscope operates on the principle of forcing each eye of the user to view a different image (of a pair of overlapping images). The optical system may consist of lenses (lens-type stereoscope), mirrors (mirror-type stereoscope) or prisms (prism-type stereoscope) or combinations of such elements. SEE ALSO anaglyph.

stereoscope, lens-type - A stereoscope in which there is a separate eyepiece and lens system for each eye, so that each eye views a different one of a pair of stereoscopic photographs.

stereoscope, mirror-type - A stereoscope in which the line of sight from each eye is directed by a train of mirrors to a particular one of a stereoscopic pair of photographs, so that each eye views a separate photograph. Also called a mirror stereoscope.

stereoscope, prism-type - A stereoscope in which the line of sight from each eye is directed by a train of prism to a particular one of a stereoscopic pair of photographs, so that each eye views a separate photograph. Also called a prism stereoscope.

stereoscopy - The art and science dealing with the principles and applications of the ability of the human visual system to create a mental image of a three-dimensional object from two separate, two-dimensional images as seen by the eyes.

stereotemplet - A composite, from a pair of slotted templets, adjustable in scale and representing the horizontal plot of a stereoscopic model. An assemblage of stereotemplets provides a means of doing aerotriangulation with a stereoscopic plotter not designed specifically for sterotriangulation.

stereotemplet areotriangulation - SEE stereotemplet method (of aerotriangulation).

stereotemplet triangulation - SEE stereotemplet method.

stereotriangulation - (1) Phototriangulation using a stereoscopic plotting instrument. Also called instrument phototriangulation. (2) Aerotriangulation using a stereoscopic plotting instrument to successively orient stereoscopic pairs of photographs into a continuous strip. Also called bridging, instrument photo-triangulation, multiplex triangulation (if done with a Multiplex), triangulation and stereoscopic aerotriangulation. Extension of horizontal and/or vertical control using these strips may be done either graphically or by calculation. (3) Aerotriangulation using stereoscopic pairs of photographs. Because most stereo-triangulation is done using aerial photographs, the preceding definitions are often applied to aerotriangulation in particular rather than to phototriangulation in general.

stereotriangulation, vertical - Stereotriangulation concerned with the establishment of vertical control or elevations. Use of vertical stereotriangulation is often limited or impossible because smaller errors are allowed in elevations than in horizontal coordinates.

stereo triplet - A set of three overlapping photographs, the photograph in the middle having a field of view in common with the overlapping photographs on each side, and arranged in such a manner as to allow stereoscopic viewing of the central photograph.

Sterneck pendulum - A pendulum made according to the original design by Sterneck: the mass is at the end of a cylindrical rod and is disk-shaped.

Sterneck pendulum-apparatus - A pendulum apparatus containing from 1 to 4 pendulums, made according to the design of Sterneck, and using the Sterneck pendulum. The pendulum's movement was monitored by light reflected from a mirror mounted on the pendulum's knife-edge. Apparatus of this kind was in common use from about 1890 to 1930.

Sterneck's method - The method of determining latitude by measuring zenith-distances of stars on the meridian, the stars being so chosen that an equal number culminate north and south of the zenith and the difference of the zenith- distances of pairs are as small as possible. Also called the method of meridional zenith distances. A universal theodolite or transit of high accuracy must be used. The method is similar to the Horrebow-Talcott method but does not require that the stars of a pair be at equal distances from the zenith. The mathematical procedure is therefore also somewhat different, in that it requires the introduction of more corrections.

stick - A length of 2 chains. The term stick is used in some surveyors' field notes and deeds in the USA in the early 1800's.

stick-up - Adhesive-backed or wax-backed film or paper on which names, symbols or descriptive terms, etc., have been printed and which is applied to map manuscripts in place of the hand-drawn names, etc. When the letters or symbols are placed in the correct position on a manuscript with the adhesive side down, they will adhere to the manuscript when rubbed or burnished. This procedure is more rapid and gives more uniform results than lettering by hand.

Stieltjes matrix - A real, symmetric, positive-definite matrix A having elements $a(i,j)$ equal to or greater than zero for all i not equal to j.

stilling device - SEE device, stilling.

stilling well - SEE well, stilling.

stipple - A pattern of fine dots, usually applied by hand, to give the approximate effect of tinted or vignetted regions in a picture.

stochastic (adj.) - SEE random.

stock, flat - Flat sheets of paper, as contrasted to rolled sheets.

Stokes' formula - (1) The formula derived by Stokes in 1849 for the height h_o of the geoid above a solid, rotational ellipsoid whose center of mass is at the center of mass of the Earth: $h_o = \int [F \Delta g \, dS] / (4\pi g_a R_a)$, in which g_a and R_a are average values of gravity and average radius of curvature, respectively, over the geoid; dS is an element of area S on the ellipsoid; $\Delta g(s, \phi)$ is the gravity anomaly at the point which is at distance s and direction ϕ from the point of interest and F is the function $F = 1 + \csc \sigma - 5 \cos 2\sigma - 6 \sin \sigma - 3 \cos 2\sigma \ln[(1 + \sin \sigma) \sin \sigma]$, in which σ is $s/(2R_a)$. The formula is based on the assumptions that (a) all matter lies inside the geoid; (b) the rotational ellipsoid to which heights are referred as the same volume as the geoid and has its center at the Earth's center of masss; and (c) Δg is known over the entire surface of the Earth. F is known as Stokes' function.

Stokes' function - The factor F in Stokes' formula. SEE Stokes' formula.

Stokes parameter - One of four quantities describing the intensity and polarization of an electromagnetic wave.

stone bound - A substantial stone post set into the ground with its top usually flush with or a short distance above the surface, to mark, accurately and permanently, an important corner of a land survey. Stone bounds with tops 6-9 cm above the ground are common and distances of 90-120 cm having been reported.

stool, taping - SEE taping stool.

stop - (1) Any part of an optical system limiting the angular size of the bundle of rays coming from the object or the amount of light passing through the system. The stop is often considered the theoretical analog of the diaphragm, which is always a material object. There are three principal kinds of angle-limiting stops: the aperture stop, the field stop and the vignetting stop. SEE ALSO glare stop. (2) SEE aperture stop. (3) SEE diaphragm, iris. SEE ALSO aperture stop; field stop.

stop, leveling-rod - A small platform, fixed to a tide staff or to the support of a tide gage, on which the foot of a leveling rod is placed in leveling from the tide gage to the tide-gage bench-mark. The leveling-rod stop itself is placed so that its upper surface, on which the leveling rod rests, is exactly at some integrally-numbered mark on the tide staff or corresponds to some integral umber on the recorder.

stop, vignetting - A stop introduced into an optical system to cut off rays passing through parts of the system which cause large aberrations. SEE ALSO aperture, vignetting.

stope - Any underground excavation in which ore is mined.

stop number - SEE aperture, relative.

storage factor - The quantity: 2π [(average energy stored per cycle) / (energy dissipated per half-cycle)], characterizing the ability of a system to store energy. The storage factor is usually denoted by Q and is therefore called the Q or Q factor of the system. It is also called specific dissipation factor by some geophysicists.

storm, magnetic - SEE magnetic disturbance.

storm drain - A drain which carries storm water, surface water and wash waters or drainage, but excludes sewage.

storm sewer - A conduit designed specifically to carry storm-water, the wash from streets and other surficial waters to points of dispersal.

storm surge - A departure from the normal elevation of the sea's surface because the water is piled up against a coast by strong winds such as those accompanying a hurricane or other intense storm.

storm tide - The increase in water level caused by a storm.

strain gage - (1) An electromagnetic device which tranforms small changes in size to changes in resistance which are proportional to the change. The strain gage is often incorporated in instruments for measuring small, dimensional changes in an object. Such an instrument may also be called a strain gage but may also be called an extensometer, strainmeter or strainometer. In particular, such an instrument used for measuring the effect of torsion, compression or similar forces on a specimen. (2) SEE extensometer.

strainmeter - SEE extensometer.

strainometer - SEE strain gage.

strait - A relatively narrow waterway connecting two larger bodies of water.

stratosphere - That portion of the atmosphere, lying between the troposphere (which ends at about 8 to 11 km) and the mesosphere (which begins at about 50 km), in which the temperature remains nearly constant or increases with altitude. The bottom of the mesosphere, the region in which the temperature begins decreasing again, is at about the same altitude as the bottom of the ionosphere. The stratosphere is too rarified to refract substantially either light or radio waves at frequencies higher than about 1 GHz; because it is not ionized, it does not greatly affect radio waves at frequencies lower than about 1 GHz.

stratum - A layer of sedimentary rock or earth.

stream - (1) A body of water moving under the influence of gravity to lower levels in a narrow, clearly defined channel. Stream is often used in a general sense, for any such body of water regardless of the width of its channel. Technically, it is more often used in the sense of a body of water moving in a channel narrower than that of a river but larger than that of a creek or brook. (2) SEE current.

stream, braided - A watercourse not filled by the normal flow of water, which subdivides into an interlaced pattern of channels.

stream gage - A graduated staff placed vertically on the bed of a stream or river to measure changes in the height of the surface, or to measure the height above a specified surface. Also called a river gage.

streamline - A line whose tangent at any point in a fluid is parallel to the instantaneous velocity of the fluid at that point.

street - Any public street, avenue or boulevard dedicated to public used generally in a city, town or village. The street includes the traveled way, parkway and sidewalks.

street, cul-de sac - A street with a dead-end that widens sufficiently at the end to permit an automobile to make a U-turn.

street line - (1) A line dividing a lot or other piece of land from a street. (2) The boundary of the side (or boundary of the end, of a dead end) of a street, defined by the document creating that street as having a stated width. Street lines may be created inside a lot and not be coincident with lot lines.

strength of current - (1) The speed of a current; the rate at which a fluid passes through a plane transverse to the direction in which the fluid is moving. In particular, the speed with which a stream or a current in the ocean is flowing. (2) The greatest speed of a tidal current. Beginning with the time of slack before flood in the cycle of a reversing tidal current (or of maximum before flood in a rotary current), the speed gradually increases to a maximum at flood and then decreases to slack before ebb (or a minimum before ebb in a rotary current), after which the current turns in direction, its speed increases to that at ebb, and then diminishes to slack before flood and completing the cycle. (3) That phase, of the cycle of tidal current, in which the speed is a maximum.

strength of ebb - The greatest speed of a (tidal) current during ebb. SEE ALSO strength of current (2).

strength of figure - A number relating the precision with which lengths of sides in a triangulation network can be determined, to the sizes of the angles, the number of conditions to be satisfied and the distribution of base lines and points of fixed location. Strength of figure, in triangulation, is not based on an absolute scale but rather is an expression of relative precision. The number is really a measure of a network's weakness, because the number increases in size as the strength decreases. The strength of figure, N_f, derived from that part of the formula for probable error of a triangle's side which is independent of the accuracy of the observations is $N_f = [(N_d - N_c)/ N_d] \Sigma (\delta_a^2 + \delta_a \delta_b + \delta_b^2)$, in which N_d and N_c are the numbers of directions observed and of conditions to be satisfied, respectively, and δ_a and δ_b are the rates of change of the sines of the distance angles A and B (usually expressed by the differences of the logarithms of the sines for a difference of 1" in the angles, the sixth decimal place being the unit place). By summing the values obtained for the simple figures composing a triangulation network, the strength of figure of the network is obtained. Because a triangulation network is usually composed of several different sets of simple figures, comparable numbers for the different sets can be obtained, and the route giving the strongest total network can then be selected for calculating lengths. Reconnaissance for a proposed triangulation network is usually done under instructions that specify limiting values of the strength of figure for the best and second-best chains of triangles between adjacent base lines; the sites for stations and for base lines are selected accordingly. Where desirable, the length of a section may be reduced by inserting an additional base line, and the numbers for strength of figure reduced accordingly. (2) A number expressing the precision with which the sides or angles of a triangle can be determined as a function of the triangle's shape.

stress - (1) The force, internal to a material, resulting from forces applied externally. (2) The force per unit area, internal to a material, resulting from forces applied externally. Also called unit stress. Although unit stress and pressure are both defined as force per unit area, pressure is that special form of stress in which the force is perpendicular to the area, is exerted against the area, and is usually taken to be uniform over the area.

stretching apparatus - SEE tape stretcher.

stride - SEE pace.

stride level - SEE striding level.

striding level - A spirit level so mounted that it can be placed above and parallel to the horizontal axis of a surveying or astronomical instrument, and used to measure the inclination of the horizontal axis to a horizontal plane. Sometimes referred to, in jargon, as a stride level. Generally, the striding level is of greater than average sensitivity. It is mounted on inverted wyes which rest directly upon the pivots on which the telescope rotates. It is used to make the inclination of the horizontal axis quite small and then to measure the magnitude of any remaining inclination.

strike - The direction of the line of intersection between a horizontal plane and the plane of a stratum, fault, etc. Strike is at right angles to dip.

strip - (1) A set of overlapping photographs which can be arranged in a sequence such that, except for the final photograph in the sequence, part of the object-space shown in one photograph is also shown in the succeeding photograph. (2) A set of overlapping photographs obtained sequentially from a moving aircraft or satellite. (3) A single, strip-like photograph made by sequentially exposing narrow, cross-wise sections through a moving slit to corresponding parts of a fixed object, or through a fixed slit to corresponding parts of a moving object. SEE ALSO strip camera.

strip, landing - A strip of land prepared specifically for the taking-off and landing of airplanes. It may contain several runways. Also called an airstrip.

strip adjustment (of aerotriangulation) - (1) The adjustment of all the coordinates, simultaneously, of points shown on a strip of aerial photographs. Also called the strip method of adjustment. Similar to block adjustment of aerotriangulation, but limited to a single strip of overlapping photographs. SEE ALSO strip method of adjusting aerotriangulation. (2) An adjustment of aerotriangulation in which the photographs are arranged in strips (in the order in which they were taken) and the coordinates are adjusted first for points imaged in each strip separately, and a second set of adjustments then made to reach agreement between coordinates of points imaged on different strips. The adjustment may be carried out graphically or analytically.

strip aerotriangulation - Aerotriangulation done on a whole strip of aerial photographs at one time.

strip aerotriangulation adjustment - SEE strip adjustment (of aerotriangulation).

strip camera - A camera designed for photographing a moving object, such as the ground if the camera is in an aircraft or a missile on a firing range, on a continuous strip of film. SEE ALSO camera, continuous-strip.

strip coordinate - One of the coordinates of any point in a strip, whether on the ground or actually an air station, referred to the coordinate system of the first pair of overlapping photographs.

stripfilm - A photographic film from which the emulsion can be removed as a membrane from the base after exposure and development, and the membrane then placed on a flat piece of plastic or glass to which it will adhere. Also called stripping film.

strip method of adjusting aerotriangulation - An aerotriangulation adjustment in which the photographs are arranged in strips and the corrections are determined simultaneously only for the coordinates of ground points whose images appear on each strip; a second set of computations is then made to minimize the discrepancies between coordinates obtained for points common to two or more strips. The strip adjustment of aerial photographs.

strip method of adjustment - SEE strip adjustment.

strip method of aerotriangulation - Aerotriangulation done on a whole strip photographs at one time.

strip mosaic - An aerial mosaic consisting of one strip of aerial photographs taken serially on a single flight.

strip photography - Photography done using a strip camera. The image of an object moving with respect to the camera is focussed onto a surface containing a narrow slit. The film is moved along behind the slit, or the slit is moved over the surface of the film, at a rate compensating for the movement of the object. This method of image-motion compensation is much used in aerial photograph.

stripping - (1) The removal of an emulsion, with its image, from individual photographic negatives and combining them by cementing them to a plate of glass or plastic. (2) The operations of cutting, attachment, etc., which are used in assembling cut sections of film into flats. (3) A method of enabling solidly-colored regions on a map to be reproduced. A transparent thin, plastic sheet carrying an opaque film has the outlines of the regions etched on it photochemically and the opaqaue film in the interiors is then peeled off (stripped) using a sharp blade. The printing plate is then prepared by appropriate photochemical methods using the stripped sheet as a mask.

stripping film - SEE stripfilm.

strip radial plot - SEE strip radial aerotriangulation.

strip radial triangulation - SEE strip radial aerotriangulation.

strip width - The average width, measured perpendicularly to the line of flight, of a series of neat models in a strip. Strip width is generally considered equal to the distance between flights.

Struve ellipsoid - A rotational ellipsoid characterized by the constants:
length of semi-major axis (a)   6 378 298.3 meters;
flattening (f)   1/294.73 .
Also called the Struve spheroid.

Struve spheroid - SEE Struve ellipsoid.

stub traverse - SEE spur traverse.

style sheet - A graphic guide to the way the grid and marginal information on a map are to be shown. Also called a mock-up.

subchord - (1) Any chord, of a circular chord, whose length is less than that of the chord adopted for laying out the curve. In a railroad curve, for example, a subchord is a chord less than 100 feet in length. (2) Any chord, of a circular curve, which is less than the long chord between the extremities of the curve.

subdivision - (1) An unimproved tract of land surveyed and divided into lots to be sold. In some localities, it is distinguished from a development, upon which improvements are made before sale; in other localities, the terms are synonymous. (2) A tract of land created by dividing a township into smaller pieces of land such as a section, half-section, quarter-section, quarter-quarter section or lotting, including the lot, section, township and range numbers, and the description of the principal meridian to which referred, all according to the approved plat of the township. (USPLS) (3) The process or method of dividing a piece of land into smaller pieces.

subdivision ordinances - SEE subdivision regulations.

subdivision plat - A plat (map) of a subdivision of land, usually prepared in accordance with a State's statutes on plats or local regulations on subdivisions, or both.

subdivision regulation - A locally adopted law (ordinance) governing the process of converting unimproved land into sites upon which buildings are to be erected. Also called subdivision ordinances.

subdivision survey - A land survey consisting of the following: locations of legal boundaries of a piece of land; division of the piece into smaller pieces such as lots, streets, rights of way and other necessary pieces; marking or monumenting all necessary corners or dividing lines; maps or plats for record showing all information regarding the boundaries, divisions, and adjoining land affecting boundaries, and including all necessary certificates; and all other items needed to correlate the survey with the record of the survey.

subgrade (adj.) - Designating or pertaining to a layer next under the uppermost, principal one.

subsidence - A rapid lowering of the surface, caused by the caving in of caves and mines, or by pumping out water or oil.

substitute center - SEE center, substitute.

subtangent - The length of a line tangent to the arc of a circle, from an extremity of that arc to its intersection with a similar line tangent to the other extremity of the arc. Also called a semi-tangent.

subtense bar - SEE bar, subtense.

subtense-bar traverse - SEE traverse, subtense-bar.

sum, algebraic - The quantity obtained, from a sequence of signed numbers, by adding together all the positive numbers, changing the sign of all the negative numbers and adding them together, and then subtracting the second sum from the first. Unless the contrary is stated, a sum is understood to be an algebraic sum. Also called algebraic addition.

sum, arithmetic - SEE sum of absolute values.

summation convention - The convention that if any index (subscript or superscript) appears more than once in a product of indexed symbols, the product is to be evaluated for each value of the repeated index and the sum of all these products taken. Also called the Einstein convention or Einstein summation convention. A frequent variant of the summation convention is to require summation only if the repeated index occurs as once as a subscript and once as a superscript.

summer solstice - (1) That point, on the ecliptic, occupied by the Sun at its greatest northerly declination (about + 23.5°). Equivalently, that point, on the ecliptic, which is north of the Equator and is midway between the equinoxes, or, that solstice which is north of the Equator. The Sun attains its greatest northerly declination at the summer solstice. Also called First Point of Cancer. (2) That instant at which the Sun reaches the point of greatest northerly declination. This is about 21 June. It marks the beginning of summer in the Northern Hemisphere.

summit level, accordant - A horizontal plane drawn over a large region and connecting mountain-summits of similar elevation.

Sumner line - A short segment of a circle of position, represented as a straight line, as obtained by the measurement of angular elevations of some celestial object. An observed angular elevation is reduced, together with an assumed latitude, to obtain a longitude. The assumed longitude is changed by small amounts and further longitudes calculated. The calculated longitudes and assumed latitudes lie on a straight line called the Sumner line. Also called a line of position and position line, although those terms are properly applied to a more general category of lines of which the Sumner line is just one special type. The Sumner line is named for Captain Thomas J. Sumner, an American sea captain, who invented its use in 1843. The St. Hilaire method is a special adaptation of the Sumner line to navigation. SEE ALSO line of position; Sumner method.

Sumner method - A method of obtaining one's location on a map from two or more measurements of the angular elevation of one star or from one measurement of angular elevation of each of two or more stars. The observations on one star are made sufficiently far apart in time that the azimuth (bearing) has changed by at least 45°; observations on two or more stars are made on stars of at least 45° difference in azimuth. The Sumner line corresponding to each observation is plotted and the intersection is the point corresponding to the observer's location. If three or more Sumner lines are drawn, the lines will in general not intersect at one point and a suitable method used for fixing on one point. SEE ALSO Sumner line.

sum of absolute values - The sum obtained, from a sequence of signed numbers, by changing the signs of all negative numbers to make them positive, and then adding all the numbers in the sequence together. Also called an arithmetic sum. If the numbers in the sequence are unsigned, they are assumed to be positive.

sump - An excavation made at the bottom of a shaft to collect water.

Sun - That star about which the Earth revolves. The prefix helio- is used to create terms pertaining to the Sun alone: e.g., heliocentric, heliotrope. The principal physical characteristics of the Sun are given in the Appendix. SEE star.

Sun, apparent - The actual Sun as it appears in the sky. Also called the true Sun.

sun, dynamical mean - A fictitious Sun which moves westward along the ecliptic at the average rate of the apparent (actual) Sun.

sun, fictitious - SEE sun, mean.

sun, fictitious mean - (1) An imaginary body moving at such a constant rate along the celestial equator that its location is never more that 16m from the true Sun in hour angle and such that the rate approximates the average rate of the annual motion of the Sun. (2) The mathematical formula giving the motion of the fictitious mean sun as defined in (1).

sun, mean - A point which moves at such a uniform rate along the celestial equator that it makes one apparent revolution around the Earth in the same length of time as does the Sun that is, in one year. Also called a fictitious sun.

sun, true - (1) The actual (real) Sun; the Sun on which astronomical observations are made. (2) An imaginary body whose center coincides with the barycenter of the actual Sun as observed from the Earth. The concept is in contrast to that of the mean sun, whose center moves uniformly along the path of the actual Sun but with uniform motion.

sun compass - A navigating instrument for determining the direction of the astronomic meridian mechanically and instantaneously from an observation on the Sun. One such instrument was designed by A. Bumstead of the National Geographic Society and used by R.E. Byrd for aerial navigation in the polar regions. It consists of a clock with a 24-hour dial giving mean solar time, an hour-hand with a gnomon at one end and with graduated circles on which latitude and azimuth are set. In use, the clock is set for latitude so that the plane of its face is parallel to the plane of the equator, the hour hand is set to show local time, and the whole instrument is oriented until the gnomon casts a shadow down the middle of the hand. In this position, the lubber line will be in the local meridian. The term sun compass is preferred for this type of instrument, to distinguish it from a solar compass.

sundial - An instrument indicating local, apparent solar time by the shadow cast by some suitable object onto a calibrated dial. The shadow's position indicates the hour angle of the Sun. The object casting the shadow is called gnomon.

sundial, analemmatic - A sundial which has on its dial an analemma for converting apparent solar time, as marked by the real Sun, to mean solar time as marked by the mean sun.

sunrise (sunset) - (1) The apparent rising (setting) of the Sun above the horizon. (2) The time at which the apparent upper limb of the Sun is on the astronomical horizon when the Sun is rising (setting); i.e., when the true zenith distance, referred to the center of the Earth, is 90° 50', based on adopted values of 34' for horizontal refraction and 16' for the Sun's semidiameter. SEE ALSO twilight.

sunset - SEE sunrise (sunset).

superelevation - The amount the outer edge of a curving road or track is raised above the inner edge to counteract the effect of centrifugal force on vehicles using the road or track.

Super-Invar - An alloy consisting of iron (63%), nickel (34%) and cobalt (3%). Super-Invar has been used for making very precise surveyor's tapes. Its coefficient of dilation is 2×10^{-7}. A base line can be measured, using the alloy, with a precision of $0.5 - 1.3 \times 10^{-6}$.

supplement (of an angle) - SEE angle, supplementary.

surface, aspheric - A surface, of an element of an optical system, which is not shaped like part of a sphere. Most astronomical telescopes incorporate mirrors having paraboloidal surfaces. Some lenses have cylindrical or toroidal shapes, in part.

surface, developable - A mathematical surface which can, in concept, be flattened without tearing or stretching, i.e., without changing the metrical properties of the surface. The cone and the cylinder, after being cut along a generating element, are developable surfaces. So is any surface obtained by bending a flat surface provided that surface is not stretched, shrunk or torn by the bending. Developable surfaces occur in geodesy as intermediaries in the mapping of the Earth or other body onto the plane, and as surfaces drawn to intersect the Earth's surface and so create a profile of the surface or a section of the surface and interior.

surface, equigeopotential - An equipotential surface representing the Earth's surface. SEE ALSO geopotential surface; surface, level.

surface, equipotential - A surface at every point of which a potential is defined and has the same value. Because the potential is the same at every point, no work is done in moving a point-mass, point-charge or other appropriate type of point about on the surface. When pointlike bodies of the same type and having the same value of mass, etc., are moved from one equipotential surface to another, the same amount of work is developed or expended, regardless of the route followed. SEE ALSO surface, geopotential; surface, level.

surface, equiscalar - A surface on which a scalar quantity is defined and has the same value at every point.

surface, geopotential - SEE geopotential surface.

surface, grained - The roughened or irregular surface of a printing plate used in an offset press.

surface, hypsometric - A surface whose geodetic height above a terrestrial ellipsoid is such that, at each point on it, standard gravity for the geodetic longitude, geodetic latitude and geodetic height is equal to the measured magnitude of the force of gravity at those coordinates.

surface, isobaric - A surface over which the pressure is constant. In oceanographic leveling, the surface to which leveling is to be referred is that at which an isobaric surface and a level surface coincide over the region of interest. That surface is commonly identified as the one lying at the depth at which there is no movement of the water. An aircraft flying at constant altitude follows an isobaric surface. That surface is not, however, parallel to the geoid not, in general, is it parallel to or related in any simple way to sea level or to mean sea level. Therefore elevations of points on the ground cannot be obtained by subtracting measured distances of the aircraft above the ground from the altitudes indicated by the aircraft's barometric altimeter.

surface, level - (1) An equipotential surface on which the potential is that of gravity. Also called a geop or geopotential surface if the surface is associated with the Earth's gravity-field and equipotential surface. The last term is inappropriate unless it is clear that gravity potential is meant. This is not always the case, e.g., in situations where both the magnetic field and the gravity field are being studied. Geopotential surface properly applies only to an equipotential surface of the Earth's gravity field. (2) A surface everywhere perpendicular to the direction of gravity. The two definitions (1) and (2) are equivalent insofar as they denote the same surface. However, the connotations of the two definitions are quite different and are inherent in the methods by which the surface is determined. E.g., the surface of a body of still water, under suitable conditions, is a level surface. The surface of the geoid is a level surface. Any line lying in a level surface is a level line. Level surfaces are approximately spheroidal, i.e., have the shape of a rotational ellipsoid. The distance between any two level surfaces decreases with increase in latitude. For example, a level surface which is 1000 meters above the average surface of the ocean at the equator is 995 meters above that surface at the poles.

surface, multiquadric - A surface which can be represented as the sum of a set of functions of the form $q_j = k_j \sqrt{[(x - x_j)^2 + (y - y_j)^2 + B]}$.

surface, principal - The locus of those points in which straight lines representing rays entering an optical system parallel to the optical axis intersect straight lines representing the same rays after they leave the optical system. Also called a principal plane. However, that term should be applied only to that part of the principal surface lying near the optical axis (i.e., in the paraxial region). An optical system has two, possibly coincident, principal surfaces. One is created by rays entering from the left and leaving at the right; the other is created by rays entering from the right and leaving at the left. The intersection of a principal surface with the optical axis is called a principal point.

surface, ruled - A surface generated as the locus of successive positions of a moving straight line. Alternatively, a surface through every point of which passes at least one straight line. All developable surfaces are ruled surfaces; the converse is not true in general. SEE ALSO surface, developable; surface, undevelopable.

surface, spheropotential - SEE spheropotential surface.

surface, undevelopable - A surface which cannot be mapped onto a plane so that all distances on the surface are kept unchanged on the plane. The sphere and paraboloid are undevelopable surfaces. SEE ALSO surface, developable; surface, ruled.

surface anomaly - Any irregularity of or close under the Earth's surface that produces anomalous results in geophysical measurements.

surface chart - SEE weather map.

surface correction - The amount applied to a measurement of gravity or geomagnetism to account for the effects of anomalies near the surface and for elevation of the surface.

surface density layer - SEE density layer.

surface easement - The right to use only the surface of the land. E.g., for easments of access, flowage or for rights of way.

surface fitting - The process or technique of determining that function which best represents a function of two variables, given the value of the second function at a number of points. Alternatively, the process or technique of determining a surface such that some function of the distances of a set of given points from the surface is less than the corresponding function of distances from any other, similar surface.

surface gravity anomaly - The difference between the value of gravity on the Earth's surface and the value of gravity calculated from a standard gravity formula.

surface harmonic function - A harmonic function defined on a surface. E.g., a spherical harmonic function in which the radial distance is held constant. Also called a surface harmonic.

surface of revolution - The surface generated by rotating a planar curve about an axis in the plane of the curve. Also called a rotational surface.

surface plate - A type of plate used in lithography, in which the exposed, sensitized coating is treated so that it will become the foundation of the printed design. Negative flats are used for preparing a surface plate. SEE ALSO deep-etch.

surf zone - SEE breaker zone.

surprint - SEE overprint.

surrender (v.) - (1) In law, yield, render or deliver up; give up. (2) Yield a particular estate to him who has an immediate estate in remainder or reversion, merging the yielded estate in the other.

surround - SEE margin (3).

survey (n.) - (1) The orderly process of determining data relating to any physical, chemical or geometric characteristics of the Earth. The list of such processes which may properly be called surveys is long. Surveys may be divided into classes according to the types of data obtained, the methods and instruments used, the purposes to be served, etc. For example, there are the geodetic survey, topographic survey, hydrographic survey, land survey, geologic survey, geophysical survey, soil survey, mine survey and engineering survey, among others. (2) The data obtained by a survey as defined in (1). The data obtained in a particular project may be designated by the name of the project, for example, the topographic survey of the District of Columbia or Cape Canaveral Survey. SEE ALSO station, oceanographic. (3) An organization engaged in making a survey as defined in (1). Such an organization is often given an official name which includes the word survey; for example, The United States Geological Survey or The Massachusetts Geodetic Survey. SEE ALSO artillery survey; beach survey; boundary survey; city survey; compass survey; completion survey; construction survey; Control Survey, Airborne; engineering survey; extension survey; ground survey; inventory survey; land survey; location survey; magnetometer survey; mine survey; mineral survey; plane survey; reconnaissance survey; resurvey survey; route survey; snow survey; soil survey; stadia survey; subdivision survey.

survey (v.) - (1) Measure distances, angles, heights, etc., to determine the relative locations of points under, on or above the Earth. (2) To inspect or investigate something in order to determine the state or condition.

survey, ABC - SEE Airborne Control Survey.

survey, aerial - (1) A survey made from an aircraft. Also called an aero-survey. (2) A survey in which aerial cameras, radar, infra-red detectors and/or other equipment aboard an aircraft are the primary source of the data collected. Also called an aerosurvey or air survey.

survey, areal - (1) A survey of a region extensive enough to require a network of control stations, i.e., loops of control. Also called an area survey. (2) An extension and densification of control. Also called an area survey.

survey, as-built - (1) A survey done after a construction project has been completed, to re-establish the principal horizontal and vertical control-points and to locate all structures and improvements. (2) A survey made to determine the locations and/or positions of points in finished structures and the dimensions of such structures.

survey, basic - A hydrographic survey so complete and thorough that it does not need to be supplemented by other surveys and is adequate to supersede, for charting, all earlier hydrographic surveys of the region.

survey, cadastral - (1) A survey relating to boundaries and subdivisions of land, and made to create regions suitable for transfer of or to define the limitations of title. The term is derived from cadastre, meaning register of the real property of a political subdivision with details of area, ownership and value. (2) A survey of the public lands of the USA, including retracement surveys for identification, and resurveys for restoration, of property lines. The term can also be applied properly to corresponding surveys made outside the public lands, although such surveys are usually called land surveys.

survey, dependent - A survey which derives the coordinates of some of the points in the survey from the coordinates of points determined by a previous survey, accepting these earlier coordinates without change. The previous survey is sometimes called a higher-order survey. While the previous survey usually is of a higher order of accuracy than the dependent survey, it need not be and sometimes is not. Using the terms higher-order survey and lower-order survey for the previous and dependent surveys is therefore misleading.

survey, detailed - A survey which takes account of the smaller features in the region surveyed. The term is relative, but is usually applied only to surveys concerned with distances less than those specified for third-order control. Control surveys are not detailed surveys unless they treat of engineering control.

survey, double-run - SEE leveling, double-run.

survey, elder - A survey which antedates the one being performed. The term was used in dividing crown lands or province lands in the American colonies.

survey, electrical-resistivity - A survey made to determine the electrical resistivity of the ground, as an indication of the presence or absence of useful minerals, ores, petroleum, etc.

survey, exploratory - A survey made to obtain general information about regions for which no general information was on record. The U.S. Geological Survey makes exploratory surveys in remote regions of Alaska. These surveys are used in making maps on scales of 1:500 000 or smaller.

survey, field-completion - A survey or inspection made in the field to obtain the additional information needed to complete a map based in part on information obtained from aerial photographs. Also called field completion or completion survey. The first of these terms is more often used, however, for the process of making such a survey. A field-completion survey may, in spite of its name, be done before or after completing the photogrammetric part of the survey. Among the items checked or added are names of places, boundaries, elevations, distances, types of roads and the nature of vegetation.

survey, first-order - (1) A geodetic survey of the highest prescribed order of precision and accuracy. Such surveys were formerly called primary surveys. In 1921, representatives of the various map-making and map-using organizations in the USA changed the name to precise survey. (2) A geodetic survey which established first-order control. In 1925, the Federal Board of Surveys and Maps adopted the designation first-order control for control of the highest order of precision and accuracy; established standards for the accuracy of such control; and specified essential elements of the procedures to be used in establishing such control. Hence, a first-order survey is defined in terms of the order of the control established by it.

survey, geodetic - (1) A survey in which account is taken of the size and shape of the Earth. This is in contrast to a plane survey, in which the reference surface for the survey is assumed to be flat. (2) A survey for determining the size and shape of a portion of the Earth, in which account is taken of the figure and size of the Earth. (3) An organization engaged in making surveys in which the size and shape of the Earth are taken into account.

survey, geo-electric - A survey made to determine the electrostatic potential or the resistivity of rocks. Also written geoelectric survey.

survey, geographic - A general term, not susceptible of precise definition, for a wide variety of surveys lying between and merging into exploratory surveys at one extreme and basic topographic surveys at the other. Geographic surveys usually cover large regions, are based on coordinated control, and are used to record physical and statistical characteristics of the region surveyed.

survey, geologic - A survey or investigation of the character and structure of the Earth, of the physical changes which the Earth's crust has undergone or is undergoing and of the causes producing these changes.

survey, geological - An organization making geologic surveys and investigations.

survey, gravimetric - A survey made to determine the force or acceleration of gravity at various places on the Earth.

survey, higher-order - SEE survey, dependent.

survey, hydrographic - (1) A survey having for its principal purpose the determination of geometric and dynamic characteristics of bodies of water. A hydrographic survey may consist of the determination of one or several of the following classes of data: depth of water and configuration of the bottom; velocities of currents; heights and times of tides and water stages; and the location of fixed objects used in surveying and navigation. (2) A record of a survey, of a given date, of a water-covered region, with particular attention to the relief of the bottom as shown by soundings and depth contours. A hydrographic survey is the authority for all data on features below the surface of high water, including the names of hydrographic features.

survey, inshore - A hydrographic survey conducted adjacent to the shoreline and, in general, for depths of about 40 meters (20 fathoms) or less.

survey, joint - The running, marking and establishing of a new boundary, or the retracement or resurvey of lines fixed by an earlier survey, in conformity with an agreement between adjoining owners or in accord with a judicial decree stipulating the manner of the survey. In the decree all parties in interest are duly represented, are charged with responsibility for identifying and accepting evidence, and are responsible for the correct technical execution of the decree. The parties must be present during the field operations.

survey, judicial - A survey ordered by a court or made expressly for a judicial action to resolve a dispute over a boundary. Also called a legal survey.

survey, legal - SEE survey, judicial.

survey, lineal - (1) A survey made in a straight line. (2) SEE alinement.

survey, lower-order - (1) A survey whose accuracy is less than that of the survey to which it is connected. (2) SEE survey, dependent.

survey, magnetic - A survey made to measure the direction and/or intensity of the Earth's magnetic field at specific points on the surface.

survey, marine - SEE survey, oceanographic.

survey, metes-and-bounds - A survey describing the boundary of a tract of land by giving the bearing and length of each successive line of the boundary. Much of the land other than public lands of the USA has been surveyed and described as a metes-and-bounds survey. The survey is also used in surveys of public lands to defined the boundaries of irregular tracts such claims, grants, and reservations not easily subdivided into rectangles of uniform size. SEE ALSO metes and bounds; description, metes and bounds.

survey, new - SEE resurvey.

survey, oceanographic - A survey of conditions in an ocean or a part thereof, with reference to animals or plants in it, chemical elements present, temperatures, etc. Also called a marine survey.

survey, original - SEE survey.

survey, photogrammetric - (1) An original geodetic survey in which monuments are placed at points which have been determined photogrammetrically. (2) A geodetic survey done photogrammetrically. (3) A survey done using either ground photographs or aerial photographs.

survey, photographic - A survey made either from aerial photographs or from photographs taken on the ground.

survey, plane-table - A topographic survey done using a plane table.

survey, polygonometric - SEE traverse. The term occurs in English literature as a close translation from European literature.

survey, preliminary - (1) A survey collecting data on which to base studies for a proposed project. (2) A survey done to obtain data from which to plan a geodetic survey. Also called a reconnaissance survey. (3) A survey made in the detail, to the accuracy, and of the scope necessary for adequately designing and preparing detailed plans for construction. SEE ALSO survey, reconnaissance.

survey, rectangular - A geodetic survey in which survey stations are established in a pattern such that the connecting lines form a rectangular grid. Also called a rectangular survey system and a rectangular land survey. Rectangular surveys were made by the Etruscans and after them, by the Romans. The scheme used for the rectangular survey of the public lands of the USA was established by the Continental Congress in 1785. It provided for lines to be run north-south and east-west at right angles to each other so that the land could be divided into squares 1 mile on a side. SEE ALSO centuratio.

survey, rural - A survey of land such as farms and other undeveloped land lying outside cities or towns and not used exclusively for single-family residences or as residential subdivisions.

survey, second-order - (1) A geodetic survey resulting in second-order control. (2) A geodetic survey of next-to-the highest order of accuracy and precision.

survey, seismic - A geophysical survey that creates seismic shocks by setting of explosives or by high-energy tamping, recording the resulting echoes and analyzing the records.

survey, single-run - SEE leveling, single-run.

survey, standard - A survey which, in scale, accuracy, and content satisfies criteria prescribed for such a survey by competent authority or general agreement.

survey, suburban - A survey of land lying outside a city or town and devoted almost exclusively to single-family residences or residential subdivisions.

survey, topographic - (1) A geodetic survey which has, for its major purpose, the determination of the configuration (relief) of the surface of the land and the location of natural and artificial objects on the surface. The information gathered by a topographic survey is usually published as a topographic map. (2) An organization making a topographic survey in the sense of the preceding definition.

survey, town-site - A survey resulting in the marking of lines and corners, within one or more regular units of the subdivision within a township, by which the land is divided into blocks, streets, and alleys as a basis for disposal of title in parcels of land.

survey, transit-and-stadia - A survey in which horizontal and vertical angles are measured with a transit and distances are measured with a transit and stadia rod. The results of a transit-and-stadia survey are usually processed in the office; the result of a survey made with alidade and planetable is the map made in the field as the measurements are taken. (An exception is found in the topographic survey of Baltimore, made in the 1890's, in which transit and planetable were used together in the field, and the map was constructed as the measurements were made.)

survey, urban - (1) A survey of a city and its environment, in which the following work is done. (a) Horizontal and vertical control are established. (b) A topographic survey is made and a topographic map, at a scale between 1:1000 and 1:2500 is prepared. (c) Survey markers are implaced at suitable locations to serve as referents in subsequent surveys. (d) A map is made showing lengths and directions of all street lines and boundaries of public property, coordinates of important points, survey markers, important structures, natural features and so on, all with appropriate legends and notes. (e) A map is made showing essentially the same information as the topographic map but at a smaller scale (e.g., 1:25 000). (f) A map is made showing underground utilities. Also called a city survey. (2) A survey of land lying within of adjoining a city or town.

survey, wire-drag - A hydrographic survey made to determine the locations of obstacles to navigation, by towing, at a set depth, a cable carried between two vessels which sweep back and forth through the region to be surveyed. The cable is suspended from floats on the surface by wires of fixed lengths. When the cable encounters an obstruction, the floats are forced into a V shape with its vertex above the obstacle. The obstacle's location is then determined by taking azimuths to the vertex from the two vessels.

survey adjustment - (1) The process of determining small changes to measured values for angles and distances in a horizontal network or traverse, or for elevations in a vertical network, in such a way that the changed values fit the geometric conditions on the network better than do the measured values. The criterion for fitting better is usually that the sum of the squares of the differences between the changed values and the measured values be a minimum and that the changed values exactly satisfy the conditions. (2) The process of determining small changes to the previously established coordinates in a horizontal, vertical or 3-dimensional network in such a way that the sum of the squares of the differences between the changed values and the previous values shall be a minimum. Conditions are often imposed on the coordinates.

survey class - A category into which surveys meeting certain specified criteria are placed. E.g., the American Congress on Surveying and mapping has established four survey classes for classifying cadastral surveys: Class A - Urban surveys; Class B - Suburban Surveys; Class C - Rural Surveys; Class D - Mountain and Marshland Surveys. For a survey to be included in a particular class, the survey must meet all of 8 distinct criteria for accuracy and method of surveying.

survey control, airborne electronic - The airborne part of a very accurate locating system used in making a survey from an aircraft.

survey coordinates - SEE coordinates, space rectangular.

survey foot - The unit of length defined by the relationship 1 foot = (1/3)(3600/3937) meters, established by the U.S. Coast and Geodetic Survey as published in its Bulletin No.26 (5 April 1893). Although a particular meter was not specified in the definition, the International Prototype Meter in Paris was probably meant. Practically, the meter used was that derived from Meter Bar 27 (primary standard) and Meter Bar 21 (auxiliary standard) of the the U.S. Coast and Geodetic Survey and later of the U.S. Bureau of Standards. Also called U.S. Survey Foot.

surveying - (1) The art and science of making a survey. (2) The process of making a survey. Surveying is an ancient profession, of which there is evidence in the clay tablets of Babylon, the hieroglyphics of Egypt and the epics of Homer. (3) The process of acquiring or accumulating qualitative information and quantitative data by observing, counting, classifying and recording according to need. Examples are traffic surveying and soil surveying. SEE ALSO city surveying; engineering surveying; land surveying; occultation surveying; plane surveying; planetable surveying; planetable surveying, topographic; route surveying; satellite surveying; soil surveying; stadia surveying; traffic surveying; transit surveying.

surveying, advanced - SEE geodesy, higher.

surveying, astronomical - Surveying in which longitude and latitude are determined astronomically; distances are calculated on the reference ellipsoid using the astronomical longitude and latitude.

surveying, cadastral - Surveying concerned with values, specifications and the mapping and recording of boundaries of property; the process of making a cadastral survey.

surveying, electromagnetic - Surveying done using an surveying instrument which uses electromagnetic radiation for measuring distance.

surveying, electronic - Surveying done using electronic instruments.

surveying, geodetic - (1) The art or process of making a geodetic survey. (2) That branch of the art of surveying in which account is taken of the figure and size of the earth. Also called geodetic engineering.

surveying, horizontal-stadia - SEE stadia surveying.

surveying, hydrographic - The process of showing, upon a sheet, all that portion of the Earth's surface which lies beneath the water in a particular region, including a delineation of the contours of submerged channels, banks and shoals and a collection of samples of water and specimens of the bottom. It also includes that part of physical oceanography which takes account of tides and currents and, in modern surveys, the temperature and salinity of the water insofar as they affect accurate measurement of depth by echo sounding.

surveying, photogrammetric - Surveying done using photographs of the Earth's surface for preparation of the maps.

surveying, plane-table - SEE plane-table surveying.

surveying, stadia-intercept - SEE stadia surveying.

surveying, stadia-subtense-bar - SEE stadia surveying.

surveying, tachymetric - Surveying done using an instrument (called a tachymeter) which allow both distance and direction to be determined more or less simultaneously. Differences of elevation (i.e., height) may also be determinable.

surveying, topographic - (1) Surveying done to obtain data for mapping. (2) Geodetic surveying in which data sufficient for determining both horizontal and vertical coordinates are determined simultaneously. Some kinds of stadia surveying are of this type (e.g., tachymetric surveying). So are aerial phototriangulation and photomapping. Surveying done merely to establish the coordinates of a few points, as in satellite surveying, does not constitute topographic surveying. (3) SEE plane-table surveying, topographic.

surveying, trilinear - The determination of the location of a point from which observations are made, by measuring the angles at that point between lines to three points of known location. Trilinear surveying involves solving the three-point problem. The solution may be obtained graphically, as by using a mechanical device such as the chorograph or the three-arm protractor, or algebraically. SEE ALSO resection.

surveying, urban - (1) Surveying done within the boundaries of a city to prepare maps and data for municipal construction; i.e., the process of making a city survey. (2) The process of making a survey of city or of a substantial part of a city. Also called city surveying.

surveying, vertical-stadia - SEE stadia surveying.

surveying altimeter - An aneroid barometer with a dial graduated in meters or feet, and used to determine approximate differences of altitude or elevation between points.
surveying arrow - SEE taping pin.

surveying camera - SEE mapping camera.

surveying computer - A computer designed for solving problems of the type commonly encountered in surveying.
surveying instrument - Any instrument used for measuring angles, distances, heights and/or elevations, or for determining geographic coordinates. SEE ALSO alidade; altimeter; angle-measuring instrument; compass; distance- measuring instrument; gyro-theodolite; level; leveling instrument; leveling rod; tape, surveyors; telescope; theodolite; transit.

surveying sextant - A sextant designed primarily for use in hydrographic surveying. Also called a sounding sextant.

surveying system - (1) The complete collection of everything needed: equipment, material, personnel and procedures for a particular type of surveying. Also called a survey instrument. (2) Any complicated apparatus or equipment used for surveying. Also called a survey instrument.

surveying system, electronic - A surveying system depending on electromagnetic distance-measuring instruments for the measurements.

surveying system, inertial - A surveying system capable of determining the geodetic coordinates or rectangular Cartesian coordinates of a sequence of points by measuring the acceleration or velocity with which a component of the system is moved about the region being surveyed. An inertial surveying system typically consists of a set of three accelerometers (or integrating accelerometers) whose axes are mutually perpendicular a set of two or three gyroscopes determining the orientation of the accelerometers' axes, a computer which accepts the measured data from the accelerometers and gyroscopes and computes the coordinates of a reference point in the unit containing the accelerometers and gyroscopes, and a vehicle in which the equipment is carried. In operation, the equipment is moved from marker to marker, stopping at intermediate points if the time needed to move between markers is greater than 10 minutes (or greater than 3 minutes if high accuracy is wanted). At each marker, the markers' coordinates are recorded. The vehicle is usually a small truck or a helicopter. However, one version developed for the U.S. Geological Survey has the equipment mounted in an airplane and no stops are made at markers or at intermediate points. Laser-type distance- measuring equipment is used to measure distances to corner-cube reflectors placed on the markers and a laser- type altimeter is used to measure altitude. Most inertial surveying systems can determine not only the coordinates of points of interest but also the force of gravity in the region surveyed.

surveying telescope - SEE surveyor's telescope.

survey instrument - (1) A portable instrument for detecting and measuring radiation under various physical conditions. (2) SEE surveying instrument.

survey marker - (1) An object (marker) placed at the site of a station (a) to identify the location of that station and (b) having itself a location determined by surveying. Also called a mark, marker or monument. However, mark should be used for the point or other symbol placed on the marker to identify the point to which the coordinates of the station refer. Monument should be used if the entire structure including the mark, the marker and the structure to which the marker is fastened or of which it is a part. A station having a survey marker should be prominent, permanent, stable and should have a definite location. (2) In particular, such an object whose coordinates are used for control in a geodetic network.

survey navigation - Aerial navigation done as part of an aerial survey - particularly for aerotriangulation.

survey navigation system - A navigation system designed for aerial navigation done for aerial surveying in particular, for aerotriangulation.

survey net - SEE survey network.

survey network - (1) A network in which the points are survey stations and the lines represent adjusted distances and directions, or show the sequence in which measurements were made or survey stations were occupied. The two principal types of survey network are the horizontal survey-network and the leveling network. (2) The set of points and lines, on an ellipsoid, representing survey stations and the lines of sight along which distances or directions were measured. The lengths of the lines and the angles between the lines represent adjusted angles and distances, or show the sequence in which measurements were made or survey stations were occupied.

survey network, horizontal-control - A survey network in which the survey stations are control stations for horizontal control.

survey network, vertical-control - A survey network in which the survey stations are control stations for elevations.

Survey-of-India method - SEE principal-plane method.

surveyor - (1) A person who surveys. (2) A person who determines the relative locations of terrestrial points by measuring or otherwise determining distances and heights or elevations, etc. In particular, a person who engages in such pursuits (surveying) as a profession. SEE ALSO land surveyor; quantity surveyor.

Surveyor General - (1) An officer in charge of the surveying and surveys of public lands. In particular, but not necessarily, public lands of a State or of the USA. (2) SEE surveyor general.

surveyor general - A principal surveyor; as, e.g., the surveyor general of the queen's manors. Also spelled Surveyor General. (Brit.)

surveyor's arrow - SEE pin.

surveyor's certificate - A document furnished by a surveyor to indicate his findings to a client.

surveyor's chain - SEE Gunter's chain.

surveyor's compass - A magnetic compass for determining the magnetic azimuth of a line of sight by means of a sighting device, a graduated horizontal circle and a pivoted, magnetic needle. Older versions almost universally read bearings, not azimuths. The surveyor's compass used for the early land surveys in the USA contained a pair of peep sights to define the line of sight and was usually mounted on a single leg, called a Jacob's staff. This instrument has been completely displaced by the surveyor's transit and the solar transit.

surveyor's tape - A ribbon of metal, specially reinforced cloth or other material, on which graduations are marked, and used for measuring distances or lengths. Usually referred to within the surveying profession simply as a tape. A common type used for geodetic surveying is of metal having a low coefficient of expansion, with two marks a convenient distance (e.g. 50 m or 100 feet) apart and usually with the one unit at either or both ends subdivided into finer units. For precise surveying, surveyor's tapes are made of invar and are 50 m or 100 m long. European practice is to use wire instead of ribbon.

surveyor's tape, metallic - A graduated, waterproof ribbon of fabric into which are interwoven, longitudinally, wires of brass or bronze to prevent stretching. Also called metallic tape.

surveyor's telescope - An optical telescope consisting of an objective lens-system, a reticle placed at the focus of the objective lens-system, and an eyepiece which magnifies the image on the reticle together with the markings on the reticle. These components are mounted in a thick, rigid tube. Also called a surveying telescope. The type of telescope is used in the majority of leveling instruments, transits and theodolites. Modern telescopes of this type also contain an optical component between the objective lens-system and the reticle to allow focusing the telescope without moving either of the other optical components.

surveyor's transit - A transit used primarily for surveying, as distinguished from the astronomical transit, which is used solely for astronomy.

surveyor's wire - A metallic wire, usually of invar or similar alloy, having a graduated scale attached to each end and used for measuring distances. A surveyor's wire is less affected by wind than is a surveyor's tape. On the other hand, tape is less likely to twist or curl and is sturdier. The corrections for slope and catenary (sag) are the same as those for distances measured by tape; the temperature, tension and calibration corrections to wire length differ from the corresponding corrections to tape length only in the values of the constants used in the formulae. However, wire undergoes significant changes of length when reeled up and unreeled many times. For geodetic surveys, a surveyor's wire usually has a diameter of about 1.65 mm and a length of 24 meters, with the scale at each end marked in millimeters from 0 (exactly 24 meters from the 0 on the scale at the other end) to 80 mm.

survey photography - SEE mapping photography.

survey plat - (1) A map on which land surveys are recorded to a scale of 2 inches to the mile. Each map usually shows one township. (2) Any survey map.

survey point - Any point to or from which differences of elevation have been determined.

survey signal - A natural or artificial object or structure whose location horizontally (and sometimes vertically) is determined by surveying and which is selected or constructed to be easily and precisely observed. Also called a signal, although that term should be used only if no confusion with other meanings of signal is likely. Survey signals are given special designations according to the kind of survey in which they are used: e.g., triangulation signal and hydrographic signal.

survey station - (1) A point at which or from which observations have been made for a geodetic survey. (2) A definite point, on the Earth, whose location has been determined by geodetic surveying. (3) A point, on a traverse, over which an instrument is placed (also called a set-up). In these three instances, a survey station is often referred to simply as a station. (4) A length, in a traverse, of 100 feet measured on a given broken, straight, or curved line. (5) A stake indicating one end of a 100-foot long interval on a traverse. The nature of a survey station is usually indicated by adding a word describing the station's origin or purpose: for example, triangulation station, topographic station or magnetic station. A survey station may or may not be marked on the ground. If it is marked, a geodetic marker (called a monument of special construction, or a natural or already-present, artificial structure is often used to mark the station.

survey system, rectangular - SEE survey, rectangular.

survey traverse - (1) A route and a sequence of points (stations) for which the location of each point has been determined by measuring that point's direction (or angle) and distance from the preceding point in the sequence. Commonly referred to as a traverse if the meaning is unmistakable. Also called a polygonometric survey or goniometric survey. Those terms usually occur in English literature as translations from European literature. If the last point in the sequence is identical with the first point or with a point whose location is already known, the sequence is called a closed traverse. Otherwise, the traverse is called an open traverse. (2) A route, a sequence of points on that route for which the location of each point has been determined as a distance and direction from the preceding point and the associated distances and directions. (3) The process by which a traverse of the type defined in (1) or (2) is established. Unless specifically stated otherwise, a survey traverse is a horizontal traverse i.e., a traverse for determining only the horizontal coordinates of the points in the traverse. (4) A sequence of distances and directions between points on the Earth or on an ellipsoid representing the Earth, obtained by or from measurements made in the field and used in determining the locations of the points. A survey traverse may determine the relative locations of the points which it connects in sequence. If the locations are determined using coordinates of control stations on an adopted datum, the locations may be referred to that datum.

swale - (1) A slight depression, often wet and covered with rank vegetation. (2) A wide, shallow ditch, usually covered with grass or paved.

swamp - (1) Low-lying land saturated with moisture and overgrown with vegetation, but not covered by water. The vegetation characteristically includes trees or shrubs. (2) Land, at elevations below that of upland, such as would be wet and unfit for agriculture without artificial drainage. (3) A wet or moist region with water standing on or just below the surface of the ground, and usually covered by a heavy and dense growth of vegetation. The term is usually applied to extensive regions in which the water is fresh.
(4) A region of spongy land usually saturated with water.

swamp, submerged - A region under water and partly or wholly covered or filled with vegetation.

swath - The region or area illuminated by a transmitter or viewed by a detector carried on an aircraft or satellite over an extended period of time. The swath is usually specified by giving the width and the locus of the centers of the instantaneous ground coverage.

sweep bar - A heavy section of steel rail or steel beam suspended at a predetermined depth by two vertical cables and towed by a vessel to find any obstructions to navigation which extend above that depth. A sweep bar is often used in hydrographic surveys.

swing - (1) The angle of rotation of a photograph about the photograph perpendicular, from some reference direction such as the direction of flight of the aircraft from which the photograph was taken. Usually denoted by κ. (2) The angle, at the principal point of a photograph, measured clockwise from the positive y-axis to the principal line at the nadir point. Usually denoted by κ but also sometimes denoted by π. (3) On oblique photographs taken by a trimetrogon camera, the angle between the principal line and the y-axis or between the isometric parallel and the z-axis. (4) The angle of rotation of a projector, in a stereoscopic plotting instrument, about the z-axis of the instrument. Usually denoted by κ. (5) (of a pendulum) The total movement of a pendulum between its passage from one end of its arc to the other, or from the point of greatest velocity in one direction to the point of greatest velocity in the other. Also called a vibration. Two successive swings or vibrations make up one oscillation. SEE ALSO ground swing.

swing, relative - The angle of rotation of an oblique camera about its own optical axis and with respect to the plane of the photograph in an adjacent vertical camera, measured on the oblique photograph by the angle between the isoline and a line joining the fore and aft fiducial marks.

swinger - SEE revolver.

swing offset - The perpendicular distance from a point to a transit line, found by holding the zero point of a tape at the given point and swinging the tape in an arc until the smallest horizontal distance is measured.

swing-swing method (of relative orientation) - The relative orientation of two projectors by merely rotating the projectors. Identical rotations (swings) applied about the z-axes have the same effect as translation of one projector in the y-direction. Identical rotations (tilts) about the y-axes have the same effect as translating one projector in the z- direction. Hence, y-parallax can be eliminated by this method without translating either projector.

switch - A device for diverting rolling stock from one track to another.

swivel graver - A graver with a swivel that allows changes in direction of scribing.

symbol - A mark or marks (diagram, design, letter or abbreviation, etc.) placed on charts or maps, which by convention, usage or reference to a legend is understood to stand for or represent a specific real characteristic or object. A symbol is not a conventional sign, which is more general and may refer to a concept or give a command. Standard symbols for charts and maps of the USA were adopted by the Federal Board of Surveys and Maps in 1938. SEE ALSO area symbol; line symbol; point symbol.

symbol, cartographic - A symbol representing some feature, equality or characteristic on a map.

symbol, proportional - A symbol whose size is proportional to the value of a quantity.

symmetry - The property possessed by some objects (bodies and geometric figures) of having two or more orientations showing no perceptible differences in the appearance of the object. Alternatively, the property that to every point in the object there corresponds a second point indistinguishable from the first and such that a line joining the two points is bisected by a point, line, or plane common to all such pairs. Or, that property of a body of remaining unchanged in appearance upon rotation about a suitably chosen point or line or upon reflection in a suitably chosen plane. An object whose appearance changes under any rotation or reflection about a line or plane is said to be asymmetric with respect to" that line or plane; the property is called asymmetry with respect to the line or plane. E.g. an arbitrary planar curve rotated about a line generates a surface which is symmetrical with that line but which is in general not symmetric with respect to any plane perpendicular to that line. Another example is the representation of the geoid by a series of zonal harmonic functions. If only terms of even degree are present, the representation is symmetrical both about the polar axis and about the equatorial plane. If terms of odd degree are admitted, axial symmetry still exists but the figure is no longer symmetrical with respect to the equator. If sectorial harmonic functions are also present, even the rotational symmetry disappears.

symmetry, axial - The property possessed by some objects or figures of being symmetrical with respect to a suitably chosen straight line. Also called polar symmetry and rotational symmetry. A distinction is sometimes made between the case in which axial symmetry exists for all orientations about the line and the case in which it exists only for a finite number of orientations. The first case is sometimes called cylindrical symmetry.

symmetry, central - The property possessed by some objects or figures of being symmetrical with respect to a point i.e., any line drawn from one side of the figure to another and passing through that point is bisected, by the point.

symmetry, cylindrical - SEE symmetry, axial.

symmetry, polar - SEE symmetry, axial.

symmetry, rotational - SEE symmetry, axial.

synergistic (adj.) - SEE system.

system - Any combination of things which work together so as to produce a purposeful effect or an effect considered purposeful. The individual components of the system are said to be synergistic. In geodetic usage, the terms system, equipment, instrument (or apparatus or device), component and element form a hierarchy, with each term denoting a less complex aggregate than the term preceding it.

system, accuracy-control - Any method attempting to detect and control errors, as by making measurements at random (i.e., by random sampling) and checking deviations of these measurements from arbitrary or standard values.

system, achromatic optical - An optical system producing the same definite effect for light of two different wavelengths, no matter what that definite effect may be.

system, active - A system which transmits or relays an electromagnetic signal.

system, amphidromic - The set of co-tidal lines radiating from a point at which the range of tides is zero (the amphidromic point). I.e., the set of cotidal lines in an amphidromic region. Note that an amphidromic point usually exists for only one constituent of the tide. It will not be the amphidromic point for another constituent.

system, amphidromic degenerate - A system of co-tidal lines whose amphidromic point appears to be located on land rather than on the open ocean.

system, anamorphic optical - An optical system producing an image which is magnified differently in two orthogonal directions.

system, binary coded-decimal - A code (set of rules) for representing each digit of a decimal number by a binary number (a sequence of 0's and 1's).

system, catadioptric optical - An optical system containing both refractive and reflective elements.

system, catoptric optical - An optical system containing only reflecting elements.

system, coherent optical - An optical system in which the relative phase of light waves in various parts of the system is invariant with time.

system, diffraction-limited optical - An optical system free from aberrations, the quality of the resulting image being determined by diffraction alone. No real optical system is diffraction-limited, but many systems come close to this ideal within a narrow field of view and a limited range of wavelengths.

system, digitizing - SEE digitizer.

system, dioptric optical - An optical system containing only lenses and flat mirrors. It is more usually called a lens system.

system, folded optical - An optical system containing reflecting components reducing the physical length of the system from what it would otherwise have been, or changing the path of the optical axis. Also called folded optics. The most common type of folded optical system is the prism binocular. Some cameras made for aerial reconnaissance contain folded optical systems producing focal lengths of over 5 meters within a physical length of about 1 meter.

system, geographic-information - SEE information system, geographic.

system, holonomic - A system of N equations (or a physical system represented by such equations) in N + M independent variables, plus M condition equations giving enough conditions on the independent variables, or on the differentials of these variables, that the system can be solved. If the condition equations give relations between differentials, these equations must be integrable. If they are not, the system is termed non-holonomic.

system, imaging - SEE imaging system.

system, isostatic - A theory of isostasy. The term usually covers also the formulas or tables needed for calculating the effect of isostasy on gravity or the deflection of the vertical.

system, metric - (1) A system of units based on the meter as the unit of length. Examples are the c.g.s. system, the M.K.S.A. system, and the S.I. (2) A system of units based on the meter as the unit of length, the gram as the unit of mass and the second as the unit of time. Also called more precisely, the c.g.s. system or the C.G.S. system. An international conference was called by France in 1870 to work out an internationally acceptable metric system. The results were; a set of official definitions incorporated in the Treaty of the Meter of Paris (1875), the establishment of an International Bureau of Weights and Measures for preserving the standards, and the establishment of a General Conference on Weights and measures to act on changes as the need arose. The General Conference in 1960 adopted the Systéme International d'Unitès (S.I. or SI) as an improved version of the previous metric system. It added to the fundamental units of length, mass and time the units of temperature, electric current and luminous intensity.

system, operating - SEE operating system.

system, optical - Any assemblage of optical elements designed to force light rays to follow a specified path for a specific purpose.

system, pancratic optical - SEE system, varifocal optical.

system, passive - A system which records energy emitted or reflected but does not produce or transmit energy of its own.

system, position-finding - SEE position-finding system.

system, position-fixing - SEE position-fixing system.

system, right-ascension - SEE coordinate system, equatorial.

system, sexagesimal - A system of counting by increments of 60; as, e.g., the division of the circle into 360°, each degree into 60 minutes and each minute into 60 seconds.

system, varifocal optical - An optical system whose magnifying power can be varied. Also called a pancratic optical system, varifocal lens, zoom lens and zoom optical system. Varifocal optical systems are much used in television cameras and copying cameras, as well as in some kinds of stereoscopic plotters.

system, zoom optical - SEE system, varifocal optical.

Système International d'Unités - A self-consistent system of units adopted by the general Conference on Weights and Measures in 1960 as a modification of the then-existing metric system, changed in 1983, and consisting of the ollowing units considered fundamental:

quantity	unit	symbol
speed of light in vacuum		c
mass	kilogram	kg
time	second	s
electric current	ampere	A
temperature	kelvin	K
quantity of matter	mole	mol
luminous intensity	candela	cd

In 1983, the International Conference on Weights and Measures removed the unit of length as a fundamental unit and replaced it with the speed of light in vacuum, defined to be exactly 299 792 458 meters per second. The meter is now a derived quantity, being defined as the speed c of light times 1 second, divided by 299 792 458. Also called the International System of Units. Abbreviated as SI. Some derived units with special names are:

frequency	hertz	Hz cycles/s
length	meter	m
force	newton	N $kg(m/s^2)$
work	joule	J $kg(m^2/s^2)$
pressure	pascal	P a $(kg/m^2)(m/s^2)$

Prefixed denoting multiples, by powers of 10, of the units are the same as in the earlier (c.g.s.) system. A few have been added to cover the anticipated range of values for often-used quantities.

system of astronomical constants - SEE constant, astronomical.

system of galactic coordinates - SEE coordinate, galactic; coordinate system, galactic.

system of units, centimeter-gram-second - SEE system of units, c.g.s.

system of units, c.g.s. - A metric system of units for physical measurement in which the fundamental units of length, mass and time are the centimeter, gram and the mean solar second. It is sometimes referred to simply as the metric system. The dyne, liter and calorie are derived units in this system. Although the c.g.s. system is particularly useful for expressing quantities of the size met with in a physical laboratory, its units are too small to be convenient for expressing quantities met with outside the laboratory in ordinary work. It has therefore been replace by the Système International d'Unités (S.I.), which uses the speed of light, kilogram and second as fundamental units of speed, mass and time. (The recent adoption of the speed of light as a fundamental constant does not affect the status of the second as a fundamental unit of time.)

System of Units, International - SEE Système International d'Unités.

system uncertainty - The root-mean-square value of the unknown constant and random errors in a quantity as inferred from a set of measurements and computations. An example of system uncertainty is a rate of motion inferred from measurements between two points at different times.

syzygy - (1) One of the two points, in the orbit of a planet or satellite, at which the body, the Earth and the Sun have the same longitude or differ in longitude by 180°. In particular, the location of the Moon when it is new or full.

T

table - SEE digitizing table; plotting table; reference table; tracing table.

table, digitizing - A measuring engine connected to a circuit which converts the location of the measuring head to coordinates in a rectangular coordinate system and makes a record of the coordinates in a form readable by a computer.

table, fundamental - SEE gravity table, fundamental.

table, supporting - The flat surface supporting the frame to which the projectors in a stereoscopic plotting instrument are clamped and to which the stereoscopic model's coordinates are referred.

table of meridional parts - A table listing distances along the geodetic meridian from the equator to various parallels of geodetic latitude, on a specified, rotational ellipsoid, and the ratio, for each of the various parallels of latitude, of the length of one minute of latitude to the length of one minute of longitude at the equator. This definition is given in many texts but does not correspond to what is actually found in tables of proportional parts. A table of meridional parts contains the meridional parts (1) for parallels of latitude at constant intervals and the meridional differences corresponding to these latitudes. On a sphere, at the equator, the length of a minute of longitude is equal to the length of a minute of latitude, but on receding from the equator and approaching the poles, the length of a minute of longitude steadily decreases. (Because the Earth is usually represented by a rotational ellipsoid rather than by a sphere, the variation with latitude is slightly different.) However, in the graticule constructed using a Mercator map projection, the minutes of longitude are made to appear of the same length at all latitudes. To preserve existing proportions between lengths along parallels of latitude and lengths along the meridian at different latitudes, it becomes necessary to increase the lengths along the lines representing meridional arcs from equator to parallels of latitude, such increase becoming greater and greater the higher the latitude. The length of that line, expressed in units corresponding to the one minute of arc along the equator, is given as a meridional part (meridional distance) in the table. The lengths are usually given for each minute of latitude, so that the differences between successive meridional parts are the meridional differences and are also given.

tablemount - SEE guyot.

tacheographometry - SEE tachmetry.

tacheometer - SEE tachymeter.

tacheometer, electronic - A tachymeter using electromagnetic radiation for determining the distance.

tacheometer traverse - SEE traverse, tacheometric.

tacheometry - SEE tachymetry.

tachometer - An instrument for measuring rates of rotation. The word is sometimes used erroneously for tachymeter.

tachometry - The measurement of rates of rotation. The word is sometimes used erroneously for tachymetry.

tachymeter - (1) A surveying instrument for rapidly determining direction and distance, using a short base line which may be an integral part of the instrument or may be separate and placed at the point to which observations are made. Also called a tacheometer and tacnygraphometer. There are two basic types of tachymeter. Both determine direction by measuring the angle, on a graduated horizontal circle, between a reference direction and the line of sight to the target. Both determine distance by measuring the angle subtended by a short line of known length and solving for the altitude of the triangle of which this line is the base and the measured angle the vertex. They differ only in how the length of the base line is determined. In one type, the length is fixed. It may be the distance between two fixed marks near the ends of a short bar (subtense bar) or the distance between two predetermined graduations on a vertical or horizontal graduated rod. The angle at the vertex is determined by changing the distance between two lines on the reticle until the fixed length is seen to be just included within the lines. The angle is indicated on the instrument. Distance is calculated or is given directly by the instrument on a dial connected by gear train to the reticle. In the other type, the reticle has on it two lines a fixed distance apart. The observer reads off the length of graduated rod seen between the two lines. The distance is calculated by the same formula as that used with the first, fixed-length type of base line. Both types also allow the difference in height of observer and target to be determined at the same time if the vertical angle can be measured. The so-called rangefinder, consisting of a short rod with a prism mounted at each end and a known distance apart, transmitting the separate images to a telescope with beam-splitting prism mounted halfway between the two prisms, is also of the second type but is not usually referred to as a tachymeter. SEE ALSO rangefinder; stadia method; subtense method. (2) A surveying instrument for rapidly determining direction and distance. Also called a tacheometer and in advertisements, a total station. This definition covers also the type of instrument included in the preceding definition. In addition, it includes the type which measures the time it takes electro-magnetic pulses to travel from the instrument to the target and back, while simultaneously measuring the direction and (usually) vertical angle to the target. Difference of height is calculated by the instrument from the measured quantities. (3) A surveying instrument for rapidly determining direction, distance and difference of height from a single measurement.

tachymeter, auto-reducing - A tachymeter which provides distance and difference of height directly from the readings on a graduated stadia rod to which the lines of sight through three lines on the tachymeter's reticle are directed. Also called a self-reducing tachymeter. In a common form, the reticle is geared to the vertical motion of the telescope so that as the telescope is elevated or depressed, the reticle is rotated and the graduated lines on the reticle appear to come loser together or move farther apart. (Separate lines are used for reading horizontal distance and differences in height.). The horizontal distance to the target is 100 times the distance between the intercepts of the top and bottom lines on the reticle. The vertical distance is the value at which the central line intercepts the rod, multiplied by a factor which appears in the field of view.

tachymeter, self-reducing - SEE tachymeter, auto-reducing.

tachymetry - (1) The measurement of distance by a method which provides rapid, though not necessarily accurate, results. Also called tacheometry. (2) Surveying in which direction and distance, and sometimes differences of height, are determined more or less simultaneously by the same instrument. Also called tacheometry. There are three principal forms of tachymetry: (a) the stadia method, in which the length subtended at a leveling rod by a known angle at the instrument is measured; (b) the subtense method, in which the angle which a fixed and known length on the rod subtends at the instrument is measured; and
(c) the distance is determined directly by measuring the amount of time it takes an electromagnetic pulse to travel the distance from instrument to target and back. There are two kinds of stadia methods: the kind in which the stadia rod is held horizontally and the kind in which the stadia rod is held vertically. Direction or angle is measured the same way in all forms: a horizontal circle is read. Difference of height is determined by reading the vertical angle and multiplying the cosine of this angle by the distance determined. SEE ALSO tachymeter.

taking, partial - The taking of only a part of a property for public use under the power of eminent domain, and for which compensation must be paid, taking into consideration the damages and/or benefits to the remaindered property.

Talcott method (of determining latitude) - SEE Horrebow-Talcott method (of determining latitude).

talus, insular - The declivty from the lower edge of an insular shelf into the oceanic depths. It is characterized by a marked change in slope and usually begins at a depth of about 100 fathoms (200 meters).

tangent - (1) A straight line, in the plane, which meets a given curve in that plane in one and only one point within a small neighborhood of the point and which, if rotated by any amount, however small, will then intersect the curve in two or more nearby points. (2) In general, the limiting position taken up by a straight line which intersects a curve in two nearby points as the one point approaches the other (fixed) point. This definition applies to curves in a space of two or more dimensions. Also called a tangent line. (3) That part of a traverse or alinement included between the point of tangency of one curve and the point of curvature of the next curve.

(4) (USPLS) A great-circle line tangent to a parallel of latitude at a township corner. (5) A long, straight line of a traverse, particularly on a route survey, whether the termini of the line are points of curve or not. (6) SEE plane, tangent.

tangent, principal - SEE direction, principal.

tangent distance - SEE distance, tangent.

tangential method (of determining distance) - A method in which the interval, on a stadia rod, intercepted by those two lines which would make an angle $\frac{1}{2}(\beta - 2\alpha)$ at the center of the instrument is read. β is the angle determined by lines of sight through the upper and lower cross-hairs of the reticle; α is the angle the line of sight through the central cross-hair makes with the horizontal. Called the tangential method of distance measurement if the calculations involved are done automatically by the instrument.

tangent method (of establishing a parallel) (USPLS) - A method of establishing a geographic parallel, for surveying a base line or standard parallel, by measuring offsets from a great-circle line started at an established township-corner and tangent to the base line or standard parallel at that corner. The tangent great-circle is projected at an angle of 90° from the meridian at the township corner from which it starts, and proper offsets remeasured north from the tangent to the parallel upon which the corners are established.

tangent plane - SEE plane, tangent.

tape - SEE surveyor's tape; base tape; cut tape; Invar tape; linen tape; Lovar tape; surveyor's tape, metallic.

tape, adding - A surveyor's tape which has, beyond the zero mark near the end of the tape, an additional meter (or foot) of tape graduated in tenths or hundredths of a meter or foot. SEE ALSO tape, subtracting.

tape, breaking - SEE breaking tape.

tape cartridge - SEE cassette (2).

tape cassette - SEE cassette (3).

tape correction - An amount applied to a distance measured with a surveyor's tape, to eliminate or reduce errors caused by differences between the conditions under which the tape was used and those under which it was calibrated. Tape corrections are of two kinds: corrections to the length of the tape, necessary because distances between marks on the tape were not correct when the tape was being used; and corrections to the distance as measured (taped length), necessary because the tape was used in a way that did not give the true distance. There are commonly corrections made, to the length of the tape, for temperature, tension, length and calibration. There are commonly corrections made, to the measured distance, for alinement, grade and sag. (The amount subtracted from a distance, after the corrections to tape length and measured distance have been applied, to obtain the corresponding distance on the geoid or the rotational ellipsoid is not considered a tape correction.) SEE ALSO alinement correction; calibration correction; gravity correction; length correction; temperature correction (1); tension correction.

tape, instantaneous-reading - A surveyor's tape on which the foot mark is repeated at each subdivision. Thus, such a tape divided into tenths of a foot would have the foot mark imprinted at every tenth of a foot subdivision.

tape, metallic - SEE surveyor's tape, metallic.

tape, piano-wire - Piano wire used instead of a metallic tape when it is advisable to control hydrography by precise traverse rather than by a weak extension of triangulation.

tape, subtracting - A surveyor's tape with the first and last meter or foot divided into tenths or hundredths of a meter or foot. Also called cut tape.

tape gage - (1) A device consisting of a tagged or marked chain, tape or other line attached to a weight which is lowered until it just touches the water's surface, whereupon the height is read on a graduated staff. Also called a chain gage. (2) SEE tape gage, electric.

tape gage, electric - A tide gage consisting of a metallic tape on a metallic reel (with supporting frame), voltmeter and battery. The tape is graduated with numbers increasing toward the free end. The free end is lowered into the stilling well until movement of the voltmeter's needle indicates that contact has been made. At that moment, the length of tape unreeled is read against an index mark.

tape length - The nominal, assumed or exact length of a surveyor's tape. Tape length should not be confused with taped length. Small amounts (corrections) are added to the nominal or assumed lengths between graduations on the tape to arrive at what the lengths would have been at the times and under the conditions in which the tape was used. SEE ALSO calibration correction; gravity correction; length correction; temperature correction (1); tension correction.

tapeperson - A person who assists in measuring distances with a surveyor's tape by carrying one end of the tape and inserting or removing taping pins to mark the distance measured. Also called a tape man, chainman and contactman. There are usually only two tapeperson's - a rear tapeperson and a head tapeperson. The rear tapeperson places the zero mark on the tape at the proper mark or pin on the ground and lines the head tapeperson, who carries the other end of the tape, with the following mark or pin. The head tapeperson pulls the tape taut and inserts a pin at the last graduation on the tape. The rear tapeperson removes the rear pin after every measurement and keeps count of the number of measurements made. (In using some kinds of surveyor's tape, the head tapeperson carries the end with the zero mark.) SEE ALSO taping.

tapeperson, rear - SEE tapeperson.

tape measurement, broken - The short distances measured and accumulated to total the full length of the tape when a standard, 100-foot or 50 meter tape cannot be held horizontally without plumbing from above shoulder level.

tape rod - A leveling rod consisting of a frame with rollers at both ends, over which an endless, graduated tape moves. Also called an automatic rod. The tape rod is designed to permit the observer to read the graduations directly, eliminating all additions and subtractions required when using other types of rods.

tape stretcher - Mechanical device for holding a tape at a prescribed tension and in a prescribed position. A tape stretcher generally consists of a spring balance to which the tape or wire is attached and a lever for applying tension to the tape through the spring balance. Also called a stretching apparatus.

tape thermometer - A precise thermometer fitted into a case designed so that it can be clipped against a surveyor's tape.

tape tripod - SEE taping tripod.

taping - The operation of measuring a distance on the Earth, using a surveyor's tape. The persons who mark the two ends in taping are called contact men or, particularly when measuring base lines, tape men or tapemen. The term taping is used in this sense in all surveys except in those of the public lands of the USA; in those surveys, for historical and legal reasons, the terms chaining and chainmen are preferred. SEE ALSO chaining; slope taping.

taping, standard-tension - (1) Taping with the tape at the tension at which it was calibrated. (2) SEE tension, standard.

taping and plumbing, horizontal - A method of taping whereby the tape is held horizontally and its pertinent graduations are projected to corresponding locations on the ground by plumb bobs.

taping arrow - SEE chaining pin.

taping buck - SEE taping stool.

taping correction - An amount added to a measured distance to account for an error introduced by a particular kind of difference between the way the distance was actually measured and a distance measured in proper alinement along a straight line with the tape held horizontally throughout its length. SEE alinement correction; grade correction; sag correction.

taping pin - SEE chaining pin.

taping station - The stake marking each interval (one tape length) along a traverse from the initial point. In particular, each stake placed at 100-foot intervals along a route survey. SEE ALSO station, plus.

taping stool - A stool, usually of metal, placed under the graduated end of a surveyor's tape, with a surface on which the position of the graduation can be marked. Also called a tape tripod, taping tripod, tripod support, taping buck, etc.

taping tripod - A tripod carrying a smooth flat or slightly curved support on which a surveyor's tape is supported at intermediate or end point when distances are to be measured accurately. SEE ALSO buck; taping stool.

tare (of a gravimeter) - An abrupt change in the measurements made with a gravimeter and caused by a change in the mechanism of the gravimeter. It is the equivalent of a shift in the value of gravity taken as standard at a base.
(2) A correction for a sudden change in the calibrated value of a gravimeter during the course of a survey.

target - (1) Any object at which a telescope is pointed for measurement of angle, direction, distance, height or elevation. The salient feature of a target is its marked visibility. Targets are selected or constructed to be clearly visible even when the lighting is poor. This definition makes signal and target almost synonymous. See, however, the next definition. (2) That point, on a signal, at which a telescope is pointed for measurement of angle, direction, distance, elevation and/or height. (3) A brightly painted, round disk which can be moved up and down a leveling rod until a mark on the disk is on the line of sight of the telescope in a leveling instrument. The rod person moves the disk up and down until the mark (e.g., a horizontal line) on it coincides, according to the instrument person, with the horizontal line in the telescope's reticle. The corresponding number on the leveling rod is then recorded. Targets are no longer used for precise leveling. In precise leveling, the instrument man himself reads the number at which the horizontal line on the reticle intersects the image of the rod (or interpolates to get that number). (4) SEE target, photographic.

target, natural - An existing natural or cultural feature, on the ground, which serves as a photographic target. Also called a substitute point or photocontrol point.

target, photographic - A specially-made, natural or already existing, man-made mark or feature, on the ground, which can be identified on an aerial photograph and which indicates the location of a station such as a control point. Also called a signal or target. A photographic target made by assembling large strips of cloth or plastic into a geometric figure with the control point at the center of symmetry is called a panel or paneling.

target, resolving-power - A chart on which are depicted bars or other kinds of figure in various sizes, separation and/or degrees of shade and photographed or observed to determine the resolving poser of the camera or optical system employed. The target may be of a size to be pinned to the wall of an optical laboratory, or of a size to be laid on the ground and photographed from an aircraft.

target, standard - A target used by radar to produce an echo of known power under various conditions. Smooth metallic spheres or cube-corner reflectors of known dimensions are such targets.

target column - A post to which targets are attached when leveling by the river-crossing method.

target leveling rod - A leveling rod to which is attached a distinctively marked disk, the target which can be moved up and down according to commands from the observer. The disk is moved by the rodman until its image in the telescope is exactly bisected, horizontally, by the central line of the reticle. The location of the target on the rod is then read by the rodman. Some targets have a vernier attached to permit more accurate reading. The self-reading leveling-rod, in contrast, does not have a target and is read by the observer. Target leveling-rods are seldom used today in geodetic leveling except when leveling across wide rivers or steep valleys.

target rod - SEE target leveling rod.

targeting - The process of marking or otherwise distinguishing ground-control points so that they are identifiable on aerial photographs.

taut-wire method - The method of measuring the distance between two points in shallow water by measuring the length of wire deposited along a course between the two points. A heavy weight is attached to the free end of a reel of wire and tossed overboard while the craft is nearing the closer of the two points. The wire is then reeled off under tension as the craft proceeds, and the amount reeled out is noted as the craft passes over the first point and then the second.

Tavistock theodolite, geodetic - A theodolite of high precision, designed for geodetic surveying according to a design by Tavistock; the average of readings at both ends of a diameter of a circle is indicated automatically and both circles are observed from a position near the eyepiece end of the telescope. One division of the micrometer for the horizontal circle is 0.5" and of the micrometer for the vertical circle, 1.0".

tax sale - (1) A sale of land for unpaid taxes. (2) A sale of property, by authority of law, for the collection of a tax assessed upon it, or upon its owner, which remains unpaid.

tax title - (1) The title by which a person holds land which he purchased at a tax sale. (2) That species of title which is inaugurated by a successful bid for land at a tax-collector's sale of the same for non-payment of taxes, completed by the failure of those entitled to redeem within the specified time, and evidenced by the deed executed to the purchaser, or his assignee, by the proper officer.

technique, gain-riding - A method of contgrolling the rise time of a signal from a Hiran station (or of a pulse from any similar transmitter) by keeping the amplitude of the receiver's output at the same reference level when zeroing and measuring distance.

tectonics - That part of geology which deals with large-scale changes, such as folding and faulting, in the crust. SEE ALSO plate tectonics.

tectonosphere - The crust and upper mantle of the Earth.

teledetection - The detection of objects or events at great distances to obtain information about the object or event. Teledetection is has about the same meaning as remote sensing. It differs from telemetry in that, in the latter, the detecting device is placed close to the object or event in question and also does the measuring; the data are then sent to the observer, who is far away. In teledection, the device is far from the object or event and measurements may or may not be made on the output of the device. SEE ALSO sensing, remote; telemetry.

telegraphic method (of determining longitude) - Time signals are sent by telegraph from a station whose astronomical longitude is known, to the station whose astronomic longitude is wanted. The time sent is compared to the time determined locally at the second station, and the difference is converted to difference in longitude. The method was originated by the U.S. Coast Survey in 1846, two years after the first transmission of telegraphic messages over wires. It became obsolete when time signals began to be sent by wireless.

telemeter (n.) - (1) Any instrument which can make measurements at one place and send the data to another place some distance away. The data are usually transmitted by wire or radio. However, some telemeters transmit the data pneumatically or hydraulically through tubes. (2) SEE distance-measuring instrument. (3) SEE tachymeter. (4) SEE rangefinder.

telemeter (v.) - Measure at one place and send the data to another place some distance away. The instrument is usually unattended.

telemeter, electronic - SEE distance-measuring instrument, electromagnetic; distance-measuring instrument, electronic.

telemetering - Transmitting the readings of instruments to a remote location. Also called telemetry and remote metering.

telemetry - (1) The technique or art of making measurements at one place and sending the data to another place far away for use or analysis. (2) SEE distance measurement. (3) SEE telemetering.

telephoto lens - SEE lens, telephoto.

teleprinter - A printing machine operating on instructions from a computer which may be some distance away.

telescope - (1) An instrument collecting electromagnetic radiation from a distant source to create an enlarged (with respect to the size in the unaided eye) image of that source. Telescopes may be designated according to (a) the frequency of the radiation they collect (e.g., radio telescope, infrared telescope, optical telescope); (b) the arrangement for collecting the radiation (reflecting telescope, catadioptric telescope, refracting telescope), or according to the kind of structure (mounting) to which the telescope is fastened while observing (equatorial telescope, meridian telescope, fixed-mount telescope). Unless specifically stated otherwise, telescope is assumed to mean optical telescope. (2) SEE telescope, optical. SEE ALSO finder telescope; guide telescope; meridian telescope; radar telescope; radio telescope; Ritchey-Chretien telescope; Schmidt telescope; Schupman telescope; surveyor's telescope; zenith telescope; zenith telescope, photographic.

telescope, achromatic - An optical telescope having an achromatic lens system for its objective (i.e., for collecting the light).

telescope, anallactic - An optical telescope having the anallactic center at or near the axis about which the telescope rotates. Also called an anallatic telescope.

telescope, anallatic - SEE telescope, anallactic.

telescope, astrographic - A telescope used primarily for photographing stellar fields to determine the locations of the stars with respect to each other. Also called an astrograph and astrographic camera. The usual astrographic telescope consists of two similar telescopes fastened together; one is used for photographing the stars, the other is used for keeping the first telescope aimed constantly at the same point in the sky. A normal astrographic telescope has a focal length of about 3.650 meters; however, a focal length of 2.000 meters has commonly been used for taking photographs for photographic star-catalogs. Telescopes of this kind were used in preparing the Astrographic Catalogue.

telescope, astronomical - An optical telescope consisting simply of a positive objective-lens and a positive eyepiece which produce, in combination, a final image which is inverted and reversed.

telescope, catadioptric - A telescope having a catadioptric optical system.

telescope, catoptric - A telescope having a catoptric optical system.

telescope, equatorial - A telescope movable independently about two perpendicular axes of rotation: an axis parallel to the Earth's axis of rotation and an axis perpendicular to it. A telescope on an equatorial mounting.

telescope, erecting - An optical telescope presenting, to an observer, an image having the same orientation as the object viewed directly. I.e., a down-to-up motion at the object corresponds to a down-to-up motion at the image. An erecting telescope introduces an additional component into the optical system of an inverting telescope. It therefore pays for the erectness of the image by being somewhat longer and losing a bit more light.

telescope, externally-focusing - An optical telescope focused by moving the eyepiece (or, more rarely, the objective lens) in or out (i.e., closer to or farther from the optical component at the other end).

telescope, finding - SEE finder telescope.

telescope, folded - An optical telescope whose physical length is made shorter than the focal length by introducing a mirror to fold back the optical path of the rays.

telescope, internal-focusing - SEE telescope, internally-focusing.

telescope, internally-focusing - An optical telescope containing a component between the objective and the eyepiece, for changing the location of the focus by moving the component back and forth along the tube. Also called an internal focusing telescope. An internally-focusing telescope is distinguished from an externally-focusing telescope in that it allows the observer to change the focal length without having to move the eyepiece or objective lens. The type was introduced into Europe at the turn of the century; it was not manufactured in America until many years later. Almost all telescopes used in modern theodolites and leveling instruments are of this type.

telescope, inverting - Properly, an optical telescope presenting an image such that a down-to-up motion at the object corresponds to an up-to-down motion at the image. The image appears upside down to the observer. However, all inverting telescopes used in surveying not only invert the image but reverse it also i.e., a left-to-right motion at the object causes a right-to-left motion at the image. Inverting telescopes contain fewer components than an erecting telescope and are usually shorter. Furthermore, the lose less of the light entering the telescope and hence are preferred for astronomical work.

telescope, meridian-seeking - An optical telescope with a component, placed in front of the objective lens system, consisting of two rigidly connected prisms so constructed that light from two circumpolar stars can be sent through the telescope simultaneously parallel to the line of sight. The angle between the direction of the line of sight and the direction to the first (second) star is the co-declination of the first (second) star. When the telescope is pointed so that the images of the two stars coincide, the line of sight is directed towards astronomic north.

telescope, optical - A telescope which collects light from a source to produce an enlarged (with respect to the size of the image produced in the unaided eye) image of the source. Usually referred to simply as a telescope if the meaning optical telescope is obvious. Optical telescope are constructed using lenses, mirrors (or prisms) or combinations of these elements. Most telescopes used in theodolites and leveling instruments are refracting telescopes, using only lenses. They consist essentially of two lens-systems: an objecting lens (-system) which brings rays of light from a distant object to a focus within the telescope's tube; and an eyepiece which magnifies the image formed by the objective lens. By means of a reticle placed at the principal focus of the objective lens, a definite line of sight is established, and the telescope thereby made into a precise device for astronomical observation and surveying. The magnifying power of a telescope depends on the ratio of the focal lengths of the objective lens and the eyepiece.

telescope, photographic zenith - SEE zenith telescope, photographic.

telescope, reflecting - A telescope which uses a reflecting surface or structure for collecting and focusing the radiation. Also called a reflector or, in the case of an optical or infrared telescope, a catoptric telescope.

telescope, refracting - A telescope which uses a large lens or lens-system for collecting and focusing the radiation and one or more smaller lens systems or mirrors for directing the radiation to the observer or detector. Also called a refractor or dioptric telescope.

telescope, terrestrial - (1) An optical telescope consisting of a positive objective lens-system, a positive eyepiece and an internal component which produces a final, erect image. SEE ALSO telescope, surveyor's. (2) SEE telescope, Galilean.

telescope, universal - A transit which has a very precise level, and a telescope which has a micrometer for determining the instant of a star's passage through a given vertical circle and a micrometer for measuring very small differences of zenith distance. A universal telescope may have only a single micrometer which can be rotated or replaced to allow measurements both in elevation and azimuth.

telescope level - A spirit level attached to a telescope in such a way that the axes of the spirit level and of the telescope are parallel at all times.

telluroid - (1) That surface whose distance below the surface of the Earth at each point is Molodenski's normal height at that point; the distance is taken in the direction of the gradient of a standard formula for the gravity potential. Also called a terroid or Telluroid. (2) That surface at every point of which the potential, according to some standard formula, is equal to the actual potential at a point P above it on the Earth's surface, both points lying on the same normal to the equipotential ellipsoid used as reference. (3) A surface defined with respect top an equipotential ellipsoid by the requirement that if the gravity potential U is given as a function of geodetic longitude λ, geodetic latitude ϕ and geodetic height h by a particular formula $U = U(\lambda, \phi, h)$, so that gravity potential U_e on a particular ellipsoid is obtained by setting geodetic height equal to zero in the formula, then the gravity potential on the telluroid shall be given by $U = U_e + (W_s - W_g)$, in which W_s is the actual gravity potential on the Earth's surface and W_g is the actual gravity potential on the geoid.

Telluroid - SEE telluroid.

temperature - A quantity measuring the degree of hotness or coldness of matter in particular, as measured by the expansion or contraction of a narrow column of liquid such as mercury or alcohol. On a submicroscopic scale, temperature is a measure of the average speed of the randomly moving elementary particles composing a body. On a macroscopic scale, it is a measure of the rate at which heat (thermal energy) is transferred from one body to another. Heat moves in the direction in which temperature is decreasing, and at a rate which is proportional to the difference in temperature. A body to which heat will move from any other body at a different temperature is said to be have a temperature of, or to be at, absolute zero. The temperatures of all other bodies are therefore positive. (The concept of negative temperatures has been used in the theory of masers to simplify some aspects of the theory.) SEE ALSO temperature scale; noise temperature.

temperature, absolute scale of - SEE temperature scale, absolute.

temperature, ambient - The temperature of the gas or liquid immediately in contact with a specified object (such as an instrument). Unless specifically stated otherwise, the average temperature of the fluid is meant. If the instrument is large and very sensitive to differences of temperature at various points, an average value may not be adequate.

temperature, approximate absolute - A temperature scale having the triple point of ice at 273° and the boiling point of water at 373°.

temperature, differential - A small, variable and undetectable difference between the temperature of a surveying instrument and the temperature indicated by a thermometer. The difference causes errors in reading that cannot be detected. It is usually caused by direct sunlight and can be made small by shading the instrument while observing.

temperature, dry-bulb - Temperature as measured with a dry- bulb thermometer.

temperature, potential - The temperature of a parcel of water raised adiabatically to the surface.

temperature, standard - (1) The ice point, 0° C. (2) The temperature at which the temperature of pure water is a maximum - 4° C. These two definitions are used in physics, the first more frequently than the second. In meteorology, the term has no generally accepted meaning.

temperature, wet-bulb - Temperature measured using a wet-bulb thermometer. The difference between dry-bulb and wet-bulb temperatures is an indication of the relative humidity of the air, and hence is very important for those measurements, such as electromagnetic distance-measurement, whose values depend on the humidity of the air.

temperature and pressure, standard - SEE standard temperature and pressure.

temperature correction - (1) The amount applied to the nominal length of a surveyor's tape to account for a change in its effective length if the tape is used at a temperature other than that at which it was calibrated. If the coefficient of thermal expansion is k_t and the tape was calibrated at temperature T_o, the expansion or contraction ΔL caused by measuring at a temperature T is $\Delta L = k_t L (T - T_o)$. (2) The amount added to the period of a pendulum (or to the pendulum period) to allow for the difference between the length of the pendulum at the temperature at which measurements were made and the length at which the pendulum was calibrated or at which other measurements were made. A temperature of 15° C is commonly adopted for calibration or reference. (3) The amount which is added to a difference of elevation determined directly from measurements to correct for the error introduced when the temperature at which the leveling rods were used is different from the temperature at which the rods were calibrated.

temperature correction, dynamic - The amount added to the observed period or half-period of a pendulum to account for the rate of change of the pendulum's temperature. The dynamic temperature correction is definitely related to the type of pendulum apparatus used, and depends especially on the way the thermometer is mounted and on the shape of the pendulum. It may change not only in magnitude but even in sign when the the type of apparatus is changed.

temperature scale - A continuous, single-valued function having the same value for all bodies between which heat is not transferred when the bodies are brought into contact, and having different values for bodies between which heat is transferred. This definition is so general that a large number of different temperature-scales can be devised to fit it. Most such scales are linear: the rate of transfer of heat from one body to another is, under ideal conditions, directly proportional to the difference in the values of the function for the two bodies. SEE ALSO Celsius temperature scale; Centigrade temperature scale; Fahrenheit temperature scale; International Temperature Scale; Kelvin temperature scale; Rankine temperature scale; Réumer temperature scale.

temperature scale, absolute - (1) A temperature scale assigning the value zero to the temperature of a body to which heat will flow from any other body in contact at a different temperature. (2) A general name for a temperature scale related to absolute zero and based on the performance of an ideal heat-engine i.e., a frictionless engine working by transferring heat from a source to a gas, letting the gas expand to do work (and cooling in the process), and letting the gas subsequently contract, without losing energy because of friction, leakage of heat, etc. Because absolute zero is unattainable and an ideal heat-engine does not exist, an absolute scale of temperature cannot be realized. Note that absolute scale of temperature and absolute temperature scale have different meanings. (2) SEE Kelvin temperature scale.

temperature scale, approximate absolute - A temperature scale on which the ice point (temperature at which water and ice are in equilibrium) at 270° and the boiling point of water at 373°. This temperature scale is intended to approximate the Kelvin temperature scale with sufficient accuracy for many sciences and is widely used in meteorology.

temperature scale, centigrade - A temperature scale in which the temperature at the triple-point of water is given the value of 0° and the temperature at the boiling point of water (at standard pressure) is given the value 100°. Now called the Celsius temperature-scale (q.v.).

temperature scale, thermodynamic - A temperature scale in which the temperature is taken to be proportional to the energy contained in a given volume of a perfect gas. The thermodynamic temperature scale in which the unit is the degree Kelvin is called the Kelvin temperature scale; that in which the unit is the degree Fahrenheit is called the Rankine temperature scale.

template - (1) A guide, gage or pattern designed to guide certain work or a tool used in the work. Also called a templet. A template is commonly a thin sheet of material cut to a certain shape and carrying suitable guide line or markings. An example is the ordinary protractor as used for drawing angles. In obtaining elevations from maps, for topographic and isostatic gravity-reductions, templates of plastic are used. (2) SEE templet (1). SEE ALSO Hammer template; Hayford template; Hayford deflection-of-the-vertical template; Hayford gravity template; Rice's template.

templet - (1) A slotted plate, assemblage of rods or marked sheet of transparent plastic to record the directions from the radial center of an aerial photograph to images of other points, and which can be assembled with similar plates, sheets, etc., to represent the connections between points on an assemblage of aerial photographs. Also called less often, a template. A transparent sheet used as templet shows the center and all radial lines from that center through images of control points and other important points, as well as lines connecting the center to images of points which show on the photograph and are the centers of other photographs. Plates are usually slotted, the slots radiating from a round hole representing the principal point, nadir point, etc., and going in the directions of pass points, corresponding image- points, etc. SEE ALSO templet, hand; templet, mechanical; templet, slotted; templet, spider. (2) SEE template. SEE ALSO calibration templet; hand templet; spider templet.

templet, mechanical - A templet which is manipulated mechanically in laying out and forming the radial aerotriangulation.

templet, mechanical-arm - SEE spider templet.

templet, radial-arm - SEE spider templet.

templet, radial-line - A templet, made from the photograph on which the radial-control points have been marked, making use of the fact that angles measured radially from the nadir point on a photograph remain constant regardless of tilt and relief. The templet contains all the marked radial-control points for that photograph. Rays from the center of the templet pass through each of the wing points, centers of adjacent photographs and horizontal-control points. There are four kinds of radial-line templets in common use: slotted templets, mechanical templets, hand templets and stereotemplets.

templet, slotted - A sheet of metal, thick cardboard or plastic in which a hole is punched to represent the location of the radial center on a photograph and slots are punched in the radial directions of (and containing) pass points, corresponding image-points, etc. Slotted templets are assembled by placing one sliding pin in each pair of slots representing directions to corresponding image-points.

templet, stereo - SEE stereotemplet.

templet cutter - A device used in photogrammetric adjustment for cutting the central hole and slots in templets. Also called a slot cutter. The slots are centered on points transferred from aerial photographs and are radial to the central hole.

templet method - Any method of aerotriangulation or phototriangulation in which templets are used to establish radial directions. SEE ALSO aerotriangulation, radial; slotted-templet method; etc.

templet ratiograph - A device for determining the ratio between two distances. One distance is that betgween the principal point and another, designated point on the aerial photograph. The other distance is the distance betwewen the corresponding principal point on a templet and the marked center of the stud for the designated point when all the templets have been assembled. The device is designed for a specific templet cutter. Also called a ratiograph. SEE ALSO ratiometer.

tenancy, joint - The holding of property by two or more persons in such a manner that, upon the death of one joint owner, the survivor or survivors take the entire property. This is to be distinguished from tenancy in common and tenancy by the entirety.

tenancy by the entirety - The holding, of an estate, created by a conveyance to husband and wife whereupon each becomes seized and possessed of the entire estate and after the death of one, the survivor takes the whole. SEE ALSO tenancy, joint; tenancy in common.

tenancy in common - The holding of property by two or more persons, each of whom has an undivided interest that, upon his death, passes to his heirs and not to the survivor or survivors.

ten-mile rule - SEE rule, ten-mile.

tension, normal - (1) That tension which must be applied to a surveyor's tape to compensate for the shortening effect of the sag (catenary) and to bring the tape to standard length. (2) That tension which must be applied to a surveyor's tape to make the tension correction and the sag correction exactly equal.

tension, standard - That tension at which a surveyro's tape was calibrated. The term standard tension taping is given in one place as a synonym, but the validity of this is doubtful.

tension correction - The correction applied to the nominal length of a surveyor's tape to allow for a change in effective length when the tape is used at a tension other than that at which it was calibrated. A surveyor's tape L meters long, with coefficient of elasticity k_e and cross-sectional area a and under a tension $(T + \Delta T)$ compared to the tension T at which it was calibrated, experiences a change in length given by $\Delta T L / (a k_e)$.

tension handle - A spring balance having a handle at one end and a spring-closed hook at the other, for attaching to the ring at the end of a surveyor's tape and applying appropriate tension to the tape.

term, absolute - A term, in an equation, which has a known, absolute, numerical value and does not contain any unknowns or variables.

term, generic - That part of a geographic name which describes the kind of feature to which the name is applied and which has the same meaning in the current local usage. For example, the generic term wan in Tokyo-wan means bay.

term, periodic - (1) A term, in the representation of a function by a series, which is a circular or elliptical function of the independent variable (argument) in particular, of time or of a function of time. The presence of periodic terms in a series does not guarantee that the series itself will show corresponding periodic motions, because some periodic terms may cancel others. (2) SEE function, periodic.

term, secular - A term, in the representation of a function by series, which is a constant times a power of the time. The presence of secular terms in a series does not mean that there is necessarily a secular change in the function, because the sum of all the secular terms may be periodic or zero.

terminal, marine - Thar part of a port or harbor which provides dockin, cargo-handling and storage.

terminator - The boundary between the illuminated and dark regions of the apparent disk of the Moon, a planet or a planetary satellite.

termini at headlands - Those points on shore (the low-water mark, in the international law of the sea) between which the closing line at indentations is drawn to mark the seaward limits of inland waters.

terrace - (1) Sloping ground re-formed into a level or gently sloping band separated by steep inclines from similar lower or higher bands on either side. The inclines are often made very steep and protected by riprap from erosion. Retaining walls may be substituted for the riprap, thus giving an even greater width for the band. Terraces are made to permit cultivation of land otherwise unusable because of its slope.
(2) A low ridge of earth constructed across sloping ground to control the rate at which rain runs over the surface, to retain some of the water and to reduce the rate of erosion. (3) A region, on cultivated land, bordered by low, broad ridges constructed of such height, alinement and spacing as to conform to the topography and permit agricultural machinery to travel along it. The purpose is to reduce erosion. (4) SEE terrace, river. (5) SEE terrace, continental.

terrace, continental - (1) A zone, around a continent, extending from the low-water line to the base of the continental slope. (2) The sediment and rock underlying the coastal plain, the continental shelf and the continental slope.

terrain - (1) A ground region considered as to its extent and topography. (2) A ground region considered as a physical feature, an ecological environment or a site of some planned, human activity.

terrain analysis - The process of interpreting data on a geographical region to determine the effect of natural and man-made features on military operations.

terrain base, stepped - The representation, made of acetate sheet in stepped form, of the contours appearing on the base map.

terrain correction - SEE gravity correction, terrain.

terrain emboss - A technique for portraying relief on a chart; a photograph is used to produce the effect of shaded relief from a relief map made by embossing.

terrain factor - Landforms, drainage features, ground, vegetation and cultural features or man-made changes in the surface of the Earth.

terrain gravity correction - SEE gravity correction, topographic.

terrain model - SEE map, relief.

terrain profile photography - SEE photography, terrain-profile.

Terrain Profile Recorder - SEE Airborne Profile Recorder.

terroid - SEE telluroid.

tesla - The name, under the Système International d'Unités, for weber per square meter.

test chart - A design for testing the performance of optical systems. The design usually consists of ruled lines, squares or disks of various sizes so arranged that the quality of a lens system can be determined by examining the image of the design.

test glass - An optical element having a curvature equal and opposite to that of the lens to be tested. When the two surfaces are placed in contact and viewed in monochromatic light, an interference pattern (Newton's rings) is formed. The rings are a contour map of the spacing between the two glasses; the contour interval is one-half the wavelength of the light being used.

test mass - A body of accurately known mass, used for testing an instrument sensitive to mass or used as a standard within such an instrument. Also called a proof mass.

test pile - (1) A pile driven to ascertain conditions affecting driving at a particular locality and the lengths probably required. (2) A pile on which a loading test also can be made to determine the extent to which it will settle under load and the carrying capacity of the soil. It is a guide in designing foundations based on piles.

test pit - An open excavation large enough to permit a person to enter and examine (geological) formations in their natural condition. Test pits are by far the most accurate method of determining the character of matgerials, but it is also the most costly. Samples in the disturbed or undisturbed condition can be taken readily.

test, side-equation - A side equation modified to make it simple and easy to use by survey parties in the field for readily checking observational accuracy and locating the points where there may be errors in the measured directions.

test, three-peg - A procedure for determining the level collimation error, by taking readings on three leveling rods, two placed about 200 meters apart and a third half way between those two. Also called three-peg adjustment.

test, two-peg - A method for determining the collimation error of a leveling instrument by taking readings on two leveling rods held on two intervisible turning points, the leveling instrument being placed considerably closer to one of the leveling rods than the other. Also called the C-test.

testator - A person who leaves a will or testament in force at his (her) death, disposing of his (her) property.

testimony - A declaration or affirmation made by a witness, under oath or affirmation. Testimony is a kind of evidence.

testimony, expert - Testimony of a person or persons skilled in some art, science, profession or business, which skill or knowledge is not common to their fellow men and which has come to such experts by reason of special study and experience in such art, science, profession or business.(Culver v. Prudetial Insurance Company, 6 W.W. Harr. 582, 179 A. 400).

testing function - SEE function, testing.

tetrabrach - SEE tetrapod.

tetrapod - A supporting structure having four legs. Also called a quadripod or tetrabrach.

thalweg - (1) The line following the lowest part of a valley, whether under water or not. The intricacy of detail in ordinary terrain often makes it difficult to locate a thalweg in practice; in a survey of a political boundary-line, this difficulty may be considerable. (2) The line down the center of the main channel of a stream. (3) The line following the deepest part of the bed of a river. (4) The line of greatest slope, cutting all contours at right angles. (5) The line joining the lowest points of a valley or a submarine valley. (6) The central or chief navigable channel of a waterway. (7) An underground stream of ground-water percolating beneath and in the general direction of a stream's course or valley on the surface. SEE ALSO rule of thalweg.

thalweg doctrine - SEE rule of thalweg.

thence (adv.) - In surveying, and in descriptions of land by courses and distances, this word, preceding each course given, imports that the following course is continuous with the one before it.

theodolite - (1) A precise surveying instrument consisting of an alidade with a telescope so mounted that it can be rotated about a vertical axis, the amount of rotation being measured on an accurately graduated, stationary horizontal circle. There are two major categories of theodolites: direction theodolites, often referred to as direction instruments and repeating theodolites. (2) A precise surveying instrument consisting of an alidade with a telescope so mounted that it can be rotated independently about a vertical axis and about a horizontal axis, the amount of rotation about either axis being measured on an accurately graduated, stationary, horizontal (or vertical) circle. The principal components of a theodolite are the telescope, standard, base, tribrach, foot-plate, horizontal circle, vertical circle, circular level and cylindrical level. (3) A transit of high precision. SEE ALSO cinetheodolite; construction theodolite; direction theodolite; gyrotheodolite; Parkhurst theodolite; phototheodolite; Tavistock geodetic theodolite; transit; transit theodolite.

theodolite, acoustic - An instrument using sound waves to provide a continuous, vertical profile of oceanic currents at a specific location.

theodolite, astronomical - SEE instrument, alt-azimuth.

theodolite, cine - SEE cine-theodolite.

theodolite, direction-instrument - A theodolite in which the graduated horizontal-circle remains fixed during a series of observations. SEE direction theodolite. SEE ALSO theodolite, repeating.

theodolite, double-center - SEE theodolite, repeating.

theodolite, gyro - SEE gyro-theodolite.

theodolite, gyroazimuth - SEE gyrotheodolite.

theodolite, gyroscopic - SEE gyro-theodolite.

theodolite, hanging - A theodolite designed so that it can be suspended from a supporting arm rather than supported on a tripod. The hanging theodolite is intended for work in restricted places such as mines.

theodolite, optical-reading - A theodolite in which the value of the desired angle is visible at the eyepiece of the telescope, rather than having to be read through a separate microscope looking directly at the circle.

theodolite, pilot-balloon - A theodolite designed specifically for measuring horizontal and vertical directions to a pilot balloon.

theodolite, repeating - (1) A theodolite so designed that the sum of successive measurements of an angle can be read directly on the graduated horizontal circle. The vertical axis of rotation is represented physically by two concentric spindles. One of these lets the alidade be rotated independently of the horizontal circle and the other, by a clamp, lets the alidade and the horizontal circle rotate together. Also called a double-center theodolite, reiteration theodolite, engineer's transit and repeating instrument. The value of the angle is obtained by dividing the total arc passed through (the final reading on the circle, plus an appropriate multiple of 360°) in making the series of measurements by the number of times the angle has been measured. In theory, the repeating theodolite is an instrument capable of giving very precise results; in practice, it does not give results as satisfactory as those obtained with a direction theodolite. (2) SEE engineer's transit.

theodolite, universal - (1) A theodolite which can be used for determining first-order astronomic coordinates as well as for determining first-order horizontal geodetic coordinates. The telescope has a micrometer attached for following a star's motion during transit over a given vertical circle or for measuring small differences in zenith distances, and with precisely graduated horizontal and vertical circles. (2) A theodolite which, because of its construction, can be used for engineering surveying as well as for geodetic surveying.

theodolite-magnetometer - A combination of theodolite and magnetometer modified to fit onto a common base. The astronomical meridian and the magnetic meridian can then be observed in a single observation.

theory, concentric-circle - The theory that cities tend to expand in concentric circles (or approximately similar shapes) from their point of origin if there are no barriers to such expansion.

theory, fermenting-dough - SEE Pratt-Hayford theory of isostasy.

thermistor - A resistor made of semi-conducting, metallic oxides such as NiO and Mn_2O_3, and exhibiting a very large change of resistance with temperature. The thermistor is therefore very widely used for measuring changes of temperature.

thermometer - Any instrument used for measuring or indicating temperature. The most common type of thermometer is that called the liquid-in-glass thermometer, which depends on the expansion and consequent rise of a liquid in a narrow tube on which a temperature scale is marked. Such a thermometer is capable, when properly designed, of measuring temperature with an error of 0.001° or less. Almost as common is that which depends on the expansion of a metal or combination of metals. This kind is usually not as accurate as the liquid-in-glass kind. Thermistors, thermocouples and similar devices which cause changes in current or voltage proportional to changes in the temperature being measured are widely used for making measurements of temperature far from the observer or recorder, particularly when the temperatures are to be recorded automatically. Thermometers for measuring extremely high temperatures usually measure the amount of radiation from the hot body; the temperature of the body is calculated from the intensity of the radiation. Such thermometers are called pyrometers or bolometers. The thermometers used most by surveyors are of the liquid-in-glass type having mercury as the fluid. However, thermistors are supplanting such thermometers form many kinds of work. SEE ALSO maximum thermometer; minimum thermometer.

thermometer, bimetallic - A thermometer using the difference in the coefficients of expansion of two different strips of metal to determine the ambient temperature. The measuring rods used for measuring the base lines of the Dunkirk-Barcelona triangulation (1792-1808) were also bimetallic thermometers. Each rod was composed of a strip of copper lying on a strip of platinum. These strips were fastened together at one end, leaving their other end free to move in response to changes in temperature. A scale on the free end of the strip of copper was read by means of a vernier on the free end of the strip of platinum, thus affording a means of determining the temperature of the bars. This apparatus was designed by Borda. SEE ALSO Borda scale.

thermometer, Borda-scale - SEE Borda scale.

thermometer, dry-bulb - A liquid-in-glass thermometer which is uncovered and open directly to the atmosphere. Contrasted to the wet-bulb thermometer.

thermometer, liquid-in-glass - A thermometer consisting of a thick, glass tube closed at both ends and marked along its length with a temperature scale; the interior is enlarged at one end to form a bulb which contains a liquid such as alcohol or mercury filling the bulb and extending a short distance further when the temperature is at its lowest. When the temperature rises, the liquid in the bulb expands and forces a thread up along the interior to a distance proportional to the rise in temperature.

thermometer, mercurial - A liquid-in-glass thermometer consisting of a graduated glass tube containing mercury and closed at both ends. Also called a mercury thermometer. A large reservoir of mercury is contained in a bulbular cavity at one end of the tube, and the bore of the tube is very narrow. The mercury expands, with increasing temperature, much more rapidly than does the glass; the mercury therefore rises and falls almost linearly with temperature as measured by the temperature scale on the glass.

thermometer, mercury - SEE thermometer, mercurial.

thermometer, mercury-in-glass - A liquid-in-glass thermometer having mercury as the liquid.

thermometer, metallic - A thermometer using the difference in the coefficients of thermal expansion of the metals of which it is made to determine the temperature of these metals and the temperature of the ambient fluid. A metallic thermometer uses two or more different metals; a bimetallic thermometer uses only two different metals and is the most common form of the metallic thermometer.

thermometer, protected - A thermometer protected against the pressure of water at great depths by being inclosed in a heavy glass tube which is partially evacuated except for the portion containing the reservoir; that part is filled with mercury to conduct heat from the surroundings to the reservoir.

thermometer, reversing - A mercury-in-glass thermometer which, when turned upside down, registers the temperature of the thermometer at that time. Reversing thermometers are usually sent into the water in pairs. Each pair is fastened to a special mounting. One of the thermometers is left exposed to the full pressure of the surrounding water; the other is protected from the pressure by a shield of metal. At a suitable moment, a weight is dropped from the ship to slide down the cable to which the thermometers are attached. When the weight reaches a pair of thermometers, it sets off a mechanism which inverts the thermometers. The protected thermometer indicates only the temperature of the water; the temperature indicated by the other thermometer results from the combined effects of temperature and pressure on the thermometer. The pressure of the water can be calculated from a formula involving the readings of both thermometers of the pair.

thermosphere - That region lying above the minimal temperature at 80 km, where absorption of ultraviolet radiation (of shorter wavelength than that absorbed by ozone) causes the temperature to increase with altitude up to 1 500° C or 2 000 ° C at 300 to 400 km.

theta angle - (1) In the Lambert conformal conic map-projection, the angle of convergence, on the developed surface, between the central meridian and the meridian through the point. (2) The central angle of the Eulerian spiral (clothoid), measured at the point of intersection of the tangents passing thru the tangent-spiral and spiral-tangent points of the spiral. Also known as the Δs or θ-angle.

Th function - The ratio of the length of a chord of a circle to the length of the corresponding arc.

Thompson prism - A composite prism formed by cementing together two similar pentagonal prisms having faces of different widths in such a manner that light entering perpendicular to a face of one component leaves in the same direction from the opposite face of the other component after having undergone five internal reflections. Also called a Malmros prism, Schmidt prism, Pechan prism and Z-prism. It is used, for instance, in prism binoculars.

thread of the river - The thread of the stream when the stream is a river.

thread of the river, middle - The line equidistant between low-water lines on the two sides of a river, extending from headland to headland without considering arms, inlets, creeks and affluents as parts of the river.

thread of the stream - The center of the main channel of a stream. If there are two prominent channels, the thread of the stream is the center of the channel used for navigation. SEE ALSO thalweg.

three-point method of radial aerotriangulation - A method of determining the horizontal coordinates of the ground points corresponding to the principal points of overlapping aerial photographs, by resecting on three horizontal-control points appearing in the overlap.

three-point problem - The problem of finding the location of a point by measuring the angles between three other points of known location. The procedure used in solving the problem is called resection.

three-wire method - SEE leveling, three-wire.

throw - (1) (mining) The vertical distance between those parts of a vein which have been separated by a fault, measured at right angles to the strike of the fault. SEE ALSO heave (mining). A horizontal fault can not have a throw, and a vertical fault can not have a heave. (2) The perpindicular distance from the offset point of curve of a circular curve to the back tangent of a spiral curve.

throwing the chain - The process of changing the figure-eight shape in which a surveyor's tape is commonly gathered up into a circular coil. Also called throwing the tape. After making a measurement, the surveyor commonly gathers the tape up into a figure-eight held in one hand at the place where the loops cross. The figure-eight can be changed into a circular coil by a dextrous movement of the hand and wrist called throwing the chain. The ability to do this was considered to distinguish the experienced chainman from an inexperienced one.

throwing the tape - SEE throwing the chain.

tick - (1) A very short, usually thin, line drawn to mark a subdivision of a scale, to guide the placement of another, similarly marked object, etc. SEE, IN PARTICULAR, grid tick; register mark.

tick, marginal - SEE grid tick.

tide - (1) A periodic change in one celestial body caused by another celestial body or bodies. (2) A periodic change in the Earth or other planet, moon or celestial body, related to the locations of the Sun, Moon and other members of the Solar System. E.g., thermal tides are changes caused by heating and cooling of a body such as the Earth's atmosphere; gravitational tides are changes in figure and are caused by the attraction of other bodies. (3) A periodic change in the size and shape of the Earth or other planet, moon or celestial body caused by movement through the gravitational field of another body or bodies. Also called astronomic tide and astronomical tide. This should not be confused with gravitational tide. The word tide is most commonly used to refer to changes in size and shape of the Earth in response to the gravitational attractions of the other members of the Solar System, the Moon and Sun in particular. In this instance, three different tides are usually distinguished: the atmospheric tide - acting on the atmosphere; the earth tide - acting on the lithosphere; and the oceanic tide (usually called, simply, the tide) - acting on the hydrosphere. SEE ALSO tide, astronomical; tide, atmospheric; tide, radiational; tide, thermal. (4) The height or elevation of the land or a body of water at some instant and place as compared to the height or elevation averaged over a specified time interval (longer than one day and preferably over a period of at least 19 years). Also called water when referring to a body of water. (5) The surface of the land or a body of water, at some instant and place, as compared to the position of that surface averaged over a specified interval of time (longer than one day and preferably over a period of at least 19 years). Also called water when referring to a body of water. (6) SEE tide, oceanic. (7) SEE tide, gravitational. (8) SEE tide, thermal. (9) SEE water (1). SEE ALSO age of the tide; amplitude of the tide; analysis of the tide, harmonic; cut tide; constituent (of the tide); constituent, diurnal; earth tide; ebb tide; equilibrium tide; equilibrium theory of the tide; flood tide; height of the tide; neap tide; line of the tide, nodal; range (4); periodicity of the tide, pole tide; range of the tide; range of the tide, great diurnal; range of the tide, great tropic; range of the tide, small diurnal; range of the tide, small tropic; rise of the tide, mean; spring tide; spring tide, perigean; stand of tide; storm tide; type of tide.

tide, anomalistic - That component of a tide which can be ascribed to changes in the Moon's distance.

tide, apogean - A tide occurring near the time when the Moon is farthest from the Earth.

tide, astronomic - (1) The periodic change in magnitude and direction of gravity, as caused by the attraction of the Moon, Sun and other members of the Solar System. (2) SEE tide, gravitational. (3) SEE tide. (4) SEE tide, equilibrium.

tide, astronomical - (1) The height of the tidal waters at a given place, as predicted from astronomical considerations only. Also called astronomic tide level. (2) SEE tide. (3) SEE tide, equilibrium.

tide, atmospheric - (1) Those periodic changes in the Earth's atmosphere which are caused by the gravitational attraction of the Moon, Sun and other planets. Atmospheric tides are always so called; the word tide alone refers to oceanic tides specifically or to any one of the three basic kinds of tides: atmospheric, earth and oceanic. The most easily observed changes are changes in barometric pressure. (2) The periodic rising and falling of the Earth's atmosphere, regardless of the cause. This usage is uncommon.
(3) Periodic changes in the atmosphere of any celestial body, caused by the effect of another body or bodies on it.

tide, compound - A tidal constituent whose phase changes at a rate equal to the sum or difference of the rates at which the phases of two or more elementary constituents change.

tide, declinational - That component of the tide which can be ascribed to a change in declination.

tide, double - (1) A high water consisting of two maxima of nearly the same height, separated by a relatively small depression. (2) A low water consisting of two minima of nearly the same depth, separated by a relatively small elevation. Sometimes called an agger.

tide, ebbing - SEE ebb tide.

tide, equatorial - A tide occurring semi-monthly as a result of the Moon's being over the equator.

tide, equinoctial - A tide occurring at about the time of the equinoxes.

tide, falling - SEE tide, ebb.

tide, gravitational - (1) A periodic motion of a level surface in response to the gravitational attraction of the Moon, Sun and other members of the Solar System. Also called an equilibrium tide. In particular, such a periodic motion of the figure of the earth. The gravitational tide is different from the earth tide and the oceanic tide; those tides are movements of the lithosphere and hydrosphere. It is almost the same as the astronomic tide. Spirit leveling gives the elevation of the Earth's surface with respect to the geoid and hence is affected by both earth tide and gravitational tide. (2) SEE tide, equilibrium. (3) SEE tide, astronomical.

tide, high - SEE water, high.

tide, highest (lowest) astronomical - The highest (lowest) height of the oceanic tide which can be predicted to occur under average meteorological conditions and under any combination of astronomical conditions.

tide, internal - A tidal wave moving along a sharp discontinuity in the water, such as at a thermocline or in a region of gradual change in density vertically.

tide, leeward - A tidal current moving in the direction opposite to that from which the wind is blowing. Also called a lee tide.

tide, lithospheric - A periodic motion of the lithosphere, caused by the movement of the Earth in the gravitational and other field of other members of the Solar System. SEE ALSO earth tide.

tide, low - SEE water, low.

tide, lowest astronomical - SEE tide, highest (lowest) astronomical.

tide, lowest normal - SEE tides, lowest normal.

tide, lunar - That portion of the tide which can be attributed directly to the Moon's attraction. The term can, if used in a general sense, include both direct and indirect effects of the Moon's attraction.

tide, mean high - SEE water, mean high.

tide, mean low - SEE water, mean low.

tide, meteorological - A tidal constituent having its origin in the local daily or seasonal variations in weather conditions. These conditions may occur with some degree of periodicity. The principal tidal constituents having a meteorological origin are Sa, Ssa and S_1.

tide, normal - SEE tide (2).

tide, oceanic - A periodic change in size and shape of the hydrosphere caused by the Earth's movement in the gravitational fields of the other members of the Solar System, particularly the Moon and Sun. For historic reasons, and because the methods used for measuring are quite different, oceanographers refer only to the vertical component of the tide as tide; the horizontal component is referred to as a tidal current.

tide, ordinary - SEE tide, mean. In legal language, ordinary is usually taken to mean average. Thus, ordinary high water line may be assumed to be the same as mean high water line.

tide, over - SEE overtide.

tide, partial - SEE constituent, tidal.

tide, perigean - The tide which occurs when the Moon is closest to the Earth during a lunar month.

tide, radiational - Periodic variations in water level related primarily to meteorological changes such as the semi-daily cycle in barometric pressure, daily breezes from the land and sea and seasonal changes in temperature. The term may be synonymous with meteorological tide.

tide, rip - SEE current, rip.

tide, rising - SEE flood tide.

tide, semi-diurnal - A tide having a period of approximately one-half tidal day. The predominating type of oceanic tide throughout the world is semi-diurnal, with two high waters and two low waters each tidal day. A semi-diurnal constituent has two maxima and two minima each constituent day, and its symbol is distinguished by the subscript 2.

tide, solar - (1) That part of the oceanic tide which is caused by the Sun's attraction. (2) The oceanic tide observed in regions where the Sun's attraction is dominant.

tide, synodic - That component of the tide which can be ascribed to changes in the phases of the Moon.

tide, thermal - A change in the size and shape of the Earth resulting from the heating of the Earth by the Sun and the subsequent re-radiating of this heat into space. At the present rate of rotation of the Earth, the lithosphere and hydrosphere are not affected much below a few tens of meters and so do not show appreciable thermal tides. The atmosphere, however, is heated throughout its depth and displays marked thermal tides. Thermal tides are difficult to distinguish from tides caused by gravitation.

tide, tropic - An oceanic tide occurring semi-monthly when the Moon's declination is at its greatest or least and the tidal effect is then greatest. At such times, there is a tendency for the diurnal range to increase. The tidal datums pertaining to the tropic tides are designated as tropic higher high water, tropic lower high water, tropic higher low water and tropic lower low water.

tide, vanishing - A semi-diurnal tide whose presence is hidden by the diurnal tide. In a mixed tide with very large diurnal inequality, the lower high water (or higher low water) frequently becomes indistinct or vanishes at times of extreme declination. During these times, the diurnal tide has such an over-riding dominance that the semi-diurnal tide, although still present, cannot readily be seen on the curve of the tide.

tide, windward - A tidal current moving in the direction from which the wind is blowing.

tide amplitude - SEE amplitude of the tide.

tide correction - Any correction made necessary by a tidal effect. In particular, the amount added to an angular elevation determined by sextant, because the sea's surface is tilted by the tide.

tide curve - A graphic representation of the rise and fall of the tide. Time is usually represented as the abscissa and height or elevation as the ordinate. For normal tides, the curve is approximately sinusoidal. SEE ALSO marigram.

tide cycle - SEE cycle, tidal.

tide gage - An instrument for measuring the height of the tide. The simplest form is a graduated staff placed vertically in the water in a sheltered location. The height of the water is read on the staff or may be recorded on paper through linkages. Still another type, called a gas-purged pressure gage or bubbler tide-gage, determines the height by relating it to the pressure needed to force a constant stream of bubbles out of the end of a duct open near the bottom of the well. SEE ALSO bubbler tide gage; tide staff; tide station; stilling well; float well.

tide gage, automatic - An instrument which automatically records the tide curve, usually on gridded paper.

tide gage, electronic - SEE tide gage, sonic.

tide gage, mechanical - A tide gage which indicates the height of the water by means of a float attached to a graduated staff and which possibly records the height by means of a mechanical linkage between the float and the recorder. A common type has a float with pointer attached to the staff in such a way that the float is free to move up and down with the tide. Another common type passes a cable from the float over a pulley which turns a worm screw on which a stylus is mounted. As the stylus moves back and forth with the rise and fall of the float, it marks its location on a strip of chart advanced at a uniform rate by a clock-controlled motor.

tide gage, sonic - A tide gage consisting of a sound generator and receiver placed together at the top of a stilling well. The generator sends pulses of sound downward to the surface of the water in the well. The receiver measures the amount of time elapsed between transmission and reception of each pulse and converts this interval to the distance of the surface below the top of the stilling well. Also called an electronic tide gage.

tide gage, standard automatic - A tide gage consisting of a of a float connected by a wire to a worm-screw that moves a pen on a chronograph. The tide gage is used where recordings are wanted of tidal movment over a long period of time. It is, of course, only one of a large variety of automatic tide gages.

tide-gage zero - SEE zero, tide-gage.

tide indicator - That part of a tide gage which indicates the height of the tide at any instant. The indicator may be in the immediate vicinity of the tidal water or at some distance.

tideland - (1) Any coastal region situated above mean low tide and below mean high tide; in particular, a region which is ordinarily covered and uncovered by the ebb and flow of the ordinary, daily tides. (2) Land which is under water at high tide and uncovered at low tide. Beach, seashore, strand and tideland have nearly the same meanings. Beach refers to the land sometimes covered by tidal water. Seashore is a loose term referring to the general region close to the sea. Strand refers to the strip of sand or pebbles along the edge of the sea.

tide level - (1) The height of the tide halfway between the heights of mean high water and mean low water. Also called ordinary tide level. (2) The height of the tide above a specified level. SEE ALSO height (3). (3) The tide when it is at a specified level.

tide level, mean - (1) The height of the tide half way between mean high water and mean low water. Also called ordinary tide level and half-tide level. (2) The reference surface half way between surfaces of mean high water and mean low water. Because the tide curve is asymmetric, mean tide level is not exactly the same as mean sea level.

tide level, ordinary - SEE tide level, mean.

tidemark - (1) A high-water mark left by tidal water. (2) The highest point reach by a high tide. (3) A mark placed to indicate the highest point reached by a high tide or, occasionally, by any specified state of tide.

tidemark, rule of the - SEE rule of the tidemark.

tide plane - SEE plane, tidal; datum, tidal.

tide plane, harmonic - SEE plane, Indian tidal.

tide plane, Indian - SEE Indian tidal plane.

tide pole - SEE tide staff.

tide prediction - The predicted time and height of high water or low water at a reference station.

tide range - SEE range (of the tide).

tide records - SEE records, tidal.

tides - SEE sequence of tides; types of tide.

tides, lowest normal - A surface of reference lower than mean sea level by half the greatest range, not taking into account the effects of wind or barometric pressure.

tide staff - A graduated staff of sufficient length to extend above the highest tide when held vertically, and on which the height of the water is measured. Also called a tide pole. Tide staffs are either fixed in place and not readily removable, or they are portable and are fastened in place in such a way that, while held securely, they can be removed easily for installation elsewhere. The tide staff is quite accurate in protected waters where the waves are small, but on the open coast, waves and swells make accurate measurement difficult.

tide staff, multiple - A succession of tide staffs on a sloping shore, so placed that the vertical graduations on the several staffs will form a continuous scale when referred to the same datum.

tide station - (1) The geographic location at which tidal measurements are made. (2) The equipment used to make tidal measurements and the housing for such equipment. A tide station may include a house, a tide gage and tidal bench-marks. According to the importance of the measurements and the period over which they have been made continuously, a tide station is classified as a primary control tide-station, subordinate tide-station, secondary control tide-station or tertiary tide-station.

tide station, primary - A tide station at which measurements have been made continuously for at least 19 years.

tide station, secondary - A tide station at which measurements have been made continuously for more than 1 year but for less than 19 years.

tide station, subordinate - (1) A tide station at which a relatively short series of measurements have been made and the measurements have been reduced by comparing them with records from a nearby tide-station with a long series of measurements. (2) A tide station listed in the Tide Tables and for which predictions are to be obtained by means of differences and ratios applied to the full predictions of a nearby tide-station.

tide station, tertiary - A tide station at which measurements have been made for at least 30 days but for less than 1 year.

tide table - An annual list of daily predictions of the times and heights of high water and low water at various places. Tide tables are assembled from astronomical data and from the results of harmonic analysis of the tides at the desired places. They are compiled and issued by national organizations such as the British Admiralty. The heights given in the tables are usually measured from chart datum rather than from mean sea level.

tie - A survey made to determine, from a point of known location, the location of another point. The location of a supplementary point may be needed for mapping or as a referent, or to close a survey on a previously determined point.

tie (v.) - Used in the past tense, connected; e.g., monuments are tied together by measurements. The corner of a property is tied to offset monuments or to other corners of property.

tie in (v.) - Make a survey to determine the location of a point with respect to the location of a point of known location.

tie point - (1) That point to which a survey is made from a point of known location, to determine the location of the former. (2) One of a set of image-points identified on photographs in the overlap between two or more adjacent strips of photographs. Tie points serve to connect the individual strips of photographs into a single unit and to connect points on photographs from adjacent flights into a common network. (3) One of a set of points used (a) to establish common scale between two adjacent photogrammetric models from two adjacent strips of photographs and (b) to link the two models together.

tier - (1) Any series of contiguous townships situated east and west of each other. (2) (USPLS) Any series of sections situated east and west of each other in a township.

tie strip - (1) An overlay showing all planimetric and topographic features in the regions along the edge of a map or chart. Also called a match strip. The tie strip is used to ensure the matching of these features on adjoining sheets. (2) SEE control strip.

till - The rocks and flour composed of finely-ground rock dragged along in the lower part of the ice of a glacier and left behind when the ice melts. Also called boulder clay.

tilt - (1) The dihedral angle between the plane of a photograph and a horizontal plane. (2) The angle, at the center of perspective, between the photograph perpendicular and the vertical (or other external direction used as reference). Also calle tilt angle and tip. The line of intersection of the two planes is called the axis of tilt. The direction of tilt is expressed by the swing, when referred to the axes of a coordinate system on the photograph, or by azimuth, when referred to an external system of coordinates. In aerial photography, tilt is commonly resolved into two components: x-tilt and y-tilt, with x-tilt being the component more nearly in the direction of the line of flight. A positive x-tilt results from the left wing of the aircraft being lowered, displacing the nadir point in the positive direction of y. Similarly, a positive y-tilt results from the nose of the aircraft being lowered, displacing the nadir point in the positive direction of x.
(3) SEE x-tilt. (4) The failure of the optical axis of a lens to be parallel to the optical axis of a preceding or following lens in a lens system. The term is applied particularly to the objective lens-system of a camera. SEE ALSO direction of tilt.

tilt, cross - An error introduced into stereotriangulation when the exact locations of camera stations for successive pairs of photographs cannot be determined. Cross tilt is usually caused by variations in equipment, materials or by imperfect relative orientation.

tilt, lateral - SEE roll.

tilt, longitudinal - SEE pitch.

tilt, relative - (1) The dihedral angle between the plane of a photograph and some other plane (not necessarily horizontal) such as that of the adjacent photograph in a strip. (2) The angle between the photograph perpendicular and a reference direction such as the photograph perpendicular of the preceding or following photograph.

tilt, x - SEE x-tilt.

tilt, y - SEE y-tilt.

tilt circle - A circle, in a tilted aerial photograph, passing through the isocenter and having a diameter lying along the principal line. When this diameter is drawn to a convenient linear scale, any chord through the isocenter give the component of tilt for that particular direction.

tilt determination - SEE Church's method; displacement method; Morse method; scale-point method.

tilt displacement - Displacement of images, on a tilted photograph, radially outward or inward with respect to the isocenter according as the images are, respectively on the low side or the high side of the isometric parallel (the low side is the one closer to the object plane).

tiltmeter - (1) An instrument which measures and records small deviations of a reference line in the instrument from the horizontal. (2) An instrument for measuring changes in differences of elevation between points on or in the Earth. The term has been applied by some writers to the comparison of successive levelings. This usage is pretentious.

timber rights - Ownership of standing timber without ownership of the land.

time - (1) The instant when an event occurs. Time is one of the four coordinates which are necessary and sufficient to completely identify the location of a particle or event. It is that one of the four which always increases during a change, regardless of whether or how the other three coordinates change. In Newtonian mechanics, there is a clear distinction between time and the other three coordinates. Time is called the temporal coordinate of an event and is usually denoted by t, T or τ. The other three coordinates are called spatial coordinates and are usually denoted by x, y, z or a similar set of three letters. In relativistic mechanics, time and the other three coordinates are not as clearly distinguished. For this reason, relativistic mechanics often uses x_i i = 1 to 4, for the set of coordinates, making the equation symmetrical in all four coordinates. The x_4 coordinate is then defined as j_t, where j is $\sqrt{-1}$. t or x_4 is referred to as the time-like coordinate and the other three coordinates as space-like coordinates. Time is not an essential concept in physics. For example, the Ludovici system of units does not contain time as a fundamental unit. The time associated with a particular event is determined by the location of the origin (called the epoch); it is usually called the date of that event. The origin commonly used by astronomers is that for which the Julian Date is 0. The origin used by geologists is the present instant, time before the present being referred to as time before the present or time B.P. The fact that the dates of past events keep changing is immaterial because the errors in dates of past events are very much greater than the constant shift in the origin. The origin commonly used for civil purposes and many scientific purposes is defined by assigning the date 15 October 1582 A.D. to the day on which the Gregorian calendar replaced the Julian calendar then in use. This day has the date 5 October 1582 A.D. in the Julian calendar. This definition puts the origin at 0h of 1 January of the year nominally taken as that of the birth of Christ. There is no year zero in the civil calendar. The year immediately preceding the origin is the year 1 B.C.; that following the origin is the year 1 A.D. In astronomical practice, the year immediately preceding this origin is designated year 0 (zero); the other years B.C. are designated by negative numbers. The position of the origin within a day depends on how the day is defined. The civil day usually begins at midnight. Before 1925, astronomers reckoned mean solar time from noon, with the mean solar day beginning at noon 12 hours after the midnight at the beginning of the same civil date. In 1925, the national ephemerides began reckoning the day from midnight, so that 31.5 December 1924 was designated 1.0 January 1925 in the new ephemerides. SEE ALSO calendar; date; time scale. (2) The time-like interval or distance between the instants at which two events E_1 and E_2 occur; it is usually taken as the simple difference $(t_2 - t_1)$ between the times denoting the instants. Also called preferably, a time interval. When used in this sense, it is usually in a phrase such as the time taken for, if the interval between the beginning and end (i.e., the duration) of one event is in question, or the time between if the time-like interval between two events is in question. A time interval in Newtonian physics is independent of the spatial coordinates of the two events. In relativistic mechanics, it is related to the distances between space-like coordinates by the equation $ds^2 = dx^2 + dy^2 + dz^2 + (jdt)^2$, or by the complete homogeneous quadratic equation in all four coordinates. The unit of time, in any system, is always a time interval. (3) The hour of the day, reckoned by the location of a specific point, such as the center of the Sun, on the celestial sphere relative to a celestial meridian used as reference. Also called the time of day. The time may be solar, lunar or sidereal time, according as the specified point is the center of the Sun, the center of the Moon, or the vernal equinox. SEE ALSO chronometer time; clock time; ephemeris time; equation of time; equation of ephemeris time; equation of Universal time; Greenwich time; Greenwich apparent sidereal Time; Greenwich Civil time; Greenwich lunar time; Greenwich mean time; Greenwich mean sidereal time; Greenwich sidereal time; Newcomb sidereal time; War Time; watch time.

time, apparent - Time determined directly from measurements i.e., based on the true place of the Sun. Apparent time is usually adjusted to remove effects of refraction and aberration from the measurements. Two kinds of apparent time are in common use: apparent sidereal time and apparent solar time.

time, apparent sidereal - Time measured by the hour angle of the intersection of the true equator of date with the ecliptic of date. Also called true sidereal time.

time, apparent solar - Time measured by the observable (apparent) diurnal motion of the Sun. Also called apparent time and true solar time. At any given instant at a point on the Earth, the apparent solar time is the hour angle of the Sun. In civil usage, apparent solar time is measured along the equator from the two branches (the two arcs separated by the North and South Poles) of the meridian through 12 hours. The hours from the upper branch (that adjacent to the point) are marked P.M. (post meridian); those from the lower branch are marked A.M. (ante meridian). Before 1925, astronomers measured apparent solar time from the upper branch of the meridian through 24 hours; since 1925, they have measured from the lower branch, and the civil day has taken the place of the astronomical day. To completely identify the time, the meridian used as reference must be named. For example, 75th-meridian apparent solar time; Greenwich apparent solar time; local apparent solar time (at the meridian of the observer). Apparent solar time is determined by observations on the Sun; e.g., it is the time shown by a sundial. SEE ALSO time; time, apparent; equation of time.

time, astrographic mean - A form of mean time used in setting an astrograph; a mean time set at 1200 occurs when the local hour angle of the First Point of Aries is 0°.

time, astronomic - (1) Time measured by the motion of celestial bodies. Also called astronomical time. (2) SEE time, astronomical (1).

time, astronomical - (1) That part of a solar day which begins at noon (0h) and continues for 24 hours, ending at noon of the next day. Also called astronomic time. By convention, the second noon belongs to the next day. Astronomical time may be either apparent solar time or mean solar time, depending on whether the apparent Sun is used or the mean sun is used. Before 1925, national ephemerides and almanacs tabulated events as functions of astronomical time; after that date, Universal time, which is the same as Greenwich mean time, was used as the argument. (2) SEE time, astronomic.

time, atomic - Time determined from a specified epoch and in terms of the atomic second. Many nations determine their own atomic times, making the measurements with their own atomic clocks. These times are, however, subordinate to International Atomic Time which, according to a definition approved by the 14th Federal Conference on Weights and Measures in 1971, is the atomic time determined by the Bureau International de l'Heure.

time, civil - Solar time in a civil day beginning at midnight. Civil time may be either apparent solar time or mean solar time. It may be counted in two series of 12 hours each, beginning at midnight and marked A.M., and at noon and marked P.M.; or it may be counted in a single series of 24 hours beginning at midnight.

Time, Coordinated Universal - SEE Coordinated Universal Time.

time, daylight-saving - A time indicated by a clock so set that the time it indicates is advanced by 1 hour from the usual standard time. In the USA, daylight saving time has usually been adopted by local (e.g., State) ordinance through the summer to make greater use of daylight hours (assuming that the round of daily activities will take place according to the indicated times and will not according to the indicated time minus 1 hour). Where daylight saving time is adopted throughout a particular time zone, the time in that time zone is given the same designation as the zone: e.g., Central Daylight-saving Time. During World War II, daylight-saving time was put into effect nation-wide and was called War Time. In 1966, daylight-saving time was established for the nation as a whole by an act of Congress providing that during the period beginning at 2 A.M. on the last Sunday of April of each year and ending at 2 A.M. on the last Sunday of October of each year, 1 hour be added to the standard time of each time zone. (A State may exempt itself by law from observing daylight-saving time.)

time, dynamical - SEE ephemeris time.

Time, International Atomic - SEE International Atomic Time.

time, local - (1) Time in which noon is defined by the transit of the Sun over the local meridian. This is in distinction from a time based upon the meridian of Greenwich or the meridian of a time zone. Local time may be either mean or apparent, according to whether the mean Sun or the actual Sun is used. Local time was in general use in the USA until 1883, when standard time was adopted. (2) Any time kept locally.

time, local apparent - The apparent solar time for the meridian of the observer.

time, local astronomical - Mean time calculated from the upper branch of the local meridian.

time, local civil - SEE time, local mean.

time, local lunar - The arc of the celestial equator, or the angle at the celestial pole, between the lower branch of the local celestial meridian and the hour circle of the Moon, measured westward from the lower branch of the local celestial meridian through 24 hours. Local lunar time is equivalent to the local hour angle of the Moon, expressed in time units, plus 12 hours. SEE ALSO time, Greenwich lunar.

time, local mean - The mean solar time at the meridian of the observer. Also called local civil time.

time, local sidereal - The sidereal time at the meridian of the observer.

time, lunar - (1) Time based upon the rotation of the Earth relative to the Moon. (2) Time based on the rotation of the Moon with respect to the Sun.

time, mean - (1) An epoch, in the system of Universal Time, referred to any meridian other than that of Greenwich. (2) SEE time, mean solar.

time, mean sidereal - Time measured from the intersection of the mean equator of date with the ecliptic of date. Mean sidereal time is affected only by those secular changes caused by precession.

time, mean solar - Time measured by the diurnal motion of a fictitious body, called the mean Sun, which is supposed to move at a constant angular speed in the celestial equator, completing the circuit in one tropical year. Often called mean time. The mean Sun may be considered as moving in the celestial equator and having a right ascension equal to the average celestial longitude of the true Sun. At any given instant, mean solar time is the hour angle of the mean Sun. In civil affairs in the USA, mean solar time is counted from the two branches of the meridian through 12 hours; the hours from the lower branch are marked A.M. (ante meridian); those from the upper branch are marked P.M. (post meridian). Before 1925, astronomers counted mean solar time from the upper branch of the meridian through 24 hours; from then on, the count has been from the lower branch. Naming the meridian of reference is essential to completely identify the time (e.g., Greenwich mean solar time). By using the same meridian as reference over a belt or zone of the Earth, watches and clocks can be adjusted to show the same mean solar time throughout the region. SEE ALSO time, standard. The Greenwich meridian is the reference for the world-wide standard of mean solar time called Universal time. Because the mean Sun is a fictitious body, mean solar time cannot be determined directly by observation.

time, observed - Time as determined by measurement, without any change or correction for personal error, aberrational or refractive effects.

time, real - (1) Time represented as a real quantity rather than as an imaginary quantity. This is usually the case in relativistic mechanics; the other three, space-like coordinates are then imaginary quantities. However, many scientists treat the space-like coordinates as real and the time as imaginary. (2) A time very close to the instant of observation or other activity. (3) The situation in which events are reported or recorded at the same time as they are happening. (4) The absence of delay in getting, sending, and receiving data.

time, rise - SEE rise time.

time, sidereal - The hour angle of the vernal equinox - the measure of time defined by the apparent diurnal motion of the vernal equinox; hence, a measure of the rotation of the Earth with respect to a coordinate system defined by reference to the stars rather than the Sun. The vernal equinox itself is not observable. Sidereal time is therefore actually defined by the right ascensions of the stars used in determining the time. Note that although the sidereal day and sidereal year are determined by timing the transits of fixed stars, the coordinates of these stars are then referred to the true equator and equinox at that instant. Sidereal time is then determined by conversion to the mean equator and mean equinox. If the apparent vernal equinox is used, we have apparent sidereal time; if the mean equinox is used, we have mean sidereal time. If the hour angle is referred to the meridian of Greenwich rather than to the local meridian, we have Greenwich mean sidereal time.

time, solar - Time determined by reference to the hour angle of the Sun. Solar time involves both the rotation and revolution of the Earth. Depending on whether time is determined by the hour angle of the apparent Sun or by the hour angle of a fictitious Sun moving at a steady rate, we speak of apparent solar time or mean solar time, respectively. Ephemeris time, which is measured in terms of a specific fraction of a particular year, is therefore also a kind of solar time.

time, standard - Mean solar time for a particular meridian and specified to be used throughout a zone (time zone) including that meridian. In the continental USA, the meridians to which standard times are referred are located at multiples of 15° (1 hour) from the initial meridian (Greenwich meridian). The standard time for each zone is identified by the number of its meridian and also by some name of geographic significance: 75th-meridian Standard Time or Eastern Standard Time; 90th-meridian Standard Time or Central Standard Time; 105th-meridian Standard Time or Mountain Standard Time; and 120th-meridian Standard Time or Pacific Standard Time. The meridians for standard time in Alaska are at 150° and 165° west longitude; in the Hawaiian Islands, at 150° west longitude; in the Philippine Islands, at 120° east longitude. Standard time was established in 1883 to correlate schedules of trains of various railroads serving the same regions. At noon on 18 November 1883, the telegraphic time-signals from the U.S. Naval Observatory in Washington were changed to this system. The time zones were planned to be roughly symmetrical about the meridians of reference and to extend 7½° to either side thereof. Practical considerations such as the need to correlate time in cities outside the original boundaries of a zone with time in cities within the zone, have caused a gradual shifting of these boundaries until some of them now exhibit large irregularities. SEE ALSO time, daylight-saving.

time, stepped atomic - The time kept by an atomic clock whose rate is correct but whose epoch is reset periodically to keep the time within 0.1s of Universal Time 2.

time, true sidereal - SEE time, apparent sidereal.

time, true solar - SEE time, apparent solar.

Time, Universal - SEE Universal Time.

time belt - SEE time zone.

time constant - The time required for an instrument to indicate a given percentage of the final reading resulting from a signal arriving at the instrument. SEE ALSO rise time.

Time Coordinated, Universal - SEE Coordinated Universal Time.

time determination - The determination of time by any method, but particularly by astronomical methods e.g., by measurements using the Moon, Sun or other celestial bodies such as stars. Determining time astronomically is a process related to that of determining longitude and employs much the same kind of equipment and instruments. If the observer knows his longitude, he can determine the time (referred to some meridian other than his own) by essentially the same kind of observational method he would use to determine longitude if the time, referred to some meridian other than his own, were known. SEE ALSO Bessel's formula; DÖllen's method; equal-altitude method (of determining time); Mayer's formula; Tsinger's method (of determining time)

time diagram - A diagram in which the celestial equator appears as a circle, and the celestial meridians and hour circles appear as radial lines. Also called a diagram on the plane of the equinoctial. A time diagram is used to make it easier to solve problems involving time, arcs of the celestial equator or angles at the poles by showing graphically how the quantities involved are related.

time dial - A disk, graduated in units of time (0 hours to 24 hours), attached to a globe with the center of the disk coincident with a pole on the globe, and adjustable to show the relationship between local time and the time at any other meridian.

time interval - The difference between two times. Also called in relativistic mechanics, a time-like interval. SEE time (2).

time interval, sidereal - The difference between two times measured in units of sidereal time. The basic time-interval in sidereal time is the mean sidereal day, defined as the time interval between two successive upper transits of the mean vernal equinox over some meridian and corrected for polar motion and short periodic irregularities in the rate of rotation of the Earth. SEE ALSO day, sidereal.

time interval, solar - The difference between two times measured in units of mean solar time. The basic time-interval in mean solar time is the mean solar day, defined as the interval between two consecutive transits of the fictitious Sun over a meridian and corrected for motion of the pole. SEE ALSO day, mean solar; year, mean solar.

time meridian - Any meridian used as a referent for reckoning time. In particular, a standard meridian or other meridian at which the time for an entire zone is reckoned.

Time 1, Universal - SEE Universal Time 1.

time scale - The system of units into which that axis (of a coordinate system) showing the progress of time is divided. The location of the origin is immaterial. Time scales are usually classified according to the method used for measuring time as: astronomic time, in which the scale is determined by the quasi-periodic rotation or revolution of celestial bodies; clock time, which is determined by the indications of more or less periodic movements of a mechanism; and atomic time, in which the scale is set by the frequency (assumed constant) of molecular or atomic phenomena (as displayed, for example, by an atomic clock). The terms time scale and time are frequently used as synonyms, and there is in fact little real difference in the meanings.

time-scale, astronomic - A time scale in which the units are those of astronomic time. There are three different types of astronomic time-scale in general use. They involve, respectively, the day, using one rotation of the Earth with respect to a celestial body as the unit; the month, using one revolution of the Moon about the Earth; and the year, using one revolution of the Earth about the Sun.

time scale, atomic - A time scale in which the unit is the atomic second.

time series - A sequence, discrete or continuous, of data assigned to specific times. A function in which the argument is time and the dependent variable an observed or measured quantity. There is little if any difference between a time series and a stochastic process, and the two terms may be considered synonyms.

time series, multiple - A time series consisting of one or more given observations or measurements at each instant (time).

time series, simple - A time series consisting of a single, given observation or measurement at each instant (time).

time service - An organization which is responsible for establishing standards of time and for providing accurate times to users.

time signal - A signal associated with a definite, accurately known time and readily observable by many users. Some of the former ways of indicating time were firing a cannon; rapidly raising a flag; and dropping a large, shiny ball from the top of a mast. These method were replaced, for practical purposes, in the mid-nineteenth century, by electrical pulses sent along telegraph wires. This method was in turn replace, in the early 20th century, by pulsed radio waves originating at radio stations associated with astronomical observatories.

time transfer - The synchronization of clocks.

Time 2, Universal - SEE Universal Time 2.

time unit, canonical - The time needed by a hypothetical satellite of the Earth to move one radian in a circular orbit having the same radius as the Earth's equatorial radius. It is approximately 13.447 052 minutes, and is $1/2\pi$ of the period of a Schuler pendulum.

Time 0, Universal - SEE Universal Time 0.

time zone - A geographic region in which the time used in civil affairs is everywhere the same, as prescribed by law, and differs by an integral number of hours from Greenwich mean time. Also called a time belt.

timing correction - An amount added to a distance measured by radar to account for the amount the pulse is delayed as it passes through the transponder.

tint - SEE tint, hypsomertric.

tint, hypsometric - One of a succession of shades or graduated colors used to depict ranges of elevation. Also called a gradient tint, elevation limits, altitude tint and layer tint.

tinting, hypsometric - The technique of showing relief on maps by coloring, in different shades and colors, those parts which lie within specified ranges of elevation. Also called altitude tinting.

tint scale, hypsometric - A graphic scale, in the margin of a map, indicating elevations or depths by means of graduated shades of colors. SEE ALSO tinting, hypsometric.

tip - SEE tilt (2).

Tissot's indicatrix - That ellipse, on a plane, into which an very small infinitesimally small) circle on a curved surface is transformed when the curved surface is mapped onto the plane. Also called the ellipse of distortion. The size and shape of the indicatrix depend on the nature of the curved surface and on the kind of mapping used.

title - (1) The union of all the elements which constitute ownership. Title is divided, in common law, into possession, right of possession and right of property. The last two of these should, however, be considered the same. (2) That which constitutes a just cause of exclusive possession; the facts or events which, collectively, give rise to the ownership of real or personal property. SEE ALSO cloud on title; color of title; evidencing of title; guarantee title; tax title.

title, allodial - Absolute ownership, not subject to any feudal duties or burdens.

title, clear - (1) Good title. (2) Marketable title. (3) Title free from encumbrance, obstruction, burden or limitation.

title, clouded - A title on which there is a cloud. SEE cloud on title.

title, defective - (1) An unmarketable title. (2) A clouded title.

title, insured - A title, the validity of which is insured by an abstract, title or indemnty compay. Also called a guarantee of title and guarantee title.

title, paramount - A title giving its owner a right of possession over all those claiming to have title.

title, quiet - SEE quiet title.

title, quitclaim - (1) A title to property that extends no further than the title released by the grantor. (2) A claim one may have in property without professing that the title is valid.

title block - A space, on a nonstandard picture such as a mosaic, photograph, of plan, reserved for title, reference and information about scale.

title certificate - SEE abstract of title; certificate of title; title insurance; title policy; Torrens title system.

title examiner - A person who analyzes a chain of title to land and passes on the validity of various (legal) instruments and then renders his opinion.

title insurance - Insurance against financial loss resulting from claims arising out of defects in the title to real property, which were existant but undisclosed at the time the policy was issued by the title cvompany. SEE ALSO abstract of title; certificate of title; title policy; Torrens title system.

title opinion - An analysis and interpretation of a title search concerning present ownership, encumbrances, clouds on title and other infirmities.

title plat, master - SEE master title plat.

title policy - A policy insuring the title to real property, issued for the protection of peresons acquiring interests in real property either as owner, lender or lessee. It insures against forgery, incompetents, insanities and other matters that are not shown by public records. It insures the actual title to property as distinguished from the record title, such as is guaranteed in a guarantee of title. It usually does not insure location. SEE ALSO abstract of title; ceertificate of title; title insurance; Torrens title system.

title search - A search of public records and documents to ascertain the history and present status of title to property, including ownership, liens, encumbrances, charges and other interests.

t-minus-T correction - SEE correction, arc-to-chord; grid convergence.

toise - A French unit of length based on a French legal standard of length of the same name and equal to about 6.4 English feet. The actual length varies according to the standard used in the definition. It was used in early geodetic surveys and was the basis, in many countries, for the standard of length. Excluding the Rhineland Rod used by Snell, the toise was the first geodetic unit of length. SEE ALSO toise of Peru.

Toise of Peru - The length of that iron bar which was used as a standard in measuring the base lines controlling lengths in the Peruvian Arc of triangulation surveyed in 1736-1743 to determine the figure and size of the Earth. The Toise of Peru became the legal standard of length in France in 1776. From it, the French legal meter was derived as follows. The toise was divided into 6 pieds (feet); each pied was divided into 12 pouces (inches); each pouce was divided into 12 lignes (lines); and one French legal meter was defined as equal to exactly 443.296 lignes at a temperature of 13° Reumer. (The Peruvian Arc does not lie in modern Peru but in Ecuador, which was included in the old Spanish presidency of Peru.) In the first decades of the 1800's, direct copies of the Toise of Peru were made and used by various states as base-line apparatus for their triangulation. Carelessness in handling the bar embodying the Toise of Peru resulted in loss of this bar as a primary standard and in use of the set of primary and secondary copies of the bar as reference for the true value of the Toise.

Tokyo Datum - A horizontal-control datum defined by the following coordinates of the origin and by the azimuth of station Kano-Yama, on the Bessel ellipsoid of 1841; the origin is at the center of the transit circle at the old Tokyo Observatory:

longitude of origin 139° 44' 40.50" E
latitude of origin 35° 39' 17.51" N

azimuth from the origin to Kano-Yama 156° 25' 28.443"

tolerance - (1) The permissible deviation from an assigned, observed or nominal value. In surveying, tolerances are used primarily to state the permissible deviation of (a) a measurements for a specified value or the average value of a set, or (b) the deviation of a dimension of a part of an instrument from a specified value. A tolerance is not the same as a standard deviation or other measure of variation and should not be used as such. I.e., it is a number adopted a priori and not one determined a posteriori. A variable which has a tolerance imposed upon it does not have a Gaussian distribution. (2) The allowable variation from standard or from specified conditions.

tolerance, standard - A number, or set of numbers, established within an industry, science or technology, setting limits on the precision or accuracy with which operations, measurements, or products are to be made. Equivalently, a tolerance accepted throughout an industry, science or technology. Also referred to as a standard if it is clear that standard tolerance is meant. However, the possibility of misunderstanding are so great and the consequences possibly so serious that use of the abbreviated form should be discouraged.

tolerance factor - A number by which the square root of the length of a section of leveling is multiplied to find the tolerance for the differences of elevation between the ends of that section.

tomography - The technique of reconstructing a function of distance and angle from the projections of that function onto certain planes. The term originated in X-ray radiography, where the technique was used to determine the three-dimensional shapes of various organs in the patient by taking radiographs at many different angles. The technique has now been extended to seismology and other sciences.

tomography, triangular - A tomography-like method of determining vorticities in the oceans by using long-range acoustical soundings.

ton, metric - SEE tonne.

tone (of an image) - Each distinguishable variation of shade from black to white.

tone, continuous - The gradual change of shading, in an image, from black to white, rather than abrupt changes between different shades of black or white. The source of this definition uses the term to denote the image itself; this appears to be an error.

tonne - A unit of mass equal to 1000 kilograms. Also called a metric ton, megagram and millies.

tooling, optical - The use of optical instruments such as telescopic sights, optical micrometers, laser beams, etc., in setting up tools for precise manufacture - usually of other tools. Also called optical alinement, although that term is somewhat more limited in scope.

topangulator - An instrument for measuring the vertical angles in the principal plane of an oblique photograph.

topocentric (adj.) - Centered at or referred to the location of the observer.

Topographic Map of the United States - The topographic map of the USA prepared as an atlas by the U.S. Geological Survey with assistance from other organization.

topography - (1) The configuration of the contours of any surface: i.e., the detailed shape of a surface. The term is used in geology for the natural contours of the Earth's surface; in meteorology, for the contours of an isobaric surface; in metallurgy, for the detailed variations of a surface from a flat surface; in physical oceanography, for the shape of the oceanic bottom or the shape of a surface of given characteristics within the water; etc. The following definitions are particular applications of the general definition to particular sciences or technologies. (2) The features of the actual (natural) surface of the Earth, considered collectively as to form. (3) A detailed, accurate description of such features. (4) The shape of the features of the Earth's surface, including both natural and artificial features. Also called relief. Topography in this sense includes: hypsography, the shape of the land; hydrography, the shape of the water surfaces and drainage features; culture, the shape of man-made features; and vegetation. SEE ALSO relief. (5) Those masses above sea level of continents and islands, and the deficiency of mass in spaces occupied by the waters of oceans, seas and gulfs. SEE ALSO geopotential topography; geopotential topography, absolute.

topography, dynamic - (1) The topography of a surface (usually an isobaric surface) specified in terms of the dynamic height of that surface above a specified surface of reference. (2) Isopleths depicting, on a map, topography corresponding to the intersection of the Earth's surface with a sequence of level surfaces. (3) The configuration formed by the dynamic height (measured in dynamic meters) between a given surface of constant pressure and a reference surface (for example, in the ocean, the 2000- decibar surface).

topography, sea-surface - The elevation of the surface of the ocean (above the geoid).

topography crest - SEE crest, topographic.

toponym - A name applied to a natural or man-made topographic feature. Also called a place name.

toponymy - That branch of cartography dealing with the names of places, natural features or man-made features.

topple - (1) The vertical component of precession or polar motion. (2) The algebraic sum of the vertical components of precession and polar motion. The term is used almost exclusively in the technology of gyroscopes.

topple axis - That horizontal axis, perpendicular to the spin axis of a gyroscope, around which there is a vertical component of precession.

torr - A unit of pressure exactly equal to 1/760th of an atmosphere. One torr is equal to 133.322 pascals, approximately.

Torrens Acts - Statutes adopting the Torrens System of compensatory registration of titles to land.

Torrens Registration System - A method of recording the ownership of land, in which the title to the land is registered instead of, as in other methods, registering the evidence of such title. The State then insures the owner against loss because of defective title. Also called the Torrens Title System. Upon the land-owner's application, the court may, after appropriate proceedings, direct the issuance of a certificate of title. With some exceptions, this certificate is conclusive as to applicant's estate in land. After registration, all deeds and documents affecting the property are duly registered. The system is named after its inventor, Sir Robert Torrens, who introduced it in South Australia in 1857. SEE ALSO abstract of title.

Torrens Title System - SEE Torrens Registration System.

Torrid Zone - The region, on the Earth, between 23° 27' north and 23° 27' south.

torsion balance - An instrument for measuring very small differences of attraction or repulsion caused by gravitational, electrical or magnetic forces, etc.. It consists basically of a bar suspended horizontally by an elastic filament. One end of the bar is subjected, because of the variation of the force from place to place in the field, to a greater force than the other end. The bar therefore rotates until the difference in forces is balanced by the torque on the filament. The difference in forces is measured by the angle through which the bar is rotated. Also called a gradiometer or variometer, depending on the purpose for which the torsion balance is being used. In geophysics, a special form of torsion balance is used to determine the rate at which gravity changes from point to point. In geophysical prospecting over a large region, a large mass of abnormal density may be detected using the torsion balance. For example, the oil fields of the Gulf Coast are associated with salt domes - solid, saline, plug-like masses of such size and density that they are easily discovered and located by geophysical methods. Among the instruments used in making such discoveries, the torsion balance has been prominent. SEE ALSO gravity gradiometer.

touch plate - SEE kiss plate.

tower - In surveying, a structure used to raise the line of sight above intervening obstacles. SEE ALSO Bilby tower; Bilby steel tower; Lambert twin tower; observing tower; signal tower; triangulation tower.

township (USPLS) - (1) The unit of land into which the public lands of the USA are officially divided by surveys. Normally, it is a quadrangle approximately 6 miles on a side with boundaries conforming to meridians and parallels of latitude, and located with respect to the initial point of a principal meridian and baseline. Townships are numbered consecutively north and south from a baseline; thus, township 14 north designates a township in the 14th tier north of a baseline. The word range is used together with the appropriate number and direction to indicate the coordinates of a particular township with respect to the initial point; thus township 14 north, range 3 east indicates that township which is the 14th township north of the baseline and the 3rd township east of the principal meridian controlling the surveys in that region. The plural form townships (abbreviated tps.) is used whenever more than one unit is to be indicated; thus, townships l4 north, ranges 3, 4 and 5 east and townships 14 and 15 north, range 3 east. A township contains 36 sections, some of which are designed to take up the convergence of the eastern and western boundary lines or range lines of the township. (2) SEE township, civil.

township, civil - A unit of local government; a subdivision of a county or a region. Usually referred to, simply, as a township if the meaning is clear. In the public-land States, boundaries of civil townships may coincide in whole or in part with lines of the public lands.

township, fractional (USPLS) - A township containing appreciably fewer than 36 sections of normal size, usually because the township was invaded by a segregated body of water or by other land not properly surveyed as part of that township. Townships may be rendered fractional in closing the public-land surveys on State boundaries or other limiting lines.

township, unorganized - Tracts of land, principally in Maine, New Hampshire and Vermont, laid out by State authority for administratite purposes but having no local govenrent.

township corner - A corner of a township.

township corner, closing - (1) The point of intersection of a guide meridian or a range line with a previously-fixed, standard parallel, or a base line. (2) The point of intersection of any township line or range line with a previously-fixed boundary between previously established corners.

township line - An external boundary, of a township, extending in an east-west direction. Township lines were intended by law to lie on parallels of latitude.

townsite - A region in one or more townships and layed out into streets, alleys and blocks of lots.

trace - The sum of the elements along the principal diagonal of a matrix. Also called the spur.

traceability - The extent to which the accuracy of a given instrument can be traced back to a primary standard.

tracing table - A device consisting of a small stand supporting a horizontal disk with matte surface. The disk can be moved up and down so that an illuminated reference mark on the disk can be placed at any point in a stereoscopic image projected onto the table an which the stand moves about. The stand also carries a pencil or similar tool for drawing the map. The stereoscopic image is created by a pair of projectors mounted above the surface on which the map is to be drawn, and a small segment of that image is intercepted by the matte surface of the tracing table. The operator moves the device about the plotting table (reference table) in such a way that the mark on the tracing table follows the desired outline or contour in the image. The device then traces out the corresponding line on paper placed on the plotting table. SEE ALSO light table.

track - (1) The actual path of an aircraft over the surface of the Earth. The astronomical azimuth is usually used for the direction of the path. (2) The actual path of an aircraft above, or of a vessel on, the surface of the Earth. The course is the path which is planned; the track is the path actually taken.

track, lead - An extended railorad track connecting one end of a railroad yard with the main track.

track adjustment - Adjustment made to the plotted track of a ship, to take account of the ship's longitudinal and lateral movement under the influence of current, wind, etc.

track chart - A chart showing recommended, required or established tracks, and usually indicating turning points, courses and distances.

tracker, optical - SEE tracking system, optical.

tracking - (1) A motion given to the main lobe of a radar or radio system so that that lobe is kept pointed in a desired, changing direction usually the direction of a moving object. (2) The process of following the motion of a moving object. (3) The process of following the frequency, phase and/or amplitude of a changing signal. SEE ALSO tracking filter.

tracking, optical - (1) The technique of using an optical system to photograph a moving object, measure distance to it, or other wise obtaining information about the object by optical means, by mounting the optical system so that it can be kept pointed at the object. The optical system may be coupled to an electronic circuit in such a way that the pointing is done automatically. (2) The technique of keeping a line of sight directed at a moving object by a device depending on an optical system coupled to an electrical circuit. Note that the first and second definitions denote completely different techniques.

tracking camera - SEE camera, tracking.

tracking filter - An electrical circuit which eliminates most of the noise contaminating a received signal by generating, internally, a signal adjustable in frequency or phase and forcing the generated signal to follow gradual changes in the received signal; the generated signal is then used for information instead of the received signal. Also called a tracking system. This process eliminates most of the noise in the received signal but also eliminates high-frequency components of the signal itself.

tracking station - Any complete set of equipment set up for observing a moving object such as an aircraft, satellite, meteor or ship, by following the motion of the object. In particular, such a set of equipment which can also measure one or more components of the object's location and/or velocity in a local coordinate system. The tracking station may use photographic, photoelectric or radio equipment or may do the tracking visually. SEE ALSO tracking system.

tracking system - (1) An optical system, radio system, or radar which is kept pointed, usually automatically at a moving object in such a manner that the power received by the system from the object or received by the object from the system is great as possible. A tracking systems may used as part of a communications system or it may be a complete system used for taking photographs of or determining distance and/or direction to an object. There are some radars and radio systems which send out signals in a fixed direction only, or which rotate at a fixed rate while sending out signals. Such systems are not tracking systems. On the other hand, there are radar and radio systems which contain no moving parts but which direct the main lobe electronically in the desired direction. Such systems are tracking systems. The term is also used, erroneously, for any equipment which gathers information about an artificial satellite, whether or not that equipment actually tracks the satellite. SEE ALSO station, tracking. (2) SEE tracking filter.

tracking system, active - (1) A tracking system which transmits signals to a moving object and receives data about the object in return. A radar which tracks an aircraft is an example of an active tracking system. (2) A tracking system in which a transmitter or transponder placed aboard the object to be tracked transmits or receives and retransmits information to the tracking equipment e.g., DOVAP, SECOR, MINITRACK.

tracking system, optical - A tracking system depending on light or infrared radiation to determine distance and/or direction. Also called an optical tracker. The term covers automatically-tracking theodolites and cameras as well as systems making use of beams of light or infrared radiation.

tracking system, passive - A tracking system which receives data about or from a moving object without itself sending signals to the object. Cameras such as the Baker-Nunn camera are passive tracking systems. Cameras such as the BC-4 camera used for photographing artificial satellites is not a tracking system. Distance-measuring equipment of the laser type, as mounted for measuring distances to artificial satellites, are active tracking systems.

track level - SEE track sector.

track sector - An instrument (clinometer) consisting of a half-circle protractor with a single movable arm carrying a spirit level, and so constructed that when the base of the protractor is parallel to a rail and the bubble of the spirit level is in the center of its tube, the movable arm will indicate the angle (slope) the rail makes with the horizontal. Also called a track level. The track sector was designed and used for determining the slope of a surveyor's tape when the tape was used in traverses along a railway. It proved to be unsatisfactory.

tract - (1) In general, a metes-and-bounds survey of a region within a township. (USA) (2) (USPLS) A parcel of land lying in more than one section, or a parcel which cannot be identified completely as part of a particular section.

tract book - A narrataive, journal-like recod that is an index to and digest of all essential actions affecting public lands.

traffic island - A small region in the center of a road and intended to provide temporary refuge to pedestrians from vehicles passing on the rest of the road. Also called traffic island. Although islands are sometimes separated from the traffic merely by strips painted on the road, they are more often raised and paved surfaces separated from the traffic by curbs.

traffic surveying - Surveying to determine the type, number, speed, relative positions and origin and destinations of vehicles.

trajectory - (1) Those points in space and time occupied by a body (or by a specified point in that body) moving from one specified place to another specified place. The body may be acted upon by internal as well as by external forces. The terms orbit, path and trajectory have similar but distinct meanings and should not be used as synonyms. Path denotes the set of points in space only and not in time, and no conditions need be put on the place of starting or ending of the motion. Orbit denotes those points in space and time occupied by a (point in a) body moving under natural forces only, usually under conservative forces. It is often implied that the path is a closed figure. A trajectory usually results from the application of man-made forces. The geometric history of a body launched into orbit about the Earth therefore consists of an initial trajectory, from the instant when the object was launched to the time when rocket-like motion ceased and free-fall began, and of an orbit from that instant until the object again hit the ground because its energy was dissipated by drag. (Some workers consider the last stage - that of rapid descent within the atmosphere, to be a trajectory). The entire geometric history, from launching to hitting the surface, constitutes the path. (2) The path followed by a mass point or massive body in a gravitational field. This definition is not in accord with general usage. E.g., the set of points and times occupied by an electron moving in a magnetic or electrostatic field is usually called the trajectory of the electron.

trajectory, cislunar - A trajectory lying wholly whithin a sphere including the Moon's orbit.

trajectory, translunar - A trajectory lying partially or wholly outside as sphere including the Moon's orbit.

TRANET - A set of 13 (in 1977) radio receiving stations established world-wide to receive signals from the TRANSIT satellites.

Transcontinental Triangulation Datum - The horizontal-control datum which is defined by the following coordinates of the origin and by the azimuth (clockwise from South) of triangulation station La Crosse, on the Clarke spheroid of 1866; the origin is at triangulation station Hays:

longitude of origin 99° 16' 16.73" W.
latitude of origin 38° 54' 50.18" N.
azimuth from origin to La Crosse 359° 44' 19.00"

The datum was adopted and used for scientific studies. It served no other purpose and no formal designation was given it.

transcriber - SEE device, point-transfer.

transducer - (1) Any device converting one kind of physical phenomenon (energy, power, force, motion, etc.) into another: e.g., converting heat into motion or pressure into displacement. Almost any instrument is a transducer. In electrical engineering, the output of a transducer is assumed to be voltage or current. In hydrography, the term is applied in particular to a device which converts mechanical vibrations as from a diaphragm to sound in water or for converting sound in water to voltages in a receiver. (2) A device accepting power from one source and transferring that power to another device. (3) An instrument for indicating the magnitude of one physical quantity (such as temperature or hydraulic pressure) by measuring a second physical quantity (such as pressure or voltage) related to the first.

transfer, water-level - The extension of leveling across a lake or between large bodies of water by assuming that the average position of the surface of the lake or of each body, averaged over a suitable length of time, is a level surface. Water-level transfer was used in 1875 to extend leveling from bench mark GRISTMILL on the Hudson River in New York to the water-level gage at Escanaba, on Lake Michigan. The level was averaged over 3 to 4 months.

transfer function - The ratio between the Laplace or Fourier transform of the output function of a system to the Laplace or transform of the input function, subject to specified boundary conditions. A transfer function is the analog, in the frequency domain, of the quantity per unit input in the time domain. It is used when an input function can be expressed, as is usually possible, as a Fourier series, so that the system's performance can be specified as a sum of independent, periodic terms. SEE ALSO modulation transfer function; phase transfer function.

transfer function, optical - The ratio of the complex Fourier transform of the distribution of illumination over an image to the complex Fourier transform of the distribution of luminance over the corresponding parts of the object. Commonly denoted by OTF. This is the same as the complex Fourier transform the the line-spread function. The absolute value of the OTF is called the modulation transfer function; the phase of the OTF is called the phase transfer function.

transfer instrument - A device used to transfer detail from a photograph or photographs to a manuscript. Transfer instruments differ from plotting instruments in that the latter are commonly used for transferring the complete picture. There are two principal types of transfer instrument: the single-point transfer instrument and the stereoscopic transfer instrument.

transfer instrument, single-point - A device which transfers planimetric detail from a single photograph to the manuscript. A reflecting projector or a camera lucida is conveniently used as a single-point transfer-instrument, although a pantograph is often used instead.

transfer instrument, stereoscopic - A transfer instrument used to transfer detail from a stereoscopic pair of photographs to the manuscript.

transfer of water level - SEE transfer, water-level.

transform (n.) - An integral $F(k) = \int G(x,k) f(x) dx$, in which x and k are vectors. Also called an integral transform or integral transformation, although the latter term is more properly used for the process of determining the transform. The function $G(x,k)$ is called the kernel.

transform (v.) - Project an image of a photograph onto a plane making a given angle with the plane of the photograph.

transform, integral - SEE transform.

transformation - (1) An equation relating a set of l variables $\{x_i\}$ to a set of variable $\{y_j\}$ in such a manner that to a particular set of values of the x_i there corresponds one and only one set of values of y_j. Also called a mapping. (2) An equation of the form $y = A x$, in which x and y are vectors and A is a matrix. he matrix A may be square or rectangular. Its elements may be constants or variables. The elements of x and y may be finite quantities or infinitesimals. (3) A one-to-one function between members Γ and τ of a group, such that to each element τ corresponds an element $\Gamma = \alpha^{-1} \tau \alpha$, where α is a member of the group. (4) SEE rectification. (5) SEE function. (6) A function relating coordinates in one coordinate system to coordinates in another coordinate system.

transformation, affine - (1) In general, any transformation from a set $\{x_i\}$ of coordinates to a set $\{y_j\}$ of coordinates, which changes finite values into finite values. In particular, a linear equation of the form $y_j = \Sigma a_{ji} x_i + b_j$, in which the $\{a_{ji}\}$ and $\{b_j\}$ are sets of constants. Affine transformations used in geodesy are usually two- dimensional or three-dimensional. The transformation may be affine only within a small region about a particular point; e.g., in the situation represented by $dy_j = \Sigma (\delta y_j / \delta x_i) dx_i$. (2) A transformation such that, in the geometric representation, straight lines remain straight and parallel lines remain parallel. Angles and scale may change. (3) A function which transforms a linear space to itself and is the the sum of a linear transformation and a fixed vector.

transformation, bilinear - SEE transformation, linear fractional.

transformation, linear fractional - The function $y = (ax + b) / (cx + d)$, in which a, b, c, d are constants and x is the independent variable. Also called a bilinear transformation and Möbius transformation.

transformation, orthogonal - The transformation of one vector into another by multiplication by orthogonal matrices. In particular, the conversion of one set of geodetic coordinates into another by multiplication by orthogonal matrices.

transformation, projective - An equation of the form $y_j = (\Sigma a_{ji} x_i) / (\Sigma b_{ji} x_i)$ relating a set of J variables $\{y_j\}$ to a set of I variables $\{x_i\}$; the $\{a_{ji}\}$ and $\{b_{ji}\}$ are constants.

transformer - (1) A photogrammetric rectifier in which the angle or angles of tilt are fixed, as for the rectification of the eight oblique photographs taken by a nine-lens camera into the plane of the central photograph. (2) SEE printer, transforming.

transform fault - A strike-slip fault along which the displacement suddenly stops or changes form. Many transform faults are associated with mid-oceanic ridges, where the actual slip is opposite in direction from the apparent displacement across the fault.

transit - (1) The apparent passage of a star, other celestial body, or point on the celestial sphere across a specified line or a specified plane. The line may be a meridian, primer vertical or almucanter on the celestial sphere, or it may be in the reticle of a telescope, or it may be some marked line of sight, or it may mark the limb or edge of the apparent disk of a larger celestial body. When no particular line is specified, a passage across the meridian is usually meant. The transit of a star across the meridian occurs at the moment of its culmination, and the two terms transit and culmination are sometimes used as if they were synonyms; such usage is not correct, even when the instrument used for observing is in perfect adjustment. At the poles, a star may have no culmination, but it will transit the meridians. The transit of a small celestial body across a larger one is exemplified by the transit of the planet Mercury across the disk of the Sun, or of a satellite of Jupiter across the disk of that planet. Both types of transit have been used geodetically. (2) An astronomical or surveying instrument consisting primarily of a telescope mounted so that it can be rotated about a horizontal axis to describe an arc of about 180° from horizon to horizon. The term transit is sometimes used as a synonym for theodolite, or to mean a surveying instrument similar to a theodolite but of insufficient accuracy to permit it to be used for establishing geodetic control. The first usage is definitely undesirable, because some theodolites do not permit the telescope to be rotated through 180° vertically. The second usage is undesirable for the same reason and because some instruments referred to as theodolites are not sufficiently accuracy for geodetic use. SEE ALSO transit, astronomical; transit, surveyor's. (3) SEE transit, astronomical. (4) SEE TRANET; TRANSIT. SEE ALSO Bamberg transit; Brunton pocket transit; engineer's transit; ephemeris transit; jig transit; micrometer transit; mining transit; surveyor's transit; vernier transit.

TRANSIT - A general name for a system of satellites and radio receivers devised by the Applied Physics Laboratory (Johns Hopkins University) for determining the location of vehicles and vessels carrying the receivers by measuring the Doppler shift in continuous-wave signals received from the satellites. The term Navy Navigation Satellite System applies to the satellites of TRANSIT together with those fixed radio stations which are used to determine the orbits of the satellites.

transit, astronomical - A telescope used for observing transits and mounted so that its motion is limited to a vertical circle. Also called a transit instrument. Sometimes called a transit. An astronomical transit is usually designed to have its horizontal axis fixed (except for small adjustments to it) and perpendicular to the plane of the local meridian. Portable or quasi-portable transits, such as the broken-telescope transit, can be rotated about a vertical axis also, but this capability is for establishing the direction of the horizontal axis and is not for measuring horizontal angles.

transit, bent - SEE transit, broken-telescope.

transit, broken-telescope - An astronomical transit having its telescope so constructed that a ray of light entering the objective lens is reflected at right angles by a prism within the telescope and directed along the horizontal axis of the telescope to the eyepiece. Also called a bent transit. Sometimes erroneously referred to as a Bamberg transit, Bamberg instrument or Bamberg broken-telescope transit. The Bamberg transit is a particular make of broken-telescope transit, albeit a very common one. The eyepiece of a broken-telescope transit lies outside the supports of the telescope. The design permits the instrument to be used to determine latitude by the zenith-telescope method. The instrument is heavy but without undesirable flexure. This and the unusually long distance between the wyes supporting the horizontal axis make for stability. The striding level used with instruments having straight telescopes is replaced by a hanging level. The position of the eyepiece is independent of the angular elevation of the object observed; this makes observing easier. The instrument has been used in determining, with first-order accuracy, astronomic longitude, astronomic latitude and astronomic azimuth at stations in high latitudes. SEE ALSO micrometer method (for determining azimuth.

transit, inferior - SEE transit, lower.

transit, lower - Transit of the lower branch of the celestial meridian. Also called inferior transit.

transit, optical - SEE transit, optically-read.

transit, optically-read - A theodolite having graduated horizontal and vertical circles of glass which are read through a microscope connected optically to the circles by prisms and mirrors. Also called an optical transit.

transit, plain - An engineer's transit without vertical circle or level on the telescope.

transit, solar - (1) A transit theodolite designed for use in place of a solar compass. (2) A transit to which has been added an attachment which solves the astronomical triangle and allows an astronomic meridian or astronomic parallel of latitude to be established directly by observation.

transit, superior - SEE transit, upper.

transit, upper - Transit of a celestial body thru the upper branch of the celestial meridian.

transit and stadia survey - SEE survey, transit and stadia.

transit bearing - A bearing taken of two objects when in line with them.

transit circle - SEE meridian circle.

transit instrument - SEE transit (2).

transition curve - (1) A curve joining the adjacent end point of two other lines (usually circular arcs) in such a way that the tangents to the lines and joining curve coincide at the points of junction and the curvature of the joining curve changes in some suitable manner. Also called easement curve or spiral curve (this latter term should properly be used only for curves that are actual spirals.) Transition curves commonly used are the cubic parabola, the clothoid (Cornu spiral) and the spiral of Archimedes. SEE ALSO curve, vertical. (2) SEE curve, spiral.

transition spiral - A spiral curve used as a transition curve.

transit level - A surveying instrument combining the major features of a transit and a leveling instrument. Transit-levels are usually not as precise or accurate as theodolites and leveling instruments separately but are more convenient for some kinds of surveying.

transit line - Any line, of a traverse, which is projected (i.e., extended), either with or without measurement, by the use of a transit or similar device. Also called a traverse line. It is not necessarily an actual line of a final survey, but may be an accessory line.

transit method - SEE traverse adjustment, transit method of.

transit micrometer - An impersonal micrometer having the moving filament placed in the focal plane of an astronomic transit and at right angles to the direction of motion of a star observed at or near culmination. Certain contact-points on the head of the screw complete an electrical circuit as the pass a fixed contact. The circuit moves a pen or other marker so that a record is made, on a sheet of paper wrapped around the drum of a chronograph, at each separate instant when the moving line reaches a position corresponding to a contact.

transit rule (for adjusting a traverse) - The correction to be applied to the departure (or latitude) of any course has the same ratio to the total misclosure in departure (or latitude) as the departure (latitude) of the course has to the arithmetical sum of all the departures (latitudes) in the traverse. The transit rule is often used when it is believed that the misclosure is caused less by errors in the measured angles than by errors in the measured distances. It meets the assumptions on which it is based only when the courses are parallel to the axes of the coordinate system used. Care must be taken in applying the rule because there is not universal agreement on what sign should be used for the misclosure. SEE ALSO misclosure.

transit surveying - Surveying done using a transit.

transit theodolite - A theodolite with a telescope which can be rotated through 180°, from horizon to horizon, about a horizontal axis.

transit traverse - A survey traverse in which the angles are measured with an engineer's transit or a theodolite and the distances are measured with a surveyor's tape. A transit traverse is usually made to establish control for local surveys and is of second-order or third-order quality.

transit time - (1) The time interval needed for a signal to pass through a circuit or mechanical device or through a component thereof. Transit time through a transponder must be added to the time needed to travel from transmitter totransponder and from transponder to receiver. (2) The time interval between the instant the leading edge (limb) of a celestial object passes the meridian to the instant the trailing edge passes the meridian.

translation - Movement in a straight line, without rotation.

transliteration - The process of recording the graphic symbols of one language in terms of the corresponding symbols of another language. (2) A set of symbols, of a language, which has undergone this process.

translocation - The process or method of determining the location of one point relative to another by observing continuous-wave signals from a satellite simultaneously from both points. The term is used, specifically, for the method of determining relative locations by counting cycles of continuous-wave radiation from TRANSIT satellites. The method is most useful if the observations are made when the satellite is at a height of the same order of magnitude as the distance between points.

transmission - SEE transmittance.

transmittance - The ratio of the intensity of light transmitted through a unit area of a translucent substance to the intensity of the light incident on that area. Also called transmission. The angle of incidence should be specified if it is not zero.

transparency - (1) A photographic print on a clear base, particularly adaptable for viewing by transmitted light.
(2) The degree to which a material is able to transmit light.

transparency, factored - SEE plate, land-mass-simulator.

transponder - A device which, upon receiving a signal, transmits a second signal. The second signal may be merely an amplified version of the first, in which case the transponder acts as a rely, or it may be different in frequency and/or content. The transponder is used, in geodesy, principally in distance-measuring instruments particularly in those constructed for measuring distances to satellites. The transponder is placed in the satellite and retransmits the received signal, usually at a different frequency, back to the ground.

traverse (n.) - (1) A route and the sequence of points on it at which observations or measurements are made. (2) A route, the sequence of points on it at which measurements are made, and the measurements themselves. (3) The process by which a route and a sequence of points of measurement on it are established. (4) SEE survey traverse. (5) SEE polygon, directed. SEE ALSO azimuth traverse; classification of a traverse; control traverse; specifications (for a traverse); stadia traverse; survey traverse; transit traverse.

traverse (v.) - Proceed along a definite route, making measurements or observations at specific points along the route.

traverse, altimetric - A traverse in which elevations are determined by measuring air pressure at points along the traverse and converting these pressures to elevations. The barometer used is usually an altimeter which provides the conversion from pressure to altitude directly. Elevations are then calculated from the altitudes.

traverse, angle-to-left (right) - A technique used in making a survey traverse, wherein all angles are measured in a counterclockwise (clockwise) direction after the surveying instrument has been oriented by a backsight to the preceding station. The technique is applicable to either closed or open traverses.

traverse, angle-to-right - SEE traverse, angle-to-left (right).

traverse, astronomical - A survey traverse in which the geographic coordinates of the points (stations) are obtained from astronomic observations, and lengths and azimuths of lines are obtained by computation.

traverse, closed - A survey traverse starting and ending upon the same station, or upon stations whose relative locations have been determined by other surveys. Also called a loop traverse.

traverse, deflection-angle - A survey traverse in which the direction of each course is measured as an angle from the direction of the preceding course. A deflection-angle traverse is often an open traverse.

traverse, double-run - A traverse made by an inertial surveying system from a starting point to an end point an back again, all stations visited during the journey out being visited again on the return.

traverse, first-order - A survey traverse which by itself forms a closed traverse, or which extends between adjusted locations of first-order control, and the points of which are first-order control. The standard set in 1973 by the Federal Geodetic Control Commission required that the ratio of the standard deviation in distance between directly connected, adjacent points in the traverse to the distance between the points should not be greater than 1:100 000. The procedures used in making the traverse should also meet certain numerical and procedural criteria called specifications. The standards and criteria superseded those recommended earlier by the U.S. Bureau of the Budget in 1958. The earlier category for first-order traverse was roughly equivalent to the second-order class II category of 1974 and to the category of first-order traverse in the classification of 1925. Before 1925, the category was called precise traverse.

traverse, fourth-order - A survey traverse establishing control less accurate than than that of a third-order traverse. In a fourth-order traverse, angles are observed with a transit or sextant, or are determined graphically, and distances are measured with tape, stadia or wheel.

traverse, framed - A traverse with complete location and bearing control.

traverse, geodetic - A survey traverse for the establishment of geodetic control.

traverse, geographical-exploration - A route along which approximate locations are determined by surveying or by navigational methods.

traverse, goniometric - SEE survey traverse.

traverse, interior-angle - A traverse whose distances are measured and only measurements of interior angles are used.

traverse, open - A survey traverse which starts from a station of known or adopted location but does not end upon such a station. Also called an open-end traverse.

traverse, open-end - SEE traverse, open.

traverse, phototrig - A set of points whose elevations are obtained either by trigonometric leveling or from measurements on terrestrial photographs, and whose horizontal coordinates are determined photogrammetrically. This is not, properly speaking, a traverse; the term is inappropriate and artificial. It may be peculiar to a few organizations.

traverse, precise - A category roughly equivalent to the category second-order, class II traverse in the USA classification of 1974 and used in the U.S. Coast and Geodetic Survey before 1925. After May 1925, the name was changed to first-order traverse.

traverse, random - A survey traverse extending from one station to a survey station not visible from the first station, in order to determine the relative locations of the two stations. SEE line, random.

traverse, second-order - A survey traverse which by itself forms a closed traverse, or which extends between control points of the same or higher category and consists of second-order control. There are two classes of second-order traverse, corresponding to the two classes of second-order control. SEE traverse, second-order class I; traverse, second-order class II. The U.S. Federal Board of Surveys and Maps established a category of second-order traverse in May 1925. Because the bases for classification were different from those used for the 1974 and later classifications, the category second-order traverse as established in 1925 cannot be placed in on of the categories of the later classifications. It was probably somewhere between third-order class I and second-order class II of the 1974 scheme.

traverse, subtense-bar - A survey traverse in which distances are determined using a subtense bar. Both horizontal and vertical control can be established by a subtense-bar traverse. The method has some advantages over a traverse in which distances are measured with a surveyor's tape if the traverse is in steep regions or in places where lines must be carried over broken country or bushes.

traverse, tacheometric - A survey traverse made with a tacheometer i.e., a survey traverse made with an instrument allowing both distance and elevation or height to be determined simultaneously.

traverse, third-order - A survey traverse which by itself forms a closed loop or which extends between control points of the same or higher category, and which consists of control of third-order or higher category. There are two classes of third-order traverse, corresponding to the two classes of third-order control: third-order class I traverse and third- order class II traverse. In addition to standards of accuracy for this category, specifications and criteria for the procedures to be used also apply. The earlier category third-order traverse of the the U.S. Federal Board of Surveys and Maps recommended in May 1925 was roughly equivalent to the same category in the 1974 classification. However, the bases of classification in the two schemes are so different that the categories cannot be exactly equivalent.

traverse, three-dimensional - A survey traverse in which all three coordinates of each point in the traverse (except possibly for the first point) are determined.

traverse, trigonometric - A modified form of survey traverse in which distances are not measured but each is calculated as the unknown side of a triangle having measured base line and measured angles. Also called a trig traverse (jargon). A base line is usually put in at every second station along the traverse so as to bisect the internal angle there, the station being at one end of the base line. Angles to the other two, adjacent stations are then measured at both ends of the base line.

traverse adjustment - (1) Those coordinates, of stations, directions (or angles) and/or distances between the stations, determined by changing measured values so as to arrive at a best set of values for these quantities. I.e., the results of a balancing a traverse (making a traverse adjustment). (2) SEE balancing a traverse. (3) The process of applying the method of least squares to adjusting a traverse. SEE ALSO angle method; Bowditch's method; compass rule; Crandall's rule; direction method (of adjusting a traverse); rule, distance-prorate; Jolly's method; least-squares method; transit method.

traverse angle - A measurement of the horizontal angle from a preceding, adjacent station to the following, adjacent station.

traverse classification - SEE classification (of a traverse).

traverse error of closure - SEE misclosure of traverse.

traverse line - SEE transit line.

traverse net - SEE traverse network.

traverse network - A geodetic network in which the stations have been located by a survey traverse and the lines connecting points on the representational ellipsoid represent distances and directions between stations. Also called a traverse net.

traverse planetable - A planetable consisting of a small drawing board mounted on a light tripod in such a manner that the board can be rotated and clamped in any position. A compass is contained in a recess in the board.

traverse specifications - SEE specifications (for a traverse).

traverse station - (1) A point, in a traverse, over which an instrument is placed for measuring (set up). (2) A measured length of 100 feet, in a traverse, whether measured in a straight, broken or curved line. (3) A point which has had its location determined by traverse.

traverse table - A mathematical table listing the lengths of the sides opposite the oblique angles for each of a series of right angles, as functions of the length and azimuth (or bearing) of the hypotenuse. Traverse tables are used in calculating departures and latitudes in surveying and in courses on computing for navigation. One argument of such a table is the angle which the line or course makes with the meridian (its azimuth of bearing), and the other argument is a distance. In tables used in land surveying, the distance argument is usually a series of integers, from 1 to 9, with which lines of greater length can be composed. In navigation, the distance argument may run to several hundred kilometers. The general availability of small, inexpensive calculators has made the traverse table almost obsolete.

traverse the instrument (v.) - Rotate a surveying instrument about its upright axis i.e., turning the instrument in azimuth.

traversing - The process of making a traverse. Also called polygonization.

traversing, three-dimensional - The process of making a traverse in which elevations of height are determined in the same process as horizontal distances.

Treaty of Ghent, 1814 - The treaty of peace concluded between USA and Great Britain at Ghent on 24 December 1814. The treaty provided for a final adjustment of those boundary lines, described in the Treaty with Great Britain, 1783, that had not yet been ascertained and determined. The adjustment embraced certain islands in the Bay of Fundy and the whole of the boundary line from the source of the River St. Corix to the most northwestern point of the Lake of the Woods.

Treaty with Great Britain, 1782 - A provisional treaty concluded in 1782 between USA and Great Britain, in which the original limits of the USA were first definitely described.

Treaty with Great Britain, 1783 - The definite treaty of peace between USA and Great Britain, concluded on 3 September 1783. The treaty defines the boundary lines of the USA in terms similar to those of the provisional Treaty with Great Britain, 1782. The northern boundary line became at once a fruitful source of dissension between the two countries. From the time of the conclusion of peace almost to the present day, the definite location of this line has been subject to a series of treaties, commissions and surveys.

Treaty with Spain, 1795 - A treaty between USA and Spain, concluded on 27 October 1795, in which the southern boundary line of the USA was settled with Spain. The southern boundary-line was defined in the Treaty with Great Britain, 1782, and in the Treatty with Great Britain, 1783, but that location was not accepted by Spain and was disputed by that country until settled by the Treaty of 1795.

trench - A long, narrow depression with relatively steep sides. Also called a trough. SEE ALSO trench, deep-sea.

trench, deep-sea - A narrow, elongation of very great depth, occurring near the edges of oceanic basins. Also called a trench, marginal trench or trough, although some oceanographers use trough for a trench-like form with more gradual walls and somewhat wider.

trench, marginal - A deep-sea trench occurring near the edge of an oceanic basin. Most deep-sea trenches are marginal trenches. They occur in the margins of the Pacific Ocean, in the Caribbean, the middle-Atlantic Basin and elsewhere. A particularly well-explored one lies along the edge of Puerto Rico and is known as the Puerto Rico trench.

trespass (v.) - (1) Encroach, as on another's privileges, rights, privacy, etc.; intrude. (2) Enter unlawfully upon the land of another.

trestle - (1) A braced framework of timbers, piles or steelwork built to carry a road accross a depression such as a gully, canyon or the valley of a stream. (2) SEE buck; taping stool.

triangle - A geometric figure consisting of three points and the three uniquely defined lines joining them. There are five kinds of triangle of survey interest and commonly referred to simply as triangles: (a) a set of three points in a plane and the three straight lines joining them; (b) a set of three points on a sphere and the three great-circle arcs joining them; (c) a set of three points on an ellipsoid and the set of three geodesics joining them; and (d) a set of three points on an ellipsoid and a set of three normal-section arcs joining them. Small-circle arcs and normal-section arcs on an ellipsoid are not uniquely defined by two points, so an additional specification is needed to make the triangle unique. To use the term plane triangle for triangle is is to use two words where one is entirely sufficient. SEE ALSO plane triangle; spherical excess.

triangle, acute - A triangle whose angles are less than 90°.

triangle, astronomical - (1) The spherical triangle formed, on the celestial sphere, by arcs of great circles connecting the celestial pole, the zenith and a celestial body. The angles of an astronomical triangle are: the hour angle, with vertex at the pole; the azimuth angle, with vertex at the zenith; and the parallactic angle, with vertex at the celestial object. The sides (arcs) are: the co-latitude, from pole to zenith; the zenith distance, from the zenith to the celestial body; and the polar distance, from the celestial body to the pole. (2) SEE triangle, navigational.

triangle, celestial - (1) A spherical triangle formed by three points on the celestial sphere and the arcs of great circles joining these points. (2) The spherical triangle formed by a pole, a celestial body and a third point, and the arcs of three great circles joining the three points. (3) The spherical triangle formed by the elevated pole, a celestial body and the zenith of the assumed location of the observer, and by the three arcs of great circles joining those points. This form of celestial triangle is one of the two forms of the navigational triangle.

triangle, center-point - A triangle divided into three smaller triangles by lines to an internal point, the angles in all the smaller triangles having been measured.

triangle, ellipsoidal - The figure formed by three points on an ellipsoid and by three geodesics or normal-section arcs joining those points.

triangle, geodetic - A geometric figure, on a given surface, whose three edges (sides) are geodesics. Also called a geodesic triangle. In particular, such a triangle on an ellipsoid representing the Earth.

triangle, misclosure - SEE triangle of error.

triangle, missing - A triangle representing the failure of the two sides of the knife-edge (the pivot of a pendulum) to intersect in a perfectly straight line.

triangle, navigational - An astronomical triangle with the elevated pole specified or a terrestrial triangle. The navigational triangle is used in calculating angular elevation or azimuth or in solving great-circle problems in sailing.

triangle, preliminary - An ellipsoidal triangle derived from selected measurements and used in forming equations imposing conditions on longitude and latitude.

triangle, spherical - A triangle formed by three points on a sphere and the three arcs of great circles joining them. To each set of three points correspond eight different spherical triangles. The conditions of the problem usually specify implicitly which of the eight is meant.

triangle, spheroidal - A triangle formed, on a rotational ellipsoid (spheroid) by three points and the arcs of the three geodesics joining those points.

triangle, tachymetric - (1) The triangle formed by (a) the two lines of sight drawn from the observer's eye through two lines, on a reticle, whose distance apart is known, to two marks on an object and (b) the line joining those two marks. (2) The triangle formed by (a) the two lines drawn from the observer's eye through two lines on a reticle to two marks of known distance apart on an object and (b) the line joining those two marks.

triangle, terrestrial - The triangle formed, on a sphere representing the Earth, by (a) a pole and two points representing the assumed location of the observer and the geographic location of the nadir point of the observed body, together with the arcs of great circles joining those points; or (b) a pole and two points representing, in great-circle sailing problems, the points of departure and destination.

triangle equation - SEE angle equation.

triangle of doubt - That triangle resulting, in solving a simple two-point problem, when a third line drawn as a check fails to pass through the point of intersection of the other two intersecting lines. SEE ALSO triangle of error.

triangle error of closure - SEE misclosure.

triangle of error - The triangle formed when three plotted lines fail to intersect at a common point. Also called a triangle of misclosure and a cocked hat (jargon). The center of the triangle may be considered to be the adjusted location i.e., the true or correct location. SEE ALSO resection.

triangle-of-error method - Any technique for solving the three-point problem graphically, using the triangle of error. Among these are Bessel's method, the Coast-Survey method and Lehmann's method.

triangulation - (1) A method of surveying in which the points whose locations are to be determined, together with a suitable number (at least two) of points of known location, are connected in such a way as to form the vertices of a network of triangles. The angles in the network are measured and the lengths of the sides (i.e., the distances between points) are either measured (at least one distance must be measured or calculated from known points) or are calculated from measured angles and measured or calculated distances. The sides having measured lengths are called base lines. Classically, only a very few, short base-lines are in the network; those are connected by the sides of triangles of normal size by a sequence of triangles of increasing size. Triangulation permits sites for stations and base lines favorable for use both from topographic and geometric considerations to be selected. It is well adapted to the use of precise instruments and methods, and can yield results of great accuracy and precision. It has been used generally where the region to be surveyed was large. The term can be considered as including not only the actual operations of measuring angles and base lines, and the reduction of the data, but also the reconnaissance which precedes those operations and any astronomic observations which are required. (2) The process using the method as defined in (1). (3) The survey network resulting from triangulation in sense (1) preceding. (4) The set of those points, on an ellipsoid representing the Earth, which correspond to stations established by triangulation, and the lines (geodesics or normal-section lines) joining them. Surveyors seldom bother stating whether they are discussing distances and angles measured between stations on the Earth's surface or the corresponding lengths and angles between points on the ellipsoid. The context must be studied to determine what is meant. When the lengths of lines in a triangulation are referred to, it is usually the lengths of lines on the ellipsoid that are meant. (5) Any method by which the location of one point is determined from the known locations of two or more other points and from measurements of the angles in the triangles formed by the points. The term is usually applied to the case in which all the triangles lie on a common surface. (6) SEE aerotriangulation; stereo-triangulation; phototriangulation. SEE ALSO arc triangulation; arc of triangulation; area triangulation; classification of triangulation; flare triangulation; flash triangulation; satellite triangulation.

triangulation, aerial - SEE aerotriangulation.

triangulation, analytical - SEE aerotriangulation, analytical.

triangulation, analytical nadir-point - SEE aerotriangulation, analytical nadir-point.

triangulation, areal - A triangulation designed to progress in every direction from a central control-point. Also called area triangulation. An areal triangulation provides geodetic control over a region such as a city or county, or fills in the regions between arcs of triangulation which form a network extending over a large region such as a county or State. SEE ALSO triangulation, arc.

triangulation, base-extension - SEE triangulation network, base-extension.

triangulation, direct radial - SEE aerotriangulation, radial; radial method, direct.

triangulation, first-order - (1) A triangulation network consisting of first-order control i.e., composed of control having a relative accuracy not worse than 1:100 000. (2) A triangulation creating a triangulation network consisting of first-order control. The category first-order triangulation of the February 1974 classification does not correspond exactly to any category of earlier classifications. The closest is the first-order, class I triangulation of the March 1957 classification. First-order triangulation of the classification of 1925 is roughly equivalent to second-order class I of the 1974 classification. First-order triangulation is not, according to the classification of 1974, subdivided into classes. That of the 1957 classification was divided into special, class I and class II first-order triangulation.

triangulation, first-order class II - In the classification used by the U.S. Government from 1957 to 1974, first-order class II triangulation corresponds approximately to second-order class I triangulation of the 1974 classification; first-order class II and second-order class I triangulation of the 1957 classification correspond approximately to second-order class II of the 1974 classification.

triangulation, geodetic - First-order triangulation together with such second-order triangulation as may be suitable for extending the network of geodetic control or for contributing to its strength.

triangulation, graphic radial - SEE aerotriangulation, radial; radial method, graphic.

triangulation, hand-templet - SEE hand-templet method.

triangulation, mechanical templet - SEE mechanical-templet method (of aerotriangulation).

triangulation, nadir-point - SEE aerotriangulation, nadir-point.

triangulation, photo - SEE photo-triangulation.

triangulation, photogrammetric - SEE phototriangulation.

triangulation, precise - In the classification used by the U.S. government from 1921 to 1925, triangulation producing control of sufficient accuracy to be classified as precise control. The term was dropped in 1925 and replaced by the term first-order triangulation.

triangulation, primary - In the classification used by the U.S. government before 1921, triangulation producing control of the highest accuracy. In the classification used from 1921 to 1925, triangulation producing control of an accuracy greater than that produced by secondary triangulation and tertiary triangulation. The term was dropped after 1925, when the terms first-order triangulation, second-order triangulation, third-order triangulation and fourth-order triangulation were introduced.

triangulation, principal-point - SEE triangulation, radial.

triangulation, radial - (1) Triangulation in which the figure or figures consist of triangles completely surrounding a common apex or apices. (2) SEE aerotriangulation, radial. Also called principal-point triangulation, radial plot and minor-control plot.

triangulation, secondary - In the classification used by the U.S. Government before 1925, triangulation which produced control of higher accuracy than that of tertiary control (which was the lowest category). The revision of 1925 added another category, that of precise triangulation, to the top of the scheme.

triangulation, second-order - In the classification used by the U.S. government from 1974 onwards, triangulation consisting of second-order or first-order control. In the classification of 1974, there are two sets of standards for second-order control, leading to the corresponding categories for triangulation of second-order class I triangulation and second-order class II triangulation. The category second-order triangulation does not correspond exactly to a particular category in any earlier classification. Second-order triangulation of the 1925 classification was roughly equivalent to third-order triangulation of the 1974 classification.

triangulation, second-order class I - In the classification used by the U.S. government from 1974 onwards, triangulation producing control having a relative accuracy not less than 1:50 000. Second-order class I and first-order class II triangulation of the 1957 classification correspond approximately to second-order class II triangulation of the 1974 classification.

triangulation, second-order class II - In the classification used by the U.S. government from 1974 onwards, triangulation producing control having a relative accuracy not less than 1:20 000.

triangulation, semi-analytical - SEE aerotriangulation, semi-analytical.

triangulation, ship-shore - SEE triangulation, ship-to-shore.

triangulation, ship-to-shore - A method for extending triangulation along a coast by making simultaneous measurements from three or more stations on shore to a target mounted on an anchored ship. Also called ship-shore triangulation. Ship-to-shore triangulation is used only when it is impractical to establish a chain of triangles or quadrilaterals entirely on land. SEE ALSO two-transit method.

triangulation, special first-order - In the classification used by the U.S. government from 1957 to 1974, the category of most accurate triangulation.

triangulation, stellar - A method of obtaining directions between points on the ground by photographing, from three or more of the points simultaneously, a beacon or lighted object against a stellar background. Photographing the light in at least two different parts of the sky, as seen from each pair of points on the ground, establishes a direction from the one point to the other. If three or more points are involved at the same time, a network of directions is established to determine, except for scale, the relative locations of the points. An extensive region can be surveyed by photographing from a large number of pairs or triplets of points. Väisälä's method of stellar triangulation originally used lights carried aloft by balloons. It has been used to establish a very precise network of first-order horizontal control in Finland. Light-carrying or Sun-illuminated artificial satellites have been used to establish vertical and horizontal control across or between continents. The term is used to cover all varieties of stellar triangulation except that in which the Moon is photographed against a stellar background. SEE ALSO flare triangulation; flash triangulation; satellite triangulation; Väisälä's method.

triangulation, stereo - SEE stereo-triangulation.

triangulation, stereoscopic - SEE aerotriangulation, stereoscopic.

triangulation, stereotemplet - SEE stereotemplet method (of aerotriangulation).

triangulation, tertiary - In the classification used by the U.S. Government before 1921, a category of triangulation third in order of decreasing accuracy of the control obtained by it. In the classification used between 1921 and 1925, it was fourth in order, following precise, primary and secondary triangulation. After 1925, it was no longer used.

triangulation, third-order - In the system of classification used by the U.S. government from 1974 onwards, a triangulation producing control of a relative accuracy not less than 1:5 000. The category is divided into two classes: third-order class I triangulation and third-order class II triangulation. The term was also used for certain categories in the classification schemes of 1925 and 1957. Those categories are not equivalent to the category third-order triangulation of the 1974 classification, although there is rough equivalence between the 1957 and 1974 categories.

triangulation, third-order class I - In the classification used by the U.S. government from 1974 onwards, a triangulation producing control having a relative accuracy not less than 1:10 000.

triangulation, third-order class II - In the classification used by the U.S. government from 1974 onwards, a triangulation producing control having a relative accuracy not less than 1:5 000.

triangulation, three-dimensional - Triangulation in which the network is treated as a three-dimensional figure, with measurements and computations being done on that basis.

triangulation adjustment - The process of determining, from a set of measured distances and angles, another set of distances and angles between the corresponding points of a triangulation network, or the coordinates of the points of the network, in such a way that (a) the network is fully determined by the set of values obtained; (b) all geometric conditions are satisfied exactly; and (c) values calculated from the derived set of distances and angles agree best with the original set of measurements. The qualification best is usually understood to mean producing the smallest sum of the differences between the derived values and the measured values. The method of least squares is therefore used for most triangulation adjustments. It is customary to distinguish between the method of least squares as used for deriving values for distances and angles and the method of least squares as used for deriving coordinates. The first variant is called adjustment by conditions or adjustment of observations; the second is called adjustment by observations equations or adjustment of coordinates. The two variants are mathematically identical and differ only in details of the process of calculation. SEE ALSO angle method of triangulation adjustment; arc method of triangulation adjustment; bearing method of triangulation adjustment; Bowie method of triangulation adjustment; direction method of triangulation adjustment.

triangulation adjustment, Helmert block-method of - SEE Helmert blocking.

triangulation adjustment, local - The satisfying of conditions existing among angles measured at a survey station. Also called station adjustment. In making a local triangulation-adjustment, two kinds of discrepancy may be removed: the failure of the sum of a set of angle measured around the horizon to equal exactly 360° or the failure of the sum of several contiguous angles comprising a larger angle and measured separately to equal the larger angle measured directly. The term measured angles or observed angles is also applied to the angles actually measured, after they have been corrected by local triangulation adjustment.

triangulation adjustment by variation of coordinates - A method of adjusting a triangulation network by using the the difference between measured values and calculated values as observables and corrections to assumed values of the coordinates as the unknowns in a set of simultaneous linear equations. The equations together with the condition equations and Lagrangian multipliers as unknowns, are solved using the method of least squares.

triangulation arc - SEE triangulation, arc of.

triangulation base line - SEE base line.

triangulation classification - SEE classification (of triangulation).

triangulation development - The process of determining those points, lines, lengths and angles on an ellipsoid which represent the corresponding stations, lines of sight, distances and angles on the Earth's surface, using the development method. SEE ALSO development method.

triangle equation - SEE angle equation.

triangulation net - SEE triangulation network.

triangulation network - A survey network in which the survey stations are triangulation stations and the lines represent adjusted distances or directions. Also called a triangulation net or triangulation system. If it consists of a connected sequence of single, simple triangles or quadrilaterals, it is often called an arc of triangulation, chain of triangulation or triangulation chain.

triangulation network, base-extension - That part of a triangulation network which consists of a measured base-line and the sequence of triangles by which the measured length of the base line is transformed into the length of the side of one of the triangles of average size in the network. Also called a base net, base network, base-extension triangulation (Brit.) and baseline-extension network.

triangulation network, three-dimensional - A set of points and connecting lines representing survey stations or control points and lines of sight or reduced distances and directions in three dimensions i.e., representing the points in their correct relative locations in three-dimensional space.

triangulation reconnaissance - SEE reconnaissance.

triangulation signal - (1) A rigid structure erected over or close to a triangulation station and used for supporting instrument and observer, or target or all three, in triangulation. (2) Any object, natural or artificial, whose location is determined by triangulation. The term may be applied to a structure whose location is determined by triangulation, but whose primary purpose is to serve later in a hydrographic or topographic survey. It would then be called a hydrographic signal or a topographic signal.

triangulation station - A point which has had its location determined by triangulation. Also called a trig point (jargon).

triangulation system - SEE triangulation network.

triangulation theodolite - SEE direction instrument.

triangulation tower - A tower used in triangulation and consisting of two separate structures, independent of one another: an inner structure which supports the theodolite (and sometimes the target or signal lamps), and an outer structure which supports the observer and his assistants (and sometimes the target or signal lamps). Before 1927, towers of wood were used by the U.S. Coast and Geodetic Survey. The inner structure was a tripod and the outer structure was a four-legged scaffold. In 1927, the Bilby tower was put into use. Triangulation towers more than 100 feet in height were not unusual.

tribrach - The three-armed base, of a surveying instrument, in which the foot screws used in leveling the instrument are placed at the ends of the arms. Also called a leveling base or leveling head. Some surveying instruments have a four-armed base or quadribrach in which are the foot screws. Tribrachs are used instead of quadribrachs for control surveys because they do not introduce strains into the base of the instrument; quadribrachs do. Some strains tend to change the instrument's orientation in azimuth during the measuring.

trier, level - SEE level trier.

trig list - A publication containing all available data on locations of control points in a particular region, with descriptions of horizontal and/or vertical control and usually organized according to the location of the control points within the limits of large-scale maps of the region.

trig point - SEE station, triangulation.

trilateration - That method or procedure of extending horizontal control or establishing horizontal control which depends on measuring a sufficient number of distances between survey stations that the locations of those stations are determined thereby. Although trilateration does not depend upon angles or directions for determining the locations, directions or angles may be measured and used as additional data. (2) The stations, distances between stations, and other data which are created in determining the locations of points by measuring distances between stations. (3) Any method of surveying in which the location of one point with respect to two others is determined by measuring the distances between all three points. (4) SEE trilateration network. SEE ALSO classification of trilateration; Hiran trilateration; Shoran trilateration.

trilateration, first-order - A trilateration producing first-order control or a trilateration network consisting of first-order control.

trilateration, second-order - Trilateration producing second-order control or a triangulation network consisting of second-order control. Second-order trilateration and second-order trilateration networks are divided into two classes: second-order class I and second-order class II trilateration, according to the category of the corresponding horizontal control.

trilateration, second-order class I - Trilateration producing or consisting of second-order class I horizontal control.

trilateration, second-order class II - Trilateration producing or consisting of second-order class II horizontal control.

trilateration, third-order - Trilateration producing or consisting of third-order horizontal control. Third-order trilateration is divided into two categories: third-order class I trilateration and third-order class II trilateration.

trilateration, third-order class I - Trilateration producing or consisting third-order class I horizontal control.

trilateration, third-order class II - Trilateration producing or consisting of third-order class II horizontal control.

trilateration classification - SEE classification (of trilateration).

trilateration net - SEE trilateration network.

trilateration network - A survey network in which the survey stations have been located by trilateration. Also called trilateration. The term applies, strictly, only to networks consisting solely of point representing survey stations located by trilateration. A network constructed partly from trilateration and partly from triangulation is often referred to as a triangulation network. However, the terms control network or survey network would be preferable. The lengths of lines connecting points represent distances between survey stations. SEE ALSO classification of trilateration network.

trilateration specifications - SEE specifications (for trilateration).

trimetrogon camera - SEE camera, trimetrogon.

trimetrogon method - A method of getting photographs for mapping by using an assemblage of three cameras: one pointing vertically downwards and two mounted to point obliquely downwards and as nearly perpendicular to the line of flight as possible.

trimming and mounting diagram - SEE diagram, trimming and mounting.

triple point - SEE point, triple (of water).

tripod - (1) Any three-legged structure used to hold or support something. (2) In particular, a three-legged structure used for supporting an instrument or survey signal and of a height convenient for the observer. Tripods designed for supporting instruments usually consist of a disk-like platform (the head) on which the instrument is placed, three legs (usually of wood) fastened to the platform by hinges and three shoes, one on each leg at the unhinged end. Just above each shoe, a small spur projects outward. Stepping on the spur pushes the shoe below it into the ground. By spreading the legs farther apart or closer together, the height of the platform can be adjusted so that the eyepiece of the instrument is at a convenient height for observing. Some legs can be adjusted in length so that the platform can be placed at a suitable height for observing and the legs can be placed at the best distance apart for stability. SEE ALSO Bumstead tripod; Johnson tripod; shoe; taping tripod.

tripod, centering - A tripod allowing the platform on which the instrument rests to be centered over the mark on the ground.

trivet - A device used instead of a tripod for mounting a theodolite or leveling instrument, and consisting essentially of a platform with three very short legs, the whole device being cast as a single piece of metal. A trivet is used for placing an instrument in a position where a regular tripod could not be used conveniently, or where greater stability is needed. When used to support a theodolite, a trivet may include a foot-plate in which are V-shaped grooves cut to receive the feet of the leveling screws.

Tropic of Cancer - (1) That parallel of declination which passes through the summer solstice. (2) That geographic parallel whose latitude corresponds to the declination of the summer solstice. Although the obliquity of the ecliptic is steadily changing so that the summer solstice is not a point of fixed declination, the Tropic of Cancer is shown on terrestrial maps as a line of fixed latitude. The value 23° 27' north latitude is generally used for this purpose. While the term tropics is sometimes applied, in popular usage, to the zone, on the Earth's surface, bounded by the Tropic of Cancer and the Tropic of Capricorn, in technical language the term designates the lines themselves while the included region is known as the Torrid Zone. Tropics also has a climatic connotation.

Tropic of Capricorn - (1) That parallel of declination which passes through the winter solstice. (2) That geographic parallel whose latitude corresponds to the declination of the winter solstice. Although the obliquity of the ecliptic is steadily changing, so that the winter solstice is not a point of fixed declination, the Tropic of Capricorn is shown on terrestrial maps as a line of fixed latitude. The value 23° 27' south latitude is generally used for this purpose. SEE ALSO Tropic of Cancer.

tropopause - A hypothetical layer separating the troposphere, within which the temperature gradually decreases with altitude, from the stratosphere, within which the temperature remains nearly constant or increases slightly. The tropopause normally occurs at an altitude of about 7 to 14 km in polar and temperate zones, and at 17 km in the tropics.

troposphere - That layer of the atmosphere within which the temperature normally decreases regularly (at a rate of about 6° C /km) from the temperature at the surface to the level above which the temperature remains approximately constant with increase in altitude. On the Earth, the troposphere normally reaches an altitude of from 7 to 14 km in polar and temperate zones and up to 17 km in the tropics, depending on the season and the latitude. Almost all clouds, precipitation and storms occur in this layer, which contains about 3/4 of the total mass of the atmosphere.

trough - (1) A long, broad depression with gently sloping sides. (2) SEE trench. (3) SEE trench, deep-sea.

trough compass - SEE box compass.

Troughton Bar - A graduated, brass bar 82 inches long made by Troughton of England and brought to the USA by Hassler in 1813 for use as a standard of length. It was supposed to be a copy of the British Imperial Yard (1760). It was replaced in 1856 by two copies of the 1855 British Imperial Yard.

Troughton level - An English leveling instrument, no longer used, having the spirit level permanently attached to the top of the telescope. This is said to be the original dumpy level, a distinction also claimed for the Gravatt level.

true (adj.) - (1) Actual, existing or observed, as contrasted to ideal, imaginary, or theoretical. (2) Corresponding to what would be predicted by a theory accounting for all systematic effects. (3) Average for all possible cases. (4) A term applied to a particular concept to distinguish it from a number of similar concepts e.g., true anomaly from mean anomaly and eccentric anomaly. In this sense, true neither denotes or connotes an actual existence of the concept or thing called true. The terms true bearing, true meridian, true north, etc., occur frequently in reports of land surveys, distinguishing the quantities so named from similar quantities derived using a magnetic compass. In descriptions of boundaries, the term true has legal significance and, except in rare instances, refers to values based directly on astronomical observations. True should never be used to indicate geodetic quantities, although when the difference between astronomical and geodetic values is insignificant, the term has been applied equally to both as a means of distinguishing them from corresponding magnetically derived values. The use of true to indicate astronomically derived values should be avoided. An exact meaning is conveyed by the terms celestial meridian, geodetic meridian, magnetic meridian, etc.

true to scale (adj. phrase) - Having the property that measurements on the map convert exacty to measurements between corresponding points on the spheroid when using the scale of the map. Because no map projection can keep a constant ratio between distances on the map and corresponding distances on the spheroid, no map is true to scale at all points.

truncation error - (1) An error caused by using only a finite number of terms in an infinite series or a finite number of steps in an infinite process. (2) An error caused by ignoring all terms following a selected term in a finite or infinite series.

trunk sewer - A sewer which receives many tributary branches and which serves as an outlet for a large region.

trust deed - SEE deed in trust; deed of trust.

Tsinger method (of determining longitude) - The method which determines astronomic longitude by recording the times at which each of a pair of stars reaches the same zenith distance, the pair selected being near the prime vertical and symmetrically placed with respect to the local meridian. The difference in time is converted to a difference in longitude. Corrections are applied for diurnal aberration and for the inclination of the horizontal axis of the telescope (this affects the vertical angle when the telescope is rotated in azimuth).

Tsinger pair - A pair of stars selected to meet the conditions for use in the Tsinger method of determining time or longitude. The principal criteria are that the stars should be near the prime vertical and symmetrically placed with respect to the local meridian.

Tsingtao Datum - A vertical-control datum based on the bench mark at Tsingtao and assigning to that bench mark an elevation of 72.289 meters above mean sea level of the Yellow Sea.

TsNIIGAik map projection with oval isolines - A map projection defined by the formulas $r = 3 k \sin (S/3)$ and $\theta = A - 3 (S/k)^2 \sin 2A$, in which r and θ are the polar coordinates, in the plane, of that point, on the sphere, at an angular distance S from the origin and at an azimuth A. k is the scale factor. It was devised by G.A. Ginsburg for cartography in Russia.

tsunami - The wave or set of consecutive waves caused in bodies of water by an earthquake or volcanic explosion. Sometimes called, in the earlier literature, a tidal wave. But tsunamis are not caused by the tides and the term tidal wave is misleading. In the open oceans, tsunamis are normally a meter or less high and have a wavelength of 100 to 200 km, with speeds of up to 200 m/s and periods of 20 to 60 minutes or more. On nearing coast-lines, the wavelength of a tsunami becomes less and the height increases correspondingly, with waves reaching a height of 30 to 50 meters or more. Landslides, particularly submarine landslides, also cause tsunamis.

tunnel - A nearly horizontal, underground passage open to the atmosphere at both ends.

tube, cathode-ray - An evacuated tube having an electron-emitting cathode at one end and a fluorescent screen, on which the electons impinge, at the other. Near the cathode are electrostatic plates or electrostatic coils (or both) that form the emitted electrons into a beam that can be focused on the screen and deflected in any desired direction by any desired amount according to voltages impressed on the plates or currents flowing in the coils. Frequently or usually written as CRT. CRTs are the imaging part of radar, television, computing machinery, etc.

Turner's method - A method of determining the standard coordinates of stars from the measured, rectangular Cartesian coordinates of those stars on a photograph, by assuming that the differences between the standard and easured coordinates are linear functions of the measured coordinates, and determining the six coefficients by measuring the coordinates of three stars whose standard coordinates are known. Standard coordinates are related in a simple manner to the right ascension and declination of the stars. Turner's method gives a good approximation if the field of view is less than about 5° and the zenith distance is not more than 60°. It has therefore been much used in the reduction of astrographic data.

Turner's plate constant - One of the set of six constants in that pair of equations devised by Turner for relating standard coordinates to rectangular, Cartesian coordinates of a photographic image of a celestial body.

turning pin - A stake, of wood or metal, on which the leveling rod rests when in use. Also called a pin or turning-point pin.

turning plate - SEE footplate.

turning point - (1) A point on which a leveling rod is held and to which a foresight is taken from one station in a line of levels and a backsight is taken from the next station. A turning point is established to allow the leveling instrument to be moved forward along the level line without a break in the sequence of measured differences of elevation. (2) A point, the elevation of which is determined determined before the leveling instrument is moved, used in determining the height of the instrument after the resetting. Commonly abbreviated T.P. (3) The pin, plate or other object on which a leveling rod rests. A turning point may be a steel pin driven into the ground, the head of a spike on a railroad track, a nail driven into a tie of a railroad track, a ball bearing set in a small dent in a concrete pavement. SEE ALSO footplate.

turnout - The arrangement of a switch and frog for diverting traffic from one railroad track to another.

turret graver - A scriber that keeps scribing points of different sizes in a rotatable holder. By rotating the holder to the desired position, a scribing point of the desired size can be brought into position to scribe a line of desired width without having to remove and replace points.

turtle - SEE footplate.

tusche - An ink used for drafting in those parts, on lithographic plates, which will do the printing.

tusching (v.) - Using tusche for adding artwork, correcting lines and lettering, and adding completely inked-in parts to the image on a pressplate. Also called lithographic drafting.

twilight - The interval of incomplete darkness following sunset (evening twilight) or preceding sunrise (morning twilight). The time at which evening twilight ends or morning twilight begins is determined by convention. Several kinds of twilight have been defined and used: civil twilight, nautical twilight and astronomical twilight; the limiting angle of the Sun below the horizon is, respectively, 6°, 12° and 18°. The duration of twilight varies with latitude and date because the Sun's diurnal path across the part of the sky visible to the observer meets his horizon at quite different angles and declinations at different latitudes and different times of the year. When the Sun approaches the horizon obliquely and is at a higher declination, the Sun takes longer to reach the limiting angle below the horizon. At very high latitudes north or south, any of the above three types of latitude may last as long as twenty-four hours, or may not occur at all.

twilight, astronomical - (1) The interval of incomplete darkness between sunrise or sunset and the instant when the center of the Sun's disk is 18° below the horizon. At this instant, there is no discernible glow at the horizon at the Sun's azimuth, and sixth-magnitude stars can be seen near the horizon. (2) The interval when the central point of the Sun's disk is between 6° and 18° below the horizon.

twilight, civil - (1) The interval of incomplete darkness between sunrise or sunset and the instant when the center of the Sun's disk is 6° below the horizon. (2) The interval when the center of the disk is between 50' and 6° below the horizon. The angle 6° was chosen to correspond, approximately, to stage at which the amount of daylight is the least needed to carry on normal work out of doors. It also represents the time at which stars of the first magnitude are just discernible near the zenith.

twilight, nautical - (1) The interval of incomplete darkness between sunrise or sunset and the instant when the center of the Sun's disk is 12° below the celestial horizon. (2) The interval when the central point of the Sun's disk is between 6° and 12° below the horizon.

twilight map projection - SEE Clarke's map projection.

twinkling - SEE scintillation.

twin low-oblique photography - Aerial photography using two cameras whose optical axes are inclined at small angles (typically about 20°) to the vertical and lie in a plane parallel to the direction of the line of flight.

two-base method (of barometric altimetry) - Altimetry in which variation of pressure with time is observed at two fixed stations near the highest and lowest elevations in a project area, and used to correct observations made by a third, roving altimeter. SEE ALSO fly-by method.

two-transit method - A method of ship-to-shore triangulation in which the location of the vessel is determined by measuring angles from two intervisible transits set up on shore over points previously located.

Tyler's standard scale - A scale for grading sediment on the basis of the sizes of the particles; the size increases from grade to grade as $\sqrt{2}$ and the values at the midpoint of each class become whole numbers of fractions.

type of tide - SEE periodicity of the tide.

typography - The art of setting type, composing and printing from raised type.

U

umbra - That portion of a shadow in which none of the light from an extended source of light (ignoring refraction and diffraction) can be observed.

Unalaska Datum - The horizontal-control datum which is defined by the following coordinates of the origin and by the azimuth (clockwise from South) to Observatory station, on the Clarke spheroid of 1866; the origin is at Unalaska astronomic station:

longitude of origin	166° 32' 05.55" W
latitude of origin	53° 52' 36.45" N
azimuth from origin to Observatory	180° 00' 00.00"

This datum is based on astronomical observations made in 1896, the longitude having been determined by the chronometric method.

uncertainty - (1) Any numerical representation of the amount of error of a value. (2) The standard deviation or probable error of a value. (3) The complement of the probability (i.e., 1 minus the probability) of an event, or the complement of the probability associated with 1 standard deviation.

undevelopable (adj.) - Not able to be flattened without tearing.

undulation - (1) A rise and fall with time; e.g., the undulation of the surface of the ocean. (2) A rise and fall with distance; e.g., the undulating hills of Oklahoma. (3) SEE height, geoidal.

undulation, astrogeodetic - SEE height, astrogeodetic.

undulation, geoidal - (1) The hilliness or rise and fall of the geoid with distance. (2) SEE height, geoidal.

undulation, gravimetric - SEE height, gravimetric geoidal.

undulation of the geoid - SEE height, geoidal; undulation.

unit - (1) Any physical quantity to which the number 1 is assigned by definition. (2) A group of people who have one or more common interests and who therefore can be considered a single social or political body for certain purposes. (3) A part, of a structure, intended for any type of independent use and with an exit to a public street or corridor. SEE ALSO base unit.

unit, absolute - Any unit based directly upon associated fundamental units of length (or the speed of light), mass and time, or length (or the speed of light), mass, time and electrical charge.

unit, astronomical - A conventional unit of distance in astronomy, equal roughly to half the length of the diameter of the Earth's orbit, and determined by assigning defined values to the constants in Kepler's third law. It is commonly abbreviated to A.U. or used as the acronym au (preferred) or AU. It is determined from the equations 1 A.U. = r and (Kepler's equation) $r^3 = k^2 T^2 M_s (1 + m) / 4\pi^2$, using the values T, the period of revolution in mean solar days; M_s, the mass of the Sun, taken as 1; m, the ratio of the mass of the Earth to the mass of the Sun; and k, the Gaussian gravitational constant 0.017 202 098 950 000. Initially, Gauss used r, m and T do determine k. Although better values of r, m, and T are now available, k has been used in so many computations and in so many tables that it has been found easier to use the equation for determining the astronomical unit than for determining k. The distance in kilometers equivalent to the astronomical unit is 149 597 870, as calculated from the constants adopted by the International Astronomical Union in 1976. The diameter of the Earth's orbit is 2.000 000 031 A.U.

unit, basic - SEE unit, fundamental.

unit, derived - Any unit, the size of which is determined directly or indirectly from the fundamental units of a system by means of an empirical law or defining equation.

unit, dwelling - A part of a building, intended to house one person or family, and with an exit to a public street or corridor.

unit, fundamental - Any one of a set of units necessary and sufficient for expressing all physical quantities. Also called a basic unit.

unit, inertial measuring - SEE measuring unit, inertial.

unit, lowest common cartographic - The smallest unit of land identified for a given geographic region and on which information is to be collected and analyzed.

unit, metric - A unit which belongs to or is derived from the c.g.s. or SI system of units.

unit, supplemental - (1) A unit whose size can be specified in a totally theoretical manner because the quantity it measures is free from any physical nature (i.e., is dimensionless) and is the ratio of like physical quantities. Also called a supplementary unit. The degree (of angle) is a supplemental unit.

unit, supplementary - SEE unit, supplemental.

unit development, planned - The same as planned community development, with unit substituted for community.

unit development ordinance, planned - An ordinance providing for and regulating development within a planned unit development locality.

U.S. Coast and Geodetic Survey first-order leveling rod - A self-reading leveling-rod adopted for first-order leveling in 1916 and having a graduated strip of invar attached. The strip is fastened at the bottom to the shoe of the rod and held in place at the top by a spring. The strip is graduated in centimeters, the intervals being alternately black and white, and is read by estimation to millimetres. The back side of the rod is graduated in feet and inches. The rod carries a thermometer and circular level. Also called Fisher leveling rod, meter rod and precise leveling rod. It was in use until about 1967 for first-order and second-order leveling.

U.S. Coast and Geodetic Survey leveling rod - A target leveling-rod having graduations on a metallic strip about 3 meters long set in a wooden rod of cross-shaped (+) cross-section. The lower end of the strip was firmly attached to the wooden rod; the upper end was free to respond to changes of temperature. Pointings were made to a target to which a small, metallic scale was fitted. The leveling rod was used by the U.S. Coast and Geodetic Survey for doing precise (now called first-order) and primary (now called second-order) leveling before 1895. The first model was constructed about 1894. It had graduations painted to show 1-centimeter intervals and had plugs set at 2-centimeter intervals. The rod carried a target and was read to millimeters by means of a 2-centimeter-long scale. It carried a thermometer and a circular level. This model was later replaced by a self-reading leveling-rod having painted on it graduations at intervals of 1 centimeter, read by estimation to millimeters. Two metallic plugs set in the rod 3 meters apart were used in its calibration. In 1916, it was replaced by the U.S. Coast and Geodetic Survey first-order leveling-rod.

U.S. Coast and Geodetic Survey pendulum-apparatus - SEE Brown pendulum apparatus.

U.S. Engineer precise leveling rod - A self-reading leveling-rod of T-shaped cross-section, 12 feet long and graduated in centimeters.

U.S. Geological Survey level - A dumpy level, constructed of stainless steel, having an internally-focusing telescope. The bubble of the spirit level is centered by making the ends of two images of the bubble coincide using a prism-and-mirror arrangement adjustable by the observer. Also called a prism level. This instrument has been used by the U.S. Geological Survey for second-order and third-order leveling. The interval between stadia wires is such that the half-interval intercept on the leveling rod, multiplied by 1000, will give the distance to the rod in feet.

U.S. Geological Survey precise leveling rod - (1) A self-reading leveling-rod graduated in yards and fractions of a yard. (2) A target leveling-rod of cross-shaped (+) cross-section, a little over 12 feet in length, used by the U..S. Geological Survey for precise leveling. There are two forms of this rod: the single-target leveling rod, having graduations on one face only and the double-target leveling rod, having graduations on two opposite faces.

U.S. National Map Accuracy Standards - A set of standards set by the USA, specifying the accuracy required of topographic maps published by the USA at various scales. Commonly referred, in the USA, as Map Accuracy Standards or National Map Accuracy Standards. The standards for horizontal accuracy specify that for maps at scales larger than 1:20 000, 90% of all well-defined features (with the exception of those unavoidably displaced by exaggerated symbolization) shall be located within 1/30 inch (0.85 mm) of their geographic locations as referred to the graticule, while for maps published at scales of 1:20 000 or smaller, 1/50 inch (0.50 mm) is the criterion. The standards for vertical accuracy specify that 90% of all contours and elevations determined from contour lines shall be accurate to within one-half of the basic contour interval. Errors (of contours and elevations) greater than this may be decreased by assuming a horizontal displacement within 1/50 inch (0.50 mm).

U.S. Public Land Survey - The survey carried out by the Bureau of Land Management and its predecessors for establishing the boundaries and subdivisions of the public lands of the USA, using the rules embodied in the U.S. Public Land System.

U.S. Public Land Survey System - SEE U.S. Public Land System.

U.S. Public Land System - The set of rules and procedures for setting boundaries of public lands and parts, as established and used by the Bureau of Land Management and its predecessors. Also called the U.S. Public Land Survey system. In particular, the rules according to which subdivisions of the public lands were classified by size and location. The U.S. Public Land System is frequently used for designating the location of a parcel of land. SEE ALSO description; range; section; tier; township.

United States Standard Datum - The name given on 13 March 1901 to the New England Datum. Adjustment of geodetic networks in the USA had been done using a datum with origin at Meades Ranch. The final, adjusted coordinates differed so little from those in the earlier, New England Datum that the latter was retained but given a different name. This name was later changed to North American Datum.

unit matrix - A matrix having all elements equal to zero except those having the same number for their row as for their column; these are equal to 1. Also called the identity matrix.

unit matrix, extended - A square matrix having 1's and zeros along the main diagonal and zeros everywhere else.

unit of length - A nominal length fixed by definition and independent of temperature, pressure and other physical conditions.

unit stress - SEE stress (2).

Universal Polar Stereographic coordinate - SEE coordinate, Universal Polar Stereographic.

Universal Polar Stereographic Grid - (1) A plane rectangular grid associated with the Universal Polar Grid System. (2) SEE Universal Polar Stereographic Grid System.

Universal Polar Stereographic Grid system - A grid system consisting of a pair of grids (one for the north polar region and one for the south polar region) with the following features. (a) They are on the polar stereographic map projection. (b) The International Ellipsoid is used. (c) The meter is the unit of length. (d) The scale factor at the origin is 0.994 . (e) The approximate secant line of unit scale is at 81° 06' 52" North and South. (f) The north (south) polar grid covers the region from the pole to 80° 30' North (79° 30' South). (g) The origin is at longitudes 0° and 180° . (h) The false eastings and false northings are both 2 000 000 meters. It is often, but improperly, referred to as the Universal Polar Stereographic Grid.

Universal Polar Stereographic Map Projection - (1) The stereographic map-projection used for the Universal Polar Stereographic Grid System. (2) SEE Universal Polar Stereographic Grid.

Universal Space Rectangular Coordinate System - A coordinate system having three mutually perpendicular axes intersecting at the center of an ellipsoid representing the Earth. The z-axis coincides with the least axis of the ellipsoid, its positive direction being northwards and either the x-axis or the y-axis lies in the plane of reference through the least axis. Scale is the same along the three axes. The x-axis is usually placed in the plane of reference and forms, with the other two axes, a right-handed coordinate system. However, if the y-axis is placed in the plane of reference, the three axes usually form a left-handed system.

Universal Time - Any one of three different times related to mean solar time at Greenwich and distinguished from each other by the amount of correction made for irregularities in the Earth's rotation. Also called Greenwich mean time; Greenwich mean solar time. The three different times are designated Universal Time 0, Universal Time 1 and Universal Time 2. However, the term Greenwich mean time is used, properly (except for special purposes in Britain) for Greenwich mean solar time before 1 January 1925 and reckoned from noon to noon; from 1 January 1925 onward, the term Universal Time is used and the time is reckoned from midnight to midnight. A fourth variety, Universal Time Coordinated, was introduced on 1 January 1972. It is determined by atomic clocks and not by the rotation of the Earth.

Universal Time Coordinated - SEE Coordinated Universal Time.

Universal Time 1 - Universal Time 0 corrected for the difference between the instantaneous and the average longitude of the observer. Commonly denoted by UT1. This correction is derived from measurements of polar motion. It gives the angular rotation of the Earth about an average or standard (e.g., the Conventional International Origin) axis. Because Universal Time 1 is related directly to the instantaneous rotation of the Earth, which is not perfectly uniform, it does not progress uniformly as compared with time measured by an atomic clock.

Universal Time 2 - Universal Time derived from Universal Time 0 by correcting that time for seasonal, periodic variations in the rate of rotation of the Earth. Commonly denoted by UT2. Universal Time 2 is a measure of the average angular rotation of the Earth and is free of predictable periodic variations. It is still affected by irregular and secular variations. The correction is not greater than 0.03 seconds. It is used for investigations of the Earth's rotation and in the reduction of data from various artificial satellites such as those of TRANSIT.

Universal Time 0 - The Greenwich hour-angle, plus 12 hours, of a point on the celestial equator whose right ascension, measured from the mean equinox to date is αU = 18h 38m 45.836s + 8 640 184.542s TU + 0.0929s TU^2, in which TU is the number of Julian centuries of 36 525 days of Universal Time elapsed since the epoch of Greenwich mean noon (12h U.T.) on 1900 January 0. Commonly denoted by UT0. Also called observed time. This, however, is a misnomer. Time is determined from observed times of transit of selected stars to give Greenwich apparent sidereal time which is then corrected, using the equation of the equinoxes, to give Greenwich mean sidereal time, which is then substituted into the equation for Universal Time 0. Universal Time 0 is therefore equivalent to the Greenwich hour angle of the vernal equinox of date plus 12 hours.

Universal Transverse Mercator coordinate - SEE coordinate, Uiversal Tansverse Mercator.

Universal Transverse Mercator Grid - (1) A plane rectangular grid associated with the Universal Transverse Mercator Grid System. (2) SEE Universal Transverse Mercator Grid System.

Universal Transverse Mercator Grid System - A grid system having the following features. (a) Maps and grids are on the transverse Mercator map projection (Gauss-Krüger type) in zones 6° wide longitudinally. (b) The Clarke spheroid of 1866 is used for maps of North America. (c) For North America, the correct datum is the North American datum 1927. (d) The longitude of the origin is on the central meridian of each zone. (e) The latitude of the origin is 0° . (f) The unit of length is the metre. (g) The false northing is 0° meter for the northern hemisphere and 10 000 000 meters for the southern hemisphere. (h) The false easting is 500 000 meters. (i) The scale factor at the central meridian is 0.9996. (j) The zones are numbered beginning with 1 on the zone from 180° W to 174° W, and increasing westward to 60° on the zone from 174° E to 180° E. The grids in all zones are identical except for the displacement in longitude. (k) The limits of latitude are 80° N and 80° S. (l) The zones are bounded by meridians whose longitudes are multiples of 6° west or east of Greenwich. It is sometimes referred to, improperly, as a Universal Transverse Mercator Grid. This term properly applies only to one grid, not to the collection of grids, and not to the system of numbering and lettering used as coordinates. On large-scale maps and in tables and overlap of approximately 40 km (25 miles) on either side of the junction is provided for the convenience of surveyors and for surveying and firing by artillery. This overlap is never used, however, in giving a reference from the grid. The transverse Mercator map projection is conformal: angles measured on the map or computed from coordinates on the grid closely approximate their true values. At any point, corrections to distances are the same in all directions. Conformality is an important characteristic for all users of numerical values of the grid. The same name may be applied to similar grid systems defined differently.

Universal Transverse Mercator Map Projection - (1) The transverse Mercator map-projection used for the Universal Transverse Mercator Grid System. (2) SEE Universal Transverse Mercator Grid.

Universal Water Chart - A series of blank charts at a scale of 1: 1000000, published for each 4° of latitude. Universal Water Charts are used for aerial navigation over water or for plotting locations, distances and courses in travel over unmapped lands such as Antarctica.

update, zero-velocity - The process of controlling the growth of velocity-related errors, in an inertial surveying system, by bringing the system to rest (reducing the external velocity to zero) and recalibrating.

upland - (1) Land situated above ordinary high water or mean high water. (2) Land situated at a higher elevation than riparian land or land adjacent to riparian regions but remote from the body of water and having no riparian rights.

uplands - (1) Land above ordinary high water or mean high water and subject to private ownership, as distinguished from tidelands, the ownership of which is prima facie in the State but is also subject to divestment under statutes of the State. (2) The same as the preceding definition, but specifying high water line instead of high water mark.

uplift - The slow rising of a portion of the Earth's crust. In some regions (e.g., Finland and Scandinavia), the uplift is believed to be a return to earlier elevations after a prolonged pushing down by glaciers. In such instances, the uplift is called glacial rebound.

Urnormal-1937 - A primary standard for the chlorinity of Normal Water.

use plat - A copy of the master-title plat and any supplemental plats of a township, showing, in addition to the status shown on the master-title plat, information about the use to which the land is put, as indicated by applications, leases, permits and so on.

usque ad filum aquae - As far as the central line, or middle, of the stream. Literally, as far as the thread of the stream. This phrase occurs frequently in modern law; ad medium filum aquae (q.v.) is another and more etymologically exact form.

V

vacancy - A strip of unsurveyed and unsold public land.

vacate (v.) - (1) Annul. (2) Set aside. (3) Cancel or resecind. (4) Render an act void. (5) Put an end to e.g., to vacate a street. (6) Move out. (7) Make vacant of empty. (8) Leave especially, surrender possession by removal. (9) Cease from occupancy.

vacation of highway - The formal determination that a section of highway is no longer needed for the purpose of a public highway. This results in the reversion of title to such highway to the owners of the underlying fee.

vacuum back - A locating back which holds the photographic film by applying a vacuum to the rear surface of the film.

vacuum box - The fram, containing its own vaccuum pump, inclosing the mold for forming plastic relief-maps.

Väisälä base line - A base line established by using the Väisälä base-line apparatus. Also written Väisälä baseline. Such base lines were intended to be used both to enforce a uniform scale on national, horizontal geodetic networks and to calibrate other distance-measuring instruments and apparatus. SEE ALSO baseline, standard.

Väisälä base line apparatus - A Fabry-Perot interferometer for setting up a base-line of accurately known length. It splits a collimated beam of light into two parts. One part travels from a base mirror M_1 at one end of the base line to a mirror M_3 at the end of a segment of the base line and thence back to the observer at a telescope. The second part goes from mirror M_1 to a reference mirror M_2 (at an accurately- known, previously measured distance from M1) and then back to the observer after reflection between M_1 and M_2 enough times that the total distance traveled is nearly equal to the distance from M_1 to M_3 and back. That is, the two parts travel approximately the same distance, but one of these distances is already known accurately. The two parts, on being recombined in the telescope, interfere to form a pattern of light and dark lines called fringes. The distance between fringes depends on the amount (in fractions of a wavelength, by which the distance traveled by one part exceeds the distance traveled by the other. If white light is used as a source, the fringes are colored and disappear altogether if the two distances differ by more than a few wavelengths (or if they are exactly the same. Because the distance between M_1 and M_2 is known exactly, the distance between M_1 and M_3 is then also known exactly. In practice, the distance between M_1 and M_2 is first determined by spacing the mirrors exactly 1 meter apart, an accurately calibrated bar of quartz being used for this purpose. Also called a Väisälä light-interference comparator. This term, however is a misnomer because two established lengths are not compared. Instead, one known length is used to construct a distance. Refractive anomalies in the atmosphere limit the measurable distance between mirrors to between 500 and 1000 meters. Under good conditions, an accuracy of a few tens of micrometers can be obtained.

Väisälä comparator - An interferometer consisting basically of three mirrors in line, of which the first two are semi-transparent and placed a known distance apart (they are placed in contact with the ends of a quartz bar exactly 1 meter long); the third mirror, fully silvered, is placed at the distance to be measured from the first mirror in line. A very powerful source of white light sends a part of the beam through the first two mirrors to the third and thence back to the observer. The other part of the beam is reflected back and forth between the first two mirrors and finally back to the observer where it interferes with the light from the third mirror. The separation between interference fringes indicates the difference between some multiple of the distance between the first two mirrors and the distance between the first and third mirrors. Also called a Väisälä light-interference comparator. It can measure differences between distances up to 100 m with an accuracy of about 0.1 micrometer. SEE ALSO Väisälä base line apparatus.

Väisälä interferometer - SEE Väisälä baseline apparatus; Väisälä comparator.

Väisälä light-interference comparator - SEE Väisälä comparator.

Väisälä's method - A method of triangulation in which a light carried on a balloon drifting slowly at a high altitude is used as the target and is photographed against a stellar background by cameras at the triangulation stations. Also called stellar triangulation although that term has now taken on a more general meaning also. Väisälä's method has been used to establish a very precise triangulation network of first-order control in Finland.

Valdez Datum - The horizontal-control datum defined by the following coordinates of the origin and by the azimuth (clockwise from South) to the azimuth mark, on the Clarke spheroid of 1866; the origin is at station Pete:
longitude of origin 145° 23' 48.560" W.
latitude of origin 60° 22' 44.440" N.
azimuth from origin to azimuth mark 179° 44' 35.4"
The coordinates of Pete were derived from the astronomic longitude of Valdez longitude station, determined by telegraph in 1905, and through triangulation from the astronomic latitude of Orca astronomic station, 1898. The azimuth, Pete to azimuth mark, is an astronomic azimuth.

valley crossing - The technique used, in leveling, to determine differences of elevation between points on opposite sides of a valley into which leveling cannot be carried. It involves simultaneous and reciprocal sightings from points on opposite sides of the valley. SEE ALSO water crossing. SEE ALSO attachment, valley-crossing.

value - (1) The number associated with something: for example, the value of a length. Commonly, the thing and its value are considered together and called a quantity; the number is then called the value of the quantity or thing. Two kinds of value are commonly distinguished: values determined by measurement and called values of observed quantities, measured values or observed values; and values determined by calculation and called calculated values. (2) The amount of money or its equivalent associated with something; the monetary worth of that thing. (3) The ratio of exchange of one commodity for another. For example, one bushel of wheat in terms of a given number of bushels of corn. Money is the common denominator by which value is measured.

value, absolute - (1) Of a real number, that number with its original sign replaced by a plus sign. Also called the magnitude of the number or numerical value. For example, the absolute value of +2 is +2; so is the absolute value of -2 An equivalent definition is: the positive square root of the square of the number. The absolute value of a number x is denoted by |x|. (2) Of a complex number, $z = x + iy$, in which x and y are real numbers and i is $\sqrt{-1}$, the positive square root of the product of z by its conjugate $x - iy$. The absolute value is denoted by |z| and is called the magnitude or modulus of the number. If y is zero, z is a real number, and its absolute value by this definition is the same as that by the previous definition.

value, adjusted - A value derived from measurements by some orderly process which eliminates of minimizes, in some specified sense, discrepancies between the measured and adjusted values or between the measured values and values of a function of the adjusted value. The process of deriving an adjusted value is called an adjustment. It may be a graphical method or analytical (arithmetical). SEE ALSO adjustment; adjustment, analytical.

value, apparent - SEE value, observed.

value, measured - A value obtained by measurement. Also called an observed value. However, the term observed value is often used, in surveying, for a value from which systematic errors have been removed.

value, most probable - (1) That value which, as a function of measured quantities containing random errors, has the greatest probability of being the errorless value, given a number of actual measurements. To determine the most probable value, the frequency function or the cumulative distribution function of the errors affecting the measurements must be known. (2) That value of a quantity which is mathematically determined from a set of measurements and is more nearly free from the effects of errors than any other value which can be derived from the same set of measurements. The most probable value is calculated after blunders and systematic errors have been removed from the data.
(3) That value of a measurable quantity affected by random error which maximizes the probability of a function of that quantity. An adjusted value determined by the method of least squares is a most probable value if the measured quantity has a Gaussian distribution.

value, normal - SEE value, standard.

value, observed - A value obtained by measurement or observation. Also called an apparent value or true value. The term observed value is often applied to that value which is derived from a measurement after systematic errors have been removed but before random errors are minimized by adjustment. An angle obtained with a repeating theodolite, after correction for closure of horizon has been applied, is considered an observed angle. SEE ALSO value, adjusted; value, measured; value, most probable.

value, standard - A number accepted as the value to be used for a particular constant. Also called a normal value.

value, true - (1) That value of a quantity which is completely free from errors of all kinds. Because the errors to which physical measurements are subject cannot be known exactly, the true value of a quantity cannot be known exactly. In surveying, the most probable value as determined by some plausible method is used as best representing the true value of the quantity. (2) The average value of a set of measurements. (3) The average value of a quantity, the average being taken over all possible cases or instances. Average values of this type occur in problems associated with sampling. (3) SEE value, observed.

van der Grinten's map projection - A map projection mapping the sphere onto a disk in such a way that the central meridian and the equator are represented by mutually perpendicular, straight lines while the other meridians and parallels of latitude are represented by arcs of non-concentric circles having centers on the x-axis and the y-axis, respectively. The map projection is neither conformal nor equal-area. Scale is exact along the line representing the equator but there is great areal exaggeration in polar regions. The map projection was invented by Grinten in 1904. He used a geometric construction and gave the map projection as a transformation from geodetic coordinates to bipolar coordinates.

VanDyke process - Printing by direct contact from a drawing on a translucent sheet onto a sensitized plate. Also called the brown-line process and the sepia process. It can be used to print maps in colors, but is most suitable for monochromatic reproduction. Reproduction can be done only at the scale of the drawing, and a high standard of draftsmanship is required. On the other hand, the process tends to make lines narrower.

vanishing-point method - A graphical method of producing a map from an oblique by locating the geometric horizon, nadir point and isocenter on the photograph, and using these to determine, for each desired point on the photograph, the direction and distance to the corresponding point on the map.

vaporware - A program (computing) that has been advertised, long before its actual completion, as completed or soon to be completed.

vara - An obsolete Spanish unit of length having various values in terms of metric or English units of length, according to where and when it was used. The vara was used in southwestern USA, Mexico and some Latin American countries. Values for some States and countries are as follows:

Arizona	33.00 inches
California	32.953 inches to 33.372 inches.
Cuba	33.38 inches
Florida	33.372 inches
Mexico	839.16 mm [Humbolt 1803; 839.16 mm.]
	837.377 mm [U.S. Coast Survey 1850]
New Mexico	33.00 inches [surveys of private claims; official recognition by court of private land claims]
Portugal:	legalized 1.1 m [Merriam/Webster 1954]
Texas	33.40 inches [Seth Ingram, Colonial Surveyor, in Document 18 August 1825]
	39.60 inches [Seth Ingram's chain used for measurement]
	33-1/3 inches [Texas law 1919].

variable - A quantity to which any number from a specified set of numbers can be assigned, either freely or according to a definite set of rules. Also called, principally by statisticians, a variate. The specified set is called the domain of the variable. A variable is classified as either an independent variable or a dependent variable.

variable, dependent - A quantity whose value depends on the value or values of another set of quantities.

variable, hidden - SEE parameter, latent.

variable, independent - A quantity whose values may be assigned arbitrarily within the domain of the variable. Also called an argument when it appears in a function.

variable, random - (1) A quantity which takes on a sequence of values such that the value of the quantity at any point in the sequence is not predictable from a knowledge of the values elsewhere in the sequence. (2) A quantity to each of whose values is assigned a number, from 0 to +1, called the probability of that value.

variable, stochastic - Any observed quantity associated with a set of repetitions of a given process which can vary from one set to another.

variable, systematic - (1) A dependent variable whose value is determined when the values of the independent variables (the arguments) are known. (2) A variable whose sequence of values is not random i.e., whose values can be predicted as a function of time or of order in the sequence.

variance - (1) The sum of the squares of the differences between the values in a set and the average value of that set, the sum being divided by one less than the number of values in the set. The positive square root of the variance is called the standard deviation. (2) The sum of the squares of the differences between the values in a set and the values of some function of the set, the sum being divided by some function of the number of values in the set. (3) The range of uncertainty in the value of a random variable. SEE ALSO Allen variance.

variate - SEE variable.

variation - (1) An inequality in the Moon's longitude which has a period of one-half of a synodical month, or about 14-3/4 days, and an amplitude of 39' 29.9". Also called the lunar variation. The variation was unknown to the Greek astronomers because it does not affect eclipses, from which the Greeks obtained most of their information about variations in the Moon's motions. It may have been discovered by Abul Wafa (10th century A.D.), an astronomer of Baghdad, but it was certainly discovered by Tycho Brahe. (2) SEE variation, magnetic.

variation, abnormal magnetic - An anomalous value, in the reading of a magnetic compass, made in some locality where there are unknown sources deflecting the needle from the magnetic meridian.

variation, annual - (1) That periodic perturbation in the Moon's longitude caused by the eccentricity of the Earth's orbit. Also called the annual equation. (2) The value of the annual rate of change of right ascension (or declination) because of precession, for a specific epoch, together with the effect of proper motion.

variation, annual magnetic - A regular component of the variation in the Earth's magnetic field, having a period of one year.

variation, daily magnetic - The transient change in the Earth's magnetic field associated with the apparent daily motions of the Moon and Sun, and having a corresponding period of about one day. Also called diurnal magnetic variation and ambiguously, a magnetic disturbance. In most places, the Sun's contribution to the daily magnetic variation follows a fairly consistent pattern, although appreciable and unpredictable changes in form and amplitude do occur.

variation, diurnal - A component, of a periodically varying quantity, which passes through a complete cycle in one day. Such a component is said to have a periodicity of one day. Magnetic declination, for example has a diurnal variation.

variation, diurnal magnetic - (1) That simple, harmonic component of the daily magnetic variation which has a period of 24 hours. (2) SEE variation, daily magnetic.

variation, elliptic - Those variations in the Moon's longitude caused by the eccentricity of the Moon's orbit. The evection is one such variation.

variation, geomagnetic - SEE variation, magnetic.

variation, lunar daily magnetic - That periodic, daily variation of the Earth's magnetic field which is in phase with the Moon's hour angle.

variation, magnetic - A regular or irregular change, with time, of magnetic declination, dip or intensity. Also referred to as variation if only geomagnetism is being discussed. In aeronautical and nautical navigation, and sometimes in surveying, the term is used for magnetic declination. Also called unnecessarily, geomagnetic variation. The regular magnetic variations are: annual, daily and secular (q.v.). Irregular variation, when severe and world-wide, are known as magnetic disturbances or magnetic storms. The Earth's magnetic field may be affected within small regions by direct-current electricity and other artificial sources. It was once a common practice of surveyors to denote as magnetic variation the net amount the compass departed from the direction taken as north in the description of a particular line, even when this was known to be slightly at variance with the celestial meridian.

variation, sea-level - SEE variation of sea level.

variation, secular - (1) The rate of change of right ascension (or declination) per century. (2) SEE variation, secular magnetic.

variation, secular magnetic - The slow change of magnetic declination, inclination and intensity of the main geomagnetic field at a place. Also called secular variation.

variation, solar daily magnetic - A periodic variation of that component of the Earth's magnetic field which is in phase with the Sun's hour-angle.

variation of compass - SEE declination of compass. Navigators use variation; land surveyors use declination.

variation-of-coordinates method - That method of adjusting a survey network in which the known quantities are the discrepancies between measurements of distance, angle, etc., and values calculated from assumed coordinates of the points in the network, and the unknowns are corrections to assumed values of the coordinates. This method and that of solving directly for corrections to measured distances, angles, etc. by the use of condition equations are the two principal methods used in surveying for adjusting survey networks.

variation of latitude - The measured variation of the astronomic latitude of a particular point on the Earth. The variation amounts to about 0.25". It is caused by rotation of the Earth's axis of figure about the axis of rotation. A secular change of 0.001" or less per annum may also be present.

variation of sea level - That variation of the daily, monthly or yearly average level of the sea from mean sea level which is caused by meteorological conditions and not by luni-solar attraction. Those variations which are of short duration, such as waves, surges, tsunamis, etc., are of course expected to average out over periods of a day or longer.

variations (of the magnetic field) - SEE variation, magnetic.

variation of the compass - SEE declination, magnetic (2).

variation of the poles - SEE motion, polar.

variometer - A torsion balance designed particularly for comparing the intensities of magnetic fields.

vectograph - A stereoscopic picture composed of two superimposed images which polarize light in planes 90° apart. When these images are viewed through spectacles having filters whose axes of polarization are at right angles to each other, an impression of depth is obtained.

vector - (1) A physical quantity having both length and direction. Velocity, acceleration and force are the most common instances of vectors in physics. (2) A number in the complex plane; a complex number. (3) A matrix having N rows and 1 column or 1 row and N columns; N is an integer. (4) A mathematical entity representable as a directed line-segment.

vector, absolute - A vector, the distances to the endpoints of which are masured in absolute units from a specified point called the origin.

vegetation - Plants, in general. The term includes trees and shrubs. It is sometimes used erroneously (as in some definitions of, e.g., marsh) for plants other than trees.

vehicle, ground-effect - A vehicle which is kept above the surface of the ground or water by a constant blast of air forced downward at the edges of the vehicle by large fans. Also called a hovercraft. Such a vehicle may be particularly suitable for use in inertial surveying and similar projects.

velocity - The rate of change of location of a moving point with time. If the location of a moving point is given by the vector r, the velocity of the point is dr/dt, in which t is the time. Speed is given by dr/dt, in which r is the absolute value of r. The velocity of a point moving along a planar curve is often resolved into two components: radial velocity and transverse velocity, or into components along the tangent, normal and binormal. If the point is moving along a non-planar curve, a third component, such as that along the normal to the osculating plane, must be specified also. In popular usage, velocity is often used for speed and speed for velocity. This usage should be avoided. SEE ALSO escape velocity; group velocity; phase velocity.

velocity (of light) - (1) The speed of light traveling in a specified direction. The velocity of light is a vector; the speed of light is a scalar. (2) SEE speed (of light).

velocity (of sound) - (1) The speed of sound traveling in a specified direction. The velocity of sound is a vector, the speed of sound is a scalar. (2) SEE speed (of sound).

velocity, absolute - (1) The lowest speed enabling a body to leave the Solar System, starting from the surface of a body of the System. Absolute velocity with respect to the Earth is 16.7 km/s. (2) The rate of change of location with respect to time, measured with respect to a coordinate system whose axes are fixed in space.

velocity, angular - A vector giving the rate of change of angle with respect to time. The quantity is a vector pointing upward from the plane of the angle if the motion is counterclockwise and pointing downward if the motion is counterclockwise.

velocity, areal - (1) A vector giving the rate of change of the area between a given straight line, a fixed curve, and the segment of a straight line passing through a fixed point on the given straight line and a moving point on the curve. The quantity is a vector having the same direction as the angular velocity of the point. (2) In particular, a vector giving the rate of change of the area included between the line of apsides of an orbit, the path and the radius vector to a point moving in that orbit.

velocity, circular - (1) The speed of a satellite in the smallest possible circular orbit about a body. The circular velocity of a body in orbit about the Earth is about 8 km/s. (2) The speed a body in a elliptical orbit would have if the orbit were circular with the same diameter.

velocity, cross-track - That component of a body's velocitgy which is perpendicular to the tangential (along-track) velocity and to the radial velocity. SEE ALSO velocity, transverse.

velocity, normal - (1) That component of a body's velocity which is in the direction of the normal (in the plane) to the path or of the principal normal (of a curve in space). SEE ALSO velocity, radial. (2) SEE velocity, transverse.

velocity, orbital - (1) That velocity having the smallest magnitude (speed) which a body must be given in the gravitational field of another body to put it into orbit about that body. Orbital velocity is a function of the length of the major axis of the orbit, according to the expression $\sqrt{(GM/a)}$, in which G is the gravitational constant, M is the mass of the primary body and 2a is the length of the major axis (the satellite is assumed to have negligible mass). It is applied in the direction perpendicular to the direction of the force of gravity. Note that orbital velocity decreases with increasing distance from the primary body; it is not the same as the speed needed to go into orbit, which increases with distance from the primary. (2) The speed corresponding to the orbital velocity as defined in (1) above. (3) The velocity of a body in its orbit. The term is inflated when applied to a satellite. Orbital velocity and velocity are the same thing.

velocity, radial - That component of the velocity of a moving point which lies on the line joining the moving point to another fixed and specified point. Called range rate by radar engineers. The component is usually taken to be positive in the direction away from the specified point. However, velocities caused by gravitational forces are often taken to be positive in the direction towards the center of attraction.

velocity, tangential - That component of the velocity of a moving point which is in the direction of the tangent to the path followed by that point. Also called velocity along the path or along-track velocity.

velocity, transverse - That component of the velocity of a moving point which is in the direction of the binormal to the path followed by the moving point. Also called cross-track velocity in the case of an orbit. Also called normal velocity.

velocity correction - An amount applied to the speed of light to obtain the speed under standard or other specified conditions of humidity, temperature and altitude.

velocity potential - A scalar the negative of whose gradient is a velocity.

vendee - The purchaser of real estate under an agreement.

vendor - The seller of real estate, usually referred to as the party of the first part in an agreement of sale.

Vening Meinesz formula - (1) The function F' in Vening Meinesz' formula for deflection of the vertical. (2) SEE Vening Meinesz formulas for deflection of the vertical.

Vening Meinesz gravity reduction - A modification of the Airy- Heiskanen gravity reduction in which the compensation or contribution of each prism is made to vary according to a bending curve representing the flexure of the Earth's crust under the weight of mountains and other high regions.

Vening Meinesz isostatic gravity anomaly - The gravity anomaly introduced by Vening Meinesz in 1931 and calculated on the assumption of certain radii for regions of different degrees of isostatic compensation, rather than on the assumption of Airy and Heiskanen that compensation is local. The Vening Meinesz isostatic gravity anomaly is little used. It is difficult to calculate and the advantages of using it are not great. SEE ALSO gravity reduction, Vening Meinesz.

Vening Meinesz pendulum-apparatus - A pendulum apparatus containing 5 pendulums and designed by Vening Meinesz for measuring the intensity of gravity at sea. The apparatus was of the Sterneck type, with three half-second pendulums of brass swinging in the same plane. This configuration allowed a correction to be made for the motion of the vessel. Motions of two of the pendulums were highly dampened and used as references for the motions of the three other pendulums.

verification - SEE calibration.

vernier - A scale attached as an auxiliary to a primary scale along which it slides and graduated so that the total length of a certain number of divisions on the auxiliary scale is equal to the total length of one more or one fewer of the same number of divisions on the primary scale. The vernier makes it possible to read a primary scale (such as a graduated circle) to an interval much less than one division of the scale. Depending on the relationship between the graduations on the vernier and on the primary scale, one has the direct vernier, retrograde vernier, double vernier and folding vernier. SEE ALSO contact vernier.

vernier, direct - A vernier on which the least interval is shorter than the least interval on the primary scale.

vernier, doubled - A pair of verniers extending and numbered in opposite directions from the same initial mark. It is used in reading a circle having graduations numbered in both directions.

vernier, folding - A vernier so constructed and numbered that it may be read in either direction.

vernier, optical - A microscope designed for viewing a graduated line (primary scale), having a vernier ruled on a glass slide located in the focal plane common to the objective lens system and the eyepiece, so that it can be compared with the image of the primary scale. The microscope is adjusted so that the image of the primary scale falls in the plane of the vernier and the scale of the vernier subtends the proper number of divisions of the primary scale. The optical vernier is superior to the contact vernier in that no errors are caused by parallax because of wear of the edges in contact. Optical verniers are, however, more liable to breakage and loss of adjustment than are contact verniers.

vernier, retrograde - A vernier on which the least division is longer than the least division on the primary scale.

vernier closure - The difference between the initial and final readings on the vernier when closing the horizon.

vernier transit - A transit having graduated, metallic, horizontal and vertical circles with verniers.

verst - An obsolete Russian unit of length or distance, equal to 1.067 km.

vertex - (1) The highest point of an object. (2) One of the two points, on a great circle, closest to the poles. (3) SEE point of intersection of curve.

vertex of curve - SEE point of intersection of curve.

vertical (adj.) - Alined with the direction of gravity.

vertical (n.) - (1) A line along which the downward-pointing tangent has the direction of gravity at the point of tangency. Equivalently, a line everywhere perpendicular to surfaces of constant gravity-potential; or, a line whose equation is given by the gradient of the potential (or of the negative of the gradient, depending on the convention used for the sign of the potential). Plumb line and vertical are often used as synonyms. The directed plumb line is usually taken as positive downwards, however. Furthermore, vertical has two closely allied but similar meanings, while plumb line has two very different meanings. If the vertical is considered a directed line, the positive direction may be taken either downwards or upwards. Older textbooks take the downward direction as positive; present practice seems to be to take the positive direction as upwards. (2) The direction, at a point, opposite to (or the same as, depending on convention) that of gravity at that point. Modern usage seems to favor the first convention, older usage the second. Rectangular, Cartesian coordinate systems associated with the older usage usually have the z-axis pointing downwards and the y-axis positive eastwards, making a right-handed coordinate system. However, left-handed coordinate systems with the y-axis positive northwards have also been used. Modern practice is to have the z-axis positive upwards and the y-axis positive northwards. Note that in geodesy the vertical is the perpendicular to an equipotential surface; the normal is the perpendicular to an ellipsoid.

vertical, deflection anomaly of the - SEE deflection anomaly.

vertical, deflection of the - SEE deflection of the vertical; deflection anomaly.

vertical, geographic - The direction of the vertical at the geoid.

vertical, geometric - (1) The direction of the radius vector drawn from the center of an ellipsoid representing the Earth through the location of the observer. (2) The direction of the radius vector drawn from the center of the Earth through the location of the observer.

vertical, local - The direction of the force of gravity at a point, as distinct from the normal to an ellipsoid.

vertical, magnetic prime - The vertical circle through those points on the horizon which lie to the magnetic east and magnetic west.

vertical, prime - (1) Either the prime-vertical circle or the prime-vertical plane. The latter is usually meant. (2) A vertical circle whose plane is perpendicular to the plane of the celestial meridian. I.e., a great circle in a plane through the vertical at a point and perpendicular to the celestial meridian. Also called the prime vertical circle. The plane of the prime vertical cuts the horizon in points due east and west. (3) The plane of the prime vertical as defined above.

vertical, principal - That vertical circle which passes through the points on the horizon which lie due north and south of the observer, coinciding with the celestial meridian.

vertical, true prime - The vertical circle through astronomical east and astronomical west points of the horizon.

vertical-circle left - SEE circle left.

vertical-circle right - SEE circle right.

vest (v.) - Give title to or pass ownership of property.

V/H computer - A computer carried aboard an aircraft doing aerial photography, to develop a voltage proportional to the speed of the aircraft and inversely proportional to the altitude. The voltage is used to control the image-motion compensation, the time interval for exposure, and the rate of travel of the moving grid in a view finder.

vibration - (1) A motion to and fro. (2) Of a pendulum, a single movement of the pendulum in either direction, to and fro. SEE ALSO oscillation.

vicinage - Neighborhood; vicinity.

vicinal - Neighboring.

Vienna level - SEE Stampfer level.

Vienna system of gravity - A gravity system using Oppolzer's value (corrected by Helmert) of 9.808 66 m/s² for gravity acceleration at the Kaiserliche Sternwarte in Vienna (1884) as the standard to which other values of gravity acceleration were referred. The Vienna system (V.s.) was used in F. Helmert's report to the International Association of Geodesy in 1900. It is related to the Potsdam system of gravity (P.s.) which succeeded it by the equation: (gravity acceleration of V.S.) = (gravity acceleration of P.s.) + 0.016 cm/s².

viewfinder - A device similar to a camera but having a ground glass plate in the focal plane, so that the scene can be viewed as it would appear in the camera. Viewfinders are commonly used when taking aerial photographs. The device is mounted vertically in the floor of the aircraft and is marked with graduations to determine what spacing between photographs is necessary to obtain the desired overlap. Because the viewfinder is alined with the true direction of flight, the angle of crab between the longitudinal axis of the aircraft and the direction of flight can be determined.

Vignal elevation - (1) An approximation H^V_N to the elevation at a point P_N by dividing the potential number by $\tau_\phi + \delta g_f / 2$. δg_f is the free-air correction to the value τ_ϕ at P_n and ϕ is the latitude of P_n. τ_ϕ is calculated from a gravity formula. Also called a Vignal height. (2) The value H^V_N calculated for the orthometric elevation at a point P_N as given by the formula $H^V_N = [\int_{P_0}^{PN} g\, dH] / \tau_V$, in which the quantities, have the same meanings as in the preceding definition and τ_V is either 980.529 cm/s² or, as in French practice, is calculated from the International Gravity Formula. Also called a Vignal height and Vignal normal height. The Vignal elevation is used in Western Europe.

Vignal height - SEE Vignal elevation.

Vignal normal height - SEE Vignal elevation.

Vignal's formulas - A set of formulas developed by J. Vignal, describing the growth of error with distance in spirit leveling and adopted in 1936, with some changes, by the International Association of Geodesy (IAU). Vignal's formulas replaced Lallemand's formulas. They were slightly modified, and the notation changed, by a resolution adopted by the IAU in 1948.

vignetting - (1) The diminution in brightness at the edges of an image, caused by obstructions in the optical system which cut off light coming in from the edges of the field of view. (2) A gradual reduction in the amount of exposure of parts of a photograph because some of the light arriving at the camera is prevented from reaching the film. The mounting of a lens may interfere with extremely oblique rays. An anti-vignetting filter is a glass plate made so that its transmissivity gradually decreases from the edges toward the center. It does not decrease vignetting but merely removes enough light from the otherwise unvignetted portion of the focal plane to ensure a uniform exposure over the entire surface. The anti-vignetting filter is used with many wide-angle lens systems to produce uniformly photographs. (3) A photographic process portraying a solid color as shading off gradually into the unprinted paper. Open water is often shown by this process.

vignetting filter - An optical filter whose density increases gradually from the center toward the edges. A vignetting filter is sometimes used in photograph or in printing to produce a positive of uniform (optical) density by compensating for gradual decrease in density of the negative with distance from the center.

vinculum - A short, horizontal bar placed under the seconds of a numerically expressed angle or direction to indicate that the seconds are used in connections with a value in minutes 1 less than is recorded.

vinculum, double - A pair of short, horizontal bars placed under the seconds of a numerically expressed angle or direction to indicate that the seconds are used in connection with a value of minutes 2 less than is recorded.

visibility chart - A map or diagram showing which regions can be seen and which can not be seen from a given point.

vision - The ability to form mental pictures from signals received by the eyes and sent to the brain. Vision is sometimes taken to imply the brain's active and essential participation in forming the pictures, while sight is taken to involve principally the eyes and the optic nerve.

vision, binocular - That form of vision which depends on the reception, by the brain, of signals from two eyes rather than one. Binocular vision should not be confused with stereoscopic vision, which is a particular kind of binocular vision.

vision, stereoscopic - That particular application of binocular vision which enables the observer to obtain the impression of depth by means of two different perspectives of an object (such as two photographs of the object taken from two different camera stations).

volume, in site specific - The volume of sea water per unit mass, as a function of salinity, temperature and pressure.

volume, specific - (1) The volume per unit mass, divided by the volume per unit mass of pure water at standard

temperature and pressure. Also called standard volume, in physical oceanography. (2) SEE density.

volume, standard - SEE volume, specific.

volume anomaly, specific - The departure of the water of the real ocean from the water found in a standard ocean because of different temperatures and salinities at different locations. Also called a steric anomaly.

Vondrak's method - A method of fitting a smooth curve (a sequence of third-order Lagrangian polynomials) to observed data by requiring that (a) the weighted sum of the differences between the observed values and the corresponding points on the curve, plus (b) a constant times the integral over the curve of the third-order derivatives along the curve, be (c) a minimum. This method does not require that the observations be at equal intervals.

W

wadi - A ravine or water course which is dry except during the rainy season. Also spelled ouady and wady. Some wadis are permanently dry.

wading rod - A rod graduated in feet and tenths of feet or in centimeters, used for gaging depths in shallow water particularly in streams.

wady - SEE wadi.

walk, random - (1) A sequence of movements whose distances and/or directions are random quantities. (2) The path taken by a point moving with a velocity which is constant within discrete intervals of time but changes randomly from interval to interval of time. (3) A succession of movements along segments of straight lines, the direction of length of each segment being determined in a random manner i.e., by chance. The Brownian motion of a particle is an example of a random walk. So is, to a lesser extent, the path taken by the image of a star when seeing is bad. The sequence of errors in leveling networks and traverses displays some of the characteristics of a random walk. Note that some random walks are restricted to take within a limited region (i.e., the point cannot proceed off to infinity), while others take place on a closed surface. The theory of random walks has been applied to inertial surveying.

wall, hanging - The wall or rock on the upper side of steeply inclined deposits. It is called a roof in bedded deposits.

Walsh function - One of the set of functions $W_n(x)$ defined as
$W_0(x) = +1 \quad 0 < x < 1$
$W_1(x) = +1 \quad 0 < x < \frac{1}{2}$
$W_1(x) = -1 \quad \frac{1}{2} < x < 1$, etc.
with $W_n(x) = 0$ at all end points. The Walsh function takes on the values +1 or -1 almost everywhere between 0 and 1 and resemble square waves of various wavelengths. They are orthonormal and can be generalized to two or more dimensions. They are used in the same way as Fourier and Laplace functions have been used but are simpler and are more suitable for use on automatic computing machinery.

wander - (1) Polar motion other than Chandlerian motion. (2) A secular or long-period component of polar motion. (3) SEE motion, Chandlerian. (4) SEE polar motion. (5) SEE precession, apparent.

wander, apparent - SEE precession, apparent.

wander, polar - SEE motion, Chandlerian.

Wansdorf zenith telescope - SEE zenith telescope.

want of correspondence - SEE y-parallax.

warping, crustal - Deformation of the land over extensive regions because of differential movement within the crust, probably in response to substantial movement of magmatic material.

warping, endogenic - Crustal warping caused by internal forces (in the mantle, for example).

warping, isostatic - Crustal warping caused by external forces such as loading by glaciers.

warranty deed - A legal instrument by which title to real property is conveyed and wherein the freehold is guaranteed by the grantor, his heirs or successors.

warranty deed, general - SEE deed, general-warranty.

warranty deed, special - SEE deed, special-warranty.

War Time - Daylight-saving time as prescribed in 1942 by Act of Congress and inaugurated throughout the USA on 9 February 1942; it was made effective by setting clocks ahead 1 hour at 2 A.M. standard time on that date (Public Law 403, 77th Congress). The use of War Time was discontinued on 30 September 1945 at 2 A.M.

Washington Meridian - (1) The astronomic meridian through the center of the dome of the old Naval Observatory in Washington, D.C. Also called the American Meridian. The Washington Meridian was used in defining the meridional boundaries of several western States. An Act of Congress, 28 September 1850, provided that "hereafter the meridian of the observatory at Washington shall be adopted and used as the American meridian for astronomic purposes and Greenwich for nautical purposes." This Act was repealed 22 August 1912 and the Greenwich Meridian adopted. Many early surveys in individual states are referred to the Washington Meridian. Its astronomic longitude was 77° 03' 06.276" west of Greenwich on 31 December 1960. Since that date, the astronomic longitude has changed by small amounts caused by improved methods of observation, different methods of calculation, and so on. The geodetic longitude is 77° 04' 02.24" west of Greenwich (referred to North American Datum 1927.) (2) The principal meridian, adopted in 1803, governing surveys in the southwestern portion of the State of Mississippi.

waste bank - A place where excess material from an excavation is put. Sometimes called a dump or spoil bank.

waste water - The liquid effluent of a community; that waste, of a group of people, which is carried off by water.

watch - SEE hack watch.

watch time - Time determined by reference to the inclination of a watch. I.e., clock time for which a watch is used.

water - (1) The height or elevation of the surface of the still water (or of a mathematical surface approximating that surface) at a particular instant and place, referred to the average height or elevation there. Also called the tide. The average is taken over a period longer than one day and preferably over a period of 19 years or longer. (2) The surface of still water (or of a mathematical surface approximating that surface) at a particular instant and place. Also called the tide. (3) A state of horizontal motion of the water, or some quality having to do with motion e.g., slack water; dead water. (4) A body of water such as a lake or river. Often used in the plural, as waters, with the same meaning or meaning all the bodies of water in a particular region. (5) SEE tide. SEE ALSO spring low water, Indian; interior water; neap high water, mean; sea water.

water, extreme high (low) - The greatest (least) height reached by the sea, during a given period, as recorded by a tide gage.

water, extreme low - (1) The least height of the water observed or estimated for the limits of a nautical chart, as part of the tide note included on that chart. Extreme low tide may be based on the least height observed at a tide station over a short period or a long period, or it may be an estimated value based on the best available information. It is not a recognized tidal plane and should not be confused with the lowest water resulting primarily from astronomical causes. (2) SEE water, extreme high (low).

water, high - (1) The greatest height reached by a rising tide. Use of the synonym high tide is discouraged. (2) The surface of the water at the instant when the height reached by a rising tide is greatest.

water, higher high (low) - (1) The highest of the high (low) waters occurring on a particular tidal day. (2) The highest of the high (low) waters (or single high water) of any specific tidal day due to the declinational effects of the Moon and Sun.

water, higher low - SEE water, higher high (low).

water, inland - A body of water, tidal or non-tidal, lying landward of the low-water mark or of the seaward limits of ports, harbors, bays and rivers. Also called interior water, internal water or national water. SEE ALSO waters, inland.

water, internal - SEE inland water.

water, international low - (1) A plane so low that the tide will but seldom fall below it. This definition was originally suggested for international use at the International Hydrographic Conference in London in 1919 and discussed at the Monaco Conference in 1926. The definition has not yet been generally adopted and, because of the presence of the word plane, is not above criticism. (2) A plane of reference below mean sea level by 1½ times half the distance between mean lower low water and mean higher high water. This definition has been suggested as an alternative to the one preceding.

water, low - (1) The least height reached by a falling tide. Use of the term low tide as a synonym is discouraged. (2) The position of the water's surface at the time when the least height is reached by the falling tide.

water, lower high (low) - (1) The lower of two high (low) waters of any tidal day. (2) The lowest of the high (low) waters of any specified tidal day due to the declinational effects of the Moon and Sun. (3) The lower of two high (low) waters of any tidal day where the tide exhibits mixed characteristics.

water, lower low - SEE water, lower high (low)

water, lowest low - (1) A reference surface whose distance below mean sea level corresponds to the average level of lowest low water of any tide. Also called lowest normal low water. (2) A reference surface whose distance below mean sea level corresponds to the average level of lowest low water of any normal tide. Also called lowest normal low water.

water, lowest normal low - (1) The height of the lowest water normally occurring at a place. (2) SEE water, lowest low.

water, mean high (low) - (1) The average height of all high (low) waters recorded at a particular point or station over a particular period of time such as 19 years. For tidal waters, the cycle of change covers a period of about 18.6 years, and mean high (low) water is the average of all high (low) waters for that period. For any body of water, it is the average of all high (low) waters over a period of time so long that increasing its length does not appreciably change the average. (2) The average of all the heights of high (low) water observed over the National Tidal Datum Epoch. By the word epoch used here, the National Ocean Survey means period. (3) The seaward limit of growth of a particular kind of grass (genus Spartina). This definition is a legal one accepted in some courts. (4) SEE neaps, mean high- (low-)water.

water, mean higher high - (1) The average height of the higher high waters of a mixed tide during a specific interval in particular, over a specified period of 19 years. Only the higher high water of each pair of high waters of a tidal day is included in the average. For stations with series of observations shorter than 19 years, comparisons are made of observations made simultaneously with those at a primary-control tide-station to derive the equivalent of 19 years of observation. (2) The arithmetic average of the higher high waters of each tidal day observed during a National Tidal Datum Epoch.

water, mean low - SEE water, mean high (low).

water, mean lower low - (1) The average height of all the lower low waters recorded during a 19-year period. It is usually associated with a mixed tide. Only the lower low water of each pair of low waters of a tidal day is included in the average. (2) The average of the lower low waters of each tidal day observed during the National Tidal Datum Epoch. (3) The average height of all lower low waters recorded during syzygy during a 19-year period, or a computed equivalent period.

water, navigable - (1) A body of water used or susceptible of being used, in its ordinary condition or after reasonable improvements have been made, as a highway for commerce in the customary means of travel on water. Often occurring in the plural to denote all such bodies of water in a particular region i.e., navigable waters. (2) A body of water or river which allows the passage of craft engaged upon useful commerce or travel. Under the laws of the USA, navigable waters have been and now are common highways. They include all tidewater streams and any other important, permanent bodies of water whose natural and normal condition at the date of admission of the contiguous State into the Union was such as to classify it as navigable water. The question of navigability in law, where there may be controversy, is to be decided by the courts, based on the facts and conditions in each case as these prevailed at the dates of statehood.

water, neap high - SEE neap high water.

water, neap low - SEE neap low water.

water, normal - A prepared standard of sea water, the chlorinity of which lies between 19.30 and 19.50 per mille and has been determined to 0.001 per mille. Normal water is used as a convenient standard for comparing measurements of the chlorinity of sea water by titration. It is prepared by the Hydrographical Laboratories, Copenhagen.

water, slack - (1) The state of a tidal current when its speed is near zero. (2) The moment when a reversing tidal current changes direction. (3) That period of low speed, near the time of turning of a tidal current, when the current is too weak to be of any practical importance in navigation.

water, still - Water whose level is not significantly affected by wind waves or other disturbances of short duration. Still water does not usually occur naturally on the open oceans or regions adjacent to them. It is created artificially at tide gages by placing a tube (called a stilling well of large diameter upright in the water; the tube is perforated near its lower end by many small holes around the circumference. The holes act both as inlets for the water and as filters preventing short-period disturbances from being transmitted to the interior.

water, tidal - Any water, the level of which changes periodically because of the attraction of the Moon, Sun and other bodies of the Solar System.

water, triple-point of - SEE triple-point of water.

water, tropic higher high (low) - The mean higher high (low) water of tropic tides.

water, tropic higher low - SEE water, tropic higher high (low).

water, tropic lower high (low) - The mean lower high (low) water of tropic tides.

water, tropic lower low - SEE water, tropic lower high (low).

water boundary - SEE boundary, riparian.

Water Chart, Universal - SEE Universal Water Chart.

water course - A stream flowing in a definite channel, having a definite bed and banks, and discharging itself into some other stream or body of water.

water crossing - A special set of measurements taken when extending a level line across a stream or other body of water when no suitable bridge is available and the width of the body of water is greater than the greatest distance allowed between leveling instrument and leveling rod. Also called river crossing, in the older literature. The same techniques are used in leveling across valleys. SEE ALSO valley crossing.

water datum - SEE datum.

water full and change, high - SEE establishment of the port.

water full and change, low - The average interval of time between the upper or lower transit of the full or new moon and the next low water.

water gap - A pass, in a mountain ridge, through which a stream flows.

water inequality - SEE inequality.

water interval - SEE interval.

water level - (1) The elevation, or the height used as an approximation to the elevation, of a particular point or small patch on the surface of a body of water above a specific point or surface, averaged over a period of time long enough to remove the effects of disturbances of short duration. Water level should not be confused with sea level. Water level deals with a surface of small area; sea level deals with the general surface of an entire sea. (2) A leveling instrument consisting of water contained in a long U-tube. The line of sight is established by sighting across the surfaces of the water at the two ends of the tube. The U-tube may be entirely of glass or it may consist of two straight, upright tubes of glass connected at their bottoms by a plastic tube. This kind of leveling instrument suffices for coarse measurements only.

water level, half-tide - (1) The average of the heights of mean high and mean low waters. Also called half-tide level and mean tide level. (2) That ideal surface half way between mean high water and mean low water. (3) The plane lying exactly midway between the points of high water and low water.

water level, legal - That stage of a body of water at which the shore line defines the riparian boundaries e.g., the normal high-water line on a lake.

water level, mean - (1) The average position of the water's free surface at a particular place. (2) The height determined by averaging the heights of the surface of water at equal intervals of time (usually hourly) over a considerable period of time. Mean sea level is distinguished from mean water level in relating to the entire body of water and not just to that part whose height is actually measured.

water level, mean-tide - SEE water level, half-tide.

water level, normal - The water level most prevalent in a water course, reservoir, lake or pond, and generally indicated by a shore line of permanent vegetation characteristic of that growing on land. Along the shores of large bodies of water, the action of waves may retard the growth of of such vegetation above normal water level.

water level, still - SEE level, still-water.

water leveling - Determining differences of elevation between points by measuring the relative elevation of each point with respect to the surface of a body of still water such as a lake. The surface of a body of still water is very nearly a level surface and provides a common surface of reference for all the points. The term water levels is generally used with this meaning.

water level line, mean - (1) The line formed by the intersection of the land with the water's surface at the height of meanwater-level. Also called the normal water-level line. (2) That line, on a map, which represents the intersection of the land with the water's surface at the height of mean water level.

water level line, normal - SEE water level line, mean (1).

water level transfer - SEE transfer of water level.

water line - (1) The line formed by the intersection of the land with the surface of a contiguous body of water. (2) That line, on a map, which represents the intersection of the land with the surface of a contiguous body of water.

water luni-tidal interval - SEE luni-tidal interval.

water mark - A line or mark left on the shores of a body of water by the water as an indication of the water's former elevation or height. Because marks below the usual level of the water tend to be removed by the action of waves, currents or tides, water marks usually indicate the highest levels reached by the water. SEE ALSO mark, high-water; mark, low-water.

water parting - The high land forming the division between two contiguous river-basins.

water rights - SEE rights, riparian.

waters - A general term for the hydrosphere or some part of it.

waters, inland - Those waters, both tidal and non-tidal, of a nation which lie landward of the marginal sea (baseline), as well as the waters, such as lakes and rivers, within its lands and over which the nation exercises complete sovereignty. Also referred to as national waters.

waters, national - SEE waters, inland.

waters, non-navigable - Waters which are not navigable waters (q.v.).

waters, navigable - SEE water, navigable.

waters, territorial - SEE sea, marginal.

watershed - (1) The whole region supplying water to a river or lake, catchment or basin. (2) The boundary between one drainage and another. Also called a divide or water parting. (3) The region contained within a line above a specified point on a stream. In water-supply engineering, it is called a watershed. In river-control engineering, it is called a drainage area, drainage basin or catchment area.

water springs - Tides (waters) of increased range occurring semi-monthly as the result of the Moon's having approximately the same right ascension (or longitude) as the sun.

water supply - A volume of water treated (or otherwise safe for drinking) and ready for distribution.

water table - (1) The upper surface of a zone, in the ground, which is saturated with water except where that surface is formed by an impermeable body. (2) The locus of those points, in water in the soil, at which the pressure is equal to atmospheric pressure.

wave, forced - A wave generated and maintained by a continuous force, in contrast to a free wave.

wave, free - A wave which, once started, continues without the application of further application of force.

wave, myriametric - A very-low-frequency radio-wave approximately 10 km or more in length (less than 30 kHz in frequency).

wave, standing - A wave whose points of zero amplitude are fixed in place and do not move with time. Also called a stationary wave. The points of zero amplitude are called the nodes or nodal points of the wave. One theory of the tides postulates that the tidal waves are a system of standing waves, any progressive movement of the waves being of secondary importance except as the tide advances into tributary waters.

wave, tidal - (1) That wave, in the oceans and seas, which evidences itself as tides and tidal currents. The term is used by some oceanographers to denote such a wave in shallow waters only. Because all waters are shallow compared to the wavelength of the tidal wave in the open oceans, such usage is ambiguous. (2) SEE tsunami.

wavelength - (1) The distance, on a wave, from one maximum or minimum to the succeeding or preceding maximum or minimum. (2) The speed with which a wave propagates, divided by the frequency of the wave. The two definitions can be taken as equivalent. Which is used depends on custom and application. E.g. Light is usually specified directly as wavelength because this is more easily measured than frequency. Wavelengths of sound are usually determined from the frequencies.

wave number - The reciprocal of the wavelength.

wave number, effective - The wave number of such monochromatic radiation as would have the same effect as the radiation actually present.

way - (1) A passage, path, road or street. (2) A right of passage over land.

wayland - The land abutting a right-of-way.

weather map - A map of the weather at present, or is predicted to be, over a considerable area. The map is usually based on observations taken at the same time at several weather stations. Also called a synoptic chart and surface chart.

weber - The unit of magnetic flux equal to 1 joule/ampere.

weber per square meter - The unit of magnetic-flux density B in the Système International d'Unités, in which system the magnitude is given the name tesla. B is related to force F and current J by the equation $F = J \times B$. The geomagnetic field is measured in terms of b and varies from about 0.000025 to 0.00007 tesla in magnitude.

Weddle's rule - The formula
$$\int_{x_k}^{x_k + 6\Delta x} f(x)\, dx = (3\,\Delta x/10)(y_k + 5y_{k+1} + y_{k+2} + 6y_{k+3} + y_{k+4} + 5y_{k+5} + y_{k+6}).$$
The fraction 3/10 was obtained by changing the coefficient 41/140 of the sixth-degree difference in the Newton-Cotes formula to 42/140. Weddle's rule is more accurate than Simpson's rule but requires 7 consecutive, known values of y rather than 3 as in Simpson's rule.

wedge - (1) SEE wedge, optical. (2) SEE gray scale. SEE ALSO distance wedge; step wedge.

wedge, continuous - SEE gray-scale, continuous-tone.

wedge, optical - A refracting prism of very small deviation, which can be placed in an optical system to introduce a small bend in the path of the rays. By placing a pair of such wedges, rotatable equally in opposite directions about a common axis, the deflection of the ray can be controlled very precisely. An optical wedge may be placed, for example, in the eyepiece of a stereoscope or as an attachment to the eyepiece of a theodolite, and used for determining distances. Also called a wedge and wedge prism. SEE ALSO distance wedge.

wedge distortion - Distortion caused, in a lens system, by the failure of an element or component to have its optical axis alined with the optical axis of the rest of the system. Also called decentering distortion. The distortion can be simulated by placing a wedge-shaped prism on the optical axis of a lens system.

wedge prism - SEE wedge, optical.

wedge reticle - A reticle carrying a cross-shaped design, one arm of the cross being split in half longitudinally and the two halves spread apart to form a V (wedge); the apex of the V is at the center of the cross. The wedge reticle is common in leveling instruments because it permits the horizontal line to be set more precisely on the center of a graduation of the leveling rod.

wedge unit - The interval between two graduations on the scale of the rotary wedge attached to an Ni2 leveling instrument for use in river crossings.

Wegener's theory - The theory of continental drift as proposed by A. Wegener (1915), according to which the continents were originally part of one, original Urcontinent from which the continents as we know them today split off and drifted to their present positions. The split occurred, according to present theory, about 200 000 000 years ago. Wegener's theory is the original of the present-day theory of plate tectonics - a theory differing from Wegener's theory principally in having a more detailed explanation of the physical processes involved and in having the entire crust divided into segments called plates which may be of continental size or smaller and which drift about.

weight - (1) The force with which a planet or satellite attracts a body on its surface. In particular, the force with which the Earth attracts a body on its surface. The weight is equal to the product of the mass of the body by the acceleration imparted by gravity. It should not be confused with the mass of the body (as the U.S. Bureau of Standards does). Note that the body must be on the surface of the planet or satellite or attached to it in some manner. Otherwise the body does not participate in the rotation of the attracting body. (2) SEE weight,statistical. (3) SEE factor, weighting.

weight, statistical - A factor by which a quantity is multiplied to increase or decrease the effect of that quantity on the results of an adjustment. Usually referred to as weight when the meaning is clear. Also called a weighting factor. When statistical weights are applied, each quantity of a similar nature in the adjustment is assigned a statistical weight; this weight indicates the relative importance of the quantity (or combination of quantities) it multiplies. In geodesy, the most common form of statistical weighting is the reciprocal of the variance of a quantity. It is used most frequently in adjustments by the method of least squares, where the equations are linear.

weighting - The process or method of applying statistical weights.

weighting factor - SEE weight, statistical.

weightman - The person responsible for applying tension to a surveyor's tape in use, so that the tape is under correct tension and in correct position when measuring a base line.

weight matrix - (1) A matrix containing the weights (adjusting coefficients) assiogned to the terms in a system of simultaneous linear equations. (2) The reciprocal (inverse) of a covariance matrix.

weir - An artificial barrier placed in an open channel and over which the water flows.

Weissbach's method - A method of carrying direction from a line of known azimuth on the surface, or at one level, to a horizontal line of unknown azimuth in another level by measuring directions and distances at the surface (or in the first level), using the line of known azimuth, to two plumb lines in a vertical shaft extending to the second level in question, and repeating the measurements at that level, using the line of unknown azimuth. The azimuth of the unknown line can then be calculated by solving for the unknowns in the two triangles created by the measurements.

Weissbach's triangle method - SEE Weisbach's method.

Weisse method (of connecting surveys) - SEE Hause method.

well, stilling - A large enclosure placed about a tide gage to prevent waves from affecting the level of water at the gage itself. The enclosure reaches from the bottom to above the highest expected waves. Small holes a short distance above the bottom let water flow slowly in or out. SEE ALSO float well.

well hydrograph - A graphical representation of the fluctuations of the water-surface in a well, plotted as ordinates, against time as abscissas.

Werner's map projection - An equal-area map projection from the sphere to the plane as defined by the formulas $r = \frac{1}{2} k R (\pi - 2\phi)$ and $\theta = (2 \Delta\lambda \cos \phi) / (\pi - 2\phi)$, in which r and θ are the polar coordinates of that point, in the plane, which corresponds to the point, on the sphere, at longitude $\lambda_o + \Delta\lambda$ and latitude φ; λ_o is the longitude of the central meridian. R is the radius of the sphere and k is the scale factor. Also called the Stab-Werner map-projection. It is a particular case of the Bonne map projection, with the standard parallel of latitude at the pole and the cone becoming a plane tangent at the pole. It was used by Goode for his polar equal-area map.

westa - An unit of area equal to 60 acres or half a hide of land.

westerly - A direction within 22.5° of west.

westing - (1) The distance westward (positive) or eastward (negative) from a meridian to which the distances are referred. Negative, eastward westings are usually given a positive sign and called eastings. (2) The distance measured westward (positive) or eastward (negative) from a central line (central meridian or y-axis) on a graticule or grid and multiplied by a scale factor to give the corresponding distance on the ellipsoid. (3) SEE departure.

west point - The western intersection of the plane through the vertical and perpendicular to the astronomical meridian with the horizon.

wetland - A region permanently or intermittently covered by shallow water e.g., a marsh, swamp or bayou.

wharf - (1) A structure of timber, stone, concrete or other material running parallel to a contiguous to the shoreline of a navigable body of water, and providing a place on one side for vessels to berth and dischage and receive cargo, passengers, stores and/or fuel. Also called a quay. (2) Any structure, on a waterfront, designed to make it possible for vessels to lie alongside and take on or unload cargo, pasasengers, etc.

whirler - A disk rotating about an axis through its center and carrying a sheet of plastic to be coated for scribing. The disk is usually large enough to carry four sheets at a time. It rotates at about 50 r.p.m. in a nearly vertical position. Paint is poured on at the center and moves slowly outward until the entire surface is covered uniformly.

Wiechel's map projection - A pseudo-azimuthal map projection mapping the rotational ellipsoid onto the plane according to the formulas $r = 2 k R \sin \frac{1}{4} (\pi - 2\phi)$ and $\theta = \lambda + \frac{1}{4} (\pi - 2\phi)$, in which r and θ are the polar coordinates of that point, in the plane which corresponds to the point, on the ellipsoid, at geodetic longitude λ and geodetic latitude φ. It was first introduced by Wiechel in 1879.

Wiechert-Gutenberg discontinuity - A seismic discontinuity between mantle and core, in which the speed of P-waves decreases and the speed of S-waves drops to zero. The discontinuity occurs at a depth of about 2 900 km and appears to lie within a region about 1 km thick. Also called the Gutenberg discontinuity and the Gutenberg-Wiechert discontinuity.

wiggling in - The technique of setting an instrument on a direct line between two specified points by successively sighting first on the one point and then on the other and moving the instrument until the two sightings differ by exactly 180°. Also called ranging-in. It is used in surveying when neither of the two points can be occupied.

wiggling-in on line - SEE centering, double.

will - The written instrument, legally executed, by which one provides for distribution of one's estate after one's death.The instrument, completely written in the handwriting of the testate, by which one provides for the distribution of ones estate after one's death.

wind (of a spirit level) - The lack of parallelism between the axis of the vial of a spirit level and the line joining the centers of the supports of the vial. Also spelled wynd. Also called spirit-level wind. When wind (pronounced to rhyme with find) is present and the spirit level is rocked on its supports, the bubble will respond with a longitudinal motion.

wind, solar - The flux of particles emitted by the Sun and sweeping through interplanetary space. This flux, predicted by Bierman in 1951 as the cause of acceleration and ionization of the tails of comets, occurs as a thin cloud (about 5 particles per cubic centimetre at the distance of Earth) moving with a speed of about 400 km per second. That part of the solar wind reaching the Earth is deflected by the Earth's magnetic field to form a many-layered sheath bending the lines of the magnetic field. The solar wind also causes auroras and magnetic storms.

window - A region of the spectrum to which the cloudless atmosphere is nearly or completely transparent. SEE ALSO entrance window; exit window.

wind rose - (1) A diagram showing the relative frequency with which winds blow from different directions. A wind rose may also show average speed or the frequency with which various speeds occur in different directions. (2) A diagram showing the average relation between the occurrence of winds from different directions and the occurrence of other meteorological phenomena.

windshield survey (USPLS) - SEE reconnaissance survey.

windward - The direction from which the prevailing littoral drift of sand moves.

wing photograph - A photograph taken through one of the side (wing) lenses of a multi-lens camera.

wing point - One of three easily-identifiable points, one near each end and one near the middle of each side of an aerial photograph.

winter solstice - (1) That point, on the ecliptic, occupied by the Sun at its greatest southerly declination (about - 23.5°). Also called December solstice and First Point of Capricornus. (2) That instant at which the Sun reaches the point of greatest southerly declination. This is about
22 December and marks the beginning of winter in the Northern Hemisphere.

wintze - A vertical or inclined passage driven downwards from a point inside a mine, to connect with a lower level or to explore the earth for a limited distance below a level. The only difference bvetween an wintze and a raise is that the former is driven downwards, the latter upwards.

wire - SEE surveyor's wire.

wire, sounding - SEE sounding wire.

wire, surveyor's - SEE surveyor's wire.

wire drag - SEE drag (3).

wire length, correction to - SEE the various corrections for tape length. Those corrections are in general also applicable to the length of surveyor's wire.

wireless method (of determining longitude) - SEE radio method (of determining longitude).

witness, expert - A witness in a case who testifies in regard to some technical or professional matter and who is permitted to give their opinion on such a matter because of their special training, skill or familiarity with the matter.

witness corner - (1) A marked point established on firm ground at a measured distance and direction from a corner which may itself be so situated that it cannot be marked permanently. A witness corner in land surveying corresponds to a reference mark in control surveying. When the true point for a corner falls upon an inaccessible place such as an unmeandered stream, lake, or pond, or in a marsh or upon a precipitous slope or cliff, a witness corner may be established at some suitable point where the monument may be permanently constructed, but preferably on a line of the survey. Usually only one witness corner will be established in each instance, and it will be located upon any one of the surveyed lines leading to a corner it a secure place within a distance of 10 chains is available. (2) (USPLS) A monumented, surveyed point near a corner, established as a reference mark when the corner is so situated as to render its monumentation or ready use impracticable. A witness corner is marked in a prescribed manner comparable to that for the true corner.

witness mark - A mark placed at a known distance and direction from a property corner or survey station to aid in the recovery and identification of that station. Also called a witness post or witness stake. A mark which is established with such precision and accuracy that it may be used to restore or take the place of the original station is more properly called a reference mark in control surveys and a witness mark in land surveys.

witness point (USPLS) - A monumented station on a line of a survey, used to perpetuate an important location remote from, and without any special reference to, any regular corner.

witness post - SEE witness mark.

witness stake - SEE witness mark.

witness tree - SEE bearing tree.

wobble - (1) Polar motion other than Chandlerian motion. (2) An erratic perturbation of a regular motion. (3) SEE motion, Chandlerian. (4) SEE motion, polar. The term wobble is vague and should defined precisely if it must be used.

wobble, Chandlerian - SEE Chandlerian motion.

Wollaston prism - A pentagonal prism having, between faces, three dihedral angles each equal to 135° and two dihedral angles each equal to 67.5°. The Wollaston prism has been used, for example in the Danjon astrolabe.

Woodward base line apparatus - SEE base-line apparatus, iced-bar.

words of exclusion - In a surveying description, the words between, by, from, on and to are words of exclusion unless there is something in the phrase that makes it apparent that the words were used in a different sense. by the river, on Brown's land and to a stone mound exclude other terms.

work - (1) The product of the force exerted upon a body times the distance the body moves in the direction of the force. Denoting the force by F and the location of the body by s, the work W done in moving the body from point P to point Q is given by $W = \int_P^Q F ds$. Work is a measure of the amount of energy expended in transforming a system from one physical state to another by accelerating masses in the system. An agent doing positive work is losing energy to the body on which work is being done, and one doing negative work is gaining energy from that body. (2) The design (map or chart) to be reproduced. SEE ALSO line work. (3) The procedures and operations of geodesy or surveying. SEE, FOR EXAMPLE leveling; survey; traverse; triangulation; trilateration.

work, continuous-tone - A design which is to be reproduced and which contains continuous shades or tones of gray or a color, down to the limit of resolution of the optical system and process being used.

working-in on line - SEE centering, double.

work unit - A single project, in leveling.

World, International Map of the - SEE International Map of the World.

World Aeronautical Chart - One of a standard series of charts at a scale of 1:1 000 000 and designed for aerial navigation throughout the world. For regions outside the USA, these charts have been superseded by the Operational Navigation Charts.

world datum - A datum specifying a coordinate system to which points throughout the world are or can be referred.

world geodetic system - (1) A set of points located in a coordinate system having its origin at the Earth's center of mass. (2) A datum to which control points throughout the world are referred. (3) A consistent set of quantities describing the size and shape of the Earth, the locations of a number of points with respect to the center of mass of the Earth, transformations from major geodetic datums to the datum of the system, and the gravity potential of the Earth. SEE ALSO World Geodetic System 72. (4) A datum so defined that it is suitable for world-wide use and specifying that the origin be at the Earth's center of mass, together with a set of transformations from other geodetic datums,

coordinates of particular points, and a representation of the gravity field.

World Geodetic System 72 - A set of constants specified by the U.S. Department of Defense in 1972 and consisting of the following constants: a datum, the coordinate system of which has its origin at the Earth's center of mass; the coordinates, in that datum, of a number of points around the world; constants for transforming from other datums to the System's datum; and values determining the Earth's gravity-potential. The system has been replaced by World Geodetic System 84.

World Geodetic System 84 - A set of constants specified by the U.S. Department of Defense in 1984 and consisting of the following constants: a datum, the coordinate system of which has its origin at the Earth's center of mass; the coordinates, in that datum, of a number of points around the world; constants for transforming from other datums to the System's datum; and values determining the Earth's gravity- potential.

World Geographic Reference System - SEE GEOREF grid.

world polyconic grid - A grid derived mathematically from the elements of a polyconic map projection.

World Polyconic Grid - A military-grid system in which a grid is applied to polyconic projections of zones of the Earth's surface covering 9° of longitude with 1° of overlap between zones and extending to 72° north and south latitudes.

Wright-Hayford rule - The rule that a measurement should be rejected if its residual exceeds five times the probable error of a single measurement.

writings, ancient - Documents bearing on their face every evidence of age and authenticity, of age of 30 years, and coming from a natural and reasonable custody. These are presumed to be genuine without express proof when coming from the proper custody. Only the original of a deed, not the record copy, can be considered as an ancient document.

writ of ejectment - A legal document issued by a court and ordering a person to vacate designated premises.

writ of entry - A real action to recover the possession of land where the tenant (or owner) has been disseised or otherwise wrongfully dispossessed.

writ of execution - A court order authorizing and directing the proper officer of the court (usually the sheriff) to carry into effect the judgment or decree of the court.

wrong-reading (adj.) - Of an image, reversed in appearance; having left-to-right in the object appearing as right-to-left in the image. I.e., appearing as a mirror image. Also called reverse reading, but this term is not recommended.

WWV time - Time determined by reference to the signals from radio station WWV or WWVH. The signals originate at the radio stations operated by, and were derived from clocks run by, the National Institute of Standards and Technology (NIST), formerly the National Bureau of Standards (NBS). The signals are set to Universal Time Coordinated (UTC). SEE ALSO time service; time signal.

WWV time signals - Accurately cvontrolled time signals brtoadcast from radio stations WWV in Boulder, Colorado and WWVH in Hawaii. SEE ALSO time service; time signal.

WWVH time signals - SEE WWV time signals.

wye - A Y-shaped fixture for supporting or holding a cylindrical or disk-like object. A single or composite lens is commonly held, on an optical bench, in a single wye. The telescope in a wye-level is supported on a pair of wyes.

wye level - A type of leveling instrument having the telescope and attached spirit level supported in wyes (Y's) in which it can be rotated about its longitudinal axis (collimation axis) and from which it can be lifted and reversed, end for end. Also called a Y-level and wye-type leveling instrument. The wye level was invented in 1740 by J. Sissons. The adjustments made possible by this mounting are peculiar to the instrument. The wye level is one of two general classes of leveling instruments, the other being represented by the dumpy level.

wynd - SEE wind.

X

x axis - (1) That axis, in a coordinate system, along which the values the variable labelled x are plotted. (2) SEE axis, fiducial.

x-correction - (1) The correction to an x direction. (2) The correction at a component along the x-axis.

x direction - An observed direction, in a figure in triangulation, for which an approximate value is obtained and treated like an observed direction in adjusting the figure. Adjustment of a figure in triangulation by the method of least squares sometimes requires the use of an x direction, for which an approximate value is obtained by computing distance and direction between two given points, by solving the three-point problem, or by other means, and then using this x direction in the adjustment and obtaining a correction (the x correction) for it to make it consistent with the adjusted values of the observed directions.

x-displacement - A movement of an image in that direction which, in the machine holding the image, is labeled x.

xerography - A copying process which projects an image of the original onto a sheet of paper on which an electrostatic charge following the pattern of the original is created. A powder is deposited on the charged portions and is fixed to the paper by heat.

x-motion - That linear motion made approximately parallel to a line connecting two projector stations in the adjustment of a stereoscopic plotter; the path of the motion corresponds, in effect, to the flight line between the two corresponding exposure stations.

x-parallax - SEE parallax, absolute stereoscopic.

x-scale - In an oblique photograph, the scale along a line parallel to the true horizon.

x-tilt - Given a rectangular, Cartesian coordinate system in the plane of an aerial photograph, with the x-axis alined approximately in the direction of flight, the x-tilt is the angle between the y-axis and a horizontal plane or between the y-axis and the horizontal reference plane of the stereoscopic plotter. Also called list or roll when related to the orientation of the aircraft from which the photograph was taken. x-tilt is positive if the y-coordinate of the nadir point is positive.

Y

Yakutat Datum - The horizontal-control datum defined by the following coordinates of the origin and by the azimuth (clockwise from South) of the azimuth mark, on the Clarke spheroid of 1866; the origin is at the Yakutat astronomic station:

longitude of origin	139° 47' 15.60" W
latitude of origin	59° 33' 50.50" N.
azimuth from origin to azimuth mark	0° 00' 19.5"

The coordinates and azimuth were determined astronomically in 1892, the longitude being determined by the chronometric method.

yard (length) - A unit of length equal to exactly 36 inches and 3600/3937 meter. The yard has been the basic unit of length in the English system of measure since at least 1742, when a brass bar inscribed with a 3-foot scale was made by the Royal Society. A copy made in 1760 was adopted by Act of Parliament, 1 January 1826, as embodying the legal definition of the yard. This copy was called the Imperial Standard Yard and was stored in the Houses of Parliament. It was destroyed by fire in 1834. A new standard of bronze and gold was constructed by comparing existing copies; its length was, in 1855, designated legally to be the Imperial Standard Yard. The 1894 value remained the legal value until 1963. In 1959 the International Yard was adopted in the British Commonwealth and the USA for scientific purposes. In 1963 the British yard was redefined officially as exactly 0.9144 meters. Until 1836 there was no standard unit of measure in the USA. In that year, an Act of Congress established the yard as a standard, defining it as the distance between the 27th and 63rd inches of the Troughton Bar, an 82-inch long, graduated brass bar brought from England by F. Hassler. On 28 July 1866, Congress passed a law making use of the metric system legal in the USA and defining the yard as 3600/3937 of the meter. In 5 April 1893, the Secretary of the Treasury approved an order by the Office of Weights and Measures officially establishing the definition given in the act of 28 July 1866. This order applied specifically to weights and measured used officially by the U.S. Government and by the separate States. For measurements of base lines made by the U.S. Coast and Geodetic Survey, the meter had been in use since Hassler brought to the USA in 1805 a copy (in iron) of the French meter of 1799. SEE ALSO Dominion Standard Yard; Imperial Standard Yard; International Yard.

yard (railway) - A set of tracks devoted to such work as making up trains, storing cars and sorting cars, and over which movements are not (in general), subjected to timetables or train orders.

yard, Canadian - SEE Dominion Standard Yard.

yard, gravity - SEE gravity yard.

yard, hump - SEE hump yard.

yard leveling rod - (1) A leveling rod graduated in yards and hundredths of a yard so that the sum of readings on the three horizontal lines of the reticle in the leveling instrument equals the length of the foresight or backsight in feet. (2) A leveling rod graduated in yards and hundredths of a yard.

yaw - (1) The variation, or amount of variation, of the longitudinal axis of a craft from the direction in which the craft is moving; i.e., the difference between the direction in which a craft is pointed and the direction in which it is moving. Also called crab. (2) The rotation, or amount of rotation, of a camera or photograph about either the photograph's z-axis or the z-axis in object space. Also called angle of yaw. If, as often happens, there may be ambiguity as to whether the movement or the amount of movement is meant, use yaw for the movement and angle of yaw for its amount. SEE ALSO angle of yaw.

yaw angle - SEE angle of yaw.

y-axis - (1) That axis, in a coordinate system, which is labeled y-axis and along which the values of the variable y are plotted or scaled.

y-displacement - A movement of a point in image space in that direction which, in the machine holding the iumage, is labeled y.

year - The interval of time needed for the Sun to make two successive passages through the same, designated plane. This is equivalent to specifying the year as the interval between two successive passages of the Earth through some designated plane. The direction of the plane is usually fixed with respect to (a) the stars, resulting in the sidereal year; (b) the vernal equinox, resulting in the tropical year; or perihelion, resulting in the anomalistic year. There is also a calendar year, a conventional interval whose value is based on the tropical year. SEE ALSO calendar year; eclipse year; leap year; light year.

year, anomalistic - The period of one revolution of the Earth around the Sun, from perihelion to perihelion. The anomalistic year contained 365 days, 6 hours, 13 minutes and 53.16 seconds in 1955 and was increasing at the rate of 0.002627 seconds annually.

year, astronomical - (1) The period of time between two successive passages of the Sun through the same right ascension (the Besselian year) or longitude (the tropical year). The length of the day and the length of the astronomical year are incommensurable. (2) SEE year, tropical.

year, Besselian - SEE Besselian year.

year, Besselian solar - SEE Besselian solar year.

year, climatic - An interval one year long and containing the times for which measured or computed values of the amount of water supply, precipitation, etc., are given.

year, embolismic - SEE leap year.

year, equinoctial - SEE year, tropical.

year, fictitious - SEE year, Besselian.

year, fiscal - An interval of time usually equal in length to the calendar year but beginning and ending on dates convenient for accounting purposes.

year, great - The interval of time required for one complete cycle of the movement of the equinoxes around the ecliptic. Also called a Platonic year. It is about 25 800 ordinary years long.

year, Julian - SEE Julian year.

year, lunar - An interval of time equal to 12 synodic months. It is approximately 354.367 d.

year, natural - SEE year, tropical.

year, Platonic - SEE year, great.

year, sidereal - (1) The interval of time from the instant when the Sun moves from a given celestial longitude to the instant when it returns to the same longitude, completing one circuit of the ecliptic. (2) The average interval of time needed for the Sun to move from a given celestial longitude around the ecliptic back to the same longitude. The average length of the sidereal year is approximately 365.25636 mean solar days.

year, solar - SEE year, tropical.

year, tropical - (1) The interval of time during which the Sun's mean longitude, referred to the mean equinox of date, increases by 360°. Also called an astronomical year, equinoctial year, natural year and solar year. The tropical year is the basis for the conventional calendar-year used in chronology and civil reckoning. It is, to a close approximation, the average length of time required for the Sun to pass from vernal equinox to vernal equinox. (2) The average interval between two successive passages of the Sun through the vernal equinox.

Y-level - SEE level, wye.

Yof Astro Datum 1967 - A horizontal-control datum defined by the following coordinates of the origin and by the deflection of the vertical at the origin, which is at station Yof Astro 1967; the ellipsoid is the Clarke spheroid of 1880:
longitude of origin 342° 30' 52.98" E
latitude of origin 14° 44' 41.62" N
deflection of the vertical 0.0

Yolo Base Datum - SEE Standard (California) Astronomical Datum 1885.

y-parallax - The difference between the perpendicular distances of the two images of a point, on a pair of photographs, from the vertical plane containing the air base. Also called vertical parallax, instrumental parallax and want of correspondence. The existence of y-parallax is an indication of tilt in either or both photographs, or of a difference between the scales of the photographs. It interferes with stereoscopic examination of the pair. Also called "want of correspondence".

y-scale - On an oblique photograph, the scale along the line of the principal vertical or any other line, inherent or plotted, which, on the surface photographed, is parallel to the principal vertical.

y-swing method - SEE one-swing method.

y-tilt - Given a rectangular, Cartesian coordinate system on a photograph, with the x-axis being more nearly in the line of flight than the y-axis, y-tilt is the angle between the x-axis and a horizontal plane or the angle between the x-axis and the reference plane of the stereoscopic plotting instrument. Also called pitch when considered in relation to the orientation of the aircraft from which the photograph was taken.

Yukon Datum - The horizontal-control datum which is defined by the following coordinates of the origin and by the azimuth (clockwise from South) to triangulation station Bald, on the Clarke spheroid of 1886; the origin is at triangulation station Boundary:
longitude of origin 141° 00' 00.00" W
latitude of origin 4° 40' 51.42" N
azimuth from origin to Bald 270° 00' 00.0"
The datum is based on a single astronomic station near the crossing of the 141st meridian and the Yukon River, and was adopted for use in computing the triangulation along that part of the Alaska-Canada boundary defined as being along the 141st meridian.

Z

z-axis - That axis, in a coordinate system, which is labeled z and along which the values of the variable z are measured or plotted.

Zeiss parallelogram - SEE parallelogram of Zeiss.

zenith - That point at which a directed line opposite in direction from the direction of the plumb line at a specified point on the Earth's surface meets the celestial sphere. Also called the astronomical zenith. The zenith and nadir are poles of the horizon. The plumb line is perpendicular to the surface of the geoid but not, except in rare cases, to the surface of the ellipsoid representing the geoid. The angle between the plumb line (the vertical) and the perpendicular to the surface of the ellipsoid (the normal) is the deflection of the vertical (also called the deflection of the plumb line). The terms geodetic zenith and geodetic nadir, and the terms geocentric zenith and geocentric nadir are sometimes used with meanings different from those given.

zenith, astronomical - SEE zenith.

zenith, geocentric - That point at which a line from the center of the Earth through a point on the Earth's surface meets the celestial sphere. This term is sometimes used in astronomy, but it seldom appears in geodesy. It should be used only in its entirety; the single word zenith should be reserved to designate the point determined by the direction of the vertical.

zenith, geodetic - That point at which the normal to an ellipsoid representing the geoid, extended upward, meets the celestial sphere. The term has some use in geodesy but should be used only in its entirety. The single word zenith is reserved for the point determined by the direction of the vertical.

zenithal (adj.) - SEE azimuthal.

zenith angle - (1) The angle, measured in a positive direction downwards, from the observer's zenith to the object observed. Also called zenith distance and co-latitude. It is usually denoted by ζ or by z. The former is preferred except when it may be confused with the deflection of the vertical (also usually denoted by ζ). (2) SEE azimuth, astronomical.

zenith camera - (1) A portable camera designed for photographing the stars in the neighborhood of the zenith, these photographs to be used for determining the astronomical longitude and altitude of the observing station. The camera has a small field of view. It has leveling screws and a very sensitive level so that the optical axis may be set precisely in the vertical. The time of each observation is recorded. The focal length is less than 1 meter. (2) A camera designed particularly for photographing stars in the immediate neighborhood of the zenith. A zenith camera (2) differs from a zenith camera(1) in that it is not portable, has usually a smaller field of view, and is used for other purposes than determining longitude and latitude e.g., determining time or corrections to a stellar catalog. One type is called a photographic zenith telescope. The focal length is typically several meters.

zenith distance - SEE zenith angle.

zenith distance, double - SEE angle, double-zenith.

zenith sector - (1) An astronomical instrument for measuring zenith distances, in which the position of a plumb line is read against a short, vertical, graduated arc. In its early form, before the invention of the telescope, the zenith sector consisted of a quadrant and a plumb line. With the introduction of the telescope, zenith sectors used in astronomical geodesy became short, vertical arcs attached to telescopes up to 1.5 or 2 meters long, the vertical angle being read by means of the plumb line suspended from the horizontal axis of the instrument. (2) An astronomical instrument for measuring zenith distances, in which the position of a short, vertical, graduated arc is read against an index mark on a spirit level suspended from the horizontal axis. A telescope equipped with an ocular micrometer was attached to the graduated arc. Zenith sectors were used for measuring zenith distances and differences of zenith distances of stars close to the zenith. For determining astronomic latitude, the zenith sector has been replaced by the zenith telescope.

zenith telescope - Any optical telescope designed particularly for making observations on stars near the zenith. In particular, a portable optical telescope adapted for measuring small differences of zenith distance, and used in determining astronomic latitude. (Also called a Wansdorf zenith telescope.) This type of instrument consists of a telescope having an ocular micrometer and a spirit level attached, and so mounted on a vertical support that it can be placed in the plane of the meridian for observations on a star culminating north (or south) of the zenith, and then rotated 180° in azimuth and a second star then observed as it culminates south (or north) of the zenith. The difference of the zenith distances of the two stars is measured with the micrometer; the spirit level is used to determine any change occurring in the direction of the telescope's axis of rotation between the two observations. The present form of this instrument is essentially the invention, in 1834, of Captain Andrew Talcott of the U.S. Engineers. It was adopted by the U.S. Coast and Geodetic Survey for determining latitude, and some improvements were made. SEE ALSO zenith-telescope method (of determining latitude).

zenith telescope, photographic - An optical telescope designed for photographing stars in the immediate vicinity of the zenith. The photographic zenith telescope may be portable, in which case it is used primarily for determining longitude and latitude; or fixed, in which case it is used for determining time and variations in latitude (and may be called a photographic zenith tube). In the portable version, the optical axis is pointed vertically by leveling against very sensitive levels; photographs are taken with the photographic negative in one position only. In fixed versions, the photographic plate is commonly rotated through 180° after each exposure.

zenith tube, photographic - SEE zenith telescope, photographic.

zero - A point taken as the origin of coordinates. Also called the zero point or null. These usages is common in surveying.

zero, absolute - That state (temperature) of a body at which no thermal energy can be transferred from it to any other body. A temperature scale in which absolute zero is assigned the value zero is called an absolute temperature scale. Absolute zero is at -373.15° on the new Celsius (Centigrade) scale. By definition, the triple-point of water is at +273.16° on the absolute Kelvin temperature scale. In the classical theory of heat, absolute zero was supposed to be the temperature at which the atoms of a substance are at rest with respect to one another. Quantum mechanics shows that, in the lowest obtainable state, atoms retain a definite amount of motion. The third law of thermodynamics states, however, that absolute zero cannot be attained under either classical or quantum-mechanical theory.

zero, tide-gage - A level surface established at a tide gage by reference to the contact mark, and used as a practical definition of tide-gage datum. Heights of the tide at a particular station are measured from this surface in the day-to-day operation of the tide gage. Adjustments may be required in the relationship between tide-gage zero and the contact mark as a result of levelings to tide-gage bench-mark.

zero correction - The negative of zero error. Also called index correction for some instruments.

zero error - The amount measured by an instrument when the actual amount should be zero. Also called index error for some instruments. The zero error is assumed to be constant and to affect all measurements by the instrument, regardless of size. It can therefore be determined by comparing the true (known) value of a measured quantity to the measured value and does not have to be determined actually at zero. However, other errors which are not constant with size or time of measurement are usually present and have to be determined also.

zero lot line - The positioning of a structure on a lot in such a way that one side of the strucxture rests directly on the boundary line of the lot.

zero meridian - (1) That plane which is parallel to the least axis of an ellipsoid, or to a specified axis of a sphere or rectangular, Cartesian coordinate system in three dimensions and from which angles to all planes similarly parallel to this axis are reckoned. (2) SEE meridian, prime.

zero point - SEE zero.

zero-velocity update - SEE update, zero-velocity.

zincography - Lithography using zinc plates in place of stones. The term is obsolete.

zinc plate, etched - An etched copy, on a zinc plate, of the contour lines drawn on the base map and used as a guide in cutting the stepped terrain base of a model for making relief models.

Zinger method - SEE Tsinger method (of determining longitude).

z-motion - Movement of a streoplotter's projector in the z-direction.

Zöllner pendulum - SEE pendulum, horizontal.

zone - (1) The region between any two concentric circles in the plane. In gravimetric geodesy, the term is applied to regions lying between circles concentric about the point at which gravity or geodetic height is to be calculated and having diameters specified by convention. (2) The portion of a sphere or of an ellipsoid included between to parallel planes perpendicular to the major axis. (3) In particular, a region bounded by two parallels of latitude. Also called a band, latitude band and latitudinal band. For example, A zonal harmonic is a function of latitude which goes through one complete cycle of values when the latitude varies from one zonal boundary to the other but which does not change when the longitude is changed. (4) A narrow region between two lines. SEE ALSO breaker zone; confluence zone, coastal; fracture zone; Hayford zone; littoral zone; offshore zone.

Zone, Antarctic - SEE Antarctic Circle.

zone, clear - That portion of the ground directly under the inner portions of a runway approach-surface.

zone, climatic - A region, on the Earth, bounded by specific parallels of latitude and showing having a climate more or less determined by the latitudes bounding it. For example, the Northern Temperate and Southern Temperate Zones are climatic zones bounded respectively by the arctic circle and the Tropic of Cancer, and by the Antarctic Circle and the Tropic of Capricorn.

zone, contiguous - A zone of the high seas, contiguous to the marginal sea of a coastal State, in which the State may exercise control to prevent infringement of its customs, regulations, etc.

zone, equatorial - (1) A narrow zone containing the equator. (2) SEE Zone, Torrid.

zone, intertidal - The region between mean high water and mean low water. Also called the littoral, foreshore and littoral zone. Although that last term also has other meanings.

zone, littoral - (1) The region extending from the shoreline top where the depth is 200 meters. This is practically equivalent to the zone of the continental shelf. (2) SEE breaker zone. (3) SEE zone, intertidal.

zone, nearshore - The region extending seaward from the shore to an indefinite distance beyond the breaker zone.

zone, neritic - (1) That region on the ocean floor which extends from the line of low tide to a depth of 200 meters i.e., approximately to the edge of the continental shelf.
(2) That part of the oceans lying at depths less than 200 meters.

Zone, Torrid - SEE Torrid Zone.

zone catalog - SEE zone star catalog.

zone plate - A transparent plate on which has been drawn a set of opaque, concentric rings whose width and spacing decrease with distance from the center. Zone plates have been used instead of a lens for special purposes such as photography with X-rays or for establishing a line of sight. The Fresnel zone plate is the most commonly used variety.

zone star catalog - A star catalog devoted to stars within a narrow band of declinations. It is usually simply referred as a zone catalog. The Yale Zone Catalogs and the Astrographic Catalogs are outstanding examples.

zone time - The time in a time-zone.

zoning - A means of regulating land-use, by which a community exercises the police power to ensure that the community's uses for the land do not conflict with each other and that adequate space is provided for all types of development. in the interest of protecting the public health, welfare, safety and other interests. SEE ALSO airport zoning; cluster zoning; density zoning; map, zoning.

zoning, exclusionary - Zoning that excludes certain classes of people, such as poor people or members of a minority, from a certain region.

zoning, inclusionary - Zoning that requires residential builders to include a certain percentage of dwelling units for households of low- and moderate income, as a condition for approval, by the government, of building.

zoning, large-lot - Zoning intended to reduce residential density by requiring large lots for building on. Also called acreage zoning.

zoning map - SEE map, zoning.

zoom lens - SEE system, pancratic optical.

zoom optical system - SEE system, varifocal optical.

zoom system - SEE system, varifocal optical.

z-term - SEE Kimura term.

z-time - SEE Greenwich mean time.

zulu time - SEE Greenwich mean time.